S0-CPD-364

HUMAN PHYSIOLOGY

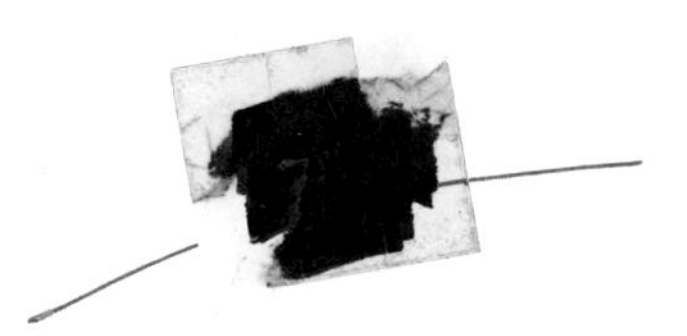

HUMAN PHYSIOLOGY

Joseph J. Previte, Ph.D.

Professor of Biology
Framingham State College

McGRAW-HILL BOOK COMPANY

New York St. Louis San Francisco Auckland Bogotá Hamburg
Johannesburg London Madrid Mexico Montreal New Delhi
Panama Paris São Paulo Singapore Sydney Tokyo Toronto

To my wife, Rachel, and children,
whose sacrifices made this book possible;

To my mom, Sarah, and dad, Albert,
who made me possible.

This book was set in Palatino by Progressive Typographers.
The editors were Jim Vastyan, Jay Ricci, David T. Horvath,
and Stephen Wagley;
the designer was Joan E. O'Connor;
the production supervisor was Charles Hess.
The drawings were done by ANCO/Boston.
The cover photograph was taken by Globus Bros.
Von Hoffmann Press, Inc., was printer and binder.

HUMAN PHYSIOLOGY

2 3 4 5 6 7 8 9 0 VHVH 8 9 8 7 6 5 4 3

ISBN 0-07-050786-4

See Acknowledgments on pages 737–739. Copyrights included on this page by reference.

Library of Congress Cataloging in Publication Data

Previte, Joseph J.
Human physiology.

Bibliography: p.
Includes index.
1. Human physiology. I. Title. [DNLM:
1. Physiology. QT 104 P944h]
QP34.5.P7 1983 612 82-14052
ISBN 0-07-050786-4

CONTENTS IN BRIEF

PREFACE		xix
TO THE STUDENT		xxiii
1	INTRODUCTION TO PHYSIOLOGY: Organization and Integration of the Human Body	1
2	THE CHEMICAL AND BIOLOGICAL BUILDING BLOCKS OF THE BODY	12
3	PHYSIOLOGY OF TISSUES	41
4	MOVEMENT OF SUBSTANCES ACROSS CELL MEMBRANES	79
5	THE ENDOCRINE SYSTEM: Hormones	102
6	PHYSIOLOGY OF THE NERVOUS SYSTEM: Structural and Functional Elements of Nervous Tissue	134
7	PHYSIOLOGY OF THE NERVOUS SYSTEM: Transmission of Information from Neurons	154
8	PHYSIOLOGY OF THE NERVOUS SYSTEM: Sensory Receptors, Codes, and Special Sense Organs	181
9	PHYSIOLOGY OF THE NERVOUS SYSTEM: The Spinal Cord and Autonomic Nervous System	205
10	THE NERVOUS SYSTEM: General Structure and Function of the Brain	236
11	PHYSIOLOGY OF THE NERVOUS SYSTEM: Structure and Function of Specific Areas of the Brain	256
12	THE PHYSIOLOGY OF MUSCLE: The Properties of Muscle Cells	282
13	THE PHYSIOLOGY OF SKELETAL MUSCLE: Muscle Mechanics, Energy Metabolism, and Exercise	307
14	THE PHYSIOLOGY OF BONE	334
15	THE CIRCULATORY SYSTEM: The Structure and Function of the Heart	359
16	THE CIRCULATORY SYSTEM: Effects of Chemical and Physical Factors on Cardiac Function	378
17	THE CIRCULATORY SYSTEM: The Functions of Blood and Lymphatic Vessels	401

18 THE CIRCULATORY SYSTEM: The Physiology of Blood, Part I: The Properties of Whole Blood and Erythrocytes (Red Blood Cells) 430

19 THE CIRCULATORY SYSTEM: The Physiology of Blood, Part II: The Properties of Leukocytes (White Blood Cells) and Thrombocytes (Platelets) 453

20 PHYSIOLOGY OF THE RESPIRATORY SYSTEM 472

21 THE EXCRETORY SYSTEM 505

22 THE DIGESTIVE SYSTEM: Regulatory Mechanisms and Defenses 542

23 THE DIGESTIVE SYSTEM: Food Processing 571

24 THE PHYSIOLOGY OF REPRODUCTION 619

APPENDIX: NORMAL REFERENCE LABORATORY VALUES 659

ADDITIONAL READINGS 672

GLOSSARY 683

ACKNOWLEDGMENTS 737

INDEX 740

CONTENTS

PREFACE xix
TO THE STUDENT xxiii

CHAPTER 1 INTRODUCTION TO PHYSIOLOGY: Organization and Integration of the Human Body 1

The Human Body: An Incredible Machine 2
Relationship of Physiology to Biological and Health Sciences 2
Organization of the Body 5
The Cell 5
Tissues 5
Organs 5
Systems 5
Integration of Body Function: Homeostasis and Biological Regulation 6
Role of the Circulatory and Lymphatic Systems 6
Role of the Excretory System 7
Role of the Nervous System 7
Role of the Endocrine System 7
Feedback Mechanisms 8
Summary 10

CHAPTER 2 THE CHEMICAL AND BIOLOGICAL BUILDING BLOCKS OF THE BODY 12

Physical Measures 13
Length, Volume, and Weight 13
Temperature and Heat 13
Chemical Building Blocks 15
Atoms 15
Molecules 15
Measurement of Quantities of Chemicals: Concentration, Equivalent Weight, and pH 18

Types of Complex Molecules (Compounds) 19
The Biological Building Block: The Cell 29
The Cell Membrane 30
The Nucleus 30
The Cytoplasm 30
Direction of Cellular Activity: The Role of DNA and RNA 34
Summary 37

CHAPTER 3 PHYSIOLOGY OF TISSUES 41

Epithelial Tissue 42
General Functions 42
Specific Characteristics 42
Functions in Body Cavities, Vessels, and Glands 47
Dynamics of Structure and Function of Epithelium 53
Connective Tissue 53
Loose (Areolar) Tissue 55
Reticular Tissue 65
Dense Connective Tissue 66
Cartilage 67
Bone (Osseous) Tissue 69
Dentin 72
Diseases of Connective Tissue 72
Muscular Tissue 73
Skeletal Muscle 73
Cardiac Muscle 73
Smooth Muscle 75
Nervous Tissue 75
Summary 76

CHAPTER 4 MOVEMENT OF SUBSTANCES ACROSS CELL MEMBRANES 79

Water and the Cellular Environment 80
Intracellular Compartments 80
Extracellular Compartments 80
Chemical Differences between Intracellular and Extracellular Compartments 81
Regulation of Total Body Water 82
The Cell Membrane 82
Role of Proteins 83
Role of Lipids 87
Transport Mechanisms 87
Nonmediated Transport 87
Carrier-Mediated Transport 92
Vesicle-Mediated Transport 95
Summary 99

CHAPTER 5 THE ENDOCRINE SYSTEM: Hormones 102

Introduction 103
Types of Chemical Control Agents 103
Hormones 103
Parahormones 103
Neurohormones 104
Neurotransmitters 104
Pheromones 104
Drugs 104
The Endocrine System 104
Nature of Hormones 105
Control of Hormone Production 105
General Regulatory Mechanisms 105
Feedback Mechanisms 107
Stress 109
Levels of Hormone Activity 111
Rates of Synthesis 111
Mechanisms of Secretion, Transport, Degradation, and Excretion 113
Role of Receptors 114
Interaction of Hormones 116
Cascade Effects 116
Physiological and Pharmacological Effects 116
Radioimmunoassay: A Sensitive Technique for Measuring Hormone Levels 117
Mechanisms of Hormone Action 119
Effects on Nuclear Receptors: Modification of Gene Expression (Transcription) 120
Effects on Cell Membrane Receptors: Modification of Posttranscriptional Processes 120
Summary 123

CHAPTER 6 PHYSIOLOGY OF THE NERVOUS SYSTEM: Structural and Functional Elements of Nervous Tissue 134

Role of the Neuron 135
Neuron Structure and Function 135
Shapes of Neurons 140
The Macroscopic Structure of Nerves 141
Supportive Tissues: The Glia 142
Glia of Peripheral Nerves 142
Glia of the Central Nervous System (Brain and Spinal Cord) 145
Demyelination of Fibers: Multiple Sclerosis 147
Types of Nerves 147
Origin of Nerves 147

Direction of Impulses 147
Destination of Impulses 147
Size and Myelin Content of Nerve Fibers 147
The Role of Nutrients in the Health of Nervous Tissue 148
Carbohydrates 149
Lipids 149
Proteins 150
Vitamins 151
Summary 151

CHAPTER 7 PHYSIOLOGY OF THE NERVOUS SYSTEM: Transmission of Information from Neurons 154

Development of Impulses by Neurons 155
The Resting Membrane Potential: An Electrochemical Equilibrium 156
The Action Potential 160
Relationship between Different Types of Potentials 167
Junctions between Neurons: The Synapse 168
General Structure and Function of Boutons and Synaptic Vesicles 169
Role of Calcium 169
Properties of Synaptic Membranes 169
Types of Synaptic Circuits 169
Function of Neurotransmitters 171
Development of Postsynaptic Potentials 172
Fate of Neurotransmitters 175
Interactions between Neurotransmitters and Receptor Molecules on the Cell Membrane 175
Examples of Neurotransmitters 175
Summary 178

CHAPTER 8 PHYSIOLOGY OF THE NERVOUS SYSTEM: Sensory Receptors, Codes, and Special Sense Organs 181

General Types of Sensory Receptors 182
Specific Examples of Sensory Receptors 182
Codes Generated by Impulses 185
Pattern Codes 185
Pathway Codes 186
Special Sense Organs 187
The Eye 187
The Ear 194
Taste (Gustation) 201
Smell (Olfaction) 202
Summary 202

CHAPTER 9 PHYSIOLOGY OF THE NERVOUS SYSTEM: The Spinal Cord and Autonomic Nervous System 205

The Spinal Cord 206
Origins and Arrangements of Spinal Nerves 206
Relationship of Cord Structure to Nerve Function 208
Reflexes 211
Nerve Pathways 214
Pain 220
Effects of Cord Damage 222
The Autonomic Nervous System (ANS) 225
Types of Neurotransmitters Secreted by ANS Fibers 227
Variations in Responses to ANS Transmitters 228
Examples of Adrenergic and Cholinergic Effects 228
Interactions between the Brain and the ANS 229
Summary 231

CHAPTER 10 THE NERVOUS SYSTEM: General Structure and Function of the Brain 236

General Properties 237
Development of the Brain and Cerebrospinal Fluid 237
Embryonic Origin of the Brain and Its Subdivisions 237
Cavities in the CNS 241
Tissues Covering the Brain 242
Cerebrospinal Fluid (CSF) 242
Diagnostic Tests and Brain Function 244
Chemical Analysis of CSF 245
X-ray Analysis 245
Immunological Studies 246
Electrical Studies 246
Brain Neurotransmitters and the Influence of Drugs 249
Endorphins 249
Dynorphin 250
Drug Effects on Neurotransmitter Function 251
Summary 253

CHAPTER 11 THE NERVOUS SYSTEM: Structure and Function of Specific Areas of the Brain 256

The Cranial Nerves 257
The Hindbrain 259
The Medulla Oblongata 259

The Pons 261
The Cerebellum 261
The Midbrain 265
The Forebrain: Its Diencephalic Subdivision 265
The Hypothalamus 267
The Thalamus 269
The Pineal Body 269
The Forebrain: Its Telencephalic Subdivision, the Cerebral Hemispheres 270
Location of Lobes 270
Functions of Lobes 271
Somatotopic Organization 274
The Split Brain 274
Learning and Conditioned Reflexes 278
Summary 279

CHAPTER 12 THE PHYSIOLOGY OF MUSCLE: The Properties of Muscle Cells 282

Structure of Muscle Tissue 283
Skeletal Muscle 283
Gross Anatomy 283
Cell Structure 285
Molecular Architecture 286
The Neuromuscular Junction 291
Molecular Sequence of Events in Contraction 295
Inhibitory Effects at the Neuromuscular Junction 295
Types of Skeletal Muscle Cells 297
Relation between Nerve Fiber and Type of Muscle 298
Cardiac Muscle 299
General Properties 299
Cell Membrane Structure 299
Cellular and Molecular Architecture 299
Metabolic Properties 300
Electrical Activity 300
Smooth Muscle 300
General Properties 300
Cell Membrane Structure 300
Cellular and Molecular Architecture 300
Electrical Activity 301
Summary 304

CHAPTER 13 THE PHYSIOLOGY OF SKELETAL MUSCLE: Muscle Mechanics, Energy Metabolism, and Exercise 307

The Mechanics of Contraction 308
Development of Tension 308

Muscles and Levers 315
Muscles and Work 316
Energy Metabolism 318
Anaerobic Glycolysis and Lactic Acid Formation 318
Aerobic Oxidative Phosphorylation 320
High-Energy Phosphate Metabolism 322
Energy Production and Muscle Contraction 323
The Effects of Exercise 325
Exercise and Physical Fitness 325
Training and Adaptation 326
Calculation of Target (Training) Heart Rate 328
Effects of Endurance Training on Muscle 328
Summary 330

CHAPTER 14 THE PHYSIOLOGY OF BONE 334

Major Functions 335
Bone as an Organ 335
General Organization 335
Chemical Composition 338
Dynamic Exchange of Constituents 341
Formation and Growth of Bone 341
Cellular Elements 341
Bone Formation (Osteogenesis) 341
Intracartilaginous Bone Formation 343
Factors Influencing Bone Formation 345
Acidity of ECF 345
Vitamins 345
Role of Hormones 348
Aging 351
Mechanical Stress and Exercise 351
Disorders of Bones and Joints 352
Summary 355

CHAPTER 15 THE CIRCULATORY SYSTEM: The Structure and Function of the Heart 359

Structure of the Heart 360
Lining and Walls of the Heart 360
Chambers and Route of Circulation 361
Blood Supply to Cardiac Muscle 363
Cardiac Physiology 364
The Cardiac Cycle 364
Cardiac Output 365
Electrical Conduction 367
Summary 375

CHAPTER 16 THE CIRCULATORY SYSTEM: Effects of Chemical and Physical Factors on Cardiac Function 378

Factors Influencing Cardiac Physiology 379
Chemical Agents 379
Blood Pressure 381
Sex 381
Body Size 381
Age 381
Body Temperature 381
Exercise 382
Cardiac Circulation during Development 385
Circulation in the Fetus 385
Tetralogy of Fallot: A Congenital Abnormality 387
Cardiac Abnormalities in Adults 387
Congestive Heart Failure 387
Atherosclerosis and Coronary Artery Disease 389
Myocardial Infarction and the Risk of Coronary Artery Disease 389
Diagnosis of Cardiovascular Function 390
Summary 398

CHAPTER 17 THE CIRCULATORY SYSTEM: The Functions of Blood and Lymphatic Vessels 401

Structure and Functions of Blood and Lymphatic Vessels 402
Regulation of Fluid Movement 404
Pulse 404
Reflex Control of Blood Vessels 407
Alternative Routes of Circulation 407
Flow of Blood through Capillaries 407
The Exchange of Materials between Blood, Interstitial Fluid, and Lymph 412
Regulation of Blood Flow through Veins 414
Other Factors Related to Blood Flow 414
Disorders Related to the Walls of Blood Vessels 421
Hypertension 422
Atherosclerosis 425
Summary 426

CHAPTER 18 THE CIRCULATORY SYSTEM: The Physiology of Blood, Part I: The Properties of Whole Blood and Erythrocytes (Red Blood Cells) 430

General Functions and Major Constituents 431
General Properties of Blood as a Fluid 432
Physical Properties of Whole Blood 432
Role of Buffers 433

Origins of Blood Cells (Hematopoiesis) 435
Properties of Erythrocytes 435
Development and Structure 435
Normal Variations in Number 435
Red Blood Cell Function and Hemoglobin 436
Blood Group Antigens 439
Major Blood Groups 440
Determination of Compatibility 441
Other Red Blood Cell Antigens 442
Pathophysiology of Red Blood Cells 443
Effects of Blood Loss: Hemorrhage and Shock 443
Hemolytic Disease of the Newborn 445
General Types of Anemia 446
Hereditary Anemias 446
Summary 450

CHAPTER 19 THE CIRCULATORY SYSTEM: The Physiology of Blood, Part II: The Properties of Leukocytes (White Blood Cells) and Thrombocytes (Platelets) 453

Properties of Leukocytes 454
Development and Appearance 454
Numbers of Leukocytes 455
General Functions 455
Defense Functions 459
Properties of Thrombocytes 461
Development and Structure 461
Numbers of Platelets 461
General Functions 462
Pathophysiology of Leukocytes 467
General Disorders 467
Leukemias 468
Pathophysiology of Thrombocytes (Platelets) 469
Summary 469

CHAPTER 20 PHYSIOLOGY OF THE RESPIRATORY SYSTEM 472

Architecture of the Respiratory System 473
Gross Anatomy 473
Cellular Anatomy and Physiology 476
The Physics of Breathing 479
Gas Laws and Variations in Pressure 479
Pressure Gradients in Inspiration and Expiration 480
Ventilation of the Lungs 480
Transport and Exchange of Gases 484
Oxygen Transport 484
Carbon Dioxide Transport 486

Exchange of Gases 486
Regulation of Respiration 488
Nervous Control Mechanisms 488
Chemical Regulators 491
Effects of Exercise 492
Environmental Effects 493
Exposure to High Altitude 493
Underwater Diving 494
Pollutants 496
Artificial Respiration 497
Mechanical Devices: Advanced Life-Support Techniques 497
Mouth-to-Mouth and Mouth-to-Nose Resuscitation—Basic Life-Support Techniques 497
Summary 501

CHAPTER 21 THE EXCRETORY SYSTEM 505

Organs and Functions of the Excretory System 506
The Skin 506
The Epidermis 507
The Dermis (*Corneum*) 511
The Kidneys 511
Anatomy 511
General Functions 512
The Nephrons: Functional Units of the Kidney 513
Physiology of the Nephrons 516
Effects of Hormones 524
Renin-Angiotensin System 525
Countercurrent Theory of Nephron Function 525
Regulation of Acid-Base Balance 527
Measurement of Kidney Function 530
Effects of Disease on Kidney Function 533
Kidney Transplants and the Artificial Kidney 535
Summary 537

CHAPTER 22 THE DIGESTIVE SYSTEM: Regulatory Mechanisms and Defenses 542

Structure 543
General Gross Anatomy and Physiology 543
Microanatomy 546
Accessory Digestive Organs 549
Secretions 554
Gastrointestinal Hormones 554
Exocrine Secretions 555
Regulatory Mechanisms 560
Control of Secretions 561

Control of Motility 561
Defenses and Inflammatory Responses 564
Inflammatory Reactions 565
Ulcers 565
Liver Disease 566
Summary 567

CHAPTER 23 THE DIGESTIVE SYSTEM: Food Processing 571

Digestion 572
Absorption 572
Role of Various Segments of the Digestive Tract 572
Absorption of Specific Substances 577
Digestive Tract Excretions 585
Malabsorption 586
Intermediary Metabolism 590
Carbohydrates 592
Proteins 592
Lipids 592
Metabolism and Nutrition 592
Basal Metabolic Rate 593
Metabolism and Body Temperature 594
Effects of Food and Exercise 595
Summary 596

CHAPTER 24 THE PHYSIOLOGY OF REPRODUCTION 619

The Male Reproductive System 620
The Testes 620
Accessory Sex Organs 622
The Production and Release of Sperm 622
Orgasm 625
The Influence of Hormones 626
The Female Reproductive System 629
The Ovaries 629
Accessory Sex Organs 629
The Production of Eggs (Oogenesis) 632
Orgasm 634
The Influence of Hormones 634
Conception and Pregnancy 639
Early Embryonic Development 640
The Pattern of Development from the Third Week 642
Maternal Hormonal Changes during Pregnancy 644
Amniocentesis 644
Reproductive Capacity 645
Fertility and Infertility 645

Effects of Sexually Transmitted Diseases 646
Control of Conception 646
Childbirth (Parturition) 653
Summary 653

APPENDIX: NORMAL REFERENCE LABORATORY VALUES 659
ADDITIONAL READINGS 672
GLOSSARY 683
ACKNOWLEDGMENTS 737
INDEX 740

PREFACE

This book is written with the assumption that much of the knowledge concerning the function of the human body in health and disease can and should be understood by every educated individual. It presents concepts so that they may be grasped by those with little or no college level background in the sciences. However, the topics are developed to depths that allow those with some science training to gain fuller appreciation of the principles and mechanisms involved.

The emphasis is on that physiology which will be most helpful to those interested in the allied health sciences and especially in the medical aspects of the life sciences. The approach used has evolved over a teaching and research career spanning more than two decades. It was developed for students in premedical sciences, medical technology, nursing, dietetics, biology, and other allied health sciences, as well as those studying the human body for their general education. The author has realized (as have many others) that most people's natural curiosity concerning human body function is heightened by the body's malfunction. In order to capitalize on this observation and to interest the student in the subject, the principles of physiology are often illustrated in relation to clinical examples and diseases with which most people are generally familiar or are likely to become familiar in their lifetime.

Current concepts that represent new ideas in the scientific and clinical literature, as well as in recent medical articles, are presented in relation to widely accepted basic principles. Each chapter contains a set of broad objectives, an outline summary, and review questions. For each chapter there is a list of additional introductory references at the end of the text, prior to the Glossary. The additional readings often present detailed information on some subjects in depths beyond that which is possible for an introductory textbook.

ACKNOWLEDGMENTS

I wish to acknowlege the education provided me by all of my former teachers and especially my mentor, L. Joe Berry, now professor emeritus at the University of Texas, Austin; Ed Hart, who worked assiduously as my research assistant and who drew many original sketches under my direction; Elaine Storella and Connie Day, whose comments were extremely helpful in refining first drafts of each chapter; Sandy Knowlton Lively, who typed drafts of the text from its inception, and JoAnne Cattani Newton, who joined in this effort shortly thereafter; and Michelle Murphy Carrier, Sue Hunter, Debbie Berkovich, and Sharon Murphy, whose general assistance shortened the time needed for completion of this lengthy task.

Thanks also go to the staff of McGraw-Hill, especially Jim Vastyan, Jay Ricci, David Horvath, and Stephen Wagley, whose encouragement sustained my efforts and to reviewers provided by them including William A. Cooper, West Texas State University; Darrell Davies, Kalamazoo Valley Community College; William E. Dunscombe, Union College; Patricia A. Lorenz, Penn Valley Community College; Duncan Martin, University of Arkansas; Peter Mel, El Camino College; Timothy O. Patschke, County College of Morris; Dixon L. Riggs, University of Northern Iowa; Timothy A. Stabler, Indiana University Northwest; Edith Wallace, William Paterson College; and Mary C. Welty, The University of South Dakota.

Finally, deep expressions of gratitude are due the following colleagues who freely gave of their time and expertise to improve specific chapters. They include Joseph Albert, Framingham Union Hospital; Marvin Adner, Framingham Union Hospital; Wilbert Bowers, U.S. Army Research Institute of Environmental Medicine, Natick Mass.; John Burke, Framingham Union Hospital; Roy Burlington, Central Michigan University; William Castelli, Harvard University Medical School; Carolyn Cohen, Brandeis University: JoAnne Cutter, Framingham Union Hospital; Carol Chauvin, Worcester State College; David DeNuccio, Central Connecticut State College; Gregory Eastwood, University of Massachusetts Medical School; Joel Feinblatt, University of Massachusetts Medical School: Kathy Fitzpatrick, Merrimack College; Ralph Goldman, U.S. Army Research Institute of Environmental Medicine, Natick, Mass.; Lawrence Hirsch, Framingham Union Hospital; Roger Hubbard, U.S. Army Research Institute of Environmental Medicine; Dana Jost, Framingham State College; Marlin Kreider, Worcester State College; Carl Lieberman, Framingham Union Hospital; Bruce Marcel, Framingham Union Hospital; René LeBlanc, Framingham State College; Sandy Marks, University of Massachusetts Medical School; Richard Masson, Framingham Union Hospital; Walter Morin, Bridgewater State College; Paula Parise-Hammet, Framingham Union Hospital; Stephen Previte, Lawrence General Hospital; Joel Rankin, Framingham Union Hospital; Isadore Rosenberg, Framingham Union Hospital; Chester Roskey, Framingham State College; Melvin Scher, Framingham Union Hospital; Nancy Tanner, Massachusetts General Hospital; Pér Tesche, U.S. Army Research Institute of Environmental Medicine; James Vogel, U.S. Army Research Institute of Environmental Medicine; George Wermers, Merrimack College.

The task of writing this text has strongly reinforced my appreciation of the intricacies of human body function. In view of the breadth and depth of knowledge required, and my own limitations, I will most heartily welcome suggestions which might help improve the clarity, accuracy, and organization of future editions of this book. Please send suggestions to me at the Biology Department, Hemenway Hall, Framingham State College, Framingham, MA. 01701

Joseph J. Previte

TO THE STUDENT

You are about to begin a study of an extraordinary subject, the human body. In the absence of more appropriate terminology, it has been termed an incredible machine. The human body has the awesome capacity to regulate and integrate its functions simultaneously. Trillions of cells and countless molecules, atoms, and ions are engaged in an ever changing and seemingly infinite number of interactions, and yet are maintained in a unique organization to sustain life in a single human being.

To help you to understand how your body functions and to prevent you from losing sight of the major physiological concepts, I recommend the following approach:

1. Study for one or more hours at regular intervals, preferably a few times each week.
2. Scan the outline at the beginning of each chapter. It previews the material that follows.
3. After reviewing the outline, read the objectives. They indicate the major points covered in the chapter and identify the goals you should achieve.
4. Before you study the chapter itself, look at the summary. Although some of the statements may puzzle you at first, the summary is a synopsis of the major concepts in the chapter.
5. When you read the main body of the chapter:
    a. Keep the objectives in mind.
    b. Read and note carefully, the italicized terms. They are the language of physiology. Mastery of this vocabulary is one of the most formidable tasks facing the beginning student. If a term perplexes you, check its definition in the glossary located at the back of the text, prior to the acknowledgments.
    c. Observe closely the illustrations, photomicrographs, and their legends. These visual aids display many of the major themes in the text.

6. The tables in each chapter are for further reference. The data should not frighten you. The details in the tables complement principles described in the body of the text. Unless your instructor indicates otherwise, you need not memorize this information. Do not lose sight of the major thrust of the chapter because of tabulated details.
7. After you have completed your study of a chapter, review the summary carefully. Determine whether you have met the objectives listed at the beginning of the chapter. If time permits, answer the review questions. The effort will help determine how well you understand the concepts each chapter addresses.
8. If you are interested in pursuing a topic consult the introductory reference materials (Additional Readings) listed by chapter at the end of the text, prior to the glossary. Your instructor also may be able to suggest additional reference materials.

The subject you are about to study is most fascinating. It is basically about you. Furthermore, the textbook is designed with you, the student, in mind. Each topic is developed step by step, and presupposes no college-level training in the sciences. I trust that your studies in physiology will provide valuable information to you, but moreover, lead to greater understanding and appreciation of the human body and its functions.

Joseph J. Previte

INTRODUCTION TO PHYSIOLOGY

Organization and Integration of the Human Body

1

1-1 THE HUMAN BODY: AN INCREDIBLE MACHINE

1-2 RELATIONSHIP OF PHYSIOLOGY TO BIOLOGICAL AND HEALTH SCIENCES

1-3 ORGANIZATION OF THE BODY
- The cell
- Tissues
- Organs
- Systems

1-4 INTEGRATION OF BODY FUNCTION: HOMEOSTASIS AND BIOLOGICAL REGULATION
- Role of the circulatory and lymphatic systems
- Role of the excretory system
- Role of the nervous system
- Role of the endocrine system
- Feedback mechanisms

1-5 SUMMARY

OBJECTIVES

After completing this chapter, the student should be able to:

1. Describe a few examples that illustrate that the body is an incredible machine.
2. Discuss the relationship of physiology to other scientific disciplines.
3. List several scientific sources through which one can read about the most recent discoveries in physiology.
4. Discuss the organization of the body in terms of its biological building blocks, the cells, and their relationships to tissues, organs, and systems.
5. Briefly describe the functions of each of the 10 systems in the human body.
6. Define homeostasis and describe the role of the circulatory and the lymphatic, excretory, nervous, and endocrine systems in the maintenance of homeostasis.
7. Distinguish between positive and negative feedback mechanisms and give at least one example of how each integrates body functions.

1-1 THE HUMAN BODY: AN INCREDIBLE MACHINE

The architecture and function of the human body are exquisite, even down through the molecular level. Trillions of molecules are meticulously arranged within all living cells, which are delicately but dynamically balanced. A constant ebb and flow occurs in response to change, but an equilibrium, called *homeostasis*, is maintained.

Approximately 60 trillion cells act together to form higher levels of organization as tissues, organs, and systems in the body. Their cooperative function infinitely exceeds the most distant visions of design of engineers. This exquisite organization is demonstrated in innumerable ways. For example, each unit of biological activity, i.e., each cell, synthesizes sugars, fats, proteins, acids, bases, salts, and a vast array of other complex chemicals. Yet the trillions of body cells do so in a controlled manner in cooperation with their neighbors to allow vital functions to flourish. For example, the cells of the liver store sugar as glycogen to maintain a reservoir of energy. Cells lining the joints of the body produce oils whose lubricating capacity exceeds that of the most expensive synthetic oils. Billions of molecules act in concert to generate a contractile force in a single muscle cell, which may be but one cell among hundreds of thousands within a muscle. The coordination of the activity of more than 600 muscles occurs through a maze of electrical circuits dependent on precise levels of salt solutions and of complex chemicals called *neurotransmitters* that are part of the nervous system. Functions of the nervous system are controlled by billions of cells which have awesome computational capacity and are located in its central computer, the brain. The durability of the human body is perhaps no better illustrated than in the cells of the heart. They must last a lifetime and yet pump about 55 million gal of blood through about 62,000 miles of blood vessels during that period. These are but a few of the examples by which the reliability, durability, performance, and utility of the human body and its parts are manifested. Many others will be discussed throughout the text. The human body as an entity possesses an unsurpassed beauty of design. Little wonder it has been called an incredible machine.[1]

1-2 RELATIONSHIP OF PHYSIOLOGY TO BIOLOGICAL AND HEALTH SCIENCES

In order to understand how the body works, it is useful to become at least generally aware of its structure or *anatomy* from the molecular level up through its overall general organization. *Physiology* is the study of the function of the whole organism and its parts. It is difficult if not impossible to study the body's function without some understanding of its *anatomy* (structure). The minimal amount of anatomy required is called *functional anatomy*. Again and

[1] An excellent film entitled "The Incredible Machine" is available from the National Geographic Society, 17th and M Sts. N.W., Washington, D.C. 20036

again it is often true that structure must generally be understood to comprehend function. For example, one must have some idea about the geography of a country to appreciate how its various areas and peoples interact and function. One must know a minimum about the structure of a car (for example, the location of the accelerator, ignition, brakes, and steering wheel) before one can learn how it is driven. This text, although an introduction to human body function, will therefore of necessity discuss its functional anatomy as well.

Progress in the life sciences and physiology is constantly being made. Contributions come from many physical, chemical, and biological disciplines. Apart from anatomy, biochemistry (the study of the chemistry of living organisms), genetics (the study of heredity), immunology (the study of defense mechanisms), pharmacology (the study of the action of drugs), and pathology (the study of disease), many other sciences add to our understanding of human body function. Textbooks published in these fields are listed by subject and author in *Books in Print, Medical Books and Serials in Print,* and *Scientific and Technical Books and Serials in Print.* These reference sources are usually available in libraries. Recent discoveries are published in periodicals, some of which are listed in Table 1-1. Scientists and others interested in physiology search these periodicals to keep abreast of new developments. Summaries of significant discoveries occurring over a period of time are described in reviews which are published quarterly or annually.

Few if any libraries can afford to purchase and/or store all journals relevant to the life sciences. Large university libraries subscribe to many periodicals, and regional medical libraries to even more. The latter are supported by the federal government and are located in several areas of the United States, normally on university campuses. The National Library of Medicine and the Library of Congress in Washington, D.C., are the largest depositories of such publications.

In spite of the overwhelming number of articles appearing in a vast array of periodicals, an individual may keep pace with general developments in specific areas. This is most often done through the use of one of several publications available at many college or university libraries. These include *Biological Abstracts, Current Contents, Excerpta Medica,* and *Index Medicus* (Table 1-1).

Biological Abstracts, which is printed twice monthly, publishes brief summaries of articles and citations from original sources appearing in more than 7600 international journals. *Index Medicus* abstracts articles taken from more than 2500 journals related to the field of medicine. *Excerpta Medica* publishes abstracts and citations from more than 3000 journals in the life sciences, which are found in separate sections for specialized subdisciplines (e.g., physiology, immunology, public health). *Current Contents* publishes weekly lists of titles, authors, and journal citations of articles from more than 1000 life science periodicals.

Once an article is located, an individual may obtain it or a photocopy of it through an interlibrary loan. In some cases the author may make a reprint available upon request. For a fee, publishers of many abstract or citation periodicals provide a computerized search and printout of titles and/or abstracts. They may also provide copies of the original articles. Many libraries have purchased access to such services. As a result, a student of physiology can obtain the most recent information in print. This is important, since concepts and ideas change as discoveries are made.

The object of this text is to introduce the student to the fascinating study of physiology. Most students can gain general insight to body function in health and disease. For the sake of clarity, concepts about the physiology of the human body are presented in somewhat dogmatic fashion in this and most other introductory texts. Yet, accumulation of scientific evidence sometimes buffets ideas which are simplified or overstated. Our understanding evolves as new discoveries are made. The latter are presented in the advanced texts and periodicals described above. Articles published in them are especially useful once a student has acquired some basic information. Study of the advanced references allows further pursuit of

TABLE 1-1
Sources of information in physiology

1. **Selected journals** (usually published monthly)

American Journal of Clinical Nutrition
American Journal of Gastroenterology
American Journal of Hematology
American Journal of Human Genetics
American Journal of Medicine
American Journal of Nursing
American Journal of Pathology
American Journal of Physiology
American Journal of Veterinary Research
Annals of Clinical Research
Annals of The New York Academy of Medicine
Biochemical Medicine
Clinical Chemistry
Clinical Endocrinology
Comparative Biochemistry and Physiology, A, B, C
Computer Programs in Biomedicine
Computers and Biomedical Research
Digestive Diseases and Sciences
Endocrinology
Experimental Cell Research
Federation Proceedings
Gastroenterology
Gerontology
International Journal for Vitamin and Nutrition Research
Journal de Physiologie
Journal of Animal Morphology and Physiology
Journal of Biomechanics
Journal of Cell Biology
Journal of Cellular Physiology
Journal of Comparative and Physiological Psychology
Journal of Electron Microscopy
Journal of Endocrinology
Journal of Experimental Biology
Journal of Experimental Medicine
Journal of General Physiology
Journal of Laboratory and Clinical Medicine
Journal of Neural Transmission
Journal of Neurochemistry
Journal of Neurocytology
Journal of Neurophysiology
Journal of Nutrition
Journal of Reproduction and Fertility
Journal of the American Dietetic Association
Journal of the American Medical Association
Journal of the Neurological Sciences
Journal of Ultrastructure Research
Laboratory Investigations
Lancet
Medical Laboratory Sciences
Medicine
Molecular and Cellular Endocrinology
Nature
New England Journal of Medicine
Pharmacology
Physiological Chemistry and Physics
Physiological Psychology
Physiology and Behavior
Proceedings of the Nutrition Society
Proceedings of the Society of Experimental Biology and Medicine
Psychological Review
Psychoneuroendocrinology
Psychopharmacology
Psychophysiology
Quarterly Journal of Medicine
Respiration
Respiration Physiology
Science
Scientific American

2. **Reviews** (published quarterly or annually)

Annual Review of Biochemistry
Annual Review of Medicine
Annual Review of Nutrition
Annual Review of Physiology
Nutrition Reviews
Pharmacological Reviews
Physiological Reviews
Quarterly Review of Medicine

3. **Compendia of current titles and/or abstracts**

Biological Abstracts. (Published twice monthly by Bioscience Information Service of Biological Abstracts, Philadelphia. This publication indexes and abstracts information from more than 7600 journals).

Books in Print. (Published annually by R. R. Bowker Co., New York. Each of four volumes provides by author and by title the latest bibliographic list of all books currently in print. A list of publishers is also included. Companion publications entitled *Medical Books and Serials in Print* and *Scientific and Technical Books and Serials in Print* are also published annually).

Current Contents. Life Sciences. (Published weekly by Institute for Scientific Information, Philadelphia. This publication lists the title, author(s), and journal of articles appearing in more than 1000 periodicals in the life sciences.)

Excerpta Medica. 2A Physiology. (Published monthly by the Excerpta Medica Foundation, New York. This periodical lists, indexes and abstracts information from journals all over the world which publish articles in physiology. Other sections do likewise for related disciplines.)

Index Medicus. (Published monthly by the National Library of Medicine, Washington, D.C. This reference indexes and abstracts information from more than 2500 journals related to the medical sciences.)

knowledge and refinement of concepts related to specific aspects of a subject.

1-3 ORGANIZATION OF THE BODY

Living organisms are composed of physical, chemical, and biological building blocks. Their complexity increases in the biological level as cells form tissues which unite to form organs which work together in systems. The cooperative interaction of all the components is required to create the vital and dynamic entity called a human being.

The Cell

The *cell* is the body's basic biological building block. It is the structural and functional unit of all living organisms. The cell is able to incorporate substances into its structure, grow, reproduce, and respond to stimuli. It will be described in the next chapter and others throughout the text.

Tissues

A division of labor occurs among cells. *Tissues* are groups of cells which perform a specialized function. Some locations of the four major types of tissues listed in Table 1-2 will be described further in Chapter 3.

TABLE 1-2
Some locations of the four major tissue types

Tissue	*Locations in the Body*
1. Epithelium	Interior and exterior surfaces of the body (e.g., mucous membranes of the mouth and nose, outer layers of the skin)
	Glands (e.g., salivary glands, hormone-producing glands)
2. Connective	Binds biological units together throughout the body (e.g., ligaments binding bone to bone)
3. Muscle	Associated with bones
	Body wall
	Walls of organs (e.g., heart, digestive organs, uterus)
4. Nerve	Brain, spinal cord, nerves

Organs

Organs are structures composed of many tissues interacting to perform a specific function(s). The heart is an example. It is held in place by elements of connective tissue that form an external sac or membrane. It is a hollow organ lined by smooth epithelial tissue of one type (endothelial tissue) and contained within a layer of muscle, the *myocardium*. A specialized type of connective tissue, blood, is found within the central chambers of the heart. The myocardium creates a forceful beat and propels blood through its smooth interior chambers to blood vessels.

The heart is but one organ involved with the circulation of blood. Organs that cooperate in this function are part of the *circulatory system*. They include the heart and all the blood vessels within the body. Fluid which escapes from the circulatory system is returned to it through the *lymphatic system*, its vessels, and lymph nodes.

Many other organs of the body, including the brain, kidneys, stomach, liver, spleen, and pancreas, will be described throughout the remainder of the text. The relationship of some of these organs to the systems of the body is outlined in Table 1-3 and illustrated for the digestive tract in Fig. 22-1.

Systems

Ten systems are integrated within the human body. The *circulatory system* has just been described above. The *lymphatic system* filters and recirculates fluid lost from the bloodstream and tissues and produces cells that some describe as part of a separate defense system, the *immune system*. The *respiratory system* exchanges oxygen for carbon dioxide. The *muscular system* makes movement possible. The *skeletal system* supports the body, protects it, and aids in its movements. The *excretory system* eliminates waste materials through the skin, urinary tract, and body linings, including those of the lungs and digestive system. The *endocrine system* pro-

TABLE 1-3
The ten systems of the human body and some of the organs that compose them*

System	*Organ*
Endocrine	Pituitary, pineal gland, thyroid, parathyroids, thymus, pancreas, adrenals, gonads (ovaries, testes)
Nervous	Brain and spinal cord
Skeletal	Bones
Muscular	Skeletal muscles
Respiratory	Lungs
Circulatory	Heart, blood vessels
Lymphatic	Spleen, thymus, tonsils, lymph nodes, lymph vessels
Digestive	Stomach, intestines, liver, salivary glands, pancreas
Excretory	Skin, sweat glands, lungs, kidneys, bladder, digestive organs
Reproductive	Testes, ovaries, uterus

* The number of systems listed depends on the means of classification that is adopted. For example, the circulatory and lymphatic systems contain cells (some of which are derived from bone) that defend the body against disease. These cells are often regarded as composing an immune system. Some of them ingest particles, including microorganisms, and may be considered a distinct subdivision, the reticuloendothelial system. Depending on properties of these cells, a further subdivision, the mononuclear phagocyte system, may be distinguished. So too, some authors prefer to separate the excretory functions of the kidneys and their associated functions into a urinary system.

duces hormones that help regulate body chemistry and function. The *digestive system* converts foods to molecules that can enter the body and be used by it. The *reproductive system* is the source of cells to reproduce human life. The *nervous system* integrates body function via the transmission of signals between cells often located in remote parts of the body.

1-4 INTEGRATION OF BODY FUNCTION: HOMEOSTASIS AND BIOLOGICAL REGULATION

The body can be analyzed and described in terms of its physical and chemical properties. However, a batch of chemicals does not make a human being; the materials must be assembled in a very precise and intricate manner. The variety and complexity of elements, molecules, cells, tissues, organs, and systems that must be integrated to produce a human being are awe-inspiring. Quadrillions (millions of billions, 10^{15}) of components must work independently and cooperate closely with one another.

For life to continue, the viable (living) cells must be able to respond to a continually changing environment. These changes occur internally as well as externally. *Homeostasis* is the capacity of cells and the body to maintain a relatively constant internal environment. Homeostasis is maintained in spite of internal and external changes. If the body is unable to sustain it, illness, disease, and/or death may follow. For example, when the environmental temperature drops, the body responds by increasing heat production to support body temperature around 37°C. Should the body fail to regulate temperature, tissue damage such as frostbite and even death might occur from cold exposure if body temperature drops low enough.

The cells of the body continually expend energy and materials to maintain homeostasis. There is a constant and very often intense exchange of chemicals between the fluid found inside cells and that found outside of them. The exchange occurs so rapidly as to often go undetected unless carefully scrutinized. There is a continual flow of substances in and out of cells with no apparent net change. The cells are said to be in a *steady state*.

The body needs well-developed communication networks to support homeostasis. Five systems in particular serve to preserve the steady state. They are the circulatory, lymphatic, excretory, nervous, and endocrine systems.

Role of the Circulatory and Lymphatic Systems

The circulation of fluids helps maintain homeostasis. The chemicals of the body are in constant flux, moving in and out of cells. Chemicals within the *intracellular fluid* (found within cells) move outside the cells to enter the *extracellular fluid* and between cells to become *interstitial* or *intercellular fluid*. The chemicals may enter the bloodstream from cells or move from the bloodstream in the opposite direction. The pumping action of the heart moves blood through the body. Some cells and substances may escape from the bloodstream, passing at various points through spaces between the cells of small blood vessels. The substances then are temporarily located in the interstitial fluid and move into vessels of the lymphatic system. Later, they are returned to the bloodstream. The remaining components in the blood continue to flow through the vessels of the circulatory system. In this way substances are transported by the circulatory system from their points of origin to other cells into which they may enter. The substances may be converted into energy or into a variety of different materials needed to build body parts in order to carry on body functions.

Role of the Excretory System

Excretion is the elimination of substances no longer needed by the body. Excretions are formed in the body by the cells lining the digestive tract, lungs, skin, and tubules in the kidneys. The kidneys are among the prime excretory organs and play an important role in the maintenance of homeostasis. They are paired organs which filter excesses of a variety of substances out of the bloodstream into the urine. They are found above the hips posterior to the abdominal cavity. The tubules of the kidneys control water and salt balance and regulate the acid-base balance in the body's fluids. They perform these functions by selective excretion of substances from the bloodstream and their retrieval (reabsorption) from the filtrate (potential urine) which is thereby formed. Among the substances excreted by the kidneys to control pH are excess acid (such as hydrogen ions, H^+) and excess base (such as ammonia, NH_3^+). In this manner, the pH of the blood is usually maintained within a narrow range (pH 7.35–7.45). A dynamic balance between excretion and reabsorption of substances helps the body maintain a relatively steady state in the composition of its fluids.

Role of the Nervous System

The nervous system can transmit signals between different parts of the body. The endings of neurons (nerve cells) can release chemicals that modify the activity of other cells. Chemicals may be moved in microtubules within the neurons and may travel several feet to cause changes in the flow of charged particles or ions. The changes in ion flow across nerve cell membranes are called *nerve impulses*.

Some neurons are specialized to transmit information to those in the central computer of the body, the *brain* (or to its outpost, the spinal cord). Other neurons transmit impulses from the brain (or spinal cord) to sites some distance away. The communications go two ways, each way utilizing different neurons, so that messages pass from one to another through the central nervous system. The brain is the key organ in this system, exceeding the capacity of any computer known to humans. The impulses that are transmitted by neurons allow cells to respond to change and affect physiological functions. In this manner the homeostasis of cells, tissues, organs, and systems of the body is maintained.

Role of the Endocrine System

Hormones are chemicals released into the bloodstream after their synthesis by specific groups of cells that form *endocrine* glands. After their release into the circulatory system, hormones travel to and affect cells which are receptive to them. Hormones are a major means of maintaining homeostasis. They exert powerful influences on body chemistry and function. Some endocrines are illustrated in Fig. 5-1.

Glands which produce hormones collectively constitute the *endocrine system*. The *pituitary gland* is the master endocrine and is located at the base of the brain. It secretes several hormones, which influence a variety of body functions, including growth. The *pineal gland*, located near the center of the brain, secretes melatonin and influences the rhythms of biological functions in the body and the onset of sexual maturity. The *thyroid gland*, located at the base of the neck, produces thyroxine and calcitonin. Thyroxine and calcitonin affect the oxidation of foodstuffs and the development of bone, respectively. The *parathyroids* are found on the back (dorsal surface) of the thyroid. They produce parathormone which, along with calcitonin from the thyroid, helps regulate calcium levels in the bloodstream. The *adrenal* glands are endocrines attached to the superior surface of each kidney. Their products include steroid hormones which affect carbohydrate and protein chemistry, as well as the transport of mineral elements (salts). Adrenal steroid secretion is high in response to stress. The *thymus* is found in the lower part of the neck overlying the breastbone (sternum). It forms thymic hormones which play roles in defense against disease. The *pancreas* is found lying within the curvature of the first part of the small intestine (duodenum). Its hormones especially influence glucose metabolism. The *ovaries* are found above the bladder in the female. Their hormones stimulate synthesis of compounds in most cells and influence the development of sexual characteristics. The *testes* are found in the scrotal sacs outside the abdominal cavity in the male. Their hormones generally favor synthesis of many compounds in most cells and influence the development of sexual characteristics.

FIG. 1-1
A model of negative feedback mechanisms. The biological unit may be a cell, tissue, organ, or system. The unit serves two functions. It generates some type of product(s) and acts as a sensor, receiving feedback signals (inputs or stimuli). The products of the biological unit are able to return to it and serve as negative feedback signals, which turn the unit off. Consequently, the level of the product falls. The lower level of the product then serves as a negative feedback signal to stimulate the unit to form more product and thus complete the cycle. The paths shown as 1→2 and 3→4 can be activated repeatedly.

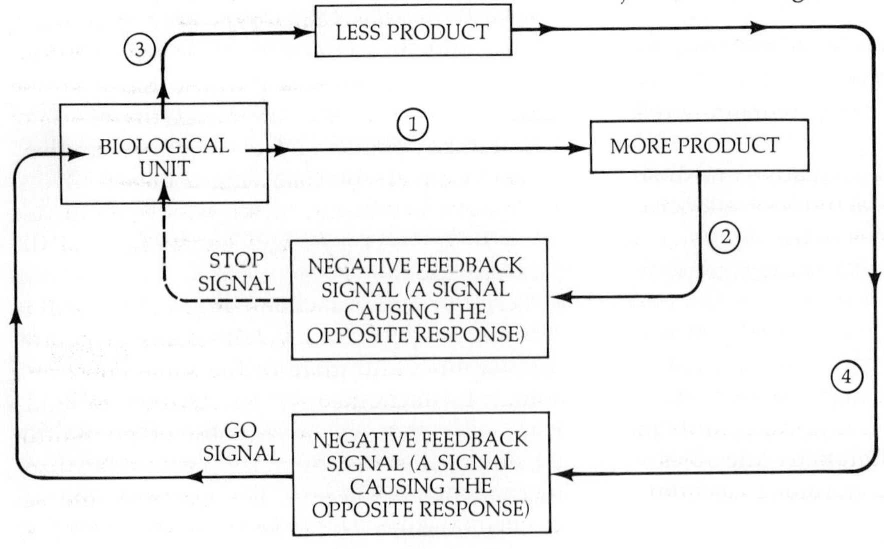

Feedback Mechanisms

The nervous and endocrine systems are among the most important biological units that integrate and regulate bodily functions. They assist in maintaining homeostasis by means of *feedback mechanisms.* Cells in these and other systems of the body may act as sensors and/or transmitters. The cells behave as *sensors* when they receive inhibitory or excitatory signals. They behave as *transmitters* when they produce a response (output) by releasing some substance or changing some activity.

Some signals cause the same activity in the unit to intensify. These are called *positive feedback* signals. Generally, however, *negative feedback* signals occur. They cause an effect *opposite* to that of the original activity and are illustrated in Fig. 1-1.

The response of neurons in the brain to a rise in blood temperature is an example of *negative feedback.* It might occur with fever or exposure

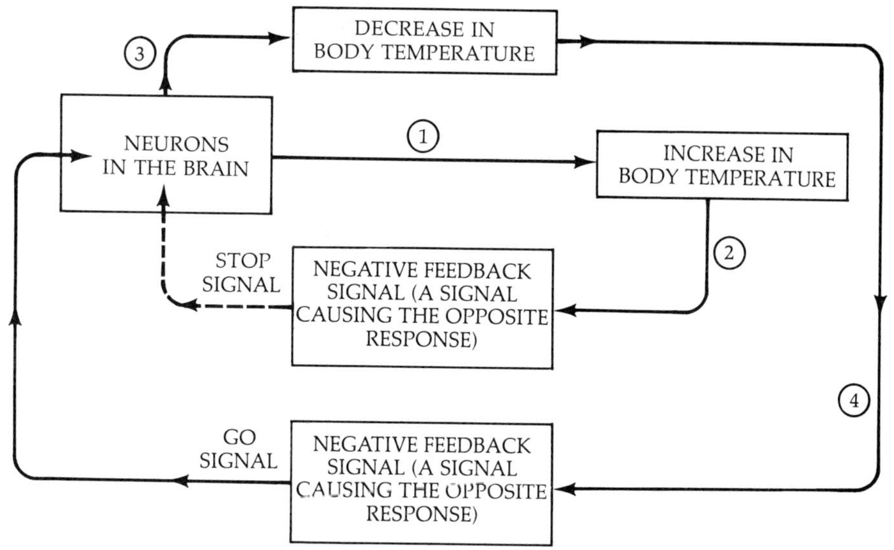

FIG. 1-2
A feedback mechanism in the brain, affecting body temperature. The paths shown as 1→2 and 3→4 can be activated repeatedly.

to high environmental temperature (Fig. 1-2). The neurons respond to elevations in blood temperature by turning off heat production and promoting heat loss through sweating (path 1 to 3, Fig. 1-2). This produces a drop in body temperature, which serves as a negative feedback signal. It stimulates the neurons to turn on heat-production mechanisms (e.g., an increase in production of heat by muscle, path 3 to 4 to 1). This response causes a rise in body temperature, which in turn serves as a negative feedback signal, repeating the cycle (path 1 to 3, Fig. 1-2). In this way, body temperature oscillates around a normal set value, remaining relatively constant.

Other illustrations of negative feedback mechanisms occur in the activity of endocrine glands. Levels of substances in the bloodstream may constitute negative feedback signals for endocrine glands. For instance, the release of calcium from bone to the blood is stimulated by parathormone secreted by the parathyroid glands (Fig. 1-3) and regulated by a negative feedback mechanism involving calcium in the blood. For the sake of simplicity the roles of other factors, such as the hormone calcitonin, are not discussed here.

An increase in the level of calcium in the blood is a negative feedback signal, which turns off production of parathormone. A drop in hormone level follows, which leads to a decrease in blood levels of calcium. The latter serves as a negative feedback signal to turn on parathormone production. This response causes an elevation in blood levels of calcium to complete the cycle.

Another example of negative feedback is in the generation of hunger pains. These pains may arise when blood sugar levels drop and stimulate strong contractions of the empty stomach. The decline in blood sugar seems to stimulate responsive neurons, called *glucostats,* in the hypothalamus of the brain. The neurons generate nerve impulses which cause strong contractions of the smooth muscle in the wall of the stomach signaling the subject of the need to eat. Conversely, following a meal and elevation of blood sugar, the glucostats are turned off and the strong hunger contractions of the stomach subside.

Very few body functions depend on positive feedback mechanisms. Such means of control lead to more and more of the same effect and could, if unchecked, be disastrous to body function, leading it to a point of no return. However, in some instances positive feedback mechanisms are useful. For example, the sequence of stages in childbirth is stimulated by

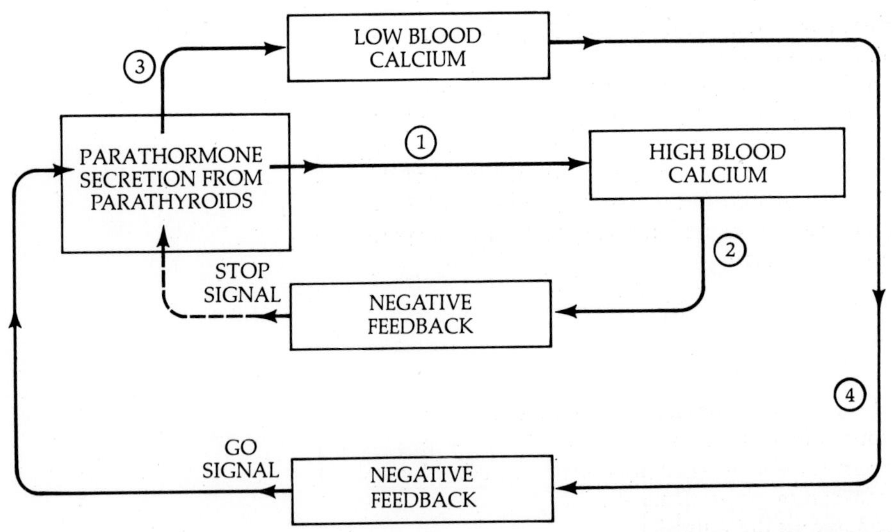

FIG. 1-3
An endocrine feedback mechanism affecting levels of calcium in the blood. The paths shown as 1→2 and 3→4 can be activated repeatedly.

positive feedback mechanisms. A hormone from the pituitary gland, oxytocin, causes muscle in the womb (uterus) to contract. The muscles are able to produce signals for the release of more oxytocin, which in turn stimulates more forceful contractions. This cycle does not subside until the baby is forced out of the birth canal and the afterbirth (placenta) is shed.

The human body possesses many mechanisms that assist in the maintenance of homeostasis and life itself. The variety of regulatory mechanisms that preserve the steady state constitute but a few of a myriad of reasons why the human body can be viewed as a miraculous machine.

1-5 SUMMARY

1 The human body is an incredible machine in which trillions of cells, each composed of trillions of molecules, interact, cooperate, and respond to change in order to maintain an equilibrium called *homeostasis*.

2 The study of function in living organisms is called *physiology*. Comprehension of it requires at least a minimal understanding of structure (functional anatomy).

A variety of disciplines contribute to physiology. They include anatomy, biochemistry, genetics, immunology, pharmacology, pathology, and many other physical, chemical, and biological sciences. Advances in these areas are published in periodicals available in many libraries.

3 The body works as a whole. It is more than the sum of its parts. Its building blocks are organized to work in integrated fashion. One approach, used to make comprehension of its functions simpler, is to first study the body's parts (building blocks) and later its more complex integrated features.

The basic biological building block of the body is the *cell*. It is the structural and functional unit of which all living organisms are composed.

Cells cooperate in groups called tissues. Distinct members of each group perform specialized functions. Tissues interact with one another to form organs, each of which performs specialized functions. The organs of the body work together in systems. The 10 systems of the body are the endocrine, nervous, skeletal, muscular, respiratory, circulatory, lymphatic, digestive, excretory, and reproductive systems.

4 The functions of the body are integrated, especially by activity of the circulatory, lymphatic, excretory, nervous, and endocrine systems. Cells, tissues, organs, and systems constitute biological units which can respond to environmental change. In health they possess a capacity to maintain equilibrium, or homeostasis. They normally do so by receiving stimuli and transmitting signals which oppose the environmental change. The usual response to change is a negative feedback signal.

REVIEW QUESTIONS

1 What is physiology and how does it compare with anatomy?

2 Why is the human body more than the sum of its parts?

3 Compare the use of journals, reviews, and compendia of current titles and abstracts to gain recent information about discoveries in physiology. When would you use one as opposed to the others?

4 Explain the levels of organization in the body from cells through tissues, organs, and systems.

5 Describe 10 systems in the body and some organs found within each of them.

6 Define homeostasis and explain its importance to life. What happens when homeostatic mechanisms fail?

7 How do the circulatory and lymphatic systems and kidneys help maintain homeostasis in the human body?

8 What part does the nervous system play in maintaining homeostasis?

9 A certain thermostat operates in such a way that when room temperature increases to a predetermined point, the furnace comes on, heating the room further. Is such a mechanism an example of positive feedback or negative feedback? Explain.

10 Define negative feedback and explain, in general terms, how it operates to regulate human body temperature.

THE CHEMICAL AND BIOLOGICAL BUILDING BLOCKS OF THE BODY

2

2-1 PHYSICAL MEASURES
Length, volume, and weight
Temperature and heat

2-2 CHEMICAL BUILDING BLOCKS
Atoms
Molecules
Measurement of quantities of chemicals: concentration, equivalents, and pH
Types of complex molecules (compounds)
Inorganic compounds
Organic compounds and their building blocks
Carbohydrates • Proteins • Lipids • Nucleic acids

2-3 THE BIOLOGICAL BUILDING BLOCK: THE CELL
The cell membrane
The nucleus
The cytoplasm
Endoplasmic reticulum and ribosomes
Golgi apparatus
Lysosomes
Mitochondria
Tubules and filaments
Centrioles, cilia, and flagella
Direction of cellular activity: The role of DNA and RNA

2-4 SUMMARY

OBJECTIVES

After completing this chapter, the student should be able to:

1. Discuss the importance of understanding some elementary chemistry to the comprehension of body function.
2. Compare the equivalent interconvertible measures of length, volume, and mass in the metric and English systems.
3. Distinguish among atoms, molecules, and compounds.
4. Define molarity, pH, and equivalent weight.
5. Describe the building blocks of carbohydrates, proteins, lipids, and nucleic acids and list some important functions of these four major compounds.
6. Describe the structure and function of the cell membrane, nucleus, and at least five components of the cytoplasm.
7. Discuss the role of DNA, mRNA, tRNA, and rRNA in protein synthesis.

The individual chemical elements in the body are valued at but a few dollars. However, many of the complex substances are worth infinitely more. If purchased from biochemical supply houses, the molecules found in the human body would cost several million dollars. Yet their true value is not measurable. For example, a diabetic's life is dependent on a hormone, insulin, which controls sugar levels in the body. What is the worth of such a compound to that individual? Normal body function is dependent on specific levels of a huge number of different life sustaining molecules. However, the human body is more than the sum of its chemicals. It is a miraculous creation possessing an organizational beauty and coordination of functions that surpasses the imagination and excites the minds of the students of physiology.

The body works as a whole. Its parts interact and affect each other. However in this text, to make physiology easier to understand, the body's physical, chemical and biological building blocks are first described as separate units and then analyzed in terms of the role they play in the whole organism. These concepts are presented below in a simplified manner so that students with little or no chemistry or physics background can comprehend some of the mechanisms involved in human body function. The necessity of understanding some basic chemical and physical concepts is evident if one notes the large number of Nobel laureates in physiology and medicine whose studies in the chemistry and physics of life have contributed so much to our understanding of human body function (Table 2-1).

2-1 PHYSICAL MEASURES

Units of the metric system are often used to describe the physical and chemical properties of the human body. The prefixes illustrated in Table 2-2 are used to give an accounting of length, volume or weight.

Numerical values are expressed as a function of the power of 10. These are exponents, or numbers of times one multiplies 10 or a fraction of 10 times itself, to attain a particular value (e.g., $10^3 = 10 \times 10 \times 10 = 1000$; $10^{-3} = 0.1 \times 0.1 \times 0.1 = 0.001$).

Length, Volume, and Weight

The basic units of length, volume, and weight are the meter (m), liter (L), and gram (g). One meter equals about 3.3 ft or 39.4 in. One liter equals about 1.06 quarts or 33.8 ounces. One gram equals about 1/454 lb or 0.035 oz. Other equivalents and subdivisions of the metric system are indicated in Table 2-3.

Temperature and Heat

A measure of temperature is the degree Celsius (°C). It is based on a scale in which water freezes at 0°C and boils at 100°C.

Temperature in °C may be converted to the Fahrenheit scale (°F) by the following formula:

$$1.8 \times {}^\circ\text{C} + 32 = \text{degrees Fahrenheit}$$

TABLE 2-1
Nobel prizes in physiology or medicine since 1955

Year	Investigator(s)	Area of Study or Discovery
1955	H. Theorell	Oxidative enzymes
1956	D. W. Richards, Jr., A. F. Cournand, and W. Forssman	Treatment of heart disease
1957	D. Bovet	Drugs to treat allergies and to relax muscles
1958	J. Lederberg, G. W. Beadle and E. L. Tatum	Biochemical genetics Genetic transmission of characteristics
1959	S. Ochoa and A. Kornberg	Enzymes to produce artificial DNA and RNA
1960	Sir McFarlane Burnet and P. B. Medawar	Immunological tolerance
1961	G. Von Bekesy	Physical principles of stimulation of cochlea (contains the organ of hearing)
1962	J. D. Watson, M. F. Wilkins, and F. H. Crick	Structure of genetic material (DNA)
1963	A. L. Hodgkin, A. F. Huxley and Sir John C. Eccles	Chemistry of nerve impulses and neuron function
1964	K. E. Bloch and F. Lynen	Regulatory mechanisms of cholesterol and fatty acid metabolism
1965	F. Jacob, A. Lwoff, and J. Monod	Regulation of cellular activity
1966	C. B. Huggins F. P. Rous	Hormonal therapy of cancer of prostate Viral cause of some tumors
1967	H. K. Hartine, G. Wald, and R. Grant	Function of the eye
1968	R. W. Holley, H. G. Khorana, and M. W. Nirenberg	Chemistry of the genetic code
1969	M. Delbruck, A. D. Hershey, and S. E. Luria	Mechanisms and chemistry of viral infection
1970	J. Axelrod, U. S. Von Euler, and Sir Bernard Katz	Chemical transmission of nerve impulses
1971	E. W. Sutherland, Jr.	Chemical effects and mechanisms of action of hormones
1972	G. M. Edelman and R. R. Porter	Chemistry and structure of antibodies
1973	K. VonFrisch, K. Lorenz, and N. Tinbergen	Patterns of individual and social behavior
1974	G. E. Palade, C. deDuve, and A. Claude	Ultrastructural organization and function of components within the cell
1975	D. Baltimore, H. M. Temin, and R. Dulbecco	Tumor virus interaction with cell's genetic material
1976	B. S. Blumberg and D. C. Gajdusek	Mechanisms of origin and spread of infectious disease
1977	R. S. Yalow R. G. Guillemin and A. V. Schally	Radioimmunoassay of hormones Chemistry and activity of hormones of the hypothalamus
1978	D. Nathans, H. Smith, and W. Arber	Chemistry of restriction enzymes and their roles in molecular genetics
1979	A. M. Cormack and G. Newbold	Computerized analysis of serial x-ray technique, CAT scan (computed axial tomography)
1980	G. Snell, J. Dausset, and B. Benacerraf	Studies on the regulation of immunological reactions by genetically determined structures of the cell surface
1981	D. Hubel and T. Wiesel R. W. Sperry	Processing of visual information by the brain Separation of functions of the hemispheres of the brain and location of many higher realms of function in the left half of the brain
1982	S. Bergstrom, J. Vane, B. Samuelsson	Biology and chemistry of prostaglandins

In this scale, the freezing point of water is recorded as 32°F and the boiling point as 212°F.

A measure of the amount of heat or energy used or produced by the body or contained in food is a kilocalorie (kcal), sometimes called a large Calorie (Cal). It is equal to 1000 calories,

TABLE 2-2
Prefixes commonly used with units of physical measurements

Prefix (and Abbreviation)	*Term*	*Exponent Form*	*Numerical Value*
mega (M)	One million times	10^{6}	(10 × 10 × 10 × 10 × 10 × 10)
kilo (k)	One thousand times	10^{3}	(10 × 10 × 10)
deci (d)	One-tenth	10^{-1}	(0.1)
centi (c)	One-hundredth	10^{-2}	(0.1 × 0.1 = 0.01)
milli (m)	One-thousandth	10^{-3}	(0.1 × 0.1 × 0.1 = 0.001)
micro (μ)	One-millionth	10^{-6}	(0.1 × 0.1 × 0.1 × 0.1 × 0.1 × 0.1 = 0.000001)
nano (n)	One-billionth	10^{-9}	
pico (p)	One-trillionth	10^{-12}	
femto (f)	One-quadrillionth	10^{-15}	

corresponding to the amount of heat needed to raise the temperature of one kilogram of water 1°C, and is the unit most often used to describe the energy value of food.[1]

2-2 CHEMICAL BUILDING BLOCKS

Atoms

Matter consists of individual elements, examples of which include carbon (C), hydrogen (H), oxygen (O), and nitrogen (N). The smallest particle of an element which cannot be subdivided without changing the nature of that substance is called an *atom*. Each atom is the smallest division of matter capable of independent prolonged existence. About 103 different elements have been discovered to date.

A number of subatomic particles have been discovered within atoms. Three of these contribute to the essential stability of all atoms. They include *electrons,* which are negatively charged and rotate in orbits about the center or *nucleus* of each atom. The nucleus is made up of positively charged *protons,* the number of which equals the number of electrons, and neutral *neutrons.* Atoms may gain or lose electrons and thereby acquire negative or positive charges. They are then referred to as *ions.* For example, when a sodium atom loses an electron, it acquires a positive charge and becomes a sodium ion (Fig. 2-1).

Molecules

Atoms may gain, donate, or share electrons. When sharing electrons, two or more atoms

[1]A calorie (cal) is the amount of heat needed to raise the temperature of one gram of water by 1°C. Since metabolic and other chemical reactions involve thousands of calories of energy, the kilocalorie is the unit most commonly used in describing these reactions. The term Calorie is its equivalent often used in discussions of nutrition.

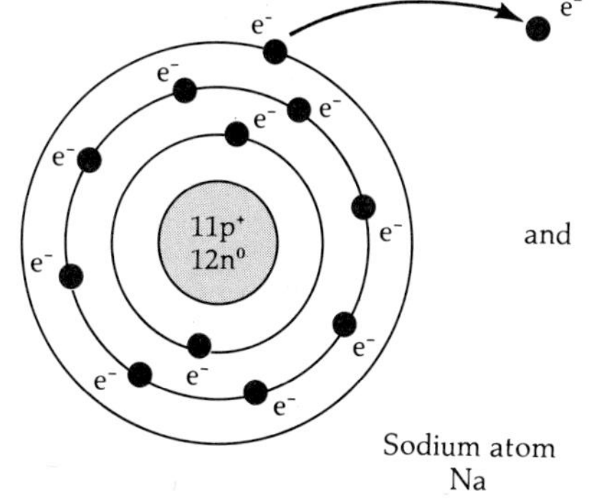

and

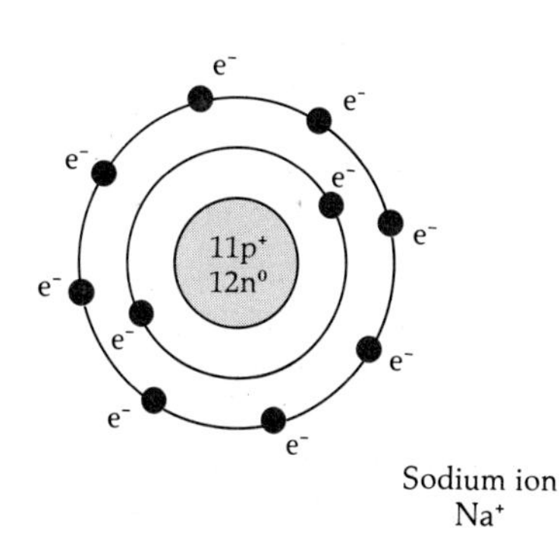

FIG. 2-1
Formation of a sodium ion from a sodium atom. The atom contains 12 neutrons (n^0), 11 protons (+), and 11 electrons (e^-). Upon loss of one electron, sodium has one more proton than it has electrons. It then has a positive charge and is called a sodium ion (Na^+).

TABLE 2-3
Units of measure often used in physiology

1. *Length*
   - Meter (m)
     - 100 centimeters
       - 39.37 in.
       - 3.281 ft.
   - *Centi*meter (cm)
     - one-hundredth (0.01) m
       - 0.3937 in.
       - 2.54 cm = 1 in.
   - *Milli*meter (mm)
     - one-thousandth (0.001) m
       - one-tenth (0.1) cm
       - 0.0394 inch ($^1/_{25.4}$ in.)
   - *Micro*meter (μ, μm)
     - (micron)
     - one-millionth (0.000001) m
       - one-thousandth (0.001) mm
       - $^1/_{25,000}$ in.
   - *nano*meter (nm)
     - one-billionth (0.000000001) m
       - one-thousandth (0.001) μm
   - Angstrom (Å)
     - one ten-billionth (10^{-10}) m
       - one hundred-millionth (10^{-8}) cm
       - one ten-thousandth (10^{-4}) μm
       - approx. $^1/_{250,000,000}$ in.
2. *Volume*
   - Liter (L)
     - 1000 mL
       - 1.057 qt
       - 33.824 oz
   - *Milli*liter (mL)
     - one-thousandth (0.001) L
       - 0.033824 oz
   - *Micro*liter (μL)
     - one-millionth (0.000001) L
       - one-thousandth (0.001) mL
   - *Pico*liter (pL)
     - one-trillionth (10^{-12}) L
       - one-millionth (10^{-6}) μL
   - *Femto*liter (fL)
     - one-quadrillionth (10^{-15}) L
3. *Weight* (mass)
   - Gram (g)
     - $^1/_{453.7}$ lb
     - 0.03527 oz
   - *Kilo*gram (kg)
     - 1000 g
       - 2.2046 lb
       - 35.27 oz
   - *Milli*gram (mg)
     - one-thousandth (0.001) g
   - *Micro*gram (μg)
     - one-millionth (0.000001) g
       - one-thousandth (0.001) mg
     - $^1/_{28,000,000}$ oz
   - *Nano*gram (ng)
     - one-billionth (10^{-9}) g
       - one-thousandth (0.001) μg
4. *Temperature and heat*
   - *Degree Fahrenheit* (°F)
     - 32°F = freezing point of water
     - 212°F = boiling point of water
     - $^5/_9 \times$ (°F − 32) = °C
   - *Degree Celsius* (°C)
     - 0°C = freezing point of water
     - 100°C = boiling point of water
     - 1.8 × °C + 32 = °F
   - *Kilocalorie (kcal)*
     - 1000 calories (the heat required to raise 1 kilogram of water 1°C)
5. *Concentration*
   - *Molar (M)*
     - A solution in which the number of grams expressed in the molecular weight of a substance is dissolved in sufficient solvent such as water to make 1 L in total volume.
   - *Equivalent Weight (Eq)*
     - Equivalent weight of a substance

$$= \frac{\text{Atomic or molecular weight in grams}}{\text{number of replaceable hydrogen ions } (H^+) \text{ or hydroxide ions } (OH^-) \text{ per formula unit}}$$

$$= \frac{\text{Atomic or molecular weight in grams}}{\text{number of electrons gained or lost by a substance in a reaction}}$$

   - *Milliequivalent (mEq)*
     - one-thousandth Eq (0.001 Eq)
   - *Acidity-Alkalinity (pH)*
     - $-\log_{10}[H^+] = \log (1/[H^+])$
     - The pH is the reciprocal of the hydrogen ion concentration expressed as an exponent of 10.
     - pH 7.0 = 10^{-7} *M* in H ions = $^1/_{10^7}$ *M* in H ions
     - pH 7.0 = neutral in chemistry (human blood's normal pH is about 7.4)
     - pH less than 7.0 = acid
     - pH greater than 7.0 = alkaline (basic)

form a union or *bond* to produce a molecule. A *molecule* is the smallest possible quantity of a substance that is composed of two or more similar or dissimilar atoms. The atoms exist independently but the properties are those of the molecule of which they form a part. Molecules may contain a variety of bonds, which can be nonpolar or polar, or covalent. Another type of bond is formed by hydrogen atoms. The bonds are used to form simple or complex molecules. For example, simple molecules consist of two atoms of the same element bonded to one another. Hydrogen gas (H_2) is a simple molecule. Furthermore, each hydrogen atom is bonded to

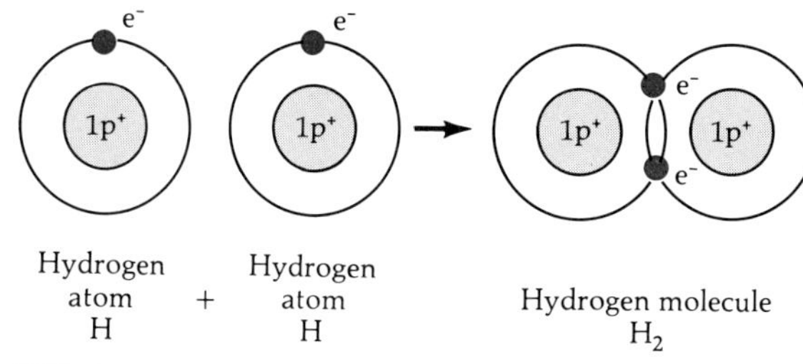

FIG. 2-2
Formation of a hydrogen molecule by means of electron sharing (nonpolar covalent bond) between two hydrogen atoms.

the other by sharing a pair of electrons. The distribution of electrons is equal between the atoms, resulting in a *nonpolar covalent bond* (Fig. 2-2). Carbon atoms also form nonpolar covalent bonds; they do so with each other as well as with hydrogen or other atoms.

The union of two or more dissimilar atoms produces a complex molecule, i.e., a molecule of a *compound*. Compounds normally have physical and chemical properties unlike the atoms of which they are composed. An example of a small molecule is that of water (H_2O), which consists of two atoms of hydrogen both bonded to one atom of oxygen. These bonds are *polar covalent bonds*. The electrons have a greater attraction to the oxygen atom; this renders the oxygen slightly more negative and the hydrogen less negative (more positive), as shown in Fig. 2-3.

The weak charges developed around the hydrogen atom as a result of polar covalent bonds can attract oppositely charged particles. The electrostatic attraction forms *hydrogen bonds*. Although they are weak individually, hydrogen bonds cumulatively can produce tremendous forces (much like the thousands of threads in the seams of a pair of jeans). They can link chemical building blocks and chains of molecules together. Many types of large molecules are built in this way. They include structural proteins, enzymes, antibodies, hemoglobin, and DNA. Their role in body function will be discussed in several chapters in this text.

Sodium chloride is a compound formed by the union of a sodium atom and a chlorine atom. In this compound, sodium donates an electron to the chlorine atom to form an *ionic* or *electrovalent bond*.

The loss of the electron in the outer orbit of sodium leaves it with an excess of protons. It becomes a *positively charged ion* and is said to be *oxidized*. The chlorine atom gains an electron and acquires a negative charge. The ion thereby formed is *less positive* than was the neutral atom and is said to be *reduced*. The oppositely charged ions are attracted to each other and may be considered to constitute a molecule of sodium chloride (Fig. 2-4).

FIG. 2-3
Formation of a polar covalent bond by the union of two hydrogen atoms with one oxygen atom to form water. The electrons are shared unequally between hydrogen and oxygen, the latter being favored. This results in association of a slight positive charge with each hydrogen and a slight negative charge with the oxygen atom.

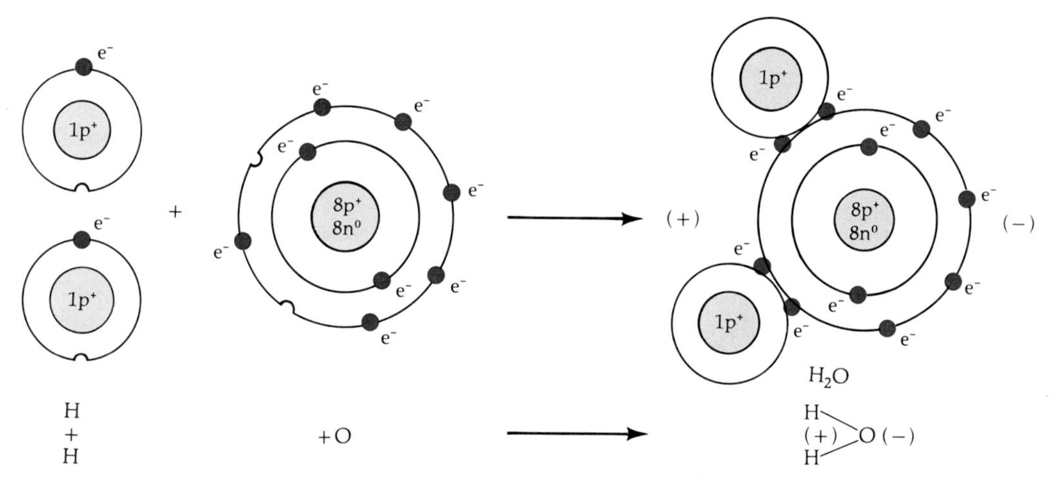

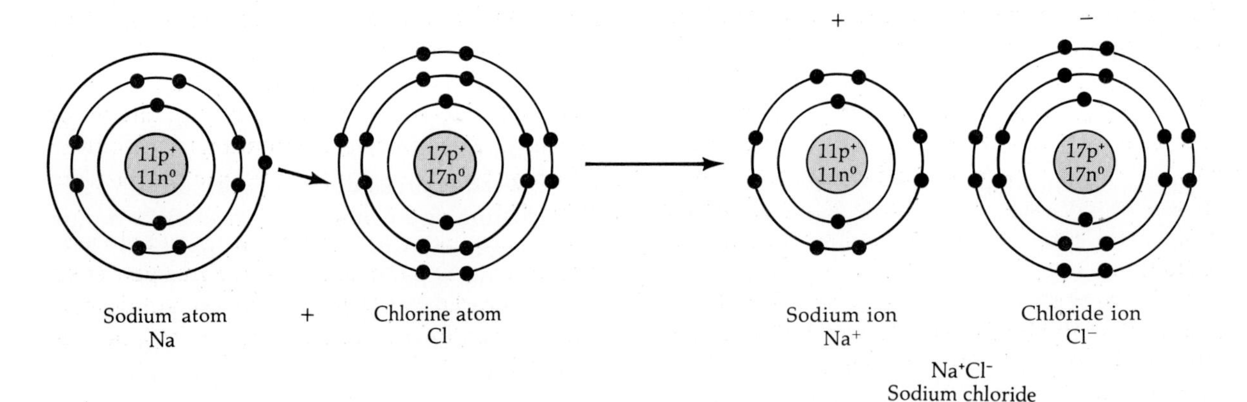

FIG. 2-4
Formation of an ionic or electrovalent bond. The sodium atom donates the sole electron in its outer orbit to that of the chlorine atom. Consequently, sodium becomes positively charged and chlorine negatively charged. The attraction of the two oppositely charged ions produces sodium chloride.

Measurement of Quantities of Chemicals: Concentration, Equivalent Weight, and pH

The *molecular weight* of any substance is approximately the number of times heavier the molecule is than the weight of an atom of hydrogen, the assigned weight of which is close to unity on an arbitrary scale. For example, one molecule of glucose is 180 times as heavy as one hydrogen atom. The molecular weight of glucose is 180.

A *mole* of a substance is the number of grams corresponding to its molecular weight (MW) in grams. A mole of glucose is 180 g of it. The concentration of chemicals dissolved in fluids may be described in *molar units*. A molar solution (1 *M*) of a substance is made by dissolving one gram-molecular-weight of the substance (1 mol) in water to make a total volume of one liter (1 L). A 1 *M* glucose solution is made by dissolving 180 g of glucose in enough water to make the final volume equal to 1 L.

Many substances in the body are dissolved in water. They constitute *solutes* in the dissolving fluid, which is called a *solvent*. Many solutes consist of ions, which are particles possessing electrical charges. The number of charges corresponds to the number of electrons gained or lost by the neutral atom in forming the ion, i.e., to the atom's *valence*. The valence may be positive or negative and in some elements it is variable. For example, the chloride ion (Cl^-) has a charge indicating a valence −1, while that for sulfate ion (SO_4^{2-}) is −2. Potassium ions (K^+) have a valence of +1 while that for calcium ions (Ca^{2+}) is +2.

In order to compare electrical activity of substances, one must determine their *equivalent* weight (Eq). An equivalent weight of an ion is equal to its atomic or molecular weight in grams divided by its charge, i.e., by its valence. A milliequivalent (mEq) equals 0.001 × Eq. A milliequivalent of any substance has the same number of charges as a milliequivalent of any other substance (see Table 2-3, No. 5)..

$$\begin{aligned}
1 \text{ Eq } K^+ &= \text{MW/valence} = 39/1 = 39 \text{ g} \\
1 \text{ mEq } K^+ &= 0.001 \times \text{Eq} = 0.001 \times 39 \\
&= 0.039 \text{ g} = 39 \text{ mg} \\
1 \text{ Eq } Ca^{2+} &= \text{MW/valence} = 40/2 = 20 \text{ g} \\
1 \text{ mEq } Ca^{2+} &= 0.001 \times \text{Eq} = 0.001 \times 20 \\
&= 0.02 \text{ g} = 20 \text{ mg}
\end{aligned}$$

Acidity is a measure of a substance's ability to donate hydrogen ions (H^+, called protons) and *alkalinity* of its ability to accept them. Acidity and alkalinity are reflected in the pH of a solution, pH being the reciprocal of the hydrogen ion concentration expressed as an exponent of 10. The pH scale ranges from 0 to 14. A pH of 7.0 is neutral, corresponding to a molar hydrogen ion concentration of 10^{-7} (0.0000001 *M* or 10^{-7} *M*). As the pH goes below 7.0, a solution becomes more acidic and its hydrogen ion concentration increases. For example, at a pH of 6.0 the hydrogen ion concentration is 10^{-6} *M* (0.000001 *M*) and is 10 times higher than at pH 7.0. As the pH goes above 7.0, the hydrogen ion concentration is lower (e.g., pH 8 = 10^{-8} *M*) and the solution is more basic (alkaline). Thus, an *acid* is defined as a substance that can donate protons (H^+) and a *base* as a substance that accepts them.

Blood, for example, normally has a slightly alkaline pH, which ranges between 7.35 and 7.45 ($10^{-7.35}$ to $10^{-7.45}$ *M*). When disease causes it to drop to 7.0, the person suffers from *acidosis*. At this point the hydrogen ion concentration, while neutral, is lower and more acidic (less basic) than normal. The hydrogen ion level is 10^{-7} *M*, or 10^{-7} Eq/L or 10^{-4} mEq/L. Conversely, diseases can also induce elevations above pH 7.45, causing *alkalosis*.

Types of Complex Molecules (Compounds)

Inorganic Compounds

Inorganic compounds consist of molecules without carbon-hydrogen bonds. They were once thought to be exclusively of mineral origin, and it was assumed they were not found in living organisms. However, modern science has proved otherwise. Carbon dioxide (CO_2), water (H_2O), sodium chloride (NaCl), and potassium chloride (KCl) are examples of such compounds in the human body, and many others exist which are extremely important to its function and will be described later.

Organic Compounds and Their Building Blocks

Organic compounds are those which contain carbon-hydrogen bonds and include *hydrocarbons* and their derivatives. Other elements, such as oxygen or sulfur, may be present. Because it was once presumed that such compounds originated only from plants or animals (i.e., from living organisms), they were named *organic* compounds. However, a number of such molecules have been produced in laboratories and they are found as products apart from living organisms.

Carbon atoms can form four single *covalent* bonds with four other atoms or their equivalent in multiple bonds (double or triple covalent bonds). They have a marked ability to form groups bonded to each other in long chains or in rings. Therefore, a tremendous number of different types of organic compounds may be produced.

The simplest hydrocarbon found in nature is methane. In this molecule, each hydrogen atom shares an electron equally with a single carbon atom in four covalent bonds. It has the formula CH_4:

```
     H
     |
  H—C—H
     |
     H
```

Methane

Individual chemical bonds may be indicated by small, straight lines between adjacent atoms. In the case of carbon, each represents a covalent union with other atoms. Methane is a gas found in swamps and bogs as a product of the action of microorganisms on organic matter. It is also produced by the breakdown of organic compounds by intestinal bacteria.

Carbon atoms may be linked to one another to form chains or rings. Glucose is an example of a 6-carbon compound found as a chain or ring structure in foods as well as in tissues. It has the empirical (symbolic) formula $C_6H_{12}O_6$. It contains 6 carbon atoms (C), 12 hydrogen atoms (H), and 6 oxygen atoms (O). The six carbon atoms form a chain to which the hydrogen and oxygen atoms are attached. In different

```
      H
      |
      C=O
      |
   H—C—OH
      |
  HO—C—H
      |
   H—C—OH
      |
   H—C—OH
      |
   H—C—OH
      |
      H
```

***d*-Glucose**

```
      H
      |
      C=O                   ┌──────────────┐
      |               1  H—C—OH          |
   H—C—OH                  |              |
      |               2  H—C—OH          |         ⁶CH₂OH
  HO—C—H                   |              |        H /5 ——— O  H
      |        =      3 HO—C—H            O  =      4 H        1
   H—C—OH                  |              |         OH      H
      |               4  H—C—OH          |      HO  3       2  OH
   H—C—OH                  |              |          H        OH
      |               5  H—C—————————————┘
   H—C—OH                  |
      |               6    CH₂OH
      H                  Glucose
```

forms of sugar molecules (*d* or *l*, D or L, + or −),[2] the H and OH groups may be on sides opposite those shown on page 19.

Illustrations of organic molecules are often depicted in linear, two-dimensional fashion. However, the bonds formed between atoms within a molecule are found in different planes so that the molecules actually have three-dimensional structures.

As mentioned above, the carbon atoms of organic compounds may also form rings. For instance, in glucose they may form a *cyclic* (ring) structure, as is illustrated above. The carbon atoms of the ring are found at the angles between any two lines in the diagram.

CARBOHYDRATES Carbohydrates are composed of building blocks of simple sugars. They generally are made up of carbon (C), hydrogen (H), and oxygen (O) in the ratio of $C_x(H_2O)_x$. There are different numbers of carbon atoms in various carbohydrates, which may be indicated in a general formula by the letter x and are linked to x atoms of oxygen and 2x atoms of hydrogen.

Many sugars contain six carbon atoms, with a ring structure similar to that shown for glucose above. These simple sugars are called *monosaccharides.* Double units (disaccharides), such as sucrose, which is table sugar, and triple units (trisaccharides) are formed by linking monosaccharides together. Ultimately, large macromolecules, called *polymers,* may be formed. In the case of carbohydrates, the polymers are called *polysaccharides.*

Glycogen (animal starch) is a common carbohydrate polymer found in high concentrations in muscle and liver, as well as in other tissue types. This polysaccharide is a polymer of glucose molecules linked so as to produce branched chains (Fig. 2-5).

This polymer exists in a variety of sizes. They include structures that contain as few as 1500 up to hundreds of thousands of glucose molecules. Their molecular weights range from 270,000 to 100,000,000.

Plant *starch* is another common polysaccharide formed by the linkage of glucose molecules. It is found in two forms, an unbranched straight-chain polysaccharide and a branched-chain polysaccharide. Its molecular weight ranges from 50,000 to 1,000,000 or more, and it is formed from hundreds to thousands of glucose molecules. Starch is commonly found in potatoes, rice, and a variety of plant products.

Another polysaccharide is *cellulose.* It is the most abundant organic compound in the world. It accounts for more than 50 percent of the carbon atoms found in plants. Its individual glucose building blocks are linked in a different manner than those in glycogen and plant starch. Cellulose is a polymer which is unbranched (straight) and ranges in molecular weight from 50,000 to 400,000 or more. Various forms of cellulose contain 300 to 2500 glucose units.

[2] The prefixes *d-* and *l-* (for *dextro-* and *levo-*, i.e., right and left) refer to the configurations of the molecules, whereas D and L and + and − refer to the direction in which they rotate the plane of polarized light.

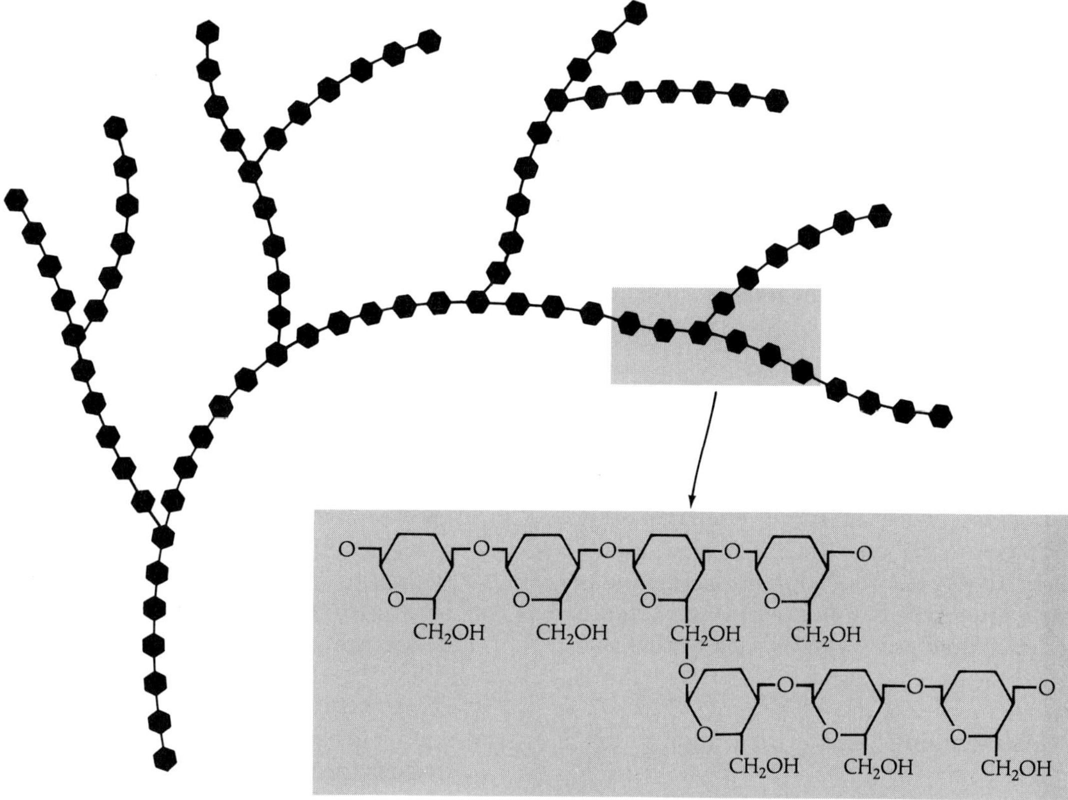

FIG. 2-5
The polysaccharide glycogen consists of glucose molecules linked to one another in chains which branch. Each glucose molecule is indicated by a hexagon in the treelike model. The structure of a small segment of that model is indicated on the lower right.

The manner in which the sugar units are linked is critical. The body contains molecules in the form of enzymes (organic catalysts[3]) which are capable of cleaving the linkages in glycogen and starch but incapable of breaking those found within cellulose. Consequently, glycogen and starch are digestible foods in humans, while cellulose is not. As a result, cellulose remains a large molecule within the intestine. It provides residue and bulk long after other molecules have been broken into their building blocks. Unlike many other molecules, it does not enter the bloodstream nor provide building blocks for the body's energy needs.

[3] A catalyst is a substance which alters the speed of a chemical reaction but is not used up in the reaction.

PROTEINS Proteins consist of polymers formed from building blocks called *amino acids.* Amino acids are organic compounds with an amino group (NH_2) bonded to a carbon atom, which is bonded to a carboxyl group (COOH), as shown below in the formula for glycine. The amino group is basic (alkaline). It can combine with (accept) positively charged particles, such as protons. A positively charged particle which is abundant in the human body is the hydrogen ion (H^+). When a H^+ combines with an amino group ($-NH_2$), it produces ammonia (NH_3^+). The carboxyl group within amino acids may

$$-\overset{\overset{\displaystyle O}{\|}}{C}-OH$$

Carboxyl group

donate hydrogen to other groups and thus function as an *acid.*

Depending on the conditions, amino acids may either accept or donate hydrogen ions (H^+). They may neutralize excess base or acid found in the body fluids. Thus, they are also referred to as *buffers.* Buffers are substances which can neutralize excess acid or base. Certain inorganic compounds also buffer body fluids (see Chaps. 20 and 21). The maintenance of specific levels of acidity or alkalinity of the body fluids by buffers is necessary to sustain the optimal activity of vital functions. For example, enzymes found in various locations work best within a specific pH range. Thus the function of the entire chemistry of the body is dependent on the maintenance of an optimal pH of the fluids bathing the cells. Diseases which cause acidosis or alkalosis can severely disrupt body chemistry and if severe and unreversed, cause death.

Variations among Proteins The length of the carbon chains in amino acids may vary. So, too, they may or may not contain side chains or ring structures. The simplest amino acid, which has a 2-carbon chain, is known as *glycine.*

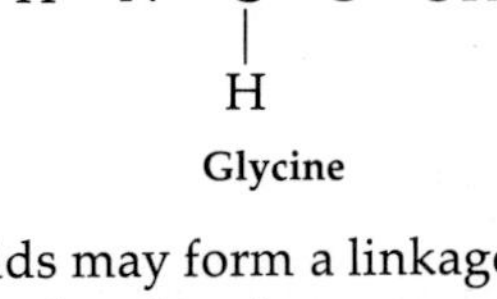

```
    H  H  O
    |  |  ‖
 H—N—C—C—OH
       |
       H
```

Glycine

Amino acids may form a linkage between adjacent carboxyl and amino groups, which is referred to as a *peptide bond.* The reaction below indicates an example of peptide bond formation between two single units of glycine. They combine to form the dipeptide glycylglycine.

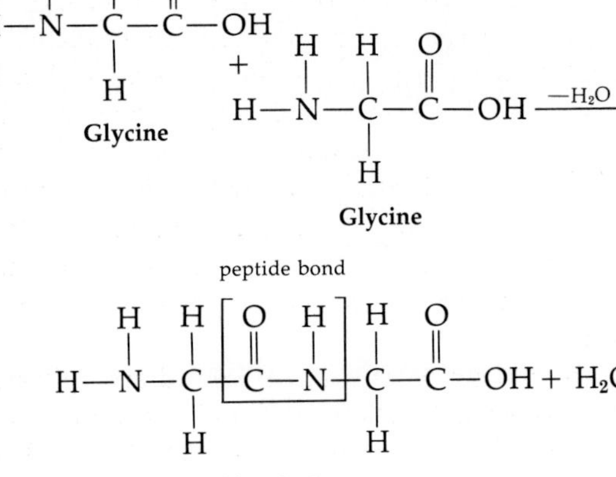

```
    H  H  O
    |  |  ‖
 H—N—C—C—OH        H  H  O
       |     +      |  |  ‖
       H         H—N—C—C—OH  —H2O→
                       |
                       H
```

Glycine **Glycine**

peptide bond

```
    H  H [O  H] H  O
    |  |  ‖  |  |  ‖
 H—N—C—[C—N]—C—C—OH + H2O
       |        |
       H        H
```

Glycyl-glycine

The linkage of many amino acids in polymers forms compounds called *polypeptides.* Proteins are composed of one or more very large polypeptides. The primary structures of proteins differ in the number and sequence of amino acids of which they are composed. These differences directly determine the functions of specific proteins. A variety of proteins with molecular weights which range from about 10,000 to 10,000,000 are found in the body.

In addition to carbon, hydrogen, oxygen, and nitrogen, proteins may contain other elements, such as sulfur, iodine, iron, or various other minerals. These elements are often essential to their function. For example, iron plays a critical role in the center of the protein chains of hemoglobin. It is essential for the transport of oxygen. Sulfur plays a very important role in the structure of many proteins. The sulfur atom often bridges and binds two or more separate chains of proteins to form multichain molecules. It does so in antibodies, which are molecules useful in defense mechanisms (Fig. 19-4).

As a group, proteins are extremely important in body function. Proteins perform several important functions as part of the limiting membranes of individual cells (see Chap. 4) and furthermore provide the architectural framework around which cells are built. Much of the body's chemistry is regulated by protein enzymes. Furthermore, it is greatly influenced by hormones, many of which are proteins or derived from them.

Enzymes are protein catalysts which alter the speed of chemical reactions but are not used up in the reaction. Although the rates vary, some enzymes repeat their function as often as 100,000 times per second or more. Each enzyme acts on a specific substance called its *substrate.* The enzyme may combine with and act upon one or more specific substrates, converting the latter into a new *product.* Various enzymes act in different ways. Some combine substances into products. For example, the two substances A and B may be converted into a product C, as shown in Fig. 2-6. Furthermore, in more complex reactions two substances W and X may be converted to products Y and Z by enzyme activity. An enzyme used to diagnose the severity of

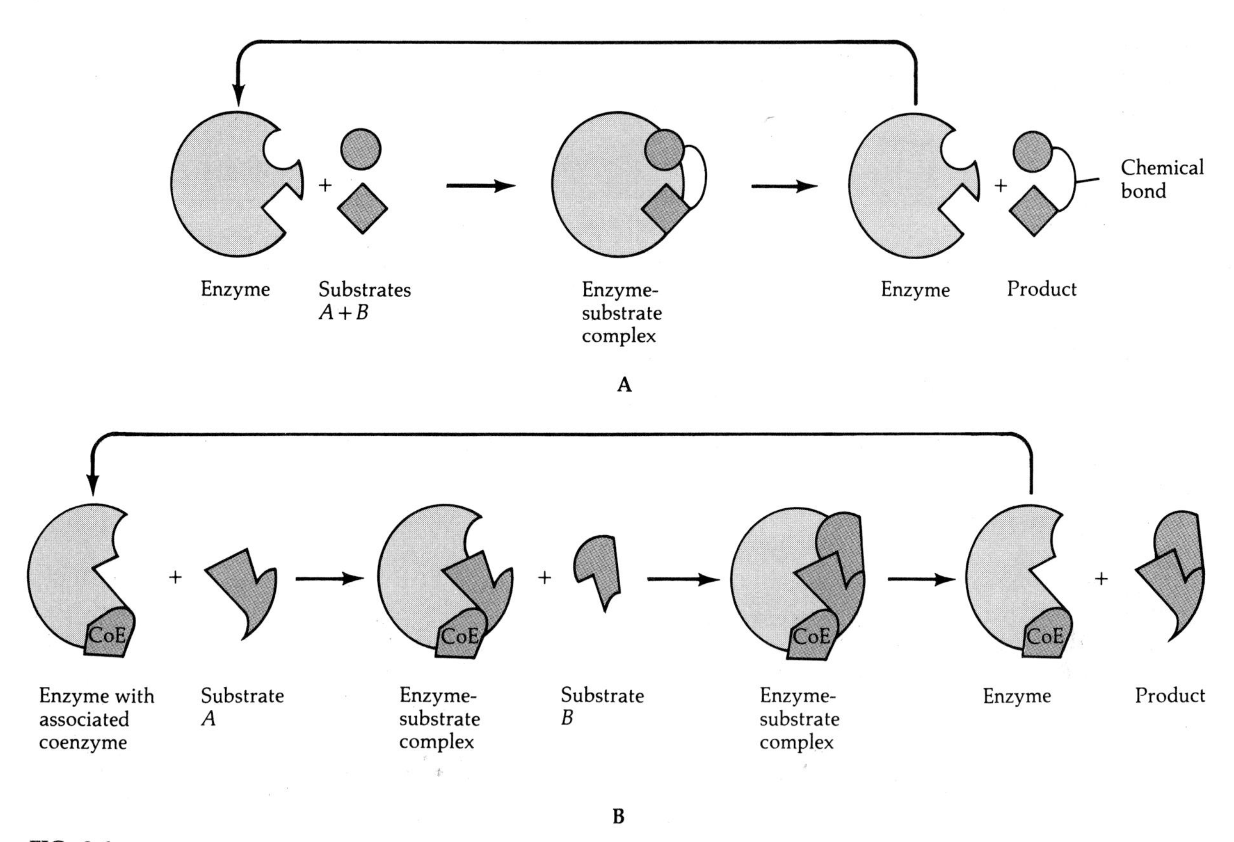

FIG. 2-6
Mechanisms of enzyme activity. (A) Lock-and-key model. (B) Participation of a coenzyme.

a heart attack, called aspartate amino transferase (AST), carries out this type of reaction:

$$\text{W} + \text{X} \underset{\text{action}}{\overset{\text{enzyme}}{\longleftrightarrow}} \text{Y} + \text{Z}$$

$$\begin{matrix}\text{Aspartic} \\ \text{acid}\end{matrix} + \begin{matrix}\alpha\text{-Ketoglutaric} \\ \text{acid}\end{matrix} \underset{\text{action}}{\overset{\text{AST}}{\longleftrightarrow}} \begin{matrix}\text{Oxalacetic} \\ \text{acid}\end{matrix} + \begin{matrix}\text{Glutamic} \\ \text{acid}\end{matrix}$$

In a reaction termed *hydrolysis,* some enzymes cleave groups from a substrate and by the addition of components of water produce a new compound(s) or product(s). For example, amylase (found in saliva and pancreatic fluid) cleaves sugar molecules from glycogen in successive steps to form smaller polysaccharides and ultimately converts all the glycogen to glucose:

$$\begin{matrix}\text{Compound A} \\ \text{(substrate)}\end{matrix} \underset{\text{action}}{\overset{\text{enzyme}}{\rightleftharpoons}} \begin{matrix}\text{compound} \\ \text{B}\end{matrix} + \begin{matrix}\text{compound} \\ \text{C}\end{matrix}$$

$$\begin{matrix}\text{Glycogen} \\ \text{(a poly-} \\ \text{saccharide)}\end{matrix} \overset{\text{amylase}}{\rightleftharpoons} \begin{matrix}\text{smaller} \\ \text{polysaccharide}\end{matrix} + \text{glucose}$$

The three-dimensional structure of an enzyme contains sequences of amino acids which possess specific shapes that are complementary to parts of the specific substrate. Those parts of the enzyme to which the substrate binds are called *active sites.* Some of them alter the substrate and are catalytic in function. They speed up the chemical reaction. The combination of an enzyme with a substrate to form a product is illustrated in Fig. 2-6A. The enzyme and substrate fit like a lock and key. However, the shape of the lock is somewhat flexible. It may be modified slightly in order to combine better with the substrate. In some instances another molecule, called a *coenzyme,* assists the enzyme and is required for its activity, as shown in Fig. 2-6B.

An enzyme molecule, like all others, vibrates at temperatures above $-273°C$. Its vibrational activity provides energy, which is used to speed up chemical reactions. An analogy of the phenomenon may be made by considering boulders which could roll down a hill with greater speed if they were given a push at locations where a slight rise occurred. In this model, the boulders are substrates. They are constantly but slowly moving and periodically colliding and combining with one another in the body fluids. (They are rolling down the hillside.) The push over the periodic rises is provided by the vibrational activity of the enzymes, which in effect tunnel out the rises and increase the rate of movement and frequency of collision of the substrates (boulders). The collision of the substrates allows chemical bonds to form, creating a new product.

Enzyme-catalyzed reactions are influenced by a number of factors. Usually an enzyme can "push along" more substrates than are available. Therefore the rate of an enzyme-catalyzed reaction is often dependent on the concentration of substrates. Most enzyme-catalyzed reactions are reversible, i.e., the products can be converted back into the substrates. In fact, the reactions usually proceed in both directions simultaneously. The net effect depends on the available concentrations of substrates and products. Hundreds of enzymatic reactions occur in the body. Many of them are part of a long, stepwise sequence, each catalyzed by a different enzyme and leading to the synthesis or degradation of specific compounds. Unless an outside source of energy is provided, the reactions move in a favored direction (downhill). Enzymes work best at an optimal pH, which varies from one enzyme to another and usually is in the range of 5.0 to 9.0. These catalysts also usually work optimally at body temperature (37°C). Within limits, the rate of their activity about doubles for each 10°C rise in temperature (from about 5° to 37°C).

Enzymes are found in specific locations or compartments in cells. Often they are associated with membranes. Some are found in the fluid within or around the cell. Enzymes may be turned on (activated) by some chemicals (e.g., hormones, or ions such as calcium ions) and poisoned or inhibited by others (some drugs). The role of these fascinating catalytic proteins in the regulation of body chemistry and energy metabolism will be discussed further in chapters throughout the text, and especially in those on cell transport mechanisms (Chap. 4), hormones (Chap. 5), energy metabolism in muscle (Chap. 13), and the digestion of foods (Chaps. 22 and 23).

LIPIDS *Lipids* are molecules which are soluble in organic solvents such as ether, chloroform, and benzene. They may be classified as *simple lipids,* which include neutral fats and waxes; *compound lipids,* which are molecules containing lipids and other groups (phospholipids, glycolipids, and lipoproteins are compound lipids); and *derived lipids,* which originate from other lipids (cholesterol, steroid hormones, bile acids, and some vitamins are derived lipids). Some examples of each of the three major types of lipids will be described below.

Neutral fats or *triglycerides* are simple lipids which possess a 3-carbon chain derived from glycerol.

```
      H
      |
  H—C—OH
      |
  H—C—OH
      |
  H—C—OH
      |
      H
```

Glycerol

Three groups of individual fatty acids may be attached to the chain to form *triglycerides.* Fatty acids consist of carbon chains which differ in length. The end of each has a carboxyl group (COOH). When the carbon atoms in the chain are bonded to one another as well as to a pair of hydrogen atoms or other such groups in single bonds, the fatty acid is said to be *saturated* or *hydrogenated.* The addition of hydrogen raises

```
         H  H  H
         |  |  |
  HOOC—C—C—C—R
         |  |  |
         H  H  H
```

Saturated fatty acid

(R = remainder of carbon chain)

the melting point of fats, converting them from liquid fats (*oils*) to solid fats.

In some instances, adjacent carbon atoms are linked by one or more double bonds. They make the fatty acids *unsaturated* (*dehydrogenated*) and lower their melting points. Therefore, as fats become unsaturated or dehydrogenated, they are more likely to be found as oils (liquid fat) at body temperature.

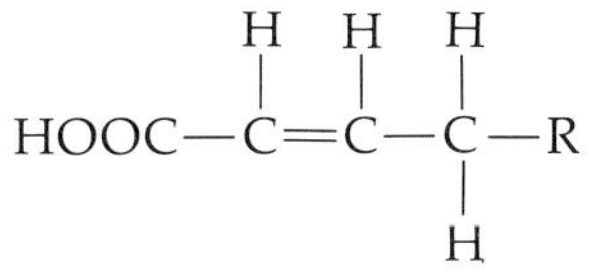

$$
\begin{array}{ccccccccc}
 & & \mathrm{H} & & \mathrm{H} & & \mathrm{H} & & \\
 & & | & & | & & | & & \\
\mathrm{HOOC} & - & \mathrm{C} & = & \mathrm{C} & - & \mathrm{C} & - & \mathrm{R} \\
 & & & & & & | & & \\
 & & & & & & \mathrm{H} & &
\end{array}
$$

Unsaturated fatty acid

A general formula for triglycerides is shown below. R_1, R_2, and R_3 symbolize fatty acids, which may be similar or different, saturated or unsaturated.

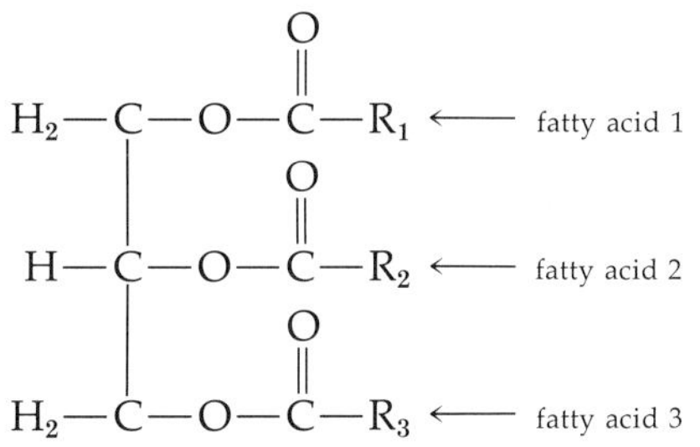

$$
\begin{array}{lcccccccl}
 & & & & & & \mathrm{O} & & \\
 & & & & & & \| & & \\
\mathrm{H_2} & - & \mathrm{C} & - & \mathrm{O} & - & \mathrm{C} & - & \mathrm{R_1} \longleftarrow \text{fatty acid 1} \\
 & & | & & & & \mathrm{O} & & \\
 & & | & & & & \| & & \\
\mathrm{H} & - & \mathrm{C} & - & \mathrm{O} & - & \mathrm{C} & - & \mathrm{R_2} \longleftarrow \text{fatty acid 2} \\
 & & | & & & & \mathrm{O} & & \\
 & & | & & & & \| & & \\
\mathrm{H_2} & - & \mathrm{C} & - & \mathrm{O} & - & \mathrm{C} & - & \mathrm{R_3} \longleftarrow \text{fatty acid 3}
\end{array}
$$

Triglyceride

A relationship has been demonstrated between dietary intake of fats containing saturated fatty acids and risk of heart attack. It will be discussed more fully in Chap. 16 on the heart and its function.

Phospholipids are compound lipids with two fatty acid molecules plus a third group linked to the 3-carbon glycerol chain.

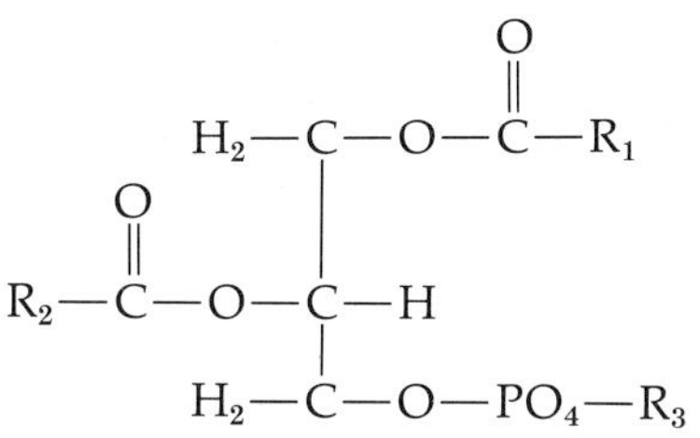

$$
\begin{array}{ccccccccccc}
 & & & & & & & & \mathrm{O} & & \\
 & & & & & & & & \| & & \\
 & & & & \mathrm{H_2} & - & \mathrm{C} & - \mathrm{O} - & \mathrm{C} & - & \mathrm{R_1} \\
 & & \mathrm{O} & & & & | & & & & \\
 & & \| & & & & | & & & & \\
\mathrm{R_2} & - & \mathrm{C} & - & \mathrm{O} & - & \mathrm{C} & - \mathrm{H} & & & \\
 & & & & & & | & & & & \\
 & & & & \mathrm{H_2} & - & \mathrm{C} & - \mathrm{O} - & \mathrm{PO_4} & - & \mathrm{R_3}
\end{array}
$$

General formula for phospholipids

The third group (R_3) is phosphate or phosphate plus a variety of other groups of differing composition. Phospholipids are abundant in cell membranes. They play major roles in determining the freedom with which molecules pass into or out of cells. This will be the subject of discussion in Chap. 4 on transport mechanisms across cell membranes.

Glycolipids are compound lipids composed of fatty acids, other lipid-derived molecules, and a sugar, as indicated below:

$$
\begin{array}{lcl}
\mathrm{L} & - & \mathrm{R} \\
| & & \\
\mathrm{S} & &
\end{array}
$$

R = fatty acid
L = lipid-derived molecule
S = sugar

Glycolipids are major components of all cell membranes but may differ in chemistry in various cells. They affect movement of substances into and out of cells. Some forms of glycolipids, such as cerebrosides and gangliosides, are abundant in the brain and play an important role in the function of the nervous system.

Lipoproteins are compound lipids which contain proteins. They are important constituents of cells and the fluid part of the blood. They may be used as an index of the risk of a heart attack (Chap. 16).

Steroids are derived lipids. They are formed in the body by repeated combination of two carbon fragments of fatty acids or other compounds, which produce four rings of carbon atoms. Three are six-membered rings and the fourth is a five-membered one. The fundamental cyclic structure is called a steroid ring. When various groups are attached to the ring, different properties result. Cells in the body are able to make such transformations. For example, the liver, by virtue of its enzymes, is able to synthesize cholesterol and degrade it. The ring structure in cholesterol is typical of that found in other steroids. Cholesterol, whose formula is shown on the next page, constitutes nearly 25 percent of cell membranes. Excesses of it may accumulate in arteries and help clog them in a disease called *atherosclerosis*.

Steroids influence a vast number of body functions and include major male sex hor-

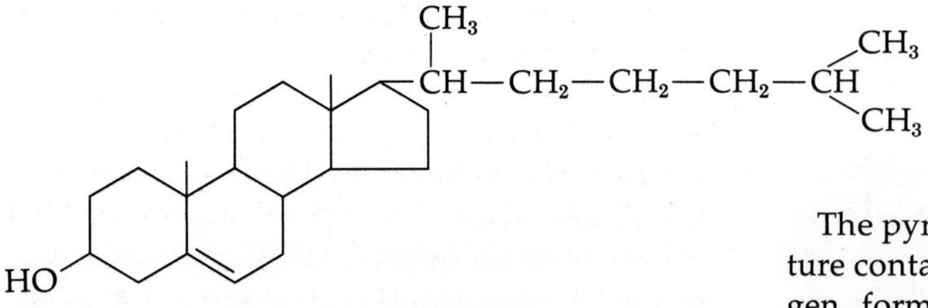

Cholesterol

mones such as testosterone, and female sex hormones such as the estrogens and progesterone. Steroids also form adrenal cortical hormones (the mineralocorticoids such as aldosterone, which influences salt metabolism, and the glucocorticoids such as cortisone, shown below, which influences protein metabolism and is used to relieve certain allergies). Vitamin D, which affects calcium metabolism and the growth of bone, also is a steroid.

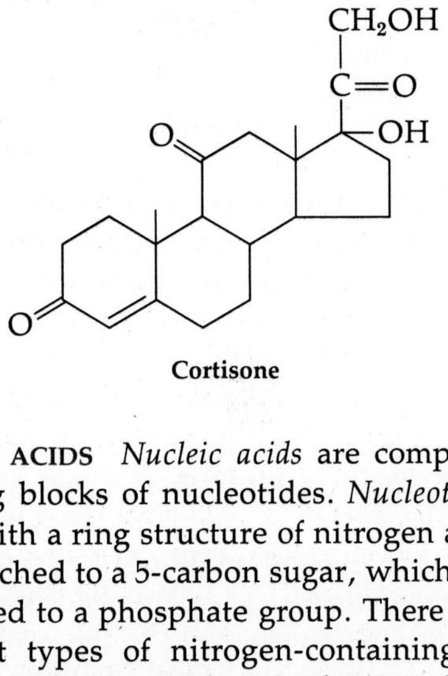

Cortisone

NUCLEIC ACIDS *Nucleic acids* are composed of building blocks of nucleotides. *Nucleotides* are made with a ring structure of nitrogen and carbon attached to a 5-carbon sugar, which in turn is bonded to a phosphate group. There are two different types of nitrogen-containing rings, namely, *pyrimidine* and *purine* bases, which are critical components of the genetic material in cells.

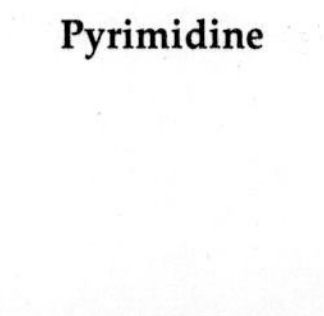

Pyrimidine

The pyrimidine ring is a six-membered structure containing two nitrogen atoms. Each nitrogen forms three covalent bonds with other atoms.

Three varieties of *pyrimidines* are *thymine, cytosine,* and *uracil.* Each contains the basic ring structure with different side groups attached to it. Thymine and cytosine are found in deoxyribonucleic acid (DNA), the material of which genes are composed. Cytosine and uracil are found in ribonucleic acid (RNA), which plays an important role in directing protein synthesis in cells.

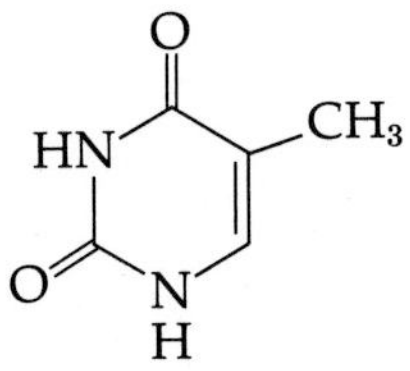

Thymine 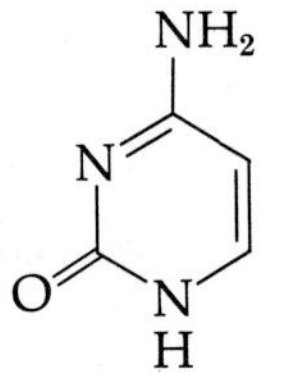 **Cytosine** 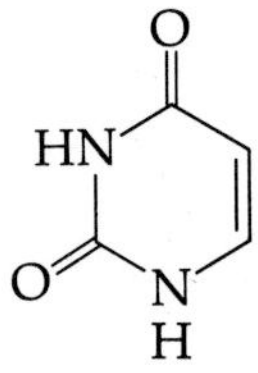 **Uracil**

The *purines* are molecules which consist of pyrimidine attached to another ring formed by two nitrogen atoms and one carbon atom. Purines such as caffeine are found in coffee, tea, cocoa, and chocolate. Theobromine is in the latter three beverages and theophylline is in tea.

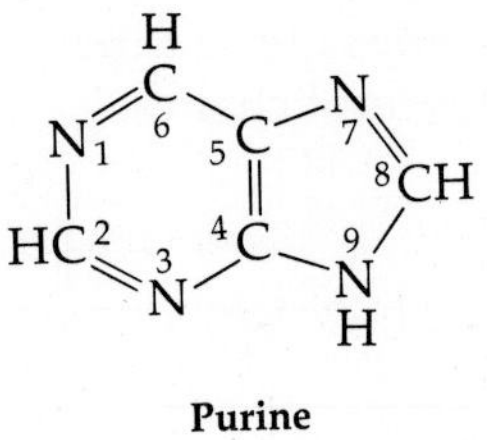

Purine

The molecules *adenine* and *guanine* are purines found in DNA and RNA. The relationship of purines to the stimulants caffeine, theobromine, and theophylline, which are found in several common beverages, is shown on page 27.

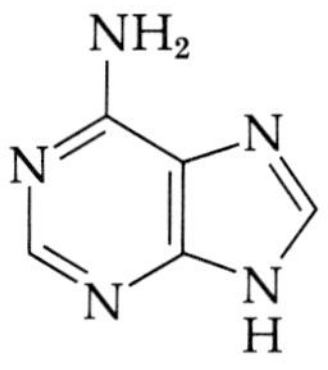

Adenine

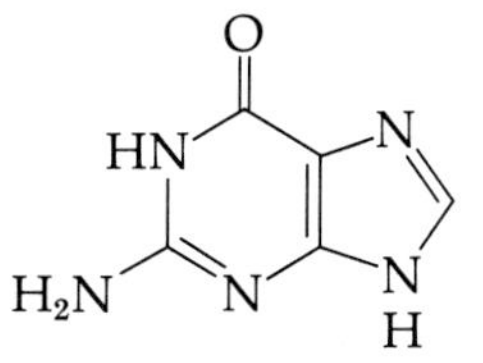

Guanine

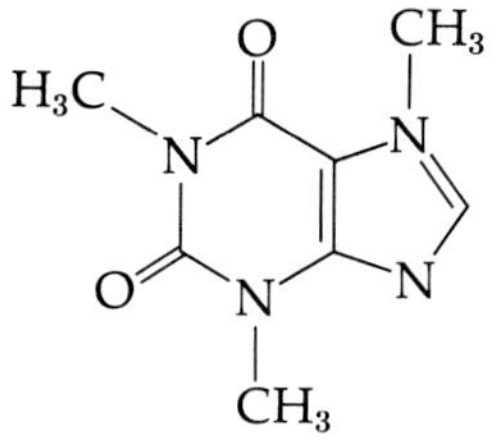

Caffeine

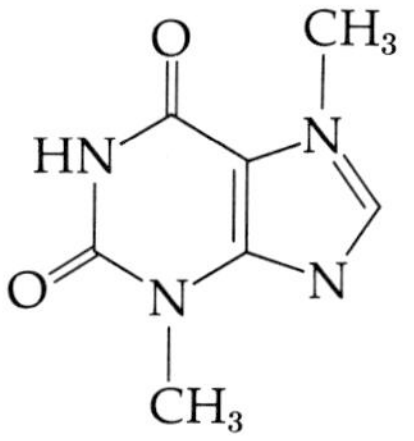

Theobromine

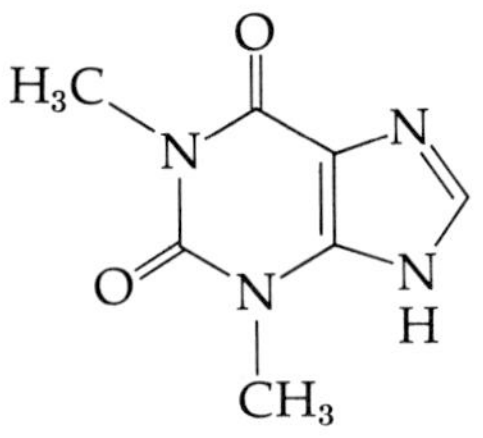

Theophylline

An example of a nucleotide is illustrated below in a molecule of *adenosine monophosphate* (AMP). AMP consists of the purine base, adenine, linked to a 5-carbon sugar, ribose, which in turn is linked to a phosphate group. When the phosphate group is missing, the molecule is

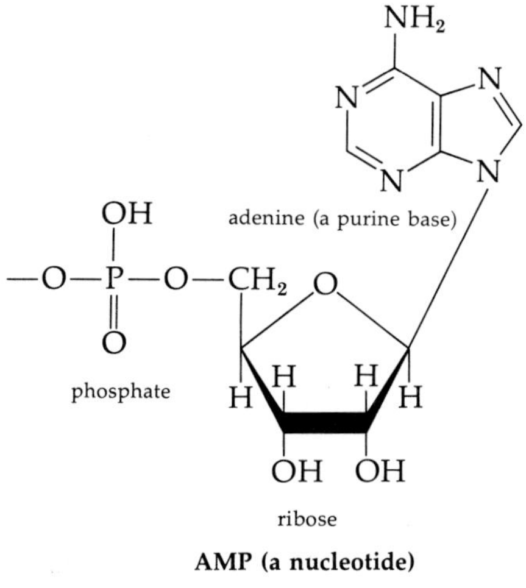

adenine (a purine base)

phosphate

ribose

AMP (a nucleotide)

called a *nucleoside*. AMP is a product of metabolism that is formed as cells run out of energy.

Polymers of nucleotides (polynucleotides) form *nucleic acids*. Two important nucleic acids are *deoxyribonucleic acid* (DNA) and *ribonucleic acid* (RNA). These complex substances are vital to the nature of life itself. DNA is a coded molecule which contains the information necessary to direct complex chemical reactions in living cells. RNA participates with it, assisting in the assembly of critical components that are vital to body structure and function.

DNA is the genetic material within the cell. It is a polymer of nucleotides that contains the sugar deoxyribose and forms a pair of coiled (helical) chains.

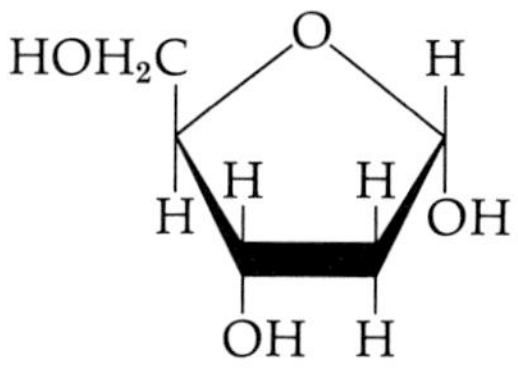

Deoxyribose

In DNA the phosphate groups are linked to the OH groups of the sugar components (deoxyribose) on adjacent nucleotides. As a result, chains are formed and linked in pairs by hydrogen bonds. The nucleotides in the pairs of chains face and combine with one another in specific patterns. They are called *complementary base pairs* (Fig. 2-7 on page 28).

There are about 125,000 base pairs in each of 46 strands called chromosomes in cells. The chromosomes are found in a part of the cell called the nucleus. In all the base pairs in DNA, a purine is joined to a complementary pyrimidine. In this fashion the purine adenine and the pyrimidine thymine unite, as do their cousins guanine and cytosine (Fig. 2-8 on page 28).

Estimates of the size of human DNA molecules indicate an average molecular weight of 82.5 million. When the molecule is extended, it is 4.1 cm (1.61 in.) long. Normally, it is so tightly coiled that it fits within a space about one ten-millionth (10^{-7}) of that length.

RNA molecules are polymers of nucleotides containing the sugar ribose (shown on page 28).

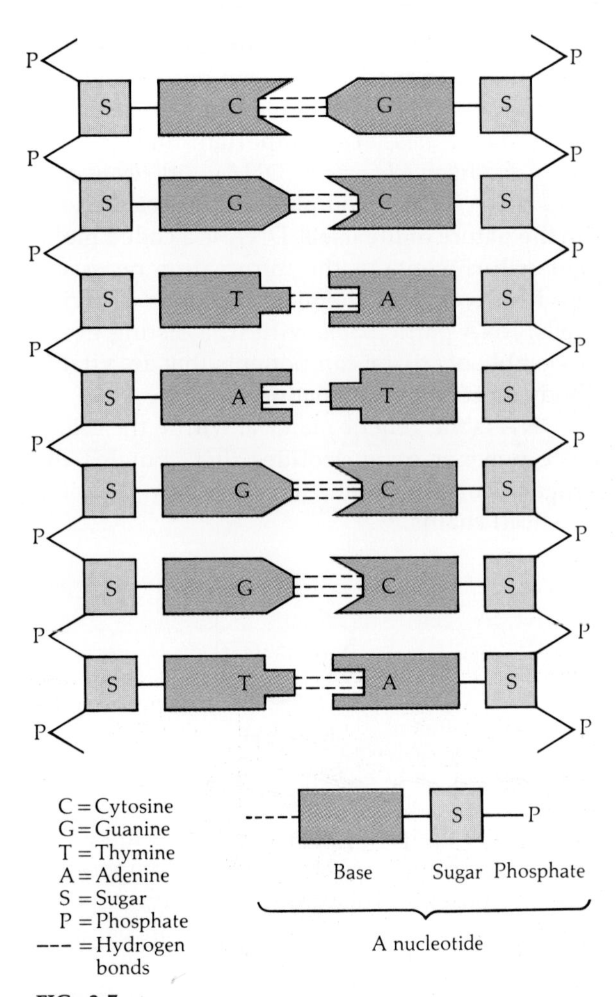

FIG. 2-7
The components of an uncoiled model of DNA. The units shown in each strand are repeated in the polynucleotides of DNA.

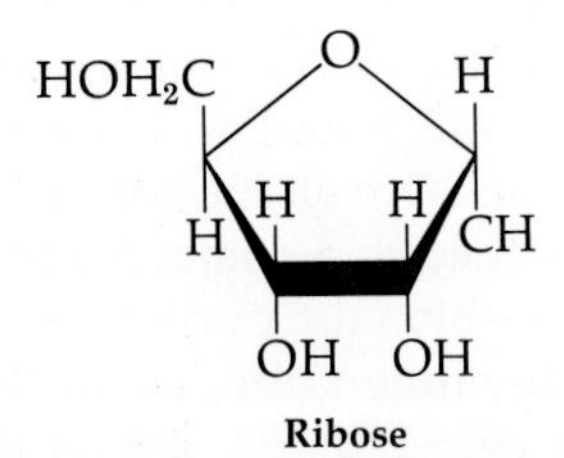

Ribose

Three of the four nitrogen-containing rings (adenine, guanine, and cytosine) found in RNA nucleotides are also found in DNA. The fourth is the pyrimidine uracil, which takes the place of thymine.

Three forms of RNA are found in cells and produced under the direction of DNA. They include messenger RNA (mRNA), transfer RNA (tRNA), and ribosomal RNA (rRNA). The mRNA is composed of hundreds of nucleotides with molecular weights ranging up to 5 million, while tRNA contains about 75 to 90 nucleotides and has a molecular weight of about 25,000. The rRNA has a molecular weight between 1 and 1.6 million. It, along with protein, forms ribosomes, found as small granules within the cell. Ribosomes function as way stations, at which amino acids are united to produce protein.

FIG. 2-8
DNA and mRNA structure. The double helix (coils) of DNA represents chains of nucleotides. The paired coils are linked by hydrogen bonds shown as horizontal lines between complementary nucleotides. mRNA is shown as a ribbon produced by DNA.

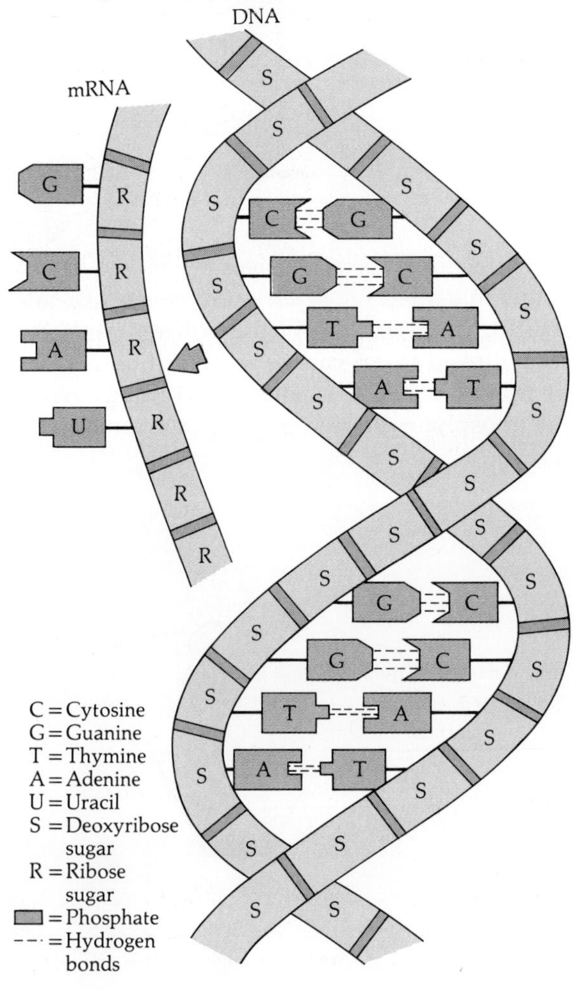

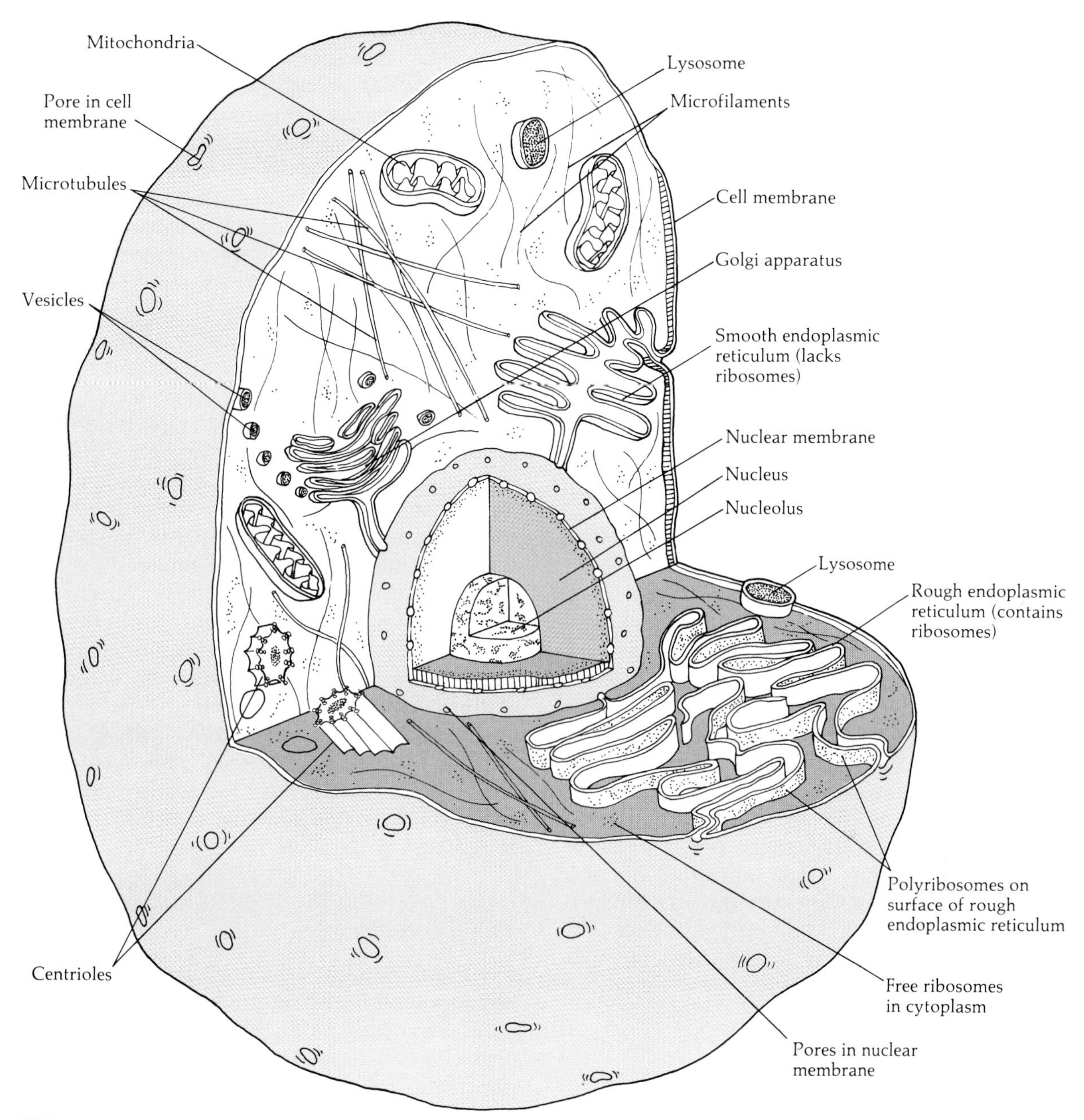

FIG. 2-9
Generalized structure of the cell.

2-3 THE BIOLOGICAL BUILDING BLOCK: THE CELL

The biological building block with which we will begin our study of the human body is the *cell.* From conception through early development, aging, and death, the structure, chemistry, and functions of cells continually change, as does their number. Different cells vary in size and shape and possess a wide range of functions. Their diversity may be observed by comparing tadpole-shaped microscopic male reproductive cells called *sperm* with huge skeletal muscle cells, whose length may be measured in inches. Their functional capacity may be appreciated by the study of a fertilized egg, a cell barely visible to the naked eye, yet one whose contribution to the universe is a unique human being. Throughout the text many cell types will be described. What follows below is a generalized description of this biological building block.

The Cell Membrane

The cell is the basic biological building block of the human body (Fig. 2-9). It contains an inner *nucleus* surrounded by *cytoplasm,* which in turn is enclosed by the outermost boundary of the cell, the *cell (plasma) membrane.* All the material within the cell is called *protoplasm.*

The cell membrane is a complex aggregate of proteins and other chemicals found in an oily (lipid) fluid. Its average thickness is 7.5 nm (approximately 1/3,400,000 in.). The cell membrane is selective in determining which substances pass across (permeate) its borders. Phospholipids form its main barrier. Tiny spaces or pores lie within and/or between protein molecules of the cell membrane. Their structure and size influence the rate of passage of molecules and ions. The composition and nature of the cell membrane and the role it plays in body function will be described in Chap. 4. The appearances it may assume will be discussed in subsequent chapters.

The Nucleus

The *nucleus* is often a conspicuous portion of the cell (Fig. 2-10). It is surrounded by a *nuclear membrane,* which consists of two layers separated by a space. Like the cell membrane, it is selectively permeable. Which molecules enter or leave the nucleus is determined by its surrounding membrane. Parts of the nuclear membrane are continuous with those within the cytoplasm known as the endoplasmic reticulum.

Within the nucleus, strands of DNA, along with RNA and protein, make up 23 pairs of chromosomes. Included among the 46 chromosomes is a unique pair, the sex chromosomes, which contain the genetic information that influences sexual development. Females have paired sex chromosomes called X chromosomes, while males have an X chromosome paired with a shorter Y chromosome. Genetic information is missing in the shorter Y chromosome of the male. This allows genes (segments of DNA) found at the end of the X chromosome to be more freely expressed, since there are none present on an equivalent portion of the Y chromosome of male cells. Traits governed by genes on this segment of the X chromosome are described as *X-linked.* Thus, red-green color blindness and hemophilia are found more often in males, since defective genes governing those traits, when present, are located on the part of the X chromosome described above.

The *nucleolus* (little nucleus) is found as a small, often spherical, structure within the nucleus (Fig. 2-9). It contains a large amount of granular material, which forms the subunits of the *ribosomes.* The nucleoli are the sites of synthesis of ribosomal RNA (rRNA). The number of nucleoli and small ribosomal granules varies with the physiology of the cell. Their presence correlates roughly with the amount of protein synthesized by the cell.

The Cytoplasm

Cytoplasm surrounds the nucleus within the cell membrane. It is also referred to as the *cytosol* or *intracellular fluid* (ICF). Many small structures, *organelles,* are located in the cytoplasm. The endoplasmic reticulum, ribosomes, Golgi apparatus, lysosomes, mitochondria, and various tubules and filaments are among the cell's major organelles. Their structure varies with the function of the cell.

Endoplasmic Reticulum and Ribosomes

Groups of paired membranes, which seem to be continuations of the nuclear membrane, are found in the cytoplasm. They wind their way in

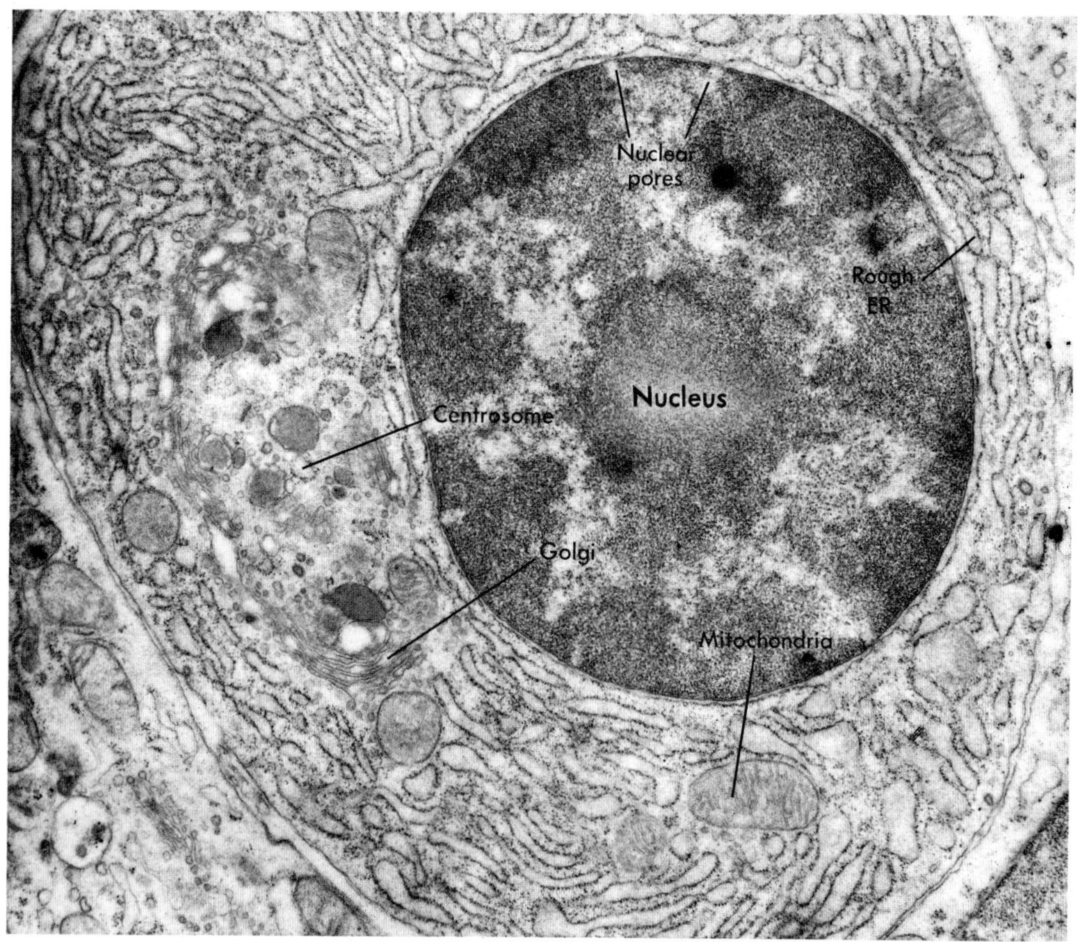

FIG. 2-10
Electron micrograph of a cell (×25,000). Rough ER = endoplasmic reticulum. Centrosome = an area of cytoplasm in which the centriole and its related structures are located. From *Histology*, 4th ed., by Weiss, L. and R. O. Greep. Copyright © 1977 McGraw-Hill Book Company. Used with permission of McGraw-Hill Book Company.

a maze throughout the cytoplasm, forming the *endoplasmic reticulum* (Figs. 2-9, 2-10). They form channels, cavities, or sacs, in which molecules may be transported or stored.

In some areas these membranes contain *ribosomal granules* attached to their surface which are flat or spherical and contain RNA and protein. Ribosomes are not thought to be active in protein synthesis when found as single granules. They may produce protein when grouped in chains as *polyribosomes* (polysomes). They do so by a process of translation, described below.

The ER that contains ribosomes is the *rough ER*. It is the site of much of the protein synthesis in the cell. Many protein hormones are synthesized on the rough ER, while other proteins are synthesized on polysomes that are not associated with the ER. These are the proteins that usually remain within the cell in which they are manufactured. An example is hemoglobin, which is formed in red blood cells and is a carrier of oxygen.

The *smooth ER* consists of paired membranes without ribosomal granules and seems to be the site of synthesis of steroids, carbohydrates, lipids, and other substances. The smooth ER in skeletal muscle is called the *sarcoplasmic reticulum* and stores and releases calcium.

Golgi Apparatus

Aggregations of tubes and saclike structures in the cytoplasm form membrane-enclosed sacs or *vesicles.* They are referred to as the *Golgi apparatus,* Golgi complex, or Golgi body (Fig. 2-10). Stacks of these structures seem to be related to the accumulation and transport of protein. In some instances carbohydrates may be added to form glycoproteins within the Golgi apparatus.

The Golgi apparatus may also give rise to *vacuoles,* which are membrane-bound storage sacs that form lysosomes (described below). Substances are assembled within the Golgi apparatus as granules. The substances are delivered to the exterior of the cell by fusion of vesicles (sacs) from the Golgi apparatus with the cell membrane to create an external opening. The stored material, in the form of granules or within vesicles, is released outside the cell by a process called *exocytosis.* Sketched in Fig. 2-9, it will be described in some detail in Chap. 4.

Lysosomes

Lysosomes are membrane-bound sacs which appear to be derived from the Golgi apparatus. They contain digestive enzymes that may degrade a variety of molecules and play a role in many body functions. Should the contents of lysosomes be released within a cell, destruction of that cell and surrounding cells and tissues may follow. This phenomenon may be responsible for the degeneration of tissues in joints which occurs in rheumatoid arthritis. The elimination of dying or dead cells and their remnants is assisted by the degradative activity of lysosomal enzymes. This function is also a useful defense mechanism in the body. For example, lysosomal enzymes of white blood cells destroy many types of microbes (bacteria and viruses).

Mitochondria

The mitochondrion (Fig. 2-10) is a membrane-bound structure within the cytoplasm which may be Y-shaped, round, or cylindrical. It is surrounded by a double membrane and contains ribosomes and DNA (Fig. 2-11). Mitochondrial DNA is equivalent to self-contained genes that allow the mitochondrion to direct its own synthetic activities. Interaction with DNA in the nucleus seems essential for mitochondria to carry on some of their functions.

Mitochondria contain a large number of respiratory enzymes, which oxidize foods to release energy in a series of separate steps. The respiratory enzymes require oxygen to break

FIG. 2-11
The mitochondrion. (A) Electron micrograph showing internal structure of mitochondrion with many cristae. Also note the double outer membrane of the mitochondrion (×14,000). (B) Diagram of mitochondrion.

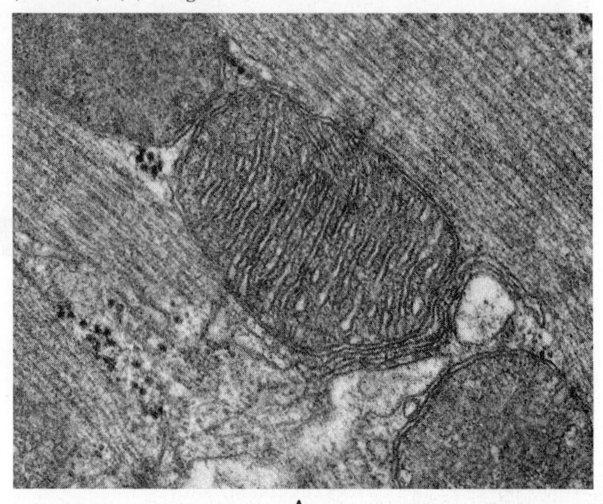

A

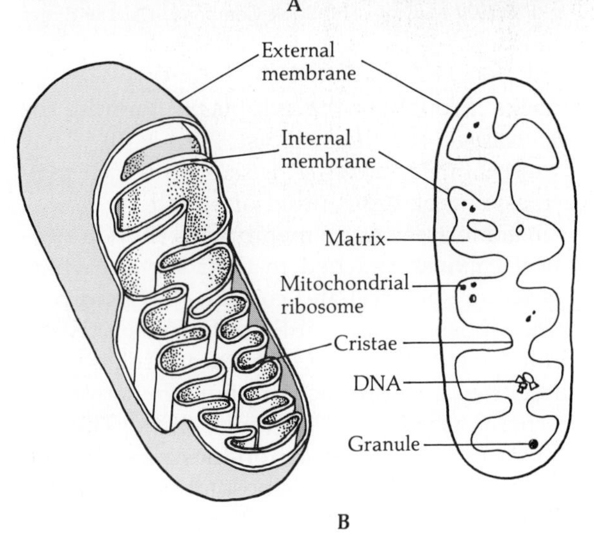

B

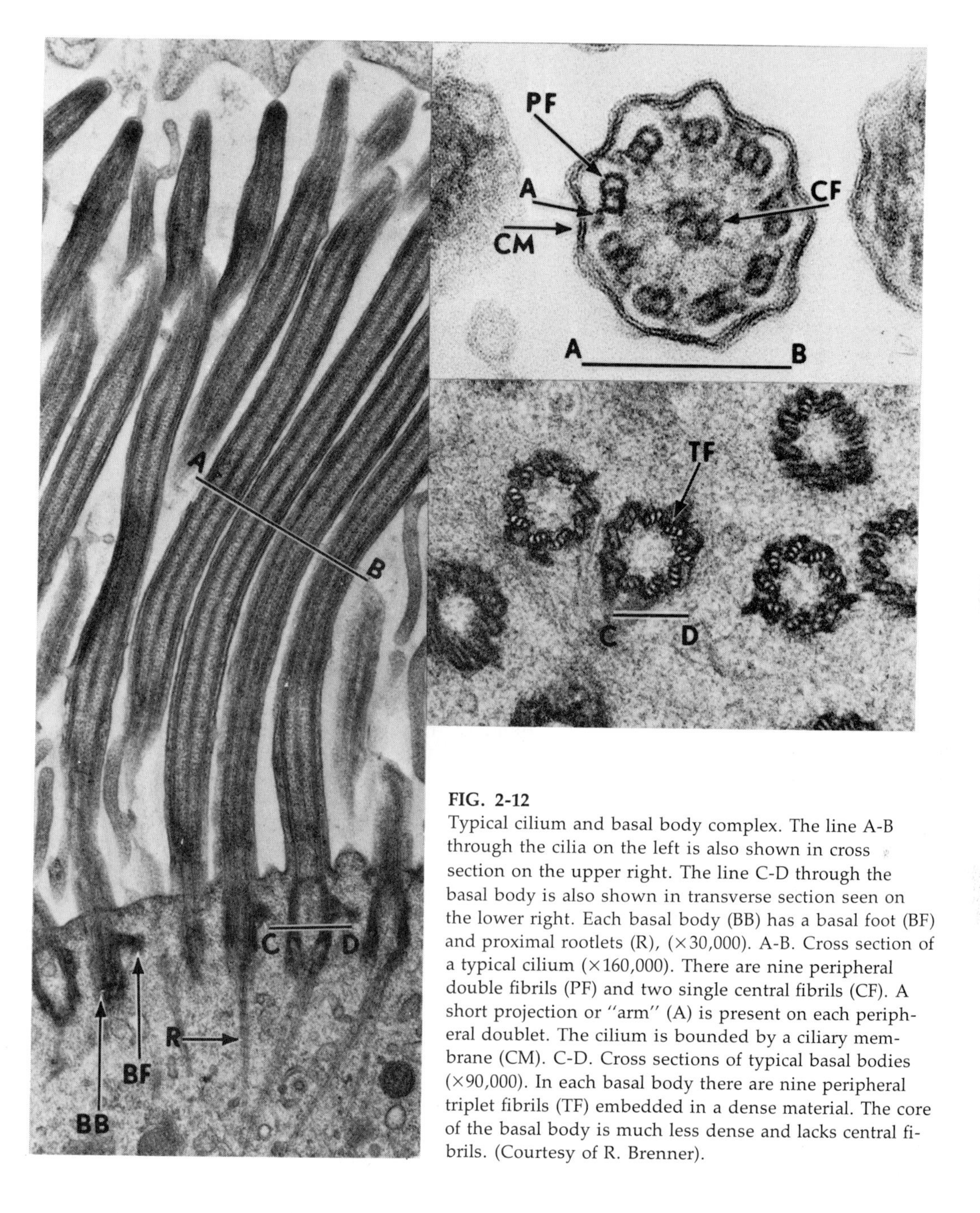

FIG. 2-12
Typical cilium and basal body complex. The line A-B through the cilia on the left is also shown in cross section on the upper right. The line C-D through the basal body is also shown in transverse section seen on the lower right. Each basal body (BB) has a basal foot (BF) and proximal rootlets (R), (×30,000). A-B. Cross section of a typical cilium (×160,000). There are nine peripheral double fibrils (PF) and two single central fibrils (CF). A short projection or "arm" (A) is present on each peripheral doublet. The cilium is bounded by a ciliary membrane (CM). C-D. Cross sections of typical basal bodies (×90,000). In each basal body there are nine peripheral triplet fibrils (TF) embedded in a dense material. The core of the basal body is much less dense and lacks central fibrils. (Courtesy of R. Brenner).

chemical bonds and release energy to drive other chemical reactions (Chap. 23). The enzymes are located on the inner folds of the inner mitochondrial membrane. These folds are known as *cristae* (singular, crista). Mitochondria are the major sites of oxygen consumption within human cells.

Tubules and Filaments

Various *tubules* and *filaments* are found in the cytoplasm. Hollow *microtubules* are those which average 25 nm in diameter. They may be used to transport materials, maintain cell shape, and assist in movement. For example, microtubules

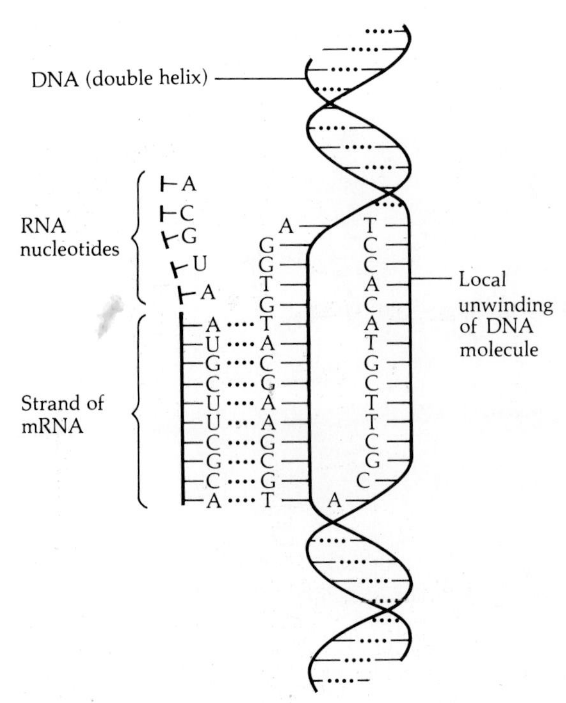

FIG. 2-13
Transcription of DNA (synthesis of mRNA). Parts of the double helix of DNA unwind. The nucleotides of a strand of DNA serve as a template with which complementary RNA nucleotides may align. The RNA nucleotides are combined (polymerized) to form a strand of mRNA.

may carry chemicals a few feet in some long nerves. *Microfilaments* are solid strands which average 4 to 6 nm in diameter. Various forms are found in different cell types. For example, muscle contains the filamentous proteins myosin and actin, which are involved in contraction. Centrioles, cilia, and flagella are other forms of filaments.

Centrioles, Cilia, and Flagella

Paired cylindrical bodies, the *centrioles,* produce the spindle apparatus, a structure composed of microtubules. The spindle apparatus gives rise to fibers, which separate chromosomes in cell division.

Cilia are hairlike extensions which are about 0.5 μm in diameter and 2 to 10 μm long (Fig. 2-12). They are found lining the upper respiratory tract and the oviducts. They consist of microtubules arranged in a circle, with nine pairs of microtubules surrounding a central pair. Their wavelike contractions propel mucus and particles away from the lungs and the egg from the ovary toward the uterus (womb).

Flagella are similar to cilia, but they are 100 to 200 μm long. They are used to propel the male reproductive cell (sperm).

Direction of Cellular Activity: The Role of DNA and RNA

A *genetic code* exists within DNA created by groups of three nucleotides (triplets) in linear sequence. Different segments of the DNA molecule may have various nucleotides arranged along a line in different orders. This produces a variety of coded triplets (*codons*). The sequences of nucleotides within the DNA form a code, which accounts for inherited characteristics (e.g., blood type A, B, AB, or O). The message within each codon, as dictated by the unique sequence of its nucleotides, is used to direct synthesis of mRNA in a process called *transcription.* This is the basis for the development of traits controlled by *genes,* the units of heredity. It is much like a language transmitted by the use of signals which constitute an equivalent of a Morse code.

Specifically, the genetic code directs the as-

FIG. 2-14
DNA direction of protein synthesis. (A) DNA directs the synthesis of three forms of RNA, namely, rRNA, mRNA, and tRNA, which are transported from the nucleus to the cytoplasm. (B1 and 2) Amino acids are activated and attached to specific molecules of tRNA. (C) Individual ribosomes, which contain rRNA, move onto a strand of mRNA. (D) The tRNAs carry individual amino acids to a coded portion of mRNA. The codon of mRNA and the anticodon of tRNA consist of complementary base pairs, which unite. (E) Ribosomes move from one codon of mRNA to the next, binding specific amino acids together by peptide bonds. (F) Ultimately, a polypeptide chain or protein may be produced. (G) The ribosome separates from the polypeptide (or protein) when it reaches the end of mRNA. The ribosome may return to the beginning of the strand of mRNA to repeat the process and be utilized repeatedly, along with the tRNAs and mRNAs, as long as energy is available (from ATP) to activate the amino acids and to drive the reactions. Several ribosomes may act in sequence and cause the synthesis of the protein molecules in assembly-line fashion.

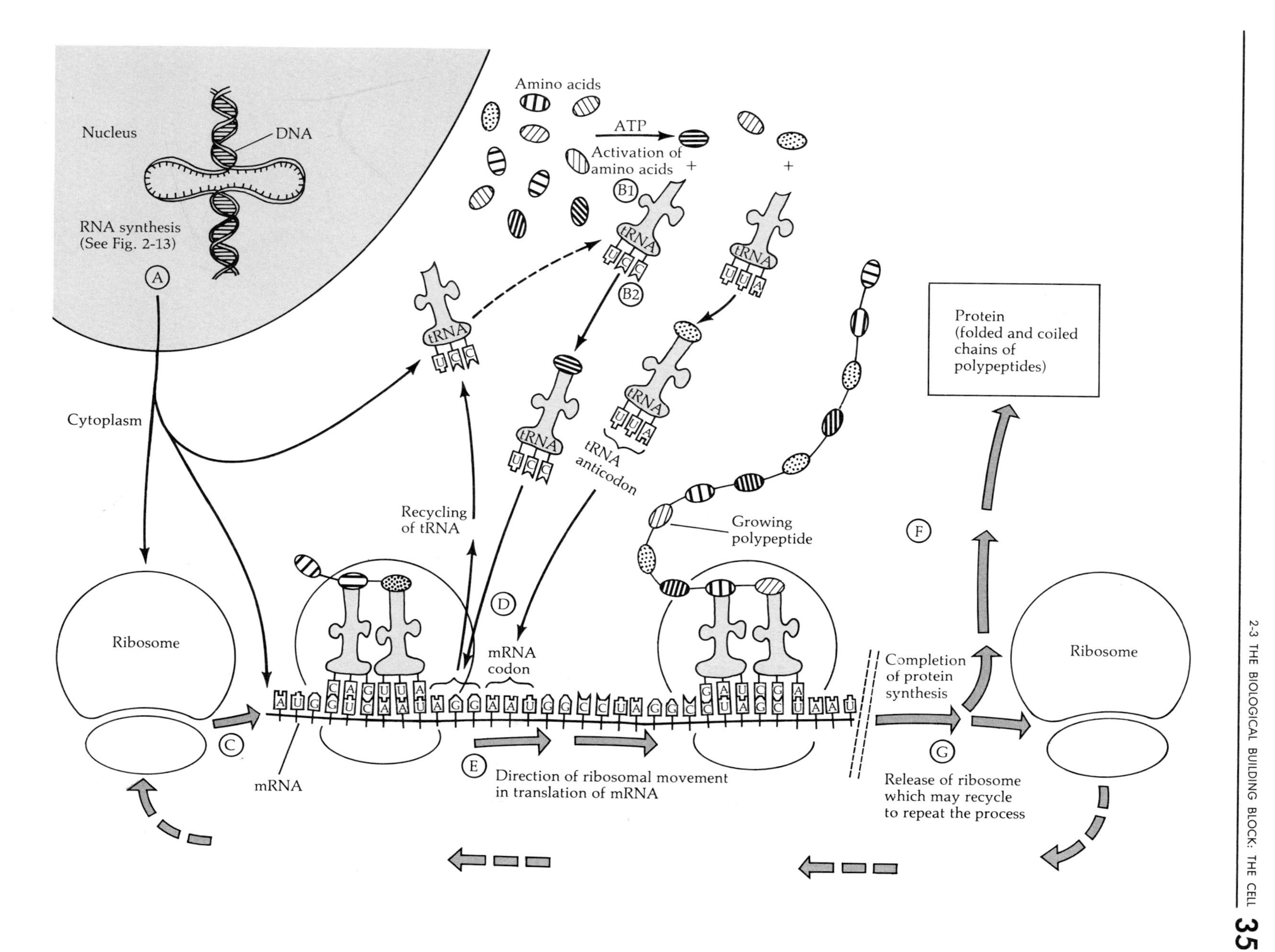
Nucleus
DNA
RNA synthesis
(See Fig. 2-13)
A
Cytoplasm
Amino acids
ATP
Activation of
amino acids
B1
B2
tRNA
tRNA
anticodon
Recycling
of tRNA
D
mRNA
codon
Growing
polypeptide
Ribosome
C
mRNA
E
Direction of ribosomal movement
in translation of mRNA
F
Protein
(folded and coiled
chains of
polypeptides)
Completion
of protein
synthesis
G
Release of ribosome
which may recycle
to repeat the process
Ribosome

sembly of many protein molecules. Nucleotides which complement those found in a section of DNA are assembled to produce a ribbon of mRNA. Transcription creates specific types of mRNA, each of which carries its own message in its specific nucleotide sequence (Fig. 2-13).

A long strand of mRNA is transferred from the nucleus of the cell to the cytoplasm. The size of the mRNA is related to that of the protein whose synthesis it will direct. Hundreds to thousands of nucleotides are required to form the necessary number of triplet codes for the transcription. The mRNA attaches to ribosomal granules which are composed of rRNA and protein and located in the cytoplasm (Fig. 2-14). In so doing, it produces a mold or template to which complementary tRNA may attach. The code in mRNA regulates the assembly of amino acids into a polypeptide or protein in the process of *translation.*

Amino acids are activated prior to attachment to the mRNA template. This process involves the expenditure of energy to combine one of the 20 individual types of amino acids found in protein with a specific type of tRNA. Each form of tRNA contains a nucleotide with its own triplet code. Each triplet in tRNA is an *anticodon,* which complements the codon in mRNA. Each complex of tRNA and its specific amino acid moves to a ribosome. Different complexes combine in succession with the complementary forms of mRNA on the ribosomes. As this occurs, the amino acids are lined up and link in a specific sequence. The tRNA thus transfers amino acids to the mRNA template. The ribosomes align the mRNA and the tRNA amino acid complexes, allowing the amino acids associated with specific tRNAs to form peptide bonds with one another and be added to a growing polypeptide chain.

A group of ribosomal granules rolling along an mRNA molecule are called *polysomes* (polyribosomes). They assist in the assembly of the amino acids. They seem to move along mRNA and bind to specific tRNA–amino acid complexes to initiate protein synthesis. The ribosome serves as a small machine that links amino acids by promoting the formation of peptide bonds. The ribosome moves off the end of the mRNA strand and separates from the resulting peptide chain or protein. Other ribosomes move onto the mRNA at the other end, repeating the process about every 60 to 90 s. The amino acid sequence dictated by the mRNA is unique for a particular protein and gives the protein its biological function.

This mechanism is used to build a variety of structural and functional proteins for each cell throughout the body. Various proteins give different cells their special identity and often confer upon them diverse and important functions. Proteins include hundreds of different enzymes which regulate body chemistry, as well as molecules as different as hemoglobin used by red blood cells to transport oxygen, hormones such as insulin used to control levels of sugar, and collagen used for structural reinforcement of bone and other tissue.

2-4 SUMMARY

In order to gain appreciation of body function one must understand the physical, chemical, and biological building blocks of the body. The importance of these concepts can be ascertained by study of the discoveries for which Nobel prizes in physiology or medicine have been awarded.

1 Scientific descriptions of the human body are often made by using the metric system. The measurements are expressed in multiples of 10. The values summarized in Table 2-3 are most easily remembered by learning the meanings of the following prefixes: mega, one million; kilo, one thousand; deci, one-tenth; centi, one-hundredth; milli, one-thousandth; micro, one-millionth; nano, one-billionth; pico, one-trillionth; and femto, one-quadrillionth.

2 The chemical building blocks of all matter are atoms and molecules (the smallest units of compounds). An atom is the smallest part of an individual element which cannot be subdivided without changing its nature. Common elements in the body include carbon, hydrogen, oxygen, and nitrogen. A molecule is composed of combinations of two or more atoms. A compound is formed from two or more dissimilar elements which produce a substance with properties different from those of each constituent element. Atoms and/or molecules may unite repeatedly to produce large compounds.

Inorganic compounds consist of elements which do not contain carbon-hydrogen bonds. Organic compounds are hydrocarbons, composed of carbon atoms bonded to hydrogen atoms, or their derivatives, which contain other elements as well. The carbon atoms link to form chains or rings (cyclic structures).

Carbohydrates are organic compounds consisting mostly of carbon, hydrogen, and oxygen. Many important carbohydrates are sugars or consist of sugar units (monosaccharides). These may be combined repeatedly to form polymers, which include macromolecules found in straight chains (e.g., cellulose), branched chains (e.g., glycogen), or a mixture of straight and branched chains (e.g., plant starch).

Proteins are polymers of amino acids united by peptide bonds. They contain basic and acidic groups, the amino acid and carboxylic acid groups, respectively. Proteins form enzymes, many hormones, and other molecules that regulate body chemistry. They also provide the supporting architecture for the body.

Lipids are molecules which dissolve in certain organic solvents. They may be classified as simple, compound, or derived.

Neutral fats, or triglycerides, are simply lipids containing a 3-carbon chain derived from glycerol. Attached to each of the carbon atoms of this chain are fatty acids. When all the fatty acid chains hold all the hydrogen atoms that they are able to bind, the fat is saturated, or hydrogenated. When they do not, the fat is unsaturated, or dehydrogenated.

Phospholipids are examples of compound lipids, which are composed of two fatty acids and phosphorous in a third group linked to the glycerol chain. Phospholipids are important constituents of the outer boundaries of cells and, like glycolipids and lipoproteins, are critical to body function.

Steroids are derived lipids formed by the combination of 2-carbon fragments in fat metabolism. Steroids are composed of three six-membered and one five-membered ring of carbon atoms. Attachment of different groups on the rings produces varieties of steroids. They include the major male and female sex hormones, as well as adrenal steroids, vitamin D, and cholesterol.

Nucleic acids are composed of building blocks of nucleotides. The latter are formed from nitrogen-carbon rings of two types (the pyrimidine and purine bases) plus sugar and phosphate. The phosphate groups attach to simple sugars in adjacent nucleotides to form long chains.

In deoxyribonucleic acid (DNA), the sugar is the 5-carbon molecule deoxyribose. Pairs of chains are bound by hydrogen atoms that are attached to the complementary nitrogen-containing bases of each nucleotide. The pairs of chains are twisted into a double coil. The complementary purine bases are linked to pyrimidines. The base pairs are adenine-thymine and guanine-cytosine, respectively. Groups of three nucleotides (triplets) along a DNA chain constitute a codon. These are the functional components of genes which determine inherited characteristics. The order of arrangement of codons on the chains constitutes the genetic code.

Ribonucleic acid (RNA) is composed of chains of the 5-carbon sugar ribose linked together by phosphate groups. The bases attached to ribose are the same as in DNA except for substitution of the pyrimidine uracil for thymine.

3 The cell is the body's biological building block. Although cells vary considerably from one another and during their individual lifetimes, they normally have a nucleus surrounded by cytoplasm and a cell membrane. The cell membrane is composed of layers of lipid and protein molecules. It selectively regulates the entry and exit of most substances. The nucleus contains 23 pairs of strands, the chromosomes, composed of protein, and DNA, the genetic information of each cell. Within the boundary of the nucleus is the nucleolus, or little nucleus, which contains rRNA and which forms subunits of ribosomes. The nucleus is surrounded by a highly selective but porous nuclear membrane.

The cytoplasm is surrounded by the cell membrane and contains a number of organelles. The endoplasmic reticulum consists of a labyrinth of paired membranes in contact with the nuclear membrane. When studded with ribosomal granules, it is called the rough ER. When it lacks ribosomes, it is called the smooth ER. Many macromolecules, including proteins, are manufactured in the ER. The Golgi apparatus consists of tube- and saclike membranous structures in which carbohydrates and proteins accumulate. These membranes are in contact with the cell membrane and may deliver materials to the outside of the cell. Lysosomes are membrane-bound sacs containing chemicals (enzymes) that can degrade many other molecules. Mitochondria are organelles surrounded by a double membrane. The inner membrane forms folds, the cristae, which contain respiratory enzymes. The latter cause oxidation of foodstuffs within cells.

Cells contain a variety of hollow microtubules and strands of microfilaments. They may be used as delivery systems and/or cause movement by their contraction. Paired cylindrical bodies known as centrioles are found in the cytoplasm. They are associated with tubular structures that form cilia and flagella in different cell types. Centrioles also produce the spindle apparatus. This consists of fibers that separate chromosomes during cell division.

DNA directs cellular activity by means of its genetic code, which consists of a linear sequence of groups of triplets of nucleotides. The coded triplets are called codons and are responsible for characteristics inherited through entities called genes. DNA directs the production of three forms of RNA and ultimately through this function is able to direct protein synthesis and govern cellular activity and traits.

DNA is capable of unwinding its double coil at one end and using that portion to assemble complementary bases and form a single-stranded ribbon of mRNA. The process is called transcription and produces a strand of nucleotides called mRNA. Twenty different forms of tRNA are also produced. Each combines specifically with one of 20 types of amino acids used to build body protein. The rRNA molecules are found with protein in ribosomal granules used as machines to combine amino acids attached to tRNA. The tRNA-amino acid complexes are arranged in an order dictated by translation of the coded nucleotide sequence on mRNA, with which the anticodon of each tRNA combines. In this manner, amino acids are lined up and transferred to mRNA to be linked within ribosomes to form proteins.

REVIEW QUESTIONS

1 What are the differences between atoms, molecules, and compounds?
2 A molecule of sodium chloride weighs 58.5 g. How would one make a 0.1 M solution of it?
3 The molecular weight of hydrogen chloride (HCl) is 36. Each molecule can release one hydrogen ion (H^+). How much HCl produces 1 Eq of H^+?
4 Using the information in the above question describe the way to make a hydrochloric acid solution with a pH of 5.
5 How are chemical bonds formed? Describe three types of chemical bonds.
6 What is the difference between an inorganic and an organic compound?
7 How do monosaccharides, disaccharides, and polysaccharides differ? Describe the differences between glycogen and cellulose.
8 What is the primary reason for differences among proteins?
9 What are enzymes and how do they function?
10 Distinguish among neutral fats, phospholipids, steroids, and glycolipids.
11 How do saturated and unsaturated fatty acids differ?
12 What are pyrimidines and purines? Name three pyrimidines and two purines found in nucleic acids. Which are found in DNA? Which are found in RNA?
13 Describe the composition of the cell (plasma) membrane.
14 Which cellular organelle contains the chromosomes, and what is their role in the body?
15 Cite the major function of the rough endoplasmic reticulum.
16 Describe the Golgi apparatus. What is its apparent function?
17 Identify the lysosomes, their functions, and the manner in which they may play a defensive role in the body.
18 Identify the structural components of a mitochondrion. How do mitochondria release energy to drive chemical reactions?

19 What roles do microfilaments, microtubules, and flagella play in human cells?

20 What is a nucleotide and how do nucleotide triplets form a genetic code?

21 Describe mRNA, tRNA, and rRNA and their roles in transcription and translation of the genetic code. Include a description of the synthesis of a polypeptide chain.

PHYSIOLOGY OF TISSUES

3

3-1 EPITHELIAL TISSUE
General functions
Specific characteristics
Cell shapes
Arrangement of cells
Functions in body cavities, vessels, and glands
Mucous membranes
Serous membranes (mesothelium)
Endothelium
Glands
Exocrines • Endocrines
Dynamics of structure and function of epithelium

3-2 CONNECTIVE TISSUE
Loose (areolar) tissue
General properties
Role of major cell types
Fibroblast • Fat cell (adipocyte) • Defense cells
Reticular tissue
Lymphoid tissue
Reticuloendothelial tissue
Hematopoietic tissue
Dense connective tissue
Regular dense connective tissue
Irregular dense connective tissue
Cartilage
Hyaline cartilage
Fibrous cartilage
Elastic cartilage
Bone (osseous) tissue
Compact bone
Spongy (cancellous) bone
Dentin
Diseases of connective tissue
Scurvy
Osteoarthritis

3-3 MUSCULAR TISSUE
Skeletal muscle
Cardiac muscle
Smooth muscle

3-4 NERVOUS TISSUE
Neurons
Neuroglia

3-5 SUMMARY

OBJECTIVES

After completing this chapter the student should be able to:

1. List four major tissue types and distinguish them with respect to structure and general function(s).
2. Compare the structure and functions of epithelial tissues.
3. Explain the role of epithelial tissue in body linings, cavities, organs, and glands.
4. Describe the general structure and function of the following connective tissues: areolar, dense, reticular, cartilage, and bone.
5. Discuss defense mechanisms and disease in relationship to the properties of connective tissues.
6. Compare the structure, function, and nerve supply of skeletal, cardiac, and smooth muscles.
7. Identify the major cell types found in nervous tissue.

Cells transport materials in both directions across their boundaries and work together in groups referred to as *tissues.* In this manner, they carry on vital functions. Four main tissue types are epithelial, connective, muscular, and nervous tissues. Their interrelationships reflect the extensive organization and variety of functions of the body.

3-1 EPITHELIAL TISSUE

General Functions

Epithelial tissues line surfaces on the inside and outside of the body. Their major functions include protection, absorption, secretion, and excretion. The outer surface of the skin, the epidermis, is an example of epithelial tissue which *protects* the body by preventing excessive loss of fluids. Healthy skin normally prevents penetration of the body by foreign materials such as bacteria, viruses, or other agents. *Absorption* is the passage of materials into cells. It is especially evident in epithelial cells lining the digestive tract. In this case, the properties of epithelial tissue determine whether foods that we eat enter the bloodstream so that they might be of nutritive value. Those foods which are not absorbed merely pass through the digestive tract and do not provide the body with energy or building blocks. *Secretion* involves the synthesis and release of useful substances. It is readily observed in different glands formed by clusters of epithelial cells. Some of the cells release enzymes, while others release acids, bases, hormones, or other products. *Excretion* is a process by which waste is eliminated. Epithelial cells lining the tubules of the kidney are actively involved in excretion as well as in reabsorption. These cells release materials that are no longer of any use to the body and also those which are potentially toxic. Through reabsorption, the kidney's tubular cells conserve substances that are still of use.

Specific Characteristics

The relative scarcity of intercellular substance distinguishes epithelium from other tissue types. The cells are generally tightly connected by proteins in any of the three forms of intercellular attachments shown in Fig. 3-1. These are: (1) *desmosomes,* in which protein firmly binds the cells together and fills the intercellular space (Fig. 3-1A); (2) *tight junctions,* which hold cells together at discrete sites and which are relatively impermeable (Fig. 3-1B); and (3) *gap junctions* (nexuses), which form intercellular pores that allow substances with molecular weights up to 1000 to pass between adjacent cells (Fig. 3-1C). Various parts of a single cell may have different types of intercellular attachments to its neighbors.

Epithelial cells characteristically are attached to underlying tissues by a dual structure known

FIG. 3-1
Models of cell junctions. The photomicrographs on the left are depicted in the diagrams on the right. (A) Desmosomes (× 115,000). (B) Tight junctions (× 115,000). (C) Gap junctions (× 275,000). Each cell's cytoplasm is located on either side of the cell junctions.

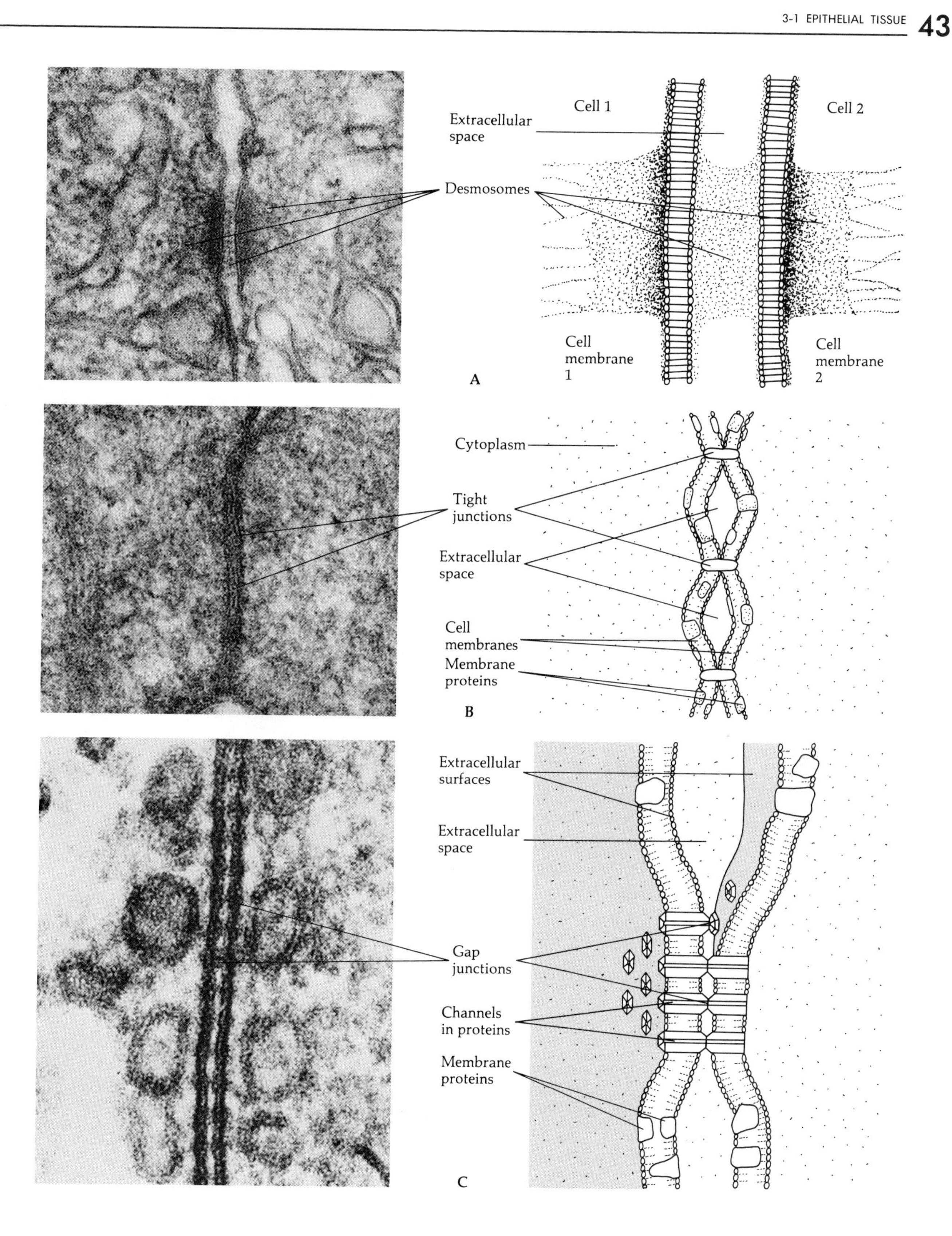

Cell 1
Cell 2
Extracellular space
Desmosomes
Cell membrane 1
Cell membrane 2
A
Cytoplasm
Tight junctions
Extracellular space
Cell membranes
Membrane proteins
B
Extracellular surfaces
Extracellular space
Gap junctions
Channels in proteins
Membrane proteins
C

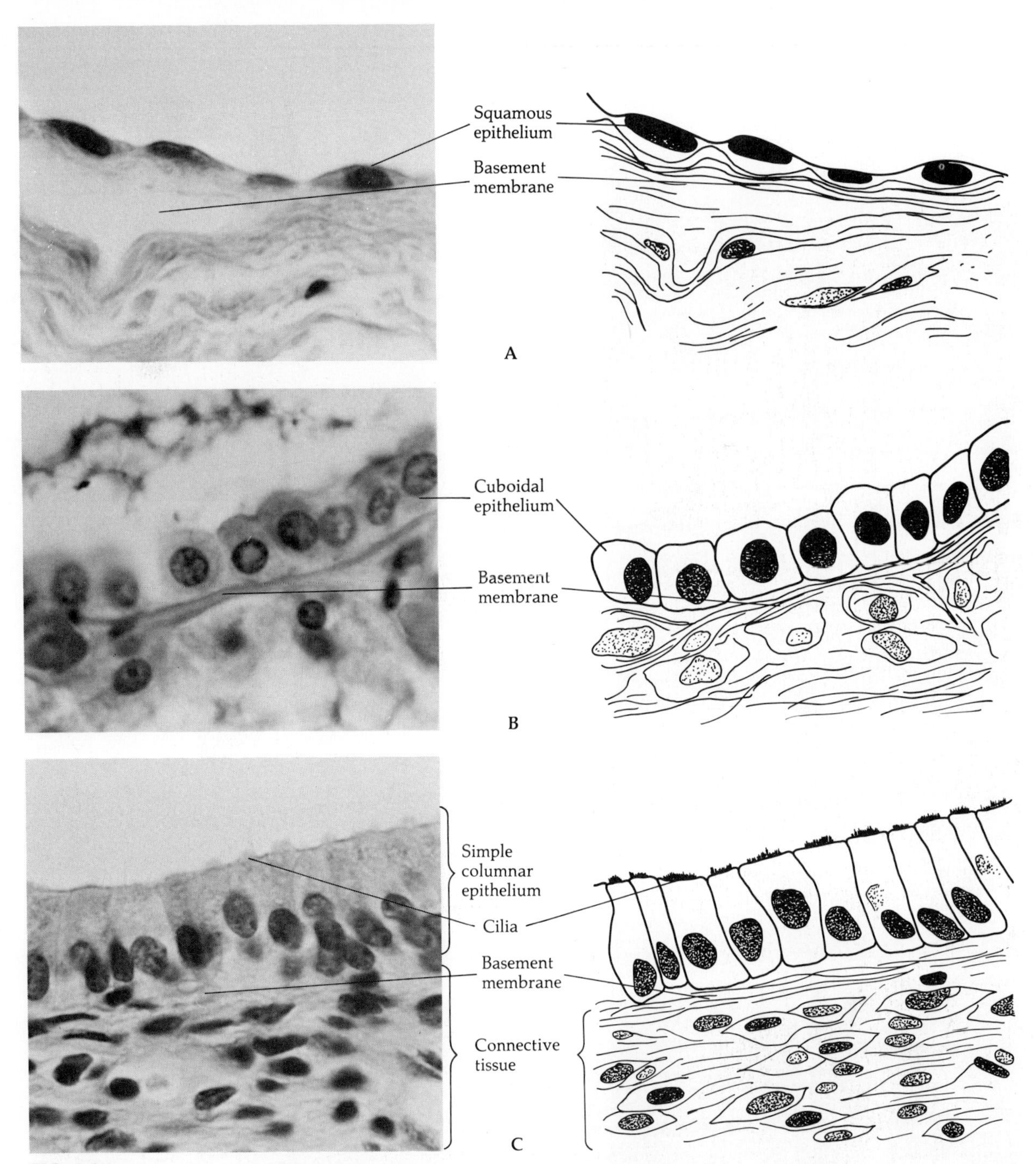

FIG. 3-2
Shapes of epithelial cells. Photomicrographs on the left are shown in diagrams on the right. (A) Simple squamous epithelium (× 1700). (B) Simple cuboidal epithelium (× 1700). (C) Simple columnar epithelium with cilia (× 1700). (D) Pseudostratified columnar epithelium with cilia and goblet cells (× 1700). (E) Transitional epithelium (× 1700).

as the *basement membrane* (basal lamina) (Fig. 3-2). The latter consists of a layer of *glycoproteins* (carbohydrate-protein complexes) produced by the epithelial cells and a reinforcing network of fibers derived from connective tissue. Nourishment for epithelial tissues usually filters through the basement membrane.

Cell Shapes

Epithelial cells assume three general shapes: flat, cubical, or column-shaped. Flat, irregularly

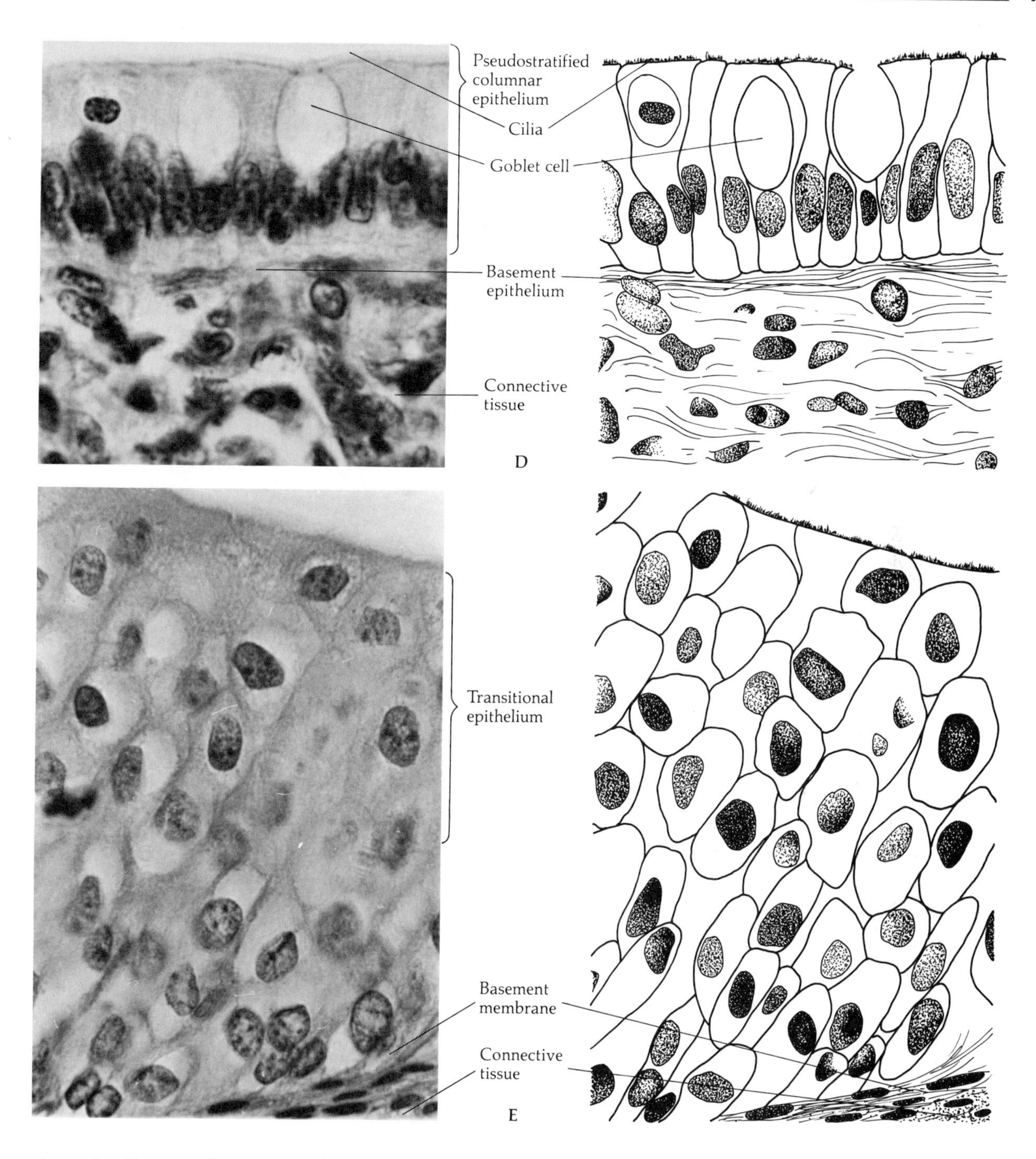

shaped cells are called *squamous epithelium* (Fig. 3-2A). They are found in large numbers at the surface of the skin. Cubical cells are referred to as *cuboidal epithelium* (Fig. 3-2B). They form the lining of hollow structures such as ducts, tubules, and secretory glands. These cells also form a layer in the retina of the eye and surround cells in the ovary that will mature into eggs (ova).

Column-shaped cells are classified as *columnar epithelium* (Fig. 3-2C and Fig. 3-3). These cells line specific ducts in the body and parts of the digestive tract (stomach and intestine), where they form a soft surface coated with

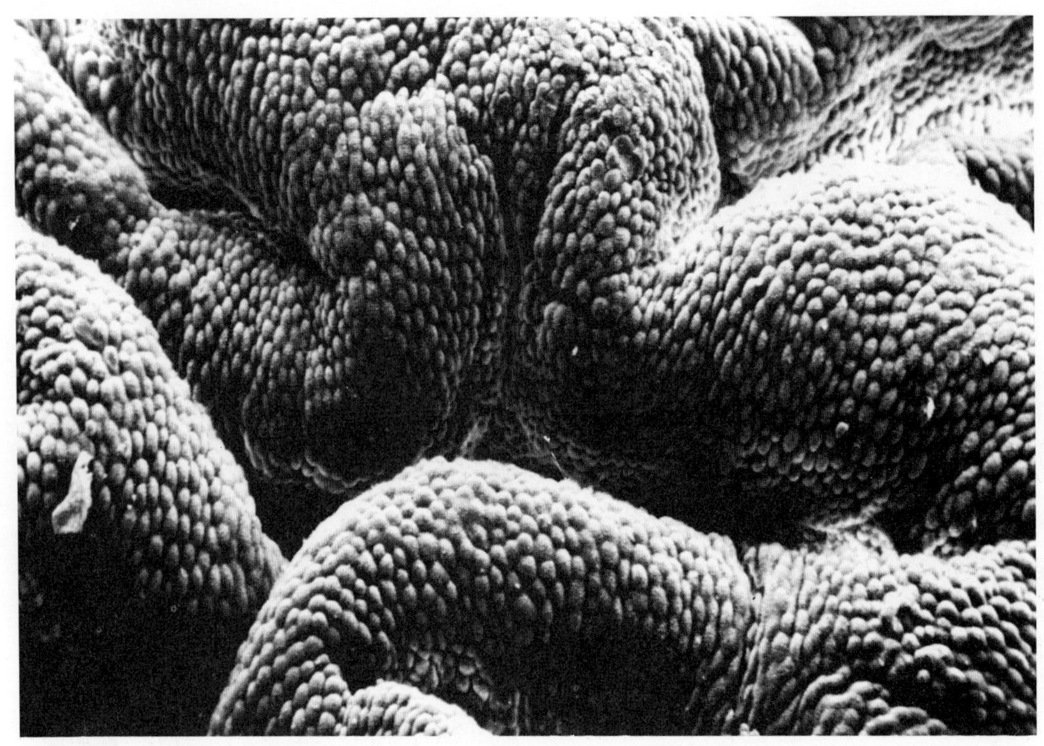

FIG. 3-3
Columnar epithelium of the gallbladder. Note the tightly packed cells which are viewed from above in this scanning electron micrograph of the inner folds of the organ (×350).

mucus. The linings of the respiratory tract and the urethra (a tube which carries urine to the exterior) are also composed of columnar epithelial tissue, as are bile ducts in the liver.

Arrangement of Cells

Epithelium may be found as one layer of cells (*simple epithelium*) or as multiple layers (*stratified epithelium*). Simple squamous cells line all the body cavities, within a membrane known as the *mesothelium*. They also line blood vessels and the heart in a similar structure, called the *endothelium*. Simple columnar epithelium lines the stomach and some parts of the respiratory tract.

Stratified epithelium is less absorptive and less secretory but more protective than simple epithelium. Its structure varies according to the location and needs of the area in which it is found. Stratified squamous cells form the epidermis, or outer layer, of the skin, the lining of the mouth, and the surface of the tongue (Fig. 3-4).

Stratified cuboidal epithelium lines the seminiferous tubules of the testes and surrounds the egg in the developing follicle within the ovary. It is also found in ducts of the sweat (sudoriferous) and oil-producing (sebaceous) glands.

Cells that are not truly stratified but only appear so are called *pseudostratified epithelium* (Fig. 3-5). All pseudostratified epithelial cells rest on a common geometric plane, the basement membrane, although their nuclei may be located at various levels. This gives a false impression of layers of cells. Pseudostratified epithelium is found in parts of the respiratory tract, including the nasal cavity, the trachea or windpipe, and the bronchi, which lead from the trachea into the lungs. In many of these areas the cells are ciliated.

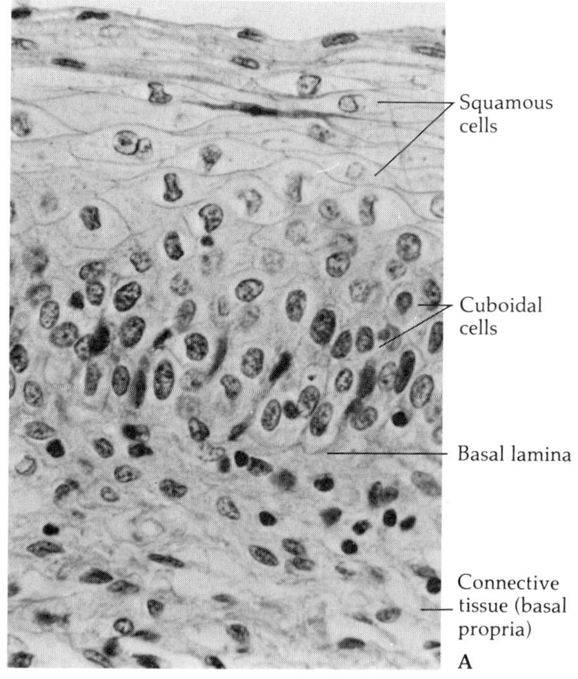

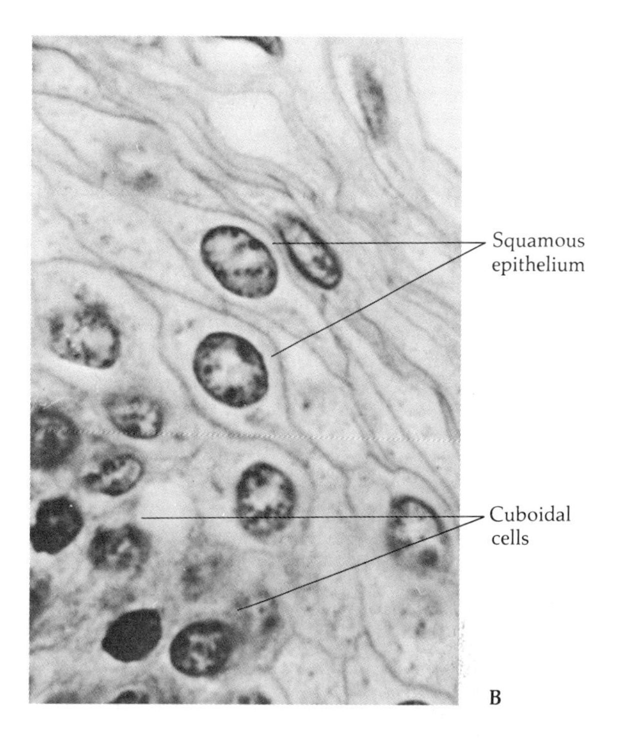

FIG. 3-4
Stratified epithelium. (A) The photomicrograph on the left illustrates squamous epithelial cells overlying cuboidal epithelium in the epidermis of the esophagus (×700). An underlying basement membrane is also evident. (B) A higher magnification of the epidermis of the esophagus illustrates cells which range from flattened squamous epithelium in its upper layers to cuboidal and columnar cells in its lower layers (×1700).

When cilia are present in this type of tissue, it is called *pseudostratified ciliated columnar epithelium.* The cilia of these cells are used to propel foreign matter clinging to mucus droplets from the lungs toward the mouth and nose.

The shape of epithelial cells often varies with the function of the organ. For example, *transitional epithelial cells* tend to be flat when an organ is stretched and sawtoothed or irregular when it is relaxed (Fig. 3-6). The cells assume these respective shapes in the lining of the bladder as urine fills or leaves that organ. Transitional epithelium is also found lining the ureter, which drains urine from each kidney, and lining the central portion of the kidney itself.

Functions in Body Cavities, Vessels, and Glands

Four functional types of epithelium can be identified. They are found lining the mucous membranes, the serous membranes, the body's vessels, and its glands.

Mucous Membranes

Mucous membranes cover the inner surfaces of cavities, tubes, and ducts in the body. They are located in parts of the digestive, respiratory, urinary, and reproductive systems. Mucous membranes consist of an epithelial lining supported by connective tissue (the lamina propria) and in some cases also by a deeper layer of smooth muscle (the muscularis). The columnar epithelial cells synthesize and secrete a viscous (thick) fluid called *mucus.* This fluid provides a wet internal barrier within passages lined by mucous membranes. The composition of mucus varies according to the needs of the area in question. A variety of different molecules secreted by the epithelial cells in the membrane are carried in mucus. For example, the mucus on the lining of the respiratory tract is coated

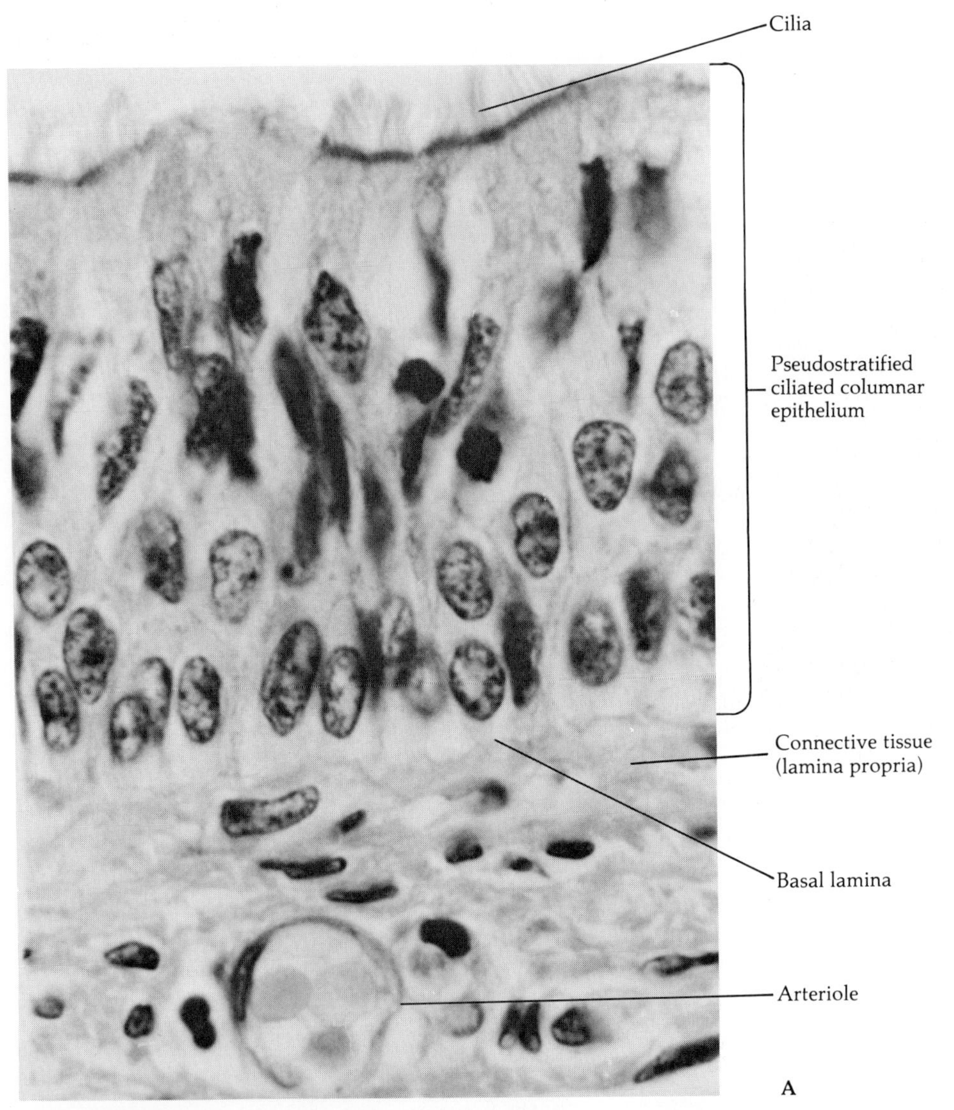

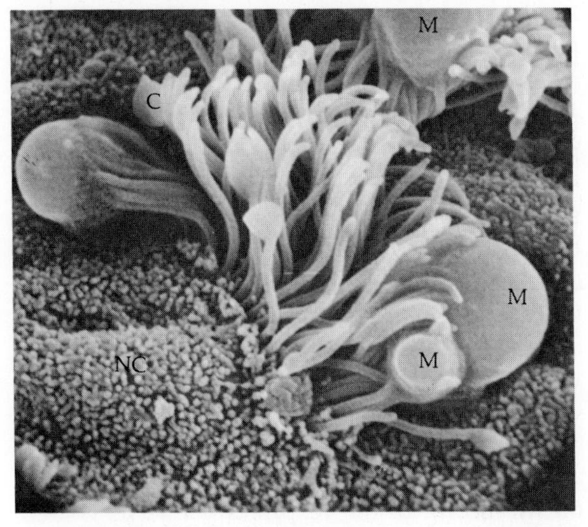

FIG. 3-5
Pseudostratified ciliated columnar epithelium. (A) The upper figure is a photomicrograph of the tissue (×1700). (B) The lower figure is a scanning electron micrograph (×7700) of the tissue. It illustrates cilia (c) on the cell, mucus droplets (m), and the surface of nonciliated cells (nc).

with an abundance of phospholipids, whose properties are much like those of detergents, while that of the stomach is laden with hydrochloric acid and the enzyme *pepsin,* which breaks down proteins.

Serous Membranes (*Mesothelium*)

Mesothelium is a serous membrane composed of simple squamous epithelium supported by

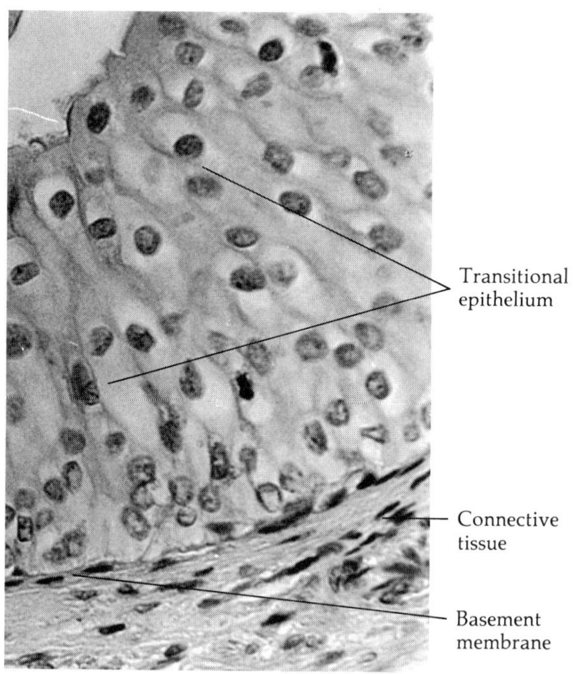

FIG. 3-6
Transitional epithelium. The upper layers are formed by transitional epithelial cells overlying a horizontally placed basement membrane (×700).

Endothelium

The inner lining of the circulatory and lymphatic systems is composed of simple squamous epithelium called *endothelium*. It is the sole tissue layer in the smallest vessels of the circulatory and lymphatic systems. These vessels are called *capillaries*. The endothelium continues into the chambers of the heart, forming the *endocardium*. Endothelium provides a smooth surface which tends to prevent spontaneous clotting of blood and minimize friction between blood cells and blood vessel walls.

The endothelium in capillaries is a site of entry and exit for fluids, ions, molecules, and cells carried into and out of the bloodstream and lymph stream. Differences in permeability (passage of substances across membranes) occur at various locations in the body. Permeation may be adversely affected by a variety of disorders, including dietary deficiency, infection, allergy, and changes in blood pressure. Such alterations can dramatically affect normal body function.

underlying connective tissue. Serous membranes form a watery (serous) secretion which reduces friction and minimizes wear and tear within the body. Serous membranes are named after the three major body cavities that they line (Fig. 3-7). The *pleural* membranes surround the lungs and the *pericardial* membrane surrounds the heart. These organs and membranes are located within a large chamber called the *thoracic cavity*. The *peritoneal* (abdominal) membrane lines the *abdominal cavity* and its contents.

In some areas of the abdominal cavity, the serous membranes not only surround the organs but also form a supporting structure called the *mesentery*. The latter is composed of a double layer of mesothelium (the peritoneal membrane) surrounding a central layer of loose connective tissue in which blood vessels, lymphatic vessels, fat, nerves, and digestive organs are suspended (Fig. 3-8).

FIG. 3-7
Body cavities.

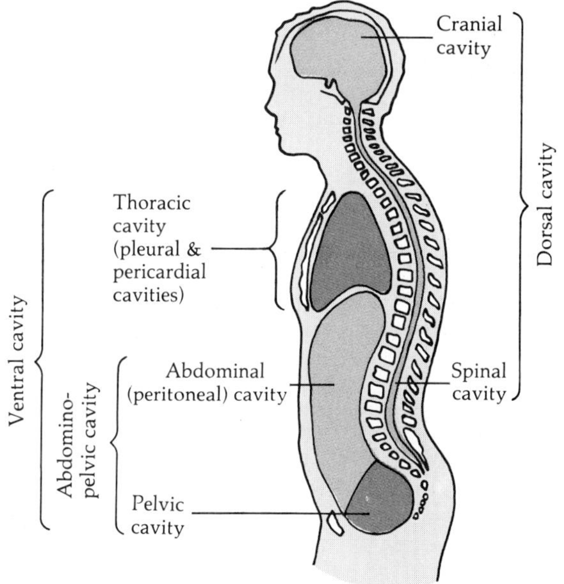

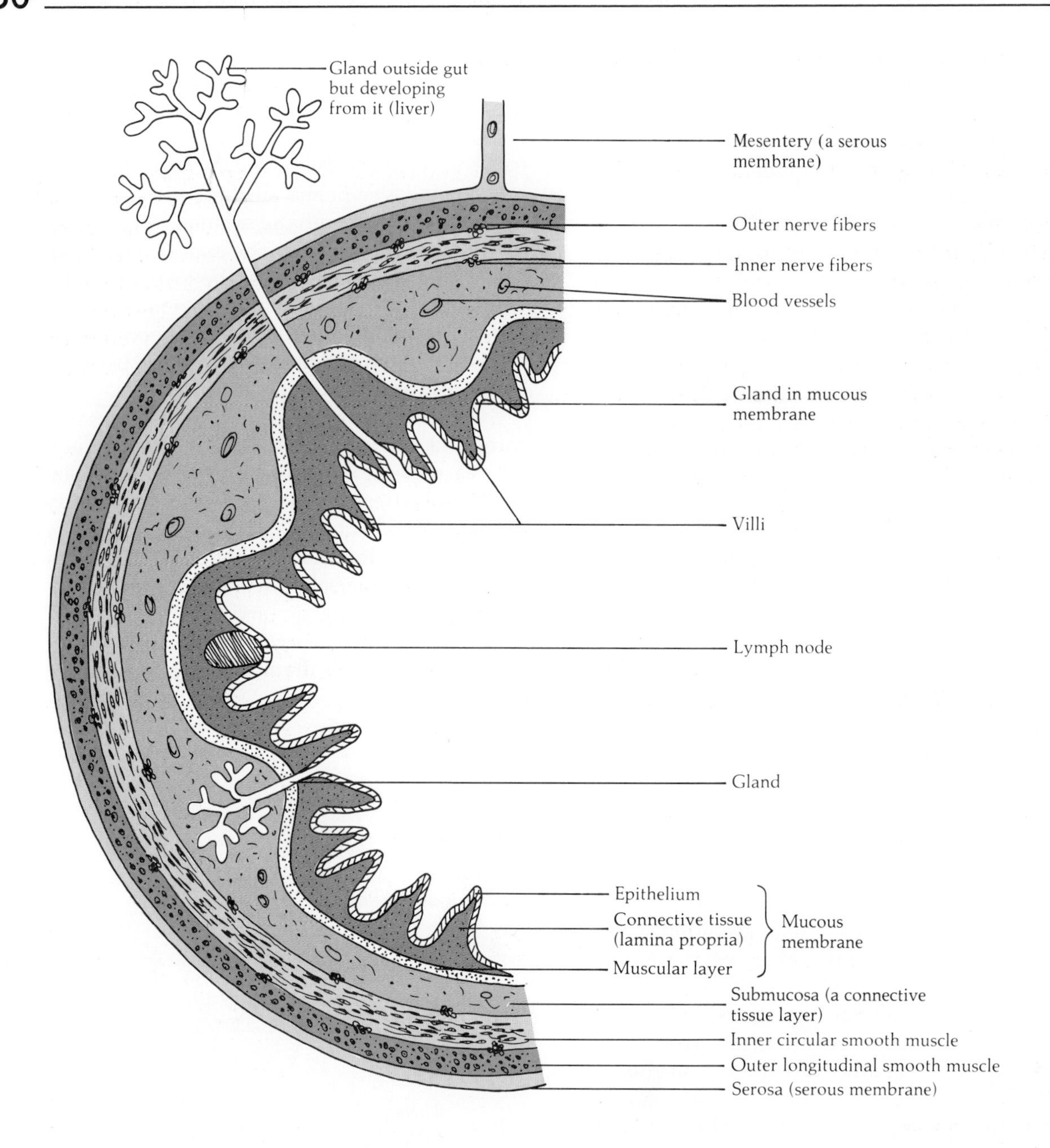

FIG. 3-8
Mucous and serous membranes of the digestive tract. The intestine is shown loosely suspended by the peritoneum, a serous membrane. The intestine is lined by a mucous membrane.

Glands

A *gland* is a cluster of epithelial cells which synthesize and secrete specific substances. The epithelium in a gland is normally mixed with connective tissue that divides the structure into lobes and lobules. Glands are surrounded and/or penetrated by blood vessels and nerves that supply nutrients and help regulate their activity. There are two general types of glands. *Exocrine glands* release their products through ducts while *endocrine* glands deliver hormonal secretions directly into the bloodstream.

EXOCRINES Exocrine glands may be distinguished by the complexity of the ducts and cells which compose the gland, as well as the nature of their secretions and the manner in which they are released (Fig. 3-9).

Simple exocrines possess a single unbranched duct. They include glands which form sweat and some sebaceous glands which produce oil in the skin. Simple exocrines also form the majority of glands along the digestive tract. Some are *unicellular* while others are multicellular. *Goblet cells,* which secrete mucus in the digestive and respiratory tracts, are examples of unicellular glands (Fig. 3-2D).

Multicellular glands may be simple or compound (Fig. 3-9). *Compound* glands have ducts which branch into several lobules. Mammary and salivary glands are examples. Multicellular glands may also be classified according to their shapes and secretions. They may be tubular, saclike (alveolar or acinar), or a combination of both. They may form viscous *mucus* or watery *serous* secretions.

Exocrines may also be distinguished by the manner in which they deliver their products. *Merocrine* glands release their secretions via the Golgi apparatus in a process called *exocytosis,* which will be described in Chap. 4. Examples of merocrines include the salivary and sweat glands. Holocrine glands shed the whole cell and its contents. This is how oil is secreted by the sebaceous glands. *Apocrine* glands seem to set their products free by releasing the tip of the cell's cytoplasm. This mechanism is debatable but thought by some to occur in the mammary glands of the breast.

FIG. 3-9
Types of exocrine glands. Basic cellular arrangments in simple and compound glands. (A) Simple gland (duct does not branch). (B) Compound gland (duct branches). (C) Classification of glands according to secretory segments: (1) tubular (if secretory portion is tubular); (2) alveolar or acinous (if secretory portion is flasklike); (3) tubuloalveolar (if secretory portion is both).

If duct doesn't branch:
it is a simple gland.

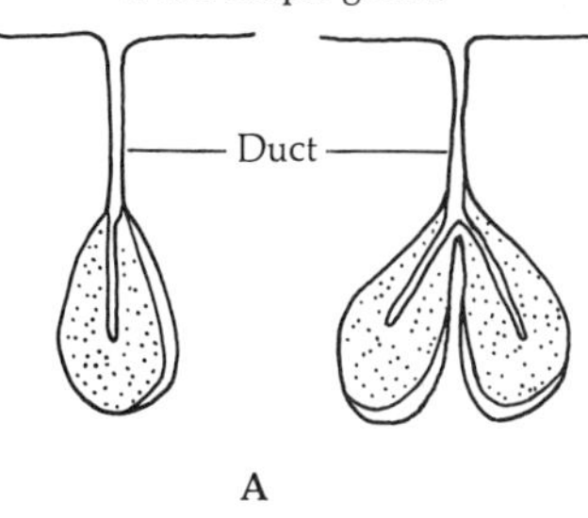

If duct branches:
it is a compound gland.

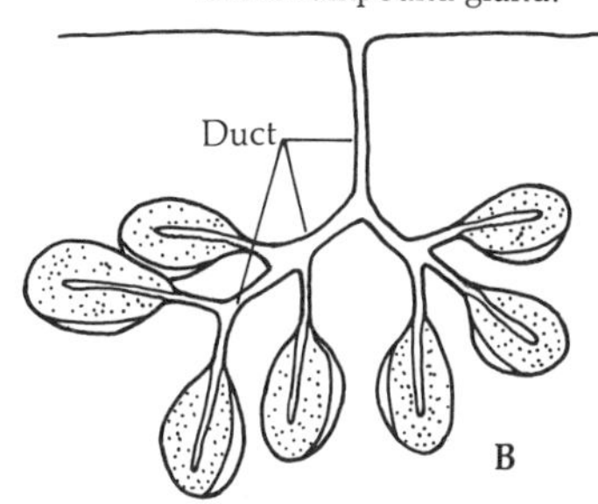

If secretory portion is:

1 tubular, it is a tubular
exocrine gland.

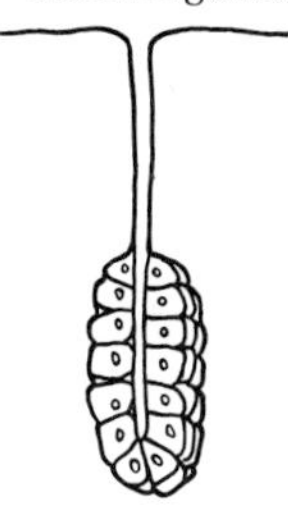

2 flasklike, it is an alveolar or
acinous gland.

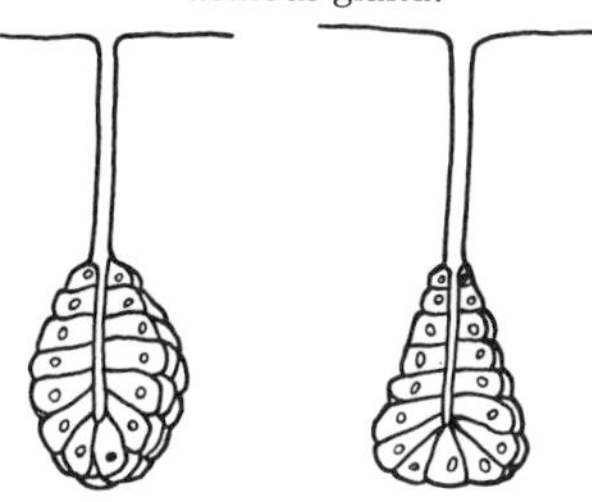

3 both, it is a tubulo-
alveolar gland.

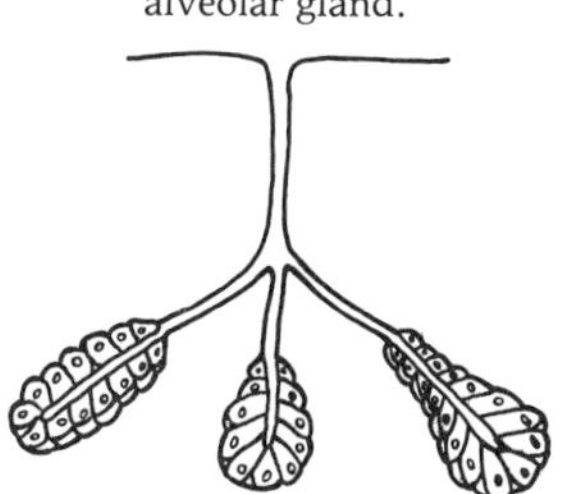

C

ENDOCRINES The endocrines are usually composed of irregular, cordlike groups of epithelial cells whose secretions are delivered directly into the blood. The endocrines consist of a variety of specialized cells found in several glands scattered throughout the body. Some of these

TABLE 3-1
Some endocrine glands and their products

Gland	*Location*	*Major Hormones*
Thyroid	Base of neck	Thyroxine Calcitonin
Parathyroids	Posterior surface of thyroid	Parathormone
Pancreas	Posterior and inferior to the stomach, in curve of first part of small intestine	Insulin Glucagon Somatostatin
Adrenal	Superior (upper) end of kidneys	
Cortex	Outer rim of gland	Glucocorticoids (e.g., cortisone) Mineralocorticoids (e.g., aldosterone) Sex steroids
Medulla	Inner central tissue	Epinephrine Norepinephrine
Pituitary	Base of center of brain	
Anterior pituitary (adenohypophysis)		Somatotropic hormone (STH), growth hormone (GH) Adrenocorticotropic hormone (ACTH) Thyroid-stimulating hormone (TSH, thyrotropin) Luteinizing hormone (LH, interstitial cell-stimulating-hormone, ICSH) Luteotropic hormone (LTH, prolactin) Follicle-stimulating hormone (FSH)
Posterior pituitary (neurohypophysis)		Oxytocin Antidiuretic hormone (ADH, vasopressin)
Hypothalamus	Superior to pituitary within brain	Releasing and inhibitory factors which regulate the secretion of anterior pituitary hormones Source of oxytocin and ADH
Thymus	Base of neck at inferior edge of thyroid	Thymic hormones, including thymosin
Pineal gland	Center of brain	Melatonin
Digestive tract	Glands along its inner surface in the abdominal cavity	Secretin Gastrin Cholecystokinin-pancreozymin (CCK-PZ) Gastric inhibitory polypeptide (GIP)
Testes	Outside of and below abdominal cavity in scrotum	Testosterone
Ovaries	In lower (inferior) abdominal cavity on each side of the uterus	Estrogens
Corpus luteum	Within ovaries	Progesterone
Placenta	Attached to fetus and wall of uterus	Estrogens Progesterone Gonadotropins
Tissues	Throughout body but especially in seminal vesicles at lower part of bladder	Prostaglandins

glands and their products are shown in Table 3-1. Their powerful effects on body function were briefly described in Chap. 1 and will be discussed in detail in later chapters, including Chap. 5 on the endocrine system.

Dynamics of Structure and Function of Epithelium

The description of epithelial structure and function given above may leave one with the impression that the differences ascribed to the epithelial tissues are absolute and unchangeable. In fact, this is true neither for epithelium nor for any other tissue. The structure and function of all cells and tissues vary during their lifetime. The changes are intimately linked to the genes of the cells, which direct a series of changes according to a programmed time-table whose speed and direction are markedly influenced by environmental factors. Thus, the shape and function of epithelial tissues are extremely variable and can change in response to chemical and physical stimuli. For example, *vitamin A deficiency* may alter the ciliated columnar epithelium of the trachea. These column-shaped cells normally secrete mucus but can be converted into irregular, squamous, non–mucus-producing cells. This change represents loss of a normal barrier to infectious agents. Since squamous cells lack mucus and cilia, they are capable neither of trapping foreign material nor of propelling it out of the respiratory tract. Such alterations are associated with variations in the cells' ribonucleic acid (RNA) and with the inability of the cells to synthesize some specific proteins. The administration of vitamin A may restore their normal structure and function.

Further investigations indicate that vitamin A is needed for the development and maintenance of differentiation (maturation) of epithelial cells. Its precise role is not clear; however, long-term deficiencies may lead to formation of epithelial tumors. The growth of tumors involves the loss of properties characteristic of normal tissue. This process is called *dedifferentiation*. Some evidence suggests that derivatives of vitamin A might inhibit the dedifferentiation of cells from normal to abnormal types. Cells might even *redifferentiate* to normal states under the influence of vitamin A. For example, recent evidence shows that vitamin A, administered orally or injected into the peritoneal cavity of mice, inhibits the development of a skin cancer known as *melanoma*. The vitamin apparently increases the animal's defense mechanisms against tumor cells, leading to destruction of the tumor.

3-2 CONNECTIVE TISSUE

Most connective tissues share three common features; besides cells, they contain cellular fibers and ground substance. There usually is more space between individual cells of connective tissue than between those of epithelium. The intercellular substance is also termed the *matrix* or *ground substance* and is composed of a variety of fibers and chemicals which produce an amorphous (formless) liquid, semisolid, or solid material. In connective tissue, the ground substance consists mostly of proteoglycans (combinations of protein and mucopolysaccharides). They are formed by the repeated linkage of many individual molecules (monomers) to make large polymers, such as hyaluronic acid and chondroitin sulfate. The ground substance characteristically traps a large amount of water to form a homogeneous viscous fluid or a thin gel. Small molecules diffuse through the ground substance to and from surrounding blood capillaries. The passage of large molecules, cells, and microorganisms is hindered.

Collagen and elastic fibers are the two main types of fibers found in the ground substance of connective tissues. *Collagen* forms groups of parallel fibers in coarse, wavy bundles (Fig. 3-10). Collagen molecules are composed of three protein chains of tropocollagen, cross-linked in a triple helical coil. The triple helical coils are bound together as cablelike structures linked side to side and along their length to form fibrils of collagen 5 to 200 nm wide. The fibrils in turn cross-link to form collagen fibers which are visible microscopically. Four major types of normal collagen have been found in different connective tissues. They are produced from

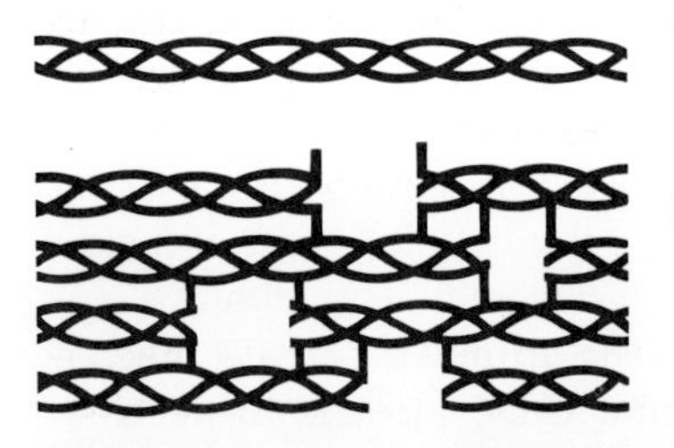

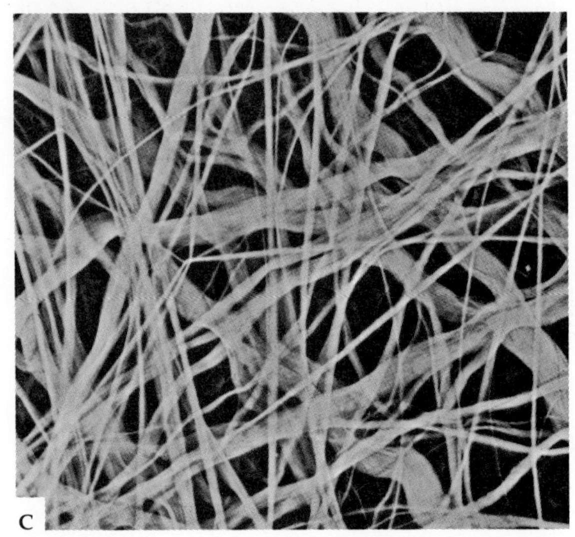

FIG. 3-10
Collagenous fibers. (A) and (B) The molecular architecture of collagen. (C) Photomicrograph of collagen fibers which vary in appearance and size as illustrated (×550). The fibers are composed of many molecules of collagen protein. From *Principles of Biochemistry* by A. White, P. Handler, C. L. Smith, R. L. Hill, and I. R. Lehman. Copyright © 1978 McGraw-Hill Book Company and *Histology*, 4th ed. by L. Weiss and R. O. Greep, copyright © 1977 McGraw-Hill Book Company; used with permission of McGraw-Hill Book Company.

chains of protein called α-1 chains. The composition of α-1 chains varies with the function of the tissue in which they are found. Type I alpha chains are found in tendon, bone and skin; type II in cartilage; type III in blood vessels, the skin, and the fetus; and type IV in the basement membrane. Boiling breaks down the collagen into an amorphous product, gelatin.

Some evidence supports the hypothesis that aging may be related to an increase in crosslinking of collagen. This stiffens the molecules and restricts movement of the body. The crosslinkages also may result in a decreased rate of diffusion and make it more difficult for vital molecules and ions to get in and out of cells. As a result, waste products may accumulate, so that the cells age more rapidly.

Elastic fibers are composed of elastin, an amorphous protein in the form of microfibrils. Elastin forms the major part of freely branching elastic fibers and imparts a high degree of pliability to them. Elastic fibers are located in areas that stretch (e.g., arteries and the trachea) and are found in bundles that are smaller and finer than collagenous fibers (Fig. 3-11).

Connective tissues are marvelously engineered for various functions. They provide support by linking and binding tissues and organs together. Cartilage and bone are special-

FIG. 3-11
Elastic fibers. The elastic fibers in this network are thinner and more homogeneous than the collagenous fibers shown in Fig. 3-10 (×550). From *Histology*, 2nd ed., by L. Weiss and R. O. Greep. Copyright © 1966 McGraw-Hill Book Company; used with permission of McGraw-Hill Book Company.

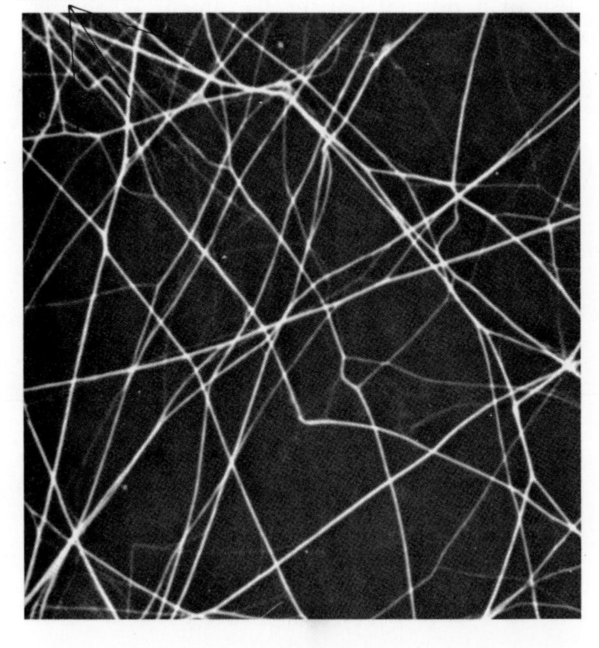

ized connective tissues that provide the body with a supporting frame. Blood is a specialized connective tissue which carries fluids through the body. With aging, variations in the proportions of fibers to ground substance occur and impart different properties to the several types of connective tissues.

Loose (Areolar) Tissue

General Properties

Areolar tissue is the most common kind of connective tissue. It is found in spaces between organs and it penetrates them. Areolar tissue contains several cell types located among a variety of pliable, elastic, and rugged woven fibers (Fig. 3-12). Its loose arrangement is responsible for the tissue's name. Large quantities of areolar tissue are found surrounding blood vessels, muscles, and nerves. Areolar tissue makes up the *superficial fascia* beneath the skin, attaching the dermis to underlying muscle.

Role of Major Cell Types

Five cell types are abundant in loose connective tissue. They are fibroblasts, fat cells, macrophages, mast cells, and plasma cells. In addition, a variety of blood cells move in and out of the intercellular spaces of this tissue.

FIBROBLAST The most common type of cell in loose connective tissue is the *fibroblast,* a small, flat, often star-shaped cell, which possesses an elongated nucleus with two to four nucleoli (Fig. 3-13). The fibroblast is thought to be a primitive cell type formed in early embryological development from mesenchyme derived from the middle (mesodermal) germ layer. The fibroblast is active in the synthesis of fibers found in connective tissue and plays an important role in the repair of injury. This function is stimulated by steroid hormones which are re-

FIG. 3-12
Loose (areolar) connective tissue. The major cell types and fibers found in loose connective tissue are depicted (×950).

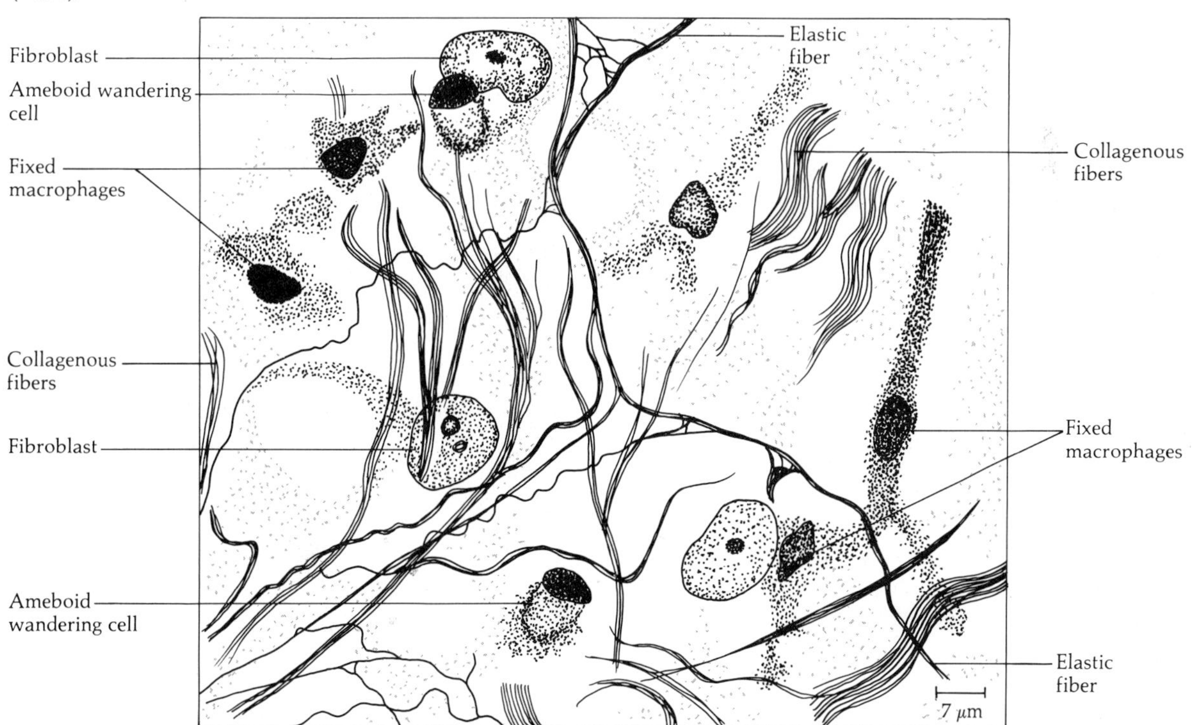

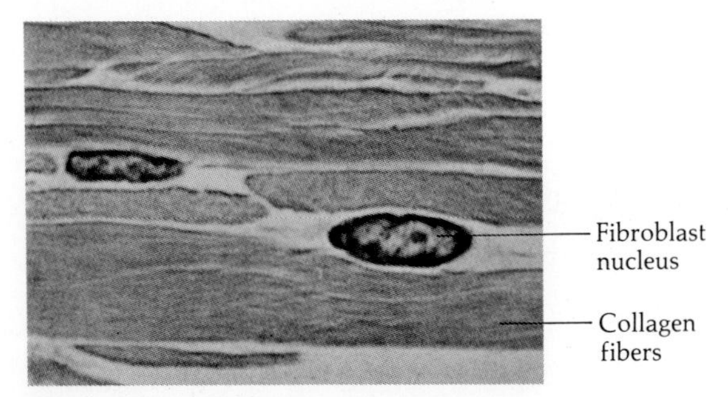

FIG. 3-13
Fibroblasts surrounded by collagen fibers in dense connective tissue (× 1500). The elliptical nuclei of two fibroblasts are apparent although the limits of their cell membranes are not. There are abundant numbers of collagen fibers surrounding the two cells. From *Histology*, 4th ed., by L. Weiss and R. O. Greep. Copyright © 1977 McGraw-Hill Book Company; used with permission of McGraw-Hill Book Company.

leased from the adrenal glands in increasing amounts during times of stress.

Mammalian cells revert to forms resembling fibroblasts when grown in tissue culture (Fig. 3-14). As a result, cells used in tissue culture generally resemble each other in structure regardless of their source. Tissue culture cells are used to grow viruses and also to synthesize molecules for laboratory and medical studies. For example, the virus for the polio vaccine is grown in tissue cultures of monkey kidney cells.

FAT CELL (ADIPOCYTE) Loose connective tissue in which fat cells (adipocytes) are the major cell type is called *adipose tissue.* Fat accounts for about 15 percent of adult body weight in the average male and 18 percent in the average female. Fat (lipid) is most often found as a single droplet, accounting for up to 85 percent of the weight of each cell. The nucleus of an adipocyte is usually inapparent, having been pushed to the side by accumulated lipids. The cell is surrounded by a fine membrane, which seems to contain empty space. Fat is found as a fluid or oil since normal body temperature is above its melting point. The transparency of the fluid makes the cell appear to be occupied by a cavity or vacuole (Figs. 3-15 and 3-16).

Fat cells are thought to be special forms of fibroblasts that store fat. Evidence indicates that the fat cell number in the body may be determined in infancy. The maximum number found in adults varies from 21 billion to 43 billion. Some data suggest that the higher number is attained as a result of excessive proliferation (hyperplasia) of the cells when overfeeding occurs during infancy. The size of fat cells varies with lipid content and increases following excessive intake of food. It decreases when energy expenditure of the body exceeds the energy content of the diet.

Fat cells provide a firm but resilient packing material. They protect the tissues and organs of the body from physical shock. Fat cells also afford a two-way barrier to heat exchange, just as insulation keeps a house cooler in summer and warmer in winter. Fat represents the largest easily usable energy reserve present in all the body's tissue types. Heat production is the primary function of brown fat, which has more blood vessels than the more common white fat. Brown fat is characteristically found in the fetus and newborn, but little of it is found in adults.

DEFENSE CELLS Three common types of cells in connective tissue are among the body's prime means of defense against disease. They are plasma cells, macrophages, and mast cells. These cells, along with those of specialized reticular connective tissues (which include blood and lymphoid tissues), are found in organs whose collective activity is often described as an immune system. The *immune system* consists of those organs whose cells defend the body against disease and which react to foreign substances by synthesizing molecules to assist body defenses. Many of its functions are car-

FIG. 3-14
Fibroblasts in tissue culture (× 500). Note the irregular shapes assumed by the different cells. From *Histology*, 4th ed., by L. Weiss and R. O. Greep. Copyright © 1977 McGraw-Hill Book Company; used with permission of McGraw-Hill Book Company.

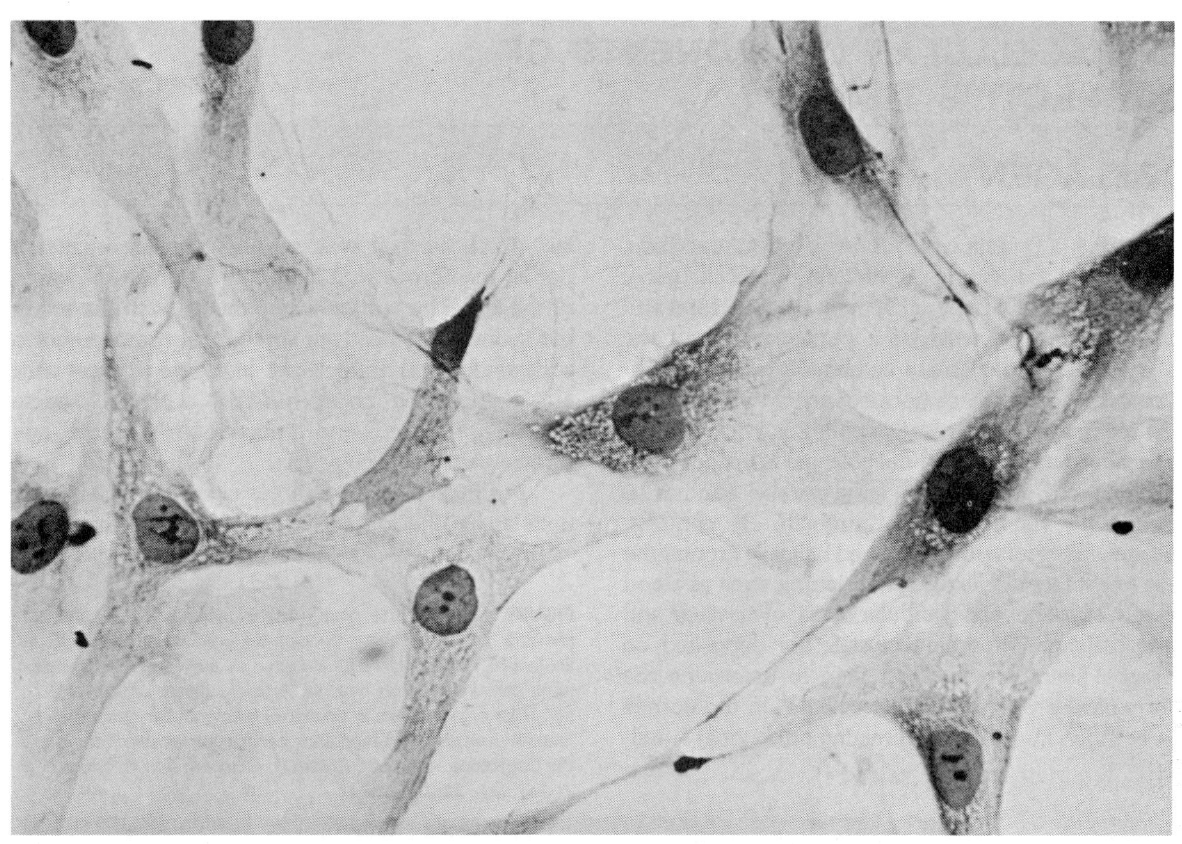

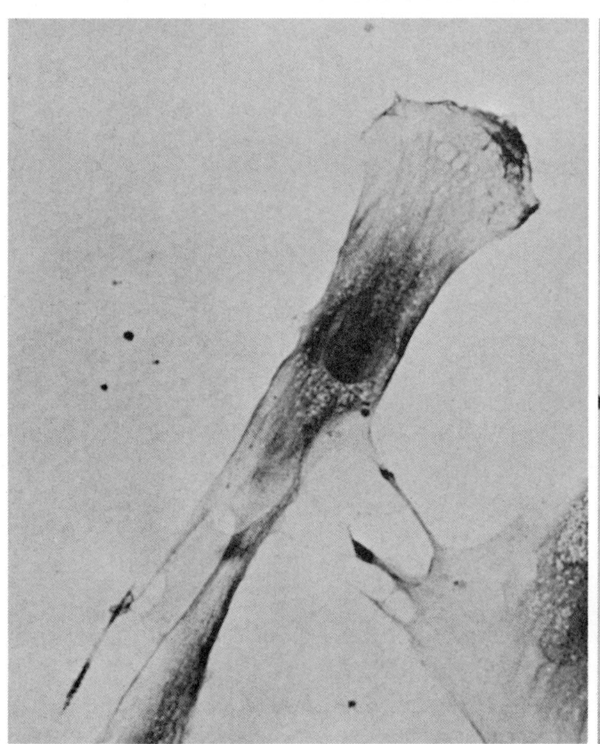

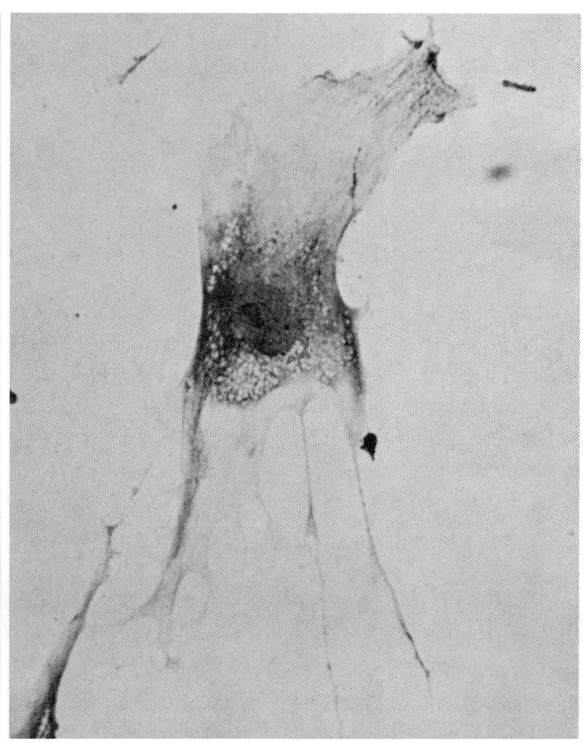

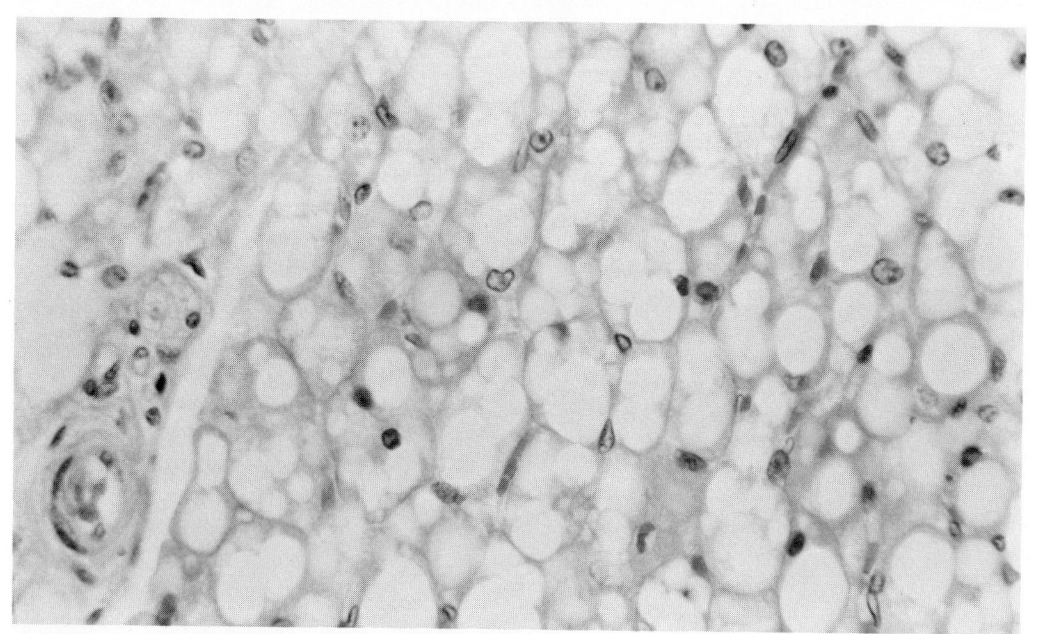

FIG. 3-15
Fat cells (adipocytes). Adipose tissue. The fat has been dissolved in the preparation of the tissue, leaving the appearance of white space surrounded by cell membranes and elliptical nuclei (×700).

FIG. 3-16
Lipid droplet in fat cell. Part of the nucleus, a lipid droplet, and the adjacent cytoplasm (at arrows) are shown at high magnification (×34,000).

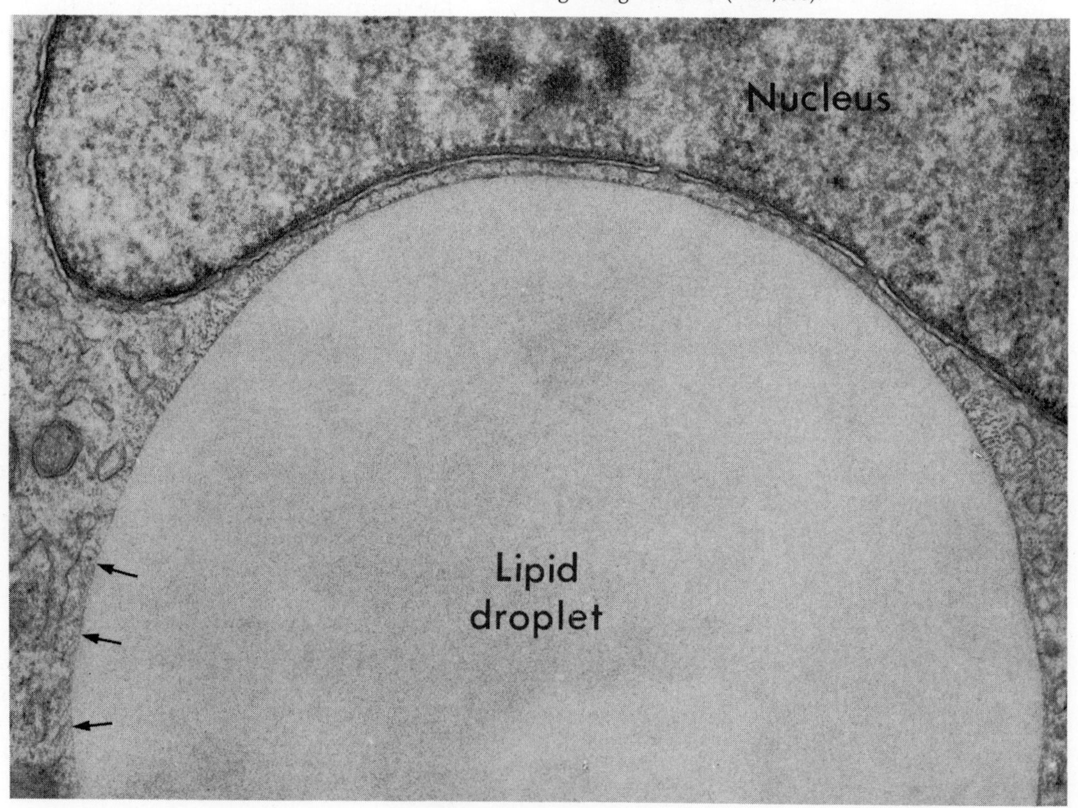

ried out by cells found in loose and reticular connective tissues, which are described below.

Plasma Cell Plasma cells are derived from white blood cells called *lymphocytes.* Lymphocytes are formed in the reticular tissue of the thymus as T cells, or may develop in the bone marrow as B cells. Reticular connective tissue will be described below. The two types of lymphocytes (T and B cells) may be distinguished under the scanning electron microscope, which provides high magnifications of the surfaces of cells (Fig. 19-3). The T cell often shows about 100 antibody-like proteins, while each B cell is more often studded with about 100,000 of them. Both cell types are derived from the multiplication of embryonic cells in their tissues of origin and migrate through the blood to settle in lymph nodes. They also may proliferate and migrate back to the blood or to other tissues. Following cell division, some B cells mature into plasma cells, which secrete protective proteins, called *antibodies,* into the body's fluids.

Plasma cells have an eccentric (off-center) nucleus, which has an affinity for pyronine dye. The cytoplasm of plasma cells often contains a large amount of *rough endoplasmic reticulum* (Fig. 3-17), which is associated with active protein synthesis, especially the synthesis of antibody. Plasma cells are located throughout loose connective tissue and mucous membranes of the gastrointestinal and respiratory tracts. The antibodies they produce provide the body with a strike force that inhibits penetration of vital tissues by potentially damaging molecules or microbes.

Role in Antibody Production The primary role of the plasma cell is to secrete antibodies which are physically and chemically complementary to the substances that cause their production. These substances, which are usually large foreign molecules, are called *antigens.* Antigens stimulate B lymphocytes to divide and form the prime antibody-producing cells, the plasma cells, and daughter lymphocytes as well.

Antibodies are globular proteins produced as part of a dual defense mechanism known as the *immune response.* Antibodies circulate through the body fluids and provide protection called *humoral immunity.* The other facet of the defense mechanism stems from the activity of T lymphocytes and phagocytes and is known as *cellular immunity.* Plasma cells produce five major classes of antibodies, each molecule of which is called an *immunoglobulin* (Ig). Antibodies are proteins often formed of units of four chains but differing in structure in molecules of IgG, IgM, IgA, IgD, and IgE. Their general form is shown in Figs. 3-20 and 19-4. IgG is the most common antibody in the bloodstream. IgM is generally produced earliest in the immune response. IgA is found in the highest amounts in body fluids outside the blood (e.g., tears and saliva). The role of IgD is not yet clear although it seems to form receptor molecules on some cells. IgE plays an important role in some types of allergic reactions.

Role of IgA Antibody in Protein Absorption IgA, or secretory antibody, is released from plasma cells to the interior and exterior surfaces of the body. Its presence in tears, saliva, and other secretions aids our body's defenses. Like other antibodies, it can neutralize and lead to the destruction of microbes and their products. Specific IgA molecules can bind to intact proteins in the digestive tract and prevent their intestinal absorption as whole molecules, thereby inhibiting the development of allergic reactions to them. Normally, digestive enzymes break down proteins into amino acids, which are not antigenic or allergenic. However, when foreign, intact proteins enter the bloodstream, an immune reaction may follow. A recent study indicates that animals that lack secretory IgA may more readily absorb whole proteins and thus be more susceptible to foodborne allergies.

Macrophage The *macrophage* is the second most common cell in loose connective tissue. It is large and *phagocytic* (cell-eating); it is referred to as a *histiocyte* when it is stationary and as a *wandering macrophage* when it moves from site to site (Figs. 3-18 and 3-19). Macrophages are irregularly shaped and best identified by their biological properties (Table 3-2) and ability to combine with special stains. For example, staining can reveal lysosomal enzymes, which accumulate in high amounts in active macrophages.

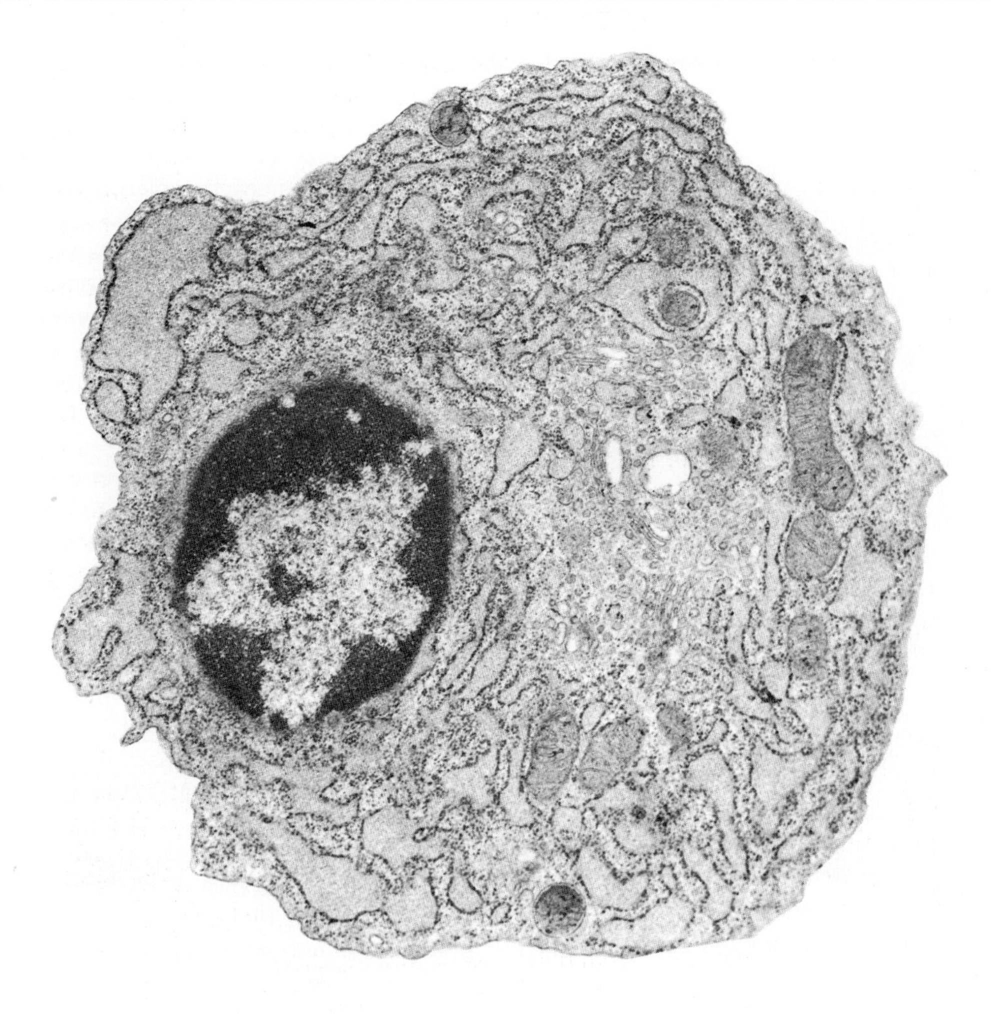

FIG. 3-17
Plasma cell. The eccentric nucleus and abundant rough endoplasmic reticulum are characteristic of intermediate-size plasma cells (× 17,160).

The macrophages originate from bone marrow cells which mature into wandering white blood cells, the *monocytes.* The monocytes are bloodborne macrophages which migrate from blood vessels to penetrate tissue spaces and mature into tissue macrophages. Collectively, the tissue macrophages form part of a group of phagocytic cells called the *reticuloendothelial system.*

Relation to Defense Mechanisms Macrophages, along with other phagocytic cells, including some white blood cells, constitute the body's primary defense. These cells can phago-

TABLE 3-2
Modification of macrophage activity

Stimulus	*Activity*
An antigen (a large, foreign, or altered agent)	Induces phagocytosis (ingestion by the macrophage)
Antibody to foreign substance	Increases efficiency of phagocytosis
Release of lymphokines from lymphocytes	May increase or decrease macrophage activities such as its ability to release lysosomal enzymes which cleave the antigen into smaller components

cytize (ingest) molecules or microorganisms (Fig. 3-19) by *endocytosis*, described in the next chapter. They also devour cells which are worn out or differ in architecture from normal healthy ones. Their targets may include aged, diseased, or cancerous (malignant) cells or others which are recognized as altered or foreign.

Relation to Antibody The efficiency of macrophage activity is increased by the presence of antibodies produced against the material that is to be ingested. Antibodies assist the macrophage in recognizing and engulfing foreign substances. The opposite ends of thousands of antibodies attach to macrophages and foreign substances that are to be ingested (Fig. 19-4). The combination allows cross-linking of molecules along the surface of the macrophage cell membrane and stimulates phagocytosis.

Activation by Lymphokines in the Inflammatory Reaction The activity of macrophages may be altered by molecules called *lymphokines* (Chap. 19). Lymphokines are produced by T cells in response to the presence of antigens. Lymphokines diffuse to the macrophage and can regulate its activity. Under their influence, macrophages may form larger quantities of lysosomal enzymes. The latter hydrolyze (digest) susceptible molecules at the site of the reaction. Under these conditions the macrophages are described as "angry" or "activated" macrophages. They are more efficient in engulfing and degrading ingested matter.

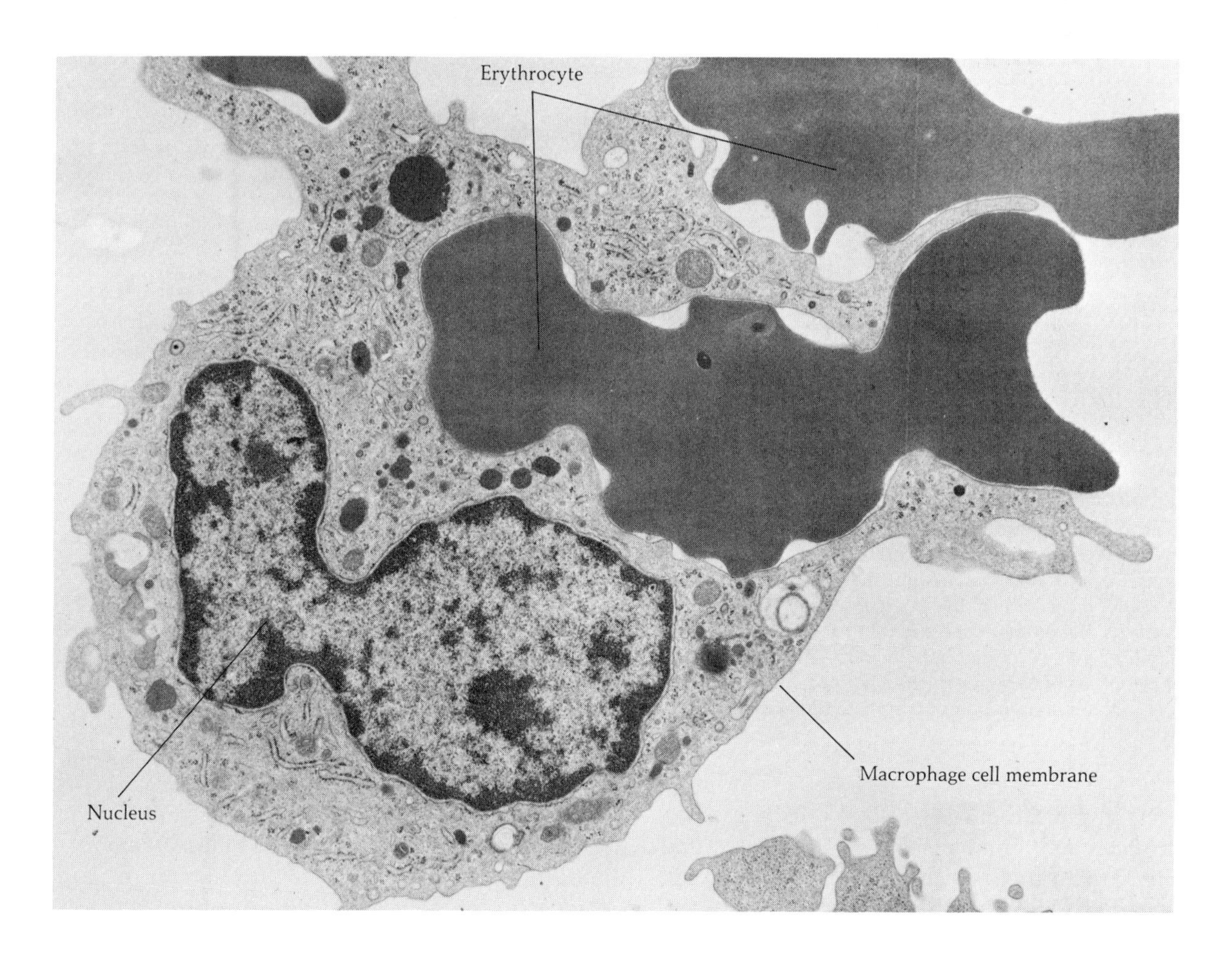

FIG. 3-18
Macrophage ingesting a red blood cell (erythrocyte) (× 9000).

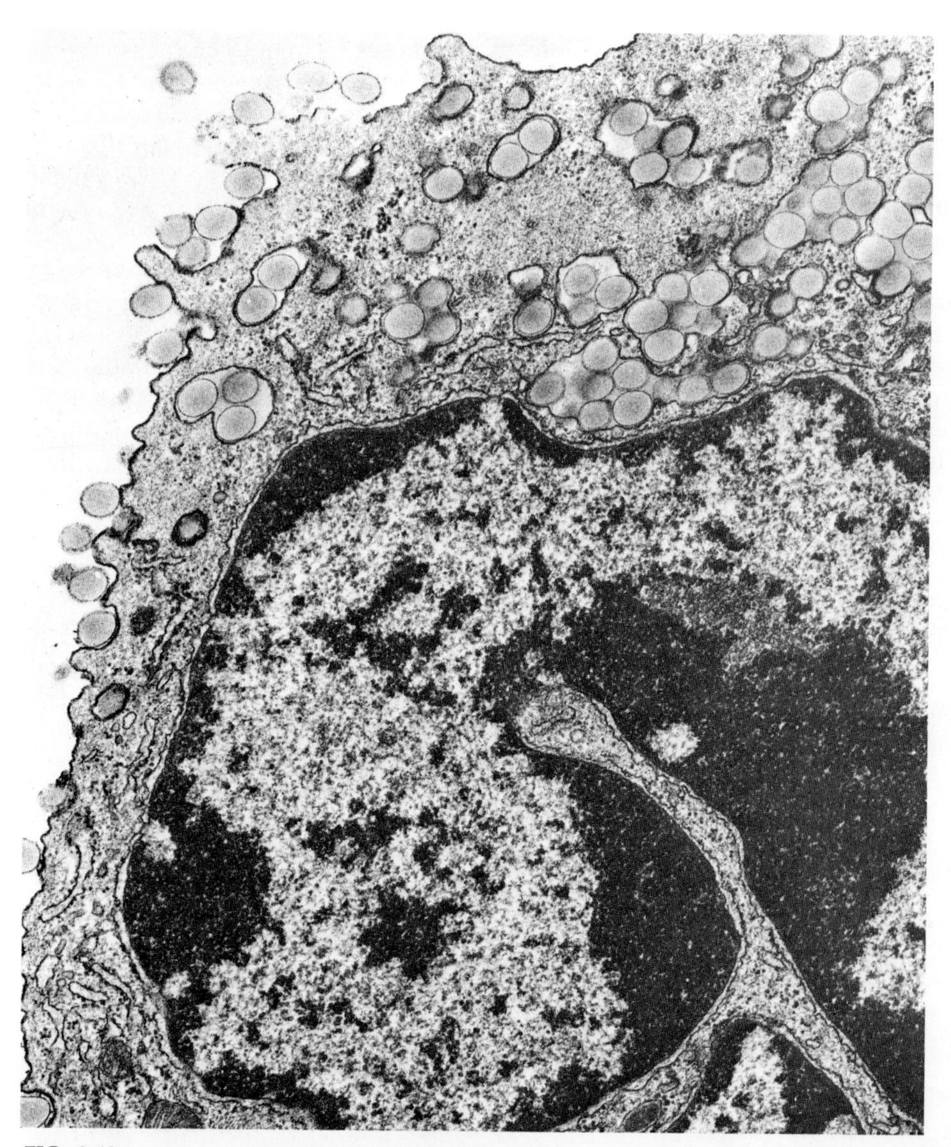

FIG. 3-19
A macrophage ingesting plastic beads (×21,000). Polystyrene beads are evident in the extracellular space in the upper left portion of the photomicrograph. Some beads are nestled in pockets at the external surface of the macrophage cell membrane. Others have been phagocytized (ingested) and are located in membrane enclosed sacs in the cytoplasm. The large structure with dark and light areas in the bottom center of the photomicrograph is the nucleus of the cell.

The migration of white blood cells, including lymphocytes, monocytes, neutrophils, and others, to an invaded area constitutes a defense mechanism known as the *inflammatory response*. Early in the response, large numbers of neutrophils migrate to the site of inflammation. They are small phagocytic cells called *microphages* and they release great quantities of lysosomal enzymes. An abundance of these cells, as well as monocytes, macrophages, and lymphocytes, can be observed under the microscope. As a result of their presence, an increase in phagocytosis and hydrolytic enzyme activity occurs during the response. The release of a number of other defense factors from the bloodstream also occurs. At a wound, the accumulation of neutrophils, cell debris, and fluid is seen as *pus*.

Role in Cancer Defense Activation of the macrophage seems to be an important feature in defense against the growth of tumor cells. Its failure may, to a large extent, determine susceptibility to various types of cancer.

Role in Rheumatoid Arthritis In some cases, a defense reaction may produce undesirable side effects as a result of elevated enzyme activity following lymphokine production. For example, for as yet unknown reasons, antibodies may be produced against components in the tissues of joints of some people. The antibodies may attract and activate cells that cause an inflammatory reaction in the surrounding connective tissue (*synovial membrane*) resulting in *rheumatoid arthritis*. *Collagenase* activity of macrophages seems to increase under these conditions. This enzyme hydrolyzes collagen and destroys connective tissue at the site of the reaction.

Mast Cell Mast cells are round or polygonal and are found in connective tissue adjacent to capillaries. Their bloodborne equivalents are called *basophils*. Both cell types often contain large numbers of *metachromatic granules*, which change the color of applied dyes. The granules represent stored materials, two common types of which are the molecules *heparin* and *histamine*.

Role of Heparin and Histamine *Heparin* is an anticoagulant. It prevents blood from clotting. *Histamine's* vasoactive properties cause blood vessels to widen and increase the permeability of capillaries and thereby promote a loss of fluid into intercellular spaces. The accumulation of fluid causes swelling and is called *edema*.

Relationship of Histamine to Allergic Reactions Foreign molecules called *allergens* can induce two types of allergic reactions, namely *immediate-type allergies* and *delayed-type allergies*. The release of histamine or other types of compounds which affect blood vessels causes immediate-type allergies to occur within minutes to hours after the second or subsequent exposure to the allergen. The release of lymphokines from T lymphocytes (Chap. 19) causes delayed-type allergies to develop hours to days after second or subsequent exposure to other types of allergens. Skin sensitivities such as poison ivy, reactions to chemicals in cosmetics, and certain allergies to drugs (such as penicillin) are examples of delayed-type allergic reactions. Allergies are examples of defense reactions with unpleasant side effects.

Examples of immediate-type allergic reactions include asthma, hay fever, and hives. In these disorders, the allergens include pollen, animal hair, household dust, drugs, foods, airborne fungi, and a variety of other substances. The localization of the reaction in different sites causes various types of disorders. When the effects of the allergen and compounds such as histamine are limited primarily to the upper respiratory membranes, the disorder is known as *hay fever;* in the skin allergens produce *hives* or *eczema;* when the reaction induces contraction of smooth muscle in the bronchioles of the lungs, the resulting disorder is called *asthma;* when it causes contraction of smooth muscle in the digestive tract, it mediates *foodborne allergies*.

In immediate-type allergies, hypersensitivity exists to allergens which cause an immune reaction and the production of IgE by plasma cells. The first reaction sensitizes the individual, that is, it causes antibody to be produced (Fig. 3-20A). The allergic reaction appears when a person is exposed to the allergen again. Sufficient IgE molecules are present to combine with the allergen and simultaneously cross-link receptors on the surface membrane of mast cells or basophils. The reaction on the cell surface triggers the release of stored vasoactive substances from granules within the cells (Fig. 3-20B). As a result, substances such as histamine pour into the local interstitial spaces. Under its influence small arteries widen and fluid is lost from the capillaries, which causes swelling in the area. The tissue becomes slightly raised if it is near the surface of the skin. The swelling may be associated with pain or itching because of the pressure it creates on nerve endings. The increased blood flow through the capillary beds produces an abnormal red color. The local allergic response is called a *wheal and flare* reaction. Histamine's release in the lungs causes

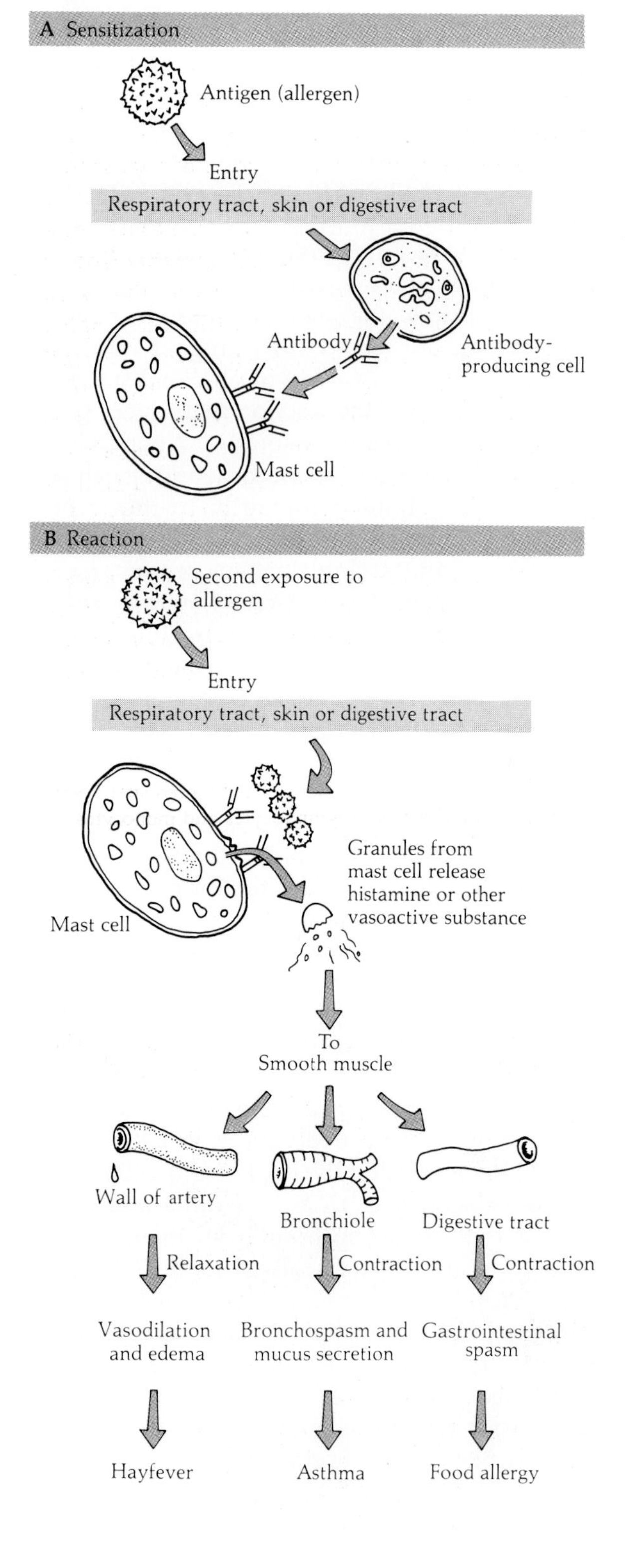

FIG. 3-20
Mast cell and immediate-type allergy. (A) Sensitization occurs when an antigen (allergen) penetrates the tissues to stimulate the production of antibodies. (B) The allergic reaction occurs with a second exposure to antigen (allergen) when sufficient antibody to the allergen is present. The antibodies bind to mast cells, and their opposite ends bind to and are cross-linked by allergen molecules. The cross-linkage of the structures stimulates the release of vasoactive molecules such as histamine from the mast cells. These molecules quickly react with cells at the tissue sites to cause the symptoms of the immediate-type allergic response. Examples illustrated include hayfever and asthma in the respiratory tract, and food allergies in the digestive tract.

the small air tubules, the bronchioles, to contract, causing *asthma,* which is characterized by labored breathing.

Immediate-type allergic reactions may be widespread when allergens gain entrance to the general circulation, producing responses in basophils. The disorder involves a systemic (systemwide) shock reaction called *anaphylaxis,* which may be fatal. Shock and death sometimes occur after allergic response to the protein in the sting from insects.

Histamine is the principal pharmacological agent involved in immediate-type allergies in humans. However, other vasoactive substances such as serotonin, leukotrienes, also called slow-reacting substance anaphylaxis (SRS-A), eosinophilic chemotactic factor, and kinins can be extremely important. Serotonin, a derivative of the amino acid tryptophan, plays a principal role in allergic reactions in rodents.

Antihistamine Therapy Only certain types of allergic reactions are mediated by histamine. It should be clear, therefore, that the administration of antihistamines (such as Benadryl, Dramamine, or Chlor-trimeton) will not be likely to alleviate symptoms of those allergies in which histamine does not play a role. Antihistamines are drugs that block the effects of histamine. They are most effective in preventing the effects of histamine and least effective in reversing them.

The release of stored products from cells can

be inhibited by elevated blood levels of specific steroid hormones in the bloodstream. The administration of cortisol, a steroid hormone, decreases the release of histamine from mast cells. Cortisol renders cell membranes more stable, causing a slower release of histamine. This allows the body a greater opportunity to degrade and eliminate the vasoactive substance without developing noticeable allergic symptoms. Thus, cortisol is an effective therapeutic agent in the treatment of many immediate-type allergic symptoms.

Reticular Tissue

Reticular tissue consists of a network of cells surrounded by fine, branched, nonelastic reticular fibers. The fibers resemble, yet differ from, pure collagen. Some of the cells retain a primitive capacity for differentiating into a variety of other types. They are able to mature into macrophages and fat cells, as well as to form more embryonic stem cells that produce lymphocytes, granular white blood cells, and red blood cells. The reticular cells, along with all phagocytic cells lining blood vessels, constitute a specialized connective tissue group referred to as the *reticuloendothelial system.* Reticular tissue forms the framework of many organs, including the liver, spleen, and kidneys. It also is found in such specialized connective tissues as lymphoid, reticuloendothelial, and blood-forming tissues (bone marrow).

Lymphoid Tissue

This is a specialized connective tissue in the lymph nodes, tonsils, spleen, and thymus. Its reticular framework contains clusters of lymphocytes that may react to specific foreign substances by actively replicating (Fig. 3-21). Clusters of actively proliferating cells are often found in areas known as *germinal centers* (Fig. 3-22). These cells differentiate into a variety of lymphocytes and plasma cells and play an important role in mediating defense mechanisms against infectious disease agents as well as causing some allergic responses.

Reticuloendothelial Tissue

Many phagocytic cells lie between and are attached to reticular fibers and are found along endothelial cells of blood vessels. The connective tissue phagocytes are poised to ingest material that is potentially harmful or of no value to the body. Cells of the *reticuloendothelial system* (*RES*) form the body's first internal line of defense beyond its external barriers (the skin, mucous membranes, and cilia). RES cells ingest infectious agents and foreign particles, as well as aging, dying, and abnormal cells. The name RES was coined to describe *all* phagocytic connective tissues cells in the body. Recently, a more limited group of phagocytic cells within the RES has been studied in detail. They are derived from specialized white blood cells, the monocytes, and are collectively called the *mononuclear phagocyte system,* or *MPS.* Phagocytosis stimulates activity within the cells, transforming the cells into "angry" (activated) macro-

FIG. 3-21
Reticular tissue from lymph node. The fine reticular fibers are laden with a variety of lymphocytes and monocytes and other cell types (× 1700).

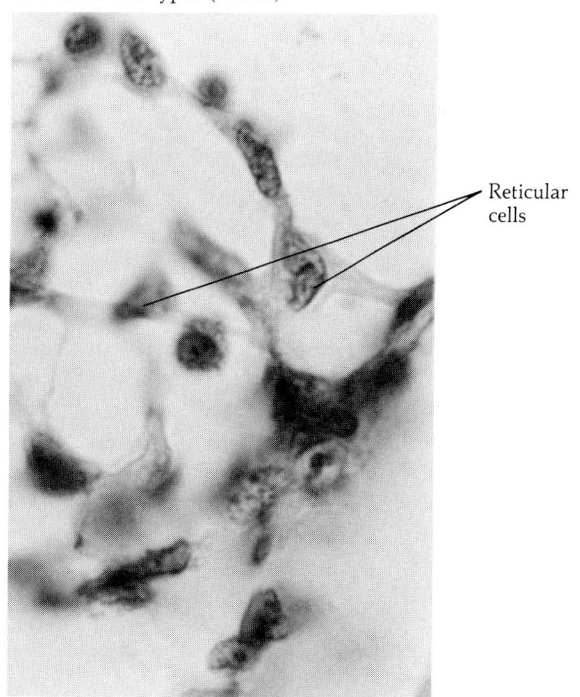

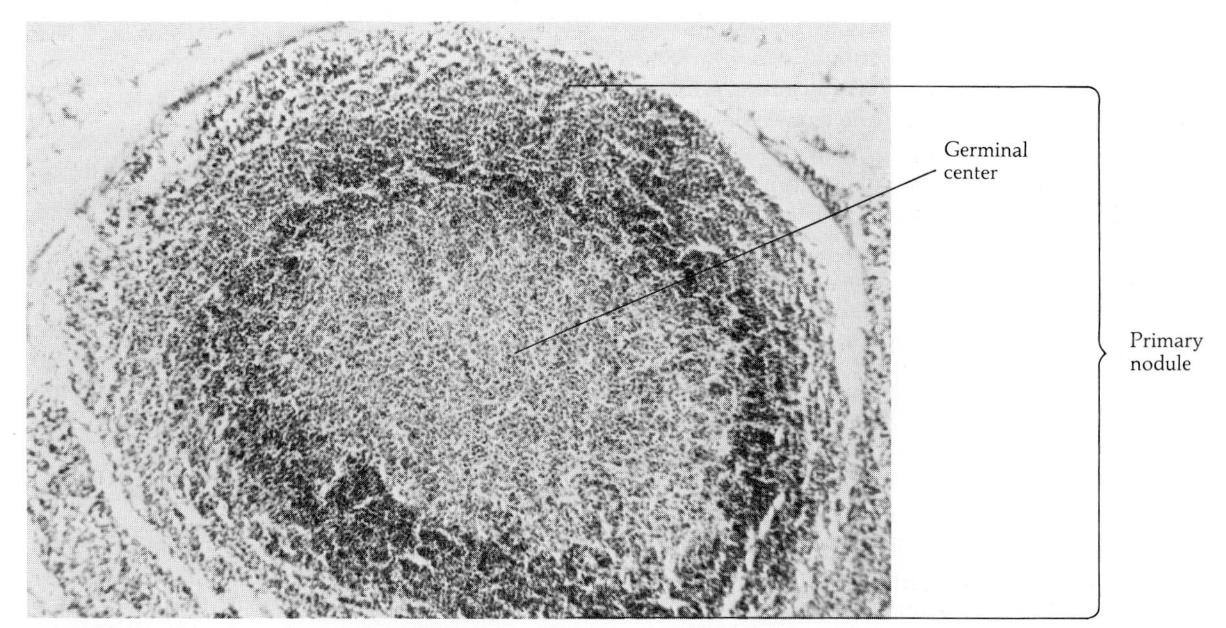

FIG. 3-22
Germinal center in a lymph node. Low-power photomicrograph of lymph node and germinal center in which cell proliferation occurs (×70).

phages. The macrophages ingest and "process" antigen, making it more effective in provoking an immune response.

The major locations of the RES/MPS cells include the liver, (Küpffer cells), the spleen, and bone marrow. MPS cells also include macrophages of lungs (alveolar macrophages), bone (osteoclasts), and the central nervous system (microglia).

Hematopoietic Tissue

Hematopoietic tissue is a specialized connective tissue with a framework of reticular fibers which support many embryonic cells called *stem cells.* This tissue is found mostly in bone marrow. It forms *blood,* which is a specialized connective tissue in which the matrix is fluid. The functions of bone marrow will be described in chapters 14 on bone and 18 and 19 on blood.

Dense Connective Tissue

In dense connective tissue, the cells are packed closely together and the intercellular space is filled with elastic or collagenous fibers. The tissue is *regular* when the arrangement of the fibers is orderly and *irregular* when the fibers are random in appearance (Fig. 3-23).

Regular Dense Connective Tissue

The proportion of elastic to collagen fibers varies in different types of regular dense connective tissue. Elastic fibers are prominent in structures that require a good deal of elasticity, such as the walls of arteries. Arteries must be able to recoil after the marked stretching that occurs as the heart pumps blood through them to tissues.

Collagen fibers are common in regular dense connective tissue, where there is a need for a tough, less elastic tissue to bind components together. This is the case in *ligaments,* which bind bone to bone, in *tendons,* which bind muscle to bone, and in *aponeuroses,* which are broad fibrous sheets of connective tissue that surround muscle. For example, aponeuroses are found binding abdominal muscles together.

Irregular Dense Connective Tissue

Irregular dense connective tissue consists of cells located between fibers that interlace irregularly with one another. The tissue is found in the dermis of the skin and in capsules surrounding many organs. It forms the sheaths which bind nerve fibers together and those which bind muscle cells in muscles. Irregular dense connective tissue also forms a tough layer, the *periosteum,* which surrounds bone.

Cartilage

Cartilage consists of groups of cells embedded in a firm matrix. Collagenous and elastic fibers run through intercellular spaces filled with large amounts of proteoglycans (proteins and mucopolysaccharides). The most abundant proteoglycan in cartilage, *chondroitin sulfate,* helps produce a flexible, smooth intercellular matrix and minimizes friction in joints.

The cells in cartilage, the *chondrocytes,* are normally large and round with spherical nuclei. Chondrocytes are often found in pairs or greater numbers within spaces called *lacunae.* Cartilage is an avascular tissue. That is, blood vessels do not usually penetrate it. Exchange of nutrients and wastes occurs by diffusion. Molecules move randomly between the cartilage and capillary beds in its surrounding connective tissue coat, the *perichondrium.*

Hyaline, fibrous, and *elastic* cartilages are the main forms of this tissue. They vary in the composition of the ground substance and the rela-

FIG. 3-23
Dense regular and irregular connective tissues. Note the flattened nuclei lying between the wavy collagenous fibers. (A) Dense regular connective tissue (×1800). (B) Dense irregular connective tissue (×1500).

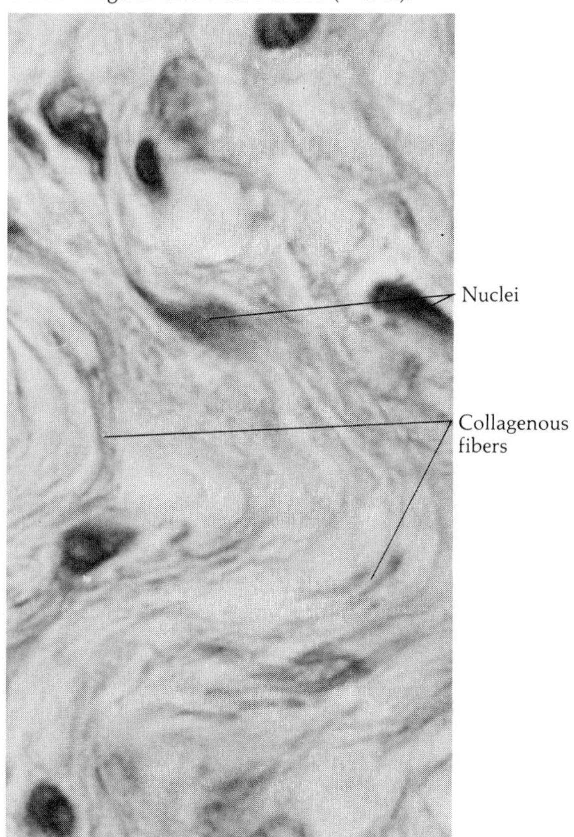

A Dense regular connective tissue

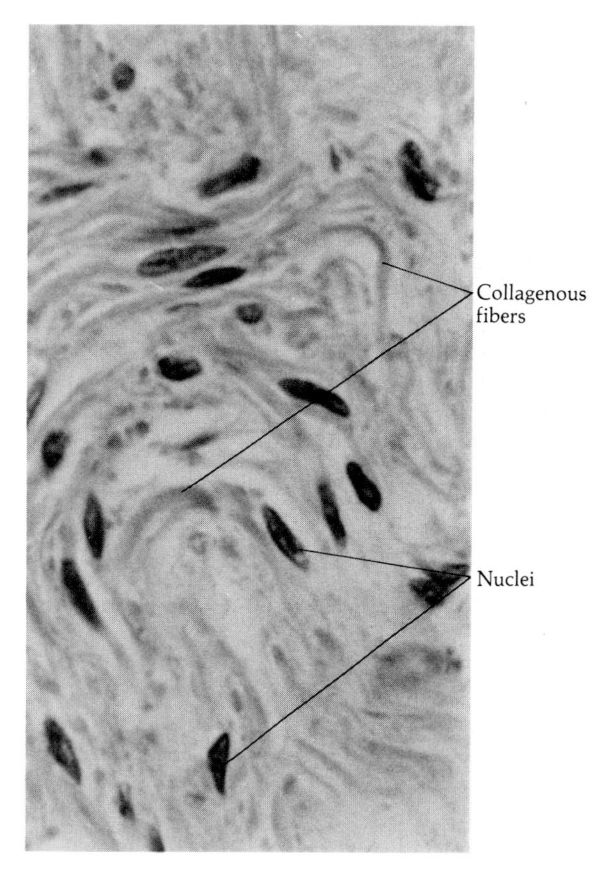

B Dense irregular connective tissue

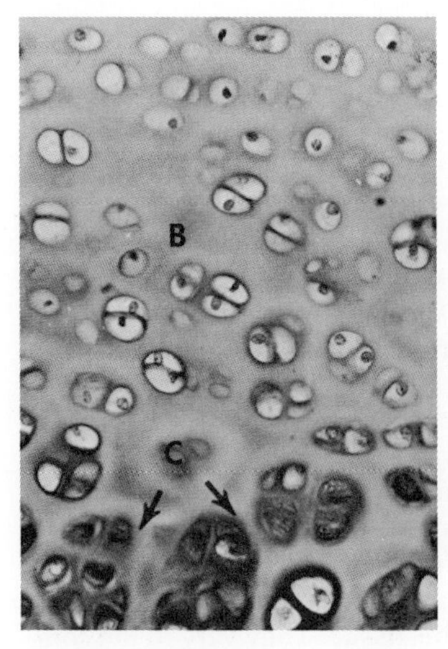

1 Hyaline cartilage

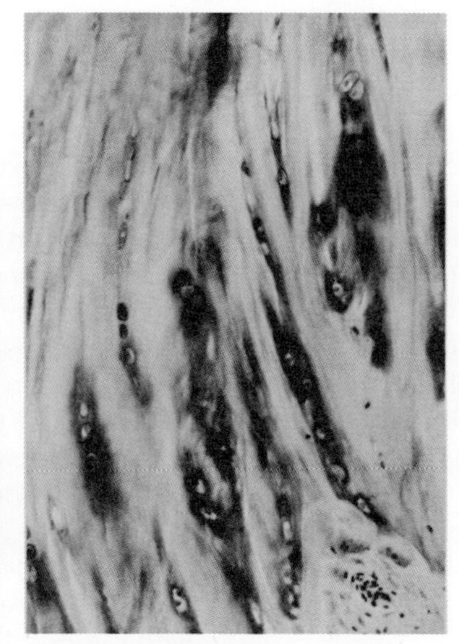

2 Fibrous cartilage

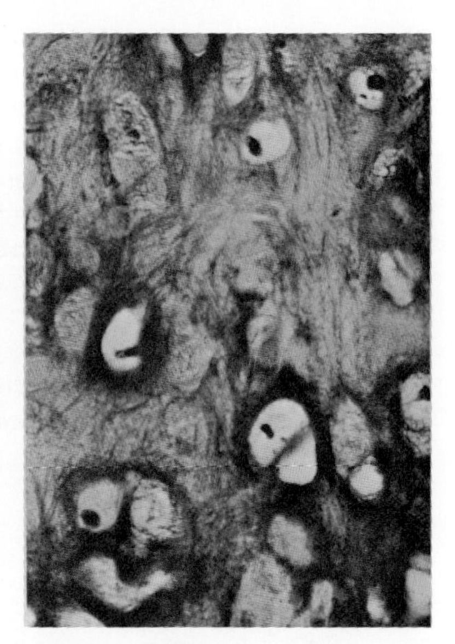

3 Elastic cartilage

FIG. 3-24
The three major types of cartilage. (1) The relatively clear matrix of hyaline cartilage is shown surrounding the chondroblasts (B), and chondrocytes (C), at the tips of the arrows (× 370). (2) The cells in fibrous cartilage are seen lined up in rows surrounded by strands of collagen fibers (× 240). (3) Note the irregular pattern to the cells in elastic cartilage (× 370). From *Histology*, 4th ed., by L. Weiss and R. O. Greep, copyright © 1977 McGraw-Hill Book Company; used with permission of McGraw-Hill Book Company.

tive proportions of different types of fibers (Fig. 3-24).

Hyaline Cartilage

Hyaline cartilage appears clear under the microscope because of the translucency of its intercellular matrix. However, an abundance of collagenous fibers is hidden within it. This form of cartilage is the precursor of much of the skeletal system. During fetal development and before birth, a large portion of the skeleton is present as hyaline cartilage, which allows compression during passage through the birth canal. After birth, much of the hyaline cartilage is gradually converted to bone. Some remains at the ends of long bones as *articular cartilage* in joints. These remnants serve as a rugged but flexible or resilient material, interposed between and covering the ends of long bones. Hyaline cartilage aids mobility by providing a surface at joints that is smooth and much more flexible than bone. This minimizes the likelihood that bits of tissue in joints will break or splinter under mechanical stress.

Hyaline cartilage is also prevalent in tissues associated with the mechanics of breathing. It makes up parts of the nose and larynx (voice box) and also joins the upper seven ribs to the sternum (breastbone) in the form of pieces called *costal cartilages.* The latter afford the *thorax,* or chest cavity, the flexibility that is necessary for breathing. The *trachea* and *bronchi* are tubes which contain C-shaped rings and pieces of hyaline cartilage. These structures keep the airways to the lungs open.

Fibrous Cartilage

Fibrous cartilage consists of cells found in rows parallel to one another lying between bundles of dense, unbranching collagenous fibers. Less flexible than hyaline cartilage, this tissue is located in areas where a tough but resilient material is needed. Fibrous cartilage is found closely associated with dense connective tissue in capsules and ligaments. It also blends with the hyaline cartilage that is normally positioned between it and the junction of bones.

Fibrous cartilage is also located within the *intervertebral discs* between the bones of adjacent vertebrae (Fig. 3-25). The middle of each intervertebral disc is composed of cellular debris and softened cartilage called the *nucleus pulposus*. It is located just anterior to the vertebral canal which extends through the center of the vertebrae. The spinal cord is located within this long channel as a bundle of nerve cells and fibers that form part of the central nervous system.

FIG. 3-25
Fibrous cartilage in the intervertebral discs. (A) The upper diagram illustrates a lateral view of the vertebral column, spinal cord, and intervertebral discs. Note that the upper disc has slipped out of place and compresses the spinal cord. (B) The lower diagram illustrates the effects of rupture of the disc and protrusion of the softer, centrally located nucleus pulposus. The tissues press on a spinal nerve as it exits from the vertebral canal.

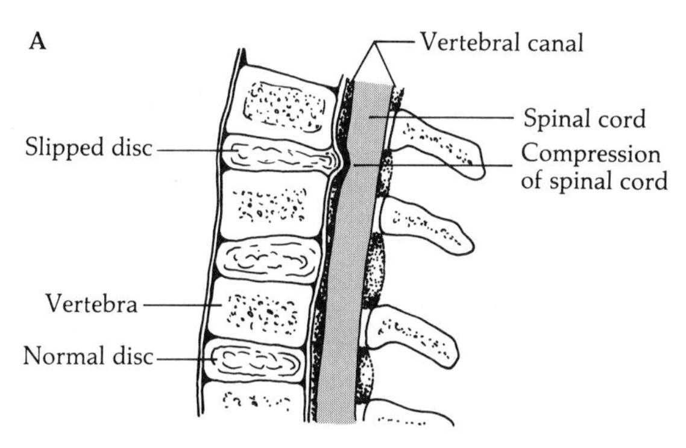

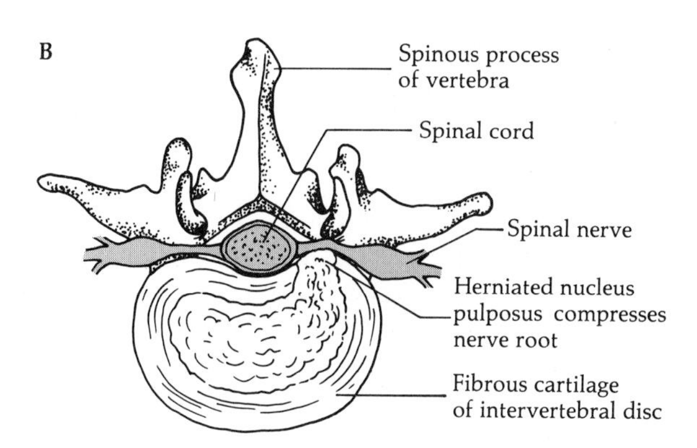

Discs provide flexible tissue between the vertebrae, thereby minimizing damage due to bending, twisting, or jumping. Even so, excess force can push the disc off-center. Dislocation can compress the nucleus pulposus, the spinal cord, and its nerves, leading to destruction of the cord and/or nerves below the affected area. Paralysis of all parts of the body supplied by nerves descending below the damaged site may follow. Dislocation (herniation) of a disc in the neck (cervical region) of the vertebral column may lead to total paralysis of the lower part of the body. Should such damage occur in the lower part of the vertebral column (lumbar region), the legs may be paralyzed.

Fibrous cartilage is also found at the junction of the anterior aspect of the hip bones (the pubis), called the *symphysis pubis*. It allows slight but extremely important alterations in the shape of the mother's pelvic cavity during the delivery of a baby.

Elastic Cartilage

Because of the predominance of elastic fibers, elastic cartilage is the most flexible of the three types of cartilage. It is found in several areas of the body, including the *external ear*, which can be easily bent but still return to its original form. This type of tissue is also located in the *eustachian tube* (the auditory tube), which leads from the middle ear to the back of the throat, and in the *epiglottis* (a structure which guards entry into the trachea, or windpipe), as well as in parts of the *larynx* (voice box).

Bone (Osseous) Tissue

About 70 percent of bone consists of inorganic salts, which give its matrix a firm structure. About 80 percent of the salt content is calcium phosphate and 14 percent is calcium carbonate. The salts and collagen fibers surround bone cells, or *osteocytes*, and entrap them in spaces called *lacunae* (Fig. 3-26). Bone cells possess membranous cytoplasmic extensions which radiate as projections through small channels, or *canaliculi*. These channels permit contact between bone cells and their extracellular matrix.

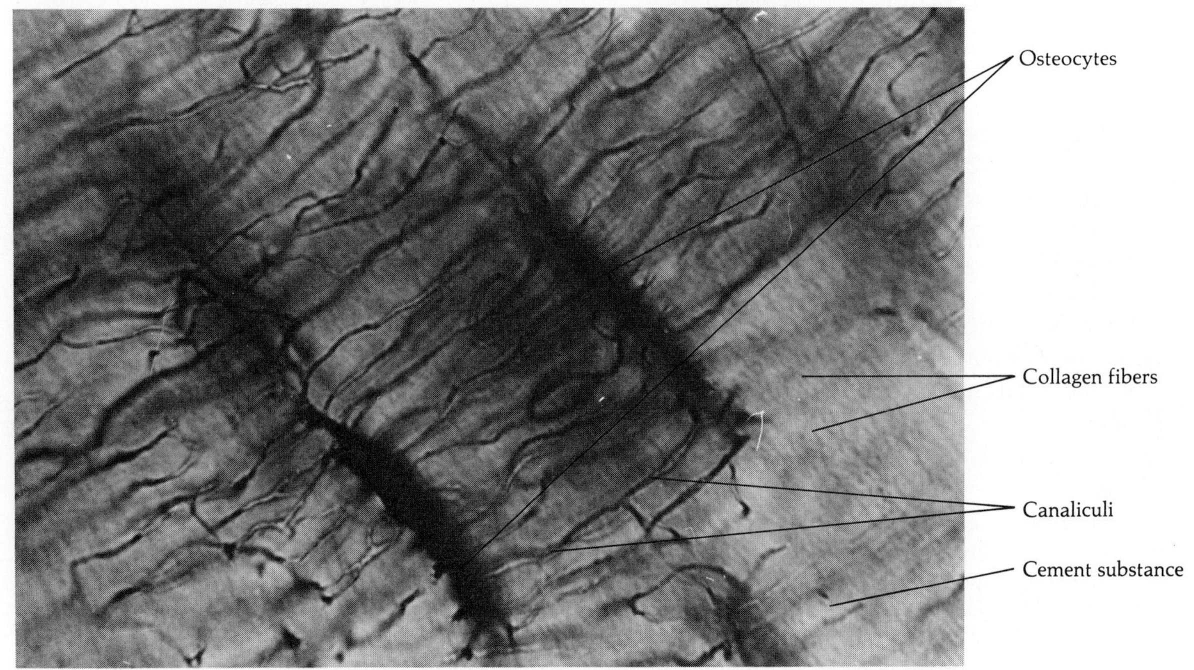

FIG. 3-26
Cellular structure of bone. Two large osteocytes are located in the dark elliptical structures labeled in this photomicrograph (× 1800). Canaliculi are seen as coarse, dark lines connecting the two areas and extending from either side of them. Collagen fibers may be noted as faint lines extending parallel to each other between the osteocytes. The tissue components are anchored together by a cement substance which is not visible in this photomicrograph but whose location is indicated.

FIG. 3-27
Haversian systems in compact bone. The cellular architecture of bone is illustrated. The arrangements of layers of bone cells (osteocytes) in Haversian systems is characteristic of compact bone found in the shaft of long bones in the body. (A) Photomicrograph of Haversian systems (× 70). (B) Enlarged photomicrograph of a single Haversian system (× 180). (C) Diagram indicating location of bone cells and collagen fibrils in Haversian systems.

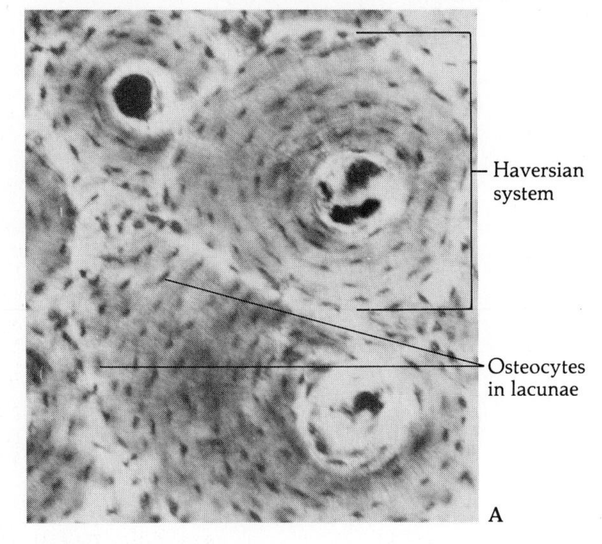

Compact Bone

Compact bone is formed by building blocks called *Haversian systems*. It is found at the outer layer of the shafts of the long bones in the body. Layers of cells form concentric circles, or *lamellae*, surrounding a central Haversian canal (Fig. 3-27). Blood vessels, lymph vessels, and nerves penetrate the Haversian canals and surround each of the cells in bone. Therefore, although they are surrounded by a rigid and rugged matrix, bone cells can constantly engage in dynamic exchange. The cells interact with one another and with substances in their extracellular matrix, which results in interchange with fluids carried from remote parts of the body in vessels

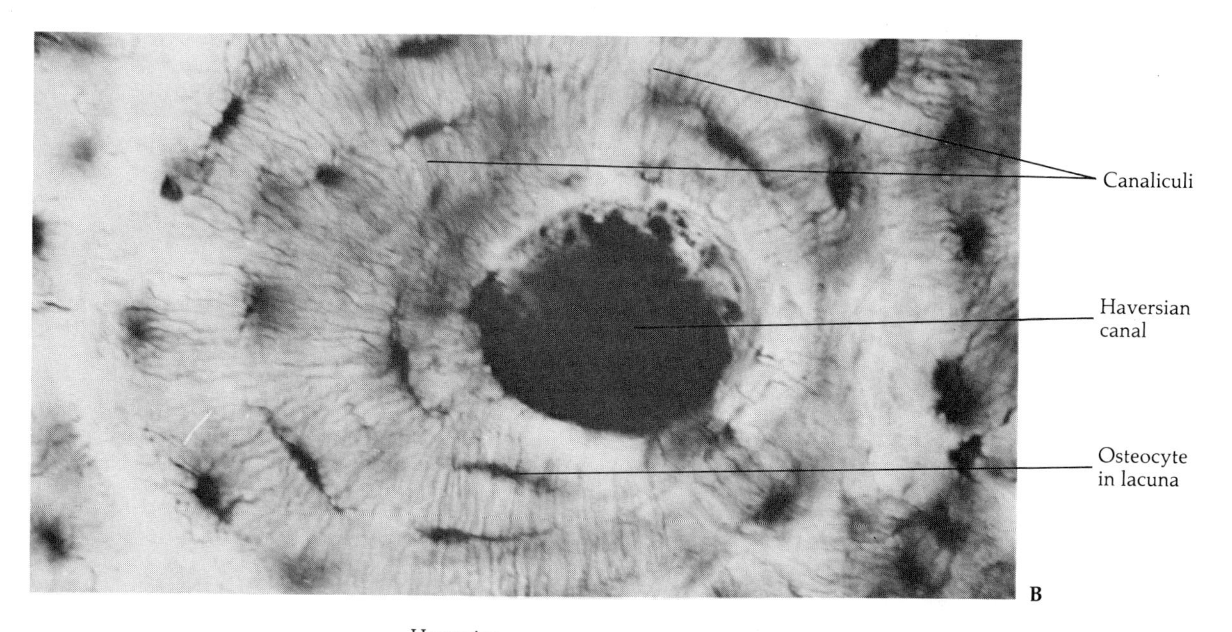
Canaliculi
Haversian canal
Osteocyte in lacuna
B

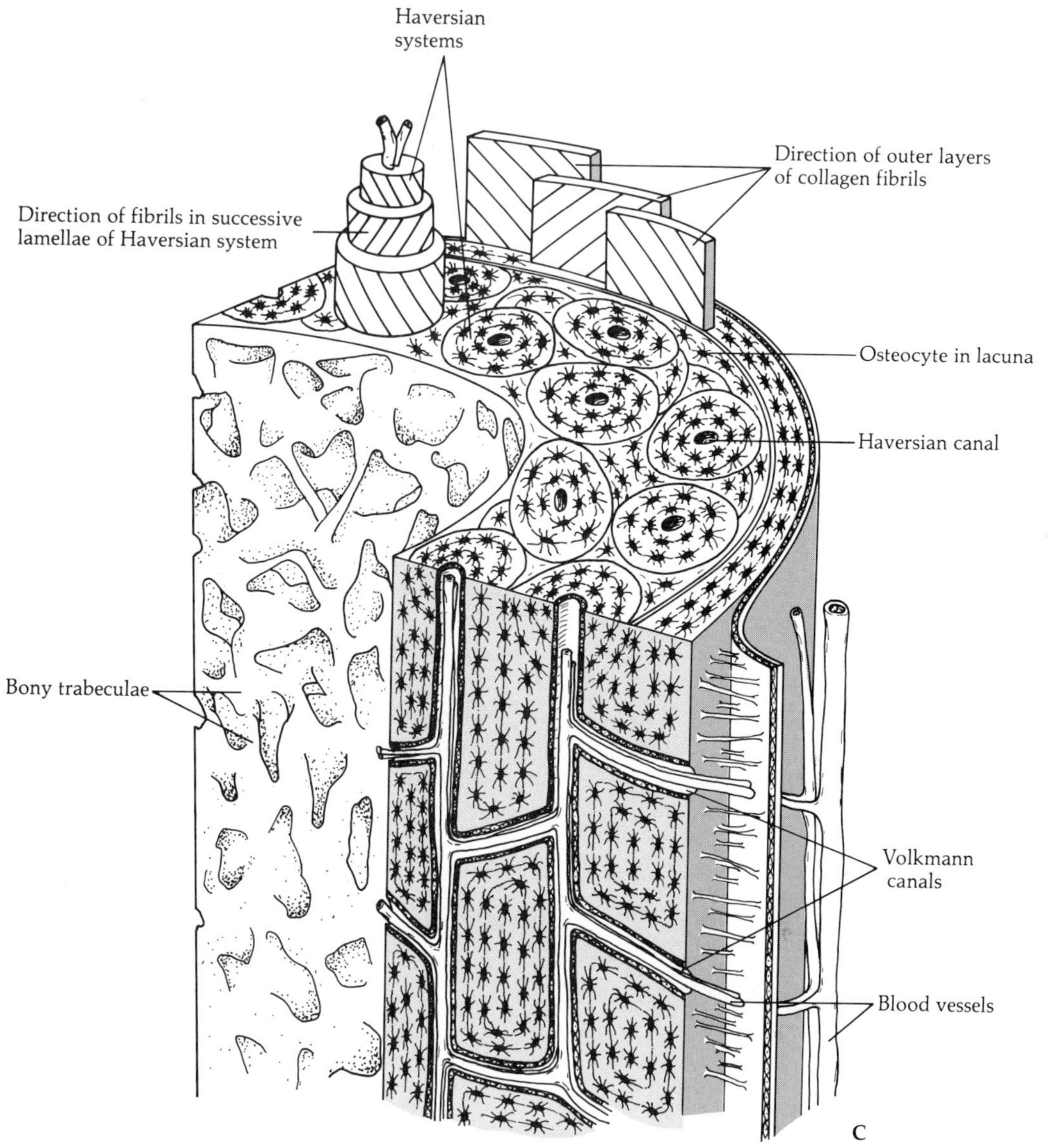
Haversian systems
Direction of outer layers of collagen fibrils
Direction of fibrils in successive lamellae of Haversian system
Osteocyte in lacuna
Haversian canal
Bony trabeculae
Volkmann canals
Blood vessels
C

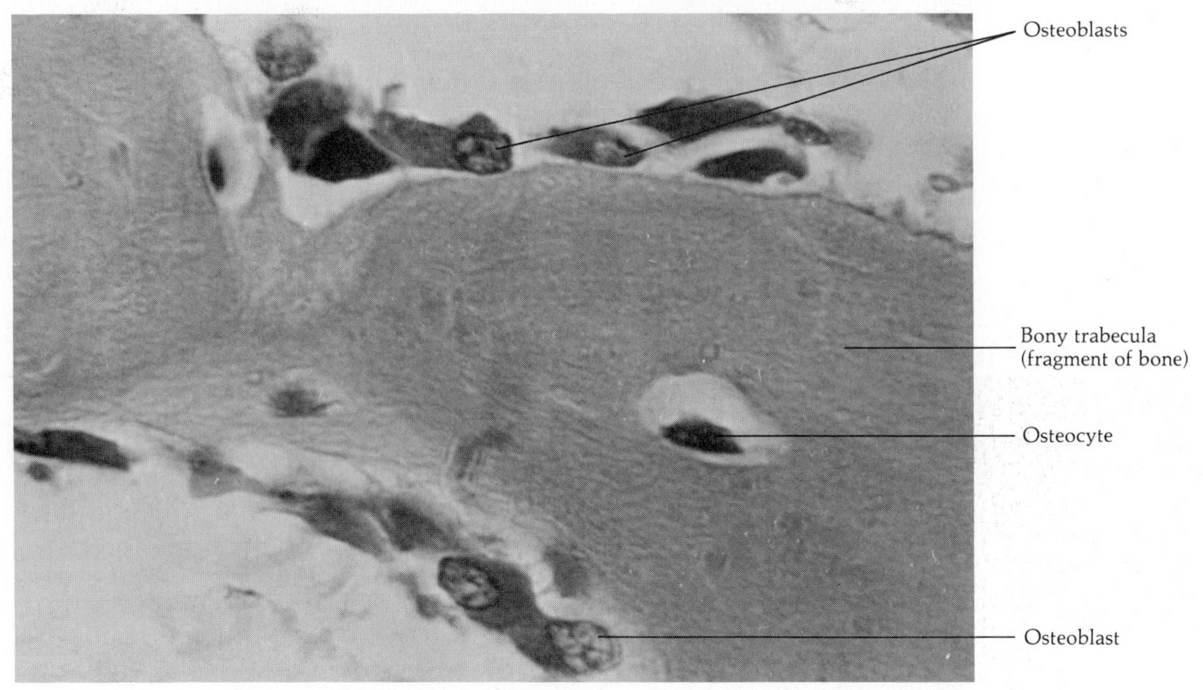

FIG. 3-28
Cancellous bone. Spongy or cancellous bone is composed of cells arranged in a lacy network and surrounded by a matrix which forms bony splinters called trabeculae. Osteoblasts are found on the surface of the trabeculae (× 1700).

that penetrate the bone obliquely in channels called *Volkmann's canals.*

Spongy (Cancellous) Bone

In many areas the osteocytes are loosely arranged in a lacy network. Large spaces between the cells make the bone spongy or more porous than compact bone. In *spongy bone* the matrix is reduced to small splinters, which are called *spicules* or *trabeculae* and which surround marrow spaces (Fig. 3-28). This type of bone resembles Swiss cheese, as it contains walls of varying thickness separating irregular spaces. The inner layers of the shaft, and medullary cavities of long bones are formed of spongy bone. This type of bone can be seen at the center of a chicken drumstick after it is broken. Active mineral and hormone metabolism occur in bone, as do remarkable exchanges with body fluids. These topics will be discussed in detail in Chap. 14.

Dentin

Dentin is acellular (it does not have cells) but is otherwise similar in structure to bone. It contains collagenous fibers embedded in a calcified ground substance. However, since there are no cells present, dentin contains no lacunae. The ground substance in dentin is harder than that in bone, makes up the bulk of teeth, and is secreted by cells that lie adjacent to the central pulp cavity in each tooth (Fig. 3-29). *Enamel* is secreted onto dentin from overlying specialized epithelial cells, comprising the enamel organ, which is formed in the early development of the tooth. Enamel is produced before the teeth erupt through the gums.

Diseases of Connective Tissue

A number of diseases affecting connective tissues result from inflammatory responses as well as from nutritional deficiencies, hereditary factors, and aging. Many disorders involve alterations in the properties of collagen. Scurvy and osteoarthritis are two examples described below.

Scurvy is a disease in which collagen in the matrix of connective tissue is poorly formed or malformed. The illness is due to a diet deficient in vitamin C (ascorbic acid). Vitamin C is present in raw fruits and vegetables, including citrus fruits, tomatoes, leafy green vegetables, cabbage, and green peppers. Prolonged vitamin C deficiency retards collagen production by fibroblasts and the formation of cross-linkages between collagen monomers. The adverse effect on collagen also hinders the repair of tissues and the healing of wounds and bone fractures. It may cause the loosening of teeth. The presence of malformed collagen makes the walls of blood vessels more susceptible to rupture and often results in hemorrhage.

Osteoarthritis is a disease in which articular cartilage breaks down. The degeneration, often called wear-and-tear arthritis, begins in areas of the body exposed to great mechanical stress. As a result, the hyaline tissue degenerates and is ultimately lost. Normal articular cartilage contains type II collagen. It is composed of three identical chains in a triple helix of protein (Fig. 3-10). In osteoarthritic cartilage, different forms of chains are produced. Their structural features seem to be responsible for some of the failure of this tissue to withstand mechanical stress. Why the cells shift to the production of a new type of collagen is not clear.

FIG. 3-29
The location of dentin in teeth. Section of a molar showing the pulp cavity, which contains blood vessels and nerves.

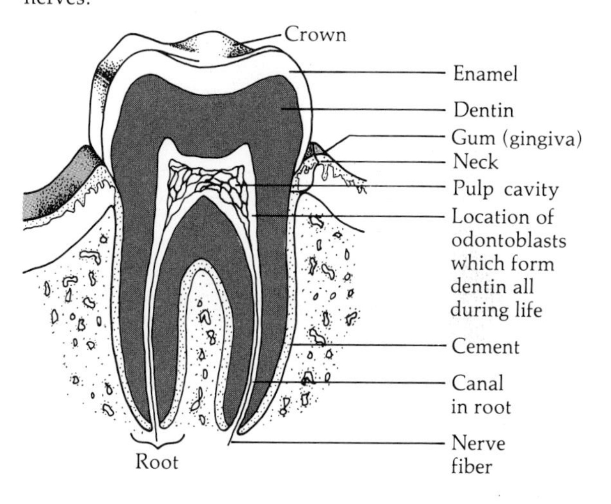

3-3 MUSCULAR TISSUE

The three types of muscle are skeletal, cardiac, and smooth muscle (Fig. 3-30). The long cells in skeletal and cardiac muscle resemble and are called fibers. Muscle is responsible for movement and the production of most of the heat in the body. The physiology of muscle will be described in detail in Chaps. 12 and 13.

Skeletal Muscle

Skeletal muscle is the largest single mass of tissue in the body. Its cells contain large amounts of protein in the form of fibrils (tiny filaments). The fibrils exhibit horizontal striations (stripes) and cross-striations (Fig. 3-30A). Individual cells (fibers) are parallel to one another. Each one contains many nuclei located near the periphery of the cell membrane. Skeletal muscle is under voluntary control, responding to neurons in the cerebral cortex of the brain. Skeletal muscle is found attached to the bony skeleton, within the tongue, and attached to the outside of the eyeball.

Cardiac Muscle

Cardiac muscle consists of striated, branching cells, each of which possesses many centrally located nuclei (Fig. 3-30B). The cells are coarser than those in skeletal muscle and are found only in the heart. Specialized junctions, the *intercalated discs,* are modifications of the cardiac cell membranes that facilitate passage of electric signals, termed action potentials, from cell to cell. In some areas the connections between cardiac cells resemble desmosomes and tight junctions (Fig. 3-1). In the areas of the intercalated discs they form gap junctions (nexuses), which are sites of low electrical resistance allowing impulses to pass more readily between cells (Fig. 3-1C).

Although the heart is not voluntarily controlled, conscious thought processes may modify the autonomic functions it displays. The in-

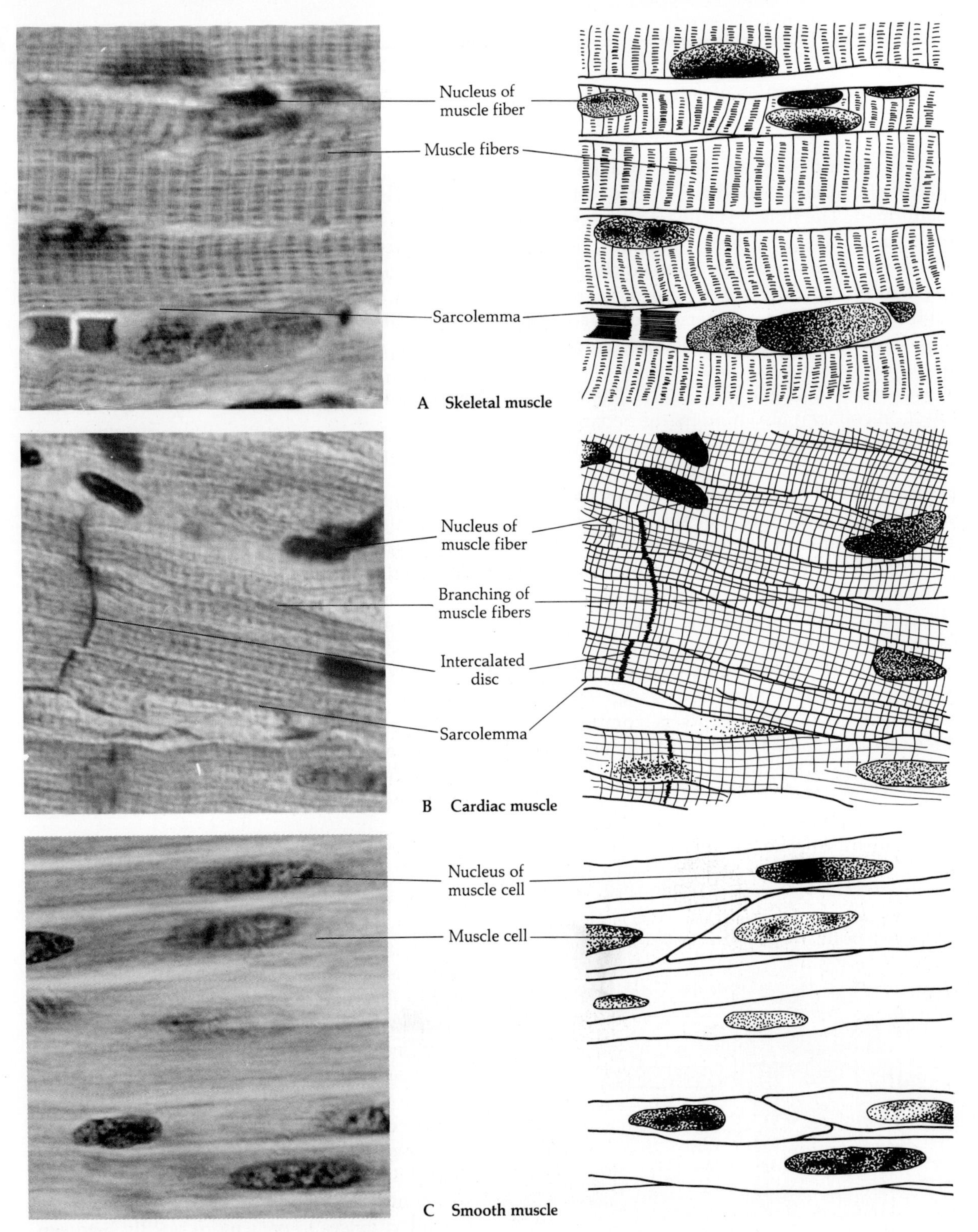

FIG. 3-30
Muscle tissue. The photomicrographs on the left are drawn on the right. (A) Skeletal muscle (× 1700). (B) Cardiac muscle (× 1700). (C) Smooth muscle (× 1700).

fluence of feedback mechanims in the nervous system, called *biofeedback,* attests to the highly integrated nature of the human body.

Smooth Muscle

Smooth muscle cells are elongated and spindle-shaped, each containing a single, long, central nucleus (Fig. 3-30C). Each cell is surrounded by cytoplasm in which the fibrils are not apparent, which gives smooth muscle a nonstriated appearance. This tissue is supplied with autonomic nerve fibers and is also called *visceral muscle* because it makes up the walls of the viscera, or internal organs. It is located in the walls of the digestive tract; the blood vessels; the bladder, ureter, and urethra of the urinary tract; and in the female, the uterus.

3-4 NERVOUS TISSUE

The physiology of nervous tissue is extremely complex since it is the most highly organized of tissues. Its functional cells, the *neurons,* respond to stimuli and transmit impulses but have lost the ability to regenerate by cell division. Therefore, neurons require the protection and help of other tissues to remain functional. For this reason, they are closely intermingled with connective tissues.

The nucleus and nucleoli of each neuron are located within its *cell body* (Fig. 3-31). The cell body is surrounded by cytoplasm from which *axons* and *dendrites* project as extensions. Axons generally carry nerve impulses away from the cell body, and dendrites usually carry them toward it. The cytoplasmic processes produce a maze of pathways, which constitute *nerve tracts.* Signals are carried along these tracts in

FIG. 3-31
Nervous tissue. Neurons from the spinal cord surrounded by connective tissue cells which appear as small granules (×700).

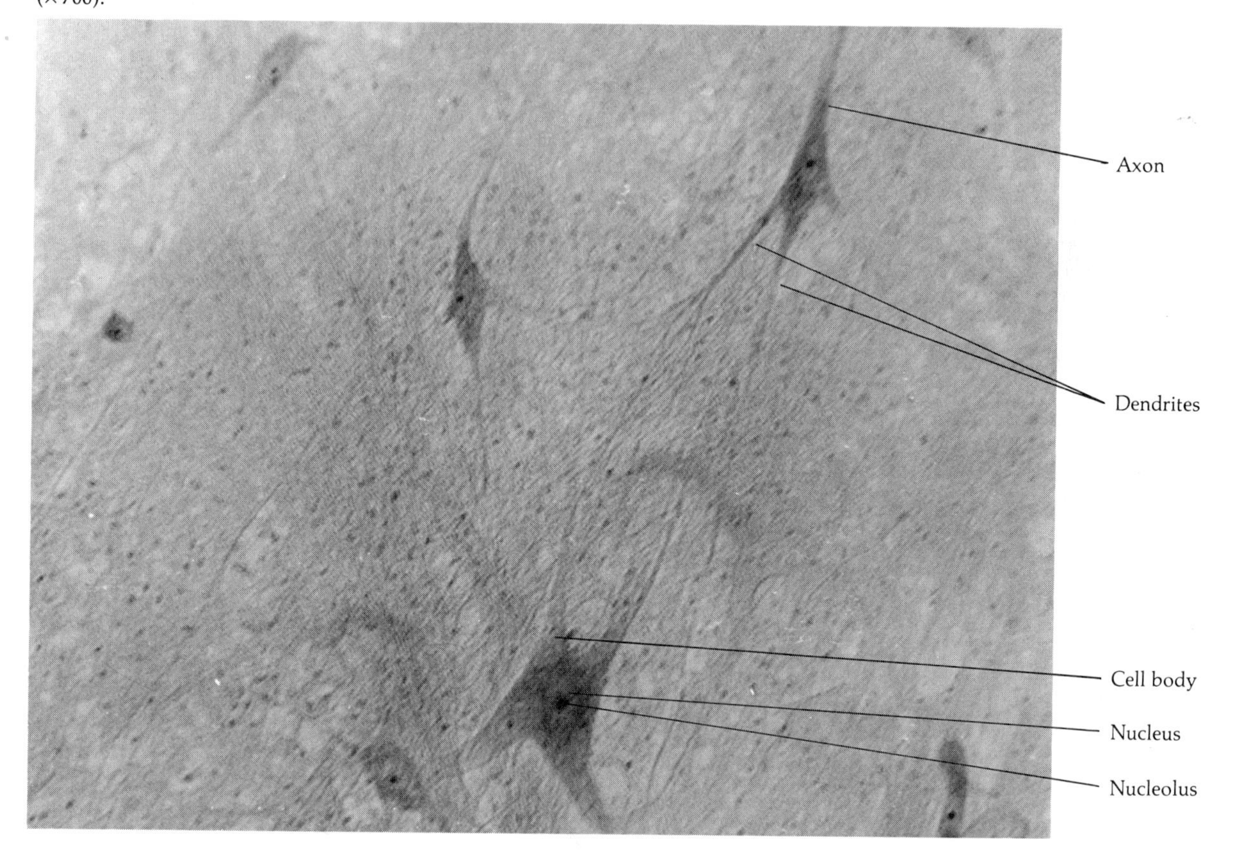

all directions between the brain and essentially all parts of the body.

Interstitial connective tissue surrounding the neurons contains cells called *neuroglia,* which literally means "nerve glue." They support, protect, and nourish neurons. A more complete discussion of nervous tissue will be presented later in Chaps. 6 to 11 on the nervous system.

3-5 SUMMARY

The four major tissue types are epithelial, connective, muscular, and nervous tissues.

1 Epithelial tissues are found lining surfaces of the body and play a role in protection, absorption, and secretion. The cells are tightly joined to one another with a minimum of intercellular space. They are supported by an underlying basement membrane composed of glycoproteins. Epithelium may be distinguished by the shape of its cells, which are flat (squamous), cube-shaped (cuboidal), or column-like (columnar). Epithelial cells may form one layer (be simple) or several layers (be stratified).

Squamous epithelium helps form the mesothelium of serous membranes and lines the pleural, pericardial, and abdominal (peritoneal) cavities. The cells secrete a watery (serous) fluid that lubricates the surfaces of tissues and organs. Endothelium contains squamous cells forming the smooth lining of the circulatory and lymphatic systems. Cuboidal cells are often found lining ducts, tubules, and secretory glands. Columnar epithelium is found in mucous membranes, which lines ducts and tubules and secretes a viscous fluid (mucus) containing a variety of molecules important to body functions.

Clusters of secretory epithelial cells surrounded by connective tissue form exocrine glands, which deliver products through ducts, and endocrine glands, which deliver secretions directly to the bloodstream.

2 Connective tissues include loose areolar tissue, reticular tissue, dense tissue, cartilage, bone, and dentin. The cells in these tissues are surrounded by varying amounts and types of intercellular fibers and chemicals. Areolar connective tissue loosely binds parts of the body together. Its intercellular spaces contain tough, wavy bundles of collagen fibers loosely interlaced with fine, freely branching elastic fibers. Fibroblasts, fat cells, plasma cells, macrophages, and mast cells predominate among the many cells found in connective tissue.

Fibroblasts are irregularly shaped cells that synthesize collagen and are active in the growth and repair of tissues.

Fat cells often appear with a fine cell membrane that seems to surround empty space. The number of fat cells found in areolar tissue seems to be determined in infancy. Fat cells (adipocytes) provide resilient packing material with high insulation value and offer a major store of energy.

Loose connective tissue cells which defend against disease include plasma cells, macrophages, and mast cells. Along with the cells of reticular tissue and blood, they are part of the immune system. This system is formed of organs whose cells defend the body against disease and which react to foreign substances by synthesizing molecules to assist body defenses.

Plasma cells may be identified by their eccentric nucleus and abundant rough endoplasmic reticulum. Plasma cells produce five classes of protective immunoglobulins, the antibodies IgG, IgM, IgA, IgD, and IgE.

Macrophages are large, phagocytic, irregularly shaped cells derived from monocytes. Macrophages ingest foreign materials as the body's first line of defense.

The inflammatory response is another defense mechanism of the body. In this response, white blood cells migrate out of blood vessels into the assaulted tissue area. Neutrophils, monocytes, and lymphocytes accumulate in large numbers in the interstitial spaces. Lymphocytes release lymphokines, which are molecules that may regulate the activity of other cells, including macrophages. Lymphokines may assist macrophages in destroying foreign or malignant cells. The inflammatory response may also lead to the destruction of tissues.

Mast cells are round or polygonal cells that contain stores of heparin (an anticoagulant) and histamine in metachromatic granules. Histamine may cause tiny blood vessels to leak and smooth muscle cells to contract. It, like other chemicals released by certain cells, may cause some types of allergic reactions. For example, IgE antibodies formed by plasma cells in reaction to some allergens bind to allergens and receptor molecules on mast cell membranes. As a result, the mast cell membrane becomes cross-linked and histamine pours out of the cell, exerting its effects in a rapid, immediate-type allergic reaction.

In contrast, delayed-type allergies are the result of activity of T cells. Their stimulation causes a response which is slower to develop after a second or subsequent exposure to specific allergens.

Reticular tissue contains star-shaped cells surrounded by fine-branched fibrils which are chemically similar to collagen. It forms the supporting framework of many organs, including the liver, lymph nodes, tonsils, thymus, spleen, and kidneys. Lymphoid tissue is a specialized form of reticular tissue. It contains lymphocytes, which proliferate within germinal centers and which may mature into plasma cells. In many areas of the body reticular tissue contains connective tissue cells which phagocytize. The entire collection of reticular cells is called the reticuloendothelial system (RES). Those derived from phagocytic monocytes form the mononuclear phagocytic system (MPS). Hematopoietic tissue is a specialized connective tissue which is also supported by a reticular framework. It is abundant in red bone marrow and contains cells which form blood cells.

Dense connective tissue contains cells and fibers packed more closely together than those in areolar tissue. Tough supporting connective tissues with regular arrangements of collagen and elastic fibers form ligaments, which bind bones together, aponeuroses, which bind muscles together, and tendons, which bind muscle to bone. Irregular arrangements of fibers are found in dense connective tissue in the dermis of the skin, in capsules that surround organs, and in sheaths around nerves and muscles.

Cartilage is composed of cells surrounded by collagen and elastic fibers embedded in a firm matrix with large amounts of proteoglycans. Three forms of cartilage exist, which vary in structure and in the proportion of collagenous and elastic fibers. They are hyaline cartilage (the precursor of much of the skeletal system), fibrous cartilage (which forms intervertebral discs), and elastic cartilage (found in the external part of the ear and the larynx).

Bone is a specialized connective tissue containing a high percentage of calcium salts. The salts crystallize around collagen fibers entrapping bone cells, the osteocytes in lacunae and their cytoplasmic extensions in small channels, the canaliculi. In this way, adjacent bone cells communicate with each other. In compact bone the cells form sheets or lamellae around a central Haversian canal. The latter is a tunnel through which blood vessels, lymph vessels, and nerves penetrate the tissue. In spongy bone the cells are more loosely arranged. The dentin and enamel of teeth are similar to bone in structure but lack cells.

Collagen is an important structural element in connective tissues. Collagen malformation occurs in a number of diseases including scurvy (due to vitamin C deficiency), osteoarthritis, and other disorders which cause a loss of cartilage in bony joints.

3 Muscle produces movement and most of the body's heat. Skeletal muscle is composed of multinucleated, striated cells aligned parallel to each other. Its function is under voluntary control and is associated with that of the bony skeleton. Cardiac muscle consists of multinucleated, branched, striated cells located in the heart. Its function is not under voluntary control. Smooth muscle consists of single nucleated, nonstriated, spindle-shaped cells found in the walls of many hollow organs. Its function is not under voluntary control.

4 Nervous tissue transmits impulses throughout the body. It consists of neurons, each of which contains a cell body housing a nucleus, and cytoplasmic extensions, called axons and dendrites, that usually convey signals away from and toward the cell body, respectively. Protective, supportive, and nourishing tissue elements are intermingled with neurons and are called neuroglia.

REVIEW QUESTIONS

1. Name the various shapes and arrangements of epithelial tissue.
2. What roles do epithelial tissues play in the digestive tract? In what two ways does epithelial tissue perform a protective function in the skin?
3. Describe the role of vitamin A in the development and function of ciliated columnar epithelium and in the resistance to melanoma.
4. What are the five major types of cells found in loose connective tissue? Describe the functions of each.
5. Compare the functions of plasma cells, macrophages, and mast cells in defense reactions.
6. Why is it that a person's second exposure to some allergens cause an immediate-type allergic reaction?
7. What is the reticuloendothelial system? How does it differ from the mononuclear phagocytic system?
8. What is the inflammatory response and how does it assist the body's defenses?
9. Why is there reason to believe that fat babies are more likely than other babies to become overweight adults?
10. Explain the differences between ligaments, tendons, and aponeuroses.
11. Discuss the importance of reticular tissues in the human body.
12. Describe hyaline cartilage and its importance (1) at birth and (2) in adulthood.
13. Where in the human body is fibrous cartilage found? Where is elastic cartilage found?
14. How are collagen structure and malformation related to scurvy and osteoarthritis?
15. What are the major differences between striated and smooth muscle?
16. What are the basic parts of the neuron and how are neurons related to neuroglia?

MOVEMENT OF SUBSTANCES ACROSS CELL MEMBRANES

4

4-1 WATER AND THE CELLULAR ENVIRONMENT
Intracellular compartments
Extracellular compartments
Plasma
Interstitial fluid and lymph
Dense connective tissue, cartilage, and bone
Transcellular fluids
Chemical differences between intracellular and extracellular compartments
Regulation of total body water

4-2 THE CELL MEMBRANE
Role of proteins
Pores
Receptors
Structural proteins
Enzymes
Pumps
Role of lipids

4-3 TRANSPORT MECHANISMS
Nonmediated transport
Diffusion
Factors involved • Role of pores
Osmosis
Osmotic pressure • Osmolarity and Avogadro's number • Effect of antidiuretic hormone
Carrier-mediated transport
Facilitated diffusion
Active transport
Carrier systems • Role of membrane-bound ATPases: the Na-K pump
Vesicle-mediated transport
Endocytosis
Exocytosis (emiocytosis)

4-4 SUMMARY

OBJECTIVES

After completing this chapter the student should be able to:

1. Compare the fluid compartments of the body and explain the basis of their existence.
2. Describe the structure of the cell membrane and explain the roles of proteins, lipids, and carbohydrates.
3. Distinguish nonmediated, carrier-mediated, and vesicle-mediated transport and discuss the role of each in cell function.
4. Identify the factors that influence the diffusion of ions, nonionized molecules, and ionized molecules through the cell membrane.
5. Compare the mechanisms of transport that require energy production by the cell and those that do not.
6. Discuss the passage of water through the cell membrane.
7. Explain the relationship between the osmolarity of the intracellular and extracellular fluids.
8. Describe the differences between facilitated diffusion and active transport, including a description of the role of carriers in each.
9. Outline the differences among pinocytosis, phagocytosis, and exocytosis.

4-1 WATER AND THE CELLULAR ENVIRONMENT

Cells exchange nutrients and waste products between their internal and external environments. In this way they maintain homeostasis. Movement of substances into and out of cells occurs through a fluid medium, of which water is the major component.

Water accounts for 65 to 90 percent of the weight of individual cells. It is responsible for an average of 50 and 60 percent of lean body weight in young adult females and males, respectively, and for about 75 percent of body weight in the newborn. The proportion of total body water decreases with aging and obesity, as other tissue elements increase. On the average, about 75 percent of all body water is located in muscle, skin, and bones, but the water content of individual tissues also varies. For example, water accounts for about 76, 72, and 31 percent of the weight of muscle, skin, and bone and 82 and 10 percent of the weight of kidneys and adipose tissue, respectively. The low percentage of water in adipose tissue causes the total body water to account for about 50 percent of body weight in obese individuals and about 70 percent in those with lean body mass.

About 55 percent of body water is *intracellular* (within cells) and 45 percent is *extracellular* (outside of cells). It serves as a *solvent,* that is, a medium in which substances are dissolved. The dissolved substances, including charged particles, moving about within water are called *solutes.* Their movement and interactions greatly influence body function. For example, the transmission of nerve impulses and the coagulation of blood are two body functions that are much affected by the movement of positively charged ions in the body's fluids.

Intracellular Compartments

The interior of the cell is divided into compartments by its internal membranes. The membranes include those of the nucleus, mitochondria, Golgi apparatus, and endoplasmic reticulum. Therefore intracellular fluid (ICF) is actually compartmentalized. However, since fluid exchange occurs rapidly between the intracellular compartments, ICF is described as though it were located within a single undivided compartment.

Extracellular Compartments

Extracellular fluid (ECF) is outside of cells. With aging, a lesser proportion of water is found in the body and the decrease is due mostly to a decline in the ECF volume.

Water passes rather freely into and out of cells. However, the rates of movement of water and its solutes vary in different tissue types, so that four ECF "compartments" may be distinguished (Fig. 4-1). They are: (1) plasma; (2) in-

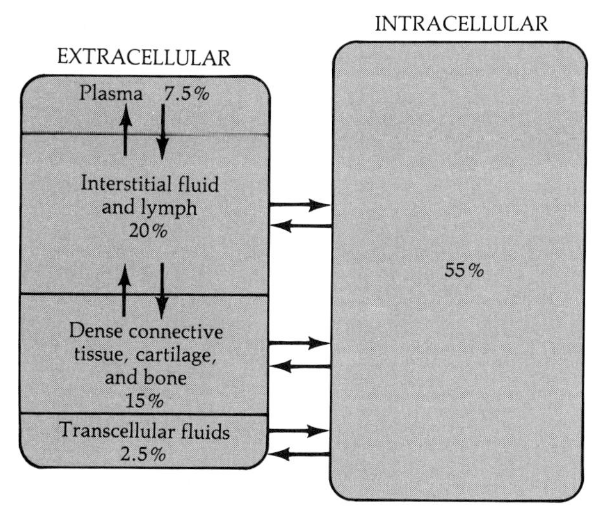

FIG. 4-1
Water content and routes of exchange among the fluid compartments of the body. The percentage of total body water located in each compartment is indicated. The arrows show the direction of exchange.

terstitial fluids and lymph; (3) certain connective tissues; and (4) transcellular fluids.

Plasma

The fluid component of blood is *plasma*. It accounts for 7.5 percent of body water. It undergoes exchange with interstitial fluid and lymph.

Interstitial Fluid and Lymph

Interstitial fluid and lymph contain 20 percent of body water. *Interstitial* (*intercellular*) *fluid* is that which immediately surrounds and bathes cells. It is formed from liquid that passes out of cells and from water, small molecules, and white blood cells which escape from blood through tiny blood vessels called *capillaries*. Interstitial fluid and plasma resemble each other because of the relatively free exchange that takes place across the walls of many capillaries. Those components of the plasma, which do not return to the bloodstream directly, enter lymph capillaries to form *lymph*. Lymph is carried by vessels of the lymphatic system, which return the fluid to the circulatory system.

Dense Connective Tissue, Cartilage, and Bone

Dense connective tissue, cartilage, and bone possess a relatively sparse supply of capillaries compared with other tissues. Consequently, less fluid exchange occurs between capillaries and these three tissue types. About 15 percent of body water is found in this compartment, and half of it is in bone.

Transcellular Fluids

Cells, membranes, glands, and organs form secretions which constitute *transcellular fluids*. These fluids include saliva, mucus, cerebrospinal fluid, and others secreted across cell membranes. They account for 2.5 percent of total body water.

Chemical Differences between Intracellular and Extracellular Compartments

Numerous ions and molecules of biological importance are contained within various body fluids. Their rates of exchange across cell membranes are governed by interrelated mechanisms. Exchanges that occur are dynamic and provide the means for healthy cells to maintain homeostasis. Continual movement of millions of particles occurs every second across most cells of the body. The movement of ions and molecules is governed to a large extent by the needs of the cell and by carrier molecules within it. Unequal concentrations of many types of ions and molecules are maintained on either side of the cell's membranes and are reflected in differences in the composition of the intracellular and extracellular fluids (Fig. 4-2). For example, sodium ions (Na^+), calcium ions (Ca^{2+}), chloride ions (Cl^-), bicarbonate ions (HCO_3^-), and glucose are among the substances which are often significantly higher in concentration in the ECF than in the ICF. Potassium ions (K^+), negatively charged organic anions, and proteins are usually in higher concentrations in the ICF. Differences of this nature account for significant variations in the composition of the ICF and ECF.

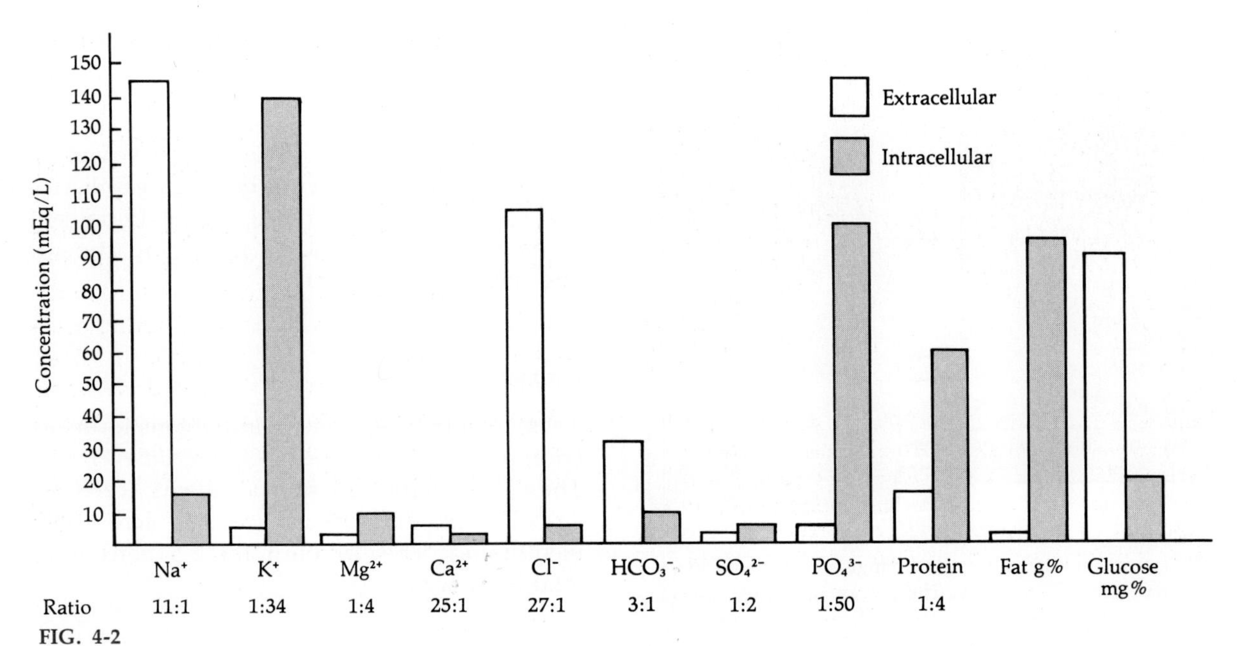

FIG. 4-2
Some differences in the composition of extracellular and intracellular fluids. Typical concentrations of substances in the body fluids are noted. They vary in different ECF and ICF compartments. Values are listed in milliequivalent per liter (mEq/L) unless noted otherwise. The normal concentrations of the substances may vary approximating the maximal values shown.

Regulation of Total Body Water

Since the bulk of body weight is due to water, factors that regulate its passage into and out of cells are extremely important. The total amount of body water is controlled by a number of mechanisms that influence its movement and that of other substances across cell membranes.

The amount of total body water is regulated by the balance of its intake, its movement among the fluid compartments of the body, and its loss by excretion. Water is excreted in the solid wastes (feces) of the digestive tract, in urine and sweat from the excretory system, and as droplets exhaled with air from the respiratory system.

The movement of water involves mechanisms that influence cell permeability. The nature of the cell membrane and the effects of molecules which carry substances across it are extremely important. So, too, are differences in the chemical composition of the ICF and ECF. A number of physiological factors,including hormones and blood pressure, also influence the water content of the body.

4-2 THE CELL MEMBRANE

The cell membrane varies considerably from one cell type to another. It consists primarily of a mixture of proteins and lipids which forms a barrier averaging 7.5 nm in thickness between the ICF and ECF. The proteins generally comprise the largest component in cell membranes by weight, while the lipid content is slightly less and the carbohydrate content is a distant third. For example, the average red blood cell membrane is about 6 nm thick and by weight contains 49 percent protein, 44 percent lipid, and 7 percent carbohydrate.

The lipids exist as a double layer of molecules about 4.5 nm thick in most cell membranes. Since normal body temperature is above their melting points, they are found in a fluid state, forming a viscous solvent in which molecules of other substances, including proteins, may move. Some of the proteins are like ships, submarines, and icebergs floating in an ocean of

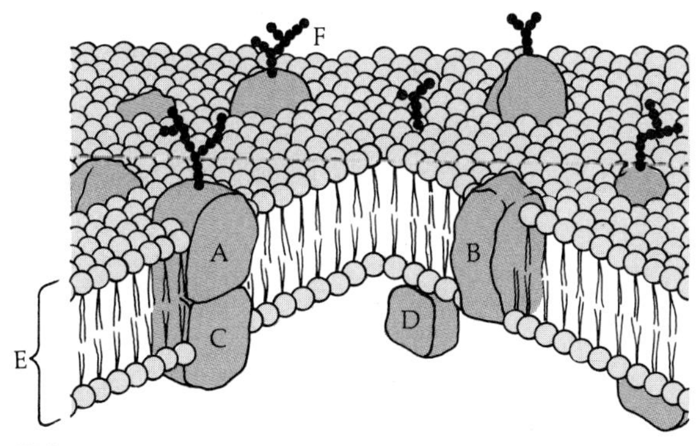

FIG. 4-3
Molecular model of the cell membrane. A, B, C, and D are proteins located on, in, and under the lipids of the cell membrane. E is the double layer of fluid lipid molecules possessing hydrophilic (water-loving) heads and hydrophobic (water-fearing, water-repelling) tails. F represents carbohydrate chains attached to proteins or lipids.

lipid. Some are located on its outer or inner surfaces while others penetrate both layers of lipids (Fig. 4-3).

Temporary spaces or *pores* may be found between or within molecules in the cell membrane. They range in size from about 4 to 20 nm (about one-millionth to one six-millionth of an inch) in different cells. Pores occupy about one hundred-thousandth to one-millionth of the total area of the average cell membrane.

There is some controversy as to the nature of pores. Much evidence indicates they are temporary channels in proteins through which ions and small molecules may pass. Since dynamic movement occurs among the molecules that make up the fluid cell membrane, the location and size of the pores may vary with the physiology of the cell. Differences in the chemical composition of cell membranes may also be responsible for variations among pores.

Role of Proteins

A variety of proteins are found as large globular molecules in cell membranes. The molecules possess charged hydrophilic groups (ionic or polar groups attracted to water), which protrude toward the extracellular fluid, and which line openings in the membrane, called *channels* or *pores*. The noncharged hydrophobic groups of proteins are nonpolar and water-repelling and project toward similar components of the lipids in the interior portions of the cell membrane (Fig. 4-4).

Proteins consist of long polypeptide chains which possess three-dimensional structure. They exist as right-handed coils (*alpha helices*) or folded sheets of paired chains of amino acids (*beta sheets*) linked by hydrogen bonds. Their overall structure is directly related to their amino acid content. Proteins in the cell membrane demonstrate several properties. They may act as (1) pores, (2) receptors, (3) structural proteins, (4) enzymes, and (5) pumps, which transport substances into and out of cells.

Pores

Some evidence suggests that the interior of proteins can function as water-filled channels or pores that extend from one side of the membrane to the other (Fig. 4-4). Proteins have been shown to move and aggregate along the plane of the cell membrane. If pores are temporary channels as previously suggested, then their nature and position could vary with that of proteins. For example, certain but not all parts of neuron cell membranes are especially suited to receive or transmit signals. These properties seem to be related to a unique distribution of proteins and pores within those cells.

Movement of specific substances through pores may be affected by the size of the channel, the distribution of charges along its inner surface, and the size and nature of the charge(s) on the substance approaching the pore for passage.

Water is found in the body as H^+OH^-, a molecule with two oppositely charged poles (a dipolar molecule). Ions are attracted to it in varying degrees, resulting in hydrated forms of ions that attain different sizes. For example, the diameter of a water molecule (H^+OH^-) is about 0.3 nm, which is almost that of hydrated Cl^- or K^+. The diameter of the hydrated Na^+ ion is 0.5 nm. Since the water molecule is polar (it has a positive and negative pole) and is small, it

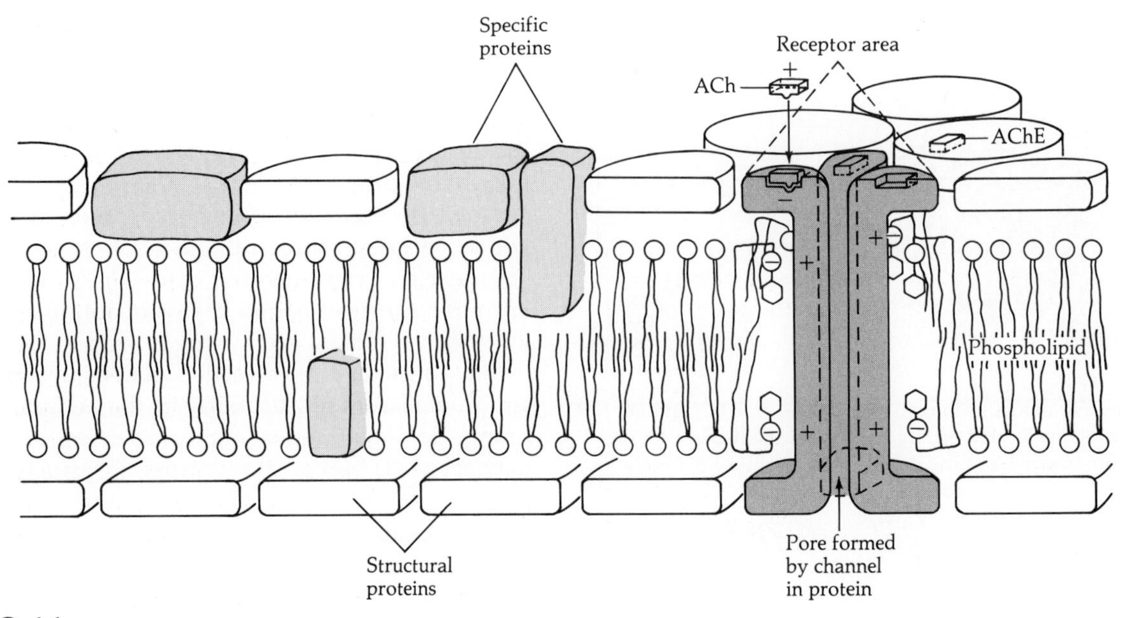

FIG. 4-4
Proteins in the structure of the cell membrane. Proteins make up the largest components (by weight) of the cell membrane. Some proteins contribute to its framework as structural protein located on its external and internal surfaces. Other specific proteins are found on and may penetrate its surface. Still others form channels or pores whose surface may contain specific sites that act as receptors for molecules such as acetylcholine (ACh) and enzymes such as acetylcholinesterase (AChE, which hydrolyzes ACh).

most readily passes through pores in cell membranes. Ions seem to pass through cells in specific channels that may differ for each ion type. The external proteins on the cell membrane have a net positive charge, which retards the passage of positively charged particles. Thus, hydrated Cl^- ions, which are equal in size to hydrated K^+ ions, pass across nerve cell membranes twice as fast as K^+ and 100 to 200 times as rapidly as the larger hydrated Na^+.

The channel in a coiled protein molecule in the cell membrane may be compared to that of a Slinky (a toy shaped like a spring, which can change shape and topple down a flight of stairs). The center of a Slinky may be large enough to allow a tennis ball, but not a softball, to move through it. Similarly, only molecules and ions of specific sizes may move through the center of proteins (the Slinkies of the cell membrane). Owing to differences in amino acid content, the interior of proteins may be charged. Accumulation of charges on the protein may lead to greater or lesser rates of movement of specific charged molecules or ions, such as Cl^- and Na^+ across the cell membrane.

A Slinky can flip-flop downstairs when provided with energy (a push from the hand or a pull from gravity). Similarly, energy changes may alter the shape of protein molecules, resulting in their movement. Such alterations may be responsible for the transport of substances across the cell membrane that can fit within the coils of a protein molecule. Parts of proteins may bend to aid the movement of specific substances within their coils. This change in protein shape will be described below in the section on carrier systems (see Fig. 4-10).

Like all analogies, this one has some shortcomings. A Slinky is rather uniform throughout its structure. Proteins are nonuniform, being composed of building blocks the sequences of which are unlikely to be repeated elsewhere within their structure.

Evidence also exists which indicates that pores may be formed by phospholipid molecules in the cell membrane. Such molecules possess chains which are flexible enough to allow small charged ions or molecules to pass between them. Their structure and role in the membrane will be discussed later.

Receptors

Carbohydrates often combine with proteins to form glycoproteins on the external surface of the cell membrane (Fig. 4-3). These external proteins may act as *receptors* and at the same time confer a unique surface chemistry on different cell tissue and organ types acting as *histocompatability (transplantation) antigens,* which are described below.

Receptors serve as sites for attachment of a variety of molecules, such as hormones, neurotransmitters, and drugs. The response of a cell or its failure to respond to these different agents often depends on the chemical makeup of surface glycoproteins (Fig. 4-5). Examples of the effects of interactions of specific receptors with a neurotransmitter, infectious microorganism, and drug will be described below, while interactions of other chemicals and cell receptors will be described in subsequent chapters.

Acetylcholine (ACh) is a neurotransmitter (molecule causing the transmission of nerve impulses) which is released from certain nerve endings. It binds to specific receptors on certain cells and causes changes in the electrical properties of the cell membranes. When it binds to receptors on skeletal muscle cells, the changes which are promoted lead to muscle contraction. The proposed location of ACh receptors adjacent to pores in a cell membrane is indicated in Fig. 4-4.

The outer protein coat of some viruses also seems to bind selectively to specific cell receptors. Viral attachment to cell membranes plays a role in susceptibility to infections and may determine which cells become the target of viral attack. Variations in binding of bacteria or their toxins to specific receptor sites may also influence the course of bacterial disease.

So, too, the response of cells to drugs depends on specific interactions between the drug and cell receptors. For example, in response to the administration of histamine, smooth muscle with H_1 receptors in the respiratory tract produce the symptoms of hay fever or asthma, while cells with H_2 receptors in the stomach release hydrochloric acid.

The composition of glycoprotein receptors is extremely important to cell function and is directed by genes within each cell. For example, the genetic control of production of a specific large-cell-surface protein [large external

FIG. 4-5
Role of some receptors on the surface of cell membranes. (A) Responsive cell has receptors that bind to the agent (⚲). Hence, the cell responds to it. (B) Unresponsive cell has receptors that do not bind to the agent (⚲). Hence, the cell is unresponsive to it.

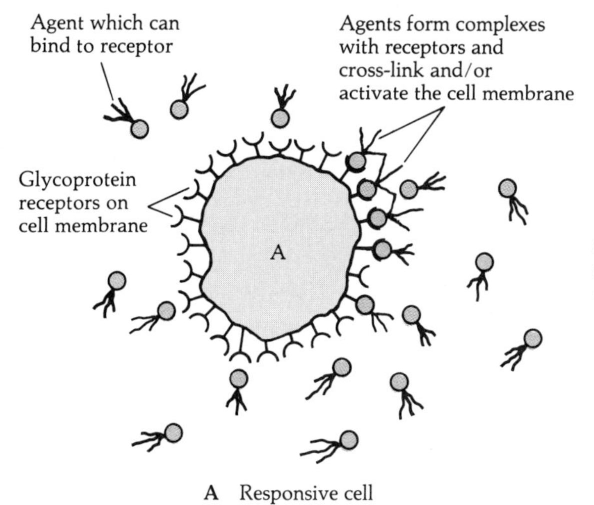

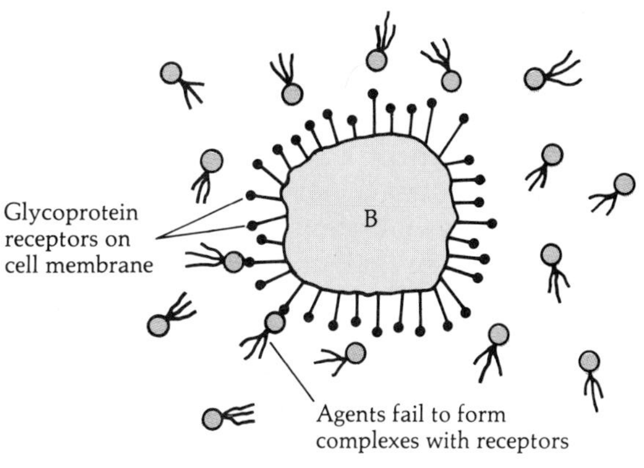

transformation sensitive protein (LETS)] in some cells has been demonstrated. Cells that fail to form LETS lose their ability to mature. Loss of LETS may be accompanied by failure of the cell to control its growth rate and biochemical properties and may thus be associated with the onset of cancer. Evidence exists that tumor viruses or chemical *carcinogens* (agents which induce cancer) may attach to and/or penetrate cells and cause loss of LETS. This may signal certain types of enhanced tumor formation following exposure to different kinds of carcinogens.

Structural Proteins

Proteins also form important structural molecules. They may: (1) serve as a supporting frame for the cell; (2) form interconnections between cells; (3) confer a unique chemistry (histocompatibility antigens) upon cells and tissues; and (4) serve as cell receptors, as described above.

A physical relationship often exists between structural proteins in the cell membrane and microfilaments and microtubules within the cytoplasm (Fig. 4-6). The presence of microfilaments and microtubules not only supports cell structure but also may facilitate passage of substances across the cell membrane and intracellular compartments.

Proteins are found as interconnections between cell membranes in the form of various types of desmosomes, tight junctions, or gap junctions, described in Chap. 3 and illustrated in Fig. 3-1.

Proteins which form histocompatibility antigens are those which are foreign to and cause an immune reaction in a recipient. The antigens are usually dominant structures on the exterior of the cell membrane. They cause a defense reaction in an individual whose major histocompatibility antigens are not matched by those on donor cells, tissues, or organs. The reaction causes rejection (destruction) of the transplanted cells. Rejection is rapid if caused by antibodies (such as rejection of blood cells in a blood transfusion reaction). Rejection proceeds more slowly when caused by cellular defense reactions involving T lymphocytes (as occurs in rejection of a skin graft). Some tissue rejection involves antibodies and T lymphocytes (for example, the rejection of a kidney transplant).

The important role of carbohydrate in defining the unique surface of different glycoproteins and cell types is clearly shown in human blood types. The difference of a single sugar molecule attached to the surfaces of large glycoproteins which project from the surfaces of red blood cells determines whether they are type A, B, AB, or O (Fig. 18-7).

Enzymes

A number of important enzyme activities occur as the result of catalytic action of proteins at the surface of the cell membrane as well as of proteins within the cell.

Within the same protein molecule some components may behave as receptors while others act as enzymes. For example, adjacent parts of

FIG. 4-6
Relation of cell membrane proteins to cytoplasmic filaments and microtubules. A, B, C, and D are proteins located on various sites of the cell membrane. E is the double layer of fluid lipid molecules possessing hydrophilic heads and hydrophobic tails. F represents carbohydrate chains attached to proteins or lipids. The globular proteins may move in the fluid lipid of the cell membrane. Microfilaments and microtubules in the cytoplasm are closely associated with and may influence the location and movement of some of the proteins.

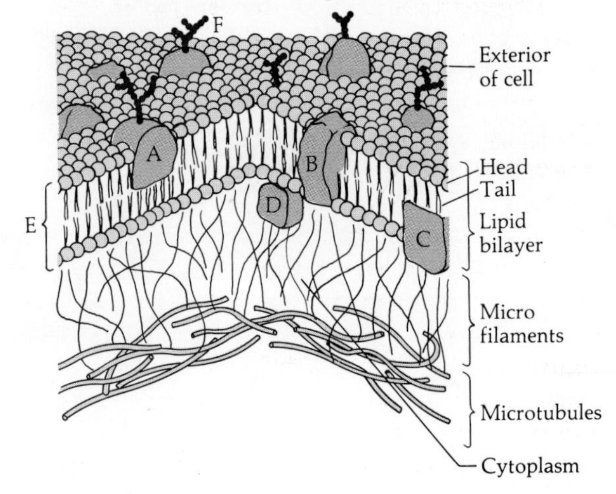

certain proteins on some cell membranes act as receptors for ACh and as enzymes (acetylcholinesterases, AChE) shown in Fig. 4-4, which can hydrolyze ACh. Some proteins bound to membranes have ATPase activity, which means they can split adenosine triphosphate (ATP) to release energy that drives chemical reactions in cells. It should be noted, furthermore, that in many instances the activity of enzymes in cells is turned on and off by specific hormones, which bind to protein receptors on or in the cells (Chap. 5).

Pumps

Some parts of proteins bind specifically to ions or molecules and transport them across the cell membranes into or out of cells. The proteins may have some components with enzyme activity as described above and other components that act as specific pumps to move substances across the cell membrane. The role of various proteins in the movement of substances across cell membranes is discussed at length below in the section on transport mechanisms.

Role of Lipids

Lipids or fats account for nearly half of the mass of the cell membrane. They exist as a double layer of molecules which form the interior matrix of the membrane (Fig. 4-6). Many lipids combine with proteins to form *lipoproteins,* or with carbohydrates to form *glycolipids* that project from the outer surface of the cell membrane.

The primary types of lipids include phospholipids, cholesterol, and glycolipids. *Phospholipids* constitute the principal barrier influencing the entry and exit of substances into and out of the cell (Fig. 4-6). The cell's capacity to control passage of substances across its borders is called *selective permeability.* In general, lipid (fat)–soluble molecules move through this part of the membrane with greater ease, whereas water-soluble molecules move through less easily or not at all. Phospholipids in the outer layer of the cell membrane may differ from those in the inner layer. So, too, the length of their fatty acid chains may vary, as well as their degree of saturation with hydrogen. The two layers of phospholipids contain hydrophilic (water-loving) heads that project toward the exterior and interior of the membrane, respectively (Fig. 4-6). The carbon chains of the phospholipids vary in length and produce water-repelling (hydrophobic) tails that project toward one another and the center of the membrane. About 75 percent of the phospholipids in most cell membranes are one of three types: *phosphatidyl choline* (*lecithin*), which is the most abundant, *sphingomyelin,* and *phosphatidyl ethanolamine.*

Cholesterol makes up slightly more than one-quarter of the total lipid in many cell membranes. Cell membranes become less permeable as their cholesterol level increases.

Glycolipids (combinations of lipid and carbohydrate) represent a small percentage of the total lipid population. Along with glycoproteins they confer unique properties (antigenicity) on the surfaces of particular cell types.

4-3 TRANSPORT MECHANISMS

The movement of substances across membranes may occur with or without the assistance of the membrane's components. Nonmediated transport, called passive transport, includes diffusion and osmosis. Carrier-mediated transport involves facilitated diffusion and active transport. Vesicle-mediated transport occurs via the formation and movement of membrane-bound sacs or vesicles.

Nonmediated Transport

Diffusion

Diffusion is random movement in gases and fluids of molecules or ions due to activity within their atomic structure. It is a consequence of the normal kinetic energy that results from electrons traveling around the nuclei of atoms. A continual random motion of substances is produced. Diffusion causes the separation of particles and results in their passage from areas of higher concentration to areas of lower concentration. It can be readily observed in the passage of molecules of perfume through a room from one corner to another.

FACTORS INVOLVED When a membrane is freely permeable to a substance, the rate of diffusion across the membrane is directly proportional to the difference between the concentration of that substance on one side of the membrane and its concentration on the other side, the temperature of the environment, and the cross-sectional area of the membrane. As the molecular weight and distance involved increase, the rate of diffusion decreases. The diffusion rate may be calculated by the following formula:

$$DR = \frac{[C_1 - C_2] \times T \times A}{\sqrt{MW} \times d}$$

where

DR = diffusion rate
C_1 = concentration of substance on one side of the cell membrane
C_2 = concentration of substance on the other side of the cell membrane
T = temperature of fluid
A = total surface area of membrane
MW = molecular weight of substance
d = distance over which diffusion occurs (e.g., thickness of cell membrane)

An analogy might be used to explain the principles governing diffusion. We may compare the cell, its contents, and its membrane to a room having holes in its walls and containing bouncing Ping-Pong or tennis balls, which are also bouncing around outside the room. In this analogy, the room is equivalent to a cell surrounded by its membrane and the holes in the wall are equivalent to pores in the membrane. The net rate of the movement of the balls into or out of the room, the *diffusion rate,* could be measured. It would depend on the difference between the number of balls on the outside of the room moving in and the number inside moving out (the *concentration gradient,* C_1–C_2). The rate would increase if energy (temperature) were applied to speed up the movement of the balls. It also would increase if the area available for passage increased (that is, the area of the wall and the number of holes in it, the total *surface area* of the membrane). The rate would be lower for the larger, heavier tennis balls than for the smaller, lighter, Ping-Pong balls (molecules of large size and high molecular weight compared with those of small size and low molecular weight). It also would be lower if the distance involved increased (that is, the distance across the cell membrane).

The potential for diffusion of ions or charged particles is also related to their electrical properties. Oppositely charged particles are attracted to each other, while those with like charges repel one another. Consequently, an electrical gradient is established when differences in net charge exist across the two sides of a selectively permeable membrane. Assuming they are free to diffuse, net movement of similarly charged particles will occur down their electrical gradient across the membrane. Cells attain an ionic equilibrium (no net movement of charged particles) when an electrochemical equilibrium is attained. The *electrochemical equilibrium* results from a balance of the electrical charges *and* concentration of charged particles across each side of a selectively permeable membrane. In fact, owing to the selective permeability of cells, the balance is uneven (like a balanced but unlevel seesaw) and creates a potential for diffusion of the charged particles (a voltage). The electrochemical equilibrium of cells is especially important in the passage of electrical signals along cell membranes. It will be discussed in detail in Chaps. 7 and 12 on nerve and muscle physiology.

ROLE OF PORES Another important variable in this analogy is the size of the holes in the walls of the room (pores in the cell membrane). If the holes are large enough, softballs (very large molecules) could pass through in addition to the smaller tennis and Ping-Pong balls. Variations in pore size can result in accommodation or exclusion of different ionic and molecular species. The effective size of cellular pores is diminished by increased levels of positively charged *calcium ions* (Ca^{2+}). As the Ca^{2+} level increases, the number of positive charges around the pore increases. The Ca^{2+} ions become associated with negatively charged components in the cell membrane, thus creating a barrier to the passage of other positively charged particles, which are repelled by the like charge. Calcium is lost from these locations following prolonged dietary calcium deficiency. This effectively increases pore size and, if pronounced, allows excessive diffusion. In this manner, calcium deficiency can lead to erratic passage of impulses across cell membranes. It may impair muscle and nerve function and may produce *tetany*, a condition associated with muscular spasms. (This is not the same as *tetanus*, which is the disease, also called *lockjaw*, that is due to poisoning by protein from the bacterium *Clostridium tetani*).

The presence of charges also affects the passage of molecules through cell membranes. Nonionized forms of molecules pass through cell membranes with greater ease than ionized forms. For example, aspirin and barbiturates are weak acids which remain largely nonionized in the acid of the stomach and are therefore readily absorbed there. When the stomach's pH increases (becomes more alkaline), the percentage of ionized forms of these molecules increases. This results in a lesser ability of the drugs to permeate the membranes of the epithelial cells lining the stomach. Consequently, less of the drug enters the bloodstream in a given amount of time. Nonionized forms of drugs are also absorbed faster than their ionized counterparts in the small intestine. The mixture of aspirin with an antacid causes aspirin to assume its ionized form and decreases its rate of absorption.

It seems clear then that the presence of charges on the surface of membranes is an important factor that influences the passage of charged particles. Owing to the accumulation of Ca^{2+}, positive ions move through pores with much more difficulty than negatively charged ions. For example, hydrated potassium ions (K^+) permeate nerve cell membranes with half the speed of hydrated chloride ions (Cl^-) although the two ions are the same size.

Osmosis

The diffusion of water across a selectively permeable cell membrane is called *osmosis*. Because of their small size, electrical neutrality, and abundance in body fluids, water molecules often move across cell membranes millions of times faster than do other molecules or ions.

OSMOTIC PRESSURE The ability of a membrane to discriminate among different ions and molecules and to control their rates of passage is termed *selective permeability*. Variations among the diffusion rates of Na^+, K^+, Cl^-, and H_2O across cell membranes illustrate this property, as does the exclusion of molecules above a certain size. Water diffuses very rapidly down its concentration *gradient* (from more dilute solutions, to less dilute ones, that is into those with less water and more solutes per unit volume). When the extracellular fluid has more water and fewer dissolved (solute) particles per unit volume than the ICF, the net diffusion of water is into the cell (Fig. 4-7B). The ECF is more dilute and is thus termed *hypoosmotic* and the ICF has a higher osmotic pressure. Water diffuses into the cell. *Osmotic pressure*, the pressure associated with osmosis, can be considered as simply a measure of a solution's tendency to draw water into it. In this situation the movement of water into the interior of the cell causes an increase in cell volume. If some of the solutes in the cell do not diffuse readily (as is the normal situation), the cell swells as a result of the net inflow of water. This creates pressure that usually resists more net water movement into the cell. Finally, an *equilibrium* may be attained in which the number of water molecules moving into the cell at any one instant equals the number moving out. At equilibrium there is no *net* diffusion of water.

OSMOLARITY AND AVOGADRO'S NUMBER The *osmotic pressure* of a solution is proportional to the total *number of solute particles* dissolved in it. Osmotic pressure is unrelated to the size of the particles. The osmotic pressure of a solution is expressed as its *osmolarity*. Solutions with the same osmolarity contain the same number of osmotically active solute particles. The osmolarity depends on the number of particles in one mole (1 mol) of the substance when it is in

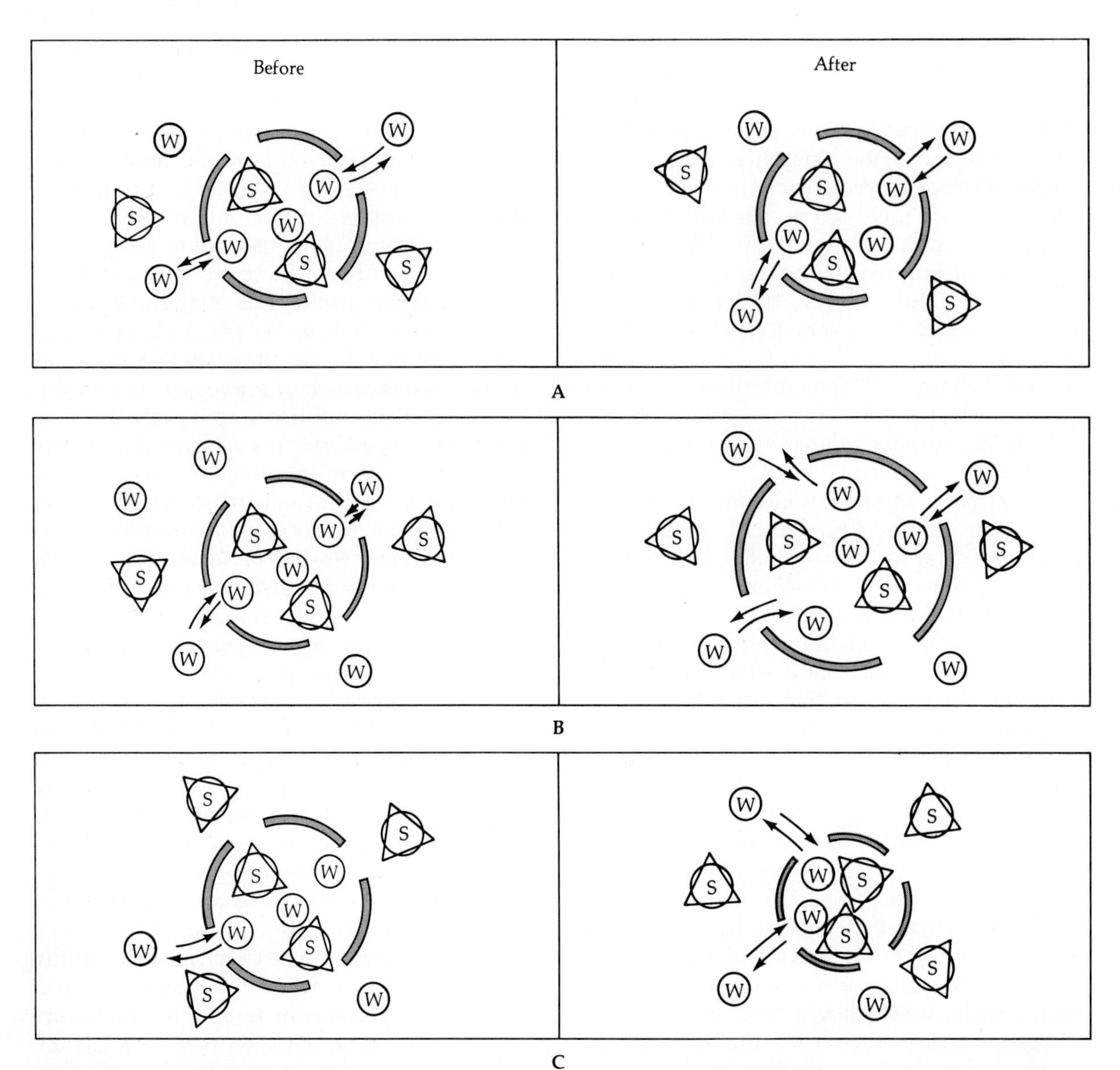

FIG. 4-7

Osmosis: The diffusion of water across a cell membrane. Count the number of water molecules (W) and nondiffusible solute molecules (S) inside and outside the cell after equilibrium has been attained. On reaching equilibrium, the ratio of the two types of molecules to one another is now the same inside and outside the cell although the total number of molecules inside the cell and the size of the cell may have changed. (A) Cell in an isotonic solution. *Before:* Theoretical cell with equal concentrations of water and nondiffusible solute inside and outside the cell. The ECF and ICF are *isotonic. After:* Water moves in and out of the cell at equal rates leaving the cell unchanged. (B) Cell in a hypotonic solution. *Before:* Theoretical cell with greater concentration of nondiffusible solute inside the cell placed in an ECF, with a lesser concentration of solute. The ECF is *hypotonic* relative to the ICF and contains relatively more water molecules. *After:* More water entered than left the cell. The net entry of water caused the cell to swell. The ICF and ECF both attain a 2:1 ratio of water to solute molecules at equilibrium. (C) Cell in a hypertonic solution. *Before:* Theoretical cell with lesser concentration of nondiffusible solute inside than outside the cell. The ECF is *hypertonic* relative to the ICF and has relatively fewer water molecules in it. *After:* The ICF and ECF have attained an equilibrium, with a net movement of water out of the cell. A 1:1 ratio of water to solute molecules is produced. However, the net movement of water out of the cell has caused it to shrink.

solution. A *mole* is the gram-molecular-weight of the substance, i.e., its molecular weight expressed in grams. A mole of any substance contains the same number of molecules (6.02×10^{23}) as a mole of any other substance. This is known as *Avogadro's number.* However, some compounds are *dissociated* in solution, which means that the particles of which they are formed are separate. Other molecules do not dissociate. For example, if 1 mol of a substance dissociates into two particles per molecule, it will produce $2 \times 6.02 \times 10^{23}$ or 12.04×10^{23} particles in solution.

The effect of dissociation on osmotic pressure can be described by comparing glucose and sodium chloride. A 1 *M* (1 molar) solution of glucose ($C_6H_{12}O_6$) can be made by dissolving 180 g (1 gram-molecular-weight) in 1 L of water. A 1 *M* sodium chloride solution (NaCl) contains 58.5 g (1 gram-molecular weight) per liter. Glucose does not dissociate when placed in solution, whereas each NaCl unit produces two particles, one Na^+ and one Cl^- ion. Therefore, a 1 *M* solution of glucose produces 6.02×10^{23} particles, and thus its osmolarity is 1. A 1 *M* solution of NaCl produces $2 \times 6.02 \times 10^{23}$ particles, so its osmolarity is 2. A 1-osmolar solution of NaCl is made by dissolving 29.25 g (half its molecular weight) per L of water. It contains a total of 6.02×10^{23} particles in the form of Na^+ and Cl^- ions, corresponding to only 3.01×10^{23} NaCl units. It is simultaneously a 1-osmolar and a 0.5 *M* solution.

Fluids which have the same osmolarity are said to be *isoosmolar* or *isotonic.* For example, the average osmolarity of human ECF and ICF is about 0.3 osmol, or 300 milliosmol (mosmol), per liter. The *tonicity* of a solution is a measure of its effective osmotic pressure relative to that of plasma. When the osmotic pressure of two fluids is the same, the rate of movement of water molecules between them is equal (Fig. 4-7A). When the osmolarity of the ECF is less than that of the ICF, the net movement of water is into the cell, which causes it to swell and, if sufficient, may cause it to burst (lyse). The extracellular fluid in this case is *hypotonic* (hypoosmotic) with respect to the cell's interior (Fig. 4-7B). *Hypertonic* (hyperosmotic) extracellular fluids cause net movement of water out of cells (Fig. 4-7C). They may cause cells to shrink (crenate). Solutions bathing cells of the body should be isotonic to prevent disruption of cellular physiology. An isotonic solution for human tissues is equivalent to 0.15 *M* (0.89%) NaCl.

A few practical examples may illustrate the importance of these concepts. For instance, the replacement of body fluids and/or injection of substances should be made with isotonic solutions. Cells may be disrupted and burst following hypotonic fluid administration. These principles must be considered in designing fluids for the storage of blood cells used in transfusions. The action of many laxatives is also based upon these factors. For example, milk of magnesia is a useful laxative because it consists of a relatively nonabsorbable hypertonic salt solution. Its presence causes osmosis of water into the large intestine. This softens solid wastes (feces) and promotes bowel movement.

EFFECT OF ANTIDIURETIC HORMONE Movement of water can also be affected by hormones. For example, the antidiuretic hormone (ADH, vasopressin) promotes the conservation of water by the body. The formation of urine is termed *diuresis,* and ADH acts as an antidiuretic. Produced in the hypothalamus of the brain, ADH seems to effectively increase the diameter of the pores of cells lining the tubules called collecting ducts within each kidney. The fluid in parts of the tubules is hypotonic as the result of the reabsorption of salts into surrounding capillaries. The increased pore size of more distant tubular cells resulting from the activity of ADH allows water molecules to diffuse more readily back down their gradient into the adjacent blood vessels and is thus antidiuretic.

Water movement is also affected by various drugs. For example, caffeine in coffee and theobromine in tea are diuretics. They are in a class of compounds, the xanthines, which include a most potent diuretic, theophylline, also found in tea. The xanthines promote movement of water from the blood into the cells of the tubules of the kidney and thus *increase* the formation of urine (promote diuresis).

Carrier-Mediated Transport

Movement of substances with the assistance of components in the cell membrane is called *mediated transport*. In one form, carrier molecules (proteins) bind specifically to the substances in question. The process is called *active transport* when energy is provided from the high-energy phosphate compound adenosine triphosphate (ATP) to assist the process. It is called *facilitated diffusion* when this energy source is not required.

Facilitated Diffusion

Facilitated diffusion is the movement of a substance through a cell membrane assisted by a carrier that passively but specifically transports the substance. At first, the substance is present at unequal levels across the membrane, i.e., a gradient exists. Random motion of the substance results in net movement from areas of higher to areas of lower pressure, concentration, or electrical activity. The carrier passively aids its transport. Only a few types of molecules (e.g., glucose, some vitamins, and some amino acids) move by facilitated diffusion through membranes at rates faster than simple diffusion would allow.

Some evidence indicates that the carrier is a protein which specifically combines with the substance on one side of the membrane and drops it off on the other. The carrier might be likened to a revolving door, which passively assists movement when it is pushed by people going in or out of a building. A model for the facilitated diffusion of glucose is shown in Fig. 4-8.

Carriers seem to be relatively specific for molecules of different shapes and sizes. Substances of similar structure may compete with each other for the same carrier (just as people might compete for a space in a revolving door). When competition occurs, the rate at which the carrier transports its normal substrate through the membrane is reduced. Evidence is available that the carrier may provide a unidirectional channel (somewhat like a water slide) favorable for passage of a specific material. In some instances facilitated diffusion may be linked (coupled) to other transport processes which produce energy. With this assistance, the carriers used in facilitated diffusion seem to be able to cause movement against concentration gradients.

FIG. 4-8
Facilitated diffusion of glucose across a cell membrane. Net movement of glucose molecules (G) is facilitated by a carrier in the cell membrane. It is shown assisting passage of the sugar along a concentration gradient from the ECF to the ICF. The gradient is temporary and can change direction depending on the availability of sugar and on cell and body function.

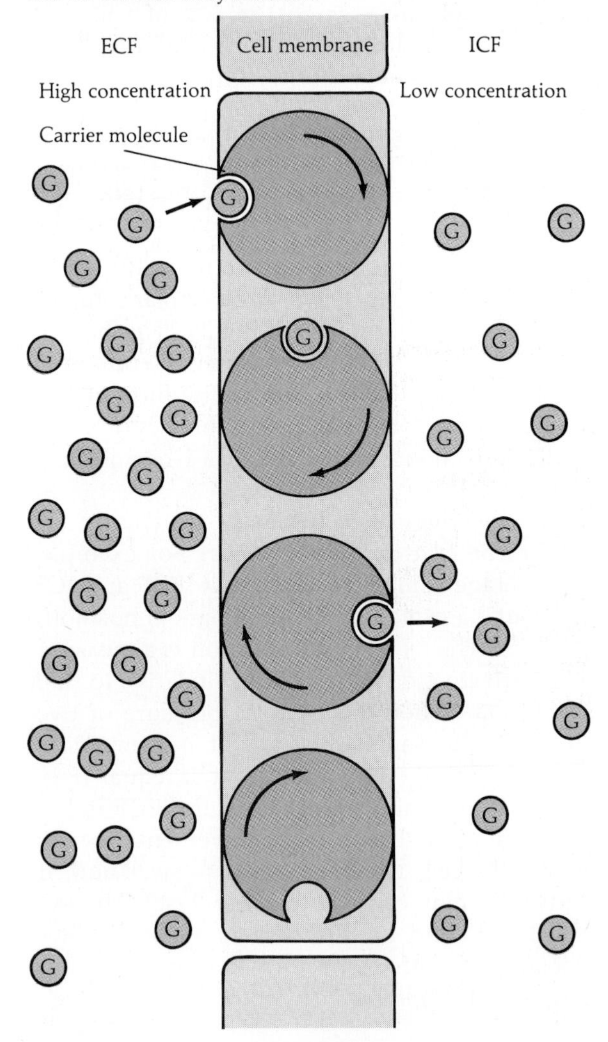

Active Transport

The ICF often contains a higher concentration of some substances than is normally found in the ECF. Yet cells may require even more of the substances intracellularly (e.g., potassium). These unequal distributions of substances across cell membranes are often produced by carriers, which require energy and are necessary to maintain vital functions. The necessary energy is usually derived from the breakdown of ATP and is used to change the shape of the carrier. The process is known as *active transport*. In this process specific carriers move molecules and ions against concentration, electrical, or pressure gradients. Active transport differs from faciliated diffusion in that net movement always occurs against a gradient and requires energy, which is supplied by ATP.

CARRIER SYSTEMS Carriers used in active transport might be likened to revolving doors with motorized moving platforms which actively move people into and out of buildings (Fig. 4-9). They do so only when energy is provided by a generator inside the building. In this case, the building is analogous to a cell and its generator to cellular metabolism of ATP. Carriers at the surface of cell membranes actively move substances into and out of cells when energy is supplied by ATP.

While analogies that compare carriers to revolving doors are easy to envision, they are not entirely correct. Proteins do not seem to revolve in the cell membrane. Rather, transport of ions and molecules may more properly be pictured as involving changes in the shape (conformation) of channels within a carrier protein, as shown in Fig. 4-10. The subunits of protein molecules can be likened to flexible walls forming a narrow corridor in a fun house. Segments of the walls can move on pivot points away from and toward each other. Individuals must be of the right shape or size and be provided with sufficient energy in order to pass readily along the corridor. Ions or molecules also must possess specific properties to pass through channels created by proteins in the cell membrane. The energy may be derived directly from reactions which accompany active transport. It

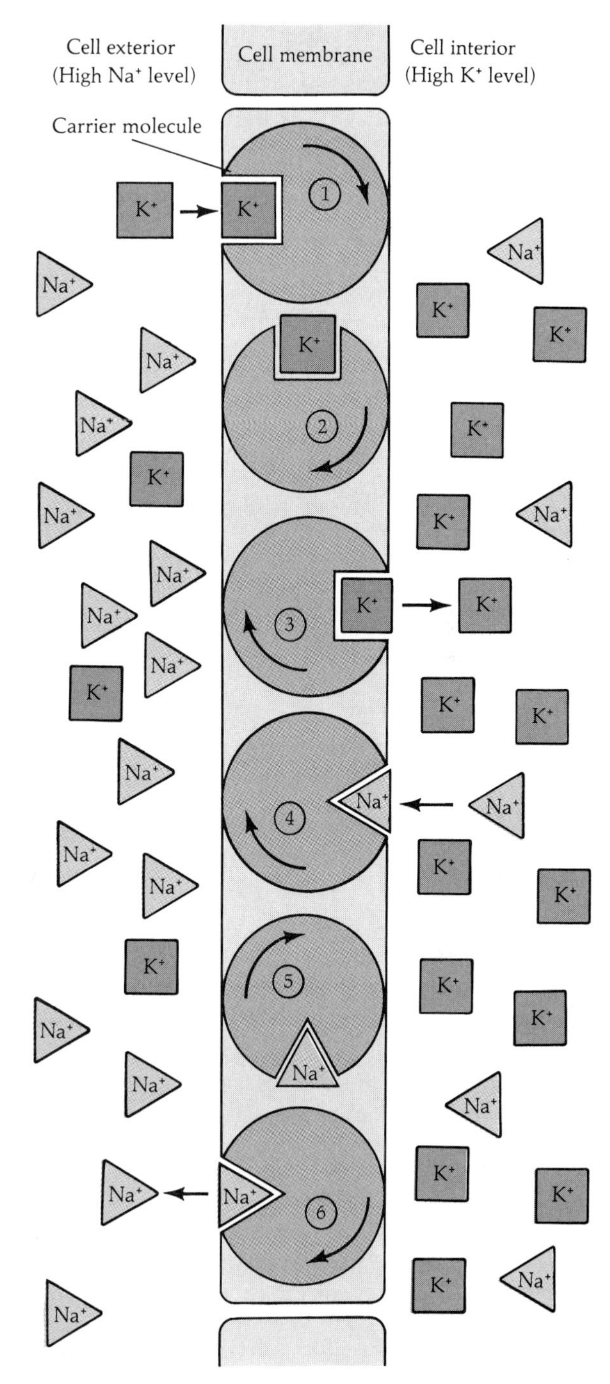

FIG. 4-9
Active transport of Na^+ and K^+. Net movement of K^+ and Na^+ ions is aided by a carrier in the cell membrane. The carrier assists movement of both types of ions in opposite directions and against their concentration gradients. It changes its shape and requires energy from ATP to do so.

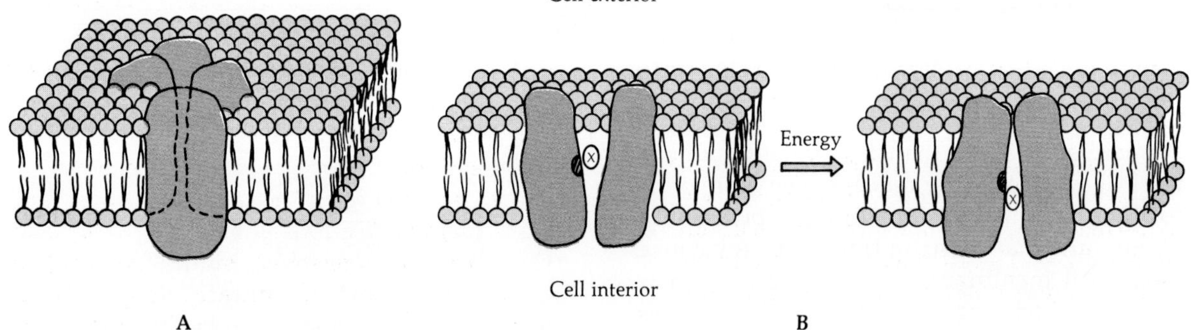

FIG. 4-10
Ion transport and change in shape (conformation) of protein channels. Proteins which span the thickness of cell membranes may form channels or pores that are the sites through which specific ions or molecules may be "carried." (A) The channel on the left is formed by four subunits which are parts of a single protein. (B) Changes in the shape of the protein may parallel alterations in its attraction for the ion or molecule that is moved across the membrane. The required energy may be derived directly from energy-producing reactions in active transport. Alternatively, a concentration gradient can be developed which stores potential energy to assist the carrier in facilitated diffusion.

also may be derived from an energy reservoir (gradient) which stores potential energy that may facilitate diffusion.

Cells are equipped with carriers to transport specific substances related to particular functions. For example, the stomach rapidly transports hydrogen ions to produce hydrochloric acid. The kidneys transport hydrogen ions from the circulatory system into the urine to maintain blood at an alkaline pH. The thyroid readily absorbs the iodine necessary for synthesis of the hormone thyroxine.

ROLE OF MEMBRANE-BOUND ATPASES: THE SODIUM-POTASSIUM PUMP A model for active transport has been studied extensively. It describes a carrier with ATPase enzyme properties located on the cell membrane. The carrier contains groups which may bind to a specific substance when facing the cell's exterior and to a different substance when facing its interior (Fig. 4-11).

One carrier that has been extensively studied actively transports Na^+ and K^+ and cleaves ATP to produce energy. It is a three-part molecule with a molecular weight of about 250,000. Two of its subunits have molecular weights of about 95,000 and the third has a molecular weight of 50,000. The larger subunits contain sites which bind to K^+ on the outside of the cell. The molecule extends to the inside of the membrane, where other sites bind to Na^+ and ATP. The action of this molecule has been likened to that of a pump, and therefore it has been called the *sodium-potassium pump* (Fig. 4-11). Regardless of the mechanism used, the sodium-potassium pump does account for most of the movement of these ions through cell membranes, while diffusion is responsible for relatively little.

The levels of Na^+ are usually high extracellularly, while those of K^+ are high intracellularly. The relative number of the symbols in Fig. 4-9 is a general reflection of the differences illustrated in Fig. 4-2 for a typical cell.

Groups on ATPase which are oriented toward the inside of the membrane (E_I) may pick up Na^+ ions and move them to the outside (Fig. 4-11). After this occurs, E_I's shape (conformation) changes. Groups which are oriented towards the outside (E_O) may pick up extracellular K^+ ions. The E_O carrier then may move the K^+ ion to the inside and change its shape to E_I.

The change in conformation of E_O to E_I requires energy. That energy is provided by the hydrolysis of ATP. In the presence of Mg^{2+} and the ATPase site of the carrier, ATP is hydrolyzed into adenosine diphosphate (ADP) and inorganic phosphate (P_i). Cleavage of the chemical bond releases energy used to drive reactions within the cell.

Three Na^+ ions are transported outside the cell for two K^+ ions moved inside. E_I can move Na^+ out of the cell, change its shape to E_O, and reorient itself to the interior of the cell without transporting K^+. That is, it can return "empty-handed." When energy is available, E_O can be converted into E_I to move more Na^+ ions out of the cell.

Some evidence indicates the E_I cannot change shape as readily without Na^+ as can E_O without K^+. That is, E_I does not move out without Na^+ as readily as E_O can move inward without K^+. Thus, the ratio of ions carried by E_I to ions carried by E_O can vary; it may be close to unity or it may be greater than unity. The latter is especially true when high levels of ATP are available. Because the ratio may be varied, a cell can increase secretion of Na^+ to the extracellular fluid while maintaining constant rates of K^+ uptake. This gives rise to relatively higher levels of Na^+ extracellularly and K^+ intracellularly. Similar active transport systems exist for other ions (e.g., Ca^{2+}), as well as for a variety of molecular species.

FIG. 4-11
ATPase carrier model of sodium-potassium pump. E_I represents an Na^+ carrier oriented toward the inside of the membrane. E_O represents a K^+ carrier oriented toward the outside of the membrane. It is a modified form of E_I. E_I transports Na^+ to the exterior. The carrier may change its shape to form E_O that associates with K^+, and transports it in the opposite direction. Reversal from E_O to E_I requires energy derived from the breakdown of ATP. The carriers may move "empty-handed." Three Na^+ ions are transported to the exterior for two K^+ ions carried to the interior. Lesser amounts of the ions may also diffuse through pores in the membrane, along the paths indicated by the broken arrows in the diagram.

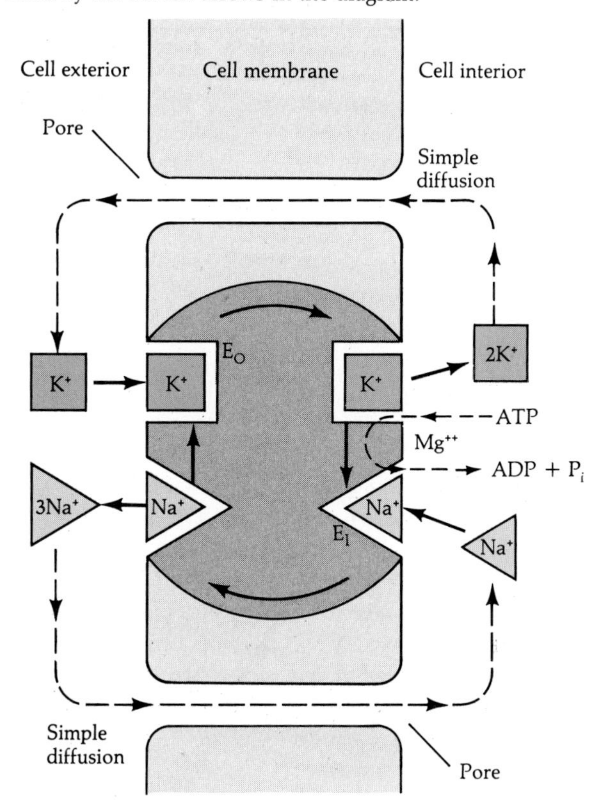

Active symport or cotransport exists for some substances. One process derives energy from and is linked to the other. For example, active symport of glucose into the blood occurs with and obtains energy from that of Na^+. The phenomenon is evident in the intestine and in the tubules of the kidney. Active symport of some amino acids and Na^+ ions also occurs. It also exists for some drugs. For example, digitalis, a drug which strengthens the heartbeat, is actively symported with Na^+.

Vesicle-Mediated Transport

Enclosure within membranes provides a mechanism whereby the cell actively participates in movement of materials into or out of its boundaries. Inward engulfment of fluid, molecules, and particles by the cell is called *endocytosis*. Extrusion of materials by the cell is called *exocytosis* or *emiocytosis* (cell vomiting).

Endocytosis

The cell membrane may be physically or chemically stimulated to fold in on itself and bring substances to its interior in *endocytosis*. When the substances that stimulate the response are ionic or molecular, the process is called *pinocytosis* (cell drinking); when they are particles or cells, the response is called *phagocytosis* (cell eating).

Pinocytosis results from interaction of strong salt solutions and proteins with molecules in the cell membrane, which can lead to its infolding (Fig. 4-12). As a result, ECF and its contents become surrounded by the membrane. Changes in surface tension cause the fluid components

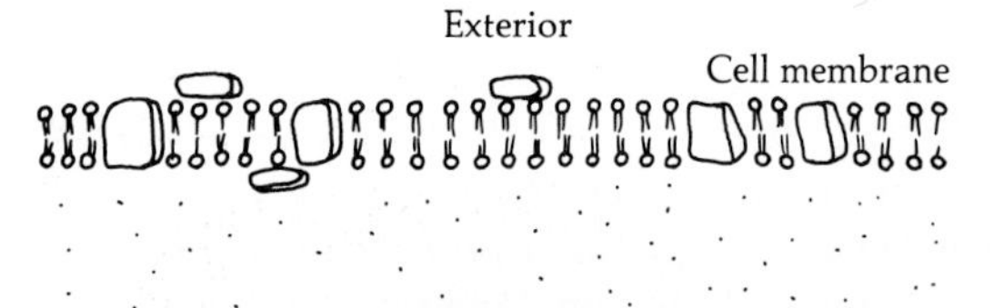
Exterior
Cell membrane
Cytoplasm
1

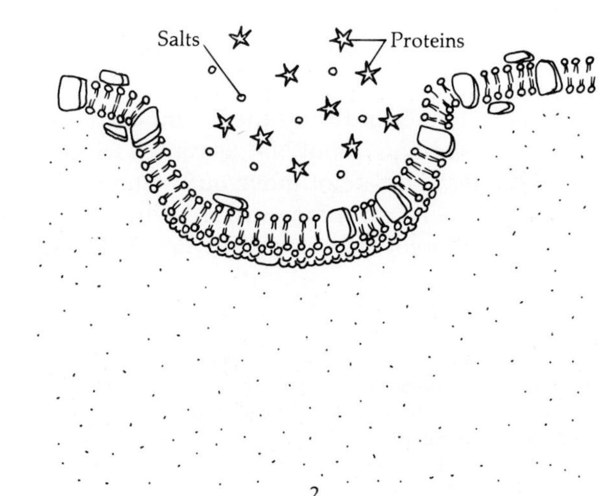
Salts
Proteins
2

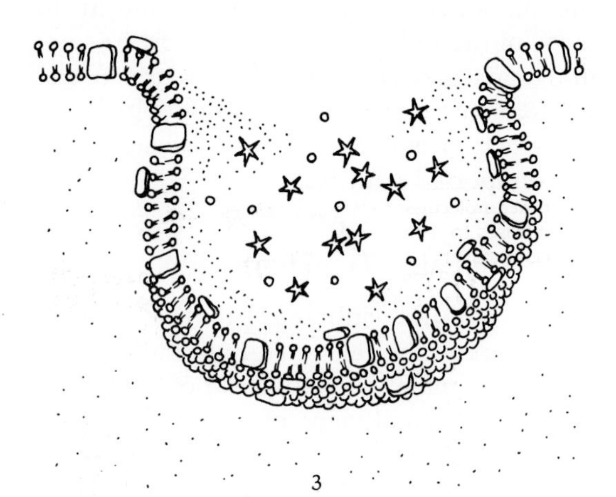
3

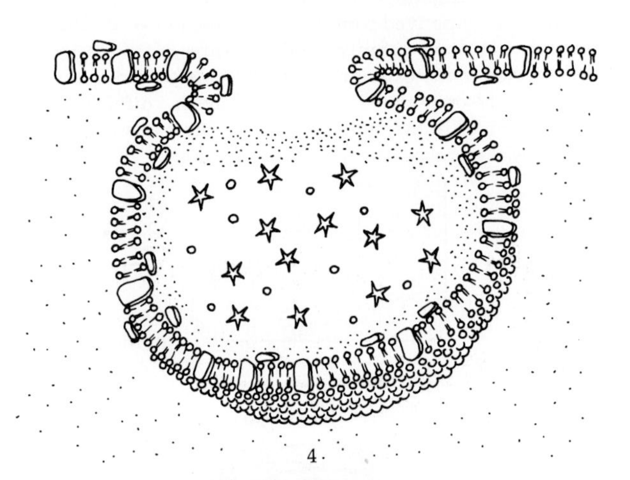
4

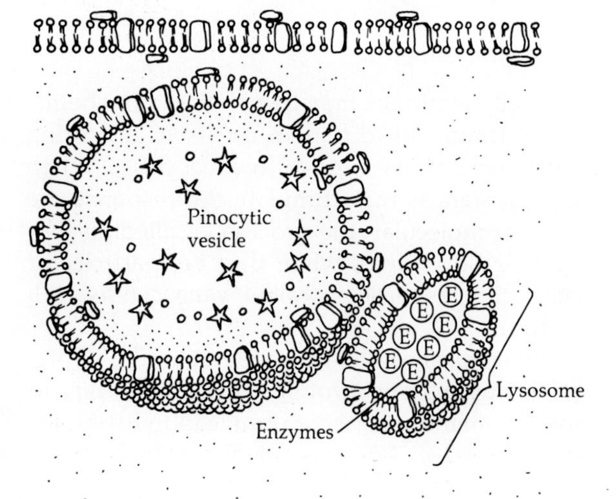
Pinocytic
vesicle
E
Lysosome
Enzymes
5

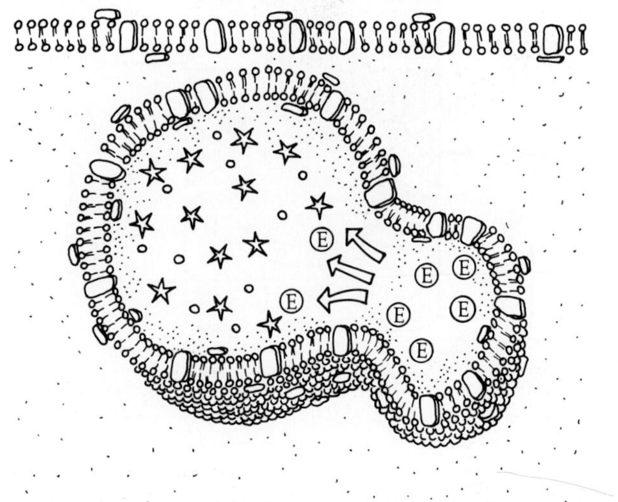
E
6

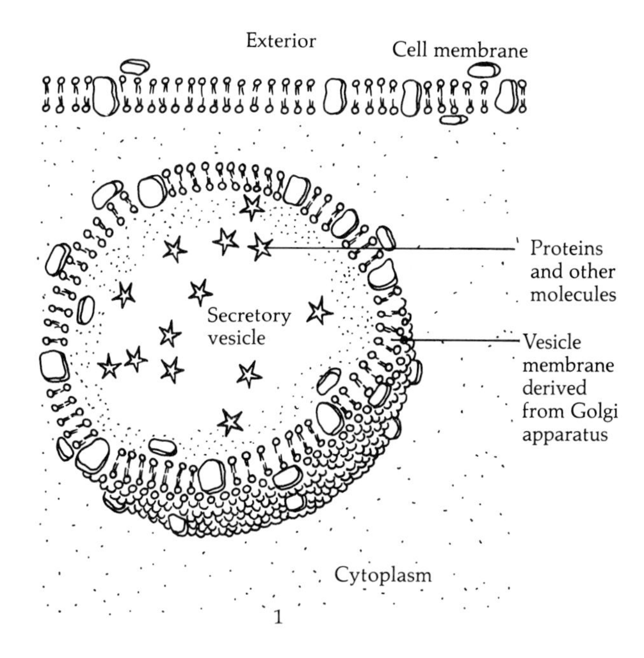

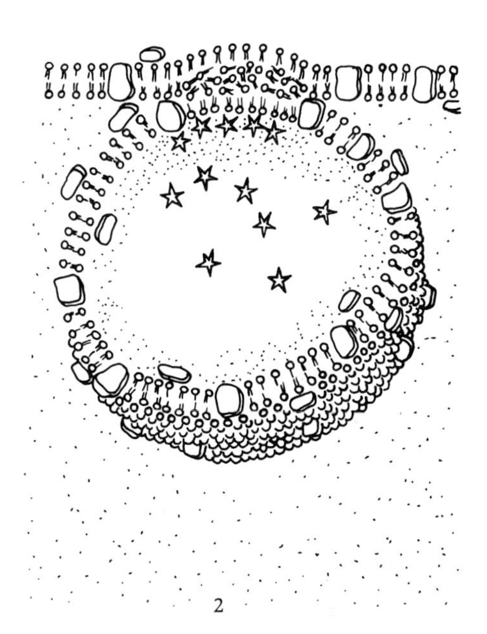

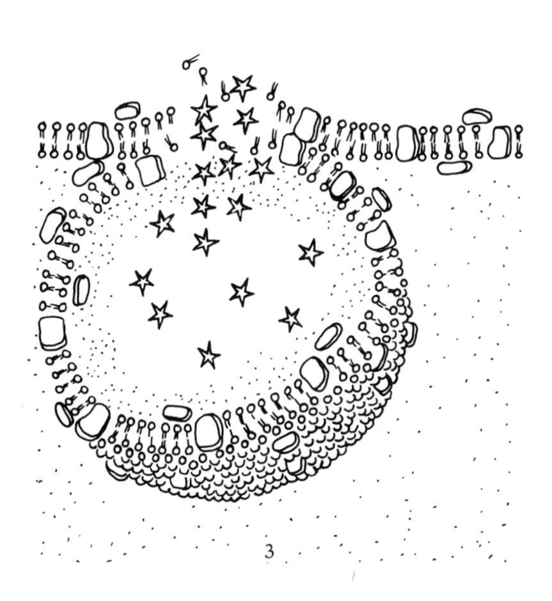

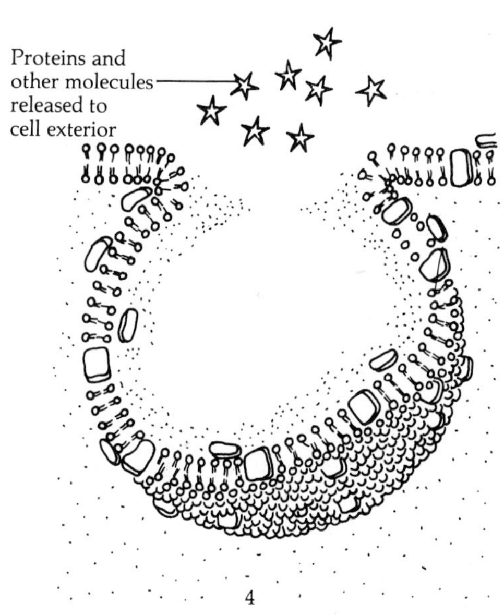

FIG. 4-12
Pinocytosis, a form of endocytosis. (1, 2 and 3) The cell membrane folds inward upon stimulation by the presence of salts or proteins in the ECF. (4) The membrane forms a deeper fold, which leads to formation of a segment that is pinched off as a pinocytic vesicle (5). (6) Enzymes (E) released from adjacent lysosomes may penetrate the vesicle and degrade its contents.

FIG. 4-13
Exocytosis. (1) The cell membrane is illustrated, along with a secretory vesicle formed by the Golgi apparatus. Products of cellular metabolism such as proteins are found within the vesicle. (2) The vesicle migrates towards and fuses with the cell membrane. (3) The phospholipids of the membranes separate as parts of the membrane are lost. (4) Molecules within the vesicle are now free to diffuse into the ECF. The integrity of the membrane is probably restored by fusion with remaining parts of the vesicle membrane and synthesis of components by the endoplasmic reticulum and Golgi apparatus within the cell.

of the cell membrane to fold in, fuse, and form a somewhat spherical sac known as a *pinocytic vesicle* (Fig. 4-12). The reaction is similar to that observed when droplets of oil are mixed in water. *Lysosomes* within the cytoplasm become attached to the pinocytic vesicle and release hydrolytic enzymes into it. The enzymes break up large molecules such as proteins and polysaccharides into their building blocks. The latter can diffuse out of the vesicle and into the cytoplasm.

Two examples that illustrate the importance of pinocytosis involve the absorption of intact proteins and vitamin B_{12} by the digestive system. Allergic reactions may develop following pinocytosis of whole proteins which are foreign to the body. Susceptibility to certain food allergies seems to be related to a greater or lesser tendency of cells to pinocytose specific molecules. The likelihood of this occurrence among people reflects differences in genetic activities that control the composition of the cell membrane. These differences can account for variations in the incidence of allergies among individuals and during the lifetime of the same person. This may account for the incidence of allergy to protein in eggs in some infants, which is often outgrown by adulthood. Pinocytosis may render an important role in the absorption of vitamin B_{12} (cyanocobalamin) by the intestinal epithelium. This vitamin is necessary for DNA synthesis and for formation of healthy red blood cell membranes. A glycoprotein, the *intrinsic factor*, is synthesized by cells lining the wall of the stomach. Its combination with vitamin B_{12} may cause pinocytic ingestion of the vitamin when it reaches the cells in the lower part of the small intestine. Lack of the glycoprotein leads to vitamin B_{12} deficiency and pernicious anemia.

Phagocytosis is a common example of endocytosis performed by the cells which defend us against an invasion of microorganisms. *Macrophages* and *microphages* are the names of large and small cells capable of phagocytosis. The microphages are white blood cells (leukocytes).

Exocytosis (Emiocytosis)

Exocytosis is essentially the reverse of endocytosis (Fig. 4-13). Molecules synthesized within the cell are surrounded by membranous components forming *secretory vesicles,* which seem to originate from the Golgi apparatus. Each vesicle fuses with the plasma membrane and makes openings in it. The molecules composing the outer limiting vesicle membrane separate, and the contents pass into the ECF. The pieces of vesicle or plasma membrane lost from the surface of the cell during this event seem to be replaced by synthetic processes involving the endoplasmic reticulum and the Golgi apparatus of the cell. Exocytosis is often linked to the accumulation of Ca^{2+} ions as described in Chap. 5. The secretion of proteins, including neurotransmitters, some enzymes and hormones, may involve exocytosis. For example, the hormones norepinephrine from the adrenal gland and insulin from beta cells in the islets of Langerhans in the pancreas are released by this process.

4-4 SUMMARY

1 Water accounts for up to 90 percent of the weight of cells in the human body and up to 60 percent of lean adult body weight. The relative proportion of water to tissues in the body decreases with aging. Water is an important solvent, about 55 percent being intracellular and about 45 percent extracellular. Water moves between cells and extracellular fluid at varying rates, which allow the following four extracellular compartments to be distinguished: (1) Plasma is an extracellular compartment of the blood containing 7.5 percent of all body water. (2) Interstitial fluid and lymph are formed by fluid leaking out of cells and the bloodstream. This compartment accounts for 20 percent of all body water. (3) Dense connective tissue, cartilage, and bone are tissues lacking a profuse supply of small blood vessels. These tissues contain about 15 percent of all body water. (4) Fluids secreted by cells, membranes, glands, and organs constitute the transcellular fluid compartment and account for about 2.5 percent of all body water.

Unequal rates of entry and exit of dissolved ions and molecules and of water account for chemical differences in the composition of intracellular and extracellular fluids. The total amount of water in the body is governed by processes which affect its intake, exchange, and excretion.

2 The cell membrane is a fluid structure consisting of two layers of lipids partially surrounded and penetrated by globular proteins. The proteins are coiled polypeptides, which may act as pores, receptors, structural supports, enzymes, and pumps. Protein channels may constitute pores, through which small ions may pass selectively. Carbohydrate molecules may be bonded to proteins on the exterior of the membrane to form glycoproteins. Some surface glycoproteins which are unique to an individual's cells are called histocompatibility antigens. Glycoproteins also may serve as specific receptors for other molecules. Neurotransmitters, components of microorganisms, and drugs are among a few of the types of chemicals which may attach to specific glycoprotein receptors.

A double layer of lipid molecules constitutes the cell's primary selective barrier. The lipids are found in the fluid state, with hydrophilic heads projecting towards the exterior and interior. They also contain hydrophobic tails of carbon chains projecting toward one another and toward the center of the double layer.

3 Movement of substances occurring with the assistance of components of the cell membrane is referred to as mediated transport. Movement of substances occurring without such assistance is called nonmediated transport. Nonmediated forms of transport include diffusion and osmosis. Diffusion is the random spreading in fluids or gases of molecules or ions due to activity within their atomic structure. Its rate is directly related to differences in concentration across the cell membrane, to the area of the membrane, and to the temperature at which the process takes place and is inversely related to the square root of the molecular weight of the substance and to the distance involved. The rate of diffusion increases as pore size increases in the cell membrane. The accumulation of Ca^{2+} effectively decreases pore size and the diffusion rates of positively charged ions.

Osmosis is the diffusion of water across a selectively permeable cell membrane. Water diffuses from more dilute to less dilute solutions. Its passage depends on a concentration gradient and is directly related to the number of particles per volume of fluid (the osmotic pressure) on either side of the membrane. The osmolarity of a solution is a measure of its osmotic pressure. It is a function of the number of molecules in one mole (gram-molecular-weight) of a substance (Avogadro's number) and of the number of particles produced from each molecule in solution.

Molecules that assist in the movement of substances across the cell membrane are called carriers. They are proteins which seem to change shape (conformation) in carrier-mediated transport. In facilitated diffusion carriers passively assist the transport of specific substances along a concentration gradient. In active transport, movement of specific substances by a carrier requires the immediate expenditure of energy supplied by ATP in the cell. Active transport occurs against a concentration gradient. ATPase enzymes that are specific active transport carriers of Na^+ and K^+ are used in the sodium-potassium pump. Energy supplied by the breakdown of ATP allows the carrier to pick up Na^+ ions inside the cell, drop them at the outside, change the carrier shape, pick up K^+ ions, and drop them off inside. Additional energy is required to repeat the process. The carrier may move "empty-handed" (change shape) in either direction. The net effect is the outward transport of three Na^+ ions for every two K^+ ions transported into the cell.

Substances are also moved into or out of cells by vesicle-mediated transport. This involves entrapment in pinched-off segments of membrane or sacs called vesicles. Movement of fluids and whole macromolecules into the cell is called pinocytosis, while that of particles and cells is called phagocytosis. Both are versions of endocytosis. Pinocytosis is an important mechanism for the absorption of large protein molecules. It may play a role in the development of food allergies. Phagocytosis is an important mechanism by which the body removes foreign or abnormal cells from circulation. It is a defense mechanism carried on by leukocytes and macrophages. Movement of substances out of the cell occurs by exocytosis, which is the reverse of the process described above. It may be responsible for secretion of proteins including neurotransmitters, some enzymes and hormones, and other important molecules.

REVIEW QUESTIONS

1 What are several major differences between the ICF and ECF?
2 Compare the location and properties of four ECF compartments.
3 Why is water important in the cell membrane?
4 How is total body water regulated?
5 Explain why the cell membrane has been described as an ocean of lipid in which ships, submarines, and icebergs of protein are found.
6 Name five roles of proteins in the cell membrane.
7 Name and describe three types of attachments between cells that are formed by proteins.
8 What are receptors and what role do they play in cell function? What role do they play in the rejection of transplanted cells, tissues, and organs?
9 How do proteins form pores or channels and what influence do such structures have in the transport of molecules and ions across the cell membrane? Use specific examples.

10 What factors influence diffusion rates and what is the effect of each? Include a description of pore size and charge in your discussion.
11 A compound ABC dissociates into three particles A^+, B^+, and C^{2-} when dissolved in water. Its molecular weight is 300. How many grams of ABC must be dissolved in 1 L of water to make a solution with an osmolarity of 1.0?
12 Why should the replacement of body fluids be done with isotonic solutions? Explain osmotic pressure and tonicity as part of your answer.
13 Explain the major differences between facilitated diffusion and active transport.
14 How do ATPases relate to the cell membrane and the sodium-potassium pump and of what importance is this model to cell function?
15 How do cells transport substances by vesicle formation? What are the names and types of the activities?

THE ENDOCRINE SYSTEM

Hormones

5

5-1 INTRODUCTION

5-2 TYPES OF CHEMICAL CONTROL AGENTS
Hormones
Parahormones
Neurohormones
Neurotransmitters
Pheromones
Drugs

5-3 THE ENDOCRINE SYSTEM

5-4 NATURE OF HORMONES

5-5 CONTROL OF HORMONE PRODUCTION
General regulatory mechanisms
Feedback mechanisms
Stress

5-6 LEVELS OF HORMONE ACTIVITY
Rates of synthesis
Mechanisms of secretion, transport, degradation and excretion
Secretion
Transport, degradation and excretion
Role of receptors
Location of receptors
Effects of alterations of receptors
Interaction of hormones
Cascade effects
Physiological and pharmacological effects
Radioimmunoassay: a sensitive technique for measuring hormone levels

5-7 MECHANISMS OF HORMONE ACTION
Effects on nuclear receptors: modification of gene expression (transcription)
Effects on cell membrane receptors: modification of posttranscriptional processes
mRNA-directed protein synthesis
Hormonal influences on cellular permeability
Stimulation of secondary messengers

5-8 SUMMARY

OBJECTIVES

After completing this chapter the student should be able to:

1. Define the term hormone; distinguish it from parahormone, neurohormone, neurotransmitters, and drugs; and identify the chemicals of which hormones are composed.
2. Describe the endocrine system and name its major glands.
3. Compare the chemical and physical properties of protein-type and lipid-type hormones.
4. Describe the two major endocrine organs that regulate hormone secretion and their interaction with each other and with the nervous system.
5. Discuss the effects of the hypothalamus and its portal system and regulatory factors on the activity of the anterior pituitary.
6. Outline the role of negative feedback mechanisms in the regulation of hormone action. Use specific examples.
7. Explain the relationship between the following factors and levels of hormone activity:

   Number of endocrine cells
   Rates of secretion and method of transport
   Interaction with cell receptors
   Interaction between hormones
8. Discuss the effects of hormones on enzymes and on cellular permeability.
9. Describe the transcriptional and posttranscriptional influences of different chemical classes of hormones.
10. List and explain the sequence of steps in the secondary messenger hypothesis of hormone action.

5-1 INTRODUCTION

Various chemicals act as control agents and regulate cells, tissues, and organs to one degree or another. The effects of these agents demonstrate a wide range of activities. Different agents may alter the cellular uptake of ions, regulate metabolism, and/or modify reproductive capacity, behavior, or personality. Their influence on a wide range of body functions begins shortly after conception and continues through early prenatal development, birth, puberty, and aging until death.

5-2 TYPES OF CHEMICAL CONTROL AGENTS

Chemical control agents fall into several categories: hormones, parahormones, neurohormones, neurotransmitters, pheromones, and drugs. The general functions of endocrine organs and their secretions are described in Chaps. 1 and 3 and are summarized in Table 3-1 and more extensively here in Table 5-5. This chapter will emphasize the cellular mechanisms of endocrine activity. The specific effects of hormones, neurotransmitters, and drugs will be detailed more fully in subsequent chapters (including those on the physiology of the nervous system, bone, cardiovascular system, digestion, excretion, and reproduction).

Hormones

Hormones are chemicals synthesized by specific cells and secreted into the blood to be delivered to specific targets, upon which they exert regulatory effects. Cells that respond to a particular hormone are called *target cells* (of that hormone). For example, follicle cell–stimulating hormone is secreted by the pituitary gland. It binds to cells in the female's ovaries to influence the production and maturation of an egg. The responsive cells in the ovaries are its target cells.

Parahormones

Parahormones are chemical control agents whose production is not limited to specific cell types. Examples include histamine, CO_2, H^+, and various nutrients. The distinction between hormones and parahormones is not always clear or approved by endocrinologists. For example, prostaglandins fit the definition of parahormones. They are derivatives of fatty acids synthesized by a number of different cell types, yet they are often called hormones.

Neurohormones

Neurohormones are chemical agents that are synthesized by special neurons that function like gland cells. Such neurons, often called *neurosecretory cells,* end near small blood vessels. Impulses traveling along these neurons cause their terminals to release the neurohormone into the bloodstream, which carries it to target cells some distance away. These agents have a distant rather than a local effect. Examples of neurohormones include oxytocin, antidiuretic hormone (ADH), and various hormone release–inhibitory factors produced by neurosecretory cells of the hypothalamus in the brain. Thus, the hypothalamus has some cells which are neural in function (part of the nervous system proper) and others which are neurosecretory and often considered part of the endocrine system.

Neurotransmitters

Neurotransmitters are chemical agents that are synthesized by neurons and stored in their terminals. Impulses traveling down the neurons cause the release of the neurotransmitter molecules, which act on cells in the immediate vicinity of the terminal (a local effect). Examples include norepinephrine, acetylcholine, dopamine, and GABA (γ-aminobutyric acid). These substances will be studied in more detail in Chaps. 7, 9, and 10.

Pheromones

Pheromones are chemical agents that affect the physiology of animals in a population. Most pheromones play some role in reproductive behavior. For example, female dogs in heat are known to release a pheromone that attracts male dogs. Pheromones also can be extracted from certain insects and used to attract and trap males. In some instances, as a means of insect control the males are sterilized and released to mate with females, which fail to form fertile eggs.

Drugs

Drugs are powerful exogenous chemicals (chemicals from outside the body) which, when administered, affect the physiology of cells, tissues, and organs. Their actions may mimic, exaggerate, or block those of normal chemical control agents found in the body. For example, neosynephrine mimics the activity of the neurotransmitter norepinephrine, while dibenzyline blocks it. Amphetamines enhance the release of the neurotransmitters norepinephrine and dopamine. Atropine blocks acetylcholine activity.

5-3 THE ENDOCRINE SYSTEM

The *endocrine system* consists of ductless glands composed of clusters of cells that secrete hormones directly into the bloodstream. The glands of the endocrine system include the hypophysis (pituitary), the pineal gland and hypothalamus, the thyroid, the parathyroids, the thymus, the islets of cells in the pancreas, the adrenals, the gonads (testes and ovaries), and the placenta (which develops during pregnancy). These glands are formed from epithelial cells bound together by connective tissue and are richly supplied with blood vessels. They are found in several areas of the body, as indicated in Fig. 5-1.

Hormones are also manufactured by endocrine tissues in organs whose functions are associated with other systems. These organs include the stomach and duodenum in the digestive system and the kidney in the excretory system.

Endocrine hormones, neurohormones, and neurotransmitters can be distinguished on the basis of their cells of origin and/or means of delivery to target cells. Endocrine hormones are synthesized by glands formed by epithelial cells, not by neurons (as are neurohormones and neurotransmitters). Endocrine hormones are delivered to distant targets by the bloodstream. Neurotransmitters are delivered by nerve fibers of a neuron which often terminate close to their target.

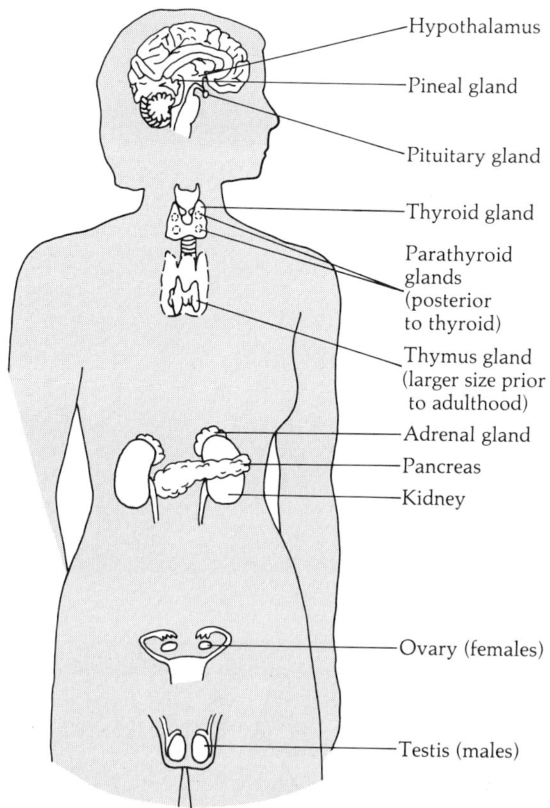

FIG. 5-1
Location of major endocrines in the adult. With aging the thymus changes markedly in size: it weighs 12.5 g at birth and 30 g at puberty; decreases to 20 g at 20 years of age; and shrinks over the next 60 years to the same size it had at birth.

Hormones alter specific physiological reactions to maintain homeostasis in a wide range of cells, tissues, and organs. For example, insulin from the pancreas alters the cellular uptake of glucose; growth hormone from the pituitary regulates body size; and steroids from the gonads (sex organs) and adrenal glands modify reproductive capacity, behavior, and personality. Overactivity (*hyperfunction*) or underactivity (*hypofunction*) of various endocrines may result in a number of different disorders (Fig. 5-2A to 5-2F). A study of some of the normal and abnormal conditions, as listed in Table 5-5 at the end of this chapter, further reveals the extensive capabilities of hormones. Table 5-5 should be noted as a useful reference for the study of specific hormones as they are described in future chapters in the text.

5-4 NATURE OF HORMONES

Endocrine hormones are of two types chemically, protein-type or lipid-type. The protein types are amino acids, short peptides, proteins, or derivatives of such molecules (e.g., glycoproteins). Many of the lipid-type molecules are steroids synthesized from the precursor cholesterol. Most hormones are of the protein type, although the sex hormones (such as testosterone, estrogen, and progesterone) and the hormones from the adrenal cortices are steroids. Generally, the protein-type hormones are water-soluble, while the steroids are fat-soluble. This distinction is critical. Protein-type hormones and other chemical control agents which are water-soluble (*hydrophilic,* or water-loving) often combine with receptors on the cell surface. Lipid-type chemical control agents are fat-soluble (*hydrophobic,* or water-fearing), and usually dissolve and diffuse through the lipid barrier of the cell membrane to combine with receptors inside the cell. Prostaglandins are lipid-type hormones which are exceptions to this principle.

Many protein-type hormones are released by ribosomes as *prohormones.* Prohormones may represent molecular species that are more readily stored and/or transported through cells. Specific enzymes in the blood or in tissues can convert these larger molecules into smaller ones which have hormone activity.

5-5 CONTROL OF HORMONE PRODUCTION

General Regulatory Mechanisms

The *hypothalamus* is composed of neurons near the center of the base of the brain. These neurons influence the secretion of endocrines

A

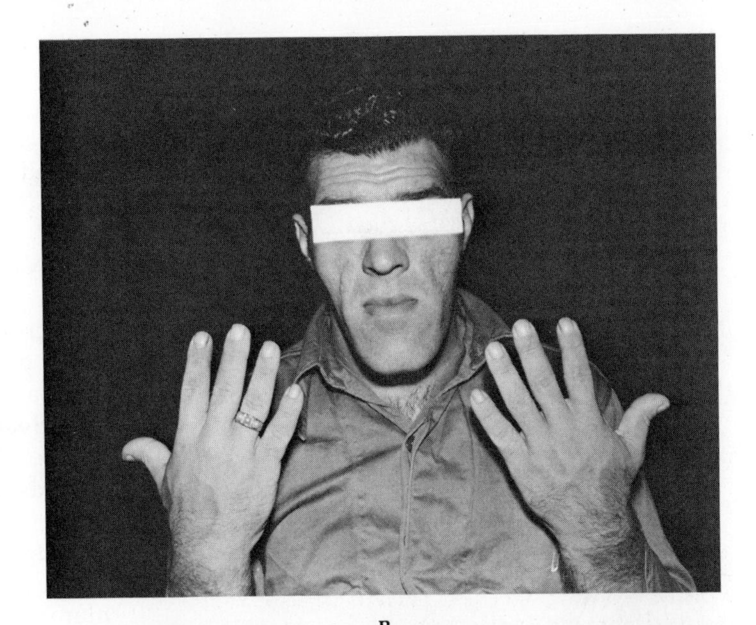

B

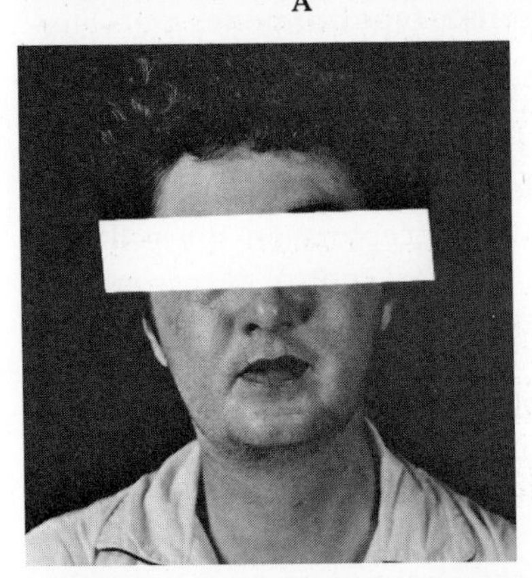

C

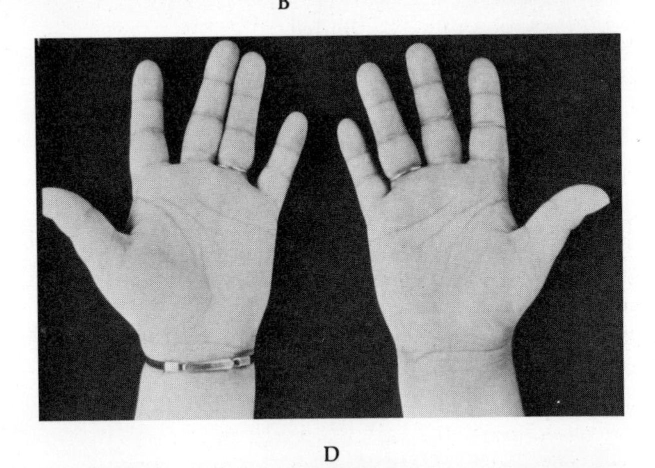

D

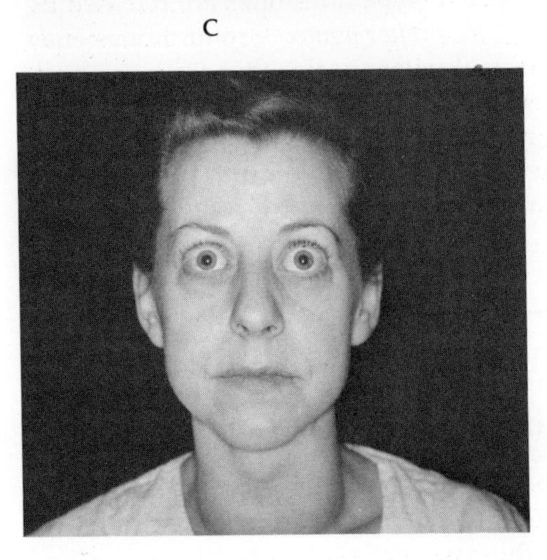

E

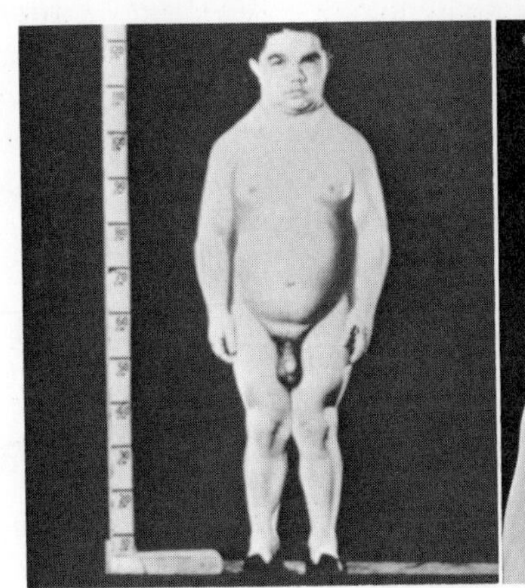

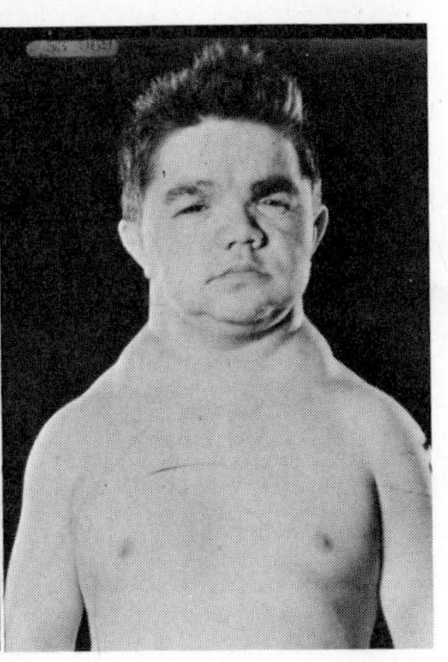

F

FIG. 5-2
Some of the effects of hyperactivity and hypoactivity of endocrines. (A) Gigantism due to overactivity of the pituitary. The normal person on the right may be compared to the giant on the left. Overactivity of the pituitary due to a tumor caused him to be 86 in tall and 338 lb at age 24. (B) Acromegaly due to overactivity of the pituitary. The features and bones in the face are coarse, as are those in the hands and feet. (C) Cushing's syndrome due to oversecretion of adrenal glucocorticoids in a young female. The subject shows the accumulation of hair and fat on her neck and face. Often the individual appears moon-faced. An increase in protein catabolism may lead to tissue rupture and pink to purple stripes on the abdomen. (D) Addison's disease due to adrenal insufficiency. Hyperpigmentation is evident in the creases of the hands of this patient whose adrenal glands were removed to treat Cushing's Syndrome. Underactivity of the adrenals also may produce excess pigment in other areas, general fatigue and a decline in blood pressure. (E) Graves disease due to hyperthyroidism. The subject's thyroid and eyes bulge (exopthalmos). Metabolic rate and nervous irritability are elevated. (F) Cretinism due to hypothyroidism in childhood (juvenile mxedema). Although sexual development is normal, the individual is dwarfed in height and is 49½ in tall and has very little body hair.

throughout the body by the production and secretion of regulatory *neurohormones.* The latter are carried by portal blood vessels (vessels which connect two capillary networks) directly to the anterior pituitary (adenohypophysis or glandular hypophysis), located below the hypothalamus. Close cooperation exists between the hypothalamus and the anterior pituitary and their vast regulatory activities. The interactions between these components of the nervous and endocrine systems serve as a major model in the study of *neuroendocrinology.*

The neurons of the hypothalamus not only manufacture regulatory substances that affect the anterior pituitary but also form two hormones, *oxytocin* and *antidiuretic hormone* (ADH). Oxytocin and ADH are neurohormones conveyed through the fibers of neurons (axons) from the hypothalamus to the *posterior pituitary* (neurohypophysis or pars nervosa). From the posterior pituitary the neurohormones enter the blood to affect target cells (Table 5-5-2b). The anatomical features of the neural and circulatory connections between the hypothalamus and pituitary are shown in Fig. 5-3.

Neurons within the hypothalamus produce several short peptides, which control the release of specific hormones from the anterior pituitary. Technically, the hypothalamic regulatory substances whose structures have been determined are *neurohormones,* while those of yet undetermined structure are called *factors.* The first part of the names of the hypothalamic releasing and inhibitory substances arises from the name of the pituitary hormone which they regulate, while the last portion describes their effects. For example, growth-hormone releasing factor (GH-RF) causes the release of growth hormone from the pituitary, while growth-hormone-release inhibiting hormone (GH-RIH) inhibits the release of growth hormone. The hypothalamic factors, hormones, and their effects are described in Table 5-1 on p. 109.

The anterior pituitary gland produces hormones, which are named after other endocrines whose secretion they regulate (Table 5-1). For this reason, the anterior pituitary hormones are called *tropic* (nourishing) hormones and the anterior pituitary is sometimes called the "master endocrine." The anterior pituitary is extremely important in the regulation of body chemistry since it controls the rates of secretion of numerous endocrines.

Feedback Mechanisms

The control of blood levels of hormones results from the homeostatic interaction of the nervous system, hypothalamus, anterior pituitary, each endocrine, and the products of their target cells. The *nervous system* has direct connections through nerve fibers to the posterior pituitary (neurohypophysis) and adrenal medullae and thereby influences their activities. Nerve impulses also influence the release of regulatory factors from the hypothalamus, which in turn control *anterior pituitary* secretions. The anterior pituitary's *tropic hormones* regulate several *endocrines.* In this sequence, the nervous system, hypothalamus, anterior pituitary, and individual endocrines interact through mechanisms which are listed in Table 5-2 and illustrated in Fig. 5-4 on p. 110.

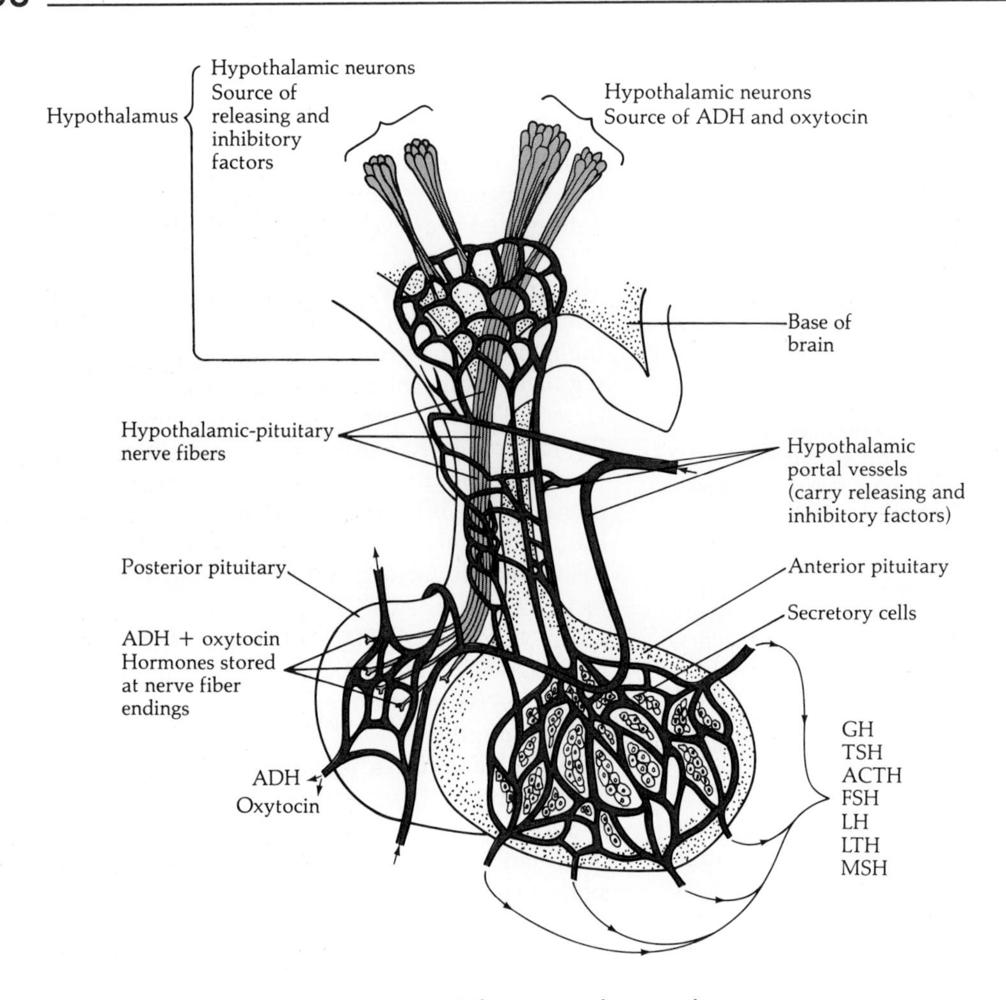

FIG. 5-3
Anatomical connections between the hypothalamus and pituitary. Releasing and inhibitory factors from the hypothalamus are carried by capillaries (the portal vessels) to the anterior pituitary to regulate the secretion of tropic hormones. Antidiuretic hormone and oxytocin formed in the hypothalamus are conveyed by transport through nerve fibers (axons) for release from the posterior pituitary.

Negative feedback signals control the production and secretion of most hormones (as described earlier in Chap. 2). Negative feedback signals could be called opposite or inverse signals because they cause an effect opposite to that of the original activity (Fig. 5-5).

Specific substances are often released to the bloodstream by cells under the influence of hormones. Elevated blood levels of the substances usually turn off further hormone production. Conversely, a decrease in the blood level of a specific substance can occur following its use, degradation, or excretion. If the quantity of a substance falls below a threshold (critical level), hormone synthesis and release are often triggered. For example, the pancreatic hormone glucagon stimulates the conversion of liver glycogen to glucose, as shown in Fig. 5-6. Increased levels of glucose in the blood serve as a negative feedback signal, turning off further glucagon secretion. Low blood glucose levels have the opposite result, triggering the release of glucagon. Thus the negative feedback signals cause effects *opposite* to those of the original endocrine activity.

TABLE 5-1
Hypothalamic factors, anterior pituitary hormones, and their effects

Hypothalamic Factor or Hormone	*Responsive Anterior Pituitary Hormone*	*Effects of Anterior Pituitary Hormone*
Corticotropin releasing factor (CRF), (adrenocorticotropin releasing factor, ACRF)	Corticotropin (adrenocorticotropic hormone, ACTH)	Release of adrenal cortical hormones
Thyrotropin releasing hormone (TRH)	Thyroid-stimulating hormone (TSH, thyrotropin)	Release of thyroid hormones
Melanocyte-stimulating hormone releasing factor (MRF)	Melanocyte-stimulating hormone (MSH)	Pigment production in skin; influences reproductive organs; influences biological rhythms
Melanocyte-stimulating-hormone-release inhibiting factor (MIF)	Melanocyte-stimulating hormone (MSH)	
Growth-hormone releasing factor (GH-RF)	Growth hormone (GH), the somatotropic hormone (STH)	Synthesis of protein and glycogen; anabolic effects on Ca^{2+}, phosphate and nitrogen metabolism
Growth-hormone-release inhibiting hormone (GH-RIH), (somatostatin)	As above	Inhibits GH
Luteinizing-hormone releasing hormone/follicle-stimulating-hormone releasing hormone (LH-RH/FSH-RH)	Luteinizing hormone (LH) and follicle-stimulating hormone (FSH)	Maturation of egg and follicle and production of estrogens in female, production of sperm and androgens in male
Prolactin releasing factor (PRF)	Prolactin (PRL), (luteotropic hormone (LTH), luteotropin)	Production of milk in mammary glands and enzymes in ovaries
Prolactin-release inhibiting factor (PIF)	As above	Inhibits above

TABLE 5-2
Regulators of hormone secretion

Regulator	*Mechanism*
Nervous system	Nervous stimulation or inhibition
Hypothalamus	Releasing or inhibitory factors
Anterior pituitary	Stimulating (tropic) hormones
Individual endocrine gland	Negative feedback inhibition by specific hormone or cell product
Target cell	Negative feedback inhibition by target cell product

Stress

The responses to stress illustrate the complexity of interrelations in the regulation of hormone activity. *Stress* may be described as any stimulus which prompts the hypothalamic-pituitary circuit to secrete corticotropin releasing factor (CRF) at higher than normal rates. Hans Selye first described the phenomenon as a general adaptation syndrome (GAS) occurring in response to a wide variety of emotional or physical stimuli.

Exposure to sounds, sights, physical factors (such as cold, heat, or altitude), and physical or psychological trauma (including emotions such as fear and anxiety) are nonspecific stimuli that can cause stress. They initiate impulses that are transmitted to the hypothalamus through the autonomic division of the central nervous system. They invoke pathways shown in Fig. 5-4 and elicit the effects outlined in Fig. 5-7 (p. 112).

Corticotropin releasing factor (CRF) is secreted by the hypothalamus and delivered by portal blood vessels to the anterior pituitary. CRF

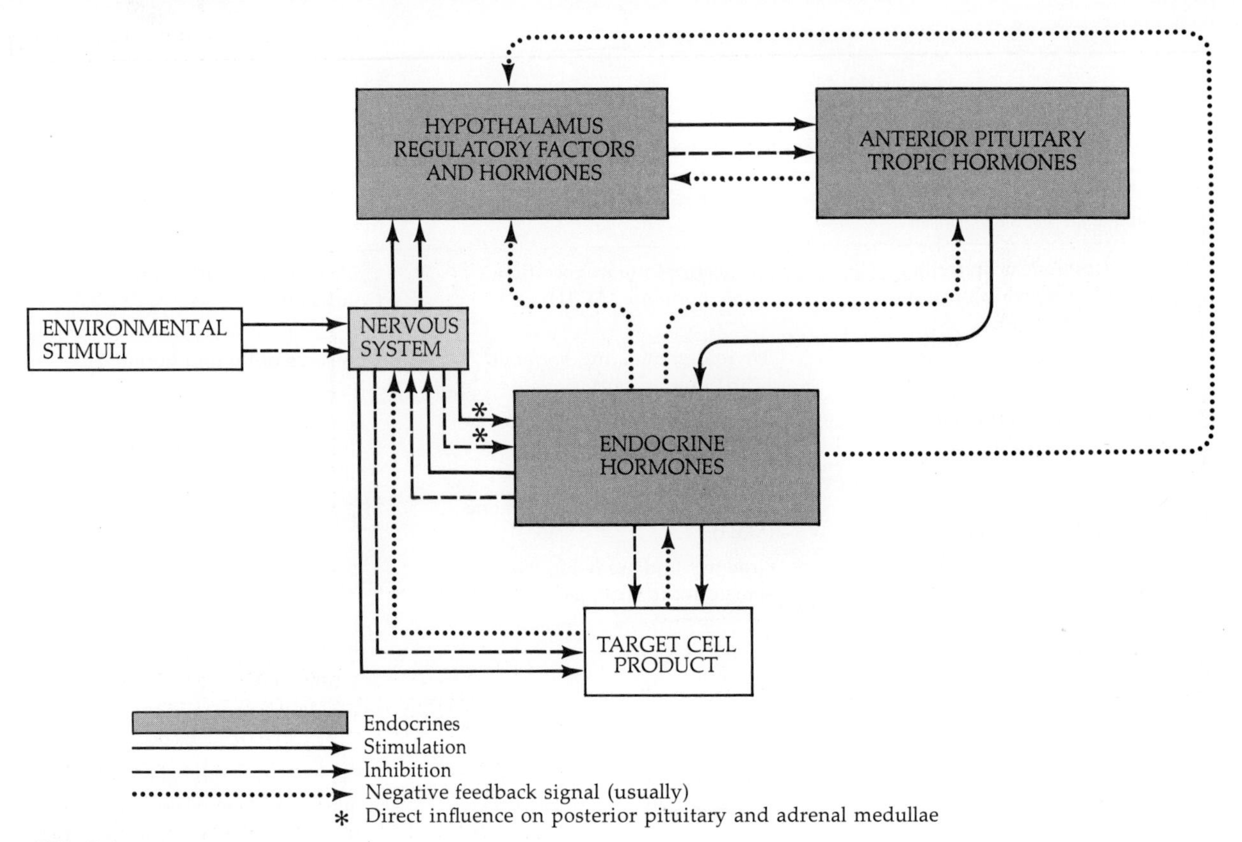

FIG. 5-4
Interactions between regulators of endocrine activity. The hypothalamus, anterior pituitary, and endocrines interact through regulatory factors (hormones) that control the activity of specific target cells. The relationships also include a circuit of reactions involving nerve impulses, which may be initiated by external or internal stimuli.

FIG. 5-5
A model of negative feedback mechanisms. The biological unit, which may consist of a cell, tissue, or organ, acts as a sensor receiving negative feedback signals (inputs or stimuli). The negative feedback signal may inhibit or stimulate a specific activity (output) of the biological unit, but (as the term negative feedback implies) the output transmitted from the biological unit tends to be the opposite of that which originally caused the feedback signal. The paths shown as 1 → 2 and 3 → 4 can be activated repeatedly.

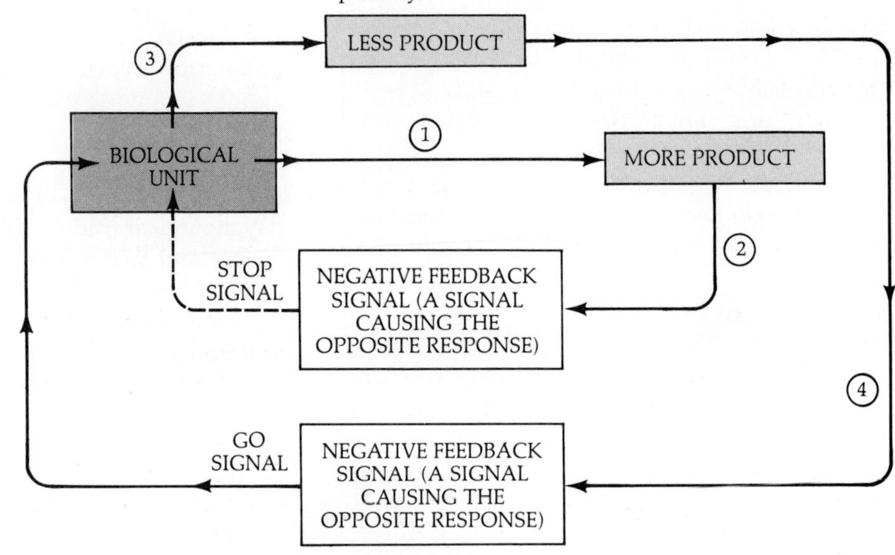

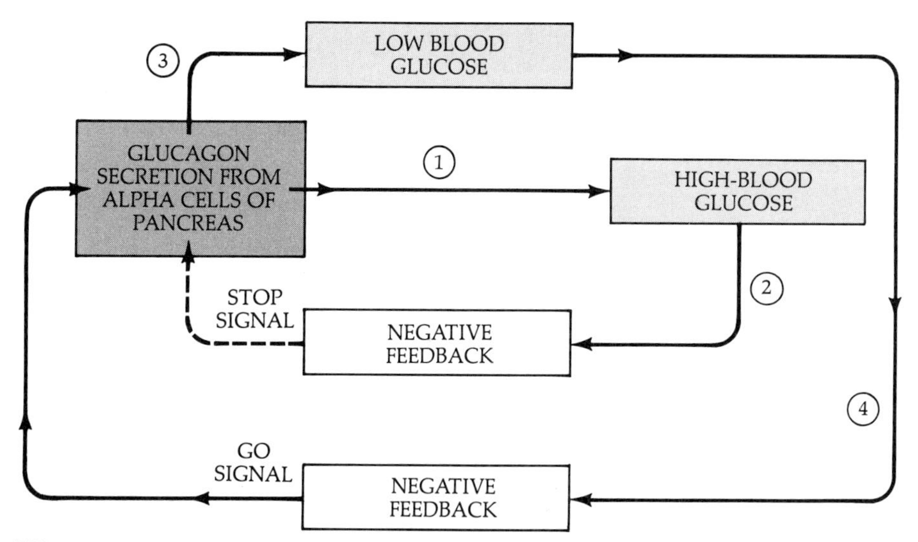

FIG. 5-6
Regulation of blood glucose involving negative feedback and pancreatic A cell (alpha cell) production of glucagon. The paths shown as 1 → 2 and 3 → 4 can be activated repeatedly. They are also influenced by the hormone insulin from B cells (beta cells) in the pancreas. Insulin lowers blood glucose, promoting its transport into cells and its conversion into glycogen.

causes the discharge of *adrenocorticotropic hormone* (*ACTH*) into the bloodstream. ACTH stimulates the release of adrenal cortical hormones called the *glucocorticoids*. The latter promote the conversion of proteins (and fats) to glucose, the levels of which increase in the blood. The conversion of noncarbohydrates to carbohydrates is called *gluconeogenesis*. The elevated blood levels of glucocorticoid hormones normally inhibit further ACTH secretion by feedback inhibition. This response results in *adaptation* and a decreased response to the stressor.

Autonomic nervous stimulation also influences the secretion of other endocrines, including cells in the adrenal glands, the hypothalamus, and the kidneys, in response to stress. The *adrenal medullae* release the hormones *norepinephrine* and *epinephrine*. The net effect of the two hormones is to elevate blood pressure and increase the availability of blood sugar for energy metabolism in skeletal muscle. The *hypothalamus* secretes *antidiuretic hormone*, which promotes water retention to help maintain blood pressure. Autonomic nerve impulses also promote the release of the enzyme renin from the kidney into the blood, and reactions follow which elevate blood pressure. Renin converts a protein in the plasma into a compound called *angiotensin II*. The latter increases blood pressure by making blood vessels constrict and by stimulating the *adrenal cortex* to secrete *aldosterone* which promotes sodium and water retention to assist the response.

The reactions to stress have been referred to as the "fight or flight" syndrome. They assist the body in stressful situations that otherwise may be harmful to health. However, evidence indicates that with repeated exposure, some of the responses to stress may result in the development of high blood pressure, atherosclerosis, and ulcers of the gastrointestinal tract.

5-6 LEVELS OF HORMONE ACTIVITY

The level of hormonal activity depends on a number of variables, which include: (1) rates of hormone synthesis; (2) mechanisms of secretion, transport, degradation, and excretion; (3) the role of receptors; (4) interactions among hormones; (5) a cascade of effects; and (6) physiological and pharmacological effects of hormones.

Rates of Synthesis

The rates of hormonal synthesis are generally affected by the level of cellular activity and in

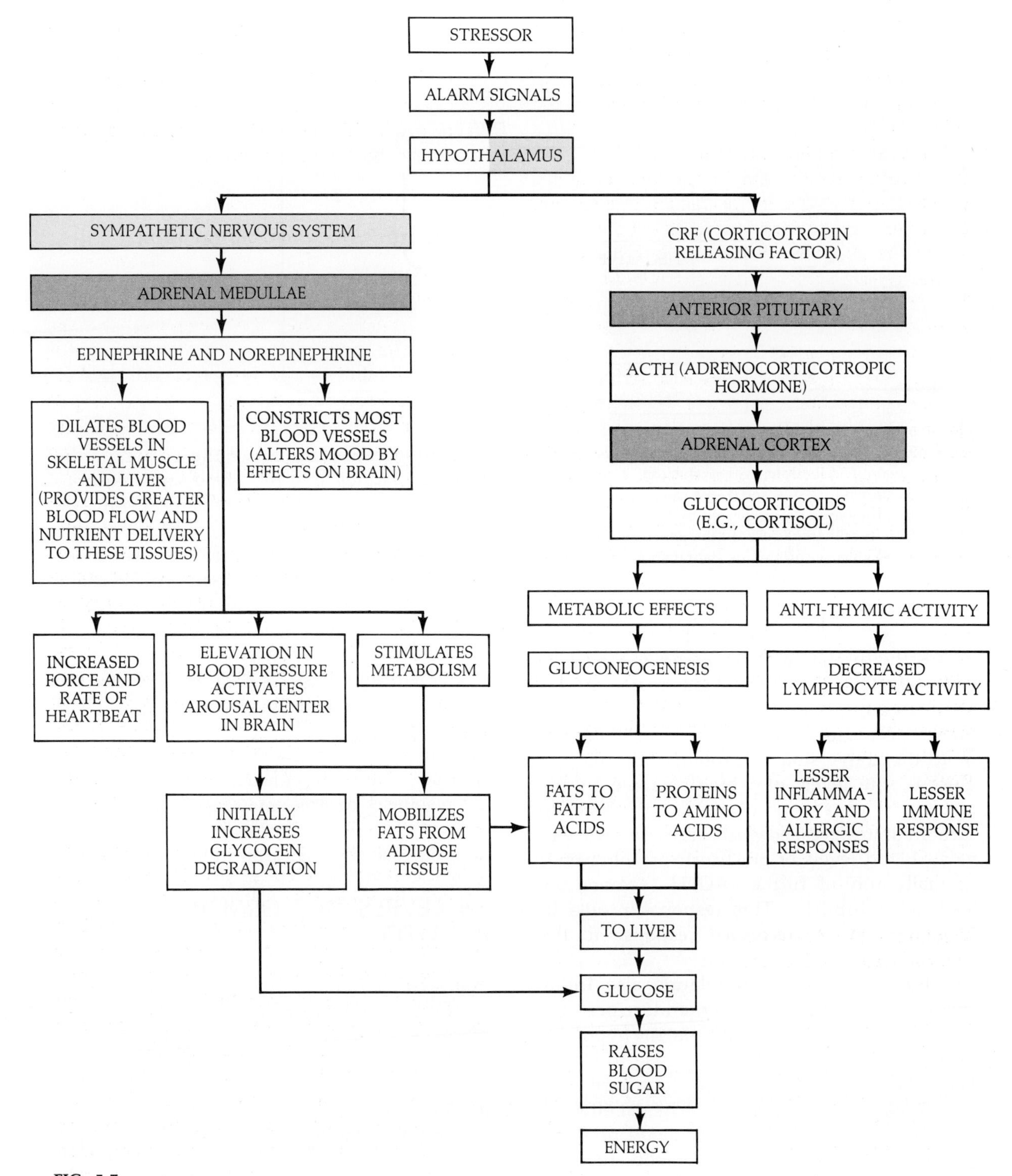

FIG. 5-7
Responses to stress, mediated by the hypothalamic-pituitary circuit and sympathetic nervous system.

unusual circumstances are related to the number of hormone-producing cells. These two factors are markedly influenced by the availability of nutrients and the genetic machinery of the cells.

Nutrients provide raw materials for the synthesis of molecules. Nutritional deficiency can lead to hypofunction of specific endocrines. For example, three to four iodine atoms are essential to the structure of the active thyroid hormones referred to as T_3 (triiodothyronine) and T_4 (tetraiodothyronine or thyroxine). Hypothyroidism can result from prolonged dietary iodine deficiency or from the inheritance of specific genetic defects. The genetic defects may impair the transport of iodine by carriers into the cells of the thyroid. The iodine deficiency and other genetic defects in thyroid hormone synthesis can lead to lack of growth and development (cretinism) in early childhood and to myxedema and goiter (enlargement of the thyroid) in adults (Table 5-5, 4 and Figs. 5-2E and 5-2F).

When genes fail to regulate cell multiplication in tissues and organs, endocrine tumors may grow. The increased number of endocrine cells within the tumor can result in an elevation of the levels of hormones circulating in the blood. An example is a tumor of the adrenal medulla which elevates norepinephrine and epinephrine levels in a disorder called *pheochromocytoma* (Table 5-5, 9b). Some basic mechanisms by which genes and nucleic acids control hormone production will be described below.

Mechanisms of Secretion, Transport, Degradation and Excretion

Secretion

The mechanisms which are responsible for the discharge of hormones into the bloodstream are not clearly understood. Agents which promote hormone release are often associated with an increase in the intracellular level of Ca^{2+} ions which somehow promotes exocytosis. Most hormones are stored in granules in the cytoplasm. In some cases the hormones seem to pass along microtubules and microfilaments in the cells (described in Chap. 2). Some hormones are stored in membrane-bound vesicles and extruded from the cell by exocytosis (see Chap. 4). For example, insulin is released by exocytosis from β cells in the pancreas and epinephrine from cells in the adrenal medullae.

Transport, Degradation and Excretion

The fat-soluble hormones (e.g., steroids) tend to combine with various blood proteins, particularly albumin, after the hormones are secreted. The combination with a specific carrier protein solubilizes the hormone and aids its transport throughout the body. The binding of hormone to protein in the blood is governed by the laws of dynamic equilibrium. Simply, this means that as the level of free hormone in the blood rises, hormone binds to protein. Conversely, as the level of free hormone falls, hormone dissociates from the protein. This mechanism helps to keep the blood level of hormones relatively constant. It is important to keep in mind that only free hormone in the blood affects cells and is susceptible to metabolic degradation and loss from the body in the urine. Target cells are unresponsive to hormone bound to and protected by protein. Fluctuations in the protein concentration of the blood can therefore significantly influence the amount of free hormone circulating in the blood and its effect on target cells (Fig. 5-8).

Hormones are subject to metabolic degradation after they are secreted. That is, they are enzymatically broken down into inactive, metabolites. Many hormones are degraded in the liver (for example, insulin and steroids such as the sex hormones, i.e., estrogens and androgens). Others are broken down in their target cells (for example, the catecholamines, epinephrine, and norepinephrine), in the blood (ACTH), and in the kidneys (parathormone). The kidneys also filter out and remove inactive metabolites derived from the hormones and excrete them in urine. Neurotransmitters are usually degraded at their site of action, and very often their metabolites are reabsorbed into the neurons and recycled into new molecules of neurotransmitters.

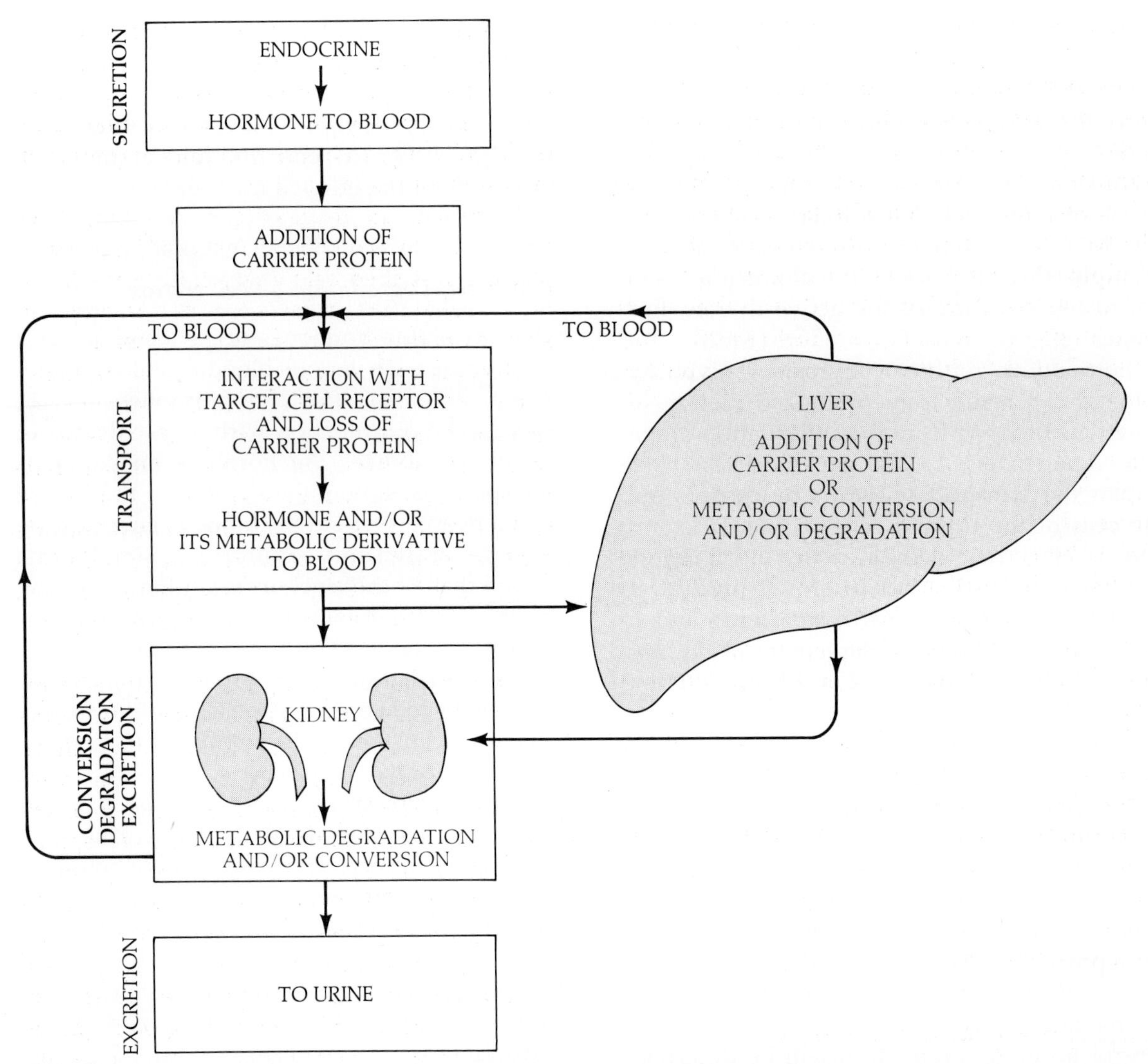

FIG. 5-8
Secretion, transport, degradation, and excretion of hormones. After their release from endocrines, some hormones are complexed with specific proteins from the liver. The proteins are shed when the hormone interacts with target-cell receptors. The hormones then may be metabolized or degraded in the liver, blood, or kidneys, prior to their excretion in the urine.

Some, however, find their way into the bloodstream and are excreted via the kidneys. Because of the degradation processes, the activity of most endocrine hormones is reduced by 50 percent in less than 60 min after secretion. Therefore, maintenance of a constant blood level of hormones requires almost continual synthesis and release of these important compounds.

Role of Receptors

Location of Receptors

Cells that are the targets of hormone activity possess specific protein or glycoprotein *receptors* that complement the unique chemical and physical properties of the hormone. When molecules of a hormone (or neurotransmitter) come into contact with a target cell, they may bind to specific receptors to trigger built-in activity in the chemical machinery of the cell.

Hormone receptors are generally located in one of three places: the cell (plasma) membrane, the cytoplasm, or the nucleus. Most protein-type hormones bind to receptors on the cell membrane. Cell membrane receptors usually have two or more functional parts. One is a binding site for the transmitter agent (hormone or neurotransmitter) and the other a channel or pore whose shape changes when the transmitter-receptor complex forms. This allows ions to move along gradients into or out of cells by diffusion (Fig. 4-4).

Steroid hormones enter the cell and complex with *cytoplasmic receptors.* The steroid hormone–receptor combination enters the nucleus and adheres to a nuclear receptor which is formed of histone protein in the chromatin of the genetic material. Such steroid hormone–cytoplasm receptor–nuclear receptor complexes influence the expression of genes, as shown in Fig. 5-10.

Other hormones, such as thyroxine, enter cells and bind directly to *nuclear receptors.* Evidence also indicates some hormones may bind to receptors at more than one site in the cell. Insulin, for example, adheres primarily to cell membrane receptors, following which the insulin-receptor complex may enter the cell to exert long-term effects.

Effects of Alterations of Receptors

The number of receptors per cell ranges from thousands to millions during the cell's lifetime. An excess receptor number probably exists so that all are not active at any given moment. Variations in the number and types of receptors present in cells can influence the cells' responsiveness to hormones. The reverse is also true —namely, that hormones can alter the number of receptors as well as their activity.

Cells with abnormally small numbers of hormone receptors are less responsive to hormones than normal cells. Decreases in the number of functional receptors can result from stressful stimuli or from the normal aging process. Changes in the nature of the receptors can also affect the responsiveness of cells. An important factor that can alter receptors is a change in the ionic composition of the extracellular fluid (ECF). Changes in its calcium ion concentration are especially notable.

Alterations in receptor function may have profound effects on body function. Three examples are described below which indicate the effects of alteration in receptors on memory, responses to drugs, and endocrine disease.

Changes in hormone receptors accompanying aging can adversely affect the nervous system and memory. A decrease in the number of functional receptors for catecholamine hormones accompanies aging and has been related to loss of memory. Those hormones include norepinephrine, epinephrine, and their derivative dopamine (dihydroxyphenylethylamine). All three are derived from the amino acid tyrosine.

The responsiveness of cells to *drugs* also relates to the nature of cellular receptors. Some drugs may mimic hormones. Others may inhibit hormonal receptors. For example 9-tetrahydrocannabinol (THC), the active agent in marijuana, binds to cytoplasmic estrogen receptors. This interaction may account for the estrogenic (feminizing) effects of heavy, long-term marijuana smoking. Some drugs which inhibit hormone activity combine with receptors to prevent hormones from binding to their receptors. Propranolol is an example of a drug with an inhibitory action which blocks the hormone epinephrine and its effects.

Antibodies may form against cell membrane receptors or against hormones to cause specific endocrine diseases. The combination of antibodies with the receptors can alter cell function and alter the normal response to a hormone. For example, failure to respond to insulin, called insulin resistance, occurs in diabetics who produce anti-insulin antibodies. Anti-insulin antibodies can bind directly to insulin, making it ineffective. Some individuals also form antibodies against insulin receptors and other molecules on the surface of the pancreatic cells. The combination of antibodies with the molecules alters or blocks them as targets for insulin or other chemical control agents.

Conversely, the adherence of antibody to hormone receptors can stimulate endocrine activity and hormone secretion. For example, Graves' disease results in hyperthyroidism. It results from combination of antibodies with the receptors or nearby sites for thyroid-stimulating hormone (TSH) in the thyroid. An increase in thyroid hormone secretion results and causes the symptoms illustrated in Fig. 5-2E and Table 5-5, 4. As more knowledge develops about receptors, a better understanding of the mechanisms underlying the effects of hormones on body functions should be attained.

Interaction of Hormones

The presence or absence of one hormone may influence the net activity of others. Two or more hormones may cause similar responses. One may create conditions that permit another to be more effective. These interactions are called *synergism, potentiation,* or *permissiveness.* For example, thyroxine is synergistic when present with epinephrine. It allows the latter to more effectively cause the release of lipids from adipose tissue. So, too, cortisol increases norepinephrine's constrictor effects on smooth muscle in the walls of blood vessels.

Conversely, hormones may produce *antagonistic* or opposing results. For instance, insulin promotes the synthesis of glycogen from glucose in the liver, while glucagon stimulates glycogen breakdown into glucose.

Cascade Effects

The initial response to a hormone is its *primary* effect. It is difficult to determine, since a series of responses rapidly follows. These occur in a *cascade,* or shower of events, spreading and multiplying the effects to influence many aspects of cell function. Since the changes often occur in fractions of a second, it is not easy to tell which comes first.

This is evident in the multiple effects of insulin. This hormone is especially active in skeletal muscle, liver, and fat cells. However, insulin also affects many other cell types and general body chemistry. The binding of insulin to cell membrane receptors promotes: (1) changes in the cell's microvilli (the microscopic folds of the cell membrane); (2) alterations in movements of small ions across the cell membrane; (3) modifications in the form and activity of carriers and in the synthesis of enzymes; (4) an increased absorption of a variety of metabolites, including glucose; (5) stimulation of cellular enzyme activity; (6) degradation of the secondary messenger cAMP (see below); (7) and the synthesis of fat, protein, and glycogen. Which of these several responses occurs first is hard to establish and illustrates the problem of identifying the primary effect of most hormones.

Physiological and Pharmacological Effects

The *physiological levels* of substances in the cells and fluids of the body are the normal quantities found in each location. The appendix at the end of this text provides tabulation of the range of physiological levels of a large number of substances found in the human body. The levels of hormones in the blood are very slight, usually below 10^{-8} *M* but ranging from 10^{-6} *M* to 10^{-12} *M*. *Pharmacological levels* are abnormally high amounts of hormones in the body. They may occur as the result of hyperfunction due to endocrine tumors or disease, or they may follow artificial administration of hormones. Responses to pharmacological doses of hormones can differ considerably from normal effects. For example, physiological levels of insulin promote the rapid entry of glucose from the blood into liver and muscle cells. Pharmacological levels of the hormone cause this transport to occur so fast that the blood can be depleted of glucose and low levels of blood sugar (*hypoglycemia*) result. However, glucose is the major energy source for the brain. Its continual supply to this organ is critical. When hypoglycemia occurs, mental confusion, dizziness, weakness, convulsions, and unconsciousness can follow. A loss of control of respiration and death may result. When these reactions are brought on by excess insulin, the reactions are termed *insulin shock*. The initial symptoms may be reversed if hypoglycemia is not maintained for a prolonged period.

Another example of the pharmacological effects of hormones is the activity of the glucocorticoids, steroid hormones secreted by the adrenal cortex. Cortisol is a glucocorticoid which exerts anti-inflammatory effects and is valuable in the treatment of some types of allergies. However, pharmacological doses of glucocorticoids may convert protein to carbohydrate at abnormally high rates, resulting in wasting of body protein and a decrease in the density of bone (osteoporosis). They also can produce severe behavioral changes (psychoses).

Great care must be taken in the artificial administration of any hormone. When attempting to interpret scientific data, one must also distinguish between physiological and pharmacological doses and effects of hormones. Furthermore, one must keep in mind that the body has built-in feedback mechanisms, which help maintain hormone levels within a narrow range. This capability is not normally available with artificially administered hormone. However, some attempts to develop slow, gradual release devices have been made. The effects of pharmacological levels of a few hormones produced by endocrine hyperfunction are listed in Table 5-5; some of these are also illustrated in Figs. 5-2A, B, C, and E.

Radioimmunoassay: A Sensitive Technique for Measuring Hormone Levels

Physiological levels of many hormones (10^{-6} *M* to 10^{-12} *M*) provide nanogram (10^{-9} g) to picogram (10^{-12} g) quantities in the plasma, cell extracts, or urine and may be measured by *radioimmunoassays*. The technique is so sensitive that its ability to detect minute amounts of a substance may be likened to locating molecules from a cube of sugar that has been dissolved in one of the Great Lakes.

Specific antibodies to human hormones often can be produced by injection of the hormones into laboratory animals. The antibodies may be extracted from the animal's serum. Hormones form a specific complex with their corresponding antibody when their solutions are mixed in a test tube. In one version of the assay, urine or serum samples containing unknown amounts of hormone are mixed with antihormone antibodies that have previously been made to cling to the walls of a test tube. Known amounts of hormone in which a radioactive substance has been artificially incorporated (i.e., labeled hormone molecules) are also added to the mixture. The labeled and unlabeled (patient's) hormones compete for and complex with antibodies, as shown in Fig. 5-9A. After incubation of the mixture, the fluid in the tube may be poured off and radioactivity of the tube measured in a radiation (scintillation) counter. The radioactivity comes from the labeled hormone, which binds to antibody clinging to the walls of the tube. When the patient's hormone level is high, the radioactivity of the tube (the labeled hormone-antibody complex) is low and vice versa, as shown in Fig. 5-9B. That is, the level of radioactivity of the complexes that form is inversely proportional to the level of hormone in the patient's serum or urine.

This technique has been applied to the measurement of many hormones. For example, it is used to assess levels of thyroid hormone and thyroid function. Measurement of the iodine bound to protein in the plasma (protein-bound iodine, PBI) was formerly used as a general indication of thyroid activity. However, more sensitive *radioimmunoassays* are now used to assess levels of T_3 and T_4, the active forms of thyroid hormone in plasma samples. The subject's plasma, containing an unknown amount of T_3 and T_4, is mixed with anti-T_3 or anti-T_4 antibodies. Each hormone forms a specific complex with its corresponding antibody. Known amounts of hormone containing radioactive iodine (i.e., labeled hormone) are also added and incubated with the mixture. The amount of radioactivity remaining on the walls of the tube after the fluid is poured out is inversely related to the amount of T_3 and T_4 in the patient's plasma.

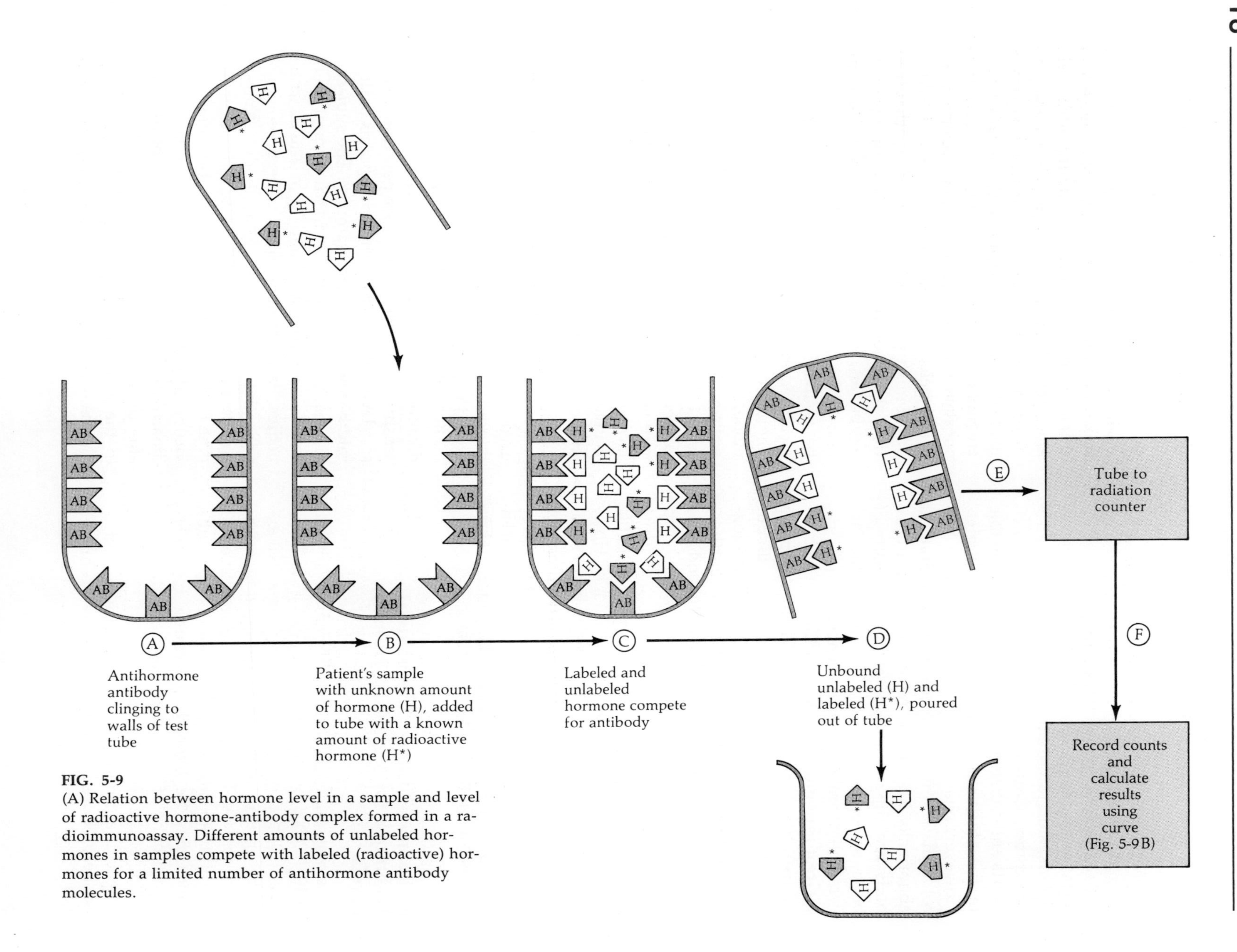

FIG. 5-9
(A) Relation between hormone level in a sample and level of radioactive hormone-antibody complex formed in a radioimmunoassay. Different amounts of unlabeled hormones in samples compete with labeled (radioactive) hormones for a limited number of antihormone antibody molecules.

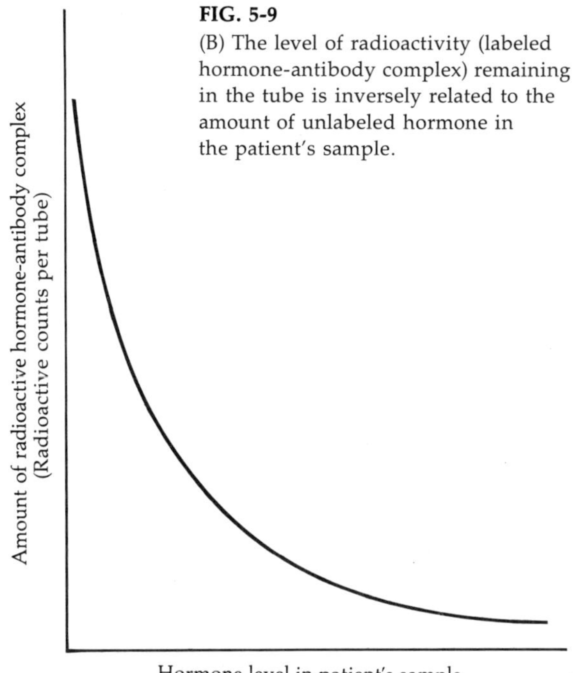

FIG. 5-9
(B) The level of radioactivity (labeled hormone-antibody complex) remaining in the tube is inversely related to the amount of unlabeled hormone in the patient's sample.

5-7 MECHANISMS OF HORMONE ACTION

Although hormones produce a wide variety of reactions, their effects result from changes in enzyme activity or in permeability of the cell. Changes in enzyme activity alter important metabolic reactions, while alterations in permeability affect the movement of key substances into and out of the cell. Hormones can influence cellular activities by modifying transcription (the synthesis of mRNA, rRNA, and tRNA) or translation (the mRNA-directed synthesis of protein, illustrated in Fig. 5-10). Both these processes were described earlier in Chap. 2 and are illustrated in Figs. 2-13 and 2-14.

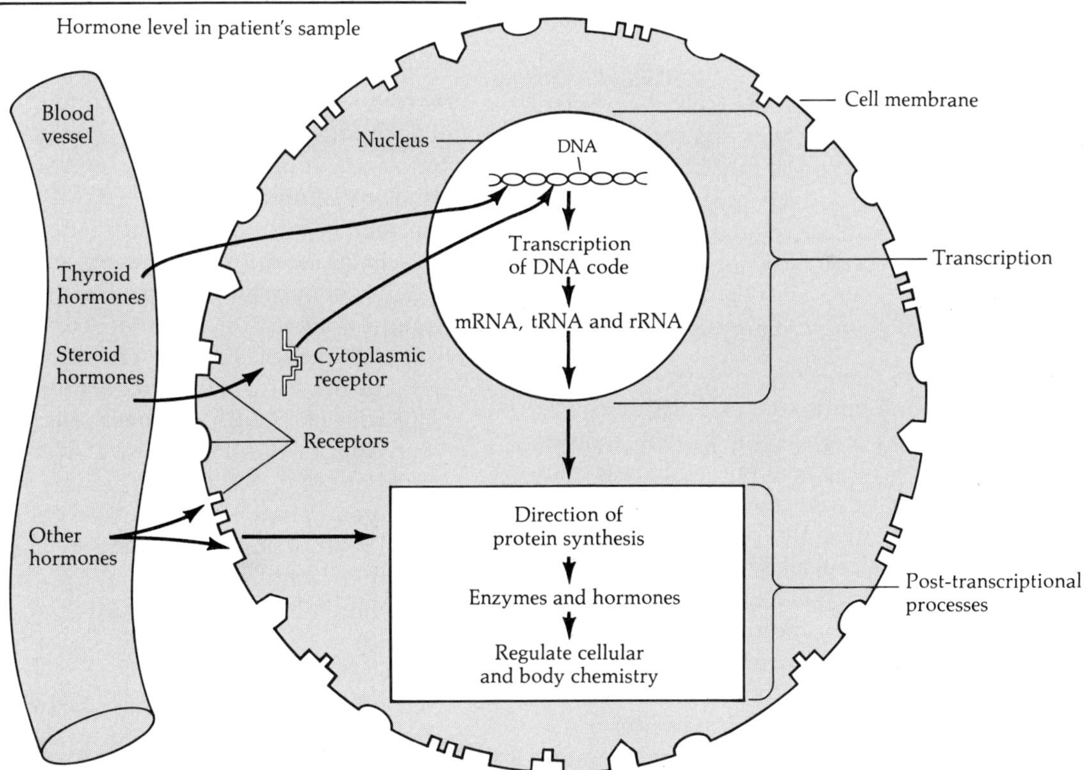

FIG. 5-10
Hormone receptors and action on transcription and posttranscriptional processes. Hormones that bind to receptors in the nucleus directly and those that do so after complexing with cytoplasmic receptors alter the expression of DNA. They regulate the transcription of specific types of RNA. Those hormones that attach to receptors on the cell membrane affect the translation of RNA and thereby regulate body chemistry.

Effects on Nuclear Receptors: Modification of Gene Expression (Transcription)

Thyroid hormones or steroids in combination with cytoplasmic receptors enter the nucleus of the cell to unite with a receptor that is a component of a chromosome. The combination of the hormone (or hormone-cytoplasmic receptor) with the nuclear receptor derepresses (turns on) part of a strand of DNA within the chromosome. This effect initiates the formation of a specific type of complementary mRNA, tRNA, or rRNA. The RNA migrates to the cytoplasm. The translation of mRNA directs the synthesis of specific protein in posttranscriptional reactions, and tRNA and rRNA participate in the reactions, as shown earlier in Fig. 2-14.

Effects on Cell Membrane Receptors: Modification of Posttranscriptional Processes

Protein-type hormones and prostaglandins unite with receptors on the cell membrane. They may: (1) direct protein synthesis by influencing the translation of mRNA; (2) alter cellular permeability; and (3) stimulate the production of regulators of cellular activity which have been termed *secondary messengers.*

mRNA-Directed Protein Synthesis

The translation of mRNA may be stimulated when a hormone binds to a cell membrane receptor. Synthesis of a specific protein follows. The protein may be an enzyme, which modifies cell metabolism, or a hormone. Thus, in some cases hormones guide the production of other hormones. This is especially true of the regulators from the hypothalamus and the tropic hormones from the anterior pituitary. The mechanisms whereby hormones alter translation are poorly understood. However, some evidence shows that hormones may alter the activity of ribosomes (the cytoplasmic granules that participate in protein synthesis).

Hormonal Influences on Cellular Permeability

Changes in membrane permeability can have a profound effect on a cell, as described in Chap. 4. Such alterations can occur when hormones combine with receptors on the cell membrane. A hormone may specifically complex with a protein receptor, which also can function as a channel or pore whose shape changes when the complex forms (Fig. 4-4). The change in shape of the receptor alters ion movement along gradients into or out of the cell by diffusion. Permeability changes can also occur as a result of alterations in the activity of membrane-bound carriers. Changes in the carriers can be brought about by hormonal modifications of transcription or translation. Alterations in receptors may result in changes in permeability of the cell membranes, followed by a wide variety of effects.

Stimulation of Secondary Messengers

Many hormones are considered *primary messengers.* They initiate changes by stimulating the activity of cellular regulators termed *secondary messengers,* which alter cellular respones. Several compounds may act as secondary messengers. Among the more widely studied are cyclic adenosine monophosphate (cAMP), cyclic guanosine monophosphate (cGMP), and calcium ions (Ca^{2+}). Their combination with specific cellular receptors may alter membrane permeability or enzyme activity. For the most part, hormones affect intercellular communication while substances such as cAMP, cGMP, and Ca^{2+} ions are intracellular regulators.

The role of cAMP has been studied extensively. That of cGMP is elusive. cGMP may act as a variant of cAMP or alternatively may modify the generation and reception of cAMP and Ca^{2+}. The functions of secondary messengers are further complicated in that they may affect each other. For example, cAMP can affect the distribution of Ca^{2+}, and Ca^{2+} can affect cAMP activity.

Secondary messengers may be produced or released by enzymes activated by the combination of protein-type hormones with their specific cell membrane receptors. For example, cAMP is produced after the combination of protein-type hormones with membrane receptors. The combination activates enzymes (known as adenyl cyclases), which are bound to

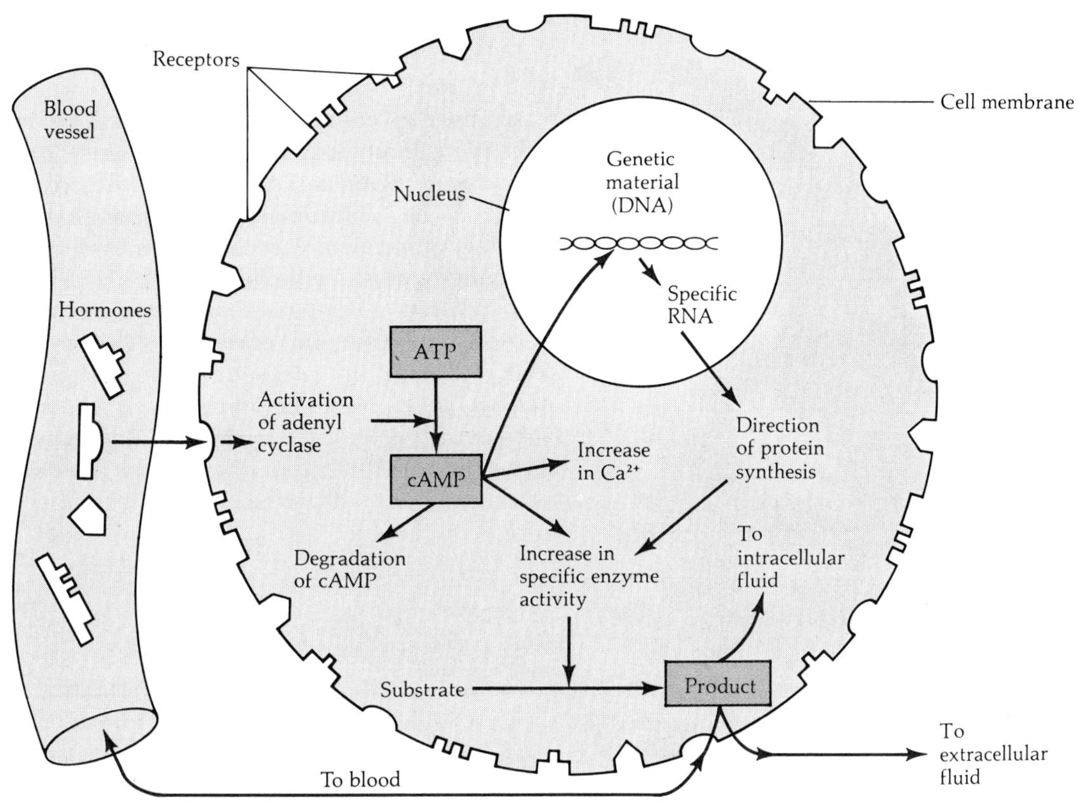

FIG. 5-11
A model of the secondary messenger hypothesis. Hormones which bind to cell membrane receptors cause the production of secondary messengers such as cAMP (cyclic adenosine monophosphate). cAMP alters the activity of enzymes which regulate cellular metabolism directly or indirectly through the transcription of genes and the translation of mRNA.

the cell membrane. The resulting activities can cause a variety of effects (Fig. 5-11).

cAMP is formed from ATP in the presence of adenyl cyclase and magnesium ions (Mg^{2+}), as shown in the following reaction:

$$\text{ATP} \xrightarrow[\text{membrane-bound adenyl cyclase}]{Mg^{2+}} \text{cAMP} + \underset{\textbf{Inorganic phosphate}}{P_i}$$

Accumulation of cAMP stimulates the activity of enzymes called protein kinases in the nucleus or cytoplasm of the cell. Different protein kinases transfer phosphate to specific acceptor proteins, converting them to active enzymes which transform substrates into products (Fig. 5-12).

The level of intracellular cAMP is influenced by its rate of production, its rate of passage out of the cell, and its rate of degradation. cAMP is cleaved by the enzyme phosphodiesterase to form inactive AMP:

$$\underset{\textbf{Active}}{\text{cAMP}} \xrightarrow[\text{phosphodiesterase}]{Mg^{2+}} \underset{\textbf{Inactive}}{\text{AMP}}$$

It is interesting to note that the purines theophylline (in tea) and caffeine (in coffee) inhibit phosphodiesterase activity, and thereby elevate cAMP levels, but also cause the release of Ca^{2+} ions from intracellular stores and block adenosine receptors. These responses, especially the latter, may account for the activity of such compounds. Some of the many effects of cAMP are noted in Table 5-3.

Not all hormones cause an increase in cAMP. Some stimulate the formation of cGMP. Its effects usually are the opposite of those of cAMP (Table 5-3). cGMP synthesis is promoted by the membrane-bound enzyme guanyl cyclase, and its degradation is caused by guanyl phosphodiesterase. cGMP often activates protein kinases which usually cause reactions opposite

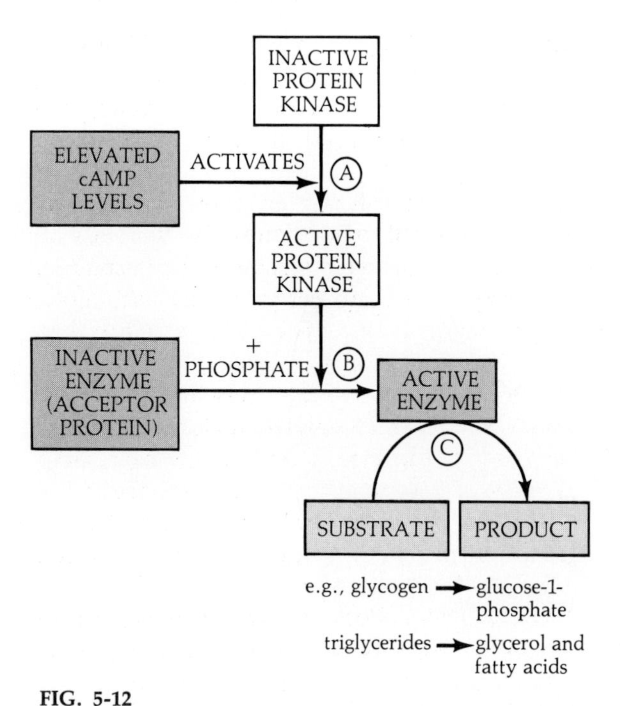

FIG. 5-12
The role of cAMP and protein kinases. (A) cAMP activates cytoplasmic or nuclear protein kinases. (B) They add inorganic phosphate to convert inactive enzymes to active ones. (C) The active enzymes alter distinct chemical reactions that cause the formation of cellular responses or products.

TABLE 5-3
Physiological effects of increased intracellular cAMP*

Stimulates	*Inhibits*
Protein degradation in liver	Protein synthesis in liver
Glycogen degradation	
Fat degradation (from adipocytes)	
Fatty acid release into blood from adipocytes	Histamine release into blood from mast cells
Contraction of skeletal muscle	Contraction of smooth muscle in the walls of bronchi, stomach, and intestine
Contraction of cardiac muscle	
Hydrochloric acid formation in stomach	
Increased urine concentration (reabsorption of water by kidney into blood)	

* cGMP usually produces effects opposite to those of cAMP.

to the ones stimulated by cAMP. For example, cGMP levels rise under the influence of the hormone insulin, which lowers blood sugar and increases liver glycogen, while cAMP levels rise under the influence of the hormone glucagon, which raises blood sugar and decreases liver glycogen. The effects of several hormones on the levels of these two secondary messengers are summarized in Table 5-4.

The contrary responses that follow elevated levels of some hormones seem to be due to opposite effects on secondary messengers. Changes in intracellular levels of cAMP and cGMP generally proceed in inverse directions. As one increases, the other decreases. The two messengers are antagonistic, their net effect

TABLE 5-4
Stimulation of increased intracellular level of second messengers

Class of Hormone	SPECIFIC HORMONES WHICH ELEVATE	
	cAMP	*cGMP*
Polypeptides	Adrenocorticotropin (ACTH)	
	Thyroid stimulating hormone (TSH)	
	Melanocyte stimulating hormone (MSH)	
	Luteinizing hormone (LH)	
	Antidiuretic hormone (ADH)	Oxytocin
	Glucagon	Insulin
	Secretin*	Secretin*
	Gastrin	
	Calcitonin*	Calcitonin*
	Parathormone	
Amino acid derived hormones	Epinephrine	Norepinephrine*
	Thyroxine	
Steroids	Corticosteroids	
Prostaglandins	Prostaglandins E_1, E_2	Prostaglandin $F_2\alpha$
Other agents	Histamine*	Histamine*
		Acetylcholine
		Serotonin

* These agents have been reported by different investigators to increase cAMP and cGMP. In some instances the effects may be due to combination with different receptors. In others they may represent differences in the experimental conditions under which the measurements were made.

being determined by the ratio of the level of one to the level of the other. Cells normally have 10 to 50 times more cAMP than cGMP.

Ca^{2+} seems to behave as a secondary messenger by binding to a protein receptor called *calmodulin* (MW ~ 16700). Calmodulin is a calcium-binding protein found in abundance in a variety of cells. Each molecule can adhere to and release four Ca^{2+} ions. Various shapes of Ca^{2+}–calmodulin complexes may be produced, which in turn can attach to different receptors to influence the activity of specific enzymes. Among them are the enzymes which regulate the metabolism of other secondary messengers (such as cAMP). Ca^{2+}–calmodulin complexes may specifically activate such enzymes as adenyl cyclase, guanyl cyclase, phosphodiesterases, or various kinases, to produce different biochemical responses.

Calcium acts more rapidly as a cellular regulator than does cAMP, with which its activities are intertwined. A number of possibilities may explain this fact; calcium is abundant in body fluids; its receptor, calmodulin, is abundant in cells; calcium does not require enzymatic synthesis as does cAMP. Stimuli which release calcium and cause an increase in its intracellular levels lead to activation of various conformations of calmodulin which may exhibit different activities toward specific cell receptors. For example, the binding of calcium to calmodulin in the cell membrane may activate adenyl cyclase to produce cAMP. Later, the calcium may diffuse to the cytoplasm to combine with calmodulin there, and activate the phosphodiesterase that degrades cAMP, diminishing the earlier response caused by elevated levels of cAMP.

As mentioned earlier in this chapter, the response of a cell to hormones also depends on the nature of the cell's receptors. For example, norepinephrine combines primarily with alpha receptors, which may cause cGMP levels to rise. The hormone causes relaxation of smooth muscle in the walls of the intestine. However, it also combines to a lesser extent with beta receptors to increase cAMP. Cells with large numbers of beta receptors, such as those in the heart, contract more forcefully under the influence of norepinephrine (an effect opposite to that in the intestine).

As indicated earlier, the interrelationship and the complexity of the hormonal reactions make it difficult to distinguish the primary effect of a hormone from the multiple effects which follow. The analysis is further complicated by differences in the nature and number of receptors, the responsiveness of different cell types, the presence of antagonists, the activity of secondary messengers, and the complex series of paths involved. Yet, hormones are generally maintained in a homeostatic balance throughout our lives. They are able to regulate growth and development from conception through old age and exert marked influences on body function.

5-8 SUMMARY

1 Body function is greatly influenced by chemical control compounds, including hormones, parahormones, neurohormones, neurotransmitters, pheromones, and drugs.

2 Hormones are chemicals synthesized by specific cells and secreted into the blood to be delivered to specific target cells upon which they exert their regulatory effects. Parahormones are formed by many cell types. Neurohormones are synthesized by neurosecretory cells such as those in the hypothalamus. Neurotransmitters are produced in neurons and released from the terminal endings of nerves. Pheromones are chemical agents which influence reproductive behavior in some animals. Drugs are exogenous chemicals (from outside the body), whose actions may mimic, block, or exaggerate those of other chemical control agents.

3 The endocrine glands are ductless and secrete hormones directly into the blood. The pituitary, hypothalamus, pineal gland, parathyroids, thymus,

pancreas, adrenals, gonads (testes and ovaries), and placenta are major endocrine glands. Some other sources of hormones include cells in the digestive tract and the kidneys. Many major hormones and their principal effects are noted in Table 5-5 and illustrated in Figs. 5-2A through 5-2F.

4 Hormones may be protein-type molecules (amino acids, peptides, proteins or their derivatives) or lipid-type molecules (steroids and derivatives of fatty acids). The former are water-soluble and usually combine with receptors on the cell membrane. The latter are lipid-soluble and usually diffuse across the cell membrane to combine with intracellular receptors.

5 The hypothalamus, anterior pituitary, and individual endocrines regulate hormone secretion. The hypothalamus produces several different polypeptides as factors which stimulate or inhibit the release of hormones from the anterior pituitary. The release-inhibitory hormones of the hypothalamus regulate secretion of the following anterior pituitary hormones: adrenocorticotropin (ACTH), thyroid-stimulating hormone (TSH), melanocyte-stimulating hormone (MSH), growth hormone (GH), luteinizing hormone (LH), follicle-stimulating hormone (FSH), and prolactin or luteotropic hormone (PRL or LTH). Each of these anterior pituitary hormones stimulates the activity of other endocrines and/or regulates critical body functions.

The hypothalamus and pituitary have direct circulatory and neural connections. Releasing and inhibitory factors from the hypothalamus are delivered by portal blood vessels to the anterior pituitary, while oxytocin and antidiuretic hormone travel down nerve fibers to the posterior pituitary.

Regulation of endocrine activity may be through stimulation or inhibition by the nervous system or hypothalamus, stimulation by the anterior pituitary, and/or negative feedback inhibition from the products of target cells that have responded to a specific hormone.

Stress activates the hypothalamus to secrete corticotropin-releasing factor (CRF), which causes the anterior pituitary to release adrenocorticotropic hormone (ACTH). The latter stimulates the adrenal cortex to secrete glucocorticoids, which elevate blood sugar and blood pressure. This prepares the body for resistance to harmful stimuli. Stress also activates the autonomic nervous system to transmit signals to several areas of the body, including the hypothalamus and the adrenal medullae. The latter release catecholamines, such as epinephrine, which further prepares the body for a "fight or flight" response. Repeated exposure to stress may cause high blood pressure, atherosclerosis, or gastric ulcers.

6 The levels of hormone activity are related to many factors. Genetic control of cell function (including cellular multiplication) and availability of nutrients play critical roles in the rate of hormone synthesis and secretion. The secretion of hormones from cells often involves their release from granules as prohormones. Hormone discharge into the blood is enhanced by increased cellular absorption of calcium ions and may involve microtubules and microfilaments of cells. Many hormones are secreted by exocytosis. In the blood they are coupled to proteins produced by the liver. The activity of most hormones declines by 50 percent within 1 h of secretion, since the molecules are rapidly degraded by the liver, kidney, and target cells.

Most hormones bind to specific protein receptors on plasma membranes of target cells. Steroids form complexes with cytoplasmic proteins, which attach to nuclear receptors. Thyroid hormones penetrate the cell to adhere directly to nuclear receptors. The structure and number of receptors per cell change during development and aging and influence responses to specific hormones.

The interaction of hormones may involve cooperation (synergism) or opposition (antagonism). The primary effect of a hormone is followed by a series of events which have a spreading or cascade effect. The rapidity of the responses makes it difficult to trace the sequence of events. Pharmacological levels of hormones are abnormally high amounts which often cause effects that do not occur with normal physiological levels.

7 Hormones may alter enzyme activity or cellular permeability. These changes result from modification of the expression of genes (transcription) or post-transcriptional processes (translation). Hormones which bind to nuclear receptors influence the direction of mRNA synthesis (transcription) and thereby exert their effects. Those which bind to cell membrane receptors may alter protein synthesis, cellular permeability, or the production of secondary messengers.

The secondary messenger cAMP is produced by activation of the membrane-bound enzyme adenyl cyclase when certain hormones bind to cell-membrane receptors. Increases in cAMP activate protein kinases in the nucleus or cytoplasm, causing further metabolic and physiologic responses. These include the degradation of protein, fat, and glycogen; the production of hydrochloric acid in the stomach; the inhibition of histamine release from mast cells; the retention of water; the contraction of skeletal and cardiac muscle; and the relaxation of smooth muscle in the walls of the bronchi, stomach, and intestine. Some hormones bind to receptors which cause an increase in cGMP to produce opposing effects. Other hormones may cause the release of Ca^{2+} ions which can form a complex with protein receptors, calmodulins, which rapidly influences cellular activities, including the metabolism of other cellular regulators such as cAMP and cGMP.

REVIEW QUESTIONS

1 Define an endocrine gland and a hormone. What is a target cell?
2 What are the categories of chemical regulators? Describe two general classes of chemicals of which hormones are composed.
3 How do the hypothalamus and pituitary influence the activity of the endocrines? What are the targets of pituitary hormones?
4 Describe the structure and function of the hypothalamic portal system.
5 Define stress in physiological terms. Describe the role of the hypothalamic-pituitary circuit.
6 What are the targets of the tropic hormones of the anterior pituitary?
7 What are the origins and functions of oxytocin and antidiuretic hormone?
8 How do the effects of hormones cascade? Describe the cascade effect of insulin.
9 What difference is there between the response of the body to the physiological and pharmacological level of a hormone?
10 Describe the principles and use of radioimmunoassays in the study of endocrine function.
11 Name three known secondary messengers. What are their general functions? Give an example of their specific effects and at least one specific target.
12 List some general effects of cAMP.
13 Describe the general relation between hormones, calcium ions, and cAMP.
14 How are cAMP levels controlled and by what means may cAMP act as a secondary messenger?
15 How does cAMP activate enzymes and influence metabolism?

TABLE 5-5
Some major sources of hormones and their principal functions

Endocrine Tissue (and Location)	*Name of Hormone*	*General Chemistry*	*Major Sites of Action*	*Principal Physiological Effects*	SOME CLINICAL SYMPTOMS OF	
					Overactivity	*Underactivity*
1. Hypothalamus (base of brain)	Hypothalamic hormones (named after pituitary hormones whose secretions they stimulate or inhibit, see Table 5-1, p. 109	Tripeptides to polypeptides	Adenohypophysis (anterior pituitary)	Enhance or inhibit secretion of anterior pituitary hormones	(See effects of pituitary hormones and discussion of hypothalamus in this chapter and table)	
2. Pituitary, the hypophysis (below the base of the brain in sella turcica of skull)						
a. Adenohypophysis (anterior or glandular pituitary)	Adrenocorticotropic hormone (ACTH), corticotropin	Polypeptide (39 amino acids MW ~ 4500)	Adrenal cortex	Production of hormones in adrenal cortex	Hyperpigmentation, Cushing's disease (see adrenal glucocorticoids and Cushing's syndrome below)	Addison's disease (see adrenal glucocorticoids)
	Thyrotropic stimulating hormone (TSH), thyrotropin	Glycoprotein (MW ~ 28,000)	Thyroid	Production of thyroid hormones	(See thyroid hormones below)	
	Growth hormone (GH), somatotropic hormone (STH) or somatotropin	Protein (191 amino acids MW ~ 21,000)	Tissues in general, especially bone and muscle and liver	Anabolism, synthesis of protein and glycogen, production of somatomedins which stimulate growth	Prior to completion of growth, increases skeletal growth, causing gigantism (see Fig. 5-2A) After growth ceases, causes burly appearance; wrinkles; enlarged facial tissues, hands, and feet; acromegaly (see Fig. 5-2B)	Hypophyseal dwarfism
	Follicle stimulating hormone (FSH), a pituitary gonadotropin	Glycoprotein (MW 25,000)	Ovaries	Maturation and release of egg from ovarian follicle (along with LH below)		Lack of sexual maturation in female and lack of menstrual periods

	(FSH, cont.)		Testes	Development of sperm		Lack of sexual maturation in male
	*Luteinizing hormone (LH) in females, a pituitary gonadotropin	Glycoprotein (MW 40,000)	Ovaries	Maturation and development of follicle, release of egg from follicle, production of progesterone		
	*Interstitial cell stimulating hormone (ICSH) in males, a pituitary gonadotropin		Testes	Development of interstitial cells, production of androgen		
	* Note: LH and ICSH are the same hormone in females and males respectively.					
	Prolactin (PRL) Luteotropic hormone (LTH), luteotropin, lactogenic hormone	Polypeptide (similar to GH; MW 21,000)	Mammary gland	Milk production after childbirth, onset of puberty	Milk production without recent birth of child	Failure to produce milk after childbirth
			Corpus luteum	Maintenance of conditions to support pregnancy; influence on ovarian enzymes		
Some cells of anterior pituitary in humans (?), intermediate pituitary lobe in lower animals	Alpha and beta lipotropins, lipotropic hormone (LPH)	Polypeptides (58 + 91 amino acids respectively	Fat cells	Lipid release, body weight effects (?), production of pigment by melanocyte-stimulating hormone; B-LPH contains endorphin and enkephalin neurotransmitters		
	Melanocyte-stimulating hormone (MSH), intermedins	Polypeptides similar to ACTH (with 13 amino acids, in α-MSH and 22 in β-MSH)	Pigment cells, (melanophores in skin)	Production of pigment	Hyperpigmentation	Lack or loss of pigmentation
b. Neurohypophysis (posterior pituitary)	Oxytocin (from hypothalamus)	Polypeptide (8 amino acid residues MW 1000)	Smooth muscle in uterus, mammary glands	Uterine contraction, childbirth, milk release		

TABLE 5-5 (*Continued*)

TABLE 5-5 *(Continued)*

Endocrine Tissue (and Location)	*Name of Hormone*	*General Chemistry*	*Major Sites of Action*	*Principal Physiological Effects*	SOME CLINICAL SYMPTOMS OF	
					Overactivity	*Underactivity*
	Antidiuretic hormone (ADH), vasopressin (from hypothalamus)	Polypeptide similar to oxytocin (8 amino acid residues; MW 1000)	Kidney tubules Smooth muscles	Increased water permeability and water retention Contraction with high doses	SIADH (syndrome of inappropriate ADH). Increased extracellular volume and blood pressure, dilution of blood, concentrated urine	Excess urine volume, extreme thirst, diabetes insipidus
3. Pineal gland (above and behind third ventricle in center of brain)	Melatonin	Derivative of amino acid tryptophan and neurotransmitter serotonin; (*N*-acetyl-5-methoxytryptamine)	Pigment cells, (melanophores in skin)	Onset of puberty, (associated with exposure to light?) establishes biological rhythms	Alteration of onset of puberty with pineal tumors (?)	
4. Thyroid follicular cells (in neck, below larynx and across upper trachea)	Thyroxine (tetraiodothyronine, T_4, and triiodothyronine, T_3)	Iodinated derivative of amino acid tyrosine, carried with two plasma glycoproteins	Body cells	Oxidative metabolism	Increased metabolism, increase in plasma protein–bonded iodine, nervous irritability, weight loss, bulging eyes (exopthalmos), bulging of thyroid, Graves' disease (see Fig. 5-2E)	In early childhood retarded growth and mental development; cretinism (see Fig. 5-2F); in adults, low mental and physical activity, skin thickening, accumulation of fat under skin, weight gain, hair loss, termed myxedema
C cells	Calcitonin (thyrocalcitonin)	Polypeptide (32 amino acids, MW ~ 3600)	Bone	Prevents loss of calcium	(See Chap. 14)	
5. Parathyroids (2–4 glands on back of thyroid)	Parathormone	Polypeptide (84 amino acids, MW 9500)	Bone Kidney tubules Intestinal epithelium	Calcium release Calcium reabsorption Calcium absorption	High blood calcium, decreased bone mass, cysts in bone, osteitis fibrosa cystica	Low blood calcium, increased muscular and nervous irritability, muscle spasms, tetany
					(See Chap. 14)	

6. Thymus (lower neck to upper sternum (breastbone) in adolescent, smaller in adult)	Thymosin (thymic hormones)	Glycopeptides, polypeptides, proteins, steroidlike molecules	Lymphoid cells (lymphocytes)	Maturation and maintenance of immunological capability, especially of cellular defense mechanisms	Autoimmune diseases (?)	Lack of cellular defense, susceptibility to cancer, infectious diseases autoimmune diseases (?)
			Other cells	Regulatory effects		
7. Digestive tract (glands in abdominal cavity)					(See Chaps 22 and 23)	
Stomach	Gastrin I and II	Polypeptides (17 amino acids MW 2100)	Stomach Pancreas	Acid production, enzyme secretion		
Duodenum (first part of small intestine)	Secretin	Polypeptide (27 amino acids, MW 3056)	Pancreas	Bicarbonate secretion		
	Cholecystokinin-pancreozymin (CCK-PZ)	Polypeptide (33 amino acids, MW 3883)	Pancreas Gallbladder	Enzyme secretion, contraction		
	Gastric inhibitory polypeptide (GIP)	Polypeptide (43 amino acids, MW 5105)	Stomach	Inhibition of acid production		
8. Pancreas (in curve of duodenum, islets of cells)						
A cells (α cells)	Glucagon	Polypeptide (29 amino acids MW 3845)	Liver	Breakdown of glycogen to glucose		
			Fat cells	release of fat		
B cells (β cells)	Insulin	Protein (2 chains with 21 and 30 amino acids respectively, MW 5700)	Liver and body cells	Uptake of glucose by cells, synthesis of glycogen from glucose, protein synthesis	Decreased blood sugar, restlessness, anxiety, tremors	Decreased entry of glucose into tissues, decreased glycogen synthesis, increased breakdown of amino acids and fats, accumulation of glucose and fatty acids in blood, decreased protein synthesis, in-

TABLE 5-5 *(Continued)*

TABLE 5-5 *(Continued)*

Endocrine Tissue (and Location)	*Name of Hormone*	*General Chemistry*	*Major Sites of Action*	*Principal Physiological Effects*	SOME CLINICAL SYMPTOMS OF	
					Overactivity	*Underactivity*
						creased urine volume, weight loss, weakness, dizziness, decrease in mental concentration; convulsions and coma may develop; disease is termed *diabetes mellitus*
D cells (δ cells)	Somatostatin (also produced by hypothalamus)	Polypeptide (14 amino acids)	Pancreas	Inhibits glucagon secretion		
			Adenohypophysis	Inhibits growth hormone (GH) secretion		
F cells	Pancreatic polypeptide (PP)	Polypeptide (36 amino acids)	Liver, digestive tract	Stimulates glycogen degradation, increases enzyme secretion, relaxes smooth muscle		
9. Adrenal glands rest on superior end of each kidney						
a. Adrenal cortex (outer section of each gland)	Aldosterone	Steroid (mineralocorticoid)	Kidney	Reabsorption of sodium (and water), excretion of potassium	Severe potassium loss, muscle weakness, increased ECF volume, increased blood pressure, kidney damage, Conn's syndrome	Excess sodium loss in urine, increased potassium retention, decreased ECF volume and blood pressure, loss of appetite
			Body cells	Elevation of sodium in ECF and net retention of water		
	Glucocorticoids	Steroids	Kidney	Increases retrieval of sodium and chloride ions from urinary filtrate		
			Sweat glands, salivary glands digestive tract	Decreases excretion of sodium and chloride ions		

	Corticosterone and cortisone	Steroid	Body cells in general and liver	Convert proteins to carbohydrate (mobilize proteins), conserve carbohydrate	Develop fat pads in face (moon face), shoulders, and abdomen; deplete protein in muscles, skin, and hair; edema, rupture of tissues under abdominal skin, producing stripes (striae); termed Cushing's syndrome (see Fig. 5-2C)	Obesity, marked pigmentation, decreased fluid volume and blood pressure, decreased heart size, Addison's disease (see Fig. 5-2D)
			Bone	Calcium loss		
			Fat cells	Inhibit synthesis of, degrade, and mobilize fats		
			Blood vessels	Maintain equilibrium		
			Thymus	Inhibit thymic function, impair protein synthesis, suppress immune and inflammatory responses		
	Androgens and estrogens	Steroids	See testes and ovaries below		Excess androgens in fetal female leads to development of male-type genitals–pseudohermaphroditism	
					Excess androgens in female, masculinization of female, beard growth, virilization	
					Early excess androgens in males produces early masculinization without testicular growth—precocious pseudopuberty	
					Accentuates sex characteristics in adults	
b. Adrenal medullae (inner section of each gland)	Norepinephrine	Derivative of tyrosine, dihydroxyphenylanine (DOPA), and dihydroxyphenylethylamine (dopamine)	Smooth muscle	Usually causes contraction of smooth muscle in blood vessels and its relaxation in the digestive tract	Adrenal tumors may produce excesses of norepinephrine and epinephrine causing high blood pressure, pulmonary edema and possibly death; termed pheochromocytoma	

TABLE 5-5 *(Continued)*

TABLE 5-5 *(Continued)*

Endocrine Tissue (and Location)	*Name of Hormone*	*General Chemistry*	*Major Sites of Action*	*Principal Physiological Effects*	SOME CLINICAL SYMPTOMS OF	
					Overactivity	*Underactivity*
	Epinephrine	Derivative of norepinephrine	Fat cells	Release of fat		
			Smooth muscle	Initially, relaxes blood vessels of skeletal muscles; contraction of blood vessels of skin and digestive tract; relaxation of wall of digestive organs, bronchioles, and bladder		
			Heart muscle	Stimulates vigorous contraction		
10. Kidney (posterior to the abdominal cavity above waistline)					(See Chaps. 17 and 21)	
	Erythropoietin	Glycoprotein (MW 36,000)	Bone marrow	Regulates red blood cell production		
	Vitamin D_3	Steroid derivative (1,25-dihydroxycholecalciferol)	Intestine	Calcium absorption	(See Chap. 14)	
			Bone	Calcium deposition		
			Kidney	Decreases phosphate reabsorption		
	Prostaglandins PGA_2, PGE_2, $PGF_2\alpha$)	Derivatives of 20-carbon-chain fatty acids, arachidonic and linoleic acids containing a 5-carbon ring	Kidney	Water and sodium loss	(See Chapter 19)	
			Smooth muscle	Relaxation and decreased blood pressure		
11. Ovaries (one on each side of uterus, above the bladder in the pelvic cavity in the female)	Estrogens (also from placenta)	Steroids	Body cells	Anabolic function	(See Chap. 24 and 9a, androgens and estrogens, above)	
		Estradiol, estrone, and estriol	Accessory sex organs, mammary glands, and secondary sex characteristics	Maturation and development		

12. Corpus luteum (group of cells in follicle of ovary after release of the egg)	Progesterone (also from placenta)	Steroid	Uterus	Readies lining and wall for implantation of embryo	(See Chap. 24 and 9a, androgens and estrogens, above)
			Mammary glands	Develops tissue for secretion of milk	
	Relaxin	Polypeptide (two chains with 22 and 32 amino acids, MW 6000)	Birth canal	Softens cervix of vagina, loosens ligaments of symphysis pubis	
13. Placenta (fetal and maternal blood vessels and adjacent tissues attached to wall of pregnant uterus)	Gonadotropins	Glycoproteins (same as FSH and LH)	Ovaries or testes	Same effects as FSH and LH (or ICSH)	(See Chap. 24)
	Human placental lactogen (HPL), human chorionic somatomammotropin (HCS)	Similar to pituitary growth hormone	Maternal placental physiology	Via permissive action produces effects similar to prolactin and growth hormone from pituitary	
	Relaxin	Polypeptide	Ovaries (corpus luteum)	Seems to prepare reproductive tract for birth; seems to relax the symphysis pubis	
	Estrogens	Steroids	As above	As in No. 11	
	Progesterone	Steroid	As above	As in No. 11	
14. Testes (in scrotal sac of male outside of and below the pelvic cavity)	Androgens, e.g., testosterone	Steroids	Body cells, accessory sex organs, and secondary sex characteristics	Protein anabolism; maturation and development	(See Chap. 24 and 9a, androgens and estrogens, above)
15. Seminal vesicles (in male, below and behind bladder and above prostate gland in lower pelvic cavity)	Prostaglandins (PGA, PGE, and PGF)	Derivatives of the 20-chain fatty acids, arachidonic and linoleic acids containing a 5-carbon ring	Digestive tract smooth muscle	Contraction with PGE and PGF	(See Chaps. 19 and 24)
			Blood vessel smooth muscle	Contraction with PGF, relaxation with PGE and PGA	
			Bronchial smooth muscle	Relaxation	
			Uterine smooth muscle	contraction (especially with PGE_2 and $PGF_2 \alpha$)	
			Fat cells	Inhibit degradation of fat	
			Kidneys	Promote sodium and water loss (?)	

PHYSIOLOGY OF THE NERVOUS SYSTEM

Structural and Functional Elements of Nervous Tissue

6-1 ROLE OF THE NEURON
Neuron structure and function
Cell body
Nucleus • Cytoplasm
Cell processes
Axons • Dendrites
Shapes of neurons

6-2 MACROSCOPIC STRUCTURE OF NERVES

6-3 SUPPORTIVE TISSUES: THE GLIA
Glia of peripheral nerves
Schwann cells
Satellite cells
Glia of the central nervous system (brain and spinal cord)
Ependymal cells
Astrocytes (astroglia)
Oligocytes (oligodendroglia, oligodendrocytes)
Microglia
Demyelination of fibers: multiple sclerosis

6-4 TYPES OF NERVES
Origin of nerves
Direction of impulses
Destination of impulses
Size and myelin content of fibers

6-5 THE ROLE OF NUTRIENTS IN THE HEALTH OF NERVOUS TISSUE
Carbohydrates
Lipids
Hereditary lipid disorders
Acquired lipid disorders
Proteins
Vitamins

6-6 SUMMARY

OBJECTIVES

After completing this chapter the student should be able to:

1. Describe the structure and function of the components of the neuron.
2. Name the layers of connective tissue that bundle nerve fibers together and describe the location of each.
3. Explain the role of glial cells in the structure and function of nerve tissue.
4. Identify the supporting tissues, the glia, in peripheral nerves and those of the central nervous system.
5. Describe three properties by which one can distinguish different types of nerves. List the three major types of nerves. Explain the relationship between their size, myelin content, and general functions. Note the differences among subdivisions of type A fibers.
6. Outline the roles of carbohydrates, lipids, proteins, and vitamins in nervous tissue metabolism. Tell why each is essential.

The awe-inspiring organization of the human body is perhaps nowhere better reflected than in the nervous system. From the moment of conception it matures more rapidly than any of the other systems. It forms a maze of interconnecting pathways infinitely more complex than those found in the largest and most sophisticated computers.

The central controls for the body's electrical circuits are located in the brain and the spinal cord, which constitute the *central nervous system* (CNS), shown in Fig. 6-1. Because its functions are vital, the CNS is protected by bone. The brain is housed in the skull and the spinal cord in the vertebral column. The nervous system is a unique part of the body, which generates information in the form of electrical signals, called *impulses.* The impulses are transmitted to and from the CNS by nerves in the peripheral nervous system (PNS). Impulses are relayed between the CNS and PNS and the tissues and organs of the body to cause appropriate responses.

The peripheral nervous system (PNS) consists of all parts of the nervous system located outside the CNS (Fig. 6-1). The peripheral nervous system includes 12 pairs of cranial nerves, 31 pairs of spinal nerves, the sensory receptors, and specialized structures known as ganglia.

This chapter will focus on the cellular organization of nervous tissue. The next three chapters, describe the generation and transmission of impulses and the role of the spinal cord and autonomic nervous system, and are followed by two chapters on the functional center of the system, the brain.

6-1 ROLE OF THE NEURON

The cell unique to nervous tissue is the *neuron.* It has properties common to other cells, as described in Chap. 2. It also possesses the capacity to generate and transmit electrical signals in response to changes. Because of its capacity to respond to electrical signals, it is termed *excitable,* a property shared with cells in muscles and exocrine glands. The neuron plays a critical role in the ability of the human body to respond to changes in its external and internal environment. This makes neuronal function a key element in the body's capacity to maintain homeostasis.

Neuron Structure and Function

Hundreds of billions of neurons are found in the nervous system. There are many different types of neurons but all contain an enlarged component, the *cell body* or *soma.* The cell body, which nourishes the neuron, gives rise to a long cytoplasmic extension called an *axon* and one or more generally shorter cytoplasmic extensions called *dendrites.* The *axon* usually carries signals away from the cell body and may divide into two or more parts called *collaterals.* The shorter *dendrites* usually carry impulses toward the cell body (Fig. 6-2).

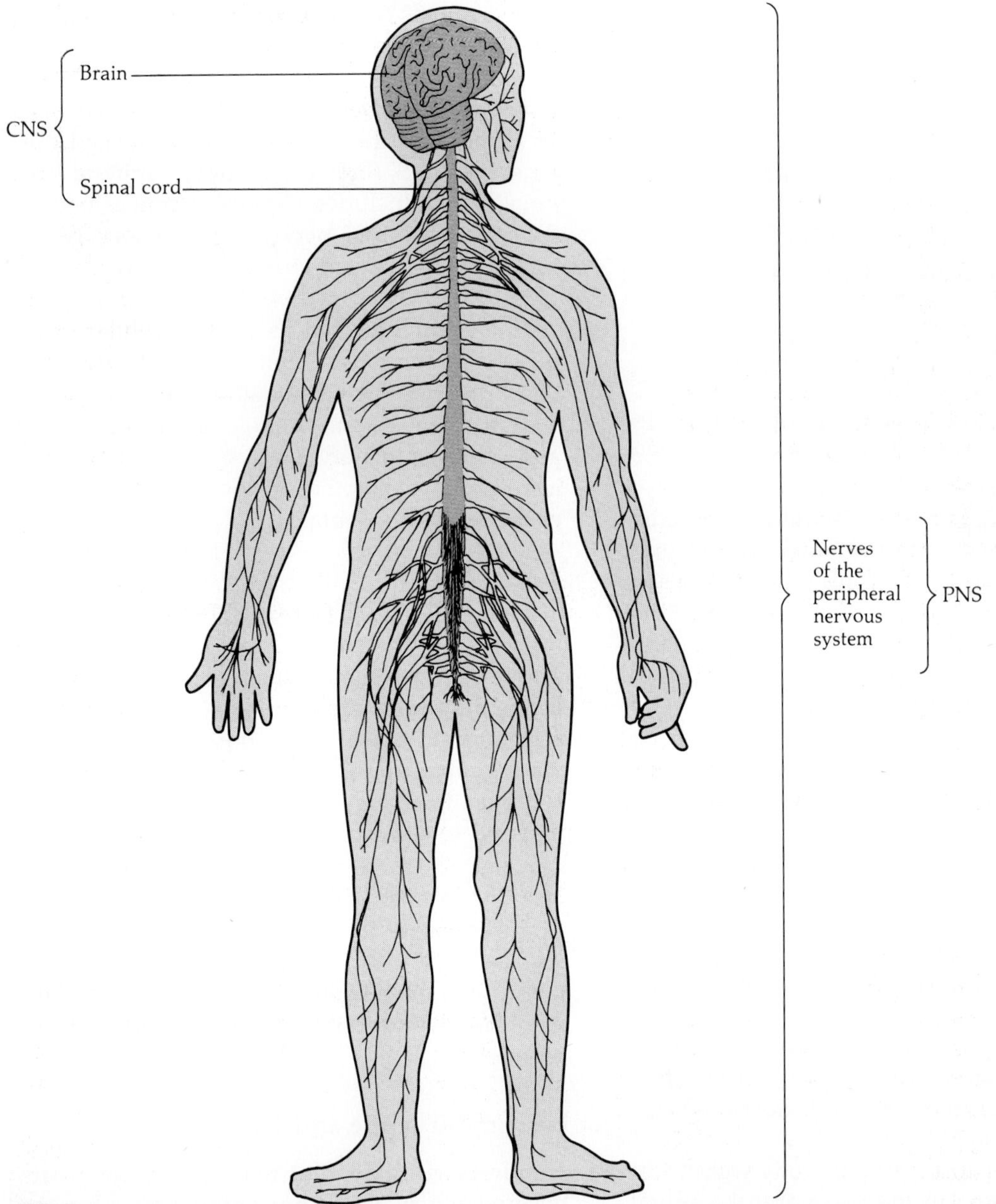

FIG. 6-1
The general organization of the nervous system. The brain and spinal cord form the central nervous system (CNS). All other structures are part of the peripheral nervous system (PNS).

Cell Body

The cell body contains a nucleus and cytoplasm, from which various cytoplasmic projections extend (Fig. 6-2). The cell body may be round, flat, or pyramidal. Its size can range from 5 μm in diameter (as in the granule cells of the cerebellum) to more than 100 μm (as in motor neurons in the spinal cord). The cell body, axons, and dendrites are capable of propagating impulses.

NUCLEUS The nucleus is a prominent part of the neuron located near the center of the cell

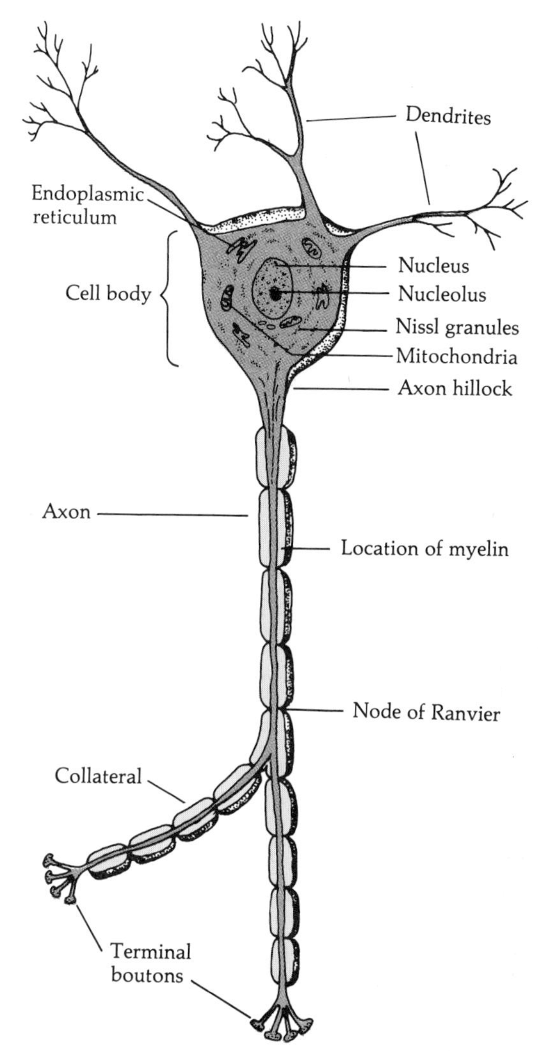

FIG. 6-2
Structure of a neuron.

body. Chromosomes are found within the nucleus, as are usually one or more large nucleoli. Ribonucleic acid (RNA) is abundant in the nucleoli.

CYTOPLASM The cytoplasm is continuous within the cell body and its cytoplasmic extensions. The cytoplasm of neurons contains many of the same organelles found in other cells. Some of these organelles are very prominent owing to their function in the generation and transmission of impulses.

Nissl Granules *Nissl granules* consist of parallel membranes coated with ribosomes and are modifications of the rough endoplasmic reticulum (RER). Nissl granules, which are active in the production of protein, may be observed under the electron microscope. They enclose cavities and have an affinity for basic dyes such as methylene blue. Application of this dye to nervous tissue may make the Nissl granules visible under the light microscope. Nissl granules are found throughout most of the cytoplasm, but their numbers decrease with distance from the cell body. None are found in the axonal extensions of the neuron.

The appearance of Nissl granules varies with the activity of the cell. They tend to disappear or disperse to the periphery of the cell body when a neuron is damaged, but they reappear on its recovery. When a neuron is lethally damaged, Nissl granules vanish permanently. This may occur following certain viral infections such as polio, which damages nerves and may lead to paralysis of the associated muscle.

Mitochondria Mitochondria are especially prevalent in the terminals of neurons which make contact with each other and with the cell membranes of other types of cells. Mitochondria contain respiratory enzymes, which oxidize nutrients to produce energy.

Golgi Apparatus The Golgi apparatus is composed of smooth, membranous structures forming sacs or vesicles in the neuron. It seems to be part of an agranular series of membranes connected to those of the Nissl bodies. The Golgi apparatus probably helps assemble products into secretory granules and packages, which are bound by the membranes in the form of vesicles. The packaged products may be transported to the terminal portion of the axon, forming *synaptic vesicles.* The role of synaptic vesicles in the transmission of nerve impulses will be described in Chap. 7.

Neurofibrils Neurofibrils are proteins which form fibrous structures in the cell body and in all parts of the axons and dendrites (Fig. 6-3). They are prominent in electron micrographs of

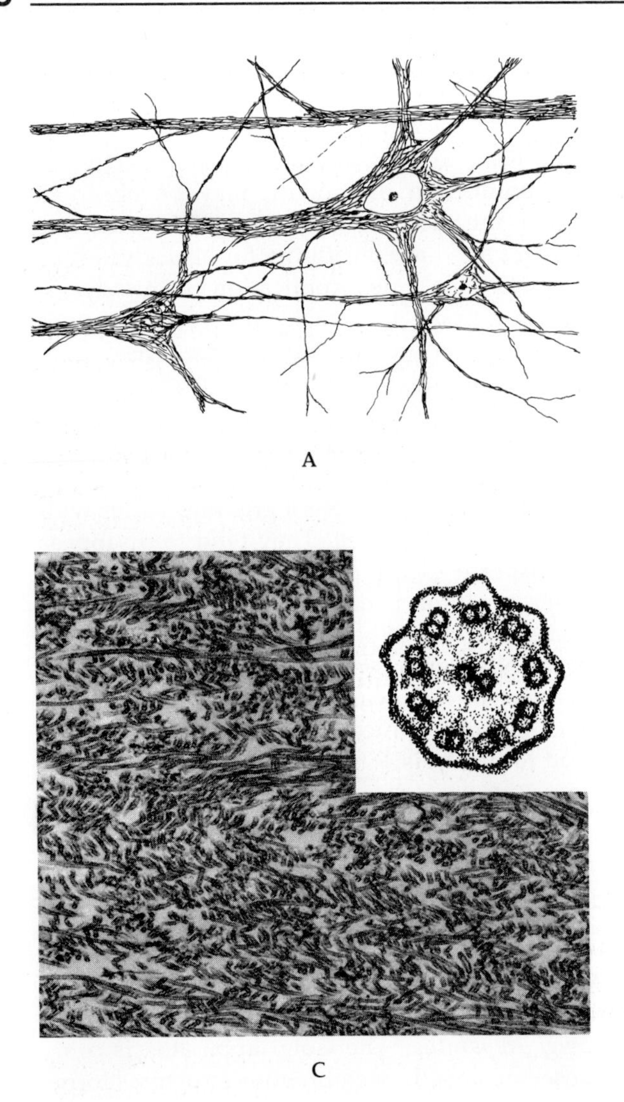

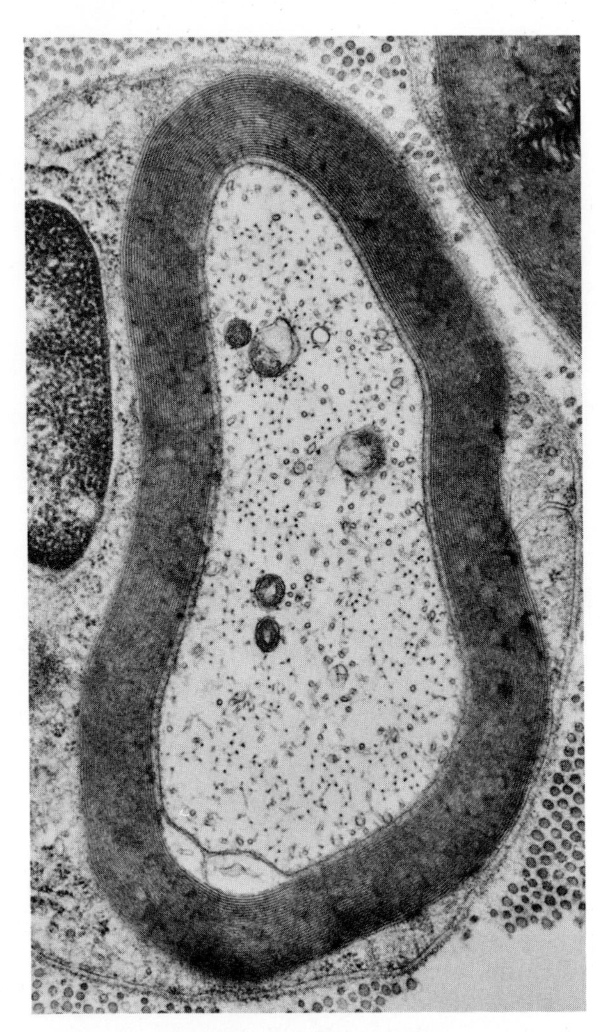

FIG. 6-3
Some types of neurofibrils. (A) Neurofibrils shown within neurons under light microscopy (×1300). (B) Neurofilaments and neurotubules surrounded by several concentric layers of myelin in a cross section of an axon viewed under electron microscopy (×33,000). (C) Neurotubules extracted from the brain and viewed in longitudinal section under electron microscopy (×22,000). Their appearance in cross section shown on the upper right resembles that of cilia in Fig. 2-12.

the growing tips of axons. With light microscopy and the application of silver stain, the neurofibrils appear as parallel bundles and threadlike structures. Neurofibrils vary in size and appearance in response to injury and with changes in temperature and in the metabolism of the neuron.

Three types of neurofibrils are: (1) *neurofilaments,* which usually run in bundles along the axon and are seen with the light microscope; (2) *neurotubules,* which are the long, straight filaments surrounding a central cavity and which correspond to microtubules in other cells; and (3) *microfilaments,* which exist as an irregular network of fibers and are evident in the growing tips of neurons.

Neurofilaments *Neurofilaments* consist of strands about 10 nm in diameter, which run the length of most axons (Fig. 6-3B). The strands are linked by cross-bridges in groups which contain spaces between their members. Neurofilaments may provide a mechanism for the flow of materials through axons. They apparently aid the movement of particles and some organelles in the cytoplasm.

Neurotubules *Neurotubules* resemble the microtubules found in cilia and flagella (Fig. 2-12). Neurotubules are cylinders about 25 nm in diameter formed by 13 parallel filaments (Fig. 6-3C) which are arranged in a circle. Each filament is about 5 nm in diameter and is composed of a distinct protein, tubulin. Neurotubules, like neurofilaments, may provide a mechanism for the rapid movement of substances in the cytoplasm. Materials may move within them at rates ranging from 1 to more than 400 mm/day. Neurotubules also offer some structural support to the cell.

Microfilaments *Microfilaments* are molecules located along and around axons. They resemble the muscle protein actin (Fig. 12-5D) and are about 5 nm (4 to 6 nm) in diameter. Microfilaments may be involved in the movement of axons during growth or regeneration of neurons.

Nerve fibers, when initially formed, grow toward those structures with which they eventually affiliate. Microfilaments may play an important part in the development of those affiliations. Cells may release a protein, the *nerve growth factor,* which attracts neurons, stimulating the growth of microfilaments in a specific direction. Evidence indicates that this may be one way in which nerves reach their exact destinations. The precise manner in which extensions of neurons reach their target receptor cells and the method by which the flow of materials occurs through neurons remain to be determined.

Cell Processes

AXONS An *axon* is single and unbranched as it emerges from the cell body. At some distance from its origin, the axon may divide into two or more stalks called *collaterals.* The axon generally has a small diameter (0.5 to 20 μm) but great length (up to 900,000 μm). Often the axon is called a *fiber.* It is bundled with other axons and other extensions of neurons to form nerves. Axons contain significant numbers of mitochondria and neurofibrils but no Nissl granules. The existence of the axon is dependent upon the cell body. When nerves are damaged, axons regenerate only if the cell body survives. The portion of the cell body to which the axon is attached is thickened and called the *axon hillock* (Fig. 6-2). Nerve impulses are generally but not always transmitted away from the cell body through the axon hillock along the axon.

The end of each axon forms *nerve terminals* with tiny swellings called *terminal boutons* (terminal knobs). Numerous membrane-bound vesicles (sacs) are often present in the terminals. The vesicles contain molecules called *neurotransmitters* which enable the neuron to relay impulses to other cells across cell junctions. A junction between the terminal of a neuron and the membrane of another cell is called a *synapse.* A synapse may occur between two neurons or between a neuron and some other cell type, such as that in a gland or muscle.

Materials continually move forward (*anterograde* transport) and backward (*retrograde* transport) through the axon, as shown in Fig. 6-4. Backward flow occurs at half the rate of forward flow. Molecules, fluids, and even cytoplasmic vesicles (membrane-bound sacs) may move at various speeds (fast and slow) and directions in an axon at the same time (Fig. 6-4). They resemble people moving up and down a sidewalk at different speeds. The rate of transport in different axons varies from a slow pace of 0.2 to 2 mm/day to a fast one around 400 mm/day.

Some evidence indicates that substances are moved along within microtubules by contractile elements of the microfilaments which behave like a ratchet mechanism (resembling the movement of the teeth on a can opener), using energy derived from ATP. In this manner, cytoplasmic components and vesicles and their contents may be transported between the cell body and distant sites. The mechanism may be similar to that of proteins which cause muscles to contract (Chap. 12).

Two types of axons, myelinated and unmyelinated, may be distinguished by the nature of

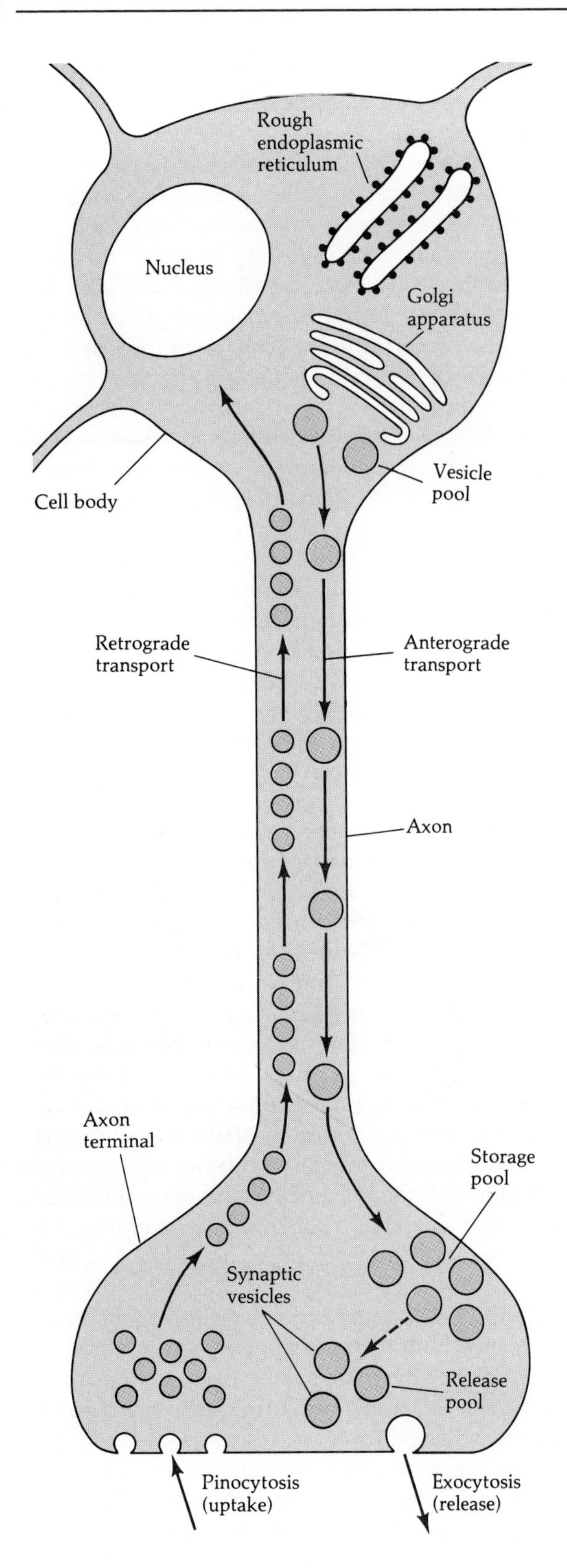

FIG. 6-4
Movement of materials between cell body and axon terminal.

their outer layers. *Myelinated axons* are surrounded by a sheath of connective tissue cells which contain the lipoprotein myelin, together with an abundance of the phospholipid sphingomyelin. *Myelin* acts much like an insulating coat surrounding an electrical cable (the axon), preventing ions from leaking across the axon membrane. Wherever it is found, myelin allows an electrical gradient of ions to be more readily maintained. *Unmyelinated axons* possess a thinner insulating coat. For this reason, the electrical properties of the two types of axons differ. Myelinated axons generally conduct impulses more rapidly than their unmyelinated counterparts. It has been estimated that if the spinal cord were to lack myelin, it would have to be several yards in diameter to provide its usual speed of impulse conduction.

DENDRITES *Dendrites* are generally short extensions of the neuron, ranging in length from less than 100 to several thousand micrometers. They contain Nissl granules, mitochondria, neurotubules, and neurofilaments. The number of neurotubules and neurofilaments decreases with distance from the cell body. The ends of dendrites branch in various patterns. Some are covered with numerous thorny projections called *spines* or *gemmules*. Dendrites and axons form macrocircuits and microcircuits among different neurons. Dendrites generally relay impulses toward the cell body. The types of circuits they make will be described in Chap. 7.

Shapes of Neurons

Neurons may be classified by their general appearance. *Unipolar* (*pseudopolar*) neurons possess a single cytoplasmic process extending from the cell body (Fig. 6-5A). The process divides into two parts, one behaving as a dendrite, the other as an axon. Unipolar neurons are found within some of the cranial nerves in the brain and all the spinal nerves. Their cell bodies make up the dorsal root ganglia of the

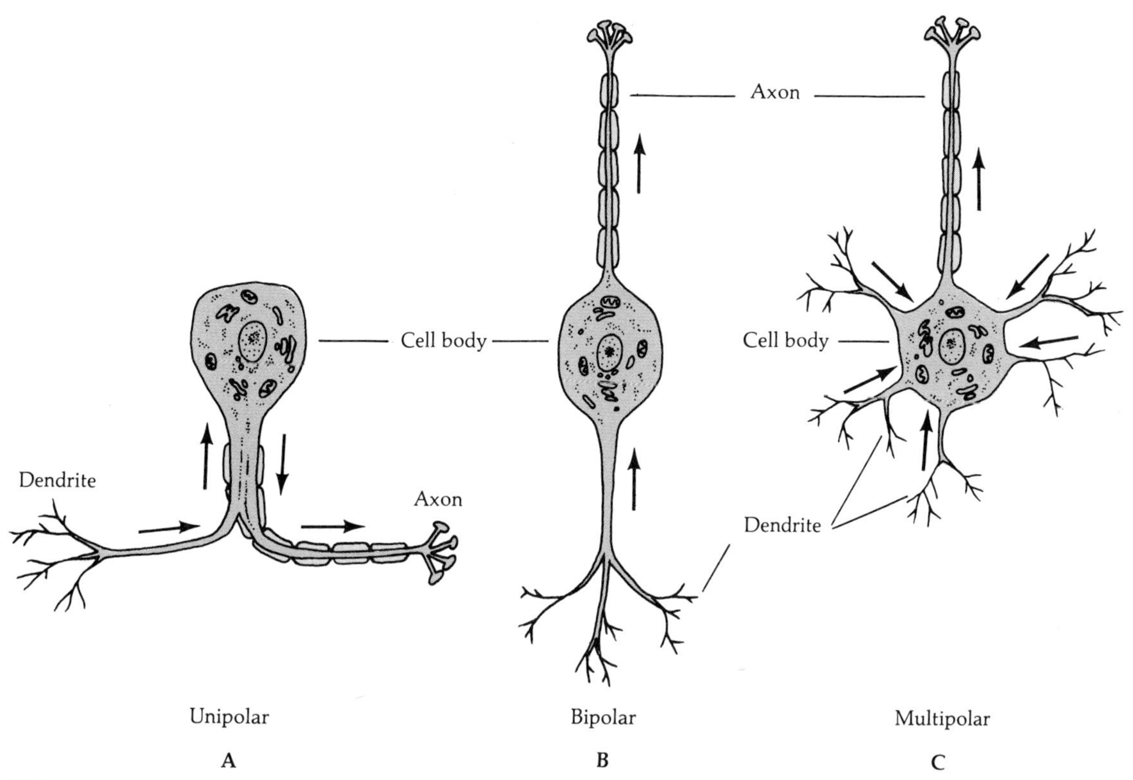

FIG. 6-5
Shapes of neurons. The impulse is conveyed in the direction noted by the broken arrows. (A) Unipolar neuron. (B) Bipolar neuron. (C) Multipolar neuron.

spinal cord. They are the primary *afferent* (or sensory) neurons since they transmit impulses from sensory receptors to the central nervous system.

The *bipolar* neuron possesses two stalks, which emerge from opposite ends of the cell body (Fig. 6-5B). One is dendritic and the other axonal. These neurons are found in the brain and spinal cord, the retina of the eye, the inner ear, and the olfactory epithelium of the nose. They are the primary afferent neurons which function in vision, hearing, smell (olfaction), and balance. Their special dendritic endings are modified to serve as receptors, which transmit information to their cell bodies and on through their axons to other neurons of the CNS.

Most neurons in the CNS are *multipolar neurons,* which possess many dendrites and a single axon stalk whose branches give off numerous collaterals and terminals (Fig. 6-5C). Multipolar neurons integrate information received in the form of impulses from various parts of the central nervous system, thus acting as complex computers. Billions are active at any instant. They include *motor* (or efferent) neurons, which carry impulses from the CNS to tissues and organs, *interneurons* (associative neurons) in the brain and spinal cord, *Purkinje* cells in the outer part of the cerebellum (cerebellar cortex), and the majority of neurons in *autonomic ganglia.*

6-2 THE MACROSCOPIC STRUCTURE OF NERVES

Nerves consist of large numbers of axons, dendrites, or both arranged in parallel bundles. They do not include the cell bodies of neurons. The cell bodies are found within the CNS and ganglia but not within nerves. A nerve, there-

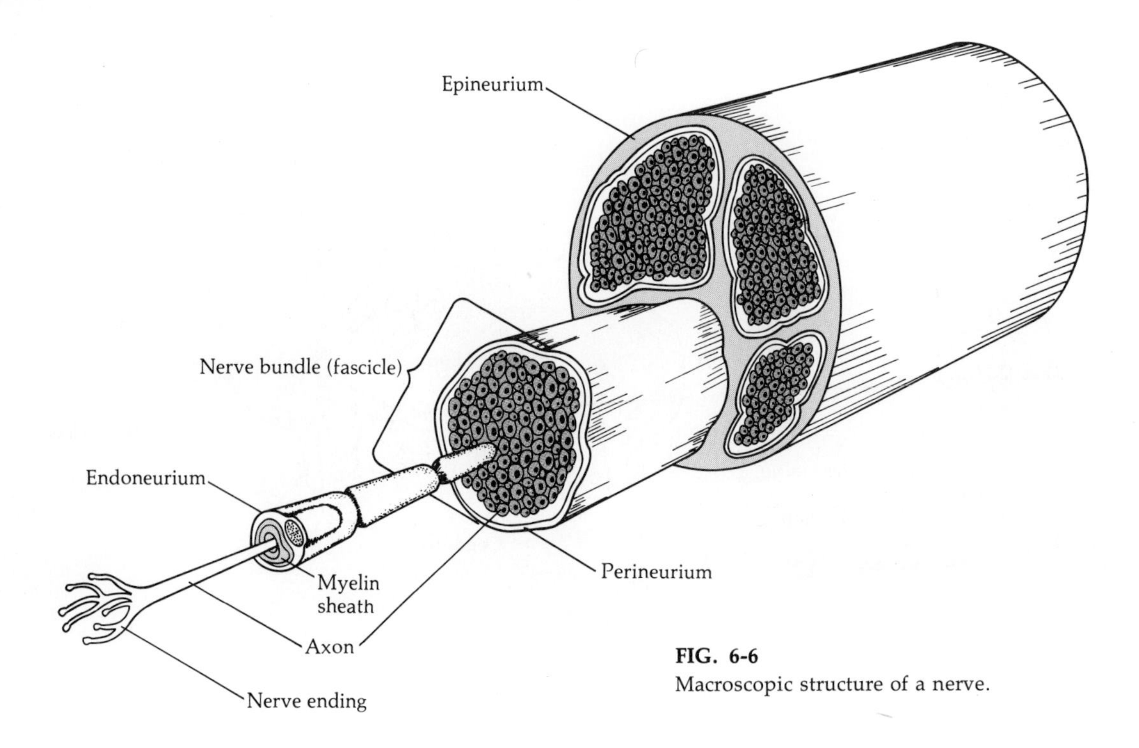

FIG. 6-6
Macroscopic structure of a nerve.

fore, is much like a cable containing microscopic telephone wires which carry impulses from one part of the body to another.

Each nerve is surrounded by three connective tissue sheaths. An outermost overcoat called the *epineurium* (Fig. 6-6) surrounds bundles (fasciculi) of axons. Each bundle is encircled by a middle layer of connective tissue called the *perineurium*. Blood vessels to the nerve are found between the epineurium and perineurium. A thin, innermost third layer is formed by reticular fibers of connective tissue called the *endoneurium*, which separately covers the surface of individual axons in the nerve.

6-3 SUPPORTIVE TISSUES: THE GLIA

Connective tissue cells, the *glia*, are located in the interstitial spaces immediately surrounding neurons and their extensions. They are 10 times more prevalent than neurons in the nervous system. They literally "glue" nerves together.

Glia of Peripheral Nerves

Peripheral nerves (those outside the brain and spinal cord) are held together, protected, and nourished by two types of glia, Schwann cells and satellite cells.

Schwann Cells

Schwann cells are wrapped concentrically around some axons in peripheral nerves as shown in Fig. 6-7. Except for interruptions which occur at 1 to 2 mm intervals, membranes of adjacent Schwann cells form a continuous sheath called the *neurilemma*, which covers most of the length

FIG. 6-7
Peripheral myelinated nerve fiber and its cell body. (A) Peripheral nerve fiber surrounded by a Schwann cell and its sheath (neurilemma). (B) Detailed drawing of the myelinated portion of an axon. The myelinated fiber is enveloped by several concentric layers of the Schwann cell membrane. (C) Photomicrograph of a longitudinal section of a myelinated nerve (×700). (D) Photomicrograph of a cross section of a myelinated nerve (×700).

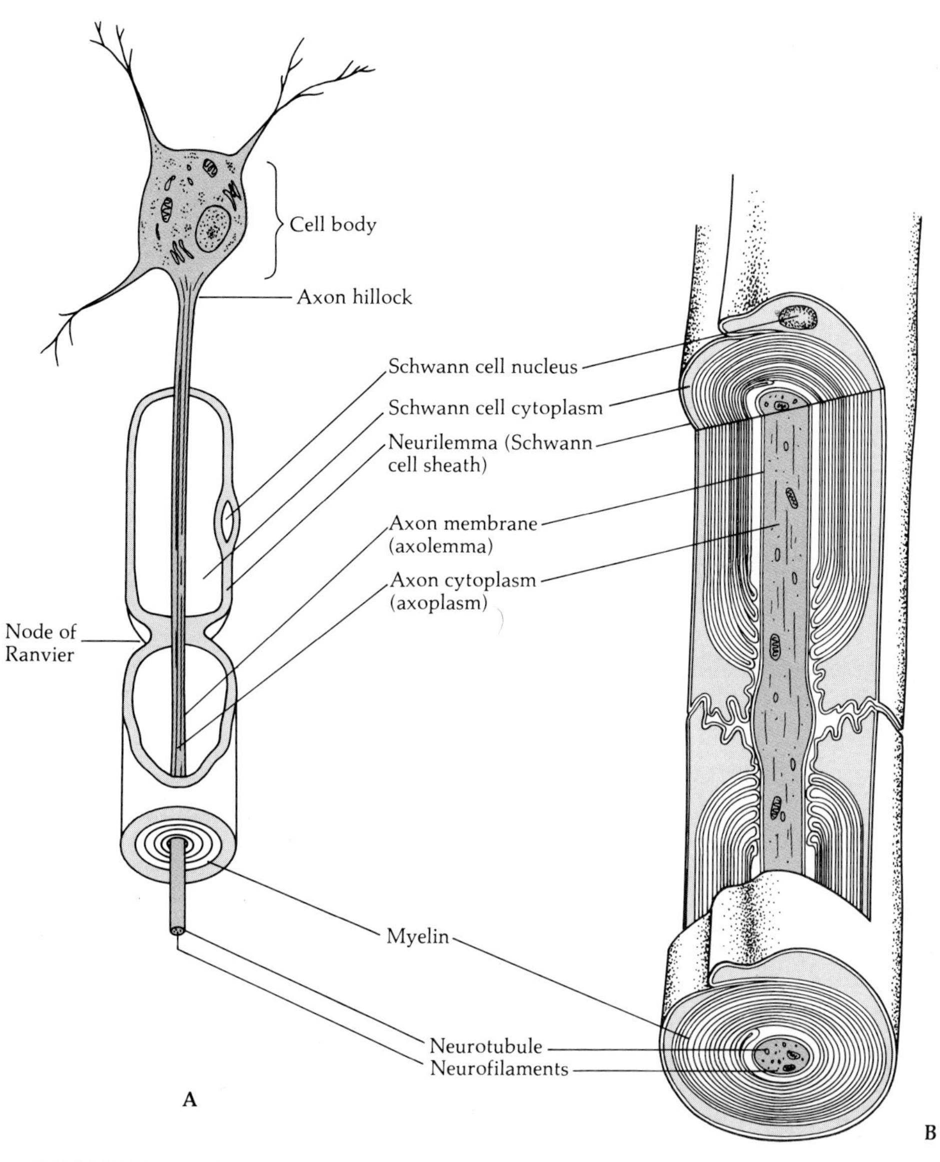
Cell body
Axon hillock
Schwann cell nucleus
Schwann cell cytoplasm
Neurilemma (Schwann cell sheath)
Axon membrane (axolemma)
Axon cytoplasm (axoplasm)
Node of Ranvier
Myelin
Neurotubule
Neurofilaments
A
B

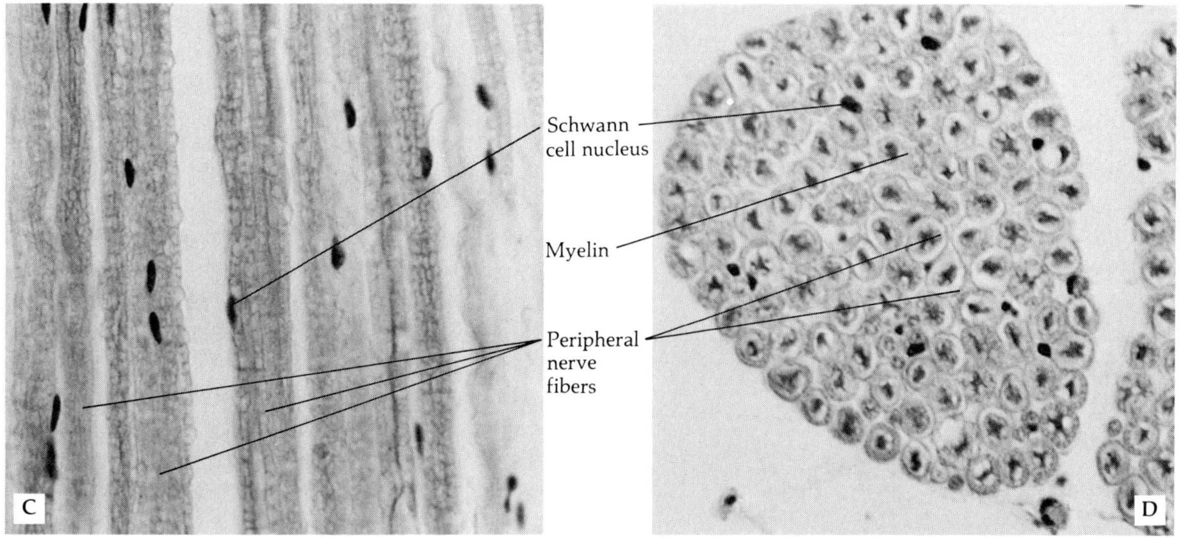
Schwann cell nucleus
Myelin
Peripheral nerve fibers
C
D

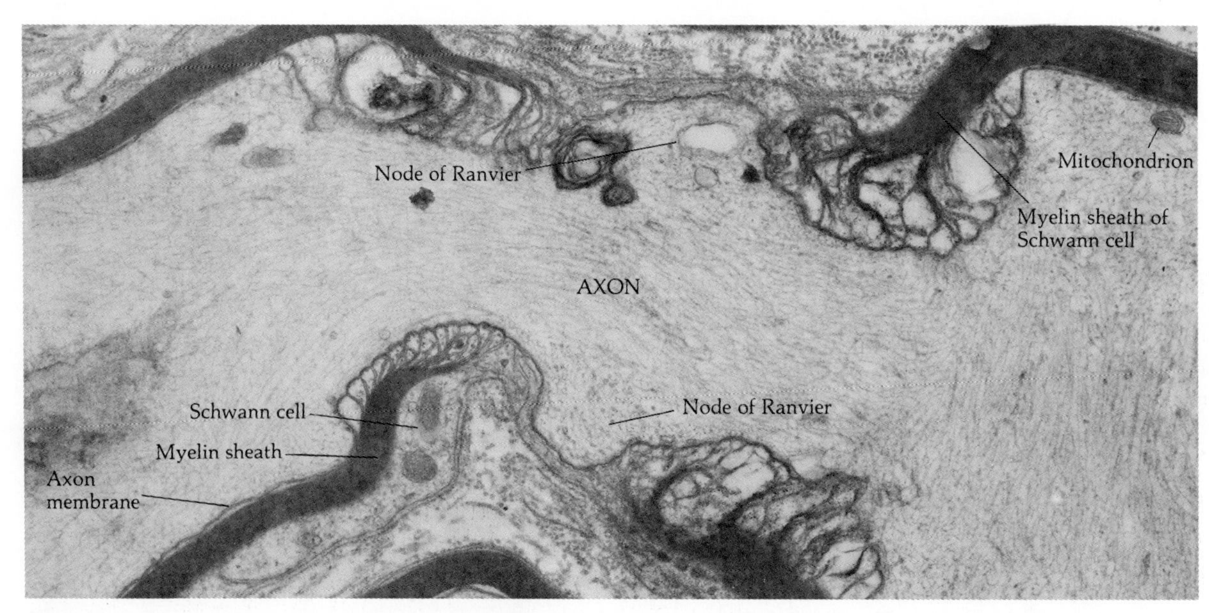

FIG. 6-8
A node of Ranvier in which the myelin coat of the nerve fiber is interrupted (×21,287).

of the axon. The intermittent gaps between Schwann cells are called the *nodes of Ranvier* (Fig. 6-8).

Schwann cell membranes contain the lipoprotein myelin, which is 70 to 80 percent lipid and 20 to 30 percent protein. The membranes may be wrapped many times around the axon (as a jelly roll may be wrapped around a central core of jelly) to form a myelinated fiber, as described above (Figs. 6-3B, 6-7B, and 6-9B).

In other instances, axons merely fit closely into indentations of Schwann cells. Such bundles of fibers contain significantly less myelin and are called *unmyelinated* fibers (Fig. 6-9A). All nerve fibers are unmyelinated in early development. The fatty myelin coat that surrounds some fibers begins its formation in about the fourteenth week of prenatal development. It proceeds as the nervous system matures. For example, a newborn child cannot walk since he or she does not yet have a complete myelin coat on motor fibers to the skeletal muscles.

Injury to a neuron may lead to deterioration

FIG. 6-9
Wrapping of Schwann cell in unmyelinated and myelinated fibers. (A) Each unmyelinated fiber merely lies in an indentation within the Schwann cell membrane. (B) The myelinated fiber is enveloped by several concentric layers of the Schwann cell membrane.

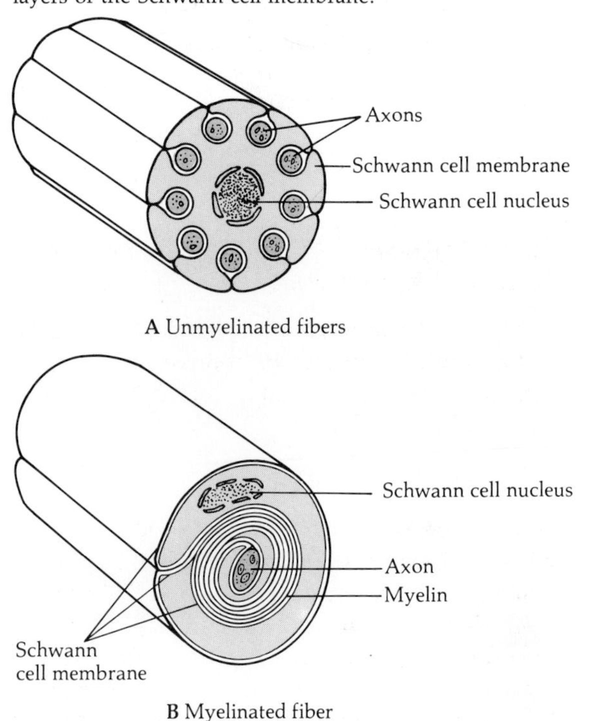

of that part of the axon distant from the cell body in a response called anterograde *Wallerian degeneration*. It may result in loss of function of the tissue or organ with which the fiber is associated. Regeneration of an axon is dependent on the presence of a healthy neurilemma. The neurilemma is the layer composed of the concentrically wrapped cell membranes of Schwann cells which form a protective shield around each myelinated axon in the PNS. The nucleus of each Schwann cell lies just underneath the neurilemma, which must be present for damaged peripheral nerve fibers to regenerate. Schwann cells actively proliferate and phagocytize (ingest) damaged components and serve as a tube that guides replacement of the axon. The equivalent supporting cells in the CNS, the *oligocytes* of the brain and spinal cord, cannot regenerate. Thus, damaged axons in the CNS are not repaired very readily.

Satellite Cells

The cell bodies of neurons are often held together in structures called *ganglia* (singular, ganglion). The connective tissue cells surrounding the cell bodies in ganglia are called *satellite cells*. They are found in ganglia of some cranial nerves, in the dorsal root of the spinal nerves, and in small numbers in autonomic ganglia.

Glia of the Central Nervous System (Brain and Spinal Cord)

Four types of supportive connective tissue cells can be identified in the central nervous system. They are ependymal cells, astrocytes, oligocytes, and microglia.

Ependymal Cells

Ependymal cells are ciliated epithelial cells that form the inner lining of interconnected cavities (ventricles) within the brain and the central canal of the spinal cord. Cerebrospinal fluid (CSF) fills these cavities and also bathes the exterior surface of the brain. The properties of CSF will be described in Chap. 10 on the brain.

Astrocytes (Astroglia)

Astrocytes are large cells possessing branched extensions (Fig. 6-10). They are phagocytic cells found between blood capillaries and neurons and act as guardians of the nervous tissue in the brain. They are supportive and help form the *blood-brain barrier*, which controls passage of materials between the blood and the extracellular fluid (ECF) around neurons in the brain. This barrier begins to form within the first three months of fetal life and continues to develop after birth through early childhood. The relative rates of passage of some common substances through the blood-brain barrier are shown in Table 6-1. Their passage is generally directly related to their solubility in lipids and inversely related to their molecular weight.

The blood-brain barrier maintains a rather constant environment for neurons in the brain. Its integrity is often crucial for survival. In denying certain substances ready access to the brain, the barrier maintains differences in the ionic composition of the blood and the ECF around its neurons. The composition of the ECF of the brain therefore resembles that of the CSF. The contents of CSF and blood plasma are compared in Table 10-2. The effects of drugs and infectious agents on the brain are also influenced by the blood-brain barrier. Certain drugs and microbial toxins do not penetrate it. This may influence the choice of drugs for therapy and the outcome of certain infections of the CNS. For example, penicillin has limited entry through the blood-brain barrier, while erythromycin gains access more readily.

Oligocytes (Oligodendroglia, Oligodendrocytes)

Oligocytes are connective tissue cells that possess long cytoplasmic extensions (about 100 μm). Oligocytes play roles in the CNS similar to those of Schwann cells in the PNS. Oligocytes are probably the most numerous cells in the CNS. Their extensions surround axons 1 to 20 times with petal-shaped processes, producing layers of myelin which insulate the axons. Because of the white appearance of myelin, regions of the CNS containing many myelinated axons are referred to as the *white matter*. Those regions which contain few myelinated fibers are relatively devoid of myelin and are referred to as the *gray matter*.

Microglia

Microglia are the smallest of glial cells. They are phagocytic and seem to enter nervous tissue by penetrating its capillaries. Although they may be macrophages which originate from monocytes of the circulatory system, some evidence indicates that they are merely a variety of oligocyte. Microglia are able to clear the central nervous system of foreign particulate material, aged cells, and cellular debris. They become very active following injury, chronic nerve degenerative diseases, cancer, and leukemia.

TABLE 6-1
Relative rates of permeation of the blood-brain barrier by some common substances

Rapid	*Intermediate*	*Extremely Limited*
Carbon dioxide	Glucose	Proteins
Oxygen	Na^+	Polysaccharides
Water	K^+	
Ethanol	Mg^{2+}	
	HPO_4^{2-}	
	HCO_3^-	
	Cl^-	

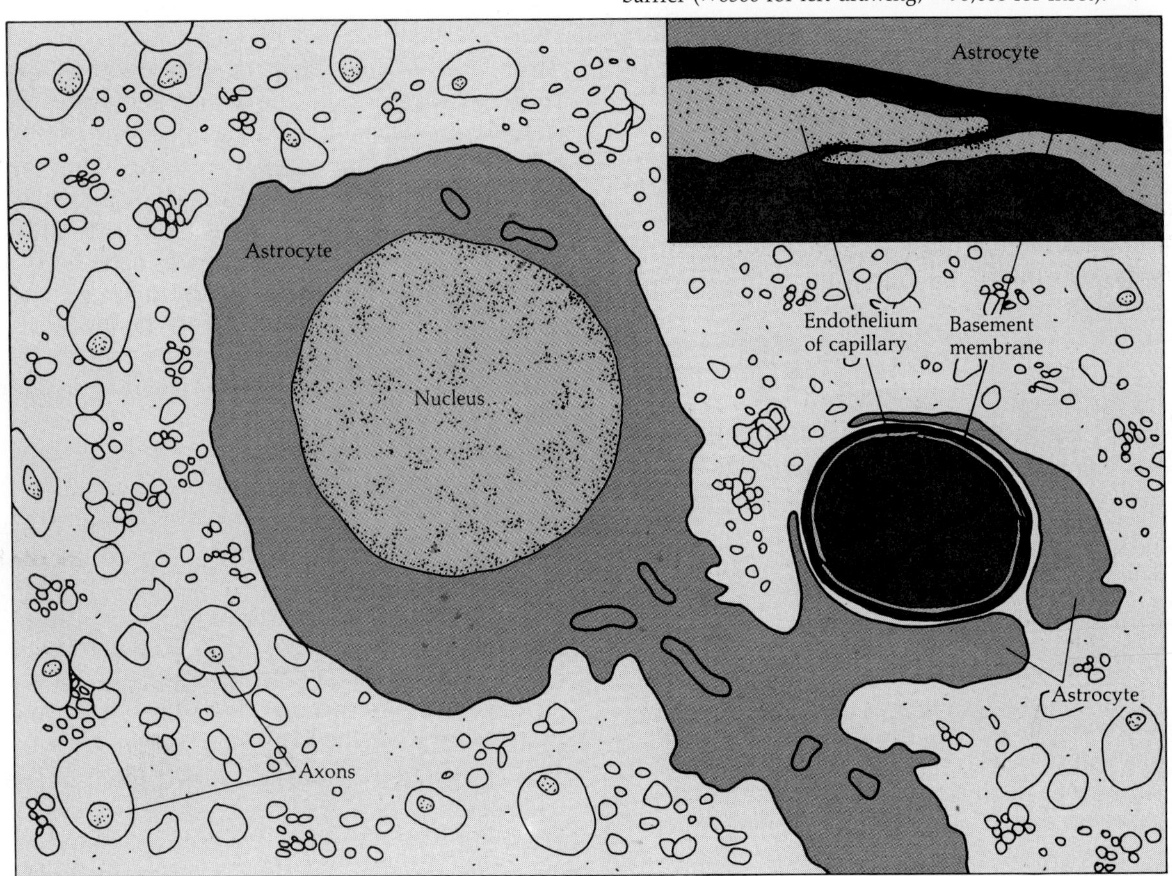

FIG. 6-10
The role of astrocytes and their basement membranes, which surround capillaries to help form the blood-brain barrier (×6500 for left drawing, ×90,000 for inset).

Demyelination of Fibers: Multiple Sclerosis

Degeneration of the myelin coat (*demyelination*) leads to diseases of the nervous system which are difficult to treat. *Multiple sclerosis* is a demyelinating disease in which defects occur in patches along the myelin sheath. Multiple sclerosis especially affects young adults, ages 18 to 40, causing degeneration of white matter in the brain and spinal cord. Over a long period of time it can progress to cause muscular weakness, loss of coordination of skeletal muscles, tremors, defective vision and speech, loss of control of the bladder or bowel, partial or complete paralysis, and other symptoms, including mental impairment and death. There is no cure for it yet. Some individuals show signs of improvement (remission), which, it is theorized, may be due to the nourishing and regenerative activities of Schwann cells.

Destruction of axons in the CNS in multiple sclerosis is followed by replacement with scar tissue composed of astrocytes and phagocytes. The cause of the disease is unknown, but some evidence indicates its onset may be due to activation of a previously latent (hidden) virus, which causes infection. Protein may be released from the fibers, leading to an immune reaction which destroys the tissue further. The susceptibility to the viral infection and allergic reaction may be inherited. In some instances damage is limited to nerves in the brain, causing *cerebral sclerosis*. Known causes of some other demyelinating diseases will be discussed in the section below on the role of nutrients in nervous tissue.

6-4 TYPES OF NERVES

Nerves may be distinguished from each other on the basis of: (1) their origin; (2) the direction in which they carry impulses; (3) the destination of the impulse; (4) their size and degree of myelination; and (5) the type of neurotransmitter released from their terminal endings. The first four properties will be described below while the fifth characteristic will be discussed in Chap. 7.

Origin of Nerves

The *cranial* nerves originate in or close to the brain and are associated primarily with tissues around the head. The *spinal* or *peripheral* nerves originate in the spinal cord and are connected to tissues and organs in more distant parts of the body.

Direction of Impulses

Efferent fibers carry impulses away from the central nervous system to tissues and organs. *Afferent* fibers relay impulses to the CNS. *Mixed* nerves contain both efferent and afferent fibers. They are a mixture of axons and dendrites.

Destination of Impulses

Efferent fibers to skeletal muscle are also called *motor or voluntary fibers,* since they control voluntary motor activities such as walking and talking. Those which extend to the body framework (for example, to the skeletal muscles) are also called *somatic* fibers. Fibers to smooth muscle, cardiac muscle, and glands are called *visceral* or *involuntary* fibers, since the activities of these structures are self-regulated by feedback mechanisms that are subconscious. Thus, the tone of the blood vessels in the body, the rate of heartbeat, and the activity of glands are not usually consciously regulated.

Many afferent fibers are *sensory* in function. That is, they deliver impulses to those centers in the brain that create conscious awareness of the transmitted information. Not all afferent fibers reach the conscious centers of the brain. Therefore, strictly speaking, not all afferent fibers are sensory in nature, although the terms are often used interchangeably.

Size and Myelin Content of Nerve Fibers

Nerve fibers can be classified according to size and degree of myelination. Group A fibers are

TABLE 6-2
Relationship between nerve fiber diameter, myelination, and function

Type of Fiber	*Diameter, μm*	*Velocity of Impulse,* m/sec	*Degree of Myelination*	*Susceptibility to Physical Trauma*	*Functions*
A	1–22	12–120	High	High	Motor and large sensory fibers
α (alpha)	12–20	70–120			Somatic (body) motor Proprioception (balance)
β (beta)		30–70			Touch, pressure
γ (gamma)	3–6	15–30			Motor to muscle spindles
δ (delta)	2–5	12–30			Pain, temperature, touch
B	3 or less	3–15	Medium	Medium	Preganglionic autonomic
C	2 or less	0.4–2.3	"Unmyelinated"	Low	Postganglionic autonomic, pain, reflexes, taste, smell

the largest and most myelinated. Group B fibers are intermediate in these properties, and those of Group C are the smallest and are "unmyelinated."

Group A fibers may be further subdivided according to size into α, β, γ, and δ types (Table 6-2). Another system of classification exists for sensory fibers which separates them into groups I, II, III, and IV.

The velocity of nerve impulses, susceptibility to physical trauma, and amount of time required for regeneration after injury all increase in direct relationship to the myelin content and diameter of the fiber. Following limited physical trauma, damage is greater to the large motor fibers in a nerve and less to its smaller δ pain fibers, which regenerate more rapidly. Because of these properties pain is felt sooner after trauma and during recovery, before motor function is regained. The relationships between diameter, myelination, and function in nerve fibers are summarized in Table 6-2.

Increased myelination and fiber size enhance the velocity of nerve impulses. Of the two factors, myelination and fiber size, the degree of myelination has more influence on the velocity of nerve impulses. As mentioned earlier, myelin acts much like insulation surrounding an electrical cable. It prevents ions from leaking across the axonal membrane and thereby makes it easier for the neuron to maintain an electrical gradient of ions. As will be seen in Chap. 7, the passage of a nerve impulse is caused by the flow of positive ions down their gradient into and along the axon. The movement of ions creates an electrical current or impulse. The current flow in myelinated fibers occurs only at the nodes between Schwann cells, where myelin is not abundant. Therefore, the impulse behaves much like an electric arc which jumps over a gap rather than like a continuous flow along the entire length of a wire. The current hops or skips rapidly from node to node in myelinated fibers in a process called *saltatory conduction*. Since an unmyelinated axon does not possess a thick coat of myelin, the current must flow along its entire length. It does not hop, skip, or jump. Therefore, the rate of movement of the impulse is slower.

6-5 THE ROLE OF NUTRIENTS IN THE HEALTH OF NERVOUS TISSUE

Metabolic pathways in neurons generally are similar to those found in other cells, but the requirements for certain nutrients are especially pronounced. Brain neurons use glucose as their primary energy source in metabolism. Fats are a secondary energy source and are also used to synthesize the myelin sheath; amino acids are used to build neurotransmitters and neurofibrils; and vitamins are used to maintain the metabolism and health of the entire nervous system.

Carbohydrates

Peripheral nerves utilize carbohydrates as well as noncarbohydrates for energy. However, glucose constitutes the main source of energy in the brain, which requires a continuous supply of the sugar. This organ, which accounts for only 2 percent of body weight, has a high rate of energy consumption. It uses 20 percent of oxygen available to the body and can consume 70 percent of the body's glucose. Thus, brain activity depends on the availability of blood glucose. Fats seem to serve as alternate sources of energy in the brain when glucose is deficient. This may occur following prolonged use of noncarbohydrate diets or starvation.

Energy is especially used to drive three types of reactions in the nervous system:

1. Reactions that lead to the synthesis of ATP. (Much of the energy released later from the hydrolysis of ATP is used for the active transport of Na^+ and K^+ ions.)
2. Synthesis of neurotransmitters (substances necessary for the transmission of impulses from a neuron to another cell).
3. Synthesis of proteins which form active transport enzymes and structural elements such as neurofibrils.

Lipids

Lipids are important components of the nerve cell membrane. The myelin sheath around many nerve fibers constitutes about 80 percent of the dry weight of lipids in nerve. Alterations in these molecules may cause noticeable changes in the CNS and PNS. Abnormalities in lipid composition may have a marked effect on conduction of impulses and the physiology of neurons. The disorders may be hereditary or acquired.

Hereditary Lipid Disorders

Heritable disorders in lipid metabolism may be passed on by recessive genes on the autosomes or on the X chromosomes (as sex-linked traits). Each disease involves a genetic defect, which causes a deficiency in a specific enzyme and accumulation of a particular lipid. The enzymes are unable to catalyze essential reactions. Consequently, the lipids are not broken down but accumulate within the lysosomes of cells. The diseases are referred to as *lipid-storage diseases, lysosomal diseases,* or *lipodoses.* The most common abnormality involves accumulation of *sphingolipids.*

Several heritable disorders affecting myelin formation have been discovered. Each can lead to mental retardation and a variety of other symptoms. Three examples are shown in Table 6-3.

Acquired Lipid Disorders

Acquired nervous disorders involving myelin may follow: (1) allergic and/or infectious diseases; (2) exposure to toxic substances; or (3) trauma. Demyelination may occur following different viral infections, e.g., measles, mumps, or influenza, or vaccinations against them. Damage to nerves from viruses is able to lead to an immune response by the body to its own nervous tissue, resulting in *allergic encephalomyelitis.* Multiple sclerosis, too, may involve an allergic response that follows some viral infections. Both of these disorders are examples of autoimmune diseases.

Accumulation of toxic levels of substances also can cause acquired myelin disorders, resulting in a variety of physical and nervous symptoms. Hexachlorophene (a bactericidal agent previously available in many soaps) may cause partial paralysis in newborn children. Carbon dioxide poisoning may cause disturbances in motor behavior and intellect. Alcoholism (coupled with malnutrition) may cause mental confusion, impairment in speech, mental retardation, and partial paralysis.

Physical trauma may cause damage to myelin by compression of nerves and/or edema. Dis-

TABLE 6-3
Examples of heritable lipid storage disorders in the nervous system

Disease	*Stored Lipid*	*Enzyme Deficiency*	*Symptoms in Addition to Mental Retardation*
Tay-Sachs	GM_2 ganglioside	GM_2 hexoseaminidase	Muscular weakness, blindness, occurs predominantly in Jews of eastern European origin
Gaucher's	Glucocerebroside	Glucocerebrosidase	Enlargement of spleen and liver, erosion of long bones in pelvis
Niemann-Pick	Sphingomyelin	Sphingomyelinase	Enlargement of spleen and liver

turbances in circulation (e.g., following application of a tourniquet) or tumors may also lead to the destruction of myelin.

Proteins

Amino acids form the building blocks of proteins in intracellular membranes, neurofibrils and other cytoplasmic organelles, and some amino acids also are building blocks of a number of neurotransmitters. Therefore, amino acid deficiencies may contribute to a variety of acquired neurological disorders. For example, norepinephrine, a compound derived from the amino acid tyrosine, is released from certain neuron terminals to affect target cells. In this setting it acts as a neurotransmitter. However, it acts as a hormone when produced in the adrenal medullae and is carried long distances through the blood to affect target cells, as described in Chap. 5. Tryptophan is another amino acid important to nerve cell physiology. It is a common building block in protein and a precursor of the pineal hormone melatonin and of serotonin, a neurotransmitter in the brain.

A heritable amino acid disorder that can cause mental retardation in newborn children is *phenylketonuria* (PKU). It is caused by a recessive gene, which occurs in about 1 in 10,000 children of northern European origin. Individuals with two recessive genes fail to produce the enzyme phenylalanine hydroxylase, causing a deficiency in the neurotransmitter dopamine as follows:

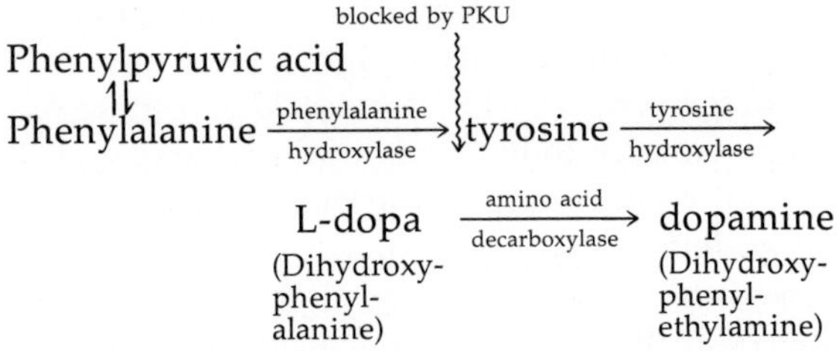

blocked by PKU

Phenylpyruvic acid
⇅
Phenylalanine —phenylalanine hydroxylase→ tyrosine —tyrosine hydroxylase→

L-dopa (Dihydroxyphenylalanine) —amino acid decarboxylase→ dopamine (Dihydroxyphenylethylamine)

In the absence of phenylalanine hydroxylase in the liver, high levels of phenylpyruvic acid accumulate and appear in the body fluids. Phenylpyruvic acid is a keto acid (an acid with a $-\overset{\overset{\displaystyle O}{\|}}{C}-$ group), whose appearance in the urine is responsible for the name of the disorder. The abnormal accumulation of catabolites of phenylalanine such as phenylpyruvic acid during the first 6 years of life is associated with inhibition of processes necessary for maturation of the brain. The specific cause of mental retardation in PKU is as yet unknown. Diets restricted in phenylalanine decrease brain injury in children with the genetic defect.

The complex interrelationships among diet, neuron function, and neurotransmitters such as serotonin are also under study. Evidence has recently been presented to indicate that diet may alter neuron activity in the brain, which in turn may influence behavior. For example, a carbohydrate-rich diet with adequate but reduced quantities of protein has been reported to favor production of high levels of serotonin in the brain. In this manner, the chemistry of foods may affect neuron biochemistry, which in turn controls the formation of neurotransmitters. The latter may complete the cycle by affecting a variety of physiological functions, including the desire for and consumption of foods.

Vitamins

Vitamins are essential nutrients required for specific metabolic activity. They play an important role, assisting enzymes as cofactors (helpers) required for specific metabolic reactions in living cells (see Appendix B, Chap. 23).

Deficiencies in specific vitamins may cause inflammation of the fibers within a nerve, a condition called *polyneuritis*. The symptoms depend on the function of the affected nerve. Seizures and diminished mental functions, including loss of memory and impaired mental ability, may result. The area supplied by the nerve can show loss of sensation, paralysis, or atrophy of muscle activity. The effects of deficiencies in vitamins B_1, B_6, B_{12}, and nicotinic acid are shown in Table 6-4. The significant effects of aberrant genes and metabolism on nervous system function are more properly the subject of study in neurophysiology, neurobiochemistry, and genetics. The effects of vitamins and their relation to nutrition will be discussed more fully in Chap. 23.

TABLE 6-4
The effects of some vitamin deficiencies on nerve function

Vitamin	*Symptoms of Deficiency*
B_1 (thiamine)	Loss of control of muscles to outside of eyeball (extraocular muscles)
	Polyneuritis
	Diminished memory (disease is called *Wernicke-Korsakoff syndrome*)
B_6 (pyridoxine)	Decreased ATP synthesis in neuron
	Adverse affects on active transport of ions
	Seizures
B_{12} (cobalamin)	Degeneration of spinal cord and peripheral nerves
	Loss of sense of balance
	Spasticity (overactive contraction of muscles)
	Dementia (decreased mental ability)
Nicotinic acid	Dermatitis
	Dementia
	Polyneuritis (disease is called *pellagra*)

6-6 SUMMARY

1 The integration of body function is achieved largely through the cooperation of the brain and spinal cord, which form the central nervous system (CNS). The CNS contains a maze of pathways that convey impulses between it and the rest of the body over the peripheral nervous system (PNS). The cell type responsible for generation of impulses is the neuron. It possesses cytoplasmic extensions bundled together to form nerve fibers.

The cell body of a neuron is an enlarged portion ranging in size from 5 to 100 μm. The nucleus is within the cell body and contains one or more prominent nucleoli.

Some cytoplasmic organelles in the neuron are highly specialized. Nissl granules, modifications of the rough ER, are active in protein synthesis. Mitochondria are especially numerous at nerve terminals. The Golgi apparatus probably packages material for transport to the terminal portion of each neuron. Neurofibrils form neurofilaments and neurotubules, which may convey chemicals through neurons at rates exceeding 400 mm/day, and microfilaments, which may be active in forming connections between neurons and other cells.

Axons and dendrites are two types of cytoplasmic processes arising from the cell body. The axon ranges from a few to 900,000 μm in length. It usually transmits signals away from the cell body. Dendrites are shorter branches, which generally convey signals toward the cell body.

The cytoplasmic processes arise from one, two, or several points of origin as stalks in unipolar, bipolar, and multipolar neurons, respectively. Unipolar and bipolar neurons are the primary afferent cells which convey signals to the CNS. Multipolar neurons include motor neurons and interneurons.

2 Nerves consist of bundles of axons and dendrites derived from various neurons, surrounded by three separate connective tissue coats, the epineurium, perineurium, and endoneurium. The latter surrounds the outer sheath, the neurilemma, of each fiber in a nerve.

3 The axons of many neurons are wrapped together by 10 times the number of connective tissue cells, called glia. The glial cells in the PNS are Schwann cells and satellite cells. Whole Schwann cells and their cytoplasmic membranes are wrapped around axons to form an overcoat containing a large amount of a lipoprotein called myelin. Such fibers are said to be myelinated. The presence of myelin acts as an insulator and facilitates a more rapid passage of nerve impulses. The portion of the Schwann cell membrane wrapped around the axon is termed the *neurilemma*. Its presence is necessary for the health of axons and their ability to regenerate when damaged. Satellite cells glue the cell bodies of neurons together in clusters called ganglia.

Four types of glial cells are found in the CNS. Ependymal cells form the inner lining of cavities in the CNS which contain CSF. Astrocytes are phagocytic cells, which form the selectively permeable blood-brain barrier between neurons and their surrounding blood vessels. Extensions of oligocytes are wrapped around axons to form myelinated fibers in the CNS. Microglia are small phagocytic cells scattered through the CNS.

Demyelination of fibers occurs in a variety of nervous disorders. In multiple sclerosis, loss of myelin occurs in patches, which become infiltrated with scar tissue formed by astrocytes and oligocytes.

4 The velocity of nerve impulses, the susceptibility to trauma, and the amount of time required for regeneration after injury increase in direct relation to the diameter of the fiber and its myelin content. The latter two properties are greatest in type A fibers, intermediate in type B fibers, and lowest in type C fibers.

5 The requirements for certain nutrients in neurons are especially pronounced. These include a disproportionate need for glucose to supply energy in the brain and the needs for fats to form myelin, for amino acids to manufacture neurofibrils and some neurotransmitters, and for vitamins to promote health of neurons. Nervous disorders can be inherited or acquired. Several heritable disorders have been established in which lipid storage and myelin synthesis are abnormal. Acquired disorders can be caused by allergic or infectious diseases, toxic substances, physical trauma, and malnutrition.

REVIEW QUESTIONS

1 Describe the cell body of the neuron, Nissl granules, and three types of neurofibrils that occur in neurons.
2 Compare axons and dendrites in terms of size, structure, and function.
3 Distinguish unipolar neurons from bipolar and multipolar neurons.
4 Name the three layers of connective tissue that bind nerve fibers together and describe the location of each.
5 What is the overall function of glial cells in nerve tissue?
6 What are the differences between Schwann cells and satellite cells? What is the neurilemma and why is it important?
7 Discuss the properties and functions of ependymal cells, astrocytes, oligocytes, and microglia.
8 What is myelin, where is it found, and what are its functions?
9 What are the major differences among A, B, and C nerve fibers? How do these properties relate to the speed of impulses and regeneration of a nerve after trauma?
10 Explain the need for glucose in brain metabolism.
11 How are lipids important in nerve physiology? Describe three hereditary defects leading to lipid storage disorders and list two causes of acquired nerve disorders.
12 What role do proteins play in nerve cell physiology?
13 Of what necessity are vitamins to the health of nerve tissue? Describe the effects on nerve function of deficiencies in choline, vitamins B_1, B_6, and B_{12}, and nicotinic acid.

PHYSIOLOGY OF THE NERVOUS SYSTEM

Transmission of Information from Neurons

7-1 DEVELOPMENT OF IMPULSES BY NEURONS
The resting membrane potential: an electrochemical equilibrium
Calculation of the equilibrium potentials of individual ions: The Nernst equation
Measured resting potential
The action potential
Flow of ions
All-or-none law
Phases in development of the action potential and their measurement
Gate model of ion flow
Requirements to develop an action potential
Velocity of nerve impulses
Relationship between different types of potentials

7-2 JUNCTIONS BETWEEN NEURONS: THE SYNAPSE
General structure and function of boutons and synaptic vesicles
Role of calcium
Properties of synaptic membranes
Types of synaptic circuits

7-3 FUNCTIONS OF NEUROTRANSMITTERS
Development of postsynaptic potentials
Fate of neurotransmitters
Interactions between neurotransmitters and receptor molecules on the cell membrane
Examples of neurotransmitters
Catecholamines
Serotonin
Glutamic acid
Gamma-aminobutyric acid

7-4 SUMMARY

OBJECTIVES

After completing this chapter the student should be able to:

1. Describe the ionic state of a nerve cell membrane when it is not conducting an impulse.
2. Discuss the roles of diffusion, selective permeability, and active transport pumps in the maintenance of an electrochemical equilibrium across nerve fibers.
3. Explain what an action potential is and how it functions in the conduction of nerve impulses. List the requirements needed to develop an action potential.
4. Name and explain the factors that influence the speed of nerve impulses, including the role of the nodes of Ranvier in the transmission of impulses.
5. Define the term synapse, describe the structure and function of boutons and synaptic vesicles, and compare type I and type II synapses.
6. Compare the events that lead to excitatory and inhibitory postsynaptic potentials.
7. Outline the general functions of neurotransmitters, including the possible role of Ca^{2+} ions in neural transmission.

The body's remarkable ability to adapt to external and internal changes and to maintain homeostasis is especially related to the function of the nervous system. This capacity is due in large measure to *excitability* of neurons, that is, to their ability to respond to stimuli. It leads to the formation of messages or signals called *impulses.* These are changes in the electrical properties of the cell membranes that are transmitted to cells, tissues, and organs of the body and that cause them to respond and to do so in predictable ways.

7-1 DEVELOPMENT OF IMPULSES BY NEURONS

Differences in the composition of the extracellular fluids (ECF) and intracellular fluids (ICF) are particularly important in the generation of impulses. As noted earlier in Chap. 4, Na^+, Ca^{2+}, Cl^-, and HCO_3^- ions are usually found in higher concentrations in the ECF, while K^+ ions and negatively charged organic (protein) ions are more concentrated in the ICF (Fig. 7-1). The unequal distribution of specific ions re-

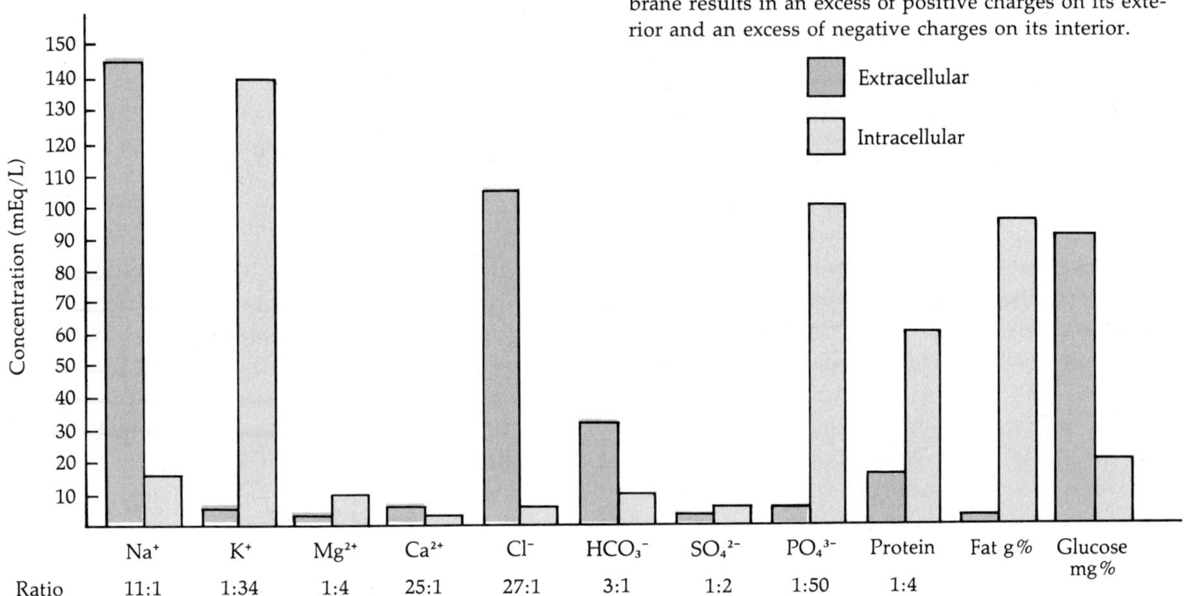

FIG. 7-1
Composition of extracellular and intracellular fluids. An unequal distribution of ions on each side of the membrane results in an excess of positive charges on its exterior and an excess of negative charges on its interior.

sults from differences in their rates of diffusion and active transport systems. Neurons, like other cells, maintain uneven concentrations of certain substances across their cell membranes. The unequal distribution results in an excess of positive charges on the exterior surface of the resting neuron cell membrane and an excess of negative charges on its interior surface.

The Resting Membrane Potential: An Electrochemical Equilibrium

The differences in the ion concentration and net (total) charge across the membrane have a significant effect on neuron function. Two types of gradients exist. One is a concentration (chemical) gradient, the other is electrical.

A *concentration gradient* exists when there is a difference in the amount of a substance per milliliter of fluid on each of two sides of a membrane. Substances tend to move along such gradients from an area of higher concentration into an area of lower concentration, as shown in Fig. 7-2A. Prior to passage of the substances, unequal concentrations produce a force, a potential for movement across the membrane. The movement is called *diffusion*. It is directly related to the size of the concentration gradient and inversely related to the square root of the molecular weight of the substance. The factors affecting diffusion were more fully explained in Chap. 4.

An *electrical gradient* exists when there is a difference in the sum total of charges across the membrane (inside and outside it). Like charges repel one another (as do two similar poles of a magnet). Opposite charges attract one another (as do opposite poles of a magnet). The sum total of electrical attraction and repulsion between charged particles may produce a force, a potential for their movement, termed *voltage*.

An excess of diffusible particles of similar charge on one side of a permeable membrane tends to cause their net movement across it as shown in Fig. 7-2B. Their movement may be assisted by the mutual attraction ot oppositely charged particles as shown in Fig. 7-2C. If the attractive forces between oppositely charged particles equals the opposing forces produced by a diffusion gradient, no net movement of ions occurs. An *electrochemical equilibrium* exists: the rate of movement of diffusible particles is equal across the two sides of the membrane. In excitable cells an electrochemical equilibrium results when a concentration gradient is opposed by an equal force of an electrical gradient. Although as many particles pass across one side of the membrane as the other in a given time, there is no net change in their distribution.

A cell usually attains a balance of electrical charges (an ionic equilibrium), as well as a balance of the concentration of particles (a chemical equilibrium), on each side its cell membrane. The *electrochemical equilibrium* is maintained through the expenditure of energy by active transport enzymes (pumps). Unlike the equilibrium shown in Fig. 7-2, that attained in a cell resembles a seesaw which is balanced unevenly as long as energy is available. The electrochemical equilibrium across a cell membrane usually maintains an excess of positive charges outside the membrane and an excess of negative charges inside the membrane. The membrane is polarized.

The voltage can be measured by placing microscopic electrodes from a voltmeter on each side of the membrane. The electrode on the outside is normally a reference electrode and that inside the cell is an active or measuring electrode. The needle on the voltmeter moves from zero (no voltage or potential) to the negative side of the scale when the electrodes are placed on opposite sides of the excitable membrane (Fig. 7-3). The measured "resting" voltage records a negative value, which averages -70 to -90 mV (millivolts) when the inside of the neuron's membrane is compared with its exterior. This measured voltage is called the cell's *resting potential*, that is, the voltage existing across two parts of a membrane of a cell that is at rest. The resting potential is equal to the attractive force that exists between opposite charges across the membrane which is not transmitting electrical signals or impulses. The term potential is used in the electrical sense, as a synonym for voltage, since opposite charges when separated have the ability to do work.

A few types of ions play major roles in the de-

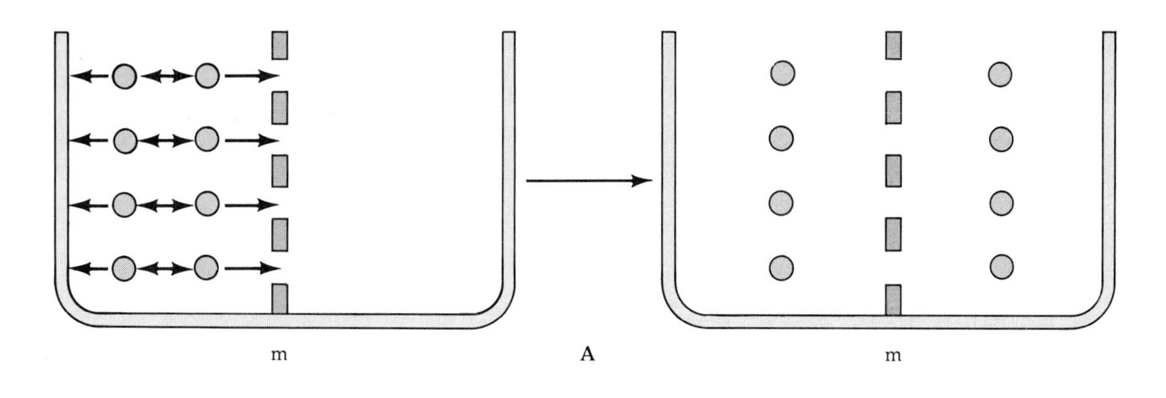

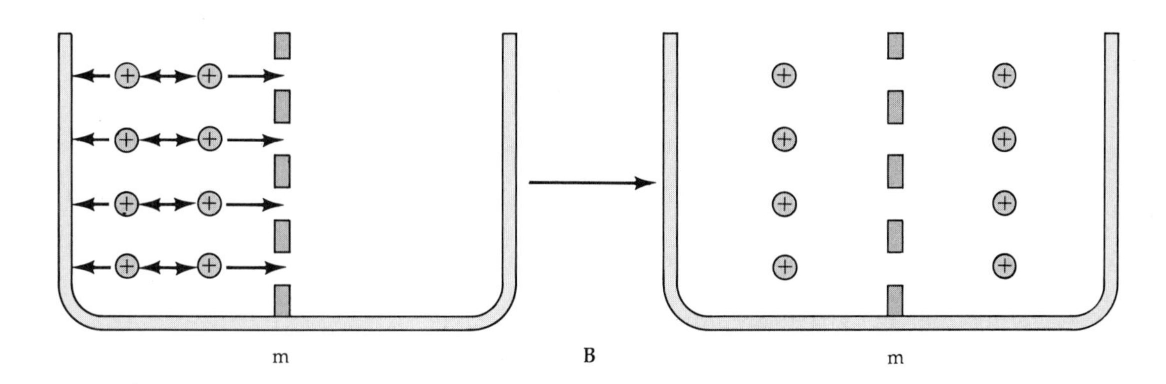

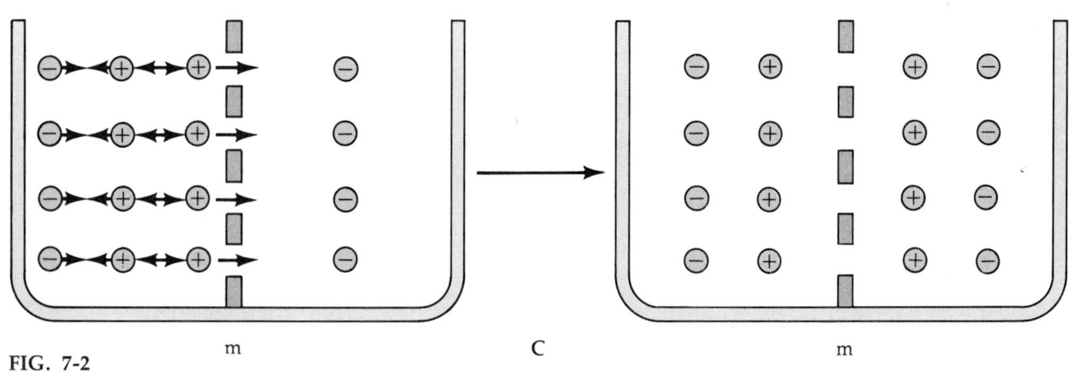

FIG. 7-2
Movement of particles down gradients to attain equilibrium. Two sides of each chamber are shown with particles in solutions separated by a selectively permeable membrane (m). The chamber on the left shows an existing gradient. That on the right illustrates the same chamber after equilibrium has been attained. Direction of movement is indicated by the arrows. (A) Movement of uncharged particles down a concentration gradient. (B) Movement of positively charged particles down an electrical gradient. (C) Movement of charged particles down a concentration gradient and electrical gradient.

velopment of the resting potential. About 90 percent of the Na^+ and Cl^- ions immediately available to the resting neuron are found on its exterior surface (Table 7-1, Fig. 7-4). Nearly 97 percent of the K^+ ions, as well as an excess of negatively charged anions such as protein and other organic anions, are found in its interior (Fig. 7-4). If all ions diffused freely, equal concentrations of them might be attained across

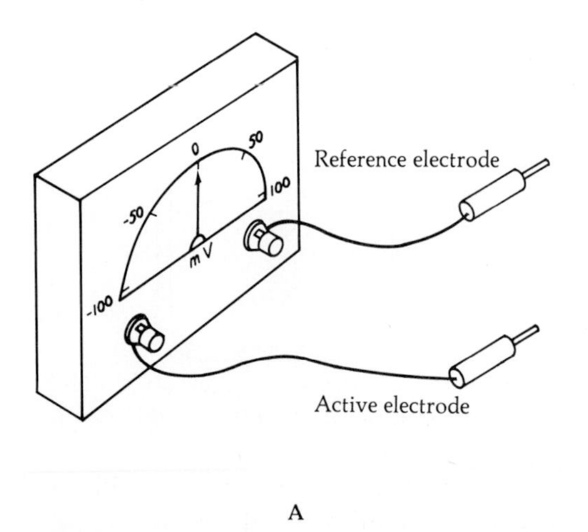

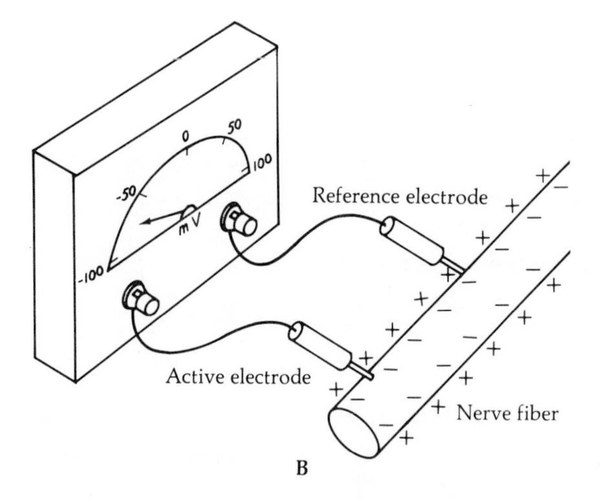

FIG. 7-3
Measurement of the resting potential. (A) A voltmeter registering no voltage (0) is shown with its reference and active (measuring) electrodes. (B) Placement of the reference electrode on the outside of the axonal membrane and the active electrode inside the membrane causes the needle to move to −70 mV. This is the measured resting potential of the cell.

the neuron's cell membrane. However, hydrated Na^+ ions are larger than hydrated K^+ ions (0.5 vs. 0.4 nm diameter, respectively). Moreover, the larger Na^+ ions diffuse inward much more slowly through pores in the membrane, their rates being about one-fiftieth to one-hundredth those of K^+ ions. Furthermore, active transport enzymes rapidly pump Na^+ ions outward to the ECF from the inside of the cell (as described in Chap. 4). Therefore, concentration *and* electrical gradients exist because of the selective permeability of the membrane and the effects of active transport enzymes.

FIG. 7-4
Electrochemical equilibrium across a theoretical neuron cell membrane. Typical concentrations of commonly observed charged particles are presented in mM/L. The equilibrium results in neutrality on each side of the membrane but inequality of net charges across it. There is an excess of sodium ions on the exterior and an excess of negative charges on the interior.

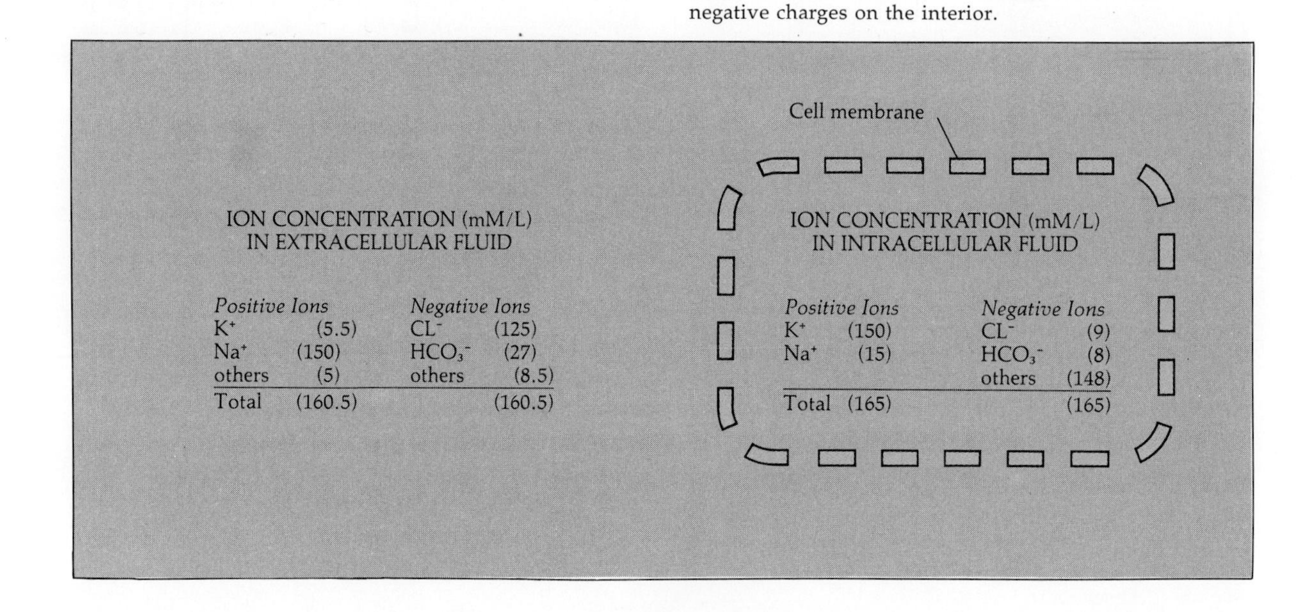

ION CONCENTRATION (mM/L) IN EXTRACELLULAR FLUID

Positive Ions		*Negative Ions*	
K^+	(5.5)	CL^-	(125)
Na^+	(150)	HCO_3^-	(27)
others	(5)	others	(8.5)
Total	(160.5)		(160.5)

ION CONCENTRATION (mM/L) IN INTRACELLULAR FLUID

Positive Ions		*Negative Ions*	
K^+	(150)	CL^-	(9)
Na^+	(15)	HCO_3^-	(8)
		others	(148)
Total	(165)		(165)

The Na^+ ions found in excess outside the cell attract the negatively charged Cl^- ions, which are able to diffuse freely across both sides of the membrane. The K^+ ions, which accumulate inside the cell by active transport, are attracted to nondiffusible negatively charged organic anions such as proteins.

The fluid on each side of the cell membrane is electrically neutral since the positive and negative charges are in balance (Fig. 7-4). However, the interior has a greater total of charges (positive and negative) than the exterior. As mentioned earlier, the K^+ ions, which diffuse more freely than Na^+ ions, accumulate in excess within the cell because of the work of active transport enzymes which carry the potassium into it. An excess of negative ions accumulate in the interior mainly through the buildup of negatively charged organic anions to which the membrane is impermeable, and to a lesser extent through the accumulation of small negative ions such as Cl^- ions. As a result, *the interior of the membrane acquires a negative potential* relative to the exterior. The membrane is *polarized.* It has a positive and a negative side (or pole).

Calculation of the Equilibrium Potentials of Individual Ions: The Nernst Equation

The net negative charge that the interior of the membrane possesses generally ranges from −70 to −90 mV compared to its exterior. The contribution to voltage or potential difference by unequal accumulation of individual univalent ions can be calculated by using the simplified Nernst equation:

$$E = 61 \log_{10} \left(\frac{C_E}{C_I}\right)$$

where

E = potential in millivolts of the exterior compared to the interior for an ion species at 37°C

C_E = concentration of the univalent ion in the extracellular fluid at equilibrium

C_I = concentration of the univalent ion in the intracellular fluid at equilibrium

The equilibrium potentials for each diffusible ion may be calculated as shown in Table 7-1.

The *equilibrium potential* is the electric force (the potential or voltage calculated in millivolts) required to oppose the natural direction and force created by the concentration gradient of the ion in question. At the voltage of the equilibrium potential, passive diffusion of the ion occurs at equal rates in both directions and equilibrium is maintained.

Measured Resting Potential

The measured voltage across the membrane is not merely the sum of the differences which can

TABLE 7-1
Ion concentrations and equilibrium potentials of a theoretical neuron at 37°C

	CONCENTRATION, mM/L			
Ion Species	*Extracellular,* C_E	*Intracellular,* C_I	$E = 61 \log_{10} \left(\frac{C_E}{C_I}\right) =$	*Equilibrium Potential,* mV
Cations				
K^+	5.5	150	= 61 log(5.5/150)	−88
Na^+	150	15	= 61 log(150/15)	+61
Others	5	——	——	——
Anions				
Cl^-	125	9	= 61 log(125/9)	−70*
HCO_3^-	27	8	= 61 log(27/8)	−32*
Others	8.5	148†	——	——

* The formula provides answers with a positive sign for Cl^- and HCO_3^-. The sign is reversed, however, since both are negative ions (anions).

† This number includes a high concentration of organic anions.

be calculated for the distribution of individual species of ions. Rather, it is a reflection of the imbalance in distribution of all ions, those that can be accurately measured individually and those that cannot. The contribution of each type of ion to the membrane's resting potential depends upon two factors: (1) the concentration of that ion on both sides of the membrane, that is, its concentration gradient; and (2) the permeability of the cell membrane to the ion.

As described above, an excess of positive ions is located outside the resting neuron and a surplus of negative ions within it. The slight but measurable unequal distribution of ions results in polarization of the membrane. The total measurable voltage across the membrane is its resting potential. The significance of the Nernst equation is illustrated by the data in Table 7-1. For example, it indicates that in order to prevent the outward diffusion of K^+ ions, the resting potential of the neuron must equal the K^+ equilibrium potential of −88 mV. Should the resting potential be less (for example, should it measure −70 mV inside compared with the outside), a net flow of K^+ ions along its concentration gradient to the exterior would occur. Active transport would be necessary to maintain an electrochemical equilibrium. As the differences between the resting potential of the cell and the equilibrium potentials of specific ions increase, neurons use more energy to maintain equilibrium through active transport mechanisms.

The Action Potential

An *action potential* or nerve impulse is a temporary change in the electrical properties of the membrane conducted along its entire length. A stimulus must attain a minimal intensity or *threshold* to cause an action potential. A *subthreshold stimulus* causes local changes which do not proceed along the entire membrane. An analogy to these conditions may be made by comparing a row of dominoes to the positive charges which move along the membrane during development of an action potential. A certain amount of energy (a threshold stimulus) is necessary to push the first domino over, so that when it topples the whole row will fall in sequence (just as the action potential, once started, spreads along the entire membrane). Less energy or push may cause the first domino to wobble back and forth or even topple sideways but not do so in a way that causes other dominoes in the row to fall. So too, a subthreshold stimulus causes local changes in the membrane which are not conducted along its entire length. Energy is required to restore the dominoes (charged particles) to their original positions. This is equivalent to the energy needed for active transport to reestablish the original electrochemical equilibrium, the resting potential. More energy is required to reset the dominoes if the whole row topples (as when an action potential occurs); less is needed when only one or a few dominoes wobble or are pushed sideways so as not to topple all their neighbors. The latter situation resembles localized changes which occur following the application of a subthreshold stimulus.

Flow of Ions

The application of a stimulus to a neuron's cell membrane increases its permeability to Na^+ ions. The Na^+ ions enter the membrane by moving through channels (pores) down the ion's electrochemical gradient. The membrane thus becomes less polarized, and when sufficient Na^+ ions have entered, it becomes *depolarized.* At this time, a comparison of the inside and outside of the membrane shows it to be electrically neutral.

The loss of polarity occurs at the point of application of the stimulus and to its left and right, as shown in Fig. 7-5. The sodium channels open at the site of the stimulus, allowing entry of Na^+ ions (Fig. 7-5B). Na^+ ion entry ultimately transforms the interior of the membrane in that area from a negative state to a positive one (Fig. 7-5C). Since like charges repel one another, the inward diffusion of Na^+ ions causes a flow of positively charged ions away from the initial point of entry, as shown sequentially in Fig. 7-5D to F. This flow of positive ions constitutes a current.

Sodium ions thus move down their electrical

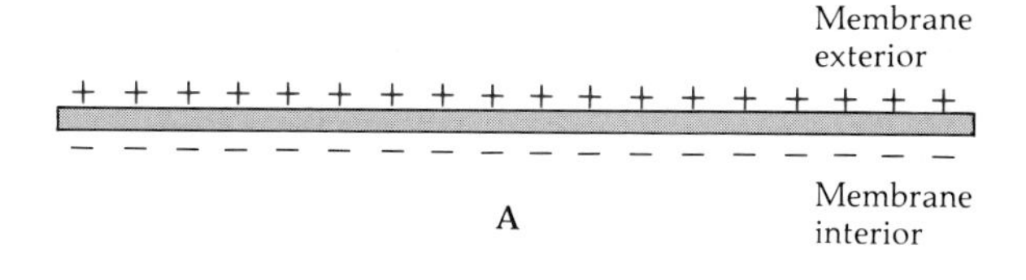

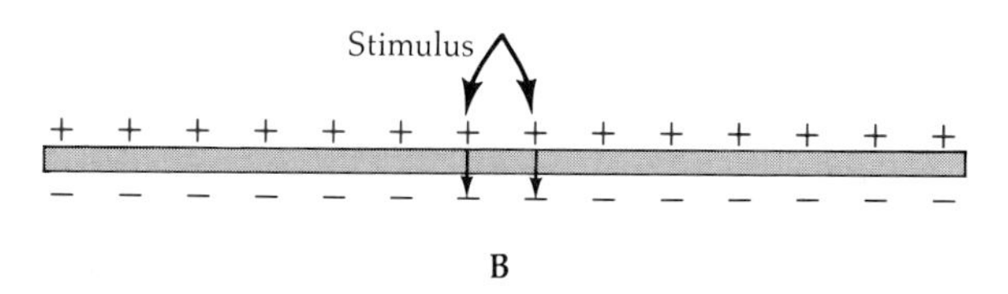

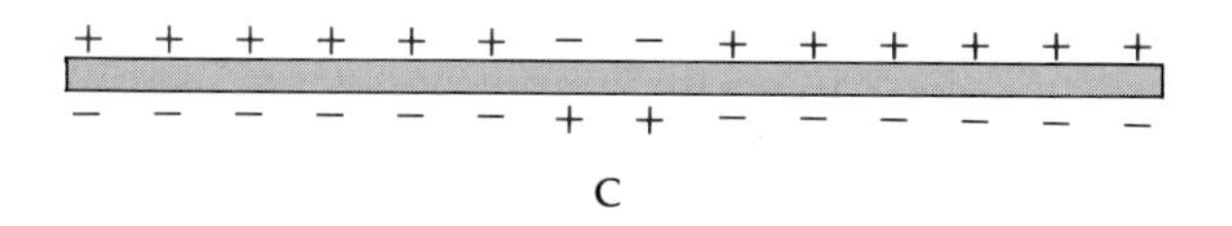

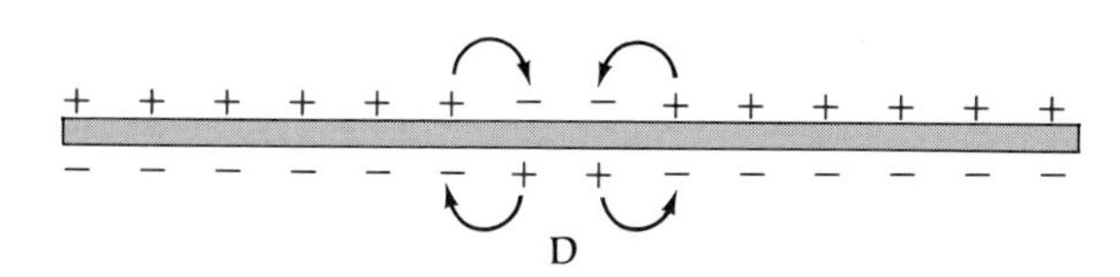

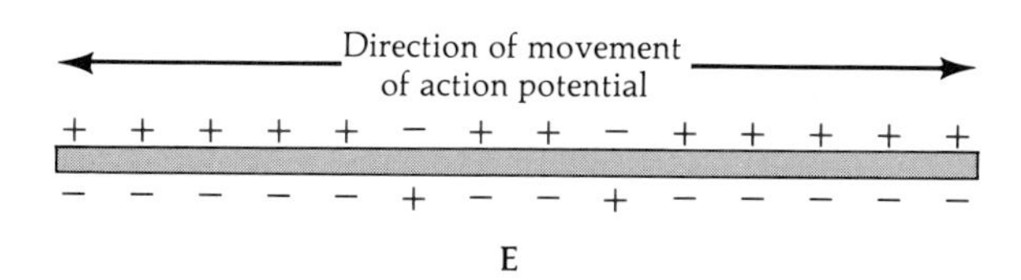

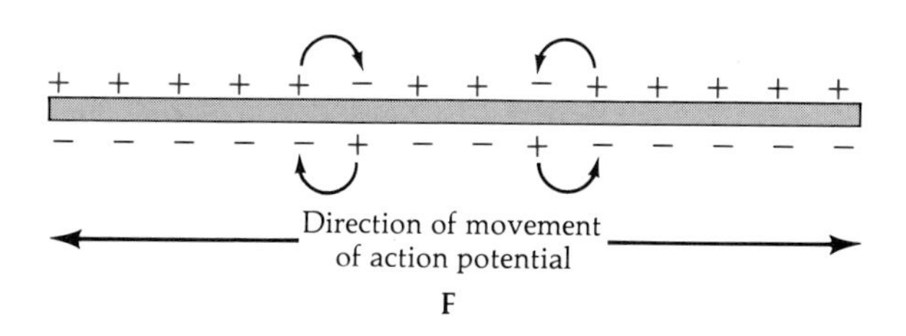

FIG. 7-5
Movement of ions in development of an action potential. The arrows indicate the flow of ions in an unmyelinated fiber. (A) Ionic equilibrium in a nonconducting (resting) membrane. (B) Application of a stimulus (↓) causes Na^+ entry. (C) Entry of positive ions makes the exterior of the membrane more negative and the interior more positive. (D and E) Oppositely charged ions are attracted to their neighbors resulting in the reversal of polarization and the action potential moving away from the initial point of stimulation. The center of the membrane becomes repolarized. (F) The attraction of oppositely charged ions occurs in a new location so that the reversal of polarization and action potential move further toward the opposite ends of the membrane.

and concentration gradients when diffusing into the cell. Consequently, the potential of the membrane approaches the equilibrium potential of sodium ions (Table 7-1). More depolarization (and Na^+ ion entry) results from a positive feedback mechanism called the *Hodgkin cycle* (Fig. 7-6). The entry of Na^+ ions stimulates greater activity in the sodium channels in the membrane, allowing even more rapid entry of the ion.

When the stimulus is sufficient, a threshold is attained so that the flow of positive ions occurs along the entire length of the membrane, depolarizing one segment at a time and creating a current which is associated with a change in voltage called an *action potential.* Evidence indicates that the action potential is caused by diffusion of ions along their gradients through specific ion channels (pores) in the cell membrane. In isolated axons in the laboratory, thousands of action potentials can traverse the fiber before recovery of the resting membrane potential is necessary. In the living animal however, active transport of ions by ATPases (the sodium-potassium pumps) usually rapidly restores the membrane to its resting potential

FIG. 7-6
The Hodgkin cycle: a positive feedback mechanism accounts for increased depolarization of a nerve fiber.

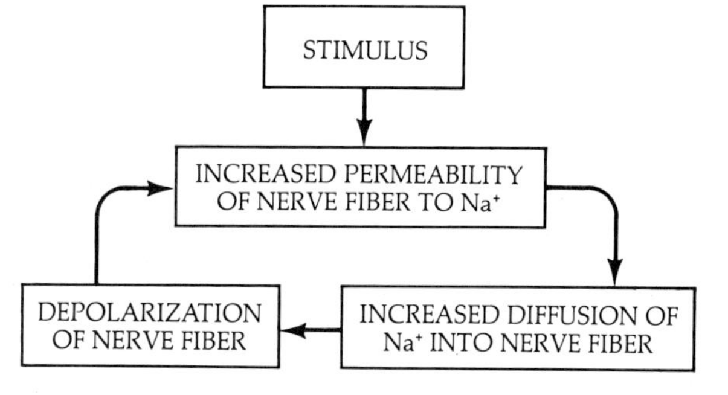

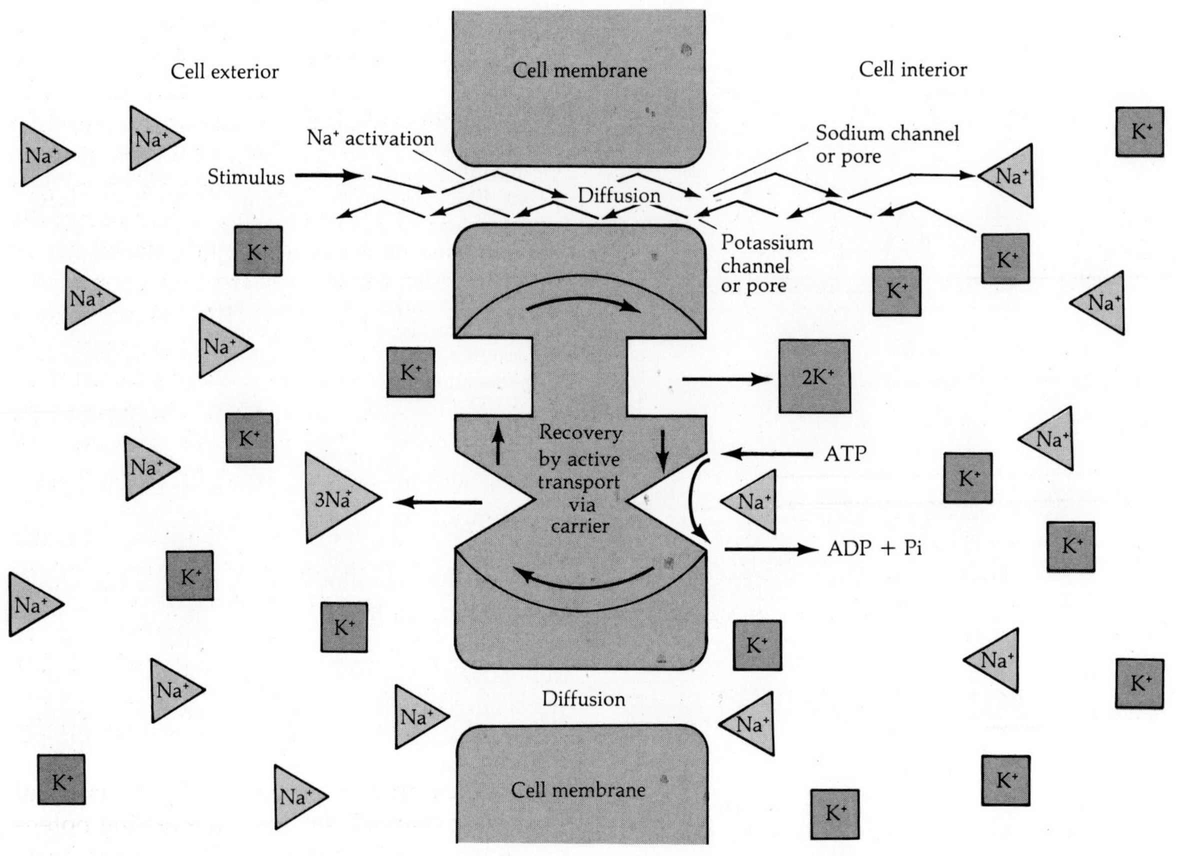

FIG. 7-7
Mechanisms of movement of positive ions in development of an action potential. Application of a stimulus leads to increased diffusion of Na^+ ions into the nerve fiber. It is quickly followed by outward diffusion of K^+ ions. Recovery of the resting condition occurs as active transport enzymes expel Na^+ from and recapture K^+ to the interior.

prior to the application of the next stimulus. The mechanisms of active movement of ions are reviewed in Fig. 7-7. The response to a stimulus as noted in steps A through D in Fig. 7-5 corresponds to the events noted in the upper portion of Fig. 7-7. Recovery of polarity of the membrane at the initial point of stimulation is noted in Fig 7-5E and F. It occurs with the participation of sodium-potassium pumps, whose activity is noted in the middle of Fig. 7-7.

All-or-None Law

According to the all-or-none law, an action potential spreads (is propagated) without a decrease in intensity along the entire membrane (like a row of falling dominoes). Action potentials either occur in response to a stimulus or do not occur at all depending on whether a threshold is reached. Action potentials are a function of the equilibrium potential of sodium in a given cell. Under normal conditions, action potentials which are formed by one neuron are of the same size. They do not develop with graded intensities along a nerve fiber under a single set of conditions. However, the size of the potential can vary under different conditions. For example, the potential decreases as it moves through the cytoplasm because that fluid conducts impulses poorly. Furthermore, graded responses can occur in nerves under natural

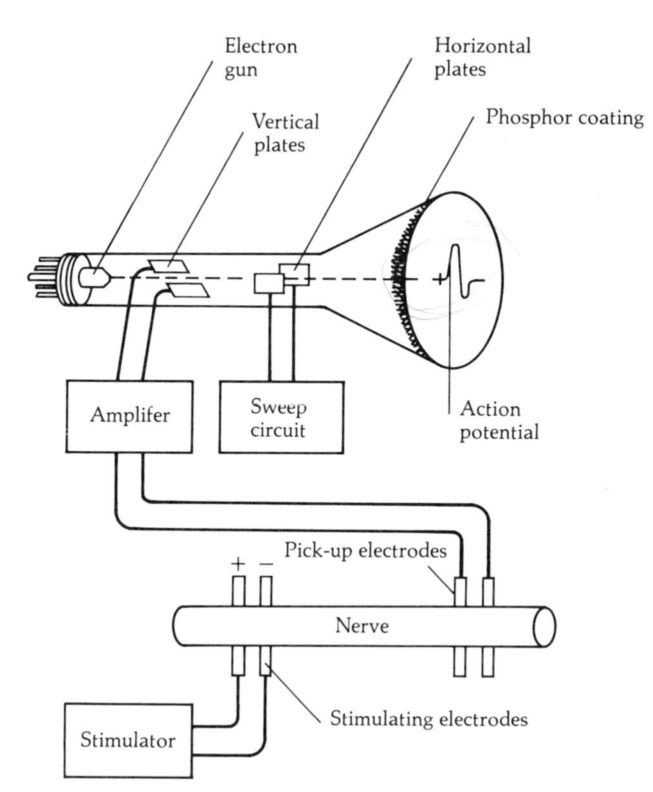

FIG. 7-8A
Recording of an action potential. Diagram of components in an oscilloscope used to measure changes in potential of excitable membranes.

conditions. Nerves are bundles of fibers from many neurons. The number of fibers that are stimulated and which respond with action potentials may vary. This is not an exception to the all-or-none law in any one fiber. These variations will be discussed below in the section on relations between different types of potentials and in that on neurotransmitters and the development of the postsynaptic potential.

Phases in Development of the Action Potential and Their Measurement

An oscilloscope can be used to record an action potential. This instrument has a screen much like that of a television (Fig. 7-8A). However, only one electron beam runs across its surface. A pair of microscopic electrodes placed on and in the fiber is used to generate an action potential. A change in voltage picked up by another pair of electrodes is carried to an amplifier and to vertical plates, which cause the electron beam generated by the cathode or electron gun to be deflected up and down. A decrease in the voltage (a decrease in potential) picked up by the electrodes causes upward deflection of the beam (for example, when the voltage moves from −70 mV towards 0 mV). An increase in voltage (potential) causes downward deflection of the beam (for example, when the voltage changes from −70 to −90 mV). The electron beam is moved left to right by voltage generated by a sweep circuit and carried to horizontal plates near the front of the tube. Thus the instrument moves the electron beam left to right continually while voltage picked up by the electrodes may deflect the beam up or down. The electron beam becomes visible when it strikes a phosphor coating, on the inside of the front of the screen, which briefly glows.

As a membrane is *depolarized* by the entry of positively charged Na^+ ions, the resting potential moves from a negative value toward zero. The membrane is depolarized; its potential has decreased. Enough Na^+ ions enter to change the potential from −70 to +40 mV. This result represents a total change of 110 mV. The entry of Na^+ ions (and decrease in voltage) causes an upward deflection of the electron beam on the oscilloscope screen. This is called a *spike* (Fig. 7-8B). Its size is a measure of the amplitude (voltage) of the action potential.

Shortly after the Na^+ entry, K^+ diffuses rapidly outward as the cell enters its *recovery* or *repolarization phase.* In this phase the potential across the membrane is returned toward its normal resting level. The greater permeability of the membrane to Na^+ following application of a stimulus is called *sodium activation* (Fig. 7-7) and is quickly lost during *sodium inactivation.* The membrane is activated and deactivated. The changes in Na^+ ion permeability are accompanied by a slower increase in permeability to K^+ ions (Fig.7-8B). The K^+ ions diffuse

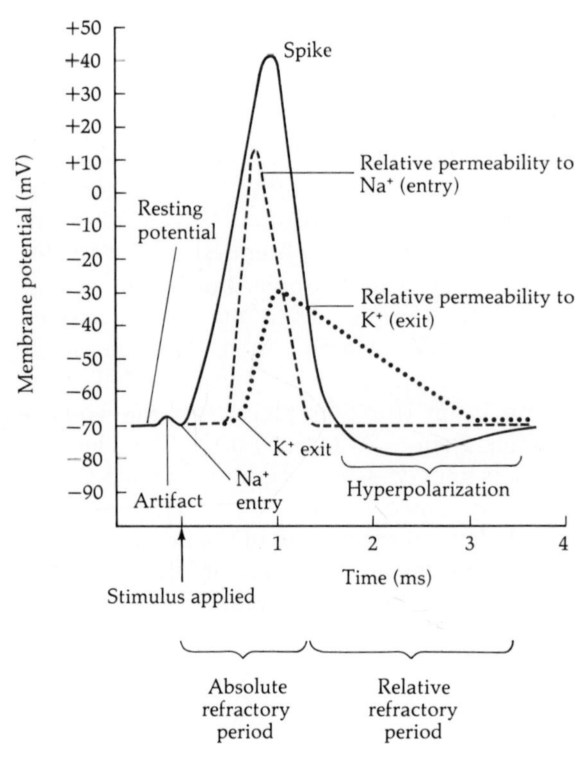

FIG. 7-8B
Diagram of events in an action potential as recorded by an oscilloscope. The action potential as recorded by an oscilloscope is shown as a thick solid line. The broken lines are not recorded by the oscilloscope but are drawn in for informational purposes. The entry of Na^+ follows the stimulation of the nerve fiber and its peak precedes the development of the spike potential and the exit of K^+. The interior of the membrane, which initially was negative, becomes positive. Upon expulsion of Na^+ and reentry of K^+ the membrane may become hyperpolarized prior to recovery of its original negative resting potential. The artifact is an unnatural defection caused by the placement of the electrodes on and in the fiber.

out of the cell down their concentration gradients. In some instances, more positive ions leave than enter the cell and the membrane briefly becomes *hyperpolarized.* The interior is now more negative than it was in its resting condition.

A period occurs in which K^+ ions leave the cell and Na^+ ions no longer enter it. From the time of application of the stimulus until repolarization is less than one-third complete, the fiber will not respond to a second stimulus. This interval is called the *absolute refractory* (unresponsive) *period* (Fig. 7-8B). It is the period surrounding activation and inactivation of the membrane to sodium permeability. It is followed by a *recovery* or *relative refractory period.* The latter occurs during the period of enhanced permeability to K^+ and lasts about 2 ms (milliseconds) (Fig. 7-8B). During the recovery phase, the membrane is restored to its original condition. A second action potential can be evoked at this time only if the second stimulus is stronger than the initial one. A fraction of a millisecond after the recovery phase, a fiber regains its original capacity to respond to another threshold stimulus. The whole cycle lasts a few milliseconds or less, so many neurons can fire action potentials at rates up to 1000 per second.

Gate Model of Ion Flow

A mechanical model, the *gate model,* may be useful as a hypothetical scheme to envision the development of an action potential (Fig. 7-9). Presumably, gates guard the channels or pores in proteins in the cell membrane. The gates open and close to permit more rapid movement of Na^+ and K^+ ions. Changes in voltage cause alterations in the structure of the channels. The extent to which gates for specific ions open is directly related to the voltage.

Under resting conditions, the gates are closed. Following a stimulus they open partially, allowing sodium ions to enter the cell in single file. However, if the stimulus is too small, the gate shuts. Closure may be due to attraction between oppositely charged ions. In any event, the channel is closed, preventing the development of an action potential.

The gates open if the stimulus reaches a threshold. The inflow of Na^+ ions opens them wider, overcoming the "pulling" effect of negative charges inside the membrane. Positive feedback causes more inward movement of Na^+ ions, which diffuse down their concentration gradient. Depolarization of the membrane continues as the gates open fully. After a slight delay and as the cell enters its recovery phase, K^+ ions flow outward, down their concentration gradient, and through the potassium channels.

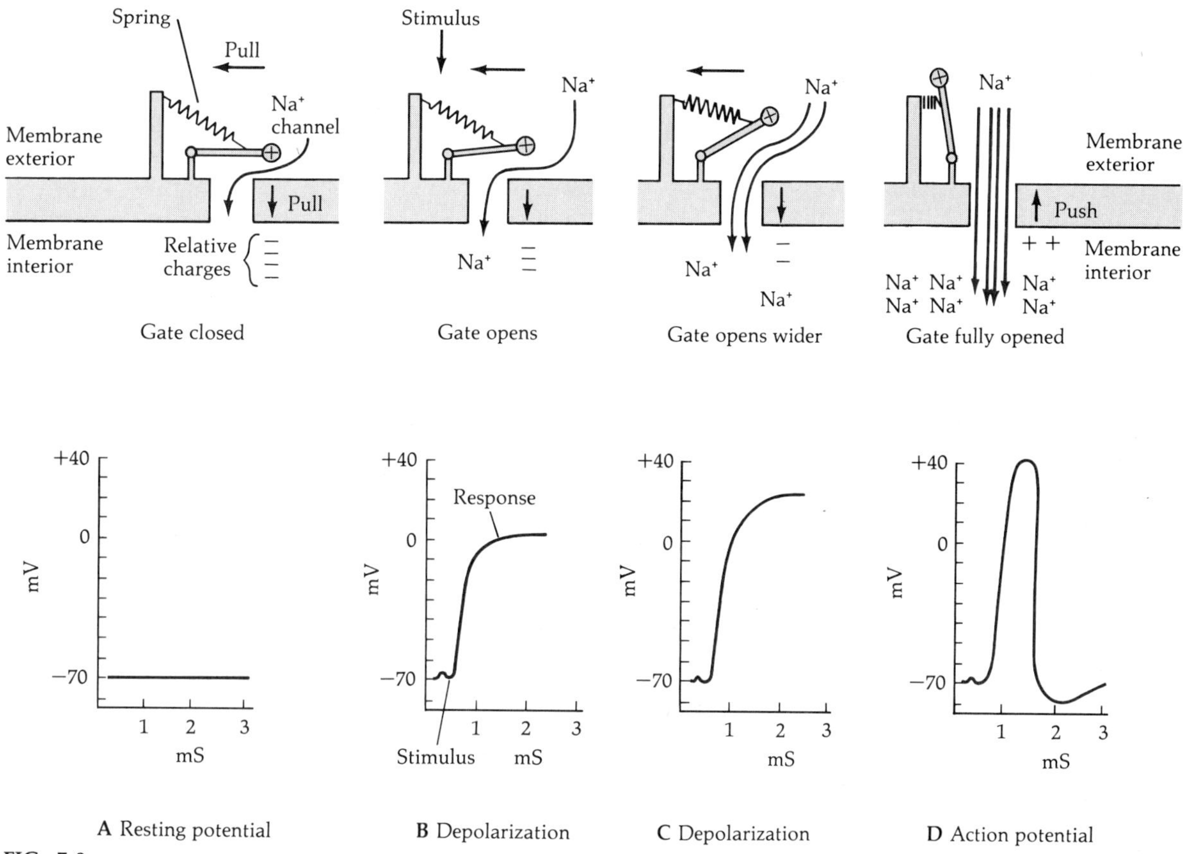

FIG. 7-9
Gate model of Na^+ entry. The upper drawings from left to right show the initial entry of Na^+, pushing open the gate to the ion channel. Continued entry opens it wider and depolarizes the interior, changing its net charge from negative to positive. The lower drawings show the corresponding patterns on an oscilloscope which record the resting potential (A), depolarization of the membrane (B) and (C), and development of the action potential (D).

Evidence on the flow of Na^+ ions has been accumulated by using tetrodoxin, a poison from the Japanese puffer fish, fugu (*Spheroides porphyreus*). This toxin specifically binds to and blocks sodium channels. Its use indicates there are an average of 10 to 50 Na^+ ion channels per square micrometer of a neuron's cell membrane. However, the number of channels ranges from 0 to 10,000 per square micrometer depending on the section of the membrane under study. The average number of channels amounts to about 1 per 300,000 phospholipid molecules, allowing Na^+ ions to diffuse into the neuron at rates of 1 million per second. This is a low rate compared with the capability of ATP pumps on the membrane to move Na^+ ions outward. Estimates indicate an average of 100 to 200 sodium pumps exist per square micrometer, for a total of 1 million pumps per neuron which together can transport Na^+ ions at rates of 200 million per second per cell. Thus, the ATPase's ability to restore the membrane to the resting condition is exceedingly rapid compared with the rate of the diffusion of ions (Na^+ in and K^+ out) that follows the application of a threshold stimulus. However, extremely little Na^+ ion entry seems needed to cause an action potential. Estimates with some neurons indicate the level of Na^+ ions within the neuron temporarily increases only 0.00033 percent during development of an action potential.

Requirements to Develop an Action Potential

A stimulus may generate an action potential when three conditions are met. The stimulus must reach a *threshold,* must *last* for a minimum time (duration), and must represent a *rapid change*.

The outward diffusion potential of K^+ ions is high (as may be deduced from the data in Table 7-1). Therefore, if a stimulus is too weak, the slow entry rate of Na^+ is exceeded by the outward diffusion rate of K^+. If the membrane is to be depolarized, a strong stimulus also must last a minimum amount of time. Only then can Na^+ move inward fast enough to exceed the exit rate of K^+. Furthermore, the change in permeability to Na^+ must occur rapidly. Otherwise, the forces of diffusion and the activity of the pumps reverse the flow of ions to prevent depolarization of the membrane.

Velocity of Nerve Impulses

The rate of conduction of impulses along fibers varies with the myelin content and size of the fiber, as mentioned in Chap. 6. Myelin insulates the nerve fiber and speeds up the conduction of impulses. It prevents ion leakage (current from flowing) across parts of the fiber. In the unmyelinated regions or nodes between Schwann cells, the movement of ions across the membrane occurs more readily. Impulses hop or skip rapidly from node to node (distances of about 1 to 2 mm) along the entire membrane. This type of flow of current is called *saltatory conduction* (Fig. 7-10). The membrane between the nodes does not allow the rapid inward movement of Na^+ because of its myelin wrapping. The primary ionic changes occur at the nodes and do not need to occur all along the fiber length. Therefore, less energy is needed for ion

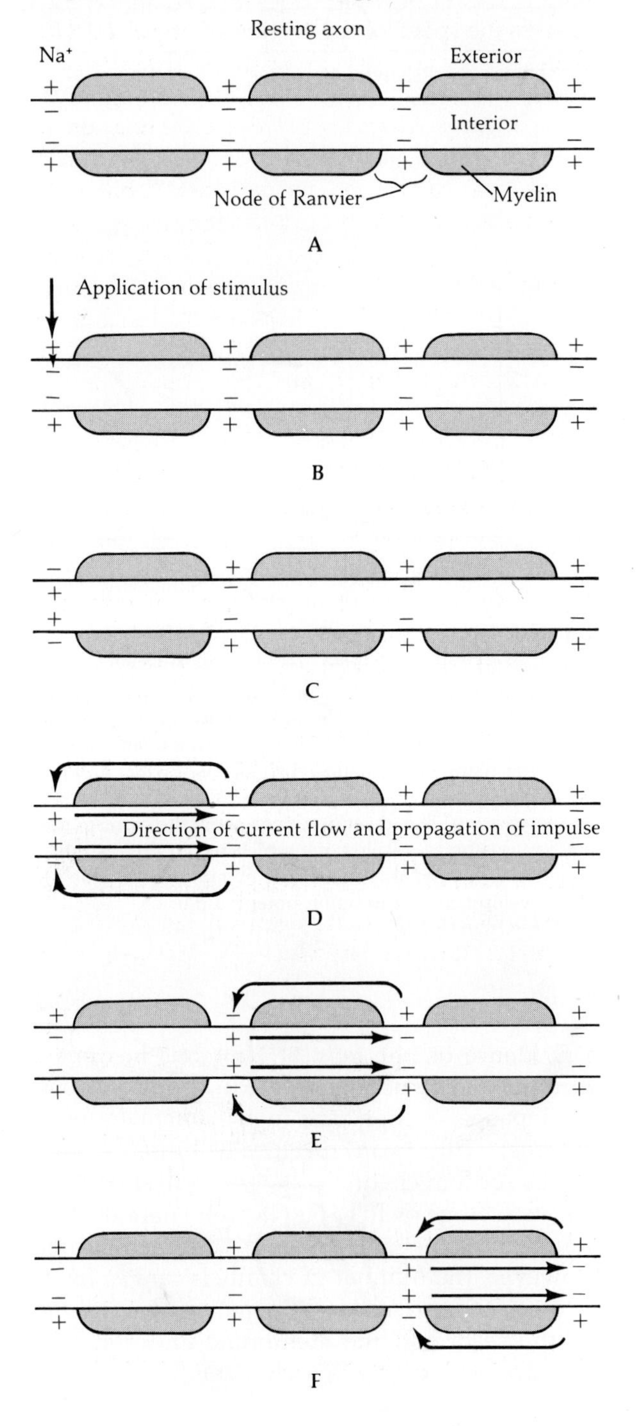

FIG. 7-10
The passage of an impulse along a myelinated fiber by saltatory conduction. The movement of ions (current flow) occurs from node to node as illustrated by the horizontal arrows. (A) Ionic equilibrium in a nonconducting (resting) membrane. (B) Application of a stimulus (↓) causes Na^+ entry. (C) Entry of positive ions makes the exterior of the membrane more negative and interior more positive. (D) Oppositely charged ions are attracted to their neighbors resulting in flow of current in ion channels from one node to the next. (E and F) The attraction of oppositely charged ions occurs in a new location so that the reversal of polarization and the action potential moves causing similar changes in the ion channels from node to node along the membrane.

pumps to restore a myelinated nerve to its resting condition than is required to restore a comparable nonmyelinated nerve. The lack of myelin at the beginning and ends of axons is associated with a slower rate of movement of impulses along these regions. Myelin, therefore, increases the speed of conduction of impulses with a lesser expenditure of energy by the neuron.

Impulse speed is not uniform among myelinated nerve fibers. It increases in relation to the diameter of the fiber as well as its myelin content, as discussed earlier (see Table 6-2). The fastest overall rate of conduction is about 120 m/s (268 mi/h) in myelinated fibers that are 20 μm in diameter. This is more than 200 times the rate of conduction observed in the slowest unmyelinated fibers, which relay impulses about 0.4 m/s (0.9 mi/h) and are about 0.1 μm in diameter.

The whole membrane of an unmyelinated fiber is uniformly excitable. However, its conduction is slow because of its large *capacitance*, that is, the number of charges needed to maintain a potential (voltage) across the membrane. Ions may leak across many parts of unmyelinated fibers, compared with the limited areas of leakage at the nodes of myelinated ones.

Therefore, a myelinated fiber has to control fewer charges to maintain polarity than does an unmyelinated fiber. In order to depolarize an unmyelinated fiber, a larger inflow of positive ions must occur in order to overcome the large number of charges along its *entire* length. The greater time needed to accomplish the change is responsible in part for the relatively slow rate of conduction of impulses in unmyelinated fibers.

Relationship between Different Types of Potentials

Potentials may be further differentiated according to the *source* of the stimulus, the *site*, and the *response* they evoke. Potentials may occur in response to external or internal stimuli. *Transduced potentials* are caused by changes in the external environment. They occur in neurons and skeletal muscle cells. *Pacemaker potentials* are elicited by changes in the internal environment of the cells. They occur as a series of slow spontaneous changes in smooth muscle and cardiac muscle.

Potentials can also be classified according to the site at which the change takes place. *Receptor potentials* occur at nerve terminals which receive signals from the external or internal environment. These usually are dendritic endings of fibers. *Synaptic potentials* develop at junctions between excitable cells.

The responses evoked vary with intensity of the stimuli and may remain as *local potentials* or develop into *action potentials.* Some stimuli cause a change in permeability which is not self-propagating and does not develop into an action potential. The stimulus is subthreshold so that only a slight flow of ions occurs and depolarization of the entire membrane does *not* follow. Transport mechanisms return the membrane to its original polarized state. The small, temporary potentials which do develop are called *local or generator potentials.* They characteristically form in sensory receptor nerve endings. By using the domino analogy described above, the local or generator potential may be compared to toppling only a few dominoes in a long row. Unlike action potentials, local or generator potentials are conducted with decreasing intensity. The voltage decreases as the distance from the point of application of the stimulus increases. The potential also lessens with time after application of the stimulus as ions leak through the permeable membrane.

The local response may be graded and vary in intensity. As the strength of the stimulus increases, greater but still localized changes in permeability may occur, thereby causing larger generator potentials (thus, more dominoes fall but not all of them do). When a number of generator potentials occur close enough in time or space or are of sufficient intensity, the total effect may be sufficient to attain a threshold and to cause an *action potential.* The sodium gates open wide, a spike potential occurs, and a self-propagating impulse flows along the membrane with maximal intensity in accordance with the all-or-none law (all the dominoes fall).

When repetitive stimuli arrive at a target membrane close enough in time (*temporal sum-*

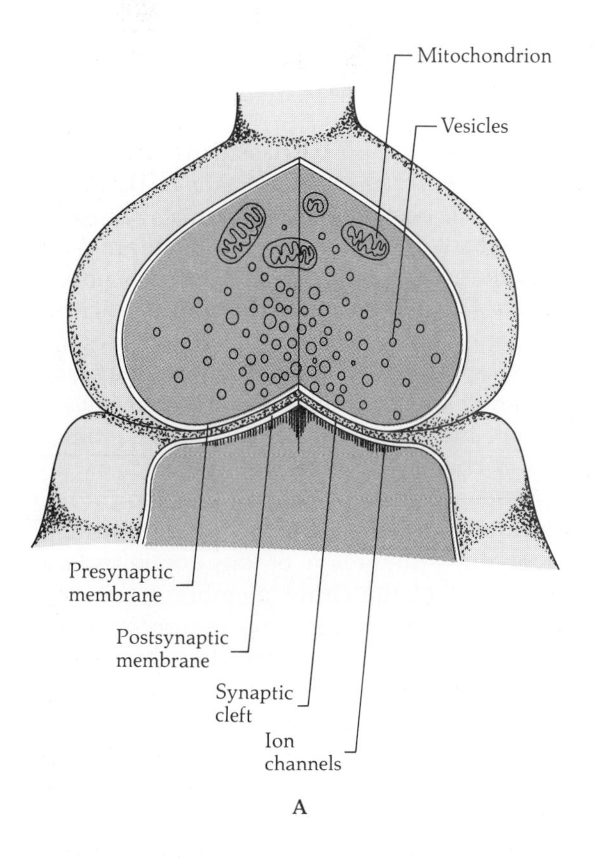

mation) or from a number of terminals (*spatial summation*), local potentials may lead to an action potential. These phenomena, which represent a few of the many ways in which variations take place in neural transmission, will be discussed further below.

7-2 JUNCTIONS BETWEEN NEURONS: THE SYNAPSE

Action potentials normally pass to the ends of the axonal fiber. At this location the terminal portions of the neuron come close to other cell

FIG. 7-11
Axon terminal, a synaptic bouton. (A) The diagram illustrates the junction (synapse) between the membranes of two neurons. One membrane is before the synapse (the presynaptic membrane), the other is after the synapse (the postsynaptic membrane). Chemicals in vesicles in the presynaptic neuron are released by exocytosis to diffuse across the synapse to join with cellular receptors on the postsynaptic membrane. (B) An electron micrograph of a junction between presynaptic and postsynaptic membranes (×43,000). Note at the arrows the presynaptic membrane of the axon terminal (A) and the postsynaptic membrane (P) of the dendritic endings (D) of a second neuron.

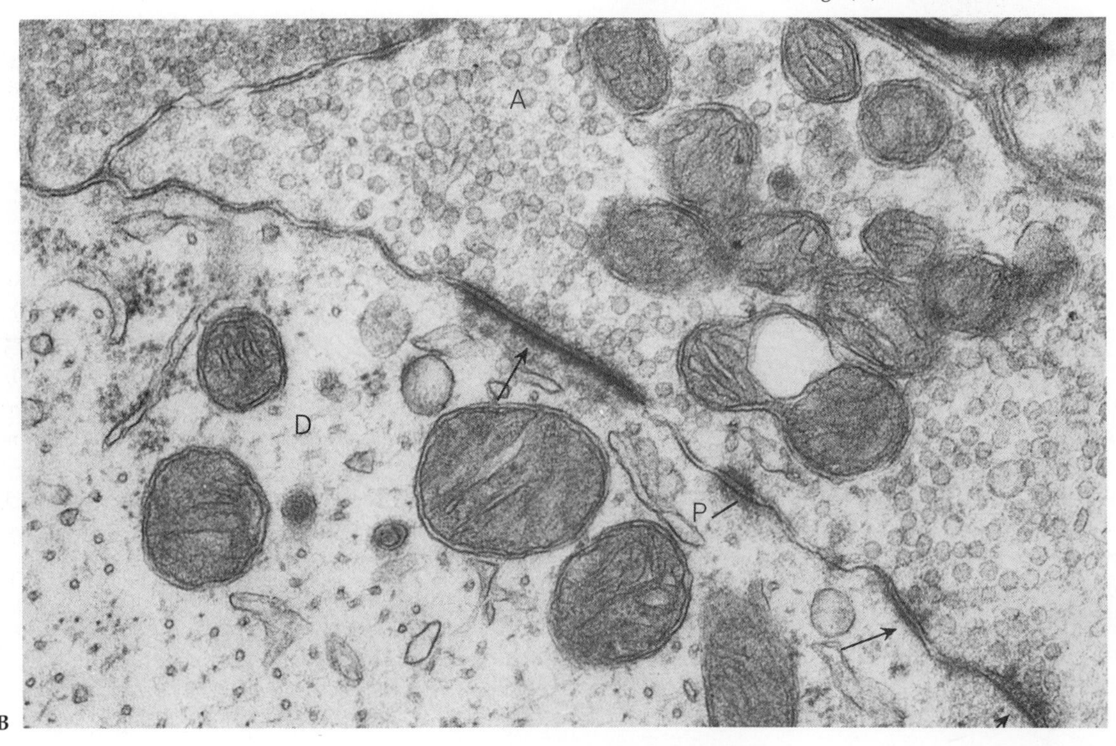

membranes. Each junction is called a *synapse*. Impulses can pass across cell junctions in different ways. One method is for the electric current to travel directly and without delay to the second cell through specialized regions shared between their membranes and known as *gap junctions* or *electrical synapses*. Gap junctions (Fig. 3-1C), which relay impulses from cell to cell, are found between smooth muscle cells and cardiac muscle cells. However, in the human body the common form of neural transmission is by the release of *neurotransmitters*. They diffuse across a synapse from a neuron to generate impulses in a second cell.

General Structure and Function of Boutons and Synaptic Vesicles

The end of the axon divides into enlargements called *terminal boutons* (Fig. 7-11). These structures contain a number of membrane-bound *synaptic vesicles*. The vesicles range in size from 10 to 60 nm. The boutons also generally contain many mitochondria, which provide energy to drive reactions at this site.

A variety of neurotransmitters are found within synaptic vesicles of different neurons, but a neuron usually has only one type of vesicle. The neurotransmitters are synthesized within the neuron's cytoplasmic organelles and are packaged in the synaptic vesicles (Fig. 6-4) for transport to the cell's terminal boutons. Neurotransmitters are released when an action potential reaches the terminals of the neuron which form the *presynaptic membrane* of the synapse.

Role of Calcium

The passage of the action potential is associated with the release of the contents of the synaptic vesicles from the terminal *presynaptic membrane* of the neuron (Fig. 7-12). The passage of the action potential is also linked or coupled with an increased rate of entry into the presynaptic membrane of Ca^{2+} ions which are found in high levels in the ECF. The role that Ca^{2+} ions play in neural transmission has not been clearly established. However, these ions can act as intracellular messengers in a number of physiological activities (Chap. 5). Their net positive charges are attracted to opposite charges on proteins. In this manner, Ca^{2+} may cause changes in the shape (conformation) of proteins, stimulating a variety of responses (such as alterations in ion channels and activation of enzymes). The entry of extracellular Ca^{2+} into the presynaptic membrane may: (1) help the Na^+ gates to open; (2) play a role in the development of action potentials; and (3) promote the release of neurotransmitter from the vesicles.

Properties of Synaptic Membranes

The junction of a neuron and another cell produces two sets of opposing surfaces, the presynaptic and postsynaptic membranes, which are separated by a space about 20 to 30 nm wide, the *synaptic cleft* (Fig. 7-12). Two types of junctions or synapses are found, which vary in properties (Table 7-2).

Type I synapses have asymmetrical membranes and large round vesicles, which primarily contain excitatory neurotransmitters that depolarize postsynaptic membranes. Type II synapses have smaller, flattened vesicles, which primarily contain inhibitory neurotransmitters that hyperpolarize postsynaptic membranes that are located across a narrower synaptic cleft.

Types of Synaptic Circuits

The terminal endings of neurons are formed by axons and dendrites which can associate in a number of different relationships. *Axodendritic* synapses, between axons and dendrites, are the most numerous type of junction. Their long cytoplasmic extensions form the *macrocircuits* of the nervous system. Axons may also, less commonly, form *axosomatic* synapses with cell bodies of other neurons or, rarely, *axoaxonic* synapses with other axons. Axosomatic and axoaxonic junctions are thought to exist as type II inhibitory synapses. Several examples of synapses are shown in Fig. 7-13.

Junctions between dendrites of neurons are called *dendrodendritic* synapses. The short length of dendrites allows them to form com-

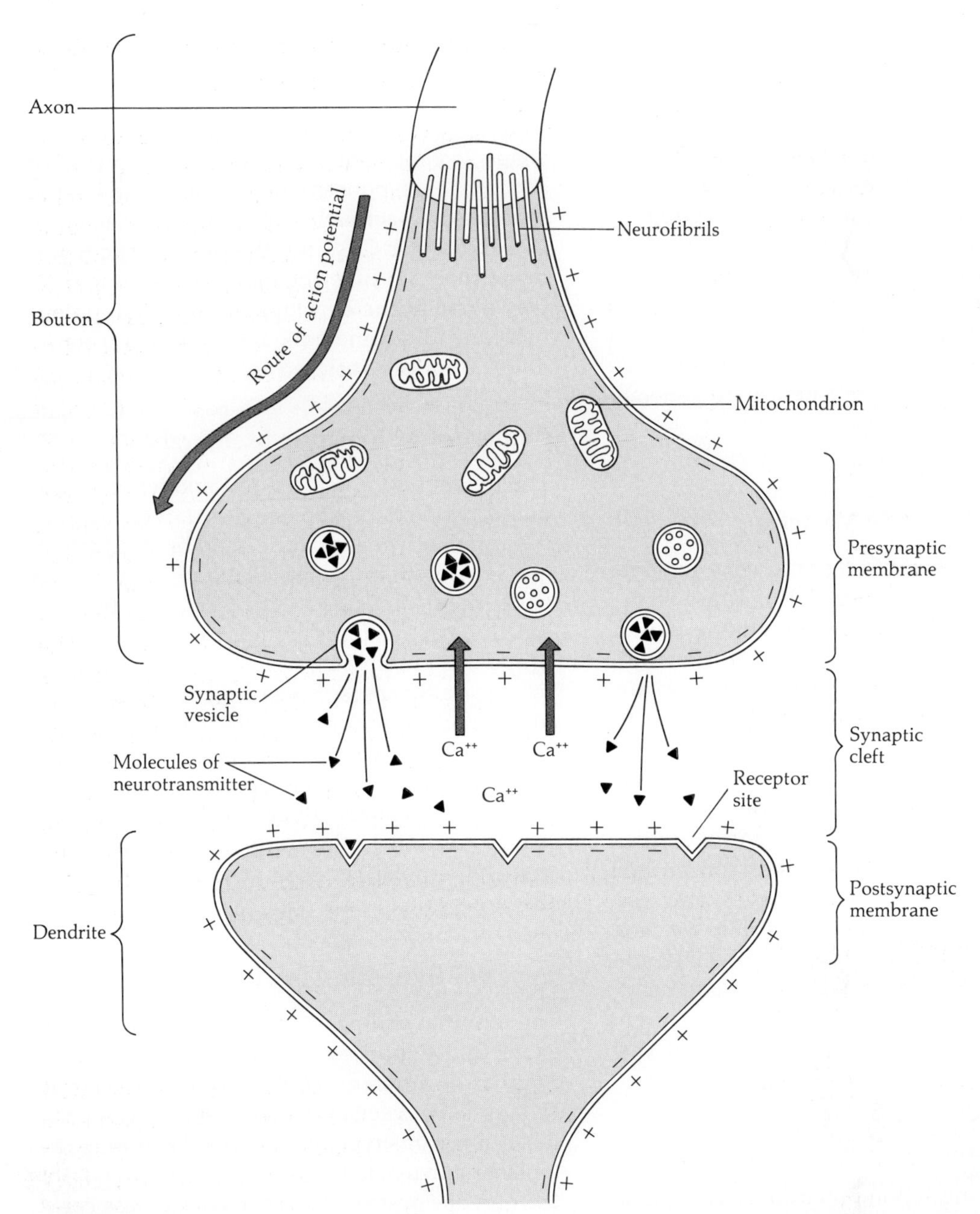

FIG. 7-12
Diagram of a bouton at a synapse. Note the membranes of the presynaptic and postsynaptic surfaces and those of the synaptic vesicles are similar in structure to the remainder of the cell membrane.

pact, local *microcircuits.* Microcircuits are common in the brain. They are evident also in the olfactory bulb, used for the sense of smell; the thalamus, a sensory relay station to the highest parts of the brain; the trigeminal nerve, which innervates the face and jaw; and the retina of the eye. Dendrites form synapses at different points along their length. Some do so by the formation of several small *dendritic spines,* which occur in the greatest number at the midpoint of each dendrite and in the least quantity at the ends nearest to and farthest from the cell

TABLE 7-2
Properties of synaptic membranes

Property	*Type I*	*Type II*
Size of vesicles	45 ± 15 nm	20 ± 10 nm
Form of vesicles	Rounded	Flattened
Primary type of transmitter released	Excitatory	Inhibitory
Symmetrical presynaptic and postsynaptic membranes	No	Yes
Relative width of synaptic cleft	Wider	Smaller

FIG. 7-13
Kinds of synapses between neurons.

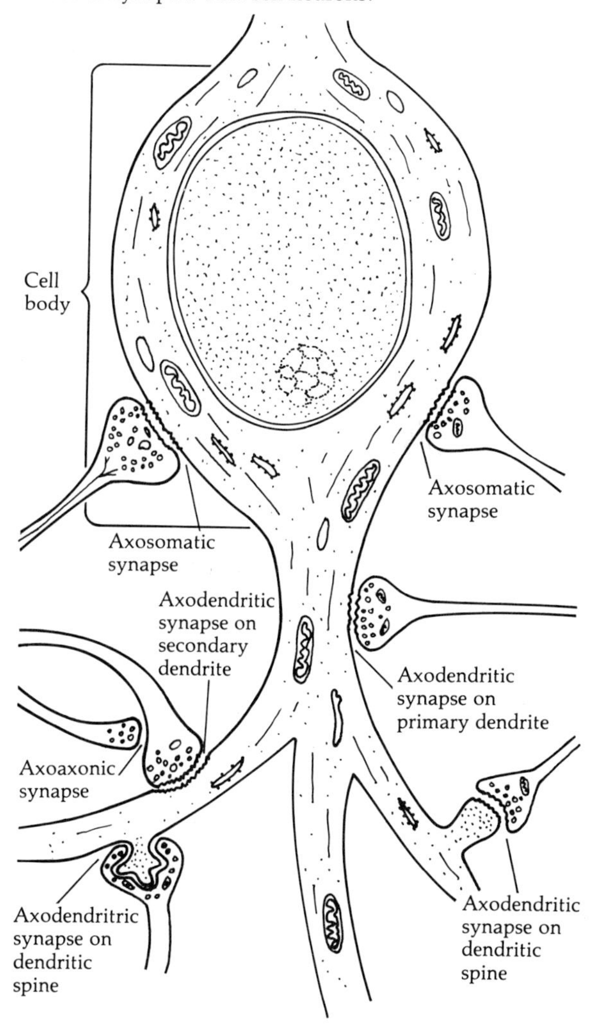

body. The number and appearance of these spines vary with the physiology and age of the neuron.

By means of a variety of circuits, neurons can form multiple synapses (polysynaptic relations) at a single location. The junctions often resemble the intersection of several roads at a traffic circle or the spokes leading to the hub of a wheel (Fig. 7-13). Analysis of the net effect of transmission of signals across synapses can be extremely complex. This is partially related to the multiplicity of junctions at a single site, as well as to variations in the quantity and type of neurotransmitters released. For example, one neuron can have 1000 to 10,000 synapses and through interneurons it may interconnect several hundred thousand cells in circuits within the brain. In this manner, the CNS is provided with functional units assembled from huge numbers of neurons. Some release excitatory neurotransmitters, others inhibitory ones. The coordination of microcircuits in the brain is multineuronal and controls the highest functions of the nervous system, including memory, learning, and behavior. Furthermore, the synapses occur not only between neurons but also between neurons and other cell types. In this manner the release of neurotransmitters regulates the function of skeletal, cardiac, and smooth muscle, as well as glandular secretions.

7-3 FUNCTIONS OF NEUROTRANSMITTERS

As described above, when an action potential reaches the terminal of an axon (its presynaptic membrane), Ca^{2+} ions move into it. The entry of Ca^{2+} is associated with the release of neurotransmitters by exocytosis from the synaptic vesicles (Figs. 7-14 and 7-15). Neurotransmitters diffuse across the synaptic cleft to bind to specific cell receptor molecules on the postsynaptic membrane. Neurotransmitters, like endocrine hormones, may alter the level of secondary messengers (such as calcium, cAMP, or cGMP), which in turn activate specific protein kinases. The kinases may cause changes in the permeability of the ion channels formed by

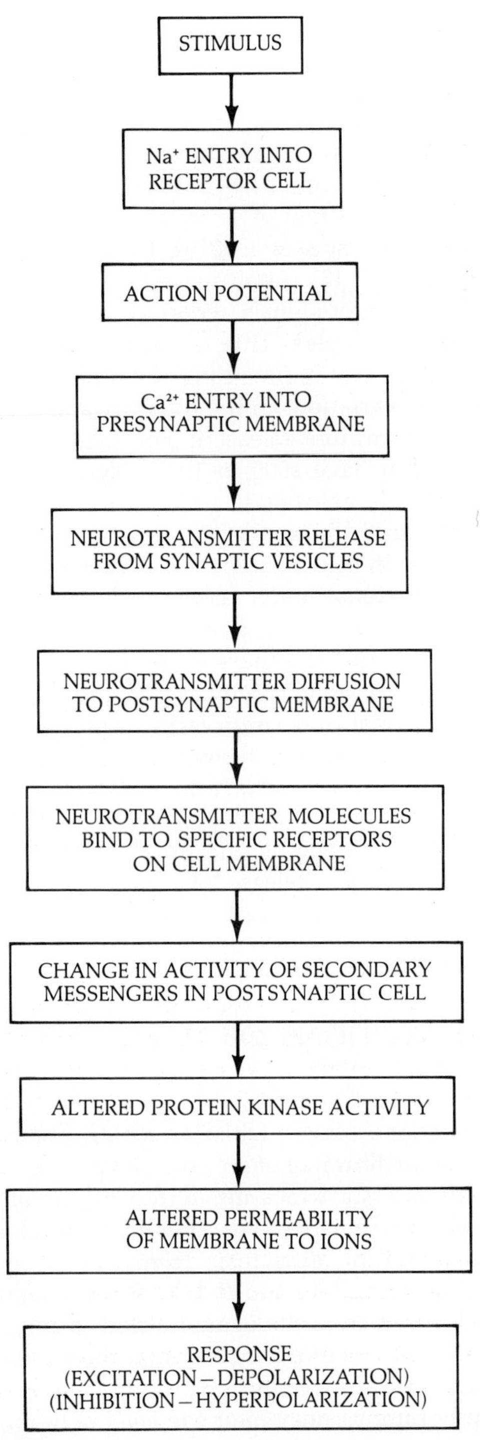

FIG. 7-14
Steps involved in excitation or inhibition of the postsynaptic membrane.

proteins. Alternatively, the binding of the neurotransmitter molecules to their receptors may directly alter the ion channel. Regardless of the mechanism, the change in conformation of the channels allows certain ions to pass across the postsynaptic membrane more freely, causing excitation or inhibition of the postsynaptic cell. The ultimate effect on the cell depends on the net flow of individual ion species which excite (depolarize) or inhibit (hyperpolarize) the postsynaptic cell membrane. The alternate responses are respectively called *excitatory postsynaptic potentials* (EPSPs) and *inhibitory postsynaptic potentials* (IPSPs).

Development of Postsynaptic Potentials

Impulses travel along nerve cell fibers to the presynaptic membrane according to the all-or-none law. The effects on the postsynaptic membrane are regulated by several factors including: (1) the type of neurotransmitters released to it; (2) the quantity of neurotransmitters; and (3) the types of receptors on the postsynaptic membrane to which the neurotransmitters bind. Therefore, the amplitude (height) of the potential which develops on the postsynaptic membrane varies.

If the net effect of the neurotransmitters is excitatory, the postsynaptic membrane is depolarized as its sodium ion channels open, producing an EPSP, as shown in Fig. 7-16A. If the stimulus reaches a threshold, an action potential flows down the membrane.

Individual boutons of one or more presynaptic neurons which converge on a single postsynaptic membrane may release subthreshold levels of neurotransmitters. The repeated formation of EPSPs in a short space of time can reach (build) a threshold; the close-timed repetition of EPSPs is called *temporal summation*. The addition of EPSPs at several synapses in one area is called *spatial summation*. Neurotransmitters may accumulate to threshold levels on the postsynaptic membrane. Strong stimuli provoke a large number of neurons to release neurotransmitters from presynaptic membranes faster than the molecules are degraded, removed, and reabsorbed. A maximal response (depolariza-

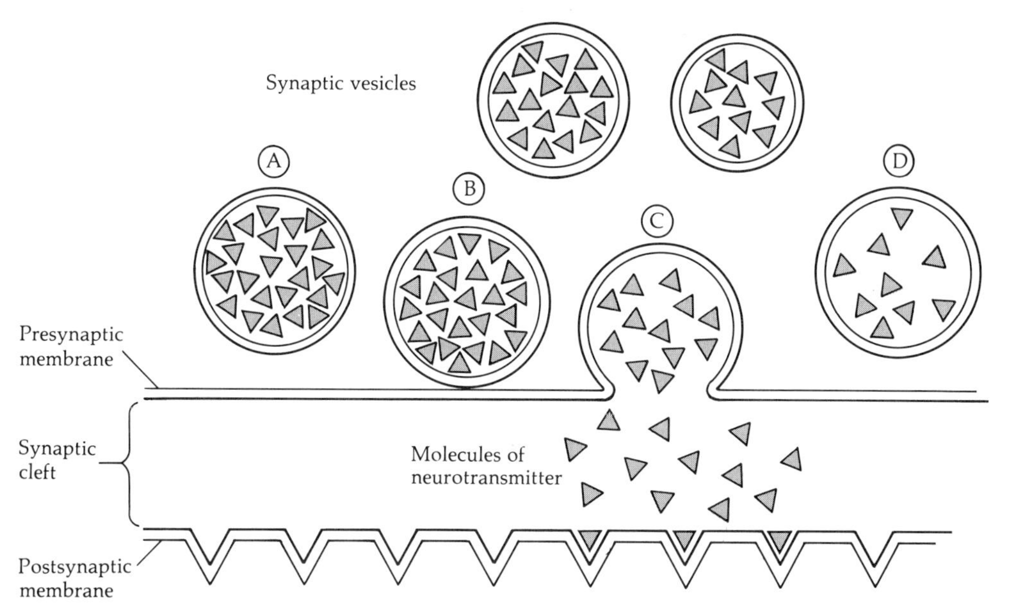

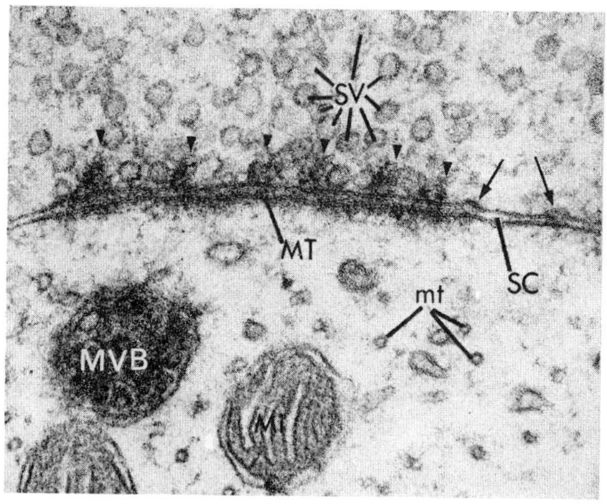

FIG. 7-15
Release of neurotransmitter from synaptic vesicles. The diagram depicts the discharge of neurotransmitter molecules by exocytosis from a single vesicle which empties its contents into the synaptic cleft and which may retrieve some of the molecules. The electron micrograph shows several vesicles at the arrows in various stages of the act of discharge of neurotransmitter by exocytosis (×61,000). SV, synaptic vesicles, MT, membrane thickening, SC, synaptic cleft, mt, microtubules, MI, mitochondria, and MVB, multivesicular body.

tion) of the postsynaptic membrane may follow. Weak stimuli provoke action potentials in fewer neurons and thus cumulatively release less neurotransmitter. The development of an action potential on a postsynaptic membrane is generally a complex function involving the net effect of multiple synaptic potentials.

Furthermore, the formation of an action potential on a postsynaptic membrane is a function of the nature of the interaction between the neurotransmitters and the receptors on the membrane. The neurotransmitters may be conceived as being analogous to keys which fit more than one type of lock. The locks are analogous to receptors on the cell membrane. Some keys fit the ignition system of an automobile used to turn on the engine. The effect of this key (neurotransmitter)–lock (receptor) interaction is similar to the generation of an EPSP. The same key may also fit another lock (a different receptor). When inserted into the lock and turned, the key fails to ignite the engine (generate an EPSP) but activates an alarm which inhibits unauthorized use of the automobile. This response resembles the formation of an IPSP. Further examples will be described below.

Some neurotransmitters hyperpolarize the postsynaptic membrane and inhibit the development of impulses. They produce an IPSP, as shown in Fig. 7-16B. IPSPs can promote the inward movement of Cl^- ions to make the interior of the cell more negative. The increased negativity may result from opening the ion gates for

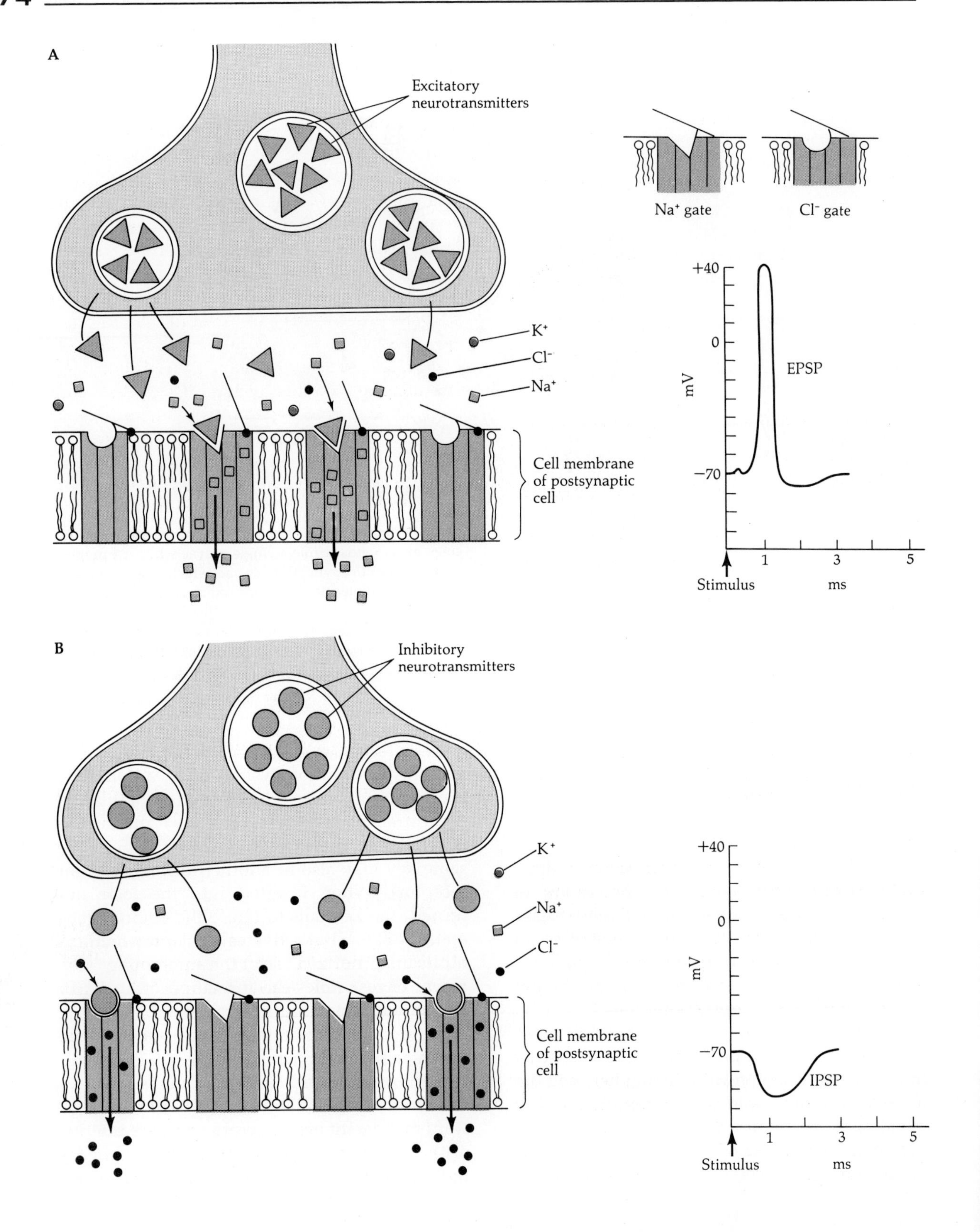

A
Excitatory neurotransmitters
Na⁺ gate
Cl⁻ gate
K⁺
Cl⁻
Na⁺
Cell membrane of postsynaptic cell
+40
0
−70
mV
EPSP
1
3
5
Stimulus
ms
B
Inhibitory neurotransmitters
K⁺
Na⁺
Cl⁻
Cell membrane of postsynaptic cell
+40
0
−70
mV
IPSP
1
3
5
Stimulus
ms

FIG. 7-16
Generation of postsynaptic potentials. The events at the synapses on the left produce the oscilloscope recordings on the right. Although more detail is shown for the postsynaptic membrane, the structures of all the membranes at the synapse are as described in earlier chapters. (A) Events during an excitatory postsynaptic potential (EPSP). (B) Events during an inhibitory postsynaptic potential (IPSP).

Cl^- by activating a chloride pump. IPSPs are promoted in heart muscle by opening ion channels for the outward diffusion of K^+ at rates greater than normal. In either case (Cl^- entry or K^+ exit), the interior of the cell becomes more negative and is more polarized (hyperpolarized). The cell's resting potential has increased (become more negative). Under such conditions the value for a typical neuron could change from -70 mV to -90 mV or more. Thus formation of an action potential often depends on the extent of the differences between the total activity of EPSPs and IPSPs. An action potential occurs when the EPSPs exceed the IPSPs by margins which attain threshold levels of depolarization.

Fate of Neurotransmitters

The level of active neurotransmitter available to the postsynaptic membrane is related to several factors. They include the rates of synthesis, diffusion, and binding to postsynaptic receptor sites, as well as recapture (reabsorption) by the presynaptic membrane, uptake by glial cells, and inactivation of the neurotransmitter at each site. After release into the synaptic cleft, large numbers of neurotransmitter molecules are recaptured by the presynaptic membrane. This represents the major route of their removal. Specific enzymes in the presynaptic and postsynaptic membranes can metabolize neurotransmitters into other products. For example, in synapses between nerve and muscle cell membranes, the neurotransmitter acetylcholine (ACh) is inactivated within about 5 ms of its release by the enzyme cholinesterase (acetylcholinesterase). Similarly, the activity of norepinephrine (noradrenaline) is diminished when it is degraded by monoamine oxidase (MAO).

Interactions between Neurotransmitters and Receptor Molecules on the Cell Membrane

Tens of thousands of receptor molecules are found on the surface of many cells. A range of molecular species, including neurotransmitters, hormones, and antigens, bind to specific receptors with stronger or weaker attraction (*affinity*). The sensitivity of receptors to specific agents varies according to the degree of molecular fit. Therefore, the effects depend on the properties of the transmitters and the receptors as well as the physiological condition of the cell. The results can be excitatory or inhibitory. As mentioned above, neurotransmitters, like endocrine hormones, seem to cause their effects in some cases by altering intracellular levels of secondary messengers and in others by the direct effect of binding to and altering the shape of the receptor sites.

Three types of receptors, α, β, and γ, have been studied extensively. Subtypes of each, whose functions differ, have been discovered. The neurotransmitter norepinephrine binds to α (and β) receptors and ACh to γ receptors. Neurons which release norepinephrine from their presynaptic membranes are termed *adrenergic*. They may decrease the intracellular levels of cAMP. Those that release ACh are called *cholinergic*. They may increase the intracellular levels of cGMP. These relations are shown in Fig. 7-17. These neurotransmitters and their receptors will be discussed more fully in Chap. 9.

Examples of Neurotransmitters

Evidence indicates a number of compounds formed by neurons act as neurotransmitters. Many have been clearly documented. A few such as the catecholamines, serotonin, glutamic acid, and gamma-aminobutyric acid will be described below.

Neurotransmitters are secreted in discrete sites and released along specific pathways. The pathways by which they are synthesized, released, recaptured, or degraded are extremely important to body function. Maps are currently

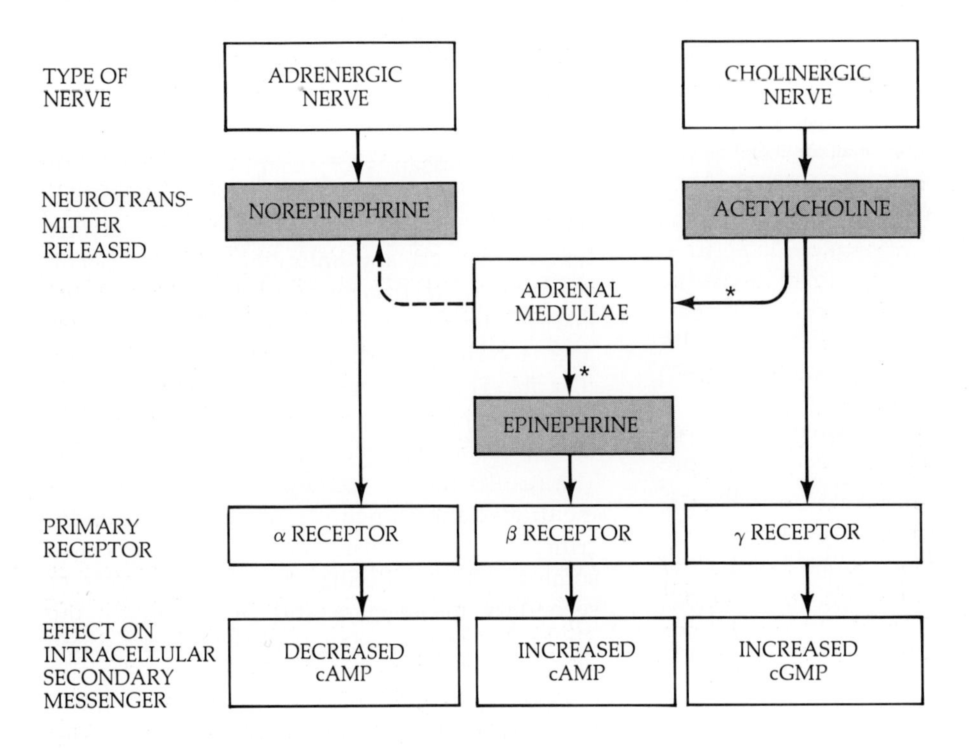

FIG. 7-17
The relationship between adrenergic and cholinergic nerve fibers, their neurotransmitters, and some of their cellular receptors.

being produced that demonstrate their cells of origin and destination. Methods of study include the administration into the brain of neurotransmitters artificially complexed to fluorescent dyes or radioactive isotopes. A fluorometer, or fluorescent microscope, or radiation (scintillation) counter is then used to determine the location and/or quantity of neurotransmitters in isolated tissues. Another approach involves the use of fluorescent or radioactive tagged antibodies formed against neurotransmitters or enzymes critical to neurotransmitter metabolism. The tagged antibody is selective for the neurotransmitter or its crucial enzyme. The adhesion of the tagged antibody to a site can be used to locate and determine the quantity of the neurotransmitter studied. These analytical approaches will be described further in Chap. 10.

Catecholamines

The catecholamines include the neurotransmitters dopamine, norepinephrine, and epinephrine, which are derived from the amino acid tyrosine. The steps in their synthesis are catalyzed by specific enzymes, as follows:

$$\text{Protein} \xrightarrow[\text{cleavage}]{\text{hydrolytic}} \text{Tyrosine} \xrightarrow[\text{tyrosine hydroxylase}]{+\text{OH}}$$

$$\text{L-dopa (dihydroxyphenylalanine)} \xrightarrow[\text{dopa decarboxylase}]{-CO_2}$$

$$\text{Dopamine (dihydroxylphenylethylamine)} \xrightarrow[\text{dopamine } \beta\text{-hydroxylase}]{+\text{OH}}$$

$$\text{Norepinephrine} \xrightarrow[\text{methylphenylethanolamine-N-transferase}]{+CH_3} \text{Epinephrine}$$

Dopamine is a neurotransmitter in the brain

that seems to especially affect movement and behavior. It does not cross the blood-brain barrier. In addition to its conversion to norepinephrine, dopamine may be inactivated through the removal of an amino group (NH_2) by the enzyme monamine oxidase (MAO). This enzyme also degrades serotonin and other neurotransmitters, including norepinephrine and epinephrine, whose effects will be discussed in Chaps. 9 and 10. Inhibitors of MAO build up catecholamine levels and have been used to treat some forms of mental illness.

Deficiencies of dopamine in a region of the forebrain known as the *basal ganglia* (Table 11-6) cause a loss of voluntary motor activity, leading to muscular tremors (involuntary body movements) in *Parkinson's disease.* Since L-dopa can cross the blood-brain barrier and be converted to dopamine, its administration may relieve the symptoms in some individuals.

Serotonin

Serotonin (5-hydroxytryptamine) is a derivative of the amino acid tryptophan, which is commonly found in many foods. Although little serotonin passes the blood-brain barrier, large amounts are found in several areas of the brain and are apparently synthesized there. Serotonin seems to be active in the transmission of impulses from the hindbrain to the forebrain as well as from the hindbrain to the spinal cord. It is found in large amounts in autonomic centers in the brain such as the hypothalamus. The specific effects of serotonin are not clearly established. Excesses of the neurotransmitter seem to stimulate the brain while deficiencies may cause a depressant effect. Serotonin is converted into melatonin in the pineal body of the forebrain and may be related to the establishment of biological rhythms governed by this section of the brain (see Chap. 11). Some evidence indicates that serotonin blocks the activities of the brain's arousal center to help induce sleep (Chaps. 9 and 10). Drugs which cause the release of serotonin (such as the tranquilizer reserpine) ultimately lead to calming effects. Conversely, the hallucinogen lysergic acid diethylamide (LSD) is an antiserotonin drug. The relationship between the effects of LSD and its antiserotonin activity remain to be clarified.

Serotonin also causes contraction of smooth muscle in some areas (such as the deep blood vessels in the skin, as well as skeletal muscle) while it has the opposite effect at other sites (such as the superficial vessels in the skin, where it causes flushing). Additionally, serotonin also may mediate some allergic responses, although it is usually of less significance in humans than in lower mammals such as rats or mice, or in invertebrates.

Glutamic Acid

Glutamic acid may be the main excitatory neurotransmitter in the brain. Its combination with ammonia (NH_3) forms glutamine. This reaction is the major means the brain has to remove ammonia, a highly toxic compound which can otherwise drastically alter acid-base balance in body fluids. The sodium salt of glutamic acid is known as *monosodium glutamate* (MSG) and is used as a flavor enhancer, especially in Chinese foods. It causes temporary neurological symptoms in many individuals and may produce the "Chinese food syndrome." These symptoms vary but may include temporary tingling, warmth, and discomfort in different areas of the skin in some individuals after consumption of foods flavored with MSG. The routine addition of MSG to baby food seems unwarranted and is on the decline, since it increases the salt content in the diet, which may cause elevation of blood pressure.

Gamma-aminobutyric Acid

Gamma-aminobutyric acid (GABA) is an inhibitory transmitter. It is formed from glutamic acid as follows:

$$\text{Glutamic acid} \xrightarrow[\text{glutamate decarboxylase}]{-CO_2} \text{gamma-aminobutyric acid}$$

This reaction is common in cells of the unmyelinated gray matter of the brain and spinal cord.

The combination of GABA with its receptors promotes the entry of Cl^- ions into postsynaptic membranes. It thereby hyperpolarizes neurons and increases the threshold required for a stimulus to cause an action potential.

Degeneration of brain cells that produce GABA may be inherited through a dominant gene and result in *Huntington's chorea*. This disorder may occur in early and later life but most often develops in middle age, usually between ages 40 and 45. Huntington's chorea produces a lack of voluntary nerve control, resulting in involuntary jerky movements and finally in progressive mental retardation and death.

7-4 SUMMARY

1 The presence of unequal levels of ions between ICF and ECF of neurons precedes and is necessary for the development of nerve messages called impulses. The resting potential is an electrochemical equilibrium maintained when a neuron is not conducting impulses. It is dependent on an electrical and chemical gradient maintained across the cell membrane. The equilibrium potential of individual univalent ions can be calculated by the simplified Nernst equation: $E = 61 \log_{10} (C_E/C_I)$. The value obtained is the voltage necessary to oppose net diffusion of the ion in question. Measurement of the resting potential with microcapillary electrodes indicates that the interior of the neuron is -70 to -90 mV compared with the exterior.

An action potential is a temporary change in the electrical properties of the membrane such that an impulse is conducted along its entire length. It is also called a nerve impulse. A threshold stimulus causes increased entry of Na^+ ions which depolarize the nerve cell membrane by a positive feedback mechanism called the Hodgkin cycle. The flow of positive ions produces a current which proceeds along the entire membrane with no decrease in intensity according to the all-or-none law. Phases in the development of an action potential include depolarization, development of peak depolarization (the spike), and restoration of the resting potential or recovery.

The gate model hypothesizes that formation of an action potential involves the flow of ions through protein channels in the membrane. The rate of entry and exit of the ions through these channels is regulated by gates which open and close.

For a stimulus to cause an action potential, it must be of sufficient strength, represent a rapid change, and last at least a minimum amount of time. Potentials can be classified according to the source of the stimulus (pacemaker or transduced), the site (receptor or synaptic), or the response evoked (local or action).

2 The junction of a neuron with another cell is called a synapse. The terminal process of an axon flares out into a bouton, which houses many membrane-bound globules, the synaptic vesicles, which store specific neurotransmitters.

Calcium in the extracellular fluid may enter the bouton upon depolarization of the presynaptic membrane and act as a secondary messenger. In this manner it may induce the release of neurotransmitters which diffuse to and bind to receptors on postsynaptic membranes to play a role in the development of an action potential.

The presynaptic and postsynaptic membranes are separated by a space or cleft in two arrangements called type I and type II synapses. Type I synapses contain excitatory neurotransmitters and possess membranes which are more asymmetrical and have larger and rounder vesicles than those of type II synapses. Transmitters in the latter are mostly inhibitory.

Synapses can form multineuronal circuits with junctions among dendrites, among axons and dendrites, or among axons themselves.

3 Neurotransmitters can cause excitation or inhibition, when they bind to the cellular receptors of the postsynaptic cell by altering the activity of secondary messengers such as Ca^{2+}, cAMP, or cGMP, or by binding to and changing the shape of ion channels. Excitatory neurotransmitters depolarize cells and open ion channels to produce excitatory postsynaptic potentials (EPSPs) while inhibitory neurotransmitters hyperpolarize cells to cause inhibitory postsynaptic potentials (IPSPs).

The level of active neurotransmitter is related to its rate of synthesis, diffusion, and binding to postsynaptic receptor sites, as well as to its rate of recapture by the presynaptic membrane, uptake by glial cells, and rate of inactivation by enzymes at each site.

Among the many types of receptors, three (α, β, and γ) have been studied thoroughly. The neurotransmitter norepinephrine released from adrenergic nerves binds to α and β receptors. Acetylcholine released from cholinergic nerves binds to γ receptors.

Among the many neurotransmitters formed are the catecholamines norepinephrine, epinephrine, and dopamine, as well as serotonin, glutamic acid, and GABA. They vary in concentration and in the functions they perform in different parts of the nervous system.

REVIEW QUESTIONS

1 State the simplified Nernst equation and explain why it does not yield results equal to the measured resting potential of a nerve fiber.
2 Trace the movement of ions in the development of an action potential.
3 What roles do active transport pumps and ion channels play in restoring a fiber to its resting potential?
4 State and discuss the all-or-none law, including its limitations.
5 Describe the phases in the development of an action potential, distinguishing among the terms polarized, depolarized, and hyperpolarized. Include a description of changes in the levels of Na^+ and K^+ ions.
6 Compare the absolute refractory phase and the relative refractory phase with respect to the response of a nerve to a stimulus.
7 What is the Hodgkin cycle and why is it significant?
8 Describe the gate model used to account for passive diffusion of ions following stimulation of a nerve cell membrane.
9 How do the nodes of Ranvier function in the transmission of impulses?
10 What is the relationship between the degree of myelination and the speed of conduction of impulses along nerve fibers?
11 Compare pacemaker and transduced potentials, receptor and synaptic potentials, and IPSPs and EPSPs.
12 How are generator potentials and action potentials related? How do they differ?
13 What are synapses and how do they differ in structure and function?
14 Discuss the possible roles of Ca^{2+} ions in neural transmission.
15 What is the general sequence of steps leading to synaptic transmission?
16 Outline the differences between neural microcircuits and macrocircuits.

17 What functions do neurotransmitters serve, and what happens to them after they have served these purposes?

18 Name six neurotransmitters and generally compare their chemistry, origins, and functions.

19 What are some differences between adrenergic and cholinergic fibers? Between α, β, and γ receptors?

PHYSIOLOGY OF THE NERVOUS SYSTEM

8

Sensory Receptors, Codes, and Special Sense Organs

8-1 GENERAL TYPES OF SENSORY RECEPTORS

8-2 SPECIFIC EXAMPLES OF SENSORY RECEPTORS

8-3 CODES GENERATED BY IMPULSES
Pattern codes
On-off codes: phasic and tonic receptors
Frequency codes
Number of receptors
Pathway codes

8-4 SPECIAL SENSE ORGANS
The eye
Physiology of visual impulses
Focusing of light on the retina
The ear
Hearing
Vestibular functions: perception of gravity and movement
Taste (gustation)
Smell (olfaction)

8-5 SUMMARY

OBJECTIVES

After completing this chapter the student should be able to:

1. Discuss the role of sensory receptors in the body.
2. Describe five general classes of sensory receptors and examples of each.
3. Outline the various means by which sensory codes are developed and interpreted.
4. Explain the structure and function of the eye in visual sensory perception.
5. Explain the structure and function of the ear in auditory sensory perception and the perception of gravity and movement.
6. Describe the mechanisms for the senses of taste and smell.

Sensory receptors detect changes in the internal and external environment to help maintain homeostasis. They are the means of our perception of the environment. They transduce (interconvert) various forms of energy and if the stimulus is of sufficient intensity, they cause the formation of action potentials, which are carried to the central nervous system (CNS) for interpretation and action.

Sensory receptors are modified neuronal endings which respond to specific stimuli. They have the capacity to relay impulses through terminal portions of afferent fibers which carry impulses to the CNS. A single afferent fiber may be undivided or may have many branches. With either arrangement, each fiber and its subdivisions is associated with a terminal receptor. The afferent fiber and its receptor(s) is called a *sensory unit.* All the receptors in a unit respond to the same type of energy for transmission of a specific type of sensation. The terms *afferent* and *sensory* are often used interchangeably although not all afferent impulses reach the level of sensory consciousness. Both types of afferent impulses are ultimately relayed across synapses to efferent fibers and then to effectors such as muscles or glands to cause a response (see Chap. 1).

8-1 GENERAL TYPES OF SENSORY RECEPTORS

Sensory receptors vary in structure and in their ability to react to different stimuli. The impulses which they generate form *codes,* which allow the CNS to interpret body feelings as a *somesthetic sense* (fr. Greek *soma,* body and *esthesis,* feeling). In this manner sensations of touch, pressure, pain, position of the body, and movement are acquired.

Various sensory receptor nerve endings may be classified according to the energy form they detect. Thus, mechanoreceptors respond to changes in pressure and tension (stretch), chemoreceptors to a variety of chemicals, osmoreceptors to osmotic pressure, thermoreceptors to temperature, photoreceptors to light, and auditory receptors (or phonoreceptors) to sound (Table 8-1). Sensory receptors also may be classified according to the source of the stimulus to which they respond. *Exteroceptors* respond to stimuli from outside the body on its surface, *interoceptors* to stimuli from internal organs, *proprioceptors* to stimuli from muscles, tendons or joints, and *teloreceptors* to stimuli from distant sources (as the eye responds to light and the ear to sound).

8-2 SPECIFIC EXAMPLES OF SENSORY RECEPTORS

The properties of several specific types of sensory receptors are listed in Table 8-1, and some are illustrated in Fig. 8-1. Receptor nerve endings encapsulated in connective tissue in the skin are Meissner's corpuscles, Pacinian corpuscles, Merkel's touch corpuscles, the corpuscles of Ruffini, and Krause's end bulbs (Fig. 8-1). Gentle touch specifically activates *Meissner's corpuscles. Pacinian corpuscles* respond to deep pressure. *Merkel's touch corpuscles* also respond to deep pressure but more slowly than Pacinian corpuscles. Merkel's touch corpuscles are formed by afferent nerve fibers surrounded by cells which resemble superficial epithelium of the outer layers of the skin. These receptors are

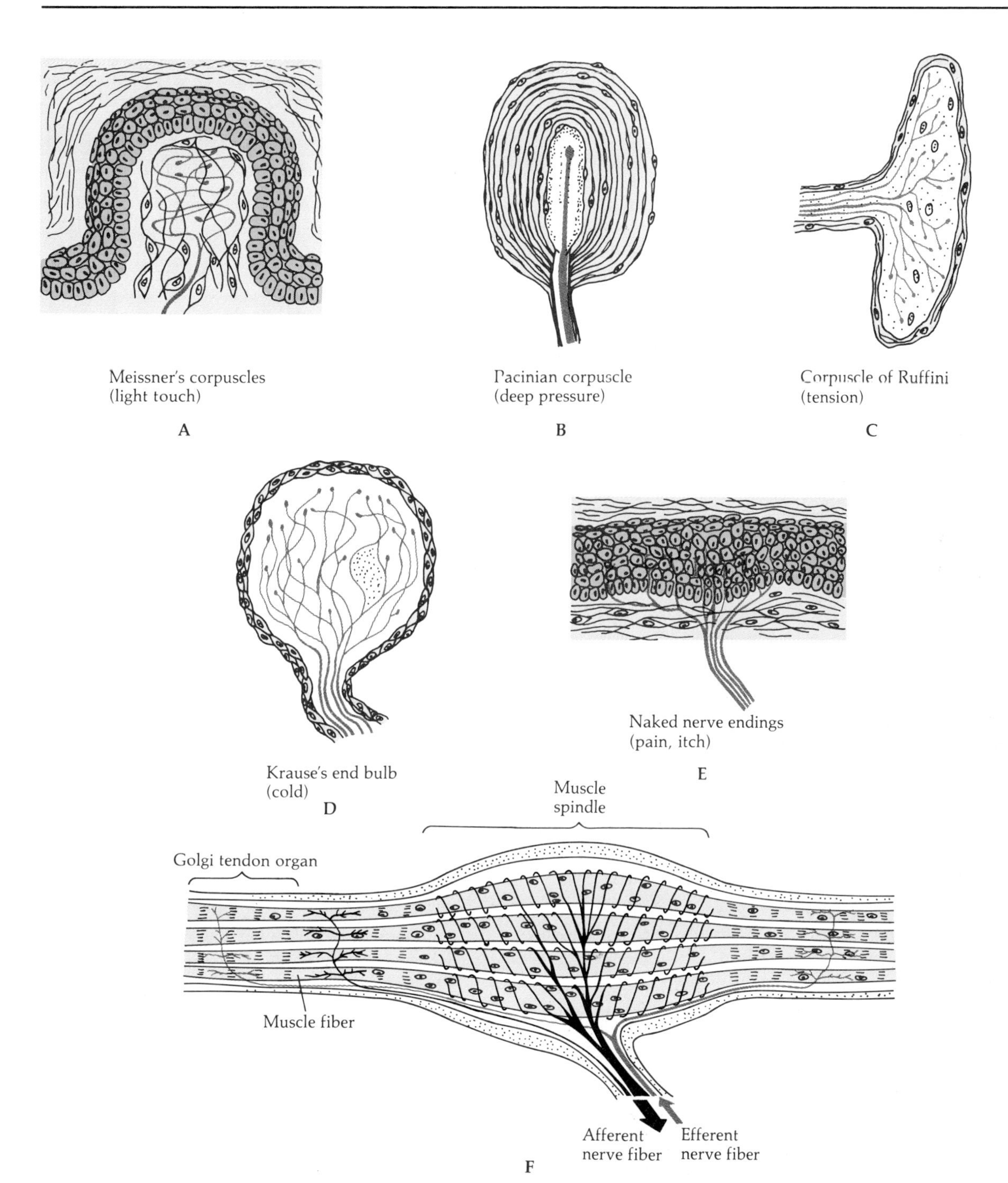

FIG. 8-1
Varieties of sensory receptors. (A to D) Encapsulated nerve endings. Connective tissues often surround nerve endings to help form the different types of receptors. (E) Naked nerve endings. (F) Sensory receptors in muscles and tendons: the muscle spindle and Golgi tendon organ. The arrows indicate the direction of movement of the impulses to or from the CNS.

stimulated when hairs from the hair follicles are deflected.

The *corpuscles of Ruffini* that are found in the skin respond to touch and tension, while those in the capsules of joints and ligaments are sensitive to changes in position of those structures. Naked nerve endings seem to be the main

TABLE 8-1
Properties of some sensory receptors

Type of Sensory Receptors	STIMULUS		RECEPTOR	
	General Name	*Specific Nature*	*Name*	*Location*
1 Mechanoreceptors	Pressure	Movement of hair in a hair follicle	Afferent nerve fiber	Base of hair follicles
		Light pressure	Meissner's corpuscle	Skin
		Deep pressure	Pacinian corpuscle	Skin
		Touch	Merkel's touch corpuscle	Skin
Baroreceptors		Blood pressure	Naked nerve endings	Blood vessels in thoracic and abdominal organs
			Aortic body	Wall of aorta
			Carotid sinus	Walls of carotid arteries
				Walls of pulmonary arteries in lungs
				Walls of venae cavae in heart
Nociceptors (algesireceptors)	Pain	Distension (stretch)	Naked nerve endings	Wall of gastrointestinal tract, pharynx, skin
Proprioceptors	Tension	Distension	Corpuscles of Ruffini	Skin and capsules in joints and ligaments
		Length changes	Muscle spindles	Skeletal muscle
		Tension changes	Golgi tendon organs	Between muscles and tendons
2 Chemoreceptors	Chemicals	Oxygen, carbon dioxide, and hydrogen ions	Aortic and carotid bodies	Wall of aorta, walls of carotid arteries
		Odors	Olfactory receptors	Respiratory epithelium
		Irritants		Respiratory epithelium
		Organic compounds, salts, and acids	Gustatory receptors	Taste buds in tongue
		Hydrogen ions, fats, acids, amino acids, other nutrients		Wall of gastrointestinal tract
		Glucose	Glucostats	Hypothalamus in brain
Osmoreceptors		Osmotic pressure	Osmostats	Hypothalamus in brain
	Itch	Vasoactive chemicals	Naked nerve endings	Skin
3 Thermoreceptors	Temperature change	Cold	Krause's end bulbs	Skin
		Heat and cold	Naked nerve endings	Peripheral body parts
4 Photoreceptors	Light	Dim light	Rods	Retina of eye
		Bright daylight and color	Cones	Retina of eye
5 Auditory receptors	Sound	Sound waves	Hair cells	Organ of Corti within inner ear
6 Vestibular receptors	Movement of head	Acceleration or deceleration	Cristae ampullaris	Swollen terminals (ampullae) of each semicircular canal in inner ear
	Gravity	Changes in force of gravity	Maculae	Two sacs (utricle and saccule) in inner ear

peripheral *thermoreceptors* while *Krause's end bulbs* are thermoreceptors in the skin which react to a decrease in temperature (exposure to cold).

Mechanoreceptors (stretch receptors) in muscles and tendons generate impulses concerning muscle tone. *Muscle spindles* (Fig. 8-1F) are wrappings of nerve fibers on the surfaces of small skeletal muscle cells called *intrafusal fibers.* Sensory fibers from muscle spindles convey afferent impulses regarding muscle length to the spinal cord and brain. Efferent fibers from the motor neurons of these two regions in turn generate impulses which control skeletal muscle contraction. Sensory endings that monitor the tension between tendons and muscle cells are called *Golgi tendon organs* (Fig. 8-1F).

Pain impulses seem to be generated by *naked nerve* fibers which are sometimes called *nociceptors* (or *algesireceptors*) since they respond to noxious or damaging stimuli (such as excess pressure, cuts, and burns). Distinct naked nerve endings mediate itch sensations, which result from stimuli of low frequency that cause release of vasoactive peptides, such as kinins or large quantities of histamines.

8-3 CODES GENERATED BY IMPULSES

Generator potentials develop as a result of stimulation at the endings of afferent neurons. They are referred to as *receptor potentials* if the sensory receptor must relay the impulse to the fiber of a second neuron for it to reach the CNS. Generator or receptor potentials may remain localized or may combine additively to produce an action potential which is carried the length of an afferent nerve fiber to the CNS (Chap. 7).

The CNS is able to interpret impulses from the internal and external environment by discriminating between stimuli on the basis of their frequency (number per second), duration (length of time), or combinations of these variables which form a pattern. The CNS also can distinguish among types of stimuli (for example, pain versus heat), their sources (such as the fingers versus the toes), and their effects (such as pleasant, neutral, or unpleasant). The effects of sensory impulses are strongly influenced by learning and cultural factors. Exactly how discrimination takes place remains unclear. However, some interpretations seem to be based on differences in the *patterns* formed by the impulses and/or the *pathways* they follow.

Pattern Codes

Pattern codes may be based on (1) on-off codes, (2) the frequency of impulses, and (3) the number of neurons involved.

On-Off Codes: Phasic and Tonic Receptors

Different sensory receptors may respond to stimuli in various ways. *Phasic* (velocity) receptors produce an impulse only when the stimulus is changing, that is, when it is applied (on), removed (off), or both, but not during a sustained stimulus. They behave in an on-off manner and become adapted (by ceasing to produce potentials) to a nonchanging stimulus. The duration of on signals and off signals forms a type of frequency code similar to variations in tones that arise from the oscillations of different tuning forks. Meissner's corpuscles (which respond to gentle touch), Pacinian corpuscles (which respond to deep pressure), and some neurons in the retina (which respond to visual impulses) are examples of phasic receptors.

Tonic (intensity) receptors produce action potentials in afferent neurons as long as the stimulus is applied. Examples include receptors that monitor arterial blood pressure in the carotid sinus, muscle spindles, pain fibers, and Krause's endbulbs. In tonic receptor function, the generator (or receptor) potential reaches a peak and then decreases to a lower, steady level (plateau), which is maintained as long as the stimulus persists. As a result, action potentials are initially produced at higher rates (frequencies) but become less frequent as the amplitude (size) of the generator potential decreases. The strength of the stimulus governs the amplitude of the generator potential, which in turn determines the frequency of formation of the action potential. Therefore, an increase in the strength of the stimulus leads to a larger generator potential, which leads to a higher frequency (rate of production) of action potentials in the afferent neuron. The opposite responses also occur.

Adaptation is the decline in the generator potential and decrease in frequency of action potentials that occur in the afferent neuron during a prolonged constant stimulus. Phasic receptors adapt rapidly, while tonic receptors do so slowly. Adaptation causes a decrease in or loss of sensation and is a means whereby the body (and brain) can adjust to stimuli. Were it not for adaptation, the brain would constantly be flooded with sensations, which might detract from its higher functions (such as concentration and judgment). For example, the body rapidly adapts to touch and pressure sensations from the hair receptors, allowing an individual to ignore their signals and concentrate on other matters (such as the study of physiology). Adaptation is also a means by which the body may distinguish constant stimuli from those that change. To illustrate, the phenomenon of adaptation occurs in the phasic photoreceptors of the eye, which rapidly adapt to light or dark or to a change in the intensity of light. Although the exact causes of adaptation are not known, some factors which may lead to it are; (1) a decrease in intensity of the stimulus as it passes from external tissues to the nerve fibers within modified receptors; (2) an unexplained decrease in responsiveness of the receptor even though the stimulus is constant and not dissipated; (3) a lesser likelihood that an action potential may develop in the afferent fiber owing to slow leakage of ions from its interior to the extracellular fluid.

Frequency Codes

The frequency with which sensory receptors develop action potentials produces information in a fashion somewhat equivalent to a Morse code. The spacing of signals can create a whole language of its own, much as a drummer can tap out distinctly different beats and rhythms. The number of action potentials produced in a specified amount of time increases in proportion to the intensity of the stimulus. The rate of formation of action potentials and their transmission along a nerve fiber might be likened to the rate of delivery of batteries by a conveyor belt running at a constant speed. The frequency of delivery of the batteries to the end of the conveyor belt depends on how close to each other they were placed at its beginning. So, too, intense stimuli produce closely spaced action potentials that move along a particular nerve fiber at a specific speed. Less intense stimuli space the action potentials further apart, resulting in their movement at the same rate of speed but in less frequent delivery at the end of a fiber. These variations alter the pattern of impulses delivered from a sensory receptor along its nerve fiber. That is, the *temporal sequence* of impulses varies as does the pattern of signals in a Morse code.

Number of Receptors

As a stimulus becomes stronger, the number of receptors that are excited increases. In addition to those receptors activated by weak stimuli, those activated by strong stimuli also respond. A stronger stimulus also causes sensory endings which overlap and mingle to fire action potentials. The CNS can distinguish between the number of afferent neurons involved as well as determine the frequency of impulses emitted. For example, heavy pressure on a finger is distinguished from light touch by stimulation of more receptors that produce action potentials as well as by the rate at which they generate them.

Pathway Codes

Impulses also may be distinguished in the CNS by the pathways over which they travel. The destination of impulses seems to play a critical role in their interpretation and the responses to them. There are specific, highly organized areas in the spinal cord and brain that contain clusters of neurons in areas which have been mapped for the receipt of specific impulses. For example, receptors in the eye transmit impulses along optic nerve pathways to the visual areas of the brain, while nerve pathways from the ear relay impulses to auditory (hearing) areas. Because of this specialization we do not *see* thunder or *hear* lightning. The structure and functions of discrete cells and pathways in the nervous system and their roles in regulating body function will be the subjects of Chaps. 9 through 11.

8-4 SPECIAL SENSE ORGANS

The Eye

Two types of photoreceptor cells, *rods* and *cones,* are located in the inner coat of the eye, the *retina,* which is surrounded by a number of other structures, as shown in Fig. 8-2. The functions of these components are summarized in Table 8-2.

The retina is an extension of the CNS and may be divided into several layers (Fig. 8-3). The rods and cones are its photosensitive end organs which respond to light. The anterior surface of the eye and retina are very transparent, so that light entering the anterior portion of the eye can travel to its posterior surface, striking cells of the *choroid coat.* The choroid coat contains blood vessels that supply the retina and contains cells with a black pigment, melanin, which absorbs some of the light and reflects the remainder to strike the rods and cones of the retina.

Because of the transparency of the tissues anterior to the choroid coat, blood vessels in this pigmented vascular layer can be observed directly with an ophthalmoscope. Such an examination may provide insight not only about diseases affecting the eye but also about those of the cardiovascular system. For example, irregularities due to atherosclerosis may appear in these vessels long before an individual is aware of onset of the disease. The lack of true intercellular connections between the pigmented layer of the choroid coat and the adjacent photoreceptor cells allows accidents or disease to separate them physically in a disorder called *retinal detachment,* which can cause blindness.

Physiology of Visual Impulses

When light is reflected from the choroid coat to the rods and cones, these cells generate impulses which are relayed to bipolar neurons and then to ganglion cells whose axons form the optic nerves (Fig. 8-3). Visual impulses then are relayed to the thalamus and on to visual centers in the brain (Tables 11-1, 11-7, 11-8; Figs. 11-7 and 11-10). Some ganglion cells behave as phasic–on receptors in response to impulses from the rods and cones. Others act as phasic–off receptors, signaling darkness in the absence of light.

The rods and cones contain various proteins called *opsins,* which are bound to different forms of colored pigments called *chromophores,* that are oxidized derivatives of vitamin A (retinal). The chromophores produce the color of rhodopsin and its sensitivity to light. Under the influence of light the chromophores are bleached and converted into a series of different shapes (conformations) and a neural dis-

FIG. 8-2
The general anatomy of the eye.

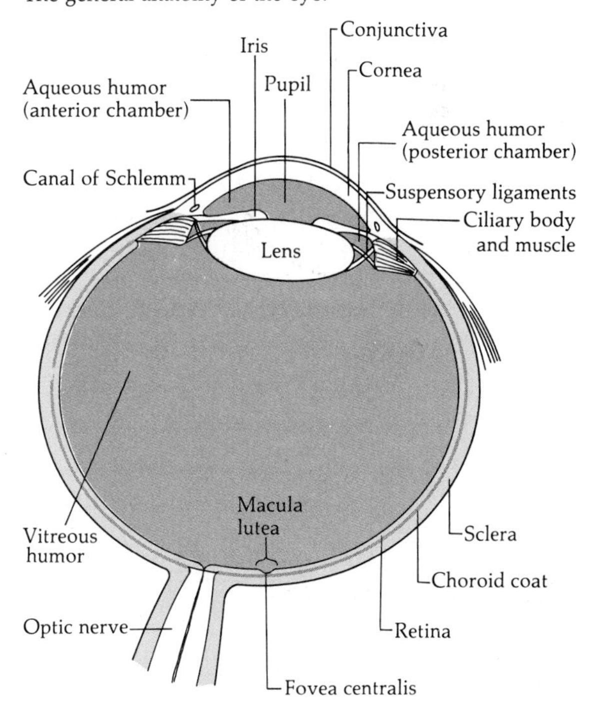

TABLE 8-2
Functions of components of the eye

Component	*Anatomy*	*Function*
Conjunctiva	Mucous membrane covering anterior surface	Moistens surface of eye; inflammation causes bloodshot eye in condition called *conjunctivitis*
Cornea	Transparent layers of epithelial cells just behind conjunctiva	Bends light, directing it toward lens
Lacrimal glands	Located on upper and lateral surface of eye Similar to salivary glands Secrete tears to lacrimal ducts	Moisten and wash the cornea with tears Contain antibacterial agents Tears moved across cornea by blinking eyelids Tears flow through lacrimal ducts into nasal corner of eye Tears move through nasolacrimal duct to nasal cavity
Aqueous humor	Transparent fluid secretion from epithelial cells of ciliary body	Fills chambers between cornea and anterior surface of lens Helps maintain shape of eye
Canal of Schlemm	A venous channel running around circumference of eye between the cornea and iris	Reabsorbs fluid (and other substances) from eye; maintains constant pressure within eye; loss of fluid regulation results in increased pressure within the eye, called *glaucoma,* which may adversely affect the blood supply and cause atrophy of the optic nerve and blindness
Iris	Ring of smooth muscle behind anterior chamber and in front of lens Pigment in cells	Reflexively regulates amount of light passing toward back of eye by widening or narrowing the size of the ring Various quantities of pigments (including melanin) produce eye color
Pupil	Opening in center of iris	Widens and narrows when iris contracts and relaxes
Ciliary body	Arises from sclera Tiny fibers attach lens to ciliary smooth muscles	Alters shape of lens (accommodation)
Lens	Epithelial cells form an elastic crystalline transparent structure	Bends light to direct and focus it on retina
Vitreous humor	Gel-like fluid with fibrils, minerals, glucose, protein, and other substances Fills the large posterior cavity of the eye	Helps maintain shape of eyeball Holds retina in place Nourishes lens and retina
Retina (inner layer of eye)	Layers of neurons in posterior and inner surface of eyeball form the innermost membrane of the eye to which all other structures of the eye are functionally subordinate	Photoreceptors in shape of rods and cones respond to black and white and to colors, respectively Synaptic neurons relay impulses through axons of ganglion cells which form optic nerve fibers
Fovea	Specialized central region in retina Contains only cones Surrounded by region with high density of cones, the macula lutea or "yellow spot"	A 1 : 1 ratio of cones and bipolar neurons in the retina characterizes the area with greatest visual acuity in sufficient light
Optic nerve	Bundle of axons from retina to visual areas of brain	Transmit inverted and reversed images from retina to brain Fibers from nasal halves of each retina cross over to opposite side of brain, others remaining uncrossed; brain receives impulses, sorts and rearranges them to produce normal visual perception

TABLE 8-2 *(Continued)*

Component	*Anatomy*	*Function*
Optic disc	Area in which optic nerve is formed by axons in the retina	Forms a blind spot (light-insensitive area)
Sclera (outer layer, the white of the eye)	Outer fibrous connective tissue coat	Modified at front of eye into cornea Helps maintain shape of eye
Choroid coat (middle layer)	Inner vascular lining between sclera and retina	Pigment absorbs light and reflects some to rods and cones Supplies retina with blood Anterior segment forms the ciliary body
Extraocular muscles	Six muscles attached to sclera	Move eye

charge is released that stimulates visual perception.

Rods and cones are *photoelectric transducers* which are designed to respond to light. The rods contain stacks of membranes shaped like discs (Fig. 8-3), which have an abundance of the pigment *rhodopsin* (visual purple). Rhodopsin dissociates into opsin and its chromophore when exposed to light (Fig. 8-4). Rhodopsin is an unstable protein which forms a channel in the cell membrane, a transmembrane protein that regulates the movement of ions. When excited by light, the membranous discs of rhodopsin release Ca^{2+} ions, which decrease the channel's permeability to Na^+ ions. Consequently, in the light, rods and cones accumulate more Na^+ ions on their external membrane surfaces and become hyperpolarized. The hyperpolarization of the cells initiates electrical activity, which develops local generator, or receptor, potentials that ultimately produce nerve impulses, which flow to synapses with other neurons in the retina. This response, in which hyperpolarization of an excitable membrane causes a nerve impulse, is opposite to that which occurs in other excitable cells. The nature of the transmitter agents released at the synapses between rods (or cones) and other cells in the retina is not certain, but they may include ACh.

In the absence of light these reactions are enzymatically reversed and the rod's membranes are less polarized. Consequently, under conditions of dim light, twilight, or night vision, when the eye undergoes *dark adaptation,* rhodopsin is more abundant in the rods; thus these cells and their associated pathways are more useful in that type of light. Because the restoration of rhodopsin occurs in the dark, twilight vision improves within a few minutes after a person enters a darkened room. The retinas of nocturnal animals such as owls and bats have a high concentration of rods, which facilitates their vision in dim light. Conversely, the sensitivity of the rods decreases in bright light in a response called *light adaptation.*

Because the pathways of the rods overlap and converge, summation of subthreshold stimuli can occur. These pathways produce less visual acuity but greater sensitivity for detection of light and dark. Consequently, twilight or night vision (*scotopic vision*), characteristically transmitted by rods, does not allow perception of the fine details and contours of an object. Although this type of vision detects low levels of light, it does not keep separate two distinct but physically close points or lines. It is for this reason that shadows in the night take on ill-defined outlines.

Cone cells are highly concentrated in a central area of the retina, the *macula lutea,* and are the only photoreceptor cells found in its central depression, the *fovea centralis.* Three types of cones can be distinguished on the basis of pigments that seem to be responsible for color vision and vision in bright light (*photopic vision*). Different opsins exist in the pigments of each of three types of cone cells that show maximal sensitivity to red, green, and violet light, respectively.

According to the Young-Helmholtz theory,

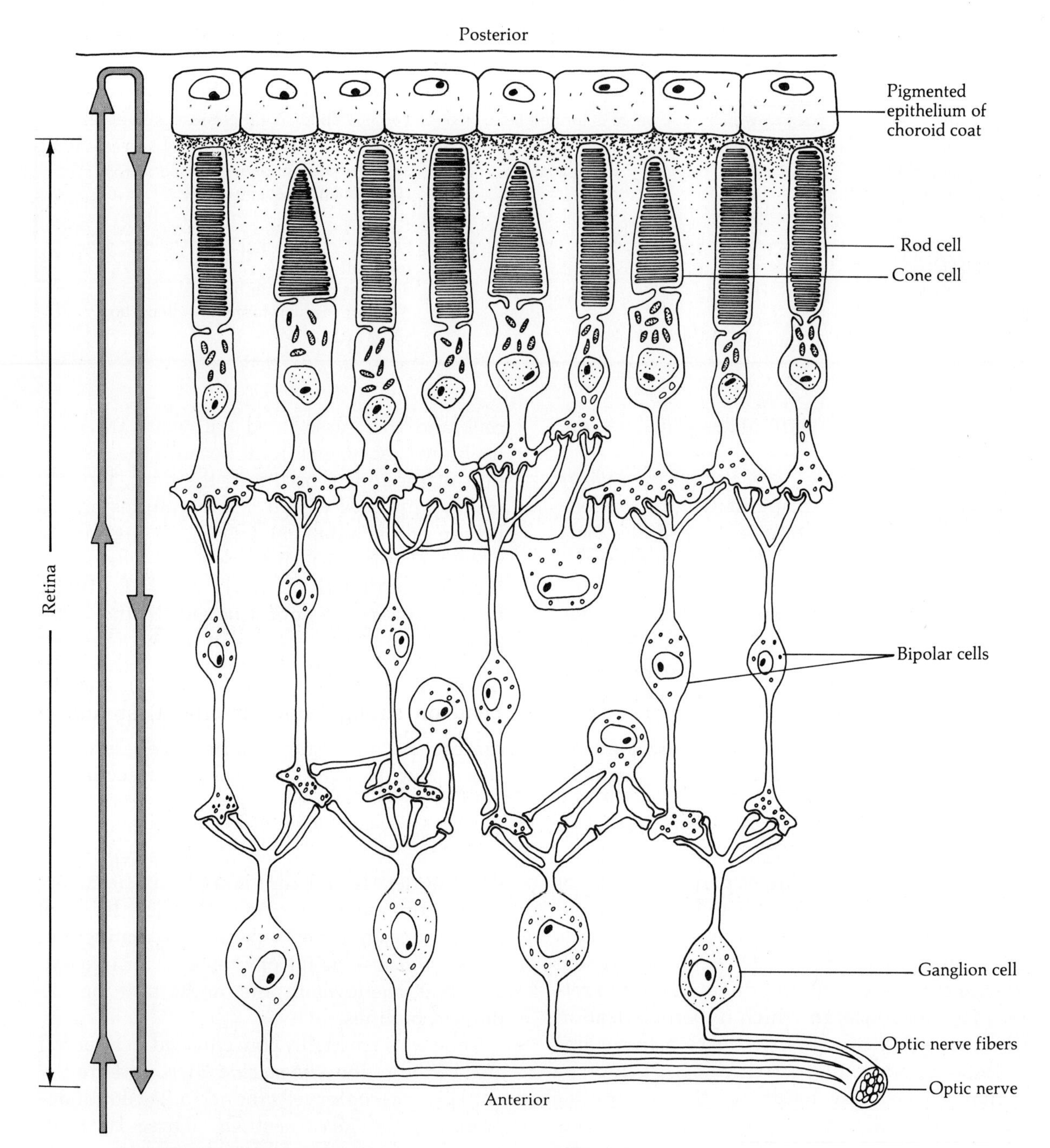

FIG. 8-3
General structure of the retina.

color is determined by the relative frequency of impulses sent to the brain by the three systems of cones. The cones are phasic receptors which cause on-off coded signals. For example, some of the neurons in the retina are turned on by red and off by green light, and vice versa. Colors reflected by objects may travel to the retina to excite the specific cone cells, whose sig-

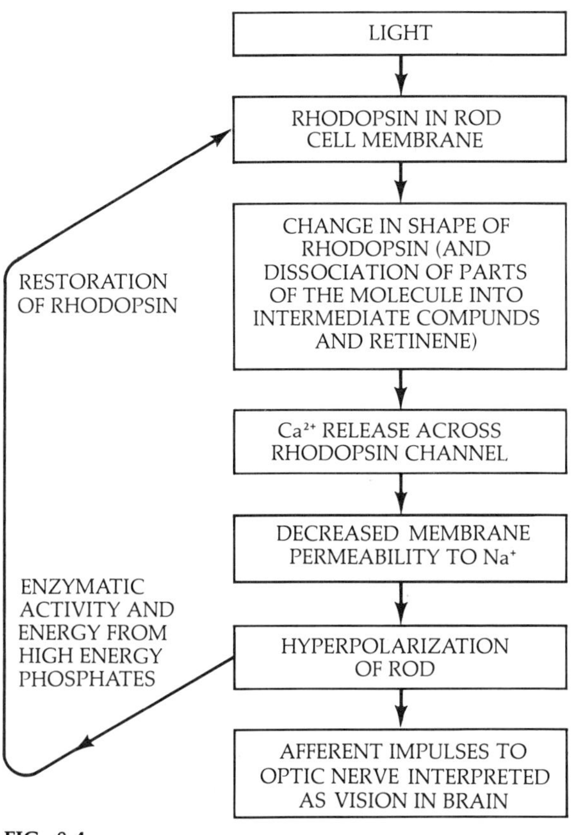

FIG. 8-4
Effects of light on rhodopsin and the formation of visual impulses.

nals are then interpreted by the brain to produce color vision.

Abnormalities in color vision may be due to inability to distinguish colors or to a weakness in one or more of the three cone systems. Some disorders are heritable and due to recessive X-linked (sex-linked) traits. The abnormalities are more common in males (who have only one X chromosome per cell) and less common in females (since they have two X chromosomes per cell and must have two recessive genes to develop the abnormality).

The difference between rods and cones is analogous to that between fine- and coarse-grained photographic film. The rods, like a coarse-grained film, provide less detail in a picture but are more sensitive to light than the cones, which are like a fine-grained film. The system of cones and their associated pathways demonstrates more direct lines to the brain, whereas impulses from the rods are carried along overlapping, converging paths. As a result, color vision is transmitted by cone cells with little overlap, which produce a greater clarity of retinal focus (visual acuity), and depend on strong (above threshold) stimuli of bright light.

Focusing of Light on the Retina

Because the eye is curved, light reaches its outer surface at different angles. When light

FIG. 8-5
Effects of cornea (prefocus) and lens (fine focus) on light entering the eye.

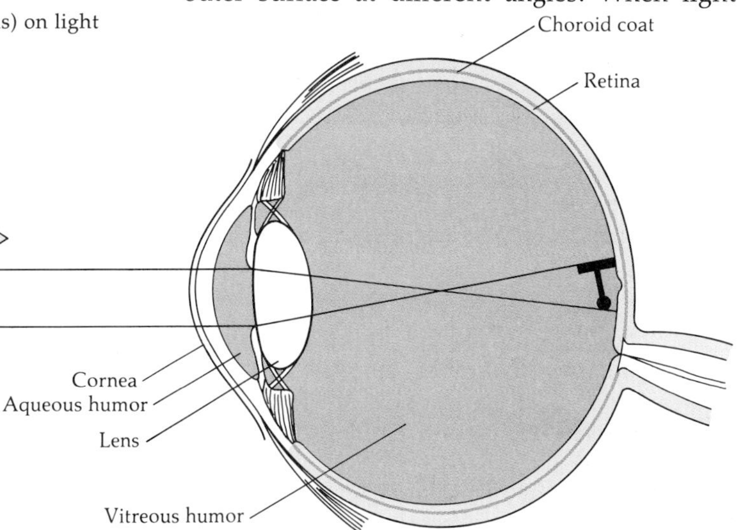

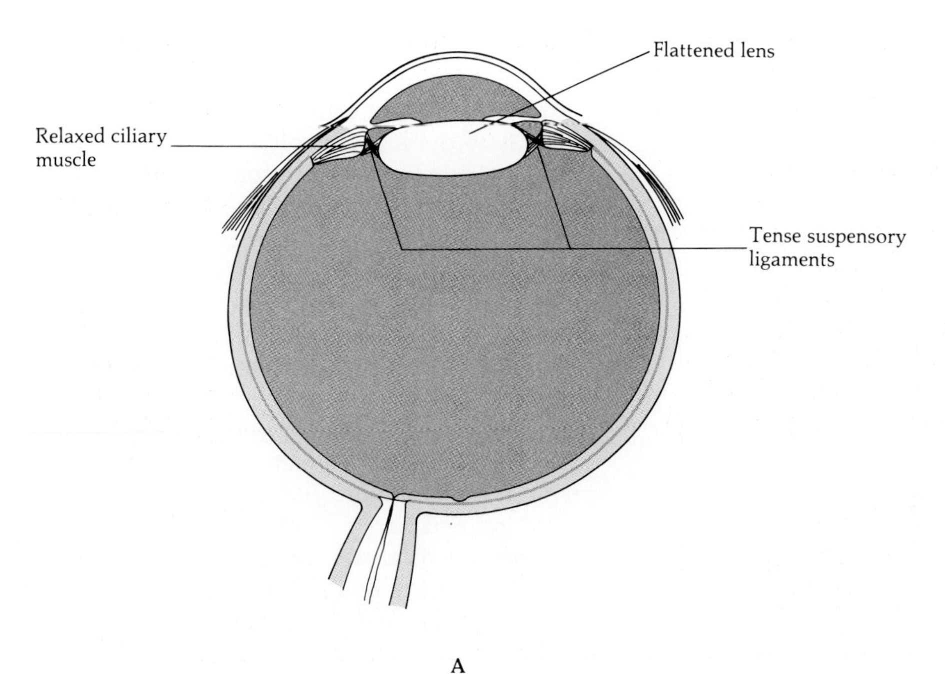

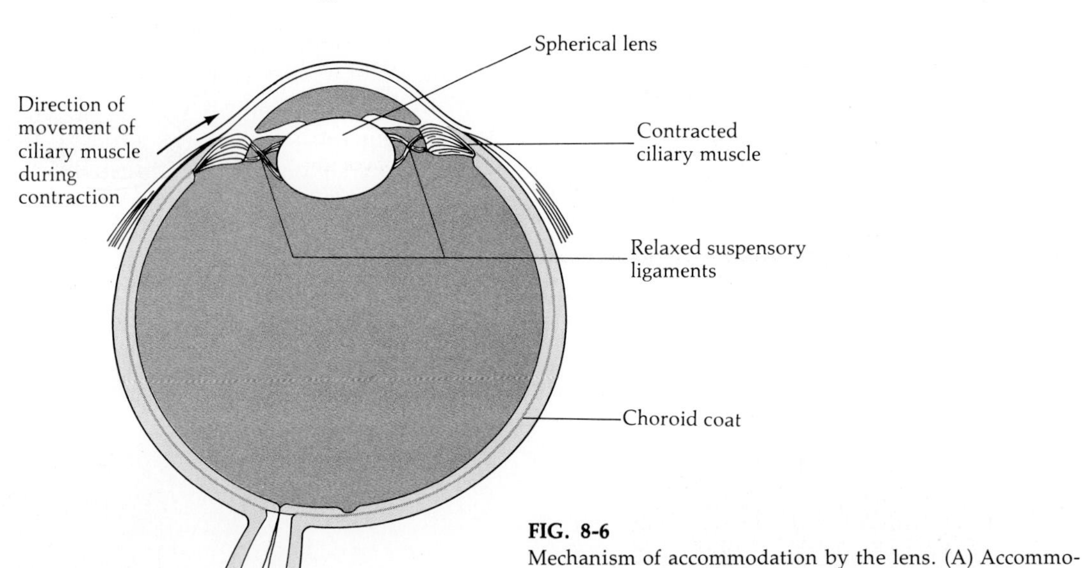

FIG. 8-6
Mechanism of accommodation by the lens. (A) Accommodation for distant vision: relaxation of ciliary muscle and flattening of lens. (B) Accommodation for near-point vision: contraction of ciliary muscle and relaxation of lens to convex shape.

passes from the air to the eye, it moves through the cornea, aqueous humor, lens, and vitreous humor, all of which are transparent. The cornea and lens, especially, bend (refract) light rays (Fig. 8-5).

The shape of the lens may be changed from a sphere to a more flattened body and may bend the light more or less in a process called *accommodation.* Ciliary muscles respond automatically to light and are attached to ligaments that hold the lens in place and maintain its shape.

Relaxation of the muscles allows forces such as the fluid pressure within the eye, to increase tension on the ligaments and stretch the lens to a flattened shape (Fig. 8-6). Contraction of the muscles eases tension on the lens and allows it to assume its natural more spherical shape. Under these opposite conditions the lens bends light rays and brings them together to a lesser or greater degree in order to focus them on the retina for distant or near vision, respectively.

Many people (especially on aging) have insufficient power of accommodation of the lens to focus light on the retina correctly (Fig. 8-7). In some individuals the eyeball is too short (front to rear) and the lens focuses light behind the retina, causing farsightedness (*hyperopia*). The person has good distant vision but poor near vision (close or near-point vision). The opposite condition is nearsightedness (*myopia*).

FIG. 8-7
Correction of optical defects by means of artificial lenses. (A) Hyperopia (farsightedness). (B) Correction of hyperopia with artificial convex lens. (C) Myopia (nearsightedness). (D) Correction of myopia with artificial concave lens.

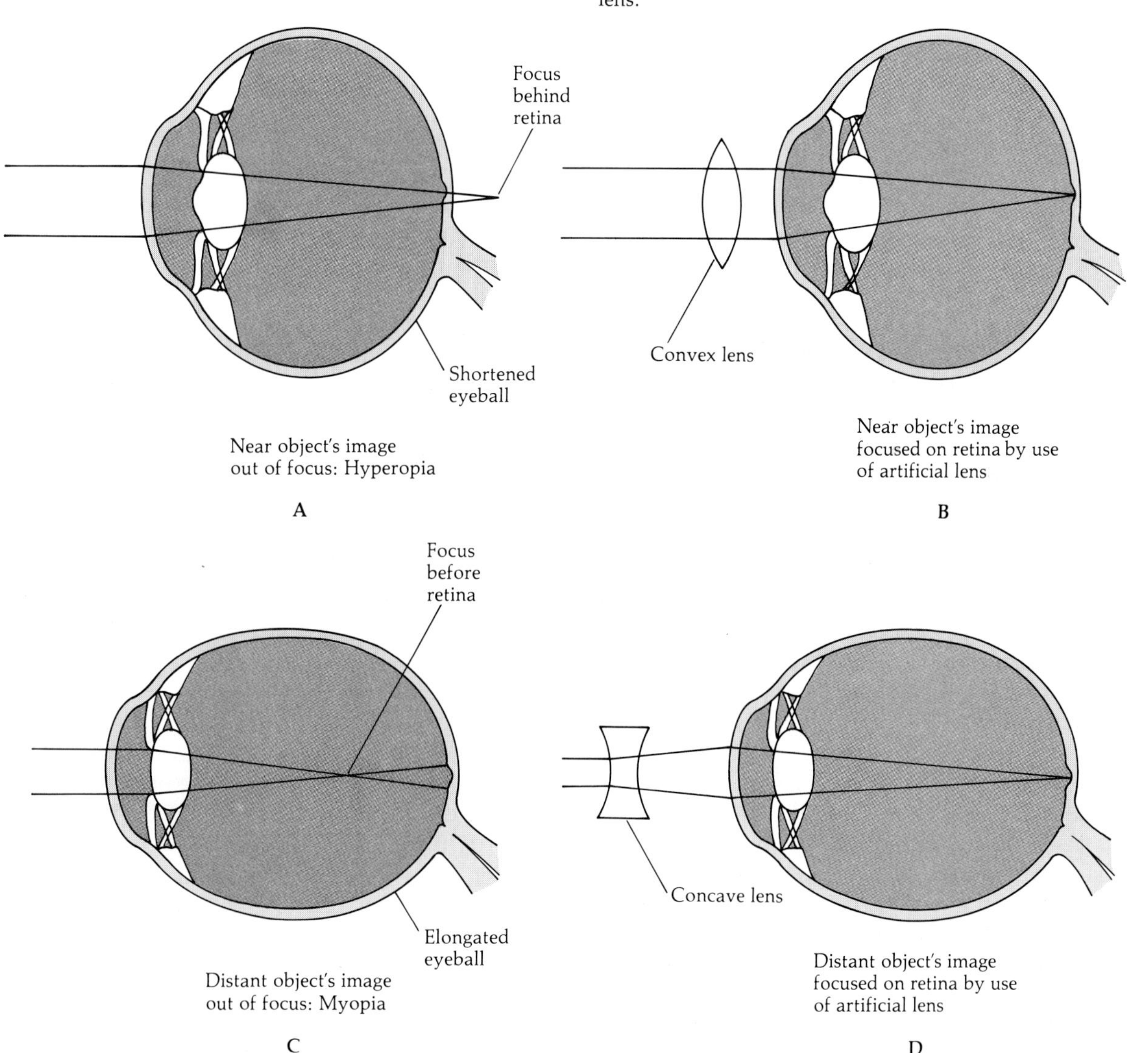

When the curvature of the cornea or lens is irregular, light is focused unevenly on the retina, causing visual blurring or *astigmatism*. An image coming from a single point is spread in various directions by the abnormal cornea or lens, causing different focal points on the retina and blurring of the image. Artificial lenses may be polished for use in bending or diverging light rays in order to focus them on the retina. Cylindrical lenses are used to correct astigmatism, convex lenses to eliminate hyperopia and concave lenses are used to eliminate myopia (Fig. 8-7).

The average person's vision allows focusing of parallel rays of light on the retina reflected from an object 20 ft away. That individual has 20/20 vision and the clarity of focus on the retina, the *visual acuity*, is used as a standard to which other people's eyesight is compared. To have 20/20 vision simply means to see an object as a normal person does when it is 20 ft away. Because each eye may vary independently of the other, visual acuity is measured separately by blocking one eye or the other (for example by covering one eye while testing the other). Thus, visual acuity is expressed as the distance from which a specific person must view an object for clear vision, divided by the distance required for an individual with standard eyesight. For example, for a myopic person with 20/40 vision, an object must be 20 ft away to be seen clearly whereas it is seen clearly by a normal person at a distance of 40 ft.

As an individual ages, the lens loses its ability to make sufficient accommodation to maintain normal vision. The loss of elasticity of the lens and a corresponding loss of its power of accommodation (*presbyopia*, or "old eye") begin about age 8 and become apparent in most individuals in their forties. As a result of the gradual hardening of the lens many adults beyond age 40 must use artificial convex lenses for near-point vision. They have hyperopia.

Aging of the lens is related to its patterns of nutrition and growth. The lens does not have its own blood vessels. Nutrients diffuse to its cells from fluids bathing its surface, and new cells which are produced at its surface are best maintained. These new cells grow over older ones, which become buried in the lens farther away from the blood vessels which provide for the exchange of nutrients and wastes. As increasing numbers of its cells age and die, the center of the lens becomes more opaque (impenetrable to light). Pigmented compounds may accumulate in sufficient quantities to impair vision significantly in a degenerative disorder called *cataract*. When vision is sufficiently impeded by cataracts, the lenses may be replaced by artificial lenses to restore some vision.

The Ear

Mechanoreceptors in the *auditory* apparatus of the ear transduce sound waves into nerve impulses that are interpreted as sounds in the brain. Receptors in the *vestibular* apparatus detect gravitational forces, movements of the head, and changes in position of the body.

The ear is structurally divided into external, middle, and inner chambers, which are illustrated in Fig. 8-8 and whose functions are summarized in Tables 8-3 and 8-4. In the inner ear mechanoreceptors formed by ciliated hair cells are located within membranous labyrinths that are filled with *endolymph* fluid and surrounded by *perilymph* fluid. Sensory signals are generated by movement of cilia that depolarize the hair cells, causing the release of transmitters. The transmitters bind to specific receptors of associated afferent nerve fibers and promote action potentials. This mechanism is common to the perception of hearing, gravity, motion, and equilibrium.

Hearing

Sound waves move through the outer ear canal to strike the tympanic membrane of the middle ear, which relays vibrations through three bones to an area near the inner ear called the *oval window* (Fig. 8-8). This area is smaller than that of the tympanic membrane. As a result, the force of the signals are amplified and then transmitted as vibrations within the perilymph of the vestibule to the *vestibular membrane*, which generates waves in the endolymph fluid

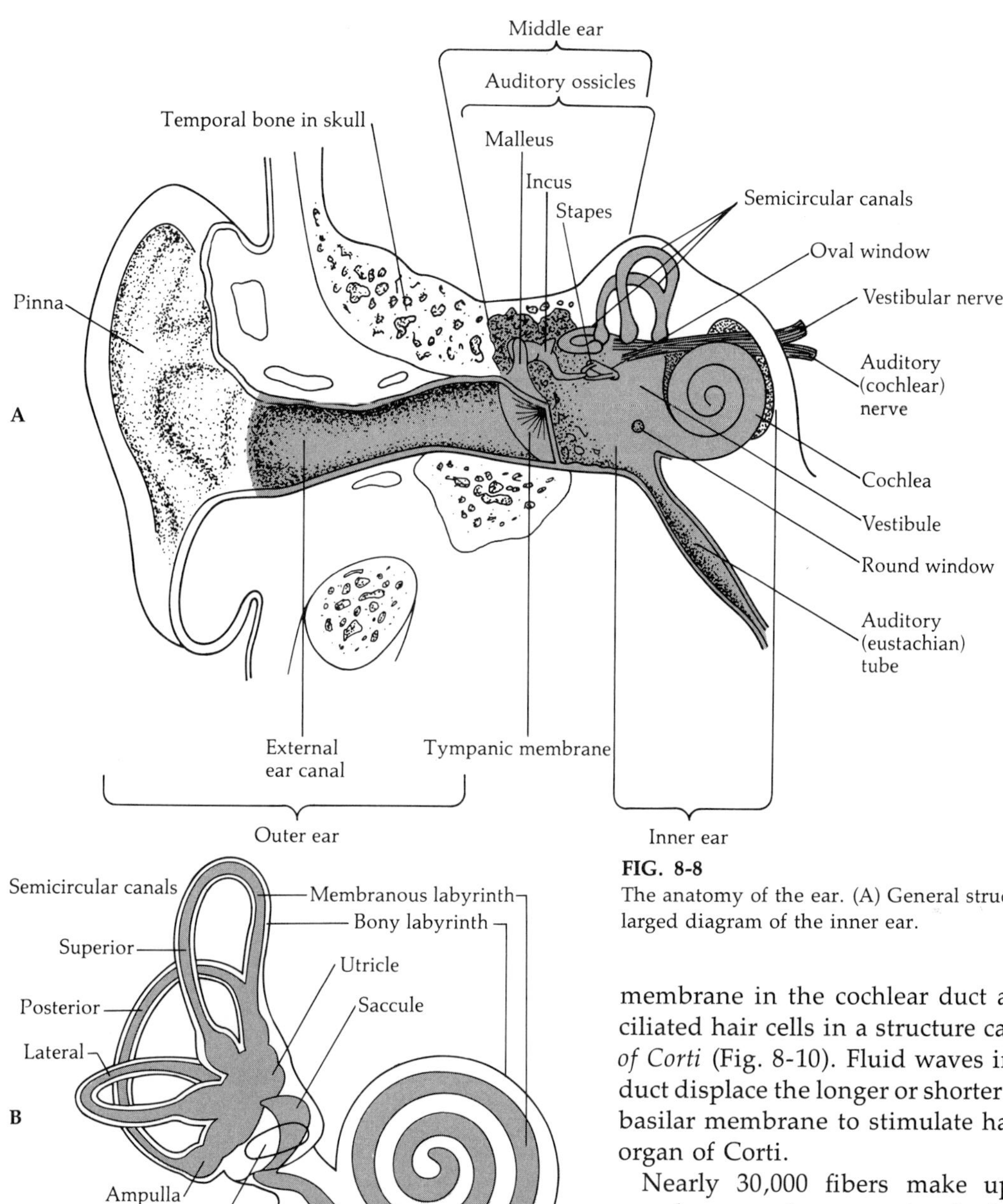

FIG. 8-8
The anatomy of the ear. (A) General structure. (B) Enlarged diagram of the inner ear.

within the *scala media* (the cochlear duct, Fig. 8-9). The *cochlear duct* is the middle membranous chamber among three located in a spiral, snail-shaped tube, the *cochlea,* which is in the inner ear (Fig. 8-10). *Auditory receptors* (phonoreceptors) which rest on the basilar membrane in the cochlear duct are formed by ciliated hair cells in a structure called the *organ of Corti* (Fig. 8-10). Fluid waves in the cochlear duct displace the longer or shorter strands in the basilar membrane to stimulate hair cells of the organ of Corti.

Nearly 30,000 fibers make up the basilar membrane and are associated with a like number of hair cells and nerve fibers of the organ of Corti. The part of the basilar membrane nearest the oval window (in the base of the cochlea) is narrow and relatively inflexible and responds best to high-pitched sound waves. Strands of the basilar membrane located at greater distances from the oval window (near the apex of the cochlea) are wider and more flexible and resonate more slowly, responding to low-pitched sound waves. In effect, a gradient of re-

TABLE 8-3
Acoustic functions of the ear

Component	*Anatomy*	*Function*
Outer ear		
Pinna (auricle), the external ear	Soft tissue (skin, cartilage, and fat)	Focuses sound for passage into ear
Ear canal (external auditory tube)	S-shaped canal lined with wax-producing (ceruminous) glands and hairs	Directs sound inward. Protects ear from entry of foreign matter
Middle ear	Cavity in temporal bone of skull	
Tympanic membrane (eardrum)	Mechanoreceptor in a membrane separating outer and middle ear	Compressed by force and velocity of air molecules by the energy from sound waves; returns to original position between waves. The frequency of vibration of the membrane is proportional to the number of waves per second, increasing as the pitch of sound becomes higher The degree of displacement increases as the sound becomes louder Moves malleus
Auditory tube (Eustachian canal)	Narrow tube between middle ear and nasopharynx	Opens during yawning, chewing, swallowing, sneezing; allows air pressure between outer and middle ear to equalize
Ear bones (auditory ossicles)	Attached to walls of tympanic cavity by ligaments; articulate with each other	Act in sequence as levers relaying vibrations of tympanic membrane to fluid in inner ear
Malleus (hammer)	Attaches to inner surface of tympanic membrane	Moves incus
Incus (anvil)	Connects malleus and stapes	Moves stapes
Stapes (stirrup or footplate)	Process, called a footplate, attaches through ligament to the oval window (opening) in the inner ear	The stapes acts as a piston in the oval window, its action generating a fluid wave, which moves through the cochlea
Skeletal muscles		Limit movement of ossicles Protect tympanic membrane from damage due to excessive vibration by high-pitched sounds; protect inner ear by damping sound waves over 90 decibels, which can impair hearing of high tones
Tensor tympani	Attaches malleus to auditory tube and bones of skull	Pulls malleus medially Increases tension on tympanic membrane
Stapedius	Attaches stapes to bones of skull	Pulls stapes so that its base pushes on the oval window, increasing pressure in the inner ear
Inner ear (labyrinth)		
Bony labyrinth	Channels in temporal bone of skull	House and protect the membranous labyrinth and the ear's sensory receptors
Cochlea	Located within bony labyrinth. Spiral tube, about 35 mm long, resembles a snail's shell with 2½ turns. Separated into three chambers (scalae) by two membranes (vestibular and basilar membranes)	Houses organ of hearing
Scala vestibuli	Upper chamber Oval window of the middle ear is located at end of scala vestibuli. Filled with perilymph	Stapes in middle ear fits into oval window, which abuts this chamber Sound stimulates movement of stapes, whose movement compresses perilymph
Scala tympani	Lower chamber. Membranous round window of the middle ear at end of scala tympani. Filled with perilymph. Communicates with upper chamber by a common opening at the tip of the cochlea	Sound waves compress the fluid in the inner ear, causing the round window to bulge outward

TABLE 8-3 *(Continued)*

Component	*Anatomy*	*Function*
Scala media (cochlear duct)	Middle chamber. Separated from upper chamber by vestibular (Reissner's) membrane Filled with endolymph Separated from lower chamber by basilar membrane	Houses organ of Corti (organ of hearing)
Basilar membrane	Up to 30,000 strands of tissue of varying lengths, forming a membrane which separates the middle and lower chambers of the cochlea	Vibrations of fluid cause selective movement of shorter and longer strands, allowing discrimination of higher and lower pitch of sounds; they occur respectively closer to and further from the oval window near the middle ear
Organ of Corti	Spiral membrane with up to 30,000 hair cells with ciliated receptors Up to 30,000 fibers of the auditory division of the statoacoustic nerve carry impulses to specific auditory regions in brain	Vibrations of 20–20,000 per second cause basilar membrane to bend cilia and depolarize the hair cells Transmitters released from the hair cells depolarize the dendrites of afferent neurons An increase in loudness of sound produces more vibration of the basilar membrane, greater depolarization of the hair cells, and a higher frequency of action potentials in the auditory nerve fibers

TABLE 8-4
Vestibular functions of the ear

Components of Vestibular Apparatus	*Anatomy*	*Function*
Semicircular canals	Membranous channels in three planes at right angles to each other and located superior to the stapes and lateral to the cochlea Surrounded by perilymph; filled with endolymph	Contains fluid which moves within channels in response to changes in the velocity of movement of the head
Ampulla	Expanded end of each of three canals each with a receptor organ, the crista ampullaris	Houses mechanoreceptors (cristae) that detect acceleration or deceleration, sense alterations in the rate of change of motion, and produce action potentials in vestibular division of the statoacoustic nerve (vestibular/auditory nerves)
Crista ampullaris	Hair cells with cilia and supporting cells are capped by a gelatinous mass within each ampulla	Movement of hairs back and forth occur with change in speed of motion of body. Cilia bent in one direction depolarize associated afferent neurons; cilia bent in the opposite direction hyperpolarize and inhibit the afferent neurons. Release of some transmitter occurs without motion, producing a low frequency of action potentials to indicate a lack of motion
Utricle and saccule	These are bulges in the membranous labyrinth of the semicircular canals and are located between the ampullae and the cochlea	House mechanoreceptors (maculae) that detect the position of the head and changes in its linear acceleration
Maculae	Sensory receptors with hair cells; hair cells are surrounded by a gelatinous covering in which calcium carbonate crystals (otoliths) are embedded	Affected by gravity or acceleration Horizontal position of hair cells in utricles senses erect position of head; near vertical position of resting hair cells in saccule poised to detect vibrations. Sensory impulses which are generated are used to cause reflex movement of the eyes to maintain them in a fixed position in spite of body movement

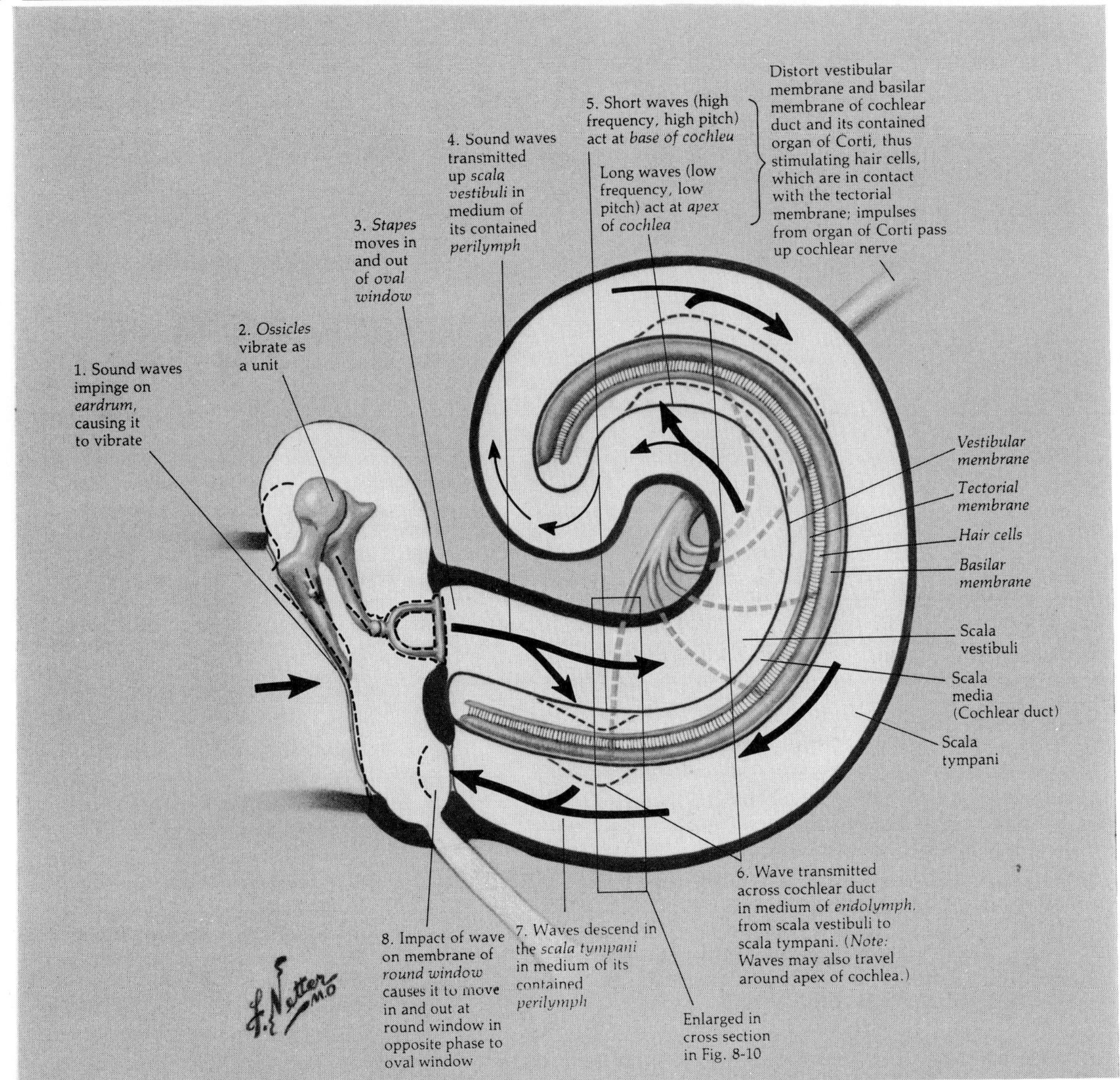

FIG. 8-9
An enlarged view of the cochlea in the inner ear. Numbers 1 through 8 depict the route over which sound waves usually generate impulses for hearing. The arrows indicate the direction of flow of fluid in the cochlea. The area which is boxed in is enlarged in Fig. 8-10. © Copyright 1969. CIBA Pharmaceutical Company Division of CIBA-GEIGY Corporation. Reprinted with permission from *Clinical Symposia*, illustrated by Frank H. Netter, M.D. All rights reserved.

sponsiveness exists for sounds of high to low frequencies along the basilar membrane and its associated hair cells and nerve fibers. The pitch of sound is distinguished by the degree of displacement of the basilar membrane and the rate of discharge (frequency) of impulses from the hair cells and their associated neurons. High-pitched sound waves displace the shorter fibers closer to the base of the cochlea (Fig. 8-9). Low-pitched sound waves displace the broader segments of the basilar membrane near the apex of the cochlea. In this fashion, the different strands of the basilar membrane are similar to the strings of a harp suited to vibrating at specific frequencies. The basilar membrane vibrates and pushes the hair cells and their cilia against the tectorial membrane. This generates

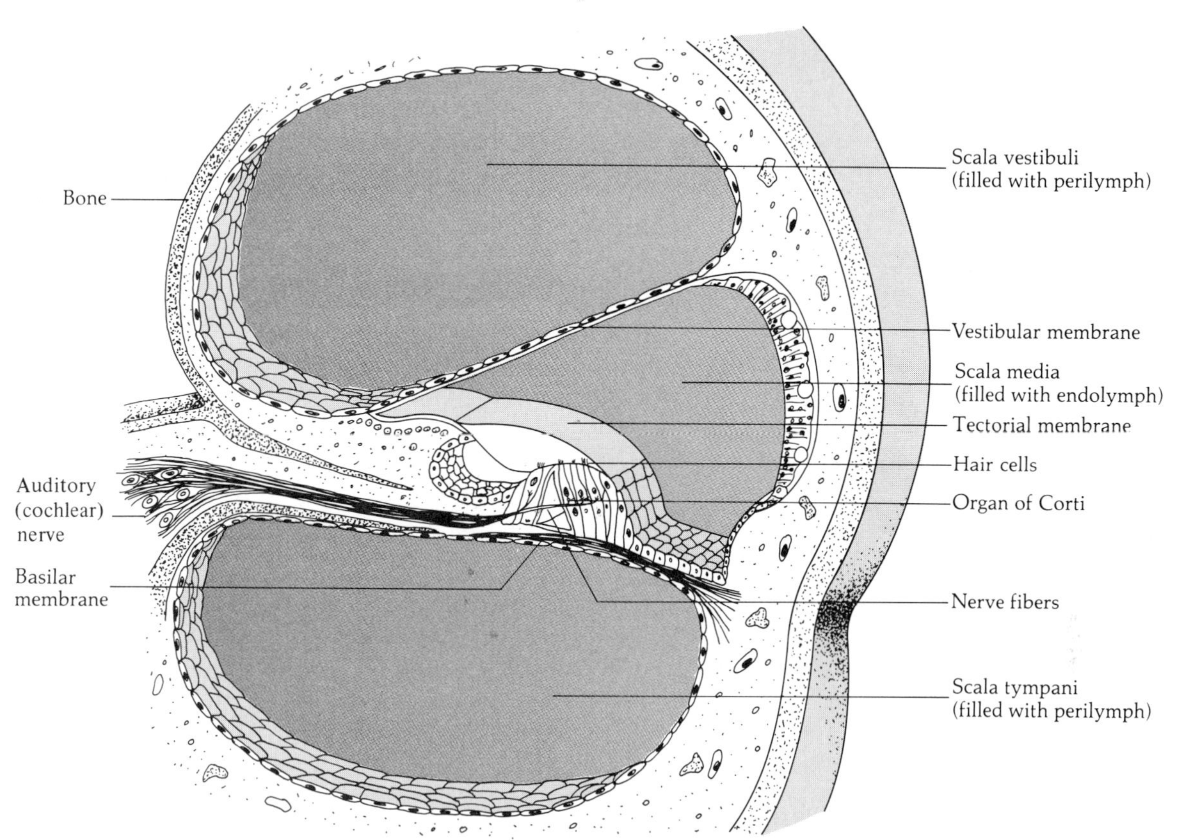

FIG. 8-10
The organ of Corti in the inner ear. A cross section of the cochlea enlarged to demonstrate the organ of Corti and the details of the sensory hair cells. The area marked with a rectangle in Fig. 8-9 is enlarged in this diagram.

nerve impulses which are carried by auditory fibers in the auditory (or cochlear) nerve (a branch of the statoacoustic nerve) to areas of the brain which perceive and discriminate sounds.

The volume (loudness) of sound is related to its intensity and is a function of the amplitude of a wave. Louder sound waves produce more pressure on auditory membranes than softer ones. The intensity of sound is measured in units called decibels and compared with a standard pressure, which is at the threshold of hearing and assigned an arbitrary value of zero. The ear can perceive without damage sounds whose intensities vary a trillionfold and do so as the tympanic membrane is displaced a distance less than the diameter of a hydrogen atom (0.1 nm).

Vestibular Functions: Perception of Gravity and Movement

The perception of gravitational changes, acceleration, and movement of the head are due to the vestibular apparatus and proprioceptors in skeletal muscle. The *vestibular apparatus* is a membranous labyrinth in the inner ear which contains sensory organs (Table 8-4). They are located at the ends of three semicircular canals (each of which occupies a different plane in space above and lateral to the cochlea) and also in two sacs, the *utricle* and *saccule,* which lie between the semicircular canals and the cochlea (Figs. 8-8B and 8-11A).

A sensory organ, the *crista ampullaris,* contains mechanoreceptors and is located in an enlargement (ampulla) at the end of each semicircular canal (Fig. 8-11B). These cristae possess hair cells whose cilia are covered with a gelatinous mass and are bent back and forth by endolymph, which moves as does fluid in a moving container. Because of the arrangements of the

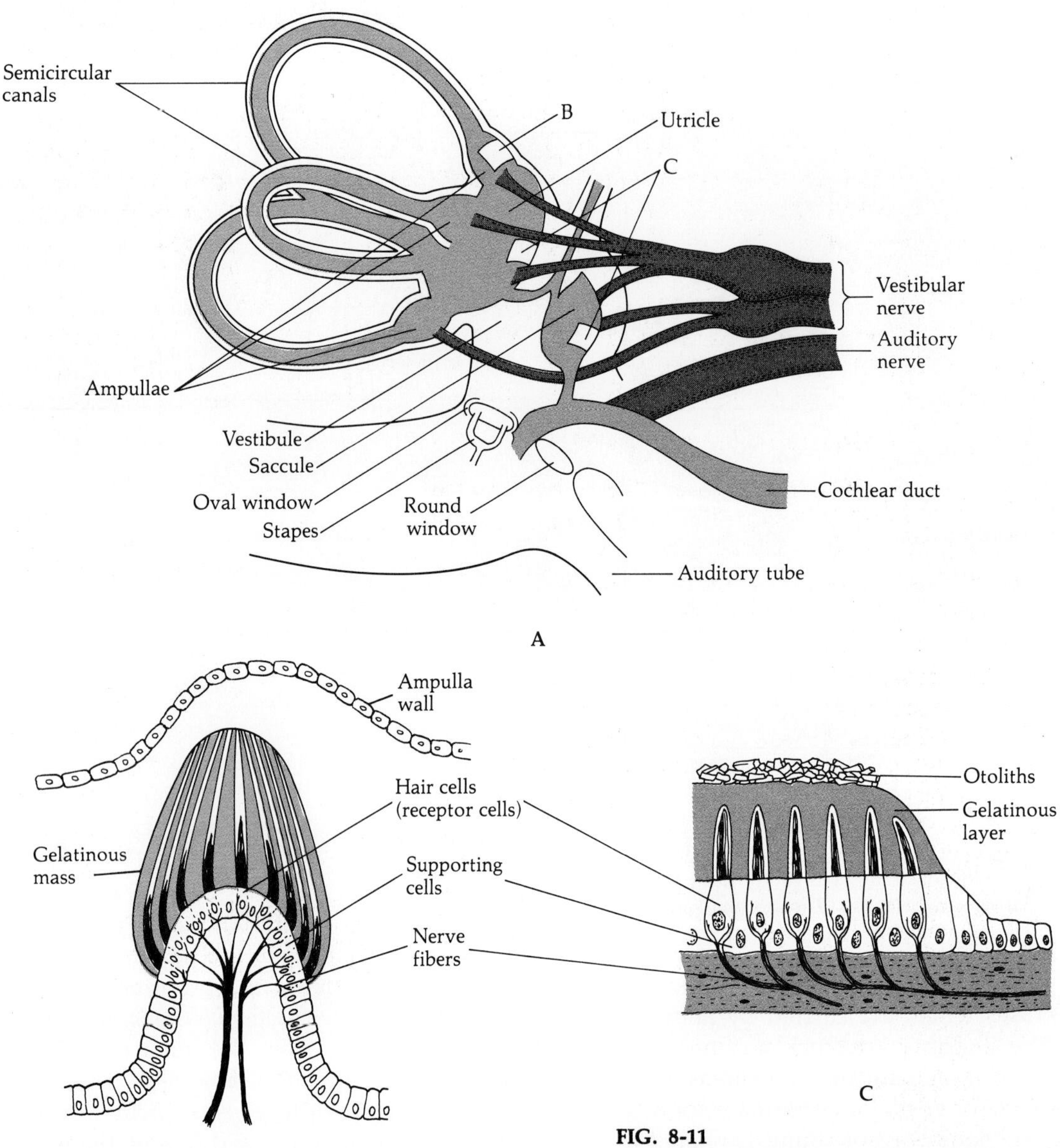

FIG. 8-11
The vestibular apparatus. (A) The vestibular apparatus. (B) A crista ampullaris; detects changes in rate of movement: a dynamic sensor. (C) A macula from a section of the wall of the utricle or saccule: a gravity (static) sensor.

semicircular canals, movement of the body in any direction affects one or more cristae, which are dynamic sensors. As a result the semicircular canals and their cristae are well suited to detect rotational acceleration, such as swiveling in a chair. The shifts of fluid are caused by acceleration or deceleration of the head (much as grocery bags shift when an automobile accelerates or decelerates). When stimulated, the hair cells release transmitters that generate impulses in the vestibular division of the *statoacoustic nerve* to provide sensations of the rate of change of movement.

Ciliated receptors, the *maculae,* are static or gravity sensors located in the utricle and saccule and covered by calcium carbonate crystals called *otoliths.* When the head is upright, the utricle is almost horizontal and the saccule almost vertical in position (Fig. 8-8), and thus are well placed to detect the erect position of the head and vibrations, respectively. Changes in the force of gravity caused by linear acceleration in the rate of movement of the head cause the endolymph to shift and displace the otoliths in the maculae. The hair cells may develop impulses, which are carried by afferent fibers in the statoacoustic nerve to the brain and indicate the position of the head and changes in the force of gravity. Movements of the head which cause displacement of otoliths are much like those which might be observed when a person carries a moist ice cube in a smooth cup. As long as the rate of movement is steady and the cup is not tipped, the ice cube remains in the same position. If there are jerky starts or stops (alterations in the rate of movement) or the cup is tipped, the ice cube slides back and forth within the confines imposed by the walls of the cup.

Taste (Gustation)

Taste buds are clusters of epithelial cells located on elevations (papillae) of the tongue's surface (Fig. 8-12). About 10,000 taste buds are located on the human tongue. Taste receptors also are located in the roof of the mouth and on the pharynx and epiglottis. A common opening (pore) leads to cilated hair receptor cells resembling those in the inner ear. However, taste receptors behave as *chemoreceptors,* not as mechanoreceptors, as do the sense organs in the ear.

The hair cells are chemoreceptors which respond to substances in fluids. Cilia of up to 18 hair cells move fluids in and out of the pore of each taste bud, in which as many as 50 nerve fibers are located. Interconnecting nerve fibers link an average of five taste buds together. It is unclear whether food molecules in fluid change the ionic composition around the receptor hair cells or whether they bind to protein receptors on the hair cells to initiate ionic changes. In either case, in the presence of specific chemicals impulses are generated, the frequency of which forms a sensory code for taste. The nerve impulses are carried to the brain by the facial (VII), glossopharyngeal (IX), and vagal (X) cranial nerves.

Four basic sensations are detected by the receptors. Those for sweet, salt, and bitter taste are found in greatest densities in an area stretching from the tip to the back of the tongue, respectively. Those for perception of sour taste are found in great numbers on the sides of the tongue.

Some receptors generate a specific taste code and others produce more than one, but no single receptor seems to relay all four types of taste. Moreover, the flavor of food is produced

FIG. 8-12
Sensory receptors in a taste bud in the tongue. (A) Diagram of papillae on surface of tongue. (B) Detail of a taste bud, pore, and surrounding cells.

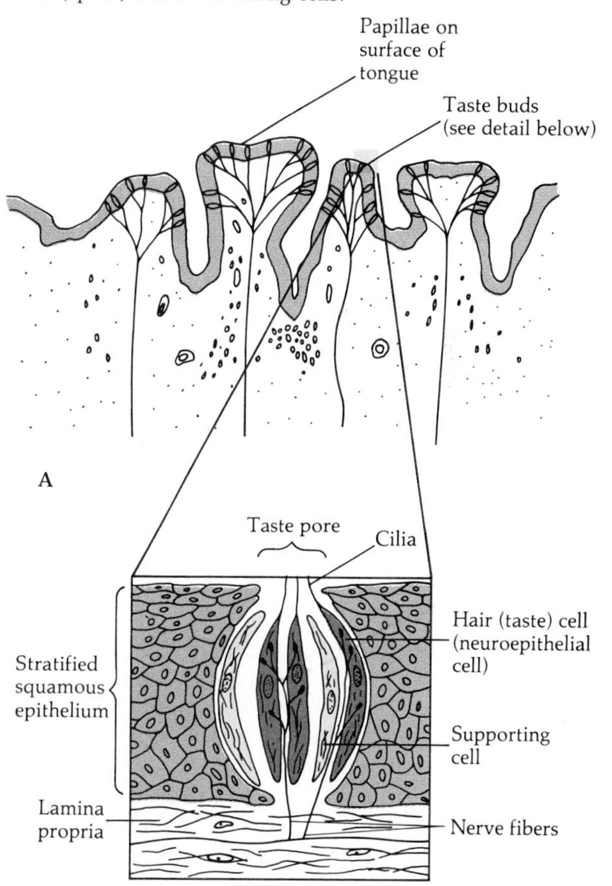

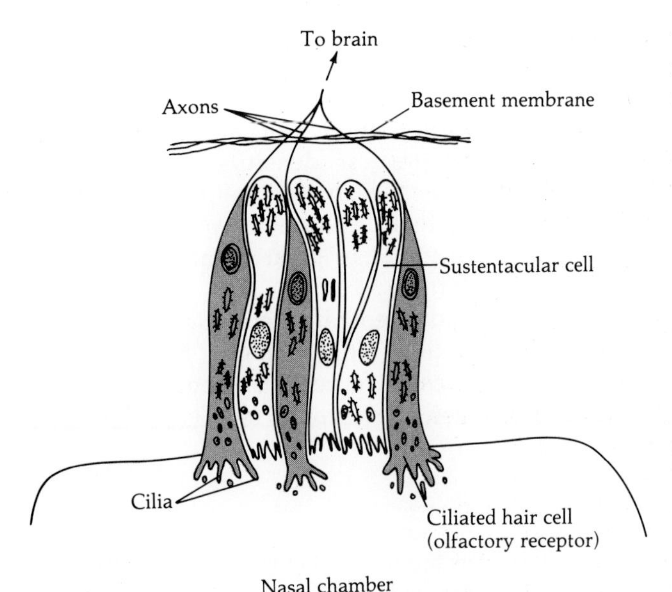

FIG. 8-13
Diagram of olfactory receptors for the sense of smell.

by combinations of responses of different taste receptors plus those of olfactory receptors in the nasal chambers, which react to odors. The combined responses are important signals that generate information which influences dietary intake of essential nutrients and warnings about potentially harmful substances.

Smell (Olfaction)

Olfactory receptors are chemoreceptors responsible for the sense of smell which are located in the mucosa of the upper posterior area of each nasal chamber. They are ciliated bipolar neurons, totaling about 10 to 20 million in number. A terminal fiber from each receptor cell merges with those of the others to produce bundles that form two olfactory nerves (Fig. 8-13). Each olfactory nerve extends to an olfactory lobe in the brain, which is located on the same side as the nostril containing the associated receptors for smell.

Molecules that generate odor must be volatile (pass into the air), water-soluble (dissolve in mucus), and perhaps somewhat lipid-soluble (penetrate lipid barriers of receptor cell membranes). It is not clear whether molecules must physically fit on sites of the receptor cells or react with them chemically in order to initiate changes that produce generator potentials. It is possible that both properties are required and involve a stereochemical interaction that deals with the spatial arrangements of atoms in the molecules.

Multiple interconnections between olfactory fibers allow fine discrimination of odors. The sense of smell is more sensitive and sophisticated in its detection of chemicals than that provided by extremely complex and expensive analytical chemical instruments. For example, the olfactory receptors not only detect odors of thousands of different chemicals but can note the presence of as little as one hundred-trillionth of a gram (10^{-14} g) of some substances in 1 L of air. In spite of this remarkable achievement in humans, the detection of odors is even more discriminating in lower animals, whose survival is more dependent on it.

8-5 SUMMARY

1 Sensory receptors are modified nerve endings which respond to specific stimuli. They may be classified into groups, which include mechanoreceptors, chemoreceptors, osmoreceptors, thermoreceptors, photoreceptors, and auditory receptors. Several types are encapsulated in connective tissue. Afferent fibers from sensory receptors carry impulses to synapses in the CNS. The impulses are ultimately relayed to neurons whose efferent fibers stimulate effector cells (in muscles or glands) capable of a response.

2 Codes for sensory impulses may be developed by means of different patterns or pathways of delivery of their signals. The pattern of impulses may vary according to the type of receptor, the frequency of impulses generated, and/or the number of receptors which are excited.

Impulses continue to come from tonic receptors as long as the stimulus is applied but arise from phasic receptors only when the stimulus changes. Impulses can be spaced closer or further apart in time, somewhat like the rhythmic beats of a drum. The number of receptors that generate impulses increases with the intensity of the stimulus, indicating its relative strength. Differences in sensation are also perceived by means of alterations in the pathways the messages follow. They can vary according to the location of receptors and their destination in the CNS.

3 Photoreceptors contain different protein pigments, which exist as channels in membranes across rod- and cone-shaped neurons in the retina of the eye. Their ionic permeability to Na^+ ions is decreased and they are hyperpolarized when individual pigment molecules are struck by specific types of light to generate visual impulses. The rods respond best in dim light, the cones to colors in bright light. Each generates impulses which can be relayed through transmitter molecules to a series of neurons whose axons form the optic nerves. Their fibers are connected to visual centers in the brain.

4 Mechanoreceptors located in the inner ear within the organ of Corti respond to sound waves, and those within the vestibular apparatus react to changes in rates of movement and gravitational forces. In both locations the mechanoreceptors are hair cells whose cilia are moved by a particular type of mechanical stimulus. As a result, transmitter molecules are released to nearby nerve endings of auditory or vestibular fibers of the statoacoustic nerve which terminate in specific areas of the brain. The receptor cells for hearing are in the organ of Corti and project from the basilar membrane. This membrane is made of nearly 30,000 fibers of different lengths, which are vibrated like strings on a harp by specific sounds of high to low pitch.

Receptor cells in the cristae ampullaris of the vestibular apparatus are moved directly by the ebb and flow of endolymph within the semicircular canals of the inner ear and convey perception of acceleration and deceleration. Others in the maculae of the utricle and saccule are covered by calcium carbonate crystals called otoliths and produce perception of the position of the head, gravity, and acceleration in linear movement.

Taste receptors are ciliated chemoreceptors which line a pore in a taste bud. The receptors respond to molecules in fluid to generate impulses in associated axons. The frequency of impulses produces a code which, along with combinations of responses from different taste receptors plus those of the olfactory receptors, causes sweet, salt, bitter, and sour sensations of taste.

Olfactory receptors are ciliated bipolar neurons which respond to chemicals in the mucosa of each nasal chamber. Substances which generate odor are volatile, water-soluble, and perhaps slightly lipid-soluble. They interact physically and/or chemically with receptors on olfactory cells to initiate impulses carried by olfactory nerves to the olfactory lobes of the brain.

REVIEW QUESTIONS

1 What name is given to sensory receptors that respond to stimuli at the surface of the body? To stimuli from internal organs? To stimuli from muscles, tendons, and joints?

2 Describe the structure of the following sensory endings, their location, and the nature of stimuli to which they respond: corpuscles of Ruffini, stretch receptors, Krause's endbulbs, naked nerve fibers.

3 Name two general variations responsible for coded nerve signals. Discuss

several ways by which codes may be generated to transmit sensory information.

4 What are phasic and tonic receptors? How do they differ?

5 What is adaptation? Of what value is it to the body? What are some of its possible causes?

6 Describe physical abnormalities of the eyes that may cause hyperopia, myopia, and astigmatism. What type of lens may be used to correct each of these conditions?

7 What responses occur in the retina and its components to generate visual impulses to the brain? Explain the cellular and chemical changes that occur.

8 How do sensory receptors in the organ of Corti generate sound impulses?

9 What structures and mechanisms allow the vestibular apparatus to relay information concerning acceleration? Position? Gravity?

10 What causes the senses of taste and smell? Describe the functions of the sensory organs involved.

PHYSIOLOGY OF THE NERVOUS SYSTEM

The Spinal Cord and Autonomic Nervous System

9-1 THE SPINAL CORD

Origins and arrangements of spinal nerves

Relationship of cord structure to nerve function

Reflexes

Nerve pathways

Types of circuits

Tracts of nerves

Ascending tracts • Descending tracts

Pain

Types of pain

Referred pain

Inhibition of pain

Effects of cord damage

Wallerian degeneration of fibers

General effects

Damage to ascending tracts

Damage to descending tracts

Effects of specific agents

Physical factors • Vitamin deficiency • Infections

9-2 THE AUTONOMIC NERVOUS SYSTEM (ANS)

Origins of ANS fibers

Types of neurotransmitters secreted by ANS fibers

Variations in responses to ANS transmitters

Examples of adrenergic and cholinergic effects

Responses of vascular smooth muscle and the heart

Responses of bronchial smooth muscle

Interactions between the brain and the ANS

9-3 SUMMARY

OBJECTIVES

After completing this chapter the student should be able to:

1. Describe the major divisions of the nervous system.
2. Identify the origins of the spinal nerves in the spinal cord and explain their general distribution.
3. Diagram the structure of the spinal cord, noting the location of all its major components.
4. Define the term *reflex* and discuss various types of reflexes.
5. List the major ascending and descending pathways of the CNS and describe their main functions.
6. Discuss pain, including its origins, variations, and inhibition.
7. Identify several factors that cause disorders in the spinal cord and account for their effects.
8. Identify the major components of the ANS; compare the origins of its sympathetic and parasympathetic subdivisions; note the relationships between preganglionic and postganglionic fibers.
9. Describe the tissues and types of organs innervated by the ANS.
10. Compare the structure and biochemistry of sympathetic and parasympathetic neurons.
11. Outline the relationship between autonomic functions and centers in the brain.

The human nervous system consists of the *central nervous system* (CNS) and the *peripheral nervous system* (PNS). The CNS is composed of the brain and the spinal cord (Fig. 9-1). All other components are parts of the PNS which include the peripheral nerves, which originate in the CNS and carry impulses to and away from it, and the *autonomic nervous system* (ANS), composed of two neuron chains which send involuntary impulses to the organs (viscera) of the body.

The different routes over which impulses are carried are highly organized and form networks of pathways in the CNS and PNS. Most often synapses (junctions) create a number of complex and alternative paths that provide the capability to generate a tremendous variety of responses. This chapter will describe the functions of some of those pathways.

9-1 THE SPINAL CORD

Efficient business organizations are set up in such a way that certain responses may occur before an incoming order reaches the main office. In this same sense the human body is extremely well organized. Several areas in the CNS can carry out rapid, automatic responses. A number of these functions, performed by the spinal cord, are described below while others, regulated by areas in the brain, are the subject of Chaps. 10 and 11.

Origins and Arrangements of Spinal Nerves

The spinal cord is a nervous tissue extension of the lower part of the brain (the medulla oblongata) and is enclosed by the *vertebral canal,* a cavity within the center of the vertebrae. Spinal nerves exit on each side through openings (vertebral foramina within or between the vertebrae (Fig. 9-1).

Generally 31 pairs of spinal nerves are in the body approximating the 33 to 34 bones that make up the adult vertebral column. The nerve pairs are named after the segments of the vertebral column from which they emerge. Eight pairs on the sides of the neck are *cervical nerves;* twelve pairs in the chest are *thoracic nerves;* five pairs in the small of the back are the *lumbar nerves;* five pairs exit through the sacrum as *sacral nerves;* and one to three nerves are located at the "tailbone" or coccyx as the *coccygeal* nerves.

During late fetal development and until shortly after birth, the vertebral column lengthens more than the spinal cord. Therefore, the spinal cord extends through only about

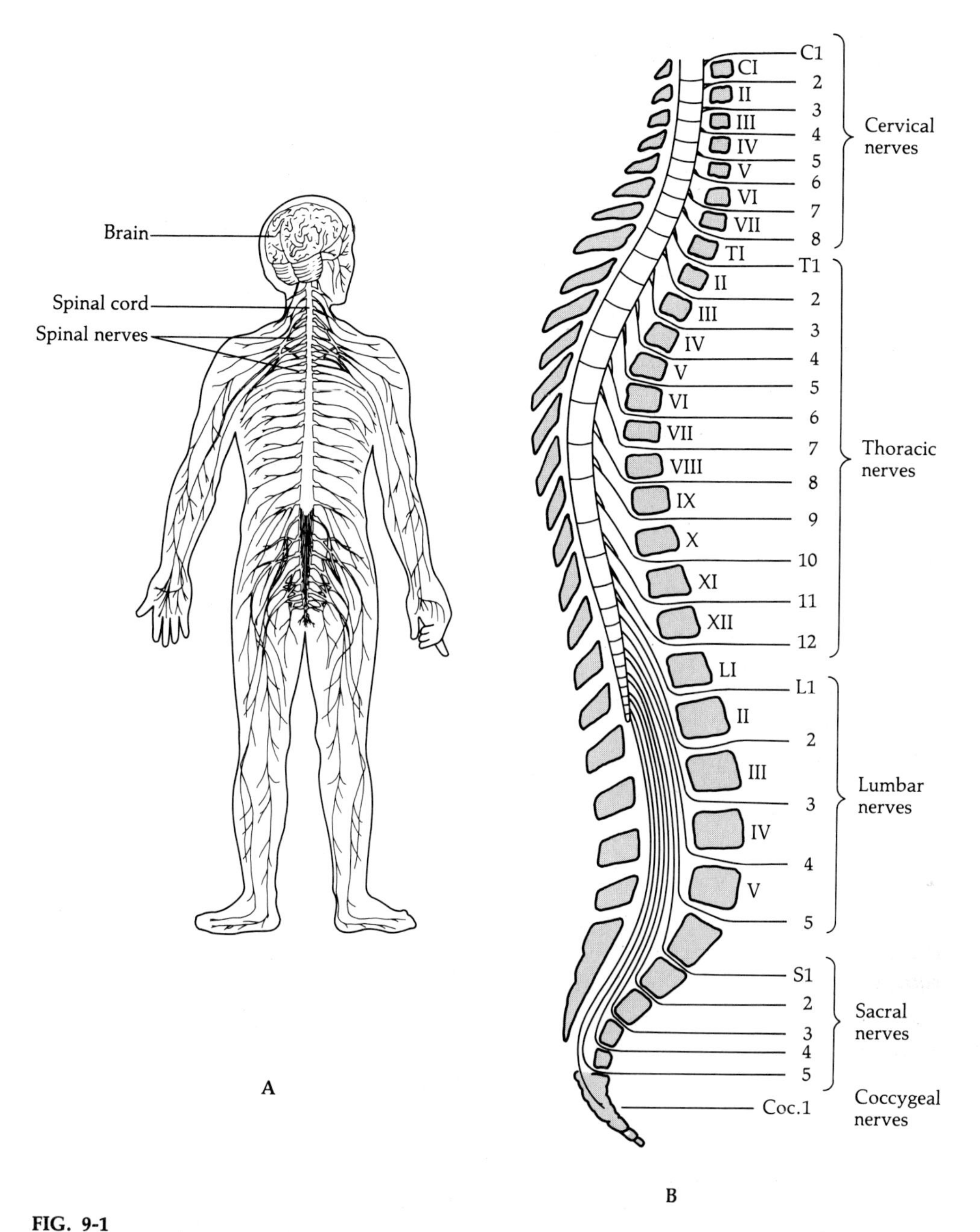

FIG. 9-1
General anatomy of the central nervous system. (A) The locations of the brain, spinal cord, and spinal nerves are shown from a dorsal (posterior) view. (B) Diagram of lateral view of the vertebral column and spinal nerves. Each vertebra is labeled. One designated member of each pair of spinal nerves emerges through the openings between the vertebrae shown at the far right. Note the end of the spinal cord about LI to LII.

two-thirds of the adult vertebral canal. The length of the spinal cord averages 43 cm in females and 45 cm in males but varies among different individuals.

The terminal portion of the adult cord divides into separate strands between the first and second lumbar vertebrae (Fig. 9-1). In newborn children, the cord may extend as far as the third

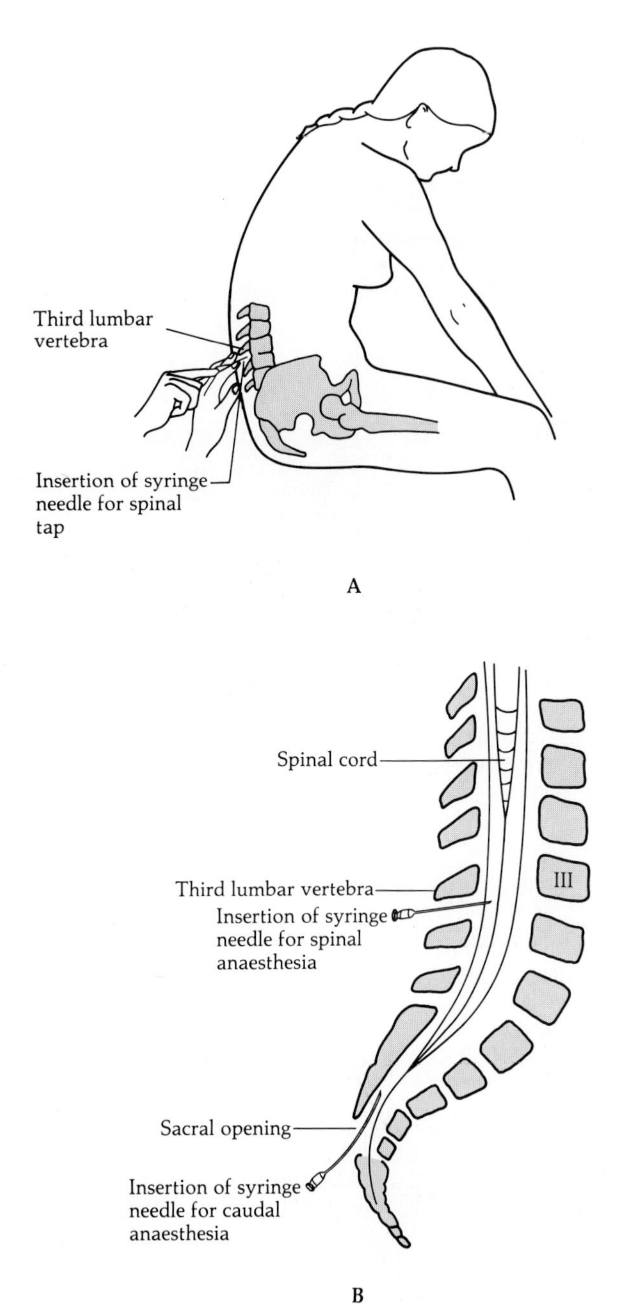

FIG. 9-2
Insertion of a syringe for withdrawal of cerebrospinal fluid. (A) A spinal (lumbar tap) made by insertion of a syringe needle between lumbar vertebrae III and IV. (B) Spinal (lumbar) and caudal (sacral) anaesthesia may be administered through openings in the vertebral column as illustrated.

lumbar vertebra. In either case the strands form a structure resembling a horse's tail and are referred to as the *cauda equina*. The cauda equina includes all the spinal nerves below the first lumbar pair.

The difference in location of the terminal segment of the cord is extremely important when one attempts to remove cerebrospinal fluid (CSF) from the nervous system for analysis in diagnostic techniques (Chap. 10) or to administer medications or anesthetics. In order to obtain the fluid a hypodermic needle is usually inserted in a spinal tap or lumbar puncture through the connective tissue covering the cord between the third and fourth lumbar vertebrae in adults (Fig. 9-2A and B). This site is beyond the terminal portion of the cord and minimizes the possibility of damage to it and to spinal nerves. In young children the insertion is usually carried out one to two vertebrae lower. The needle may be inserted in these respective areas for injection of anesthetics in spinal anesthesia. Alternatively, an opening in the sacral segment of the vertebral column may be used for caudal anesthesia (Fig. 9-2B).

Relationship of Cord Structure to Nerve Function

The spinal cord consists of myelinated white fibers which surround unmyelinated gray fibers grouped in a shape much like the letter H (Fig. 9-3). Both types of fibers carry signals up and down the cord and to other areas of the CNS. The white fibers also convey impulses to neurons in the gray matter. Although the dimensions of the cord vary in different regions, it is elliptical in cross section and averages 7 mm from anterior to posterior and 6 to 11 mm in width. It is penetrated by a *central canal*, which is continuous with cavities (ventricles) within the brain. The central canal is surrounded by the *gray matter*.

Afferent neurons are found in the *dorsal horn* (dorsal segment of the gray matter of the cord), *efferent motor neurons* in its *ventral horn*, and *interneurons* (associative neurons) in between. The sensory and motor fibers emerge from the cord as the dorsal and ventral roots, respec-

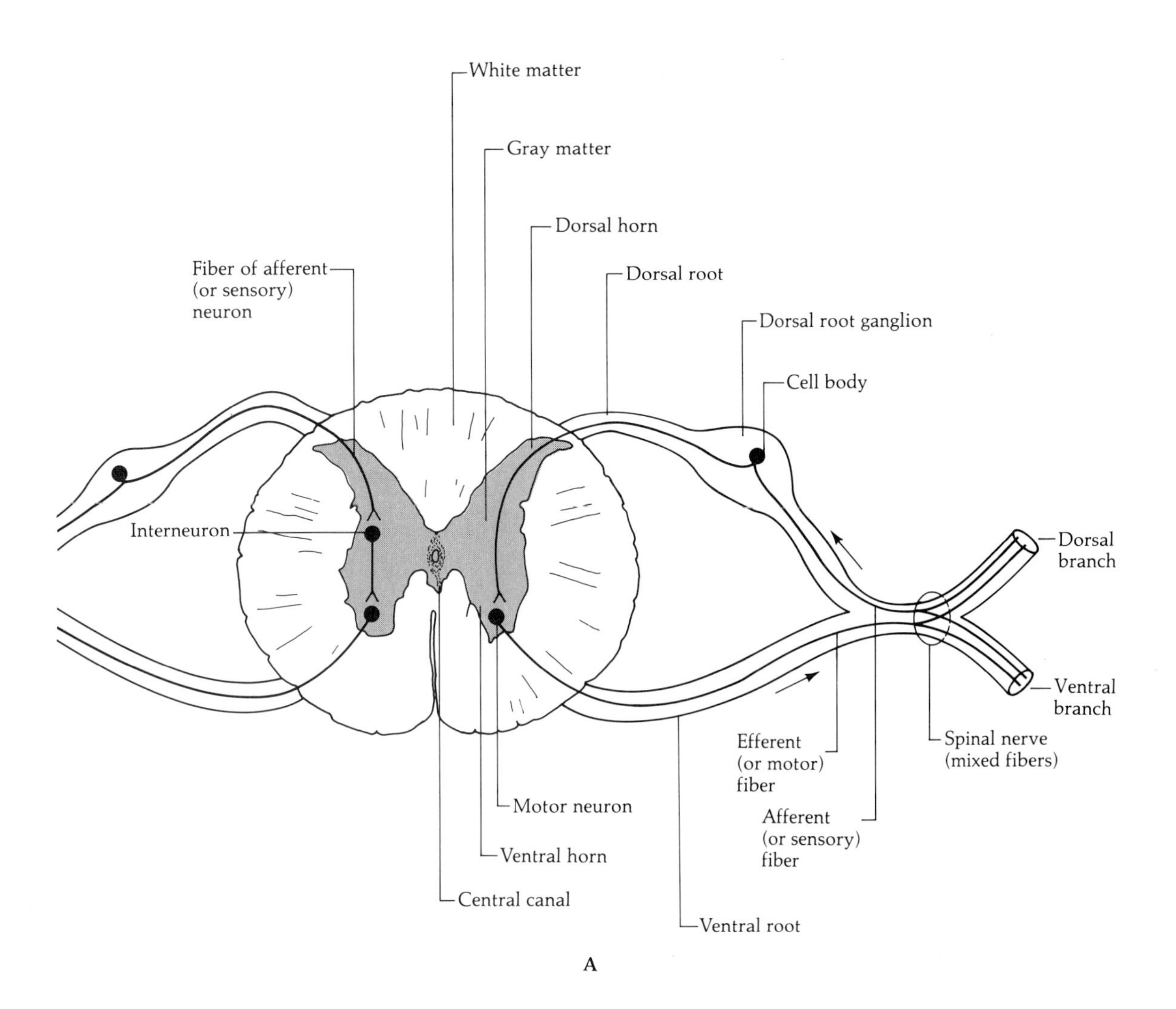

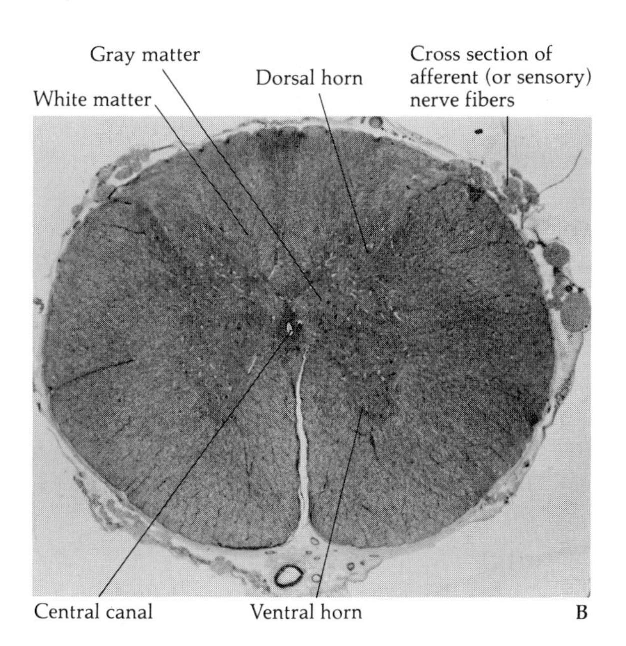

FIG. 9-3
A cross section of the spinal cord, showing spinal nerves. (A) A drawing illustrating the spinal cord in cross section. (B) A photomicrograph of the spinal cord (about ×17).

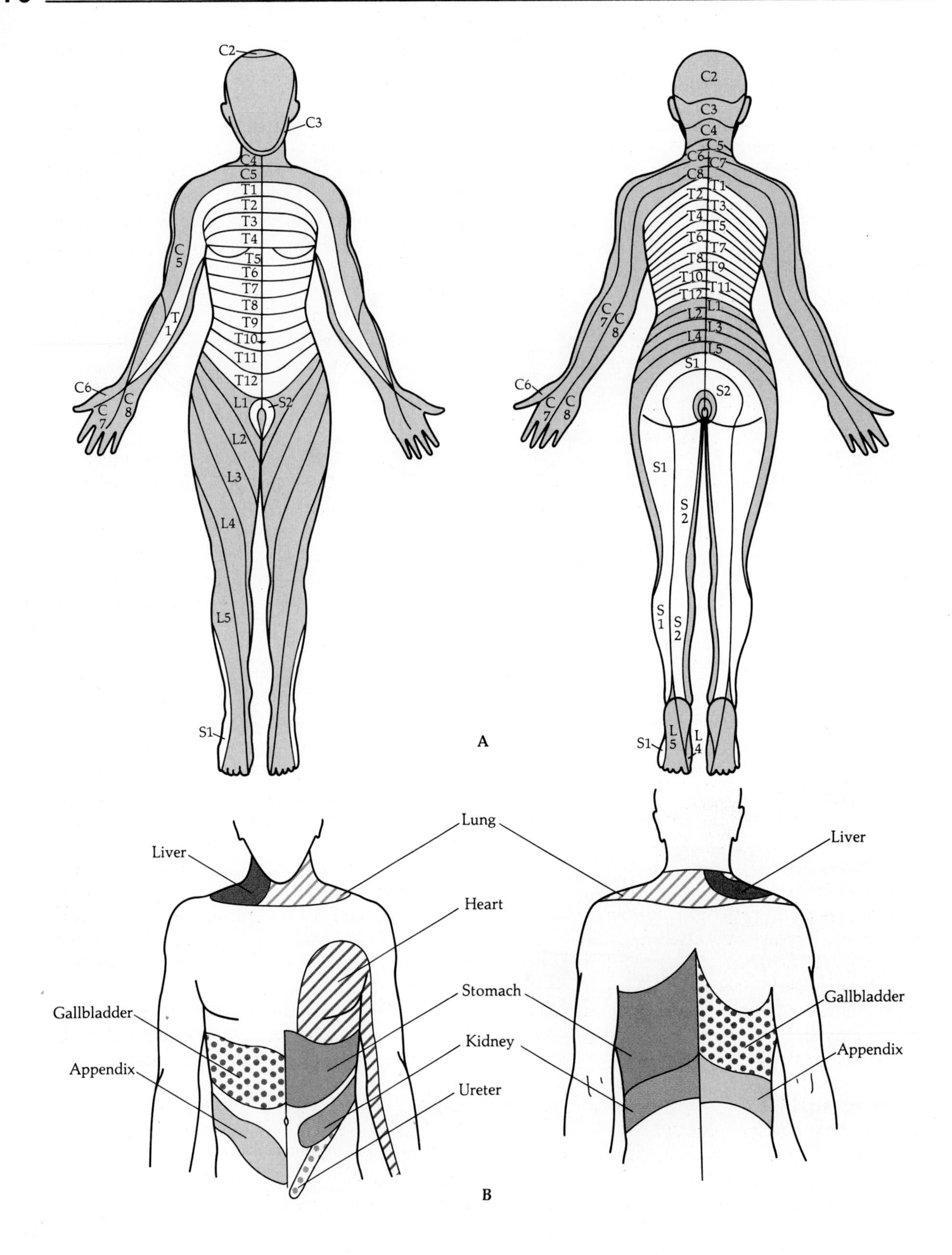
C2
C3
C4
C5
T1
T2
T3
T4
T5
T6
T7
T8
T9
T10
T11
T12
L1
S2
L2
L3
L4
L5
S1
C6
C7
C8
T1
A
S2
L5
L4
B
Lung
Liver
Heart
Stomach
Gallbladder
Appendix
Kidney
Ureter

FIG. 9-4
Segmental distribution of spinal nerves to the skin. The diagrams on the left and right are ventral and dorsal views of overlapping dermatomes which share common embryonic origins. (A) Areas between the lines are supplied by the spinal nerves noted. (B) Overlapping sensory areas are projected on (referred to) the body surface (shown as shaded areas) from internal organs. Areas marked by a specific color receive sensory fibers from a single nerve trunk. The origin of pain may be misjudged and be mistakenly identified as arising from other area(s) within the same color zone because they are supplied by the same nerve trunk. Figure 9-4B adapted from *A Systematic Approach to Neuroscience* by E. L. House, B. Pansky, and A. Siegel. Copyright © 1980, McGraw-Hill Book Co. Used with permission of McGraw-Hill Book Co.

tively, which in turn form spinal nerves on each side of the cord.

About 3000 to 5000 interneurons exist for each motor neuron in the CNS. The typical motor neuron in the spinal cord forms some 10,000 synapses (junctions), of which 80 percent are on its dendrites and 20 percent on its cell body.

The afferent neurons which transmit sensory information to the brain are called *sensory* neurons. *Afferent* fibers, both myelinated and unmyelinated, are found in the *dorsal root* (posterior root), which leads into the *dorsal root ganglion* (posterior root ganglion). The latter is located on the posterior aspect of the cord, medial to the spinal nerves. It contains cell bodies whose fibers enter the dorsal horn in the cord. *Efferent* or *motor* fibers originate from neurons in the ventral horn and emerge in the *ventral root* (anterior root) of the spinal cord.

The dorsal and ventral roots merge to form spinal nerves prior to the exit of fibers through the intervertebral openings. Consequently, *spinal nerves* are mixtures of sensory and motor fibers. During development each spinal nerve supplies dorsal root sensory fibers to an area of the skin known as a *dermatome,* and ventral motor root fibers supply an area of muscle called a *myotome.* The routes the fibers follow have been established in dermatomal maps, which relate the origin and destination of spinal nerves to specific segments of the spinal cord, as shown in Fig. 9-4A.

Muscles, vertebrae, and the spinal cord develop segmentally from a series of subdivided parts, but growth alters the initial pattern. Furthermore, nerves intermingle to form groups of fibers called *plexuses,* which are then redistributed to specific destinations. The brachial plexus supplies the shoulders and arms, while the lumbosacral plexuses innervates the pelvic region and legs. Nerve fibers from different segments of the spinal cord may overlap in the dermatomes with which they associate (Fig. 9-4B). Consequently, destruction of a single spinal nerve often results in partial rather than total loss of sensation and/or movement in a specific area of the body.

Reflexes

A *reflex* is a basic automatic involuntary response to a stimulus. The term reflex originates from the Latin prefix *re*, back, and the verb *flectere,* to bend, and thus literally means to bend back. In a reflex, impulses from receptors are carried to and from (bent back from) the spinal cord or brain, allowing the body to respond automatically. Some of the receptors were described in Chap. 8.

All reflexes involve at least one afferent and one efferent neuron, but most reflexes involve interneurons and large numbers of neurons, some of which carry impulses across reflex arcs in centers in the brain. Reflexes generally enhance survival and control automatic body responses, including heart and respiratory rates. Those which involve neurons located in the spinal cord are known as *spinal reflexes,* such as the withdrawal response to tickling of the foot. Others, which involve neurons in the brain, are *cranial reflexes,* such as the pupillary reflex which causes the pupil of the eye to constrict in response to light.

Stimuli cause impulses which are transmitted from receptor endings through afferent fibers to the cell body in the dorsal root ganglion and through its afferent nerve fiber into the spinal cord, which forms synapses with fibers of an efferent (motor) neuron in the ventral horn. Efferent motor fibers then transmit the impulse to effectors to elicit a response (Fig. 9-5). Efferent motor fibers to skeletal muscle are called *somatic*

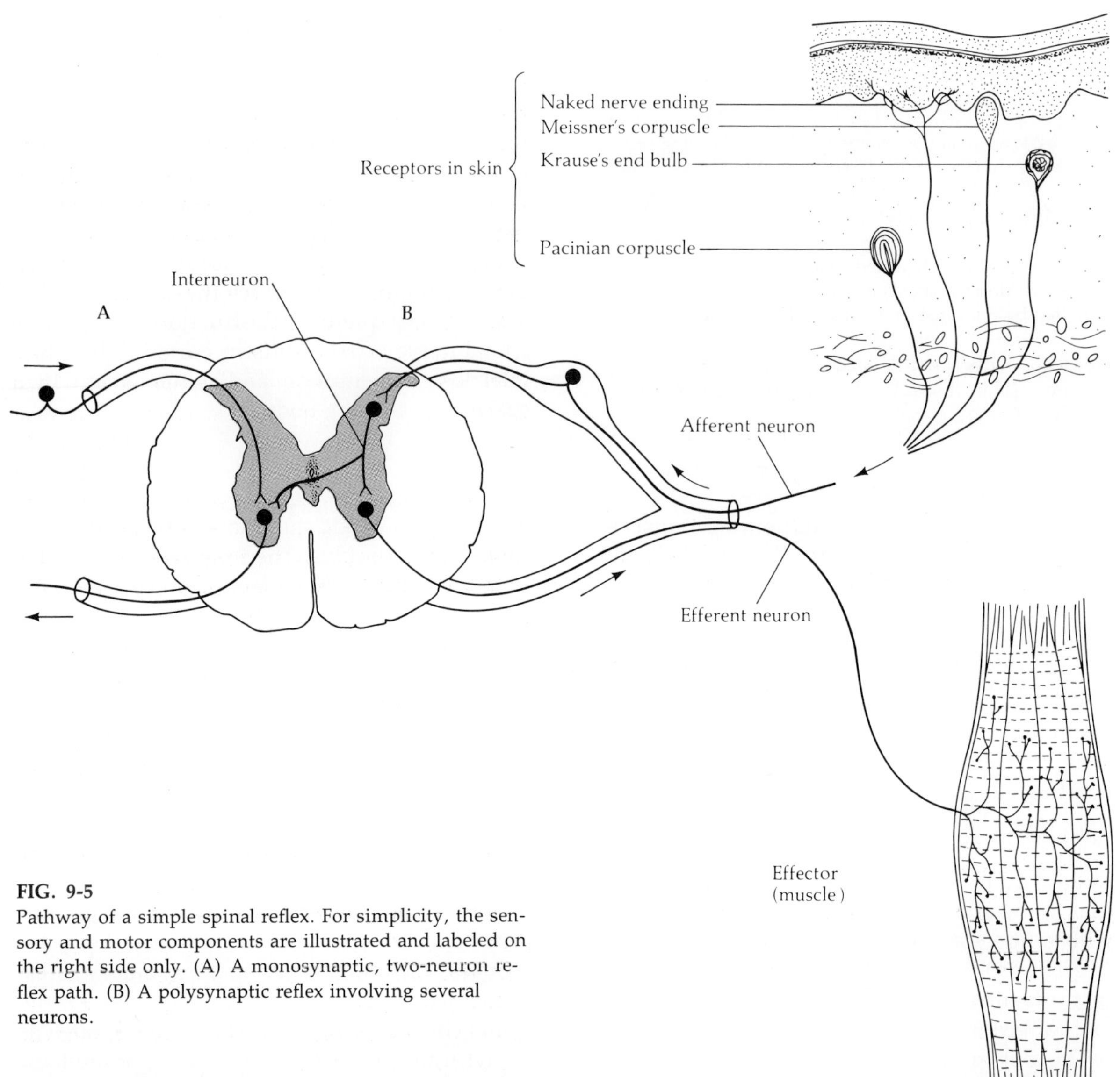

FIG. 9-5
Pathway of a simple spinal reflex. For simplicity, the sensory and motor components are illustrated and labeled on the right side only. (A) A monosynaptic, two-neuron reflex path. (B) A polysynaptic reflex involving several neurons.

motor fibers and those to smooth muscle, cardiac muscle, and exocrine glands are called *visceral* motor fibers. The path of the incoming and outgoing impulses constitutes the basic functional unit of the nervous system, the *reflex arc.*

The simplest reflex arc is *monosynaptic.* This arc is formed by two neurons, one of which generates afferent impulses that are carried across a synapse to another neuron that generates efferent impulses (Fig. 9-5A). In most instances, however, reflexes occur through *polysynaptic arcs,* in which the incoming impulse is delivered through one or more interneurons to more than one efferent neuron and its fibers (Fig. 9-5B).

The pathways of reflex arcs may be very complex (Fig. 9-6). *Ipsilateral* reflexes occur on the same side of the spinal cord, while *contralateral* reflexes cross it and occur on both sides. *Intrasegmental* reflexes take place at one level of the cord, while *intersegmental* reflexes occur at more

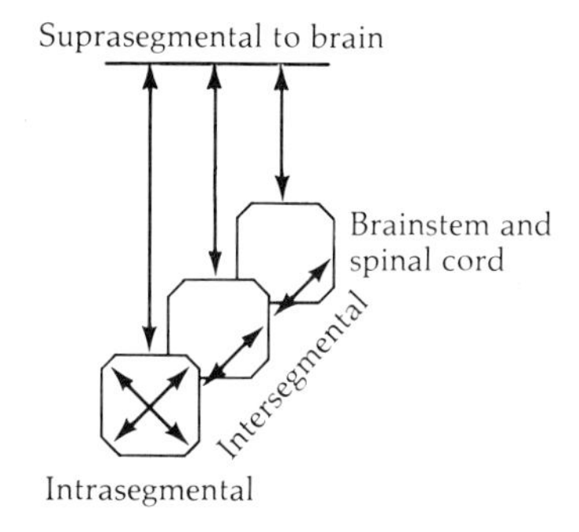

A

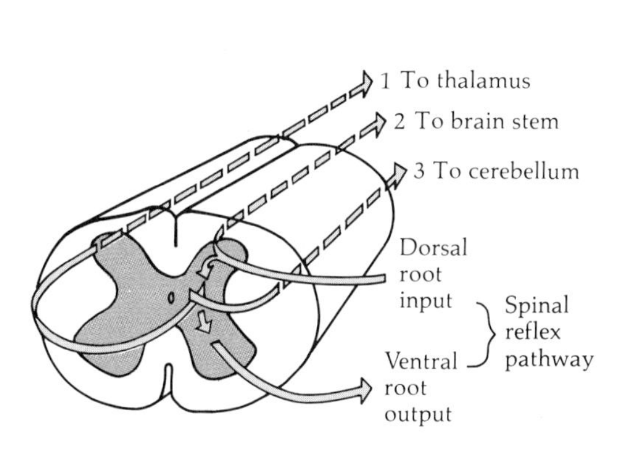

B

Alternative pathways for reflexes between the spinal cord and brain. (A) General intrasegmental, intersegmental, and suprasegmental paths. (B) Three ascending pathways: (1) The crossed spinothalamic tract to the thalamus (a contralateral path). (2) The dorsal columns to the brainstem (an ipsilateral path). (3) The spinocerebellar tract in the lateral white matter to the cerebellum (an ipsilateral path).

than one level up or down the cord. *Suprasegmental* reflexes involve paths to and from the brain.

The response to tickling of the foot (light pressure) illustrates a *simple monosynaptic reflex* involving two neurons (Fig. 9-5A). Action potentials pass from receptor endings through an afferent fiber into the dorsal root of the spinal cord, then through a synapse to a motor neuron in the ventral horn, and then to cells in an extensor muscle of the leg.

Another example of a common reflex is the *stretch reflex,* also called the *myotactic reflex* since it occurs when muscle spindles and Golgi tendon organs are stretched (Fig. 8-1F). The myotactic reflex is used as a clinical test of motor neuron function when the patellar tendon of the knee is tapped with a rubber hammer to elicit a knee-jerk response. Many other neurological tests are based on assessment of reflex responses. For example, the Babinski reflex, which curls the toes downward in response to stroking the sole of the foot, is missing when motor neurons in the brain are damaged. The Babinski sign of damage is a response in which the big toe pulls backward and the other toes spread out. In the first few weeks after birth, before the outer cortical area of the brain is fully formed, a response which is the same as the Babinski sign can be observed in infants.

Strong stimuli are carried through interneurons, which may form polysynaptic relationships (Fig. 9-5B). The interneurons may cause ipsilateral, intersegmental, contralateral, or suprasegmental connections in a variety of paths and reflexes (Fig. 9-6).

Even though the knee-jerk response to a doctor's rubber hammer is unconscious, suprasegmental paths allow an individual to become aware of the stimulus. Voluntary responses may follow the reflex response through activity of the brain, which can transmit the signals to motor neurons which are under conscious control. Afferent impulses also may be carried intersegmentally by *collateral* branches of axons. Collaterals make it possible for the spinal reflexes to travel through two or more levels of the spinal cord (Fig. 9-7).

The initial response to a stimulus occurs rapidly when the impulse passes through a simple two-neuron or three-neuron reflex arc. This route, of course, requires less time than the longer pathway to the brain and occurs automatically and prior to conscious awareness of the stimulus. Simple reflexes are natural and unlearned. They differ from *conditioned reflexes,* which are learned and more variable. The latter depend on neuron associations in the cerebral cortex in the brain and will be described in Chap. 11.

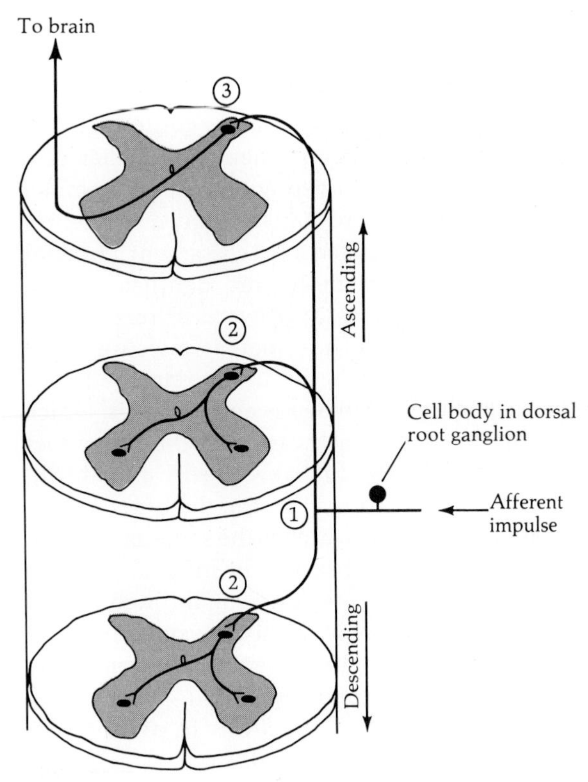

FIG. 9-7
Some alternate routes of afferent collateral fibers. The incoming afferent fiber may give off ascending and descending branches (collaterals) which: (1) travel intersegmentally in ascending and descending paths; (2) form synapses for contralateral reflexes; and (3) synapse with neurons forming contralateral, suprasegmental paths to the brain.

Nerve Pathways

Types of Circuits

Pathways in the nervous system are connections between any two points which mediate distinct functions. Those found within functional groups of white fibers, well defined and normally of a common origin and/or termination, are called *tracts.* They form trillions of interneuronal connections and create an astonishing number of pathways, which integrate the activity of the nervous system.

Amplifying, converging, parallel, and reverberating circuits are found in the CNS. *Amplifying circuits* are formed by the fibers of one neuron, which divides into two or more routes to form diverging paths involving increasing numbers of neurons (Fig. 9-8A). They can carry signals directly from one neuron to hundreds of neurons up and down the CNS and ultimately to thousands of effector cells.

Impulses also can be funneled from two or more neurons to a lesser number of neurons by *converging circuits* (Fig. 9-8B). The funneling of the information is called *convergence.* This type of pathway makes it possible for inputs from different sources to cause identical responses. For example, the thought, sight, smell, taste, or sounds associated with an appetizing meal can stimulate afferent neurons which funnel impulses to efferent neurons to regulate digestive tract activity.

Receptors may transmit impulses to presynaptic afferent fibers, several of which may then converge on the postsynaptic membrane of a single neuron in the CNS (Fig. 7-13). In this way, different sensory inputs can interact to regulate the activity of other neurons. For example, as many as 100 to 200 presynaptic afferent fibers may synapse with a single motor neuron. The neurotransmitters released by some afferent fibers are excitatory while those released by others are inhibitory. A volley of transmitter agents may be released asynchronously rather than simultaneously from an afferent fiber and last about 12 to 15 ms (milliseconds). Generation of an excitatory postsynaptic potential (EPSP) and action potential depends, then, on the net effect of the transmitters (Chap. 7).

Parallel circuits are composed of neurons each of which stimulate a series of others (Fig. 9-8C). Because the parallel paths have differing numbers of synapses, impulses pass more or less rapidly over them and reach a target at staggered intervals, causing repetitive responses. The impulses from several fibers may converge on a common output cell to govern activities that last up to 20 ms. For example, axons of granule cells in the cerebellum of the brain form parallel fibers which connect a series of Purkinje cells much as wires link telephone poles (Chap. 11). Each Purkinje cell receives fibers from 80,000 to 100,000 granule cells.

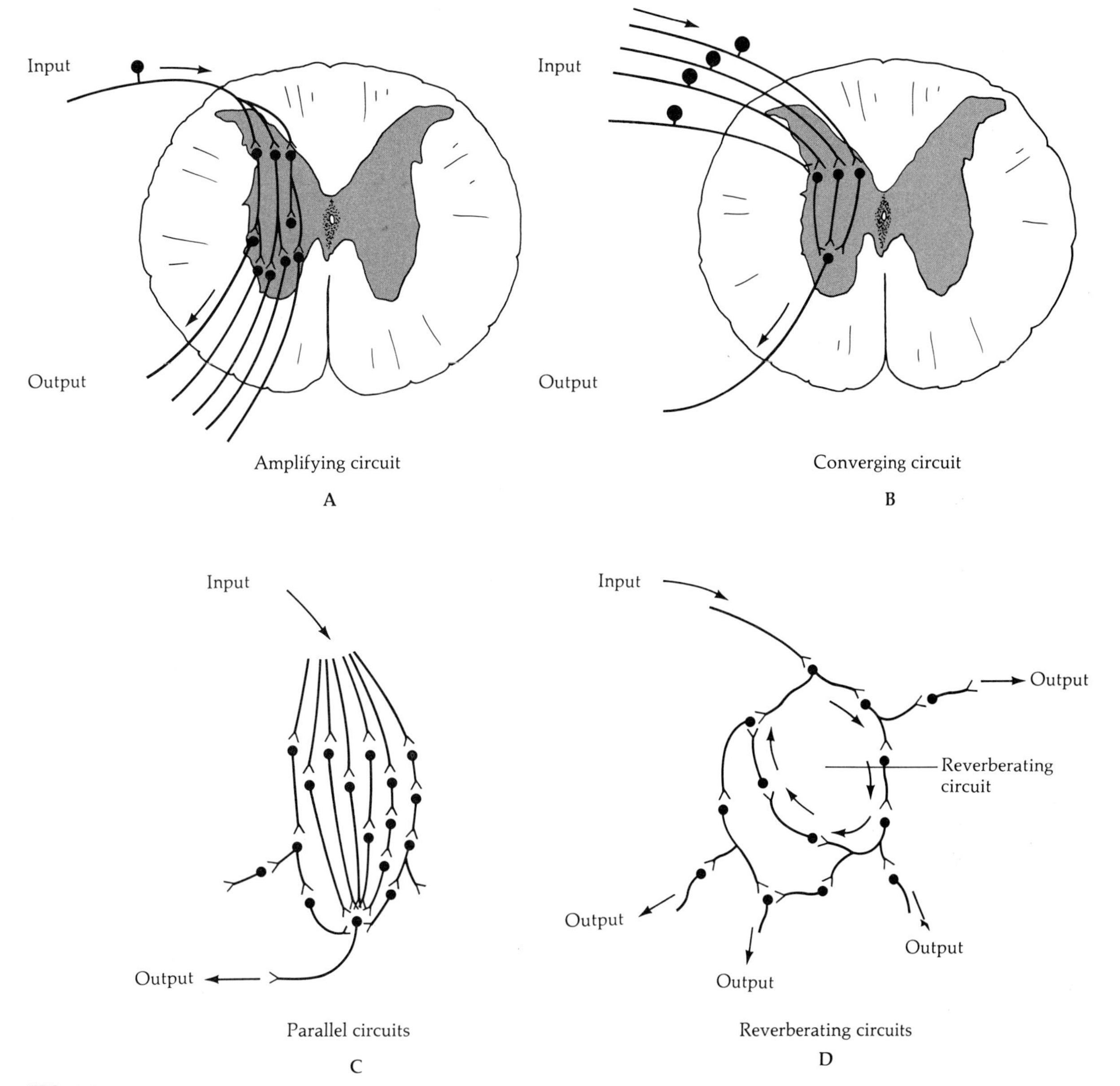

FIG. 9-8
Varieties of the arrangements between neurons forming circuits for polysynaptic reflexes. The arrows indicate the direction of movement of impulses.

Reverberating circuits are often found in the CNS and are formed by loops of neurons in which an interneuron or interneurons return the impulse to cells preceding them within the loop (Fig. 9-8D). Impulses stimulate the activity of the cells in the loop and at its termination until fatigue or inhibition occurs. Rhythmical muscular activities involved in walking and respiration are among numerous bodily functions regulated in this way. Some of the circuits utilize negative feedback signals to regulate the activity of members of the group.

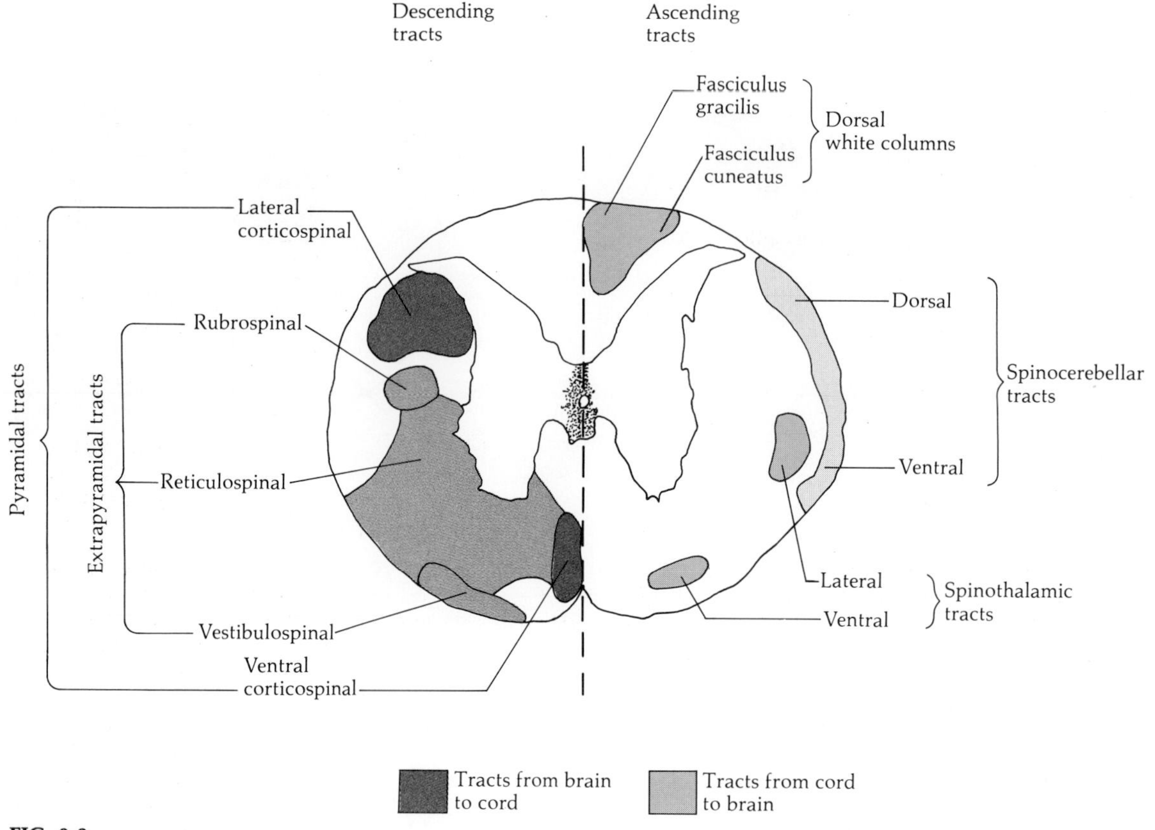

FIG. 9-9
Location of some major ascending and descending tracts in the spinal cord. Note that these tracts are found on both sides of the cord. For simplicity the diagram shows only some descending tracts on the left and some ascending tracts on the right.

Tracts of Nerves

Fibers from neurons are bundled together in discrete locations to carry impulses up and down the spinal cord. The well-defined routes they form in the white matter are called *ascending* and *descending* nerve tracts.

The spinal cord contains a complex network of these tracts, which transmit impulses to various levels of the CNS and out to peripheral tissues and organs (Fig. 9-9).

ASCENDING TRACTS Three major ascending tracts to the brain are the spinothalamic tracts, the dorsal white columns, and the spinocerebellar tracts (Fig. 9-9). Their properties are summarized in Table 9-1. The *spinothalamic tracts* and the *dorsal white columns* carry sensory impulses by means of a chain of three neurons (Fig. 9-10). The receptor endings of the primary neurons of these sensory tracts carry the impulse to a cell body in the dorsal root ganglion, then to a second neuron in the dorsal gray column of the cord or the brainstem. Fibers from the second cell body usually cross over (*decussate*) and synapse with those of a third cell body which is in the thalamus of the brain. Terminal fibers from the thalamus then carry sensory signals to the cerebral cortex of the brain. Because the cerebral cortex is responsible for conscious sensations, an individual is made aware of pain, pressure, or proprioception (balance) through impulses carried by the spinothalamic tracts.

Impulses from the *spinocerebellar tracts* travel from proprioceptors in joints and muscles over a two-neuron pathway. They pass through the

TABLE 9-1
Properties of some major ascending tracts

Name of Tract	*Receptors*	LOCATION OF CELL BODIES *No. 1*	*No. 2*	*No. 3*	*Fibers Cross Over**	*Destination of Impulses in Brain*	*Major Functions (Sensation)*
Spinothalamic		Dorsal root ganglion	Dorsal gray column of spinal cord	Thalamus	Yes†	Cerebral cortex (post central gyrus)	
Lateral spinothalamic	Naked nerve endings	"	"	"	Yes		Pain
	Krause's end bulbs	"	"	"			Temperature
Ventral spinothalamic	Pacinian corpuscles	"	"	"	Yes		Crude touch
	Meissner's corpuscles	"	"	"			Light pressure
Dorsal white columns	Proprioceptors (corpuscles of Ruffini, Golgi tendon organs, muscle spindles) Meissner's corpuscles (and Pacinian corpuscles?)	Dorsal root ganglion	Medulla	Thalamus	Yes	Cerebral cortex (post central gyrus)	Proprioception, deep sensation receptors, conscious balance and fine touch (discrimination between two points)
Lateral: fasciculus cuneatus	In upper extremities, the arms	Dorsal root ganglion in thoracic and cervical cord segments	Nucleus cuneatus in medulla	Thalamus	Yes	Cerebral cortex (post-central gyrus)	
Medial: fasciculus gracilis	In lower extremities, the legs	Dorsal root sacral and lumbar ganglia	Nucleus gracilis in medulla	"	Yes		
Spinocerebellar (dorsal and ventral tracts)	Proprioceptors (described above)	Dorsal root ganglion	Dorsal and ventral gray column	None	Generally not‡	Cerebellum	Proprioception, subconscious sensations, and balance

* Fibers from the second sensory cell bodies normally cross over to the opposite side of the CNS.
† Fibers of the lateral spinothalamic tract cross over within one segment of their entry into the cord, while those of the ventral spinothalamic tract give off ascending and descending collaterals prior to synapse with fibers that cross the cord.
‡ A few cross over.

dorsal root ganglia, ascend through the gray columns of the cord, and usually end on the same side of the cerebellum. The cerebellum functions at levels below that of consciousness.

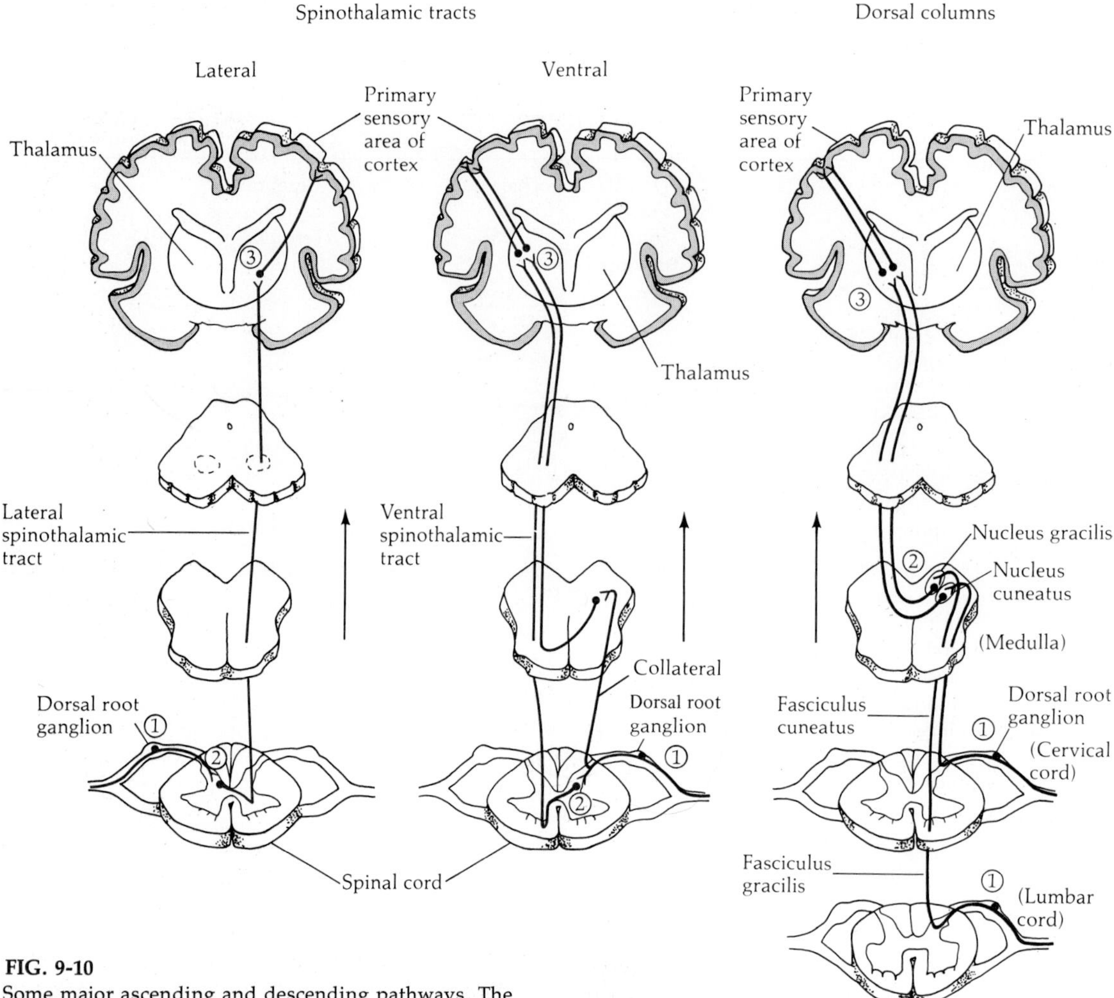

FIG. 9-10
Some major ascending and descending pathways. The figures above illustrate some ascending tracts and those on the next page show some descending tracts. The numbers indicate the locations of the cell bodies of the individual neurons in the chains in each.

Therefore, the spinocerebellar tracts subconsciously mediate muscle activity for posture and balance (proprioception).

DESCENDING TRACTS Skeletal muscle motor activity (contraction) is triggered by descending nerve fibers and may be caused by a simple spinal reflex as described above. However, higher levels of the CNS also may be used to generate voluntary movements. Neurons within the brain that cause motor activity are known as *upper motor neurons.* Their fibers synapse, usually via interneurons, with *lower motor neurons* in the ventral horn of the spinal cord to form various pathways. The lower motor neurons are involved in simple reflexes, as described above.

The pyramidal (corticospinal) and extrapyramidal tracts are descending columns that form the main motor pathways (Table 9-2 and Fig. 9-9). The *pyramidal* tracts originate in upper motor neurons in several areas of the cerebral cortex in the brain. (The structure and function of the brain will be described in Chaps. 10 and 11.) About 1 million of the fibers in this tract are grouped in triangular bundles in the pons of

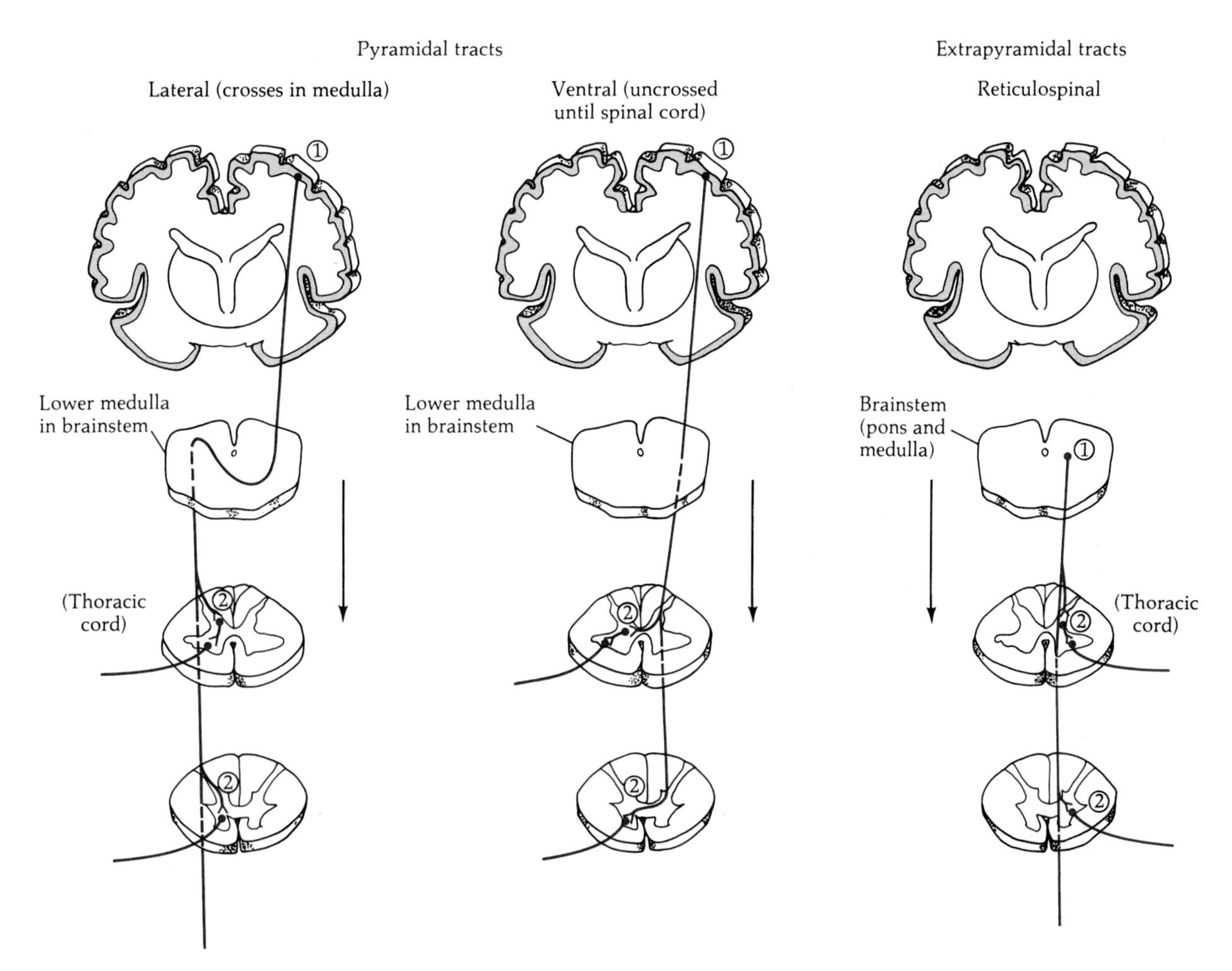

the hindbrain. They generally form synapses with interneurons in the spinal cord, which then relay excitatory or inhibitory impulses to lower motor neurons in the ventral horn which supply opposing muscles. About two-thirds of the fibers of upper motor neurons cross over in the medulla (the lowest part of the brain). These fibers form the *lateral corticospinal* tract. The remainder cross over in the spinal cord to make up the *ventral corticospinal* tract.

Since many afferent fibers cross over as they ascend in the CNS, sensory information from one side of the body is delivered to the opposite side of the brain. The information then may be transferred to nearby motor neurons. The motor neurons relay impulses through descending paths, which also cross over. Therefore, the motor neurons deliver their impulses to the side of the body from which incoming afferent or sensory signals have been received.

Additional descending motor tracts that originate in other areas of the brain (such as the basal ganglia, reticular formation, and cerebellum) are called *extrapyramidal* tracts (Fig. 9-9). The extrapyramidal system is important in the coordination of muscular activity. It may be the predominant means of motor control in mammals lower than humans, in birds, and in reptiles. In higher animals, the extrapyramidal tract is generally active in coordinating skeletal muscle activity in those responses on which survival depends. Instinctive behavior is greatly influenced by the extrapyramidal system. The development of the pyramidal tracts in humans superimposes a level of control beyond that governed by the extrapyramidal tracts. Interaction of the two systems influences motor activity and behavioral responses.

TABLE 9-2
Properties of some major descending tracts

Name of Tract	LOCATION OF CELL BODIES *No. 1*	*No. 2*	*Fibers Cross Over*	*Destination of Impulses*	*Major Functions*
Pyramidal (corticospinal) Lateral and ventral	Motor areas of cerebral cortex	Interneurons and motor neurons in gray column	Yes*	Skeletal muscle (and collaterals to brainstem, pons, cerebellum, dorsal column, and dorsal horn)	Voluntary excitation and inhibition
Extrapyramidal					Posture regulation
Reticulospinal				Skeletal muscle Visceral organs, e.g., heart and respiratory muscle	Voluntary Involuntary control of heart and respiratory rate
Lateral	Medulla, pons, and midbrain	Interneurons and motor neurons in gray column	Some	Skeletal muscle Visceral organs	Mostly excitatory
Medial	Mostly medulla	"	Some	"	Mostly inhibitory
Vestibulospinal	Nucleus of vestibular nerve in medulla	"	Some near their origin	Skeletal muscle	Involuntary control of muscle tone for balance and equilibrium
Rubrospinal	Red nucleus in midbrain	Interneurons in gray column	Yes	Skeletal muscle	Involuntary control of muscle tone and posture

* A few fibers remain uncrossed. The bulk of lateral fibers cross in the medulla, while those of the ventral tract cross in the white matter of the spinal cord.

One of the major extrapyramidal tracts, the *reticulospinal* tract, originates in the reticular formation of the medulla, pons, and midbrain. The tract derives its name from its reticular (lacy, spider web-like) network of fibers. The *medial* subdivision of this tract arises from the medulla and primarily delivers inhibitory impulses to skeletal muscle. The *lateral* subdivision originates in the medulla, pons, and midbrain and is mostly excitatory. The inhibitory and excitatory impulses can superimpose control over spinal reflexes and oppose, exaggerate, or modify reflex activity.

Other extrapyramidal pathways include the *vestibulospinal* tract from the medulla and the *rubrospinal* tract from the midbrain. Both tracts coordinate subconscious skeletal muscle activity and influence posture and balance.

In spite of the diversity of the sources and pathways of upper motor neurons and the presence or absence of interneuronal synapses, the lower motor neurons finally deliver impulses from the ventral horn of the spinal cord to regulate skeletal muscle. The efferent nerve fibers of lower motor neurons are referred to as the *final common pathway*.

Pain

Fibers which transmit pain impulses ascend from diffuse naked nerve endings (Fig. 8-1E and Table 9-1). The dorsal gray columns relay some pain impulses to the thalamus and to the cerebral cortex. Different types of pain impulses are transmitted contralaterally by the lateral spinothalamic tracts (Fig. 9-10).

Types of Pain

Pain impulses vary with the size of fibers and their degree of myelination. For instance, *first pain,* which occurs soon after stimulation of receptor endings and damage to the body, is transmitted rapidly along myelinated, A-delta fibers (Table 6-2) and experienced as definite localized sensations. *Second pain* is carried more slowly over smaller, unmyelinated C fibers (Table 6-2) and experienced as a dull, aching diffuse sensation.

The effects of pain-killing drugs also vary in relation to the properties of pain fibers. Cocaine, as an example, gains access to small, unmyelinated fibers more readily than to larger, myelinated ones. It can therefore block dull, aching, second pain more easily than first pain.

Referred Pain

The presence or loss of the sensation of pain is related to the dermatomal areas supplied by spinal nerves (Fig. 9-4). Because these areas overlap considerably, damage to individual nerves rarely produces complete anesthesia (loss of sensation). When it does, the anesthesia is confined to very small areas supplied by a single nerve trunk.

The area from which pain originates is often misjudged. *Referred pain* impulses originate in internal organs but seem to come from the body surface (Fig. 9-4B). Confusion about the source occurs because the affected dermatomes in the skin receive sensory fibers from the same dorsal nerve root as the internal organs being stimulated. The tissues of both develop from the same embryonic or dermatomal segment, as described above. For example, a heart attack often causes apparent pain in the left arm and the anterior chest wall (Fig. 9-4B). Irritation of the receptor endings in the heart transmits impulses that are misinterpreted as sensations originating in the other areas. This inaccurate assessment occurs because the sensory nerves from the heart originate in thoracic dorsal root ganglia 1 to 5, while the sensory nerves of the arm and the anterior chest wall are connected to thoracic dorsal root ganglia 1 to 2 and 1 to 5, respectively.

Convergence of afferent fibers on the primary neurons of pain tracts or *facilitation* (summation) of otherwise subthreshold signals may also produce referred pain. Pain receptors produce afferent impulses from somatic areas and from visceral organs. These dual origins of the impulses may explain the difficulty an individual may experience in correctly identifying the sources of referred pain. In addition, impulses from visceral afferent fibers may lower the pain threshold by generating excitatory postsynaptic potentials (EPSPs) in the dorsal horn of the spinal cord. The EPSPs may be additive in their effects and facilitate the generation of impulses which register the sensation of pain through the lateral spinothalamic tracts.

Inhibition of Pain

The transmission of pain impulses may be reduced by stimulation of collateral afferent fibers in the tissues near the site of pain stimuli. For example, relief of pain is often observed on squeezing or shaking a hand after it is injured. Incoming "nonpain" signals travel to neurons in an area in the dorsal horn of the spinal cord called the *substantia gelatinosa.* Interneurons in the substantia gelatinosa may release inhibitory neurotransmitters such as gamma-aminobutyric acid (GABA), which hyperpolarize and inhibit the primary pain fiber. As a result, excitation of the secondary neuron of the pain tract does not take place. Some evidence indicates that descending fibers may exert a similar effect. Cells in the substantia gelatinosa may form a limiting barrier or gate through which incoming pain signals must pass in order to ascend to the brain. They are able to block the transmission of impulses from the presynaptic membrane of the primary neuron to the postsynaptic membrane of the secondary neuron and thereby inhibit pain.

Following irritation the relief of pain by the

application of a mustard plaster or the administration of acupuncture may work by the mechanism just described. Acupuncture has been used to block pain in surgery without other methods of anesthesia. Pain-killing drugs also may block the functions of pain fibers. The activity of some of these drugs and of brain neurotransmitters called *endorphins* and *enkephalins,* which are natural opiates with pain-killing activity, will be described in Chap. 10.

Effects of Cord Damage

The spinal cord and nerves may be damaged in a number of ways, including physical injury, infection (such as poliomyelitis), immune reactions (as in multiple sclerosis), toxic poisoning (such as that caused by arsenic), and growth of tumors or formation of *cysts* (accumulations of gas, liquids, or solids surrounded by a membrane which forms a sac). In some cases the damage causes *neuritis* (inflammation of nerve fibers). The symptoms of nerve damage depend on the tracts that are adversely affected and range from loss of sensation in specific areas to partial or complete paralysis of the body and even to death.

Wallerian Degeneration of Fibers

Fibers in the spinal cord do not regenerate or do so very slightly. However, when a peripheral nerve fiber is severed, the part remaining attached to the cell body survives while the rest of it degenerates. Degeneration occurs above the cell bodies of ascending fibers and below those of descending fibers. The degeneration away from the site of damage (in an ascending or descending direction) is called *Wallerian degeneration.* The regeneration of nerve fibers requires an intact cell body and neurilemma, along with production of nerve cell growth factor, a protein released from tissues toward which the fiber grows (see Chap. 6). Following damage, a regenerating nerve fiber branches outward as sprouts grow to reestablish communications. The process is random, with growth occurring at a rate of 1 to 2 mm/day. Schwann cells proliferate in the neurilemma around the end of the fiber still attached to the cell body. They can grow and bridge a gap up to 3 mm. Larger gaps in nerve fibers may be bridged surgically by a suture that ties cut ends together and/or by a nerve graft taken from a portion of another nerve from the patient. When the gap created by the damage is too great, a dense outgrowth of nerve fibers may form a bulbous sprout called a *neuroma* (a tumor of the nervous system). Such sprouts complicate regeneration of functional nerve fibers. In some cases neuromas further impede recovery in that they may contain sensory sprouts with extreme sensitivity to pressure.

General Effects

Spinal cord damage may occur at different levels. Loss of certain body functions may follow damage to specific areas of the spinal cord (Fig. 9-11). Injury at a specific level generally affects body components below it. When damage occurs higher up the cord, more of the body is likely to be adversely affected.

After upper-level cord injuries, simple spinal reflexes carried out by intact lower levels of the cord may be depressed for several weeks. This reaction, called *spinal shock,* occurs through the loss of the net effect of suprasegmental descending motor pathways. They no longer influence the neurons involved in simple reflexes controlled by areas below the damaged section. At first all reflexes are severely depressed, then some may gradually return. Spinal shock demonstrates the importance of the whole nervous system, even in "simple" reflexes.

Many tracts cross over at different levels in the CNS. Therefore, injury to the spinal cord in different regions may cause loss of sensory or motor activity on different sides of the body. The loss of neural function occurs on the side of the body opposite the injury if damage to the spinal cord occurs prior to the point of decussation of the affected fiber (Fig. 9-12). Loss of neural function occurs on the same side of the body if the spinal injury occurs beyond the point of crossover of the affected fiber (Fig. 9-12).

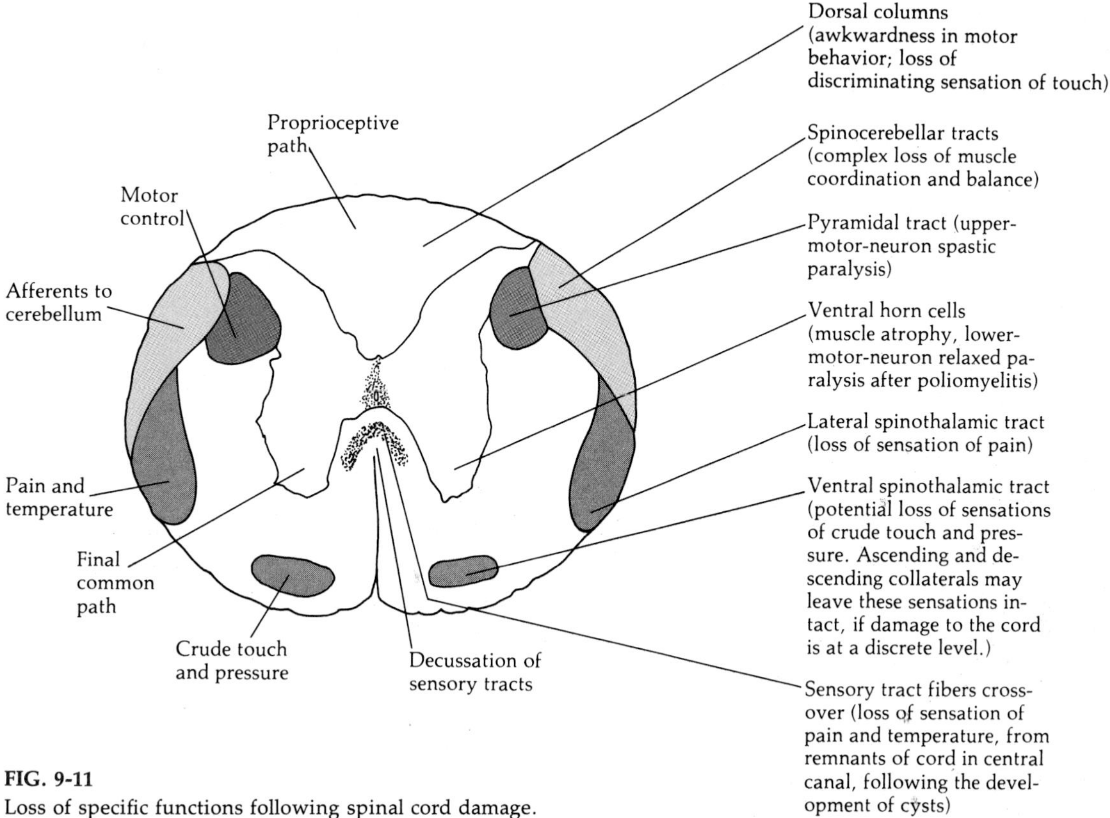

FIG. 9-11
Loss of specific functions following spinal cord damage. Damage to the areas illustrated commonly results in the loss of functions noted in parentheses.

Damage to Ascending Tracts

The spinothalamic tracts located near the outer surface of the spinal cord are most vulnerable to damage by external forces. For example, touch and pressure fibers of the ventral spinothalamic tract travel as ascending and descending collaterals for variable distances up and down the cord prior to forming synapses with fibers of the second cell body on the opposite side in the dorsal gray column (Table 9-1 and Fig. 9-10). Therefore, damage at one level of the cord may cause loss of sensations in areas associated with fibers entering it on the opposite side some three levels away. Since spinothalamic tracts cross over, damage to them often leads to loss of sensation on the opposite side of the body. The areas of sensory loss may be discrete or overlapping, depending on the location of fiber damage.

Syringomyelia (hollow cord) is a progressive degenerative disorder of the spinal cord which may be congenital or acquired. Physical trauma or physiological alterations such as a deficiency in blood supply are among the variety of causes of this disorder. As a result, neuroglial cells grow over and compress the central canal of the spinal cord. Irritation and pain are first experienced; then as the lateral spinothalamic tracts are destroyed, the sensations of pain and temperature are lost. Touch fibers of the ventral spinothalamic tract are damaged as well. The sense of touch may be left intact, however, through collaterals from this tract which ascend and descend to other segments prior to synapsing within the spinal cord.

Injury to ascending tracts also may affect some specific areas of the body. For example, a tumor growing in the dorsal columns may damage the fasciculus cuneatus, whose sensory

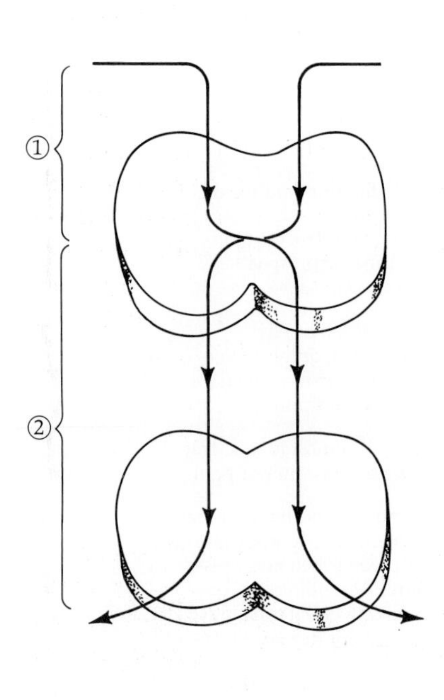

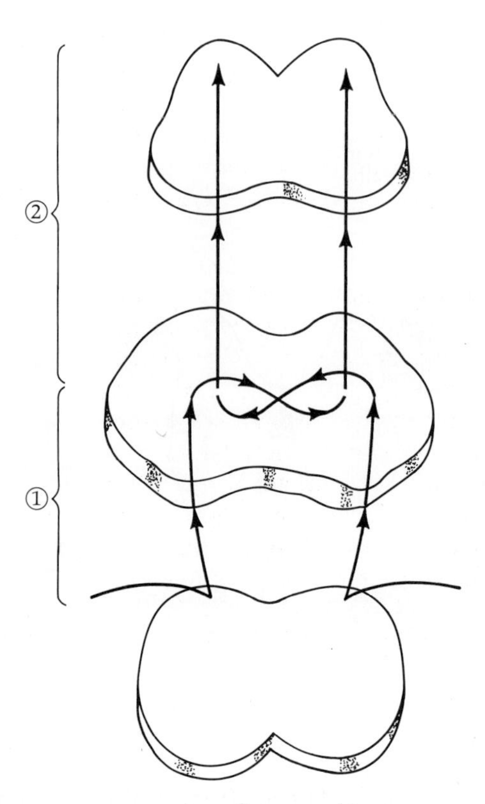

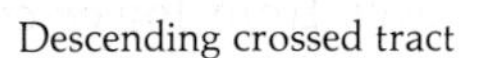

FIG. 9-12
Location of loss of function in the body related to area of spinal tract injury. In either diagram damage in region 1 causes loss of neural function on the opposite side of the body and damage in region 2 results in loss of neural function on same side of the body. (A) Pathways in descending crossed tract. (B) Pathways in ascending crossed tract.

neurons for discriminating touch are in the upper cord. The damage results in loss of sensations in the upper parts of the body such as the arm. In contrast, injury to the fasciculus gracilis, which has sensory neurons in the lower cord, affects the lower parts of the body such as the leg (see Table 9-1 and Figs. 9-9 to 9-11).

Damage to Descending Tracts

Damage to the ventral horn may produce partial or complete paralysis. Loss of motor activity occurs in areas of the body below the damaged spinal cord segments. Injury to the cervical cord may cause complete paralysis, while injury to the lumbar and sacral regions may affect only the lower part of the body and result in paralysis of one or both of the legs (see Table 9-2 and Fig. 9-12).

Effects of Specific Agents

In addition to physical injury, nutritional deficiency and infection may damage nerve tracts. The effects depend on the neurons involved and the nature of the agent (Table 6-4).

PHYSICAL FACTORS Physical injury and trauma may produce any of the results mentioned above. Excess stress of ligaments and muscles resulting from lifting too much weight, improper lifting, poor muscle conditioning, or physical trauma may lead to stress on ligaments and muscles. One rather common condition is dislocation of the intervertebral disc(s) and displacement of the connective tissue in the center of each disc, the nucleus pulposus. When dislodged, the nucleus pulposus and its softened cartilage may compress and damage neurons (Fig. 3-25). Discs in the lower back (lumbar region) are vulnerable to displacement. Their dislodgement may cause lower back and leg pain due to compression of the lateral spinothalamic tracts.

VITAMIN DEFICIENCY Vitamin deficiency may result in a wide range of diseases, with specific deficiencies causing marked effects in neuron physiology. Vitamin B_1 (thiamine) deficiency, which causes beriberi, and alcoholism, which is often accompanied by multiple vitamin deficiency, damage small nerve fibers of the lateral spinothalamic pain and temperature tracts. Vitamin B_{12} (cyanocobalamin) deficiency is associated with degeneration of large myelinated fibers in the touch and proprioceptive tracts, an effect similar to arsenic poisoning.

INFECTIONS A variety of infectious agents, including fungi, protozoa, bacteria, and viruses, may impair neuron function. A few of the more common disorders include damage from syphilitic bacteria and from herpes, polio, measles, and rabies viruses.

Syphilis is induced by a spiral-shaped bacterium normally transmitted by sexual contact with an infected individual. In later stages of the disease large myelinated fibers in the dorsal root ganglion are destroyed. Small C fibers for pain and temperature are left intact, while the touch and proprioceptive fibers are disrupted. The resulting disorder is called *tabes dorsalis.* The victim experiences pain in the limbs, loss of sensation, and loss of coordination of movement. As the infection spreads through the CNS, the individual develops general paralysis and dementia (loss of intellectual faculties).

A number of viruses attack the nervous system. Many destroy cells in the gray matter and cause accumulation of microglia in an inflammatory response termed viral encephalitis. The viruses that produce mumps, cold sores (*Herpes simplex*), and rabies are among these, as is the one that causes shingles. The latter is caused by *Herpes zoster,* which attacks the dorsal root neurons and produces pain and rashes in the dermatomes associated with the damaged segment of the spinal cord.

Poliomyelitis virus attacks motor neurons in the ventral horn. Within 10 days of infection, a sore throat, headache, stiff neck, and pains in the arms and/or legs usually appear. The muscles may be paralyzed but in a relaxed (flaccid) state. If the virus infects upper motor neurons in the brain, it may paralyze respiratory muscles and cause death. Vaccination with the Sabin vaccine prevents polio infection.

9-2 THE AUTONOMIC NERVOUS SYSTEM (ANS)

The ANS is composed of *sympathetic* and *parasympathetic* subdivisions, which relay involuntary efferent impulses from the CNS to hollow organs (viscera) through *ganglia* (clusters of neurons that act as relay stations). The pathways in the ANS are formed by visceral efferent fibers which relay impulses automatically, that is, without conscious control. The impulses travel in most instances from lower centers in the brain (areas below the cerebral cortex) through a two-neuron chain to responding tissues.

The ANS regulates the activity of exocrine tissue and organs which share a common feature, namely, the presence of smooth muscle or cardiac muscle in their structure. They include tissues in the circulatory, respiratory, excretory, digestive, and reproductive systems, as well as in parts of the eye (Fig. 9-13).

Organs affected by the ANS are generally innervated by both sympathetic and parasympathetic fibers, which usually cause opposite effects to help maintain homeostasis. For example, sympathetic stimulation speeds up the heart while parasympathetic impulses slow it down; sympathetic impulses inhibit motility of the digestive tract while parasympathetic ones accelerate it. The responses are related to differences in neurotransmitters released from the terminal fibers of the ANS and to the effects that the neurotransmitters have on cell receptors. Some organs which are solely innervated by sympathetic autonomic fibers include the

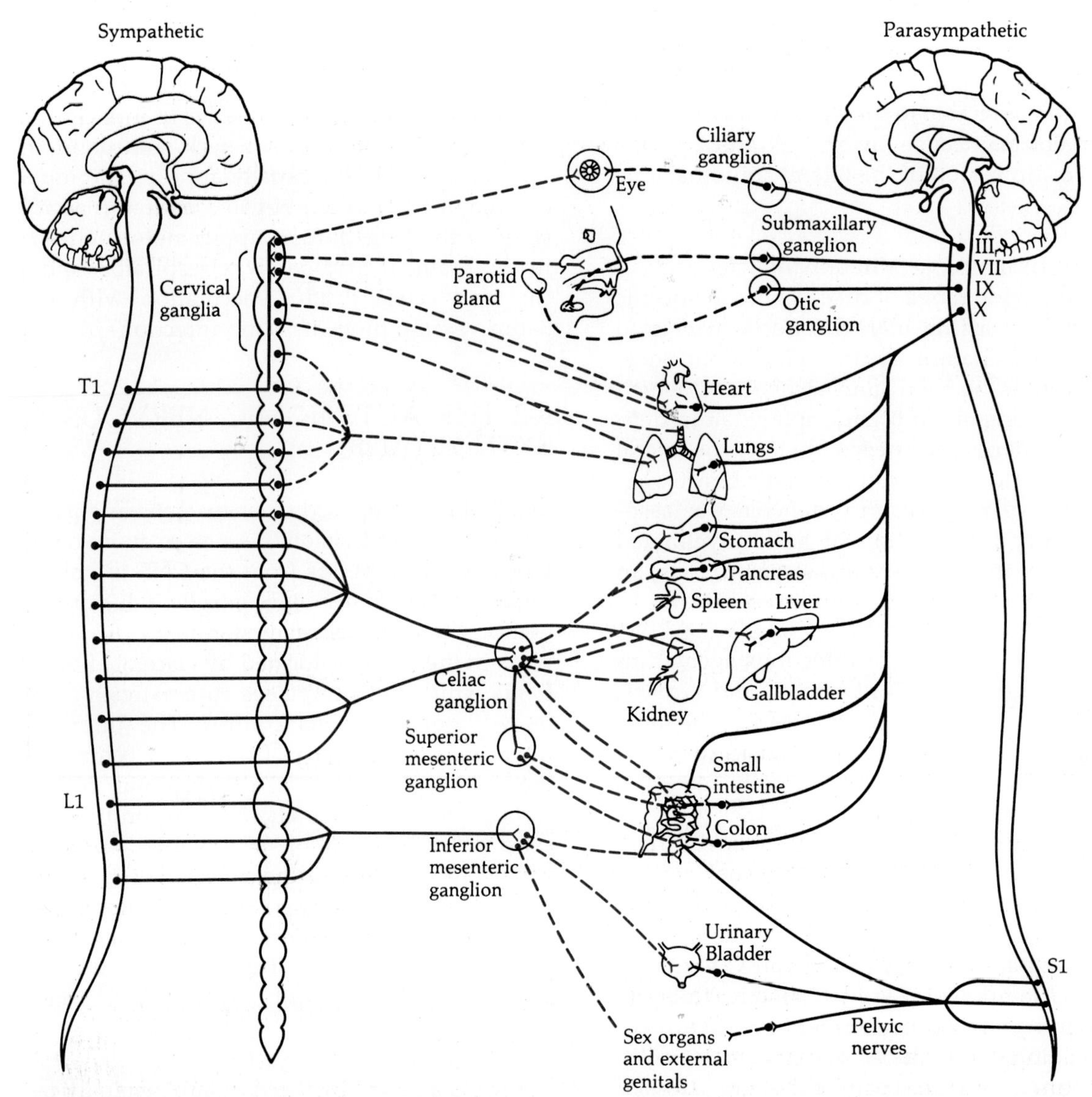

FIG. 9-13
Organs influenced by fibers of parasympathetic and sympathetic neurons. The fibers are illustrated as follows: solid lines, preganglionic; dashed lines, postganglionic; color, sympathetic; black, parasympathetic.

adrenal medullae, the sweat glands, smooth muscle around hair, and smooth muscle in blood vessels in the skin. The activity of these tissues is decreased by absence of sympathetic stimulation rather than by the opposing effects of parasympathetic fibers.

The primary cell bodies of autonomic neurons are located in the CNS and synapse with the cell bodies of second neurons clustered in ganglia outside the spinal cord. The fibers from the first cell body are located between the CNS and the ganglion and are called *preganglionic* fibers. Those between the ganglion and the innervated tissue are *postganglionic* fibers. The sympathetic and parasympathetic subdivisions of the ANS can be distinguished on the basis of (1) the origin of the fibers; (2) location of associated ganglia; and (3) type of neurotransmitter usually released from the postganglionic terminals of the fibers (Table 9-3).

TABLE 9-3
Properties of sympathetic and parasympathetic nerve fibers

Property	*Sympathetic*	*Parasympathetic*
Origin of nerves	Thoracic and lumbar segments of spinal cord	Cranial nuclei in brain and sacral segments of spinal cord
Location of ganglia	In chain beside vertebrae (paravertebral); in abdomen and thorax (prevertebral and collateral)	Near or in wall of organ innervated (terminal) (e.g., in wall of heart, digestive tract and bladder, near the eye and salivary glands)
Neurotransmitter released from preganglionic fibers	Acetylcholine	Acetylcholine
Usual neurotransmitter released from postganglionic fibers	Norepinephrine	Acetylcholine
Usual type of postganglionic fibers	Adrenergic. Some postganglionic sympathetic fibers are cholinergic; they release acetylcholine. These fibers include those which stimulate the sweat glands, inhibit smooth muscle in the coronary arteries, and relax that in blood vessels to skeletal muscle.	Cholinergic
Sole fibers to some organs	Yes	No
Terminal ganglia in or near organ	No	Yes
Diffuse effect	Yes	No
Primary receptor to usual transmitter	Alpha (α)	Gamma (γ)

Because the cell bodies of the primary neurons of the sympathetic nervous system are located in the intermediate gray matter of the thoracic and lumbar segments of the spinal cord, this system is sometimes called the *thoracolumbar* division of the ANS. Its ganglia are located alongside the spinal cord in a chain (the *paravertebral* ganglia) or at some distance from the spinal cord in the thorax and abdomen (the *prevertebral* or *collateral* ganglia).

The cell bodies of the primary neurons of the parasympathetic nervous system are located in nuclei of cranial nerves III, VII, IX, X, and XI (Tables 9-5 and 11-1) and in the sacral division of the spinal cord, which accounts for its alternate name, the *craniosacral* division of the ANS. Its ganglia are terminal in location, that is, they are within or adjacent to the tissue that is innervated.

Because of these anatomical differences, sympathetic postganglionic fibers are long, spread out considerably, and create diffuse effects, while parasympathetic postganglionic fibers are more localized. The ratio of preganglionic to postganglionic fibers approaches 1:10 or more in the sympathetic system and 1:2 in the parasympathetic system.

Types of Neurotransmitters Secreted by ANS Fibers

All preganglionic fibers of the ANS secrete acetylcholine (ACh) and are called *cholinergic nerves* (Fig. 9-14). They are mostly slow-conducting, B-type myelinated fibers. While all parasympathetic fibers are cholinergic, most but not all postganglionic sympathetic fibers release the neurotransmitter norepinephrine and are called *adrenergic nerves.* Regardless of the type of neurotransmitter released, postganglionic fibers are slower-conducting, unmyelinated (gray) C fibers.

Some postganglionic sympathetic fibers are cholinergic (Table 9-3). Furthermore, preganglionic sympathetic cholinergic fibers terminate in the adrenal medullae. The fibers release

ACh, which binds to receptor cells in those endocrine organs, causing them to secrete epinephrine (adrenaline) and norepinephrine (noradrenaline) into the blood. Although the adrenal medullae secrete 3 to 10 times more epinephrine than norepinephrine, sympathetic adrenergic neurons throughout the body secrete so much norepinephrine that the small portion of it that diffuses into the circulation (Fig. 9-15) is sufficient to keep the transmitter's level in the blood in a range of 0.2 to 0.3 mg/mL, which is nearly 4 to 5 times the level of epinephrine. The stimulation of sympathetic neurons and the adrenal medullae by stress (Fig. 5-7) elevates the quantities of both transmitters and, especially, releases large amounts of epinephrine stored in these endocrines. The secretion of neurotransmitters from cholinergic and adrenergic transmitters, their interaction with specific cellular receptors, and their general fate is illustrated in Figs. 7-17 (p. 176) and 9-15.

Variations in Responses to ANS Transmitters

Norepinephrine released from adrenergic neurons and epinephrine released from the adrenal medullae have overlapping but distinct effects (Chaps. 5 and 7). Furthermore norepinephrine can be converted to epinephrine by the addition of a methyl group under the influence of the enzyme phenylethanolamine-N-methyltransferase and both transmitters are degraded by methyltransferase (COMT) and other enzymes such as monoamine oxidase (MAO) prior to excretion in the urine:

Because of the similarity in structure, norepinephrine and epinephrine interact with α and β receptors. However, the effects of norepinephrine seem to result mainly from its stimulation of α receptors and those of epinephrine from excitation of β receptors. Different subdivisions of receptors have been discovered. For example, those in cardiac muscle are called β_1 receptors and can be distinguished from β_2 receptors in smooth muscle (Table 17-4). Although the exact nature of α, β, and γ receptors has yet to be completely unraveled, transmitters have been reported to alter the intracellular levels of specific secondary messengers (Table 9-4 and Fig. 7-17). In general, α and γ receptor stimulation usually causes opposite responses; norepinephrine binds to α receptors and decreases cAMP; epinephrine binds to β receptors and increases cAMP; and ACh binds to γ receptors and elevates cGMP.

Examples of Adrenergic and Cholinergic Effects

Response of Vascular Smooth Muscle on the Heart

The response of a tissue or organ to autonomic impulses depends not only on the types of nerve fibers and transmitters but also on the predominant effects of the different types of receptors of the responding tissue cells. For example, sympathetic adrenergic fibers release norepinephrine, which binds to α receptors and causes contraction of smooth muscle in the walls of many blood vessels in the body. However, sympathetic fibers to coronary arteries

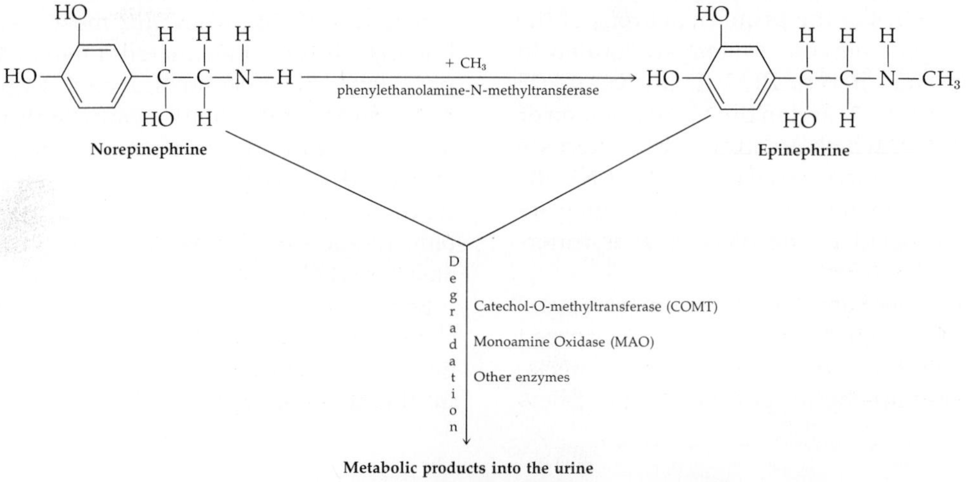

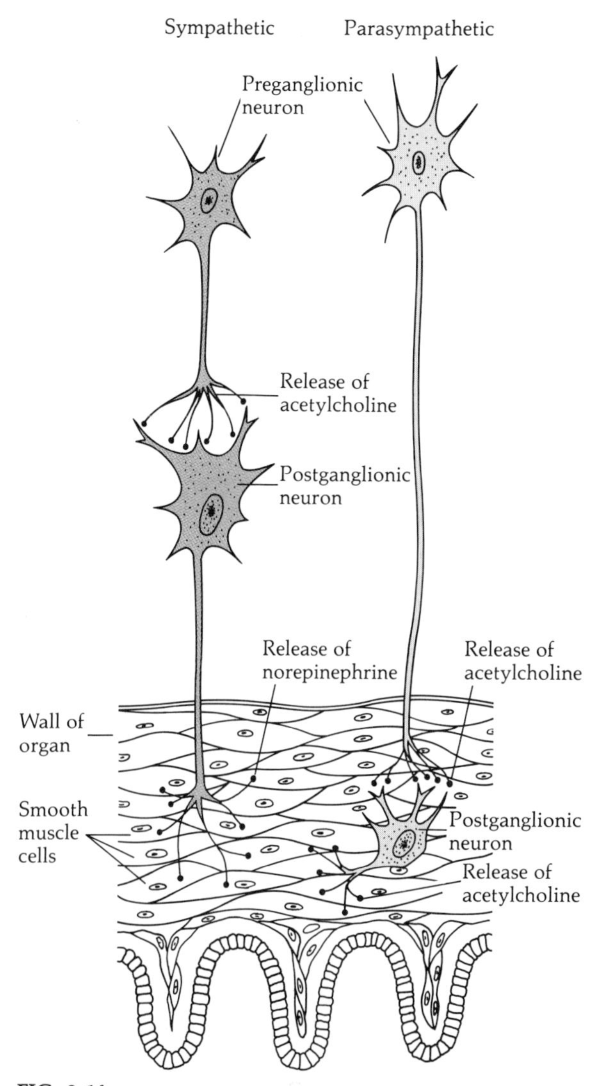

FIG. 9-14
Comparisons of autonomic nerves. Note the short preganglionic fiber and long postganglionic fiber of the sympathetic neurons. The relative lengths of the parasympathetic fibers are reversed.

and to arteries in skeletal muscles are cholinergic. They release ACh, which binds to γ receptors and relaxes the smooth muscle in the walls of the vessels. Consequently, the overall effect of sympathetic stimulation is an increase in blood pressure. This response is caused by a narrowing of many blood vessels and an increase in the rate and force of contraction of the heart due to adrenergic effects. However, more blood is simultaneously shunted through widened vessels in the wall of the heart and in skeletal muscles mainly due to B_2 adrenergic effects and some cholinergic effects (Tables 9-4 and 17-4).

Responses of Bronchial Smooth Muscle

Another example of the opposing effects of the subdivisions of the ANS can be observed in respiratory function. Constriction of smooth muscle around the bronchi (air passages to the lungs), making air exchange difficult, occurs in asthma. The changes can result from allergen-IgE complexes, which adhere to receptors on mast cells and basophils (and which decrease cAMP in those cells) and cause them to release histamine (Chap. 3).

Histamine binds to specific histamine receptors in bronchial smooth muscle (H_1 receptors), causing elevations in cGMP and contraction of the cells. Cholinergic release of ACh to γ receptors also elevates cGMP with similar results (Table 9-4 and Fig. 9-16). Adrenergic release of norepinephrine stimulates α receptors and decreases cAMP in the cells, heightening the response. Conversely, agents which elevate cAMP [e.g., epinephrine (Table 9-4) or isoproterenol] relieve symptoms of an asthmatic attack.

Interactions between the Brain and the ANS

ANS-mediated responses often are complex and may involve several critical physiological functions controlled by polysynaptic pathways (multiple synapses and their associated neurons). Many of the interconnections between afferent fibers to the brain and efferent fibers of the ANS remain to be discovered. However, the origins of several ANS routes from the brain have been demonstrated. These include parasympathetic fibers of the cranial nerves III, VII, IX, X, and XI, whose origins are indicated in Table 9-5. These fibers are associated with centers which regulate vital cardiovascular functions, respiratory functions, body temperature, and emotions. The relations between the ANS, the brain, and cranial nerves will be discussed further in Chap. 11, and their influence on cardiovascular and respiratory physiology will be considered in Chaps. 15, 17, and 20.

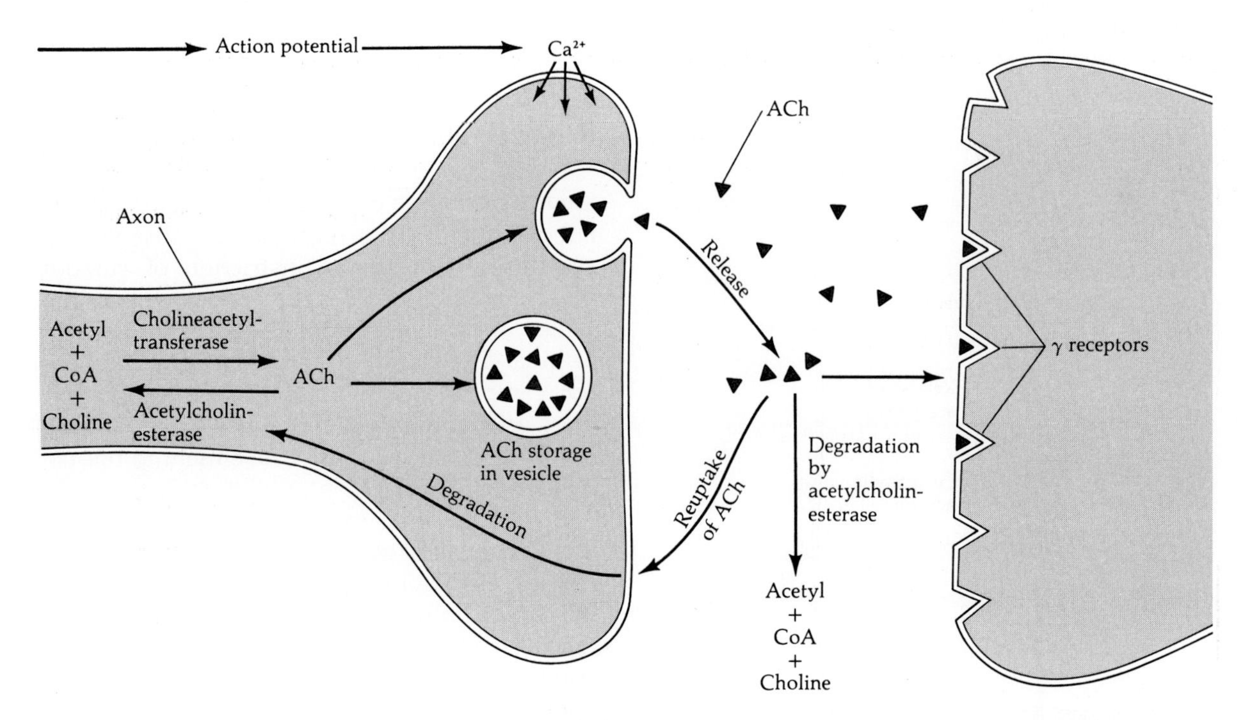

A Cholinergic synapse

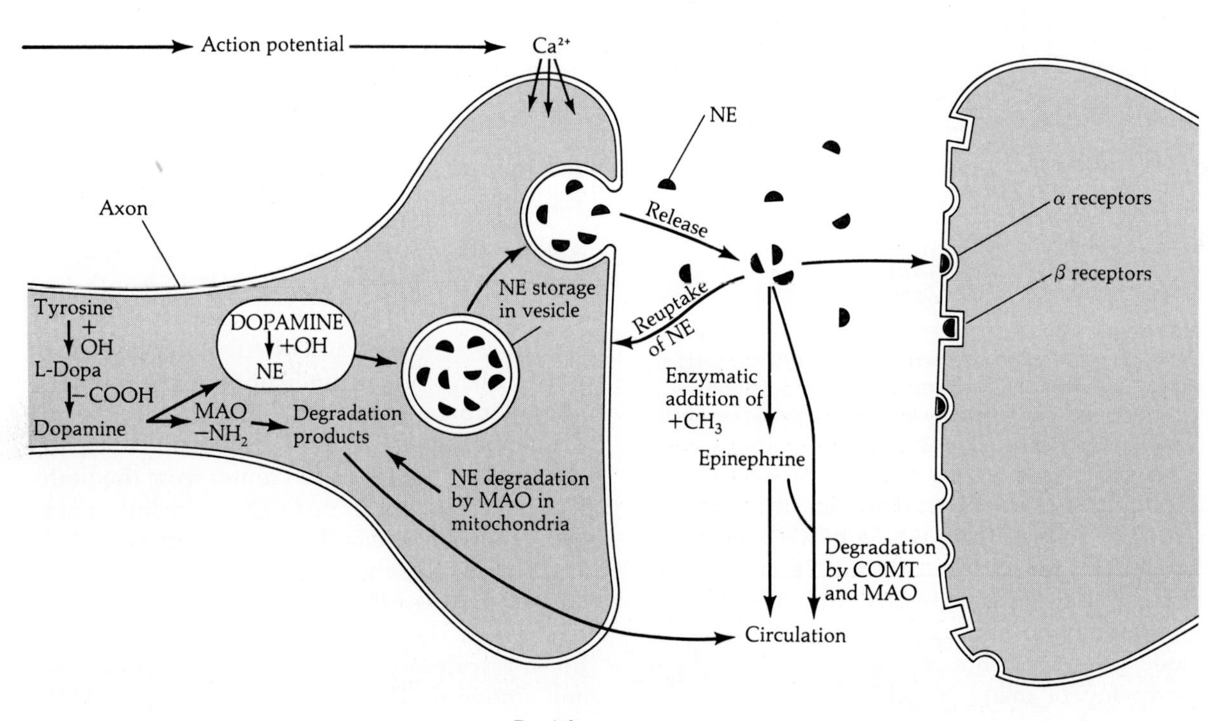

B Adrenergic synapse

FIG. 9-15
Chemical events at (A) cholinergic and (B) adrenergic synapses. Events surrounding the production and fate of acetylcholine (ACh) and norepinephrine (NE): MAO = monoamine oxidase; CoA = coenzyme A; COMT = catechol-O-methyl transferase; L-dopa = L-3,4-dihydroxyphenylalanine.

TABLE 9-4
Biological properties of adrenergic and cholinergic nerves—and comparison with usual effects of epinephrine*

Property	*Adrenergic Stimulation (by Sympathetic Fibers)*	*Hormonal Stimulation*	*Cholinergic Stimulation (Usually by Parasympathetic Fibers)*
Transmitter	Norepinephrine	Epinephrine (from adrenal medullae)	Acetylcholine
Primary receptor	α	β	γ
Effects on involuntary muscle			
Heart (heart rate and contractility)	(+ to −)†	+	−
Bronchi	+‡	−	+ (sympathetic fibers to vessels in heart and skeletal muscle)
Many blood vessels	+	−	− (sympathetic fibers to vessels in heart and skeletal muscle)
Wall of stomach and intestine (motility and tone)	−	− (usually)	+ (sympathetic fibers to vessels in heart and skeletal muscle)
Sphincters in stomach and intestine	+ (usually)		− (usually)
Effect on secondary messengers	↓ cAMP (usually)	↑ cAMP	↑ cGMP
Major effects of drugs on receptors¶			
Phenylephrine (neosynephrine)	+		
Isoproterenol		+	
Carbachol			+
Phenoxybenzamine (dibenzyline)	−		
Propranolol		−	
Atropine			−§

* (+) = stimulation, (−) = inhibition
† The stimulatory effect is much less than that of epinephrine, and leads to reflex slowing indicated by the negative (−) sign.
‡ α receptors have been demonstrated in the bronchi of guinea pigs but remain to be positively identified in humans.
§ Some types of receptors that are responsive to ACh are not inhibited by atropine.
¶ Also see Fig. 10-10.

9-3 SUMMARY

1 The two major divisions of the nervous system are the central nervous system (CNS), which contains the brain and spinal cord, and the peripheral nervous system (PNS), which consists of all of its other components.

2 The spinal cord consists of nervous tissue within the vertebral canal. Thirty-one pairs of spinal nerves emerge from the cervical, thoracic, lumbar, sacral, and coccygeal regions of the vertebral column. The spinal cord is elliptical, consisting of peripheral myelinated white fibers, which surround unmyelinated gray fibers. The latter are grouped in a shape much like the letter H.

Sensory neurons are found in the dorsal horn of the gray matter of the cord, motor neurons are found in the ventral horn, and interneurons are in between. The sensory and motor fibers emerge from the cord as the dorsal and ventral roots, respectively, and merge to form pairs of spinal nerves.

In spinal reflexes impulses move into the cord from receptors and out again to effectors without ascension to the brain. The functional units of the nervous system are reflex arcs, the simplest of which are monosynaptic. More common reflex arcs are polysynaptic. Those reflexes which are learned are called conditioned reflexes.

Ipsilateral reflexes remain on the same side of the cord, while contralateral ones cross it. Intrasegmental reflexes remain within the same segment of the cord, while intersegmental ones travel over more than one level and suprasegmental reflexes involve the brain.

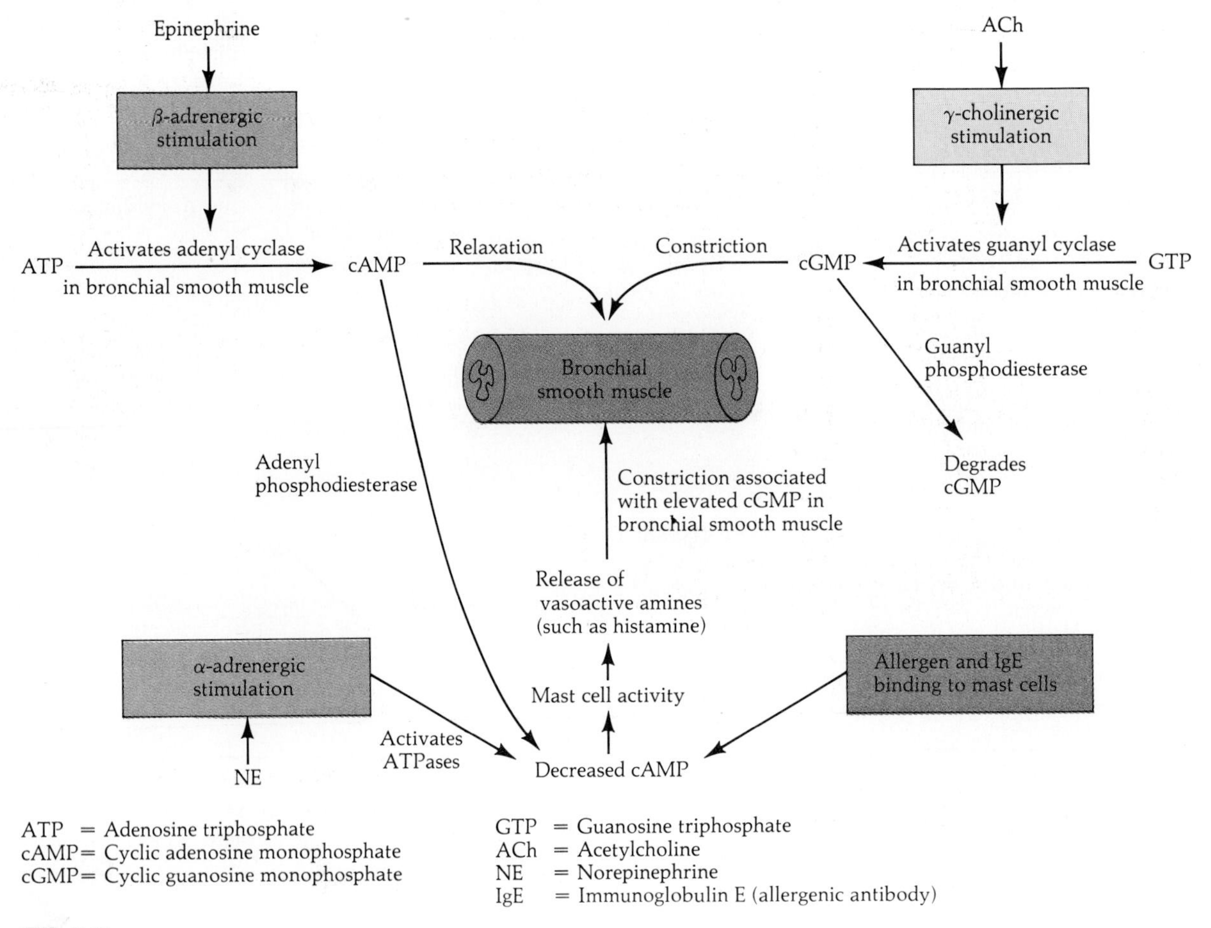

FIG. 9-16
Adrenergic, cholinergic, and allergic effects on bronchial smooth muscle.

Nerve impulses can funnel in convergence from a larger to a smaller number of neurons through converging circuits, can spread in divergence from a smaller to a larger number over amplifying circuits, can be transmitted through a series of cells in parallel circuits, or can be used to create feedback signals in reverberating circuits (feedback circuits).

Bundles of fibers carrying impulses up and down the CNS are respectively called ascending and descending nerve tracts or pathways. The spinothalamic and dorsal white columns are among the major ascending sensory tracts. They are composed of a chain of three sensory neurons with cell bodies in the dorsal root ganglion, dorsal gray column of the cord, and thalamus of the brain. Each of these tracts delivers impulses which cross the CNS as they ascend to the cerebral cortex. The major descending tracts are the pyramidal

TABLE 9-5
Components of autonomic nervous centers in the brain

Area	*Type of ANS Fibers*	*Functions*
Hindbrain		
Medulla oblongata (in lower brainstem)	Sympathetic and parasympathetic fibers to vital organs	Controls vital centers for the Heart, Respiration, Blood vessel tone
	Parasympathetic components of cranial nerves	
	IX (glossopharyngeal)	Controls production of saliva by parotid gland; assists in regulation of respiration
	X (vagus)	Regulates activity in the heart and the respiratory and digestive tracts
	XI (spinal accessory)	Distributed with vagal fibers to thoracic and abdominal viscera; also to the voice box
Pons	VII (facial)	Regulates production of saliva by submaxillary and sublingual glands Regulates production of tears by lacrimal gland
Midbrain (middle brainstem)	III (oculomotor)	Regulates all but two external muscles to the eye Fibers to the sphincter muscles of the iris reflexively alter size of the pupil Fibers to ciliary muscles attached to the lens alter its diameter reflexively in accommodation for near vision
Forebrain		
Hypothalamus	Nuclei form autonomic centers	Various centers regulate sleep, appetite, thirst, body temperature
Anterior hypothalamus	Parasympathetic fibers	Promotes heat loss by stimulating the sweat glands Decreased sympathetic activity dilates (relaxes) vessels in skin and promotes heat loss
Posterior hypothalamus	Sympathetic fibers	Promotes heat production by constriction of blood vessels in the skin and shivering of skeletal muscles
Limbic system	Interacts with hypothalamus	Influences feelings, emotions (e.g., fear, anxiety, pleasure, relaxed state), mating behavior, feeding responses, sense of smell

and extrapyramidal tracts. They are composed of neuron chains, most of which cross over in the CNS and many of which descend to skeletal muscle.

The effects of damage to the spinal cord depends on the nature of the pathways involved. Nerve fibers separated from their cell body deteriorate in a phenomenon called Wallerian degeneration. Spinal shock causes depression of spinal reflexes for periods up to several weeks and is due to the loss of descending suprasegmental pathways. Since ascending spinothalamic tracts cross over in the CNS, damage to them often leads to loss of sensation on the opposite side of the body. The areas of sensory loss may be discrete or overlapping, depending on the location of fiber damage. Loss of motor activity occurs in areas of the body below the damaged motor tract segments. Paralysis is likely to affect greater to lesser areas following damage to cervical, thoracic, lumbar, sacral, and coccygeal segments, respectively. The dislocation of an intervertebral disc, deficiency of vitamins B_1 or B_{12}, alcoholism, or infection with bacteria (as in syphilis) or with certain viruses can damage parts of the CNS and result in the loss of sensory and/or motor activity, paralysis, or death.

3 The autonomic nervous system (ANS) consists of various efferent nerves which automatically regulate the activity of smooth muscle, cardiac muscle, and exocrine glands. Ganglia of the ANS are formed by clusters of neurons, which serve as relay stations between incoming preganglionic fibers from the CNS and outgoing postganglionic fibers which become autonomic nerves to the organs of the body.

The sympathetic nerves of the ANS arise from the thoracic and lumbar segments of the CNS. Most, but not all sympathetic postganglionic fibers release norepinephrine and are called adrenergic nerves, while all fibers of the parasympathetic division release ACh and are called cholinergic nerves.

The ANS regulates the activity of cardiac muscle and of smooth muscle in the eye, as well as that in circulatory, digestive, respiratory, urinary, and reproductive organs. Both sympathetic and parasympathetic fibers innervate most organs, usually exerting opposite effects produced in response to changes in the internal or external environment in order to maintain homeostasis.

Sympathetic fibers are more diffuse than parasympathetic ones. Sympathetic adrenergic fibers release norepinephrine, which binds to α receptors and causes a decrease in cAMP. Parasympathetic cholinergic fibers liberate ACh, which binds to γ receptors and causes an increase in cGMP.

Alpha-adrenergic nerve stimulation causes contraction of vascular smooth muscle, while the stimulation of cholinergic nerves causes γ receptors to induce relaxation of this tissue. The effects of neurotransmitters on the same type of tissue in different locations can vary, owing to interaction of the neurotransmitters with different receptors. Thus, in contrast to the reactions in vascular smooth muscle, alpha-adrenergic stimulation inhibits smooth muscle motility in the digestive tract, while cholinergic excitation of gamma receptors stimulates digestive tract motility. Cholinergic nerves also cause contraction of bronchial smooth muscle, as do allergic reactions mediated by histamine. These responses are inhibited by sympathetic stimulation of the adrenal medulla, which secretes epinephrine, which binds to beta receptors to cause relaxation of bronchial smooth muscle.

4 ANS functions are especially influenced by neurons in the medulla, hypothalamus, and limbic system of the brain. Among some of the autonomic centers in these areas of the brain are those that control circulatory and respiratory functions, those that regulate body temperature, and some which influence emotions.

REVIEW QUESTIONS

1 What are the major divisions of the nervous system?

2 Describe the origins of the spinal cord and nerves and their relationship to the vertebral column. How does this change during development?

3 What is the functional significance of the dorsal and ventral horns and interneurons of the spinal cord? How do the dorsal root ganglion, ventral root, and spinal nerves relate to these structures?
4 Of what value are reflexes? How do monosynaptic and polysynaptic reflexes compare? What are conditioned reflexes?
5 Compare amplifying, converging, parallel, and reverberating neuronal circuits.
6 Describe the differences between descending and ascending tracts in the CNS and compare their general functions. What are pyramidal and extrapyramidal tracts?
7 What are the names and routes of tracts for pain and temperature, crude touch and light pressure, posture and balance, fine touch, and deep sensations?
8 Describe the origins and pathways for pain impulses. Compare first and second pain.
9 What is the anatomical basis for referred pain and what alternative explanations may be used to explain its origins?
10 Describe some ways to inhibit pain.
11 Describe Wallerian degeneration of nerve fibers and the direction of regeneration of ascending and descending fibers.
12 Account for the effects of spinal shock.
13 Explain some major differences in symptoms resulting from damage to the spinothalamic, fasciculus cuneatus, and fasciculus gracilis pathways.
14 How do the dislocation of intervertebral discs produce pain and paralysis in different areas of the body? Account for paralysis in an arm compared with a leg. Describe some causes of lower back pain.
15 Outline the effects of vitamin B_1 and B_{12}, syphilis, viral encephalitis, and poliomyelitis on nerve pathway function.
16 Explain the origins of sympathetic and parasympathetic nerves and distinguish between adrenergic and cholinergic nerve fibers.
17 Compare the sympathetic and parasympathetic subdivisions of the autonomic nervous system.
18 Compare the neurotransmitters released by the postganglionic fibers of sympathetic and parasympathetic terminals. With what receptors do they interact and what effect do they have on secondary messengers such as cGMP and cAMP?
19 What are the effects of adrenergic and cholinergic impulses on smooth muscle in the walls of blood vessels, the digestive tract, and the respiratory tract? How does an allergic attack involving histamine compare with adrenergic and cholinergic responses in the bronchi?
20 Discuss some general influences that the higher centers of the brain may have on autonomic functions.

THE NERVOUS SYSTEM

General Structure and Function of the Brain

10

10-1 GENERAL PROPERTIES

10-2 DEVELOPMENT OF THE BRAIN AND CEREBROSPINAL FLUID
Embryonic origin of the brain and its subdivisions
Cavities in the CNS
Tissues covering the brain
Cerebrospinal fluid (CSF)
Formation and circulation
Properties of CSF
Functions

10-3 DIAGNOSTIC TESTS AND BRAIN FUNCTION
Chemical analysis of CSF
X-ray analysis
Pneumoencephalography
Cerebral angiography
CAT scan
Immunological studies
Electrical studies
Electrical stimulation of the brain
Electroencephalography (EEG)
Sleep • Abnormal patterns: Epilepsy

10-4 BRAIN NEUROTRANSMITTERS AND THE INFLUENCE OF DRUGS
Endorphins
Enkephalins
Dynorphin
Drug effects on neurotransmitter function
Opiate activity
Model of drug addiction
Specific drugs and mechanisms of action

10-5 SUMMARY

OBJECTIVES

After completing this chapter the student should be able to:

1. Describe the general properties of the brain.
2. Discuss the development of the brain, identify its three major divisions, and list the major components in the four subdivisions of the brain and those of the peripheral nervous system.
3. Outline the way in which the ventricles of the brain and the central canal of the spinal cord are formed.
4. Diagram the tissues that cover the surface of the brain and spinal cord.
5. Describe the origin, circulation, properties, and functions of cerebrospinal fluid.
6. List and discuss several techniques used to diagnose brain function and dysfunction.
7. Identify and describe the opiatelike neurotransmitters in the brain and discuss the mechanisms by which drugs influence neurotransmitter function.

10-1 GENERAL PROPERTIES

The brain is the most complicated organ in the human body. It has been estimated that the neurons of the brain, if stretched end to end, would extend to the moon and back. The electrical activity and chemical communications between its hundreds of billions of cells, with their trillions of synapses, have been likened to countless flashing stars in a galaxy of infinite proportions. Yet this organ, so critical to life, weighs about 1.3 ± 0.065 kg (nearly 3 lb), which is only 2 percent of total body weight, and is generally heavier in males than females. Its weight reaches a maximum at age 20 and declines thereafter, some 100,000 cells being lost per day. However, no correlation has been established between brain size and intelligence in humans!

The brain accounts for about 20 percent of the oxygen consumption in the resting body, an amount disproportionate to its size but indicative of its importance to body function. The energy consumed is equivalent to that used by a 20-W bulb and powers hundreds of billions of neurons (estimates range from 10^{10} to 10^{12}), which represent about 70 percent of all neurons in the central nervous system (CNS). Each neuron in the brain typically forms 1000 to 10,000 synapses, to produce a total of nearly 100 trillion (10^{14}) junctions. Circuits are formed in which small groups of neurons wired to others perform specific functions.

With the exception of olfactory input, incoming information is processed in the thalamus of the brain, which acts as a "switchboard" and relays information to other areas, including the highest brain level, the cerebral cortex. The information is transmitted over a maze of pathways, which involve several kinds of fibers. Within each half of the brain are *associative neurons,* which form interconnections and which comprise about 99.98 percent of the neurons in the CNS. Fibers of neurons which carry impulses between the two halves of the cerebrum are called *commissural fibers.* Incoming afferent fibers, associative fibers, and commissural fibers form a vast number of pathways which can deliver impulses to large numbers of motor neurons.

10-2 DEVELOPMENT OF THE BRAIN AND CEREBROSPINAL FLUID

Embryonic Origin of the Brain and Its Subdivisions

The mitotic cell division which follows fertilization of the egg produces an increasing mass which separates into three distinct layers, an outer *ectoderm,* a middle *mesoderm,* and an inner *endoderm.* In the first few days after conception the ectoderm grows rapidly. It forms a thick layer in the middle of the developing embryo called the *neural plate* (Fig. 10-1). The cells continue to proliferate and produce two elevated ridges, the *neural folds.* Some cells from the neural folds merge to form a cylinder surrounding a hollow central space called the *neural tube,* which gives rise to the CNS. Other

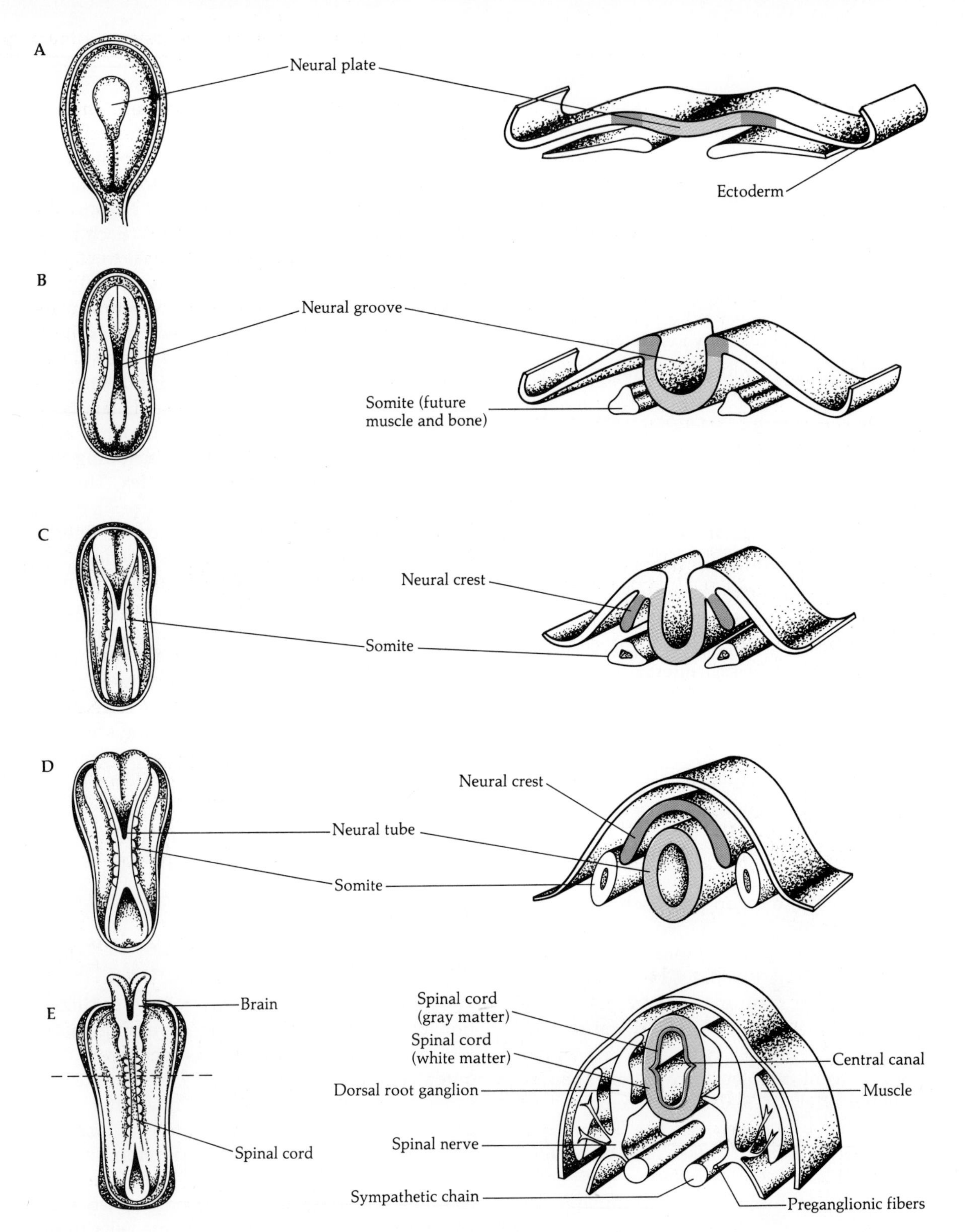

FIG. 10-1
Early development of the nervous system (first month after conception). The successive drawings on the left depict a human embryo 3 to 4 weeks after conception and viewed from above. Those on the right are cross sections taken from the approximate midregions of the respective illustrations on the left.

cells differentiate within the ectoderm from each side of the neural tube to become the *neural crest.* Neural crest cells give rise to several parts of the peripheral nervous system (PNS), including ganglia and nerves of the autonomic nervous system (ANS) as well as the nerves of the spinal cord. Early development and rates of growth are dominant at the head (anterior) end of the body and form a brain or encephalon (fr. Greek *en,* in, and *kephale,* head) disproportionately large compared with the remainder of the body. Estimates indicate that, prior to birth, an average of 250,000 neurons per minute are produced in the brain.

Growth of the neural tube is evident during the first 3 weeks after conception. By the fourth week the neural tube bends to form a C-shaped structure with three distinct regions, the *forebrain, midbrain,* and *hindbrain* (Fig. 10-2A).

FIG. 10-2
Development of the brain during the first 4 to 5 weeks after conception. (A) Dorsal view of the development of the brain's three major subdivisions. (B) through (E) Further development of subdivisions. (F) A side view of the subdivisions shown in E. From: E. L. House, B. Pansky, and A. Siegel, *A Systematic Approach to Neuroscience.* Copyright © 1979, McGraw-Hill Book Company; used with permission of McGraw-Hill Book Company.

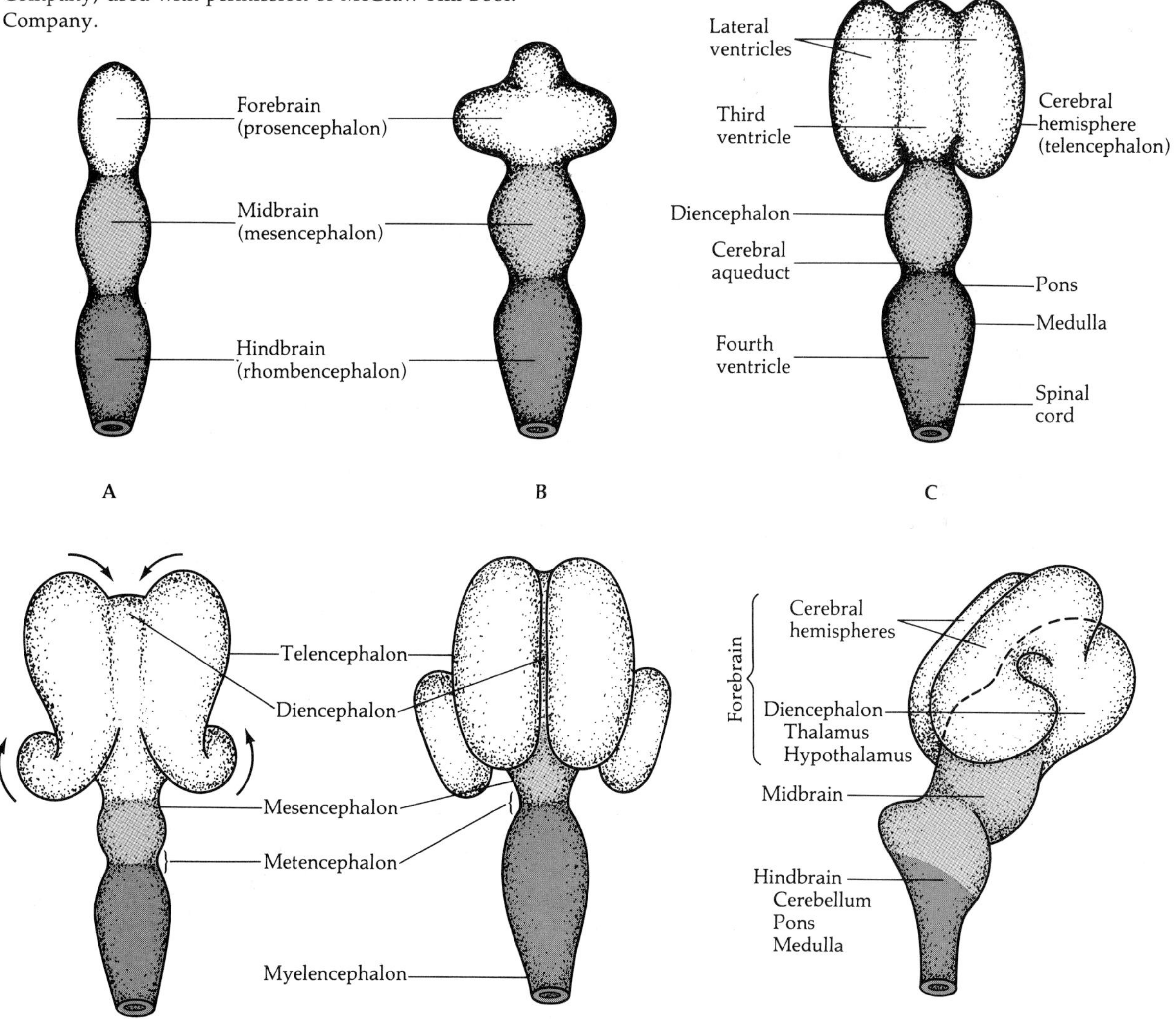

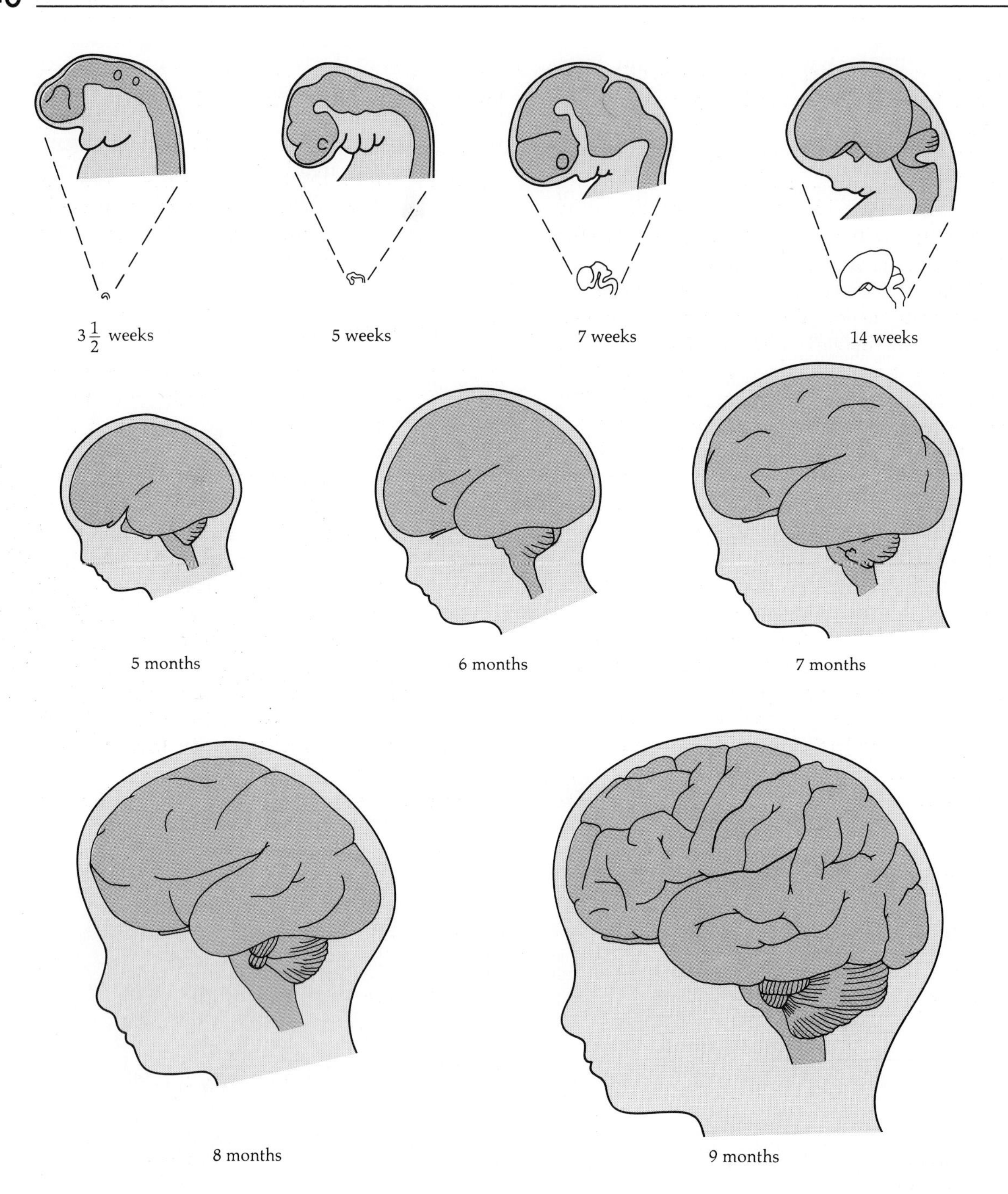

FIG. 10-3
Appearance of the brain during development, viewed from the side. The early divisions of the brain, evident at 3 weeks, bend by the fifth week. A similar bend is seen at the junction of the brain and spinal cord. Outpocketings in the forebrain develop, which later form the retina and the optic nerves to the eye. The lateral walls of the telencephalon of the forebrain grow rapidly to form large cerebral hemispheres, which are very evident by 7 weeks. About the fourteenth week, their surface starts to expand, forming hills and valleys characteristic of the human brain. (The drawings are about four-fifths life size, except for the enlarged sketches in the top row.)

The forebrain grows rapidly and bends at an angle to the midbrain. Continued development produces a large, long furrow, which separates the forebrain into the left and right *cerebral hemispheres* (Fig. 10-2C). The outer layers form the *cerebral cortex.* The anterior segment of the developing brain forms two outgrowths, the *optic vesicles,* that will form the retinas of the eyes. They are located in the second portion of the forebrain called the *diencephalon,* which also forms the *thalamus* and *hypothalamus.* The thalamus is a switchboard with a number of sensory and associative relay neurons. The hypothalamus is an important autonomic center.

Centers for hearing and vision and bundles of nerve tracts develop in the *midbrain.* This segment of the brain contains major connections between the anterior and posterior divisions of the CNS.

Growth of the *hindbrain* creates three major structures: the *pons, cerebellum,* and *medulla oblongata.* The pons consists of neurons which bridge various parts of the brain, the cerebellum coordinates and refines muscular activity, and the medulla regulates several vital activities and connects the brain to the spinal cord.

The development of the neural tube posterior to the hindbrain (toward the tail end of the body) produces the *spinal cord.* Neurons from the neural crest form the peripheral nerves and their associated cells (Fig. 10-1). Significant changes in the development of the nervous system are apparent throughout the first 9 months after conception and indicate a complex process which begins at conception and continues throughout life (Fig. 10-3).

Cavities in the CNS

Ultimately an adult brain is formed that is divided into a forebrain, midbrain, and hindbrain, each of which contains specialized subdivisions (Table 10-1). The neural tube of early embryonic development persists in modified form as five interconnected chambers lined by ciliated ependymal cells. Four are called ventricles and the fifth is named the cerebral aqueduct. The two *lateral ventricles* are the most anterior and develop side by side within each cerebral hemisphere. They are connected posteriorly with the *third ventricle,* which is found in the midline of the forebrain. The *cerebral aqueduct* extends as a channel in the midbrain leading from the third ventricle to the fourth ventricle, which is in the hindbrain.

The fourth ventricle is continuous with the

TABLE 10-1
Origins of structures in the CNS derived from the neural tube

Structure	Subdivisions	Cavity	Major Components
Brain			
Forebrain (prosencephalon)	Telencephalon	Lateral ventricles Anterior part of ventricle III	Cerebral hemispheres Olfactory bulb Corpus callosum
	Diencephalon	Most of ventricle III	Epithalamus Thalamus Hypothalamus
Brainstem: Midbrain (mesencephalon)		Cerebral aqueduct	Corpora quadrigemina (superior and inferior colliculi) Cerebral peduncles
Brainstem: Hindbrain (rhombencephalon)	Metencephalon	Ventricle IV	Cerebellum Pons
	Myelencephalon	Ventricle IV	Medulla oblongata
Spinal cord		Central canal	Cervical, thoracic, lumbar, and sacral divisions of the spinal cord

central canal in the spinal cord. By adulthood, the central canal is often filled with neurons, glial cells, and their products. The locations of the chambers and channels in the CNS are listed in Table 10-1 and illustrated in Figs. 10-2C and 10-6.

Tissues Covering the Brain

The brain is housed in the skull and, along with the spinal cord, is surrounded by three distinct membranes, the *dura mater*, the *arachnoid* layer, and the *pia mater* (Fig. 10-4). They are referred to as meninges (singular meninx). Invasion by bacteria or viruses causes an inflammation of one or more of these membranes called *meningitis*.

The outermost membrane covering the brain and spinal cord is the *dura mater*. It is composed of two sheets of tough, fibrous connective tissue with venous sinuses (cavernous structures holding pools of blood) located between the two layers. The outermost dural layer, the *periosteum*, lines the skull, and its inner layer dips into furrows on the surface of the brain. It produces folds (falx cerebri) between the left and right cerebral hemispheres, divides the cerebral cortex from the cerebellum (tentorium cerebelli), and separates the cerebellar hemispheres from each other (falx cerebelli).

The middle membrane, the *arachnoid*, is formed of a loose web of collagenous and elastic fibers. These compose a meshwork surrounding open areas, called *subarachnoid spaces*, which are filled with cerebrospinal fluid.

Fingerlike projections from the arachnoid attach to the innermost and third membrane, the *pia mater*, whose elastic and collagenous fibers adhere to the adjacent nervous tissue. In many places, the pia mater directly contacts the ciliated ependymal cells which line the ventricles of the brain and the central canal of the spinal cord. In each of the four ventricles, the pia mater is penetrated by blood vessels called the *choroid plexus* (Figs. 10-4 and 10-6).

Cerebrospinal Fluid (CSF)

Formation and Circulation

A liquid, *cerebrospinal fluid* (CSF), filters from the blood in the vessels of the choroid plexus and flows through the central canal of the spinal

FIG. 10-4
Diagram of meninges covering the brain.

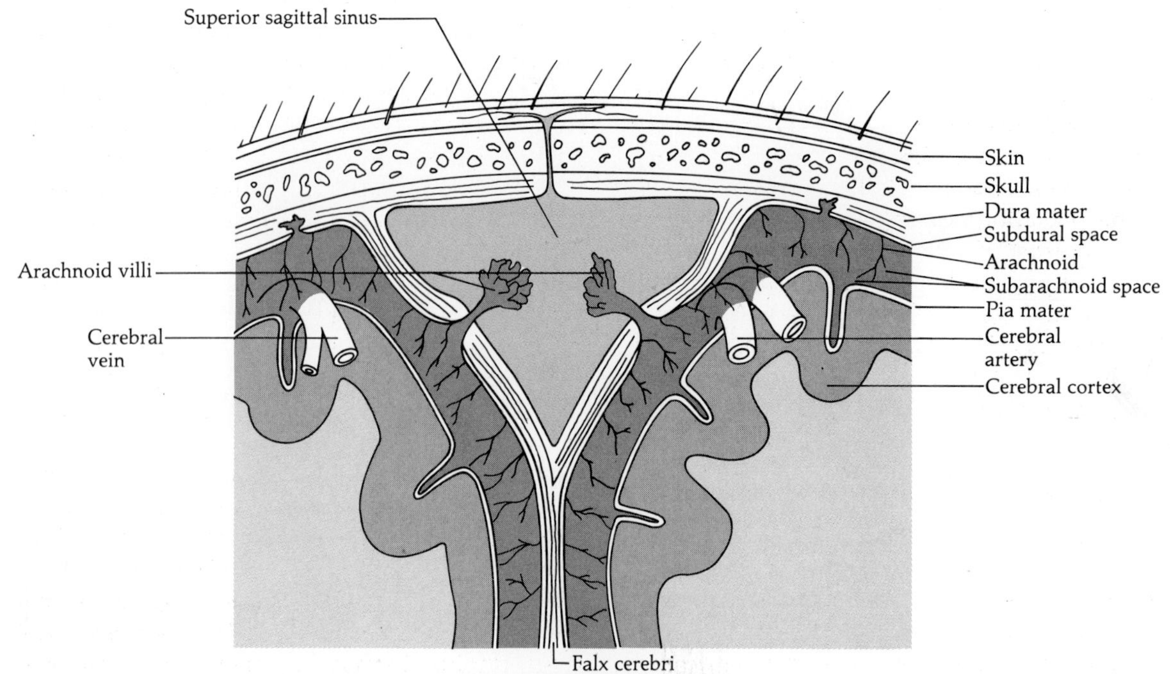

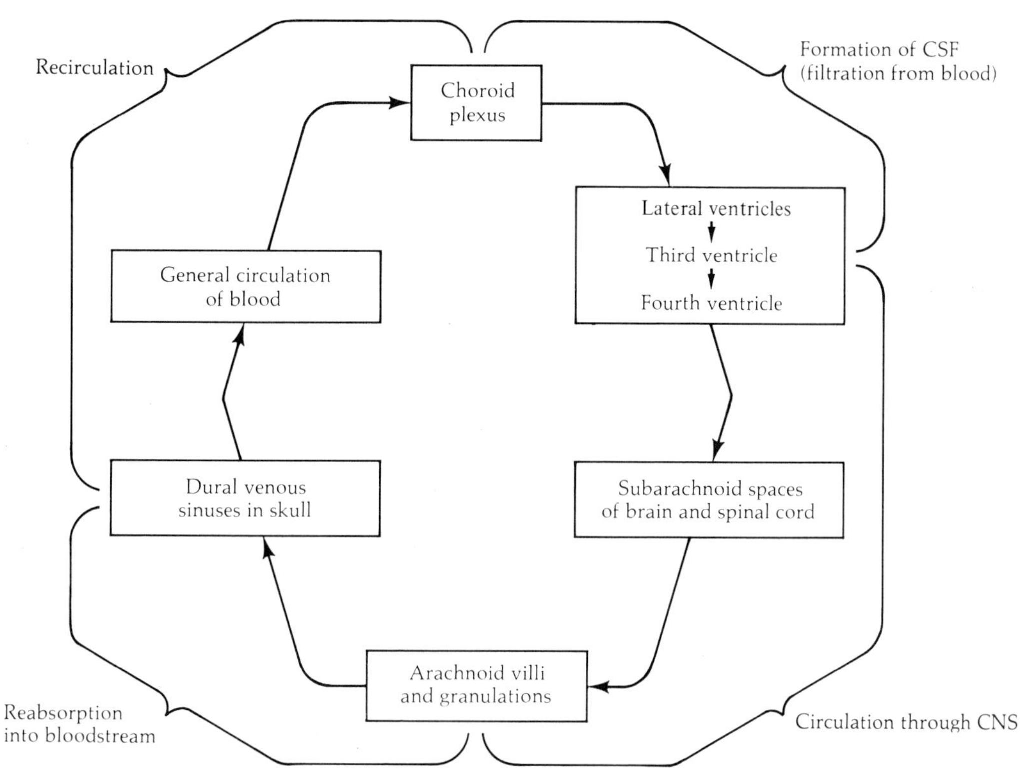

FIG. 10-5
Formation, absorption, and circulation of CSF.

cord. It then passes into the subarachnoid spaces, moves through the arachnoid villi into venous sinuses in the dura mater, and returns to the bloodstream (Fig. 10-5). The circulation of CSF (illustrated in Fig. 10-6) is promoted by the movement of cilia of the ependymal cells and the pulsations of arteries that supply the choroid plexus.

The average adult produces about 504 $\pm$ 72 mL of CSF per day, which continually replaces the 150 mL found in and around the brain and spinal cord. The CSF moves slowly through the CNS and is reabsorbed into the bloodstream (for recirculation). The rate of absorption into the blood is generally the main factor that governs the volume of CSF. Absorption ceases when the pressure of CSF within the ventricles of the brain drops below a critical level (68 mmHg). The CSF may accumulate, causing an increase in intracranial pressure during which the brain becomes compressed from the center outward against the bony lodgings of the skull. This disorder is called *hydrocephalus* and may result in brain damage. The growth of tumors and the development of meningitis can also block the circulation and absorption of CSF, causing it to accumulate.

The meninges extend to the lower lumbar area of the vertebral column, although the spinal cord does not. Therefore, CSF can be withdrawn and sampled in the lower lumbar area with less hazard to the spinal cord than when other areas are used. The CSF is usually removed with a hypodermic needle inserted through the meninges in the area between lumbar vertebrae 3 and 4 in adults (Fig. 9-2).

Properties of CSF

The permeabilities of the choroid plexus, pia mater, and ependymal cells are highly selective and are responsible for a blood-cerebrospinal fluid barrier different from the blood-brain barrier described in Chap. 6 (see Fig. 6-10 and Table 6-1). However, both barriers serve a common purpose in that they maintain the homeo-

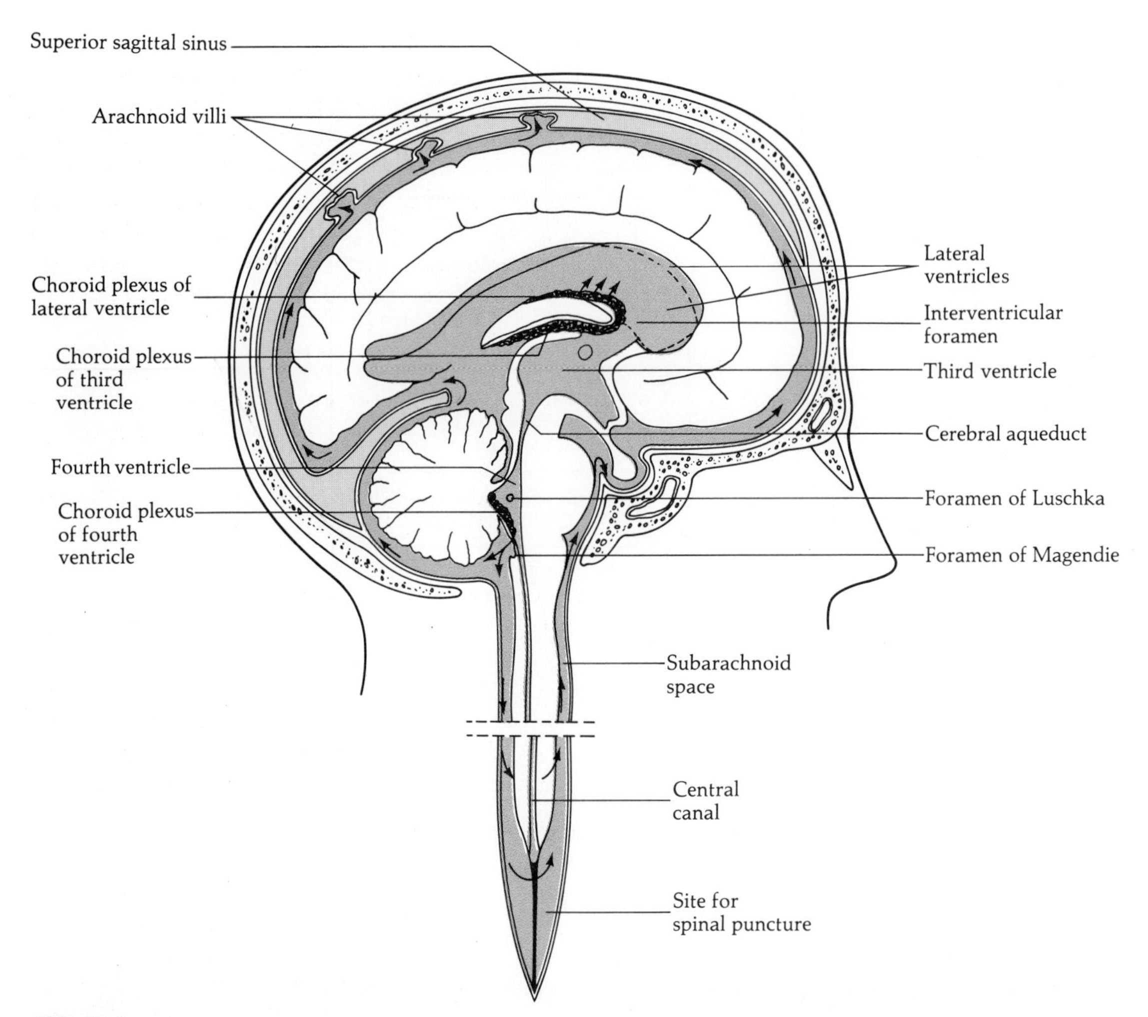

FIG. 10-6
Circulation of CSF. The segment of the spinal cord between the dotted lines is not illustrated.

static equilibrium necessary for neuronal function.

The composition of CSF is basically the same as that of the extracellular fluid around the cells of the brain and is slightly different from that of plasma (the fluid portion of blood). The general composition of CSF and that of plasma are compared in Table 10-2. CSF has much less protein, a slightly lower glucose content and pH, and a higher Cl^- ion concentration than plasma.

Functions

The CSF forms a cushion between the brain and skull and between the spinal cord and the vertebral column. It acts as a shock absorber, minimizing damage from physical trauma. The fluid also buoys the brain in the cranial cavity and reduces the amount of weight on the surrounding membranes which support the brain. For example, the average brain has a net weight of 1.3 kg when surrounded by air but only 0.05 kg when immersed in a container filled with CSF.

10-3 DIAGNOSTIC TESTS AND BRAIN FUNCTION

Much basic information about the nervous system has been derived by applying dyes to thin sections (micrometer thicknesses) of its tissues and studying the stained specimens with a microscope. A number of other techniques also

TABLE 10-2
Approximate normal values of some plasma and cerebrospinal fluid constituents*

Property	*Cerebrospinal Fluid*	*Plasma*
Cells/μL	0–5 (all are normally mononuclear white blood cells)	none†
Protein (mg/dL)‡	5–45	6000–9000
Albumin	52%	52–68%
Alpha$_1$ globulin	5%	2–5%
Alpha$_2$ globulin	14%	7–14%
Beta globulin	10%	9–15%
Gamma globulin	19%	11–21%
Glucose (mg/dL)	50–75	70–115
pH	7.30–7.40	7.35–7.45
Chloride (mEq/L)	120–130	100–106 (in serum)§

* A more complete source of these data may be found in the appendix at the back of this text. Adapted with permission from the *N. Engl. J. Med.* **302:**37–48 (1980).

† Plasma, by definition, is acellular. It is the fluid component of blood. Normal whole blood has 4.2 to 5.9 million red blood cells per microliter and 4300 to 10,800 white blood cells per microliter.

‡ A dL is a deciliter, which is equal to 0.1L or 100 mL.

§ Serum is plasma minus the clotting elements of blood.

have been developed to reveal more about the normal and abnormal function of the nervous system. They include chemical analysis of the CSF, pneumoencephalography, cerebral angiography, CAT scan (computer assisted tomography), and immunological and electrical studies.

Chemical Analysis of CSF

The CSF is normally nearly protein-free (Table 10-2). Therefore, a relatively small change in CSF protein, especially globulins, is an important index of neural disorders. Protein levels in the CSF may increase due to multiple sclerosis, development of some neural tumors, neurosyphilis, neuroencephalitis, and other diseases. Quantities of other substances also may be affected. For example, a decrease in Cl^- ion levels follows infection of the meninges by bacteria that cause tuberculosis.

X-ray Analysis

Differences in the penetration of tissues by x-rays can be recorded on film. Because x-rays penetrate soft tissues rather readily, anatomical differences may be made more apparent when some tissues are contrasted with surrounding areas following the administration of radiopaque substances (those which reflect the radiation). Three x-ray techniques currently used to assess the structure of the brain and its vessels are pneumoencephalography, angiography, and computer-assisted tomography (CAT scan).

Pneumoencephalography

Pneumoencephalography involves the removal of CSF and its replacement with air, followed by x-radiography of the brain. The lesser density of air makes it possible to contrast the emptied ventricular chambers and the surrounding tissues. The outlines of the chambers are more readily observed for abnormal growths. Unfortunately, removal of CSF eliminates its buoyant support of the brain, and the tissues, nerves, and blood vessels become stretched and may cause severe headache. The symptoms may be relieved or minimized by injection of sterile isotonic saline to replace the missing CSF.

Cerebral Angiography

Angiography is a technique in which the outlines of blood vessels are made visible without dissection. The intravenous injection of a radiopaque dye is followed by exposure to x-rays. The location and general structure of the blood vessels, such as the cerebral blood vessels on the surface of the brain, are made visible (Fig. 10-7). The diagnostic technique is often used to detect a suspected aneurysm (ballooning of the wall of a blood vessel).

CAT Scan

A *CAT scan* is a procedure in which the brain is penetrated by x-rays from several different angles. Computers are used to reconstruct the various x-ray images of the brain. Visual images can be sorted out to present a replica of the three-dimensional characteristics of the tissue layers (Fig. 10-8). Intravenous injection of iodine or another radiopaque substance may be used to improve the image of abnormal blood vessels or tissues.

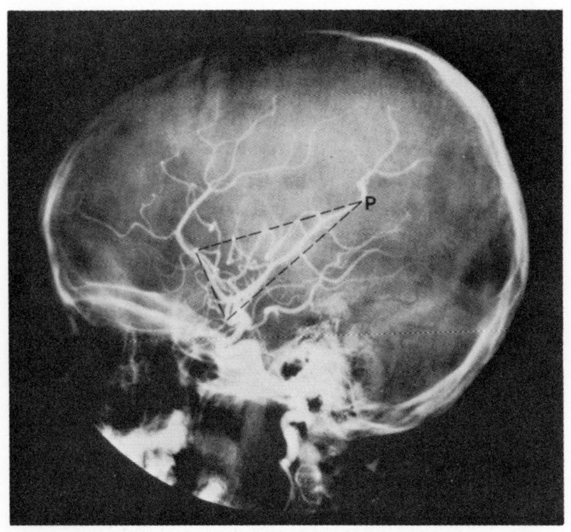

FIG. 10-7
Cerebral angiography. The branches of the middle cerebral artery are outlined in a triangle, the most posterior portion of which is labeled P.

Immunological Studies

Immunological studies have been used in research to provide information about neuron structure and function. The surfaces of neurons and their receptor sites may be analyzed, as well as the neurotransmitters which neurons secrete and to which they respond. The studies also may be applied to differentiate various neurons and to distinguish glial cells. The presence or absence of specific types of neurons in different areas of the CNS and PNS has been studied, as has the formation of neurons and functional areas of the nervous system during their development. By using these techniques the "wiring diagrams" of neuronal circuits can be traced and the cells may be studied in health and disease.

Animals such as rabbits, sheep, goats, and horses can be injected with molecules that have been extracted from human nervous tissue. The immune reaction to the foreign molecules on the cell surfaces or to molecular extracts causes the formation of antibody protein in the animal's bloodstream. The antibodies may be purified so as to react only to the molecules which originally caused their formation. In analytical tests these antibodies react specifically with components in the CNS. Immunological analyses may be made on tissues taken before surgery for a biopsy, after surgery, and during recovery or postmortem to gain basic information about neuronal function.

Radioactive elements or fluorescent dyes may be combined with antibodies. The tagged antibodies adhere to specific complementary molecules in the nervous tissue. The location and quantity of the tagged antibody (and the adhering nervous tissue molecules) can be quantitated with radiation detection equipment (scintillation counters). The technique is similar to that used for the radioimmunoassay of hormones (Fig. 5-9). Complexes of fluorescent tagged antibodies and nervous tissue components can be observed under a fluorescent microscope.

FIG. 10-8
CAT scan of tumor of the brain. Iodine was injected intravenously to provide contrast of the tumor (which appears white) with normal brain tissue (which appears dark).

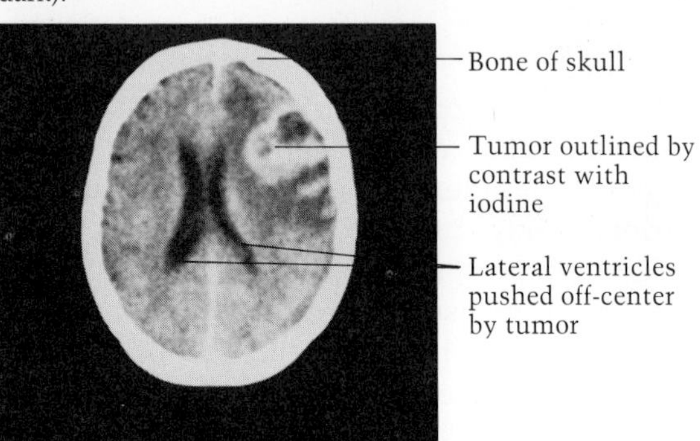

Electrical Studies

Electrical Stimulation of the Brain

Miniature electrodes can be placed on specific areas of the brain and connected to stimulators which generate current. The nature of the responses evoked in the subject indicates much about function and dysfunction in the area under study. Since brain tissue itself does not transmit pain signals, this technique may be used to assess neural activity in the brain of an individual undergoing brain surgery.

Electroencephalography (EEG)

A more commonly employed technique involves analysis of electrical activity emitted from the active brain. (The electrical activity of neurons was described in Chap. 7.) The currents generated by neurons in the brain are conducted by the surrounding fluids to the scalp and can be recorded. The totals of the recorded electrical activity produce electrical impulses called *brain waves.* They are conducted from electrodes placed on the scalp to an amplifier, which magnifies the signal and relays it to another. This second amplifier converts the signal (transduces it) into a visual display, for example, into a force to move a pen on a recording chart. The display that is produced is called an *electroencephalogram* (EEG). In a less widely used technique, the skull is opened and electrodes are placed directly on the pia mater of the brain, and a recording called an *electrocorticogram* is produced.

Brain waves vary in rate of occurrence (*frequency*) and size (*amplitude*). The size of the wave is a measure of the voltage or potential associated with it. The pattern of waves depends on the person, the position of the electrodes, and the structure and physiology of the brain. The wave patterns change from infancy to adulthood and with alterations in behavior and neural activity. Unconsciousness is accompanied by a pattern of waves of low frequency and high amplitude. Trauma or disease can destroy specific areas of the cerebral cortex and result in a localized decrease in wave frequency. The EEG becomes electrically silent with cerebral death, when there is no detectable current to provide a visual display.

Several types of waves have been identified which differ in frequency and potential (voltage). Their properties are summarized in Table 10-3. *Alpha* (α) *waves* are produced when an individual is relaxed, especially with the eyes closed, or has just awakened from a restful sleep. Alpha waves are the most evident pattern observed in an EEG and are called *dominant alpha rhythms* or *high-voltage slow waves* (HVS). They occur at rates of 8 to 12 per second and develop a potential equal to 50 μv (microvolts). Alpha waves may be detected in several areas of the brain but are more readily detected on the posterior (parietal-occipital area) because they are produced by the electrical activity of neurons in that area of the brain.

TABLE 10-3
General properties of major brain waves

Property	Alpha (α)	Beta (β)	Delta (δ)	Theta (θ)
Frequency, number/s	8–12	13–32	0.5–3.5	4–7
Potential, μv	50	5–10	20–200	10
Associated activity	Relaxed, awakening	Mental activity, arousal	Sleep	Disappointment, frustration
Appearance				

α β δ θ

Time (= 1 s)

Alpha waves disappear with visual and mental activity and during sleep. They are blocked when light is shone in a person's eyes or when a person is aroused. Arousal is associated with the activity of the reticular formation of the brainstem, which contains neurons that form the reticular activating system (RAS). This group of neurons seems to make perception possible by maintaining an alert state. Alpha wave formation is inversely related to the electrical activity of the RAS.

Beta (β) *waves* are produced during mental activity and the aroused state. They have a frequency ranging from 13 to 32 per second, are associated with a potential of 5 to 10 μv, and are called *low-voltage fast waves* (LVF).

Delta (δ) *waves* are most evident during unconsciousness and deep sleep. They occur at rates of about 0.5 to 3.5 per second and develop a potential of 20 to 200 μv.

Theta (θ) *waves* are often formed during times of disappointment and frustration in young children and adolescents. They occur at slow rates of 4 to 7 per second, with a potential of about 10 μv. Theta waves characteristically originate in the limbic areas of the brain, which are closely linked to generation of emotions.

Accurate interpretation of brain waves requires careful training. Because of the vast interconnections among different neurons within the brain and the PNS and because the waves reflect the total electrical activity of many neurons, a great number of different neuronal activities may produce similar wave patterns. Furthermore, differences in the placement of the electrodes on the brain's surface or other variations in the instrumentation may alter the appearance of the brain waves.

SLEEP *Sleep* normally produces two distinct patterns in the EEG, slow delta waves and rapid beta waves. The alpha rhythms are gradually replaced by irregular waves of lower voltage and then by slow delta waves of higher voltage. Delta waves are produced in *slow-wave sleep* which is also called *non-REM (NREM)* sleep, that is, non-rapid eye movement sleep. At intervals during *paradoxical sleep* more rapid low-voltage beta waves occur, which are characteristic of arousal and mental activity. This type of sleep often occurs during dreams and is associated with rapid eye movement, intermittent twitching of muscles of the face and limbs, and inhibition of postural muscle tone. It lasts about 5 to 30 min and is also called *rapid eye movement* (REM) *sleep* to contrast it with *NREM sleep*.

NREM and REM sleep occur in alternating periods, with NREM sleep accounting for about four-fifths of the total sleeping time. The whole cycle lasts about 1.5 to 2 h. Changes in sleep patterns proceed through four stages, during which cyclical fluctuations occur in EEG tracings, pulse rate, temperature, blood pressure, and depth of sleep.

A path between the frontal lobes of the brain and the hypothalamus, pons, and thalamus may form a "sleep" center (see Chap. 11). Its activity is overridden by sensory inputs transmitted to the RAS which cause arousal. Activation of the RAS and alertness seem to lessen with repeated stimulation. Elevations in serotonin, GABA, dopa, or other neurotransmitters may inhibit RAS activity, resulting in slow-wave (NREM) sleep. Signals also can excite the RAS (or lessen inhibition of it) to cause or to maintain the wakened state.

The purpose of sleep is not clear. The brain shows electrical activity during sleep, but specific areas may be resting. It has been theorized that biochemical and structural changes which are necessary for normal brain functions may occur during this time.

ABNORMAL PATTERNS: EPILEPSY Abnormal electrical activity of the brain may also reflect neural disorders (due to trauma or diseases involving infection, abnormal development, tumors, or abnormal genes). For example, delta waves, which are normally prominent during deep sleep, are also present in the awakened state in some children whose mental development may be retarded. *Epilepsy* is another condition in which brain wave patterns are altered. It is accompanied by a temporary abnormal increase in neural activity called a *seizure.* Epilepsy occurs in individuals predisposed to seizures because of trauma, disease, or heritable factors.

Different forms can cause excess electrical activity in specific limited areas or widespread regions of the brain. Epilepsy may be accompanied by loss of consciousness and muscular contractions. *Petit mal* epilepsy may involve such a brief period (1 min or less) that the individual or an observer is unaware of any external change. Rapid involuntary muscle contractions may go unnoticed. The individual may or may not partially or completely slump toward the ground. The EEG pattern shows alternating round and pointed (spiked) waves occurring in pairs (doublets) 3 times per second. In *grand mal* epilepsy, unconsciousness is generally preceded by a signal or warning called an *aura*. The sensation often involves sight, smell, or memory. Unconsciousness follows and lasts for over 1 min, while the skeletal muscles remain in sustained contraction (called the *tonic* phase). This stage is followed by muscle spasms causing synchronized violent muscular contractions (called *convulsions*) and by possible loss of control over the bladder and bowels (this phase is called the *clonic* phase). A deep sleep follows. The EEG pattern shows high-voltage waves occurring about 8 to 12 times per second during the tonic phase and slower waves during the clonic phase which often last through sleep and recovery.

10-4 BRAIN NEUROTRANSMITTERS AND THE INFLUENCE OF DRUGS

Over 30 compounds that exist in the brain are potential neurotransmitters. The functions of several of them were described in Chap. 7. Three types of peptide neurotransmitters with opium- or morphinelike activity formed in the brain are the endorphins, enkephalins, and dynorphin. Others formed by cells inside and outside of the CNS will be described elsewhere.

Endorphins

Endorphin (from Greek *endo*, within, and *orphin*, morphine) is a generic name for all peptides synthesized in the brain that possess morphinelike activity similar to that of opium. The polypeptides can be divided into *endorphins*, which are relatives of β-lipotropin (a hormone in the anterior and intermediate pituitary), and *enkephalins*, which may come from some other precursor compound. The endorphins may affect mood, behavior, and mental health, block pain, regulate autonomic functions such as food intake and body temperature, and modify the activity of endocrines. The narcotic effects of endorphins are like those of the opiates morphine and heroin. A *narcotic* is defined medically as a drug that induces drowsiness (acts as a sedative) and inhibits pain (acts as an analgesic). This usage differs from the legal definition, which also may include drugs such as cocaine, amphetamines, and marijuana.

The α, β, and γ endorphins range in size from 16 to 31 amino acid residues (Fig. 10-9). They are synthesized in the hypothalamus and pituitary and perhaps in some other tissues as well. The opioid (opiatelike) activity of a mole of the endorphins ranges between 18 and 33 times that of a mole of morphine. β-endorphin is the most potent of the three types. The endorphins share amino acid sequences with hormones from the anterior pituitary, including β-lipotropins, MSH, and ACTH (Fig. 10-9).

ENKEPHALINS *Enkephalins* are pentapeptides found in the brain, anterior, intermediate and posterior pituitary, and other tissues such as those of the gastrointestinal system. Two of them, methionine-enkephalin, or Met-enkephalin, and leucine enkephalin, or Leu-enkephalin, differ in only the single amino acid after which each is named. The opioid activity of 1 mol of enkephalin is slightly greater than that of 1 mol of morphine.

The enkephalins inhibit the release of the excitatory neurotransmitter, P substance, whose presence is associated with pain. P substance is an undecapeptide (11 amino acids) found in neurons in the brain, sensory neurons of peripheral nerves, and cells of the gastrointestinal system.

The sequence of five amino acids that form the enkephalins is also found within the endor-

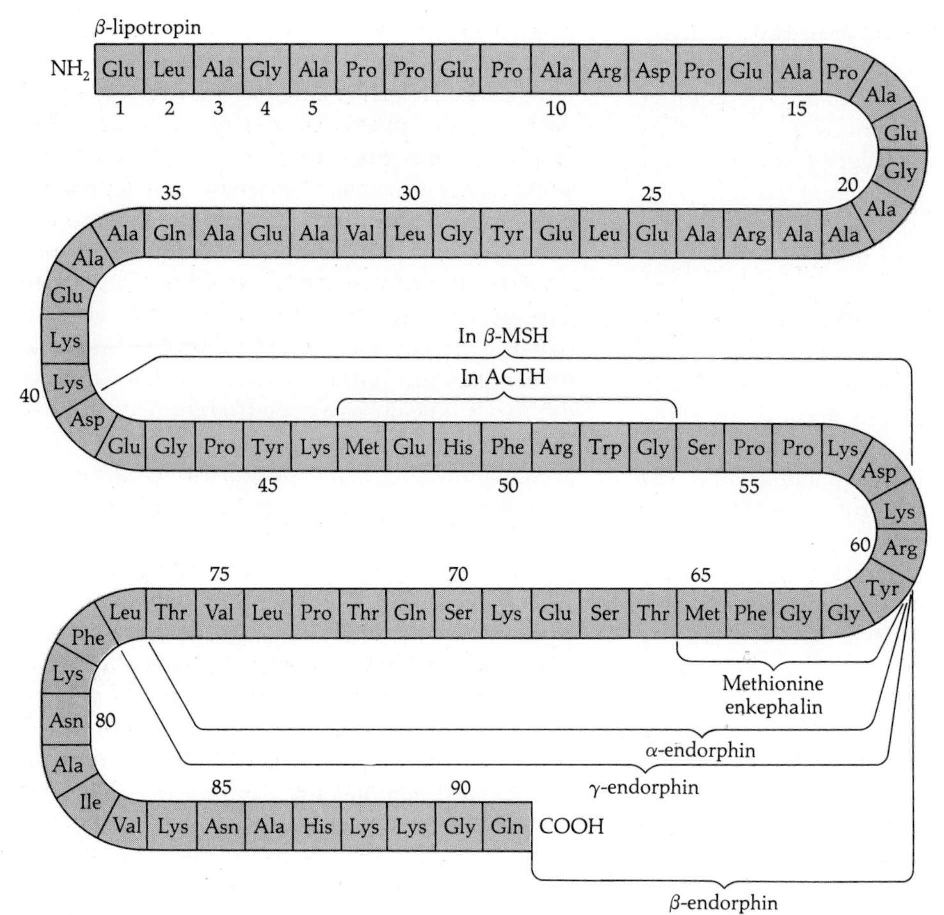

FIG. 10-9
β-Lipotropin and its relationship to polypeptide neurotransmitters. The amino acids are numbered in sequence from the free amino (NH_2) terminal end. Each group of three letters is an abbreviation for the name of a specific amino acid. β-Lipotropin is composed of a chain of 91 amino acids. The chain includes smaller sequences that are also found among 18 of the 22 amino acid residues in human β-melanocyte-stimulating hormone (β-MSH); among 7 of the 39 amino acid residues in the secreted form of human adrenocorticotropic hormone (ACTH); and in all the sequences in enkephalin and α-, β-, and γ-endorphin.

phins (Fig. 10-9). The relationship between the sizes of β-lipotropin, endorphins, and enkephalins is as follows:

β-Lipotropin —	endorphins —	enkephalins
(91 amino acids)	**(16–31 amino acids)**	**(5 amino acids)**

The enkephalins formed in the anterior pituitary may be delivered to the brain by blood vessels or by CSF through the activity of specialized ependymal cells. These two routes are opposite in direction of the usual hypothalamic to pituitary flow (Fig. 5-3).

Dynorphin

Dynorphin is a tridecapeptide (13 amino acid residues) synthesized in the brain. A sequence of five of its amino acids is also found in leucine enkephalin. In various assays, the opioid activity of 1 mol of dynorphin has been reported to range from 30 to 200 times that of 1 mol of morphine. Dynorphin may be the most potent analgesic discovered to date. Investigations regarding its uses may allow the control of pain which is currently unmanageable (intractable) except by severance of the pain tracts in the spinal cord (cordotomy).

Drug Effects on Neurotransmitter Function

Drugs are chemical compounds that can be used to treat, prevent, or diagnose disease. Many drugs alter brain cell functions. Those that change behavior, mood, and perception are called *psychotropic drugs.* Some, like amphetamines, are *stimulants;* others, such as barbiturates, are *sedatives,* which depress the CNS and in large doses can cause the following sequence of events:

Sedation ⇌ hypnosis ⇌ anesthesia ⇌ coma → death

(Drowsiness) — Sedation; **(Depression of CNS resembling sleep)** — hypnosis; **(Loss of sensation)** — anesthesia; **(Deep, prolonged unconsciousness)** — coma

Drugs may alter the levels of neurotransmitters by affecting their rates of synthesis, release, recapture, or inactivation. Drugs may interact with receptor sites on cell membranes and behave as agonists or antagonists. *Agonists* are drugs that have an affinity for and stimulatory effect on receptors. *Antagonists* block those activities.

Opiate Activity

Morphine, heroin, and opiatelike molecules, including the endorphins, enkephalins, and dynorphin, bind to opiate receptors in the brain. Activation of opiate receptors elevates cGMP, blocks sodium ion channels, and induces narcotic effects. Opiate receptors are prevalent in the midbrain, in the thalamus, and in the amygdalae of the forebrain. (The latter plays a role in neural adjustments for aggressive and defensive behavior.) Opiate receptors are also located in the substantia gelatinosa of the dorsal horn of the spinal cord. The dendrites of the small cells in the substantia gelatinosa synapse with each other and may serve as a gate through which pain impulses must pass to ascend to the brain. The adherence of opiates or opiatelike molecules to receptors in the substantia gelatinosa may block the transmission of pain.

Endorphins, enkephalins, and dynorphin are natural pain killers. They may be released following acupuncture, hypnosis, or the administration of placebos. A *placebo* is a substance containing no medication, given as a control or merely to humor a patient. Enkephalins are also elevated in certain mental disorders. High concentrations have been found in some forms of schizophrenia. The administration of antiendorphins, which oppose the opiatelike activity of the enkephalins, offers a potential remedy to certain types of schizophrenia.

Model of Drug Addiction

It has been hypothesized that drug addiction may be due to *tolerance* (a diminished response), which may develop with repeated exposure. The effects of drugs such as morphine and heroin on opiate receptors may be related to narcotic addiction. Binding of drugs to cell receptors can elevate levels of specific secondary messengers and depress others. Tolerance may arise as the cell's biochemistry adapts to the new levels of secondary messenger activity. Greater and greater amounts of the drug may be required to produce the original effect. Withdrawal of the drug may cause a decline in the level of a specific secondary messenger whose activity was originally elevated and leave behind a high level of the opposing secondary messenger. The symptoms of drug withdrawal may result from an imbalance in secondary messenger activity, leading to overactivity of some enzymes in the responsive cells. Studies on dynorphin, the endorphins, and enkephalins may lead to the development of nonaddictive pain killers which are even more potent than morphine or other addictive opiates.

Special Drugs and Mechanisms of Action

The response to a drug depends on the receptors and the specific neurotransmitters involved and on the direction of change in the activity of both. Illustrative examples of the sites where neurotransmitter synthesis may be affected are

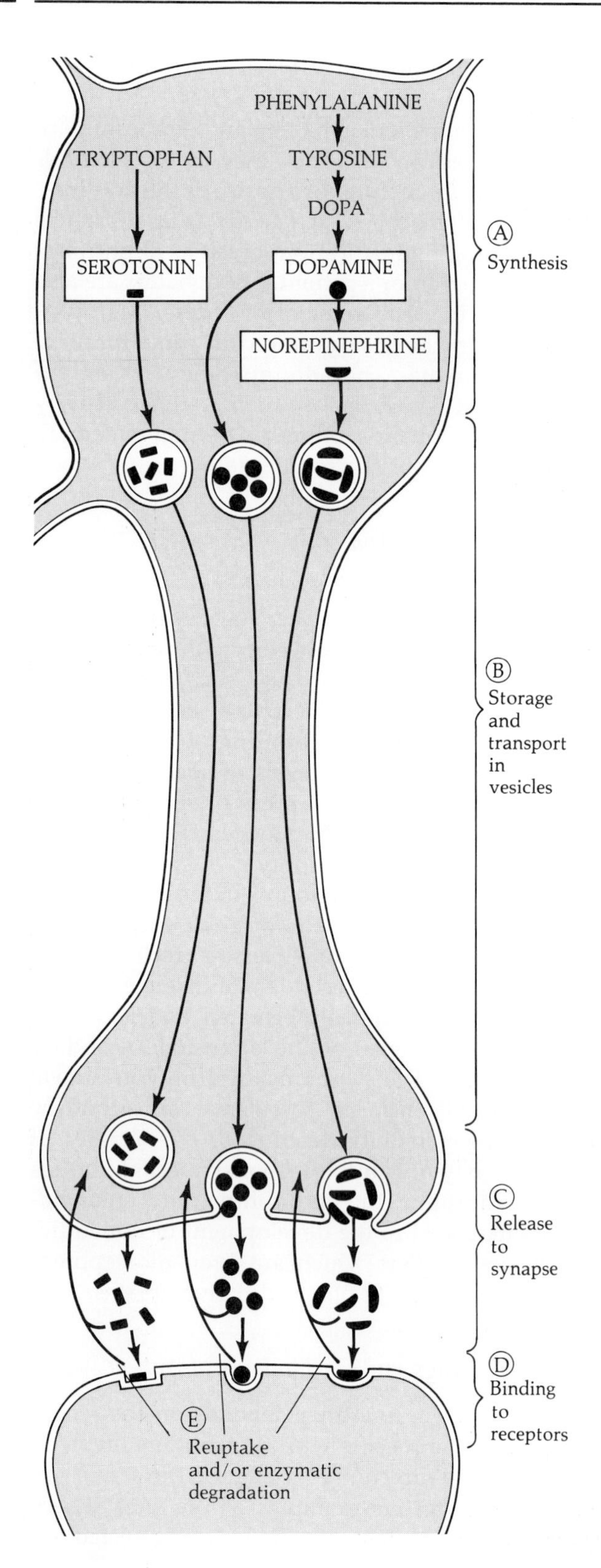

FIG. 10-10
Possible sites of action of drugs on neurotransmitter function. Five major sites (A to E) where neurotransmitter activity may be affected are indicated. As examples the illustration uses serotonin, dopamine, and norepinephrine which are formed in various neurons in the hypothalamus of the brain. Although three types of neurotransmitters are illustrated, the student should remember that evidence indicates that a single type of fiber probably releases only one form of neurotransmitter.

shown in Fig. 10-10, and the effects of certain drugs on various neurotransmitters are summarized in Table 10-4.

Some drugs are agonists to specific receptors and augment their activity. For example, the analgesics morphine, heroin, methadone, and meperidine (Demerol) stimulate opiate receptors to kill pain. Other drugs are *antagonists* to specific receptors and diminish their activity. For example, the tranquilizer chlorpromazine (Thorazine) is antiadrenergic, the hallucinogen LSD, as well as marijuana in large doses, is antiserotonergic, and the drugs scopolamine and atropine are anticholinergic.

Drugs which depress the levels of norepinephrine include the sedatives meprobamate (Miltown) and the benzodiazepines (Valium and Librium). Those which elevate the levels of norepinephrine and one or more other catecholamines act as stimulants or antidepressants. They include the amphetamines (e.g., Dexedrine), cocaine, and isoproniazid (Marsilid).

Studies of neurotransmitters and their interaction with drugs offer promise to improve neuronal functions that affect behavior, mood, sleep, memory, and learning. For example, a variety of CNS depressants have been developed which successfully control epileptic seizures. Although much research remains to be done, it appears that significant developments may lead to means of control of many disorders which are not currently responsive to existing means of therapy.

TABLE 10-4
Effects of drugs on neurotransmitter function

		MECHANISM OF ACTION		NEUROTRANSMITTER AFFECTED‡				
Drug (and Trade Name)	*Biologic Activity*	*Receptor**	*Level of Neurotransmitter†*	*Norepinephrine*	*Dopamin*	*Serotonin*	*Enkephalin*	*Acetylcholine*
Meprobamate (Miltown)	Antianxiety (sedative)		−	×				
Benzodiazepines (Valium, Librium)	Antianxiety (sedative)		−	×				
Chlorpromazine (Thorazine)	Antipsychotic (tranquilizer)	−		×	×			
Amphetamine (Dexedrine)	Stimulant		+ (enhances release)	×	×			
Cocaine	Antidepressant (arousal)		+ (blocks reuptake)	×		×		
Isoproniazid (Marsilid)	Antidepressant (arousal)		+ (inhibits monamine oxidase)	×	×	×		
Lysergic acid diethylamide (LSD)	Hallucinogen	− (in pons)				×		
Cannabis (marijuana, in large doses)	Hallucinogen	− (in pons)				×		
Apomorphine	Analgesic (inhibits pain)	+			×			
Morphine	Analgesic	+					×	
Heroin	Analgesic	+					×	
Methadone	Analgesic	+					×	
Meperidine (Demerol)	Analgesic	+					×	
Scopalamine	Anticholinergic	−						×
Atropine (belladonna)	Anticholinergic	−						×
Nicotine	Stimulant	+						×

* (+) = agonist, activates receptor
(−) = antagonist, blocks receptor
† (+) = increases activity
(−) = decreases activity
‡ = causes an effect on this neurotransmitter

10-5 SUMMARY

1 Although the brain accounts for only 2 percent of the weight of the body, it is a most complicated organ. Hundreds of billions of neurons in the brain form 100 trillion synapses and can consume 20 percent of the oxygen available to the resting body.

2 The three major areas of the brain are the forebrain, midbrain, and hindbrain. The CNS first develops as a neural tube. Some of its cells pinch off to

form the neural crest, which gives rise to the components of the PNS. The inner part of the neural tube forms the ventricles of the brain and the central canal in the spinal cord. The outer parts of the brain and spinal cord are covered by a tough double layer of fibrous connective tissue called the dura mater. This is supported by the arachnoid layer, which resembles a spider web and which surrounds the innermost pia mater. Infections of these membranes (meninges) may cause an inflammation called meningitis.

Blood vessels penetrate the pia mater and form the choroid plexus. CSF filters from the blood vessels of the choroid plexus and crosses ependymal cells, which line the cavities in the CNS. Movement of CSF is aided by cilia on the ependymal cells and pulsations of arteries in the choroid plexus. The arachnoid granulations direct the flow of CSF out of the ventricles into venous sinuses of the dura mater. CSF is similar to plasma but contains significantly less protein because of the blood CSF barrier. CSF increases the buoyancy of the brain and serves as a shock absorber.

3 A variety of tests reveal normal and abnormal brain functions. Protein analysis of CSF may be used to assess damage due to infections. Pneumoencephalography is x-ray analysis of a brain in which CSF is replaced by air in the ventricles. Cerebral angiography involves the administration of dye opaque to x-rays which makes it possible to contrast blood vessels with other tissues in the brain. A CAT scan consists of a series of x-rays taken from different angles and reconstructed by computer analysis to provide three-dimensional images. Antibodies tagged with radioactive elements or fluorescent dyes also can be used to demonstrate the quantity and location of specific molecules on cells and in precise areas of the nervous system. These studies involve analyses made with a scintillation counter or fluorescent microscope, respectively.

Electrical activity is transmitted from the brain to the surface of the scalp. Variation in its emission can be analyzed as tracings called brain waves via EEG. Three major types of waves are alpha (α) waves of low frequency and high voltage, characteristic of relaxed awakening; beta (β) waves of fast frequency and low voltage, prevalent during mental activity; and delta (δ) waves of very slow frequency and a wide range of voltage, associated with deep sleep. Abnormal patterns may be seen in various neural disorders. Epilepsy may produce temporary abnormal increases in electrical activity of the brain and may cause unconsciousness and loss of muscular control. These symptoms are less severe and last less than 1 min in petit mal epilepsy and are more severe and last more than 1 min in grand mal epilepsy.

4 Over 30 types of neurotransmitters are found in the brain. Many are peptides similar to those in other tissues. Endorphins, which are peptides produced in the hypothalamus and in the posterior pituitary, exhibit about 18 to 33 times the opioid activity of morphine. Enkephalins are pentapeptides whose sequences are found within endorphins. They are formed in the brain and anterior pituitary. They bind to opiate receptors in the brain with an efficacy slightly greater than that of morphine. Dynorphin is a tridecapeptide formed in the brain. It includes the amino acid sequence of enkephalins and is 30 to 200 times as active as morphine.

Drugs may influence brain neurotransmitter functions and may alter their effects by either activating or opposing the activity of their receptors. Drugs which bind to and activate opiate receptors mimic the effects of morphine and the enkephalins. They increase levels of cGMP, block the conductance of Na^+ ions, and inhibit neuron function. Addiction may be due to tolerance to the altered levels of secondary messengers which follow the administration of a drug.

Drugs may be classified as agonists or antagonists of specific cell receptors, or they may be classified into groups which affect the levels of specific neurotransmitters. Neurotransmitter activity may be altered at the sites of synthesis, storage, transport, release, binding to receptors or reuptake by a neuron. Agonists of opiate receptors are morphine, heroin, and methadone. Antagonists to specific receptors include the antiadrenergic tranquilizer Thorazine, the antiserotonergic hallucinogen LSD, and the anticholinergics scopalamine and atropine. The levels of norepinephrine are depressed by the sedatives Miltown, Valium, and Librium. The levels of one or more catecholamines are elevated by stimulants such as the amphetamines and the antidepressant cocaine.

REVIEW QUESTIONS

1 How are the brain and spinal cord formed?
2 What are the three general subdivisions into which the adult brain may be divided, and what structures does each area contain?
3 What three membranes surround the brain? Describe the structure of each.
4 How is CSF generated? Describe its pattern of circulation.
5 Of what is CSF composed? How does it compare with plasma? What is its function?
6 What is the significance of changes in the levels of protein and Cl^- ions in the CSF?
7 What is pneumoencephalography?
8 What is cerebral angiography?
9 Explain the basis of a CAT scan and compare its use with that of x-rays.
10 Outline some immunological approaches to the study of brain function and the types of information they may provide.
11 Describe two ways in which the electrical properties of the brain can be studied, including a description of brain waves and the changes that occur during the different phases of sleep. Describe epilepsy and the changes that occur in brain waves.
12 What are endorphins, and what are their effects on the brain? What are enkephalins, and what are their effects on the brain? What is dynorphin and how does its effects compare with those of other opioid neurotransmitters?
13 In what five general ways may neurotransmitter levels be affected? What are agonists and antagonists?
14 How do opiates cause a narcotic effect?
15 Explain a model of drug addiction based on the effects of secondary messengers.
16 Name some drugs and the specific receptors to which they are antagonists and other drugs and the specific receptors to which they are agonists.

THE NERVOUS SYSTEM

Structure and Function of Specific Areas of the Brain

11-1 THE CRANIAL NERVES

11-2 THE HINDBRAIN
The medulla oblongata
The pons
The cerebellum
Location
Functions
Cellular anatomy
Pathways
Afferent paths of the cerebellar cortex • Efferent pathways
Effects of cerebellar damage or disease

11-3 THE MIDBRAIN

11-4 THE FOREBRAIN: ITS DIENCEPHALIC SUBDIVISION
The hypothalamus
Appetite and thirst centers
Temperature regulation center
Behavioral responses
Sleep
Hormonal secretions
The thalamus
The pineal body

11-5 THE FOREBRAIN: ITS TELENCEPHALIC SUBDIVISION, THE CEREBRAL HEMISPHERES
Location of lobes
Functions of lobes
Interior of cerebral hemispheres
Exterior of cerebral hemispheres
Somatotopic organization
The split brain
Evidence from surgically treated epileptics
Control of speech
Learning and conditioned reflexes
Theories of memory
Theories on the biochemical basis of memory

11-6 SUMMARY

OBJECTIVES

After completing this chapter the student should be able to:

1. Identify the three major subdivisions of the hindbrain.
2. Discuss the origins and functions of the cranial nerves.
3. Name and explain the functions of each component of these subdivisions.
4. Describe the structure and function of the midbrain.
5. Identify four major parts of the forebrain, their general locations, and their functions.
6. Diagram the location of the lobes in the cerebral hemispheres and list the main functions of each lobe.
7. Describe the relationship between specific areas in the brain and sensory and motor function of parts of the body from head to toe.
8. Discuss the nature of the split brain.
9. Describe the control of speech.
10. Explain the relationship between learning and conditioned reflexes, describing the theories of memory and their possible biochemical bases.

The structure and function of the brain are dependent on complex interactions among cells, receptors, neurotransmitters, and the pathways described in Chaps. 6 through 10. Studies suggest that the brain initially forms a much larger number of neurons and fibers than it can use. Clusters of neuron cell bodies called *nuclei* are grouped together to perform common functions. When the boundaries of the clusters of neurons are not clearly delineated they are called *centers*. Nuclei and centers relay impulses through the central and peripheral nervous systems. The manner in which the neurons connect with one another to form these circuits is prescribed in early embryonic and postnatal development. Levels of hormones such as sex steroids, especially estradiol, seem to influence the maturation of neurons markedly. The hormones alter the activity of neurotransmitters and influence the formation of interneuronal circuits in the brain, thereby establishing paths that affect adult behavior. Furthermore, the neurons that persist appear to be those that are used. Many regress, but more are retained than are generally needed. Thus, sensory and motor stimulation in early infancy and childhood are extremely important for the attainment of optimal brain function in adulthood.

As noted earlier, the brain develops in three segments, the forebrain, midbrain, and hindbrain (Table 10-1). All the structures in the midbrain and hindbrain are collectively referred to as the *brainstem*. This chapter will detail further structural and functional relationships of specific areas in the brain.

11-1 THE CRANIAL NERVES

The cranial nerves originate in several locations in the brain (Table 11-1). These nerves form pathways in the peripheral nervous system. Twelve pairs emerge through openings in the floor of the skull and can be seen on the underside of the brain (Fig. 11-1). They have been assigned the Roman numerals I through XII, proceeding from the forebrain to the hindbrain. They are called the olfactory (I), optic (II), oculomotor (III), trochlear (IV), trigeminal (V), abducens (VI), facial (VII), statoacoustic (VIII), glossopharyngeal (IX), vagus (X), accessory (XI), and hypoglossal (XII) nerves.

With the exception of the *vagus,* the cranial nerves largely innervate structures around the head and throat. They send fibers to the eyes, ears, tongue, and some of the muscles on the upper part of the shoulder. Eighty percent of the body's parasympathetic fibers are found in the vagus. Its name is derived from the Latin word for wanderer, since it meanders from its origin in the medulla of the brain to innervate parts of the heart, respiratory tract, liver, gallbladder, pancreas, and digestive tract down through the colon.

Nerves I, II, and VIII are exclusively sensory in function, while the remaining ones are mixed and contain both sensory and motor

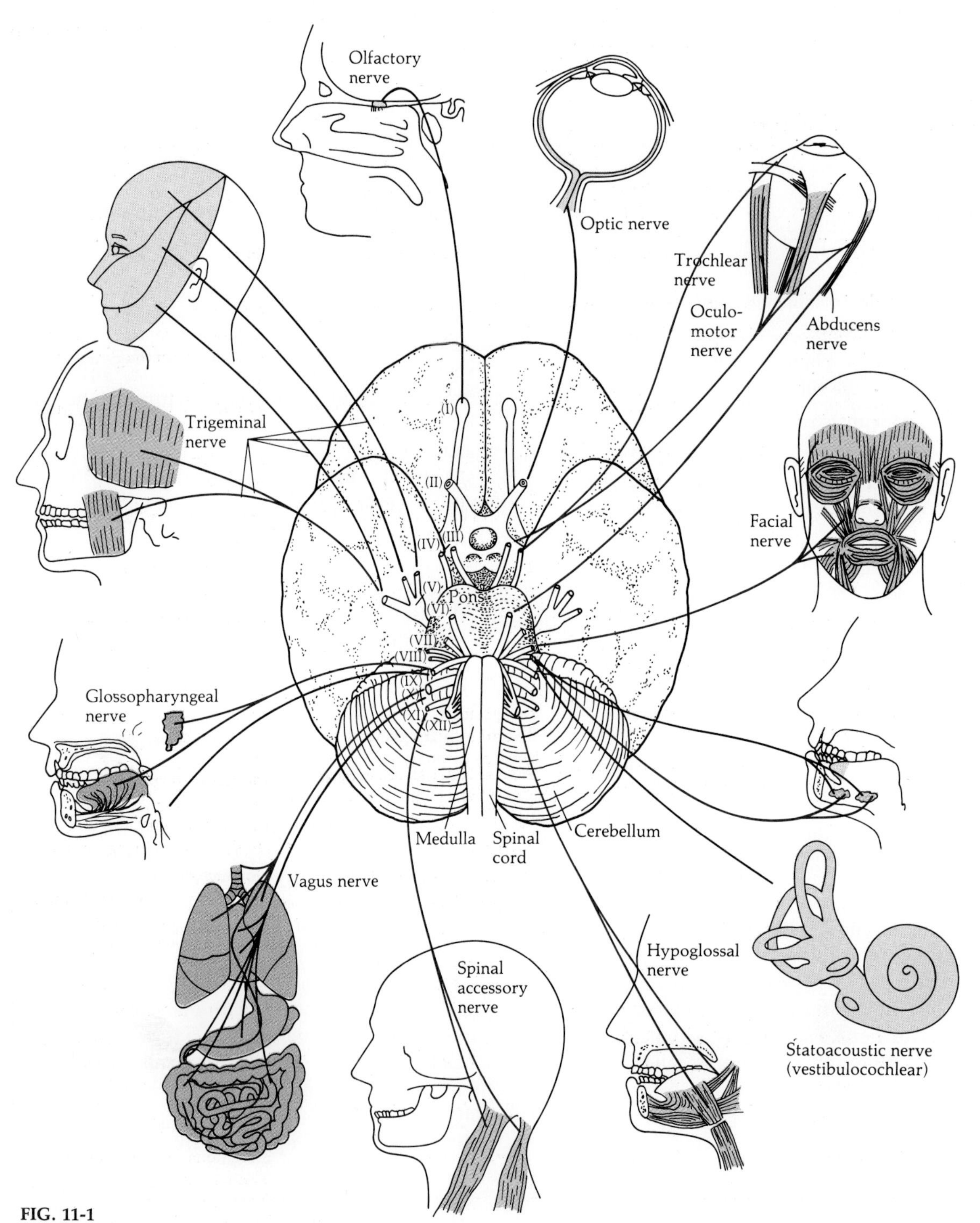

FIG. 11-1

Major routes of distribution of the cranial nerves from the base of the brain. The colored lines indicate motor fibers, the black lines show sensory fibers.

fibers. Of the mixed nerves, III, IV, VI, XI, and XII are primarily motor. Nerves III, VII, IX, and X also contain efferent fibers of the parasympathetic division of the autonomic nervous system. The properties of the cranial nerves are summarized in Table 11-1. More complete descriptions can be found in texts of anatomy and neuroanatomy as listed in *Books in Print* (Table 1-1).

TABLE 11-1
Properties of the cranial nerves

Number	*Name*	*Type of Fibers*	*Origin (Primary Cell Bodies)*	*Major Sites of Distribution*	*Major Functions*
I	Olfactory	Sensory	Olfactory lobe	Olfactory mucosa lining the nose	Smell
II	Optic	Sensory	Retina of eye	Thalamus and midbrain	Vision
III	Oculomotor	Mixed	Midbrain	Eye muscles	Autonomic control of pupil size and accommodation of lens in eye
IV	Trochlear	Mixed	Midbrain	Superior oblique eye muscle	Eye muscle movement (acts as pulley)
V	Trigeminal	Mixed	Pons	Skin of face and scalp	Sensations from face and scalp
				Jaw muscles	Chewing
VI	Abducens	Mixed	Pons	Lateral rectus eye muscles	Eye movement (outward)
VII	Facial	Mixed	Pons	Taste buds Salivary glands Facial and scalp muscles	Taste Secretion of saliva Facial expression
VIII	Statoacoustic (auditory) (vestibulo-cochlear)	Sensory	Pons, lateral wall of fourth ventricle, cerebellum	Semicircular canals and vestibule (hallway) of inner ear	Equilibrium and hearing
IX	Glosso-pharyngeal	Mixed	Medulla	Taste buds Parotid salivary gland Muscles in pharynx Carotid sinus in the wall of internal carotid artery Carotid body in wall of external carotid artery	Taste Secretion of saliva Swallowing Influences respiration and constriction of blood vessels Influences respiration and constriction of blood vessels
X	Vagus	Mixed	Medulla	Muscles of pharynx and larynx Thoracic organs	Swallowing Autonomic control of thoracic and abdominal organs; Influences heart and lung function
				Abdominal organs	Influences digestive tract motility and secretions
XI	Accessory (spinal accessory)	Mixed	Medulla	Muscles of pharynx, larynx, neck, and shoulder	Sensory fibers to pharynx and larynx; movement of neck and shoulder muscles
				Thoracic and abdominal organs	Forms cardiac branches of vagus to heart (?)
XII	Hypoglossal	Mixed	Medulla	Tongue muscles	Sensations from and movements of tongue

11-2 THE HINDBRAIN

The hindbrain consists of the medulla oblongata, pons, and cerebellum. Their major properties are listed in Table 11-2.

The Medulla Oblongata

The *medulla oblongata* is located just above the spinal cord as it enters the base of the skull through an opening called the *foramen magnum* (large opening). Situated in the lowest part of

TABLE 11-2
Major properties of the hindbrain

Area	*Location*	*Component*	*Associated Functions*
Medulla	Just above spinal cord		Contains ascending and descending tracts
		Nuclei of cranial nerves	IX –, X –, XI –, XII – See Table 11-1
		Nuclei for vital autonomic centers	
		Cardiac center	Regulates heartbeat
		Vasomotor center	Regulates constriction and dilation of blood vessels
		Respiratory center	Regulates breathing
		Crossover of pyramidal tracts	Carries motor fibers to opposite sides of body
Pons	Above medulla and anterior to cerebellum		Contains a bridge of fibers including the cerebellar peduncles, which connect the lower and upper CNS with each other and the cerebellum
		Nuclei of cranial nerves	V –, VI –, VII –, VIII – See Table 11-1
		Pneumotaxic center	Causes intermittent inhibition of breathing
		Apneustic center	Major stimulus for breathing
Cerebellum	Above medulla but behind pons	Nuclei and fibers of cerebrocerebellar circuits	Receive sensory inputs from eye, ear, muscle spindles, and tendon organs, as well as from fibers of suprasegmental descending paths; send subconscious signals to descending suprasegmental paths; modify pyramidal and extrapyramidal tract impulses to skeletal muscle, providing fine coordination of muscular activity; assist posture, equilibrium, and speech

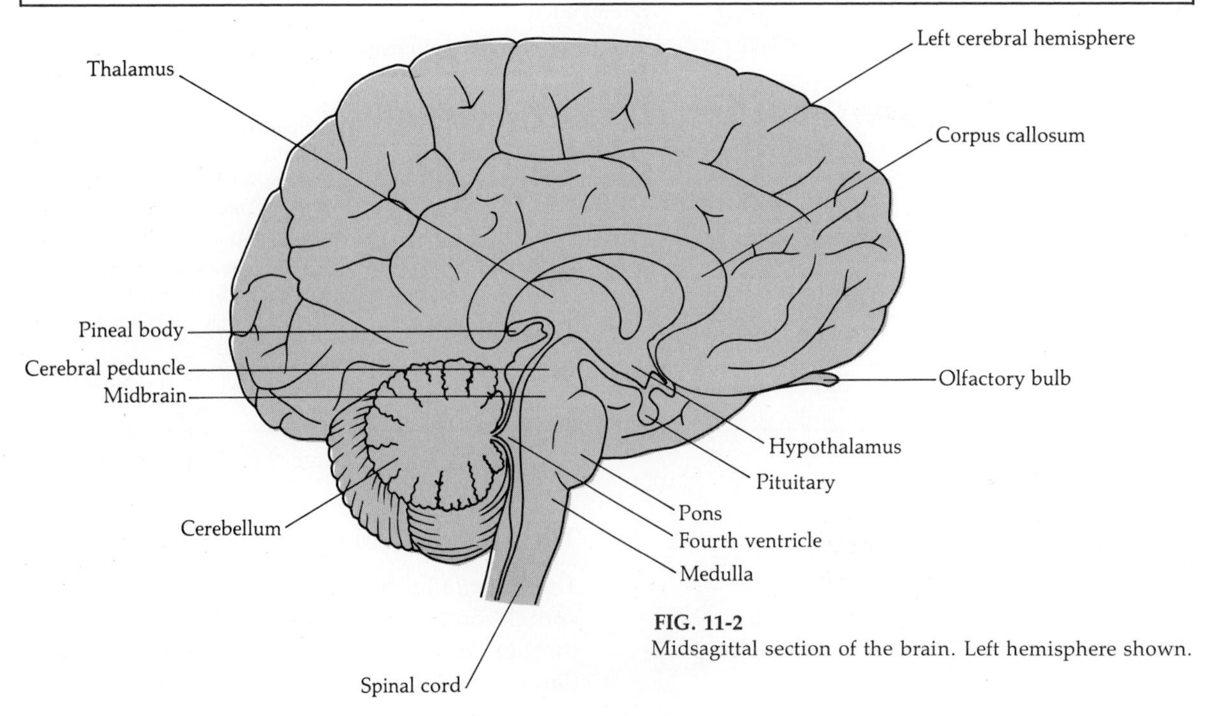

FIG. 11-2
Midsagittal section of the brain. Left hemisphere shown.

the brain, the medulla oblongata is cone-shaped, measures about 2.5 cm in diameter, and connects the spinal cord below to the pons above (Fig. 11-2). The medulla oblongata is the source of the primary cell bodies of cranial nerves IX to XII, the site of neurons that form cardiac, vasomotor (blood-vessel-tone), and respiratory centers, and is an area in which many ascending and descending nerve tracts cross over.

The Pons

The *pons* contains neurons and fibers which form a bridge that connects with the spinal cord below, the cerebral hemispheres above, and the midbrain and cerebellum behind (Fig. 11-2). The pons is the source of cranial nerves V to VII (and parts of cranial nerve VIII) and of neurons which form the *pneumotaxic* and *apneustic centers.* The two centers help maintain rhythmic ventilation of the lungs.

The Cerebellum

Location

The *cerebellum* is a three-lobed structure which accounts for nearly 10 percent of the weight of the central nervous system (CNS) and is located above and behind the pons and medulla (Fig. 11-2). Specific cerebellar neurons which control motor activity in discrete areas of the body are extremely numerous and overlapping in function. They occupy widespread but defined areas for which neurological maps have been constructed.

Functions

The cerebellum receives information from receptors throughout the body, integrates it, and generates outgoing impulses, which coordinate efferent motor impulses in other centers, including those descending from the cerebral cortex. The cerebellum may be thought of as an extremely high-speed and complex "computer," which detects and adjusts differences in the state of contraction of muscles. It does so at a subconscious level by receiving afferent impulses from receptors in muscles, tendons, and joints, as well as from centers for vision and equilibrium (whose impulses are relayed through the *corpora quadrigemina* in the midbrain). The cerebellum coordinates these incoming afferent impulses through feedback circuits. It ultimately regulates and "fine tunes" motor impulses, which are carried along descending tracts from the motor cortex, for the coordination of the contraction and relaxation of opposing muscles.

In summary, the cerebellum: (1) merely influences but does not initiate muscular contraction; (2) does so at the subconscious level; and (3) makes "computerized" adjustments through feedback activities which alter the strength of contractions of opposing muscle groups. The computer functions include: (1) adjustment of the timing of excitation and inhibition of muscular contraction of opposing muscle groups; (2) correction of errors in their activity to cause even and smooth muscular function; and (3) maintenance of posture and equilibrium.

Cellular Anatomy

The cerebellum is divided into two general areas: an outer cerebellar cortex, containing large amounts of gray matter, and an inner cerebellar medulla, which has great quantities of white fibers. A variety of neurons with multiple synapses are found in the *cerebellar cortex.* Its three distinct strata are formed by an outer *molecular* layer containing mostly axonal and dendritic fibers; a middle sheet, one cell layer in thickness, called the *Purkinje cell* layer; and an inner *granular* layer composed mostly of granule cells (Fig. 11-3A).

The *cerebellar medulla* contains neurons forming the *cerebellar nuclei* as well as fibers from the superior, middle, and inferior *cerebellar peduncles,* which relay afferent and efferent impulses between the cerebellum and other areas of the brain (Table 9-1, Figs. 11-3 and 11-4).

Pathways

Afferent impulses are transmitted over reverberating (feedback) circuits between the cere-

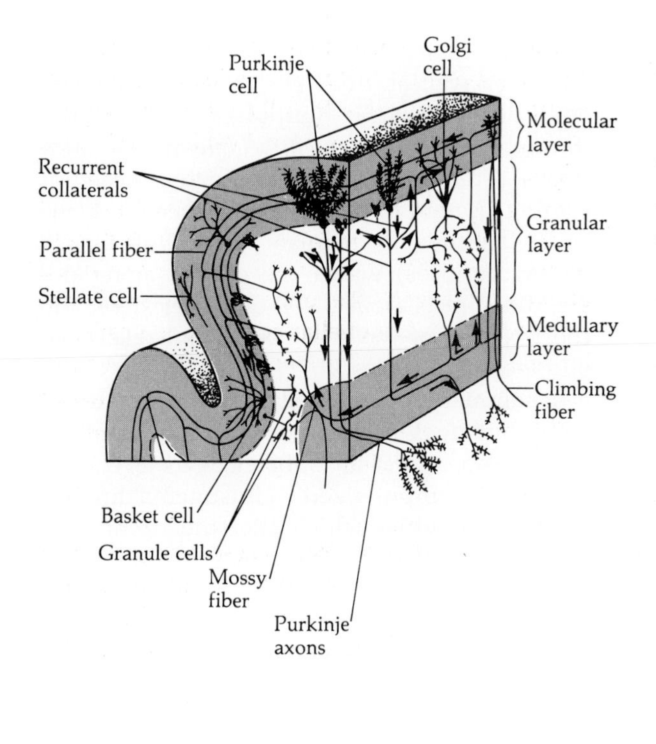

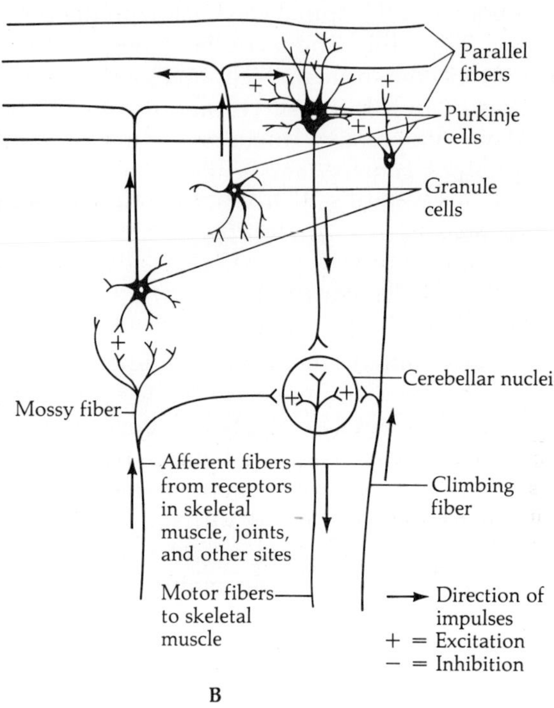

FIG. 11-3
Cellular structure of the cerebellum. (A) Arrangement of cells in a three-dimensional view. (B) Direction of impulse flow in cerebellar circuit. The arrows indicate the routes that the impulses follow.

bellum, other parts of the brainstem, and the spinal cord (Fig. 11-4). Efferent impulses are transmitted from the cerebellar nuclei to brainstem components and the thalamus to influence descending motor paths. Some of the information delivered to the thalamus is then relayed to the cerebral cortex to cause conscious awareness of skeletal muscle activity that would otherwise go unnoticed (remain subconscious).

AFFERENT PATHS OF THE CEREBELLAR CORTEX
The afferent impulses carried to the cerebellum do not directly attain the sensory level of consciousness. They must be relayed from the cerebellum to the thalamus and on to the cerebral cortex to create awareness of skeletal muscle activity. Thus, the impulses coming to the cerebellum follow several complex routes.

The cerebellar peduncles carry excitatory signals to the mossy and climbing fibers which project to the cerebellar cortex (Fig. 11-3B). The mossy fibers receive impulses from nuclei of the spinocerebellar tracts and the pons, respectively. Each mossy fiber forms excitatory synapses with hundreds of granule cells in the innermost layer of the cerebellar cortex and also sends excitatory impulses to cerebellar nuclei located in the cerebellar medulla.

The granule cells are excitatory interneurons and the most common type of neuron in the brain, totaling about 500 billion (5×10^{11}) in all, or about 7 million per cubic millimeter of tissue. Their axons form parallel fibers which project from the cell body like a letter T toward and along the molecular layer to form synapses with hundreds of Purkinje cells (Fig. 11-3A).

Each Purkinje cell, found in the middle layer of the cerebellar cortex, synapses with several types of interneurons, numbering in the tens of thousands. They include: (1) the excitatory granule cells just described; (2) excitatory climbing fibers from cerebellar nuclei (whose collaterals may link 10 or more Purkinje cells in reverberating circuits with the cerebellar nuclei); and (3) three types of inhibitory neurons in the cerebellar cortex, the basket cells, stellate cells, and Golgi cells (Fig. 11-3A).

The Purkinje cells function similar to receivers and relays on telephone poles. The parallel fibers from the granule cells serve as "wires" which connect hundreds of Purkinje cells in a series. Each Purkinje cell can receive connections from 80,000 to 100,000 different parallel fibers and from as many as 2000 climbing fibers, plus the other cell types described above. However, the only cells to send impulses from the cerebellar cortex to the cerebellar medulla are the *Purkinje cells, and their impulses are always inhibitory.* They often function additively to suppress the activity of descending motor tracts that control skeletal muscle contraction.

EFFERENT PATHWAYS The cerebellar impulses are relayed from the cerebellar nuclei to the descending motor tracts. The release of excitatory impulses from deep cerebellar nuclei to the

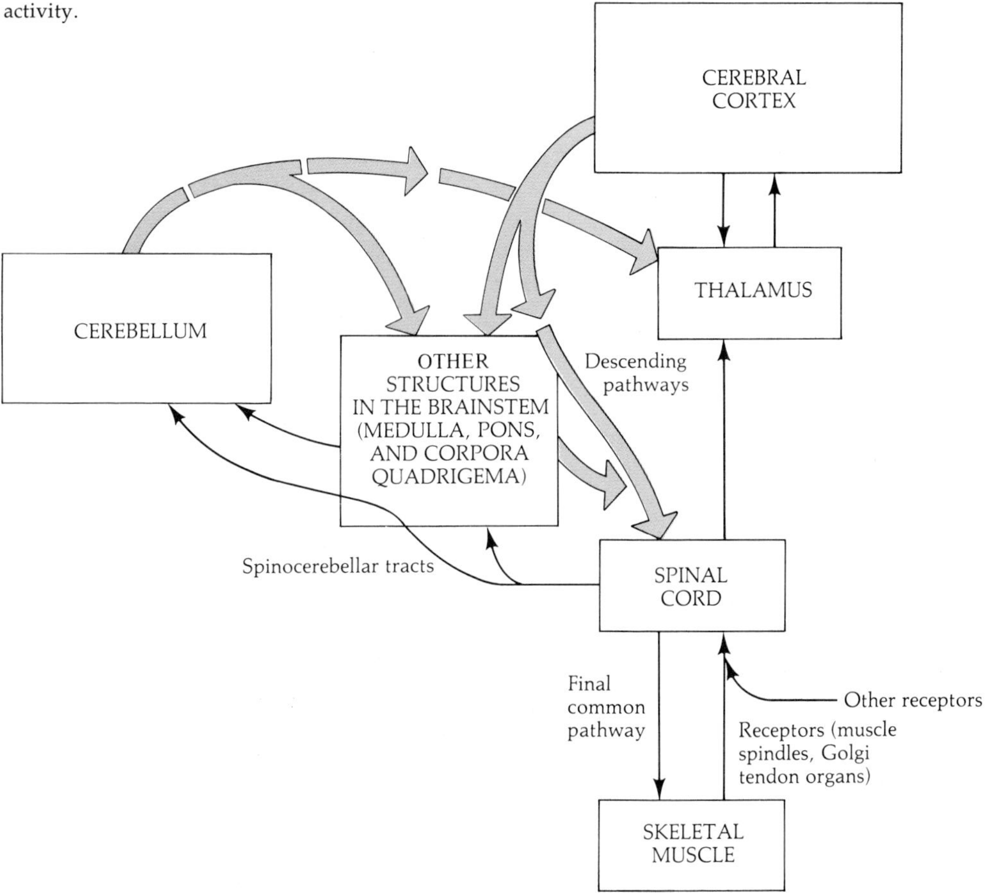

FIG. 11-4
Schematic diagram of flow of impulses between ascending tracts, the cerebellum, and descending tracts which form the motor system to skeletal muscle. The illustration is a simplification of a number of direct and indirect reciprocal connections that form circuits necessary for the cerebellum to act as a "computer" which "fine tunes" skeletal muscle contraction and coordinates motor activity.

TABLE 11-3
Some effects of cerebellar damage or disease

Condition	*Symptoms*
Ataxia (loss of coordination)	Lack of coordination of voluntary muscle movement; muscle weakness Disruption of timed movements often characterized by dizziness (such as stumbling as opposed to walking)
Decomposition of movement (by-the-numbers phenomenon, toy soldier walk, or spastic movements)	Movement does not occur smoothly; rather, it occurs in separate stages (as in walking upstairs, when a leg is lifted to the height of the next step, then in an individual movement extended forward)
Dysmetria (loss of judgment of distance)	Failure to judge distances, resulting in overmovement or undermovement (such as failure to touch one's nose with the index finger when blindfolded).
Dysdiadochokinesia (adiadochokinesia; loss of start-stop movements)	Inability to stop one movement and rapidly start another in the opposite direction (as in drumming or tapping the fingers)
Scanning speech	Clumsy, uncoordinated use of speech muscles, resulting in irregular volume and rhythm of speech
Dysarthria	Slurred, explosive speech
Asynergia	Clumsy, uncoordinated movements even in the presence of visual impulses
Dysergia	Failure of different muscle groups to cooperate (such as sliding the foot of one leg over the other lower leg)
Asthenia	Moderate weakness within a muscle group; impaired cooperation between members of the group demonstrated by a delay in starting and stopping contraction
Titubation	Jerking forward and backward while sitting or standing
Intention tremor (involuntary movements)	Limb(s) shake in jerky, coarse manner before end of intended movement (lack of tremor when the muscles are at rest)
Hypotonia (decreased muscle tone)	Flabby muscles, decreased muscle tone, weak tendon reflexes (tested for with knee-jerk reflex)
Nystagmus	Inability to coordinate lateral eye movements (spasmodic movements of eyeballs)

brainstem and thalamus is dependent on the net influence of the inhibitory impulses received from the Purkinje cells plus the excitatory ones from the collaterals of the climbing and mossy fibers (Fig. 11-3B). Most of the efferent fibers from the cerebellar nuclei are delivered to synapses in the thalamus and to the descending pathways of the extrapyramidal tracts in the medulla and the pons (Fig. 11-4), which include the reticulospinal, vestibulospinal, and rubrospinal tracts described in Table 9-2. In this manner, the cerebellum fine tunes motor activity.

Effects of Cerebellar Damage or Disease

Damage or disease in the cerebellum can result in the loss of muscular coordination. The effects become apparent when an individual attempts to maintain posture or perform skilled movements. Most of the efferent cerebellar paths cross over (that is, travel to the opposite side of the CNS), as do the descending motor tracts with which they form direct or indirect connections. As a result of the double crossover (that of the efferent cerebellar paths plus that of the descending motor tracts), damage to the cerebellum affects muscle on the same side of the body. That is, the site of damage to the cerebellum and the site of the effects are *ipsilateral.* Owing to the resulting loss of coordination, the affected skeletal muscles fatigue easily, become flabby (lose tonus), and are unable to carry out a number of motor activities (Table 11-3).

The nature and extent of a cerebellar-induced disorder depends on the site and amount of tissue affected. Since there are hundreds of billions of cerebellar neurons, limited damage does not block cerebellar function completely. Signals may be delivered but at a lesser rate. Delays in the coordination of muscular activity are in proportion to the number of neurons involved. The individual also may compensate for the damage by utilizing a number of alternative paths which transmit signals to and from the cerebellum (Fig. 11-4).

11-3 THE MIDBRAIN

The *midbrain,* the smallest division of the brain, is a short, inverted heart-shaped mass of tissue situated between the pons and cerebellum of the hindbrain and overlapped by the thalamus of the forebrain above (Fig. 11-2). The major components and functions of the midbrain are noted in Table 11-4. The midbrain delivers sensory information for sights and sounds through ascending pathways to the cerebral cortex. It also integrates incoming sensory visual and auditory impulses for reflex responses carried through the descending pyramidal and extrapyramidal motor tracts (Table 9-2). This capacity is called the *righting reflex.* Damage to the midbrain can result in the loss of this reflex, interfere with eye movements, and/or result in partial or complete paralysis of specific skeletal muscles. In summary, the midbrain is a relay area.

11-4 THE FOREBRAIN: ITS DIENCEPHALIC SUBDIVISION

The *forebrain* includes all components around the third ventricle (Fig. 10-6). It is subdivided into a posterior and centrally located section, the *diencephalon,* which includes the hypothala-

TABLE 11-4
Major functions of the midbrain

Area	*Location*	*Component*	*Associated Functions*
Midbrain	Between pons and cerebellum of hindbrain and below third ventricle and thalamus of forebrain		Coordinates descending motor paths with visual and auditory reflexes
	Roof	Corpora quadrigemina	
		Two superior colliculi	Visual reflexes
		Two inferior colliculi	Auditory reflexes; major relay nucleus in auditory pathway
	Middle	Nuclei of cranial nerves	III—Controls size of pupil and movement of eye muscles IV—Movement of eye muscles
		Red nucleus	Connects basal ganglia of cerebral hemispheres with cerebellum to control muscles and posture
		Reticular formation	Connects excitatory and inhibitory neurons in pons and medulla to control skeletal muscle activity
		Substantia nigra	Relay station for extrapyramidal fibers
	Base	Cerebral peduncles	Connects motor tracts of cerebral hemispheres in forebrain to hindbrain (including the pons) and spinal cord

TABLE 11-5
Major functions of the diencephalic subdivision of the forebrain

Area	*Location*	*Component and/or General Function*	*Associated Functions*
Hypothalamus	Lower walls and floor of third ventricle	Descending pathways	Modifies autonomic reflexes
		Nuclei and centers for autonomic functions:	
		1. Appetite center	Hunger or fullness (responses to chemoreceptors)
		2. Thirst center	Water intake (responses to osmoreceptors)
		3. Temperature regulation centers	
		Posterior hypothalamus	Responses to cold: Shivering (skeletal muscle activity); Constriction of skin's blood vessels
		Anterior hypothalamus	Responses to heat: Relaxation of skeletal muscles; Dilation of skin's blood vessels; Stimulation of sweat glands
		4. Behavioral centers (in association with limbic system of telencephalon)	Behavioral responses, emotions, responses to stress
		Posterior hypothalamus	Escape; Waking (also involves limbic system paths)
		Anterior hypothalamus	Mating; Sleep (also involves limbic paths)
		Lateral hypothalamus	Attack reactions
		Neuroendocrine functions	Modifies pituitary function: Portal vessels transport hypothalamic factors and hormones to regulate the anterior pituitary; Axons carry ADH and oxytocin to posterior pituitary
Thalamus	Walls and part of floor of third ventricle	Nuclei	Transmission of signals: From eyes, ears and skin to sensory cerebral cortex, to associative and somatosensory areas in cerebral cortex, and to motor areas of cerebral cortex; Relay of signals from cerebellum
Pineal body	Roof of third ventricle, superior to midbrain and posterior to thalamus; attached to posterior wall of third ventricle by the pineal stalk	Neurons and secretory cells	Influences rate of sexual maturation; Synthesizes melatonin; Inhibits sex organs via blocking the release of luteinizing hormone; May provide a response to light which produces a biological clock

mus, thalamus, and pineal body, and a prominent anterior section, the *telencephalon*. The telencephalon dominates the exterior surface of the brain and is formed by two halves called the cerebral hemispheres. The major functions of the forebrain are summarized in Tables 11-5, 11-6, and 11-7.

The Hypothalamus

The *hypothalamus* is located on the lower border of the forebrain and on the floor of the third ventricle (Fig. 11-2). It plays an important role in homeostasis by influencing several *autonomic* and *neuroendocrine functions* including: (1) appetite and thirst; (2) temperature regulation; (3) behavioral responses; (4) sleep; and (5) hormonal secretions (Tables 11-5, 9-5, 5-1, and 5-5).

The hypothalamus is connected by tracts to the spinal cord, midbrain, thalamus, limbic lobe, and other areas of the brain, as well as to the pituitary gland. Some afferent tracts carry impulses from the olfactory and limbic lobes of the brain to modulate smell and emotions, respectively. Some efferent tracts include those that carry impulses to the midbrain, thalamus, and posterior pituitary. Hypothalamic tracts to the posterior pituitary transport the antidiuretic hormone and oxytocin (Chap. 5). Numerous interneurons exist within areas of the hypothalamus and between it and other regions of the brain. The complexity of the circuits makes it difficult to correlate specific hypothalamic nuclei with individual functions. However, several centers which are responsive to negative feedback signals are described below.

Appetite and Thirst Centers

Neurons in the hypothalamus that respond to the level of glucose in the blood are receptors called *glucostats*. The rate at which they metabolize glucose relates to the generation of reflex responses that influence digestive tract motility and the appetite center to cause hunger or fullness. Some neurons form a *hunger* or *feeding center* responsive to low blood sugar (hypoglycemia). Others form a *satiety* (fullness) *center* responsive to a high level of blood sugar (hyperglycemia) which may cause a sensation of fullness (Chaps. 22, 23). Research indicates that certain individuals who overeat and become overweight may have faulty neurons in these circuits. Studies with animals support this concept. For example, artificial electrical stimulation of the satiety center in a starved animal causes it to avoid food unnaturally, while stimulation of the hunger center causes a fully fed animal to overeat.

Osmoreceptors are hypothalamic neurons which respond to changes in the osmotic pressure of the blood. They may generate impulses to produce a sensation of thirst or desire to drink, and the neurons involved are referred to as a *thirst center*. When the osmotic pressure of the blood is high (and its water content is low) the desire to consume water is increased in order to return the osmotic pressure to a normal range. The water level in the blood is also influenced by antidiuretic hormone (ADH), which is secreted by the hypothalamus and carried down its axons to be released from the posterior pituitary gland. ADH promotes reabsorption of water (in potential urine) from the tubules of the kidney back into the blood. The reabsorption of water minimizes urine formation and water loss. Disorders in the hypothalamus and/or posterior pituitary can result in *diabetes insipidus*, a disease in which more than 20 L of urine may be excreted each day (Table 5-5, No. 2b).

Temperature Regulation Center

Body temperature is normally maintained within a narrow range by a combination of heat production and heat loss controlled through a *temperature regulation center* in the hypothalamus.

Heat production is stimulated by fibers from neurons in the posterior hypothalamus. When the blood temperature drops, as may occur with exposure to a cold environment, body heat is conserved when the adrenergic sympathetic fibers cause the arterioles of the skin to constrict (narrow). Thus, less blood is delivered toward the surface of the body and more blood (and heat) is retained in the body. At the same time, more heat is produced as the sympathetic fibers cause the skeletal muscles to tremor about 10 to 20 times per second in weak, rapid rhythmical contractions (shivering), which produce no useful work but do generate heat (see Chap. 13).

When the blood temperature rises, the heat production center is inhibited and the heat loss center is activated. Heat loss is promoted by the stimulation of the sweat glands by cholinergic sympathetic fibers from neurons in the anterior hypothalamus. The evaporation of sweat from the skin's surface cools the skin, along with the blood passing through the vessels in it, and lowers body temperature. Inhibition of adrenergic sympathetic fibers causes arterioles in the skin to dilate (widen) and thereby increases the blood flow and heat loss to the atmosphere. A decrease in body temperature "turns off" the heat loss center through negative feedback signals to complete the cycle.

Failure of the temperature regulation center may cause an elevation of body temperature, which is called fever or *hyperthermia,* or it may cause a decrease in temperature, which is called *hypothermia* and is accompanied by chills. An elevation of 1°F in body temperature causes metabolism to speed up by about 7 percent. Conversely, a drop in body temperature slows down physiological functions. Utilization of this response has led to the use of cooling jackets (or ice baths) to slow rapidly moving body parts. For example, the induction of hypothermia is used to make open-heart surgery somewhat easier.

Normal homeostatic control mechanisms of the hypothalamus maintain body temperature of humans, other mammals, and birds within a narrow range, which is characteristic of each species and generally independent of slight changes in environmental temperature. Such animals are called *homeotherms* (warm-blooded). Animals whose body temperature rises and falls with that of the environment (such as worms, frogs, and fish) are called *poikilotherms* (cold-blooded). Humans are homeotherms with a body temperature that averages 37°C (98.6°F) orally and ranges between 36.3 and 37.1°C (97.3 and 98.8°F) in early morning. These values are about 0.5°C (1.0°F) lower when measured rectally.

Behavioral Responses

Many behavioral responses are influenced by neural connections between the hypothalamus and the limbic lobe of the brain. Their interactions will be described below in the section on the limbic system. The responses to stress are also activated by the hypothalamus, which serves as an alarm that mobilizes the body's defenses. These reactions were described earlier in Chap. 5 and illustrated in Fig. 5-7.

Sleep

The electrical activity of the brain varies from a slower to a faster pace during NREM (nonrapid eye movement) and REM (rapid eye movement) sleep, described in Chap. 10. The inducement of sleep seems to come from neurons which collectively form a *sleep center.* They are located in the hypothalamus, pons, thalamus, and frontal lobes of the brain. Conversely, excitatory stimuli activate the neurons of the *reticular activating system* (RAS, or *reticular formation*) in the medulla, pons, and midbrain. RAS impulses override those from the sleep center to maintain the wakened state. The neurons of the RAS may form a "waking center." Thus, activation of the neurons of one or the other centers can induce the sleep state or the wakened state. However, the exact location and functions of all the neurons involved remain to be determined. For example, barbiturates are drugs that are sometimes used to induce sleep. They seem to act by blocking the ascending fibers to the RAS. Barbiturates also increase the threshold and recovery time of neurons in general. The prolonged and repeated use of these sedative-hypnotic drugs to induce sleep seems undesirable. Such drugs inhibit REM sleep, which usually occurs in natural sleep and during dreaming (Chap. 10). Continual inhibition of REM sleep may have adverse consequences for the individual.

Hormonal Secretions

Perhaps no role of the hypothalamus is more encompassing than its *neuroendocrine functions.*

In addition to the secretion of ADH and oxytocin, the hypothalamus forms releasing and inhibitory hormones (Table 5-1), which are carried by portal blood vessels to the anterior pituitary gland (Fig. 5-3). The hypothalamic hormones directly or indirectly influence most other endocrines and body functions (Fig. 5-4). They regulate the responses to stress, control blood pressure and heart function, and influence body metabolism, as described in Chap. 5. Additional specific activities of hypothalamic hormones will be described in later chapters.

The Thalamus

The *thalamus* is the "switchboard" of the brain. It: (1) processes sensory information; (2) conveys a general perception of environmental stimuli; and (3) seems to be involved with the development of primitive feelings found in lower animals as well as in humans (e.g., hunger and discomfort). The thalamus is the primary means by which sensory impulses are projected to the cerebral cortex.

The thalamus consists of two lobes of tissue connected by an intermediate mass, lies just below the corpus callosum (Fig. 11-2), and is composed of more than 25 nuclei. Impulses from these nuclei are sensory, associative, or intrinsic and are relayed to the cerebral cortex, to other brain areas, and to different areas within the thalamus. Sensory information is delivered to the cerebral cortex by way of the spinothalamic tracts and dorsal columns (Table 9-1, Figs. 9-9 and 9-10). By means of many afferent tracts, impulses for sight, sound, pain, temperature, touch, and balance reach the levels of consciousness by passage through the thalamus to neurons in the cerebral cortex. However, impulses also travel in the reverse direction over corticothalamic fibers. Thus, information is transmitted in both directions so that the cerebral cortex and thalamus can modify each other's activities by feedback circuits.

The Pineal Body

The *pineal body* (*pineal gland*) is made up of secretory cells and a few neurons. It is located in the roof of the third ventricle and above the superior colliculi of the midbrain and posterior to the thalamus (Fig. 11-2) in an area called the *epithalamus*. Sympathetic nerve fibers make up a prominent neuronal part of the pineal body. The secretory cells within the pineal body contain norepinephrine and serotonin and may convert the latter into the hormone melatonin. This hormone circulates freely throughout the body since the blood vessels to the pineal gland are outside the blood-brain barrier. Some of the neurosecretions of the pineal gland can suppress the liberation of releasing factors from the hypothalamus, which in turn can influence pineal function through feedback mechanisms.

For example, melatonin inhibits the activity of melanocyte-stimulating hormone (MSH) and adrenocorticotropic hormone (ACTH), as well as the release of luteinizing hormone (LH) from the anterior pituitary. It influences sexual *maturation* by inhibiting the release of LH (Table 5-1). Children with tumors of the pineal gland often undergo sexual maturation at an early age. Some have been reported to become sexually mature by the time they were 5 years old.

The nearness of the pineal gland to the superior colliculi of the midbrain (which regulate visual reflexes, Table 11-4), provides a hint of another possible function. Exposure to light and dark seems to provide the pineal body with a timing mechanism called a *biological clock*. A number of physiological functions depend on oscillations in exposure to light. As a result, many biological activities follow a cycle that approximates one day and are referred to as *circadian rhythms* (from Latin *circa*, about, and *dies*, day). Fluctuations in the levels of many hormones, in the numbers of circulating white blood cells, and in body temperature follow circadian rhythms. The importance of such variations is considerable. They are reflected in "normal" fluctuations of levels of substances in the body each day. These alterations may change responsiveness to therapy as well as susceptibility to disease.

For example, the levels of circulating white blood cells increase and decrease each day according to an individual's circadian rhythm. These cells assist the body in bacterial destruction and removal. It has been postulated that antibiotics (and other drugs) might be more effective in assisting body defenses if their administration were timed according to the relevant circadian rhythm peaks of the individual. For example, such an approach might allow the levels of antibiotic to be coordinated with the maximal natural defenses of the body to attain optimal results. The timing of administration of the drug would be based on the circadian rhythm of the patient as opposed to the rather arbitrary use of 4- to 6-h intervals between doses that is currently used.

11-5 THE FOREBRAIN: ITS TELENCEPHALIC SUBDIVISION, THE CEREBRAL HEMISPHERES

The *cerebrum* is the uppermost part of the brain. It is composed of right and left portions, the *cerebral hemispheres,* which are partially separated by a cleft, the *longitudinal cerebral fissure.* The lower portions of the hemispheres are connected to each other by myelinated fibers (white matter) called the *commissural fibers.* The largest bundle of these fibers form a C-shaped structure, the *corpus callosum* (Figs. 11-2 and 11-6B).

The *cerebral cortex* is a thin layer of gray matter 2.5 to 4.0 mm thick which forms the cover of the cerebrum. The surface area of the cerebral cortex is significantly increased by folds called *gyri* (singular, gyrus), shallow depressions called *sulci* (singular, sulcus), and deeper ones called *fissures.* If spread out, the cortex would extend over an area of about 1600 to 2500 cm^2 (1.7 to 2.7 ft^2, or a maximum of about 3 ft^2).

The neuronal circuits of the cerebral cortex are densely packed and miniaturized to a greater extent than the modern basic circuitry chips (integrated circuits) of pocket-sized calculators. The small area of tissue which composes the cerebral cortex contains approximately 49 billion neurons with trillions of synapses that make up thousands of series, parallel, and reverberating circuits.

Six layers of cells may be distinguished in the cerebral cortex (Fig. 11-5). The majority of the cells in its outer four layers, the *outer zone,* receive impulses from the thalamus, commissural fibers, and other parts of the cortex. Star-shaped granule cells in the fourth layer are the principal receptor cells. Large pyramidal cells in the *inner zone* are the primary efferent neurons in the cerebral cortex. The largest of these pyramidal cells are called *Betz cells.*

White matter lies beneath the cerebral cortex. It contains a variety of myelinated fibers, including those of the ascending and descending paths described earlier in Chap. 9. White fibers which transmit signals within each hemisphere are *association fibers;* those which carry the impulses to the opposite side are *commissural fibers.*

The cerebral hemispheres have a multiplicity of functions. In some of their important roles they: (1) receive sensory information and bring it to the conscious level (such as that for vision, hearing, taste, and smell; (2) coordinate sensory and motor activity; (3) initiate voluntary motor activity; (4) influence behavior and emotions; and (5) are responsible for higher brain functions such as memory, judgment, language, speech, musical creativity, mathematical computations, and recognition of spatial patterns. The anatomical and physiological foundations for some of these functions will be described below.

Location of Lobes

The cerebral hemispheres are each divided into the frontal, parietal, occipital, temporal, insula, and limbic lobes (Fig. 11-6). Some of their major components and functions are listed in Tables 11-6 and 11-7.

The outer surface of the brain has the appearance of a wrinkled boxing glove lying on its thumb. Three furrows or deep grooves serve as external markings that outline the locations of several lobes. The *central sulcus* (*fissure of Rolando*) runs vertically through each hemisphere, separating the anterior *frontal lobe* from the posterior *parietal lobe* (Fig. 11-6A). The *occipital lobe,* which lies at the posterior of the hemisphere, is separated from the parietal lobe by the *occipital-parietal* fissure. The *lateral fissure* (*fissure of Sylvius*) divides the anterior and lower part of the frontal lobe from the thumb-shaped *temporal lobe* below it. The *insula* and *limbic lobes* are internal.

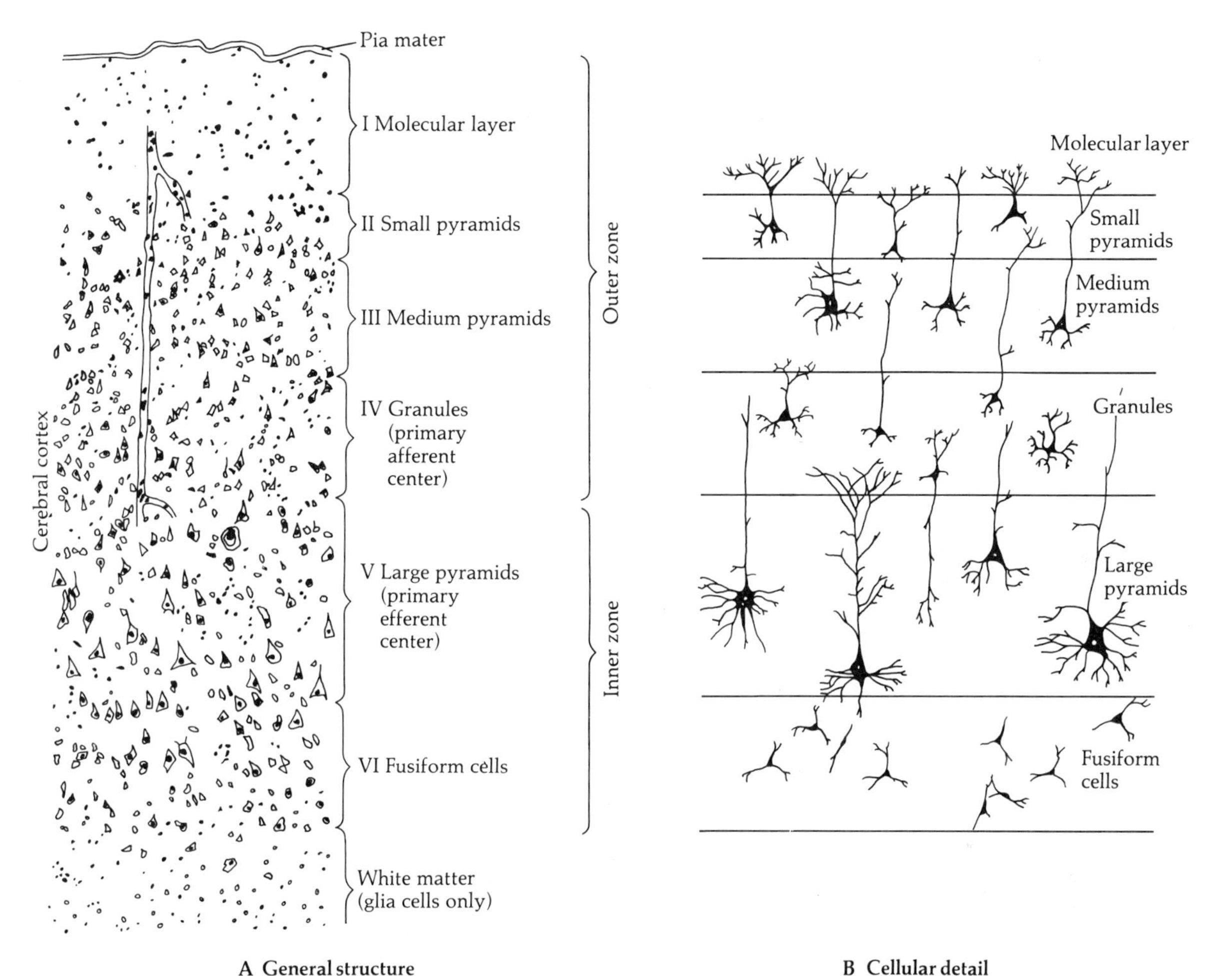

FIG. 11-5
Layers of cells in the cerebrum. (A) General structure. Six different cortical regions may be distinguished. Variations in the relative thickness of each layer may be extensive. (B) Cellular detail of neurons and neuroglia in cerebral cortex.

Functions of Lobes

Interior of Cerebral Hemispheres

The *limbic system* is composed of the limbic lobe, found within the core of the cerebral cortex (Fig. 11-6B), plus some nuclei in the thalamus, hypothalmus, and reticular formation. It influences autonomic functions by relaying impulses among several areas in the brain, as noted earlier in Table 9-5 and illustrated as follows:

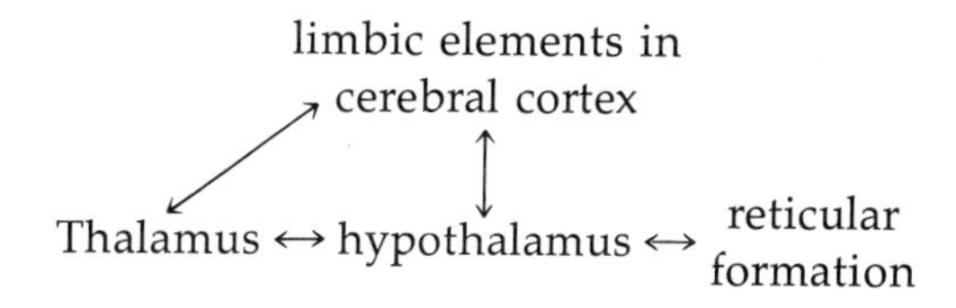

The limbic lobe receives the terminal fibers of the olfactory cranial nerves (Table 11-1), which convey signals for the sense of smell. The nuclei of these fibers originate in the olfactory bulb (Fig. 11-2) under the frontal lobe. The limbic system especially influences unlearned responses involving emotions and behavior. Its complex circuits affect feelings of fear, rage, anxiety, relaxation, pleasure, and sex. Studies have shown that the stimulation of discrete sites within the limbic system can evoke such

TABLE 11-6
Major functions of the forebrain: interior of cerebral hemispheres in the telencephalon

Area	*Location*	*Component*	*Associated Functions*
Limbic lobe	Limbic system (made up of tissue around the brainstem) plus subcortical nuclei, including some in the thalamus and hypothalamus		Interconnects sensory and autonomic pathways of thalamus and hypothalamus, plus fibers of reticular formation of brainstem, to influence behavior and emotional responses
		Hippocampus	"Recent" memory (retention of new information) Send signals to hypothalamus
		Septal nuclei (septum)	Modifies emotions
		Amygdala	Modifies emotions via efferent signals to hypothalamus
		Olfactory bulb (undersurface of frontal lobe)	Sense of smell Source of cranial nerve I, the olfactory nerve
Insula (island of Reil)	Island of tissue at center of the telencephalon; just within cerebral hemispheres and fissure of Sylvius (above middle of temporal lobe)		Receives signals of taste fibers from thalamus
Basal ganglia*	Deep within cerebral hemispheres, above and lateral to thalamus		Coordinate sensory and motor activity Interconnect with one another and lower nuclei Disorders in basal ganglia cause several illnesses, including Parkinson's disease (characterized by muscle tremors) and Huntington's chorea
		Corpus striatum	
		Caudate nucleus	Receives signals from substantia nigra of midbrain; especially abnormal in Huntington's chorea
		Lenticular nucleus	
		Putamen	Receives signals from substantia nigra
		Globus pallidus	Sends impulses to cerebral cortex via thalamus

* The *subthalamic nuclei* and *substantia nigra* of the upper brainstem also may be included in the term *basal ganglia*.

feelings in individuals undergoing neurosurgery. Specific neurotransmitters released from nerve endings in limbic centers have been associated with certain emotions. The interactions are complex. However, studies on the effects of neurotransmitters and drugs which enhance or inhibit neurotransmitter activity offer an understanding of behavior and also promise the opportunity to modify behavior.

The *insula* (island of Reil), located within the core of the hemispheres (Fig. 11-6A), receives fibers carrying taste signals from the thalamus and transmits them to the cerebral cortex.

The *basal ganglia* are clusters of nuclei located deep within the cerebral hemispheres and in the brainstem. They coordinate sensory and motor activity and form a striped mass called the *corpus striatum* (Table 11-6). Components of

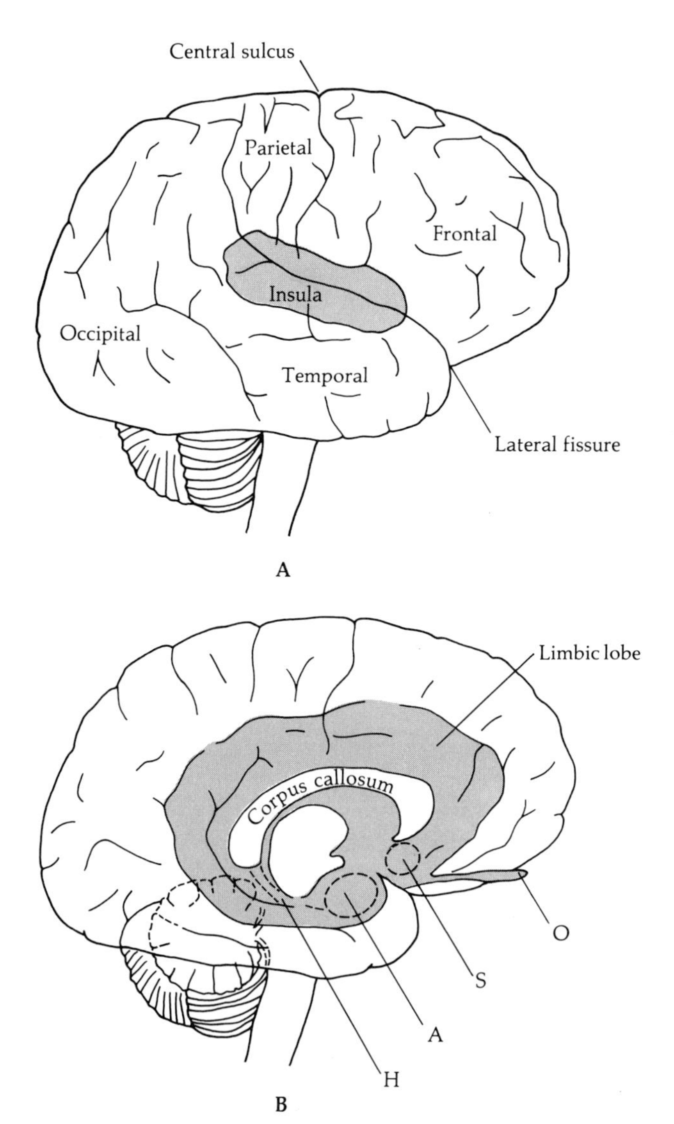

FIG. 11-6
Location of lobes of the cerebral hemispheres. Right hemisphere shown. (A) External surface of the right hemisphere is shown. The location of the insula in the midsagittal section within the hemisphere is indicated in color. (B) Interior view. The general positions of the components of the limbic lobe are drawn in color. The approximate locations of the septum (S), amygdala (A), and hippocampus (H) are indicated with dotted lines. The olfactory bulb is labeled O.

the corpus striatum include the *caudate nucleus, putamen,* and *globus pallidus.* Disorders in the basal ganglia may result in Huntington's chorea or in Parkinson's disease.

Huntington's chorea is a heritable dominant disease which appears in young adults. It results from an inability of the neurons to convert glutamic acid to the neurotransmitter gamma-aminobutyric acid (GABA, Chap. 7). The neurons degenerate and are replaced by glial cells. The individual loses voluntary control of muscles and develops progressive mental deterioration culminating in death.

Parkinson's disease is associated with elevated levels of acetylcholine (ACh) or with a deficiency in dopamine production. Its symptoms of rhythmical muscular tremor are relieved by the administration of atropine, which blocks the effects of ACh at its receptor sites, or by L-dopa, which can be converted into dopamine.

Exterior of Cerebral Hemispheres

The general functions of the lobes on the surface of the cerebral hemispheres have been determined. Some of these functions are illustrated in Fig. 11-7 and summarized in Table 11-7. The neurons of the frontal lobe send out voluntary motor impulses, some of which govern speech. Those of the temporal lobe are associated with hearing. The parietal lobe contains somatic sensory and associative neurons, and the occipital lobe receives visual signals from the optic nerves.

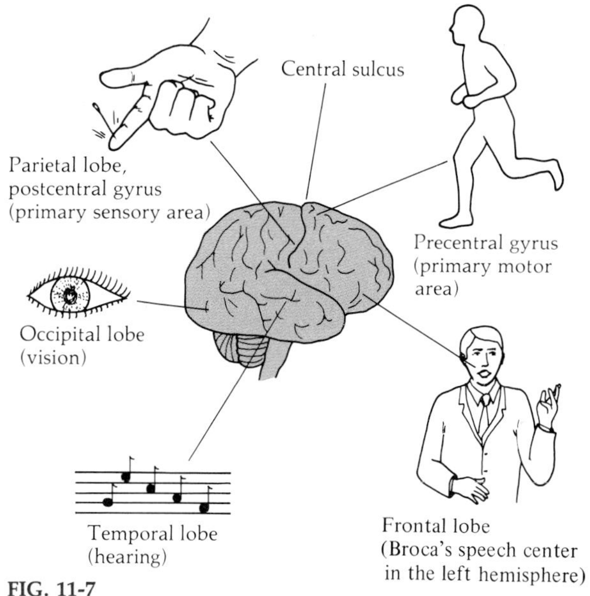

FIG. 11-7

Somatotopic Organization

The cerebral cortex exhibits *somatotopic organization*, which means that the neurons associated with parts of the body are located in orderly patterns. The approximate positions of neurons that govern specific activities have been discovered (Figs. 11-7 to 11-9). A useful landmark in locating specific functional areas is the central sulcus (fissure of Rolando). The *primary motor cortex* consists of neurons that are found anterior to the central sulcus in a fold of tissue called the *precentral gyrus*, which control voluntary motor activity (Fig. 11-8A and C). The *primary sensory cortex* consists of neurons located posterior to the central sulcus in a fold called the *postcentral gyrus* which receive impulses that cause conscious awareness of sensory information (Fig. 11-8A and B).

TABLE 11-7
Major functions of the forebrain: exterior lobes of cerebral hemispheres in the telencephalon

Area	*Location*	*Major Functions*
Frontal	Anterior telencephalon, above lateral fissure (fissure of Sylvius) and anterior to central sulcus	Precentral gyrus emits somatic motor impulses Speech area
Temporal	Below lateral fissure (under temple on side of forehead)	Hearing
Parietal	Posterior to central sulcus and anterior to occipital lobe	Postcentral gyrus receives somatic sensory impulses Associative paths within cerebral cortex
Occipital	Posterior segment of telencephalon behind posterior occipital fissure	Vision (destination of cranial nerve II, the optic nerve, which originates from ganglia in the retina and forms optic tracts which pass through the midbrain to relay impulses to this lobe)

The motor and sensory areas within each hemisphere are associated mostly with the opposite sides of the body. Specific motor functions can be related to a distorted human torso placed upside down and backward over the precentral gyrus in the primary motor cortex (Fig. 11-8C). Most of the motor neurons in each cerebral hemisphere, except those for speech, control functions on the opposite sides of the body. A similar arrangement has been noted for sensory neurons located in the primary sensory cortex (Fig. 11-8B). Most sensory inputs, except for the sense of smell, also terminate in the opposite hemisphere.

The exact locations of neurons controlling motor and sensory functions varies slightly from one individual to another. Minor variations exist in the thickness and in the composition of layers of cells in the cortex and in the fibers of the white matter. Distinct areas usually can be correlated with brain functions, but not always. Activities have been associated with spatial locations in regions which differ in cell structure. Over 200 of these areas have been assigned numbers and are called Brodmann's areas, some of which are shown in Fig. 11-9.

The Split Brain

Since each half of the cerebrum assumes a greater responsibility for certain functions, the human brain is sometimes described as a *split brain*. The left half is dominant in most individuals. It is interesting to note the correlation between the higher incidence of dominance in the left brain and right-handedness, since the left brain controls motor activity on the right side of the body. Dominance in the split brain seems to be related to the development of speech and awareness of self that distinguishes humans from lower animals. This is also true even in some nonhuman primates, such as apes, that have been trained to understand words at the level of a 2- to 3-year-old child. Animals other than humans lack the natural ability to form associations between word patterns used for names and the objects in the environment to which the words refer. Nonhuman primates nearly always communicate to express their

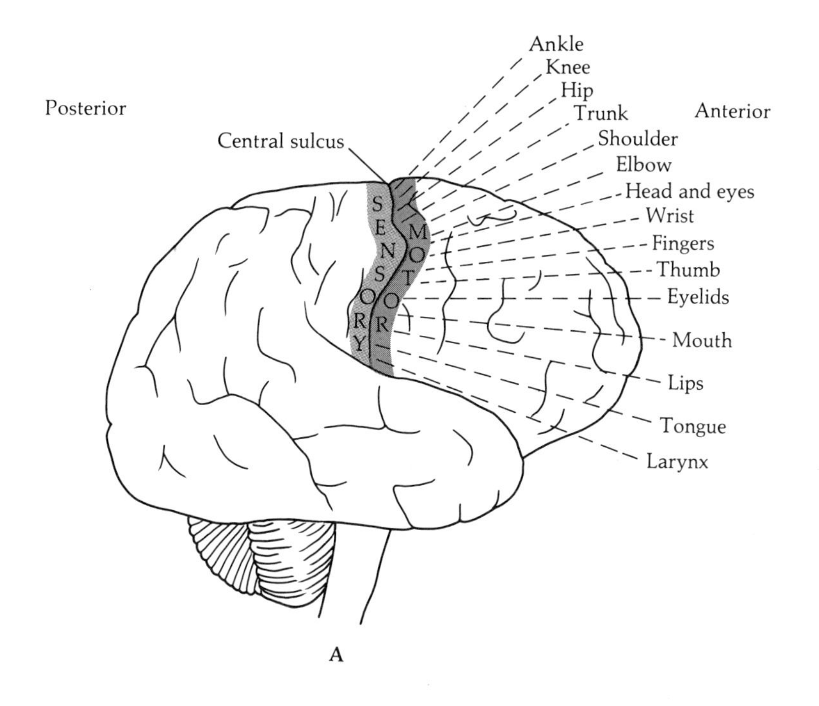

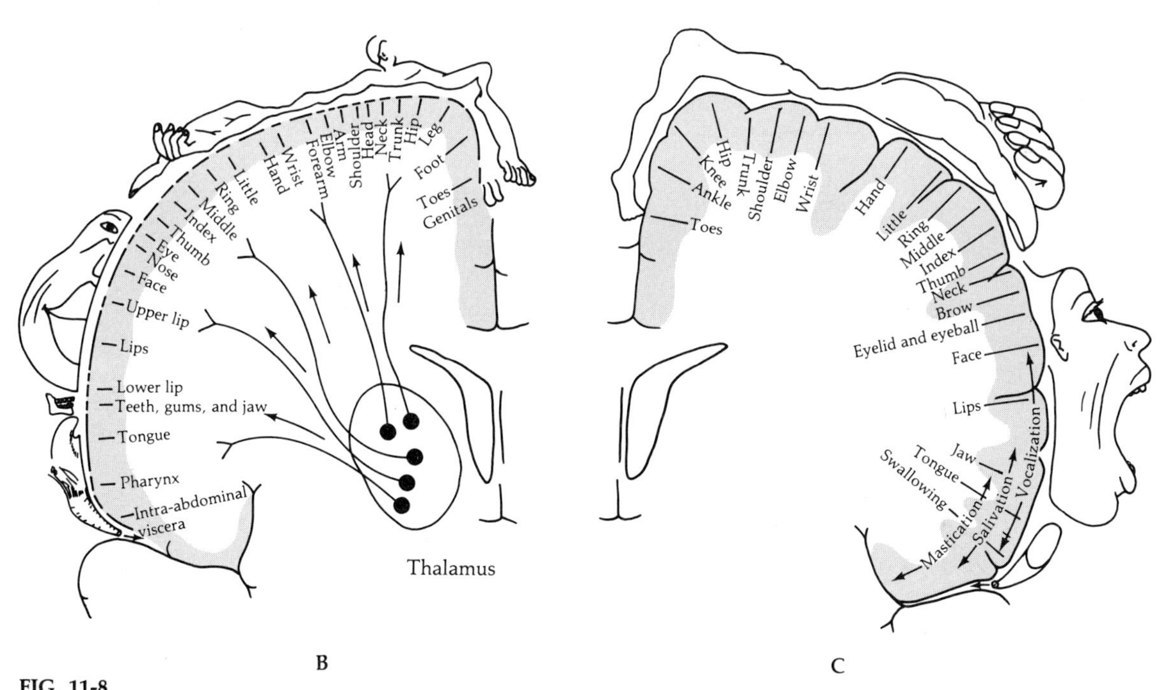

FIG. 11-8
Map of primary motor and sensory areas relating the cerebral cortex to body parts. (A) Primary motor areas anterior to the central sulcus on the side of the cerebral cortex. Sensory areas are posterior to the central sulcus. (B) Map of primary sensory reception from the body in regions of the cerebral cortex. The ornamental figure in the diagram lies over the postcentral gyrus of a cross section of the cerebral hemisphere. (C) The motor map viewed in a cross section made through the precentral gyrus. The ornamental figure illustrates areas within the primary motor cortex whose electrical stimulation causes motor activity in the illustrated body regions.

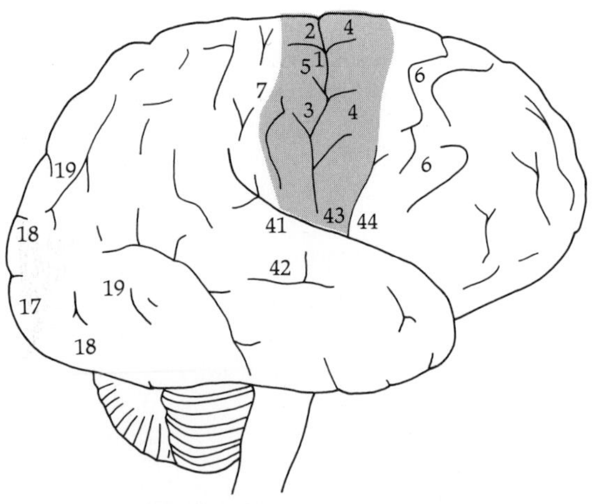

1,2,3 Primary sensory areas (bodily sensations)
4 Primary motor area (initiates movement)
5,6,7 Secondary sensory areas
17 Primary visual area
18,19 Visual associations
41 Primary auditory area
42 Auditory area (associative?)
43 Gustatory (taste) area
44 Speech, in left cortex (Broca's area)

FIG. 11-9
External surface of the right cerebral hemisphere, illustrating some numbered Brodmann's areas. The primary sensory and motor areas are colored. The association areas are not. Note that the speech center (No. 44) is located in the corresponding area of the *left* cerebral hemisphere (see Fig. 11-10).

emotional states but, unlike humans, rarely use language to communicate about their environment as well.

The left hemisphere usually governs speech, language, writing, and mathematical computations. The right hemisphere receives sensory inputs for spatial recognition (i.e., geometric patterns) and influences musical creativity, as noted in Table 11-8.

One side of the brain is dominant for some functions, even though both lobes originally seem to have had the same potential for development. In order to carry out routine motor activity, an individual normally transfers sensory inputs from one side of the brain to the other by means of commissural fibers. When one side of the cortex is damaged in young children, the corresponding neurons on the opposite side are able to assume gradually the lost function but usually do not develop the capacity to regulate it as effectively.

TABLE 11-8
Common dominant sites in the cerebral hemispheres

Left Hemisphere	*Right Hemisphere*
Speech	Spatial recognition of geometric patterns
Language	
Writing	
Mathematical computations	Musical creativity
Left nostril, smell	Right nostril, smell
Right visual field of each eye	Left visual field of each eye
Much of right ear, hearing	Much of left ear, hearing
Right-handed touch	Left-handed touch

Evidence from Surgically Treated Epileptics

Studies on patients surgically treated for epilepsy have revealed many properties regarding the split brain. Epilepsy is a disorder characterized by an abnormal increase in the firing of neurons in the brain. It may be accompanied by a decline in the inhibitory neurotransmitter GABA and alterations in the EEG (Chap. 10). Epilepsy usually can be controlled by medication, but not always. In some severe cases, the commissural fibers were cut to relieve the symptoms of the disorder. Subtle changes in motor ability were then detected by tests which reveal dominance in the split brain. For example, an epileptic so treated was requested to feel metallic numbers with the right hand, add the totals arithmetically, and indicate the results with the right hand. This ability is maintained because the fibers of sensory neurons for right-handed touch cross over in the lower brain as they carry information up to the left hemisphere, where neurons for mathematical computations are found. Association fibers transmit signals between the sensory and motor neurons within that side of the brain to direct right-handed movement. However, when metallic numbers are placed in the left hand, the sensory signals for touch which cross over in the lower brain ascend to its right side, which does not perform mathematical computations. The absence of commissural fibers prevents transfer of this information and the task cannot be completed with the left hand.

In another test, a set of blocks was arranged in a geometric pattern. The left hand of the patient but not the right hand could reconstruct the pattern. This result occurs because only in the right hemisphere (which controls movement of the left hand of the patient) can the necessary associations be made. The right side of the brain receives visual signals from the left visual field of each eye, and is the location of neurons that are used for spatial recognition of geometric patterns, and for control of motor activity of the left hand.

Control of Speech

As mentioned above, the ability to communicate through speech distinguishes humans from lower animals. Assembling sounds to produce language requires the integration of several sensory and motor areas. The auditory, visual, Wernicke's, and Broca's areas are especially involved. Their locations and interconnecting paths are illustrated in Figure 11-10.

Sensory inputs from letters, words, and sentences are transmitted from the retina of the eye through the optic nerves to the visual area of the occipital lobe. The impulses are assembled into language in the association areas of the occipital lobe and then carried to *Wernicke's* area in the left temporal lobe. Signals for sound travel from the ear through the auditory division of the statoacoustic nerve to the auditory area in the anterior and superior part of the temporal lobe. Successive sounds produce signals, which are assembled as words and sentences in auditory association areas and carried to Wernicke's area for integration with visual information to produce *thoughts.* A bundle of fibers carries these signals to *Broca's* area (sometimes called the speech center) in the left side of the frontal lobe. Impulses are transmitted from it to motor neurons to regulate respiratory muscles, as well as muscles in the mouth and larynx, to produce sound.

Neurological dysfunctions in an area may be associated with a specific deficiency in language and speech. Such disorders include: in-

FIG. 11-10
Control of speech. Signals travel over circuits as indicated by the arrows in the left hemisphere. Visual impulses from written words are sent from the primary visual area (1) to Wernicke's area (3). Auditory impulses from spoken words are sent from the primary auditory area (2) to Wernicke's area (3). Wernicke's area (3) integrates the auditory and visual information, relaying impulses to Broca's area (4). Broca's area (4) assembles words in patterns and sends impulses to coordinate respiration (5) with the control of voice (6) by the larynx (voice box). Associative neurons around the primary visual and auditory areas allow recognition of words.

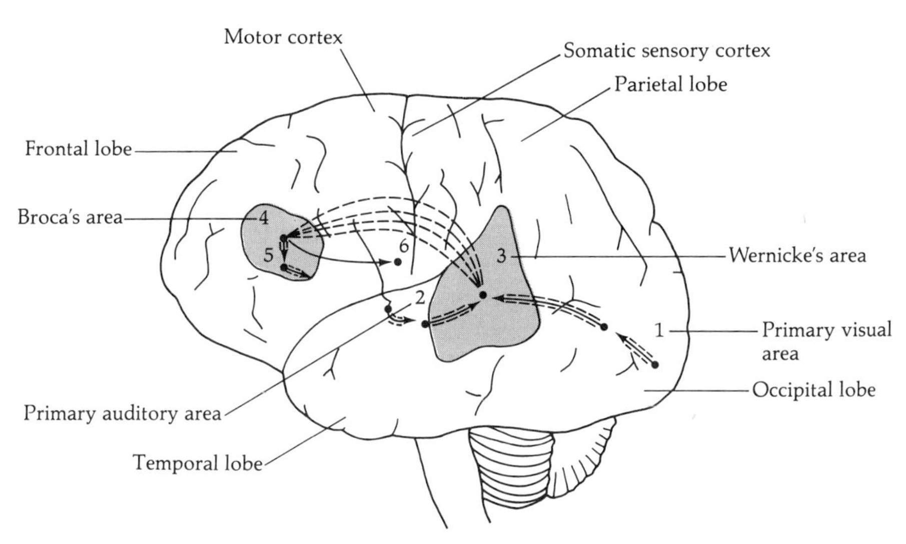

ability to understand written language and thus to read, called *alexia* or word blindness; inability to use or to understand written or spoken language, called *aphasia;* and inability to recognize familiar objects, called *agnosia.*

A type of incomplete alexia is *dyslexia,* a disorder more common in males than females but of unknown origin, and one in which the brain does not correctly organize letters and numbers. The affected person cannot read and understand more than a few lines at a time. The ability to read, write, and count may be adversely affected by letters or numbers which appear upside down, backwards, and/or transposed (such as L appearing as 7, *d* as *b,* and *tug* as *gut*).

A form of aphasia caused by disorders in Wernicke's area may produce an inappropriate choice of words and syllables. Destruction in Broca's area can result in slow, impaired, hesitant speech. Even so, individuals with hesitant speech may sing well. This apparent inconsistency is due to the fact that centers for musical creativity are on the right side of the brain, which is opposite Broca's area. Recovery from the above disorders may occur when adjacent neurons or those in nearby regions or on the opposite side of the brain assume the functions that were lost.

Learning and Conditioned Reflexes

The nervous system does more than affect muscular and glandular activity. Higher functions of judgment and language depend on learning and memory and usually are attributed to the human brain. These functions usually are poorly developed or lacking in lower animals.

Learning is the accumulation and use of neural information in the brain. It can occur without the cerebral cortex and has been observed in invertebrates, so it is not a phenomenon peculiar to humans. Advanced types of learning as reflected in language, mathematical computations, and musical creativity are associated with changes that occur mainly in the cerebral cortex and that are probably accompanied by structural and chemical alterations in its neurons.

A *conditioned reflex* is a simple example of learning. It is an acquired response that an animal may make in reacting to a stimulus. The Russian investigator Pavlov induced conditioned reflexes in dogs. A dog salivates naturally when meat is placed in its mouth. Pavlov rang a bell before feeding the animal meat. This sequence paired an artificial stimulus with one that evoked salivation naturally. He repeated this artificial stimulation sequence many times. Later, he rang the bell without feeding the dog, and the animal salivated anyway. The sound of the bell had conditioned the dog to salivate. It was a learned response evoked when the artificial stimulus was applied without the natural one. This classical experiment has been expanded by physiologists and psychologists to distinguish unconditioned reflexes (those which occur naturally in response to a normal stimulus) from conditioned reflexes (learned responses). Conditioned reflexes become an important part of human behavior and influence a tremendous variety of human body functions. Responses which are learned often result from the association of environmental signals with specific stimuli.

Pleasant or unpleasant artificial stimuli can be substituted for natural ones to elicit conditioned reflexes. A subject may *learn to respond* because of an artificial environmental signal. When the response also involves performance of a task, the reaction involves more voluntary control and is called *operant conditioning.* Experimental animals may be trained to perform tasks in this manner. Some investigators have studied the use of this principle to train monkeys or apes to perform routine tasks for people.

Theories of Memory

Memory involves retention and retrieval of information added to the brain. It may be *short-term,* lasting a few seconds; *intermediate,* lasting a matter of minutes to hours; or *long-term,* lasting for days or longer. A number of theories have been proposed to explain the development of memory. They include: (1) formation of reverberating circuits; (2) structural changes in neurons; and (3) the modification of synaptic junctions.

One theory proposes that circuits may be developed within the nervous system to mediate short-term memory. After the stimulus of learning has occurred, loops of neurons carry impulses in *reverberating neuronal circuits* (Fig. 9-8D). Information transmitted by these signals is lost when events interfere with conduction in the circuit. Thus shock, or a decrease in supply of blood to the area of the brain in which the loops are found, may cause loss of newly acquired information. For example, the kind of localized decreases in circulation associated with aging might result in the loss of short-term memory.

A second hypothesis ascribes the development of memory to several types of *structural changes* that may occur in neurons. Modifications of the surrounding neuroglia may facilitate the transmission of impulses across neuronal junctions. Repeated use of a neuron may cause new interneuronal relationships, which might result from the formation of new axonal or dendritic processes.

A third proposal theorizes that *modification of synaptic junctions* may occur as a result of learning. Functional changes may occur with or without noticeable structural alterations. The passage of signals over certain tracts may be facilitated in a phenomenon referred to as *potentiation*. An increase in transmission of impulses over a particular pathway might favor a particular response.

Theories on the Biochemical Basis of Memory

New protein and ribonucleic acid (RNA) are synthesized during the development of long-term memory in the CNS of lower and higher animals, including humans. It has been hypothesized that the nucleotide code in RNA, assembled in response to stimuli, relates to long-term memory. When animals are deprived of sensory inputs, the syntheses of RNA and protein in the brain are impeded. The neurons also may undergo other biochemical changes as well as the structural changes described above.

Stimuli may cause alterations in cyclic adenosine monophosphate (cAMP), which lead to changes in the protein structure of synaptic membranes and the development of short-term memory. Alternatively, cAMP can interact with the nuclear protein of the neurons and thereby influence the direction of new protein synthesis. These changes could result in more permanent alterations in neurons and be reflected as long-term memory.

Studies with flatworms, goldfish, and laboratory rats have produced evidence to support conflicting hypotheses of memory. Information related to human memory is needed. Which theories or combinations thereof are involved in human learning and memory remains to be established.

11-6 SUMMARY

1 Twelve pairs of cranial nerves (I through XII) originate in centers in the brain and innervate structures primarily around the head and throat. The vagus (X) sends fibers to the respiratory system, heart, and the digestive system as far down as the colon. Cranial nerves I, II, and VIII are exclusively sensory, while the others are mixed in function, including III, IV, VI, XI, and XII, which are primarily motor nerves. Cranial nerves III, VII, IX, and X contain the fibers of the cranial division of the parasympathetic nervous system.

2 The hindbrain, midbrain, and forebrain are the three major divisions of the brain. The hindbrain is the part closest to the spinal cord and consists of the medulla oblongata, pons, and cerebellum.

The medulla oblongata is connected to the spinal cord. The nuclei of cranial nerves IX to XII, as well as the cardiac, vasomotor, and respiratory centers, are located in this region of the brain and the pyramidal motor tracts cross over within it.

The pons, located just above the medulla oblongata, houses nuclei of cranial nerves V to VIII, the pneumotaxic and apneustic centers that regulate respiration, and a number of fibers that bridge other parts of the brain.

The cerebellum is posterior to the pons and contains circuits which interconnect motor centers used for speech, equilibrium, posture, and coordinated mobility. A vast number of reverberating circuits influence the neurons in the cerebellum through feedback mechanisms. The circuits include cells in cortical and medullary areas of the cerebellum. Among them are the mossy and climbing fibers, which are excitatory, as are the interneuronal granule cells and their parallel fibers which act as relays, while Purkinje fibers transmit inhibitory impulses to the cerebellar nuclei. Cerebellar impulses travel from the latter to the thalamus and brainstem to influence motor activities. In this manner, afferent signals from muscles, joints, and visual and equilibrium centers continually influence cerebellar activity. The cerebellum in turn acts as a complex, rapid computer, which precisely adjusts skeletal muscle movement. Extensive damage to large areas of the cerebellum may cause a variety of motor dysfunctions.

3 The midbrain acts as a relay area. It contains bundles of fibers running between the forebrain and hindbrain and neurons that coordinate the activity of motor paths with visual and auditory signals.

4 The main subdivisions of the forebrain are the diencephalon, which contains the hypothalamus, thalamus, and pineal body; and the telencephalon, formed by the cerebral hemispheres. The hypothalamus is situated on the floor of the third ventricle and contains centers for autonomic, neuroendocrine, and behavioral functions. Neurons of the thalamus compose a switchboard, which forms important relay stations to send sensory signals to the cerebral hemispheres. The pineal body is a neurosecretory gland, which synthesizes melatonin, regulates sexual maturation, and may provide a mechanism for biological rhythms.

5 The cerebrum is divided into halves, the cerebral hemispheres, which are linked by white commissural fibers, many of which are located in the corpus callosum. Two lobes, the insula and limbic lobes, are in the interior and four lobes, the frontal, temporal, parietal, and occipital, are on the external surface of each hemisphere. The insula receives taste signals and the limbic lobe, along with some other areas in the brain, forms a limbic system influencing emotion and behavior. The left frontal lobe governs speech, the temporal lobe controls hearing, the parietal lobe contains associative centers, and the occipital lobe contains the dominant center for visual signals.

The brain is somatotopically organized. Specific motor and sensory functions are located anterior and posterior to the central sulcus, respectively. Activities have been localized spatially in numbered regions called Brodmann's areas. Since each half of the cerebrum assumes a greater responsibility for certain functions, the human brain is called a split brain. The left half is usually dominant for control of speech, language, writing, and mathematical ability. The right half governs recognition of spatial geometry and musical creativity.

Learning is the accumulation and use of neural information in the brain. It can result in reflexes which are conditioned by environmental signals. Memory involves the retention and retrieval of information. It may result from the development of reverberating circuits, structural changes in neurons, or modifications of synaptic junctions. Changes in RNA and protein and secondary messengers within the brain may serve as the biochemical bases of memory.

REVIEW QUESTIONS

1 Name the cranial nerves and distinguish their general functions.
2 What are the major components of the medulla, and what functions do they perform?
3 What are the major components of the pons, and what functions do they perform?
4 Outline the cerebellar circuits involving mossy fibers, granule cells, parallel fibers, climbing fibers, and Purkinje cells. Explain how the cerebellum acts as a complex, rapid computer.
5 Describe the function of efferent cerebellar signals and the routes that they follow.
6 Under what conditions can the effects of cerebellar damage be observed? Describe two general types of disorders in the course of your answer.
7 What is the righting reflex, in what part of the brain is it mediated, and what impairments are caused by interference with the righting reflex?
8 What functions does the hypothalamus perform?
9 What role does the thalamus play?
10 What are the main activities of the pineal body?
11 Which lobes are located in the interior of the cerebral hemispheres, and what are their roles?
12 What are the names and major functions of the lobes located on the external surface of the cerebral hemispheres?
13 Explain what is meant by the somatotopic organization of the brain.
14 How has information gained in the study of epilepsy been used to demonstrate split-brain function?
15 How do centers for speech control interact?
16 Outline three theories that attempt to explain memory.
17 How is a conditioned reflex brought about? How does it differ from operant conditioning?
18 What biochemical changes may be involved in learning and memory?

THE PHYSIOLOGY OF MUSCLE

The Properties of Muscle Cells

12

12-1 STRUCTURE OF MUSCLE TISSUE

12-2 SKELETAL MUSCLE
- Gross anatomy
- Cell structure
- Molecular architecture
  - *Banded appearance*
  - *Nature of molecules*
    - Thin myofilament composition • Thick myofilament composition
  - *Sliding filament theory*
- The neuromuscular junction
  - *Depolarization of the muscle cell membrane*
  - *Generation of an action potential in the muscle cell*
  - *Molecular sequence of events in contraction*
- Inhibitory effects at the neuromuscular junction
  - *Botulinal toxin*
  - *Black widow spider toxin*
  - *Curare*
  - *Organophosphates*
  - *Succinylcholine*
  - *Myasthenia gravis*
- Types of skeletal muscle cells
- Relation between nerve fiber and type of muscle

12-3 CARDIAC MUSCLE
- General properties
- Cell membrane structure
- Cellular and molecular architecture
- Metabolic properties
- Electrical activity

12-4 SMOOTH MUSCLE
- General properties
- Cell membrane structure
- Cellular and molecular architecture
- Metabolic properties
- Electrical activity
  - *Type of smooth muscle*
  - *Types of potentials*
  - *External sources of stimuli*

12-5 SUMMARY

OBJECTIVES

After completing this chapter the student should be able to:

1. Compare the properties of skeletal, smooth, and cardiac muscle.
2. Describe the gross anatomy of a skeletal muscle and its connective tissue elements.
3. Describe the cellular and molecular architecture of a skeletal muscle fiber and the role of the molecular components in the sliding filament theory of contraction.
4. Discuss the properties of the neuromuscular junction and outline the biochemical and electrical events that occur during and after muscular contraction.
5. Compare the cellular and molecular architecture of skeletal, cardiac, and smooth muscle fibers.
6. Compare the metabolic and electrical properties of skeletal, cardiac, and smooth muscle.

Muscle is a contractile tissue which behaves as the body's engine, generating some movement and producing most of its heat. Its production of heat is even greater than that of the body's complex chemical factory, the liver. There are three types of muscle: skeletal, cardiac, and smooth, which contain differing amounts of contractile protein. *Skeletal muscle* is the largest mass of tissue, accounting for over 40 percent of body weight. Its ability to shorten and its attachment to bone produce voluntary body movement. *Smooth muscle* is found around hollow organs such as the digestive tract organs, blood vessels, the urinary bladder, and the uterus. In the digestive tract it moves food and undigested residues (feces) involuntarily in response to internal and external stimuli. *Cardiac muscle* is only located in the walls of the heart and is used to propel blood.

12-1 STRUCTURE OF MUSCLE TISSUE

The general structure of the three types of muscle tissue was briefly introduced in Chap. 3, and is illustrated in Fig. 12-1. Skeletal muscle cells are striated, cylindrical, multinucleated, and under voluntary nervous control. They are found parallel to each other and are anchored to the bony skeleton by collagen fibers that form tendons. Cardiac muscle cells are striated, multinucleated, branched, cylindrical, and under involuntary (autonomic) control. Smooth muscle is made up of nonstriated, spindle-shaped cells with a single nucleus and is under involuntary control. Each smooth muscle cell is wide at the center and tapered at each end. The properties of these three types of tissues are compared in Table 12-1.

12-2 SKELETAL MUSCLE

Gross Anatomy

There are over 600 skeletal muscles in the body which work in coordinated fashion directed by the nervous system to provide voluntary body movement. Skeletal muscles are composed on the average of 75 percent water, 20 percent protein, and 5 percent of a mixture of other inorganic and organic molecules. The number of cells in each muscle ranges from a few hundred to several hundred thousand.

The cells are called *muscle fibers* because they resemble the fibrous elements of connective tissue. Each muscle cell is surrounded by a cell membrane, the *sarcolemma* (from Greek *sarc*, flesh, and *lemma*, peel). The sarcolemma is enveloped by a connective tissue coat, the *endomysium* (Fig. 12-2). Clusters of 12 or more cells are grouped in variously sized bundles, called *fasciculi* (singular, fasciculus). A connective tissue layer referred to as the *perimysium* encases each bundle. Groups of fasciculi are held together by a third connective tissue coat, the *epimysium* (often called the *fascia*). The epimysium surrounds the whole muscle and is continuous with the collagenous fibers of tendons, which connect the muscle to bone.

A muscle is normally attached by tendons at each end to two different bones. Tendons are noncontractile connective tissue elements which transmit the force of muscular contraction to the bone. Tendons perform their function mainly through collagen fibers which are

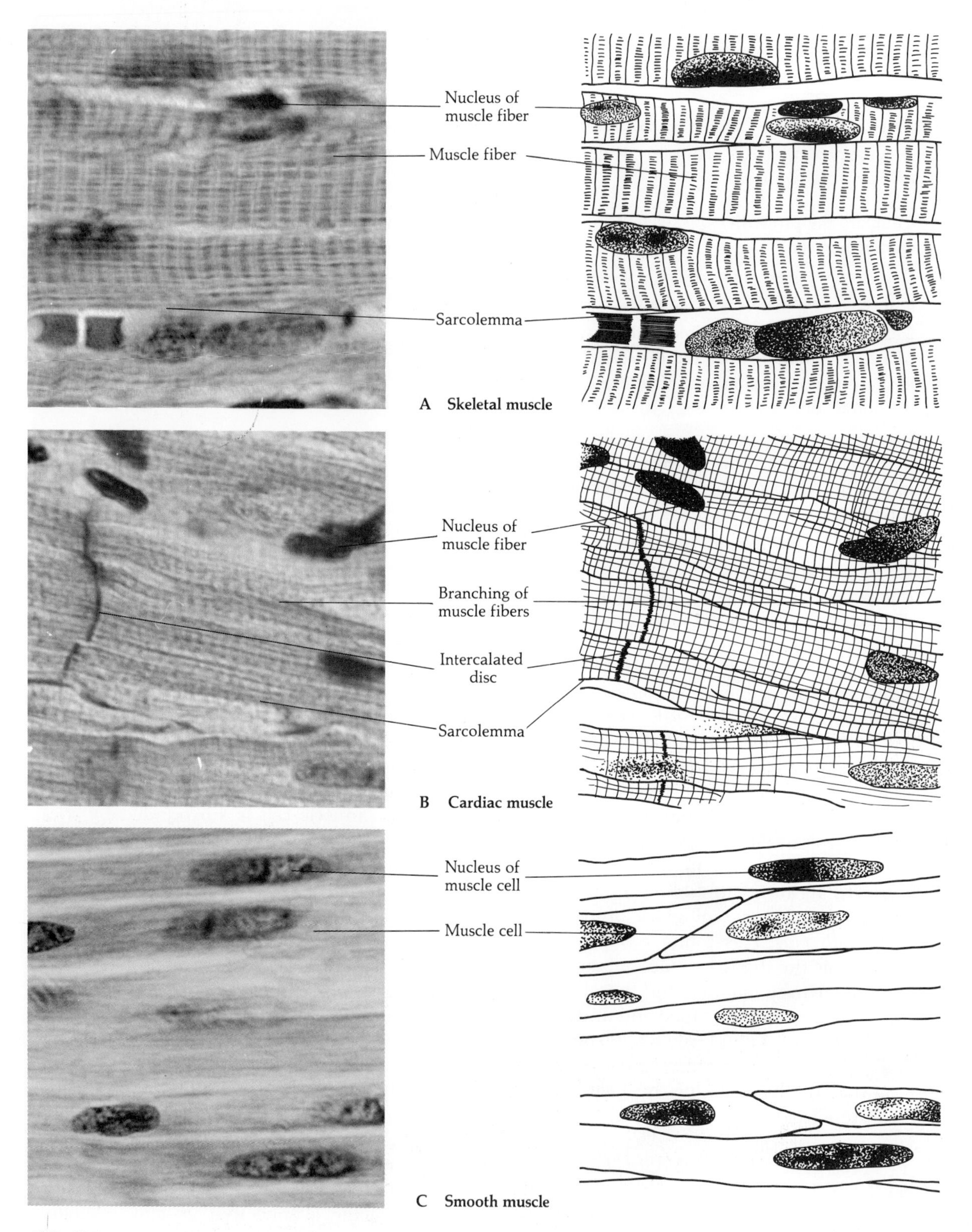

FIG. 12-1
Microscopic appearance of muscle. The photomicrographs on the left are drawn on the right. (A) Skeletal muscle (×1700). The banded pattern of skeletal muscle is apparent. (B) Cardiac muscle (×1700). Note the intercalated discs at the cell junctions. (C) Smooth muscle (×1700).

found in epimysium and are implanted in the surface of the bone. When a muscle contracts, it moves only one of two bones to which it is attached. The part of the muscle attached to the tendon and bone which does not move is called

TABLE 12-1
General properties of skeletal, cardiac, and smooth muscle

Property	*Skeletal*	*Cardiac*	*Smooth*
Location	Under skin, usually attached to bone	Heart	Walls of hollow organs (e.g., digestive tract, bronchial tubes, blood vessels, bladder, uterus)
Percent of body weight	40–50	0.5	3
Cell Structure			
Nucleus	Multinucleated	Multinucleated	Single
Location of nucleus	Peripheral	Central	Central
Shape	Cylindrical	Cylindrical	Spindle
Appearance	Striated	Striated	Smooth
Size (μm)			
Width	10–100	9–20	2–20
Length	1000–300,000	100	10–500
Amount of contractile protein (actomyosin, in mg/100 g wet weight of tissue)	100	60	6

the *origin*. The end that moves the bone to which it is attached is called the *insertion*.

Cell Structure

In early development, the fusion of embryonic cells, called *myoblasts*, produces an anatomic unit consisting of a multinucleated, cylindrical skeletal muscle cell, called a *syncytium*. Each unit is surrounded by a cell membrane, the *sarcolemma*, a portion of which is modified to form the motor end plate.

Protein molecules aggregated in groups of filaments lie parallel to the cell surface. The units, the *myofilaments*, visible only under the electron microscope, contain contractile proteins. Myofilaments are grouped together to form strands called *myofibrils* which are visible under

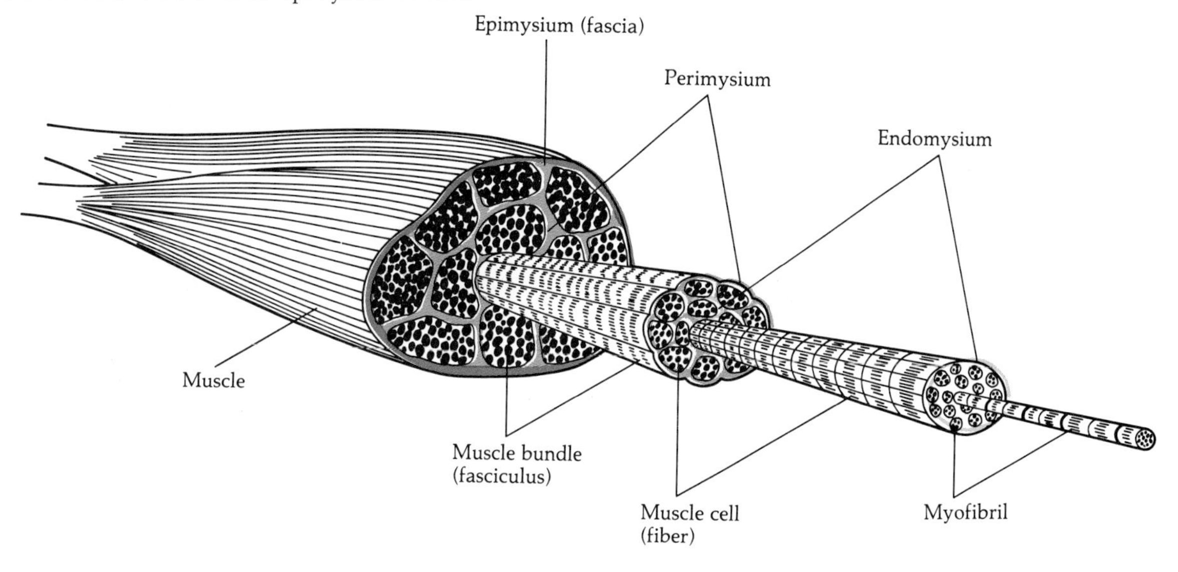

FIG. 12-2
General anatomy of a skeletal muscle. Connective tissue coats surround individual muscle cells as the endomysium, bundles of cells as the perimysium, and the muscle as a whole as the epimysium or fascia.

the light microscope, and are 1 to 3 μm in diameter, yet fill up to 80 percent of the interior of the cell. Hundreds to hundreds of thousands of myofibrils give skeletal muscle cells of varying sizes a striated (striped) appearance under the light microscope.

Skeletal muscle cytoplasm or *sarcoplasm* contains the usual cytoplasmic organelles. Large numbers of mitochondria are often present between the myofibrils. Energy is stored in the form of carbohydrate as glycogen granules, which are commonly found between the myofibrils. Lipid droplets are also found in the cytoplasm and are often near the relatively conspicuous mitochondria. Muscle cells also contain Golgi complexes, ribosomes, and other organelles that are characteristic of most types of cells.

The sarcoplasmic reticulum in skeletal muscle cells is a highly specialized and modified form of the smooth endoplasmic reticulum of other cell types. It exists as a system of tubules between and parallel to the myofibrils (Fig. 12-3). The tubules flare at right angles into enlarged chambers or cisterns at the boundary of each functional unit of contraction, the *sarcomere.* The tubules have small openings (fenestrations, or windows) near their midpoints.

FIG. 12-3
The ultrastructure of mammalian skeletal muscle.

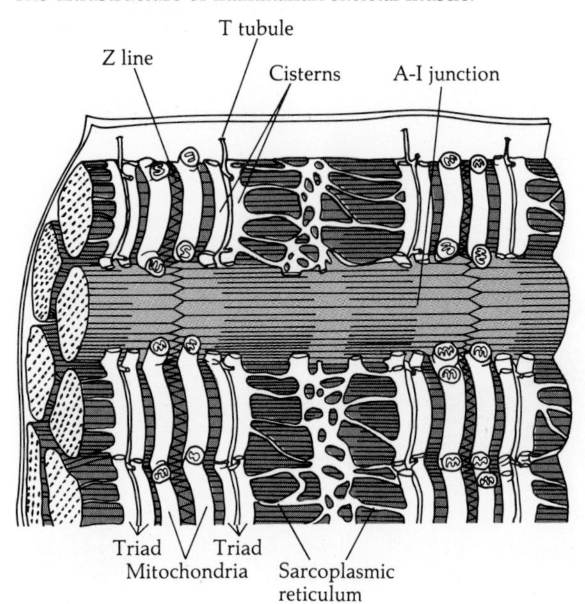

A second and separate system of transverse tubules or T tubules project inward and perpendicular as extensions of the sarcolemma. They seem to provide paths of low electrical resistance for the conduction of impulses to the interior of the muscle cell. The T tubules and the adjacent cisterns of the sarcoplasmic reticulum form repeating three-chambered structures called *triads.* A triad consists of three components, the T tubule, which extends perpendicular to the myofibrils along the A-I junction, and a cistern on each side of it (Fig. 12-3). Action potentials pass from one membrane to the other within the triad so that as each membrane is depolarized it develops an increased permeability to the flow of ions.

Molecular Architecture

Sarcomeres are the functional units of contraction in skeletal muscle. The length of a sarcomere is about 2 to 3 μm (0.002 to 0.003 mm) and there are a total of about 36 million of them in the average skeletal muscle cell. The relationship between the whole muscle and its cellular and molecular components is illustrated in Fig. 12-4.

Banded Appearance

The stained myofibrils of skeletal muscle cells have a cross-striated appearance under the microscope (Fig. 12-4, B through D). They can be viewed with polarized light, whose waves have uniform characteristics. The refraction (bending back) of polarized light indicates the homogeneity (similarity) or heterogeneity (differences) of samples viewed under the polarized light microscope. Such an instrument limits the vibrations of light waves to one plane by passing them through a quartz prism, a transparent solid which bends the rays of light. When the molecular composition of a structure being viewed is symmetric, polarized light is bent equally by it in all directions. The structure is *isotropic* and appears dark when viewed through the microscope since less light passes through the eyepiece. When the molecular composition of the structure being viewed is

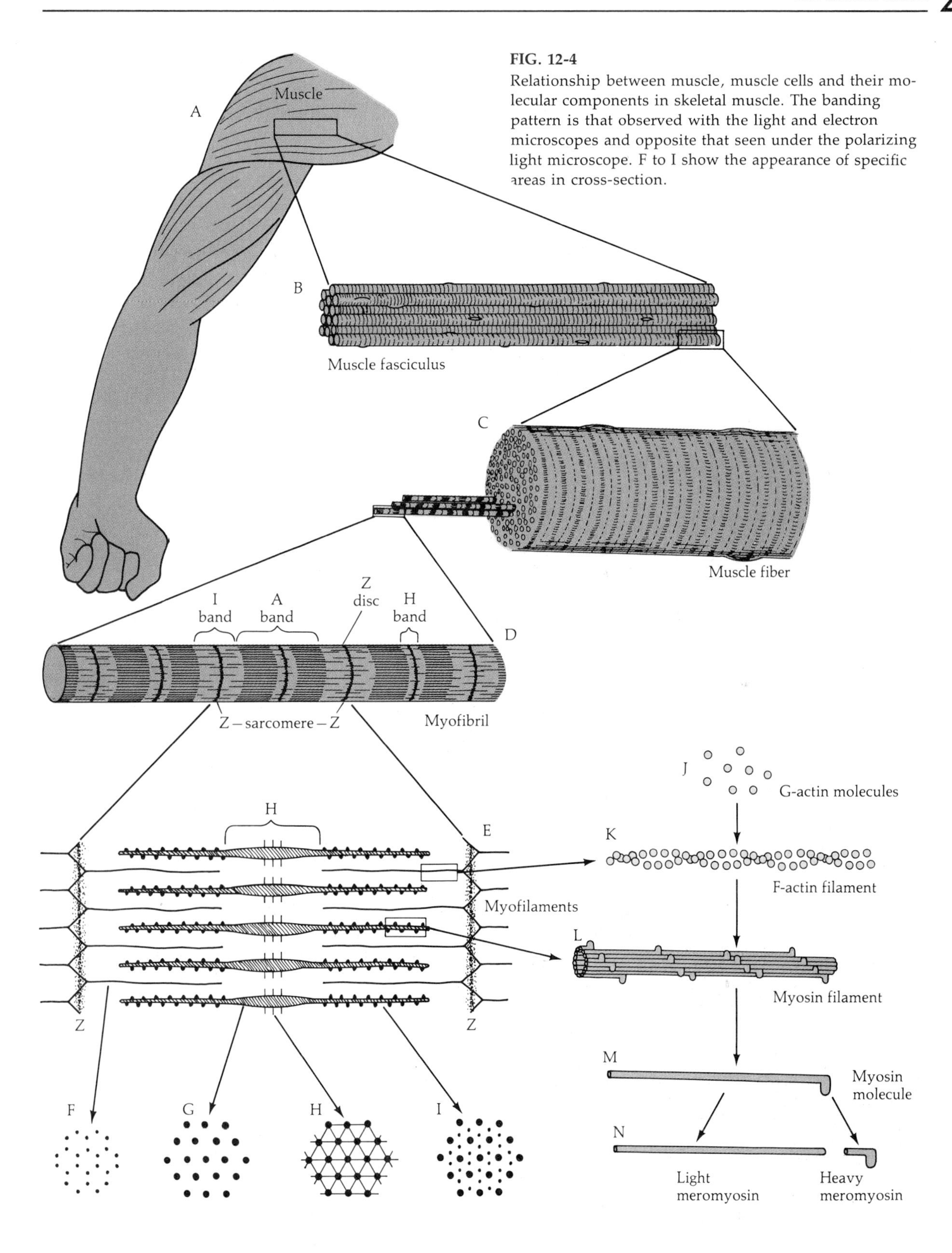

FIG. 12-4
Relationship between muscle, muscle cells and their molecular components in skeletal muscle. The banding pattern is that observed with the light and electron microscopes and opposite that seen under the polarizing light microscope. F to I show the appearance of specific areas in cross-section.

asymmetric, polarized light is bent in different directions. The structure is *anisotropic* and appears light when viewed through the microscope. It is as if a linear beam of light were split into two light rays perpendicular to each other, causing some of the light to pass through the eyepiece of the microscope to the viewer's eye.

Those bands in muscle which bend polarized light equally in all directions (isotropically) are dark and are called *I bands.* Those which do so unequally (anisotropically) are light and are called *A bands.* The appearance of the two is reversed in ordinary light or under the electron microscope. The alternating dark and light patterns are due to the regular repeating structure of molecules of varying mass and density which form filaments within the myofibrils (Fig. 12-4D).

I bands are composed of thin myofilaments which, in turn, are made up primarily of a protein, *actin*. The center of each I band is bisected by a *Z line* or disc (Fig. 12-4D and E). Each sarcomere extends from Z line to Z line.

A bands are composed of thin and thick myofilaments. The mixture of the two accounts for the unequal refraction (birefringence or double refraction) of polarized light. The thick myofilaments are made up of the protein *myosin*. The center of the A band is composed solely of myosin in an area called the *H zone* (band). A darker

FIG. 12-5
Molecular architecture of the thin myofilament.

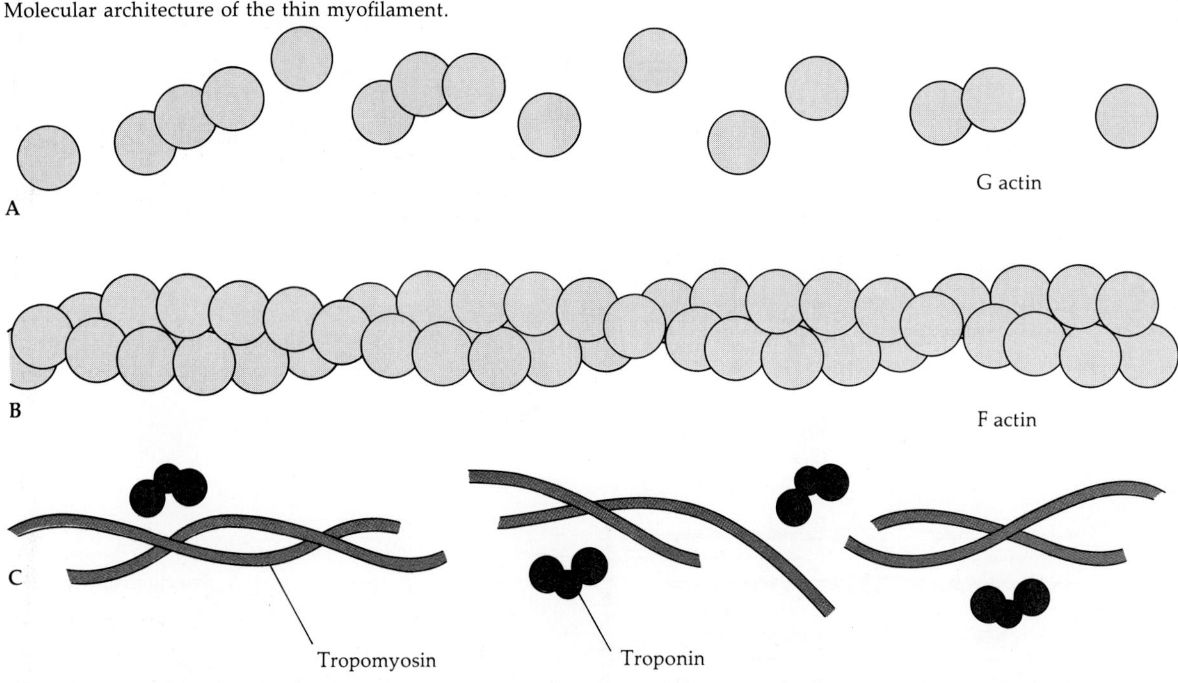

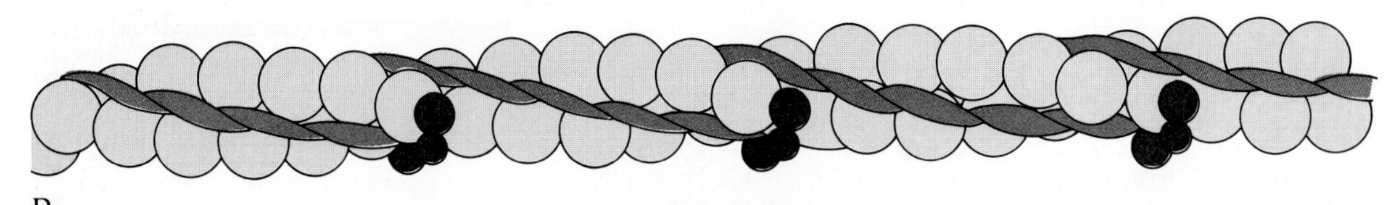

bulge formed by myosin in the middle of the H zone is referred to as the *M line* or M band (Fig. 12-4E).

Nature of Molecules

THIN MYOFILAMENT COMPOSITION A thin filament is a fibrous structure composed of about 300 to 400 molecules of globular *G-actin* polymerized to produce two coiled strands of fibrous *F-actin,* 40 to 60 cablelike strands of *tropomyosin,* and 40 to 60 spheres of *troponin.* Globular G-actin molecules, with a 42,000 MW, can be separated as individual units (monomers) in the laboratory. In intact muscle, however, they are united in repeating units of 6 to 7 spheres to form a double twisted strand of F-actin.

Tropomyosin molecules (68,000 MW) are linked end-to-end and form a long, coiled cable lying in two grooves of the twisted strands (helices) of F-actin (Fig. 12-5D).

Troponin is a spherical protein of about 69,000 MW which binds to tropomyosin at intervals of approximately every 6 to 7 G-actins (Fig. 12-5). Troponin is formed by a complex of three individual globular polypeptides which bind to tropomyosin, actin, and Ca^{2+} respectively.

THICK MYOFILAMENT COMPOSITION Each thick myofilament is composed of myosin protein molecules of 460,000 MW. A myosin molecule is 120 nm long and shaped like a lollipop stick with two globular pieces of candy bent at sharp angles at one end (Fig. 12-6A). Each of these two globular heads, connected to a long rod in the form of a double coiled tail, forms heavy meromyosin. Each globular head is associated with two smaller light polypeptide chains.

Each globular head of heavy meromyosin contains a sequence of amino acids which produces a site with ATPase activity. In the presence of its associated pair of light chains, part of each globular head also forms a site which can complex with actin. Aggregations of myosin molecules (about 400 in number) make up a thick myofilament, as illustrated in Fig. 12-6B.

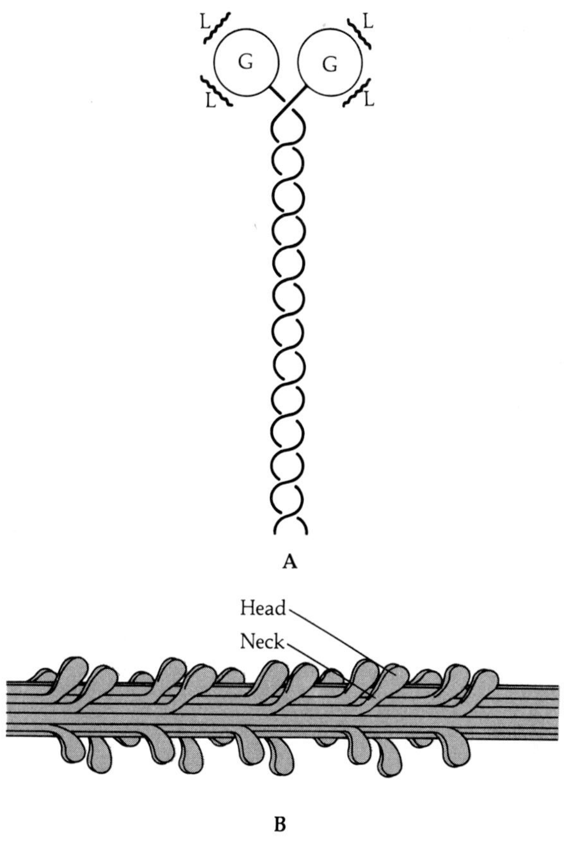

FIG. 12-6
Molecular architecture of the thick myofilament. (A) Diagram of myosin molecule: G = Globular heads; L = Light polypeptide chains. (B) Aggregations of myosin forming a thick myofilament.

Sliding Filament Theory

The overall composition of the bands of the myofibrils can now be described and related to a theory of muscle contraction. The I bands are composed of thin myofilaments which consist of actin, tropomyosin, and troponin. The A bands are composed of the thick myofilaments containing myosin and the overlapping thin myofilaments (Fig. 12-7). A bulge, the M-line, is formed by a special protein, the M-protein, which cross-links tails of adjacent myosin molecules and helps keeps them aligned (Fig. 12-4E). The area of the A band which is composed solely of the thick myofilaments is called the H zone.

When muscle contracts or stretches, microscopic changes can be observed in the banding pattern. As muscle contracts, the A band remains constant in length, but its H zone narrows and eventually disappears, as illustrated in Fig. 12-7A and B. As muscle is stretched, the H zone widens (Fig. 12-7A versus 12-7C).

The sliding filament theory is the most widely accepted hypothesis to explain skeletal muscle contraction. According to this theory, muscle shortens during contraction as the thick and thin filaments slide past each other (compare the right side of Fig. 12-7A and B).

The globular meromyosin heads are helically arranged around the thick myofilaments (Fig. 12-8). They are oppositely directed on the left and right ends of the filament. They can change their conformation (shape) and bind to actin on adjacent thin filaments. During contraction, they cause the thin filaments of adjacent I bands to slide toward one another within a sarcomere.

The sites of union between the globular heads of myosin of the thick myofilaments and actin of the thin myofilaments are called *cross-bridges*. Myosin moves from an angular or *r* (rigor) cross-bridge position to a more perpendicular or *p* position (compare Fig. 12-8A with B). As this occurs, the myosin slides the thin filaments of adjacent I bands toward the center of each sarcomere. The cross-bridges are then broken, and the globular myosin heads return to their *r* positions to bind to different sites on the actin strands. The myosin heads on each side of an A band are oriented in opposite directions and seem to rotate or swivel toward and away from each other. As a result of repetition of such movements during contraction of muscle, the H zone within the A band disappears as the thin filaments move towards the center of the sarcomere and the width of the sarcomeres (Z line to Z line) decreases (compare Fig. 12-7A and B).

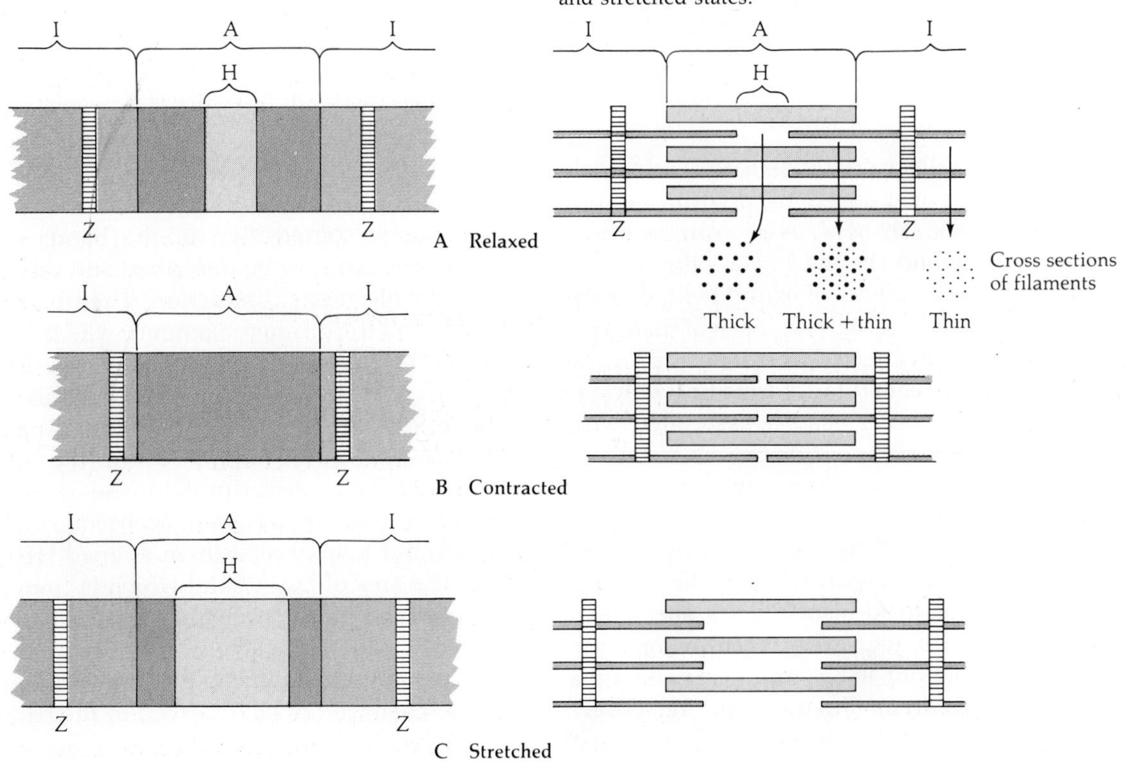

FIG. 12-7
Model of thin and thick filaments in (A) the relaxed, (B) contracted, and (C) stretched states. The diagrams on the left show the change in banding patterns. Note the disappearance of the H zone and narrowing of the width between the Z lines during contraction, and the opposite response as the muscle is stretched. The diagrams on the right illustrate the movement of the filaments when a muscle fiber changes from the relaxed to the contracted and stretched states.

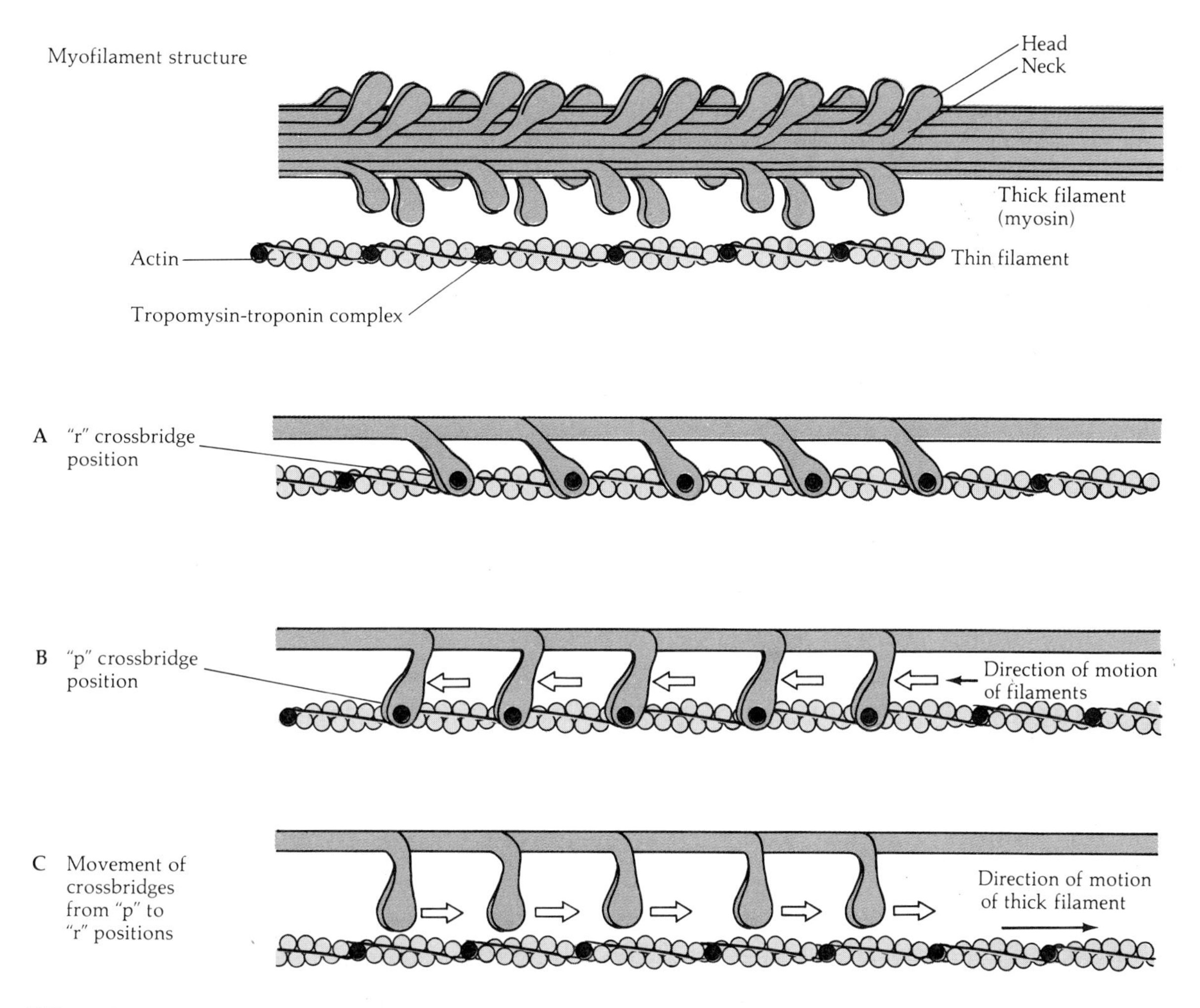

FIG. 12-8
Events in the movement of the myofilaments according to Huxley's sliding filament theory. The successive diagrams labeled A through C indicate the manner in which thick myofilaments slide thin myofilaments after forming crossbridges with them. The movement is made possible by the entry of Ca^{2+} ions into the sarcoplasm, which relieves the inhibition of myosin-ATPases. The latter hydrolyze ATP to provide energy that changes the conformation (shape) of the myosin which makes up the thick myofilaments.

The Neuromuscular Junction

The contraction of skeletal muscle is under the voluntary control of nerve fibers. Motor neurons and muscle cell membranes form synapses called *neuromuscular junctions.* The terminal endings of each axon approach, often branch out, and form junctions with several muscle cells. Each skeletal muscle cell is supplied with its own axon terminal. The localized region of the muscle cell membrane at the junction is called the *motor end plate.* The axon at this site has no myelin sheath. The basement membranes of the two cell types are separated by a synaptic cleft of about 50 nm (Fig. 12-9A, B, and C).

A single neuron, its axonal branches, and the muscle cells with which it synapses form a functional entity called a *motor unit.* Anywhere from three to a few hundred muscle cells may be included in a motor unit. Muscles whose activity is finely regulated, such as those attached to the eyeball, have a smaller number of muscle cells in their motor units than do those which are more coarsely controlled (such as the calf muscle, the gastrocnemius, in the leg). Large numbers of mitochondria and synaptic vesicles (usually 100 to 200) lie within the presynaptic membrane of each axon terminal. Many mitochondria also lie near the postsynaptic membrane of the muscle cell (Fig. 12-9C). Their presence is indicative of the need for energy to carry out functions at the synapse.

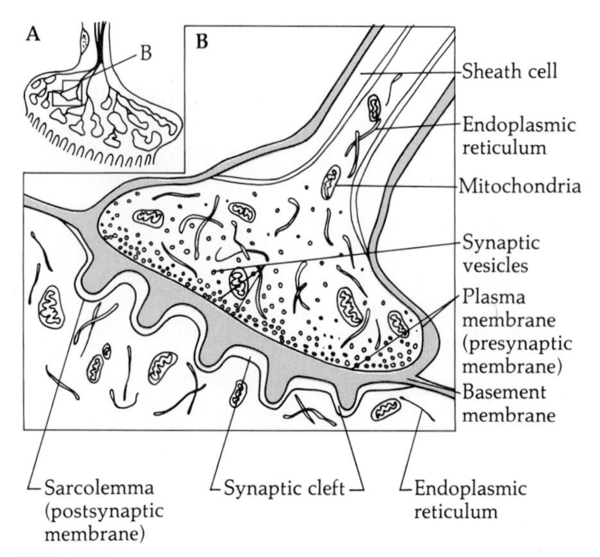

FIG. 12-9
The neuromuscular junction. (A) Diagram (×1500). (B) Enlargement of section outlined in color in A. (×22,500). (C) Electron micrograph of frog neuromuscular junction (× ~ 80,000).

Depolarization of the Muscle Membrane

As an action potential passes through the presynaptic membrane of the axon, it promotes the entry of Ca^{2+} ions into that membrane, and the secretion (exocytosis) of acetylcholine (ACh) from its synaptic vesicles. ACh diffuses across the synaptic cleft and may bind to receptors concentrated on the *motor end plate* (postsynaptic membrane) of the muscle cell. A sequence of events is initiated by depolarization of the motor end plate (Fig. 12-10).

Like neurons, muscle cells are excitable. The distribution and movement of ions across skeletal muscle cell membranes resembles that of neurons, as described in Chap. 7. ACh combines with receptors on the muscle cell membrane to open ion channels. This is followed by the entry of Na^+ ions which were initially in excess on the exterior of the resting muscle cell. Na^+ ions diffuse into the cell as directed by their equilibrium potential. The net entry of positive ions depolarizes the motor end plate and generates a change in voltage called an *end-plate potential* (*EPP*). The end-plate potential is usually sufficient to attain a threshold which depolarizes the entire length of the muscle cell membrane. The polarity of the membrane decreases as it is depolarized. The potential changes from a resting value of -90 mV to a value above 0 mV as an action potential develops.

This change in potential is called *excitation* and usually is associated with (coupled to) changes in the muscle (such as cross-bridge activity) that produce tension and *contraction*. The whole process is called *excitation-contraction* coupling. EPPs that develop at a neuromuscular junction resemble, yet differ from, potentials formed at interneuronal synapses. An action potential which develops in a motor neuron travels to all its axon terminals, causing an EPP to be produced in all skeletal muscle cells in the motor unit. Excitatory postsynaptic potentials (EPSPs), which develop on the postsynaptic membranes of neurons, may summate, whereas EPPs, which develop on muscle cell membranes, do not. A single EPP is generally greater than the threshold required to develop an action potential in the muscle cell membrane. Unlike synaptic potentials, which may be excitatory or inhibitory postsynaptic potentials (EPSPs or IPSPs), EPPs are always excitatory.

Generation of an Action Potential in the Muscle Cell

After their release and diffusion across the neuromuscular junction, neurotransmitter molecules may bind to receptors which are localized in the motor end plate of the postsynaptic membrane of the muscle cell. Ion channels are opened and EPPs are generated. The amplitude

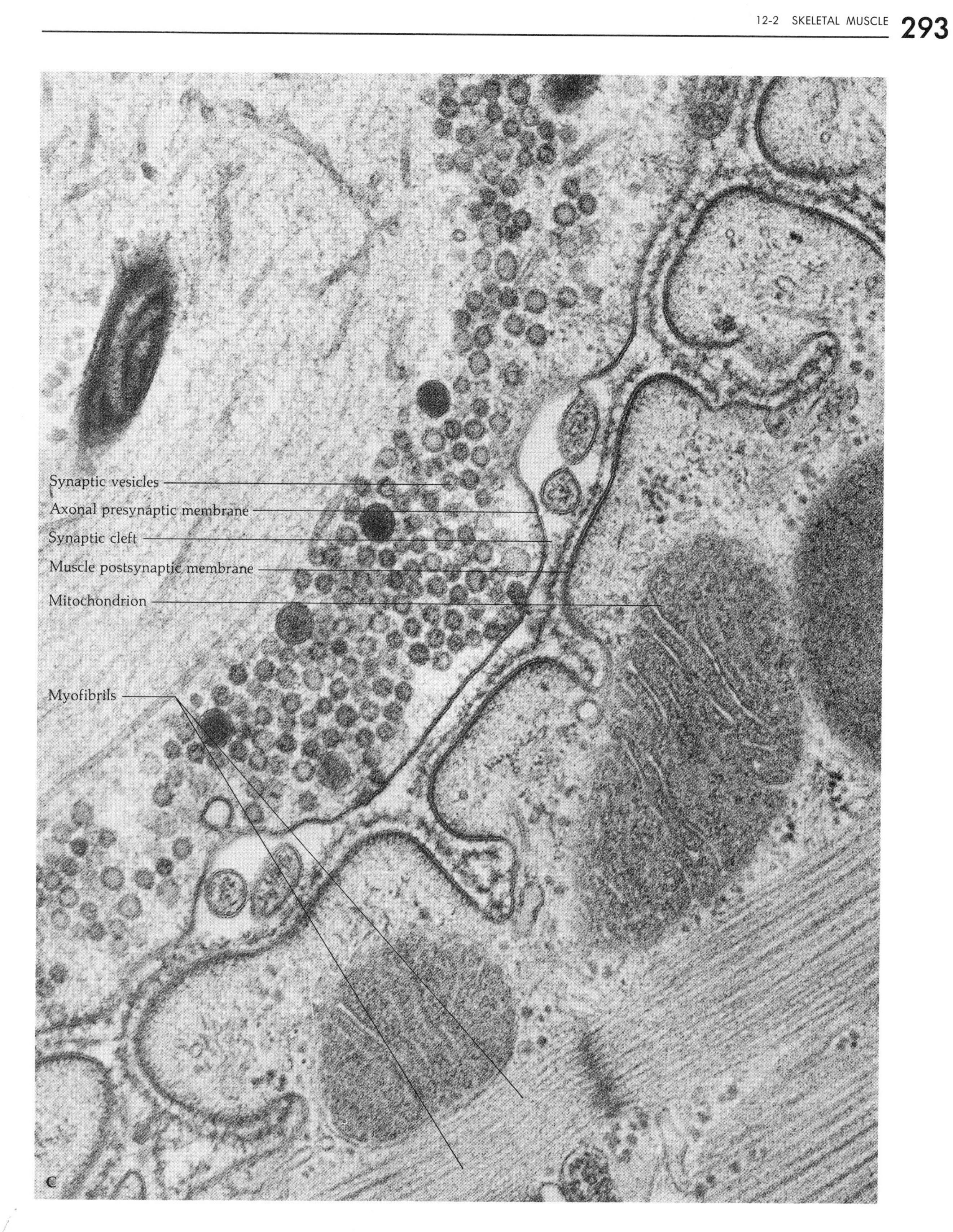
Synaptic vesicles
Axonal presynaptic membrane
Synaptic cleft
Muscle postsynaptic membrane
Mitochondrion
Myofibrils
C

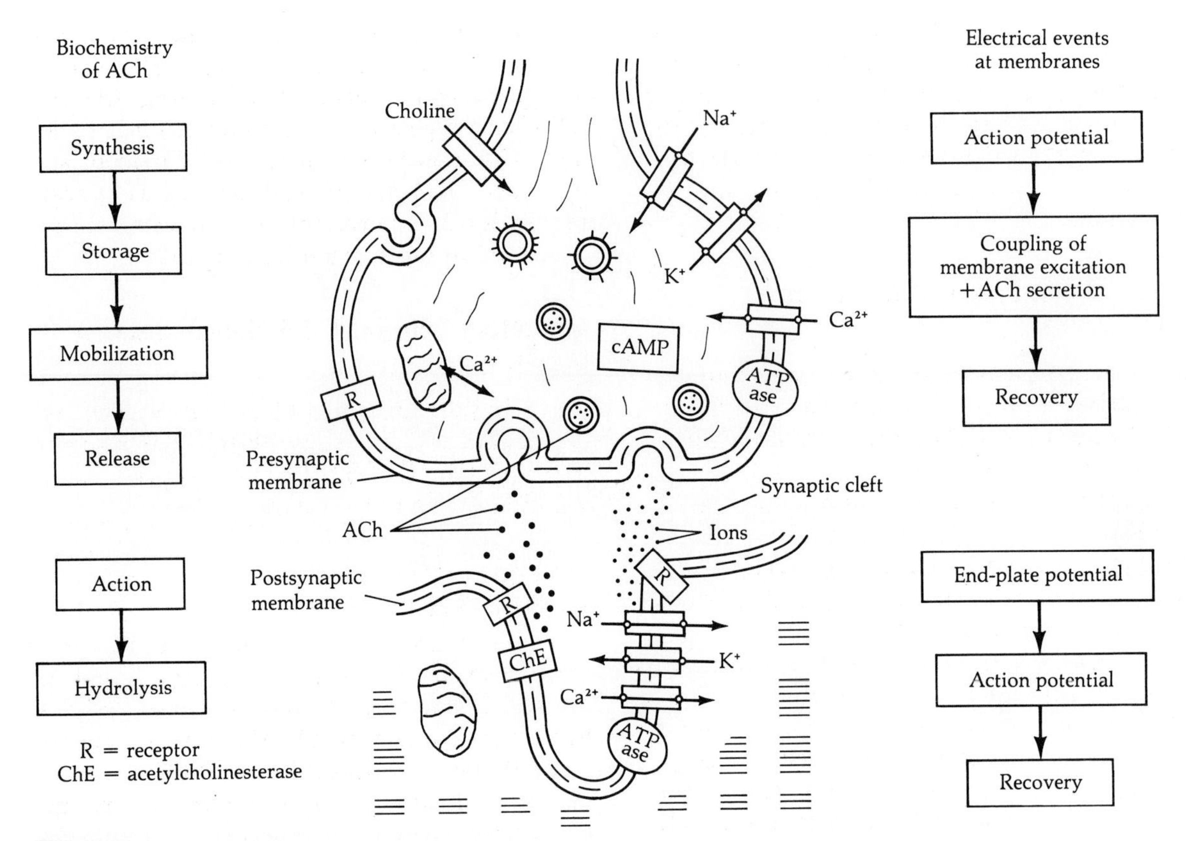

FIG. 12-10
Biochemical and electrical events in neuromuscular transmission. The arrows in the diagram indicate the net direction of transport of specific substances.

of the response of the end-plate membrane increases in proportion to the number of ACh molecules bound to the receptors. When a sufficient number of ACh molecules adheres to the muscle cell membrane, the EPP causes an action potential to develop in the muscle cell.

Apparently, two ACh molecules must bind to receptors to open a single ion channel. Prior to gaining this opportunity, some ACh molecules are destroyed by cholinesterase and others are returned to the axon terminal. However, each synaptic vesicle contains enough ACh (about 10,000 molecules) to open 2000 channels. Since about 100 to 200 vesicles are available, approximately 100,000 to 200,000 ion channels can open in the muscle cell membrane. Enough ACh opens ion channels to allow a net flow of about 3 billion ions (such as Na^+) across the muscle membrane within 0.3 ms (millisecond). This response results in the development of an EPP. Usually, each EPP is of sufficient intensity to cause an action potential. The latter travels at a speed of 5 m/s and requires about 2 to 5 ms to pass along the entire length of the muscle cell membrane. The action potential then causes changes within the cell that lead to contraction.

After its release ACh has three fates (see Fig. 10-10). Most of it is retrieved by the vesicle (this is called *reuptake*). About two-thirds of the remainder binds to receptors and later diffuses from them. It is degraded along with the balance of the ACh molecules, which remain in the synaptic cleft. ACh is hydrolyzed (degraded) by the enzyme *acetylcholinesterase* (cholinesterase, or ChE). Each molecule of ChE can hydrolyze nearly 25,000 ACh molecules per second. The connective tissue fibers near the postsynaptic membrane possess about one molecule of ChE for every ACh receptor on the membrane.

The degradation of ACh occurs within 5 ms of its release from the synaptic vesicles. The degradation and reuptake of ACh provide the muscle fiber with an opportunity to restore its resting potential (be repolarized) and be readied for another sequence of steps that lead to contraction.

Molecular Sequence of Events in Contraction

The molecular basis of skeletal muscle contraction follows the series of steps outlined in Fig. 12-11. The release of ACh from a motor neuron and its adherence to receptors at the neuromuscular junction cause an EPP to develop on the sarcolemma. The EPP generates an action potential which depolarizes the sarcolemma, the T tubules, and the sarcoplasmic reticulum in each muscle cell of the motor unit.

Depolarization of the sarcoplasmic reticulum increases its permeability and allows Ca^{2+} ions stored in the sarcoplasmic cisterns to diffuse into the sarcoplasm of the muscle cell. The accumulation of positively charged Ca^{2+} ions around the myofilaments attracts oppositely charged groups in troponin and pulls troponin's three subunits toward one another. The change in shape allows fibrous tropomyosin to roll deeper into a groove found between the coiled double strands of fibrous actin. As a result, sites on actin are exposed to which the globular heads of myosin can bind to make cross-bridges.

The Ca^{2+} ions not only unmask the sites to which myosin binds but they also activate (release from inhibition) the ATPase active sites within myosin. Myosin-ATPases rapidly hydrolyze ATP and release energy to change the conformation of myosin. The energy is also used to make cross-bridges between actin and myosin and to break and re-form them in new locations. In this way, muscles shorten in *isotonic contraction* as the globular heads of myosin slide the actin filaments within a sarcomere toward each other.

Active transport enzymes, the Ca^{2+}-ATPases, are found on the surface of the membranes of the sarcoplasmic reticulum. They actively transport Ca^{2+} ions back into the cisterns for storage. As the Ca^{2+} levels available to the myofibrils decline, troponin regains its original shape, tropomyosin again blocks the binding site on actin, and myosin returns to its original position. The muscle is restored to its precontraction state. The role of ATP, as well as energy production and consumption in muscle, will be described in more detail in Chap. 13.

Inhibitory Effects at the Neuromuscular Junction

A number of substances block the sequence of events outlined in Fig. 12-11 and thus prevent contraction of skeletal muscle. Many do so by affecting the activity of ACh.

Botulinal Toxin

Botulinal toxin is a protein poison released as an exotoxin from the bacterial cells of *Clostridium botulinum*. This microorganism is found in the soil in the form of spores, which are extremely heat-resistant. Therefore, when inadequate thermal processing is used in the preservation of foods, the spores survive. In the absence of oxygen and especially in a nonacid environment, these spores can germinate, multiply, and produce active cells which secrete toxin. Home-canned vegetables are the most common sources of botulism.

After the contaminated food is eaten, the toxin is absorbed through the intestinal epithelium and delivered through the bloodstream to the tissues of the body to cause food poisoning called *botulism*. As little as 0.1 μg of the toxin is lethal to humans. The toxin combines with the terminal endings of neurons and blocks the release of ACh (step B, Fig. 12-11). ACh accumulates in the axon but does not diffuse from the synaptic vesicles to the motor end plate of the muscle. Paralysis of the muscle results. Should the toxin affect neurons that supply the muscles of the diaphragm, respiratory paralysis and death may follow.

Black Widow Spider Toxin

The venom from the black widow spider enhances the production and/or release of ACh

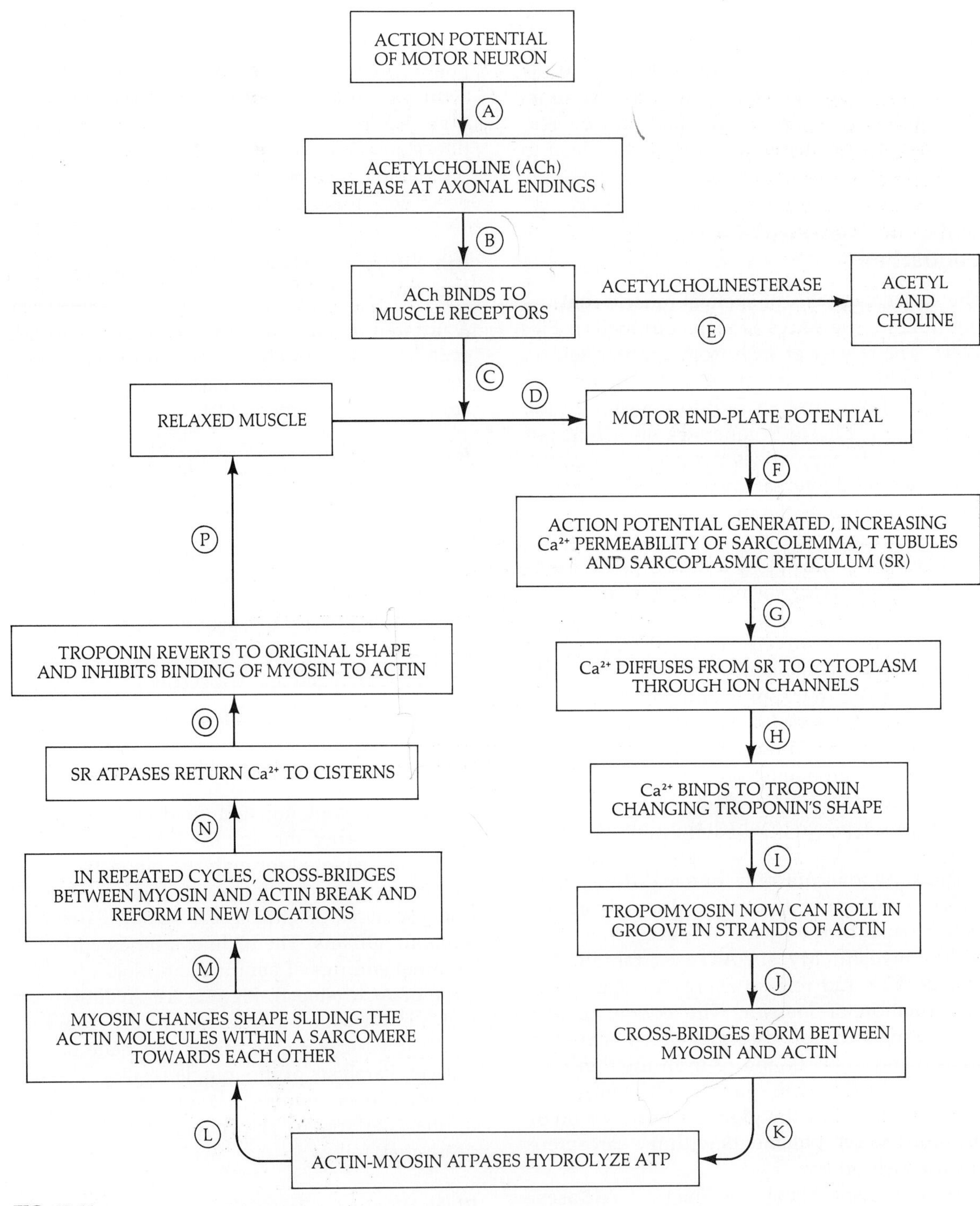

FIG. 12-11
Molecular events leading to contraction. The sequence of events which leads from contraction back to relaxation is traced by steps A through P. SR = sarcoplasmic reticulum.

(step B, Fig. 12-11). The permeability of muscle cells increases and restoration of ionic equilibrium (repolarization of the membrane or motor end plate) is difficult, if not impossible.

Curare

Curare binds to ACh receptors, competing with ACh for its binding sites on the membrane (step C, Fig. 12-11). Because this drug does not behave as does the neurotransmitter, neuromuscular activity is blocked. Curare may also make ACh resistant to degradation by ChE and may produce conditions which prevent repolarization of the muscle cell membrane (step E, Fig. 12-11).

Because the effects of curare are reversible, it may be administered with great care as a muscle relaxant during surgery. However, large amounts of the drug inhibit skeletal muscle contraction, causing respiratory paralysis and death. For preparation of weapons, South American Indians dip their blowgun darts and arrowheads in curare. The slightest scratch produced by these weapons is lethal, since it leads to paralysis of respiratory muscles and causes asphyxiation.

Organophosphates

Organophosphates are organic compounds containing phosphorus. Diisopropyl fluorophosphate (DFP), the insecticide parathion and a variety of other toxic compounds, including those used as nerve gases in chemical warfare, are examples. They act as anti-ChEs and block step E in Fig. 12-11. As a result, ACh persists longer to prevent restoration of ionic equilibrium. The response can lead to respiratory paralysis and death.

Succinylcholine

Succinylcholine is an organic compound which depolarizes muscle. It activates step F in Fig. 12-11 and causes muscle contraction for several seconds prior to paralysis. The drug is rapidly degraded metabolically. When the proper dose is administered as a muscle relaxant in preparation for surgery, paralysis lasts for only a matter of 5 to 10 min.

Myasthenia Gravis

Myasthenia gravis is an autoimmune disease characterized by muscular weakness. Its cause is unknown but it is often accompanied by a disorder of the thymus gland. (The thymus is a source of T lymphocytes, which provide defenses against disease, as noted in Chaps. 3 and 19.) In myasthenia gravis the individual forms antibodies against his or her own ACh receptors. The antibodies bind to the receptors and initiate reactions that degrade them. Normal neuromuscular function is effectively blocked at step C in Fig. 12-11. Current therapy for the disorder involves removal of the thymus and administration of anticholinesterases and anti-inflammatory corticosteroid hormones. Ninety percent of patients so treated improve.

Types of Skeletal Muscle Cells

Most skeletal muscles are composed of a mixture of three types of muscle fibers called *fast-twitch glycolytic* (*A*), *slow-twitch* (*B*), and *fast-twitch oxidative* (C) fibers (Table 12-2). Their proportions vary with the activity of the muscle. Fast-twitch glycolytic A fibers predominate near the surface of the body. They are pale (white) in color owing to a low myoglobin content. Myoglobin is a protein pigment which colors muscle and binds oxygen that diffuses to it from the extracellular fluid (ECF) and blood vessels. Muscles deep within the body usually contain large amounts of myoglobin and have abundant numbers of dark fast-twitch oxidative C fibers.

Fast-twitch glycolytic A fibers (also called *fast-white fibers*) derive most of their energy from metabolic pathways that do not require oxygen (anaerobic pathways). They contract and fatigue rapidly because anaerobic pathways do not allow complete combustion of foodstuffs. These fibers generate forces of high intensity for short durations such as those produced during weight lifting. They are also used in skilled hand movements and in movements of the eyes.

Fast-twitch oxidative C fibers are also called *fast-red fibers*. They derive most of their energy from metabolic pathways which require oxygen (aerobic pathways). Their rate of contraction is rapid; their rate of fatigue is moderate. These cells are involved in activities which demand

TABLE 12-2
Comparison of skeletal muscle fibers

Property	*Fast-Twitch Glycolytic (A)*	*Slow-Twitch (B)*	*Fast-Twitch Oxidative (C)*
Alternate name	Fast-white, low oxidative,	Slow-intermediate, high oxidative,	Fast-red, high oxidative
Size of cells	Large	Intermediate	Small
Color	Pale (white)	Dark (red)	Dark (red)
Activity	High intensity, short duration (e.g., weight lifting, skilled movements)	Endurance, long-lasting (e.g., posture)	Endurance, intermediate (e.g., swimming, long-distance running)
Speed of contraction	Fast	Slow	Fast
Rate of fatigue	Fast	Slow	Moderate
Primary metabolic pathways for energy production	Anaerobic	Aerobic	Aerobic
Number of mitochondria	Sparse	Moderate (but small in size)	Abundant
Neuromuscular junction; size and complexity of vesicles and folds of membranes	Extensive	Moderate	Slight
Major locations	Muscles close to body surface	Muscles of posture, long muscles of back	Muscles deep within body

the kind of endurance required for swimming and long-distance running.

Slow-twitch B fibers are also called *slow-intermediate, high oxidative fibers.* Aerobic metabolic pathways provide their primary sources of energy. They have the slowest rate of contraction and fatigue of the three fiber types. Slow-twitch B cells are prominent in muscles used to maintain posture. These *slow-red fibers* are the only types found in some muscles (e.g., the long muscles of the back). Such muscles are the darkest colored in the body.

Relation Between Nerve Fiber and Type of Muscle

The nature of a muscle fiber is dependent upon and can be altered by changes in neural activity at the neuromuscular junction. Experimental transplantation of a newborn lab animal's nerve endings from fast-twitch glycolytic A fibers to fast-twitch oxidative C fibers promotes conversion into the opposite cell type. Within limits imposed by heredity, specific exercises designed to alter neural activity can change the cellular and biochemical composition of whole muscle. For instance, high-frequency, high-intensity, forceful, short-term exercises seem to build up the respiratory capacity (aerobic metabolism) of the A fibers, while low-frequency, lower-intensity, less forceful, prolonged exercises increase respiratory capacity even more dramatically in C fibers. Thus, heavy weight lifting does not promote development of those fibers suitable for swimming, and vice versa. The effects of training will be discussed in Chap. 13.

All muscle cells within a motor unit are of the same muscle fiber type. However, a muscle usually is composed of cells supplied by different motor units. As a result, muscles have a checkerboard pattern of fibers. The development of individual fiber types within a muscle can be promoted by an increase in the activity of specific motor neurons. These observations demonstrate the importance of the properties of the motor neuron in relation to the structure and function of the muscle as a whole.

12-3 CARDIAC MUSCLE

General Properties

Skeletal, cardiac, and smooth muscle share the property of contractility but differ in many respects. Cardiac and skeletal muscle are multinucleated and striated, whereas smooth muscle is not. Skeletal muscle is under voluntary nervous control, while cardiac and smooth muscle are under involuntary nervous control. Some of the properties of these muscle types are noted in Table 12-1 on page 285.

Cell Membrane Structure

The sarcolemma of each cardiac muscle cell is highly developed. The junctions of these cells form *intercalated discs,* enclosing a space of 15 to 20 nm between the ends of adjacent cells (Fig. 12-1B). Narrower *gap junctions* or *nexuses,* measuring less than 2 nm, are also found between cells. Both these modifications of the cardiac muscle cell membrane aid in the transmission of impulses by providing low-resistance pathways for current flow.

Cellular and Molecular Architecture

Cardiac muscle cells have a banded appearance similar to that of skeletal muscle (Fig. 12-1). However, cardiac muscle cells and their myofibrils branch. Those of skeletal muscle do not. The actomyosin content of cardiac muscle is only 60 percent of that found in skeletal fibers.

Large mitochondria are common between cardiac myofibrils and usually extend the length of a sarcomere. The sarcoplasmic reticulum (shown in Fig. 12-12) is less extensive than

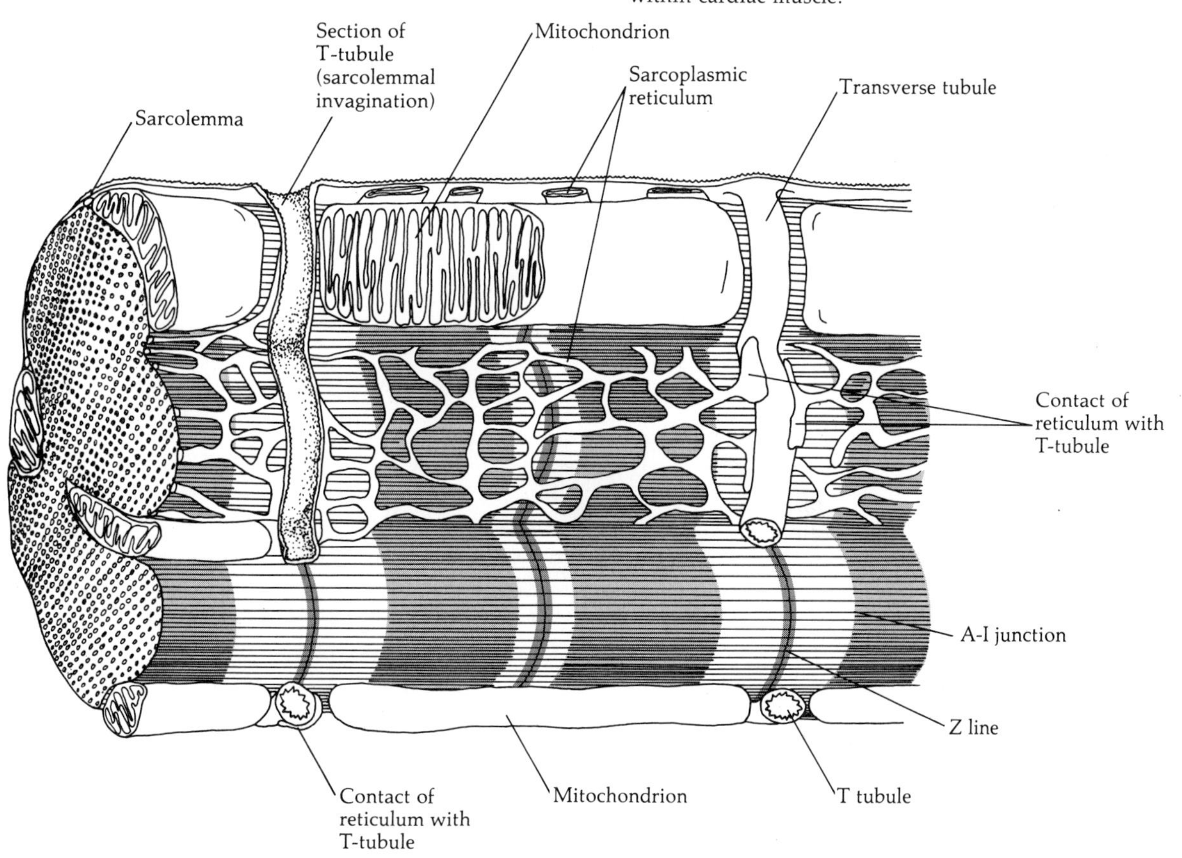

FIG. 12-12
A schematic drawing of the tubules and mitochondria within cardiac muscle.

in skeletal muscle. The cisterns are irregular and not well developed. Consequently, no triads are visible. The T tubules are fewer in number but larger than those of skeletal muscle. The T tubules cross the Z lines in human cardiac muscle. (In skeletal muscle T tubules are found at the A–I junctions; compare Figs. 12-3 and 12-12, pages 286 and 299.)

Metabolic Properties

The metabolic properties of cardiac muscle are similar to those of slow-twitch high oxidative B fibers in skeletal muscle. Cardiac muscle cells primarily use aerobic paths to generate energy and are built for endurance. Like skeletal muscle cells, they cannot regenerate. A newborn baby seems to have all the skeletal and cardiac muscle cells that he or she will ever possess. In effect, the life-span of these cells is equivalent to that of the individual. However, increases which occur in cell size during growth and with training reflect physiological hypertrophy of the cell's components.

Cardiac muscle cells generally maintain their vital capacity through life. The cytoplasmic components are degraded and replaced very rapidly. About half of the mitochondria are replaced every 5 days and the same proportion of the myofibrillar proteins are replaced every 8 to 10 days. However, the rate of replacement of cellular elements diminishes with age, and lipid content increases as myocardial cells grow old. Consequently, lipids may account for up to 25 percent of the dry weight of the aged heart.

Electrical Activity

Modified cardiac muscle cells, called pacemaker cells, initiate action potentials in the heart. The pacemaker cells are found in a discrete area, the sinoatrial (SA) node and convey impulses over a network of fibers which forms the intrinsic conduction system of the heart. (It consists of the SA and AV nodes and Purkinje fibers described in Chap. 15). Impulses are relayed from cell to cell across gap junctions (Fig. 3-1C). Each cardiac muscle cell does not have its own neuromuscular junction with external nerves. Neurotransmitters diffusing from autonomic fibers are able to modify the activity of groups of cardiac muscle cells significantly. Sympathetic nerve terminals release molecules of norepinephrine, which act as excitatory transmitters to depolarize some nearby cardiac muscle cells. The neurotransmitters released from each axon terminal bind to receptors on specific cardiac muscle cells. Excitation and depolarization of these specialized cells causes the development of an action potential. It is rapidly transmitted from cell to cell in a coordinated fashion by means of the gap junctions which facilitate the flow of current (ions) through the heart. Parasympathetic nerves release ACh, which hyperpolarizes the cardiac muscle cells, inhibiting contraction.

12-4 SMOOTH MUSCLE

General Properties

Smooth muscle forms continuous layers in the walls of hollow organs. The cells are found in layers in which the broader portion of one is sandwiched between the adjacent parts of others (Fig. 12-1C). Slight amounts of intercellular connective tissue found between the cells are penetrated by small blood vessels and nerves. The smooth muscle cells are often arranged in an inner circular layer surrounded by an outer longitudinal one (Fig. 3-8).

Cell Membrane Structure

Narrow gap junctions or nexuses are also found between the cell membranes of smooth muscle fibers (Fig. 12-13). These junctions aid in the transmission of impulses. Receptors which are spread out over most of the smooth muscle cell surface are very accessible to neurotransmitters. This is unlike the localized concentration of receptors found at the motor end plate of skeletal muscle fibers.

Cellular and Molecular Architecture

Cytoplasmic organelles such as mitochondria, Golgi apparatus, and ribosomes in smooth muscle fibers are typical of those found in most other cells. However, smooth muscle fibers do not have the highly developed T tubules and sarcoplasmic reticulum characteristic of striated muscle.

The myofibrils of smooth muscle cells lack the banded appearance of those in skeletal muscle and contain less myosin but more actin. The myofibrils average about 1 μm in width (about half the diameter of those in skeletal muscle) and have 10 to 15 thin filaments for each thick one. Troponin and T tubules are poorly developed or missing in smooth muscle. Ca^{2+} ions enter the cell from the ECF and cytoplasmic stores following depolarization of the cell membrane. The Ca^{2+} ions probably bind to and activate an enzyme which alters myosin, allowing it to form cross-bridges with actin.

Smooth muscle cells are built for endurance, like those of cardiac muscle and slow-twitch B fibers of skeletal muscle. Smooth muscle cells are high oxidative fibers, which use aerobic metabolic pathways as their prime means of obtaining energy.

Unlike other muscle cell types, those of smooth muscle can regenerate to some extent. They proliferate in *atherosclerosis* and, along with the fats which accumulate, they narrow the inner diameter of arteries in that disease (Chaps. 16 and 17). However, following injury to smooth muscle, connective tissue cells multiply rapidly to form scar tissue.

Electrical Activity

Types of Smooth Muscle

Groups of smooth muscle cells may behave as one coordinated mass (single-unit muscle) or several separate groups (multiunit muscle). *Single-unit* smooth muscle consists of a group of cells which behave as one motor unit. Action potentials pass from one cell to the other, as shown in Fig. 12-13. Single-unit muscle is found in the uterus, ureters, and walls of the gastrointestinal system (Fig. 3-8) whose contractions are localized, or *segmental,* when initiated by the inner circular layer of smooth muscle. The contractions are propagated as a wave called *peristalsis* when initiated by the outer layer of longitudinal muscle cells. Peristalsis moves food along the digestive tract.

In *multiunit* smooth muscle, fewer cells contract as a group. Since each cell is close to a nerve ending, the contractions are more localized and subject to finer control than in single-unit smooth muscle. The ratio of number of sites of neurotransmitter release to number of smooth muscle cells is much higher in multiunit than in single-unit smooth muscle. However, the nerve endings in multiunit smooth muscle do not form separate synapses with each cell. Instead, the neurotransmitters seem to diffuse from several axonal sites to groups of nearby smooth muscle cells (Fig. 12-14). Multiunit smooth muscle is located in the walls of many large blood vessels, in the bronchi in the lungs, at the base of the hair follicles in the skin, in the iris of the eye, and on the surface of the lens of the eye.

Types of Potentials

The electrical activity of single-unit muscle has been studied extensively, and the following description pertains to it. The resting potential of smooth muscle can vary from -50 to -80 mV but is usually -50 mV. This figure is considerably less than that of skeletal muscle, which averages -90 mV. Single-unit smooth muscle, like cardiac muscle, can initiate potentials independently of signals from external nerve fibers. It can contract spontaneously even if denervated. *Pacemaker cells* are present and develop slow spontaneous regular depolarizations called *pacemaker potentials.* The latter originate in different cells at various times. They may cause action potentials to form, which then cross gap junctions to depolarize and induce contraction in successive cells in the unit.

Longitudinal smooth muscle also shows slow spontaneous rhythmic depolarizations called *slow waves.* Slow waves develop over a longer period of time, often spaced seconds apart, and represent changes of only a few millivolts. The slow waves of depolarization seem to produce local circuits of current and local depolarization of the membrane. The summation of a series of slow-wave potentials can cause a spike poten-

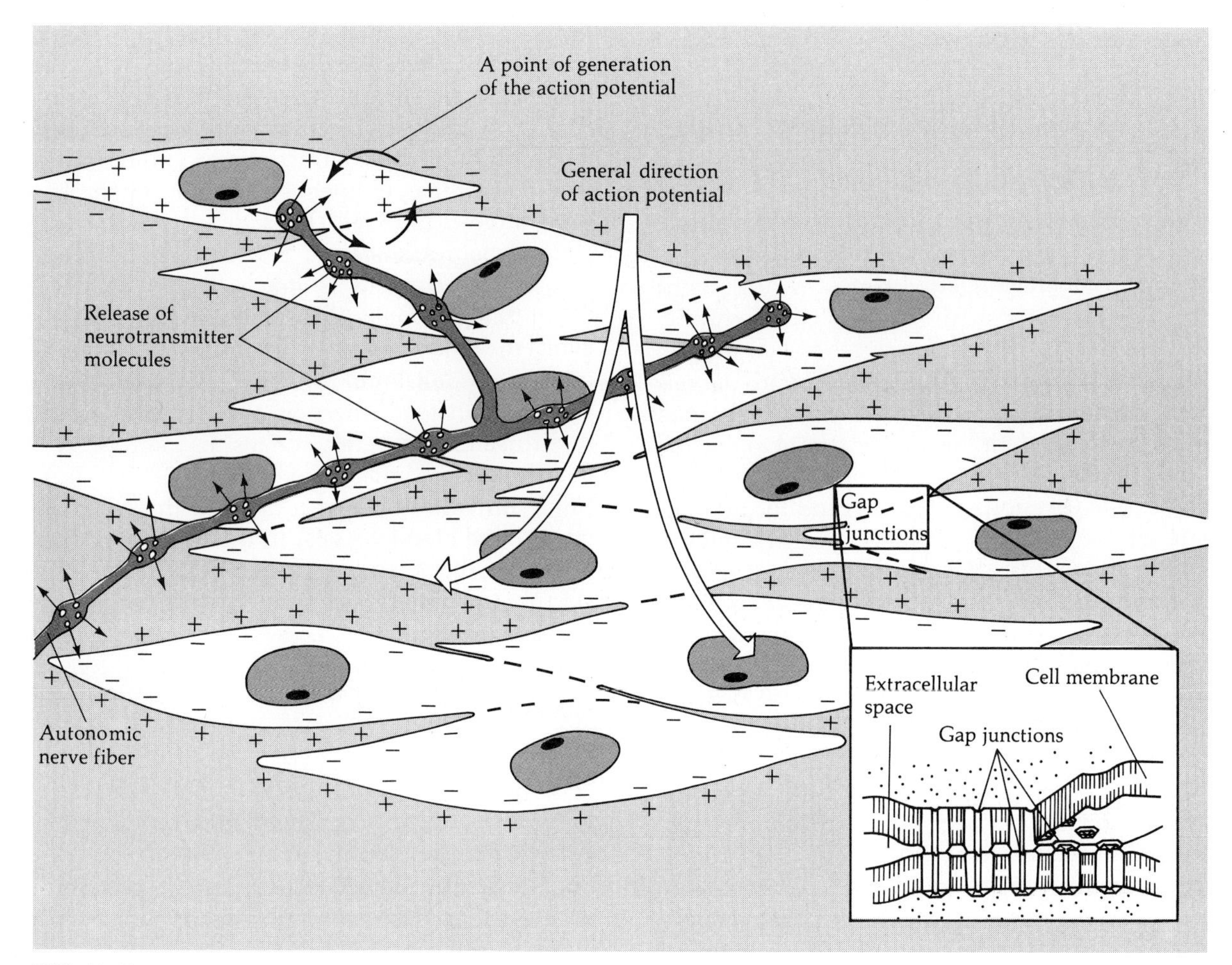

FIG. 12-13
Passage of action potentials across gap junctions in single-unit smooth muscle. Neurotransmitter is released from nerve terminals of a single nerve fiber to only a few muscle cells in the group. A current is generated that is passed through all the cells in the unit across gap junctions, by which the muscle cells are electrically united. All 11 cells in this unit ultimately respond to the nerve impulse. In a cross section of smooth muscle in an animal, each cell usually has contact with six others. As a result, many parallel pathways are cross-connected. Only one nerve fiber is shown for the sake of simplicity. However, the unit usually is innervated by a sympathetic and a parasympathetic fiber and their collaterals.

tial which leads to propagation of an action potential and smooth muscle cell contraction. Although the mechanism by which the potential develops is not clear, intestinal smooth muscle behaves like a bundle of parallel fibers rather than like single cells. Ions flow through gap junctions, causing the membrane to depolarize, and contraction proceeds from cell to cell within a group. Following stimulation, depolarization of the cell and development of a spike potential seem to be more dependent on the entry of Ca^{2+} than Na^+ ions.

External Sources of Stimuli

External autonomic nerve signals and hormones can modify potentials across smooth muscle cell membranes. A unique property of single-unit smooth muscle is that it behaves as a mechanoreceptor and is activated by stretch. The basis for this response has not been established. Regardless of the source of the stimulus, smooth muscle requires more time than skeletal muscle to develop an action potential. The pas-

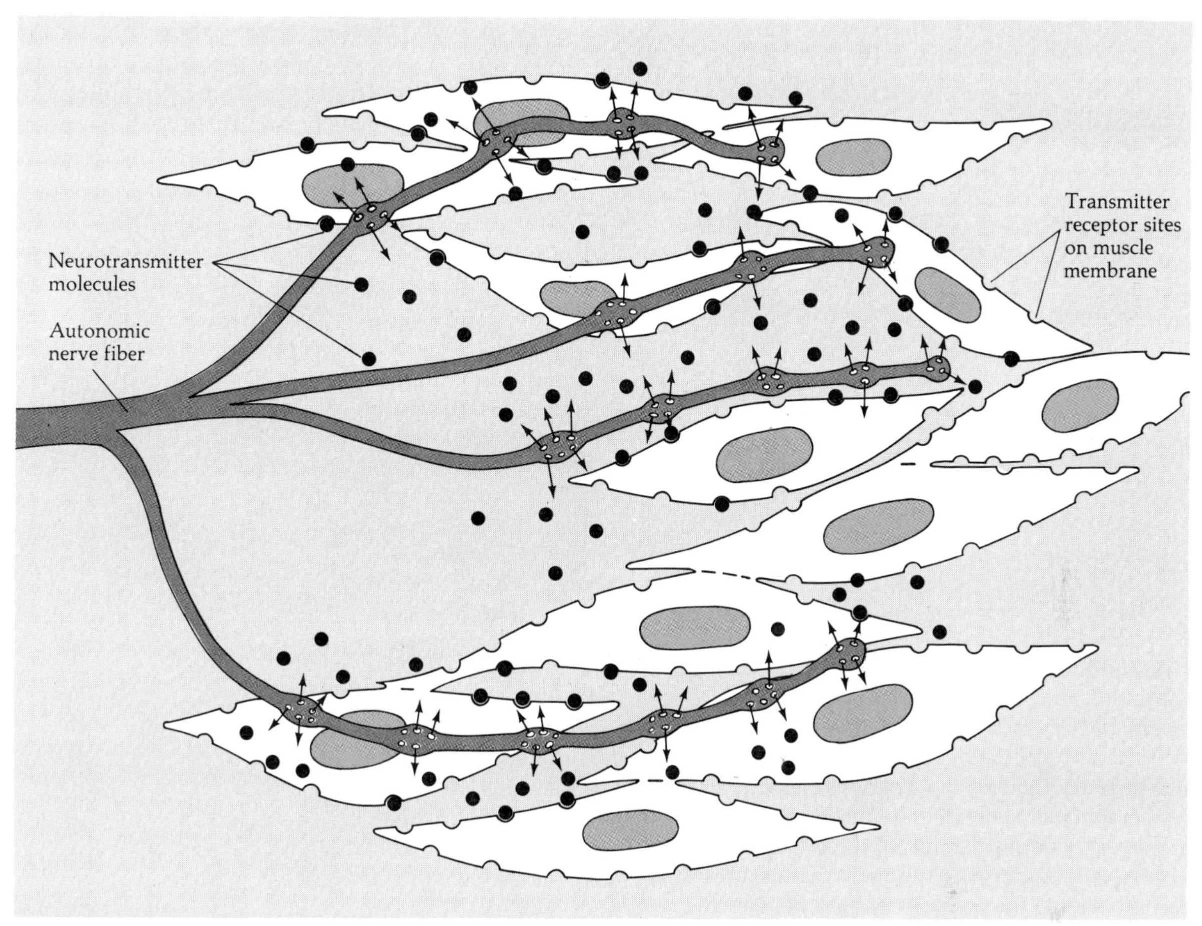

FIG. 12-14
Response of multiunit smooth muscle. Neurotransmitter is released from nerve terminals to a limited number of cells supplied by the same nerve fiber. Many nerve fiber–muscle units exist in multiunit smooth muscle. The ratio of sites of transmitter release to smooth muscle cell number is greater than that observed in single-unit smooth muscle. Consequently, nerves exert a greater degree of control over more localized clusters of smooth muscle cells. Only one nerve fiber is shown for the sake of simplicity. However, the unit usually is innervated by a sympathetic and a parasympathetic fiber and their collaterals.

sage of the impulse from cell to cell leads to asynchronous contraction of fibers rather than to the simultaneous contractions observed in skeletal muscle.

Furthermore, the response of smooth muscle cells to neurotransmitters depends on the type of neurotransmitter released by autonomic nerve endings, as well as on the nature of the receptors and responding cells. The diversity of effects illustrates the variations and fluctuations in ionic control in smooth muscle. Some of the bases for the variations were described earlier (Chap. 5; Fig. 7-17, page 176; Table 9-4, page 231) and will be reviewed here only briefly.

The density of nerve fibers to smooth muscles varies. Most sympathetic nerve fibers are adrenergic. They release norepinephrine, which interacts primarily with alpha receptors but can cause depolarization or hyperpolarization. Adrenergic stimulation usually depolarizes smooth muscle which contracts in the walls of most blood vessels, but leads to hyperpolarization and relaxation in the walls of the intestine. The latter response is identical to the beta-adrenergic effect of the hormone epinephrine (Table 9-4).

Cholinergic nerve fibers may be parasympathetic or sympathetic. They release ACh which binds to γ receptors. The effects of this neurotransmitter are generally opposite to those of norepinephrine (Table 9-4). When ACh binds to γ receptors, the smooth muscle in the walls of blood vessels relax, while that in the walls of the stomach and intestine contracts. The role of smooth muscle in the physiology of the body will be discussed further in chapters dealing with the blood vessels and the urinary, digestive, and reproductive systems.

12-5 SUMMARY

1 Skeletal muscle cells are striated, cylindrical, multinucleated, and under voluntary nervous control. Layers of connective tissue form the endomysium around each cell, the perimysium that binds cells into bundles or fasciculi, and the epimysium (fascia) that surrounds the whole muscle and forms tendons, anchoring it to bone. The part of a muscle's tendon that moves the bone to which it is attached is called the insertion. The other end, which does not move the bone to which it is attached, is called the origin.

2 The skeletal muscle cell is surrounded by a cell membrane, the sarcolemma, within which are many nuclei surrounded by muscle cytoplasm, the sarcoplasm. Eighty percent of the sarcoplasm is filled with strands of protein that form microscopic myofibrils. Cytoplasmic organelles in the form of specialized tubules, such as the sarcoplasmic reticulum and T tubules, and other organelles, including mitochondria, are found between the myofibrils. The tubules form three-chambered structures called triads, which are located at the junctions of cross-striations named the A and I bands.

3 Under polarized light myofibrils appear cross-striated with alternating dark I bands (isotropic bands), and light A bands (anisotropic bands). The I bands deflect polarized light equally in all directions, while the A bands do so unequally. The bands look the opposite color under ordinary light.

In a relaxed cell, a Z line can be observed within the center of the I band. Within the middle of the A band is an H zone, which is bisected by an M line. A sarcomere is the functional unit of skeletal muscle cell contraction and extends from Z line to Z line, a distance of about 2 to 3 μm.

Myofibrils consist of thin and thick myofilaments composed of various proteins. The thin myofilaments are formed by a double-stranded coil of actin in which lies two twisted strands of tropomyosin. A globular and trimolecular complex of troponin is embedded in the actin at intervals. Thick myofilaments are composed of myosin, which has two globular heads attached to a double coiled tail. Two light chains cap each globular head.

According to the sliding filament theory of contraction, the globular heads of myosin bind to actin, change shape, and slide the actin molecules toward each other. This results in a decrease in width of the sarcomeres (Z line to Z line) as the H zone narrows or disappears.

4 Neuromuscular junctions are formed by the axons of motor neurons, which are separated by synaptic clefts from specialized regions of skeletal muscle cell membranes called the motor end plates. The motor neuron and all the muscle cells that its terminals innervate are collectively called a motor unit.

ACh released from synaptic vesicles diffuses into the synaptic cleft and binds to receptors localized on the motor end plate, increasing the ionic permeability of the muscle fiber for a few milliseconds. The ACh is rapidly retrieved by the vesicles and degraded by cholinesterase (ChE). ACh bound to receptors on the motor end plate briefly allows net entry (diffusion) of positive ions such as Na^+ through the skeletal muscle cell membrane. This response results in generation of an EPP and depolarization of the muscle cell membrane. The EPP usually leads to the development of an action potential that passes along the entire membrane. This response is called excitation and is normally linked (coupled) to the subsequent development of tension and contraction of the muscle cell.

Contraction occurs after Ca^{2+} ions pass through depolarized sarcoplasmic reticulum membranes, bind to troponin, and relieve the inhibition of myosin-ATPases. The ATPases cleave ATP to release energy that is used to repeatedly form and break cross-bridges between myosin and actin and to slide the filaments past each other. The availability of Ca^{2+} decreases as active transport enzymes return these ions to storage sites within the cisterns of the sarcoplasmic reticulum. The depletion of Ca^{2+} in the sarcoplasm results in a loss of cross-bridge activity as the energy is used up.

5 Skeletal muscle activity may be blocked by botulinal toxin, which inhibits ACh release; black widow spider venom, which causes excess production and/or release of ACh; curare, which competes with ACh for its receptors or enhances ACh resistance to ChE and blocks repolarization; organophosphates, which block ChE activity; succinylcholine, which depolarizes the motor end plate; and autoimmune antibodies formed in myasthenia gravis, which lead to degradation of ACh receptors.

6 Most muscles are a mixture of fast-twitch low oxidative A fibers, slow-twitch high oxidative B fibers, and/or fast-twitch high oxidative C fibers. The nature of the fibers is dependent on, and can over a long period of time be changed by, alterations in motor nerve activity at the neuromuscular junction.

7 Cardiac muscle cells are striated, cylindrical, multinucleated, branched, and not usually under voluntary control. Their myofibrils are also branched. Some parts of intercellular junctions between cardiac muscle cells are found as intercalated discs; others are modified to form narrower gap junctions or nexuses. Both aid in the transmission of impulses from cell to cell.

The myofibrils of cardiac muscle are striated and similar to those of skeletal muscle but contain only 60 percent of its actomyosin. The T tubules of cardiac muscle are larger and the sarcoplasmic reticulum less extensive, so that triads are not apparent.

Like slow-twitch high oxidative B fibers of skeletal muscle, cardiac muscle fibers are built for endurance. Like skeletal muscle fibers, they cannot regenerate after birth. Some modified cardiac muscle fibers act as pacemaker cells and initiate action potentials spontaneously and rhythmically. The potentials may be modified by neurotransmitters released to groups of cardiac muscle fibers by autonomic nerves.

8 Smooth muscle fibers are nonstriated, spindle-shaped cells with a single nucleus, and are under involuntary nervous and/or hormonal control. Smooth muscle is found in layers in the walls of hollow organs. Receptors are spread over the surface of most of the smooth muscle cell membrane. The cell membranes form gap junctions or nexuses among the neighboring cells.

The small myofibrils in smooth muscle fibers are not readily apparent and contain less myosin (and thick filaments) and disproportionately more actin (and thin filaments) than skeletal muscle. The sarcoplasmic reticulum is less extensive, and the T tubules are missing or less well developed. Smooth muscle cells are built for endurance, metabolically resembling the slow-twitch B fibers of skeletal muscle. Unlike the latter, smooth muscle can regenerate slightly.

Depolarization of the smooth muscle cell membrane allows Ca^{2+} ions to enter the cell and activate enzymes, initiating a mechanism of contraction which may resemble that of striated muscle. Smooth muscle fibers can respond in small groups of finely controlled units called multiunit smooth muscle or act within larger groups behaving as single-unit smooth muscle. In the latter arrangement the muscle fibers act as one motor unit, passing the action potential from one to the other. The contractions are local (segmental) in circular smooth muscle. They also may be propagated as a wave of contraction called peristalsis when initiated in longitudinal smooth muscle.

Slow, regular, spontaneous depolarizations or pacemaker potentials are generated by smooth muscle fibers called pacemaker cells. Slow rhythmic depolarizations also may occur. They can attain a threshold and cause a spike potential, which produces an action potential that flows across all the cells in a single unit. External autonomic signals, hormones, or stretching also may cause formation of action potentials and asynchronous contraction of single-unit smooth muscle. Some smooth muscle cells behave as multiunits in which members of smaller groups respond directly and simultaneously to neurotransmitters released from nearby axonal endings.

REVIEW QUESTIONS

1. What is a muscle? What are its functions? Describe the gross anatomy of a skeletal muscle.
2. Describe the cellular structure of skeletal muscle cells and tissue.
3. Diagram and label the banded appearance of a sarcomere viewed with polarized light. Why is polarized light used to study muscle?
4. What are the molecular compositions of thin and thick myofilaments? What are their roles in muscle cell contraction?
5. Account for the generation of EPPs and action potentials in skeletal muscle.
6. What are the roles of ACh, Ca^{2+} ions, ChE, the sarcoplasmic reticulum, T tubules, troponin, and myosin-ATPase in skeletal muscle contraction?
7. How do the following substances inhibit activity at the neuromuscular junction: botulinal toxin, black widow spider venom, curare, organophosphates, succinylcholine, and antibodies in myasthenia gravis?
8. Compare the properties of fast-twitch glycolytic A, slow-twitch B, and fast-twitch oxidative C fibers.
9. How does the type of nerve fiber influence the composition of skeletal muscle?
10. Compare skeletal, cardiac, and smooth muscle.
11. What are the differences between single-unit and multiunit smooth muscle? How does single-unit smooth muscle develop and relay electrical activity?

THE PHYSIOLOGY OF SKELETAL MUSCLE

Muscle Mechanics, Energy Metabolism, and Exercise

13

13-1 THE MECHANICS OF CONTRACTION
Development of tension
Response to a stimulus
Physiological variables
Strength of impulse • Number of active muscle fibers • Coordination by the CNS
Effects of stretching
Frequency of stimulation
Summation of contractions • Tetanus • Treppe: the staircase effect or "warming up"
Atrophy and hypertrophy
Muscles and levers
Muscles and work

13-2 ENERGY METABOLISM
Anaerobic glycolysis and lactic acid formation
Aerobic oxidative phosphorylation
High-energy phosphate metabolism
Energy production and muscle contraction
Heat production and muscle activity
Formation of "extra" shortening heat
Efficiency of muscle

13-3 THE EFFECTS OF EXERCISE
Exercise and physical fitness
Fatigue
Cramps and pain
Training and adaptation
Maximal exercise
Submaximal or endurance exercise
Calculation of target (training) heart rate
Effects of endurance training on muscle

13-4 SUMMARY

OBJECTIVES

After completing this chapter the student should be able to:

1. Describe the development of tension in a muscle, note three phases of contraction, and distinguish between isotonic and isometric tension.
2. Discuss the variables of the mechanics of muscle contraction, including an account of the strength and frequency of stimuli.
3. Outline the events that lead to summation of contractions, treppe, and tetanus.
4. Explain how muscles behave as levers and the physical principles that govern work and power of a muscle.
5. Compare anaerobic and aerobic pathways and the units of energy currency that are produced and explain how the pathways are related to muscle activity.
6. Describe three phases of heat production and the events which lead to the development of muscle tension.
7. Define exercise, training, and submaximal and maximal exercise.
8. Discuss the effects of endurance exercise on skeletal muscle and its relationship to maximal oxygen utilization ($Vo_{2\,max}$) and training (target) heart rate.

Muscle is the body's engine and its activity generates force and heat. The force makes work and movement possible while the heat helps to maintain constant body temperature. Both responses are made possible by the release of energy derived from the combustion (catabolism) of organic molecules, the availability of which results ultimately from the consumption of food.

Skeletal muscle is the main source of body heat and exceeds the liver, which is second to it, in this capacity. The tissues in both sites carry on a wide array of metabolic interconversions necessary for homeostasis, one facet of which is maintenance of constant body temperature. The development of contractile force by muscle is a mechanical process which requires chemical reactions that provide energy to do work. The previous chapter described the gross and cellular anatomy of muscle, as well as its molecular architecture. This chapter discusses the mechanics of muscle contraction, the metabolic paths by which muscle cells obtain energy, and the effects of exercise on this tissue.

13-1 THE MECHANICS OF CONTRACTION

Contraction is the generation of force by muscle. It occurs when myofilaments within each muscle cell slide toward one another and their sarcomeres shorten (Chap. 12). The resulting force is transmitted through contraction of myofibrils to noncontractile elements called *elastic components*. The elastic components are stretched and transmit the force developed to the surface of muscle cells and to bone. Some elastic elements are in series and others are parallel to the myofibrils. The *series elastic elements* are the extracellular connective tissue elements around the muscle fiber and those which form tendons. The *parallel elastic elements* are the sarcolemma (the muscle cell membrane) and the sarcoplasm, owing to its natural viscosity (resistance to flow). The contractile proteins generate tension by stretching the elastic components and produce work when the elastic elements recoil (much as a stretched spring does) and the muscle shortens. Recoil of the elastic elements in tendons and shortening of the sarcomeres contribute about equally to the changes in length observed during contraction.

Development of Tension

Muscles generate tension isotonically or isometrically. *Isotonic tension* (contraction) produces a constant force as the muscle shortens (from Greek *iso*, equal, and *tonic*, tone or force). *Isometric tension* (contraction) develops increasing force while the muscle length remains essentially the same (from *iso*, equal, and *metron*, measure). A combination of both types of tension is generated in normal muscle action and

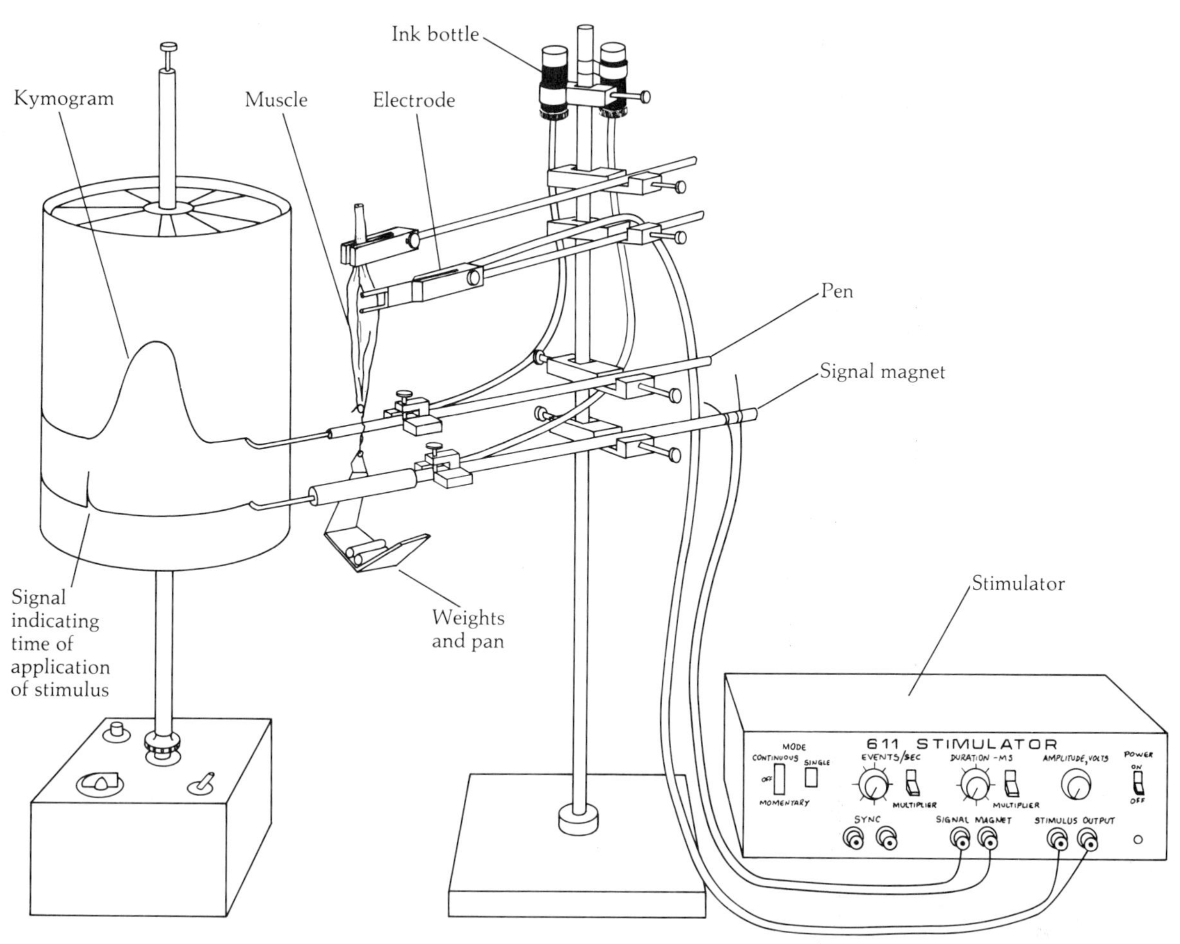

FIG. 13-1
Recording device (kymograph) to measure contraction in an isolated muscle. The stimulator generates a defined voltage carried to the muscle through the electrodes. A threshold voltage causes contraction of the muscle which lifts the pen. The pen traces a record on a sheet of paper attached to a revolving drum. The lengths of phases of the muscle responses are measured by noting the time between application of the stimulus (shown by the lines produced by the signal magnet), the change in the height of the tracing, and the speed at which the drum rotates.

is called *auxotonic contraction*. When a book is grasped, for example, muscles in the forearm generate isometric tension until the force developed overcomes the weight of the book. The muscles then shorten in isotonic contraction. If the book is held at arm's length, isometric tension is maintained. The entire action (lifting and holding the book) involves *auxotonic* contraction.

The mechanical response of muscle to a stimulus and the development of tension can be measured in isolated muscle by using the simple recording device called a *kymograph*, shown in Fig. 13-1. It consists of a lever with a weight which is attached to a pen, which moves when the muscle contracts. The muscle is artificially stimulated to contract by an electric current carried to it through electrodes from a stimulator. The pen records the movement on a piece of paper, which is rotated on a drum, producing a tracing called a kymogram.

The electrical activity of muscle can be measured by recording devices similar to those used with nerves (Chap. 7). The electrodes carry electric current to or from solutions or tissues. When placed on the skin, they transmit current to or from motor units within the underlying muscle. The activity of a single motor

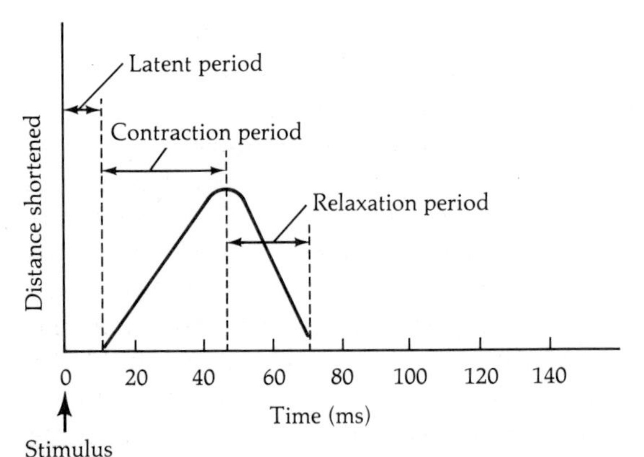

FIG. 13-2
Effects of a single stimulus on isotonic muscle contraction. The arrow indicates the time of application of a threshold stimulus. The threshold stimulus produces an action potential and membrane depolarization. The separate phases of the response to the threshold stimulus are the latent, contraction, and relaxation periods. Adapted from *Human Physiology,* by A. J. Vander, J. H. Sherman, and D. W. Luciano. Copyright © 1980. McGraw-Hill Book Co. Used with the permission of McGraw-Hill Book Co.

unit and its associated muscle fibers can be recorded by inserting fine, insulated microelectrodes shaped as needles into the muscle. The signals may be amplified and visually displayed when delivered to an oscilloscope screen (Fig. 7-8A, p. 163). The recorded pattern is an *electromyogram* (EMG) and is used to study normal and abnormal muscle activity. It is analogous to the ECG (electrocardiogram), which is a record of changes in the electric field of the heart (see Chap. 15).

Response to a Stimulus

The mechanical response of a muscle fiber to a single stimulus is called a *twitch*. The duration of a twitch varies with the type of fiber and the condition of stimulation. A twitch can last as little as 7.5 ms (milliseconds) in fast fibers or more than 100 ms in slow fibers. It develops as an action potential, flows along the whole cell membrane and involves three phases, the latent, contraction, and relaxation periods (Fig. 13-2).

The *latent period* is the time that intervenes between the application of the stimulus and the mechanical response. The *contraction period* is the time after the latent period between the start and the development of peak tension. The *relaxation period* is the time required for tension to drop from its peak back to zero. During this last phase, the return of Ca^{2+} to the sarcoplasmic reticulum consumes 30 to 50 percent of the energy used by muscle. The major source of energy, the cleavage of ATP, is used to reverse reactions that took place during contraction. After relaxation and during recovery, energy is released as myosin binds to adenosine triphosphate (ATP) and hydrolyzes it by virtue of myosin-ATPase activity. The myosin is readied for cross-bridge formation and movement when actin-binding sites become available.

In general, the time required to overcome inertia (resistance to motion) increases with increasing mass at rest. Therefore, as the load on a muscle increases, more tension must be developed prior to generation of a twitch (Fig. 13-3). The contractile components of the muscle must generate greater tension on the elastic components in order to overcome the greater inertia produced by the heavier load. This results in an increase in the latent period. As the load or weight increases, the time required to generate the tension necessary to move the load

FIG. 13-3
The effects of increasing load on isotonic contraction. Note that the increase in the latent period is coupled with decreases in the velocity of contraction and in the distance shortened as the load becomes greater. From *Human Physiology,* by A. J. Vander, J. H. Sherman, and D. W. Luciano. Copyright © 1980. McGraw-Hill Book Co. Used with permission of McGraw-Hill Book Co.

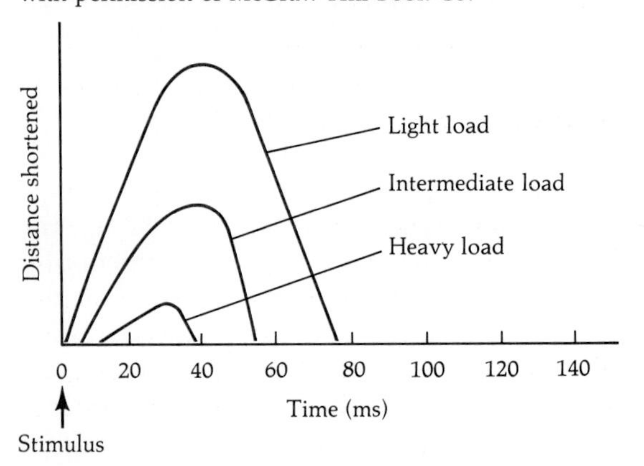

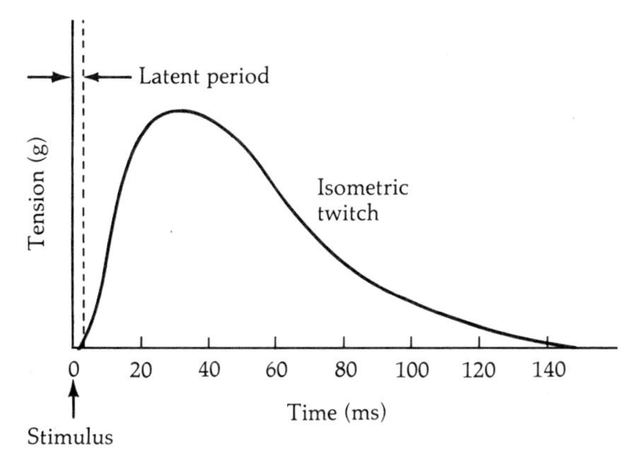

FIG. 13-4
Isometric contraction following a single stimulus. Note the short latent period and extended contraction time compared with an isotonic twitch as shown in Fig. 13-2. From *Human Physiology*, by A. J. Vander, J. H. Sherman, and D. W. Luciano. Copyright © 1980. McGraw-Hill Book Co. Used with permission of McGraw-Hill Book Co.

increases, and the velocity of contraction and the distance shortened decrease. Evidence derived from studies of rat skeletal muscle indicates that the speed of contraction is related to the levels of actomyosin and myosin-ATPase.

When a muscle cannot move a load, the myofilaments do not slide and muscle length remains constant. The contraction is isometric (Fig. 13-4); although tension and heat are generated, the speed of contraction is zero. Marked differences may be noted in comparing an isotonic twitch which moves a light load (Fig. 13-2) with an isometric twitch (Fig. 13-4). Common examples of isometric contractions are those used to maintain posture to counteract the force of gravity.

Physiological Variables

The strength of the nerve impulse and the number and nature of active muscle fibers influence the development of muscle tension.

STRENGTH OF IMPULSE As mentioned in Chap. 12, when an end-plate potential (EPP) develops, sufficient acetylcholine (ACh) is usually released by a motor neuron to cause an action potential to be generated in all the muscle fibers in the motor unit. Small doses of curare may be used to artificially block the activity of some but not all ACh released from nerve terminals to motor end-plates. This approach has been used to demonstrate that a specific number of packets of ACh must be released to generate an EPP in each muscle fiber. Subthreshold potentials develop if curare reduces the quantity of ACh that combines with end-plate receptors below a critical level. Under these conditions, repeated stimulation of the nerve fiber may cause the temporal summation and accumulation of sufficient ACh to attain a threshold. An action potential is elicited which causes contraction of the muscle fiber. However, in a living animal the motor nerve fiber normally releases enough neurotransmitter so that the EPP which develops is able to elicit an action potential in the muscle fiber and cause contraction. The response of the individual muscle fiber is governed by the *all-or-none law*. Following stimulation under a defined set of conditions, the fiber either contracts or does not contract. An increase in the strength of the stimulus above a threshold value does not produce greater tension in a single fiber. However, more forceful contractions may be elicited by varying the conditions of stimulation.

NUMBER OF ACTIVE MUSCLE FIBERS Strong stimuli cause more muscle cells in a muscle to respond than do weak stimuli, thereby increasing the force of contraction in a phenomenon called *recruitment*. The amount of tension developed by a muscle is directly related to the number of fibers and/or motor units that are simultaneously active. The number of muscle fibers in different muscles ranges from a few hundred to several hundred thousand, and the number of muscle fibers per motor unit ranges from three to a few hundred cells. A *motor unit* is composed of fibers from an individual motor neuron and the muscle fibers with which their axon terminals form synaptic junctions. Muscles whose activities are finely regulated have fewer muscle fibers per motor unit than those which are coarsely regulated. Because large muscles have more muscle fibers than small ones, they generally can develop greater tension. The central nervous system (CNS) can simultaneously excite a larger or smaller number of motor units (and muscle fibers) and thereby regulate the strength of contraction.

Most muscles contain three distinct types of fibers, whose twitch times and biochemical properties vary (see page 298). Although each motor unit consists of fibers of one type, the contractile response of the whole muscle depends on the net effect of the different motor units within it which are simultaneously active.

COORDINATION BY THE CNS Afferent signals from muscle spindles, Golgi tendon organs, and proprioceptors (Fig. 8-1) are carried to the spinal cord, cerebellum, and other motor areas in the brain. These impulses provide information that is used to coordinate the activity of individual motor units, muscles, and groups of muscles by feedback mechanisms. Excitatory and inhibitory efferent signals are relayed to various sites in the brain, and especially through the cerebellum and extrapyramidal tracts, to control the contraction and relaxation of opposing groups of muscles (see Fig. 11-4, page 263).

Muscles which generate force and act as the prime movers of the body are called *agonists*. Their activity is steadied by opposing forces produced by *antagonists,* which are muscles that return the body to its anatomical position. All motion results from contraction. Movements of the body generally involve reflex coordination of agonists and their antagonists, which provides fine gradations in the development of tension. An example of an agonist is the biceps, which flexes the forearm at the elbow. Its antagonist is the triceps, which extends the forearm. Since muscles have hundreds to hundreds of thousands of cells acting in separate motor units, extremely fine gradations of opposing tension may be produced.

Effects of Stretching

When a skeletal muscle fiber is moderately stretched, the sarcomeres (from Z line to Z line) may increase in length and the cell may develop greater tension. When stretched, the myofilaments have less overlap and have greater opportunity for the formation of bonds between actin and myosin. The resting length of muscle in the body is about the same as that attained with moderate stretching. This affords muscles optimal opportunity to generate tension. The passive stretching of muscle produces tension called *passive loading* in the series elastic elements. Muscle spindles are modified muscle cells which are arranged parallel to other cells within a muscle and which respond to the rate and extent of stretch as well as the contraction of a muscle. They do so by means of afferent nerve endings and modified muscle fibers that act as stretch receptors (Fig. 8-1F). Golgi tendon organs are stretch receptors in tendons in series with the cells of a muscle and especially suited to measure the total force exerted by a muscle during contraction. These receptors respond primarily to contraction of the muscle and secondarily to extensive stretching. Both types of receptors relay afferent impulses to the CNS. The appropriate excitatory and/or inhibitory impulses are generated in a coordinated manner to regulate opposing muscle groups.

Frequency of Stimulation

Following a response to a stimulus and in the early part of the relaxation period, skeletal muscle cells demonstrate a *refractory phase,* during which a response to another stimulus does not occur. The period is similar to, but often shorter than, that observed in nerve fibers and lasts about 1 to 3 ms. The refractory phase occurs as the spike potential is generated in the cell. This period represents the time required for the muscle cell membrane to repolarize and for Ca^{2+} ions to return to the sarcoplasmic reticulum. In the early part of this phase, the muscle is in an *absolute refractory period*. It is completely unresponsive to a stimulus. After a fraction of a millisecond, the muscle is in a *relative refractory period*. That is, it will respond to a stimulus stronger than the one which caused the initial response. Cardiac muscle exhibits a longer refractory period than skeletal muscle, thereby protecting the heart from overwork and fatigue that might occur if contractions fused, to cause a single sustained state of tension (called tetanus and discussed on page 313).

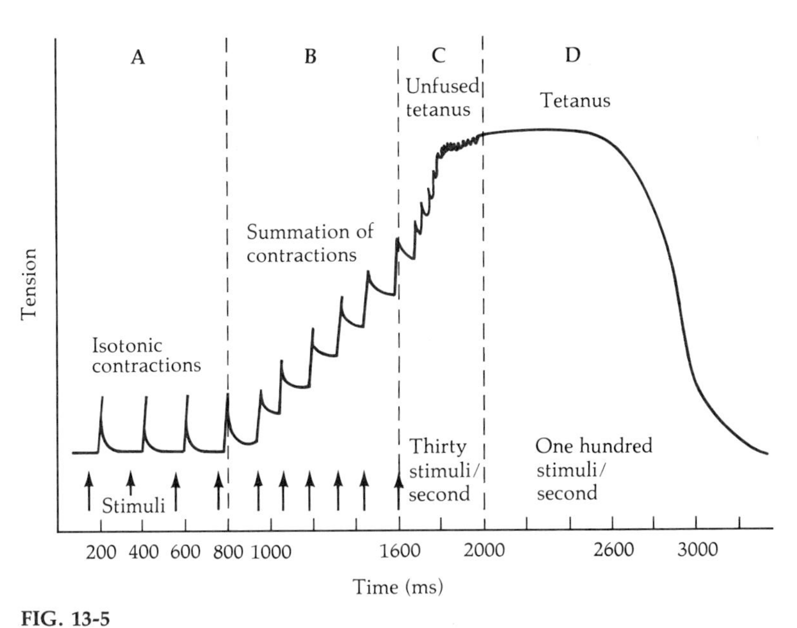

FIG. 13-5
The effects of repeated stimuli on muscle contraction. (A) Successive simple isotonic twitches. (B) Summation of contractions caused by stimuli repeated prior to complete recovery of the muscle. (C) Unfused tetanus. (D) Fused tetanus. Maximal tension is attained but tension falls as the muscle fatigues and is depleted of ATP.

SUMMATION OF CONTRACTIONS While an action potential may pass through a muscle fiber in as little as 1 to 2 ms, the mechanical response may last 100 ms or more. Therefore, repeated action potentials may be generated in a muscle fiber before it has completely relaxed.

When the muscle is partially relaxed (in a relative refractory period), a second stimulus can produce an early response. The contraction generated is greater than that following a single stimulus. The accumulation of tension due to closely repeated stimuli can cause muscle contractions to occur prior to complete relaxation; this is called *summation of contractions* and is shown in Fig. 13-5B. The frequency of stimulation necessary to cause summation varies with the type of muscle fibers involved. The stimuli must be repeated in less time than is normally required for a single twitch to occur. As the frequency increases and the relative tension produced becomes greater, only slight relaxation occurs between successive contractions; such a response is called incomplete or *unfused tetanus* (Fig. 13-5C). Since the resting state is not completely restored, the muscle does not fully relax. The partial contraction which persists between the successive stimuli, called *contracture,* warns of fatigue. Clinically, the term contracture is used in a different sense, to describe shortening of muscle which abnormally limits movement of a joint.

TETANUS Very rapid repeated stimulation may produce maximal contraction with no detectable decrease in tension and no relaxation until the muscle fatigues. A continuous sustained maximal contraction is called complete or fused *tetanus* (Fig. 13-5D). It causes muscle to generate its greatest possible force. The tension can be 4 to 10 times more than that observed after a single twitch. The number of stimuli required to cause complete tetanus varies with the twitch time of individual fibers and ranges from 30 to 100 or more stimuli per second in slow and fast fibers, respectively.

The following mechanism may help explain the summation of stimuli. When the myofilaments slide together and the sarcomeres shorten, a certain amount of time is required for stretched elastic elements within each muscle cell to recoil. This time period is analogous to that needed for a stretched spring to recoil to close a door. During this period, stimulation of the contractile elements (the myofilaments) pulls the elastic elements even further and causes greater tension. This resembles pulling the spring on an open door to cause tension and then pulling it further a second time, before the door closes. The ability of a muscle to sustain maximal tension is directly related to the availability of energy from ATP. When the ATP stores in muscle are exhausted, fatigue sets in and tension decreases.

Tetanus is produced under natural conditions when muscles must develop maximal amounts of tension. Motor units rapidly fire action potentials to sustain maximal contraction. Tetanus also may result when the electrical activity at the motor end plate continues unchecked. Sufficiently high levels of the poison strychnine have this effect. Strychnine blocks inhibitory neurons and intensifies neural activity. Reflexes are exaggerated as impulses pass unchecked from excitatory motor nerve endings. Skeletal muscles, including those used for respiration (such as those in the diaphragm and abdomen), undergo fusion of contractions (tetanus). Respiratory failure can result, which may lead to asphyxiation and death.

TREPPE: THE STAIRCASE EFFECT OR "WARMING UP" Repeated maximal stimuli at a rate below that which will induce tetanus increase muscular force in a response called *treppe* (German for "staircase"). The tension increases stepwise with each successive stimulus. The contractile force developed in an isolated muscle may double between its first and fiftieth response to maximal stimuli. (Maximal stimuli are those strong enough to excite all of the muscle's individual fibers.) The reason for this effect is unknown. It may involve enhanced actomyosin activity, or it may represent greater efficiency of chemical reactions as muscle temperature increases. Some investigators postulate that it is due to greater flow of Na^+ and K^+ ions across the membrane, as well as increased entry of Ca^{2+} ions into the sarcoplasm. The release of Ca^{2+} ions allows myosin-ATPase to bind to actin to form actomyosin-ATPase, which hydrolyzes ATP more rapidly than does its precursor, myosin-ATPase. To date, the reason for improved muscle performance after "warming up" remains unclear.

Atrophy and Hypertrophy

The size of individual cells, tissues, or organs may be changed by physiological alterations or disease. A reduction in size or bulk is called *atrophy*. (The word literally means lack of nourishment.) An increase in size is *hypertrophy*, while defective or abnormal development or degeneration is referred to as *dystrophy* (which literally means defective nourishment). Muscular atrophy is often due to a disease of the motor neurons and is not inherited, appears late in life in distal muscles (those farthest from the midline), and may be associated with spasms (involuntary muscle contractions). Muscular dystrophies are primarily direct muscle cell diseases which are heritable, usually appear in proximal muscles (those near the midline), and occur without spasms.

The degeneration of skeletal muscle resulting from nerve damage is called *denervation atrophy*. The protein and RNA content progressively decreases while DNA levels remain stable. Should restoration of nerve function be delayed for more than 3 to 4 months, muscle degeneration is unlikely to be reversed.

When the nerve supply to skeletal muscles degenerates, connective tissue gradually proliferates around the muscle fibers and the amounts of actin and myosin decrease. The ACh receptors of the muscle fibers are no longer confined to the motor end plates but are located over the entire surface of the muscle cell membrane. As a result, individual fibers become extremely sensitive to the effects of ACh. Neuromuscular control is lost and spastic paralysis can occur. Among the many causes of nerve degeneration are nutritional and hereditary deficiencies (see Chaps. 6 and 9), accidental damage, and infectious diseases (see Chap. 9), and drugs and poisons (see Chaps. 10 and 12).

Disuse atrophy follows prolonged failure to use muscles. For example, it can be observed when a cast is removed after healing of a broken bone. The muscles may shrivel and may lose half their girth when immobilized for 2 or more months. Much of the loss is due to a decrease in myofibrillar protein. Consequently, casts are used which allow as much mobility as possible while protecting the injured bone from stress and weight.

Muscular hypertrophy is a normal response to repeated, intense, forceful exercise that results in the synthesis of contractile protein and proliferation of myofibrils. About 75 percent of maximal tension must develop in a muscle for this condition to result. Adult skeletal muscle apparently can only increase in size this way. The number of skeletal muscle cells seems to remain constant from shortly after birth and the cells are thus incapable of cell division or regeneration.

Muscular dystrophies are genetic diseases in which muscle fibers are progressively replaced by fibrous tissue and fat after birth. They may be fatal later in life. As the disorders develop, muscles weaken. Among the different types of muscular dystrophy the most common is the Duchenne type, which is due to a heritable defect of genes on the X chromosome. As a result, it shows up more often in males, whose Y chromosome lacks the corresponding piece on which the gene is found on the X chromosome. A less severe type of muscular dystrophy is the X-linked Becker form. Other heritable muscular deficiencies are caused by defective genes on the autosomes (pairs of chromosomes each of which carries a gene that affects the same trait).

Muscles and Levers

The same physical principles that govern the action of levers regulate the generation of work by muscles. A *lever* is a simple mechanical device which rotates on a supporting point called a *fulcrum*. A seesaw is an example (Fig. 13-6).

If movement is to occur, the total energy expenditure (or work done) on one side of the fulcrum must exceed that on the other. *Work* is produced when a mass is moved through a distance by a force. At equilibrium, the product of force and distance on one side of the fulcrum

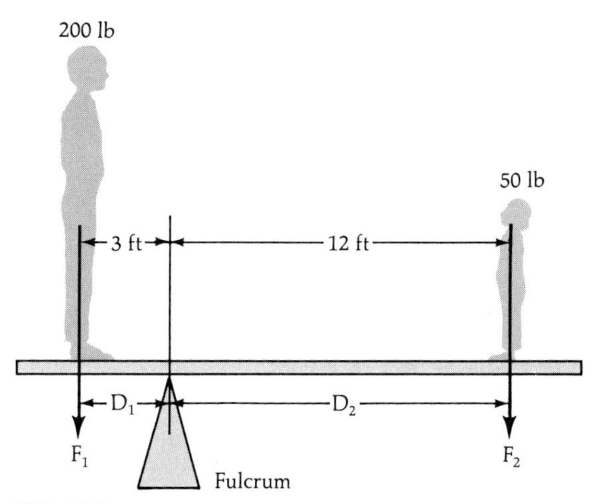

FIG. 13-6
Forces involved in leverage in a seesaw. The seesaw is balanced when the products of force-distance are equal, i.e., when $F_1 \times D_1 = F_2 \times D_2$.

equals that on the other. If the force and distance on one side are called F_1 and D_1, respectively, and those on the other F_2 and D_2, the equilibrium may be expressed by the following equation, which describes the condition under which no movement occurs:

$$F_1 \times D_1 = F_2 \times D_2$$

(Force × distance from fulcrum)
= (force × distance from fulcrum)

For example, if an adult were to sit 3 ft from the fulcrum on a seesaw and a 50-lb child were to sit on the other end some 12 ft from the fulcrum, the weight of the adult which is required to attain equilibrium can be calculated. If one ignores the weight of the plank in the seesaw itself, the calculation is completed as follows:

$$F_1 \times D_1 = F_2 \times D_2$$

$$F_1 \times 3 \text{ ft} = 50 \text{ lb} \times 12 \text{ ft}$$

$$F_1 = \frac{50 \text{ lb} \times 12 \text{ ft}}{3 \text{ ft}}$$

$$F_1 = \frac{600 \text{ ft-lb}}{3 \text{ ft}}$$

$$F_1 = 200 \text{ lb}$$

When the product on one side of the equation exceeds that on the other, movement occurs. The forces that muscles exert around their fulcrum points on bone also lead to motion. To illustrate these principles, let us assume that the biceps in the upper forearm is the only active muscle involved in the movement of the forearm and the hand toward the shoulder. The biceps is inserted an average of 2 in below the elbow. The elbow is the fulcrum in this example. The center of the hand is an average of 14 in beyond the elbow, as shown in Fig. 13-7. The force needed to support a 50-lb barbell in the outstretched hand can be calculated as follows:

$$F_1 \times D_1 = F_2 \times D_2$$

$$F_1 \times 2 \text{ in} = 50 \text{ lb} \times 14 \text{ in}$$

$$F_1 = \frac{50 \text{ lb} \times 14 \text{ in}}{2 \text{ in}}$$

$$= \frac{700 \text{ in-lb}}{2 \text{ in}}$$

$$= 350 \text{ lb}$$

In this instance, the muscle is at a significant mechanical disadvantage but has gained maneuverability (Fig. 13-7). The hand moves 14 in with a 2-in contraction of a muscle attached to the forearm, providing a 7:1 advantage in maneuverability and speed of movement at the expense of a sevenfold disadvantage mechanically to the biceps. To move the 50-lb weight the biceps had to generate 350 lb of force. This type of arrangement is the most common among muscles in the body.

Some muscles act under different circumstances. Slight movement may occur with a mechanical advantage. For example, minimal effort is required by the muscles that move the head to maintain posture or by the calf muscles in the back of the lower leg that raise the body up on the tips of the toes. The relationships described in the above equation differ for various muscles. Some of them provide greater mechanical advantage, while others afford greater maneuverability. The study of muscular motion is called *kinesiology*, a specialized subdiscipline of physiology.

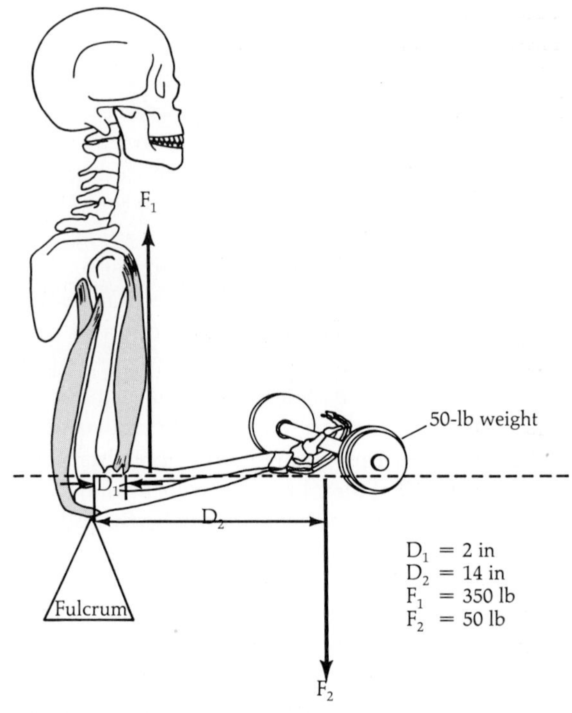

FIG. 13-7
Principles of leverage operating among the biceps muscle, hand, and elbow. A force of 350 lb (F_1) must be generated over a distance of 2 in (D_1) to move a 50-lb weight (F_2) 14 in (D_2).

Muscles and Work

The work performed by muscles is governed by additional physical principles, some of which are described in Table 13-1. These basic relationships, which incorporate the concepts of work (W), power (P), and velocity (V), may be used to derive the equations which are illustrated.

Muscles do work when they contract. By noting the distance they shorten, the load they lift, and the time required, one can use a kymograph to record the work performed by isolated muscles (Fig. 13-1). The amount of *work* is measured as a force (load or weight) moved through a distance ($W = F \times D$, Table 13-1). Greater amounts of work can be performed by increasing the force exerted or the distance through which the force moves. For example, 10 lb lifted a height of 1 ft involves 10 ft-lb of work ($W =$

TABLE 13-1
Physical principles of work and power

Equation	Manipulations to Derive Other Equations
1. $W = F \times D$	
2A. $P = W/T$	Since $W = F \times D$, one can substitute $F \times D$ for W in Eq. 2A and obtain Eq. 2B.
2B. $P = F \times D/T$	Since velocity (V) = distance/time (D/T), V may be substituted for D/T to derive Eq. 3A.
3A. $P = F \times V$	Division of both sides by V leads to Eq. 3B.
3B. $P/V = F$	

Symbols: W = work; F = force; D = distance; P = power; T = time; V = velocity.

10 lb × 1 ft). The amount of work is doubled with a twofold increase in weight or height (e.g., 20 lb × 1 ft, or 10 lb × 2 ft = 20 ft-lb).

Power is the rate of doing work. It can be calculated by the equation $P = W/T$. Powerful muscles accomplish work in less time than weaker ones. For example, a 4-year-old child might lack the power to lift a stack of books weighing 20 lb to a table 3 ft high within 2 s. The power required would be 30 ft-lb/s $\left(P = \frac{20 \text{ lb} \times 3 \text{ ft}}{2 \text{ s}} = 30 \text{ ft-lb/s}\right)$. However, if the books weighed 2 lb each, the child might lift them one at a time until 10 were raised to the table top. Assuming 20 s was needed to perform the task in this manner, the power required would be one-tenth as great as in the previous example. It would be 3 ft-lb/s $\left(P = \frac{20 \text{ lb} \times 3 \text{ ft}}{20 \text{ s}} = 3 \text{ ft-lb/s}\right)$. The amount of work is the same in both examples (20 ft-lb), but the time and therefore the power differ.

Muscles exert varying amounts of power in several ways. As described earlier, the CNS can alter the number, type, and nature of activity of the motor units involved. The CNS normally activates or recruits greater or lesser numbers of motor units within the muscle to produce more or less force. The most powerful muscles have many motor units and a preponderance of fast fibers, which are activated simultaneously. They develop maximal force and velocity to attain maximum power which can be calculated by $P = F \times V$ (Eq. 3A, Table 13-1). An optimal force exists whereby each muscle produces maximum power. To maintain constant power, muscles contract more slowly (their velocity of contraction decreases) as the force of contraction (load) increases and vice versa (Fig. 13-8). When heavier loads are lifted, more time is required for the elastic and contractile components to generate enough tension to overcome the inertia of the load, as described earlier (Figs. 13-3 and 13-4). The principles, which are illustrated mathematically in Eq. 3 of Table 13-1 and in Fig. 13-8, can be demonstrated simply if one compares the time it takes to lift a very heavy object with the time to lift a light one.

Bicycling also illustrates these physical relationships. Because of opposition due to gravity, a cyclist must exert more force to pedal uphill than on level ground. The increased requirement for force demands more strenuous (more forceful) and usually slower (lower-velocity) pedaling. Multigeared bicycles lessen the force

FIG. 13-8
The relationship between velocity of contraction and the force generated by a muscle. As increasing force is generated by a muscle, the velocity of contraction decreases.

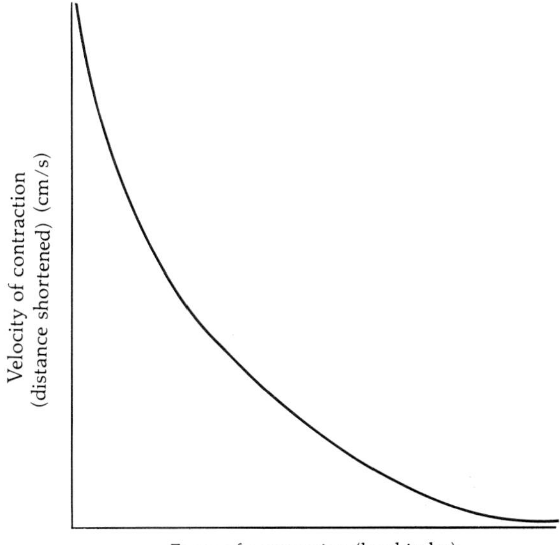

requirement. To maintain constant power (and forward movement of the bicycle) a lower (larger) gear may be used, which requires less force but faster pedaling, as described in Eq. 3A ($P = F \times V$). A higher (smaller) gear produces the opposite effect.

In summary, the work a muscle produces is the product of the force that it develops and the distance over which it contracts. The velocity of contraction varies inversely with the force developed. As the load or weight increases (i.e., as the force requirement increases), the velocity of contraction decreases until a load is reached which the muscle cannot move. Because the muscle does not shorten, the velocity of contraction is zero. The force produced is maximal and isometric contraction is attained.

13-2 ENERGY METABOLISM

Chemical reactions produce energy in muscle cells just as they do in other kinds of cells. The totality of these reactions, termed *metabolism,* may be separated into *catabolism,* which breaks down molecules, and *anabolism,* which synthesizes them. Chemical changes that take place in the absence of oxygen are termed *anaerobic reactions.* They provide much of the energy during muscle contraction and are associated with accumulation of lactic acid in muscle and blood plasma, resulting in fatigue. Chemical changes that require oxygen are called *aerobic reactions.* They generate much energy during the recovery of a muscle.

Prominent compounds metabolized in the body include the three major types of organic molecules in foods. They are carbohydrates, fats, and proteins (described in Chap. 2) and are metabolized by interconnected pathways in a series of reactions, many of which are reversible. The activities of specific enzymes regulate the rates at which the reactions occur.

Anaerobic and aerobic pathways lead to the synthesis of high-energy phosphate compounds such as adenosine triphosphate (ATP) and creatine phosphate (CP), which can be stored or used as immediate sources of energy. Exercises in which strength and power predominate rely primarily on immediate and anaerobic sources of energy. Endurance or prolonged exercises rely largely on aerobic pathways. These relationships are illustrated in Fig. 13-9.

Anaerobic Glycolysis and Lactic Acid Formation

Glycogen, a polymerized form of glucose, is stored in skeletal muscle and represents an average weight of about 0.7 percent of the tissue. The chemical bonds of glycogen and glucose contain energy which is initially released by anaerobic catabolic reactions called *glycolysis* (glycogen or glucose lysis). The reactions proceed in the cytoplasm as follows:

$$\text{Muscle glycogen} \leftrightarrow \underset{\text{(6-Carbon compound)}}{\text{glucose}} \leftrightarrow$$

$$\underset{\text{(3-Carbon compound)}}{\text{2 pyruvic acid}} \underset{-2H \; +O_2}{\overset{+2H}{\rightleftarrows}} \underset{\text{(3-Carbon compound)}}{\text{2 lactic acid}}$$

A phosphate group is added to glucose in the steps intermediate to its conversion to two mol-

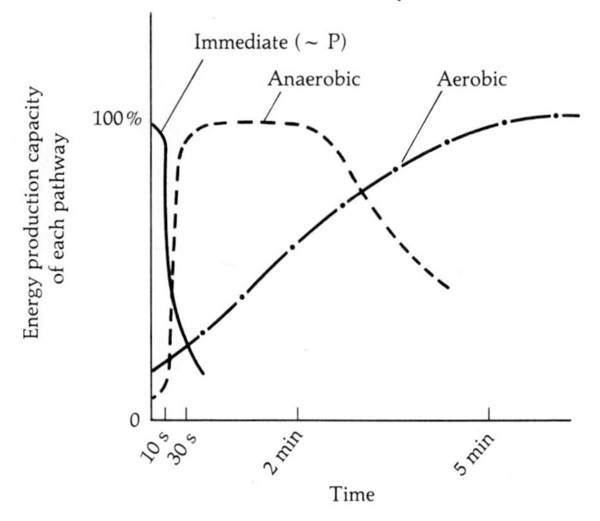

FIG. 13-9
Energy sources for muscle activity. Muscles first use energy stored in high-energy phosphate bonds, then that produced by anaerobic pathways, and finally utilize aerobic metabolism to fuel their activity.

ecules of pyruvic or lactic acid. The addition of phosphate to the carbon chain uses energy from ATP. However, the anaerobic cleavage of chemical bonds of one molecule of glucose provides enough energy to produce an excess of two molecules of ATP. Fatigue occurs when muscle consumes all the available ATP.

Energy is also locked up in the chemical bonds of lactic acid. The conversion of lactic acid to pyruvic acid and synthesis of liver glycogen require oxygen (Fig. 13-10). Accumulation of lactic acid in the muscle cell lowers the pH of its intracellular fluid. As a result, enzymes critical to energy production and sensitive to pH changes are turned off, resulting in fatigue, which may be accompanied by pain.

Lactic acid derived from the anaerobic degradation of glucose in muscle diffuses to the blood, which carries it to the liver, where it is converted to glucose, which recirculates to muscle in a series of reactions called the *Cori cycle* (Fig. 13-10). By means of aerobic reactions, liver enzymes remove hydrogen from lactic acid, converting it into pyruvic acid. About 16 percent of the lactic acid is oxidized to produce energy stored in the form of ATP. The energy is used by liver cells to reverse glycolysis, ultimately converting about 84 percent of the lactic acid into liver glycogen (Fig. 13-10). The glycogen may then be broken down into glucose, diffuse into the blood, and be carried to muscle cells. There the glucose may be polymerized to form muscle glycogen, completing the lactic acid (Cori) cycle.

Oxygen consumption rises in proportion to work but does not keep pace with the demands for energy. Therefore the cell builds up an oxygen debt as lactic acid accumulates. Because lactic acid diffuses into the blood, its level in blood may be used to assess the oxygen debt.

FIG. 13-10
Carbohydrate and lactic acid cycling (the Cori cycle) and energy production for muscle contraction. Anaerobic, oxygen-independent reactions are indicated by solid lines. Oxygen-dependent reactions are indicated by dotted lines. Glucose and lactic acid are interconverted by metabolic reactions in muscle and liver as these compounds and their intermediate, pyruvic acid, are carried to these tissues by the blood.

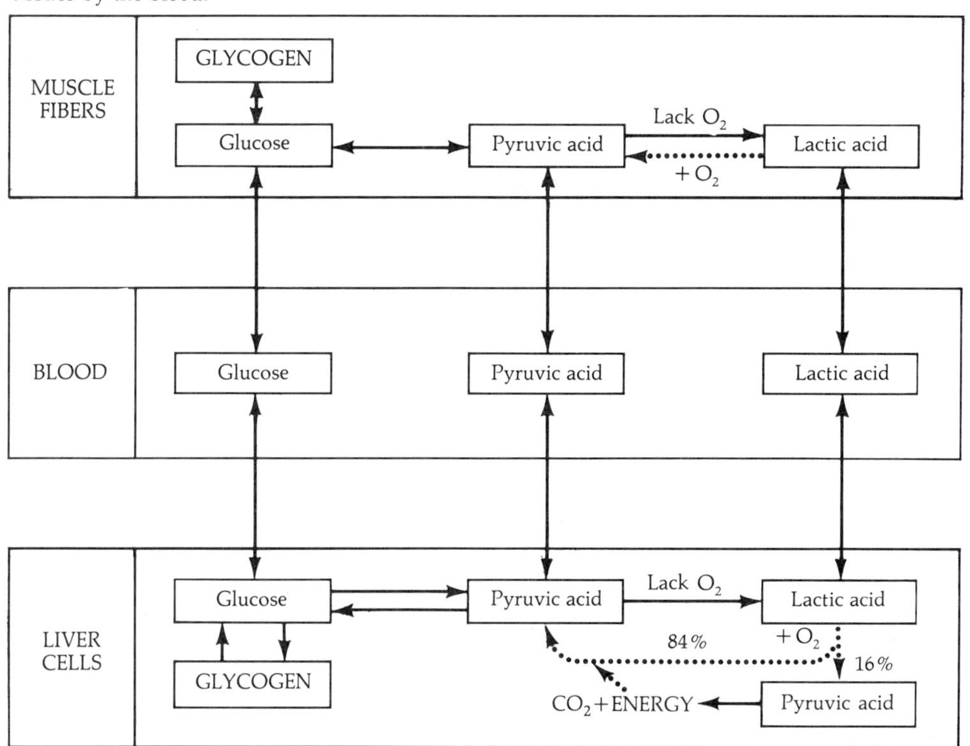

Because of work (or exercise) an individual uses up more oxygen than the resting requirement. The excess oxygen consumed during recovery is called the *oxygen debt.* The time required to repay the oxygen debt is proportional to the intensity of work, its duration, and the degree of exhaustion attained. The time decreases with training but may last 2 h or more after as little as 2 to 5 min of most exhausting work even in well-trained athletes.

Aerobic Oxidative Phosphorylation

The pathways involved in the production of ATP are shown in Fig. 13-11. Some of the principal metabolic reactions of carbohydrates, fats, and proteins are outlined. Aerobic reactions form ATP when organic molecules are oxidized, using respiratory enzymes located in the mitochondria. The process is called *oxidative phosphorylation.* Specific enzymes add phosphate to the carbon chains and break the hydrogen and carbon bonds of individual compounds, liberating energy which is used to synthesize ATP.

Complete anaerobic and aerobic oxidation of glucose produces a net gain of 38 molecules of ATP. During anaerobic glycolysis (which splits glucose anaerobically), 2 net (excess) molecules of ATP are generated, while 36 more are formed by the aerobic degradation of the two molecules of pyruvic acid derived from each molecule of glucose (aerobic oxidative phosphorylation). Thus, muscles obtain considerably more energy from glucose in the presence of oxygen. Aerobic oxidation, and thus the duration of intense exercise, are limited by failure to provide sufficient oxygen to sustain oxidative phosphorylation.

Pyruvic acid and a 2-carbon compound, acetyl coenzyme A (acetyl CoA), play pivotal roles in energy production. During anaerobic glycolysis, one molecule of glucose forms two molecules of pyruvic acid. During the first step of oxidative phosphorylation, each molecule of pyruvic acid loses CO_2 to form a 2-carbon acetyl fragment that is combined with coenzyme A to yield acetyl CoA. This reaction is the only irreversible one illustrated in Fig. 13-11.

The 2-carbon chain of acetyl CoA is then combined with the 4-carbon chain of oxalacetic acid to form citric acid. Citric acid is metabolized in a series of enzymatically catalyzed reactions in which hydrogen is stripped off its carbon chain. The 6-carbon chain of citric acid is gradually cleaved and ultimately forms more oxalacetic acid, which may combine with additional acetyl CoA fed into the cycle to form more citric acid. In this manner, a metabolic cycle occurs in which energy is released in small steps from the cleavage of chemical bonds. The energy is used to drive reactions that form ATP. The cycle is called the *citric acid cycle.* Since a few of the intermediate compounds contain three carboxylic acid groups (—COOH), it also is called the tricarboxylic acid (TCA) cycle and sometimes is called the Krebs cycle after Hans Krebs who was a pioneer in delineating this aspect of metabolism.

A single cycle produces 12 molecules of ATP for each acetyl CoA that enters it. Thus, 36 moles of ATP are derived from the aerobic phase of glucose metabolism (since the 6-carbon chain of glucose may be metabolized to form three molecules of acetyl CoA). The end products of the cycle are carbon dioxide, water, and energy. The latter is stored in the high-energy phosphate bonds of ATP and CP.

When carbohydrate intake is high and exceeds the energy demands of muscle, fats and limited amounts of glycogen can be synthesized and stored as energy reservoirs. About

FIG. 13-11
A simplified scheme of anaerobic and aerobic metabolism and ATP synthesis. ATP is produced and consumed by metabolic reactions. The net (excess) number of molecules of ATP formed by anaerobic and aerobic catabolism of one molecule of glucose is shown in blocks on the right side of the diagram. The upper half of the illustration mainly illustrates anaerobic reactions; the lower half shows aerobic reactions. (*) Specific amino acids such as aspartic acid and glutamic acid are involved in reactions in which oxidative deamination (removal of NH_2 groups) occurs as the amino acids enter the citric acid cycle. Amino acids also may enter the cycle under the influence of transaminases (which readily transfer NH_2 groups from one compound to another).

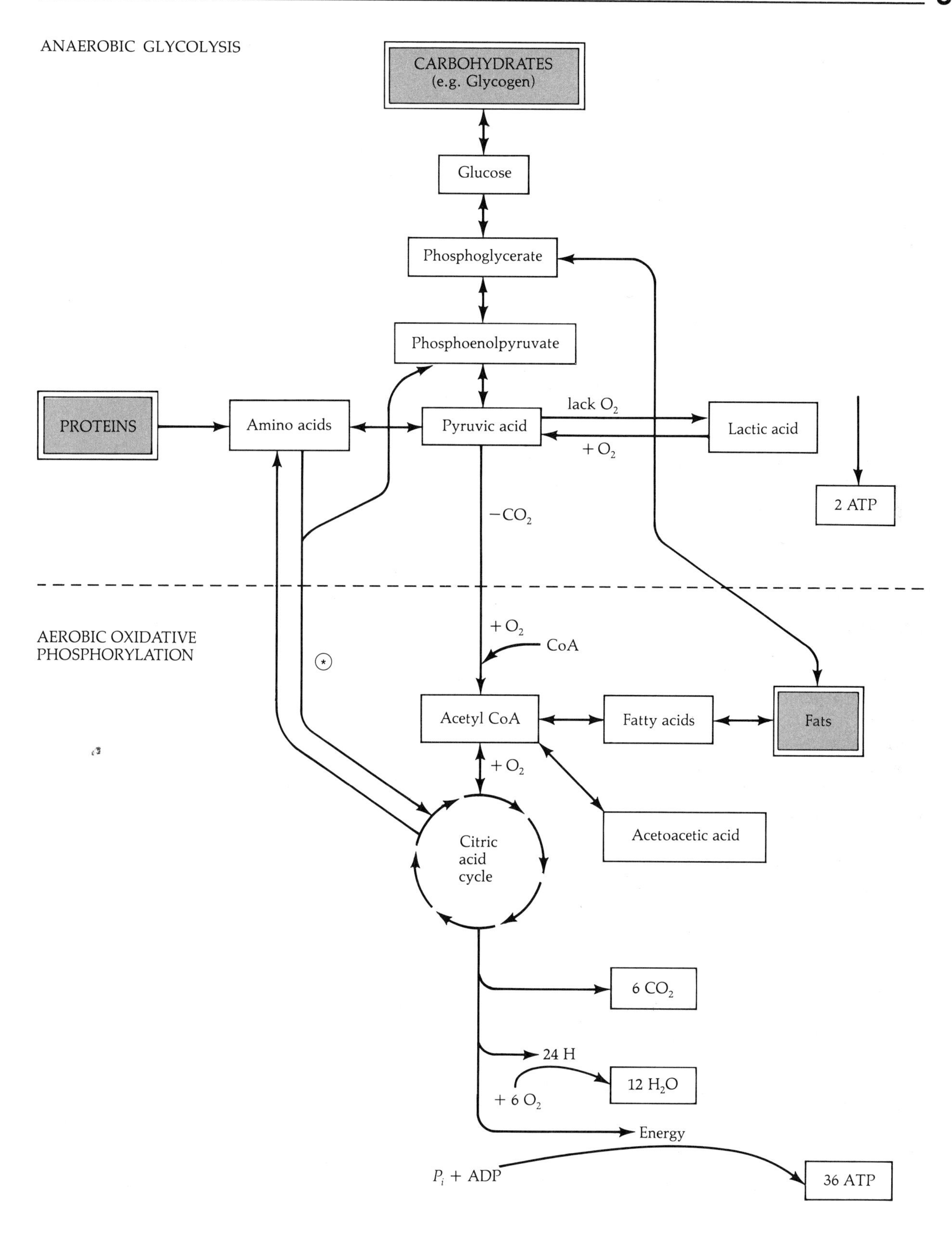

ANAEROBIC GLYCOLYSIS
CARBOHYDRATES (e.g. Glycogen)
Glucose
Phosphoglycerate
Phosphoenolpyruvate
PROTEINS
Amino acids
Pyruvic acid
lack O_2
Lactic acid
$+O_2$
2 ATP
$-CO_2$
AEROBIC OXIDATIVE PHOSPHORYLATION
$+O_2$
CoA
Acetyl CoA
Fatty acids
Fats
$+O_2$
Acetoacetic acid
Citric acid cycle
$6\ CO_2$
24 H
$12\ H_2O$
$+ 6\ O_2$
Energy
P_i + ADP
36 ATP

0.7 percent (0.5 to 1.0 percent) of the weight of muscle in an average adult is due to glycogen. In a 70 kg adult this amounts to an average of 245 g (assuming up to 50 percent of the body weight may be due to muscle). Since the complete oxidation of carbohydrate produces 4 kcal/g, muscle stores about 980 kcal of energy.[1] Up to 6 percent of the weight of the liver is due to glycogen. This amounts to 108 g or 432 kcal (assuming an average liver weight of 1.8 kg). ECF contains an average of about 0.1 percent glucose, which amounts to 10 g of carbohydrate that can provide 40 kcal of energy. These estimates correspond to an average total carbohydrate energy reserve of the body amounting to only 1452 kcal (980 + 432 + 40). Consequently, glycogen stores do not provide a sufficient energy reserve for usual daily activities. The body must use other molecules in addition to carbohydrate to provide energy. For example, fasting for 12 to 18 h almost depletes the liver of its glycogen reserves although muscle glycogen is spared.

Fats are important fuels for several types of muscle fibers. Energy is derived from the aerobic oxidation of fatty acids and ketone bodies (such as acetoacetic acid, β-hydroxybutyric acid, and acetone). Cardiac fibers, resting skeletal muscle fibers, and muscle fibers used for endurance activities can burn fat as their main caloric source. Cardiac fibers oxidize ketone bodies preferentially to glucose. Skeletal muscle fibers employed in endurance activities have a high oxidative capacity and include the slow-intermediate B fibers and the fast-red C fibers.

Proteins also may supply energy when there is insufficient carbohydrate and fat. After an individual fasts for 2 to 3 days, proteins are hydrolyzed into amino acids. Some of the amino acids are deaminated (NH_2 is removed) and enter the citric acid cycle directly (Fig. 13-11). Most of them are first converted to pyruvic acid. Should protein hydrolysis continue, myofibrils may undergo degradation and the muscle cells and muscles atrophy. This phenomenon, called *muscle wasting*, is characteristic of prolonged protein malnutrition.

[1] A calorie (cal) is the amount of heat needed to raise the temperature of 1 g of water by 1°C. Since metabolic and chemical reactions involve thousands of calories, the unit of energy used in describing these reactions is the kilocalorie (kcal), which equals 1000 cal and is also called the large calorie or Calorie. The latter term is often used in discussions of nutrition.

High-Energy Phosphate Metabolism

The catabolism of nutrients provides energy from glycolysis and oxidative phosphorylation. Energy is captured and conserved by linking (coupling) those reactions to others which synthesize *high-energy phosphate bonds.* The latter have been called the "energy currency" of the cell. The bonds are symbolized by ~P, a designation which indicates an ability to transfer energy from the phosphate bond to some acceptor (symbolized X~P). ATP is the most immediate source of energy and the common unit of "energy currency." The energy derived from glycolysis and oxidative phosphorylation is used to synthesize ATP from adenosine diphosphate (ADP) by reaction No. 1 in Fig. 13-12.

The energy produced by the hydrolysis of ATP is used to contract muscle, to drive metabolic reactions, and to produce heat. When ATP is abundant, energy stored in its high-energy phosphate bonds is converted to other units of energy currency, such as creatine phosphate (C~P, reaction 4 in Fig. 13-12). Resting muscles store about 3 to 8 times more C~P than ATP.

As ATP is consumed, C~P breaks down to provide energy for the production of more ATP. The synthesis of ATP is favored at the expense of C~P (as indicated by the thicker arrows in Eqs. 5 and 6 in Fig. 13-12). Therefore, muscle does not normally fatigue or lose all of its ATP until its C~P reservoirs are exhausted. The C~P reservoirs supply enough energy for about 30 to 100 contractions. When death occurs, muscle cells cannot replace ATP. The absence of available energy prevents the dissociation of actin-myosin bonds, leaving the muscles in the stiffened muscular condition of *rigor mortis* (Latin, stiffness of death).

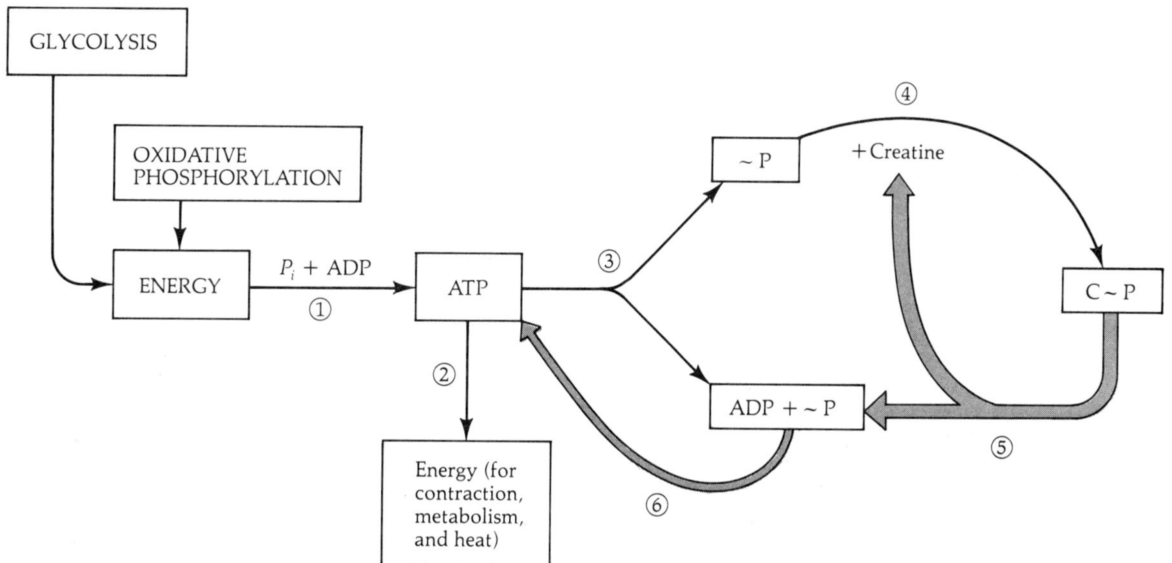

FIG. 13-12
The production of energy and high-energy phosphate compounds. When energy reserves are required by muscle, the equilibrium favors reactions in the directions indicated by the thicker arrows. ATP = adenosine triphosphate, ADP = adenosine diphosphate, P_i = inorganic phosphate, ~P = high-energy phosphate bond, and C~P = creatine phosphate.

Energy Production and Muscle Contraction

Heat Production and Muscle Activity

Muscles release heat in three phases: *activation, shortening,* and *delayed heat.* The first two phases take place in a fraction of a second, during which energy is provided from ATP and C~P reservoirs and anaerobic reactions. The third phase requires as long as 300 s and occurs during the recovery of the muscle.

During the activation heat phase about 30 to 50 percent of the energy developed by the muscle is consumed. Most of it is used to move Ca^{2+} ions back to the sarcoplasmic reticulum. Prior to entering the activation phase the globular heads of myosin are readied for movement and are in the r crossbridge (or rigor) position. Each head is at an acute angle with respect to the actin filaments (Fig. 13-13A). A threshold stimulus depolarizes the cell membranes and the Ca^{2+} ions to diffuse down their concentration gradients out of the sarcoplasmic reticulum and among the myofibrils. The Ca^{2+} ions allow myosin-ATPase to bind to actin, which produces actomyosin-ATPase, which hydrolyzes ATP rapidly. Twitch tension begins to develop during the formation of activation heat.

Shortening heat develops as the globular heads of myosin move to a p cross bridge position, more perpendicular to the actin filaments (Fig.13-13B). The filaments slide over one another, peak tension develops, and mechanical relaxation follows.

Delayed heat is formed during recovery as the muscle is readied for its next contraction. The greatest amount of ATP hydrolysis and energy production takes place at this time. Myosin-ATPase binds to and hydrolyzes ATP. The globular heads of myosin move from the p to the r state (Fig. 13-13C). They are repositioned much like the head of a pistol, cocked and ready to fire. During the next activation period, the trigger mechanism is "pulled" by a stimulus that releases Ca^{2+} ions. The Ca^{2+} ions change the shape of troponin and allow myosin to bind to actin, forming and activating actin myosin-ATPase to initiate the next cycle of contraction and heat release.

Formation of "Extra" Shortening Heat

The chemical reactions observed during contraction do not account for all the heat that is

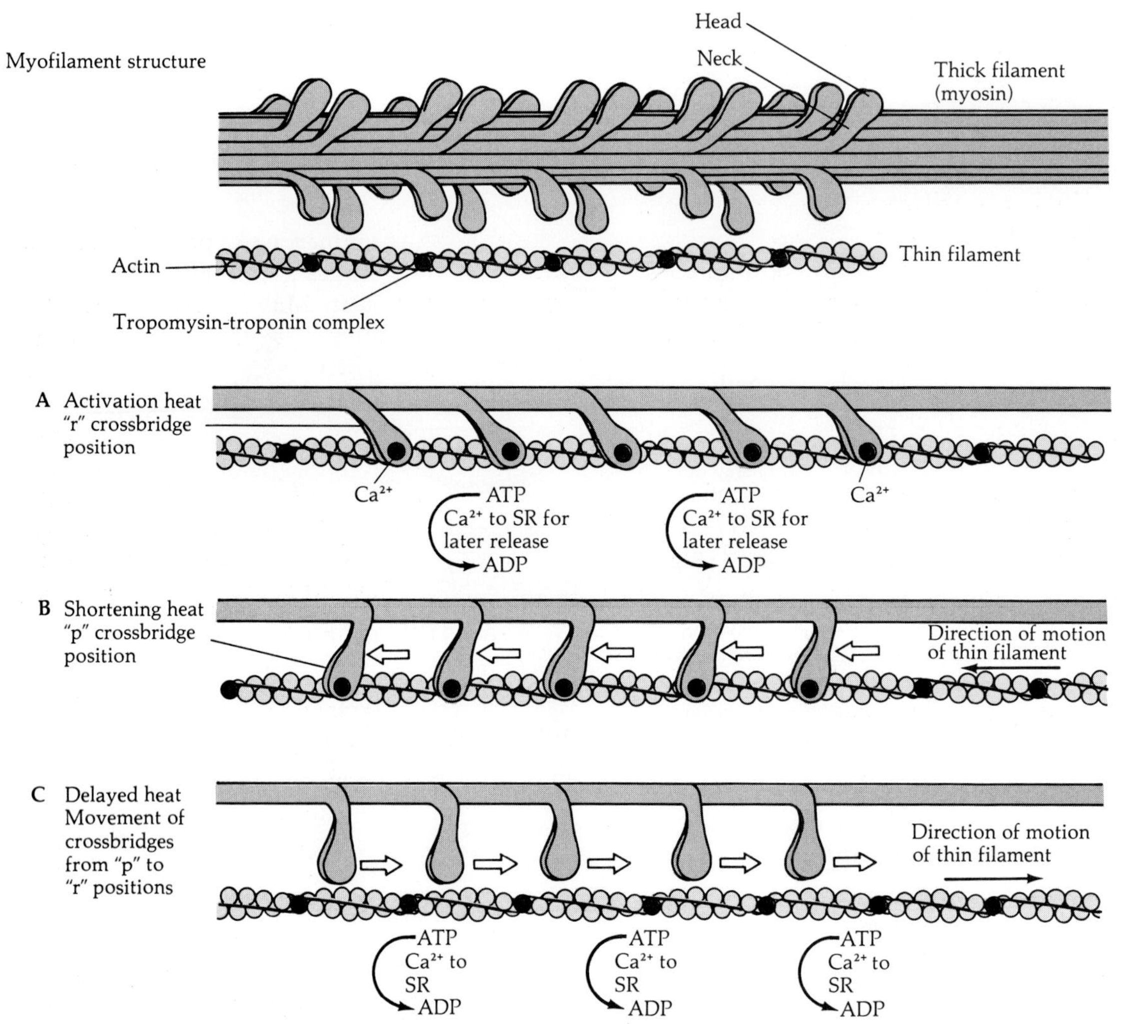

FIG. 13-13
Phases of heat production during movement of myofilaments. The arrows indicate the direction of motion of the respective myofilaments. SR = sarcoplasmic reticulum.

generated. The energy output of muscle is equal to the work it performs plus the heat it produces and is greater during isotonic than isometric (tetanic) contraction. That is, under isotonic and isometric conditions, a contracting muscle fiber releases nearly a constant amount of heat but performs more mechanical work in the first case and less in the second. The difference in utilization of energy (and of oxygen) is called the *Fenn effect*, the amount of oxygen consumed in isotonic contraction also being greater. Extra shortening heat is released in proportion to the decrease in length. No corresponding amount of metabolism has been discovered to account for the extra release of heat. Some investigators believe the Fenn effect results from the use of extra ATP in isotonic contraction. Others hypothesize that it may be due to changes that occur in proteins without the splitting of ATP or C~P. They theorize that during rapid shortening the movement of myofibrils and the making and breaking of cross-bridges may produce heat. After the muscle has shortened, these steps are apparently reversed as ATP is split.

Efficiency of Muscle

The maximum efficiency of muscle (the ratio of conversion of chemical to mechanical energy) is

25 percent. That is, 75 percent of the available energy forms heat and the remainder performs useful work during isotonic contraction. The efficiency is about the same as that of a gasoline engine. Peak efficiency is attained in muscle at an optimal speed of contraction. It is as though an energy cost is required to overcome friction in "starting up" contraction. The start-up cost is lower in red muscles than in white muscles. The relationship between energy consumption in μmoles of ATP used and velocity of contraction is shown in Fig. 13-14.

Exposure to cold causes heat production in muscle to increase several hundred percent above normal. Alternating cycles of weak rhythmic contractions of the skeletal muscles, called *tremors* or *shivering*, occur at rates of 10 to 20 times per second. They produce no useful work but generate large amounts of heat to warm the body. These responses constitute reflexes controlled by descending nerve pathways from the hypothalamus. They represent one of several mechanisms whereby the hypothalamus acts as the temperature regulation center of the body (Chap. 11, page 267).

FIG. 13-14
The relationship between energy consumption and the velocity of contraction. Note that more ATP is consumed to start up contraction. ATP consumption decreases within the optimal range of velocity of contraction.

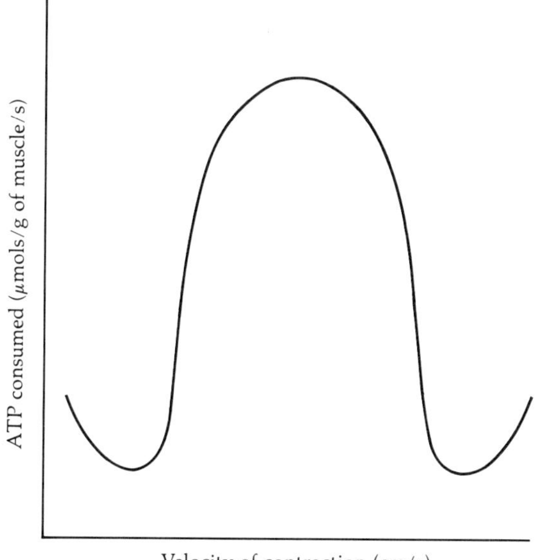

13-3 THE EFFECTS OF EXERCISE

Exercise is the performance of a physical activity during which muscles carry out useful work and/or produce heat. Muscle activity may be varied in its duration, speed, and force of contraction. Specific exercises may differ in one or more of these variables. As exercise continues, shifts occur in energy metabolism pathways from the consumption of energy stored in high-energy phosphate compounds (which is immediately ready for use) to consumption of that derived from anaerobic pathways and then to consumption of that produced by aerobic pathways, as shown in Fig. 13-9.

Exercise and Physical Fitness

Exercise and physical fitness are linked to the endurance of muscle, the flexibility of joints, and the general functions of the body, including those of the respiratory and circulatory systems. *Muscle endurance* is the ability to maintain persistent physical activity against less than maximal resistance. The respiratory and circulatory systems cooperate to supply muscle with the O_2 necessary for aerobic reactions and rid it of the excess CO_2 that results from metabolism. The following discussion emphasizes the role of muscle in exercise; later chapters (notably Chap. 16 on the heart and Chap. 20 on respiration) will describe the role of other tissues, organs, and systems.

Fatigue

Fatigue occurs in a muscle when it runs out of energy and can no longer respond to a stimulus. Although the causes are not clear, the development of fatigue has been related to several factors. Among them are the availability of glycogen, the oxidative capacity of muscle, and the nature of neural activity.

The amount of stored glycogen can be a limiting factor in skeletal muscle endurance. Depletion of glycogen, a decrease in pH, and an accumulation of lactic acid, salts, CO_2, and other metabolites parallel fatigue. Susceptibility to fatigue is inversely related to the oxidative capacity of a muscle cell and directly to its size and that of its nerve fiber. Large muscle fibers have a lower oxidative capacity and greater susceptibility to fatigue than small muscle fibers. In some instances, fatigue is neural in origin and may result from changes in the availability of critical substances or from other alterations that impede the effectiveness of neuromuscular transmission.

Cramps and Pain

Muscle *cramps* are involuntary, intermittent contractions which usually produce pain. The causes are not clear but may involve motor nerve impulses, metabolic disturbances involving salts such as those of Ca^{2+}, K^+, and/or abnormalities in actin or myosin. Cramps often can be relieved by stretching or massaging a muscle and/or replacement of salts lost through sweating.

Muscle pain and stiffness can follow vigorous exercise. The cause is not precisely known but may be due to some of the factors described above. An interruption in blood flow to muscle can result in pain. Several polypeptides have been detected and related to the onset of pain. They include kinins (polypeptides), which arise from hydrolysis of globular proteins in the blood, and the P substance produced in the cells of the brain, spinal cord, and gastrointestinal system. P substance is an excitatory neurotransmitter whose activity may be blocked by the enkephalins from the brain (Chap. 10, p. 249).

Training and Adaptation

Training is adaptation to repeated exercise. It involves the independent replication of two types of muscle proteins, those that produce tension and those that aid aerobic metabolism (oxidative phosphorylation). Both responses usually occur as a result of training. However specific exercises normally favor one or the other. Thus the muscles of a weight lifter with bulging biceps have increased amounts of myofibrillar protein although the individual is not necessarily in condition to run a long race.

Activity requiring strength (such as weight lifting) primarily builds up proteins such as actin and myosin, which are used to develop tension. Forceful, repetitive exercise in which about 75 percent of maximal muscle tension is developed leads to replication of the myofibrils. The diameter of individual fibers increases, as does that of the whole muscle. This leads to increased strength, since the force created by a muscle is directly related to its size. The force averages 4 kg/cm^2 of cross-sectional area of muscle tissue. Since the reactions leading to myofibril synthesis are reversible, cessation of exercise gradually results in loss of added myofibrillar components.

Training that demands endurance, such as swimming, bicycling, and jogging, mainly causes replication of proteins involved in aerobic metabolism. The proteins built up by such endurance exercises include myoglobin, respiratory enzymes, and others found in mitochondrial membranes. Endurance training also may increase the capillary-to-muscle fiber (cell) ratio to allow for better perfusion of muscle (movement of substances through the blood to the tissue). Thus, a trained runner may have greater amounts of proteins, which aid aerobic metabolism. The runner, in contrast to the weight lifter, is not necessarily equipped to lift heavy weights.

Maximal Exercise

Aerobic adaptations follow maximal and submaximal exercises. *Maximal exercise* is a physical activity which requires O_2 to be used at rates beyond the capacity for it to be supplied. The maximal volume of oxygen a person can deliver to muscle during the most strenuous exercise is the $Vo_{2_{max}}$. It is a measure of aerobic capacity. More O_2 is often required than can be made available immediately in spite of increases in the depth and rate or respiration. The heart rate and volume of blood pumped to the lungs for oxygenation only can increase within limits, which are determined genetically (Table 20-7).

Anaerobic reactions are dominant in highly

intense exercise (such as a sprint lasting 60 s). Energy production is limited in the absence of oxygen and the presence of carbon dioxide, causing muscle cells to derive most of their energy from anaerobic glycolysis. As discussed earlier, an oxygen debt is produced during this period. It must be repaid in order to convert the lactic acid which is formed into pyruvic acid in order to attain maximal available energy. About 80 percent of the oxygen needed to reduce the debt is made available after the exercise is over. The remainder is available for aerobic metabolism.

Submaximal or Endurance Exercise

Submaximal exercise is physical activity requiring less than $Vo_{2_{max}}$. It is also called *endurance exercise*. Examples of exercises which may be carried on as endurance activities include swimming, bicycling, walking, and jogging. During this type of activity, the Vo_2 increases linearly with heart rate, as shown in Fig. 13-15. It also increases linearly with pulse rate (since heart and pulse rates are normally identical). Therefore, the intensity of endurance exercise may be estimated by noting the pulse rate as an indirect measure of heart rate.

Optimal metabolic changes occur when exercise produces 60 to 90 percent of the maximum heart rate (equivalent to about 50 to 80 percent $Vo_{2_{max}}$), lasts 20 to 60 min, and is repeated 3 to 5 days each week. Exercise (work) is considered to be of low intensity when it produces 60 to 70 percent of maximum heart rate and of high intensity when it approaches the 90 percent level. Untrained individuals can maintain high-intensity exercise for only a few minutes, whereas long-distance runners may continue running at 80 to 90 percent of maximum heart rate for 2 to 3 h. Because stores of glycogen are limited, its availability becomes an important factor in determining the duration of high-intensity exercise.

Some athletes may improve their performance in prolonged endurance competition by increasing glycogen stores in order to make more energy immediately available. They practice *carbohydrate* or *glycogen loading*. The regimen combines a low carbohydrate diet with strenuous exercise for more than 1 to 1.5 h/day for 1 week. Three to four days before competition, the diet is altered so that carbohydrates supply most of the caloric requirements. Some reports indicate that under these conditions muscle glycogen can rise to 3.7 percent of the weight of the muscle (compared with the usual range of 0.5 to 1.0 percent). As a result, the amount of time that endurance exercise may be performed prior to exhaustion (fatigue) seems to increase. Like any other diet and exercise program, *this routine should be adopted with caution and only on the advice of a qualified physician*.

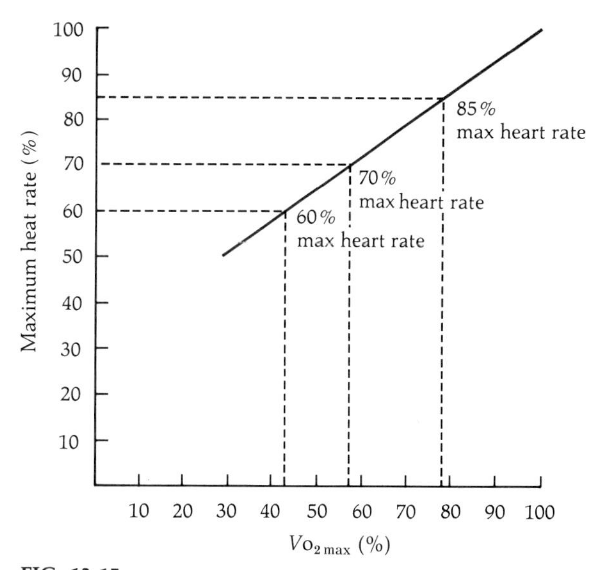

FIG. 13-15
The relationship between O_2 uptake ($Vo_{2\,max}$) and heart rate. Oxygen uptake increases linearly with heart rate.

The $Vo_{2_{max}}$ can vary among individuals of the same weight. For the same rate of work, one person may perform at a lower percentage of his or her $Vo_{2_{max}}$ than another person. Consequently, the former individual conserves glycogen by providing more oxygen and more energy from the glycogen used (since it is more completely oxidized to provide energy for muscular work).

Individuals of different weights may have the same $Vo_{2_{max}}$. When running up a hill however, the lighter individual performs less work and uses a smaller percentage of his or her $Vo_{2_{max}}$.

Under such conditions, the onset of fatigue may be delayed in lighter individuals.

Calculation of Target (Training) Heart Rate

How long physical activity can be maintained is directly related to training and heart rate associated with the exercise. The duration of endurance exercises can be set at a particular heart rate for each individual. This rate is called the individual's target or training heart rate. It may be calculated as a function of the *maximum heart rate* permissible for safe exercise by that individual. The maximum heart rate is most accurately determined as the highest heart rate recorded during a graded exercise test *under the supervision of qualified medical personnel*. It is related to the age, level of fitness, and health of the individual. It can also be estimated by measurement of the heart rate after either a 12-min or 1.5-mi run at top speed. Since maximum heart rate is a function of age, a simple means of *estimating* it is to express the value as 220 beats per minute minus the age of the individual in years.

The training (target) heart rate may be approximated by using the following equation:

$$\text{Training heart rate} = 0.60\,(\underbrace{220 - \text{age in years}}_{\substack{\text{Maximum}\\ \text{heart rate}}} - \text{resting heart rate}^{1}) + \text{resting heart rate}$$

For example, a 50-year old person's maximum heart rate is estimated at 220 − 50 = 170 beats per minute. If the resting heart rate of that individual is 70 beats per minute, the training heart rate (to provide 60 percent of maximal heart rate) is:

$$\begin{aligned} \text{Training heart rate} &= 0.60\,(220 - 50 - 70) + 70 \\ &= 0.60\,(100) + 70 \\ &= 60 + 70 \\ &= 130 \text{ beats/min} \end{aligned}$$

[1] The resting heart rate is equal to the pulse rate taken in a comfortable, quiet sitting position for 30 s immediately after rising from a restful night's sleep.

To calculate the training heart rate at 70 percent, 80 percent, or 85 percent of maximal heart rate, one merely inserts the decimals 0.70, 0.80, or 0.85 in place of 0.60 in the equation, or one may estimate the training heart rate needed to attain 70 to 85 percent maximal heart rate from Fig. 13-16.

Following a successful medical evaluation by a qualified physician, a designed exercise routine can be recommended. In the example above, the target would be to attain a heart rate of 130 beats per minute and maintain it for 20 to 60 min through exercise performed 3 to 5 days per week. Some individuals are substantially below their genetically determined maximum fitness. In these people, any activity producing sustained elevations of heart rate, even when modest, provides physical training.

Effects of Endurance Training on Muscle

The adaptive changes which occur after submaximal exercise lead to endurance that especially increases the aerobic capacity of skeletal muscle (summarized in Table 13-2).

Although endurance training affects all types of skeletal muscle cells, the changes are most prominent in fast-twitch high oxidative type C red fibers (Table 12-2, p. 298). Pronounced increases occur in fat oxidation and metabolism. The production of energy from carbohydrate metabolism also increases, but it becomes relatively less important. The metabolism of skeletal muscle comes to resemble that of cardiac muscle. There is an increase in oxygen utilization, in number and activity of mitochondria, concentration of myoglobin, and metabolism of high-energy phosphate compounds. The skeletal muscle of an endurance-trained individual is optimally prepared for long-lasting activity.

Cardiac muscle fibers have less need to increase their aerobic capacity to maintain endurance activity. Because they usually derive most of their energy from aerobic pathways, the biochemical adaptations of cardiac muscle resulting from training are not as pronounced as those of skeletal muscle. A major response, however, is the hypertrophy of cardiac myofibrils, which leads to an increase in cell and

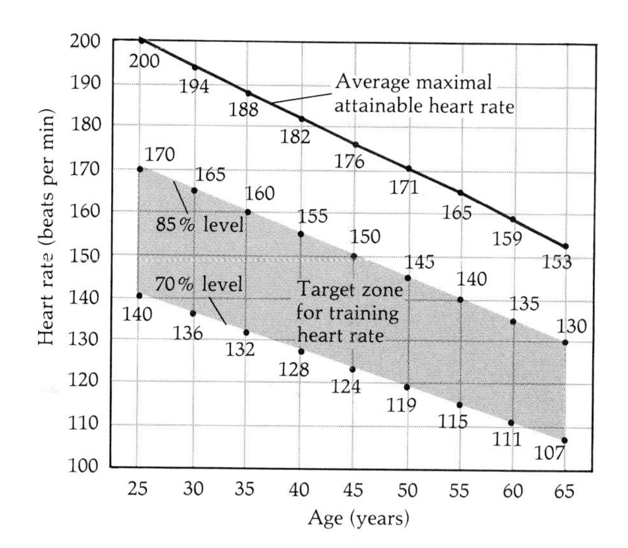

FIG. 13-16
Training (target) heart rate recommended to attain physical fitness at different ages. *Note:* A thorough medical evaluation by a qualified physician is required prior to planning any exercise program. The individual's age is noted and a program is designed to maintain the heart rate (beats/minute) within 70 to 85 percent of the average maximal attainable heart rate (the target zone) for that person during the prescribed exercise routine. The maximum heart rate equals 220 minus age in years. This chart shows that as we grow older, the highest heart rate which can be reached during all-out effort falls. These numerical values are "average" values for age. Note that one-third of the population may differ from these values. It is quite possible that a normal 50-year-old may have a maximum heart rate of 195 or that a 30-year-old might have a maximum of only 168. The same limitations apply to the 70 percent and 85 percent of maximum lines.

heart size. Effects of exercise on the heart will be described further in Chap. 16.

The capacity of the respiratory system to supply oxygen increases with training. This change, coupled with the more forceful contractions of the enlarged heart, supplies significantly more O_2 to meet the metabolic demands of aerobic oxidative reactions. Consequently,

TABLE 13-2
Effects of endurance exercise on skeletal muscle following training*

Property	*Increase*	*Decrease*	*Unchanged*
Carbohydrate metabolism	Carbohydrate oxidation (but not as much as in untrained muscle)	Lactic acid formation	
	Cellular permeability to sugar		
	Capacity for glycogen synthesis	Rate of glycogen depletion	
Fat metabolism	Fatty acid oxidation (more so than carbohydrates)	Plasma free fatty acids (during exercise)	
	Incorporation of fatty acids into triglycerides		
	Storage of triglycerides	Serum triglycerides	
Mitochondria	Number		
	Size		
	Protein content		
	Respiratory enzyme activity		
Factors in oxygen consumption	Myoglobin concentration in red fibers		
	Respiratory capacity		
	$Vo_{2\,max}$ (to larger muscle mass)	Blood flow per gram of muscle	Oxygen delivery to individual cells
High-energy phosphate metabolism	Aerobic generation of ATP		

* Endurance or submaximal exercise is that which requires less oxygen consumption than the individual's maximal capacity to supply it ($Vo_{2\,max}$).

with training, submaximal physical activity can be maintained for some time. During exercise, the heart rate of a trained individual levels off. As the individual shifts from anaerobic to aerobic pathways, the familiar "second wind" seems to occur. Oxygen consumption and demand are in equilibrium. This steady state has been reported to occur with oxygen consumption reported as high as 85 percent of $V_{O_2 MAX}$ in some trained individuals. As noted above, the effects of exercise will be explored further in Chaps. 16 and 20 on the heart and respiratory system.

13-4 SUMMARY

1 Contraction is the generation of force by muscle. It is caused when the myofilaments slide toward each other and stretch elastic elements, which then recoil. When a resting muscle fiber is moderately stretched, its recoil produces greater tension. The initial greater length caused by stretching generates more opportunity for the formation of bonds between actin and myosin. This arrangement allows the filaments to slide over an increased distance toward each other, producing more tension on the elastic elements in the muscle fiber. Tension can develop when the muscle length remains constant during isometric contraction, as well as when the muscle shortens during isotonic contraction. Isometric tension also develops when the load is too great to be moved by the muscle.

The mechanical response of a muscle fiber to a stimulus is called a twitch and occurs in three phases: the latent, contraction, and relaxation periods. The mechanical activity generated during the response can be recorded in a visual display called a kymograph, and the electrical activity displayed in an electromyogram (EMG).

Muscle tension is influenced by the strength and frequency of the nerve impulses and by the number and nature of active muscle fibers. Individual muscle fibers obey the all-or-none law. Following the application of a stimulus under a defined set of conditions, they either do or do not contract. The amount of muscle tension produced by a whole muscle generally parallels an increase in number of simultaneously active muscle fibers and/or motor units. An increase in the strength of the stimulus can cause more muscle fibers in the muscle to contract. The phenomenon is called recruitment.

Afferent signals are carried from receptors in muscles, tendons, and joints. They are relayed to motor areas of the CNS and generate excitatory and inhibitory efferent signals. In this manner, the CNS regulates contraction and relaxation of muscle groups.

Repetition of closely spaced stimuli causes maximal tension in a response called tetanus. The stimuli must be repeated at rates of about 30 to 100 times per second for slow and fast fibers, respectively. Relaxation occurs after fatigue. Stimulation at rates below those which induce tetanus also causes tension to increase in a steplike response called treppe or "warming up." Treppe may be caused by greater actomyosin activity.

The time during which a muscle is completely unresponsive to a second stimulus is the absolute refractory period. It is followed by an interval called the relative refractory period, during which the muscle will respond only when the strength of the second stimulus is significantly greater than that of the first. Closely spaced stimuli may cause more forceful contraction owing to summation of contractions.

Although skeletal muscle cells in the adult are incapable of regeneration, components within muscle cells can be replicated. Forceful, repetitive, prolonged exercise causes such changes and leads to an increase in fiber size, causing muscular hypertrophy. When skeletal muscles degenerate in varying forms of atrophy, they are replaced by connective tissue elements. Atrophy results from degeneration of nerves and/or disuse. Muscular dystrophies are inheritable disorders that primarily affect muscle cells directly.

Muscles act as levers and produce forces which rotate about a supporting point, or fulcrum, on bone. The generation of work by a muscle is regulated by the same laws that govern the function of levers. Muscles perform work according to physical principles expressed in the following formulas:

(1) Work = force × distance
(2) Power = work/time
(3) Power = force × velocity

The velocity of muscular contraction varies inversely with the force developed. At maximal force the velocity is zero, muscle does not shorten, and isometric tension is produced.

2 Muscles depend on metabolism for energy production. Metabolism is the sum total of chemical reactions in living cells. It includes a series of individual reactions catalyzed by specific enzymes in which compounds are degraded (in catabolism) or synthesized (in anabolism).

Anaerobic glycolysis converts glycogen to glucose and lactic acid in the cytoplasm and in the absence of oxygen. It results in the net synthesis of two ATP molecules for each molecule of glucose metabolized.

Aerobic oxidative phosphorylation generates the bulk of the energy stored in high-energy phosphate compounds such as ATP. Through aerobic pathways, pyruvic acid generated by the breakdown of glucose is degraded to form CO_2 and H_2O, and release energy. Each molecule of glucose (a 6-carbon compound) is first converted anaerobically into two molecules of pyruvic acid (a 3-carbon compound), which is then transformed aerobically into three molecules of acetyl groups (2-carbon compounds), which are linked to coenzyme A to form acetyl CoA. Acetyl CoA is metabolized aerobically in the citric acid cycle. Enzymes in the citric acid cycle remove hydrogen and carbon from organic acids, breaking their chemical bonds to release energy. Organic acids (such as citric acid) are derived from fats, proteins, and carbohydrates in a series of mainly reversible reactions. Net synthesis of 36 ATP molecules occurs for each molecule of glucose catabolized by these aerobic reactions. With the two extra ATP molecules formed by anaerobic steps, a total of 38 ATP molecules is gained for each molecule of glucose that is consumed. The energy which is produced is converted to units of energy currency of the cell in the form of high-energy phosphate bonds of ATP. An excess supply is stored in creatine phosphate and used later to resynthesize ATP.

As a result of energy metabolism, muscles release heat in three phases called activation, shortening, and delayed heat. The first phase occurs as muscle is readied for contraction and actomyosin-ATPase is formed; the second develops during contraction; and the third and largest quantity of heat is formed while muscle recovers. Extra heat not accounted for by observable chemical reactions may be formed by changes in the shape of muscle proteins. About 75 percent of the energy produced in muscle metabolism forms heat. The remainder is used to perform work during muscle's contraction.

3 Exercise is the performance of physical activity. It is dependent on the en-

durance of muscle, the flexibility of joints, and other body functions, especially those of the respiratory and circulatory systems. The energy necessary for muscle to perform exercise is rapidly made available from reservoirs of ATP and C~P. Thereafter, energy is provided by anaerobic glycolysis and, later, aerobic oxidative phosphorylation. During sustained exercise, the shift from anaerobic to aerobic pathways may provide an individual with a "second wind."

Fatigue occurs when a muscle runs out of energy and it can no longer respond to a stimulus. It is related to depletion of glycogen and to the inability of oxidative reactions to meet energy demands. Muscle cramps are involuntary intermittent contractions in which pain may develop. In some instances, cramps may be due to an interruption in circulation and to the release of factors such as kinins or P substance.

Repetitive exercise, called training, may lead to replication of contractile proteins and those necessary for aerobic metabolism. Maximal exercise causes O_2 to be consumed at rates faster than the $Vo_{2_{max}}$ can supply it. The heart fails to pump blood to the lungs at a rate sufficient to maintain aerobic metabolism. An oxygen debt builds up as metabolism shifts to anaerobic pathways. The debt is repaid after the physical activity ceases. The $Vo_{2_{max}}$ is the maximal volume of oxygen a person can deliver to muscle during the most strenuous exercise.

Submaximal or endurance exercise requires less than the $Vo_{2_{max}}$. It can stimulate aerobic oxidative capacity and is most effective when performed at a training or target heart rate that is specific for each individual. Endurance training results in an increase in the aerobic capacity of skeletal muscle fibers and especially elevates the utilization of fats. The most prominent change that training promotes in cardiac muscle is the enlargement of the mass of myofibrillar protein in the heart.

REVIEW QUESTIONS

1. What are elastic series components, and how do they contribute to the development of tension during muscle contraction?
2. Compare isotonic, isometric, and auxotonic contraction.
3. What is a twitch? What are the phases of a twitch? How is it affected by moderate stretching of muscle?
4. Diagram the summation of contractions, unfused tetanus, tetanus, and contracture in the response of skeletal muscle to repeated stimuli.
5. Explain why the all-or-none law is not in conflict with variations in the development of muscle tension.
6. How does the CNS coordinate muscle activity?
7. What is a lever? How does the principle of the lever apply to muscle activity?
8. Interpret Fig. 13-8 relating force and velocity of contraction. How does this principle relate to the muscle activity of a weight lifter who presses 100 lb, then 200 lb, then 300 lb? How does this exercise relate to ATP consumed, as indicated in Fig. 13-9?
9. Interpret Fig. 13-11 in terms of energy currency units used during intense exercise.
10. Write three basic equations which describe the physical principles of work and power, and outline their applications to muscle function. Calculate the power of a muscle which lifts 25 lb 1 in. in 0.25 s.

11 Describe anaerobic glycolysis, lactic acid production, and the role of glycolysis in the formation of liver glycogen.

12 Discuss aerobic oxidative phosphorylation, the pivotal roles of pyruvic acid and acetyl CoA, and the interconversion of proteins, fats, and carbohydrates. Compare the net amount of ATP formed in aerobic oxidative phosphorylation with that evolved by anaerobic glycolysis.

13 How do ATP and C~P compare as units of energy currency?

14 Describe the efficiency of muscle and the relationship of heat to muscle activity.

15 What are some factors associated with muscle fatigue and cramps that may account for muscle pain?

16 Define training and discuss the two primary types of changes it may cause in skeletal muscle.

17 How is exercise related to oxygen consumption, anaerobic glycolysis, and carbohydrate stores?

18 Frank Bee is training for a track team's 10 km race. A physical checkup by qualified medical personnel indicates he is in excellent physical condition. He is 20 years old and has a resting heart rate of 65 beats/min. Calculate his training (target) heart rate.

19 Compare the major changes that occur in skeletal and cardiac muscle in response to endurance exercise.

THE PHYSIOLOGY OF BONE

14

14-1 MAJOR FUNCTIONS

14-2 BONE AS AN ORGAN
- General organization
  - *Anatomy*
  - *Marrow*
- Chemical composition
  - *Organic molecules*
  - *Inorganic molecules*
- Dynamic exchange of constituents

14-3 FORMATION AND GROWTH OF BONE
- Cellular elements
- Bone formation (osteogenesis)
  - *Intramembranous bone formation*
  - *Intracartilaginous bone formation*

14-4 FACTORS INFLUENCING BONE FORMATION
- Acidity of Extracellular Fluid (ECF)
- Vitamins
  - *Vitamin D*
  - *Vitamin A*
  - *Vitamin C*
- Role of Hormones
  - *Polypeptide Hormones*
    - Parathormone • Calcitonin • Growth hormone
  - *Steroid Hormones*
    - Vitamin D • Androgens • Estrogens • Adrenal steroids
  - *Hormonal interactions*
- Aging
- Mechanical Stress and Exercise

14-5 DISORDERS OF BONES AND JOINTS
- Metabolic Disorders

14-6 SUMMARY

OBJECTIVES

After completing this chapter the student should be able to:

1. Discuss the major functions of bone and joints.
2. Give a description of bone as an organ, including its general organization and chemical composition.
3. Compare intramembranous and intracartilaginous bone formation and outline the origins of cells in bone.
4. Describe the effects of vitamins, hormones, and aging on the properties of bone.
5. Explain how mechanical stress and exercise may influence bone and joints.
6. Name two general metabolic types of bone disorders and list some examples of each.
7. Describe the effects of inflammation on bone and joints.

There are about 206 bones in the body, which account for nearly 18 percent of its weight. They form the skeletal system, that provides the movement which results when the force of contraction of muscle is transferred through tendons to bone. Bones perform several vital body functions, including support, protection, anchorage, storage, and blood cell formation.

14-1 MAJOR FUNCTIONS

Bone offers considerable support to the body. As a result of its high percentage of crystalline calcium salts, bone withstands great compressive force. It also is dependent on its collagen content for tensile strength (ability to withstand force that tends to tear it apart). For example, the bones in a runner's leg absorb about 5000 lb of force per square inch of tissue each time a foot strikes the ground. It requires about 20,000 psi (lb per square in) to break the average bone, compared with 10,000 to 20,000 psi to crack various types of cast iron and 14,000 psi to break a piece of spruce. On a weight basis, bone's resistance to compression and tension is comparable with that of reinforced concrete and approaches that of some forms of steel. Like the cement and the steel rods of reinforced concrete, bone is a two-phased system composed of crystalline minerals reinforced with fibrous proteins.

The remarkable strength of bone affords protection to vital organs housed within the skeletal system. Clearly, the bony skull protects the brain as the rib cage does the heart and lungs.

Muscles are anchored to bone by tendons which penetrate the *periosteum*, the fibrous connective tissue covering of bone. The contraction of muscle produces tension that not only moves bone but stimulates the growth of bony ridges and protuberances which serve as anchor points for tendons and which also are useful anatomical landmarks.

Bone is a storage site or reservoir for many mineral elements (inorganic salts) including about 99 percent of the body's calcium, 80 percent of its phosphorus, 25 percent of its sodium, and lesser amounts of a number of other minerals. The minerals are released into the blood or removed from it as dictated by the body's needs and regulated by feedback mechanisms.

Bone *marrow* is a specialized connective tissue found in hollow spaces in bone which carries out blood cell formation (hematopoiesis). Marrow contains a variety of immature cells, blood cells, and fat. This specialized hematopoietic tissue is described briefly here and more fully in Chap. 18.

14-2 BONE AS AN ORGAN

General Organization

Anatomy

Bone tissue, in conjunction with other connective tissues, forms macroscopic organs called bones (Figs. 3-26 to 3-28). An example is the humerus of the upper arm, a long bone (Fig. 14-1A). Like all long bones, it is composed of

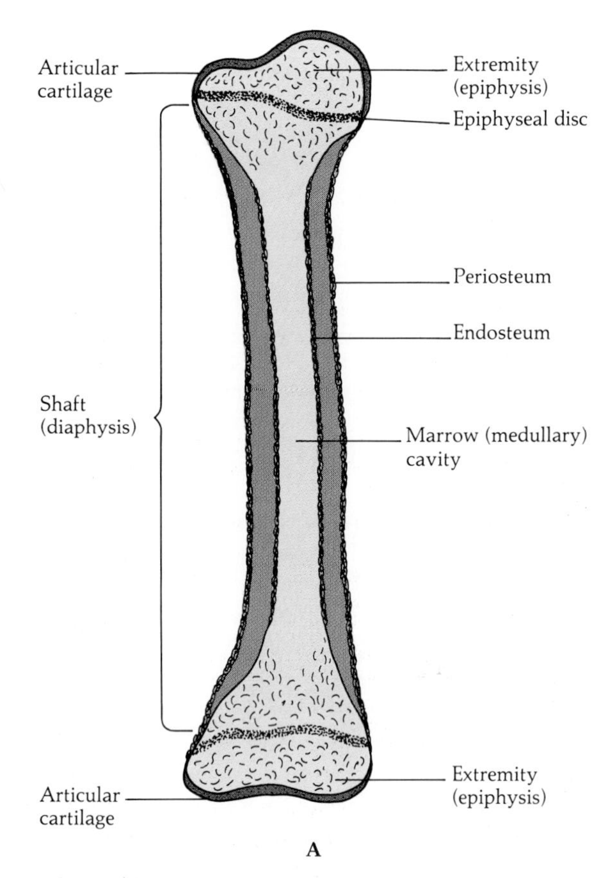

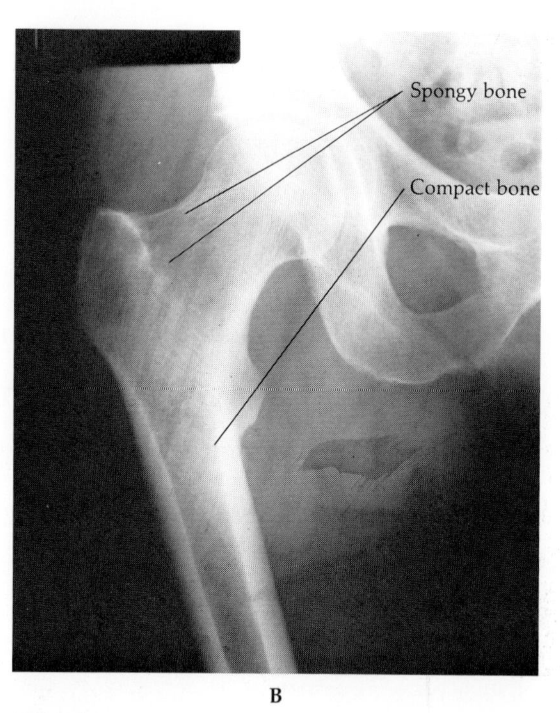

FIG. 14-1
Anatomy of a long bone. (A) General structure (a simplified diagram). (B) X-ray of the upper end of the femur in the hip joint illustrating compact and spongy bone.

spongy and compact bony tissues located in the shaft (*diaphysis*) as well as in the two extremities (*epiphyses;* singular, epiphysis).

Fibers of the periosteum adhere to and penetrate the outer surface of bone (Fig. 3-27C). The inner surface of bone is lined with a condensed layer of marrow called the *endosteum* (Fig. 14-1A). It lines the central cavity of the shaft, the marrow cavity (medullary cavity), which is filled with marrow. During growth the outer layer of the endosteum merges with a sheet of bone-forming cells, the *osteoblasts.*

Bones form junctions called *joints* or *articulations* which allow varying degrees of movement. *Diarthroses* (*synovial joints*) move freely. *Synarthroses* allow little or no movement. Several categories of joints have been distinguished by means of the arrangements of connective tissues which bind the bones together (Table 14-1).

Hyaline cartilage covers the epiphyses of bones in freely movable joints. In these joints, a synovial membrane composed of outer fibrous and inner areolar tissues forms a cavity filled with a viscous fluid, called synovial fluid. The fluid originates from blood vessels from which it is transported through the synovial membrane. Synovial fluid is rich in a proteoglycan (a complex of protein and mucopolysaccharide) hyaluronic acid, which is secreted by cartilage cells adjacent to the synovial membrane in the joint. The fluid and the smooth surface of the cartilage minimize friction at mobile bony junctions (Fig. 14-2).

A *capsule* of dense fibrous connective tissue, which is a continuation of the periosteum, surrounds the synovial cavity and merges with ligaments that bind bones together. The connective tissues are often modified according to the function of the joint. In some locations, the capsule produces crescent-shaped discs of fibrocartilage called *menisci* (singular, meniscus). When present, menisci separate the synovial cavity into two chambers and remain attached to the

TABLE 14-1
Types of joints and their major connective tissue components

Name	*Major Junctional Tissues*	*Examples (Locations)*
Diarthroses (Synovial joints) (freely movable)	Dense and areolar	Finger joints, knee joints, elbow, shoulder, hip joints
Synarthroses (slight movement or none)		
Syndesmosis (no movement)	Dense-regular	Fontanelles (soft spot) in baby's skull Sutures of skull (with age, dense-regular tissue is converted to bone and the junction between sutures become a synostosis)
Synchondrosis (no movement ordinarily)	Hyaline cartilage	Between immature facial bones Between ribs and chest plate (sternum) Connects epiphysis to diaphysis in growing long bones
Symphysis (Amphiarthrosis) (slight movement)	Dense-regular and fibrous cartilage	Between mandibles in midline of lower jaw Pubic bones in pelvic girdle Between cartilage plates of the intervertebral discs
Synostosis (no movement)	Bone	Conversion of synchondrosis to synostosis (e.g., fusion of sutures; maturation of facial bones)

FIG. 14-2
A simplified diagram of a synovial (diarthrotic) joint.

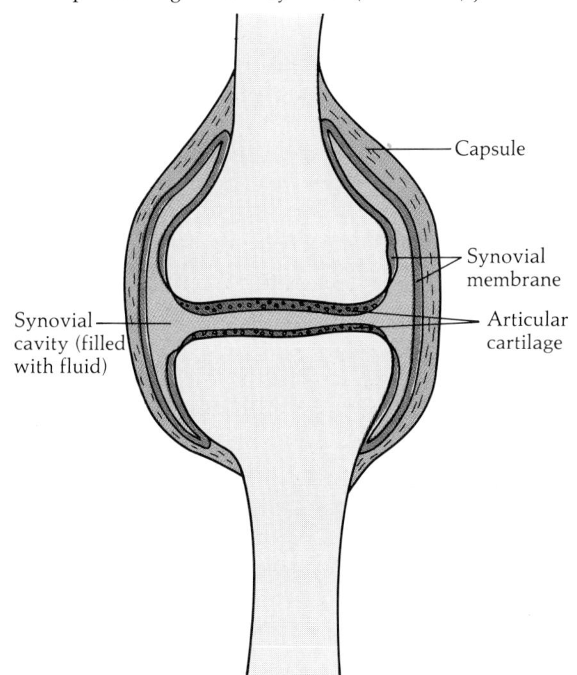

capsule at its periphery. *Meniscal cartilages* in the knees of athletes are often damaged by mechanical stress. In other locations *bursae* are formed as sacs which act as cushions to relieve pressure between moving parts. Bursae, which are similar to the pouches formed in joints by synovial membranes, surround muscles, tendons, ligaments, and bones (Fig. 14-3A). *Tendon sheaths* are sleeves of connective tissue similar to synovial membranes which form tunnels in which tendons glide over bone or ligaments (Fig. 14-3B). Some, like the metacarpal tunnel, are located in the wrist and hand. Some of these tendon sheaths are adjacent to major blood vessels and nerves to the hand and can compress them, interfere with their function, and cause the hand to "fall asleep" in a disorder called the *metacarpal tunnel syndrome.*

Marrow

There are two types of marrow: young active marrow, which is red, and older, less active yellow marrow. Marrow contains stem cells, which are precursors of blood cells, and other cells in various stages of maturation. The stem cells are derived from an embryonic tissue

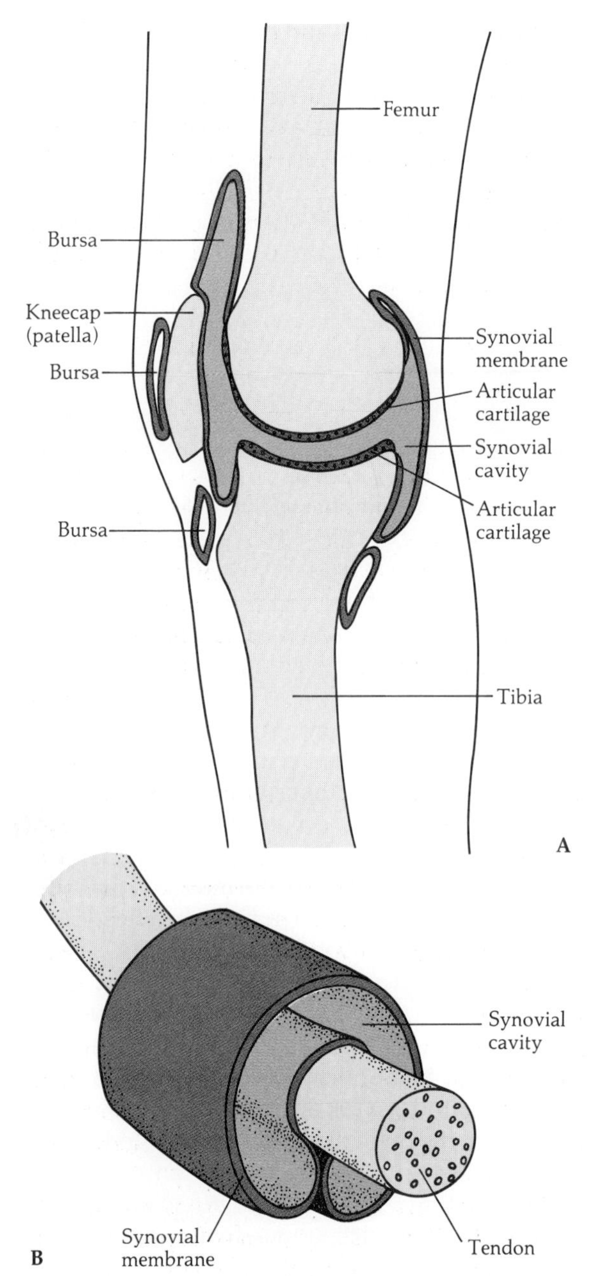

FIG. 14-3
Simplified diagrams of (A) bursae in the knee joint and (B) a tendon sheath.

called *mesenchyme*. Red marrow's color comes from its high content of red blood cells, which contain hemoglobin. It is most active in blood cell formation and is the only form of marrow present at birth. This tissue is gradually converted to yellow marrow with the accumulation of fat, beginning at 5 to 7 years of age. Yet red marrow persists in large quantities, accounting for nearly 5 percent of body weight (4.7 ± 1.3 percent) in adults. It is found at the ends of long bones and in the skull, collar bones (clavicles), vertebrae, ribs, chest plate (sternum), and hip bones (pelvis). Yellow marrow fills most of the shafts of long bones by age 18. It is located mainly in the central canal of the long shafts of bones and extends into its spongy areas.

Chemical Composition

Bone contains about 8 percent water, 20 percent organic molecules, and 72 percent inorganic salts (Table 14-2).

Organic Molecules

Collagen accounts for about 18 to 19 percent of the weight of bone and 90 percent to 95 percent of its organic material. It is a triple-stranded, cross-linked, coiled protein of about 300,000 MW and was described in Chap. 3 and shown in Fig. 3-10, p. 54. Collagen acts as a reinforcing material much like steel rods in concrete. Collagen molecules occupy nearly 40 percent of the space in bone and are responsible for its great tensile strength.

TABLE 14-2
Approximate composition of bone by weight*

Substance	*Approximate Percentage*
Water	8
Organic molecules	20
Collagen	(18–19)
Proteoglycans	(<1)
Inorganic salts	72
Calcium phosphate, $Ca_3(PO_4)_2$	(57)
Calcium carbonate, $CaCO_3$	(10)
Magnesium phosphate, $Mg_3(PO_4)_2$	(3)
Others (e.g., sodium, potassium salts)	(1)

* The composition of bone varies among different samples.

Proteoglycans form less than 5 percent of the organic molecules in bone. However, they keep collagen soft and pliable (they act as plasticizers), reduce its natural tensile strength, and cause it to swell. The proteoglycans are composed of at least 95 percent carbohydrate. The carbohydrate is built of repeating disaccharide units, which produce macromolecules with molecular weights ranging from tens of thousands to millions. Chondroitin sulfate and keratin sulfate are examples of proteoglycans common to bone and other connective tissues.

Inorganic Molecules

About 99 percent of the body's Ca^{2+} ions are found in the skeleton. Calcium ions account for about 1.7 percent of adult body weight and two-thirds of the weight of bone (Table 14-2). Most calcium is in the form of calcium phosphate, $Ca_3(PO_4)_2$, while less than 10 percent exists as calcium carbonate, $CaCO_3$.

Although only 1 percent of total body Ca^{2+} ions are found outside of bone, the mineral performs several vital functions in addition to those it serves in the skeletal system. Ca^{2+} ions strongly influence the permeability of cell membranes (Chap. 4) and constitute an important secondary messenger to hormones (Chap. 5). They also have critical functions in the physiology of excitable tissues such as nerve and muscle (Chaps. 7 and 12), as well as in the coagulation of blood (Chap. 19).

In bone, calcium phosphate is found in three different types of crystals, which are highly insoluble: hydroxyapatite, octacalcium phosphate, and amorphous calcium crystals. Two other varieties of the salt, which have increasing degrees of solubility, are calcium monohydrogen phosphate, $CaHPO_4$, and calcium dihydrogen phosphate, $Ca(H_2PO_4)_2$. These two forms become more abundant in acid fluids. As a result, changes in acid-base balance of body fluids can promote the solubility of Ca^{2+} salts and facilitate or impede their transport by precipitating the salts in tissues. For example, the absorption of dietary calcium is enhanced by acidity of the digestive juices. Precipitation of the salt is favored by alterations in pH associated with disorders such as hardening of the arteries (arteriosclerosis) and the formation of gallstones and kidney stones.

Insoluble calcium phosphate crystals are located between collagen fibers and form the bulk of the intercellular matrix of bone. *Hydroxyapatite*, $Ca(OH)_2 \cdot 3Ca_3(PO_4)_2$ or $Ca_{10}(PO_4)_6(OH)_2$, exists in an ionized state as indicated by an alternative formula:

$$Ca^{2+}\ 10^{-x}(H_3O^+)_{2x} \cdot (PO_4)_6(OH^-)_2$$

The formula illustrates the charges of the hydroxyapatite crystals. Consequently, a wide variety of ions binds to oppositely charged parts of the crystal's surface and may be exchanged with those in the blood. Ions may be *adsorbed* (adhere) to the surface of the crystals and become part of their structure or be *resorbed* (be removed from them to return to the blood). In the common form of hydroxyapatite the ratio of calcium to phosphate is 10 : 6 or 1.67 : 1. The ratio increases with age. Conversely, the crystals are considered calcium-deficient when the ratio drops below 1.5 : 1. For example, sodium and potassium ions can replace calcium ions in hydroxyapatite and cause a decline in the ratio.

Other alterations in the composition of the crystals also may change the properties of bone. A notable example is the replacement of hydroxyl ions (OH^-) by fluoride ions (F^-), which increases the strength of bone and makes it less susceptible to fracture. The same type of exchange occurs in teeth and makes them less susceptible to decay. However, excesses of fluoride ions may cause variable coloration or mottling of teeth.

Since each gram of bone contains about 10^{16} crystals, if all crystals in bone throughout the body were spread out, they would occupy a floor space of about 100 m^2, or 1100 ft^2. This large, charged surface area effectively adsorbs, stores, and releases a number of substances, including heavy metals such as lead. After its entrance from the digestive or respiratory tract, about 95 percent of lead is concentrated in bone. The transport of lead resembles that of calcium and its accumulation may be rapid or long-term. Lead poisoning may cause symptoms of intoxication in the gastrointestinal and neuromuscular systems and the CNS, as well as in blood cells, the kidneys, and other tissues and organs.

Radioactive minerals (radionuclides) from an atomic explosion may enter the body by various routes to be adsorbed on bone. For example, the radionuclides may settle on grass and be incorporated into hay eaten by cows. Some of the radioactive elements pass into the animal's blood and milk. After humans drink the milk, the radionuclides may cross the intestine to enter the person's blood and be carried to bones. Charged elements such as radioactive strontium ion (Sr^{2+}) can be adsorbed at the site of opposite charges on the surface of hydroxyapatite and/or be exchanged with ions in it. Energy released from radionuclides readily damages rapidly growing, dividing cells, such as stem cells in the bone marrow. Permanent changes in the genetic material of cells may occur as heritable alterations called *mutations.* Some stem cell mutations may lead to abnormal patterns of growth and cell division, causing leukemia (cancer of white blood cells, Chap. 19).

About 80 to 85 percent of the body's phosphorus is found in bone. The remainder is located in a variety of tissues. Phosphorus is an essential ingredient of deoxyribonucleic acid (DNA) and ribonucleic acid (RNA) and has a number of other important functions, including the phosphorylation of sugars, the formation of high-energy phosphate compounds, and the formation of inorganic buffers such as the inorganic phosphate ions HPO_4^{2-} and $H_2PO_4^-$, which are formed by the dissociation of calcium phosphate. Because it has a net negative charge, inorganic phosphate (P_i) readily moves through epithelial cells across the small intestine into the blood. Some P_i may be "dragged" passively across the cells with the active transport of Ca^{2+}. The main route of excretion of P_i is through the kidneys into the urine (see Chap. 21).

Although both calcium and phosphorus are required for the formation of calcium phosphate, dietary deficiencies in Ca^{2+} ion result more often than those of P_i. This occurs because less than 50 percent of the Ca^{2+} ion consumed usually passes across the epithelial cells into the blood and the amount absorbed decreases with age. P_i deficiency is rare because P_i is common in foods and 70 to 80 percent of that which is consumed is absorbed into the blood.

The plasma levels of the two elements influence their metabolism. Their addition (*deposition*) to bone is favored by a high Ca/P ratio, whereas their removal from bone is favored by a low one. The Ca/P ratio in blood is expressed as the product of the concentrations of the two ions in serum. (Serum is blood minus its cells and clotting elements.) The ratio ranges from 26 to 47 mg/dL (with blood levels of Ca^{2+} ions varying from 8.5 to 10.5 mg/dL and those of P_i from 3.0 to 4.5 mg/dL). The Ca/P ratio in the blood is calculated as follows:

Ca/P blood ratio =	calcium ion concentration	×	inorganic phosphate concentration
Low range:	8.5	×	3.0 = 25.5 mg/dL
High range:	10.5	×	4.5 = 47.25 mg/dL

The relatively low percentages of minerals other than Ca^{2+} ions in bone may tend to misrepresent their importance. For example, nearly 25 percent of the body's Na^+ ion supply is found in bone although it makes up less than 1 percent of this tissue. Thus, bone is an important Na^+ ion reservoir. It releases Na^+ to the blood to neutralize acid in body fluids and may also remove Na^+ from the blood. These activities help maintain the pH of body fluids within a narrow range. This close control of pH is essential to the function of critical enzymes in vital organs such as the kidney, heart, brain, and lungs. In this manner, bone assists the kidneys in the maintenance of an acid-base balance necessary for survival. The role of the kidneys in mineral metabolism and acid-base balance will be discussed in Chap. 21.

Minerals found in trace amounts in bone can be extremely important to body function (App. C and D, Chap. 23). Some act as partners with enzymes (coenzymes), Fig. 2-6, p. 23. The mechanisms of action of most trace elements are not yet established, but their importance has been demonstrated. For example, the role of fluoride ions was described above. Among some other examples are copper, cobalt, and iron, all of which are required for red blood cell formation.

Dynamic Exchange of Constituents

Bone is dynamic, constantly undergoing turnover (breakdown, exchange, and resynthesis of its components), although at rates slower than those of other tissues. Substances are continually deposited in bone from blood or returned to the blood by resorption. The rate of movement between the intracellular fluid (ICF) and extracellular fluid (ECF) compartments of bone resembles that of dense connective tissue and cartilage (Fig. 4-1, page 81). Since bone makes up about 18 percent of body weight, the exchange of materials is significant. For example, about 800 mg of calcium is exchanged between tissue and fluid compartments of the body each day. Ca^{2+} ion turnover is sufficient to replace all the mineral in the adult skeleton about every 4.7 years. Ca^{2+} ions in food are absorbed largely in the first part of the small intestine (in the duodenum, where stomach acid is delivered), crossing the epithelial cells to enter the blood. Ca^{2+} ions are lost largely by excretion with undigested residues (feces) from the large intestine. A small amount of the mineral is excreted with urine. The kidneys typically reabsorb and conserve most of the calcium from their urinary filtrate, returning it to the blood.

The rate of exchange between bone and the body's compartments decreases with age. Calcium salts accumulate in bone and the replacement of the organic matrix slows down. As a result, the ratio of inorganic to organic matter increases as bone matures. Bone begins to resemble concrete in which the number of reinforcing rods (organic molecules) diminishes. With aging, bone becomes more suceptible to fracture, mostly because of a decrease in its tissue mass and partly because of the elevation in ratio of inorganic to organic matter.

The rate of bone resorption also becomes greater as the blood levels of substances such as calcium and phosphate decrease. Thus, specific dietary deficiencies, starvation, or the demand for nutrients caused by pregnancy can accelerate losses. An increase in the dietary intake of the depleted material(s) usually prevents or reverses such losses.

14-3 FORMATION AND GROWTH OF BONE

Cellular Elements

The genesis of bone cells can be traced back to early development from the middle embryonic germ layer, the *mesoderm,* which also produces muscle and other connective tissues (see page 237 and Chap. 24). Some mesodermal cells retain their capacity for division and maturation into a variety of cell types. They become *mesenchymal cells,* which mature into a number of different connective tissue cell types.

Bone formation, or *osteogenesis,* involves the progressive transformation of mesenchymal cells into *osteoblasts* and *osteoclasts. Osteoblasts* secrete protein, laying down collagen fibers in different directions and also deposit calcium phosphate crystals between the fibers. They become trapped in their own secretions and are then called *osteocytes* (bone maintenance cells). Until about 40 years of age the cellular activity of osteoblasts predominates, as does net bone formation. After age 40, the effects of *osteoclasts,* large multinucleated phagocytes which resorb bone, become dominant. They seem to originate from mononuclear blood cells formed by embryonic mesenchyme. As a result, skeletal mass decreases very gradually and bone becomes more brittle with aging.

Bone Formation (Osteogenesis)

Bone is produced by intramembranous or intracartilaginous bone formation. *Intramembranous (mesenchymal)* bone formation occurs within connective tissue membranes in flat bones of the skull, jaw, and face and in all bones after the skeleton has completely matured (between the ages of 20 and 25). It also accounts for the increase in diameter of long bones. *Intracartilag-*

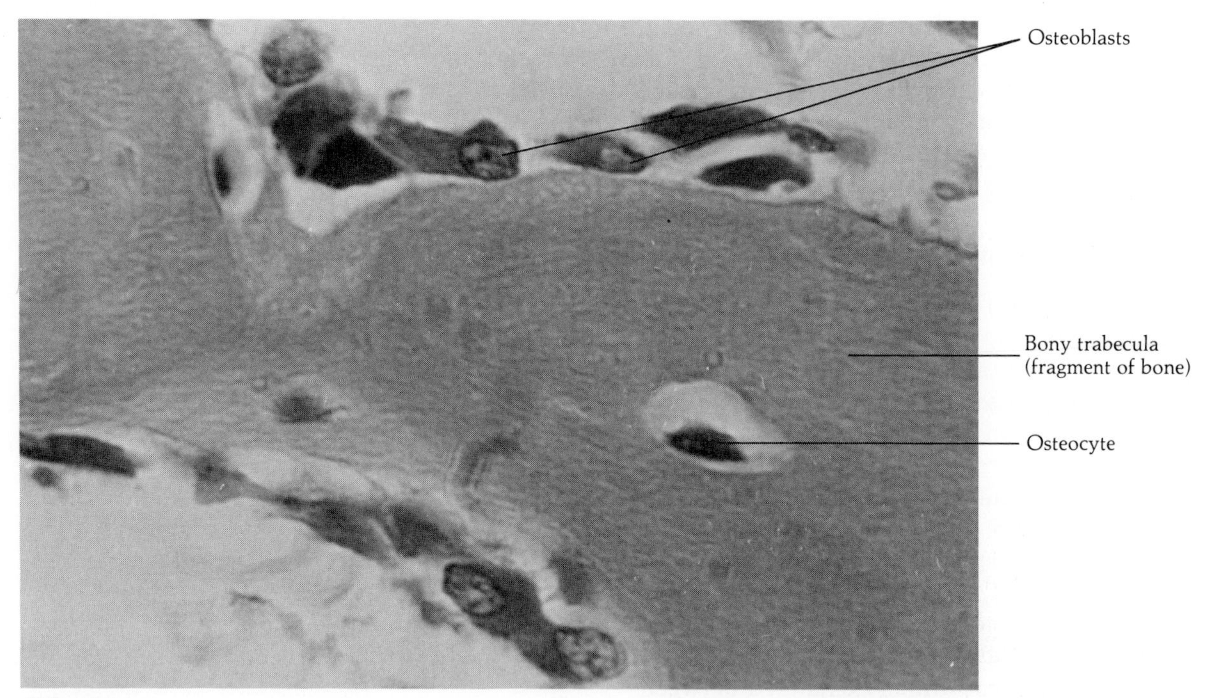

FIG. 14-4
Photomicrograph of mesenchymal (intramembranous) bone formation (× 1700).

inous (*endochondral*) bone formation replaces cartilage with bone to form the long bones of the body and increase bone length. In both processes mesenchymal cells proliferate and mature to form osteoblasts and then osteocytes. Where necessary, osteoclasts reconstruct the tissue. The cellular activity may be stimulated by stress. For example, they become active when bone is fractured.

Intramembranous Bone Formation

Intramembranous bone formation is illustrated in Fig. 14-4. Osteoblasts laden with endoplasmic reticulum secrete collagen and the Golgi apparatus releases small quantities of proteoglycans. The proteoglycans serve as plasticizers and as cement substances in the bony matrix. Ca^{2+} ions, P_i, and other minerals accumulate and crystallize around the collagen fibrils under the control of the osteoblasts. Threadlike splinters of bone called *spicules* or *trabeculae* are produced. The trabeculae are the units of *spongy* (*cancellous*) bone (Fig. 14-1B).

Osteoclasts reconstruct bone by slow, continual activity. These large, phagocytic cells contain nuclei derived from the fusion of several cells and therefore possess a variety of capabilities. The resorptive function of osteoclasts is stimulated by parathyroid hormone. Osteoclasts also may phagocytize partially degraded bony particles. Osteoclasts produce lactic acid and citric acid, which dissolve salt crystals, and *acid phosphatases*, lysosomal enzymes whose hydrolytic activity digests calcium phosphate in the matrix of bone. Evidence indicates that the acid phosphatases are most active in an acid environment. *Alkaline phosphatases*, found on the plasma membrane of bone cells, favor the formation of calcium phosphate and the mineralization of bone. All these activities allow mesenchymal bone to be reconstructed between membranes, leading to the formation of inner and outer compact layers within a spongy central mass called the *diploë*. The diploë in flat bones is comparable with the marrow cavity of long bones.

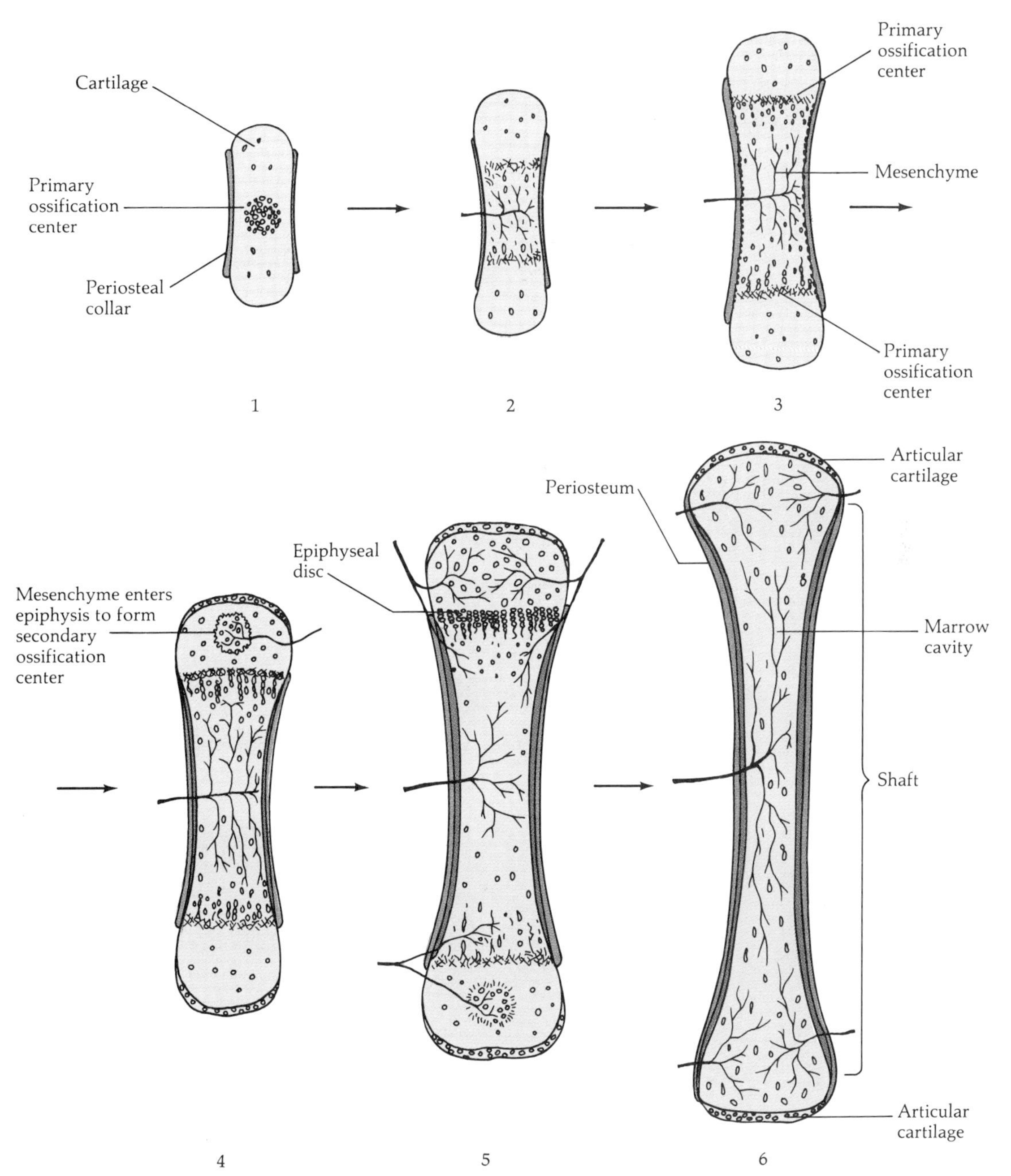

FIG. 14-5
Intracartilaginous (endochondral) bone formation.

Intracartilaginous Bone Formation

Intracartilaginous bone formation produces the bulk of the skeletal system, including all its long bones. It lengthens bones, as shown in Fig. 14-5. The cellular activities are similar to those in intramembranous osteogenesis, except that hyaline cartilage is formed first and converted to bone. Hyaline cartilage is synthesized during early development and is gradually transformed into bone in a process beginning in the second or third month of fetal life and

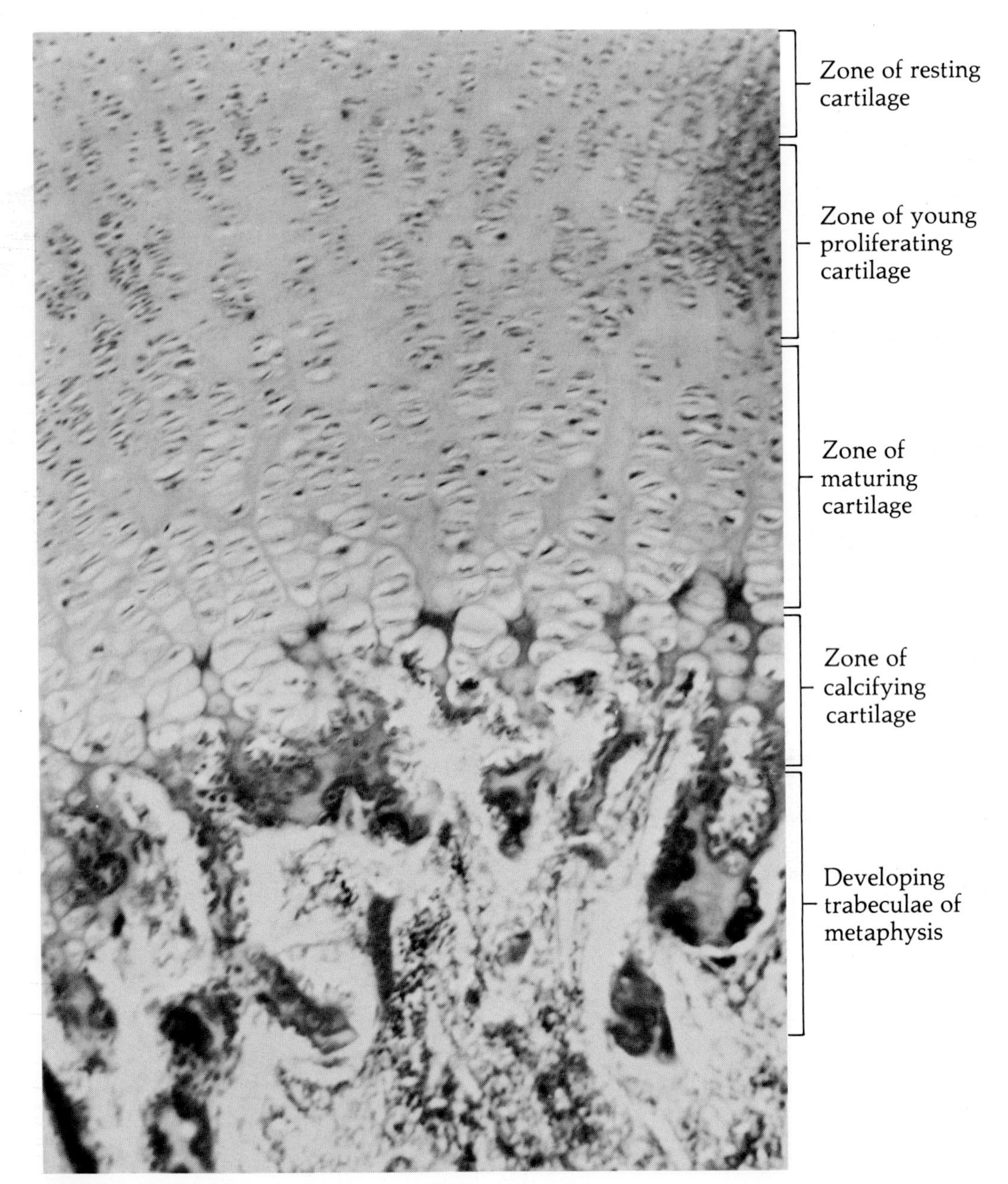

FIG. 14-6
Ossification in the epiphysis. Photomicrograph of endochondral bone formation in the epiphysis (× 170).

continuing until the individual is about 20 to 25 years of age.

In intracartilaginous bone formation, the skeletal structure is first laid down as cartilage. The sequence of steps shown in Fig. 14-5 illustrates the series of changes. The cartilage cells (chondrocytes) multiply spreading from an area in the center of the shaft, the *primary ossification center* (1 through 3). The cells in this area multiply, enlarge, mature into osteoblasts, secrete collagen, and become surrounded by calcium crystals. Long rows of cells are formd, most of which ultimately dissolve and are incorporated into bony trabeculae. Continued growth separates the center into two parts located on both sides of the middle of the shaft (3). As these events occur, mesenchymal cells form a ring of tissue, the *periosteal band* or *collar*. Blood vessels penetrate the collar, conveying embryonic cells which form marrow. Other cells in the collar lay down bone around the shaft while osteoclasts enter it to resorb bone, create channels, and produce marrow spaces. The marrow spaces become filled with hematopoietic tissue derived from mesenchymal cells, which penetrate the spaces. Later, the mesenchymal cells enter the epiphyses to form secondary ossification centers (4 through 6). In the epiphyses the cartilage cells proliferate, enlarge, and are replaced by bone in steps similar to those which take place in the primary ossification centers (Fig. 14-6). Some cartilage cells retain their capacity for growth, forming an *epiphyseal disc* or *plate* (Fig. 14-5), which connects the epiphysis and diaphysis with a transitional area of spongy bone called the *metaphysis.* The length of long bones increases as a result of cellular activity in the epiphyseal plate and metaphysis, which are collectively referred to as the *future growth zone.* The epiphyseal plate, which is less dense to x-rays than fully developed bone, may be observed in an x-ray as a thin line between the diaphysis and epiphysis. Growth stops and the line disappears when the epiphysis and diaphysis fuse.

14-4 FACTORS INFLUENCING BONE FORMATION

The structure of the skeletal system is constantly changing, as described above. Bone cell proliferation and maturation and protein and mineral metabolism are adjusted in order to maintain homeostasis. Among the factors which markedly influence the growth and composition of bone are pH, nutrients, hormones, enzymes, the age of the individual, and the mechanical stress to which the bone is subjected.

The development of bone is promoted by an increase in the alkalinity of its ECF, the secretion of protein by bone cells, the absorption of Ca^{2+} and P_i ions across the intestinal epithelium, and the reabsorption of Ca^{2+} and P_i by the kidney from the urine back into the blood.

Acidity of Extracellular Fluid (ECF)

When the acidity of bone ECF increases, more soluble types of calcium phosphate are formed in the bone and Ca^{2+} is lost to blood. Agents which stimulate the synthesis of organic acids and the accumulation of H^+ ions may demineralize bone and elevate blood levels of Ca^{2+}. As described above, substances which increase the acidity of the intestinal contents increase the solubility of calcium salts and their absorption and also result in higher blood levels of Ca^{2+}. For example, eating a high-protein meal enhances hydrochloric acid production by the stomach and stimulates the entry of the acid into the intestine. Acidity of the intestine and enhanced Ca^{2+} absorption also result from the fermentation of lactose (milk sugar) and the production of organic acids by *Lactobacillus* bacteria. These normal inhabitants of the intestine are found in high numbers in fermented dairy products such as yogurt. The presence of lactose in the diet promotes the growth of these bacteria. Conversely, decreased Ca^{2+} absorption occurs with a low-protein diet, alkalinity of the intestinal contents, or other factors which favor the formation of insoluble calcium salts. The latter form in the presence of phytic acid (found in cereal grains such as bran), oxalic acid (found in spinach and rhubarb), phosphate, and fatty acids.

Vitamins

A number of different nutrients including specific vitamins have prominent effects on bone. Among those with notable effects are vitamins D, A, and C. Their general properties are described in App. B in Chap. 23.

Vitamin D

Vitamin D promotes the intestinal absorption of Ca^{2+} and P_i and the mobilization of Ca^{2+} from old to new bone. Vitamin D not only is an essential nutrient but behaves as a hormone. Its molecular structure consists of a steroid ring

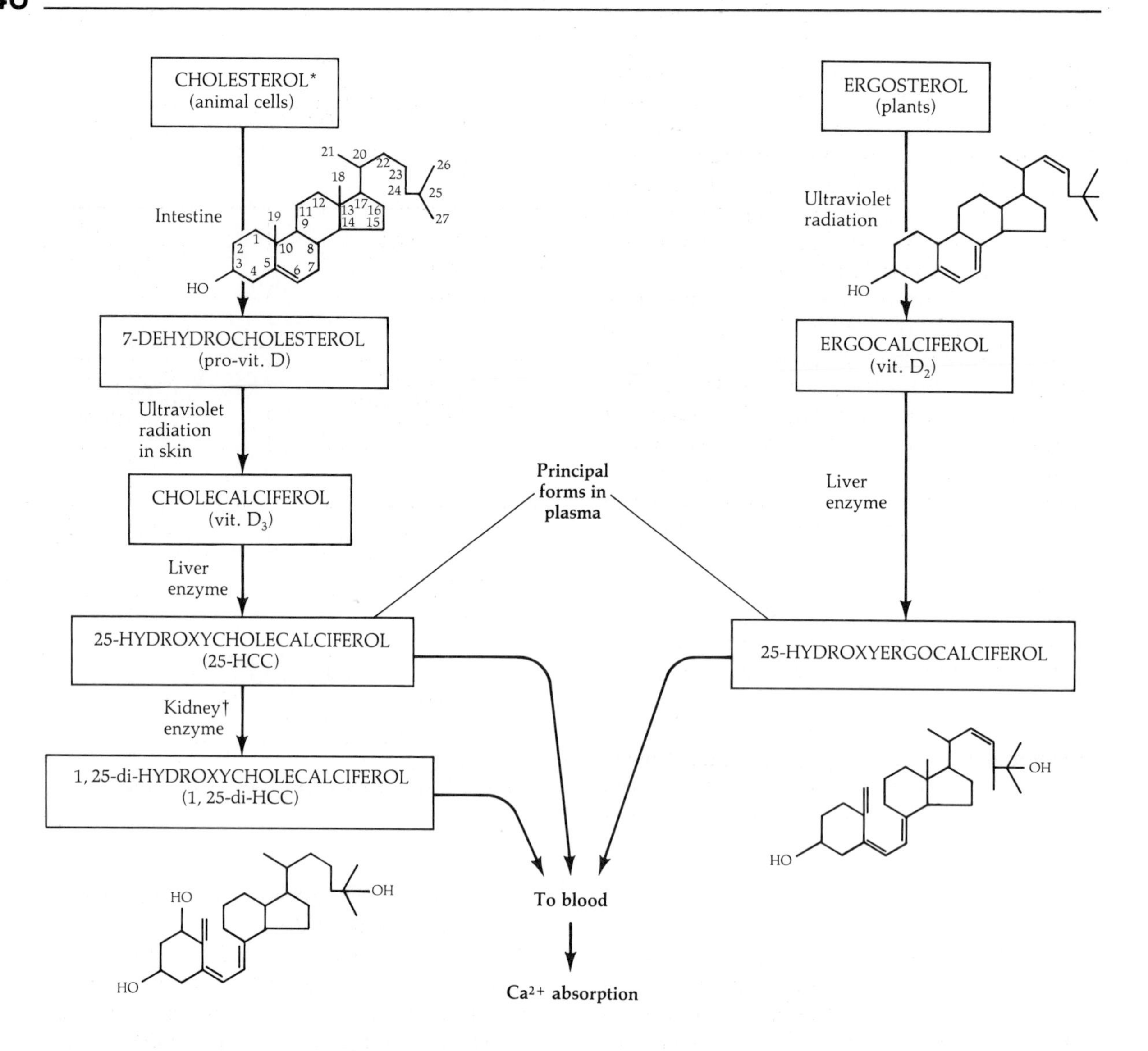

*Each carbon atom in steroids is numbered as shown in the formula for cholesterol.
†Stimulated by parathormone and low blood P_i, inhibited by high blood Ca^{2+}.

FIG. 14-7
The formation of active forms of vitamin D.

which is also found in cholesterol and cortisone, as described in Chap. 2. Active vitamin D is produced from the inactive precursors *ergocalciferol* in plants and *7-dehydrocholesterol* in animal cells (Fig. 14-7).

Ergocalciferol is present in yeast and olive oil. Exposure of yeast to ultraviolet radiation causes conversion of ergosterol into active vitamin D_2, *ergocalciferol.* Continuous exposure inactivates the vitamin. Halibut and cod liver fish oils contain provitamin D or 7-dehydrocholesterol. It is converted by ultraviolet radiation exposure of the skin into active Vitamin D_3 or *cholecalciferol.* The latter enters a series of reactions to form even more active derivatives, as shown in Fig. 14-7. The steps involve the addition of hydroxyl (–OH) groups first by liver and then by kidney enzymes to produce increasingly effective forms of vitamin D_3 called *25-HCC*

(25-hydroxycholecalciferol) and *1,25-diHCC* (1,25-dihydroxycholecalciferol). Vitamin D_3 is secreted by the liver and then the kidney into the blood for circulation as a hormone, which exerts profound effects on bone and mineral metabolism. Consequently, drugs or diseases which damage the liver or kidneys interfere adversely with these functions.

Most Ca^{2+} absorption takes place in the intestine under the influence 1,25-diHCC whose synthesis is promoted by parathormone (parathyroid hormone) secretion. Active forms of vitamin D help maintain high Ca/P ratios by stimulating Ca^{2+}-ATPases in the intestine and kidney to actively transfer Ca^{2+} to epithelial cells for transport into the blood. Ca^{2+} ion levels in the blood are regulated by negative feedback mechanisms involving parathormone and 1,25-diHCC (Fig. 14-8).

An excess of ingested vitamin D may accumulate in the liver and be released in toxic levels into the blood. Large doses stimulate osteoclasts to resorb Ca^{2+} from bone, increase susceptibility to fractures, and cause the accumulation of Ca^{2+} in soft tissues. Elevated levels of Ca^{2+} may appear in the tubules of the kidney and precipitate to form stones and cause kidney failure.

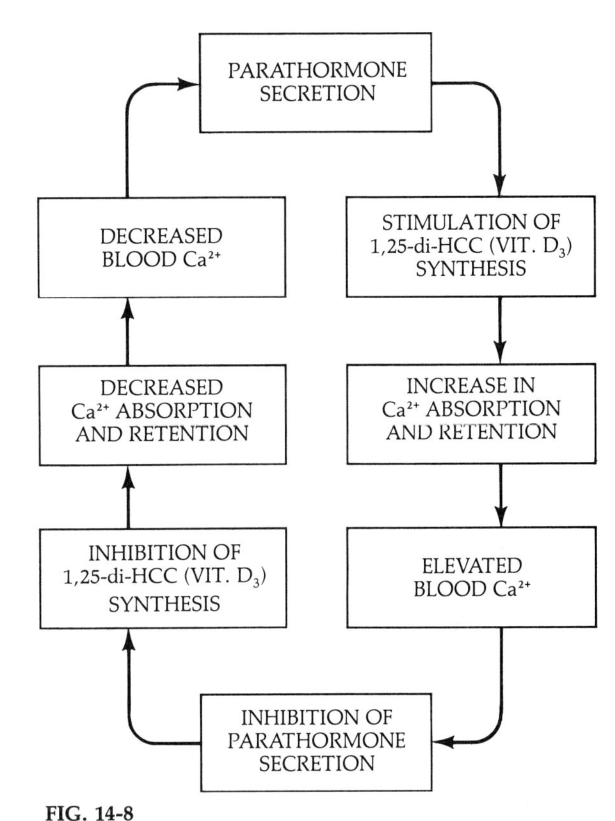

FIG. 14-8
Homeostatic regulation of Ca^{2+} levels in the blood.

Vitamin A

Vitamin A (retinol, a reduced form of retinal) is required for intracartilaginous bone growth. It is necessary for the synthesis of chondroitin sulfate, a vital proteoglycan that acts as a plasticizer in the matrix of bone. Vitamin A is also required for the maturation of epithelial cells and plays a role in resistance to epithelial tumors (Chap. 3). The visual cells in the retina are dependent on vitamin A to synthesize the visual pigment rhodopsin (Chap. 8, Fig. 8-4).

Vitamin A is an essential nutrient composed of multiples of five carbon atoms linked in open chains. Precursors to the active vitamin are found in carotenes, which form yellow to yellow-orange pigments in fruits and vegetables. The precursors can be converted into active forms of vitamins A_1 and A_2. Vitamin A_1 is found in ocean fish oils, butter, egg yolk, and liver. Vitamin A_2 is found in freshwater fish (as well as in birds which consume them).

Vitamin A is easily oxidized, thereby losing its activity. Vitamin E protects it from oxidation. Administration of large doses of vitamin A leads to storage of toxic quantities in the liver. Excesses appear in the blood and cause a number of symptoms, including thickening of the bones, pain in the joints, loss of hair, cracking of the skin, blurred vision, and loss of appetite.

Vitamin C

Vitamin C (ascorbic acid) is required for the synthesis of a healthy matrix in connective tissue including that of bone and dentin. It is necessary for the metabolism of some amino acids, such as hydroxyproline, which forms cross-linkages in collagen. Absence of the cross-

TABLE 14-3
Effects of hormones on bone maturation and degradation

	BONE MATURATION			BONE DEGRADATION			
Hormone	*Cell Proliferation*	*Synthesis of Matrix by Osteoblasts*	*Ca^{2+} Increases in Bone*	*Resorption by Osteoclasts*	*Lysis by Osteocytes*	*Ca^{2+} Loss from Bone*	*Protein Loss in Bone*
Growth hormone	+	+	+				
Calcitonin				–	–	–	
Estrogens	?*	+	+	–			
Testosterone	–	+	+	–			
Vitamin D_3			±†	+			
Parathormone	+	–		+	+	+	
Glucocorticoids	–	–		+			+

A plus sign (+) indicates stimulation of the activity, a minus sign (–) indicates its inhibition.
* Stimulates the conversion of osteoclasts to osteoblasts?
† Vitamin D_3 causes the removal and addition of Ca^{2+} from old to new bone in remodeling.

TABLE 14-4
The effects of hormones on protein and mineral metabolism in bone formation*

			Ca^{2+}			P_i
Hormone	*Protein*	*Acid in ECF*	*Bone*	*Intestinal Absorption*	*Reabsorption by Kidney*	*Reabsorption by Kidney*
Parathormone		+	–	+	+	–
Calcitonin		–				–
Growth hormone	+			+	–	
Vitamin D_3			±†	+	+	+
Estrogens	+	–	+			
Testosterone	+		+		+	
Glucocorticoids	–		–			

* *Symbols:* Generally a plus sign (+) indicates conditions which favor bone formation and a minus sign (–) those conditions which inhibit it. However, in the case of acid, the effects are the opposite. † Vitamin D_3 causes Ca^{2+} to be removed from old bone and added to new bone in remodeling.

linkages reduces the tensile strength of collagen.

Vitamin C is a 6-carbon compound similar to simple sugar. Both l-ascorbic acid and l-dehydroascorbic acid, its oxidized and reduced forms respectively, are biologically active. Vitamin C is found in citrus fruits and in a variety of vegetables, including broccoli, brussel sprouts, and spinach. Vitamin C deficiency is associated with diseases of connective tissue such as scurvy (Chap. 3). Because vitamin C is readily excreted by the kidneys it is usually not toxic. In some individuals however, it may accumulate in the kidneys and cause the precipitation of salts and the formation of kidney stones.

Role of Hormones

Various polypeptide and steroid hormones stimulate the proliferation of bone cells, their capacity for mineral transport, and/or mineral resorption (Tables 14-3 and 14-4). The general properties of the hormones were summarized in Table 5-5. Polypeptide hormones with notable effects on bone are parathormone, calcitonin, and growth hormone. Among the effective steroids are vitamin D_3, the androgens, estrogens, and adrenal steroids.

Polypeptide Hormones

PARATHORMONE The parathyroid hormone, *parathormone,* is a polypeptide which promotes the synthesis of 1,25-diHCC (vitamin D_3); it also elevates the plasma levels of calcium and makes the mineral generally available to cells throughout the body. Negative feedback mechanisms regulate the blood levels of this hormone (Fig. 14-8).

Parathormone, like 1,25-diHCC, elevates plasma levels of calcium by causing the release of Ca^{2+} from bone, its absorption across the intestine, and its reabsorption (retention) by the kidneys. Parathormone, unlike 1,25-diHCC, whose synthesis it stimulates, also causes the excretion of P_i by the kidneys.

The precise mechanism of parathormone action has not been established. Reports indicate that it increases cAMP in osteoclasts and causes the cells to proliferate and form lactic, citric, and carbonic acids. The ECF becomes acidic (Table 14-4), and collagen and proteoglycans are depolymerized and degraded to form acidic glycoproteins. These changes cause bone to release Ca^{2+} to the blood. Furthermore, the degradation of proteoglycans releases large quantities of minerals such as Ca^{2+} ions which are normally bound to these macromolecules. The depolymerization of bone also renders its amino acid building blocks more accessible to the degradative action of organic acids and proteolytic enzymes. The overall effect of parathormone is the formation of depolymerized, demineralized bone with markedly less tensile strength.

CALCITONIN *Calcitonin* (thyrocalcitonin), the calcium-lowering hormone, is a polypeptide hormone which protects the body from excess calcium (hypercalcemia) and the bone from calcium loss. Calcitonin is synthesized by the C cells of the thyroid gland, which are adjacent to the cells of the thyroid follicles which synthesize thyroxine. The effects of calcitonin in humans have not been clearly demonstrated. In animals, the most significant effects occur during growth. Calcitonin's net activities mostly oppose those of parathormone and 1,25-diHCC (Table 14-3). Calcitonin inhibits the resorption of Ca^{2+} and P_i from bone. It may do so by decreasing the formation of acid in the ECF compartment around bone. It reduces the levels of Ca^{2+} in the cytoplasm and also inhibits P_i reabsorption by the kidneys, resulting in its greater excretion in the urine (Table 14-4).

GROWTH HORMONE *Growth hormone* (GH or somatotropin) is a polypeptide whose general effects are anabolic. It promotes Ca^{2+} absorption across the intestine, protein synthesis, and the multiplication of cartilage, bone, and blood cells (Table 14-3). One of its major roles is to promote protein synthesis (Table 14-4). Growth hormone increases the uptake of amino acids and rapidly inhibits protein catabolism and the catabolic effects of glucocorticoids. It causes osteoblasts to secrete collagen and proteoglycans. Many of these effects may be mediated by *somatomedins,* which are polypeptide hormones of about 7000 MW. Their secretion is regulated by growth hormone. Somatomedins are formed in the liver and perhaps in the kidney and muscle. They stimulate proteoglycan synthesis and are directly involved in the growth of bone.

Growth hormone can spare glucose, elevate plasma glucose levels, and increase glycogen storage. It can promote the conversion of proteins into carbohydrates and in adipose tissues can promote the release of fatty acids which may be converted into carbohydrates and used as energy sources (Fig. 13-11, p. 321).

Steroid Hormones

Vitamin D_3, the androgens, estrogens, and adrenal glucocorticoids are steroids which significantly affect bone metabolism. Vitamin D_3 promotes conditions which assist the remodeling of bone; the androgens and estrogens aid its development; and the adrenal steroids generally promote its degradation.

The effects of the steroids on mineral metabolism will be described in detail in Chap. 21 and their influences on sexual development and reproductive function will be discussed in Chap. 24.

VITAMIN D Active forms of vitamin D are considered hormones since they are synthesized by specific cells, circulate in the blood, and bind to and affect target cells. The hormonal effects of these steroids on bone described above, are summarized in Tables 14-3 and 14-4.

ANDROGENS *Testosterone,* the major androgen (or male sex hormone), is an anabolic hormone synthesized in the testes. It enhances Ca^{2+} retention, the mineralization of bone, and protein synthesis during the development of bone and muscle (Tables 14-3 and 14-4).

Castration (removal of the testes in males) lowers testosterone levels and significantly affects bone growth in immature males. Lack of sufficient testosterone inhibits the development of the secondary sex characteristics and retards the rate of epiphyseal closure. Intracartilaginous bone formation proceeds for a longer period of time, and greater adult heights result. These effects are the basis in animal husbandry for the castration of a rooster to produce a capon and of a bull to produce a steer. Larger and plumper bodies result. Conversely, lesser adult heights result when abnormally high levels of testosterone are present during the same growth period.

ESTROGENS β-Estradiol, the most active form of the female sex hormones, is one of three principal forms of active *estrogens,* which exert anabolic effects on most tissues. The other major estrogens are estrone and estriol, which are derived from β-estradiol in the liver. Estrogens promote protein and nucleic acid synthesis in most cells. Because they inhibit lactate production and lower its plasma levels, the estrogens cause Ca^{2+} to accumulate and increase the density of bone (Tables 14-3 and 14-4).

Estrogens may cause the conversion of osteoclasts to osteoblasts, induce protein synthesis by osteoblasts, and promote mineralization in long bones but stimulate demineralization in the pelvis. In general, these responses lead to early closure of the epiphyses. The earlier advent of puberty in females than in males and the accompanying increases in sex hormone production may partially explain early epiphyseal closure in females and the fact that women attain a lesser average height than men.

ADRENAL STEROIDS The adrenal cortex secretes three classes of hormones: mineralocorticoids, glucocorticoids, and sex steroids. The influences of the *sex steroids* were just described. *Mineralocorticoids* (such as aldosterone) are salt-retaining hormones. They promote the retention of salts (such as sodium) and stimulate salt reabsorption from the urinary filtrate in the tubules of the kidney. They thereby maintain or elevate salt levels in the blood and the availability of minerals to tissues, including bone. The functions of mineralocorticoids will be described more completely in Chap. 21.

The main responses to the glucocorticoids are catabolic (Table 14-5). Because the hormones elevate plasma glucose and originate in the cortex of the adrenal gland, they are called *glucocorticoids.* Glucocorticoids stimulate gluconeogenesis (the formation of glucose from noncarbohydrates). Proteoglycans and collagen are degraded under their influence, leading to loss of Ca^{2+} ion from the bone and resorption.

Hormonal Interactions

The ways in which hormones influence one another were discussed at length in Chap. 5. Many of these interactions affect bone. For example, growth hormone seems to cooperate with thyroxine and insulin to promote collagen synthesis in bone. The more effective collective responses to these hormones may be due in part to somatomedins, whose secretions are stimulated by growth hormone and which compete with insulin for cellular receptors. Thyroxine may cooperate by its general stimulation of aerobic metabolism.

TABLE 14-5
Principal effects of major adrenal cortical steroids

Hormone	*Major Effects*	
Glucocorticoids		
Cortisol (hydrocortisone) and corticosterone	Protein, carbohydrate, and fat catabolism	Elevate blood glucose levels by gluconeogenesis
Mineralocorticoids		
Aldosterone	Metabolism of electrolytes (retention of Na^+, Cl^- and HCO_3^- ions) Osmotic retention of water	
Sex Steroids		
Androgens and estrogens	Anabolism and development of secondary sex characteristics	

Aging

Skeletal mass is reduced by an average of 15 percent between youth and old age by the loss of bone. The extremely complex bases for these changes have yet to be clearly defined. The individual's sex, genetics, and alterations in the circulatory system, thymus, cell receptors, and hormone secretions are involved.

The circulatory system becomes less resilient with aging. The flow of blood and exchange of metabolites decrease and the dynamic flux of materials necessary for the maintenance of bone is adversely affected. The accumulation of calcium around pores may further impede ion exchange (pp. 88–89).

The thymus shrinks after adolescence. With its regression, antibodies may be produced which may attack the body's own tissues and cause autoimmune diseases. Such diseases are more prevalent in the aged. Changes in the cellular activities of T cells from the thymus may lead to degradation of connective tissue elements, including those in bone. The functions of T cells were introduced in Chap. 3 and will be described further in this chapter and in Chap. 19.

The number of cellular receptors decreases and their structure changes with aging (page 115). The quality and quantity of receptors are influenced significantly by genes carried in each cell. The responsiveness of receptors to some hormones is adversely affected by aging, as is the body's capacity to maintain homeostasis.

The secretion of sex steroids diminishes with aging, resulting in a decrease in the anabolic effects of these hormones relative to the catabolic effect of glucocorticoids on bone. The gradual resorption of bone results.

Mechanical Stress and Exercise

Physical exercise and mechanical stress promote bone metabolism. As a result of exercise the rate of circulation increases and the dynamic exchange of substances is accelerated. Mechanical stress also seems to induce physical changes in the crystalline structure of bone which stimulate cellular activity. When crystals are put under pressure, their convex surface (outward bulge) acquires a negative charge, while their concave surface (inward curve) becomes positively charged. When bone is put under pressure, electronegative charges accumulate on the convex surface and cause effects similar to those of an increase in pH. The increased alkalinity of the ECF favors the activity of alkaline phosphatases, promoting osteogenesis. Conversely, elevations in electropositive charges on the concave surface increase the acidity of the ECF. The activity of acid phosphatases of osteoclasts is promoted, as is the solubility of calcium phosphate. These changes lead to the resorption of bone.

This principle has application in the use of braces to straighten bones. Bone growth is favored on the outward side when pressure is applied to it. Resorption occurs on the opposite side. In this manner bones (such as those holding teeth in the jaw or those in deformed limbs) are reshaped.

Inactivity or failure to put mechanical stress on bone results in resorption. For example, when teeth are lost and not replaced, chewing and mechanical stress on the jawbone decrease. Over many years gradual resorption may cause the jaw to recede by as much as 50 percent. Therefore, infants and children should eat foods which require chewing in order to make room in the jawbones for their permanent teeth. Adults should do so to maintain normal jaw structure. Whenever possible individuals should exercise regularly to promote and sustain the development of a healthy skeletal system.

The structure of the joints also may change in ways that limit movement or make it painful. Like bone, the joints of the body can be adversely affected by aging, lack of exercise, excess stretching, and a number of changes which cause metabolic responses that alter the properties of connective tissues. Exercise and moderate mechanical stress are necessary for the maintenance of healthy joints. The activity of osteoblasts decreases after several months of joint disuse, as occurs with limbs immobilized in a cast or in bedridden patients. Fat and fibrous connective tissues penetrate the joint capsule and may even replace cartilage in the immobilized joint. Conversely, excess stretching and physical trauma may damage joints. A *sprain* occurs when the tissues in a joint are stretched or torn. It often causes the temporary accumulation of interstitial fluid within the joint, thereby producing swelling and pain. A *dislocation* occurs when the tissues are stretched and often torn, allowing the bones to move out of the joint.

14-5 DISORDERS OF BONES AND JOINTS

A number of disorders can adversely affect the development and/or function of bones and joints. Among the causes are metabolically induced diseases and inflammatory reactions. Some of these were briefly mentioned earlier in Chaps. 3 and 5. Others, which are due to disorders in the bone marrow, will be described further in Chaps. 18 and 19.

Bone diseases due to alterations in metabolism can be separated into those primarily due to defects in calcium metabolism (such as rickets and osteomalacia) and those which alter the ratio of osteoblast to osteoclast activity (such as osteoporosis, hyperparathyroidism, excess vitamin D, and cancer of the bone). Several of these disorders are summarized in Table 14-6.

Rickets commonly occurs as a result of vitamin D deficiency, usually within the first 18 months after birth. Failure to absorb Ca^{2+} may lead to a decrease in the Ca/P ratio. Cartilage continues to grow but is not converted into calcified bone. As a result, the skeletal system bends when put under mechanical stress. Cartilaginous knobs on the rib cage (rachitic rosary), bowed legs, or knocked knees are signs of rickets.

Even though the disorder is not caused by vitamin D deficiency, similar symptoms occur in *familial hypophosphatemic,* or *vitamin D–resistant, rickets.* An X-chromosomal heritable defect in P_i transport enzymes causes an abnormal Ca/P ratio. This form of rickets cannot be corrected by administration of vitamin D. Another heritable form of vitamin D–resistant rickets, due to recessive genes on the autosomes, blocks synthesis of vitamin D_3. A defect in the enzyme needed to add the -OH group to the first carbon atom of cholecalciferol causes this disorder.

Osteomalacia (sick bones) is an adult variation of rickets due to Ca^{2+} deficiency. It is more common in women than in men. Lack of vitamin D, kidney disease, or prolonged lack of activity, such as may occur in bedridden, chronically ill patients, can cause osteomalacia. The bones gradually become less dense, soften, and bend. Pain results and the bone is more susceptible to fracture. Osteomalacia may begin during pregnancy, when it is perhaps due to the resorption of the mother's bone which supply Ca^{2+} for fetal growth.

Osteoporosis (porous bones) often begins around age 50. Its precise cause is unknown but may be related to inadequate Ca^{2+} intake or to abnormal metabolism caused by hyperthyroidism, vitamin C deficiency, or an imbalance in the ratio of glucocorticoids to sex steroids. In osteoporosis the osteoblasts are inhibited and the density of bone therefore decreases as its matrix becomes malformed. The Haversian canals widen, producing a porous effect.

Osteoporosis progresses twice as fast in women as in men between ages 50 and 65. Thereafter the rate of loss is equal in both sexes.

TABLE 14-6
Metabolically induced disorders in bone formation

Disorder	*Characteristics*	*Possible Causes*
1. *Major effects on calcium-phosphate balance*		
Rickets	Poor Ca^{2+} absorption Matrix formation occurs in bone without calcification, within 18 months of age	Vitamin D deficiency Ca^{2+} deficiency
Vitamin D–resistant rickets		
Familial hypophosphatemia (hypophosphatemic rickets or vitamin D–resistant rickets)	Soft ricketlike skeleton Low alkaline phosphatase activity Low plasma P_i	Genetic defect on X chromosome, causing defective transport of P_i
Vitamin D–resistant rickets	Low plasma Ca^{2+}	Recessive genetic defect on autosome Lack of kidney enzyme to form 1,25-di-HCC
Osteomalacia (adult rickets)	Accumulation of uncalcified bone matrix	Kidney disease Prolonged immobility Prolonged vitamin D or Ca^{2+} deficiency
2. *Major effects on ratio of osteoblast to osteoclast activity*		
Osteoporosis	Malformation of matrix Wide Haversian canals and porous bone Inhibition of osteoblasts	Hyperthyroidism Decrease in sex steroids Elevated activity of glucocorticoids Malnutrition (long-term calcium deficiency or vitamin C deficiency)
Osteitis fibrosa cystica	Progressive bone loss Fibrous tissue deposits in bone Appearance of discrete, "punched-out" areas Overactivity of osteoclasts	Kidney disease Hyperparathyroidism*
Osteosclerosis	Increased amount of bone	Secondary (metastatic) cancer, lead poisoning, hypoparathyroidism*
Dwarfism*	Short skeletal structure Underactive osteoblasts	Undersecretion of growth hormone
Dwarfism-cretinism*	Short skeletal structure Underactive osteoblasts	Undersecretion of thyroxine
Gigantism*	Overgrown skeletal structure Overactivity of osteoblasts	Oversecretion of growth hormone during preadult years
Acromegaly*	Abnormal growth of bones in face, hands, and feet Overactivity of osteoblasts	Oversecretion of growth hormone in adults

* Also see Tables 5-5, 14-3, and 14-4.

The early more rapid loss in females may be related to the abrupt decrease in circulating levels of estrogens that occurs at menopause (the cessation of reproductive capacity in women). In aging males, circulatory levels of the primary sex hormone, testosterone, decrease gradually. Other hormones also may influence the progression of osteoporosis. For example, parathormone causes greater bone resorption when the activity of estrogens is low.

Bone disorders also may affect nearby tissues. For example, physical trauma to bone can cause the accumulation of Ca^{2+} in adjacent soft tissues such as muscle in a disease called *myo-*

TABLE 14-7
Inflammatory disorders of bones, joints and surrounding tissues

Disorder	*Cause*	*Comments*
Osteomyelitis	Bacterial infections in bone, causing inflammatory response	Causative agents include staphylococci, tubercle bacilli, and syphilitic bacteria
Arthritis	Mechanical stress, infection, metabolic disorders, genetic factors, autoimmune reactions	Progressive degeneration of joints; over 100 types; common in elderly
Osteoarthritis ("wear and tear" arthritis) (also see Chap. 3, p. 73)	Mechanical stress on joints	Occurs more often in later life and, more often in males than females Degrades collagen Fibrocartilage replaces hyaline cartilage in joints Localized osteoporosis may develop Bony projections may develop to limit mobility and cause pain
Rheumatoid arthritis (also see Chap. 3, p. 63)	Chronic inflammation of synovial membrane from physical trauma and/or slow virus infection Rheumatoid factors (IgM and IgG antibodies) produce autoimmune responses against tissue in joints	Often appears between ages 20 and 40 Occurs 3 times more often in females than males Stiffness and pain tend to lessen during the day
Gingivitis	Inflammation of the gums	May be initiated by dental plaque
Pyorrhea	Inflammation causing resorption of bony socket of jaw around teeth	Major cause of tooth loss in adults
Bursitis	Inflammation of bursae, often due to physical trauma	Common examples are "tennis elbow" and "housemaids knee" Common site is in the shoulder
Rheumatic fever	Inflammation of synovial membrane, tendons, and other joint tissues and heart valves	May be caused by antibodies formed against unchecked streptococcal infection; antibodies react with tissues
Fibrositis	Inflammation of fibrous connective tissue May migrate from site to site	Commonly called *rheumatism* Often occurs in small of back, when it is commonly called *lumbago*
Gout	A genetic defect causing accumulation of uric acid crystals from purine metabolism results in inflammation in joints	Common sites affected are toes, knees, fingers

sitis ossificans. In addition to physical damage, disease and antigenic stimulation can cause inflammatory responses which disrupt bone and/or its surrounding tissues (Table 14-7). The basis of the inflammatory response and its role in the degradation of collagen and connective tissue disease was described in Chap. 3. To review it briefly, T lymphocytes produced in the thymus and circulating in the blood are attracted to the altered tissue site, along with other white blood cells, including monocytes, which mature into macrophages. Different lymphocytes release various lymphokines, which act as regulatory molecules influencing cellular activity, including that of macrophages. In chronic, prolonged disorders, lymphokines

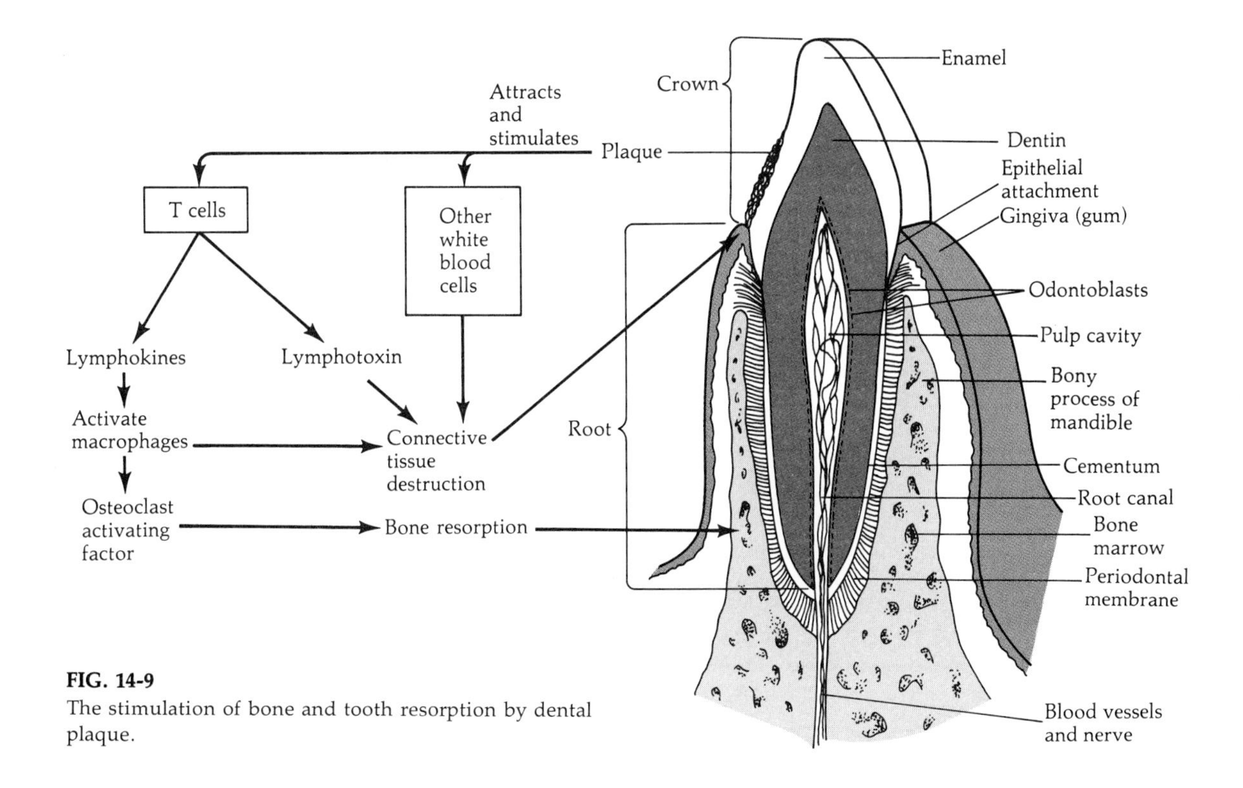

FIG. 14-9
The stimulation of bone and tooth resorption by dental plaque.

can stimulate the remodeling of bone, the formation of small bony projections, and the binding of soft tissues to bone.

Some of the lymphokines mimic the effects of parathormone (perhaps because of the common endodermal origin of the thymus and the parathyroids). Lymphokines also can antagonize the effects of growth hormone. The net result of their activity is the inhibition of bone growth coupled with resorption of the tissue.

A common example of the effects of an inflammatory response on bone is the series of reactions which can occur as the result of the accumulation of dental plaque on the teeth and gums (Fig. 14-9). Dental plaque consists of bacterial masses that produce polymers of dextran from sucrose. Plaque may interact with white blood cells as they move through the blood in the gums surrounding the teeth. Following stimulation by dental plaque, reactive lymphocytes release lymphokines which may behave as *osteoclast-activating factors* to promote bone degradation. Other lymphokines activate blood cells and macrophages to form enzymes such as collagenases that hydrolyze the components of bone. "Killer" T cells release lymphotoxins which can destroy fibroblasts, causing *periodontal disease.* Gingivitis, an inflammation of the gums, may result. *Pyorrhea* (loosening of the teeth) due to resorption of the bone in the jaw can also occur.

14-6 SUMMARY

1 Bone supports and protects the body, anchors muscles, stores minerals, and is a source of blood cells. The approximately 206 bones in the body form the skeletal system.
2 Each bone is an organ composed of compact and spongy bone tissues plus others which complete its structure. The long bones of the body have a shaft

called the diaphysis and two extremities called epiphyses. Fibrous connective tissue, the periosteum, covers and adheres to the exterior of bones and is penetrated by blood vessels, lymph vessels, and nerves located in channels in thin bone. A condensed layer of specialized connective tissue, named the endosteum, lines the central cavity in the shaft. Bones may form freely movable junctions, termed diarthroses or synovial joints, or may form junctions with little or no movement, called synarthroses. There are several types of joints formed by different kinds of connective tissues. Synovial membranes near joints form bursae or pouches. Tendon sheaths form tunnels in which tendons glide over bones or ligaments.

Marrow fills the center of the shaft. It is derived from embryonic mesenchyme and contains stem cells, which are precursors of blood cells and other cell types in various stages or maturation. Mesenchyme also produces fat cells, which accumulate with aging, converting red marrow to less active yellow marrow.

Bone contains about 8 percent water, 20 percent organic molecules, and 72 percent inorganic salts. About 90 to 95 percent of the organic molecules consist of collagen, a triple-stranded, cross-linked, coiled protein. Proteoglycans are complexes of proteins and mucopolysaccharides which keep collagen soft and pliable. Calcium phosphate and calcium carbonate account for about 57 percent and 10 percent, respectively, of the weight of bone. Many minerals found in small amounts in bone are extremely important to body functions. Examples are Na^+ and Mg^{2+} ions. Calcium phosphate, $Ca_3(PO_4)_2$, is a crystalline salt whose solubility improves in an acid environment. The product of the concentration of calcium and phosphate in the blood, the Ca/P ratio, is an important indication of mineral metabolism in the body. Three crystalline forms of the salt found in bone are hydroxyapatite, octacalcium phosphate, and amorphous calcium phosphate. Hydroxyapatite is an ionic crystal which can exchange ionized elements with the blood and/or adsorb oppositely charged particles. Among these are F^- ions, whose incorporation strengthens the crystalline structure of hydroxyapatite and that of bone.

In response to the body's needs, mineral elements are constantly exchanged between bone and blood. Bone is formed by the addition of minerals to its matrix and degraded by their resorption. Bone growth is favored by the proliferation and maturation of mesenchymal cells, the synthesis of protein, the absorption of Ca^{2+} and P_i from the intestine, their reabsorption from the kidney back into the blood, and the formation of a less acidic extracellular fluid (ECF). Resorption is generally promoted by the opposite conditions.

3 The process of bone formation and growth is called osteogenesis. The mesenchyme forms osteoblasts, which synthesize bone and then become osteocytes, entrapped residents within the bone. Bloodborne cells mature into osteoclasts, which resorb bone. Intramembranous or mesenchymal bone formation occurs within membranes in flat bones, widens the diameter of bones, and maintains all bones of the adult skeleton after 20 to 25 years of age. Intracartilaginous bone formation produces the long bones of the body from cartilage cells which proliferate in growth centers. The cartilage is dissolved and replaced by bone through the activity of cells derived from the mesenchyme.

4 Acidity, nutrients, hormones, and mechanical stress affect the proliferation and/or maturation of bone cells and the metabolism of protein and minerals. Acidity of the ECF leads to demineralization of bone and loss of calcium to the blood. Conversely, acid in the digestive tract promotes the absorption of Ca^{2+} ions and their passage from the lumen of the tract into the blood.

Vitamins D, A, and C strongly influence osteogenesis. Vitamin D is converted to active forms, 25-hydroxycholecalciferol (25-HCC) and 1,25-dihydroxycholecalciferol (1,25-diHCC) in the liver and kidney, respectively. The vitamin promotes the intestinal absorption of Ca^{2+}, its mobilization from old to new bone, and the reabsorption of P_i by the kidney. It maintains high Ca/P ratios that facilitate bone remodeling. Vitamin A (retinol) is required for the synthesis of chondroitin sulfate, for intracartilaginous bone growth, and for epithelial cell differentiation. Vitamin C (ascorbic acid) is necessary for the metabolism of hydroxyproline, which forms cross-linkages in collagen in the matrix (intercellular substance) of connective tissue.

Many hormones directly influence bone physiology. Growth hormone stimulates Ca^{2+} absorption across the intestine, protein synthesis, and bone growth; calcitonin inhibits the resorption of bone in animals. Estrogens and androgens are anabolic steroid hormones that stimulate protein synthesis and Ca^{2+} incorporation into bone, resulting in its maturation. Parathormone and glucocorticoids accelerate bone degradation. Parathormone stimulates the synthesis of active vitamin D_3 (1,25-diHCC), Ca^{2+} resorption, and depolymerization of proteoglycans in the bony matrix. Calcitonin may oppose its effects. Glucocorticoids convert protein to carbohydrate, adversely affecting collagen, proteoglycans, and bone formation.

With aging, bone becomes more susceptible to fracture because its mass decreases. The changes may result from diminished circulation, alterations in cellular receptors, elevations in the ratio of glucocorticoids to sex steroids, and lack of cellular protection due to shriveling of the thymus.

Mechanical stress and physical exercise promote dynamic exchange of the constituents of bone and retard its deterioration. Disorders in bone and joints result from aging, lack of exercise, excess stretching, and metabolic changes in connective tissue. Lack of mechanical stress may lead to accumulation of extra fat and fibrous tissue in joints. Sprains result from excess stretching and may cause accumulation of interstitial fluid in joints. Dislocations move bones out of joints.

5 Abnormal bone structure may result from metabolic disorders and inflammatory responses. The most common disorder, rickets, involves the formation of soft uncalcified bones. It is often due to a decrease in Ca^{2+} absorption caused by vitamin D deficiency. Osteomalacia is a type of adult rickets due to Ca^{2+} deficiency and may be caused by kidney disease, vitamin D deficiency, and/or prolonged immobility. Osteoporosis is the development of porous bone with a malformed matrix. It may be due to hyperthyroidism or to a high ratio of glucocorticoids to sex steroids. It may also be caused by vitamin C deficiency.

Inflammatory reactions in which T cells release lymphokines and lymphotoxins may cause disorders of the bones and joints. Periodontal diseases include inflammation of the gums (gingivitis) and resorption of the sockets of bone around the teeth (pyorrhea). Osteomylitis can occur as a result of inflammation following bacterial infections. There are over 100 forms of arthritis (inflammation of tissues of the joints). Inflammation is caused by wear and tear in osteoarthritis and by an autoimmune reaction in rheumatoid arthritis. Inflammation of the synovial sacs near a joint is called bursitis.

REVIEW QUESTIONS

1 What are the major functions of bone?
2 Draw and label a sketch of a long bone and of a synovial joint. Describe the

functions of each of their components, including bone marrow. What are bursae, tendon sheaths, menisci?

3 What are the major chemicals in bone, and what is the function of each?
4 Compare the functions of collagen and proteoglycans.
5 How are hydroxyapatite crystals important in bone physiology?
6 If the Ca^{2+} level in the blood is 8.0 mg/dL and that of P_i is 2.0 mg/dL, what is the Ca/P ratio? What might be the resulting effects on bone?
7 Describe the dynamic exchange of constituents in bone and the significance of the Ca/P ratio.
8 What are the embryonic origins of bone? Describe the origins and functions of osteocytes and osteoclasts and their roles in intramembranous and intracartilaginous bone formation.
9 What is vitamin D? Describe its conversion to the active form. What is its role in bone physiology?
10 Discuss the factors that influence the remodeling of bone, including vitamin D, parathormone, osteoclasts, phosphatases, and physical stress.
11 What is vitamin A, and what influence does it have on bone function?
12 How does vitamin C affect the formation and maintenance of bone?
13 Compare the effects of parathormone, calcitonin, and growth hormone on bone.
14 How may the thymus, glucocorticoids, plaque, and osteoclasts be related to periodontal diseases?
15 How do the glucocorticoids and sex steroids (androgens and estrogens) affect bone physiology? How are they related to aging?
16 What are the consequences of aging on bone?
17 Explain the physical, chemical, and physiological principles that underlie reshaping of bone with braces.
18 How do rickets, osteomalacia, and osteoporosis differ?
19 How do inflammatory reactions influence bone metabolism?
20 Compare osteomyelitis, bursitis, rheumatoid arthritis, and osteoarthritis with respect to their underlying causes.

THE CIRCULATORY SYSTEM

The Structure and Function of the Heart

15

15-1 STRUCTURE OF THE HEART
Lining and walls of the heart
Outer coverings
Inner lining
Wall of the heart
Chambers and route of circulation
Valves
Effects of abnormal valves
Blood supply to cardiac muscle

15-2 CARDIAC PHYSIOLOGY
The cardiac cycle
Timing of systole and diastole
Blood flow through the chambers
Heart sounds
Cardiac output
Work of the heart
Energy production
Starling's law
Electrical conduction
Intrinsic (internal) regulation
Extrinsic (external) regulation
Efferent nerves • Afferent nerves and cardiac reflexes
Changes in electrical conductivity: The ECG
Principles • Wave patterns in a normal ECG • Unipolar recordings • Bipolar recordings • Effects of heart disease
Fibrillation

15-3 SUMMARY

OBJECTIVES

After completing this chapter the student should be able to:

1. Describe the structure of the heart and its pump-like action.
2. Outline the route of circulation of blood to and from the heart and the blood supply through its muscular walls.
3. Describe the rhythmical changes that occur in the cardiac cycle.
4. Discuss cardiac output, Starling's law, and the all-or-none law.
5. Describe the electrical activity of the heart.
6. Compare the intrinsic and extrinsic mechanisms of regulation of the heart.
7. Explain the principles and applications of the electrocardiogram (ECG).
8. Describe a normal ECG and relate its phases to the cardiac cycle.

The *circulatory,* or *cardiovascular,* system is composed of the heart and blood vessels (arteries, veins, and capillaries). They form, respectively, a pump and series of flexible tubes through which blood moves nutrients to cells, tissues, and organs and carries metabolic products away from them for use or disposal. The circulatory system, like the endocrine and nervous systems, is critical to the maintenance of homeostasis (Chap. 1). It provides communication between the intracellular and extracellular fluids of remote parts of the body. Various forms of cardiovascular disease affect more than 28 million people and account for nearly 55 percent of all deaths in the United States. Thus, cardiovascular physiology is extremely important to human health.

The heart is the functional center of the circulatory system. It pumps nearly 2000 gal (7571 L) of blood each day and between 51 and 55 millions gal (193 to 208 million L) in a lifetime through more than 60,000 mi of blood vessels. Anatomically, the heart is about the size of a closed fist, weighs nearly 0.66 lb (300 g), and accounts for 0.5 percent of total body weight of the average adult. Repeated, prolonged vigorous exercise causes proliferation (*hypertrophy*) of the myofibrils in cells (fibers) of the heart wall, enlarging the organ. In trained athletes the heart may weigh up to 1.32 lb (600 g). This larger size allows the heart to contract and expel a greater volume of blood more forcefully. No pathological effects have been noted when the heart enlarges as a result of repeated vigorous exercise. However, with circulatory abnormalities or disease infections may be lethal to cardiac muscle cells. They may then be replaced by noncontractile connective tissue cells, which impair cardiac function.

15-1 STRUCTURE OF THE HEART

Lining and Walls of the Heart

Outer Coverings

The heart is located in the middle of the chest cavity (the mediastinum) slightly to the left of center and is surrounded by a continuous, double-layered serous membrane, the *pericardium* (Fig. 15-1). The heart appears much like a fist that has been pushed into a partially inflated balloon. The inner serous layer is the *visceral pericardium* (or *epicardium*), which closely adheres to the heart wall and to the large blood vessels of the heart. The outer serous layer is the *parietal pericardium.* Serous fluid, a watery lubricant secreted by the two membranes, moistens their adjacent surfaces and is contained between them in the *pericardial cavity.* An external covering, the *fibrous pericardium,* is located outside the two serous membranes and forms a loose-fitting sac in which the heart is contained. This external membrane also adheres to the large blood vessels attached to the heart and to the diaphragm and sternum (breastbone). Damage or infection to any of the membranes may cause an inflammatory response called *pericarditis.* It results in infiltration of the membranes by white blood cells and molecules such as lymphokines and histamine. Fluid accumulates and the membranes may

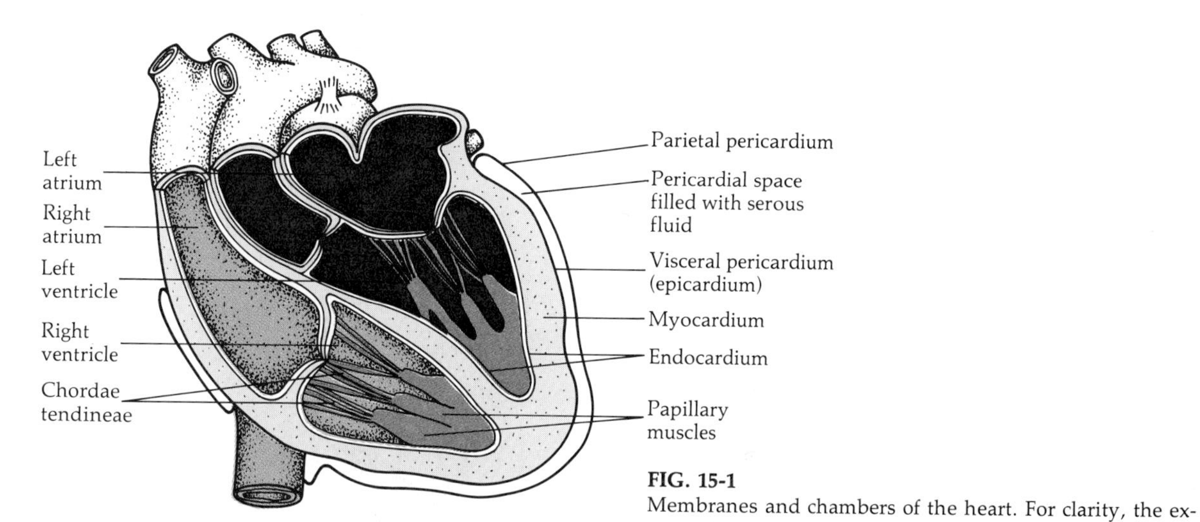

FIG. 15-1
Membranes and chambers of the heart. For clarity, the external fibrous pericardium is not shown. The lighter colored areas indicate blood low in O_2 content, while the darker colored areas indicate a high level of O_2.

form adhesions (stick together). These painful reactions may limit cardiac mobility and lead to severe cardiovascular problems.

Inner Lining

The chambers of the heart are lined with simple squamous epithelial cells which help form a smooth inner surface, the *endocardium* (Fig. 15-1). It is part of a continuous lining throughout the cardiovascular system and also covers the heart valves. In blood vessels this same lining is referred to as the *endothelium.*

Streptococcal bacterial infection can cause an immune response that damages the heart valves. In a small percentage of people, prolonged streptococcal sore throat may elicit a defense reaction leading to significant antibody production. It has been discovered that parts of the bacteria are composed of molecules which resemble some on the surfaces of the heart valves. Therefore, certain antibodies produced in reaction to streptococcal bacteria tend, because of this resemblance, to adhere specifically to the valves. The antibodies then attract white blood cells to the reactive sites, promoting an inflammatory reaction. Some of the white blood cells release pyrogens, that is, molecules that cause fever, which accounts for the name of the disease, *rheumatic fever.* This disease kills nearly 13,000 people annually in the United States. Penicillin minimizes the complications of the infection by blocking the synthesis of bacterial cell walls and the growth of the organisms. With any streptococcal infection penicillin should be given over its entire prescribed course. Otherwise, not all the bacteria will be killed and those left (the more virulent ones) may simply multiply and cause infection with a population of bacteria that is more resistant to the drug.

Wall of the Heart

The bulk of the heart wall, the *myocardium,* is composed of cardiac muscle described earlier in Chaps. 12 and 13. Trauma or infection of the myocardium can result in an inflammatory disorder called *myocarditis,* which may cause adhesions of the pericardium, accumulation of interstitial fluid, and interference with normal cardiac function. The myocardium performs the work of the heart, which is described below.

Chambers and Route of Circulation

The wall of the heart encloses a hollow space divided into four chambers, the right and left *atria* and the right and left *ventricles* (Fig. 15-1). Most of the blood delivered to the right atrium

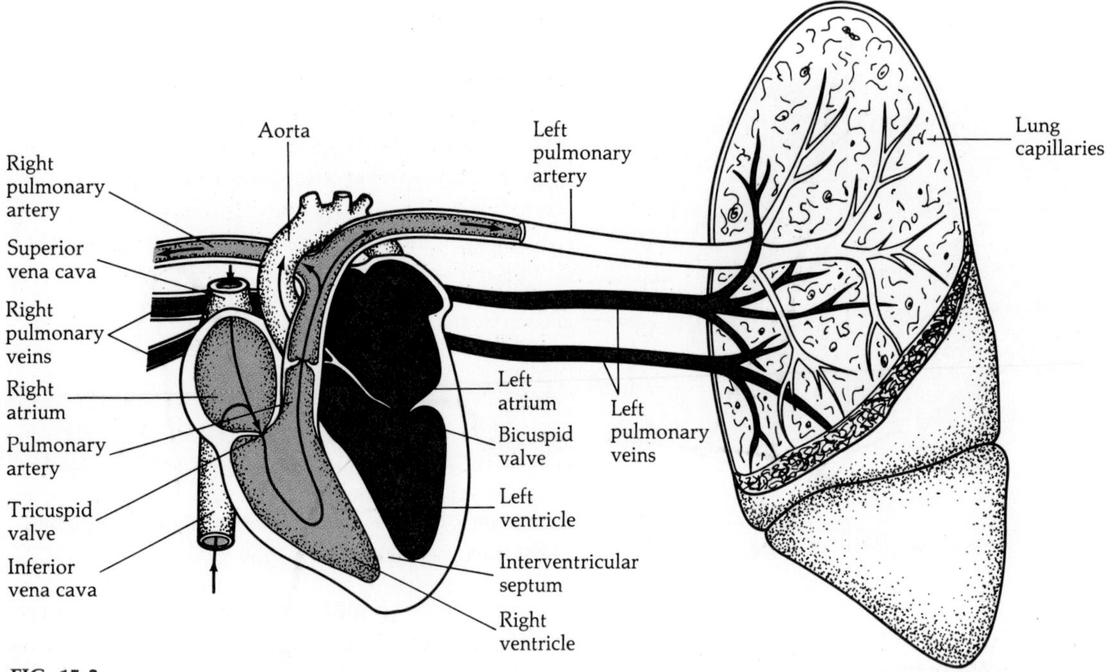

FIG. 15-2
Route of circulation of blood through the heart. The arrows indicate the route that blood follows. For simplicity only one lung (the left one) is shown. The lighter areas indicate blood low in O_2 content, while the darker colored areas indicate a high level of O_2.

comes from the large veins in the body, the *venae cavae,* and flows through them into the right ventricle to enter the pulmonary artery and the lungs (Fig. 15-2). From the lungs, it travels through the pulmonary veins back to the left atrium to enter the left ventricle and exit through the aorta to the body.

Alternate contraction (*systole*) and relaxation (*diastole*) of the heart propels blood through the body. Ventricular systole propels blood through the pulmonary arteries to the lungs and through the aorta to the rest of the body.

Valves

Valves normally restrict blood flow to one direction. They are found in the major vessels and chambers of the heart as well as in veins and lymph vessels throughout the body. A three-flap valve, the *tricuspid valve,* allows blood to flow from the right atrium to the right ventricle but not in the reverse direction (Figs. 15-2 and 15-3). The blood passes from the right ventricle through the *pulmonary semilunar valves* of the pulmonary arteries to the lungs. The three half-moon-shaped flaps account for the valve's name. The blood returns from the lungs to the left atrium through four pulmonary veins. It then flows by way of a two-flap *mitral* or *bicuspid valve* into the left ventricle and out to the body through the *aortic semilunar valve.* Because the tricuspid and bicuspid valves are located between the atria and the ventricles, they are often referred to as the right and left *atrioventricular* (A-V) *valves,* respectively.

Small papillary muscles attached to the atrioventricular valves by tendinous strands, the chordae tendineae, hold the valves in place to maintain unidirectional flow of blood through the atria into the ventricles (Fig. 15-1). The attachments prevent the flaps of the valves from opening in the opposite direction when pressure builds up in the ventricles during systole.

Effects of Abnormal Valves

Stenosis, the narrowing of any valve or blood vessel, causes a decrease in blood flow and a

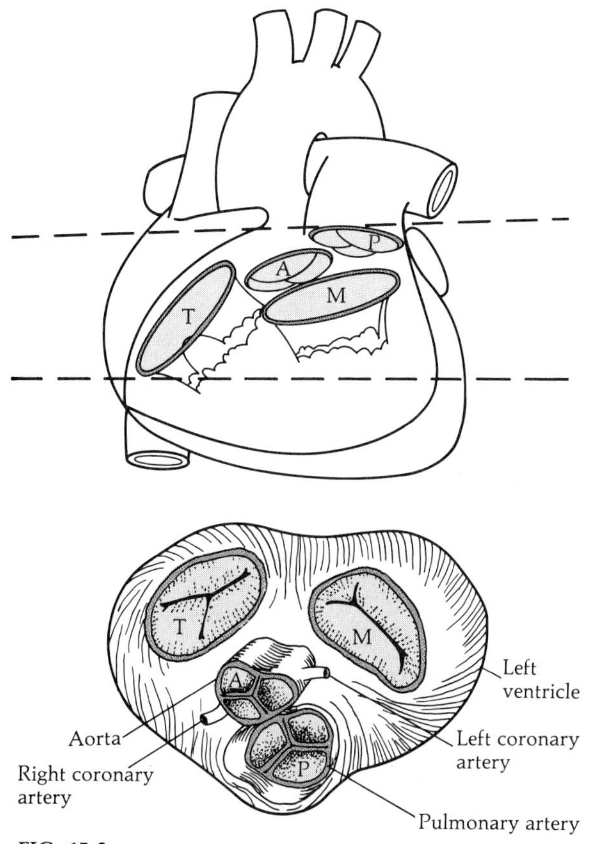

FIG. 15-3
Valves of the heart. The upper figure shows the location of the valves in the heart. The lower figure is an anterior-superior view that demonstrates their detail in a section taken from the area between the dotted lines that separates the upper and lower chambers. T, tricuspid valve; A, aortic valve; P, pulmonary valve; M, mitral (bicuspid) valve. Adapted from *The Heart,* 2d ed., J. S. Hurst and R. B. Logue (eds.). Copyright © 1970, McGraw-Hill Book Co. Used with permission of McGraw-Hill Book Co.

change in blood pressure in the vessel it serves. Stenosis is often associated with the development of a faulty or incompetent (inadequate) valve in which the flow becomes turbulent. The severity of the turbulence depends upon the degree of stenosis. Circulation through the heart is affected by tricuspid, mitral, pulmonary, or aortic stenosis.

The inflammatory response which accompanies rheumatic fever may also damage A-V heart valves and cause them to become incompetent. When this occurs, blood leaks back into the atria during ventricular systole, resulting in turbulent flow, or *regurgitation.* This alteration in flow may decrease the volume of blood moving into a given chamber and result in a gurgling sound. Backward blood flow decreases cardiac output and may lead to cardiac insufficiency. The abnormal sounds (*heart murmurs* or clicks) caused by stenosis or incompetent valves may be heard with a stethoscope.

Blood Supply to Cardiac Muscle

The right and left coronary arteries supply the muscular walls of the heart itself with blood. The left artery carries about 85 percent of the blood to the myocardium and the right one transports the remainder. During rest, the coronary arteries deliver about 200 to 250 mL of blood per minute to the heart muscle. However, during strenuous exercise, they may carry as much as 1000 mL-min.

The first branches of the aorta are the right and left *coronary arteries,* which originate just above the aortic semilunar valve and divide into smaller vessels which encircle the heart (Fig. 15-4). Arterioles (small arteries) are formed which lead to capillary beds (the smallest of blood vessels) within the heart muscle.

The coronary capillaries allow the exchange of nutrients and wastes. Oxygen in the capillaries diffuses into the heart muscle cells, and carbon dioxide, built up by metabolism in cardiac muscle cells, diffuses into the vessels. The capillaries merge to form *coronary veins,* which deliver most of the blood to the *coronary sinus,* a large vein that empties into the posterior surface of the right atrium. A few coronary veins discharge blood directly into the chambers of the heart.

Nearly 50 percent of the mass of the heart consists of blood vessels which penetrate the myocardium and are vital to its health. Heart muscle cells depend on oxygen for aerobic metabolism and are rapidly damaged by oxygen deficiency. Blockade (occlusion) or spasms that narrow coronary arteries cause *coronary insufficiency.* The disorder is often the result of atherosclerosis, which is caused by accumulation of fatty deposits in the lining of the coronary ar-

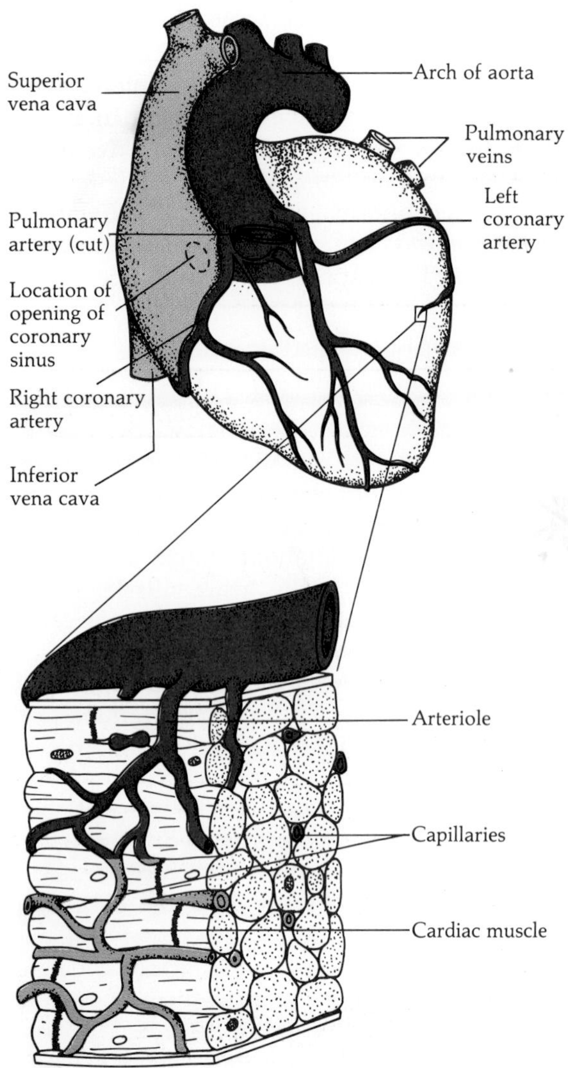

FIG. 15-4
Coronary arterial blood supply to the heart. The lower figure illustrates an enlargement of the vascular supply of cardiac muscle from the area of the heart indicated in the box in the drawing of the heart.

teries and proliferation of smooth muscle in their walls, both of which narrow the inner diameter of the vessels. If the obstruction occurs near the origin of the coronary arteries, the area of the heart deprived of oxygen is large and the damage (infarct) to the heart muscle is more extensive than occurs with an obstruction closer to the terminal branches of the vessel. The effects of coronary artery disease and atherosclerosis will be discussed in greater detail in Chaps. 16 and 17.

Prolonged and repeated exercise may lead to proliferation of capillary beds in the heart (Chap. 13). However, this increase may merely be sufficient to supply the greater muscle mass formed by the accompanying hypertrophy of the myofibrils. It is not clear whether alternate (collateral) routes of coronary circulation are produced in response to exercise. Formation of such natural bypasses would offer the advantage of alternate emergency routes of supply to the myocardium should blockade occur.

15-2 CARDIAC PHYSIOLOGY

The Cardiac Cycle

As the heart pumps blood in resting adults, it alternately contracts in systole and relaxes in diastole in rhythmical beats, which average 75 per minute. One complete heartbeat (systole and diastole) lasts about 800 ms (milliseconds) (0.8 s, Fig. 15-5).

Timing of Systole and Diastole

At the start of the cycle, both atria undergo systole almost simultaneously for about 100 ms and then relax in diastole for 700 ms. Ventricular systole follows atrial systole almost immediately, lasts for 300 ms, and occurs nearly simultaneously in both ventricles. Ventricular diastole lasts for 500 ms, 100 of which overlap the entire period of atrial systole. During a total of 400 ms all four chambers are simultaneously in diastole. This overlap provides all chambers of the heart with simultaneous rest between contractions.

Blood Flow through the Chambers

The entrance and ejection of blood from the heart are timed with the cardiac cycle. Blood flows into the atria during atrial diastole, and about 70 percent of the blood that enters the ventricles does so while the atria and ventricles are in diastole (Fig. 15-5). The remaining 30 percent of blood that flows into the ventricles does so at the beginning of the cardiac cycle

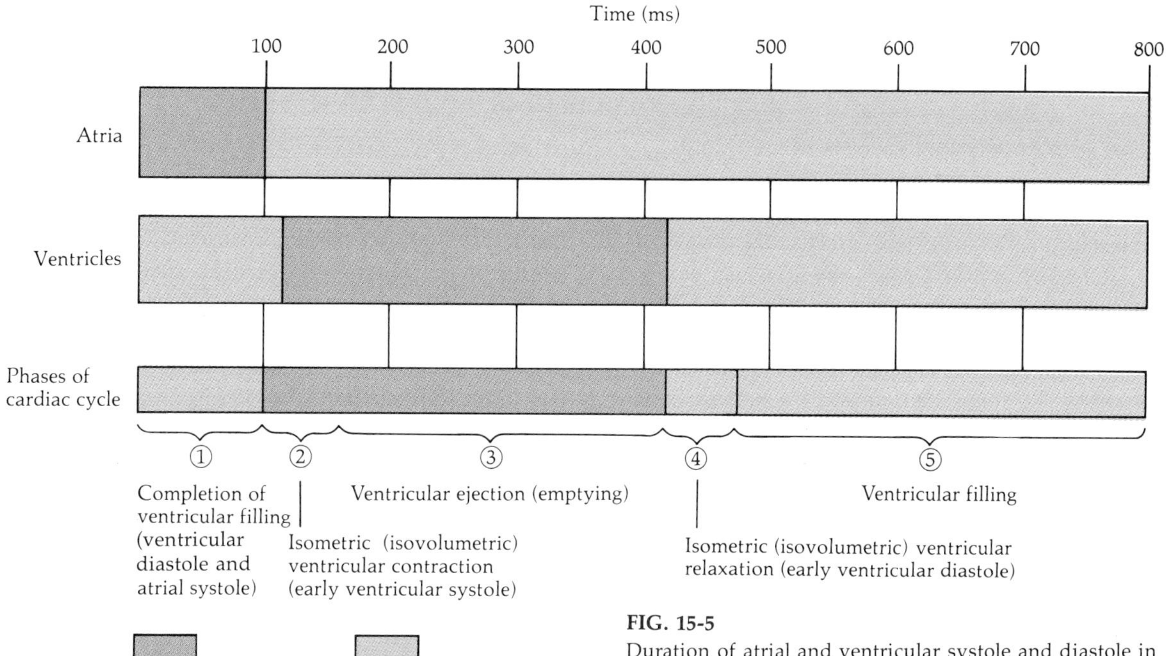

FIG. 15-5
Duration of atrial and ventricular systole and diastole in the cardiac cycle. Phases of the cardiac cycle, (1) to (5). The atrioventricular valves close in early ventricular systole (phase 2) and open in early ventricular diastole (phase 4). These respective changes precede the opening of the semilunar valves of the aorta and pulmonary artery and follow their closure. In the short intervals when one set of valves closes and the other opens heart sounds 1 and 2 are heard during the phases of the cardiac cycle which are described.

when the ventricles are relaxed but the atria are in systole. Then blood is ejected from the heart during ventricular systole. Therefore the bulk of blood enters the ventricles in the last phases of atrial diastole. Because of this phenomenon, a rapid heart rate does not initially have a noticeable effect on cardiac output (because only the duration of the early phases of systole and diastole in the cardiac cycle are reduced and the ventricles still are able to fill). As a result, cardiac output often can be maintained when the heart rate increases during exercise, stress, or in response to other stimuli. However, when the heart rate exceeds 180 beats per minute and the time for venticular filling in late diastole (Fig. 15-5) becomes inadequate, cardiac output declines (Chap. 16).

Heart Sounds

The activity of the cardiac cycle produces sounds similar to *lubb* and *dup,* which can be heard through a stethoscope placed against the anterior chest wall. The first and louder noise occurs at the beginning of ventricular systole. It is due primarily to the turbulence of the blood caused by vibrations and closure of the A-V valves and vibrations of the heart muscle. The second sound, softer and higher-pitched, results from closure of the aortic and pulmonary semilunar valves at the beginning of ventricular diastole. The heart continues through its cycle, causing alternating sounds of *lubb dup, lubb dup.* Other softer heart sounds are heard in some young adults after exercise as blood rushes into the atria and ventricles. The duration, clarity, and distinctness of heart sounds vary with the physiology of the heart.

Cardiac Output

Work of the Heart

Cardiac output, a measure of the work of the heart, is the amount of blood pumped out of each ventricle each minute. In the average adult heart it measures about 5.25 L/min. The ejection of blood from the left ventricle is opposed by pressure within the aorta produced by blood already in the vessel and by the recoil of elastic fibers in its walls. As arterial pressure increases, the heart must expend more energy to accomplish the same cardiac output. The energy expenditure decreases when blood pressure declines. During each ventricular contraction, the left ventricle normally delivers about 45 percent of its contents to the aorta. This quantity of blood, called the *stroke volume,* amounts to about 70 to 90 mL per beat, is increased by exercise, and may be decreased by certain heart diseases.

Cardiac output is also called the *minute volume* of the heart and may be calculated as follows:

Cardiac output (minute volume)
= heart rate (beats/min) × stroke volume (mL/beat/ventricle)

Cardiac output
= 75 × 70
= 5250 mL/min/ventricle
= 5.25 L/min/ventricle

Energy Production

The efficiency of cardiac muscle is similar to that of skeletal muscle. The heart converts only 20 to 25 percent of the energy made available into work (blood pumped). Cardiac output is closely related to the energy metabolism of heart muscle cells. Although they are very rich in oxidative enzymes, the cells are relatively poor in anaerobic enzymes and thus require very high oxygen levels (Chap. 13).

The heart usually obtains more than 99 percent of its energy from aerobic metabolism. It produces the bulk of its energy (about 60 percent) from the oxidation of fats, a lesser amount (about 35 percent) from carbohydrates, and the remainder (about 5 percent) from ketones and amino acids. Because of the high demands for O_2 by aerobic metabolism in cardiac muscle, nearly 65 percent of oxygen is removed from hemoglobin as blood flows through the coronary circulation, compared with 25 percent in most other tissues. The greater demand for oxygen can only be met by an increase in the rate of blood flow. During exercise, when the demand for oxygen by cardiac muscle exceeds the supply, the cardiac muscle becomes *hypoxic,* i.e., its level of oxygen is diminished. Since a maximum of only 10 percent of the energy requirements of the heart can be obtained from anaerobic pathways, an almost continuous supply of oxygen to cardiac fibers is vital. Unlike skeletal muscle fibers, cardiac muscle cells cannot build up an oxygen debt and are rapidly and irreversibly damaged when totally deprived of oxygen for short periods of time (30 s or more).

Starling's Law

According to Starling's law of the heart (also called Frank-Starling law of the heart), cardiac fibers generate greater energy during contraction if they are stretched moderately prior to shortening. (A similar response of skeletal muscle to moderate stretching was described in Chap. 13.) The length of each sarcomere in cardiac muscle cells can be extended from 1.5 to 2.5 μm, affording the myofilaments greater opportunity to form actin-myosin bonds. As a result, the extent of contraction possible becomes greater, and the myofilaments and the cells of which they are a part can contract with greater force.

Cardiac fibers are stretched when the blood volume in the ventricles increases. The ventricles are filled with blood at the end of ventricular diastole by the additional volume provided by atrial systole. The stretched ventricular fibers respond with contractions that produce a more forceful heartbeat. In these terms Starling's law translates to mean that as more blood is put into the heart, more is pumped out. However, overdistention can damage the fibers and disrupt cardiac function. Cardiac output is regulated not only by venous return to the heart but by other internal and external factors described below.

The heart produces the strongest contractions possible under a specific set of conditions. Because all the cardiac cells of the heart are coordinated as one unit, no gradations in tension develop during a single set of conditions. Thus, the all-or-none law applies to the cardiac muscle of each chamber of the heart, as it does to skeletal muscle fibers (see Chap. 13, p. 311).

Electrical Conduction

Cardiac muscle, like skeletal muscle and nerve tissue, is excitable. The resting membrane potential of its fibers ranges from −80 to −95 mV, with the interior negative compared to the exterior of the membrane. An action potential is normally initiated by pacemaker cells in the right atrium of the heart and spreads among cardiac cell membranes across their intercalated discs and even narrower gap junctions or nexuses, as noted in Chaps. 3 and 12 and illustrated in Figs. 3-1C and 12-1B.

When cardiac muscle is depolarized, it enters an *absolute refractory period,* during which it will not respond to further stimulation. This unresponsive phase, which may last 150 to 250 ms, is much longer than the same one in skeletal muscle contraction. Because of it the heart is protected from tetanic fusion of contractions, that occur in skeletal muscle, and the heart is provided with a rest. This relaxation period, in which the heart is nonexcitable, is essential for cardiac function and takes place as the organ refills with blood and recovers its resting potential during diastole.

Cardiac muscle continues to recover its resting potential during the *relative refractory period,* which follows the absolute refractory period and which lasts 30 ms. By the end of the relative refractory period the heart has returned to its normal excitability as its resting membrane potential has been fully restored.

During the relative refractory period, a stronger stimulus occurring sooner than the usual one may cause premature contraction, resulting in less than normal tension. Occasionally the premature beat may be followed by a longer relaxation period called the *compensatory pause,* which gives the appearance of the heart skipping a beat (Fig. 15-10D). However, observations of the heart rate made over a long time indicate that it usually has not done so. The cause of the apparent skipped beat is not romantic, in spite of the many song lyrics which imply otherwise. In some instances the premature beat may be due to excess caffeine, smoking, or lack of sleep. It also may follow cardiac damage, which may make the heart unusually excitable.

The ability of the heart to set its own rate independently of external stimuli is called *autorhythmicity.* Because the heart generates action potentials spontaneously, it independently establishes its own rhythmical beat. This commences as early as 3.5 weeks of embryonic development, which is prior to the formation of anatomical connections between the heart and external nerves. This capacity is so well developed that small pieces of the heart continue to beat rhythmically when the organ is dissected. Cells from the atria beat faster than those from the ventricles; however, the whole heart responds at a rate set by its fastest depolarizing cells.

Intrinsic (Internal) Regulation

The autorhythmicity of the heart originates in a cluster of modified cardiac muscle cells located in the right atrium adjacent to the entrance of the superior vena cava and called the *sinoatrial*

FIG. 15-6
Electrical conduction system in the heart.

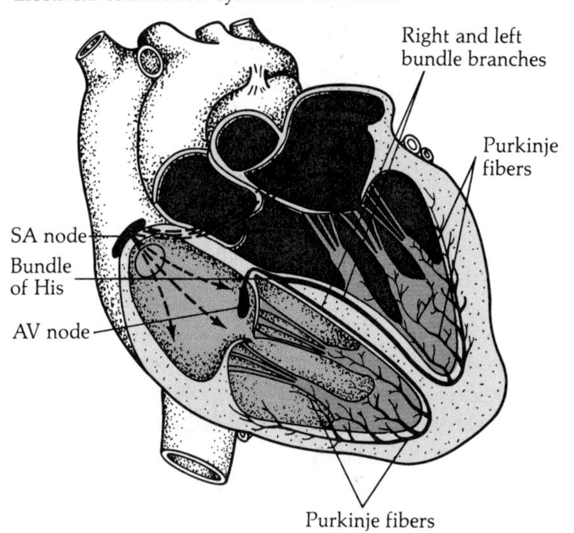

(*S-A*) *node* (Fig. 15-6). A second cluster of modified cardiac muscle cells, the *atrioventricular* (*A-V*) *node,* is located in the lower wall of the right atrium near the right ventricle. Extending from this node is the *bundle of His,* which divides into right and left bundle branches located in each side of the wall (the interventricular septum) that divides the ventricles. The terminal portions of the branches form a network of Purkinje fibers in the walls of each ventricle. The S-A node, A-V node, bundle of His, its right and left branches, and the Purkinje fibers make up electrical bridges which convey impulses in sequence from the atria to the ventricles. Within the atrium the impulses generated by the S-A node are transmitted from cell to cell across intercalated discs. The pacemaker cells in the S-A node generate the dominant rhythm of the heart, producing about 70 to 80 impulses per minute; this rhythm overrides the slower pace of artificially isolated A-V node cells, which generate a pace of about 50 impulses per minute, and of the Purkinje fibers, which are even slower and produce about 20 to 40 impulses per minute. The conduction of impulses along the various electrical bridges forms the *cardiac conducting system* which regulates the heart rate. An external artificial electric generator (pacemaker) attached to the heart can overcome this intrinsic system and set the rate at which the heart beats.

Extrinisic (External) Regulation

Nerves carry impulses in both directions between the heart and the CNS (Fig. 15-7). Their messages are integrated by the *cardiac center* in the medulla of the brain and affect both the rate and the strength of the heart's contraction. (See Chap. 11, pp. 260 and 261 for a brief description of the cardiac center.)

EFFERENT NERVES *Autonomic fibers* of the parasympathetic and sympathetic nervous system transmit impulses to the heart from the cardiac inhibitory and accelerator centers in the medulla, respectively. They form groups of nerves called the *cardiac plexus,* with fibers surrounding the arch of the aorta and following the course of the coronary arteries to the S-A node, the A-V node, the atrial myocardium, and to some extent the ventricular myocardium.

The vagus nerve carries parasympathetic cardiac inhibitory fibers primarily to the atria and releases acetylcholine (ACh). The permeability of the pacemaker cells becomes greater, and more readily allows potassium ions to diffuse out of the cell and chloride ions to move inward. As a result the resting potential increases and the interior of the cells becomes hyperpolarized (more negative). Consequently the cells require more time to depolarize when stimulated. Because of vagal activity, the pacemaker cells are hyperpolarized and generate impulses less frequently and the heartbeat slows. The hyperpolarizing effects of the vagus help maintain a slow, steady heart rate. When vagal function is blocked (if the nerves are severed or vagal stimulation is decreased), a marked increase in heart rate follows.

The *sympathetic fibers* which regulate the heart originate from thoracic spinal nerves 1 to 4 or 5 (Fig. 15-7) which also form the *superior, middle,* and *inferior cardiac nerves.* Their postganglionic fibers release norepinephrine, which accelerates and strengthens the heartbeat. Norepinephrine enhances the inward flow of sodium and calcium ions and decreases the permeability of cardiac muscle cells to potassium ions. The entry of sodium and calcium ions rapidly depolarizes the pacemaker cells and initiates systole. The rate of heartbeat increases in a positive response called the *chronotropic effect.* Some stimuli cause the strength of the heartbeat to increase, in a response called the *inotropic effect* which may be related to greater rates of entry of calcium into cardiac fibers.

Stress may cause both chronotropic and inotropic effects on the heart (Fig. 5-7). Sympathetic stimulation also causes epinephrine to be secreted from the adrenal glands and its influence on the heart resembles, but is more effective, than that of norepinephrine released from sympathetic postganglionic fibers. Norepinephrine also may be converted metabolically into epinephrine to further boost the response (Chap. 9, Table 9-4, p. 231).

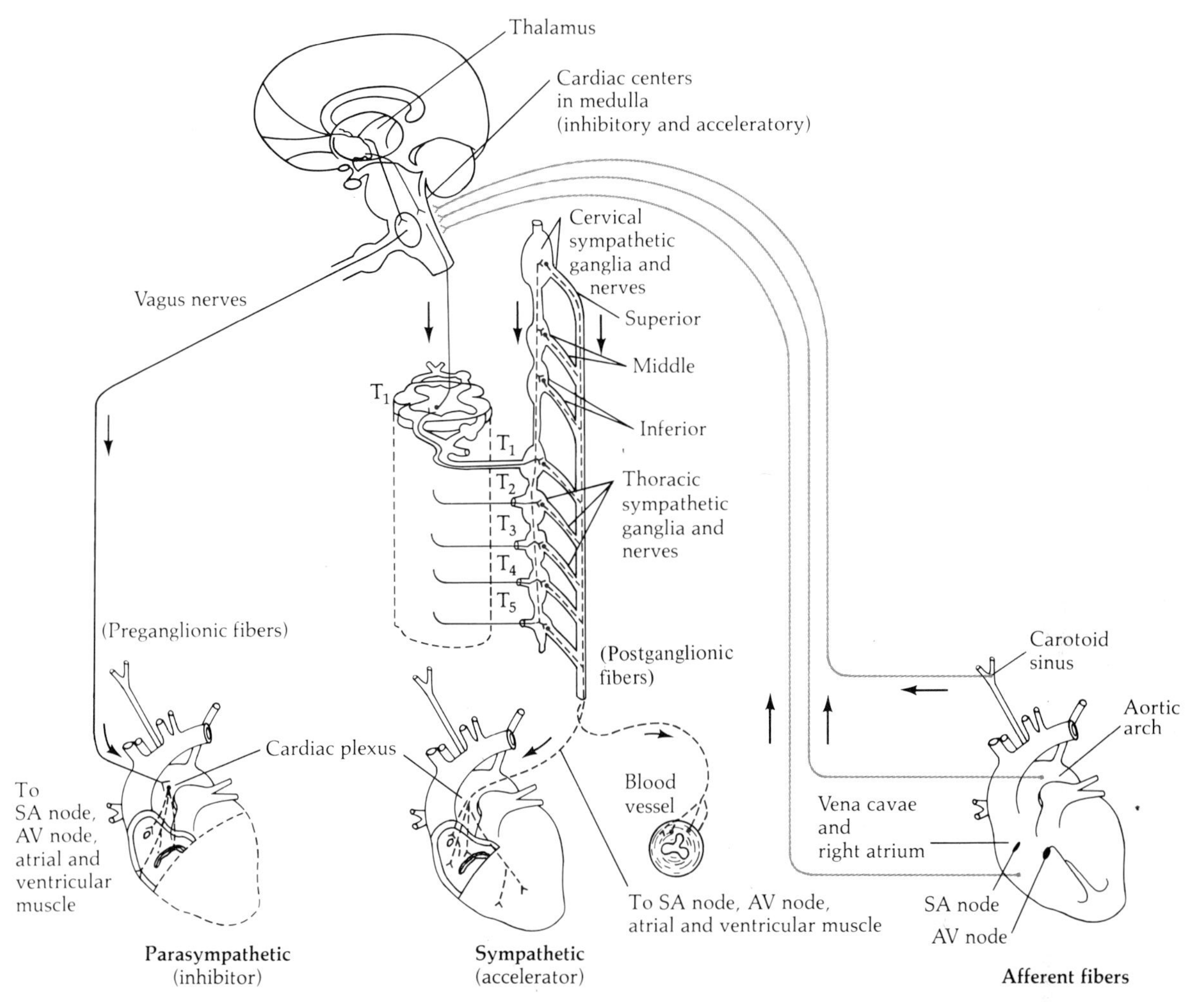

FIG. 15-7
The cardiac center in the medulla and its relation to efferent and afferent nerves of the heart. Adapted from *The Heart,* 2d ed., J. S. Hurst and R. B. Logue (eds.). Copyright © 1970, McGraw-Hill Book Co. Used with permission of McGraw-Hill Book Co.

AFFERENT NERVES AND CARDIAC REFLEXES *Response to Arterial Pressure Change* When the walls of some blood vessels containing pressure receptors are stretched, strained, or deformed, they generate afferent impulses which are carried to the cardiac center. Alterations in the level of blood pressure as well as its rate of change may cause such impulses. The receptors for these changes are *baroreceptors* (pressoreceptors), of which the most important are the aortic bodies in the wall of the aorta and those in the carotid sinuses in the walls of the carotid arteries (Table 8-1, p. 184 and Fig. 15-7). Similar but functionally less important receptors exist in the walls of the pulmonary arteries and in the heart.

When arterial blood pressure increases, impulses from the aortic bodies and carotid sinus are relayed to the cardiac centers in the CNS to efferent parasympathetic fibers of the vagus nerve. Reflex inhibition of the heart rate occurs (Fig. 15-7). When arterial blood pressure decreases, the reflex responses are the reverse, and acceleration follows.

Response to Venous Pressure Change Baroreceptors (stretch receptors) in the walls of the atria and venae cavae are associated with afferent fibers which terminate in the cardiac center

(Table 8-1 and Fig. 15-7). Two sets of atrial baroreceptors seem to generate impulses; one as the atria relax and the other as they contract. The two types of receptors may initiate reflex responses which lead to discharge of impulses by sympathetic fibers, which accelerates the rate of heartbeat, or to inhibition of the fibers, which slows the rate of heartbeat. In general then, stimulation of venous baroreceptors initiates sympathetic responses which result in an increase in heart rate and an increase in cardiac output on the venous side of the heart, whereas stimulation of the arterial baroreceptors initiates parasympathetic responses which decrease the heart rate and cardiac output. The reflex control of blood pressure is described further in Chap. 17 and illustrated in Fig. 17-12.

An increase in blood pressure in the venae cavae may stimulate the baroreceptors. The impulses, relayed to sympathetic neurons in the cardiac center, may promote reflex acceleration of the heart. This response, the *Bainbridge reflex,* occurs following intravenous infusion of large quantities of blood or salt solutions and increases the rate of blood flow through the circulatory system. It was proposed that this response also occurred as a reaction to the increased metabolic demands of exercise. It was reasoned that the Bainbridge reflex causes the heart to pump more blood to the lungs for oxygenation, to deliver more oxygen and nutrients to skeletal muscle, and makes a greater amount of energy available for muscular activity. However, the proposal seems doubtful, since extremely high artificial atrial pressures are required in order to elicit the reflex.

Response to Oxygen Deficiency Elevation of the CO_2 level in the blood (coupled with deficiency of oxygen) stimulates chemoreceptors in the aortic and carotid bodies, as does a decrease in blood flow. These receptors closely monitor blood gases to regulate respiration carefully and rapidly, and they may alter cardiovascular responses in an emergency type of control. This effect is unlike the normal, rather continuous influence of baroreceptors on cardiovascular function. Stimulation of the chemoreceptors leads to an increase in the rate and depth of breathing, sympathetic reflex acceleration of heart rate, and an increase in blood pressure due to constriction of many arterioles (small arteries), while those in the heart and brain widen. The influences of chemoreceptors will be described more completely in Chap. 20 on respiration.

A significant oxygen deficiency in the vessels in the walls of the heart also may cause afferent impulses to be generated by chemoreceptors which are transmitted to the cerebral cortex as pain messages. They warn of oxygen deficiency and elicit reflex widening of the coronary arteries to provide an increase in blood flow (and oxygen) to prevent damage to heart muscle. If the reflex response fails to provide the cardiac muscle with sufficient oxygen, tissue damage in the heart may result. The chest and arm pain resulting from an insufficient blood supply is called *angina pectoris.* The terms literally mean that there is pain radiating from the area near the pectoral or chest muscles. The pain often travels toward the left arm and the origin of the impulses is often misinterpreted. This error occurs because nerves to the heart and those of the brachial plexus to the arm contain sensory fibers from the same dorsal root of the spinal cord and represent overlapping dermatomes (described in Chap. 9 and in Fig. 9-4B on p. 210). Therefore, pain originating in the heart is often mistakenly perceived as being from the left arm, shoulder, and chest wall.

The pain may be due to the release of *P substance,* a polypeptide transmitter agent released from the dorsal roots of the spinal nerves to the brain. (The origins of pain and the nature of its transmitters were more fully described in Chaps. 9 and 10.)

Changes in Electrical Conductivity: The ECG

PRINCIPLES An *electrocardiogram* (ECG) measures the normal electrical activity of the heart as well as the rate of the heartbeat. It is a recording of differences in voltage between excited and unexcited parts of the heart. The impulses generated by cardiac muscle produce electrical activity, which is conveyed through tissues to the surface of the skin by salts dissolved in the body fluids. Metal electrodes are placed on the skin to measure the voltage that develops. One electrode actively measures the voltage and is compared with a second one, which is used for reference. An electrical current is generated when a difference in voltage exists between the two. This current is conveyed to an amplifier, which enlarges the difference, and then to a transducer, which converts the signal into a visual display. An oscilloscope is an example of such an instrument (Fig. 7-8A, p. 163). The electrical signal may also be used to move a pen to create a graphic tracing. Either mechanism produces a visual display of the electrical event.

The recording is proportional to the magnitude and type of change (positive or negative) in the electrical potentials. The action potentials transmitted by the electrodes represent the algebraic sum of all action potentials at the recording electrode at the surface of the body. As a wave of depolarization approaches an active electrode, the voltage transmitted is positive when compared with that of the reference electrode. Consequently, the tracing is deflected upward. As the wave of depolarization moves away from the active electrode, the negative voltage deflects the tracing downward.

WAVE PATTERNS IN A NORMAL ECG In the usual ECG, three wave deflections are formed (Fig. 15-8). Each reflects the function of particular parts of the conduction system of the heart. They are the P wave (caused by atrial depolarization), the QRS complex (produced by ventricular depolarization), and T wave (generated during ventricular repolarization). A U wave is sometimes formed and may be due to slow repolarization of the papillary muscles.

The *P wave* occurs as the atria depolarize with an impulse that originates in the S-A node. The number of P waves per minute corresponds to the atrial rate and rhythm, which vary with the heart rate and age of the subject. The *QRS complex* is formed as depolarization travels from the A-V node, along the bundle of His, and through its branches and spreads out among Purkinje fibers to the cardiac muscle in the walls of the ventricles. The number of QRS complexes per minute is equal to the ventricular rate.

The *P-R interval* is the time intervening between the beginning of the P wave and that of the QRS complex. It is a measure of the time required for the impulse to travel from the S-A node to the ventricles. Atrial repolarization occurs during the QRS period. A prolonged P-R interval suggests a delay in conduction in the A-V node.

During the *T wave,* the ventricles repolarize as the heart recovers following contraction. The time between the end of the QRS complex and the beginning of the T wave (the S-T segment) represents the period between depolarization and the start of repolarization of the ventricles.

UNIPOLAR RECORDINGS Various locations may be used from which electrodes may record the ECG. A *unipolar recording* measures the potential at a single active location. The potential of one active (exploring) electrode is compared with that of an indifferent one at zero potential (that is, its potential is zero and shows no major changes during the cardiac cycle). The readings from the active electrode are called V-lead recordings (V for voltage) and are taken from six

FIG. 15-8

An electrocardiogram (ECG). The U wave is not always observed.

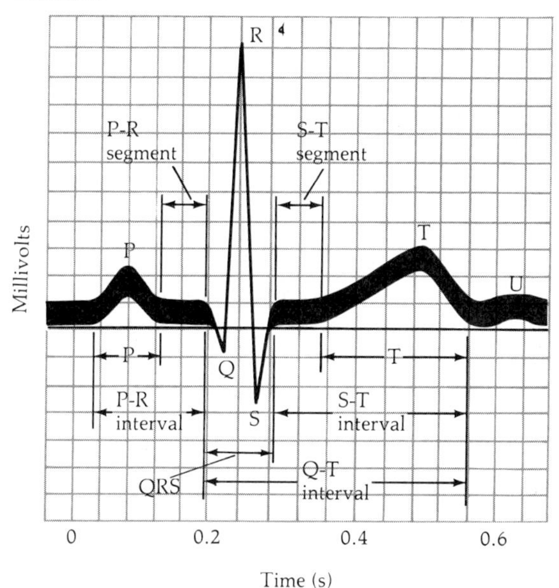

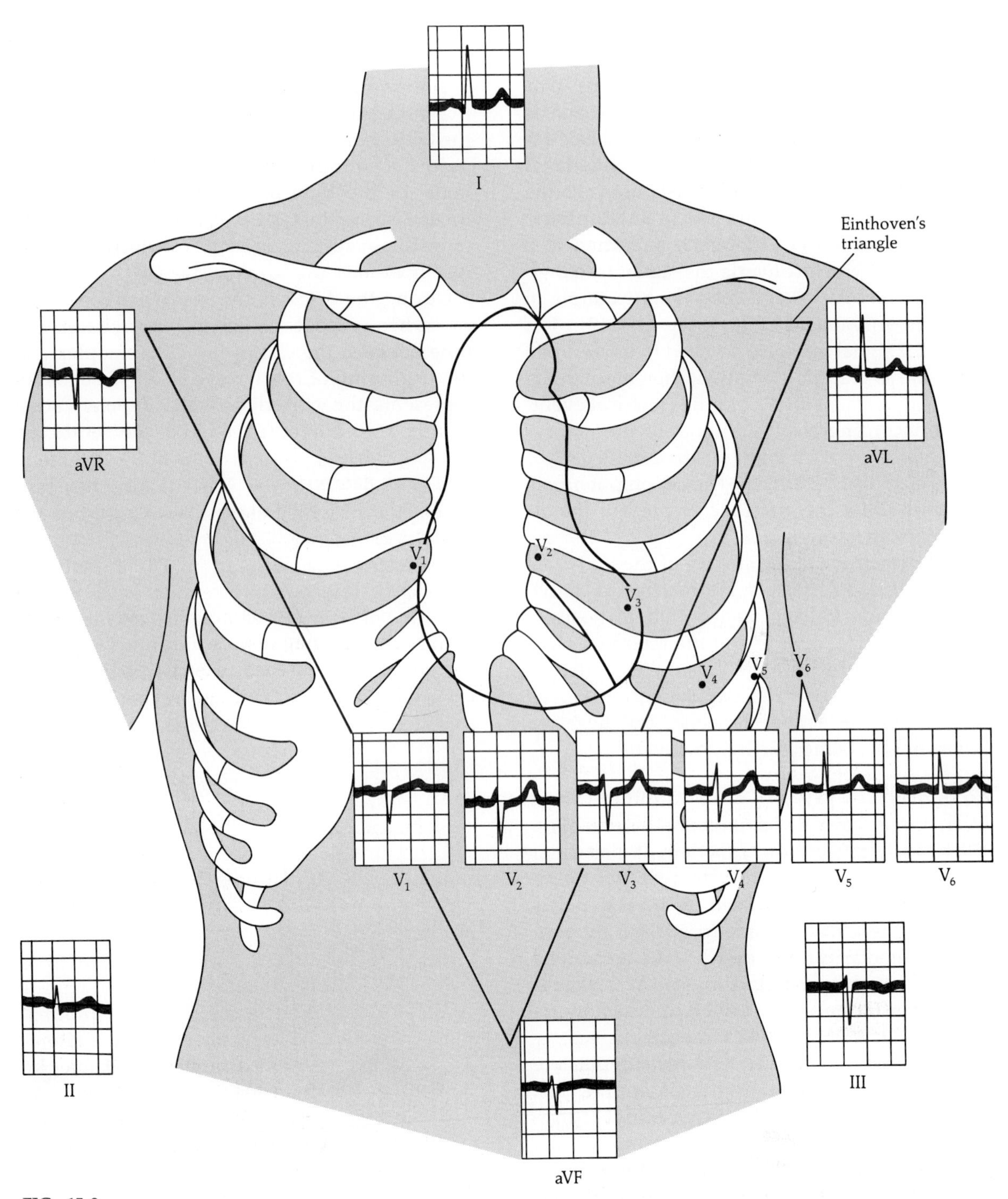

FIG. 15-9
Electrocardiograms recorded with unipolar electrodes placed in various standard positions. The position of Einthoven's triangle is noted. Bipolar recordings from leads I, II, and III and unipolar recordings from leads in six standard positions in the chest (V_1 to V_6) are shown, as well as augmented recordings from the arms (aVR, aVL) and leg (aVF).

standard positions on the chest, called V_1 to V_6 (Fig.15-9).

Unipolar recordings also may be derived by comparing the active electrode placed on one limb with the voltages on the other two limbs; these are called *augmented* (a) *limb lead recordings.* When the active electrode is on the right arm, the recording is called aVR; on the left arm the recording is called aVL; and on the left leg, the recording is called aVF.

BIPOLAR RECORDINGS *Bipolar recordings* measure differences in potentials between two sites in the body. The leads are usually placed on the left wrist, right wrist, and left ankle. Often, a fourth electrode is placed on the right leg and connected to an electrical ground to minimize electrical interference. Potentials may be compared with the recording electrodes in three standard positions (Fig. 15-9). The location of the standard limb leads for recordings are:

Lead I: compares the right arm to the left arm

Lead II: compares the right arm to the left leg (see also Fig. 15-8)

Lead III: compares the left arm to the left leg

The patterns formed in each type of recording are similar but the heights of the waves differ because the difference in potential between the pairs of electrodes varies in each of the three positions (Fig. 15-9). The differences may be used to assess cardiac functions.

The three electrodes form a triangle (*Einthoven's triangle,* Fig. 15-9) with the heart at its center. The triangle approximately parallels the frontal axis of the organ. Each electrode conveys its signal to a common terminal in the instrument, which at a given instant measures the difference in potential between two active electrodes. The recorded potential is altered by the electrical activity of the heart. The sum of the potentials at the points of an equilateral triangle is zero when the source of current is at the center of the triangle. When a wave of depolarization is conducted through the heart, some of the current is conveyed to the surface of the body. It will cause a positive tracing as it approaches an active electrode and a negative one as it moves away.

The product of the direction and magnitude of the tracing, the *electrical vector,* varies with the positions at which the electrodes are placed. A cardiac vector is related to the electromagnetic force generated by the heart. A *cardiac vector* from an ECG tracing may be recorded by an oscilloscope and may be drawn as an arrow that indicates the size of the electromagnetic force and the angle associated with each wave. (An electromagnetic force is an attraction due to the motion of an electric charge.) The angle of the cardiac vector is the direction of the force within Einthoven's triangle. If one draws a line connecting the tops of the arrows of the vectors of the P wave, QRS complex, and T wave, then a loop called a *vectorcardiogram* is attained. Such tracings may be made electronically from simultaneous measurements in more than two planes of the heart (such as the side and front of the heart). This procedure generates two- to three-dimensional vectorcardiograms, which may be used to diagnose cardiac function. They are affected by disease, exercise, or even the position of the heart during the recording.

EFFECTS OF HEART DISEASE Cardiovascular disease may affect the ECG (Fig. 15-10) and sometimes warn of heart malfunction. Therefore, it is useful for a physician to record an ECG for a patient prior to the advent of the disorder. This makes it easier to distinguish individual variations and changes due to aging from pathological alterations.

Abnormalities in an ECG which may follow a heart attack appear and disappear at various times. The QRS complex may become abnormal immediately after a heart attack and the alteration may take years to disappear or it may be permanent. An abnormal ST segment also may result immediately after heart damage. It usually disappears within 3 to 6 weeks. Irregularities in T waves may occur within seconds to 24 h of a heart attack and disappear within months to years. Changes in T waves are sometimes observed in healthy subjects. Needless to say, accurate interpretation of th ECG requires careful and extensive training.

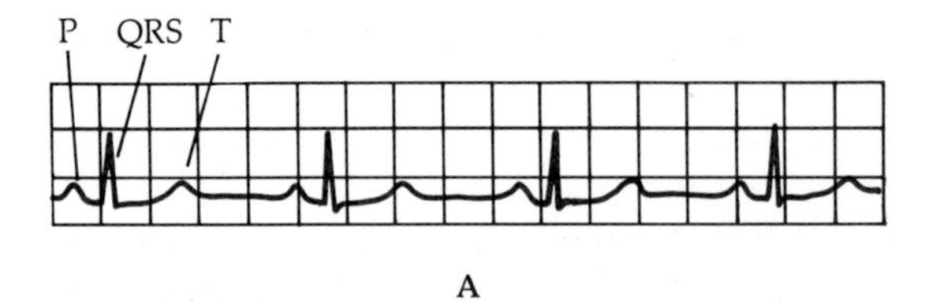

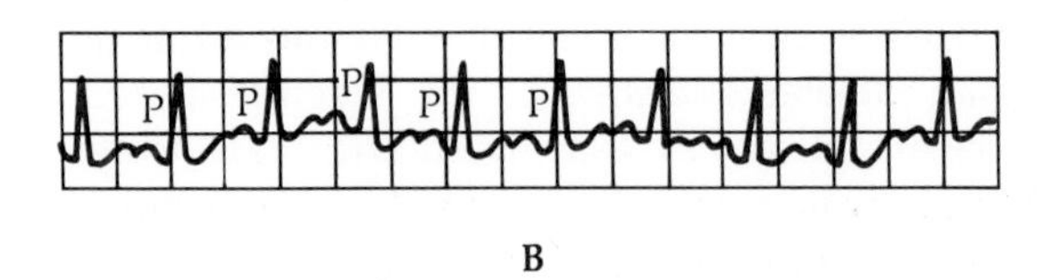

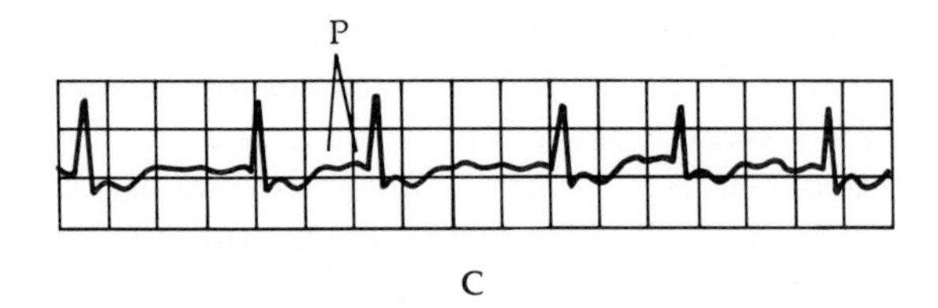

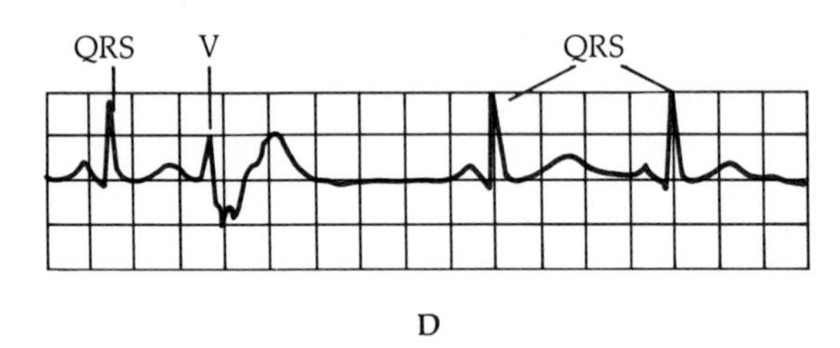

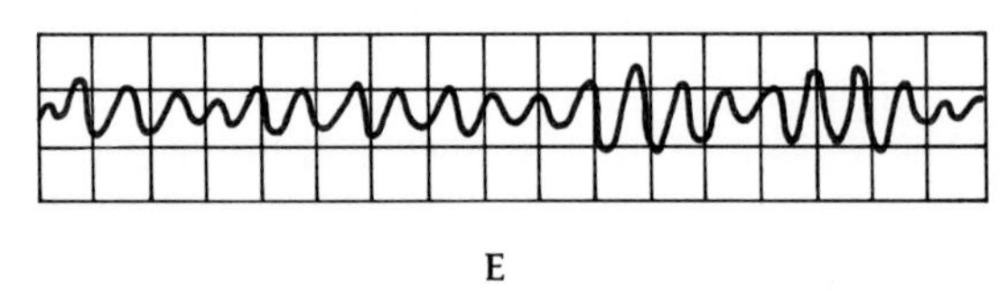

FIG. 15-10
Changes in the ECG with alterations in heart function. (A) Normal ECG. (B) Rapid atrial beat. Note the close proximity of the P waves (as well as the remaining waves) in the ECG. (C) Atrial fibrillation. Note that the P waves occur more rapidly than in A, their height is markedly diminished, and the QRS complexes are irregular. (D) A premature ventricular beat (V) followed by a compensatory pause. Note the increased distance (time) between the premature beat (V) and the next QRS complex compared with the distance between other QRS complexes. (E) Ventricular fibrillation. Note the rapid succession of weak beats (of low amplitude) and loss of the characteristic waves of a normal ECG.

Fibrillation

Fibrillation is the lack of coordination of contraction of fibers within a muscle. Cardiac fibrillation is an uncoordinated electrical and mechanical activity of the chambers of the heart. Several hypotheses have been advanced to explain cardiac fibrillation (either atrial or ventricular, as shown in Fig. 15-10C and E). The various hypotheses propose that fibrillation may be due to alterations in signals, which include their recirculation through the heart (the *circus movement*); their production in several independent sites (called *ectopic* or *displaced foci*); and an imbalance in them among different parts of the heart (*electrotonic current flow*). The heart may fibrillate following damage caused by a diminished blood supply, physical agents such as a catheter, or chemical agents. Irreversible damage follows if fibrillation is unchecked. Although fibrillation of the atria often ceases spontaneously, that of the ventricles usually requires a strong artificial electric shock which completely stops cardiac muscle activity. This procedure is called *defibrillation*.

During atrial fibrillation the heart continues to pump blood, but blood flow ceases when ventricular fibrillation occurs. External compression of the heart may maintain blood flow. The sternum (breastbone) is compressed at a point about 1 to 1.5 in above its tip near the abdomen and then allowed to return to its normal resting position. This procedure is repeated at a rate of 60 compressions per minute. The technique may revive the natural rhythm of the heart and is often used in conjunction with artificial respiration (Chap. 20, Fig. 20-16). The combination of the two techniques is called *cardiopulmonary resuscitation* (*CPR*). In some instances CPR does not revive the heart's rhythmic beat and a strong artificial electric shock is used to arrest ventricular fibrillation prior to attempts to revive the heart's natural rhythm.

An abnormal atrial beat may produce small rapid P waves or no P waves. As the heart rate increases above normal levels, the P waves become smaller and finally disappear completely (compare Fig. 15-10B and C). If through a disorder part of the atrium becomes functionally independent of the S-A node, the atrial rate becomes extremely rapid and forms an ectopic focus. An atrial rate of 200 to 350 beats per minute causes depolarization that moves in circles around the junction of the right atrium and the venae cavae and is called *atrial flutter.* When there are 300 to 500 atrial beats per minute, depolarization is completely irregular and disorganized and referred to as *atrial fibrillation* (Fig. 15-10C). The ventricles fail to keep pace with the atria, receive occasional impulses, and respond at irregular intervals often ranging from 80 to 160 beats per minute.

Digitalis slows the rapid, weak beat of a fibrillating heart. Digoxin is a form commonly used to slow ventricular rate in patients with atrial fibrillation. This drug decreases the speed of conduction between the atria and ventricles. It increases the interval between P and R waves, retarding a rapid ventricular rate and in so doing increasing cardiac output. These actions probably occur by blocking the sodium and potassium ATPases, slowing reentry of potassium into the cell and prolonging the recovery period during each cycle. Digitalis also has been shown to exert a vagal effect. That is, its activity accentuates a slow, steady heart rate and may even inhibit the reflex acceleration of the heart mediated by the carotid sinus, as described above. An excess of digitalis may inhibit A-V conduction completely and result in a *heart block* and death.

An abnormal heart rhythm also may be corrected by an artificial pacemaker. This device consists of an electrical signal generator powered by small, long-lived batteries and implanted under the skin of the chest. The pacemaker is connected to electrodes which are attached to the myocardium or threaded into the right ventricle. The artificial rhythmical signal produced by its circuits is strong enough to override the weak, irregular impulses of the natural pacemaker.

15-3 SUMMARY

1 The average heart accounts for about 0.5 percent of body weight and pumps 2000 gallons of blood daily through more than 60,000 mi of blood vessels. It is loosely held in place by an external sac, the fibrous pericardium. The external sac surrounds a double-layered serous pericardium (epicardium). The inner layer of the epicardium adheres to the muscular myocardium, or heart wall. The inner endothelial lining of the heart is called the endocardium and also covers the heart valves. Streptococcal infections may cause rheumatic fever and an immune reaction that damages the heart valves.

The route of blood flow is from the venae cavae to the right atrium, into the right ventricle, to the pulmonary artery, out to the lungs and the pulmonary veins, and back to the heart at the left atrium. Blood then passes into the left ventricle, from which it is propelled out of the heart through the aorta to the body.

Valves restrict blood flow to one direction. The atrioventricular (A-V) valves are the tricuspid and bicuspid (mitral) valves. Semilunar valves are found in the pulmonary artery and aorta. Stenosis, the narrowing of a valve or blood vessel, causes blood flow to be turbulent. Damage to a valve causes valvular incompetence, allowing blood to flow backwards. Both stenosis and valvular incompetence may produce abnormal heart sounds called murmurs.

The right and left coronary arteries supply the walls of the heart with blood. Subdivisions of the coronary arteries lead to capillary beds that supply cardiac muscle fibers with oxygen. The blood flows from the capillaries into subdivisions of veins and enters the coronary sinus to be delivered to the right atrium. Blockade (occlusion) of the coronary arteries causes coronary insufficiency, diminishing the supply of oxygen vital for cardiac muscle function.

2 The heart rhythmically beats an average of 75 times per minute when a person is resting. One beat lasts about 800 ms. During this cycle the two atria contract almost simultaneously, followed by contraction of the two ventricles. Each contraction (systole) is followed by relaxation (diastole). Atrial and ventricular diastole overlap for a period of 400 ms, providing the heart with rest between beats. Two major sounds in the cardiac cycle are associated with closure of the A-V valves followed by closure of the semilunar valves.

Cardiac output is a measure of work performed by the heart. It is equal to the stroke volume multiplied by the heart rate and averages 5.25 L/min in the adult resting heart. The work performed by the heart overcomes the resistance to blood pressure in the aorta.

The heart usually obtains 99 percent of its energy from aerobic metabolism. However, a maximum of only 25 percent of this energy is converted to useful work. About 60 percent of the energy is derived from fats and 35 percent from carbohydrates.

Moderate stretching of cardiac muscle fibers occurs as venous return to the heart increases, swelling its chambers with more blood. The greater length of the fibers allows more forceful contraction according to Starling's law of the heart. Heart muscle follows the all-or-none law and provides the strongest contraction possible under a specific set of conditions.

Action potentials are conducted from one cardiac muscle fiber to another across intercalated discs. Depolarization of each cardiac fiber is followed by an absolute refractory period. During this time, each cell is provided with rest and protected from tetanic fusion of contractions.

The rhythmical heartbeat is initiated internally by pacemaker cells in the S-A node in the right atrium. Impulses are conveyed from the S-A node to atrial fibers and the A-V node and through the bundle of His in the wall between the ventricles and the two branches of the bundle to Purkinje fibers that spread out to ventricular muscle cells.

Afferent and efferent nerve fibers carry impulses between the heart and accelerator and inhibitory segments of the cardiac center of the medulla of the brain. Sympathetic fibers convey efferent excitatory impulses to the conduction system of the heart to accelerate heart rate. Parasympathetic fibers in the vagus nerve transmit efferent inhibitory signals that slow the heart.

Cardiac reflexes are initiated by afferent impulses generated from baroreceptors in the aortic body and carotid sinus in response to changes in arterial blood pressure. Increased arterial pressure generates afferent impulses transmitted to the cardiac center which elicit efferent parasympathetic impulses that slow the heart. Decreases in arterial pressure cause opposite responses. Stimulation of venous baroreceptors causes sympathetic responses which result in an increase in cardiac output on the venous side of the heart.

Elevated levels of carbon dioxide coupled with oxygen deficiency cause chemoreceptors in aortic and carotid bodies to reflexively increase respiration and elevate heart rate and blood pressure while improving circulation to the brain and coronary arteries. Sufficient deprivation of oxygen causes afferent signals to be generated from the heart which elicit sensory perception of pain (angina). The perception of pain may be due to release of P substance from associated neurons in the spinal cord.

Action potentials developed by heart cells are conducted through extracellular fluids to the surface of the body. Metal electrodes placed on the skin can transmit impulses generated from the heart. The electrodes may be connected to an instrument that converts the electrical signal into a visual display or recording called an electrocardiogram or ECG.

The major tracings produced in a normal ECG are called P, QRS, and T waves. They are associated with atrial depolarization, ventricular depolarization, and ventricular repolarization (recovery) of the heart, respectively. Heart disease may alter the ECG. Changes can persist for varying periods of time, depending on the type and extent of heart damage.

Unipolar recordings measure the potential at a single active location. Bipolar recordings measure the difference at any moment between two of three active electrodes at different locations. Their placement on the right and left arms and left leg form Einthoven's triangle, which approximately parallels the frontal axis of the heart.

Cardiac fibrillation is due to lack of coordination among individual heart muscle fibers. Hypothetical causes are recirculation of electrical impulses, their production at several independent sites, and an imbalance among them. Strong electric shock may stop fibrillation. The administration of digitalis decreases electrical conductivity over the A-V node. It can slow the rate of heartbeat and is often used to steady and slow an abnormally fast heartbeat caused by the too rapid generation of impulses from the S-A node to the A-V node.

REVIEW QUESTIONS

1. Draw a sketch of the heart and label the chambers and tissues that make up the heart and its internal and external membranes. Include the valves, major vessels, and intrinsic conducting system.
2. Trace the flow of blood through the heart and lungs.
3. Describe the structure and function of the heart valves.
4. What is the distinction between stenosis and incompetence? How do streptococcal infections lead to heart murmurs?
5. What is meant by coronary insufficiency? How is the heart muscle supplied with blood?
6. Account for the major heart sounds in the cardiac cycle.
7. Define cardiac output and identify the factors that increase or decrease the work of the heart.
8. What are the sources of energy in heart muscle, and how effectively is energy used?
9. How do Starling's law and the all-or-none law govern heart function?
10. Describe the heart's electrical conducting system.
11. What intrinsic factors regulate rhythmical activity of the heart?
12. Describe the relationship among parasympathetic and sympathetic impulses, the cardiac center, and heart rate.
13. How do cardiac reflexes alter heart function in response to increases in arterial and venous pressure? What is the Bainbridge reflex?
14. What is an electrocardiogram? Draw, label, and explain the nature of P, QRS, T, and U waves.
15. How do unipolar and bipolar recordings differ? What is the importance and meaning of Einthoven's triangle? What is the relationship between a cardiac vector and heart function?
16. Define fibrillation and discuss its effect on heart function. How may digitalis affect this disorder?

THE CIRCULATORY SYSTEM

16

Effects of Chemical and Physical Factors on Cardiac Function

16-1 FACTORS INFLUENCING CARDIAC PHYSIOLOGY
Chemical agents
Ions
Neurotransmitters, hormones, and stress
Alcohol
Nicotine
Caffeine
Blood pressure
Sex
Body size
Age
Body temperature
Exercise
Isotonic exercise
Isometric exercise
Physical training

16-2 CARDIAC CIRCULATION DURING DEVELOPMENT
Circulation in the fetus
The foramen ovale
The ductus arteriosus
Tetralogy of Fallot: A congenital abnormality

16-3 CARDIAC ABNORMALITIES IN ADULTS
Congestive heart failure
Atherosclerosis and coronary artery disease
Myocardial infarction and the risk of coronary artery disease
Diagnosis of cardiovascular function
Physical examination
Analyses of blood samples
Images of the heart and cardiac catheterization

16-4 SUMMARY

OBJECTIVES

After completing this chapter the student should be able to:

1. Describe the effects of neurotransmitters and other chemical agents on heart function.
2. Discuss the effects of exercise on cardiovascular activity.
3. Compare the cardiac circulation of a fetus with that of an adult, noting the effects of incomplete development of the heart.
4. Describe congestive heart failure, atherosclerosis, coronary artery disease, and myocardial infarction.
5. Discuss the analysis of serum enzyme levels by electrophoresis and spectrophotometry and explain the value of such analyses in the diagnosis of cardiovascular disease.
6. Name the major lipids and lipoproteins involved in atherosclerosis, and describe the role each may possibly have in the disorder.
7. Compare the use of fluoroscopy, ultrasound, scintillation images, cardiac catheterization, and angiograms in assessing heart function.

16-1 FACTORS INFLUENCING CARDIAC PHYSIOLOGY

As noted in Chap. 15, the heart is an amazing organ and life makes strenuous demands upon it. This chapter deals with some of the chemical, physical, and anatomical changes that alter cardiac physiology. It also considers modifications which adversely affect the heart and certain diagnostic techniques available to identify and to assess its function.

Chemical Agents

A number of chemicals may influence the heart indirectly by altering blood flow through vessels leading to and from the heart or directly by delivery to the tissues of the heart through the coronary arteries (Table 16-1).

Ions

The maintenance of normal heart function, like that of the function of other excitable tissues, is influenced by the composition of the extracellular and intracellular fluids (Table 16-1 and Chaps. 4, 7, 12, and 13). The accumulation of extracellular Ca^{2+} ions hyperpolarizes the cells, but Ca^{2+} ion entry into cardiac fibers increases the force of the muscle's contractile mechanism. Ca^{2+} accumulation in the extracellular fluid, like that of K^+ ions, decreases the heart rate. Changes in the Na^+ ion level normally are not sufficient to exert a significant effect on cardiac function.

Neurotransmitters, Hormones, and Stress

Certain neurotransmitters and hormones significantly influence heart rate. As mentioned in the previous chapter, acetylcholine (ACh), from postganglionic parasympathetic vagal fibers, decreases conduction time and heart rate and slows the heart to a regular rhythmical beat. Norepinephrine, from postganglionic sympathetic fibers, increases the rate and strength of contraction, resulting in elevation of cardiac output and blood pressure. Norepinephrine also constricts small arteries (arterioles), especially those within the skin and abdomen, and diverts more blood back toward the heart. Stress causes sympathetic activity, cardiac output, and blood pressure to rise as the adrenal glands are stimulated to release norepinephrine and epinephrine (Fig. 5-7).

Alcohol

Moderate levels of alcohol in the blood dilate small arteries, allowing blood to flow more readily into peripheral vessels, especially in the skin, fingers, and toes. As a result of the increased flow into these peripheral sites, blood pressure is lowered, as is the work of the heart, which pumps against the lesser resistance, and the heart rate reflexively decreases. Since alcohol causes a greater supply of blood to be delivered to the surface of the body, heat is lost more rapidly from the skin and central (core) body temperature drops. The increase in peripheral blood flow protects the remote parts of the body

TABLE 16-1
Major effects of some chemical agents on the heart

Chemical	EFFECT* Heart Rate	Force of Contraction	Mechanism
Extracellular level of ions			
Ca^{2+} increase	↓		Ca^{2+} hyperpolarizes heart muscle
		↑	Ca^{2+} entry enhances muscle contractile mechanism
K^+ increase	↓		Lowers resting potential (owing to K^+ entry into cardiac fibers); decreases conduction in A-V node; arrhythmic beat and fibrillation may occur
Na^+ decrease (artificial 10–20% decrease and effect on isolated heart)†	↓		Permits increased binding of Ca^{2+}; slows conduction; longer period needed to attain threshold for depolarization
Na^+ increase (increase with congestive heart disease; an indirect effect)†	↑		Increased edema causes elevated venous return
Transmitters			
Acetylcholine	↓		Increases K^+ ion permeability; hyperpolarizes heart muscle and suppresses the
		↓	development of pacemaker potentials
Catecholamines (norepinephrine and epinephrine)	↑		Promote Na^+ ion entry and depolarization
		↑	Promote Ca^{2+} ion entry
Drugs			
Digitalis (effects on patients with atrial fibrillation and flutter)	↓	↑	Promotes slow depolarization of cardiac muscle coupled with decreased conduction at A-V node
Nitroglycerin and amyl nitrite	↓		Inhibit contraction of smooth muscle in veins and arteries, leading to decrease in workload of the heart and diminishing its oxygen consumption; may widen the coronary arteries and improve blood supply to the myocardium
Alcohol	↓		Dilates peripheral vessels; blood pressure decreases
Nicotine	↑	↑	Releases norepinephrine from autonomic ganglia
Caffeine	↑	↑	Stimulates cardiac center in medulla

* (↑ = increase); (↓ = decrease)
† Under *normal* conditions, changes in extracellular Na^+ ion have little effect on the heart.

against cold exposure. As a result, it has been noted that alcoholics are less susceptible to frostbite but more prone to hypothermia (lowered body temperature).

Nicotine

Nicotine is a drug derived from tobacco. It enters the bloodstream through the capillaries in the nasal passages and lungs. Small doses of nicotine stimulate autonomic ganglia, causing the release of norepinephrine, which results in an increase in systolic and diastolic blood pressure and heart rate. These changes are measurable within a few minutes or less of smoking a single cigarette. The cardiovascular system of a smoker works much harder than that of a nonsmoker.

Caffeine

Caffeine is a stimulant with a purine structure similar to that of cyclic secondary messengers described earlier (Chaps. 2 and 5). It acts on the cardiac center in the medulla to speed up the heart rate. Even though caffeine elevates cardiac output, consumption of the drug has not been related to myocardial disease.

Blood Pressure

An increase in arterial pressure generally causes reflex deceleration of the heart. The response is due to stimulation of baroreceptors in the aortic arch and carotid sinus, which leads to vagal inhibition of the heart (Fig. 15-7). When blood pressure falls, vagal inhibition is relieved and the heart rate and blood pressure rise again to complete the cycle.

Sex

The average heart rate of an adult is 75 beats per minute, ranging from 65 to 80. Males have a lower heart rate than females of the same age. The disparity may be related to differences in body size as well as to influences of the sex hormones.

Body Size

In general, the rate of metabolism and the heart rate decrease as body size (body surface area) increases. For example, the average heart rate of small laboratory mice is 700 beats per minute, that of rabbits 150, that of humans 75, and that of elephants 20.

Age

In general, the heart deteriorates with age, and the resulting diseases are responsible for over 650,000 deaths each year in the United States. The majority of deaths due to cardiovascular disease occur among people over 65. However, nearly one-third of deaths from all causes in individuals between 35 and 64 years of age are due to coronary artery disease.

The size, shape, and appearance of the heart change with age as its inner lining, myofibrils, ion channels, and responsiveness to drugs undergo adverse alterations. The cardiac output, heart rate, and response to exercise also decline with aging, while the incidence of abnormal beats (arrhythmias) increases (Table 16-2).

Body Temperature

Under normal circumstances the body reflexively adjusts heat production and heat loss to maintain a rather constant body temperature (Fig. 1-2). The adjustments for temperature homeostasis are regulated by the hypothalamus and influenced by changes in metabolism (Chaps. 11 and 23).

A rise in body temperature (*hyperthermia*) leads to an increase in heart rate. A 1°F elevation in body temperature leads to about 7 to 10 more heartbeats per minute. An increase in body temperature after exercise or ingestion of a meal (Chap. 23) is accompanied by a higher

TABLE 16-2
Some effects of aging on the heart

Property	*Alteration**
Cardiac output, L/min	↓
Males: age 22	6
age 66	5
Females: age 22	5
age 66	4
Heart rate beats/min	↓
Unborn	120–160
Newborn	110–130
Young children	72–92
Adults	65–80
Response to exercise	
Heart rate	(Less marked increase in those over 65 years of age compared with young adults)
Relaxation period	↑
Catecholamine receptor activity	↓
Protein content	↓
Cardiovascular disease	↑
Arrhythmic beats	↑

* ↑ = increase; ↓ = decrease

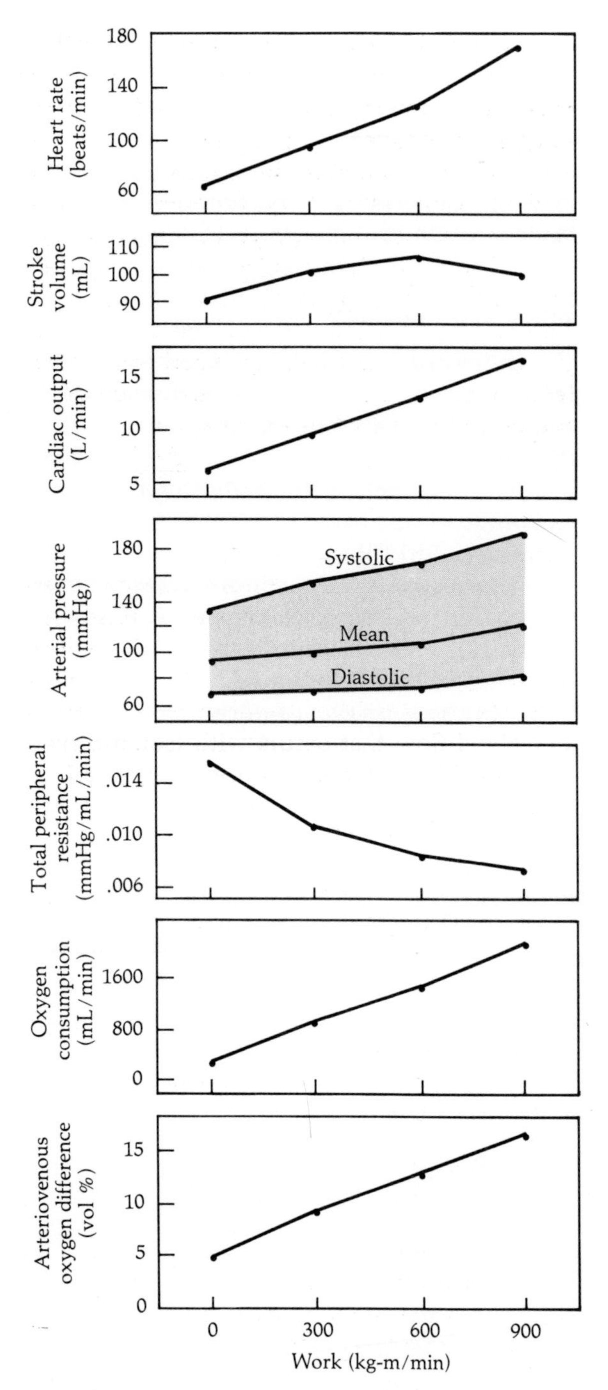

FIG. 16-1
Responses of the cardiovascular system to different levels of isotonic exercise. The physical activity or workload was assessed by the use of a stationary exercise cycle (an ergometer). Subjects of the experiment lay on their backs and pedaled against a load measured in kilograms and at a rate measured in meters per minute. The work was then calculated as kilogram-meters per minute. SOURCE: R. M. Berne and M. N. Levy, *Cardiovascular Physiology,* 4th ed., C. V. Mosby, St. Louis, 1981. Data from A. Carlsten and G. Grimby, *The Circulatory Response to Muscular Exercise in Man,* Chas. C. Thomas, Springfield, Ill., 1966.

heart rate. A decrease in body temperature (*hypothermia*) produces the opposite effect.

The induction of hypothermia slows the heart and facilitates open-heart and coronary bypass surgery. The latter procedure involves the insertion of segments of healthy veins to provide blood flow to areas of the myocardium supplied by diseased coronary arteries. In order to slow the heart during surgery, the patient may be placed in a cooling jacket.

Exercise

Responses to exercise may improve blood flow to skeletal and cardiac muscle. Thus, delivery of oxygen and nutrients and their exchange for metabolic wastes, such as carbon dioxide, are speeded up. Increases occur in cardiac output (heart rate and stroke volume), arterial pressure, oxygen consumption, and the extraction of oxygen by tissues from arterial blood (Fig. 16-1). The decline in resistance to blood flow in peripheral tissues accelerates the rate of exchange between blood and tissues. Therefore, the extent of the alterations is related to the nature and intensity of the exercise. The exercise may be isotonic (that in which skeletal muscles shorten) or isometric (that which develops tension without shortening of muscle cells).

Isotonic Exercise

Strenuous isotonic exercise causes a rise in cardiac output. The normal heart rate may increase from 75 to 180 beats per minute and the stroke volume from 70 to 100 mL per beat or more. During intense exercise cardiac output may rise to 35 L/min in some individuals.

Exercise also moderately raises systolic arterial pressure. As striated muscle contracts, it squeezes local blood vessels and increases the resistance to blood flow. However, during relaxation of muscle the number of open (dilated) capillaries increases and resistance to blood flow decreases. A major stimulus for dilation of smooth muscle in the walls of arterioles is the elevation in carbon dioxide and hydrogen ions resulting from striated muscle metabolism. In this manner blood flow is related to tissue activity, a phenomenon called *autoregulation* (Chaps. 17 and 21). Consequently, isotonic exercise is accompanied by a *drop in total peripheral resistance* (Table 16-1) and an increase in blood flow to active muscle.

At maximal oxygen utilization during exercise, about 4 times more blood may be carried to the heart, while nearly 20 to 30 times more blood flows per minute to active skeletal muscle. A few minutes after exercise, blood flow to the skin increases to promote the loss of heat generated by skeletal muscle metabolism. Blood flow to the brain remains unchanged while that to inactive tissues (such as those of the digestive and reproductive systems) is reduced. The duration of strenuous isotonic exercise is restricted by the extent to which cardiac output can be increased and by the body's ability to direct blood to active tissues and away from less active ones.

The increase in blood volume flow through the heart is accompanied by dilation of the cardiac vessels and a rise in venous return to the organ. More blood enters the rapidly beating heart in spite of its lesser filling time. Cardiac output usually increases as the heart rate rises from 75 to 180 beats per minute. Beyond that rate the chambers of the heart do not completely fill with blood, and cardiac output drops. Heart failure may occur in most adults if the rate significantly exceeds 180 beats per minute for prolonged periods. However, within normal limits, the large venous blood return stretches the heart and results in a more forceful contractile response and increased cardiac output (Starling's law).

The number of liters of air brought into and expelled from the lungs each minute (the pulmonary ventilation rate) becomes greater with exercise, so that more oxygen can be delivered to the tissues. A larger number of lung capillaries and alveoli (air sacs) open to promote gas exchange, allowing the blood to pick up more oxygen for delivery to peripheral tissues. The acceleration in metabolism resulting from exercise may cause oxygen consumption in skeletal muscle to exceed the supply, causing an oxygen debt (Chap. 13). As a result, exercise magnifies the *arteriovenous oxygen difference,* which is the difference in oxygen between the arterial supply and venous drainage of blood from tissue beds (Fig. 16-1). Less oxygen than usual is returned to the right side of the heart.

Isometric Exercise

Isometric exercise, like isotonic exercise, elevates heart rate and systolic and diastolic pressure. However, isometric tension sustains the compression of blood vessels and reduces blood flow to active muscle. This contrasts with the intermittent compression of vessels and blood flow that occurs with isotonic exercise. The flow of blood to active muscle fails to

TABLE 16-3
Physiological effects of physical training (conditioning) on cardiovascular function

Property	*Response**
Dimensions of the heart	
Size	↑
Diameter of chambers at end of diastole	↑
Mass of left ventricle†	↑
Cardiac output	
Stroke volume	↑
Maximum and resting heart rates	↓
Blood pressure	
In older subjects and those with high blood pressure†	↓
Oxygen consumption and transport	
V_{O_2} (volume of oxygen used per minute)	↑
Vascularity of heart	↑
Diffusion capacity of lungs	↑
Difference between arterial and venous O_2 levels in younger subjects†	↑

* ↑ = increase; ↓ = decrease; the effects on skeletal muscle were noted in Table 13-2, p. 329.
† This property often (but not always) changes in this direction.

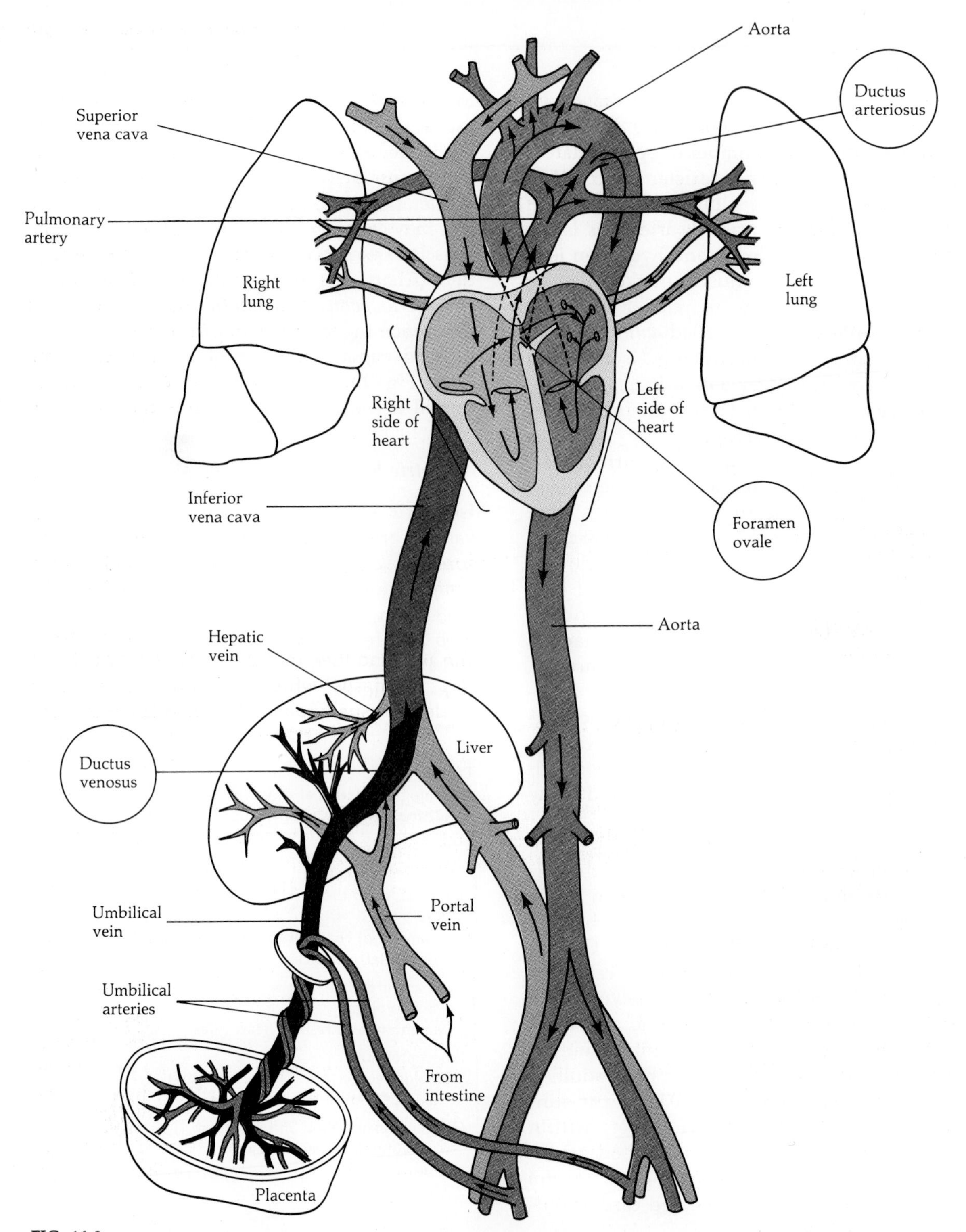

FIG. 16-2
Circulation in the fetus. Note the circled labels of three bypasses which normally close after birth: the ductus arteriosus, foramen ovale, and ductus venosus. The degree of oxygenation in the vessels is generally related to the intensity of the color. The brightest colored vessels have the highest percentage of oxygenated blood.

match metabolic needs and fatigue occurs rapidly. Because of the sustained compression of blood vessels, there is little or no opportunity for an increase in venous return from peripheral skeletal muscle to the heart and there is little or no change in the heart's stroke volume.

Physical Training

Strenuous physical activity is limited by the heart rate, which increases in direct proportion to maximum oxygen consumption (Fig. 13-15, p. 327). During and following physical training (repeated exercise), important modifications increase the capacity of the heart to meet the greater demands for oxygen, nutrients, energy, and removal of metabolites caused by strenuous exercise. Changes in the heart's size, vascularity (provision with blood vessels), stroke volume, and cardiac output are notable adaptations that occur (Table 16-3).

16-2 CARDIAC CIRCULATION DURING DEVELOPMENT

Circulation in the Fetus

After 8 weeks of development, the unborn offspring is no longer termed an *embryo* but is referred to as a *fetus.* Oxygenation of fetal blood takes place in the placenta, an organ formed by embryonic membranes and those of the wall of the uterus (womb) of the mother. In the placenta the mother's uterine arteries deliver blood to enlarged enclosed pools (blood sinuses), into which capillaries from the fetus's umbilical arteries project. The blood of mother and fetus are separated by the walls of the fetal vessels and their basement membranes formed in early development. Small molecules and particles with a molecular weight of 1000 or less diffuse readily across the placenta. Oxygen diffuses from the maternal sinuses into fetal capillaries and carbon dioxide moves in the opposite direction. Nutrients and oxygenated blood are carried to the fetus from the placenta by the umbilical vein.

Much of the fetal blood bypasses the fetal lungs because oxygenation does not occur there. Only a small proportion follows the usual adult route and the remainder flows through alternate pathways formed by two structures, the *foramen ovale* and *ductus arteriosus* (Fig. 16-2). Also, the digestive system is not the source of food as it is after birth. Therefore, a fetal vessel, the *ductus venosus,* delivers highly oxygenated blood directly from the umbilical vein to the inferior vena cava, bypassing the tissues of the liver. This route is closed after birth when the umbilical vein is tied off with the umbilical cord (Chap. 24).

The Foramen Ovale

Most of the blood in the fetal heart bypasses the right ventricle, flowing directly from the right atrium to the left atrium. It moves through a small oval opening, the *foramen ovale,* between the atria and then into the aorta. After birth two flaps of tissue which surround the foramen ovale are usually pushed against the interatrial wall to close the opening when pressure in the left atrium rises above that in the right atrium.

The difference in pressure is produced by several changes as the baby breathes and the lungs expand. The pressure within the lungs drops as the pulmonary vessels widen. Blood surges through the pulmonary veins into the left atrium, causing the pressure within the left atrium to rise. At birth the umbilical vein narrows and decreases the flow of blood from the ductus venosus to the inferior vena cava and into the right atrium (Fig. 16-2). Thus, pressure within the right atrium drops. The pressure in the right atrium declines further when the umbilical cord is tied after birth.

Should the foramen ovale remain open (patent) after birth, some blood recirculates from the left atrium, whose pressure is higher, back to the right atrium, whose pressure is lower, increasing the flow through the pulmonary artery. As a result heart murmurs sometimes develop and tissues in the walls of the vessels and chambers hypertrophy.

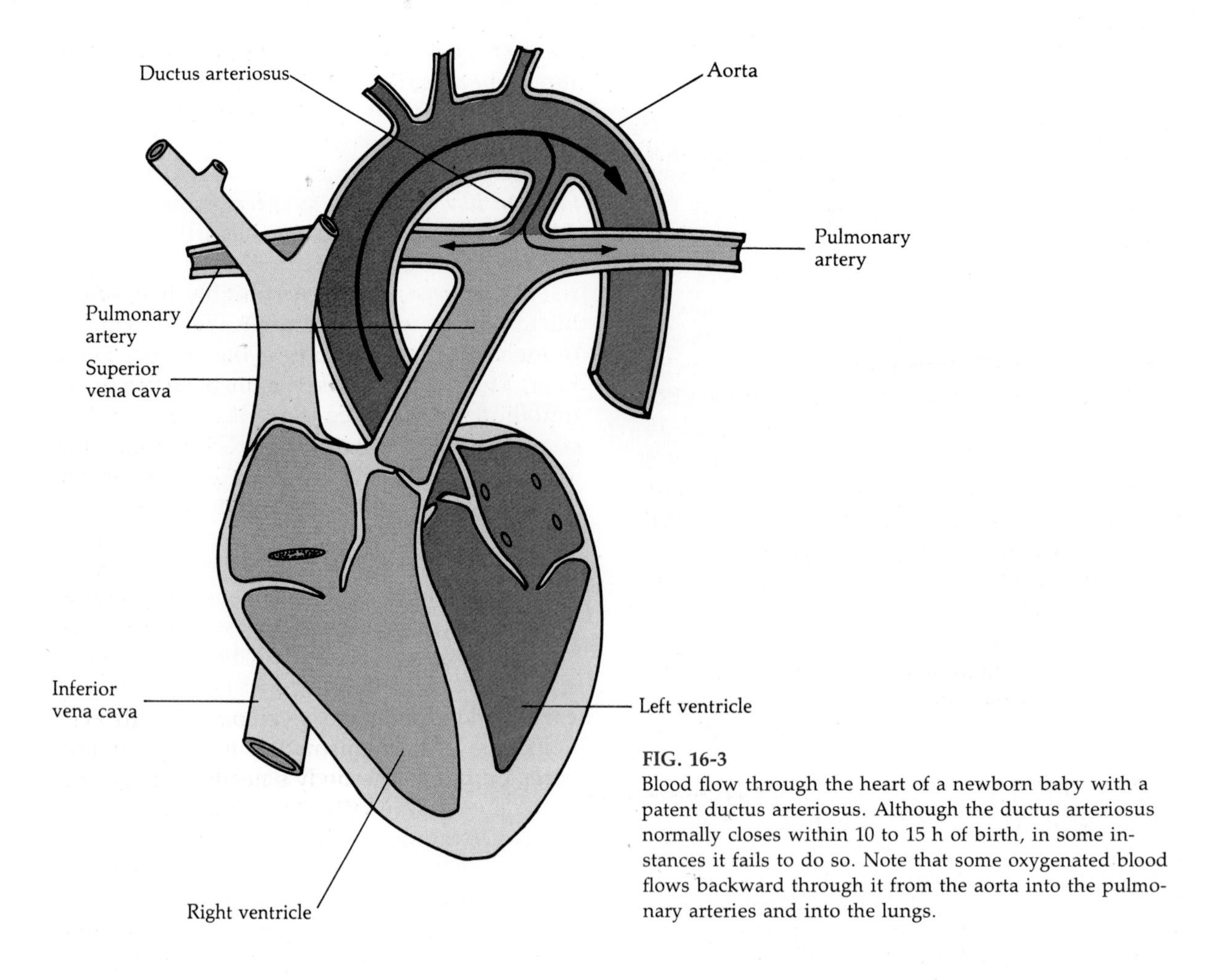

FIG. 16-3
Blood flow through the heart of a newborn baby with a patent ductus arteriosus. Although the ductus arteriosus normally closes within 10 to 15 h of birth, in some instances it fails to do so. Note that some oxygenated blood flows backward through it from the aorta into the pulmonary arteries and into the lungs.

The Ductus Arteriosus

Only a slight amount of blood flows into the right ventricle in the unborn baby's heart because of the foramen ovale. Furthermore, 80 to 90 percent of the blood in the right ventricle circumvents the lungs by flowing through the pulmonary artery and ductus arteriosus directly to the aorta (Fig. 16-2). Consequently, a disproportionately small amount of blood flows through the lungs of the unborn. Also, because of the location of the ductus arteriosus, unoxygenated and oxygenated blood mix in the left segment of the arch of the aorta, to be delivered to remote parts of the body (Fig. 16-2). Conversely, vessels to the head, which arise from earlier segments of the aorta, receive blood which is more oxygenated.

After birth when a baby breathes, the lungs and the pulmonary arteries widen allowing the blood to pass more readily toward the ductus arteriosus. However, venous return of blood from the placenta to the right atrium ceases. Pressure on the right side of the heart then declines, as does that in the pulmonary artery, which originates in the right ventricle. Blood may then flow from the aorta backward through the ductus arteriosus to the pulmonary arteries and into the capillaries of the lungs (Fig. 16-3), contrary to the patterns normally observed in adults. Although the exact cause is unknown, the ductus arteriosus usually closes 10 to 15 h after birth. Elevated oxygen levels may cause the release of vasoactive molecules, which constrict the vessel and/or dilate the pulmonary arteries, and result in closure of this fetal bypass.

The ductus arteriosus fails to close in fewer than 1 of 2000 babies. In these cases oxygenated blood flows backward through this fetal vessel and tissues receive less oxygenated blood than normal. This generates an extra load on the left ventricle to pump more blood out to satisfy oxygen demands of the tissues. The left ventricle may fail under the strain of this burden. Also, the resistance to forward blood flow through the pulmonary artery increases the work of the heart. Connective tissue deposits (fibrosis) accumulate in the pulmonary arteries to add further to the poor pulmonary circulation and congestion of the lungs. As resistance to forward flow becomes greater, more blood may be shunted backward through the pulmonary arteries. Thus, much of the blood pumped out of the heart fails to circulate through the body. If the condition is uncorrected, left ventricular failure occurs. It is a deficiency in which the heart fails to keep pace with the body's needs and can cause death within a year of birth.

Tetralogy of Fallot: A Congenital Abnormality

Fallot's tetrad is an abnormality present at birth (a congenital disorder) named after the French physician Étienne-Louis Arthur Fallot, who originally described its four major signs. The primary defect is a hole in the interventricular septum (Fig. 16-4). Less blood than normal flows from the right ventricle through the pulmonary arteries. The right ventricle reflexively responds with more rapid and vigorous contraction, resulting in muscular hypertrophy of that chamber. The aorta moves to the right into a position that allows blood to pass into it from both ventricles (rather than from the left one alone, as is normally the case). The pulmonary artery narrows (subvalvular pulmonary artery stenosis occurs). This change impedes the flow of blood to the lungs and causes more blood to flow from the right ventricle directly into the aorta. As a result, the aorta and the tissues it supplies receive a mixture of deoxygenated and oxygenated blood, which has a bluish color. The low level of oxygen in the blood is called *cyanosis*. The child is a "blue baby," with serious anatomical and physiological problems.

16-3 CARDIAC ABNORMALITIES IN ADULTS

Diseases of the cardiovascular system affect adult heart function adversely. The disorders may alter the electrical activity of the heart and be accompanied by inflammatory reactions, as described in Chap. 15. When the heart is diseased, the body attempts to compensate for abnormal cardiac output through a variety of complex mechanisms. Among them are reflex alterations of the heart rate, of its conduction of impulses, of the force of myocardial contraction, of autonomic activity to the heart, and of the dilation and constriction of arteries and/or veins in specific areas of the body. Hypertrophy of the myocardium also may occur as the heart works harder to overcome its defect. Should these mechanisms fail to keep up the heart's output to the body's needs, changes in body function gradually occur and disease and death may follow. Such alterations are likely when demands on the heart are increased in times of stress or exercise. Three of the most common cardiovascular diseases are congestive heart failure, atherosclerosis, and myocardial infarct.

Congestive Heart Failure

Congestive heart failure occurs when blood is not pumped through the heart adequately and cardiac output is low. It is called congestive heart failure because fluid accumulates in various tissue spaces. The reduction in cardiac output may lead to a decline in the filtration capacity of the kidney (Chap. 21). As a result less sodium may be excreted. Its osmotic effects can promote water retention, which causes the accumulation of more interstitial fluid, aggravating the symptoms of the disorder. There are two common causes of death due to congestive heart failure. The first is accumulation of fluid in the lungs (pulmonary edema), in which the

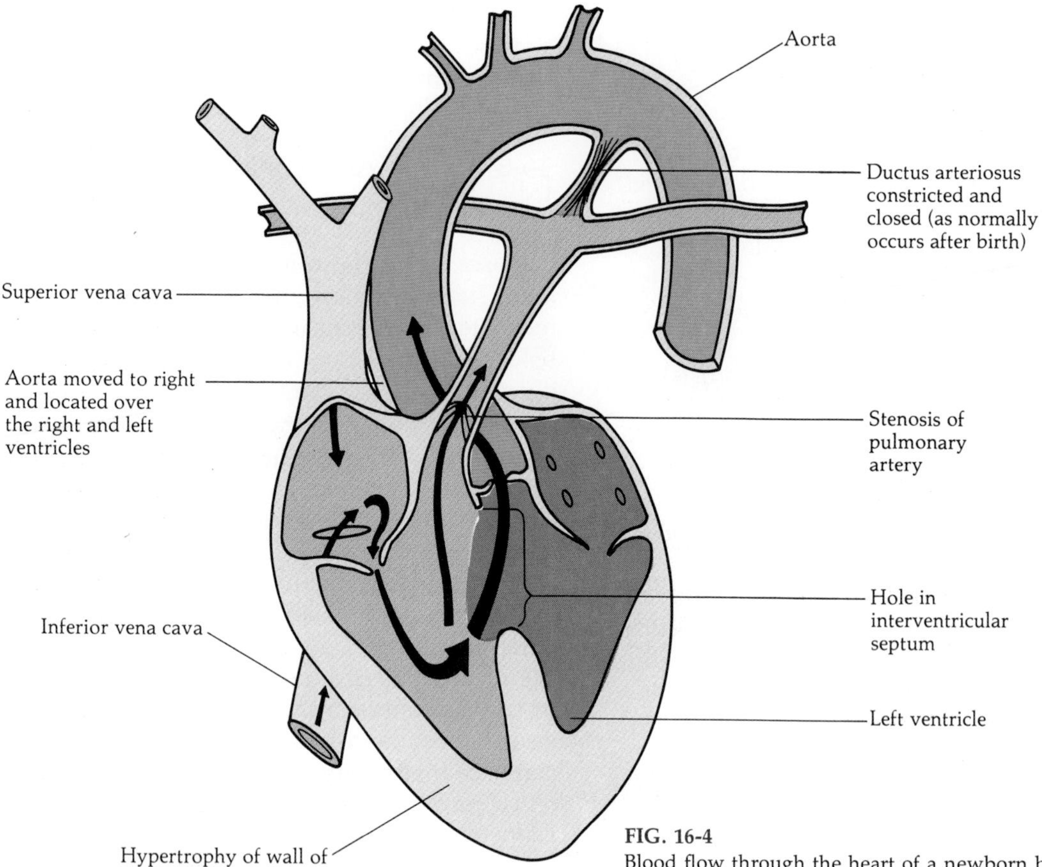

FIG. 16-4
Blood flow through the heart of a newborn baby with the tetralogy of Fallot. Note that the narrowed pulmonary artery coupled with the hole in the interventricular septum cause much of the blood to bypass the lungs and flow directly from the heart through the aorta, which overlaps both ventricles. Consequently, a mixture of oxygenated and deoxygenated blood is delivered to the aorta and the tissues and organs it supplies.

individual drowns in his own body fluid. The second is death due to oxygen deficiency in the brain produced by the diminished cardiac output, which markedly lowers blood flow to the brain. Signals between the brain and the respiratory system result initially in rapid, shallow breathing, which is followed by labored breathing, and then by a decrease and temporary cessation of respiration and a repeat of the cycle in a pattern called Cheyne-Stokes respiration (Chap. 20).

The major cause of congestive heart failure in the general population is high blood pressure. Two patterns may be described, backward and forward failure. In *backward failure* blood builds up behind a chamber or chambers. It accumulates in the veins leading to the organ, in the tissues with which those veins are associated, or in the chambers of the heart itself. Too little blood flows through the lungs for optimal gas exchange. In *forward failure* a chamber (ventricle) pumps insufficient blood forward, cardiac output is inadequate, and the effects of the lesser blood supply to peripheral tissues and to skeletal muscle become obvious during exercise. Because both ventricles are in series, one pumping blood that ultimately must flow out from the other, failure on one side ultimately affects the other.

Forward failure of the right ventricle results in too little blood flow to the lungs for normal gas exchange. Backward failure of the right ventricle reduces stroke volume. Because the large veins in the body continue to bring blood back to the heart, pressure increases in the veins and in the right side of the heart. The elevated venous pressure causes more fluid than normal to leak out of the capillary beds and to accumulate in the interstitial spaces in a condition called *peripheral edema.* The fluid buildup in the tissues may enlarge the liver and spleen and create a pool of fluid in the abdomen called *ascites.*

Forward failure of the left ventricle results in too little blood flow to peripheral tissues. Backward failure of the left ventricle increases pressure in the left atrium and in the pulmonary veins. It results in an accumulation of interstitial fluid within the lungs referred to as *pulmonary edema.* An elevation in fluid content of the thoracic cavity called *pleural effusion* also results. Both these conditions compress the air sacs in the lungs and decrease air exchange in the lungs. The right ventricle works harder to pump blood to the lungs and its capacity to sustain this activity may be overloaded, leading to right ventricular failure. Drugs such as digitalis (which increases stroke volume, Table 16-1) may be used to relieve the symptoms in some patients. However, heart muscle destroyed by the underlying disorder cannot be replaced.

Atherosclerosis and Coronary Artery Disease

Atherosclerosis is a disorder in which the walls of the arteries degenerate and are infiltrated by fatty substances forming *plaques.* Plaques are visible microscopically and/or macroscopically and are associated with the accumulation of lipids, smooth muscle cells, and other components in the artery wall. Plaques cause narrowing of and may close the inner diameter of arteries. The accumulation of plaques in arteries reduces the blood flow (causes *ischemia*) and the oxygen supply to tissues and organs. The disorder develops silently and may proceed over 20 to 50 years with no apparent signs, until it closes off more than 60 percent of a blood vessel's inner diameter. Then, within minutes, damage to the heart may occur if the coronary arteries are affected, or a stroke may result if the cerebral arteries are involved. The harmful effects of a lesser oxygen supply become apparent in times of stress and during exercise, when oxygen demand is high. When the supply of oxygen to heart muscle by the coronary arteries is reduced significantly, afferent nerve impulses to the brain are interpreted as pain. As pointed out in Chap. 15, this condition is called *angina pectoris.*

Nitroglycerin and *amyl nitrite* are drugs which are commonly used to relieve angina. The drugs decrease consumption of oxygen by the myocardium and dilate the veins in the body, increasing their storage capacity and lessening the return of blood to the heart. They also decrease resistance in arterioles, which lessens the work of the heart in ejecting blood. As a result of these changes, blood pressure drops and the heart rate decreases, as does the work of the heart (Table 16-1). The need for O_2 in the myocardium is diminished, and pain signals often decline. Some evidence indicates that these drugs also may redistribute blood flow to areas in the myocardium that are deprived of sufficient oxygen, and in angina patients they also may dilate coronary arteries to improve blood flow to the myocardium temporarily.

Myocardial Infarction and the Risk of Coronary Artery Disease

Since cardiac muscle cells primarily use aerobic metabolic pathways (Chaps. 12 and 13), they die rapidly when deprived of oxygen and are then said to be *infarcted* (killed, because of inadequate oxygen supply). In such a condition, the individual has suffered a *myocardial infarct.* The larger the infarcted area, the more severe the effects. Since the dead cardiac fibers are replaced by connective tissues, restoration of cardiac output depends on myofibrillar hypertrophy. While the size of the surviving cardiac muscle cells increases, their numbers do not. Over 1 million heart attacks occur each year and 75 percent are first attacks for the victims. The total number amounts to about 2 per minute, and heart attack is the major killer in the United States, accounting for approximately 650,000 deaths annually.

Many conditions increase the risk of coronary

TABLE 16-4
Factors associated with increased risk of coronary heart disease

High blood pressure
Type A behavior pattern (stress and tension)
Elevated blood lipoprotein, fat, and cholesterol
Obesity
Physical inactivity
Heavy cigarette smoking
Diabetes mellitus (especially in women past 65)
Familial incidence
Male sex
Increased age

TABLE 16-5
General recommendations to reduce risk of coronary heart disease*

1 Correct hypertension: with prescribed medication, blood pressure can be controlled and maintained at normal levels
2 Reduce dietary salt
3 Correct obesity; maintain ideal body weight
4 Control caloric intake
   a Reduce fat to 35 percent of caloric intake and spread through meals of the day:
      (1) Less than 10 percent should come from saturated fatty acids
      (2) Less than 10 percent should come from unsaturated fatty acids
      (3) The remainder (about 15 percent) should come from monosaturated fats
      (4) The cholesterol intake should not exceed 300 mg/day
   b Limit "empty" calories (foods that provide calories but no vitamins or other essential nutrients)
   c Assure adequate protein intake

* Any diet should be approved by one's physician. The dietary recommendations are adapted from a 1978 publication of the American Heart Association, entitled *Diet and Coronary Heart Disease.*

heart disease (Table 16-4). However, no one factor in and of itself determines the extent of the risk. Substantial evidence indicates that each of these factors may be associated with greater probability of heart attack, and the risk rises as the combination of factors, their intensity, and/or their frequency increase. Factors that may be corrected include high blood pressure (hypertension), cigarette smoking, obesity, blood levels of lipoprotein and fats, and physical inactivity. Some studies indicate that individuals with high blood pressure have as much as tenfold greater risk than those with normal pressure. Among the other risk factors, one of them alone increases the risk, two existing simultaneously nearly triple it, and three occurring at the same time increases it nearly tenfold. Genetic variations among individuals may further alter the significance of risk factors. High blood pressure, one of the most serious and accurate predictors of heart attack risk, when coupled with low blood levels of high-density lipoprotein (HDL) is an even more accurate indicator of high risk in people over age 65.

Behavior patterns of people may also be used as an index of future risk of a heart attack. Type A individuals are highly competitive, aggressive, goal-setting, and driven and seem more prone to coronary attacks than Type B individuals, who are more relaxed.

Dietary programs designed to decrease hypertension may reduce the risk of heart attack. Dietary salt should be avoided and caloric intake controlled. Attempts should be made to attain and maintain ideal body weight (Table 16-5). Furthermore, modification of individual behavior patterns are important. Included may be ways to cope better with stress, modify type A personality traits, change eating habits, eliminate smoking, and/or increase physical activity. A combination of these approaches has improved cardiovascular health in many patients. However, so many variables are involved that exact prescriptions to reduce the risk of heart attack remain somewhat hypothetical.

Diagnosis of Cardiovascular Function

Cardiovascular function may be analyzed by *noninvasive* and *invasive* techniques. Noninvasive techniques, which require no intrusion into healthy tissue, are the least aggressive means of analyses. They include a physical examination and the use of the stethoscope, electrocardiogram (ECG), x-rays, and ultrasound. Invasive techniques necessitate insertion of an instrument deep into tissues or organs and are employed when more information about the internal environment is required than can be attained through noninvasive methods.

Physical Examination

A thorough physical examination is an invaluable initial diagnostic tool. It provides a baseline or control record of the patient against which comparisons may be made. A medical history is taken, which should include a record of familial tendencies toward disease and details of behavioral habits and stresses to which the patient is exposed. The use of a stethoscope allows the physician to detect abnormal heart sounds. Blood pressure should be recorded. In individuals under 30 the checkup should include an ECG to record the normal electrical activity of the heart. Persons over 40 (especially males) should have an ECG at least once every 2 years. A record of weight should be maintained. Blood samples should be taken to analyze significant chemicals in the serum, such as lipoproteins, fats, cholesterol, and enzymes, which are extremely useful indexes of cardiovascular health.

Analyses of Blood Samples

Withdrawal of blood and analyses of the chemical constituents of plasma (its fluid portion) or serum (plasma without the clotting elements) are valuable diagnostic tools. A number of assays may be performed to assess cardiovascular health, including measurement of enzyme activity as well as the type and quantity of lipids and lipoproteins.

ELECTROPHORESIS OF SERUM SAMPLES Electrophoresis involves the movement of charged particles across a support medium placed in an electric field. It may be used to distinguish isoenzymes (isozymes) or other molecules that exhibit a difference in net electrical charge. *Isoenzymes* are slightly different forms of an enzyme that may be found in different intracellular locations and that catalyze identical chemical reactions at different rates. The activity of particular isoenzymes in the serum is significantly elevated at specific intervals after myocardial infarction. They can be analyzed electrophoretically to provide diagnostic information.

Serum contains a variety of charged molecules. A drop of serum can be applied to a cellulose acetate sheet (or agar), which acts as a support medium. Each end of the medium is submerged in separate compartments with a buffered salt solution. The salt concentration, the ionic strength, and the pH of the buffer are controlled and a cool constant temperature is maintained to minimize diffusion of charged molecules along the support medium. One compartment is connected to a positive electrode (anode), the other to a negative electrode (cathode). When the current is turned on, the direction and rate of movement toward the anode or cathode are characteristic of different species of molecules and depend on their net charge. Specific stains can be applied to the support medium after electrophoresis to make evident molecules that have separated in discrete bands (Fig. 16-5). Electrophoretic techniques can be used to identify clinically important substances, such as serum enzymes, lipoproteins, and various proteins, including those that have antibody activity (the gamma globulins).

DIAGNOSIS USING SERUM ENZYME ACTIVITY AND SPECTROPHOTOMETRY The levels of several enzymes found in serum may be elevated by cell damage, including myocardial infarct. The enzymes may leak out of cells to raise activities in proportion to the amount of tissue damaged. Several enzymes whose activities may be altered in certain diseases are aspartate aminotransferase (AST), previously called glutamic oxalacetic transaminase (GOT or SGOT, the S being an abbreviation for serum); creatine kinase (CK), also called creatine phosphokinase (CPK); lactic dehydrogenase (LD or LDH); and hydroxybutyric dehydrogenase (HBD). The diagnostic value of tests for these enzymes is briefly described in the following paragraphs.

Each enzyme catalyzes a specific reaction forming a product or products (Chap. 2) measurable under controlled conditions. The prod-

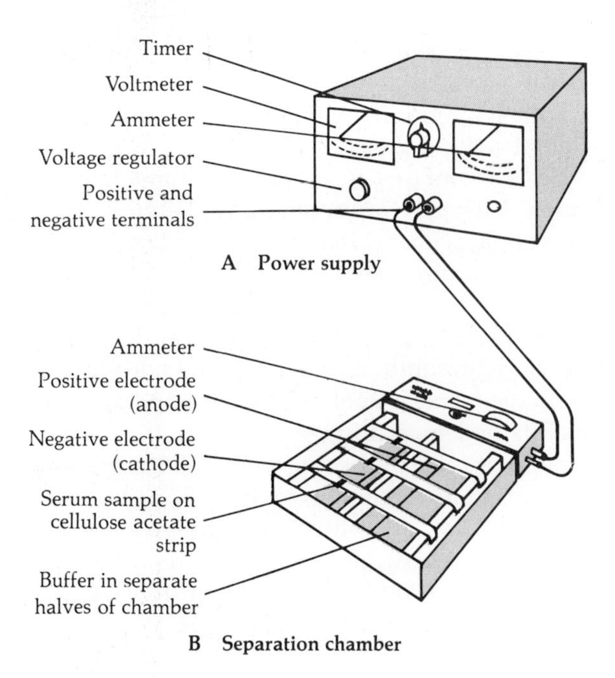

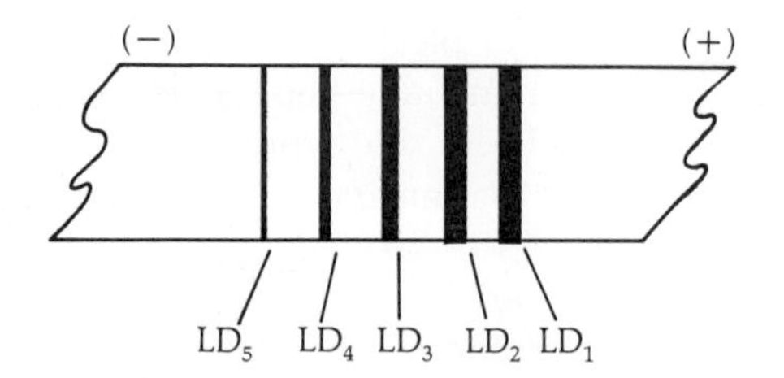

FIG. 16-5
Electrophoresis instrumentation and separation of LD isoenzymes. (A) A power supply which provides a constant current or voltage. (B) Serum samples are applied to cellulose acetate strips and placed in the chamber. After the current is turned on, molecules which differ in net charge migrate toward the anode at various speeds. (C) The separate bands of LD isoenzymes have been separated in this manner and stained to make them visible.

uct of the reaction produces a specific color when combined with an added reagent. The extent of the reaction (intensity of specific color) may be related to a difference in the absorption of light of a specific wavelength passed through the reaction mixture. The amount of light absorbed is measured in an instrument called a spectrophotometer.

The activities of AST, CK, LD, and HBD change with time after myocardial infarction (Fig. 16-6). CK shows the largest and earliest increase; the increase in AST (GOT) is somewhat slower and of lesser magnitude; LD reaches peak activity and returns to normal much more slowly than either CK or AST; and HBD is slowest in attaining its peak and in returning to normal.

AST *Aspartate aminotransferase* is especially useful as a measure of stress. The name is derived from the enzyme's capacity to transfer an amino (NH_2) group from one organic acid to another in a reversible reaction called *transamination*.

CK *Creatine kinase* is an enzyme composed of two subunits which catalyzes the metabolism of adenosine triphosphate (ATP) and creatine phosphate (C~P) in skeletal and cardiac muscle (Fig. 13-12, p. 323). Three isoenzymes, CK_1, CK_2, and CK_3, have been identified electrophoretically in various sites in the body. CK_3 is the only isoenzymic form of CK that circulates in normal serum (Table 16-6). Its activity increases in the serum after myocardial infarction. CK_2 leaks into the serum in significant amounts within 4 to 18 h after myocardial infarction. The rise and fall of serum CK_2 within 18 to 48 h of a suspected heart attack is indicative of that condition. Persistent elevated CK_2 levels indicate an increased probability of progressive injury to the heart.

FIG. 16-6
Alterations in the levels of the activity of four serum enzymes following myocardial infarction.

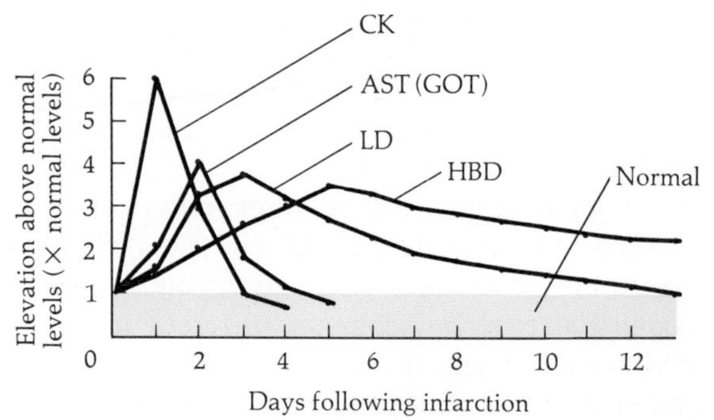

TABLE 16-6
Type and approximate percentages of CK isoenzymes normally found in various sites

		PERCENTAGES IN VARIOUS SITES			
	Subunits	*Brain*	*Heart*	*Skeletal Muscle*	*Serum*
CK_1	B,B	90			
CK_2	M,B		40		
CK_3	M,M	10	60	100	100

CK levels (CK_3) rise in response to skeletal muscle diseases such as muscular dystrophy and also are elevated after an intramuscular injection. Since CK_3 is found in heart and skeletal muscle, elevated CK levels in serum may originate from either tissue. Peripheral muscle disorders may be distinguished from myocardial infarction by calculating the ratio of CK to AST activity. A CK/AST ratio that exceeds 8.9 generally denotes peripheral muscle disease, as does a high absolute value of CK. This is so because the amount of skeletal muscle in the body and its capacity to release CK are overwhelming in comparison with that of the heart.

LD *Lactic dehydrogenase* is an enzyme composed of four subunits and found in many different types of cells in the body. It removes hydrogen from lactic acid to form pyruvic acid.

LD serum values increase within 8 to 10 h after myocardial infarction, usually peak within 2 to 3 days after the initial attack (rising 2 to 4 times above normal), and remain elevated for 10 to 14 days (Fig. 16-6). There are five LD isoenzymes (LD_1 to LD_5) whose relative abundance varies among different tissues. LD_1 and LD_2 are the predominant forms in the heart and the level of LD_1 exceeds that of LD_2 in the sera of about 80 percent of the patients with a heart attack. Furthermore, LD_1 activity persists when serum is heated at 65°C for 30 minutes, whereas LD_2 to LD_5 activities are lost. Because LD_1 is predominant in the heart, changes in LD activity of heated serum provide a useful index of myocardial infarction. Since the activity of LD remains above normal for 10 to 14 days after a heart attack (Fig. 16-6), analysis for this isoenzyme is very useful in diagnosis in patients who wait several days after the disorder occurs before being examined by a physician.

HBD In older diagnostic techniques, assay of a fourth enzyme, *hydroxybutyric dehydrogenase,* was also performed. It is associated with fatty acid degradation and synthesis in reversible reactions in cardiac and skeletal muscle that involve derivatives of acetyl CoA (page 321). However, recent studies indicate that the activity of HBD is primarily due to that of LD_1 and secondarily to that of other LD isoenzymes. Following a heart attack, levels of HBD in the serum are elevated more slowly than those of other enzymes. HBD peaks around 4 to 5 days after myocardial infarction and persists longer than other enzymes, remaining at higher than normal levels for about 2 weeks (Fig. 16-6).

In addition to muscle disorders, many other diseases affect serum enzyme activity. Among them are diseases of the liver, pancreas, prostate, and blood (anemias and leukemias) and toxemias developing during pregnancy. In many instances the pattern and extent of change in the values of several key enzymes provide insight as to the nature of the disease. Consequently, enzyme activities are measured several times over a period of days or weeks. Furthermore, other tests are usually made in addition to enzyme assays to assist in diagnosis and in differentiation of one disease from another.

PLASMA LIPIDS AND LIPOPROTEINS Products of fat digestion, absorption, and metabolism that are found in blood plasma (fluid) combine with different carrier proteins to form lipoprotein packets, whose plasma levels are useful in predicting the risk of heart attack. The complexes may contain fatty acids, triglycerides (neutral fats), phospholipids, and sterols such as cholesterol. They are transported in the blood among the intestine, liver, and other tissues. Lipoproteins may be separated into five major groups by electrophoresis or by centrifugation (Fig. 16-7).

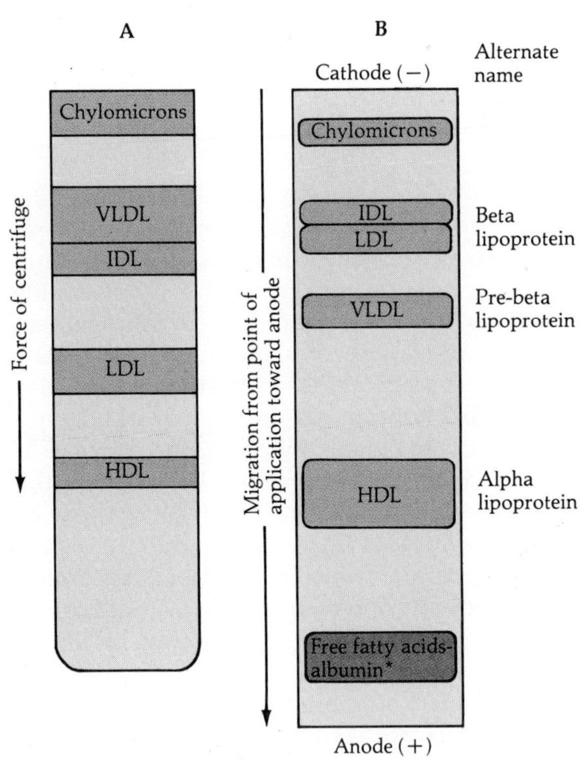

FIG. 16-7
Separation of plasma lipoproteins by density or buoyancy in ultracentrifugation (A) and by electrophoresis (B).

The five major types of lipoproteins are chylomicrons and lipoproteins of very low density (*VLDL*), intermediate density (*IDL*), low density (*LDL*), and high density (*HDL*). Their compositions vary in the percentages of lipids and proteins contained within the packets, as do their sizes, their maximum diameters being about 1.0, 0.08, 0.04, 0.02, and 0.01 μm, respectively. Chylomicrons, the largest and least dense lipoproteins, originate from dietary fat and are secreted by intestinal cells as small fat globules averaging 0.1 μm in diameter. They appear in the blood about 1 to 8 h after ingestion of a meal that contains fat. The next least dense lipoproteins, VLDL, are found in foods and secreted by the liver. They are degraded by lipoprotein lipases enzymes to IDL and LDL (Fig. 16-8). Receptors on peripheral cells, such as those lining arteries or on fat cells, have a preferential attraction for LDL (Fig. 16-8). This lipoprotein carries much of the cholesterol in the blood and deposits it in peripheral cells. The LDL-cholesterol complex enters cells by endocytosis and its cholesterol is released as a result of lysosomal enzyme activity in the cell.

Receptors on liver cells have a preferential attraction for chylomicron remnants. The liver metabolizes these remnants to cholesterol and bile acids and secretes them in bile to be reabsorbed in the intestine and recycled or excreted with intestinal wastes.

HDL, the densest of the five types of lipoprotein, is found in foods and also is synthesized by cells in the intestine and the liver. It promotes cholesterol recycling and converts lipids into chylomicron remnants, which can then be degraded by the liver and excreted (Fig. 16-8). In this manner HDL minimizes the accumulation of cholesterol in the plasma and in peripheral cells.

Individuals with above normal levels of lipoproteins in their plasma have *hyperlipoproteinemia*. They may be divided into several groups according to the types of lipoproteins found in excess in their plasma (Table 16-7). Hyperlipoproteinemias may be due to excess dietary intake of cholesterol or triglycerides and/or be associated with hereditary defects that alter metabolism (especially of enzymes that metabolize lipids), hormonal imbalances (as occurs in diabetes mellitus or hypothyroidism), and liver and kidney dysfunctions.

Hyperlipoproteinemia often can be treated successfully by diet therapy aimed at reducing the plasma lipids which are in excess. Restrictions may include the number of calories,[1] the quantity of protein, carbohydrate, or alcohol,

[1] A kilocalorie, sometimes called the large calorie or Calorie, is the amount of heat needed to raise the temperature of 1 kg (1000 g) of water by 1°C. This unit is generally used, especially in discussions of nutrition, because metabolic and other chemical reactions involve thousands of calories (the amount of heat to raise 1 g of water by 1°C).

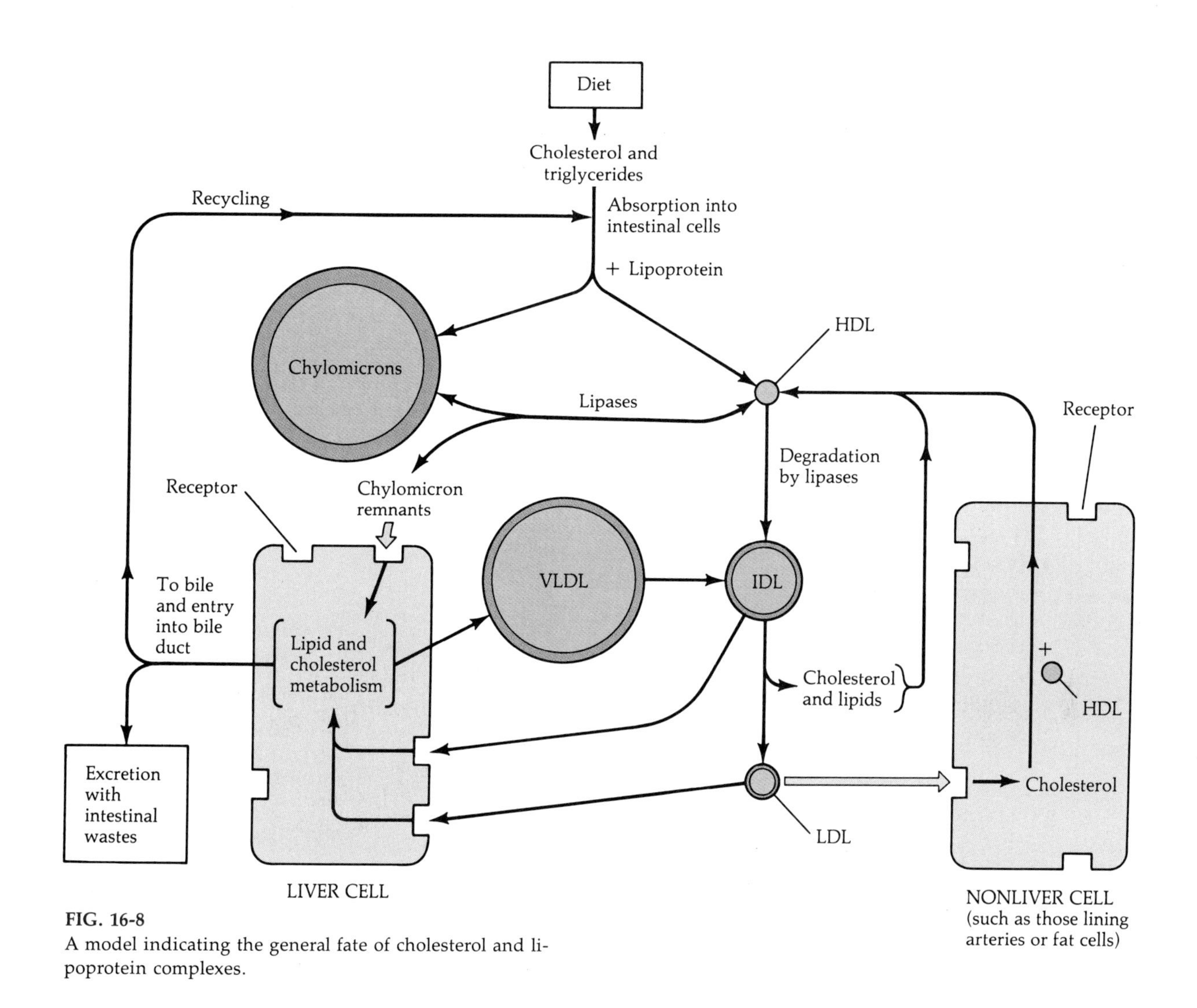

FIG. 16-8
A model indicating the general fate of cholesterol and lipoprotein complexes.

TABLE 16-7
Types of hyperlipoproteinemia

	EXCESS COMPONENT IN PLASMA							
	LIPOPROTEINS				LIPIDS			
Type	*Chylomicrons*	*LDL*	*VLDL*	*IDL*	*Triglyceride*	*Cholesterol*	*Underlying Mechanism*	*Relative Incidence*
I	+				+		Enzyme deficiency for chylomicron removal	Rare
II IIA		+				+	Excess production or inadequate clearance of LDL	Common
II IIB		+	+		+	+		Common
III	+ (sometimes)			+	+	+	Metabolic block of VLDL catabolism	Rare
IV			+		+		Excess production or inadequate clearance of VLDL	Common
V	+		+		+		Excess production of VLDL and deficiency in chylomicron removal	Rare

TABLE 16-8
Diets for types I-V hyperlipoproteinemia*

Diet Prescription	*Type I*	*Type IIa*	*Type IIb and Type III*	*Type IV*	*Type V*
Calories	Not restricted	Not restricted	Achieve and maintain "ideal" weight, reducing diet if necessary	Achieve and maintain "ideal" weight, reducing diet if necessary	Achieve and maintain "ideal" weight, reducing diet if necessary
Protein	Total protein intake not limited	Total protein intake not limited	High protein intake, 20% of kcal	Not limited other than control of patient's weight	High protein intake, 20% of kcal
Fat	Restricted to 25–35 g; type of fat not important	Saturated fat intake limited, polyunsaturated fat intake increased	Controlled to <40% of kcal (polyunsaturated fats recommended in preference to saturated fats)	Not limited other than control of patient's weight (polyunsaturated fats recommended in preference to saturated fats)	Restricted to 30% of kcal (polyunsaturated fats recommended in preference to saturated fats)
Cholesterol	Not restricted	As low as possible, meat only allowed cholesterol source	Less than 300 mg; meat only allowed cholesterol source	Moderately restricted, 300–500 mg	Moderately restricted, 300–500 mg
Carbohydrates	Not limited	Not limited	Controlled, 40% of kcal; concentrated sweets restricted	Controlled, 45% of kcal; concentrated sweets restricted	Controlled, 50% of kcal; concentrated sweets restricted
Alcohol	Not recommended	May be used with discretion	Limited to 2 servings (substituted for carbohydrate)	Limited to 2 servings (substituted for carbohydrate)	Not recommended

* From: Levy, R. I., "Hyperlipoproteinemia and its Management," *J. Cardiovasc. Med.* **5**:435–462 (1980).

and the type and amount of fat and quantity of cholesterol (Table 16-8).

Relation of Plasma Lipoprotein Levels to Risk of Heart Attack The risk of heart attack generally increases with the levels of chylomicrons, VLDL, IDL, and LDL in the plasma and decreases with increasing levels of HDL. Moreover, as the HDL levels decline (below a mean value of 45 mg/dL of plasma for males and 55 mg/dL for females), the risk of heart attack increases proportionately.

Conversely, high quantities of HDL or high HDL/cholesterol ratios are the best indicators of protection against heart attack. Nonsmokers generally have higher levels of HDL than smokers. Marathon runners also have elevated plasma levels of HDL, possibly because of utilization of lipids by skeletal muscle for energy metabolism. Weight loss through reduction in caloric intake and dietary control also may be employed to alter specific lipoprotein levels in order to diminish the risk of cardiovascular disease (Table 16-8).

Relation of Plasma Lipoprotein levels to Cell Receptors The number and shape of lipoprotein receptors influence the rate of uptake, degradation, and fate of lipoproteins. As mentioned above, liver receptors preferentially bind chylomicron remnants while receptors on peripheral cells favor LDL. The liver receptors effectively remove large amounts of cholesterol from the plasma, rendering individuals somewhat resistant to accumulations that might otherwise occur as a result of its dietary intake. If all other factors are equal, the susceptibility to heart attack diminishes as the number of lipoprotein receptors in the liver increases. The number of active receptors varies. As mentioned earlier, diet, heredity, and hormones are among variables which may influence receptor activity. Elevated dietary intake of lipids can challenge the capacity of the liver receptors and result in excess plasma lipoproteins. Conversely, plasma lipoproteins may be reduced by specific dietary restriction (Table 16-8). Specific genetic deficiencies may modify receptor number and function and result in different types of hyperlipoproteinemia. Elevated levels of the hormone thyroxine increase the number of receptors that promote the catabolism of LDL. Conversely, hypothyroidism results in elevated LDL. Hyperlipoproteinemia may therefore result from an interplay of genetic, environmental, and metabolic factors causing variations in individual responses to different diets. Theories that propose to relate these events to atherosclerosis will be discussed further in Chap. 17.

Images of the Heart and Cardiac Catheterization

Should any of the tests described above reveal an abnormality in cardiovascular function, other analyses may then be used to confirm the results. Several techniques produce images of the heart. One method, called *fluoroscopy*, projects x-rays of the moving heart on a fluorescent screen. *Ultrasound* involves beaming sound with frequencies of more than 20,000 Hz (hertz, or cycles/s) at the heart. Although most of the energy passes through the tissue, some is reflected. The time required for the reflection is related to the distance to the tissue and its thickness. The pattern of reflection of sound provides information about the movement of the structure under the probe. A transducer converts the reflected sound into electrical activity that forms a visual display or graphic tracing called an *echocardiogram*. The electrical activity may be used to deflect an electron beam on an oscilloscope screen to provide a visual display, as shown in Fig. 16-9. As the sound beam is directed at different tissues, the image changes and may reveal areas of cardiovascular disease or alterations due to hypertension, valvular defects, or congestive heart failure.

A *scintigram* (scintiscan) is an image of an organ such as the heart, recorded on film and generated from certain short-lived isotopes. These compounds selectively accumulate in specific tissues after their intravenous administration. They emit radioactivity whose level in a particular tissue site is proportional to the rate of blood flow or to the affinity (attraction) of the tissue for the radioactive substance. The sites that emit γ-rays are identified by differences in

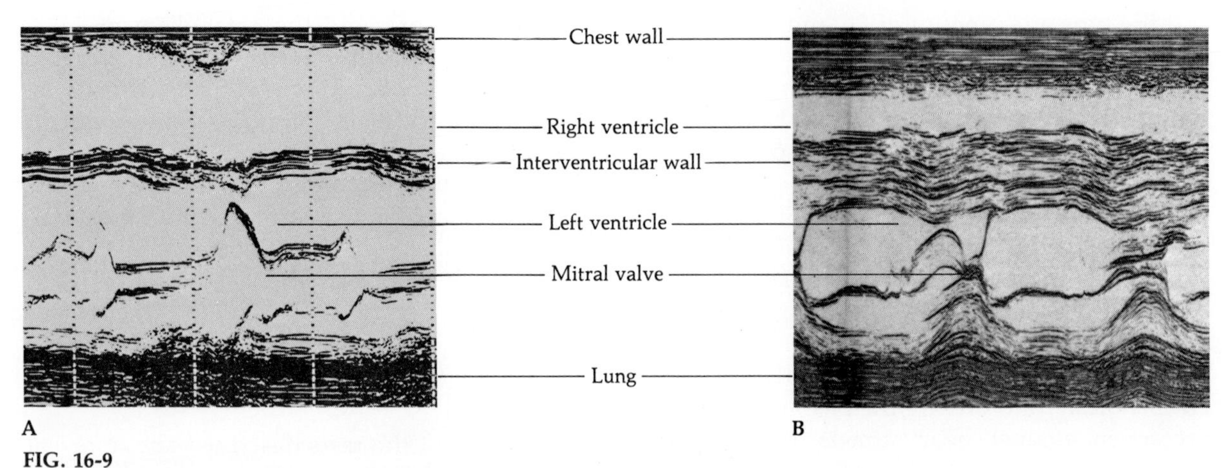

FIG. 16-9
Ultrasound analysis (echocardiogram) of (A) a normal heart and (B) one with mitral insufficiency and a thickened interventricular wall. Compare A to B.

image density. Injured areas in blood vessels or the myocardium emit less radioactivity and may thus be pinpointed.

Cardiac catheterization is another helpful, though hazardous, technique. In rare cases it may induce fibrillation. A catheter (hollow tube) may be inserted into a superficial vein in the left forearm, then moved into the superior vena cava and the right atrium. This procedure is viewed through a fluoroscope. The catheter may then be advanced through the right ventricle into the pulmonary artery. Blood samples may be withdrawn and pressure measured along the route (see Chap. 17). Similarly, an *angiogram*, a visual display of blood vessels or lymph vessels, often made by x-rays, may be employed. A radiopaque compound may be injected through the catheter and allowed to flow into the heart and its associated vessels. Because the injected material absorbs x-rays, x-radiography of the heart shows the location and structure of the radiopaque vessels. Visual analysis of the condition of the vessels may be made. (An angiogram of vessels in the brain is shown in Fig. 10-7, p. 246.)

16-4 SUMMARY

1 A variety of factors modify cardiac physiology. Heart rate is usually higher in females than males. Epinephrine, norepinephrine, moderate exercise, and consumption of caffeine or nicotine may increase heart rate and stroke volume. A rise in the level of acetylcholine, arterial blood pressure, or alcohol decreases heart rate. Digitalis decreases heart rate and increases stroke volume and may be used to treat patients with atrial fibrillation.

Strenuous isotonic exercise causes increases in heart rate, stroke volume, cardiac output, systolic pressure, oxygen consumption, and arteriovenous oxygen differences. It decreases overall resistance to blood flow in peripheral vessels. Isometric exercise elevates heart rate and systolic and diastolic pressure but reduces blood flow in active muscles. Endurance training enlarges heart size and improves cardiovascular function, vascularity of the heart, stroke volume, and the diffusion capacity of the lungs.

2 Much of the blood flow in the fetal heart bypasses the lungs by means of the foramen ovale and ductus arteriosus. Usually both bypasses close after birth.

The failure of either to do so may lead to oxygen deficiency. The tetralogy of Fallot is a serious congenital abnormality due to a hole in the wall between the ventricles and produces sufficient oxygen deficiency to cause a "blue baby" syndrome.

3 Cardiac abnormalities which adversely affect heart function include congestive heart failure, atherosclerosis, and myocardial infarction. Congestive heart failure is associated with inadequate cardiac output. Edema results from a buildup of venous pressure and loss of fluid from blood into the interstitial spaces. The major cause of congestive heart failure is high blood pressure.

Atherosclerosis is a disorder in which lipids gradually accumulate in plaques in the lining of the arteries. Smooth muscle proliferates in arterial walls and narrows the inner diameter of the vessels. Coronary insufficiency may occur and, if unrelieved, may lead to oxygen deficiency in cardiac fibers. Chest pain called angina pectoris may result. Heart cells deprived of oxygen for 30 s or more die in a condition called myocardial infarction. Nitroglycerin or amyl nitrite may temporarily decrease the workload of the heart, diminish O_2 consumption by the myocardium and alleviate coronary insufficiency.

The risk of myocardial infarction is increased by a number of factors, of which high blood pressure has the greatest effect. Type A behavior, obesity, physical inactivity, smoking, and increased age also heighten risk of a heart attack.

Diagnosis of cardiac function may be made during a physical examination. It should include a medical history, measurement of blood pressure, and an ECG and also may involve analyses of serum samples for lipids, lipoproteins, and serum enzymes. Damage to heart cells and other cell types often increases the levels of the serum enzymes creatine kinase (CK), aspartate aminotransferase (AST), and lactic dehydrogenase (LD). The magnitude and patterns of change in their activity and electrophoretic distributions help assess the severity of myocardial infarction and distinguish it from other disorders.

Plasma lipoproteins are distinguished by their density as well as by their electrophoretic properties. The accumulation of plasma chylomicron remnants, IDL, and LDL is associated with reduced protection against myocardial infarction, whereas accumulation of HDL is related to greater protection. The accumulation of HDL is associated with the production of chylomicron remnants. HDL promotes the transport of lipoproteins among the blood, nonliver cells, and liver cells. The liver metabolizes the remnants, which may be recycled or excreted in bile delivered to the intestine.

In order to diagnose cardiac abnormalities further, images of the heart may be produced by x-rays, fluoroscopy, ultrasound, scintigrams, and angiography following cardiac catheterization.

REVIEW QUESTIONS

1 What effects do alcohol, caffeine, and nicotine have on cardiac output? Why?

2 How do blood pressure, sex, body size, temperature, and aging influence heart rate and size and cardiac output?

3 Compare the effects of strenuous isotonic and isometric exercise on the heart, and describe its response to endurance training.

4 Diagram the route of circulation through the heart of the unborn, noting the roles of the foramen ovale and the ductus arteriosus.
5 What are the consequences of failure of the newborn baby's circulatory pattern to mature normally after birth?
6 What is the tetralogy of Fallot and what are its effects?
7 What is congestive heart failure, and how is it related to peripheral edema, ascites, and pulmonary edema?
8 Describe the relationship of atherosclerosis to angina pectoris and coronary artery disease.
9 What is the relationship between coronary artery disease and the risk of myocardial infarction? List several prominent risk factors and discuss their importance.
10 Describe the types of tests used in a physical examination that are related to cardiovascular health.
11 Outline the principles of serum electrophoresis and separation of LD and CPK isoenzymes.
12 Describe the principles of spectrophotometric analysis of enzymes.
13 Roland Dee is admitted to a hospital complaining of pain in the left arm, which he has had for an unspecified amount of time. He is 60 years of age. His plasma enzymes show an elevated level of LD_1 but normal levels of AST and CK. His LDL levels are high and his HDL levels are low. What is the probable diagnosis? When did the cause of his symptoms likely occur?
14 Name and describe two methods used to distinguish lipoproteins.
15 What general relationships exist between the major types of lipoproteins and heart attack risk? If Tony Francis has a high level of HDL in his plasma, low blood pressure, and Type B personality traits, how do these factors relate to his risk of heart attack?
16 In what manner does the interplay of heredity, diet, hormones, and lipoprotein receptors affect the fate of cholesterol?
17 Compare several techniques available to produce images of the heart. Include fluoroscopy, ultrasound, the use of scintigrams, and angiography following cardiac catheterization.

THE CIRCULATORY SYSTEM

The Functions of Blood and Lymphatic Vessels

17-1 STRUCTURE AND FUNCTIONS OF BLOOD AND LYMPHATIC VESSELS

17-2 REGULATION OF FLUID MOVEMENT
Pulse
Reflex control of blood vessels
Alternative routes of circulation
Flow of blood through capillaries
The exchange of materials between blood, interstitial fluid, and lymph
Return of interstitial fluid to the circulatory system
Mechanisms of lymph movement
Regulation of blood flow through veins
Other factors related to blood flow
Physical factors: Bernoulli's principle; gravity; Poiseuille's law
Measurement of blood pressure
Regulation of blood pressure

17-3 DISORDERS RELATED TO THE WALLS OF BLOOD VESSELS
Hypertension
Cardiovascular effects
Nervous factors
Hormones
Essential hypertension
Therapy for hypertension
Atherosclerosis

17-4 SUMMARY

OBJECTIVES

After completing this chapter the student should be able to:

1. Name and describe the three major layers of tissues in the walls of arteries and veins and compare the two types of vessels with respect to their walls and functions.
2. Discuss the flow of fluid through capillary walls according to Starling's equilibrium of capillary exchange and describe the role of hydrostatic and osmotic pressures in the formation of lymph.
3. Compare the factors influencing fluid movement in arteries, veins, and lymph vessels.
4. State Bernoulli's principle and explain its relationship to blood flow.
5. Define Poiseuille's law and explain its relationship to blood flow.
6. Discuss the cause of systolic and diastolic pressure and explain how both are measured.
7. Describe the relationships among cardiac output, peripheral resistance, and blood pressure.
8. Outline the manner in which nerve impulses influence blood pressure.
9. Describe the major effects of hormones on blood pressure.
10. Discuss the progression of arterial disease to atherosclerosis and compare two theories explaining the possible origins of the disorder.

The heart assists homeostasis by circulating cells, molecules, and ions in a fluid called blood which is contained within blood vessels. Some components of blood escape from the vessels and later enter a separate system of channels called lymphatic vessels which ultimately return the fluid to blood vessels. In this circulatory pattern the heart pumps blood from the right ventricle through the lungs in order to exchange gases (Chaps. 15 and 20). Upon reentering the heart from the lungs, blood is pumped from the left ventricle into the aorta to circulate throughout the body to continue the cycle.

17-1 STRUCTURE AND FUNCTIONS OF BLOOD AND LYMPHATIC VESSELS

The primary vessel to carry blood from the heart is the aorta, which subdivides into *ar-*

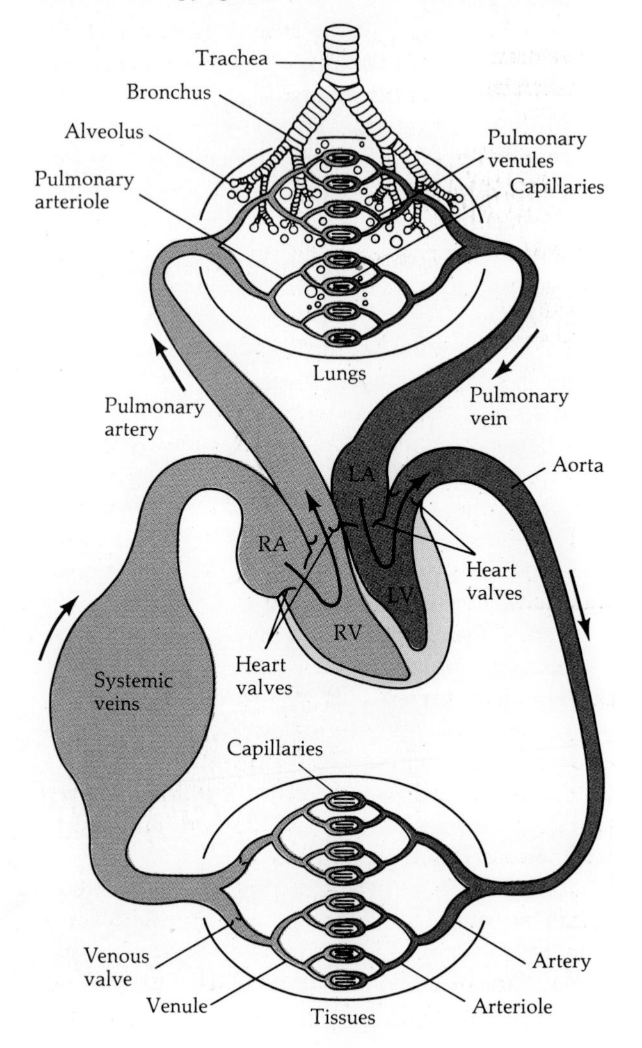

FIG. 17-1
Path of circulation of blood through the body. The arrows indicate the direction of flow of blood. RA = right atrium; RV = right ventricle; LA = left atrium; LV = left ventricle. An alveolus is a microscopic air sac in the lung across which gas exchange occurs. The segments of the circulatory system which have highly oxygenated blood are more intensely colored. Those with deoxygenated blood are lighter in color. From *The Human Cardiovascular System, Facts and Concepts,* by J. T. Shepherd and P. M. Vanhoutte. Copyright © 1979, Raven Press, New York.

TABLE 17-1
General tissue composition of the coats (tunics) of arteries and veins

Coat (Layer)	*Artery*	*Vein*	*Distinguishing Features of Veins*
1. Intima (inner)	Endothelium	Endothelium	Modified to form valves
	Connective tissue (in large arteries)		
	Elastic tissue		
2. Media (middle)	Circular smooth muscle	Circular smooth muscle	Less than in artery
	Elastic tissue	Elastic tissue	Less than in artery
	Collagenous tissue	Collagenous tissue	Less than in artery
	Reticular tissue		
3. Adventitia (outer)	Loose connective tissue	Loose connective tissue	Thinner walls but wider
	Fibrous connective tissue	Fibrous connective tissue	diameter and greater
	Some longitudinal smooth muscle		blood-retaining capacity

teries, which supply blood to the organs and tissues of the body (Fig. 17-1). The smallest of the arteries, *arterioles,* divide into the tiniest of all blood vessels, the *capillaries,* in which exchange of substances between the blood and the tissues of the body occurs. The lumen (inner passage) of a capillary is so small that red blood cells pass through it in single file. The capillaries merge into small *venules,* which transport metabolic products from cells of the body. The venules merge into *veins* and the veins into the *venae cavae,* which return blood to the heart.

As blood is transported through arteries, it is under pressure. This pressure, called *hydrostatic* (fluid) pressure, causes small molecules and ions to escape into tissue spaces through pores and gaps between the cells in the walls of the capillaries. Substances, such as proteins, which do not pass back into the capillaries are retrieved by the more permeable network of lymphatic vessels.

Arteries are the blood vessels which carry blood away from the heart while *veins* transport it toward the heart. Most arteries deliver oxygenated blood whereas most veins carry deoxygenated blood. The exceptions are the pulmonary arteries and veins; the former contain deoxygenated blood and the latter transport oxygenated blood (Fig. 17-1).

Arteries and veins are composed of three layers or coats (*tunics*) of tissues. From the inside to the outside they are the *intima,* the *media,* and the *adventitia* (Fig. 17-2). The wall of an artery is generally thicker and more complex and contains more tissue, especially smooth muscle, than that of a vein in the same vessel area (Table 17-1). As a result arteries are less likely to collapse when blood flow decreases. The intima in both types of vessels is lined by a single layer of elongated squamous epithelial cells, the *endothelium.* In veins the endothelium is periodically modified to form *valves,* which permit blood to flow in only one direction, toward the heart.

Different arteries can be distinguished as elastic or muscular according to variations in tissues. *Elastic arteries* are also referred to as *conducting arteries* because they conduct blood directly from the heart. These largest of vessels, which include the aorta and pulmonary arteries, have great amounts of elastic tissue and internal diameters between 1.0 and 2.5 cm. *Muscular arteries* are known as *distributing arteries* because they carry blood directly to organs and tissues. Most arteries of this type range in internal diameter from 0.03 to 1.0 cm and may be classified as small, medium, or large. Most arteries, excluding arterioles, are so thick that their walls must be supplied by smaller blood vessels called *vasa vasorum.* Arterioles, the smallest of arteries, average 0.003 cm (or 30 μm) in diameter (ranging from 8 to 50 μm) and branch into smaller vessels, *metarterioles,* which supply about 10 to 100 capillaries each. The circular smooth muscle in the media of arterioles is responsive to nervous impulses that alter the diameter of the vessels and the flow of blood from arterioles into capillaries.

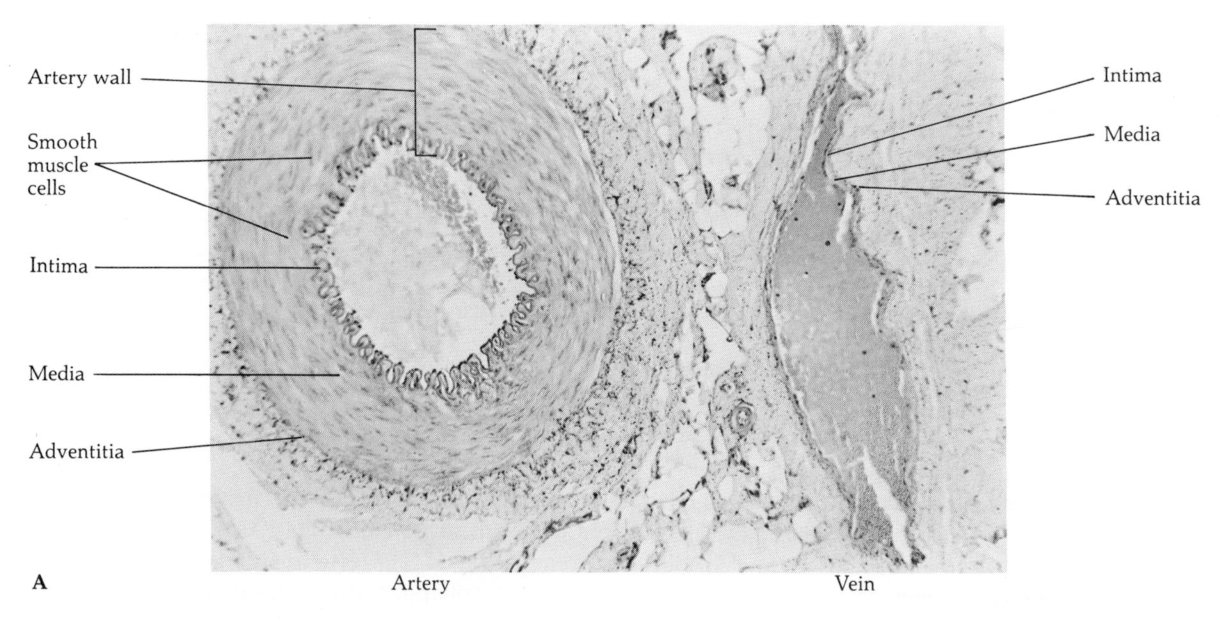

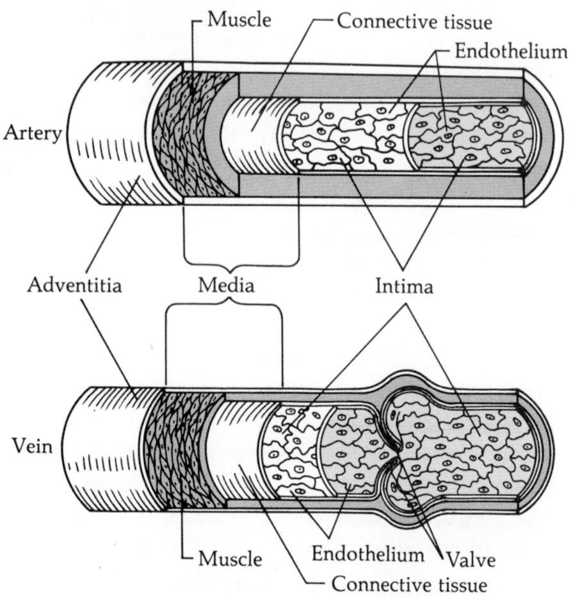

FIG. 17-2
Structure of the walls of blood vessels. (A) Microscopic view of cross section of artery and vein (× 170). Note that the wall of the vein is thinner than that of the corresponding artery. (B) Drawing of longitudinal section of artery and vein, demonstrating the adventitia, media, and intima. (C) Longitudinal section of an arteriole (× 700). The arteriole is obliquely placed in this photomicrograph. The fingerlike segment near the middle is formed by the vessel. (D) Cross section of an arteriole containing red blood cells (× 1700).

17-2 REGULATION OF FLUID MOVEMENT

Pulse

As the heart beats, the blood ejected during ventricular systole causes the expansion followed by the recoil of the elastic walls of the arteries, which allows a continuous flow between heartbeats. The rhythmic change in size of the vessels may be felt near the surface of the body as the *pulse*. The volume, tension, rhythm, and rate of pulse all provide valuable insights to cardiovascular function.

Arteries receive blood directly from the heart. Thus they are on the high-pressure side of the circulatory system while the veins are on the low-pressure side (Fig. 17-3).

Blood is relatively noncompressible. As it is ejected from the heart, it pushes against the blood ahead of it and causes the walls of the arteries to expand in a wave, which moves along the length of the vessel at a rate much faster than the flow of blood. The sequence is analogous to a model in which balls are squeezed into a rubber stocking filled with water. The

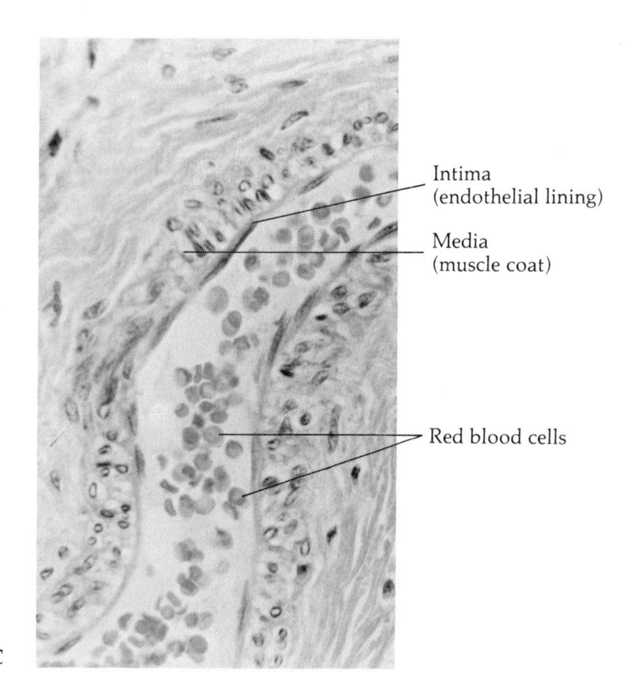

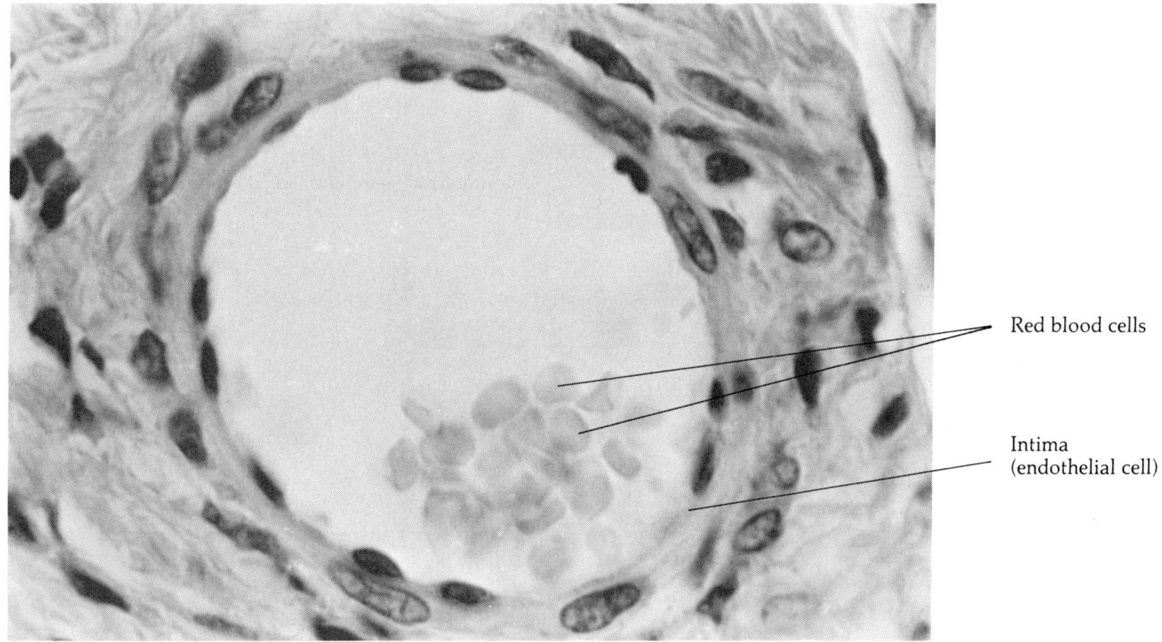

balls and water are analogous to blood cells and plasma and the stocking to a blood vessel. If a ball at one end is pushed forward, the force is transmitted through the other balls to the end of the stocking. The stocking's diameter widens to accommodate forward movement of the balls and water, and it does so before the balls move down its length. In a similar way, the pulse has a high *velocity,* traveling at about 6 to 9 m/s, while the maximum flow rate of blood is only about 40 cm/s.

The *volume* of the pulse may be large or small, depending on the volume of blood pushed through a vessel. The *tension* of the pulse may

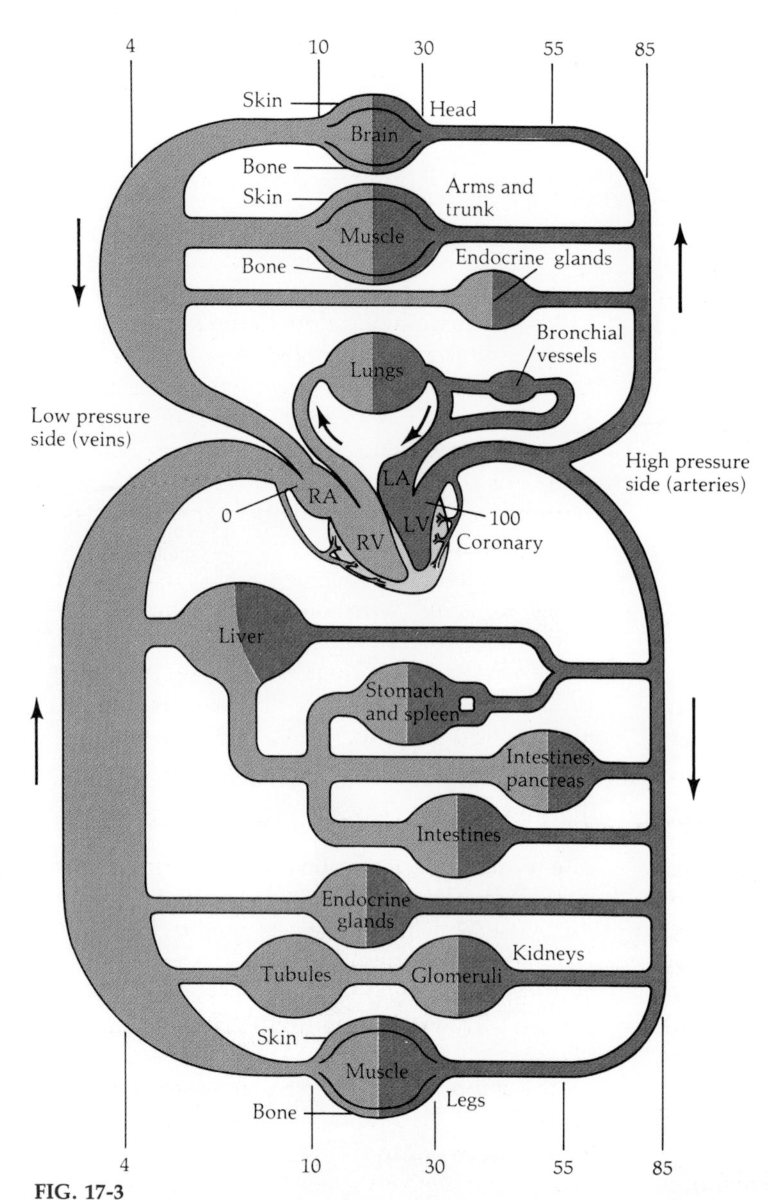

FIG. 17-3
The distribution of blood in the body. Note that arteries, which deliver blood to tissues and organs, normally have a higher pressure than the corresponding veins which drain the same tissues and organs. The approximate mean blood pressures in mmHg are indicated for different segments of the circulatory system. Areas with highly oxygenated blood are intensely colored. Those with deoxygenated blood are lighter in color. From *The Human Cardiovascular System, Facts and Concepts,* by J. T. Shepherd and P. M. Vanhoutte. Copyright © 1979, Raven Press, New York.

be hard or soft, depending on blood pressure within the vessel. Pulse rate may be *rapid* or *slow,* according to heart rate. Heart and pulse rates usually match. The pulse is rapid when the heart rate is high (*tachycardia*) and lower when the heart rate is slow (*bradycardia*). The normal pulse and heart may attain rapid rates during strenuous exercise (Fig. 13-16 and Figs. 15-10 and 16-1). However, when the atria beat very rapidly and weakly, as in atrial flutter, the ventricular beats occur so closely together that the ventricles do not fill (the stroke volume is decreased). As a result the pulse is so weak that it cannot be felt although the heartbeat can be heard with a stethoscope. For example, with atrial flutter at 240 beats per minute the pulse rate may only be 90 per minute and a *pulse deficit* of 150 exists (240 − 90 = 150). The normal pulse deficit is zero.

In addition to the force exerted by the heart, several other agents affect the circulation of body fluids. They include such factors as nervous reflex control of the diameter of the lumen of blood vessels, the use of alternative routes of circulation, the presence of valves to produce one-way flow, the movement of fluid from the capillaries to interstitial spaces, and its entrance into lymph vessels for return to the circulatory system.

Reflex Control of Blood Vessels

The rate of heartbeat and blood flow through the circulatory system is regulated directly by the autonomic nervous system and by hormones. Impulses from sensory receptors sensitive to pressure changes and to blood chemicals initiate alterations in the diameter of blood vessels (Table 8-1 and Fig. 15-7). The changes then may be mediated by autonomic nerves or by local *autoregulatory mechanisms* in which chemical changes in the blood vessels initiate the responses (Chaps. 16 and 21). The contraction of circularly arranged smooth muscle narrows the vessels in the response of *vasoconstriction.* Inhibition of smooth muscle relaxes these vessels in the response of *vasodilation* (vasodilatation).

Impulses from baroreceptors and chemoreceptors are carried to the cardiac and vasomotor centers in the medulla to initiate autonomic reflexes which bring about changes in the rate and force of the heartbeat, as well as in the size of peripheral arterioles. Together these autonomic reflexes regulate arterial blood pressure and the flow rate of blood. The flow of blood through a particular vascular bed is also influenced by local autoregulatory responses mentioned above, due to activity of the tissues or organs it supplies. Local responses to chemical and physical stimuli in a specific vascular bed cause reflex dilation or constriction of arterioles and usually maintain a flow rate in proportion to the needs of particular tissues and organs in the face of changes in blood pressure.

Alternative Routes of Circulation

Alternative circulatory routes may supply tissues with blood. Should one route of vessels to a tissue be closed, another route normally provides adequate blood flow to the particular site. These routes are designated as *collateral circulatory routes.* Arteries without collateral circulation, such as the coronary arteries and segments of the middle cerebral artery, are referred to as *end arteries.* Blockade of vessels to tissues without collateral circulation causes tissue death (necrosis).

Direct arteriovenous anastomoses connect arteries to veins with no intervening capillary beds and allow increased blood flow through the area. Arteriovenous anastomoses are important in the regulation of body temperature and are found in the distal parts of the limbs (in the fingers and toes) as well as in the penis.

Flow of Blood through Capillaries

The term *capillary* was originally ascribed to the microscopic blood vessels of the body because they resemble thin hairs or *capilli* (Fig. 17-4). On the average they are about 8 μm wide internally and hold about 2300 blood cells. (However, they range from 5 to 10 μm in width (Fig. 17-4) and 0.3 to 1.0 mm in length, depending on the activity of the particular vessel.) Billions of capillaries, called *capillary beds,* form an extensive network in tissues. The density of beds varies with the physiology of the tissue. Some reports indicate a range of capillary density from 7000 to 12,500 cm^2 per 100 g of tissue. If all the capillaries from one individual were stretched end to end, they would measure nearly 60,000 mi. Their surface area has been estimated to be about 175 m^2, which is nearly 100 times the total surface area of the body. This immense group of tiny vessels allows blood to come into very close approximation to the cells and tissues with which exchange of nutrients, gases, wastes, and other substances must occur.

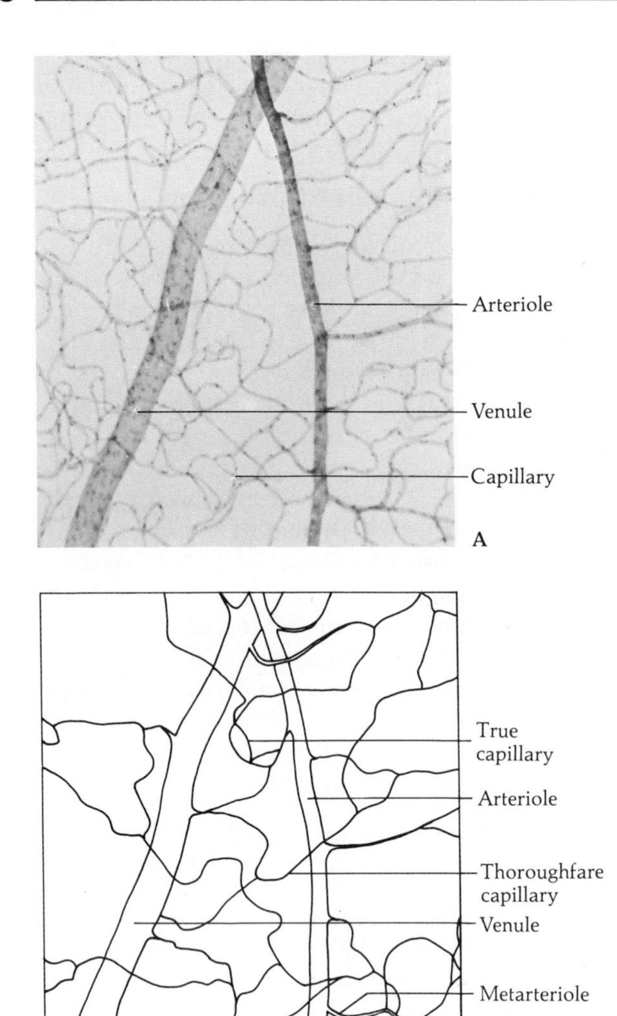

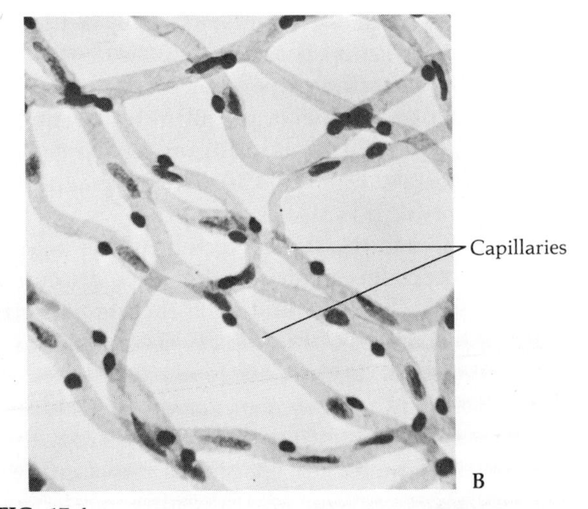

FIG. 17-4
Longitudinal section of a capillary bed. (A) A photomicrograph of the capillary bed found between arterioles and venules in the retina of the eye. The nucleus of each endothelial cell in the wall of the capillary bed appears as a dark, oval to elliptical structure (×40). (B) Section of capillary bed at higher magnification (×250). (C) A diagram of Fig. A showing the relationships between arterioles, metarterioles, capillaries, and venules.

Metarterioles lead into two types of capillaries, *thoroughfare capillaries*, which connect arterioles directly to venules, and side branches of metarterioles, called *true capillaries* (Fig. 17-4). At the point of origin of true capillaries, metarterioles have circularly arranged smooth muscle that forms *precapillary* sphincters. By means of autoregulation these sphincters open and close to control blood flow through true capillaries according to a tissue's activity and needs.

The inner walls of capillaries are formed by a circular arrangement of endothelial cells (Fig. 17-5). They are supported by a basement membrane and connective tissue cells (pericytes), which constitute the middle and outer layers that anchor the capillaries to nearby tissues (Fig. 17-5). The thickness and nature of the basement membrane varies according to the functions of the area serviced by capillaries. Variations in fibrous proteins alter the porosity of capillary pores, the structure of the intercellular cement substance, and the permeability of the capillaries. Various endothelial cells are interconnected by three types of intercellular

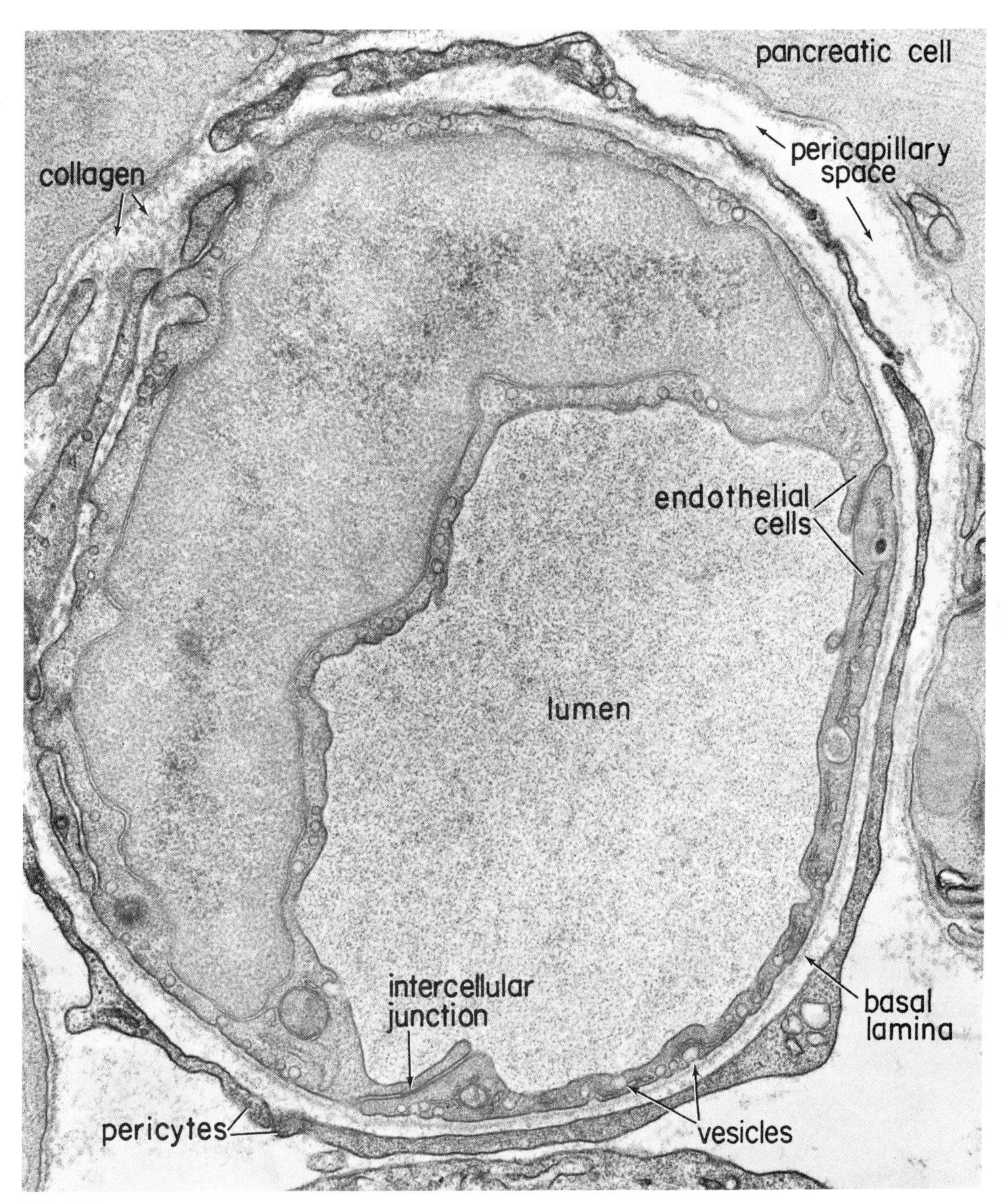

FIG. 17-5
Electron micrograph of cross sections of endothelial cells forming a capillary in the pancreas (×40,000). A tight intercellular junction is demonstrated between the endothelial cells. The basement membrane or basal lamina is located between the cells and the surrounding connective tissue cells, the pericytes. The pericapillary or extracellular space outside of the pericytes is evident in the upper right corner of the figure. From *Histology*, 4th ed., by L. Weiss and R. O. Greep. Copyright © 1977, McGraw-Hill Book Company. Used with permission of McGraw-Hill Book Company.

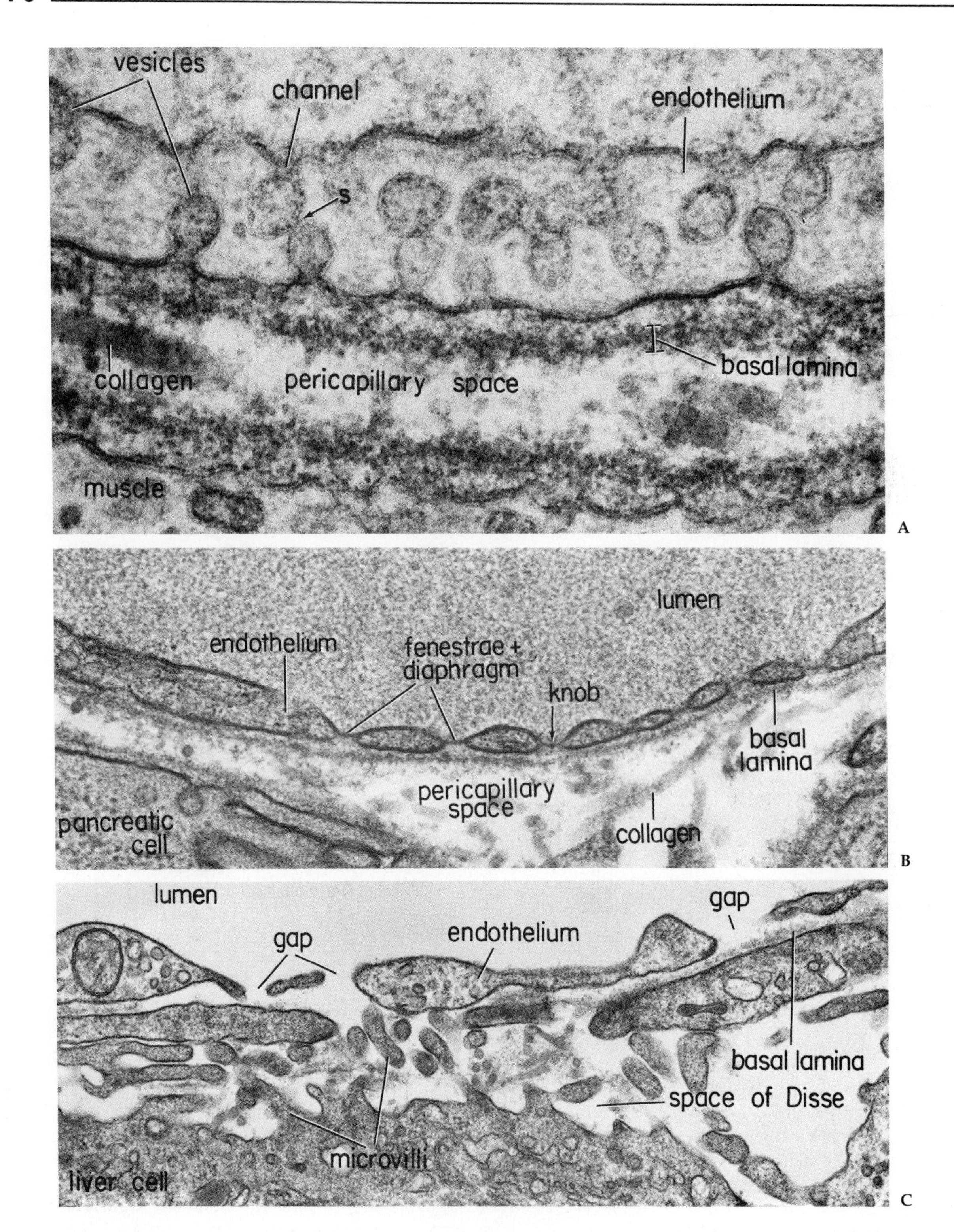

vesicles
channel
endothelium
s
collagen
pericapillary space
basal lamina
muscle
A
lumen
endothelium
fenestrae + diaphragm
knob
basal lamina
pericapillary space
pancreatic cell
collagen
B
lumen
gap
endothelium
gap
basal lamina
space of Disse
microvilli
liver cell
C

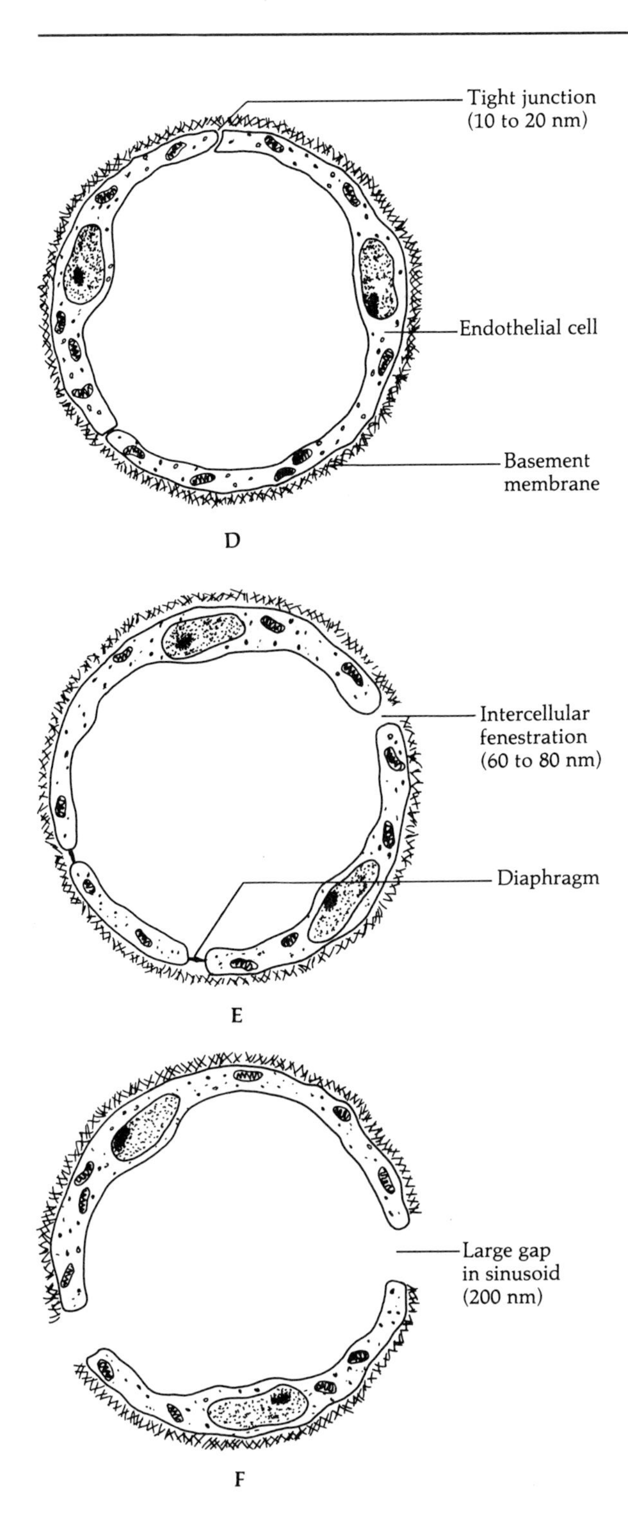

FIG. 17-6
Types of intercellular junctions between capillaries. The electron micrographs on the left are diagrammatically illustrated as gross cross-sections of capillaries on the right. Continuous capillaries, A and D. The basement membrane or basal lamina is apparent in the figures (A × 120,000). Tight junctions with channels about 10 to 20 nm in width and narrower segments labeled s, are evident. Fenestrated capillaries, B and E. The large circular spaces 60–80 nm in width are often bridged by a diaphragm which possesses a central knob (B × 70,000). Sinusoids (discontinuous capillaries, C and F). Large gaps hundreds of nanometers wide are found between endothelial cells of this type of capillary which allows an extensive exchange of components including plasma proteins. The spaces around the sinusoids in the liver are called the spaces of Disse (C × 30,000). From *Histology*, 4th ed., by L. Weiss and R. O. Greep. Copyright © 1977, McGraw-Hill Book Company. Used with permission of McGraw-Hill Book Company.

junctions, which form capillaries that are continuous, fenestrated, or sinusoidal (discontinuous capillaries, Fig. 17-6).

Continuous capillaries have tight junctions only 10 to 20 nm wide, through which few substances can pass. They are found in muscle, connective tissue, and the blood-brain barrier (p. 145). *Fenestrated* capillaries have large circular openings (fenestrations) about 60 to 80 nm wide (Fig. 17-6B). Those located in the gastrointestinal and endocrine systems are bridged by a thin diaphragm of unknown chemistry and porosity. Fenestrated capillaries in the kidney which lack a diaphragm are called *glomerular capillaries* and are highly penetrable (Chap. 21). The third kind of intercellular capillary junction, the *sinusoid,* lacks a continuous basement membrane and has large gaps that are up to hundreds of nanometers in width. There is free movement of large molecules such as plasma proteins and even of some cells between the blood and the tissues that these capillaries supply. Sinusoids are found in the liver, spleen, and bone marrow.

Small molecules may move through all three of these intercellular endothelial junctions and through temporary channels or pores formed by contractile protein in the endothelial cell membranes (Fig. 4-10). The modes of molecular and ionic movement across capillaries are identical to those observed in other cell types (Chap.

4). It would be helpful to the student to review these concepts at this time.

The Exchange of Materials between Blood, Interstitial Fluid, and Lymph

Interstitial fluid is the liquid between and surrounding cells. It is also called *intercellular fluid.* Some of it enters lymph capillaries and becomes *lymph.* Together, interstitial fluid and lymph represent an extracellular fluid (ECF) compartment that accounts for about 20 percent of body water (Fig. 4-1). Diffusion and bulk flow (hydraulic flow) are mechanisms which cause the formation of lymph. Diffusion is rapid, while bulk flow is slower and depends on the differences between hydrostatic and osmotic pressure on the two sides of the capillary wall which force fluid and dissolved solutes across the wall. Lymph formed as a result of these forces is similar in composition to blood plasma except that lymph contains less protein.

Hydrostatic and osmotic forces are the major influences of fluid transport across capillary walls as described in *Starling's equilibrium of capillary exchange.* According to this principle, the difference between the amount of fluid leaving the arteriolar end of a capillary and that reabsorbed in the venous end of a capillary is the *net filtration rate.* Hydrostatic pressure is the blood pressure in the vessels that creates a lateral force that pushes fluid through intercellular gaps between the endothelial cells of a capillary. Molecules with diameters of 9.0 nm or less readily pass through the pores of endothelial cells or through their endothelial intercellular spaces. Thus, water molecules pass easily into the interstitial spaces. Large molecules, such as plasma proteins, generally do not. The loss of water leads to an increase in the osmotic pressure of blood as it flows from the arterial end to the venous end of a capillary (Fig. 17-7). Osmotic pressure determines the tendency of water to move into a solution (Chap. 4). The osmotic force is produced mainly by nondiffusible blood proteins such as albumin, which is the most abundant of them. These proteins were originally mistaken for suspended particles that behaved as colloids, and as a result the pressure was referred to as colloidal osmotic pressure (oncotic pressure).

The hydrostatic and osmotic pressures of blood seem to be more important determinants of fluid exchange than are the lower hydrostatic and osmotic pressures of interstitial fluid. The hydrostatic pressure at the arteriolar end of a capillary exceeds the osmotic pressure, but at the venous end the reverse is true (Fig. 17-7). Therefore, hydrostatic pressure is the dominant force that causes filtration of fluid *out* of arterial capillaries (and moves blood forward), and osmotic pressure is the force that promotes some *reentry* (reabsorption) of fluid into venous capillaries. Since it requires about 1 to 2.5 s for blood to travel from one end of a capillary to the other, enough time is provided for exchange of nutrients and wastes between capillaries and the tissues which they service. Fluid not immediately returned to circulatory capillaries may be taken up by lymphatic vessels as *lymph.* Lymph formation averages 1.5 to 1.7 mL/h per kilogram of body weight and about 2 to 4 L/day. This volume nearly equals that of the blood plasma and represents about 0.1 to 1.0 percent of the fluid entering the capillary beds.

Hydrostatic pressure varies in capillaries of different organs. It is high in the kidneys and causes a greater than average filtration rate from their fenestrated capillaries, the *glomeruli.* The low hydrostatic pressure in the continuous capillaries of the lungs normally prevents the accumulation of fluid (pulmonary edema) and the buildup of pressure on the alveoli (air sacs) in the lungs by the interstitial fluid. As a result, the alveoli may function optimally for gas exchange.

Return of Interstitial Fluid to the Circulatory System

Lymph capillaries are constructed of endothelial cells with fenestrations that lack tight intercellular junctions and possess little or no basement membrane. As a result, they are more permeable than blood capillaries and accommodate passage of particles up to 25 μm in diameter. Interstitial fluid and dissolved molecules readily enter the lymph vessels and flow toward

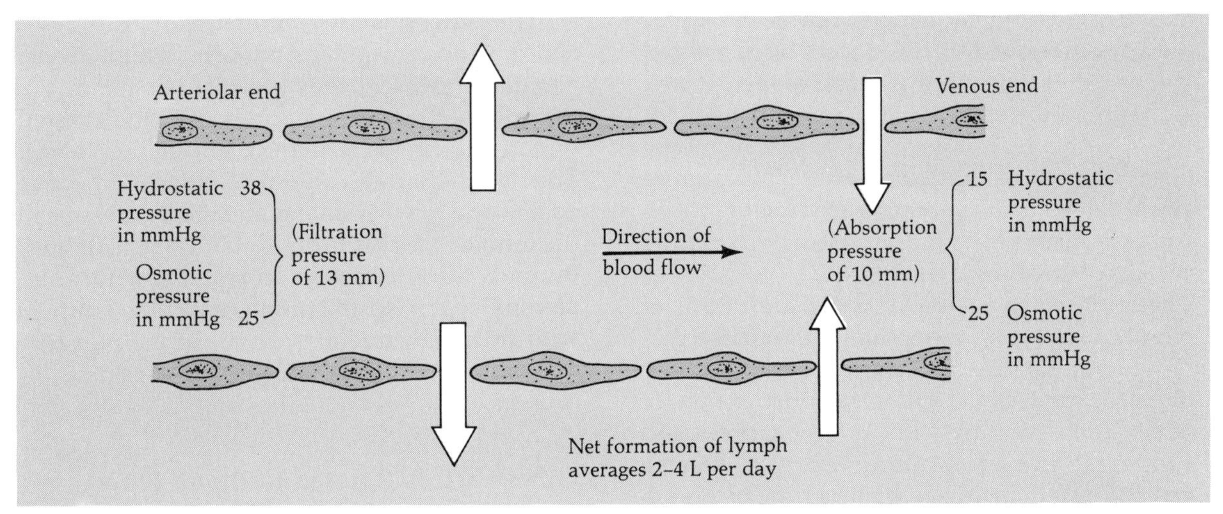

FIG. 17-7
Example of the major forces involved in the net movement of fluid across walls of a capillary. The arrows indicate the direction of movement of fluid.

the heart as directed by numerous one-way valves (Fig. 17-8).

Lymph flows in vessels passing through the reticular tissue of the lymphatic organs, which filters the lymph (Fig. 3-21, p. 65). The responses of lymphocytes and phagocytic cells in the reticular tissue to substances in the lymph may involve humoral and cellular defenses (Chaps. 3 and 19).

Lymph returns to the circulatory system by flowing into veins. The major routes of reentry are through the right lymphatic duct and the thoracic duct, which merge with the brachiocephalic veins (innominate veins); these veins empty their contents into the superior vena cava, which delivers blood into the right atrium. The movement of lymph into the circulatory system tends to maintain blood at a constant volume, replacing fluid lost from the capillaries. Valves at the junctions of the lymphatic and venous systems prevent the entry of blood into the lymph system.

Mechanisms of Lymph Movement

The flow of lymph increases as interstitial fluid pressure becomes greater and is markedly elevated by the *lymphatic pump*. The valves in lymph vessels permit only one-way flow, and a pumping action is produced by mechanical compression of the vessels. The contractions of nearby skeletal muscles and, to a lesser extent, the contraction of smooth muscle in blood vessels and the compression of tissues increase the flow of lymph. Active contraction of smooth

FIG. 17-8
Photomicrograph of longitudinal structure of a lymph vessel and a valve (× 170). Note that the endothelial cells are modified internally to form a unidirectional valve looking like an upside-down V in this photomicrograph. Fluid would normally move upward through this valve.

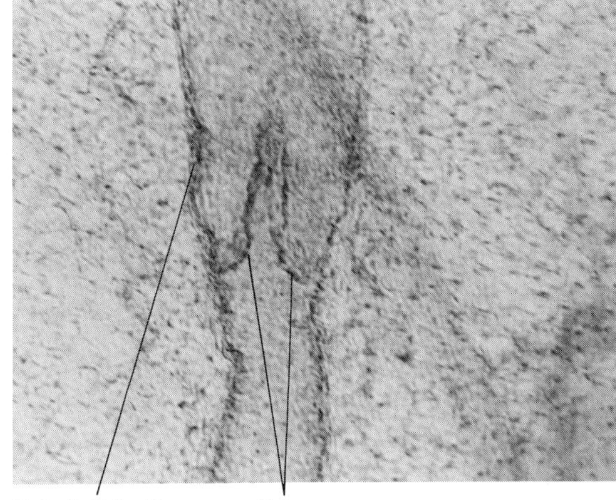

muscle in the walls of lymph vessels and negative pressure (suction) produced by the rapid flow of blood through the venae cavae, as well as changes in pressure that occur with ventilation of the lungs in respiration, assist the movement of lymph. A "milking" effect that propels lymph can be artificially produced by the recoil of elastic stockings as individuals who wear the stockings move their legs.

Factors which block or slow the flow of lymph cause *edema*, the accumulation of fluid in the interstitial (intercellular) spaces. For example, edema can result from obstruction of veins, from standing still for a long time, and from congestive heart failure (Chap. 16). Edema also can occur owing to obstruction of lymph vessels by a filarial parasite (a nematode, or roundworm) which produces a disease called *elephantiasis*. The edema stretches tissues to many times their normal size, especially in the legs and scrotum.

Regulation of Blood Flow through Veins

Blood collected by the venous capillaries flows into small venules, which converge into the larger veins of the body. The elasticity of large veins and the ability of the smooth muscle in their walls to relax allows them to act as blood reservoirs. Veins when nearly empty may increase their capacity to hold blood about sixfold, with a rise in pressure of only 1 mmHg. At any given moment the veins of a resting individual contain about 85 percent of the body's blood. The flow of blood through them is enhanced by one-way valves, by active contraction of smooth muscle in their walls, and by passive movement caused by the contraction of adjacent skeletal muscles, as described for lymph flow above. Increased pressure caused by deep abdominal breathing and the negative pressure produced by a deep breath also produce passive movement of blood in veins of the abdominal and thoracic cavities.

Gravity increases pressure on the walls of vessels in proportion to their distance below the heart. When in a vertical position, the long veins of the legs support a lengthy column of blood, which exerts great pressure. If the blood remains stagnant in the veins owing to a lack of activity, the increased pressure in the associated venous capillaries causes lymph to accumulate, which may result in edema in the ankles. Over long periods of time repeated buildup of pressure may wear out the valves. The uninterrupted column of blood may cause edema and may twist and enlarge the vessels in the form of *varicose veins* (Table 17-5). This condition is fairly common in persons with occupations that require much standing coupled with little movement, such as security guards or salespersons stationed behind a register or counter.

Other Factors Related to Blood Flow

Physical Factors: Bernoulli's Principle; Gravity; Poiseuille's Law

Although the internal diameter of individual vessels becomes smaller as arteries divide into arterioles and as these separate into capillaries, the number of vessels multiplies so much that the total cross-sectional area of the vessels increases. The reverse occurs as capillaries merge to form venules, which in turn join to form veins (Fig. 17-9).

These changes are related to physical principles which govern the movement of fluids in tubes. *Bernoulli's principle* dictates that as a blood vessel divides, the cross-sectional area of its vessel bed (its subdivisions) should increase, the rate of flow (velocity) should decrease and the pressure should increase. That is, as the total cross-sectional area of a vascular bed increases, the total lateral pressure on the walls of all its vessels should also increase while the velocity of blood flow in the individual vessels decreases.

Although the rate of flow does in fact decrease as blood vessels divide, the same volume of blood moves through the larger area in the same time. The flow rate in the individual vessels is reduced, but the total volume moved in a given period is identical. This is much like the changes in flow which occur as a riverbed widens.

According to Bernoulli's principle the total energy of blood in motion in a vessel is the sum of the energy of flow (which is related directly to the velocity of flow) plus that of pressure and

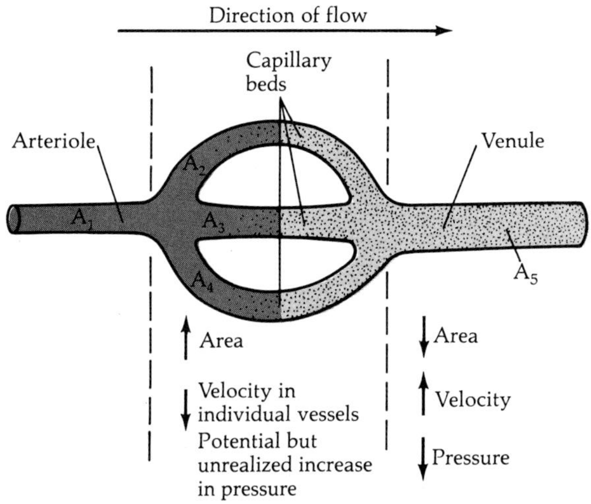

FIG. 17-9
Comparison of cross-sectional areas, velocity of flow, and pressure in arterioles to capillaries, and capillaries to venules. The cross-sectional area: A_1 is less than the sum of A_2, A_3, and A_4; A_5 is less than the sum of A_2, A_3, and A_4; A_1 is less than A_5. ↑ = increase; ↓ = decrease.

gravity. Therefore, if the total energy remains the same, as the velocity declines, the pressure should rise. However, the narrower inner diameter of the subdivisions of arteries offer more resistance to flow (more frictional forces), which generates a loss of energy as heat. Furthermore, arteries subdivide so extensively in most cases that the large increase in total surface area of the vessels offsets potential increases in blood pressure. As a result, the pressure in each of the subdivisions of an artery does not increase but rather decreases (Fig. 17-10C).

The flow rate declines further when blood moves from arteries to capillaries (Fig. 17-10B). This relationship affords an extremely slow rate of blood flow averaging about 0.07 cm/s through capillaries (Table 17-2) which usually provides optimal time for exchange of substances between the blood and tissues.

As capillaries join to form venules and these merge to form veins, the velocity of flow increases, flow becomes more turbulent, and pressure decreases. This pattern is analogous to that observed when several streams join to form a river whose bed is narrower than that of all of the streams. Because veins are so distensible (capable of dilation) pressure continues to drop as blood flows in them back toward the heart. Some of the energy of flow is released as heat due to friction caused by the movement of blood against the vessel walls.

Contraction and relaxation of the heart in the cardiac cycle and the accompanying expansion and recoil of the arteries circulates blood whose

FIG. 17-10
Approximate relationship between cross-sectional area, velocity of blood flow, and pressure in blood vessels. From *The Human Cardiovascular System, Facts and Concepts,* by J. T. Shepherd and P. M. Vanhoutte. Copyright © 1979, Raven Press, New York; and from *Circulatory Physiology—The Essentials,* by J. J. Smith and J. P. Kampine. Copyright © 1980, The Williams and Wilkins Company, Baltimore.

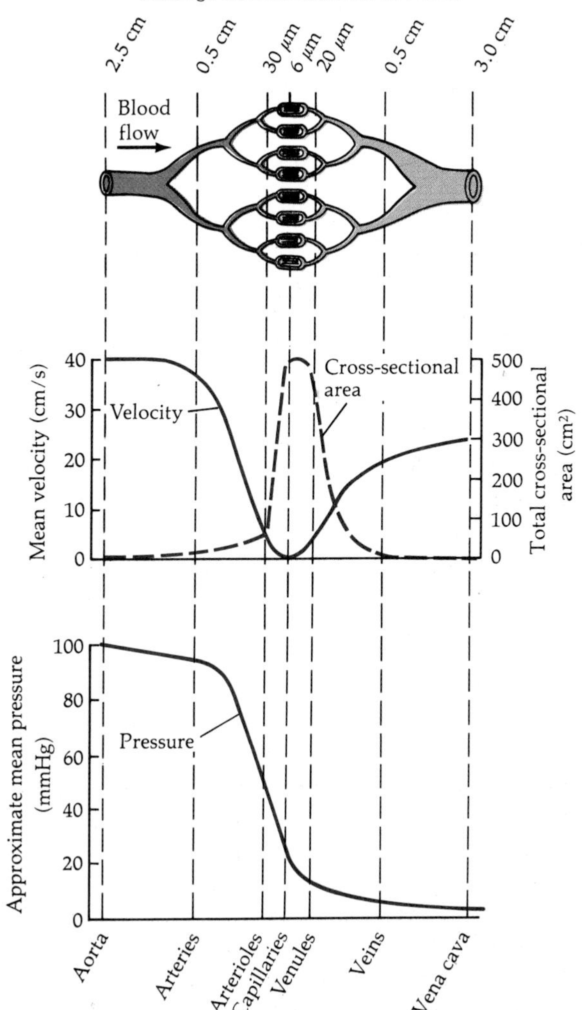

TABLE 17-2
Average velocity of blood in various vessels

Vessel	*Average Internal Diameter, cm*	*Flow Rate, cm/s*
Aorta	2.5	40
Arteries	0.5	37
Capillary beds	0.0006	0.07
Veins	0.5	20
Vena cava	3.0	24

average velocity varies according to the principles described above (Table 17-2, Fig. 17-10). The time it takes for blood to circulate through the body can be estimated by various methods following injection of substances which can be traced through the circulatory system. For example, when bile salts are injected into the arm, their movement through the heart and to the blood vessels of the tongue requires about 15 s before the person notices a bitter taste. It has been estimated by similar techniques that it takes about 23 s, or 27 heartbeats (at a heart rate of 70 beats per minute), to propel blood from the right atrium through the distal parts of the body and back. The velocity of blood varies, averaging as much as 40 cm/s in the aorta and about 0.07 cm/s in open capillaries.

As noted above, gravity also contributes to the total energy in the bloodstream. When a vessel is in a vertical position, the force of gravity elevates the pressure on a column of blood. When a person moves from a horizontal to a vertical position, the pressure in the vessels below the heart becomes greater owing to gravity, and pressure in the leg veins is increased as the blood volume shifts to the lower parts of the body. Fainting may result if stagnation of the blood in the extremities decreases flow to the brain. Low blood pressure enhances this possibility. The opposition of gravity to blood flow diminishes when a person faints and assumes a horizontal position. Gravity may cause a decrease or increase in blood pressure of 0.77 mmHg for each centimeter that the vessels are located above or below the heart, respectively.

Changes in the diameter of the blood vessels most markedly affect the rate of flow. The physical principles described in *Poiseuille's law* were established for fluid movement in nondistensible tubes rather than in blood vessels. However, Poiseuille's equation can be used in a qualitative way to analyze the factors that influence blood flow.

$$F = \frac{(P_1 - P_2)\pi r^4}{8Lv} \tag{17-1}$$

where
F = flow rate (volume of fluid circulated in mL/min)
$P_1 - P_2$ = difference in pressure between opposite ends of the vessel
π = 3.1416
r = radius of vessel
L = length of vessel
v = viscosity of fluid

According to this equation the volume of blood flow depends most importantly on the radius of the vessel. Therefore, the ability to constrict or dilate blood vessels is extremely important in the regulation of blood flow and pressure.

Measurement of Blood Pressure

Blood pressure depends on the volume of blood flow, on the resistance the wall of the vessel offers to flow, and on how far the vessel is from the heart. The contraction of the left ventricle of the heart forces blood into the aorta, thus producing pressure on the walls of that vessel which is transferred to the large arteries of the body as the blood flows through them. *Blood pressure* is that force exerted by the blood against the walls of blood vessels.

The pressure exerted when the ventricles contract is *systolic pressure*, which when measured in the upper arm of a young adult, averages nearly 120 mmHg. After elastic fibers in the walls of the vessels have stretched, they recoil to maintain *diastolic pressure* during ventricular disastole. At this time the average diastolic pressure in a young adult is nearly 80 mmHg. *Pulse pressure* is the difference between systolic and disastolic pressure and averages about 40 mmHg. The mean blood pressure is the arithmetic average of systolic and diastolic pressures and is approximately 100 mmHg.

TABLE 17-3
Mean blood pressure and standard deviations in apparently healthy persons 20 to 106 years of age*

Age Group	MALES		FEMALES	
	Systolic†	*Diastolic*†	*Systolic*†	*Diastolic*†
20–24	123 ± 13.7	76 ± 9.9	116 ± 11.8	72 ± 9.7
25–29	125 ± 12.6	78 ± 9.0	117 ± 11.4	74 ± 9.1
30–34	126 ± 13.6	79 ± 9.7	120 ± 14.0	75 ± 10.8
35–39	127 ± 14.2	80 ± 10.4	124 ± 13.9	78 ± 10.0
40–44	129 ± 15.1	81 ± 9.5	127 ± 17.1	80 ± 10.6
45–49	130 ± 16.9	82 ± 10.8	131 ± 19.5	82 ± 11.6
50–54	135 ± 19.2	83 ± 11.3	137 ± 21.3	84 ± 12.4
55–59	138 ± 18.8	84 ± 11.4	139 ± 21.4	84 ± 11.8
60–64	142 ± 21.1	85 ± 12.4	144 ± 22.3	85 ± 13.0
65–69	143 ± 26.0	83 ± 9.9	154 ± 29.0	85 ± 13.8
70–74	145 ± 26.3	82 ± 15.3	159 ± 25.8	85 ± 15.3
75–79	146 ± 21.6	81 ± 12.9	158 ± 26.3	84 ± 13.1
80–84	145 ± 25.6	82 ± 9.9	157 ± 28.0	83 ± 13.1
85–89	145 ± 24.2	79 ± 14.9	154 ± 27.9	82 ± 17.3
90–94	145 ± 23.4	78 ± 12.1	150 ± 23.6	79 ± 12.1
95–106	146 ± 27.5	78 ± 12.7	149 ± 23.5	81 ± 12.5

* The blood pressure was measured in individuals in a sitting position with a sphygmomanometer.
† ±1 Standard deviation.
Source: R. P. Lasser and A. M. Master, *Geriatrics* **14**:345–360 (1959).

In general, systolic and diastolic blood pressures are higher in young men than in young women. However, between the ages of 20 and 74 systolic pressure increases rising by about 20 mmHg in males and 43 mmHg in females, while diastolic pressure is relatively more stable. From about age 45 onward, the average blood pressure of females usually exceeds that of males of the same age. In our society greater variations in blood pressure occur among members of both sexes as they age (Table 17-3). The increasing dispersion of values becomes noticeable among individuals beyond 30 years of age.

The mean blood pressure declines from 100 in the aorta to 55 in arterioles to 30 in capillaries to 4 in the venae cavae and 0 as blood flows into the right atrium (Fig. 17-10). In veins within the thoracic cavity it may even be negative (less than atmospheric pressure) during inspiration as the chest cavity expands.

Blood pressure may be measured with a sphygmomanometer. This instrument consists of an inflatable cuff connected by rubber tubes to a mercury reservoir with a rubber bulb to provide pressure (Fig. 17-11). The bulb contains a valve to regulate air during inflation and deflation of the cuff. The cuff is wrapped around the patient's upper arm near the elbow and held in place by metal buckles or a velcro-type closure. As the cuff is inflated, it squeezes against the brachial artery near the elbow. Air pressure within the cuff is transmitted to the mercury reservoir and the mercury is pushed up a glass tube calibrated in millimeters of mercury.

The *palpatory* (feeling) *method* is the most sensitive and precise noninvasive technique used to measure systolic pressure with a sphygmomanometer. The radial artery, extending along the forearm and approaching the surface on the thumb side near the wrist, is most often used to find the pulse. In adults the radial artery is best felt on the forearm about 1 in from its junction with the wrist, and about 0.5 to 0.75 in from its bony projection on the thumb side of the wrist. To feel the pulse, the index and second finger of the observer are gently placed on the artery. Then to determine blood pressure the cuff is inflated until the radial pulse disappears. Inflation of the cuff is continued until the column of mercury rises about an additional 30 mm. The valve on the sphygmomanometer bulb is slowly opened so that the cuff deflates and the column of mercury drops at a rate of about 2 to 3 mm/s. The height of the column of mercury is recorded the moment the radial pulse reappears and can be felt and is a measure of systolic pressure.

However, the *auscultatory* (listening) *method* is employed most often for blood pressure measurement with a sphygmomanometer. Heart sounds (Korotkoff's sounds) may be heard through a stethoscope as they are transmitted through the blood to the brachial artery. The sphygmomanometer cuff is inflated to about 30 mmHg above the systolic pressure recorded earlier via the palpatory method or above the point at which no heart sounds can be heard. When air is slowly released from the cuff, the pressure within it decreases, blood surges through the artery, and sounds can be heard again. The first heart sound that is heard is

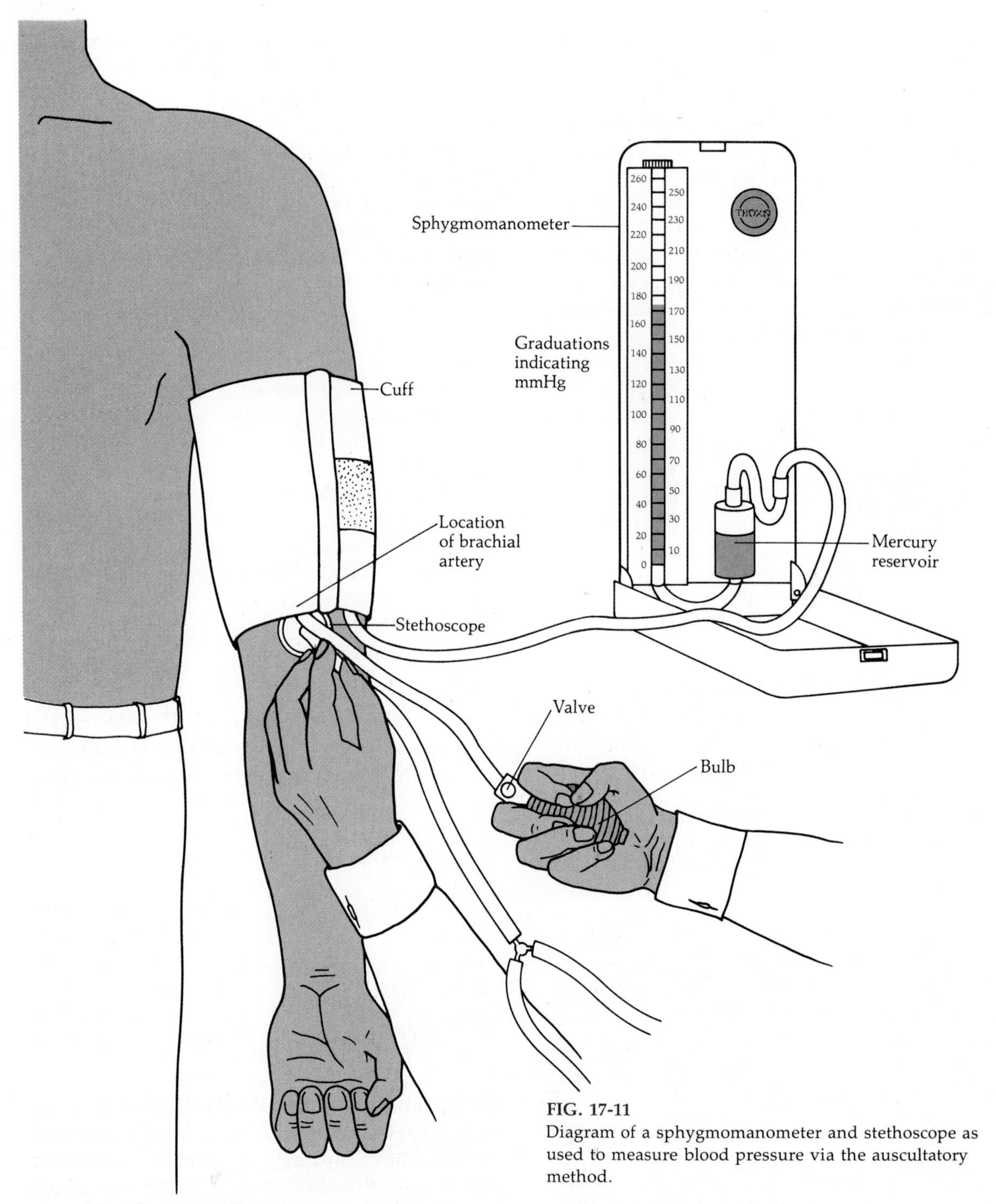

FIG. 17-11
Diagram of a sphygmomanometer and stethoscope as used to measure blood pressure via the auscultatory method.

caused by closure of the atrioventricular valves, turbulence of blood flow from the ventricles, and cardiac vibrations (page 365). The *second sound,* which is softer, is produced by the closure of the aortic and pulmonary valves. *Systolic* pressure is recorded as the height of the column of mercury when the first heart sound is heard. As pressure in the cuff continues to fall, the sounds become louder and then muffled and finally disappear. *Diastolic* pressure is noted as the height of the column of mercury when the sounds become muffled just prior to their disappearance.

A *direct method* of measuring blood pressure involves insertion of a catheter attached to a manometer (instrument for measuring pressures of liquids and gases). This technique is used mainly in laboratory animals, but may be used in humans when direct pressure measurements are critical, as in the diagnosis of severe cardiovascular disorders. The manometer may be filled with sterile isotonic salt solution and connected to the catheter. The catheter is inserted through a syringe needle into a blood vessel and may even be threaded to the heart. Blood pressure is transmitted to the saline in the reservoir, causing the fluid to rise. It is possible to calculate the equivalent blood pressure in millimeters of mercury because the density of the salt solution relative to that of mercury is known.

Regulation of Blood Pressure

Arterial blood pressure is the product of cardiac output and *peripheral resistance* (the opposition to flow in the blood vessels). When either variable is altered, blood pressure may rise or fall. Such adjustments are often made to meet the needs of the body and may cause significant effects on individual organs, on their functions, and on the body as a whole.

THE ROLES OF CARDIAC OUTPUT AND PERIPHERAL RESISTANCE *Cardiac output,* the amount of blood pumped out of each ventricle each minute, has a direct effect on blood pressure. As the volume pumped increases, blood pressure tends to change in the same direction. Factors that influence cardiac output were described earlier (Chaps. 15 and 16).

Peripheral resistance, exerted primarily by the arterioles, may alter blood pressure significantly, as indicated in the following equation:

$$R = \frac{P_1 - P_2}{F} \qquad (17\text{-}2)$$

where R = resistance to flow

$P_1 - P_2$ = difference in pressure between opposite ends of the vessel

F = flow rate

Equation (17-2) indicates that the resistance to blood flow and blood pressure are directly related to each other and inversely related to the flow rate. According to Poiseuille's law, presented above (Eq. 17-1), the flow rate is described by

$$F = \frac{(P_1 - P_2)\pi r^4}{8Lv}$$

If one substitutes this value of F for that of F in Eq. (17-2), one obtains the following:

$$R = \frac{P_1 - P_2}{\frac{(P_1 - P_2)\pi r^4}{8Lv}} = \frac{(P_1 - P_2)8Lv}{(P_1 - P_2)\pi r^4} = \frac{8Lv}{\pi r^4} \qquad (17\text{-}3)$$

The form of Eq. (17-3) on the far right $\left(R = \frac{8Lv}{\pi r^4}\right)$ shows that flow resistance R and blood pressure increase directly in proportion to the length of the vessel L and the viscosity of blood v and decrease in proportion to the fourth power of the vessel's radius r^4. The viscosity of blood is normally 3 to 5 times that of water and is directly related to the number of red blood cells and the quantity of proteins in the plasma. A decrease in red blood cell count results in less viscous blood, and as indicated by Eq. 17-3, is accompanied by decreases in resistance to flow and blood pressure. An increase in red blood cell count causes the opposite responses. However, as noted earlier in this chapter, the most notable effects on resistance to flow and blood pressure are produced in an inverse relationship to changes in the radius of the blood vessels in the body.

REFLEX CONTROL MECHANISMS As mentioned earlier in this chapter, blood vessel diameter and cardiac output are regulated by autonomic reflex control mechanisms which are governed by vasomotor and cardiac centers in the medulla oblongata. These activities are coordinated with those of the respiratory center in the medulla to provide vital organs such as the brain, heart, and kidneys with a continuous

TABLE 17-4
The properties of vasodilator and vasoconstrictor nerves to various blood vessels

	TYPE OF FIBER (AND RECEPTOR)*	
Effect	*Parasympathetic* (γ)	*Sympathetic*
Vasoconstriction:		
Most blood vessels		+ (α)
Vasodilation:		
Arteries in head and external genitals in males	+	
Most other arteries	+	+ (β_2)
Veins		+ (β_2)

* Although alpha (α) and beta (β) receptors are primarily responsive to norepinephrine and epinephrine, respectively, they will react to a lesser extent to the alternative neurotransmitter. Beta receptors in cardiac muscle are termed β_1 while all other beta receptors are called β_2. Receptors responsive to acetylcholine are called gamma (γ) receptors.

supply of oxygenated blood (Chaps. 9, 11, and 20).

The diameter of blood vessels is regulated by neurotransmitter molecules released from nerve fibers and the receptors with which they interact (Table 17-4). *Adrenergic sympathetic* fibers cause *vasoconstriction* of most blood vessels but induce vasodilation in those in which β_2 receptors predominate. *Cholinergic parasympathetic* fibers cause *vasodilation* in arterioles which they innervate.

The adjustments of vessels to these neurotransmitters are analogous to those of a nozzle at the end of a garden hose. As the nozzle is narrowed, greater pressure in the hose results if the water supply is constant. This resembles vasoconstriction of the walls of peripheral vessels, which elevates blood pressure. In some cases *hypertension* (high blood pressure) results. Alternatively, the nozzle may open and the pressure drop. This response resembles vasodilation. It is important to note that the body does not have enough blood to fill all its blood vessels were they to be simultaneously vasodilated. Vasodilation, therefore, reduces systemic pressure and may lead to *hypotension* (low blood pressure). With hypotension more blood remains in peripheral tissues than returns to the heart, and cardiac output decreases. The pooling of blood in peripheral vessels may result from the activity of vasodilator fibers or may occur passively as a result of the inhibition of vasoconstrictor nerves.

An elevation in systemic pressure stimulates *baroreceptors* in the walls of the *aortic arch, carotid sinuses,* and *pulmonary artery* (Table 8-1 and Fig. 15-7). Impulses carried to cardioinhibitory autonomic centers in the medulla oblongata generate reflex responses that inhibit cardiac output and dilate peripheral arterioles, resulting in a drop in systemic blood pressure (Fig. 17-12A). The fall in pressure produces negative feedback signals, which are carried from the baroreceptors back to cardioaccelerator autonomic centers. Then, cardiac output is accelerated, peripheral arterioles are constricted, and an increase in systemic blood pressure results (Fig. 17-12B). In this manner the circulatory system compensates for a sudden increase or decrease in systemic arterial pressure.

The reflexes which cause blood pressure to drop are relayed through nuclei in the vasomotor center in the medulla. Some are transmitted through interneurons to the nuclei of the vagus nerve and the reticular formation. The vagal nucleus forms the cardioinhibitory center, whose impulses diminish cardiac output and cause vasodilation and a drop in blood pressure (Fig. 17-12A). Inhibitory signals carried down the vagas turn off excitatory sympathetic stimuli to the heart; those carried by the reticulospinal tract turn off sympathetic vasoconstriction of arterioles in general. Slowing of the heart, vasodilation of blood vessels, and a passive reduction in blood pressure result. A decrease in arterial pressure causes feedback inhibition of these impulses. Consequently, heart rate increases, arterioles constrict, and blood pressure increases (Fig. 17-12B).

It should also be noted that when blood pools in veins, venous pressure rises, and venous baroreceptors in the terminal portions of the venae cavae and the adjacent walls of the right atrium generate cardioaccelerator impulses and depress cardioinhibitory ones. The subsequent reflex acceleration of heart rate was described earlier on page 370 and illustrated in Fig. 15-7.

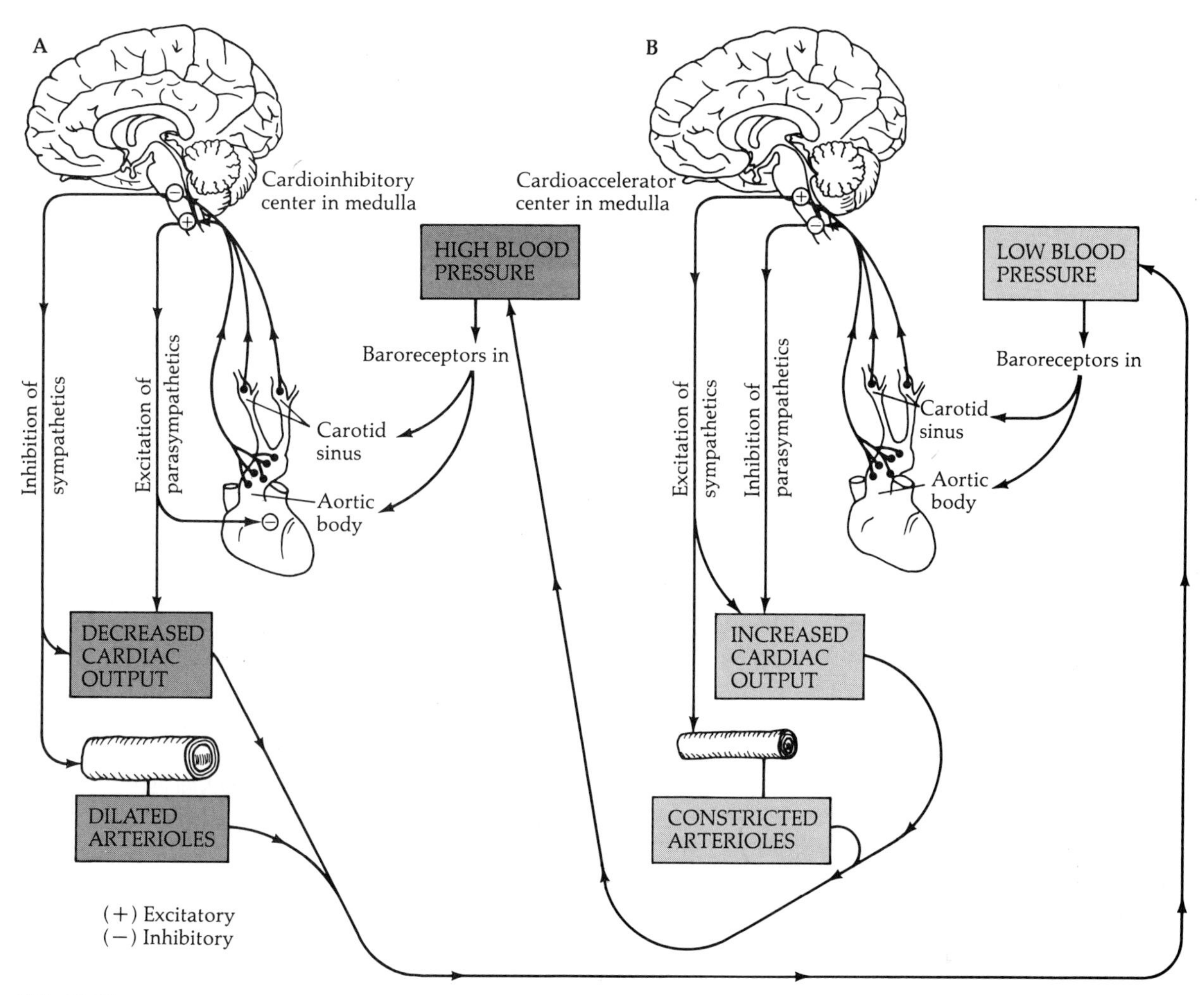

FIG. 17-12
Reflex regulation of blood pressure. (A) High blood pressure stimulates mechanisms that lower the blood pressure. (B) Low blood pressure stimulates mechanisms that raise blood pressure.

Similar pulmonary vessel baroreceptors respond to elevations in pulmonary pressure between the left atrium and pulmonary veins. The accelerated heart rate prevents blood from pooling in the pulmonary veins. This response enhances gas exchange in the lungs and protects them from a buildup of blood pressure and the adverse effects of pulmonary edema.

Interneurons in the central nervous system (CNS) from the cerebral cortex, limbic system, hypothalamus, cerebellum, and medulla may affect each other. Thus thoughts, sights, sounds, emotions, changes in temperature, and posture can also influence blood pressure. For example, the emotion of embarrassment causes vasodilation of the vessels in the skin of the face, producing a surge of blood which reddens the skin in the reaction of blushing.

17-3 DISORDERS RELATED TO THE WALLS OF BLOOD VESSELS

Although a number of disorders occur in blood vessels (Table 17-5), two recognized as the major causes of cardiovascular disease are hypertension and atherosclerosis. Their general relationship to coronary artery disease and myocardial infarction was noted in Chap. 16.

TABLE 17-5
Disorders in the walls of blood vessels

Disorder	*Symptoms*
Hypertension	Blood pressure exceeding 160/95 · Borderline cases exceed 140/90
Arteriosclerosis	A variety of arterial disorders involving degeneration of the arterial wall, proliferation of cells, and in some cases loss of resiliency due to increased deposits of calcium salts; causes susceptibility to hemorrhage when blood pressure increases with stress or exercise
Atherosclerosis	A specialized type of arteriosclerosis with accumulation of fats in the inner wall of an artery, coupled with narrowing of the vessel due to smooth muscle proliferation; increased susceptibility to clotting and hemorrhage in the vessel
Headaches	Pain, which may be due to alterations, such as spasms of smooth muscle, in arterial vessels in brain; increased pressure in brain and skull (increased intracranial pressure) and constriction causing decreased blood flow (ischemia) may occur
Aneurysm	Weakening of walls and formation of outward bulge, which may burst with increased blood pressure, leading to paralysis or death
Thrombosis	Formation of a clot (thrombus) in a vessel. It becomes an embolus when it leaves the vessel wall and floats freely in the circulation. Coronary thrombosis occurs in the coronary arteries of the heart, cerebral thrombosis may occur in the middle cerebral artery of the brain; blockade of vessels in the brain may cause a cerebrovascular accident, or stroke
Embolism	The physical blockade of flow in a blood vessel by an undissolved mass or embolus. Any bit of matter foreign to the blood which may obstruct blood flow is an embolus. It can be a blood clot (thrombus), tumor, air, tissue cells, clumps of bacteria, or other foreign bodies
Hemorrhage	Loss of blood due to rupture of vessel wall; bleeding may be internal or external
Phlebitis	Inflammation of tissues in wall of veins; may lead to formation of a thrombus
Varicose veins	Twisting, widening, and stretching of veins due to pressure of long columns of blood; weakened walls and valves allow blood to pool in lower segments of the vessels; increased pressure on the vessel produces structural changes; increased pressure in abdominal cavity during pregnancy often induces these changes; thrombosis, hemorrhage, and phlebitis may accompany serious cases of varicose veins

However, since cardiovascular disease is responsible for over 900,000 deaths annually in the United States, it seems appropriate to discuss hypertension and atherosclerosis in some detail now that the physiology of the heart and blood vessels has been described.

Hypertension

Although blood pressure generally increases with age (Table 17-3), *hypertension* is defined as a systolic pressure above 160 mmHg or a diastolic pressure above 95 mmHg. People whose respective values slightly exceed 140 and 90 mmHg are referred to as *borderline* hypertensives. Some causes of hypertension are cardiovascular, nervous, hormonal, and renal (kidney-related) in origin.

Cardiovascular Effects

Blood pressure is elevated by an increase in cardiac output and/or peripheral resistance. Higher levels of hormones (norepinephrine, epinephrine, thyroxine), nerve stimuli (from sympathetic fibers), and exercise increase cardiac output. A rise in peripheral resistance often results from a loss of elasticity in the walls of arteries and is associated with atherosclerosis, a form of arteriosclerosis (Table 17-5).

Nervous Factors

Alterations in vasomotor activity in the medulla may accelerate the activity of sympathetic neurons (Fig. 17-12B). As described earlier, the elevated activity may produce an increase in car-

TABLE 17-6
Some hormones which elevate blood pressure*

Endocrine	*Hormone*	*Possible Stimuli*	*Responses*
Adrenal Medulla	Norepinephrine	Stress	Vasoconstriction; increased cardiac output
	Epinephrine	Stress	Increased cardiac output
		Tumor of adrenal medulla, causing pheochromocytoma	Vasoconstriction; increased cardiac output
Cortex		Tumor of adrenal cortex, causing Conn's syndrome	Retention of salts; fluid retention; increased blood volume
	Glucocorticoids	Stress; overactivity of adrenal cortex	Fluid retention; increased blood volume
	Aldosterone	Stress (the effect on aldosterone release is not as great as on glucocorticoid release)	Retention of salts; fluid retention; increased blood volume
Thyroid	Thyroxine	Overactive thyroid and/or cold exposure	Elevated metabolism; increased cardiac output
Kidney	Angiotensin II	Low blood pressure; damage or disease in kidney; chronic dietary salt deprivation	Renin release from the kidney causes angiotensin II formation, which stimulates the secretion of aldosterone and constriction of blood vessels; increases the force of the heartbeat; angiotensin II may also stimulate thirst and drinking to elevate blood volume

* The overall characteristics of these hormones were noted in Table 5-5.

diac output and peripheral vasoconstriction, leading to higher blood pressure.

Hormones

Many hormones markedly elevate blood pressure (Table 17-6). Prominent among them are thyroxine, which stimulates metabolism in general; the adrenal hormones such as norepinephrine and epinephrine, whose secretion increases with stress; and aldosterone and the glucocorticoids, which promote sodium and water retention. When blood pressure drops or the kidneys become damaged or diseased, the kidneys also form *renin,* a molecule with enzyme activity. Renin promotes the formation of the most potent vasoconstrictor substance in the body, *angiotensin II* (Chap. 21). Angiotensin II causes vasoconstriction, an increse in aldosterone output, an increase in cardiac output, and a rise in blood pressure.

Essential Hypertension

In 9 of 10 cases, the cause of hypertension is not readily apparent and the condition is called *essential hypertension.* Patients may have low, moderate, or elevated blood renin levels with no apparent kidney damage. On the other hand, essential hypertension may progress to a state which may lead to kidney malfunction, atherosclerosis, heart disease, or stroke.

Therapy for Hypertension

In the United States less than one-third of those with high blood pressure adequately control it by diet or drugs. Weight loss reduces the mass of body tissue that must be supplied with blood, and thereby may decrease cardiac output and blood pressure. The consumption of less salt also may lower cardiac output and blood pressure by causing a decrease in the ECF volume. A number of drugs may be used to control high blood pressure (Table 17-7). Some act as *diuretics,* which promote salt and water loss from the kidney so that blood volume, blood pressure, and cardiac output diminish. Other drugs lower blood pressure by affecting the CNS or peripheral nervous system, or by inhibiting vasoconstriction or the synthesis and activity of angiotensin II. Individual drugs or

TABLE 17-7
Activities of some antihypertensive drugs

Activity	*Drug*	*Mechanism of Action*
Diuretic action	Thiazides β enzothiadiazides	Promote salt and water loss
	Spironolactone	Antagonizes aldosterone
Alteration of sympathetic activity in the CNS	Methyldopa	Depletes catecholamines Metabolic products of methyldopa act as antagonists of receptors in the CNS and may inhibit sympathetic outflow
Inhibit sympathetic nerves to promote vasodilation	Propranolol	Blocks β_1 and β_2 receptors
	Reserpine	Depletes catecholamine stores
	Guanethidine	Depletes norepinephrine stores and impairs its release from α-adrenergic endings
	Hydralazine Minoxidil	Direct effect, relaxing arteriolar smooth muscle
Block angiotensin II	Captopril	Blocks converting enzyme needed for the synthesis of angiotensin II
	Saralasin	Competes with angiotensin II receptors

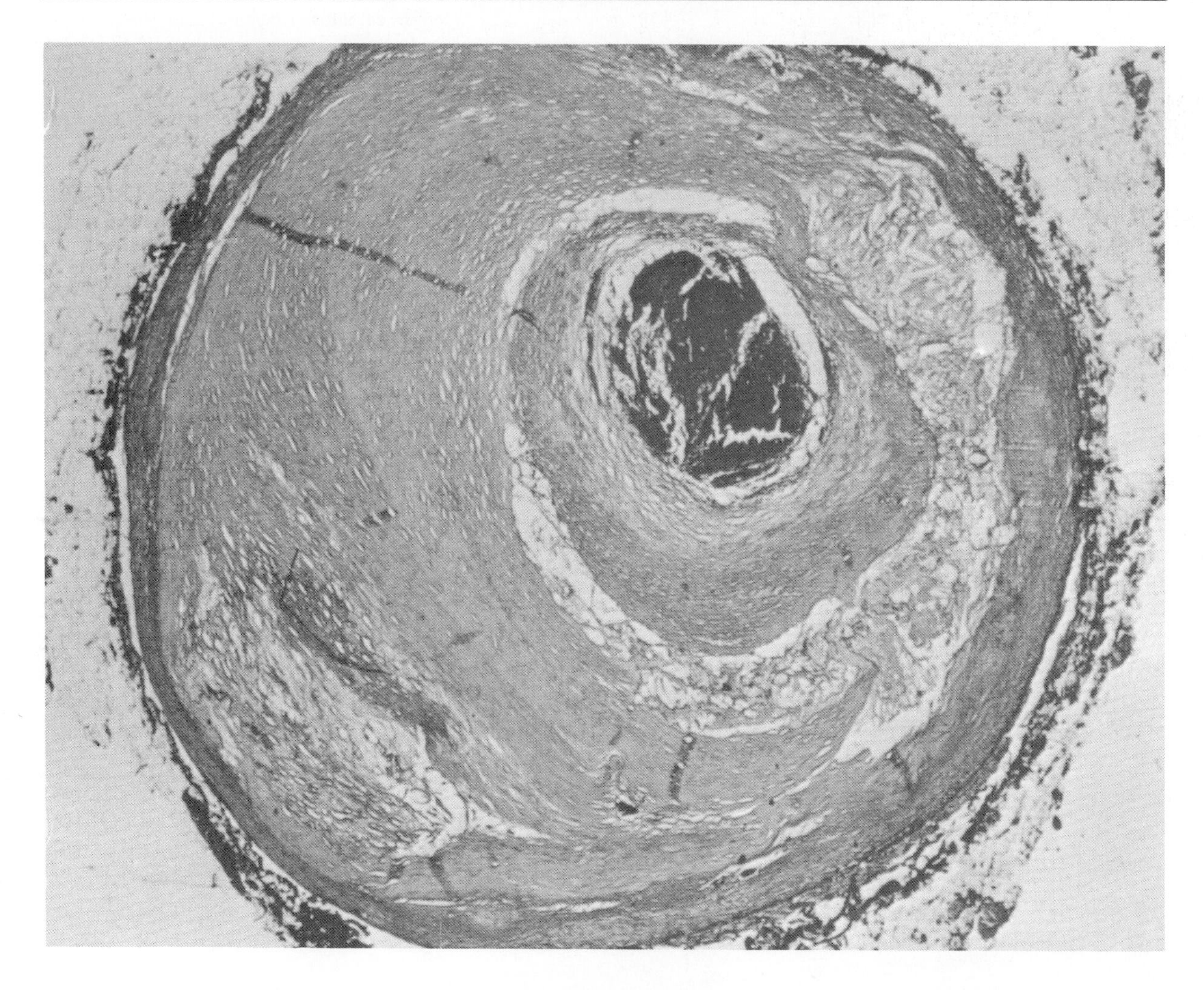

combinations of them may be used to control hypertension, depending on the needs of the patient.

Atherosclerosis

Atherosclerosis is a specialized form of arteriosclerosis in which arterial wall degeneration occurs as fats accumulate in abnormal, flat patches or *plaques* in the inner lining of arteries. Smooth muscle multiplies in the middle coat and narrows the vessels' opening (Fig. 17-13). Blood flow decreases and the vessels become more susceptible to blockade by a clot (thrombosis) and to hemorrhage (bleeding).

Two hypotheses that have been formulated to explain the origins of atherosclerosis are the monoclonal theory and the response to injury theory. The *monoclonal theory* proposes that the accumulation of materials on the inner lining of arteries is related to cellular mutation and involves the formation of benign tumors of smooth muscle. These tumors are benign only in the sense that they do not spread from organ to organ as do malignant tumors. However, they may spread gradually through the arteries of the body. The tumors are thought to arise from a single mutated smooth muscle cell, which forms a clone of genetically identical cells. According to this theory, a stimulus provokes clonal smooth muscle cell division, which narrows the opening, and the accompanying cellular metabolism then forms deposits of fats and proteins in the lining, resulting in plaque formation and atherosclerotic disease.

This theory further proposes that the presence of fats in the bloodstream stimulates the proliferation of smooth muscle. Numerous agents, including risk factors for coronary artery disease (as discussed in Chap. 16 and noted in Table 16-4), may increase the probability of formation of these tumors. For instance, the cardiovascular system of a cigarette smoker must contend with hydrocarbons and nicotine, which can have devastating long-term effects. Some of the hydrocarbons in cigarette smoke cause greater risks of mutation. Smokers' lungs also may accumulate greater amounts of airborne radioactive particles, which increase mutation rates in smooth muscle cells. Nicotine stimulates the heart to work harder (Table 16-1) and, along with the other factors, may account for the high rate of cardiovascular disease in cigarette smokers.

FIG. 17-13
Atherosclerotic artery. Note the extremely narrow opening in the center of the deteriorated artery. By proliferation, the smooth muscle has pushed inward toward the intima, narrowing the lumen of the vessel. Fat has accumulated in the intima to form plaques.

A second and widely accepted theory of atherosclerosis is the *response to injury theory*. It asserts that atherosclerosis develops in blood vessels following injury to the endothelial lining (Fig. 17-14). Presumably, chemical or mechanical agents damage endothelial cells and cause them to separate from the remainder of the inner lining. Blood platelets are attracted to the damaged tissues and release molecules that cause smooth muscle cells to multiply and push inward from the middle coat of the vessel. Collagen, elastin, and other molecules then accumulate at the site of the damage. These activities narrow the arterial lumen, and lipids also accumulate to form plaques on the vessel's inner lining. The presence of excess fats such as cholesterol surrounded by low-density lipoproteins in the bloodstream causes hyperlipoproteinemia (Chap. 16 and Fig. 16-8), which may initiate injury to endothelial cells and lead to atherosclerosis. Although the process can be reversed, chronic hyperlipidemia (a long-lasting excess of fatty substances in the blood) may lead to more widespread atherosclerosis.

As atherosclerosis develops, calcium deposits may accumulate and harden the arterial walls. This progression is responsible for the common phrase *hardening of the arteries*, as applied to this form of arteriosclerosis. The loss of elasticity in the arterial wall causes greater pressure fluctuations as blood pulses through the vessels. The surge of blood that occurs during systole may rupture the vessel and in some instances lead to enough blood loss to cause shock and death.

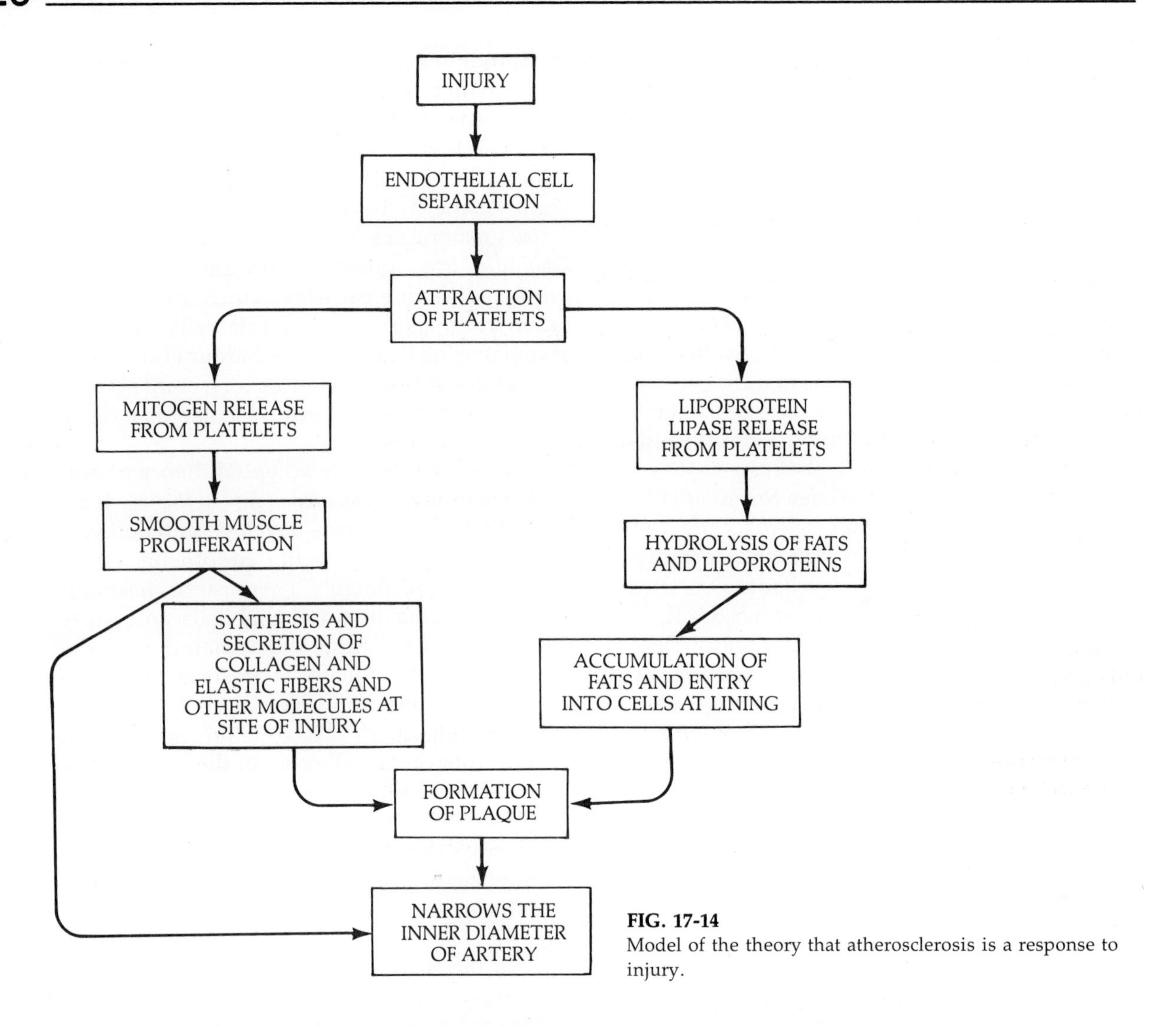

FIG. 17-14
Model of the theory that atherosclerosis is a response to injury.

Arteriosclerotic vessels are more susceptible than normal ones to *cerebrovascular accidents* (CVA). Among the causes of such disorders are thrombosis, embolism, or hemorrhage of the middle cerebral artery, an end artery in the brain (Table 17-5). In any of these three instances, blood flow becomes inadequate to a part of the cerebrum and may cause a CVA. The disorder is commonly called a *stroke* and was referred to in the past as *apoplexy*. Stroke is characterized by mental confusion, vomiting, convulsions, and severe headache and may lead to loss of sensation, loss of speech, and loss of motor activity (paralysis). The symptoms depend on the area of the brain that is deprived of its normal blood supply.

17-4 SUMMARY

1 Arteries conduct blood away from the heart to arterioles, subdivisions which lead to microscopic capillaries. The capillaries anastamose to form venules, which merge to give rise to veins, which carry blood back to the heart. Ex-

cept for capillaries, blood vessels are composed of three layers of tissues, each of which has varying amounts of connective tissues. The intima has an endothelial lining, the media has circular smooth muscle, and the adventitia (in a few instances) has longitudinal smooth muscle. Veins generally have less tissue in each coat, much less muscle and connective tissue in the media, and larger internal diameters than corresponding arteries. The endothelial lining of veins is modified to form unidirectional valves.

Precapillary sphincters, in the metarterioles, help regulate blood flow into capillary beds. Some molecules and ions escape through spaces ranging, in order of increasing size, from intracellular pores to intercellular tight junctions, fenestrated capillaries, and sinusoids. Blood pressure produces hydrostatic pressure in vessels, which pushes fluid and small molecules out of capillaries across capillary walls to form interstitial fluid. Most of the fluid moves back into the capillaries by osmotic forces in accordance with Starling's law of capillary exchange.

Lymph is excess interstitial fluid which enters the very permeable lymph capillaries. Lymph capillaries merge to form lymph vessels, the largest of which, the thoracic duct, returns lymph to the blood through the innominate vein. Lymph is mainly moved by passive mechanical compression due to skeletal muscle contractions. Unidirectional valves in lymph vessels cause the passive milking of lymph vessels. Lymph is filtered as it moves through lymph nodes and lymphatic organs, whose lymphocytes and phagocytes form part of the body's defense system.

Several physical factors influence the movement of body fluids. As vessels subdivide, flow rate in each subdivision decreases, as explained by Bernoulli's principle. However, although pressure should rise at this time, arteries subdivide so extensively that in most cases overall blood pressure does not rise; rather, much energy is lost as heat. Gravity usually increases blood pressure in vessels below the heart and decreases it in vessels above the heart. The flow rate of blood rises as a function of the fourth power of the radius of the vessel (Poiseuille's law).

Blood pressure is the lateral force of blood against the walls of vessels. Systolic pressure is the force created as a result of ventricular contraction of the heart. Diastolic pressure is produced as the expanded walls of the aorta and other arteries recoil during relaxation of the ventricles. Blood pressure is normally measured on the upper arm by a sphygmomanometer. Heart sounds can be heard through a stethoscope placed on the brachial artery. As pressure within the inflated sphygmomanometer cuff is allowed to decrease, systolic pressure is recorded as the height of the column of mercury in millimeters when the sounds first reappear. Diastolic pressure is recorded as the level of mercury just before the disappearance of the sounds. Blood pressure measured this way in young adults averages nearly 120/80 mmHg.

Blood pressure varies directly with cardiac output and peripheral resistance. The latter is related directly to the length of a blood vessel and the viscosity of blood and inversely to the fourth power of the radius of the vessel. Pressure increases as vessels narrow in vasoconstriction and decreases as they widen in vasodilation. These changes are regulated by reflex control of smooth muscle in the vessel walls.

Impulses are generated by baroreceptors when changes in blood pressure occur in the carotid sinus and aortic arch and the vasomotor and cardiac centers in the medulla respond. Cardioinhibitory neurons decrease cardiac output and cause peripheral vasodilation and a drop in blood pressure. The fall in pressure produces negative feedback signals, which turn off these responses and stimulate cardioaccelerator neurons to promote a more rapid heartbeat and vasoconstriction and cause blood pressure to rise.

2 A number of physiological disorders are related to changes in the walls of blood vessels. They include the two major causes of cardiovascular disease, hypertension and atherosclerosis, as well as other forms of arteriosclerosis, and headaches, aneurysm, thrombosis, embolism, hemorrhage, phlebitis, and varicose veins. Hypertension (high blood pressure) is defined as systolic pressure above 160 mmHg or diastolic pressure above 95 mmHg. It is borderline if the values exceed 140 and 90 mmHg, respectively.

Elevated levels of adrenal hormones (including aldosterone, norepinephrine, epinephrine, and glucocorticoids) and thyroxine can contribute to hypertension. In response to low blood pressure, low Na^+ ion content, or kidney disorders, the kidneys produce renin, which causes the formation of a powerful vasoconstrictor, angiotensin II. Angiotensin II stimulates the secretion of aldosterone and also may directly stimulate the thirst center in the brain. All these responses help to elevate blood pressure.

Atherosclerosis, a disorder often associated with high blood pressure, produces a high risk of heart attack. Atherosclerosis involves the addition of fat, collagen, and elastic fibers to the lining of blood vessels. The proliferation of smooth muscle in the middle coat pushes inward, narrowing the vessel's inner opening. The accumulations form a visible anatomical change called plaque, which becomes more widespread as atherosclerosis progresses. The monoclonal theory proposes that the changes result from a mutation in a smooth muscle cell which forms a clone of smooth cells that produce the effect. The response to injury theory holds that the changes are due to injury of endothelial cells. This theory also proposes that the release of substances from the platelets in the blood causes smooth muscle cells to proliferate, which leads to other associated changes in the vessel wall.

REVIEW QUESTIONS

1 Sketch a cross section of an artery, a vein and a capillary, labeling the tissues found in each type of vessel.
2 Compare the three types of intercellular junctions found in various capillaries.
3 How does the body regulate blood flow to tissues? Include a description of vasoconstriction, vasodilation, and precapillary sphincter control of capillaries.
4 What are the physiological factors that regulate the formation of lymph as blood moves through a capillary bed?
5 What is the fate of lymph and what causes its movement?
6 Discuss Bernoulli's principle and explain its relation to blood flow.
7 Name the law exemplified by the following formula and discuss its relation to blood flow:

$$F = \frac{(P_1 - P_2)\pi r^4}{8Lv}$$

8 What is blood pressure? When is it considered normal? How is it usually measured? How does it vary between males and females and between the young and old?
9 How are cardiac output and blood vessel diameter related to blood pressure?

10 Illustrate by means of a labeled diagram the influence of the major nervous control centers on blood pressure.
11 Compare arteriosclerosis and atherosclerosis; thrombosis, embolism, and hemorrhage; aneurysm, phlebitis, and varicose veins.
12 What is hypertension? How do the adrenals and kidneys influence blood pressure?
13 Discuss four general ways in which different drugs may control high blood pressure.
14 Define atherosclerosis and outline two theories advanced to explain its origin.

THE CIRCULATORY SYSTEM

18

The Physiology of Blood, Part I: The Properties of Whole Blood and Erythrocytes (Red Blood Cells)

18-1 GENERAL FUNCTIONS AND MAJOR CONSTITUENTS

18-2 GENERAL PROPERTIES OF BLOOD AS A FLUID
Physical properties of whole blood
Volume, viscosity, and specific gravity
Plasma
Hematocrit
Role of buffers

18-3 ORIGINS OF BLOOD CELLS (HEMATOPOIESIS)

18-4 PROPERTIES OF ERYTHROCYTES
Development and structure
Normal variations in number
Red blood cell function and hemoglobin
General properties of Hb
Types of Hb

18-5 BLOOD GROUP ANTIGENS
Major blood groups
Determination of compatibility
Other red blood cell antigens

18-6 PATHOPHYSIOLOGY OF RED BLOOD CELLS
Effects of blood loss: Hemorrhage and shock
Hemolytic disease of the newborn
General types of anemia
Hereditary anemias
Sickle-cell anemia
Thalassemia

18-7 SUMMARY

OBJECTIVES

After completing this chapter the student should be able to:

1. Distinguish among blood, its formed elements, plasma, and serum.
2. Discuss the role of transport and defense reactions of blood.
3. Describe whole blood and plasma, explaining their physical properties and buffering capacities.
4. Define the hematocrit and relate changes in it to alterations in body function.
5. Describe the development of blood cells and platelets.
6. Describe the properties of red blood cells and discuss normal variations that occur in their numbers.
7. Describe the structure and genetics of hemoglobin.
8. Explain the genetic basis of A, B, O, and Rh blood types and discuss their physiological significance.
9. Compare three major forms of anemia, identifying their causes and effects.

Homeostasis, the maintenance of the steady state, depends on the interaction of cells, tissues, and organs in distant areas of the body. It is a function essential to life and involves the delivery of metabolic products between remote sites and the generation of feedback signals that stimulate or inhibit specific cellular activities. The ability to peform such functions is dependent on many systems and is closely associated with the properties of the blood pumped by the heart through a communication network of more than 60,000 mi of blood vessels.

18-1 GENERAL FUNCTIONS AND MAJOR CONSTITUENTS

Blood conveys oxygen, nutrients, metabolites, secretions, excretions, and heat between cells, tissues, and organs; buffers the body fluids; and offers protection against disease. *Blood* is a specialized connective tissue composed of a fluid intercellular matrix, *plasma,* and cells or cell-derived components, the *formed elements* (Table 18-1).

The *formed elements* include *erythrocytes* (red blood cells, RBCs), which carry oxygen to tissues; *leukocytes* (white blood cells), which aid the body's defenses; and *thrombocytes* (platelets), which release factors that aid the formation of blood clots and that plug holes in injured vessels. Molecules and ions in the plasma may promote the aggregation of the formed elements and their adherence to tissues at the site of clot formation. A yellowish fluid, serum, escapes from the site of the clot, leaving behind the formed elements and substances that cause clotting. *Serum* may be described as plasma without its clotting elements. This chapter will emphasize the properties of erythrocytes and Chap. 19 will describe further those of leukocytes and platelets.

TABLE 18-1
The average composition of blood

				*Cells/μL**
Formed elements (40–45%)	Red blood cells (erythrocytes)			4.2×10^6–5.9×10^6
	White blood cells (leukocytes)			4.3×10^3–10.8×10^3
	Platelets (thrombocytes)			1.4×10^5–4.4×10^5
Plasma (50–55%)	Water, 90% (solvent)			
	Dissolved substances, 10%	Proteins (6–8%)	Albumin, 53%	
			Globulins, 43%	
			Fibrinogen, 4%	
		Others (2–4%)	Carbohydrates	
			Fats	
			Proteins (enzymes, hormones)	
			Amino acids	
			Salts	
			Gases	
			Excretory products	

* The number of cells per μL equals the number per mm^3.
† Further details on the composition of plasma and a comparison with cerebrospinal fluid are given in Table 10-2, p. 245. Its ionic composition resembles that of ECF as described in Fig. 4-2, p. 82. Normal reference lab values of blood and plasma are presented in the appendix of the text.

TABLE 18-2
Physical properties of whole blood*

Property		*Comments*
Percent of body weight	8.5–9.0	Decreased by fluid loss, hemorrhage; altered by disease.
Volume in liters		Decreased by fluid loss, hemorrhage; altered by disease.
Males	5.0–6.0	
Females	4.5–5.5	
Viscosity (compared with water)	3.0–5.5	Decreased by anemia; increased by malaria; increased by elevated red blood cell count in *polycythemia* caused by exposure to high altitude or by tumors in marrow (*polycythemia vera*)
Sedimentation rate (mm/h)		Often increases as result of infection and inflammation. [Blood cells with normal Hb levels fall to the bottom of a tube filled with copper sulfate solution with a specific gravity of 1.055 within 5 to 15 s. The specific gravity of blood (its density compared with that of water) is normally 1.055 to 1.065.]
Males	1–13	
Average	4	
Females	1–20	
Average	10	
Children	1–15	
Average	5–10	
Hematocrit (packed red blood cell volume, % of whole blood volume)		Elevated by altitude exposure Depressed in anemia
Males	45–52	
Females	37–48	
Osmotic fragility of cells (ability to withstand immersion in hypotonic salt solutions without lysing)	No lysis normally occurs in solutions with more than 0.5% sodium chloride while lysis occurs in those with less than 0.3%	This test is a measure of how nearly spherical are the red blood cells. Osmotic fragility decreases when cells tend to be flatter; this is true in iron deficiency anemia, sickle-cell anemia, thalassemia, and some other anemias; osmotic fragility increases in some forms of acquired hemolytic anemia

* Data adapted from R. E. Scully, J. J. Galdabini, and B. U. McNeely (eds.), *N. Engl. J. Med.* **302**(1):37–48 (1980).

18-2 GENERAL PROPERTIES OF BLOOD AS A FLUID

Physical Properties of Whole Blood

Volume, Viscosity, and Specific Gravity

Blood accounts for about 8.5 to 9.0 percent of total body weight and has a volume of 4.5 to 6.0 L, the average volume being higher in males than in females (Table 18-2). The viscosity (resistance to flow) and sedimentation rate (the rate at which cells settle out in a tube) may be used as indexes of some diseases and of recovery from them (Table 18-2).

Plasma

Plasma is an extracellular fluid (ECF) which engages in exchange with other fluid compartments and which accounts for about 7.5 percent of total body water (Chap. 4, Fig. 4-1). Plasma makes up 50 to 55 percent of the volume of blood, the other 40 to 45 percent being due to formed elements.

Plasma is composed of about 90 percent water and 10 percent dissolved molecules and ions (Tables 18-1 and 10-2). It is normally about 4.5 times more viscous than water and may be altered by diseases that change the composition of body fluids. Therefore, plasma chemistry is an important diagnostic tool. For example, levels of AST (aspartate aminotransferase) disclose the severity of a heart attack (Fig. 16-6). Similarly the quantity and types of blood lipoproteins reveal the risk of coronary artery disease and the amount of blood glucose indicates the severity of diabetes. Yet, changes in plasma composition may also reflect a normal range in values as noted in the appendix of the text.

Hematocrit

The *hematocrit,* or packed cell volume, is the percent by volume of red blood cells in whole blood after centrifugation. The hematocrit reading varies with age, sex, altitude exposure, and disease. It is highest in the newborn, decreases during the first 12 weeks after birth, and then gradually increases until adult levels are attained at about age 25, but normally does not return to the high level of the newborn infant (Table 18-3). In young adults the hematocrit averages 47 percent in males and 42 percent in females.

The hematocrit is measured after centrifugation of blood to which an anticoagulant has been added. The blood is placed in a capillary tube, or in a *Wintrobe tube* which is about 10 cm long and graduated in millimeters. After centrifugation the plasma forms an upper straw-color (orange) layer, the white blood cells and platelets make up a gray-white, "buffy" middle layer, and the red cells comprise the lowest layer. Because plasma is trapped among the red blood cells, 4 percent of the packed cell volume is subtracted from the measured value to calculate the correct hematocrit.

TABLE 18-3
Changes in hematocrit and hemoglobin with age*

Age	*Hematocrit,* %	*Erythrocyte† Count,* $\times 10^6/\mu L$	*Hemoglobin,* *g/100 mL*
1 day	61.0	5.14	19.0
7 days	56.0	4.86	17.9
12 weeks	33.0	3.70	11.3
1 year	34.5	4.76	11.1
10 years	38.5	4.80	13.2
17 years:			
Males	45.8	5.14	14.9
Females	40.9	4.56	13.4
25 years or more:			
Males	47.0	5.40	16.0
Females	42.0	4.80	14.0
60 years or more:			
Males	42.1	4.75	14.1
Females	40.8	4.71	13.7

* Arithmetic means are shown.
† The number of cells per μL equals the number per mm^3.

Role of Buffers

Buffers are organic and/or inorganic compounds that combine with excess acid or base. They maintain body fluids at a pH that is optimal for the activity of specific enzymes. The pH of blood is slightly alkaline, usually varying from 7.35 to 7.45. It must be maintained in that range so that specific enzymes can perform their functions. Otherwise severe derangements in metabolism may produce reactions similar to those of irreversible shock. When the pH of blood falls below 7.35, the condition of *acidosis* occurs, and when it rises above 7.45, the condition is referred to as *alkalosis.*

Proteins buffer both excess acid and base. For example, the amino groups (NH_2) of proteins are basic and may combine with hydrogen ions from acidic compounds and neutralize them. The carboxyl groups (COOH) of proteins act as acids and can dissociate to donate hydrogen ions to neutralize basic compounds.

The major one of the many inorganic buffers in the body is sodium bicarbonate ($NaHCO_3$). An example of its buffering action is the neutralization of hydrochloric acid. The sodium ion (Na^+) combines with the chloride ion (Cl^-) from the hydrochloric acid, forming sodium chloride (NaCl). The remaining bicarbonate ion (HCO_3^-) unites with the free hydrogen ion (H^+) to form carbonic acid (H_2CO_3), which is a weaker acid than hydrochloric acid. Formulations of this buffering activity are:

$$\underset{\text{Hydrochloric acid}}{HCl} + \underset{\text{Sodium bicarbonate}}{NaHCO_3} \rightarrow \underset{\text{Sodium chloride}}{NaCl} + \underset{\text{Carbonic acid}}{H_2CO_3}$$

Carbonic acid, in turn, neutralizes excess base to produce more $NaHCO_3$ buffer, as shown in the following equation:

$$\underset{\text{Carbonic acid}}{H_2CO_3} + \underset{\text{Sodium hydroxide}}{NaOH} \rightarrow \underset{\text{Sodium bicarbonate}}{NaHCO_3} + \underset{\text{Water}}{H_2O}$$

In this cycle excess acid (HCl) and base (NaOH) are neutralized. Potassium bicarbonate ($KHCO_3$) has similar properties. These buffers are regulated by the kidneys to maintain homeostasis of body fluids (Chap. 21), and by the respiratory tract in a series of reactions that

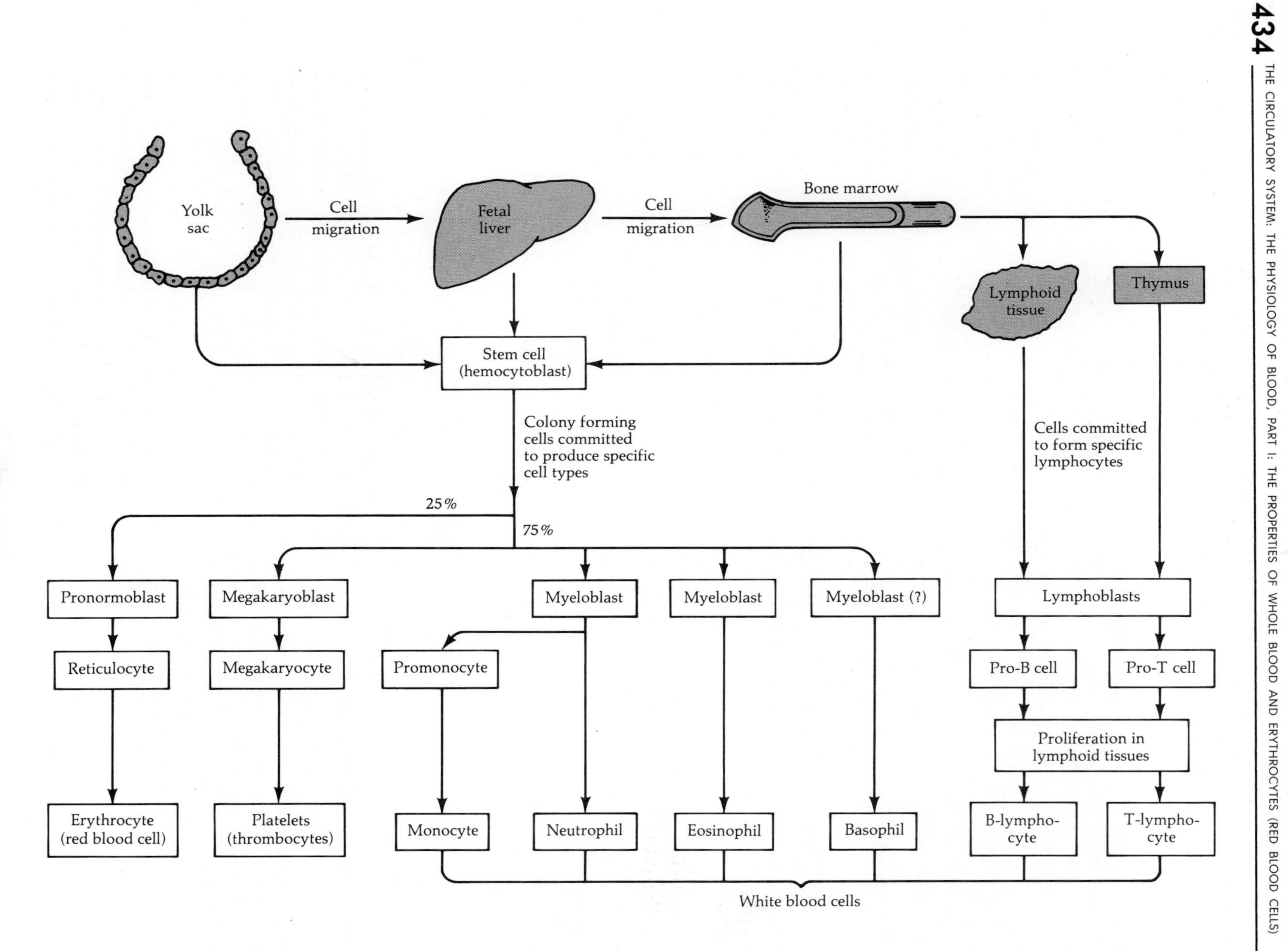
Yolk sac
Cell migration
Fetal liver
Cell migration
Bone marrow
Lymphoid tissue
Thymus
Stem cell (hemocytoblast)
Colony forming cells committed to produce specific cell types
25%
75%
Cells committed to form specific lymphocytes
Pronormoblast
Megakaryoblast
Myeloblast
Myeloblast
Myeloblast (?)
Lymphoblasts
Reticulocyte
Megakaryocyte
Promonocyte
Pro-B cell
Pro-T cell
Proliferation in lymphoid tissues
Erythrocyte (red blood cell)
Platelets (thrombocytes)
Monocyte
Neutrophil
Eosinophil
Basophil
B-lymphocyte
T-lymphocyte
White blood cells

eliminate CO_2 with exhaled air and help regulate the pH of blood (Chap. 20).

18-3 ORIGINS OF BLOOD CELLS (HEMATOPOIESIS)

In early embryonic development cells migrate from the yolk sac to the fetal liver and then to bone marrow, where colonies of *stem cells* are formed (Fig. 18-1). Young bone marrow, which is primarily red, is the source of most of the blood cells and is called *hematopoietic* (hemopoietic) tissue (Chap. 14). Various stem cells proliferate and produce unique colony-forming cells, which cannot be distinguished in appearance but which mature to produce specific cells that develop in stages to yield individual types of blood cells. Certain precursors form erythrocytes, while others undergo nuclear replication without cell division to create giant cells, *megakaryocytes,* whose cytoplasmic pieces are later pinched off to produce platelets. Still other colony-forming cells multiply, migrate to lymphoid tissues, develop, and proliferate as B and T lymphocytes. Granular white blood cells develop from specific myeloblasts which mature in a series of stages (in which the cells are called promyelocytes, myelocytes, and metamyelocytes) to form neutrophils, eosinophils, and basophils.

18-4 PROPERTIES OF ERYTHROCYTES

Development and Structure

Pronormoblast cells in bone marrow mature to normoblasts, which lose their nuclei and form *reticulocytes* (Figs. 18-1, 18-2). The *reticulocytes* are young red blood cells that possess an abundance of ribosomal granules which when stained may aggregrate into reticular strands (which accounts for the cells' name). This endoplasmic reticulum disappears within 2 to 3 days as each cell matures into an erythrocyte. Each cell loses its nucleus as it squeezes from the bone marrow to enter the blood (Fig. 18-2). Reticulocytes make up about 1.0 percent of the normal circulating red blood population entering it at a rate of about 1 to 1.5 million/s.

FIG. 18-1
Embryonic origins of blood cells.

The loss of the nucleus may explain why erthrocytes acquire indentations on each side and look like biconcave discs after they squeeze out of the marrow and enter blood vessels (Fig. 18-3). The shape is due to a cytoskeleton made up of a protein called *spectrin.* The biconcave disc shape is advantageous to the cell because it provides a greater surface/volume ratio, requires a minimum expenditure of energy to be maintained, and can fold to aid the passage of red blood cells through capillaries. The average diameter of erythrocytes is about 7.5 μm and ranges from 6 to 9 μm. As a result of diseases which cause an excess of globulins or fibrinogen protein to accumulate in the blood, erythrocytes often stack upon one another. This phenomenon is called the *Rouleau (stacking) effect* and may be reversed *in vitro* (Latin, in glass) by dilution of blood with isotonic salt solution. Diseases which cause Rouleau formations include multiple myeloma (cancer of the bone cells), cryoglobulinemia (a disorder in the globular proteins), leukemia, rheumatoid arthritis, infectious mononucleosis, cirrhosis of the liver, and hepatitis.

Normal Variations in Number

There are about 20 to 30 trillion (10^{12}) erythrocytes in the human body, which if placed side by side, would encircle the earth 4 times (a distance of 100,000 mi). In adults, the number of erythrocytes averages 5.4 million per microliter (per mm^3) of blood in males and 4.8 million in females (Table 18-1). Like the hematocrit, the quantity of red blood cells decreases between birth and 8 to 12 weeks of age, reaches adult levels by age 25, and declines with further aging (Table 18-3). The hormone *erythropoietin* is formed by the action of erythropoietic factor, produced in the kidneys, upon a globulin in the plasma. Erythropoietin stimulates the bone

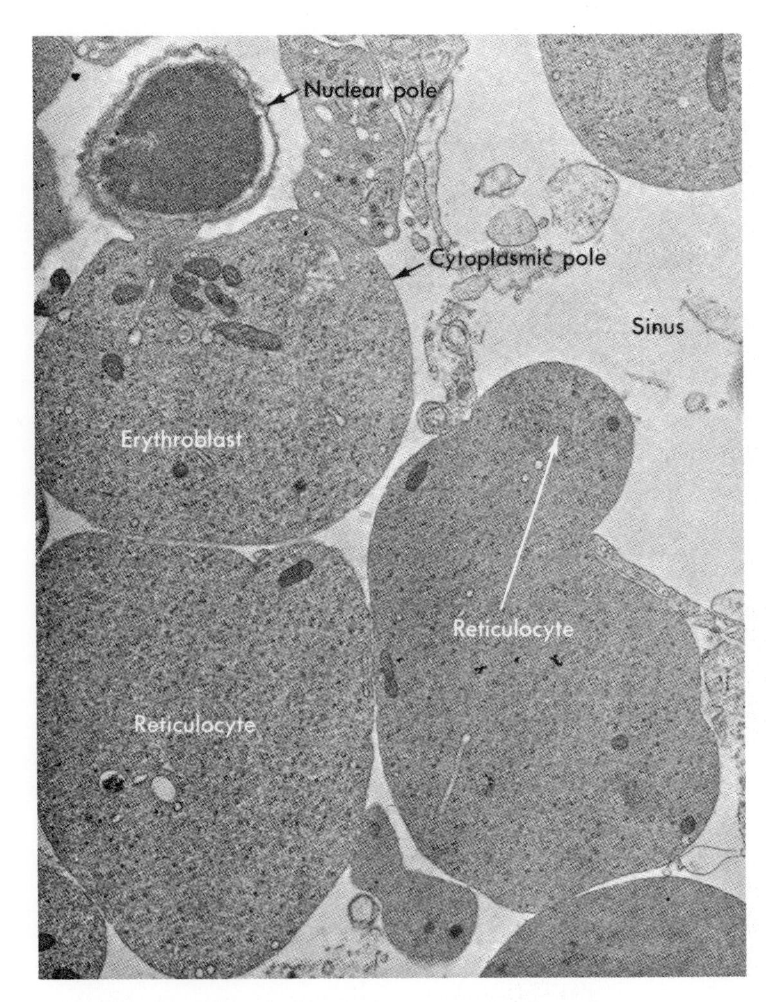

FIG. 18-2
Migration of a normoblast (erythroblast) to form a reticulocyte in bone marrow (× 12,000). As the erythroblast on the upper left matures into a reticulocyte its nucleus is pinched off and left behind. The cytoplasmic pole of the cell squeezes from the bone marrow through an intercellular junction or sinus of a blood vessel to form a reticulocyte. Two reticulocytes are illustrated in the lower portion of the figure. From *Histology*, 4th ed., by L. Weiss and R. O. Greep. Copyright © 1977, McGraw-Hill Book Company. Used with permission of McGraw-Hill Book Company and L. Weiss. From the work of J. Chamberlain, R. Weed, and L. Weiss.

marrow to produce red blood cells (Fig. 18-4). Prolonged exposure to high altitude (low levels of oxygen) causes more hormone to be secreted and the red blood cell count to increase to levels which may be above normal within 4 days of ascent from sea level (Chap. 20).

The life-span of the average red blood cell is about 120 days, a sufficient period for the cell to circulate throughout the body about 120,000 times. As aged erthrocytes pass through lymphatic organs, especially the spleen and the liver, they are ingested by macrophages. In this process, about 200×10^9 red blood cells are lost each day and an equal number replaced through hematopoiesis in the red marrow.

Red Blood Cell Function and Hemoglobin

One of the major functions of red blood cells is the transport of oxygen (O_2), which depends mostly on a globular protein, *hemoglobin* (Hb). When Hb loosely binds O_2, it is called *oxyhemoglobin* (HbO_2). When oxygen diffuses from HbO_2 to tissues, *reduced hemoglobin* is produced. The presence of excess hydrogen ions favors this reaction, and the combination of them with Hb is symbolized HHb. A decline in blood pH promotes the release of oxygen from Hb (the dissociation of oxyhemoglobin) and the formation of HHb (Figs. 20-10 and 20-11). The significance of this reaction in respiration will be discussed further in Chap. 20. Carbon diox-

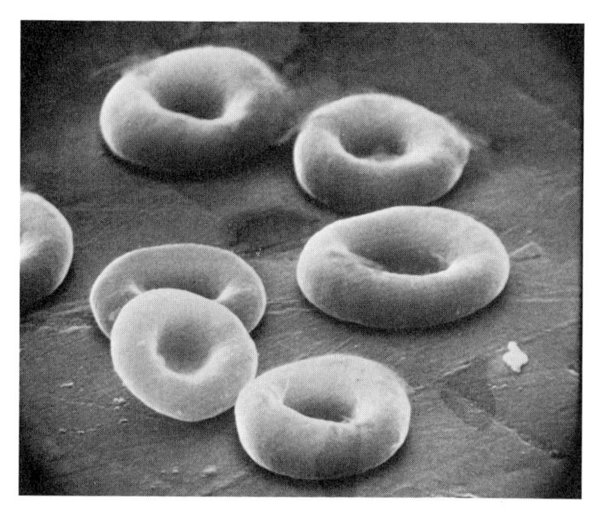

FIG. 18-3
Erythrocytes viewed in a scanning electron micrograph (× 10,000).

ide (CO_2) formed by tissue metabolism also may combine with Hb to form *carbaminohemoglobin* ($HbCO_2$). Blood is scarlet (vivid red) when HbO_2 is high and bluish red when HHb and $HbCO_2$ levels are elevated.

General Properties of Hb

The average adult level of Hb is 14 to 16 g per 100 mL of blood (ranging from 11.1 to 19 g per 100 mL) and totals about 600 g in the body. Age, sex, and exposure to high altitude alter the level of Hb, as they do the number of red blood cells (Table 18-3).

There are about 280 million globular protein molecules of Hb, each with a molecular weight of 65,000, in each red blood cell. The Hb molecule is a tetramer composed of four polypeptide chains, each of which has an organic molecule with a cyclic structure (a porphyrin ring) that holds a single ferrous iron, Fe^{2+} (Fig. 18-5). The same cyclic structure exists in other pigments that assist aerobic metabolism. For example, it occurs in oxidative enzymes (cytochromes, Chap. 23) in animal and plant cells and in chlorophyll in plant cells. The iron loosely and reversibly combines with oxygen, allowing the average adult to carry 1.34 mL O_2 per gram Hb, or a total of about 800 mL O_2 in all the red blood cells in the body.

When the iron in Hb is oxidized to the ferric state (Fe^{3+}) by exposure to drugs or oxidizing agents, methemoglobin is produced. Methemoglobin cannot carry O_2 or CO_2 and its accumulation causes the skin to darken. Although a slight amount of methemoglobin is formed normally, it is reconverted to Hb by enzymes. However, certain hereditary disorders may cause sufficient methemoglobin to be produced to discolor the skin although such individuals do not show signs of an oxygen deficit.

Normal Hb readily releases O_2 to the respiratory pigment *myoglobin,* which is found in skel-

FIG. 18-4
Effects of high-altitude exposure (hypoxia) on red blood cell formation.

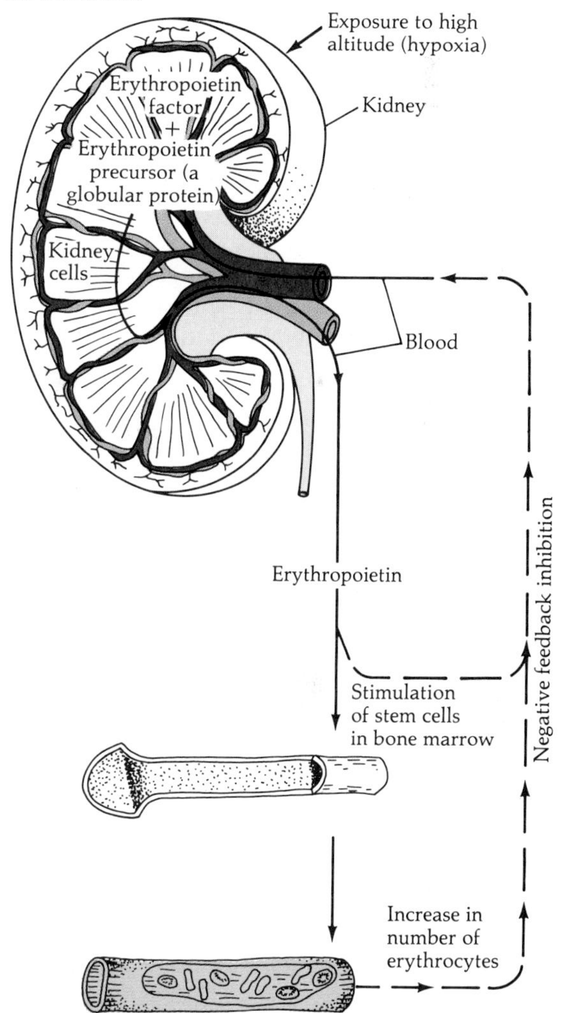

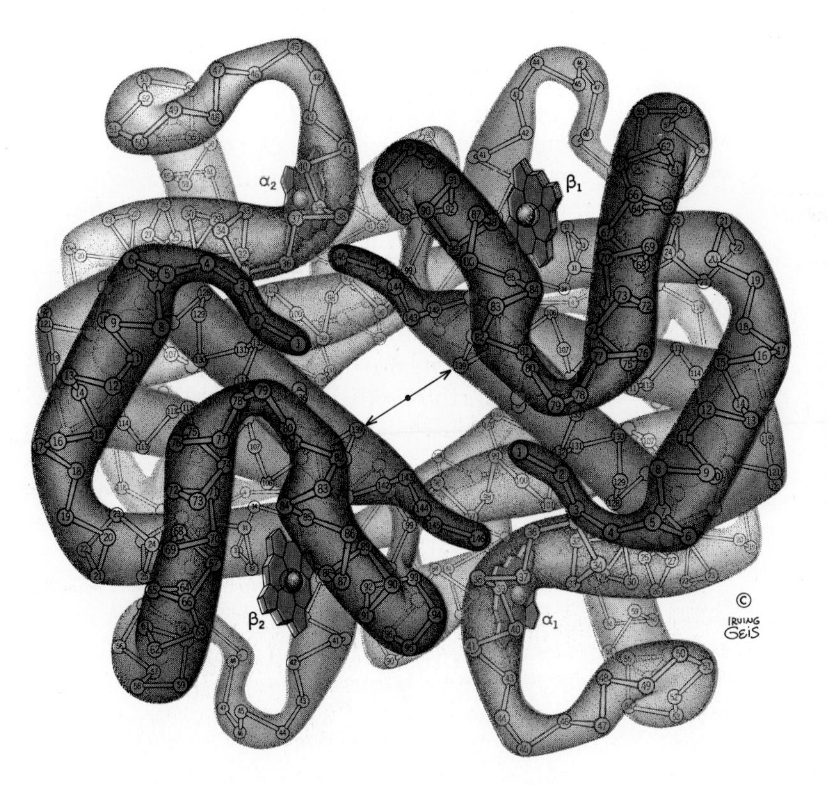

FIG. 18-5
Structure of oxyhemoglobin. Note that two α and two β chains are linked to form hemoglobin. Each contains a porphyrin ring at the center of which is an atom of iron (Fe) drawn as a circle. Each of the four iron sites loosely binds O_2 when the gas is abundant and releases it when the hemoglobin and the red blood cell in which it is found are in an environment in which the level of O_2 is low. The β chains move apart, in the direction indicated by the arrows when O_2 is released and reduced hemoglobin is formed.

etal muscle. Myoglobin is a single-chain polypeptide with a single oxygen-binding iron complex and a molecular weight of 17,000. Myoglobin exceeds hemoglobin in its affinity for O_2 and provides skeletal muscle cells with greater efficiency in removing O_2 from the blood than is true of other tissue types.

If the blood level of HHb decreases to about 5 g per 100 mL, the skin appears blue. In this condition, *cyanosis* (blue of the skin), aerobic metabolism is seriously disrupted. In another reaction that may discolor the skin, and limit the ability of Hb to carry O_2, carbon monoxide (CO) combines with Hb. It does so 210 times more readily than oxygen and produces *carboxyhemoglobin* (HbCO). As little as 0.02 to 0.04 percent of CO in the air may cause carbon monoxide poisoning. The blood becomes cherry red, and the lack of O_2 may be fatal.

Under normal conditions hemoglobin is recycled when phagocytes in the spleen and liver destroy aged and worn-out red blood cells. In the liver, Hb is gradually converted into a pigment, *bilirubin,* and incorporated into bile, which is either released to the gallbladder for storage or delivered directly to the small intestine for metabolism, reabsorption, or excretion in the feces and urine (Fig. 18-6). The iron removed from hemoglobin and other proteins in the liver may also be recycled, which reduces the dietary requirement for this mineral. Should too much bilirubin accumulate, *jaundice* (yellow skin) may occur, which often reflects hemolytic diseases and disorders of the liver and gallbladder (Chap. 22). Hemolytic diseases are disorders in which red blood cells burst and Hb leaks out of the cells during an activity called *hemolysis* (lysis of blood cells).

Types of Hb

Normal human blood contains three major types of Hb: HbA, HbF, and HbA_2. Their chain compositions and frequencies are shown in Table 18-4. Normal Hb consists of two alpha (α) chains, each of which is 141 amino acids in length paired with two beta (β), delta (δ), gamma (γ), or embryonic epsilon (ϵ) chains. The non-alpha chains are 146 amino acids in length. Specific genes govern the synthesis of the various types of polypeptide chains.

More than 150 variants of hemoglobin have

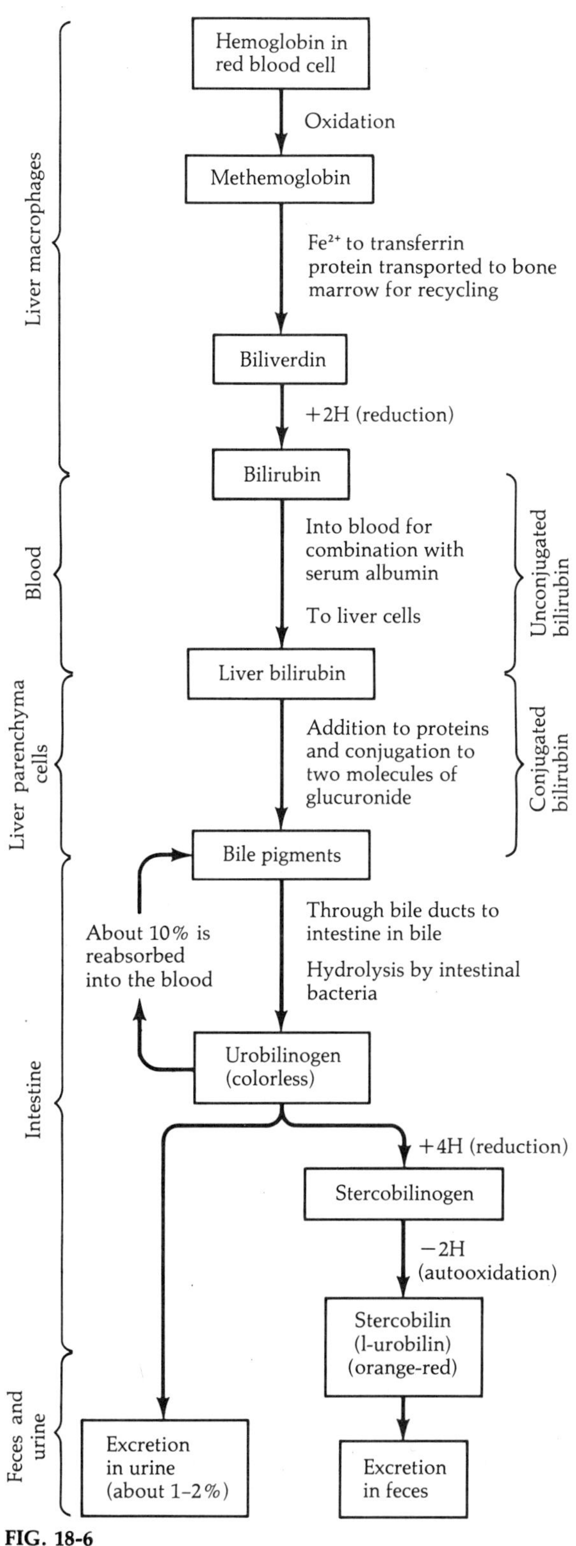

FIG. 18-6
The fate of hemoglobin: the production and destiny of bile pigments and their derivatives.

TABLE 18-4
Three major types of normal human hemoglobin

Type Hb	*Pairs of Chains*	*% of Adult Hb*
A	α,β	91–95
A_1	α,β^*	4.6–6.8
A_2	α,δ	1.5–3.5
F	α,γ	<0.5

* HbA_1 is derived as a modification of the HbA molecule.

been discovered in which single amino acid substitutions result in no apparent functional change. Only a few forms are associated with disease. In fact, under normal circumstances the activities of the genes for chain production vary, so that distinct types of hemoglobin are produced at different developmental stages in an individual's lifetime. For example, Hb with epsilon chains (Hb-Gower) is produced in the embryo during the first 2 months after conception and decreases to negligible levels 3 months later. HbF (fetal Hb) is the most prominent form of Hb in this early fetal stage and declines sharply within 6 months after birth. It has a higher affinity for O_2 than HbA (adult Hb) which is advantageous to the transport of oxygen across the placenta from mother to the fetus. HbF levels drop from 80 percent of all Hb in the fetus to less than 0.5 percent in adults. HbA levels increase slowly from less than 5 percent of Hb between the third month of development and birth to nearly adult levels (91 to 95 percent of Hb) between 3 and 6 months of age.

A variant, HbA_1, is normally formed by the modification of HbA and exists in three forms (HbA_{1a}, HbA_{1b}, and HbA_{1c}), which usually account for 4.6 to 6.8 percent of the total Hb. In diabetics an increase in HbA_{1c} to 12 percent of Hb is correlated with an inability to transport glucose from the blood into tissue cells.

18-5 BLOOD GROUP ANTIGENS

Genetic differences may not only alter Hb but also may cause alterations in other structures of the membranes of the formed elements. The variations may be significant when blood is transfused from donor to recipient. The foreign molecules on the cell surfaces are antigenic and

TABLE 18-5
Major blood groups

Blood Group	*Major Antigen (Agglutinogen)*	*Antibodies in Serum (Agglutinins)*	*Hereditary Makeup (Genotype)*	APPROXIMATE FREQUENCY, %*		
				White	*Black*	*Oriental*
O	None	anti-A and anti-B	OO	43	50	30
A	A	anti-B	AA or AO	45	29	36
B	B	anti-A	BB or BO	8	17	23
AB	A and B	none	AB	4	4	13

* The approximate frequencies of blood groups in various races are taken from *Hematology*, 2d ed., by W. Williams *et al.* Copyright © 1977, McGraw-Hill Book Company. Used with permission of McGraw-Hill Book Company.

may cause antibody to be formed in the recipient. The transfused cells will *agglutinate* (clump) when they are mixed with the corresponding antibodies. Because of this reaction major blood cell antigens are called *agglutinogens* and the antibodies are referred to as *agglutinins*.

Similar antigens on the surface of other tissues are called *histocompatibility* antigens (page 86). They are responsible for immune reactions that lead to tissue rejection when a donor and recipient's major tissue antigens do not match. Such reactions are important in skin, kidney, heart, and bone marrow transplants. Studies of the similarities between the antigens on the leukocytes of the donor and recipient are often performed to predict the likelihood of a successful transplant. The closer the antigenic identity of the two individuals' tissues, the less the likelihood of rejection of the transplant.

More than 60 blood group antigens have been identified on human erythrocytes. However, many are buried deep within the red blood cell membrane and do not elicit a strong immune response. Those which project from the surface are able to stimulate strong immune responses when administered to individuals who lack them.

Major Blood Groups

The major blood group antigens are designated as the *A* and *B antigens* and are responsible for type A and B blood, respectively. Both types of antigens are found on group AB cells (Table 18-5), and individuals with neither A or B antigens have type O blood. Low levels of antibodies against the specific antigens are normally found in the serum of individuals lacking one or both antigens. The natural presence of these antibodies is probably due to exposure to other antigens, similar to those on red blood cells. For example, such antigens are found in the cell walls of some bacteria.

Anti-A antibodies agglutinate cells with A antigens; anti-B antibodies agglutinate those with B antigens. The production of antibodies follows exposure of an individual to the antigens which are not present on that person's own red blood cells.

The presence of the blood group antigens is determined by pairs of genes. There are three main types of alternative forms (alleles)—A, B, and O genes. The A gene exists in two different forms, A_1 and A_2, which are found in nearly 80 percent and 20 percent, respectively, of type A individuals. The presence of an A or a B gene masks an O gene when paired with it. The different individuals are heterozygous and have A blood and an AO hereditary makeup or B blood and a BO hereditary makeup, respectively (Table 18-5). Persons with A or B blood also may be homozygous and have pairs of A genes or B genes, respectively, in their cells. When an A and a B are present, they are codominant, both are expressed, and AB blood is formed. The incidence of the genes and the major blood groups varies among whites, blacks, and Orientals, as may be seen in Table 18-5.

When the red blood cell membrane is being formed, certain genes of nearly all persons produce a glycoprotein called the H substance. As a result of its common occurrence it is rarely antigenic in humans. The addition of another sugar on the H substance is caused by an A or B

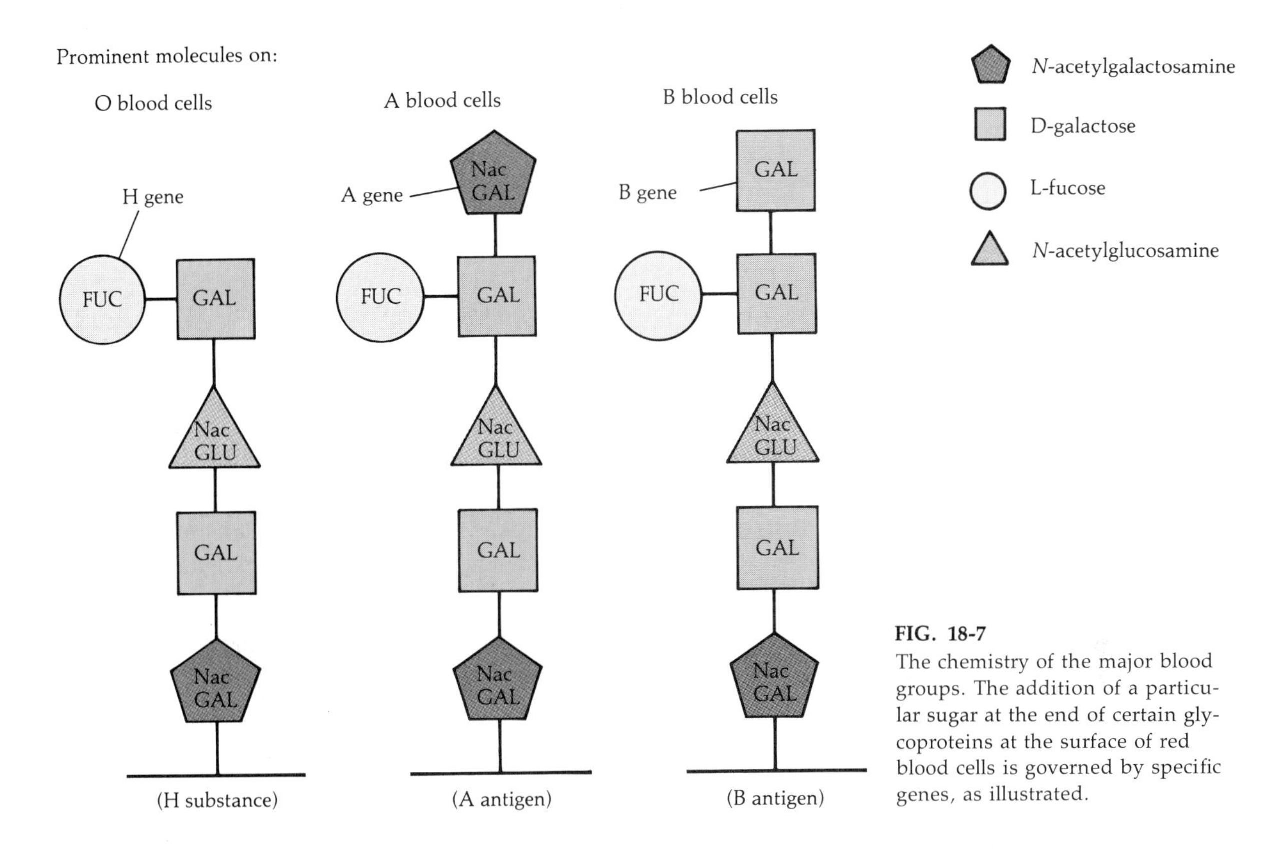

FIG. 18-7
The chemistry of the major blood groups. The addition of a particular sugar at the end of certain glycoproteins at the surface of red blood cells is governed by specific genes, as illustrated.

gene and determines the major blood groups (Fig. 18-7). The O gene fails to add a sugar to the H substance so that the presence of two O genes results in O type blood. Persons with an A gene and a B gene add two different sugars to H glycoproteins, producing type AB blood.

Determination of Compatibility

Blood transfused from donor to recipient must be *compatible*, that is, it must not agglutinate. Even a slight potential for agglutination should be avoided since such a reaction may block vital blood vessels and often is lethal. Both the recipient and the donor must be considered. Blood must not be given to a recipient who lacks its major red blood cell antigens or is likely to form antibodies against them. A recipient will form antibodies against the major blood group antigens that he or she lacks (see Table 18-5). The donor's blood also should not contain antibodies that might agglutinate the recipient's red blood cells.

The compatibility of donor and recipient are determined by *ABO grouping* and *cross matching* (Table 18-6). In ABO grouping, known types of cells or sera are used to assay the properties of the donor's blood. In cross matching, combinations of the donor's and recipient's cells and sera are mixed to measure their compatibility. In an emergency requiring immediate transfusion, cross matching may be used to determine compatibility between donor and recipient even though their blood types are not known.

In forward ABO grouping (cell grouping) the donor's erythrocytes are mixed with known types of antibodies. In reverse ABO grouping (serum grouping) the presence of unexpected antibodies in the donor's serum is measured by mixing it with known red blood cell types.

In major cross matching, the donor's cells and recipient's serum are mixed to ascertain that the recipient's antibodies will not react with the donor's cells. In another assay, the donor's serum is screened by immunological tests for the presence of unusual antibodies that might otherwise go undetected and be capable of

TABLE 18-6
Combinations of assays used in ABO grouping and cross-matching

Test	*Red Blood Cells (Antigens)*	*Source of Serum (Antibodies)*
ABO grouping:		
Forward	Donor	Known antibody types
Reverse	Known types	Donor
Cross matching:		
Major	Donor	Recipient
Minor	Recipient	Donor

causing agglutination. The latter test makes minor cross matching (the testing of the donor's serum or plasma against the recipient's red blood cells) redundant.

As a rule, recipients possessing anti-A antibodies should not receive cells containing A antigens; those with anti-B antibodies should not receive type B cells; and those who have anti-A and anti-B antibodies should not receive either type A or type B cells.

Individuals with type AB blood cells have neither of the antibodies present and are referred to as universal recipients. *In an emergency* they may receive transfusions of "washed" erythrocytes (cells minus the serum) from persons of groups A, B, or O. Persons with AB blood usually are not given a transfusion of whole blood from these groups since the respective plasmas contain anti-B, anti-A, and anti-A and anti-B antibodies, which may react with the cells of the AB recipient.

Individuals with group O blood lack both A and B antigens and are called *universal donors.* Although in a transfusion the anti-A and anti-B antibodies in type O blood are naturally diluted in the recipient's bloodstream, they still may be a source of some reaction. In order to avoid this possibility transfusions of group O blood to recipients with other blood types are performed with cells "washed" and suspended in isotonic saline (salt solution). This procedure is used only in emergencies when the ABO blood type identical to the recipient's is not available for transfusion.

Other Red Blood Cell Antigens

Among other antigens on red blood cells which may stimulate immune reactions are the Kell, Lewis, Lutheran, Kidd, Duffy, and M, N, and S antigens. Some occasionally cause transfusion incompatibility but are of less importance than the ABO and Rh antigens.

Rh antigens, which may provoke strong immune reactions, were first discovered when the red blood cells of a rhesus monkey were injected into a rabbit and caused the formation of anti-rhesus antibodies. Antibodies in the rabbit serum that reacted with the ABO human blood group antigens were removed by mixing the serum with known human ABO blood types. Still, remaining antibodies agglutinated human red blood cells from a majority of the donors tested. The antigen on the cells responsible for this reaction was named the Rh antigen, and the corresponding antibody, the anti-Rh antibody. Individuals whose erythrocytes have the antigen are called Rh-positive (Rh^+), while those who lack it are Rh-negative (Rh^-). Currently blood is routinely typed for major blood groups (ABO) and also for the presence or absence of the Rh antigen. Nearly 85 percent of whites of western European origin, 72 percent of blacks, 86 percent of American Indians, and 95 percent of Orientals are Rh^+.

Several different Rh antigens have been discovered. According to one explanation, each of three dominant genes, C, D, and E, produces a specific antigen. Among their recessive counterparts, the c and e genes produce weak Rh^- antigens, but no antigen has been discovered for the d gene. Individuals have three pairs of Rh genes but the synthesis of a particular antigen merely requires the presence of only one gene of a given type (Table 18-7). The D antigen, also called the Rh_0 antigen in another system of nomenclature, is the strongest of the Rh antigens and is responsible for nearly 90 percent or more of Rh incompatibilities. The C antigen is responsible for most of the remaining incompatibilities. Routine Rh typing involves testing for the D antigen. Erythrocytes that have it are considered Rh^+ and must never be

TABLE 18-7
Some common Rh antigens

Type	*Antigens*	*Hereditary Makeup (Genotype)*
Rh-positive (Rh^+)	c,D,e C,D,e c,D,E C,D,E	The production of an antigen requires the presence of only one specific gene (which is indicated by the same letter as its antigen). Individuals possess pairs of each of the three genes. In a person homozygous for an antigen, members of the gene pair are identical. Both are dominant or recessive, as indicated by capital or lowercase letters, respectively. Persons are heterozygous for a given gene pair when one is dominant and the other is recessive.
Rh-negative (Rh^-)	c,e C,e c,E C,E	

given to an Rh^- recipient. One variant, the D^u type is a weak D-positive antigen. Although it does so rarely, it can stimulate anti-D (anti-Rh) antibody formation, and individuals with D^u blood cells are designated Rh^+.

If one merely considers the inheritance of D antigens, then the probability of inheritance of a D gene and Rh positivity is assured if either parent is homozygous (DD). It is 3:4 if both parents are heterozygous (Dd), since one gene from each pair is contributed from each parent at conception. When both parents are heterozygous, the chances are 1:4 that their offspring will be Rh-negative. When both parents are homozygous and recessive (dd), they are Rh-negative and all their offspring will be Rh-negative for the D antigen. Similar probabilities can be computed for the C or E antigens once the parental genotypes are determined by study of the appearance of the Rh antigen in family members.

18-6 PATHOPHYSIOLOGY OF RED BLOOD CELLS

A variety of factors may alter the rate of production, release, and destruction of specific blood cells. They include physical agents, genetic, nutritional, infectious, and pharmacological factors, and many diseases. In some instances changes in the erythrocyte number and/or structure provide a clue to the cause of the ailment. The discussion below will focus on disorders related to abnormalities in whole blood and red blood cells. Dysfunction in white blood cells and platelets will be discussed in Chap. 19.

Effects of Blood Loss: Hemorrhage and Shock

The loss of blood, called *hemorrhage,* is external when blood escapes to the outside of the body and internal when blood empties into tissue spaces or body cavities. Either loss may cause *shock,* bringing about changes that decrease blood pressure (Chap. 17). The decline in blood volume and pressure stimulates the cardiac vasomotor and respiratory centers in the medulla (Fig. 18-8). Cardiac output increases, the pulse becomes rapid and "thready" to the touch, and sympathetic adrenergic nerves cause vasoconstriction of vessels in the skin. The latter reaction, coupled with the fluid loss and decreased blood volume, causes the skin to become pale and cool to the touch. Sympathetic cholinergic nerves activate the sweat glands, and fluid loss is further aggravated and body temperature drops. The diminished volume of blood to the brain results in a lesser supply of oxygen to neurons in that organ. The respiratory centers in the medulla and pons are stimulated and breathing becomes rapid and shallow. The diminished oxygen supply to the brain may cause dizziness and fainting.

The symptoms of *primary shock,* the initial reaction to blood loss, resemble those of fainting. The drop in blood pressure initiates negative feedback responses which tend to restore blood pressure (Figs. 18-8 and 17-12). *Secondary shock* results when blood volume is not restored and the level of body fluids is not maintained. A marked decrease in blood pressure and prolonged vasoconstricion may cause *tissue hypoxia,* a decline in oxygen supply to the tissues which causes a shift to anaerobic metabolism. Organic acids are formed at rates that may exceed the buffering capacity of the body fluids, and the change in pH adversely affects the ac-

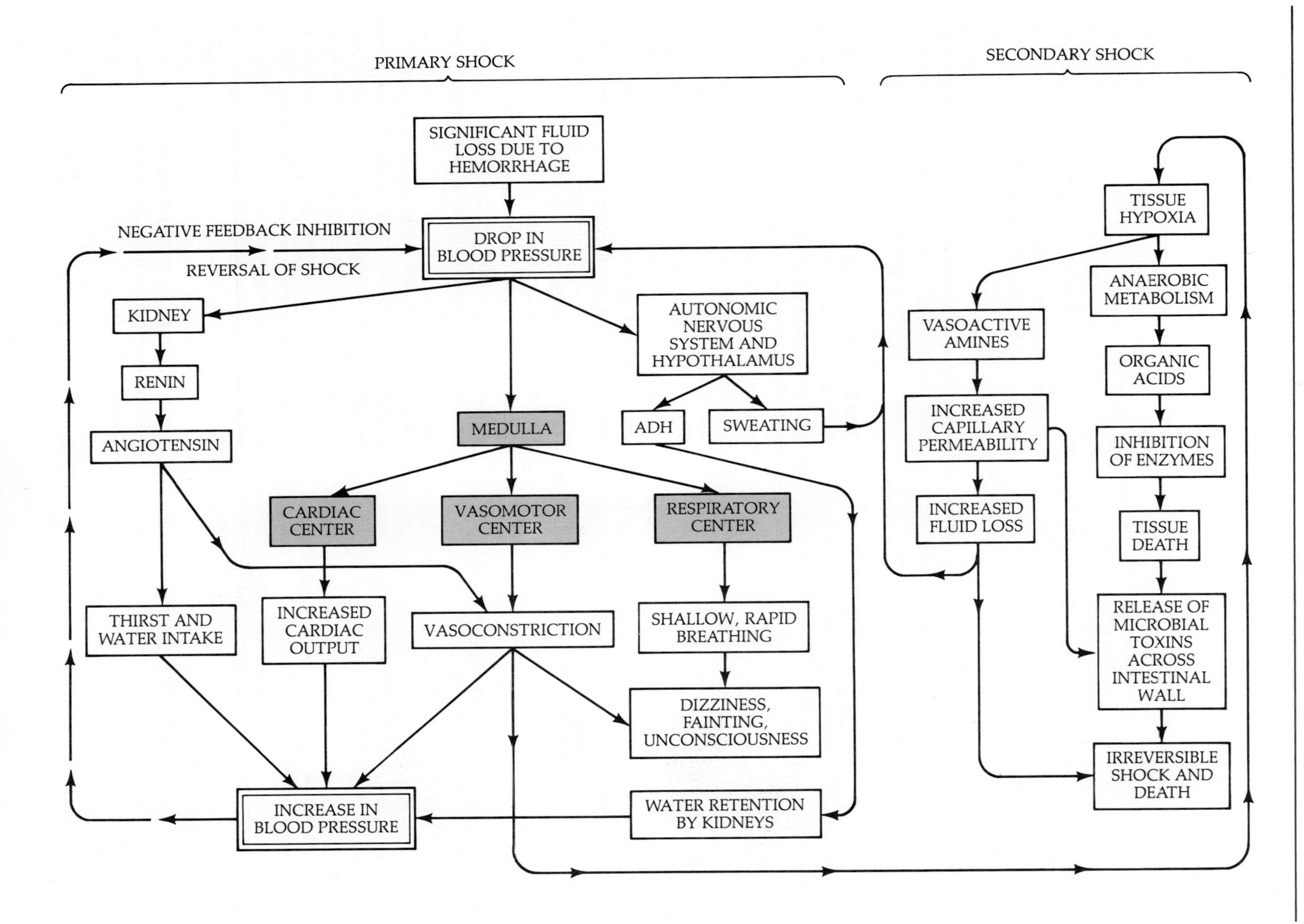

PRIMARY SHOCK
SECONDARY SHOCK
SIGNIFICANT FLUID LOSS DUE TO HEMORRHAGE
NEGATIVE FEEDBACK INHIBITION
REVERSAL OF SHOCK
DROP IN BLOOD PRESSURE
KIDNEY
RENIN
ANGIOTENSIN
AUTONOMIC NERVOUS SYSTEM AND HYPOTHALAMUS
ADH
SWEATING
MEDULLA
CARDIAC CENTER
VASOMOTOR CENTER
RESPIRATORY CENTER
THIRST AND WATER INTAKE
INCREASED CARDIAC OUTPUT
VASOCONSTRICTION
SHALLOW, RAPID BREATHING
DIZZINESS, FAINTING, UNCONSCIOUSNESS
INCREASE IN BLOOD PRESSURE
WATER RETENTION BY KIDNEYS
TISSUE HYPOXIA
VASOACTIVE AMINES
INCREASED CAPILLARY PERMEABILITY
INCREASED FLUID LOSS
ANAEROBIC METABOLISM
ORGANIC ACIDS
INHIBITION OF ENZYMES
TISSUE DEATH
RELEASE OF MICROBIAL TOXINS ACROSS INTESTINAL WALL
IRREVERSIBLE SHOCK AND DEATH

tivity of enzymes. Vasoactive amines such as histamine are released and increase the permeability of the capillary beds, which leads to further fluid loss. The increased permeability of the capillaries may allow toxins from bacterial cells in the intestine to enter the blood. Anything which reduces cardiac output may contribute to shock. Regardless of the initial cause, the mechanisms of shock are similar. In the earlier stages of the response, shock is *reversible*. Restoration of the fluid volume and correction of metabolic acidosis should be the primary treatment procedures. However, if the fluid loss approximates 30 percent or more of blood volume, *irreversible shock* and death become likely results.

Hemolytic Disease of the Newborn

The placenta is a selective barrier between the mother and the fetus which allows the delivery of nutrients to the baby's bloodstream and the removal of waste products from it but normally prevents the exchange of cells. The placenta is formed jointly by adjacent tissues, including separate pools of maternal blood vessels of the mother's uterine lining and the fetus's developing circulatory system (Fig. 24-14). As pregnancy continues through the third trimester (sixth through ninth months), the size of the baby and the limits of expansion of the uterus increase the chances that some placental blood vessels may become faulty. In one specific set of conditions, namely when an Rh-positive baby is conceived by an Rh-negative mother and when a faulty placenta develops, sufficient numbers of the baby's Rh-positive red blood cells may move into the Rh-negative mother's bloodstream to stimulate the production of anti-Rh antibodies by the mother.

During the first such pregnancy there are usually too few antibodies formed to cause harm. However, in subsequent pregnancies in which the same conditions prevail, sufficient anti-Rh antibodies may be formed by the mother and cross the placenta to agglutinate the baby's red blood cells. The antibodies combine with Rh antigens on the surface of the baby's red blood cells and enhance their attractiveness to phagocytes in the reticuloendothelial system. The macrophages in the liver (the Kupffer cells) ingest Rh-positive cells with their adhering antibody. The blood cells are destroyed (hemolyzed) and the hemoglobin is converted into bile pigments. The stem cells of the baby's marrow produce new red blood cells and release them as immature erythroblasts (precursors of erythrocytes) in the bloodsteam. Thus the disorder is called *erythroblastosis fetalis* (or hemolytic disease of the newborn, HDN, since it results in lysed red blood cells in the baby).

While the baby is in the uterus and the placenta is intact, bile pigments from the degradation of Hb enter the mother's bloodstream for disposal. Shortly after birth the newborn cannot yet efficiently excrete the bile pigments, which may accumulate and interfere with liver metabolism and produce a jaundiced skin color. Bilirubin, a pigment derived from hemoglobin, may accumulate in the gray matter of the central nervous system and be associated with neuron degeneration in a disorder called *kernicterus*. The basal ganglia deep within the cerebral hemispheres (Table 11-6) are especially affected, and varying degrees of neurological symptoms and even death may follow.

An infant born with erythroblastosis has an insufficient number of normal red blood cells and suffers from hemolytic anemia. This reaction occurs in about l percent of the pregnancies in which the baby is Rh-positive and the mother Rh-negative. Yet, some Rh-negative women have given birth to 12 or more Rh-positive babies without the development of incompatibilities. The passage of significant quantities of blood cell antigens apparently occurs only in a small percentage of pregnancies. In spite of our knowledge and the availability of therapy, nearly 7 of every 100,000 babies born alive each year in the United States die of the disease (which accounts for about 200 or more deaths annually). About 90 percent of the severe cases of erythroblastosis are due to fetal-maternal incompatibility of the D factor. Of the remaining 10 percent , the majority are due to a

FIG. 18-8
Effects of hemorrhage and fluid loss.

mismatch of ABO antigens, while about 2 percent occur because of reactions involving Kell, Duffy, Lutheran, Kidd, C, E, or other antigens.

Three modes of therapy have been developed to treat Rh-induced hemolytic disease. These include the administration of anti-Rh antibody, phototherapy, and exchange transfusion. Preventive therapy consists of administration of purified *anti-Rh antibody* (an anti-Rh_0 immunoglobulin, called Rhogam). It is injected into an Rh-negative mother just after the birth of her first Rh-positive child and subsequent ones. The antibody neutralizes any Rh_0 antigen that may have leaked into the mother's blood at the baby's birth, so that her immune system is not sensitized to Rh_0 and does not produce anti-Rh_0 antibodies. This preventive therapy markedly reduces the risk of hemolytic disease of the newborn in future pregnancies.

In some instances erythroblastosis fetalis can be corrected by exposure of the affected baby to *phototherapy*. After being blindfolded to prevent damage to the retina the infant is exposed to ultraviolet light. The ultraviolet light converts bile pigment derivatives of hemoglobin, such as bilirubin, in the skin, into compounds which do not interfere with liver metabolism. Since the mother's anti-Rh antibodies have a limited lifespan, they gradually decrease in quantity in the child's blood after birth and the symptoms of the disease disappear.

The baby also may be treated by receiving two or more complete *exchange transfusions* of blood to dilute any anti-Rh antibodies. The blood administered is the baby's major ABO blood type and is Rh-negative. This prevents possible agglutination of the transfused cells by any anti-Rh antibody remaining in the baby's blood. If the baby's major blood group is also incompatible with the mother's blood, O negative cells are administered to the baby.

General Types of Anemia

Anemia is a decrease in the quantity and/or function of red blood cells and/or hemoglobin below normal levels. Many of its symptoms, such as physical and mental fatigue, shortness of breath, and palpitations of the heart, are due to insufficient O_2 in the tissues. Anemia involves a decrease in hematocrit or hemoglobin content below normal limits for an individual at a given altitude, age, and sex (Table 18-3). For adult men at sea level, anemia exists when the hematocrit drops below 42 percent and the Hb below 14 g per 100 mL; for women the respective thresholds are 36 percent and 12 g per 100 mL. Anemias may be classified as to their cause and subdivided into those due to decreased red blood cell production, increased red blood cell destruction, and acute blood loss.

Different forms of anemia also can be described in terms of the average diameter of the red blood cells and/or their shape and by their Hb content (Table 18-8). The cells are small in *microcytic anemia*, large in *macrocytic anemia*, and of normal size in *normocytic anemia*. The hemoglobin levels may be low (*hypochromic*) or normal (*normochromic*) or may appear high (*hyperchromic*), causing the red blood cells to be pale, moderately red, or very red. In addition to nutritional deficiencies, diseases of the liver and gallbladder, and incompatibility reactions which cause *hemolytic anemias*, the disorders may develop as a result of hereditary defects (Table 18-8).

Hereditary Anemias

A variety of hereditary anemias result from specific genes which direct the synthesis of abnormal forms of Hb. One type, *sickle-cell anemia*, occurs especially in blacks (and is due to replacement of a single amino acid among the 146 in the beta chain); another is *thalassemia*, Cooley's anemia, or Mediterranean anemia, which occurs most often in people of Mediterranean, African, or Asian ancestry.

Sickle-Cell Anemia

Sickle-cell anemia is usually a normocytic, slightly hypochromic type of anemia. It kills about 80,000 people annually worldwide. When oxygen levels are low, the abnormal red blood cells transform from a biconcave disc to a more rigid crescent shape, which resembles a farmer's sickle (Fig. 18-9).

Sickle-cell Hb (HbS) is synthesized under the

TABLE 18-8
Some major forms of anemia

Type	*Characteristics of Red Blood Cells*	*Cause(s)*
Microcytic, hypochromic	Less than 6 μm in diameter, low level of Hb	Vitamin B_6 deficiency
		Iron deficiency; may be due to deficiency of iron-carrying serum protein, transferrin, or to increased iron demand during pregnancy
		Infection
		Heredity: thalassemia and some forms of sickle-cell anemia
Macrocytic, normochromic (may be hypochromic)	More than 8 μm in diameter, high level of Hb	Folid acid deficiency
		Vitamin B_{12} (extrinsic factor) deficiency
		Intrinsic factor deficiency; needed for absorption of vitamin B_{12}
		Failure to absorb B_{12} is called pernicious (Addison's) anemia and may be caused by autoimmune destruction of stomach cells which produce the intrinsic factor
Normocytic, normochromic	Normal size and Hb content	Blood loss (hemorrhage)
		Burns
		Poisons: arsenic, DDT, benzene, radiation
		Immunological disease
		Heredity, including some forms of sickle-cell anemia

direction of a recessive gene which substitutes the neutral amino acid valine for the acidic glutamine in position 6 of the 146 amino acids in the β chains. Valine is water-repelling (hydrophobic), whereas glutamine is water-attracting (hydrophilic). The substitution of valine adversely affects the ability of Hb to transport oxygen so that oxygen levels decrease in the blood. The effect is noticeable during physical activity, with respiratory infection, under conditions which alter acid-base balance, and with high-altitude exposure. When oxygen demand is increased, the cells become distorted, cling to each other, and clog capillaries. The sickle shape may be due to stacking of reduced sickle-cell hemoglobin molecules, which form rigid fluid crystals or aggregates of HbS molecules that form long fibers. The aggregated HbS decreases the flexibility of the sickle cells, making their movement within capillaries difficult. The disruption of the cells causes them to hemolyze, hemoglobin to leak out, and anemia to result.

FIG. 18-9
Appearance of red blood cells in sickle-cell anemia (×1000). Note the crescent-shaped sickle cells in the photomicrograph.

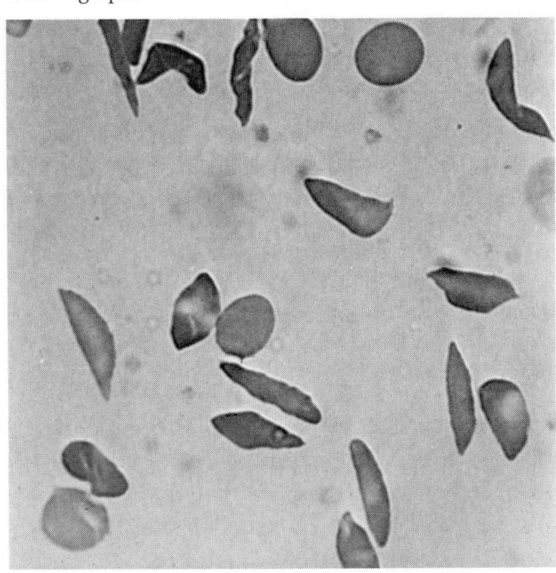

About 0.1 to 0.2 percent of blacks born in the United States have *sickle-cell disease* (sickle-cell anemia; Table 18-9), a serious malady which is often fatal before 30 years of age. These individuals have two recessive genes for the formation of sickle-cell hemoglobin, and their genetic constitution is designated as HbSS. In such adults, 80 to 85 percent of their hemoglobin is HbS,

TABLE 18-9
Possible reactions in sickle-cell disease

Fatigue, pallor weakness, headache
Phagocytosis of sickle cells in spleen
Enlargement of spleen (splenomegaly) in childhood
Sudden pooling of blood in spleen
Normocytic normochromic anemia
Impeded flow in specific capillaries and pain in the area and destruction of tissue (occurs in spleen); severe joint pain is common
Thrombosis may impede blood flow to tissue sites, causing ulcerations and gangrene
Failure to regenerate red blood cells (aplastic anemia)
Increased susceptibility to bacterial infections, such as salmonella and pneumococcal infections (which cause typhoid fever and bacterial pneumonia, respectively)
Death

1 to 20 percent is HbF, and 2 to 4.5 percent is HbA_2.

People with sickle-cell disease are protected during the first few months of life by high levels of fetal hemoglobin (HbF). Thereafter, they are in reasonably good health but may suffer from one or more types of crises. They may periodically have a painful crisis (due to blockade of the blood supply to a tissue and its death or infarction), an aplastic crisis (due to depression of the bone marrow, which causes a marked decrease in Hb levels), a sequestration crisis (in which red blood cells accumulate in pools in the spleen, which becomes enlarged), and hemolytic anemia (due to destruction of red blood cells or hemoglobin). (See Table 18-9.)

Approximately 9 percent of the blacks in the United States have *sickle-cell trait* and carry a single recessive gene for the condition and one normal dominant gene. Their genetic makeup is designated as HbAS and their hemoglobin is 60 to 70 percent HbA, 30 to 45 percent HbS, HbF less than 0.5 percent, and HbA_2 1.5 to 4.5 percent. Because a large percent of their Hb is normal, no clinical signs of blood abnormalities are evident. Sickling may occur, however, when blood levels of oxygen decrease, which may result from a number of conditions, including exposure to high altitude, respiratory infection (Chap. 20), or congestive heart failure.

Several tests exist for HbS. When cells with HbS are mixed with a 2 percent metabisulfite solution in the well of a sealed glass slide at 37°C, they assume the sickle shape within 2 h. HbS can also be detected by its precipitation when sickle cells are mixed in a test tube with inorganic buffers and reducing agent. Additionally, electrophoresis may be used to separate different forms of Hb and to identify HbS. At an alkaline pH (8.6) Hb molecules have a net negative charge and various forms (including HbS) migrate to the positive electrode (anode) at different rates (Fig. 18-10).

Although there is yet no cure, a number of therapies for sickle-cell anemia are currently under study. The most practical one is prevention by avoidance of those situations which may induce crisis. Preventive approaches include the provision of adequate nutrition, including folic acid supplements to promote red blood cell formation; prompt treatment of infections; avoidance of strenuous work and vigorous exercise to minimize oxygen demand; and avoidance of cold to minimize vasoconstriction and the clogging of blood vessels. Experimental approaches to modify HbS have been attempted with varying degrees of success (Table 18-10). Some investigators have proposed genetic engineering as a cure, in which normal Hb gene nucleotides would be inserted into stem cells in bone marrow. Then, should the new genes survive, they would direct the synthesis of normal Hb. The ethical implications of genetic engineering, its possible unexpected genetic effects, and technical difficulties have impeded this approach.

Thalassemia

Thalassemias originally were discovered in people near the Mediterranean Sea and were named after the Greek word *thalassa,* sea. *Thalassemias* are microcytic, hypochromic, hemolytic anemias caused by decreased rates of synthesis or failure of synthesis of one or more α or β chains. The disorders are accompanied by jaundice, enlargement of the spleen, and stunted growth and can be fatal.

TABLE 18-10
Agents proposed to treat sickle-cell anemia

Agent	*Mechanism*	*Disadvantage*
Urea	Ruptures hydrophobic bonds; desickles cells	Strong diuretic, causes fluid loss
Cyanate	Combines with valine; antisickling	Potent toxin of aerobic enzymes; irreversible combination with Hb
Vitamin B_{17} (nitrilosides)	Found in fruit seeds, bitter almonds, cassava, and dark lima beans; contains cyanide; metabolizes to form cyanate and other antisickling compounds	Induces goiter (thyroid growth) in absence of adequate dietary iodine; depletes amino acid levels with diets marginal in protein
Mg^{2+} administration	Inhibits prostaglandin hormone activity; antisickling	Not always effective
Bone marrow transplant	Normal stem cells are seeded in the subject to form normal red blood cells and Hb	Tissue must be matched between donor and recipient; otherwise, tissue rejection occurs owing to immune reaction

The more severe forms, *thalassemia major,* are due to a homozygous dominant defective condition. The less severe disorders, *thalassemia minor,* are due to the heterozygous genetic state, in which one defective gene is paired with a normal gene. Four genes which govern hemoglobin α chain synthesis are found in whites and three in blacks. Deficiencies in α chain synthesis and the resulting α-thalassemia becomes more severe as the proportion of normal α genes is reduced, and replaced by defective ones or are deleted entirely.

Babies that lack all four α genes die in the fetal stage (are stillborn). *α-Thalassemia major* is a homozogous recessive condition in which no α genes are present and all four Hb chains are γ chains. This produces Bart's Hb, and the fatal disorder results in stillborn infants and is known as *hydrops fetalis. α-Thalassemia minor* is a heterozygous condition in which the individual is missing three of the four alpha genes and forms Hb with four β chains, called HbH (or B_4) along with some Hb Bart's (γ_4). The anemia is chronic and its severity varies. When two α genes are absent, the cells are just slightly hypochromic, the anemia is very mild, and no clinical symptoms are evident.

β-Thalassemia is most serious when both β genes are defective in thalassemia major, or *Cooley's anemia.* Messenger RNA is either produced inefficiently or is degraded rapidly. In these instances γ chains are usually synthesized, resulting in higher HbF levels, but at a rate too slow to make up for the missing β chains. As a result, α chains are in excess, and they precipitate and damage the cells. In heterozygous individuals with one defective β gene, the disorder varies from moderately severe thalassemia intermedia to thalassemia minor, or *Cooley's trait,* in which there are no obvious symptoms.

FIG. 18-10
Electrophoretic separation of hemoglobin molecules. Two drops of lysed blood cells are deposited with a capillary tube across a cellulose acetate strip. The strip, which had been previously saturated with buffer at an alkaline pH, is placed across positive and negative electrodes in a chamber. The current is turned on to attain 350 V for 30 min. Hb proteins move from the point of application toward the anode (positive electrode) at a rate dependent on their net negative charge. After they migrate, the strip is removed from the chamber and immersed in a stain specific for protein. Separated bands of hemoglobin are made visible. The sickle-cell hemoglobin (HbS) band is drawn with a broken line since HbS is not normally present. It migrates at a rate slower than those of HbA and HbF but faster than those of HbA_1 and HbA_2.

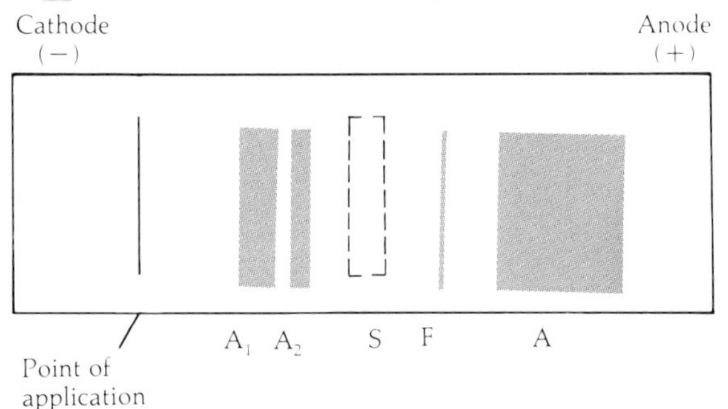

18-7 SUMMARY

1 Blood transports substances, buffers the pH of body fluids by means of inorganic salts and proteins, and defends the body against disease.

2 About 55 percent of blood consists of a fluid called plasma and 45 percent is made of formed elements (cells and platelets). When blood clots, a yellowish fluid, serum, separates from the cells and the clotting elements in plasma.

The volume, viscosity, and sedimentation rate of blood usually vary within narrow limits, which may be exceeded in specific disorders. The hematocrit or packed cell volume reflects the red blood cell number and hemoglobin content. The hematocrit varies with age, sex, and exposure to high altitude. Anemia is a decrease in the hematocrit involving a less than normal quantity and/or function of red blood cells and/or of hemoglobin.

Blood is usually maintained within a pH range of 7.35 to 7.45 by the presence of buffers, which are organic and inorganic compounds that neutralize excess acid or base. The major inorganic buffer, sodium bicarbonate, and the analogous compound potassium bicarbonate are regulated primarily by activities of the kidneys. The lungs also influence buffer levels and the pH of blood by the elimination of CO_2 with exhaled air.

3 The origin of blood cells is the yolk sac, from which embryonic cells migrate to the fetal liver and bone marrow. In those areas colonies of stem cells, which mature along separate lines, form erythrocytes (red blood cells) and leukocytes (white blood cells). Other white blood cells, lymphocytes, are formed by stem cells which originate in the marrow and migrate to lymphoid tissue to form ultimately T and B lymphocytes. Still other stem cells produce giant multinucleated megakaryocytes, from which cytoplasmic pieces are pinched off to form platelets (thrombocytes).

4 The average numbers of erythrocytes, leukocytes, and thrombocytes are 4.2×10^6 to 5.9×10^6, 4.3×10^3 to 10.8×10^3, and 1.4×10^5 to 4.4×10^5 per microliter of human blood, respectively. The hemoglobin (Hb) content of red blood cells averages 14 to 16 g per 100 mL of blood. Red blood cells are biconcave discs which lack a nucleus and range in size from 6 to 9 μm. They are derived from reticulocytes, young nucleated red blood cells that have an abundance of ribosomes and endoplasmic reticulum. Hemoglobin (Hb) loosely binds to O_2 to produce oxyhemoglobin (HbO_2), which imparts a vivid red color to blood. When it gives up O_2, it can carry CO_2 in the form of carboxyhemoglobin ($HbCO_2$) and also can carry hydrogen in place of oxygen as reduced hemoglobin (HHb), which make blood dark red or blue.

The body produces a variety of hemoglobin molecules, each of which has two pairs of polypeptide chains. Specific gene pairs active at various stages of life, from early development through adulthood, direct synthesis of different polypeptide chains. Two α chains are normally paired with other types of chains to form each molecule of Hb. In the first 2 months after conception, α and ϵ chains are produced to form Gower Hb. Then increased amounts of fetal hemoglobin, HbF (with α and γ chains), are synthesized. HbF, whose formation decreases after birth, makes up less than 0.5 percent of adult Hb, while HbA (with α with β chains) and HbA_2 (with α and δ chains) compose 91 to 95 percent and 1.5 to 3.5 percent, respectively, of adult Hb.

5 The major blood groups are distinguished by the presence of A (type A), B (type B), A and B (type AB), or neither (type O) antigen on the surfaces of red blood cells. Antibodies against A and B antigens are naturally present in low levels in individuals who lack the antigens. When the antigens are transfused to a person lacking them, large amounts of anti-A or anti-B antibodies

are formed against them. A fatal reaction may result when sufficient antibodies are formed.

Erythrocytes with the Rh antigen (Rh-positive cells) can elicit similar responses. Rh antigen on the cells of an unborn child may cross a defective placenta in sufficient quantities to cause the mother to synthesize anti-Rh antibodies. In a second or later such pregnancy enough antibody may pass from the mother into the unborn baby's blood to clump its cells. This disorder is called erythroblastosis fetalis or hemolytic disease of the newborn, HDN.

6 Hemorrhage (blood loss), which may be internal or external, may lead to shock, which at first may be reversible but later may be irreversible and fatal. Shock becomes irreversible when tissue hypoxia stimulates the excess production of organic acids by anaerobic metabolism; these acids shift the pH of body fluids beyond normal limits. The buffering capacity of inorganic salts and proteins is exceeded. As a result, the individual's heartbeat, respiration, and pulse become rapid, blood pressure declines, the pulse weakens, and the skin gets cold. Vital organs such as the brain become deprived of oxygen and unconsciousness results. Damage to the kidneys and further fluid loss due to sweating compounds the problem. Large amounts of bacterial toxins from the intestines may seep into the blood, often with lethal results.

Anemia results when the hematocrit or hemoglobin level decreases below the normal limits for a person's age and sex. Anemia may be caused by physical, genetic, nutritional, infectious, or pharmacological factors.

Sickle-cell disease (sickle-cell anemia) is a hemolytic condition which is often fatal before age 30. It results from two recessive genes that form abnormal β chains by substituting valine for glutamic acid in the 6th position of each chain. When oxygen levels are low, the sickle-cell hemoglobin (HbS) molecules may aggregate and cause red blood cells to change from biconcave discs to crescent shapes. Oxygen transport is impeded, capillaries may clog, the spleen may enlarge, and pain may result. Sickle-cell trait is the heterozygous counterpart, in which no clinical blood abnormalities are evident but which also may produce sickling when blood levels of oxygen are low.

Thalassemias (Mediterranean or Cooley's anemias) are recessive genetic anemias in which α-chain synthesis is abnormal and/or β-chain production is retarded. Thalassemias are hemolytic anemias in which the spleen may enlarge and growth may be stunted; they may be fatal.

REVIEW QUESTIONS

1 List the formed elements in blood. Describe their characteristics.
2 Name the main components of plasma.
3 How does one determine a hematocrit and under what conditions does it vary?
4 How do inorganic salts and proteins buffer the blood?
5 Give an account of the role of bone marrow in the production of formed elements.
6 Describe the formation of red blood cells.
7 What is the Rouleau effect?
8 Explain the role of the kidneys in regulating red blood cell division.
9 Describe the physical and chemical properties of hemoglobin.
10 Outline the genetics of normal hemoglobin formation. How are types of Hb identified chemically and electrophoretically?

11 Describe the antigens, antibodies, and genotypes in blood types O, A, B, and AB. What chemical differences distinguish these antigens? How can their presence be verified?
12 Define hemorrhage and discuss its effects.
13 Discuss the importance of the Rh antigen and its genetics.
14 When is the Rh factor a problem between mother and baby? Discuss the consequences of the problem and describe means of preventive and corrective therapy.
15 Compare microcytic, macrocytic, and normocytic anemias.
16 Compare hypochromic, hyperchromic, and normochromic anemias.
17 Describe sickle-cell disease and sickle-cell trait. Discuss some types of experimental therapy for sickle-cell anemia.
18 What are thalassemia, α-thalassemia, and β-thalassemia? Describe the genetic basis and consequences of each.

THE CIRCULATORY SYSTEM

The Physiology of Blood, Part II: The Properties of Leukocytes (White Blood Cells) and Thrombocytes (Platelets)

19-1 PROPERTIES OF LEUKOCYTES
Development and appearance
Numbers of leukocytes
General functions
Neutrophils
Eosinophils
Basophils
Lymphocytes
Monocytes
Defense functions

19-2 PROPERTIES OF THROMBOCYTES
Development and structure
Numbers of platelets
General functions
Clotting sequence
Inhibition of clotting
Tests for clotting elements

19-3 PATHOPHYSIOLOGY OF LEUKOCYTES
General disorders
Leukemias

19-4 PATHOPHYSIOLOGY OF THROMBOCYTES (PLATELETS)

19-5 SUMMARY

OBJECTIVES

After completing this chapter the student should be able to:

1. Describe the development of leukocytes.
2. Compare the structure of neutrophils, eosinophils, basophils, lymphocytes, and monocytes. Give the relative frequency of each in normal blood.
3. Describe the functions of each of the cells listed in objective 2.
4. Discuss the development of platelets and their structure.
5. Outline the three major stages in blood clotting, noting the roles of platelets and calcium ions.
6. Discuss the relationship of white blood cells to infectious diseases and leukemias.
7. Explain the relationships among disorders in coagulation, the liver, and platelets.

As previously described, whole blood is composed of a fluid, plasma, in which erythrocytes, leukocytes, and thrombocytes are suspended. Erythrocytes were described in Chap. 18, and the properties of leukocytes and thrombocytes will be discussed now.

Although outnumbered 100- to 1000-fold by erythrocytes, the leukocytes, which exist in several different forms, exert profound effects on body functions. They have a wide range of capabilities, including the ability to ingest cells and particles, secrete proteins which regulate cellular activities, release hydrolytic enzymes, and produce specific molecules which assist the body's defenses.

Thrombocytes are remarkable granular derivatives formed from pinched-off segments of megakaryocytes wrapped in pieces of cell membranes. Thrombocytes, commonly called platelets, influence many activities by releasing a variety of molecules from their storage reservoirs. Their effects range from involvement in blood coagulation to stimulation of smooth muscle cell proliferation in atherosclerosis.

19-1 PROPERTIES OF LEUKOCYTES

The number of leukocytes averages 6700 per microliter in whole blood (Table 18-1). The count differs between whites and blacks, males and females, and young and old and also during the course of many diseases. The size of the cells ranges from 6 to 30 μm (compared with red blood cells, which average 7.5 μm). Although there are several types of leukocytes, each of which has different functions, the various types can be broadly divided into two groups: those that contain coarse cytoplasmic granules are called *granulocytes,* and those without them are *agranulocytes* (Table 19-1). The granulocytes are the neutrophils, eosinophils, and basophils, and the agranulocytes are the monocytes and the lymphocytes (Fig. 19-1).

Development and Appearance

Stem cells in the bone marrow produce colony-forming cells, which mature along specific lines to produce various white blood cells (Fig. 18-1). Their estimated life-spans vary under normal conditions and may range from 6 to 10 h for granular leukocytes to about 40 h for monocytes to 90 to 100 days for large lymphocytes, and up to 15 to 20 years for small lymphocytes, which are called *memory cells* and are formed during an immune response. During infection the life-span of some leukocytes may be as short as 2 h. In general, leukocytes are more flexible than erythrocytes, and unlike them can squeeze between endothelial cell junctions in blood vessels and lymph vessels without losing their nuclei to enter or leave the blood (compare Figs. 18-2 and 19-2). Movement of cells between blood vessels and tissue spaces is called *diapedesis* and allows cells mobility to participate in defense reactions in the body.

White blood cells may be distinguished microscopically in stained blood smears. A combination of dyes which are acid (red) and alkaline (blue) is commonly applied in Wright or Giemsa stains. Cellular components which combine with both types of dyes equally are termed *neutrophilic* (neutral-loving). The cyto-

TABLE 19-1
Properties of white blood cells

Cell Type	CHARACTERISTICS* Cytoplasm	Nucleus	Diameter, μm	Thousands/μL (range)	Percent (range)
Granulocytes:					
Neutrophil	Reddish purple to pink-lilac granules	Nonsegmented to 3 to 5 lobes	9–12	4.2 (2.2–7.6)	64 (50–70)
Eosinophil	Red-orange to red-purple	Nonsegmented to 2 oval lobes	10–14	0.13 (0–0.5)	2 (1–4)
Basophil	Purple to black granules of varied shapes	Indented, lobed, or S-shaped	8–10	0.03 (0–0.16)	0.3 (0–1.0)
Agranulocytes:					
Lymphocyte	Usually slight amount, blue	Occupies most of cell	6–30	1.8 (0.8–3.1)	27 (25–40)
Small	Sky blue to dark blue, agranular rim of cytoplasm	Large, dark blue-purple	6–12		
Large	Indented cell membrane, more abundant cytoplasm	Large, dark blue-purple, eccentric, clumped, condensed chromatin	20–30		
Monocyte	Gray-blue, very fine granules, abundant cytoplasm	Oval to kidney-shaped	12–15	0.4 (0.1–0.8)	5 (3–8)

* Appearance with Wright's stain.

plasmic granules of certain leukocytes stain reddish purple to light pink-lilac. Those granules of cells which attract the red dye are *eosinophilic* (eosin-loving). Those that stain blue with the alkaline dye are *basophilic* (base-loving). On the basis of these reactions, granulocytes can be identified as *neutrophils, eosinophils,* and *basophils* (Table 19-1 and Fig. 19-1). They are further differentiated by variations in cell size and by the appearance of nuclear and cytoplasmic components.

Numbers of Leukocytes

The white blood cell count shows wide normal ranges (Table 19-1). The relative frequency of each type may be noted in a stained blood smear in which 100 to 200 cells are counted and the percentage of each type determined in a *differential count.* The average percentages of observed leukocytes are neutrophils 64, lymphocytes 27, monocytes 5, eosinophils, 2, and basophils 0.3 (Table 19-1). The absolute numbers and percentages vary with the physiology of the individual and are often altered dramatically by disease. Various types of leukocytes are produced and released from their tissues of origin at different times to combat particular infections.

General Functions

Neutrophils

Neutrophils are granulocytes that usually are the most abundant type of white blood cell, forming 50 to 70 percent of the leukocyte population. As the cells mature from stem cells in bone marrow, the nucleus of the neutrophil is transformed from a single round form to a band shape to a multilobed segmented structure. A band or stab cell has a nucleus formed of rudimentary lobes connected by a wider band (rather than a narrow thread) or has one lobe that is U-shaped. The nucleus of a mature neutrophil has two to five lobes connected by narrow threads of chromatin. (Chromatin comprises the colored strands of protein and nucleic acid of which chromosomes are composed.) Owing to this variable appearance, neutrophils

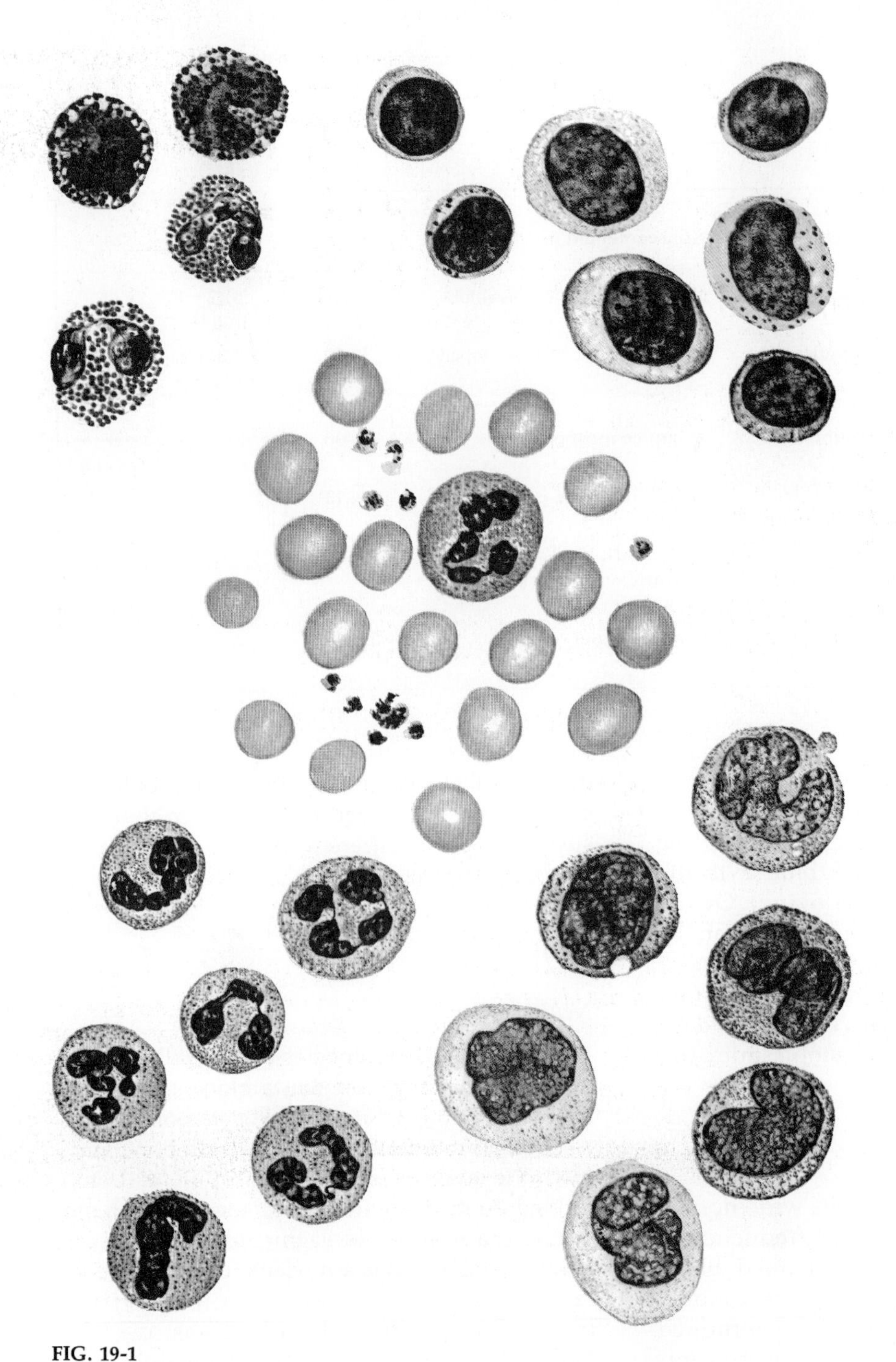

FIG. 19-1
Microscopic appearance of cells and platelets in a blood smear prepared with Wright's stain ($\sim \times 1000$). Eighteen mature red blood cells, several platelets and a neutrophil are shown in the center. At the top left, are two basophils, below which are two eosinophils. At the top right, are three large and four small lymphocytes. At the bottom left, are six neutrophils. At the bottom right, are six monocytes. From *Histology*, 4th ed., by Weiss and R. D. Greep. Copyright © 1977, McGraw-Hill Book Company. Used with permission of McGraw-Hill Book Company. (See inside back cover for color plate of Fig. 19-1.)

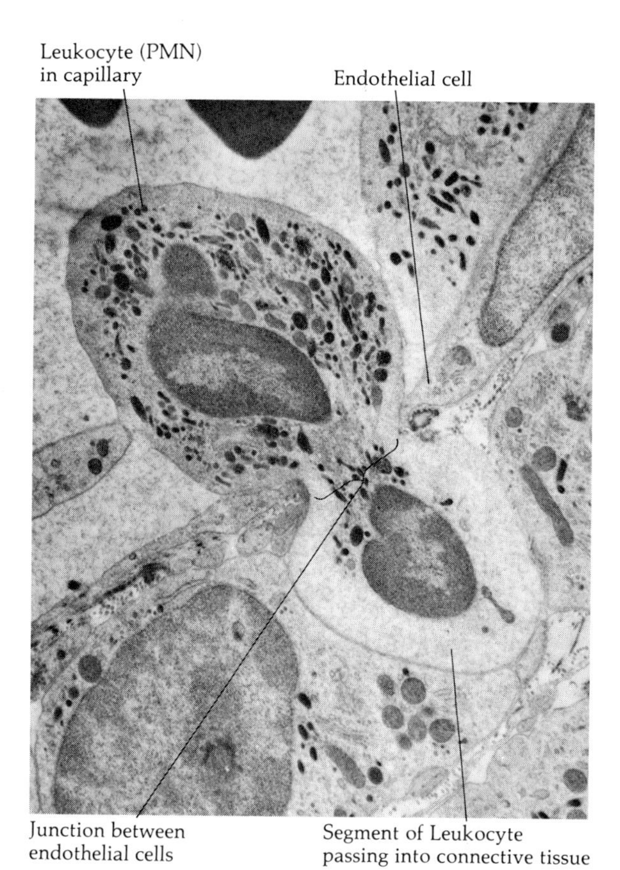

FIG. 19-2
Diapedesis: electron micrograph showing passageway of white blood cell from inflamed venule to connective tissue (× about 10,000).

and other granulocytes with segmented nuclei are called *polymorphonuclear leukocytes* or PMNs (Fig. 19-1). PMNs are small phagocytes (*microphages*) capable of ingesting and killing large numbers of other cells such as bacteria. In a process called *chemotaxis* the PMNs are attracted by chemicals to tissues that have been invaded or damaged. The PMNs then participate in defense reactions in which PMN-cell surface receptors interact with substances or particles at the tissue site (Fig. 4-5) and ingest them by *endocytosis* (page 95). Neutrophils are especially active in this regard.

The lysosomes of PMNs, called *phagosomes* or *phagolysosomes,* release enzymes that can degrade foreign molecules. Leukocytes in general, and neutrophils in particular, secrete a wide range of *bacteriostatic* chemicals (which inhibit bacterial growth) and/or *bactericidal* chemicals (which kill bacteria). An example is lactoferrin, a bactericidal agent that binds and withholds the iron necessary for bacterial growth. Other specific lysosomal granules release a group of bactericidal enzymes such as the *myeloperoxidases.* These enzymes catalyze reactions which form peroxides such as hydrogen peroxide, which oxidize susceptible molecules, in reactions that are fatal to microorganisms.

Among the most common of applied bacteriostatic agents are antibiotics which inhibit bacterial proliferation and thereby assist leukocyte defenses. Should natural or administered antibacterial compounds fail, microorganisms may multiply within leukocytes, kill them, rupture their membranes, and spread to other tissues. The cellular debris and fluid accumulating in the zone of combat between PMNs and microorganisms is commonly referred to as *pus.*

Eosinophils

Eosinophils are granulocytes characterized by reddish orange cytoplasmic granules and a reddish purple nucleus, which may be unsegmented or may have two lobes (Fig. 19-1). These cells, which constitute about 1 to 4 percent of the leukocyte population, have a high affinity for acids and are also called *acidophils.* Eosinophils are less phagocytic than neutrophils. However, the oxidative enzyme activity of eosinophils is important in defense reactions and may increase, as may their number, in certain allergic responses.

Basophils

Basophils are relatively rare in the blood, making up about 0 to 1 percent of the leukocytes. The shapes of their purple to black cytoplasmic granules are highly varied, and the nucleus may be indented, lobed, or S-shaped (Fig. 19-1). Basophils are found in larger relative numbers in interstitial fluids than in blood. Outside the blood these cells are called tissue basophils and are thought to form *mast cells* (Chap. 3).

Some of the granules of the basophils and mast cells are reservoirs of heparin or histamine, which are probably released by exocytosis. The release of heparin prevents blood from clotting and helps to keep it flowing, while histamine widens arteries and increases capillary permeability. As a result of these activities, greater quantities of antibodies, enzymes, and other molecules that aid body defenses are delivered into tissues from the blood. The numbers of circulating basophils and eosinophils decrease following increases of glucocorticoid hormones secreted by the adrenal cortex as a result of stress (Table 19-4, page 467), which is part of a general endocrine adaptation to stress (Fig. 5-7, page 112).

Lymphocytes

Lymphocytes (Fig. 19-1) normally are second only to neutrophils in frequency of occurrence of the white blood cells. They constitute 25 to 45 percent of the circulating leukocytes and are subdivided into *small lymphocytes* (6 to 12 μm in diameter) and *large lymphocytes* (20 to 30 μm in diameter). Lymphocytes become smaller as they mature and a whole array of sizes can be found in blood.

The cytoplasm of lymphocytes usually appears light sky blue to dark blue in color and lacks granules, but may possess small globules. The nucleus is usually large, round or slightly indented, and dark blue to purple. It occupies most of the cell, leaving a very narrow rim of cytoplasm. The chromatin of the nucleus appears clumped or lumpy.

Although small lymphocytes cannot be distinguished from each other under the light microscope, several forms exist. By using the scanning electron microscope or specific antibodies, two major types of small lymphocytes may be identified as *T cells* and *B cells* (Fig. 19-3). The surface of the T cell has fewer antibody-like proteins than that of the B cell. T cells participate in cellular defense reactions designated as *cellular immunity*, while B cells assist defenses by the secretion of antibodies into the body fluids in a response called *humoral immunity* (Chap. 3). Their activities will be described further in the section on defense functions.

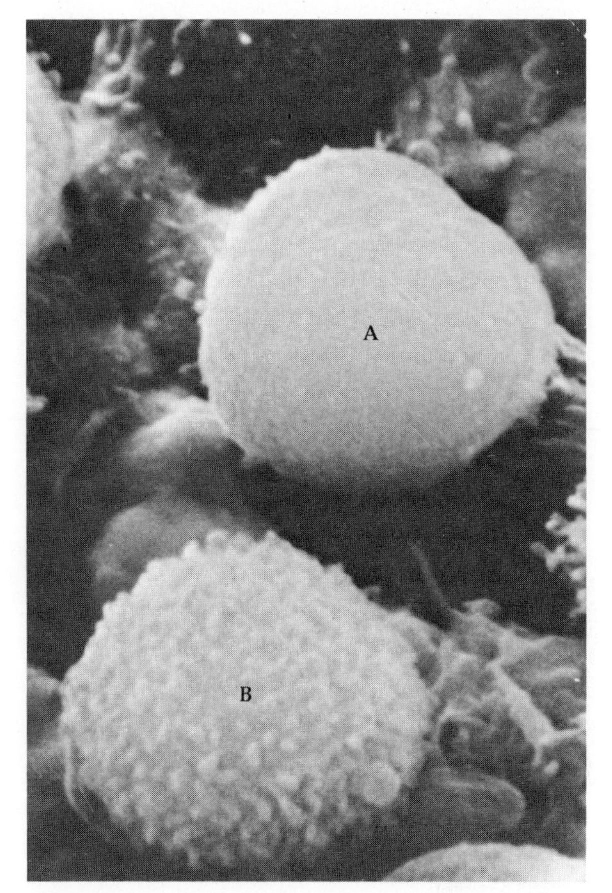

FIG. 19-3
T and B lymphocytes as viewed with a scanning electron microscope × 6300. (A) T cell. The T cell usually has a relatively smooth surface stippled by about 100 antibody-like proteins. (B) cell. The B cell has a roughened surface produced by nearly 100,000 antibody-like proteins.

Monocytes

Monocytes have a cytoplasm that is generally more abundant than that of lymphocytes, is pale, gray-blue, and dull in appearance; and

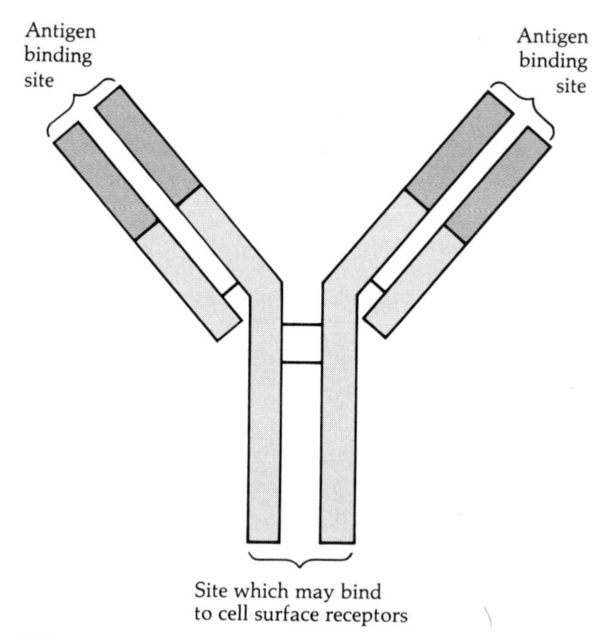

FIG. 19-4
Simplified drawing of an antibody molecule.

contains some fine granules (Fig. 19-1). Monocytes make up about 3 to 8 percent of the leukocytes and structurally resemble atypical large lymphocytes. However, monocytes often have more rounded margins to their cytoplasmic membranes than atypical lymphocytes. Furthermore, the nucleus of a monocyte is usually kidney-shaped or segmented with one or two lobes that have a lacy chromatin pattern. That of an atypical lymphocyte is usually nonsegmented and has chromatin which appears coarse, lumpy, or clumped near the periphery of the cell.

Monocytes secrete antimicrobial compounds and form myeloperoxidases but in amounts less than neutrophils. Circulating monocytes are blood-borne macrophages, which, like neutrophils and lymphocytes, are attracted by chemotaxis to areas in which various agents injure or invade tissues. Monocytes may migrate into tissue spaces and mature to produce the tissue macrophages (histiocytes) of the reticuloendothelial system (Fig. 3-18). These cells are the first line of defense against bacterial infections. Monocytes and macrophages are especially active in the ingestion and destruction of intracellular parasites (organisms which grow in cells), including many viruses and bacteria. An increase in monocyte number is generally a favorable sign of recovery from infection, except in tuberculosis, where the opposite may be true.

Defense Functions

Leukocytes are active in the body's defenses and can ingest, neutralize, and destroy foreign materials and aged and abnormal cells. In addition, they secrete a variety of molecules, including enzymes, antibodies, and lymphokines, all of which help eliminate or destroy microorganisms and toxins.

Antibodies are protein molecules produced in response to exposure to antigens (Chap. 3). Most antibodies are composed chemically of four polypeptide chains joined together, often forming a Y-shaped molecule (Fig. 19-4). Two arms of the molecule are identical and may bind to specific parts of antigens to neutralize, precipitate, or agglutinate them or prepare them for phagocytosis. The antibody molecule's base may bind to specific receptors on cells such as basophils, mast cells, macrophages, or numerous other cell types. Aggregations of sister antibody molecules may cross-link antigens. When antibodies bind to cell surface receptors, they may initiate biological responses in those cells, such as the release of histamine (Fig. 3-20), or cause a variety of other responses by mechanisms similar to those of hormones or neurotransmitters (Chaps. 5, 7, and 9).

In addition to these defense functions, the monocytes and macrophages, as well as the T and B lymphocytes, act as sentinels guarding against, and reacting to, the entry of foreign materials into the body. One of the many locations in which these cells are found is the reticular tissue of lymphatic organs through which lymph must filter (Figs. 3-21 and 3-22). At these sites monocytes mature and transform into macrophages, which actively ingest foreign particulate matter.

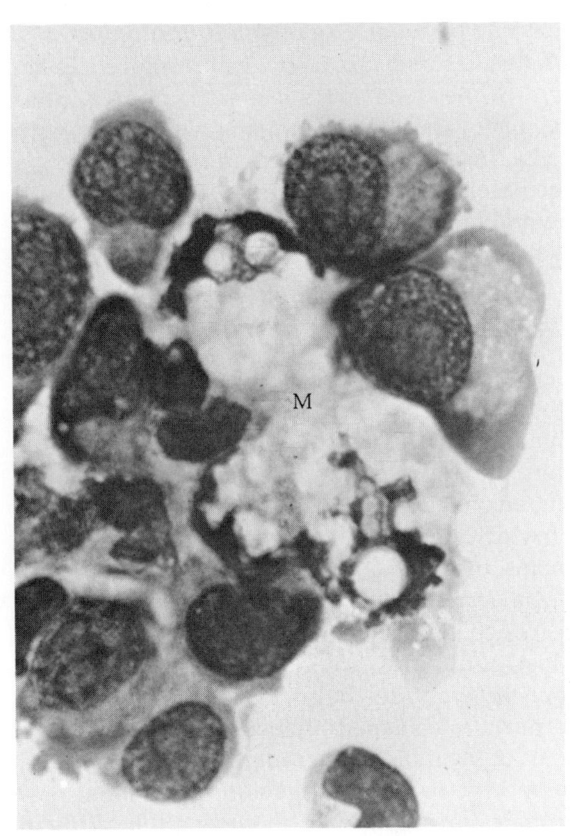

FIG. 19-5
A complex of lymphocytes surrounding a macrophage (M) which has ingested antigen (~ × 1250). The lymphocytes may be transformed into large reactive cells as a result of their interactions with the macrophage.

Upon interacting with antigen, the macrophages prepare it (process it) for T and/or B lymphocyte reaction. The three cell types then may form complexes (Fig. 19-5) whose interaction enhances defense mechanisms. The macrophages are activated by *lymphokines* released from T lymphocytes. *Lymphokines* are low-molecular-weight diffusible molecules secreted by cells that regulate the properties of other cells. Some lymphokines are chemotactic molecules that attract cells; others activate macrophages. They seem to stimulate the enzyme activity of the giant phagocytic cells, making them "angry" and improving their ability to digest and degrade undesirable materials and cells (page 61).

Perhaps the best studied lymphokines are *interferons,* low-molecular-weight proteins that may protect noninfected cells from viral infection and tumors. Different interferons are produced by lymphocytes and other cells such as fibroblasts. Some interferons enhance the production and activity of killer T cells and inhibit the replication of viruses (by preventing the translation of viral mRNA). They may also increase macrophage activity and reduce the replication of tumor cells. The clinical application of interferons offers promise as a means of therapy in some cases of cancer.

Antigens also stimulate germinal centers of lymphoid tissue (Fig. 3-22) to undergo cell division and form small and large lymphocytes. The small daughter lymphocytes are not duplicates of their parents but have acquired a new property called *immunological memory.* This memory allows them to respond more rapidly to subsequent exposures to the same and related antigens and also to multiply and thereby provide greater defenses against these antigens.

One of the two major subdivisions of small lymphocytes, the B cells, originate in bone marrow, migrate to lymphoid tissue (Fig. 18-1), and divide to form large lymphocytes, which later form plasma cells that secrete antibodies. The T cells develop in the thymus gland and migrate to other lymphoid tissues to produce a number of different functional types of T cells. Some of them are *inducer* or *helper T cells,* which enhance the activity of other cells such as macrophages, as described above. In some instances the presence of certain helper T cells improves antibody secretion by B cells. Other T cells, *suppressor* or *killer cells,* release molecules that may inhibit or destroy cell types which are recognized as foreign. Altered or foreign cells may be recognized by interaction of their cell surface receptor molecules with antireceptor molecules on the surface of the T cell. These receptor molecules form part of the cell's major histocompatibility antigens. The T suppressor cells also have been implicated in immune reactions gone awry, which cause autoimmune diseases such as multiple sclerosis (page 147). Such disorders in which normal defense mechanisms attack an individual's own tissues are called *autoimmune reactions.*

19-2 PROPERTIES OF THROMBOCYTES

Development and Structure

Thrombocytes (*platelets*) are derived from giant cells, megakaryocytes, in bone marrow (Fig. 18-1), which measure 100 μm or more in diameter. Megakaryocytes are produced as huge cells by repeated nuclear divisions in the absence of cytoplasmic division, during which up to 64 nuclei per cell may be formed. During maturation portions of megakaryocyte membranes enclose fragments of cytoplasm, which are pinched off without nuclei to form platelets. The fragments are then extruded from the bone marrow between endothelial cell junctions into capillaries (Fig. 19-6). Platelets are ultimately formed as separate oval or granular structures lacking a nucleus. They stain dark lilac to blue in a smear and range in size from 1 to 4 μm (Fig. 19-1). Platelets contain microtubules, microfilaments, myosin, and actin, which forms bundles of filaments in the presence of calcium ions. The presence of these cytoplasmic organelles and proteins is intriguing in view of the secretory capacity of platelets, their ability to bind to other cells, and their ability to cause clots to shrivel. The granular contents of platelets include a variety of important compounds, some of which are described below.

FIG. 19-6
Formation of a ribbon of platelets from a megakaryocyte (× 1980). Note the long white strands, which are aggregations of platelets. One strand is marked with an arrow. From *Histology*, 4th ed., by L. Weiss and R. O. Greep. Copyright © 1977, McGraw-Hill Book Company. Used with permission of McGraw-Hill Book Company.

Numbers of Platelets

The number of platelets varies from 1.4×10^5 to 4.4×10^5 per microliter of blood. New ones are continually released from bone marrow to replace those that are degraded and eliminated from the body. Their life-span is 5 to 10 days.

General Functions

Platelets help to: (1) maintain the integrity of blood vessels; (2) form plugs to block holes in injured vessel walls; (3) assist in the formation of blood clots; and (4) release a number of compounds which affect a variety of body functions. These compounds include serotonin and epinephrine, which may cause vasoconstriction; mitogens, which promote smooth muscle cell division and which may be related to the causes of atherosclerosis; ADP (adenosine diphosphate), which is associated with platelet aggregation; and two derivatives of prostaglandin (Chap. 5) called thromboxane A_2 (TXA_2) and prostacyclin (PGI_2). The platelets also secrete calcium ions and a variety of enzymes and factors essential to blood clotting.

Platelet aggregation is an important mechanism in the control of bleeding (*hemostasis*). When blood flows from a damaged vessel, platelets bind to exposed collagen in the damaged wall, form secretions, stick to each other, produce a plug to assist hemostasis, and release factors to help the blood clot (Figs. 19-7 and 19-8).

Thromboxane (TXA_2) causes aggregation of platelets and vasoconstriction, while prostacyclin (prostaglandin I_2, PGI_2) has the opposite effects. Both compounds have short half-lives (that of TXA_2 is about 30 s and that of PGI_2 is about 2 min) and are formed from fatty acids as indicated in the following reactions:

$$\text{Fatty acids} \rightarrow \text{arachidonic acid} \rightarrow \text{endoperoxides} \rightarrow PGH_2 \quad (\text{prostaglandin } H_2)$$

$$PGH_2 \rightarrow PGI_2 \ (\text{Prostacyclin}), \quad PGE_2, \quad PGD_2, \quad PGF_{2\alpha}, \quad TXA_2 \ (\text{Thromboxane})$$

In like manner, leukocytes have been shown to produce *leukotrienes,* short-lived, slow-acting agents that cause smooth muscle contraction in blood vessels and the respiratory tract, and edema. These extremely potent mediators of

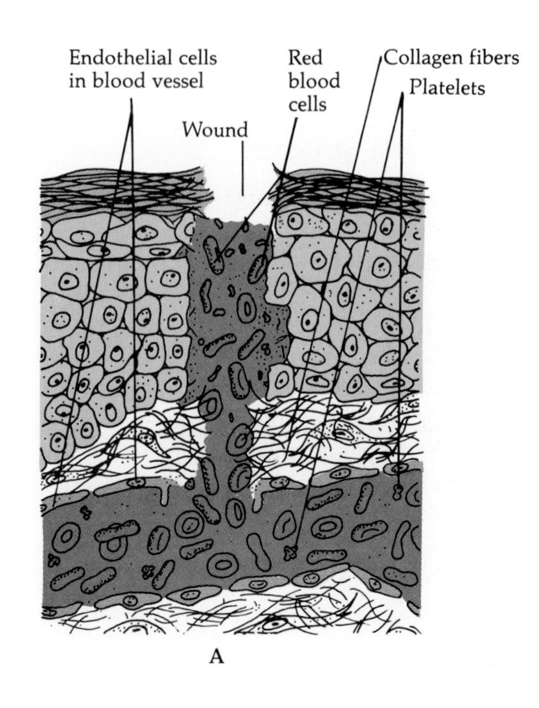

A

B

FIG. 19-7
Function of platelets in closure of a hole in a damaged vessel. (A) Movement of blood through tissue wound. (B) Platelets adhere to collagen fibers and start to plug hole. (C) Platelets adhere to each other to close the gap in the damaged vessel wall while blood clot forms to reinforce the initial platelet plug.

some allergic and inflammatory reactions are noncyclic derivatives of arachidonic acid. Various leukotrienes are 100 to 1000 times more potent on a molar basis than histamine.

Prostacyclin seems to be protective and thromboxane seems to induce a number of disorders. For example, a decrease in prostacyclin seems to be associated with atherosclerosis, heart attack, embolisms, shock caused by bacterial endotoxins, and a platelet disorder, *thrombocytopenic purpura.* Administration of prostacyclin has been proposed for use in coronary bypass surgery or other vascular surgery during which it seems advantageous to induce vasodilation and inhibit clotting temporarily. It is interesting to note that the formation of prostacyclin, thromboxane, and other products of prostaglandin metabolism are inhibited by aspirin and aspirinlike anti-inflammatory drugs.

The ability of prostacyclin to prevent platelet aggregation may also inhibit human cancers from metastasizing (spreading) through the bloodstream to other areas. Prostacyclin may prevent cancer cells from adhering to platelets. In so doing it may prevent the cells from anchoring in new sites by blocking their means of attachment to the endothelium of blood vessels.

FIG. 19-8
Platelet plug formed 10 min after a blood vessel is cut. PLT = mass of platelets; FIB = fibrin surrounding platelets; CT = connective tissue; RBC = red blood cells.

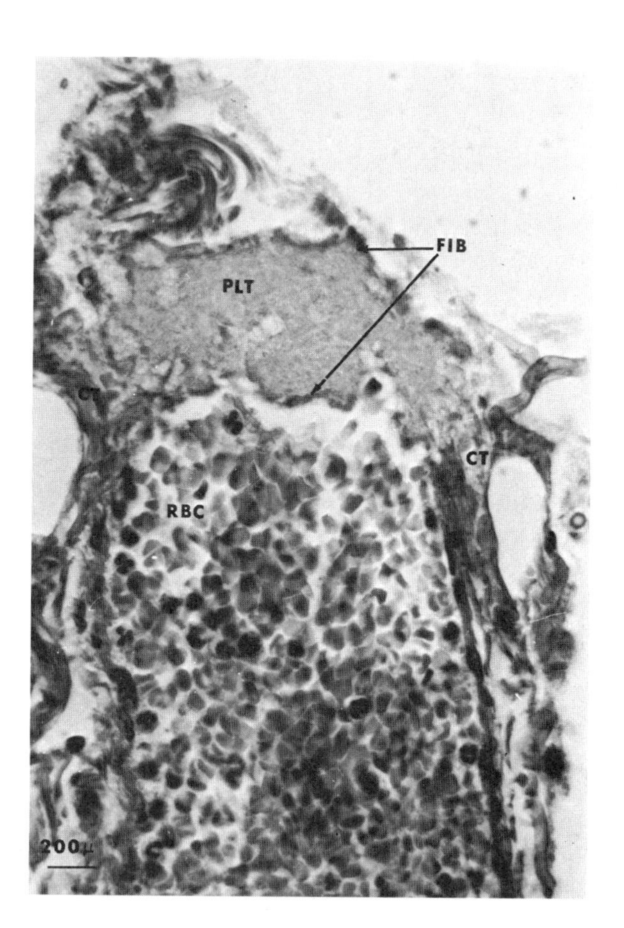

Clotting Sequence

At least 12 factors in the plasma participate in the formation of a blood clot (Table 19-2). They are numbered in the order of their discovery, with Roman numerals. Clotting factors are converted from inactive to active forms as one product activates the next in a stepwise series of reactions, which are dependent on the presence of calcium ions. The sequence flows like a waterfall or cascade and ultimately causes the conversion of soluble fibrinogen monomers to insoluble fibrin polymers (Fig. 19-9). Fibrin then forms a meshwork that adheres to tissues and to the inside of blood vessels, trapping the formed elements of blood and resulting in a clot.

Clotting results from the activation of two separate but related pathways called the extrinsic and intrinsic systems. They are initiated by release of factors from tissues and in the blood, respectively. The steps in both sequences lead to the formation of *thrombin,* a molecule with enzymatic properties (Fig. 19-9). The *extrinsic clotting system* is activated by the release of factors from tissue mixed with Ca^{2+} ions and plasma. The role of the extrinsic system is not clear but it may accelerate clotting by causing the formation of thrombin.

The *intrinsic system* initiates clotting when blood contacts a surface changed by damage or disease. The altered surface behaves as though it were foreign. Clotting begins when the Hageman or contact factor (XII) and the platelets participate in a series of reactions that require Ca^{2+} to activate prothrombin. Prothrombin may also be activated by the *extrinsic system,* which promotes the release of thromboplastin (factor III) from damaged tissues. Both systems lead to the formation of an active complex of factors X and V to complete the first stage of the clotting sequence.

In the second stage, common to both pathways, prothrombin (factor II) is converted to the active enzyme thrombin by the products of the first-stage reactions. In the third stage, active thrombin converts soluble fibrinogen (I) to soluble fibrin monomers. Fibrin-stabilizing factor (XIII) then converts the monomers to polymerized, insoluble threads of fibrin that trap blood cells to complete formation of the clot. The clot is further stabilized when *thrombosthenin,* a contractile protein released from platelets, causes it to retract (shrivel). Thrombosthenin chemically resembles actomyosin in skeletal muscle.

Clots are removed by *plasmin* (fibrinolysin), a proteolytic enzyme formed from an inactive precursor, plasminogen, found in abundance in the plasma. Plasmin hydrolyzes blood protein factors V and VIII along with fibrin polymers, which are degraded, split, and cleared by the reticuloendothelial system.

The liver synthesizes a number of coagulation factors (Table 19-3) and is critical to normal clotting. The production of the active forms of

TABLE 19-2
Factors in blood that cause clotting

Factor	*Name(s)*
I	Fibrinogen
II	Prothrombin
III	Thromboplastin, tissue factor
IV	Calcium ions
V	Proaccelerin, AC globulin (accelerator globulin, ACG), labile factor
(VI)	(Found to be the same as factor V)
VII	Serum prothrombin conversion accelerator (SPCA), proconvertin, autoprothrombin I, stable factor
VIII	Antihemophilic factor (AHF), antihemophilic factor A, antihemophilic globulin (AHG)
IX	Plasma thromboplastin component (PTC), antihemophilic factor B, autoprothrombin II, Christmas factor
X	Stuart-Prower factor, autoprothrombin III
XI	Plasma thromboplastin antecedent (PTA), antihemophilic factor C
XII	Hageman factor, contact factor, glass factor
XIII	Fibrin-stabilizing factor, Laki-Lorand factor, fibrinase, fibrin serum factor

four of these factors (II, VII, IX, and X) are also vitamin K–dependent. The influence of vitamin K in the clotting process is responsible for its German name, the *Koagulations* vitamin. Absorption of vitamin K, like that of other fat-soluble vitamins, depends on the availability of adequate bile salts formed by the liver and released to the intestine (Fig. 18-6). Therefore, liver disease can impede the synthesis of clotting elements directly and also lessen their availability indirectly by diminishing the production of bile salts and the absorption of vitamin K.

Inhibition of Clotting

Blood normally does not clot in vessels owing to a slight excess of *anticoagulants,* which are inhibitory to specific steps in the clotting sequence. One major anticoagulant, antithrombin III, is abundant in body fluids and is an alpha globulin formed in the liver. Antithrombin III may inactivate thrombin by forming a complex with it and also may inactivate other clotting factors. *Heparin* also may inactivate thrombin by combining with antithrombin III and enhancing its reactivity with thrombin. Heparin is an acidic proteoglycan formed by liver cells, basophils, and mast cells and is a natural anticoagulant. It is frequently used clinically and in laboratories to prevent clotting. Other anticoagulants used in the laboratory, such as citrate, oxalate, and ethylenediaminetetraacetic acid (EDTA), inhibit clotting by binding to Ca^{2+} ions (factor IV).

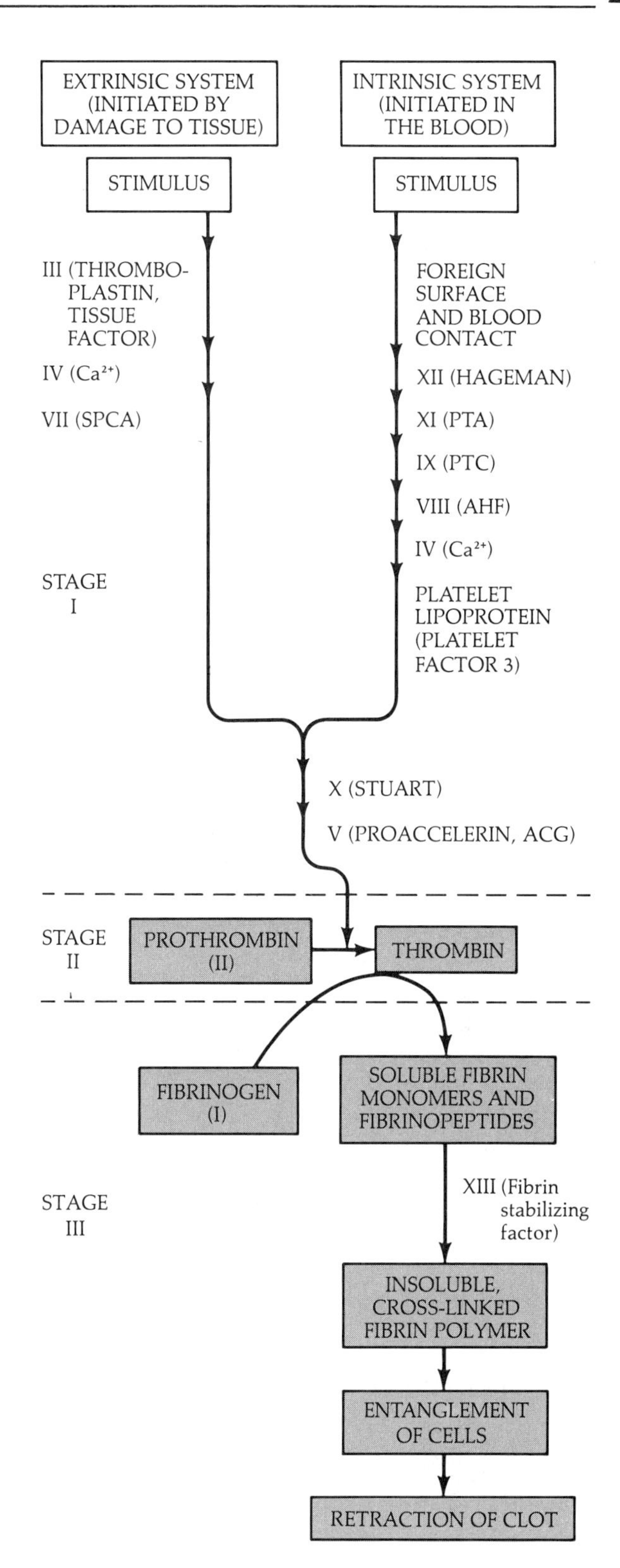

FIG. 19-9
The formation of a blood clot. The factors involved cause the stepwise conversion of inactive precursors in the blood into active molecules via two pathways. One occurs outside the vessel, the other inside. In three stages soluble fibrin monomers are converted to insoluble polymerized strands of protein that form a network in which blood cells become trapped.

TABLE 19-3
Clotting factors synthesized in the liver

Name	*Factor Number*	*Vitamin K–Dependent*
Fibrinogen	I	No
Proaccelerin	V	No
Prothrombin	II	Yes
Serum prothrombin conversion accelerator (SPCA)	VII	Yes
Plasma thromboplastin component (PTC)	IX	Yes
Stuart-Prower factor	X	Yes

Interestingly, the ingestion or administration of oral anticoagulants such as warfarin (Coumadin), which is derived from fermented sweet clover, prevents the synthesis of vitamin K–dependent factors in the liver. Warfarin acts as an antagonist of vitamin K and blocks the activation of the precursors of four clotting factors (II, VII, IX, and X). Consequently, a reduction in the activation of several clotting elements occurs, the level of plasma prothrombin is decreased (hypoprothrombinemia), and clotting is prolonged. This has led to the use of various forms of this anticoagulant (such as commercial warfarin) as rodenticides. Their consumption may lead to fatal hemorrhage of rats.

Blood also may fail to clot in the body when one or more of the essential factors is deficient or lacking. The most common hereditary deficiency is absence of antihemophilic factor VIII, which allows prolonged bleeding in the disease *hemophilia A*. Because the deficiency is due to a recessive gene on the X chromosome, when present it is always expressed in males, since they have only one X chromosome per cell. Because females have two X chromosomes per cell, the presence of a normal dominant gene on one of the two sex chromosomes masks the defective gene. Such a female carries, but does not manifest, the trait. Females must have two defective genes (be homozygous) to develop hemophilia. However, this condition is usually lethal in females. The hereditary pattern is similar for *hemophilia B* (Christmas disease) in which factor IX (Christmas factor) is deficient.

A low platelet count also may inhibit clotting. This condition may be hereditary, may be acquired by exposure to drugs, infectious agents (such as measles virus), or ionizing radiation, or may be due to nutritional deficiencies (of vitamin B_{12}, iron, or perhaps folic acid). Spontaneous bleeding usually occurs when the platelet count falls below 1×10^5 per microliter.

Clotting also may be prevented *in vitro* by the addition of oxalate or citrate, which bind calcium ions. A solution of sodium *citrate*, sodium *phosphate*, and *dextrose* (CPD) is often mixed with whole blood (about 14 mL CPD per 100 mL of blood) to preserve it under refrigeration at 4°C in blood banks. The citrate binds calcium ions to prevent clotting, the phosphate helps to maintain adenosine diphosphate levels and improve the viability of the red blood cells, and the dextrose provides a source of energy. Because red blood cells deteriorate rather quickly, whole blood stored in CPD at 4°C is considered fresh for only 7 days, when 98 percent of the cells are still viable, but may be used for transfusions for up to 21 days, when about 80 percent of the cells remain viable. A solution of *acid, citrate,* and *dextrose* (ACD), used until recently in blood banks in the United States, is less effective in preserving the cells.

Tests for Clotting Elements

The time required for blood to clot varies not only among individuals but also with the type of coagulation test employed. The *bleeding time* is measured from the time a sharp standardized incision is made on the palm surface of the forearm until blood stops flowing from the wound, and is about 1 to 3 min. The *clotting time* is assayed by placing blood in a capillary tube or a test tube at 37°C; usually 3 to 6 or 7 to 15 min, respectively, are required for clot formation. The *prothrombin time* test (PT) measures the presence of clotting elements in the blood.

The number of seconds required for plasma mixed with 3.8 percent citrate to clot is measured when it is also mixed with standard amounts of factors III and IV (thromboplastin and Ca^{2+}). The PT for normal plasma ranges from 12 to 17 s. Prolonged prothrombin times indicate deficiencies in one or more clotting elements.

19-3 PATHOPHYSIOLOGY OF LEUKOCYTES

General Disorders

Various diseases have different effects on the production of certain kinds of white blood cells (Table 19-4). Many diseases stimulate stem cells in the bone marrow and in other hematopoietic organs to proliferate and to produce excess leukocytes. The cells migrate from the sites to enter the blood and then exit into interstitial spaces. They then may be attracted by chemotaxis to the focal points of the disease.

TABLE 19-4
Some factors altering leukocyte number and/or activity

	CELL TYPES				
Effects	*Neutrophils*	*Eosinophils*	*Basophils*	*Lymphocytes*	*Monocytes*
Increased number and/or activity	Tissue damage due to various infections or drugs	Allergies such as hay fever and bronchial asthma Parasitic infections such as beef tapeworm (*Taenia saginata*), pork roundworm (*Trichinella spiralis*), blood flukes (*Schistosoma*)	Allergies Worm infections Chickenpox	Immune response to viral infections (such as hepatitis and infectious mononucleosis) and to bacterial infections (such as whooping cough) Autoimmune diseases such as rheumatoid arthritis and pernicious anemia Some stages of certain types of cancer; leukemias; early stages of Hodgkin's disease	Intracellular infections with protozoa, viruses, or bacteria (such as tuberculosis or typhoid fever)
Decreased number and/or activity	Exposure to certain chemotherapeutic drugs, irradiation, and during specific stages of some infections	Stress Cushing's syndrome (due to excess production of adrenal glucocorticoids)	Stress Acute infection Some cases of hyperthyroidism	Hereditary immunologic deficiency diseases Certain chemotherapeutic drugs Late stages of Hodgkin's disease	Treatment with prednisone (an anti-inflammatory glucocorticoid)

Many leukocytes are excreted in the feces following their migration through the walls of the intestine.

Leukocytosis is a temporary increase in the total white blood cell count above 11,000 per microliter. It often develops after infectious agents stimulate the hematopoietic tissue. When the leukocytes are killed in combat or the count falls below 4000 per microliter for any reason, the condition *leukopenia* results. This response occurs in some stages of infectious disease and after exposure to radiation or drugs that destroy or inhibit the activity of bone marrow cells.

Leukemias

Leukemias are proliferative diseases in which the white blood cell numbers increase owing to cancerous (neoplastic) multiplication of specific hematopoietic cells (Table 19-5). Leukemias tend to spread easily throughout the reticuloendothelial system. They may be distinguished from *lymphomas,* which are cancers of lymphoid tissue that initially tend to remain more localized but later may progress to a leukemic phase. Hodgkin's disease is an example of a lymphoma which rarely becomes leukemic. It is associated with a defect in T cell function and occurs most often in individuals between 15 and 35 years of age and those over 50. A variety of non-Hodgkin's lymphomas exist, including those in which the primary cell types involved are well-differentiated lymphocytes, histiocytes, or undifferentiated lymphocytes (stem cells) as in Burkitt's lymphoma.

About 60 percent of leukemias are acute and 40 percent are chronic. *Acute leukemias* are sudden in onset and are often fatal within 3 to 6 months without treatment, but with therapy, they may go into remission (lessening in intensity) for years. *Chronic leukemias* develop more gradually, causing no obvious symptoms at first. They are more responsive to drugs and/or radiation therapy; half the patients survive more than 4 years and the remainder less than that time. About 14,000 people die of leukemia each year in the United States.

Leukemic individuals frequently exhibit symptoms of fever, weakness, pain in the joints resembling rheumatoid arthritis, ulcerations of the mucous membrane of the mouth and throat, and enlargement of the lymph nodes and other organs of the reticuloendothelial system.

Three major complications of leukemias are anemia, bleeding due to a decrease in platelet count, and infection. In acute leukemias greater numbers of stem cells are often found in peripheral blood, while chronic leukemias are often accompanied by elevated numbers of specific types of leukocytes (Table 19-5). Metabolites released from leukemic cells may adversely affect a number of critical functions. The liver and kidneys usually become infiltrated with leukemic cells and develop insufficiencies which, with hemorrhage, are among the most common causes of fatality among leukemic patients.

TABLE 19-5
Characteristics of some types of leukemia

Type	*Cell Type Frequently Involved*	*Primary Susceptible Age Groups in Years*
Acute		
Lymphocytic	Lymphoblasts (of specific cell lines)	Most common in young children 3 to 4
		Rare under 40, occurs more often in adults over 60
Myelogenous (myelocytic, granulocytic)	Myeloblasts (of different cell lines)	Frequent in infants Most common in youth to middle age
Monocytic (usually acute, may be chronic)	Myeloblasts (monoblasts) or sometimes reticuloendothelial cells	Favors middle age Rare before 30
Chronic		
Lymphocytic	Abnormal lymphocytes	Usually over 60 Rare under 40
Granulocytic (myelogenous)	Granulocytes in all stages	Age 20 to 50 Often after 40

19-4 PATHOPHYSIOLOGY OF THROMBOCYTES (PLATELETS)

Disorders in platelets may seriously affect the clotting of blood. As mentioned earlier, bleeding becomes more prolonged when the platelet count falls below 1.0×10^5 per microliter and may occur as a result of disorders of the marrow caused by deficiencies in vitamin B_{12}, folic acid, or iron or by some infections, drugs, ionizing radiation, or cancer. A reduction in platelet number, *thrombocytopenia,* also may be caused by the production of antiplatelet antibodies which destroy platelets in a condition known as *thrombocytopenic purpura.* Sometimes the antibodies are formed when the donor's platelets in transfused blood stimulate an immune reaction in the recipient. In other instances antibodies are formed against an individual's own platelets in an autoimmune reaction in which the antibody-platelet reaction causes the platelet count to be reduced. Purple patches (petechiae) appear in the mucous membranes and skin, and bleeding time may be prolonged. Viral infections which damage tissues and expose hidden antigens or which lead to alterations in synthesis and structure of antigens may induce autoimmune reactions. Such responses also may follow infectious mononucleosis, some forms of leukemia, or the administration of certain drugs. Conversely, *thrombocythemia,* a rare proliferative disease of the marrow, increases the number of platelets, some of which behave abnormally. Thrombosis and hemorrhage may follow the increase in platelet number.

19-5 SUMMARY

1 Leukocytes, white blood cells with important defense functions, have counts that range from 4.3×10^3 to 10.8×10^3 per microliter of blood. Leukocytes may be identified, distinguished, and their relative percentages noted by means of a differential WBC count. Cells with coarse granular cytoplasm are classified as granulocytes and are separated into three types, neutrophils, eosinophils, and basophils. Those that lack coarse granules are agranulocytes and are distinguished as lymphocytes and monocytes.

Most of the leukocytes are formed in bone marrow, whereas lymphocytes proliferate in lymphoid tissue. Leukocytes may escape from their tissues of origin to be transported to interstitial spaces and tissues. They squeeze between gaps in the endothelial cells lining capillaries in a phenomenon called diapedesis. This migratory pattern plays an important part in the body's defensive capacity.

Neutrophils have lilac cytoplasmic granules and are very active phagocytes. Eosinophils have reddish to orange cytoplasmic granules. Basophils, which have dark blue to purple cytoplasmic granules, migrate to tissues to form mast cells (tissue basophils). Basophils and mast cells release heparin and histamine in defense reactions. Lymphocytes normally have a scant rim of smooth, sky-blue cytoplasm but can be of larger size with more cytoplasm. Lymphocytes may be classified functionally as T cells, which secrete lymphokines and are active in cellular immunity, and B cells, which secrete antibodies active in humoral immunity. Monocytes have blue-gray, relatively smooth cytoplasm and large nuclei. They are bloodborne phagocytes which develop into macrophages in tissues.

Various types of leukocytes are active in the body's defenses, behave as phagocytes, and secrete antibodies, enzymes, and lymphokines to participate in defense reactions that destroy and eliminate abnormal or foreign cells, including microorganisms and their toxins. The granules in granulocytic leukocytes contain lysosomal enzymes, including myeloperoxidases, which assist defense reactions. Myeloperoxidases form hydrogen peroxide, which helps kill bacteria. The release of lymphokines may attract leukocytes to tissues, activate macrophages, prevent viral infection, produce conditions which are toxic to invading microorganisms, or alter other host defenses. Leukocytes also can form leukotrienes that may cause some allergic responses.

2 Platelets are oval to granular, nonnucleated elements which stain dark lilac to blue. They release a number of substances essential to clotting and aggregate among collagen fibers of damaged tissues to plug holes in injured blood vessels. Platelets also produce prostacyclin, which promotes blood flow, and thromboxane, which has the opposite effects.

Blood coagulates when inactive molecules are activated in a sequence of three stages. The first stage involves molecules that activate prothrombin. During the second stage prothrombin is converted to enzymatically active thrombin. In the third stage, thrombin in the presence of fibrin-stabilizing factor enzymatically converts soluble fibrinogen protein into insoluble fibrin polymers, which trap platelets and blood cells to form the clot. Clot removal is normally initiated by a proteolytic enyzme, plasmin, whose precursor, plasminogen, is abundant in plasma. The synthesis of clotting factors I, II, V, VII, IX, and X occurs in the liver. Vitamin K is required for the formation of factors II, VII, IX, and X.

3 Several diseases cause the numbers and frequency of occurrence of white blood cell types to vary. When the WBC count is low, the condition is leukopenia. When it is temporarily high, the disorder is called leukocytosis. Exposure to infectious agents, drugs, radiation, and stress alter white blood cell physiology. In leukemias excess numbers of various specific white blood cells are produced by cancer in bone marrow or lymphoid tissue. Acute forms may be rapidly fatal but when treated may undergo long-term remission. Chronic conditions are often treatable and half the patients with such disorders survive more than 4 years.

4 Decreases in platelet number (thrombocytopenia) cause prolonged bleeding time. Antibodies which lead to platelet destruction in thrombocytopenic purpura may result from transfusion of blood containing platelets with incompatible antigens. Antibodies may also form from autoimmune reactions which sometimes follow viral infections. Conversely, platelet diseases which cause intravascular clots result from excessively high platelet counts in a disorder called thrombocythemia.

REVIEW QUESTIONS

1 What are the major types of leukocytes?
2 Where are the five major types of leukocytes produced?
3 What is a differential count?
4 What are the normal frequencies of occurrence and characteristics of each type of leukocyte?
5 Compare the major functions of the five main types of white blood cells.
6 Contrast lymphocytes and monocytes.
7 Describe the general defense functions of leukocytes.
8 Compare T cells and B cells with respect to structure and function.
9 Account for the structure and general functions of platelets, including the formation of a platelet plug in a damaged blood vessel.

10 Compare the origins and functions of thromboxane, prostacyclin and leukotrienes.
11 What are the three major stages in the coagulation of blood? Explain the extrinsic and intrinsic pathways.
12 What normally prevents blood from clotting within blood vessels? Name and explain three tests that indicate the presence of clotting elements.
13 Why is the liver important to blood clotting? What is its relationship to vitamin K?
14 What are the leukemias? How do they differ from one another and from lymphomas?
15 Describe two contrasting types of platelet disorders and their effects.

PHYSIOLOGY OF THE RESPIRATORY SYSTEM

20

20-1 ARCHITECTURE OF THE RESPIRATORY SYSTEM
Gross anatomy
Cellular anatomy and physiology
Cell types lining the respiratory tract
Respiratory membranes
Surfactants and respiratory distress syndrome

20-2 THE PHYSICS OF BREATHING
Gas laws and variations in pressure
Pressure gradients in inspiration and expiration
Ventilation of the lungs
Spirometry
Lung volumes
Types of respiration
Normal breathing • Abnormal breathing and effects of disease

20-3 TRANSPORT AND EXCHANGE OF GASES
Oxygen transport
Carbon dioxide transport
Exchange of gases
Exchange in tissues and the lungs
Factors affecting gas exchange

20-4 REGULATION OF RESPIRATION
Nervous control mechanisms
Chemoreceptors
Baroreceptors
Medulla
Pons
Cerebral cortex
Chemical regulators
Carbon dioxide and hydrogen ions
Oxygen

20-5 EFFECTS OF EXERCISE

20-6 ENVIRONMENTAL EFFECTS
Exposure to high altitude
Underwater diving
Nitrogen
Oxygen
Carbon dioxide
Decompression
Pollutants
Carbon monoxide
Cigarette smoke
Other pollutants

20-7 ARTIFICIAL RESPIRATION
Mechanical devices—Advanced life support techniques
Mouth-to-mouth and mouth-to-nose resuscitation—Basic life support techniques

20-8 SUMMARY

OBJECTIVES

After completing this chapter the student should be able to:

1. Describe the gross and the microscopic anatomy of the lung, including the alveoli, and the nature of the respiratory membranes.
2. Discuss the physical factors involved in breathing.
3. Explain the functions of a spirometer and how various lung volumes can be measured.
4. Distinguish among external, internal, and cellular respiration and between inspiration and expiration.
5. Diagram the exchange of oxygen and carbon dioxide in the lungs and in the tissues and discuss the factors affecting the exchanges.
6. Identify the effects of carbon dioxide, hydrogen ions, and oxygen as chemical regulators of respiration.
7. Outline the influences of chemoreceptors and baroreceptors of the carotid and aortic bodies on the nervous regulation of respiration.
8. Compare the effects of altitude and of diving on respiratory function.
9. Discuss the consequences of carbon monoxide and smoking on respiratory function.
10. Describe the responses of the cardiopulmonary organs to exercise.
11. Explain the purpose of artificial respiration, comparing mechanical devices with mouth-to-mouth and mouth-to-nose resuscitation; describe cardiopulmonary resuscitation.

The recognition of the importance of a "breath of fresh air" reflects an essential function of the respiratory system, namely to provide the body with a means of gas exchange (oxygen for carbon dioxide). Without oxygen the cells of the body cannot burn (oxidize) foods on a sustained basis to provide energy, propel metabolism, maintain body temperature, and sustain life. Unlike single-cell organisms such as amoebas, higher animals have many cells, too many to depend solely on diffusion to transport gases deep within the body. Instead, they possess special organs designed to facilitate gas exchange, which vary according to the natural environment of higher animals. For example, fish have gills and land animals have lungs. Lungs form the functional components within a group of structures called the *respiratory system.*

The organs of the respiratory system collectively participate in the exchange of gases between tissues of the body and the blood in a process called *respiration.* It is brought about by breathing actions (inhalation and exhalation) termed *lung ventilation.* The phase of gas exchange that occurs between the lungs and the blood is *external respiration* and is a prelude to *internal respiration,* which occurs between blood and tissues. Tissue cells use oxygen released from the blood and release carbon dioxide in aerobic metabolism or *cellular respiration* (Chap. 13).

20-1 ARCHITECTURE OF THE RESPIRATORY SYSTEM

Gross Anatomy

Air moves through the nostrils (external nares) and nasal chambers into a passageway called the *nasopharynx,* as well as posteriorly from the mouth into an area called the *oropharynx* (Fig. 20-1). The oropharynx leads to the *laryngopharynx* and then through an opening, the *glottis,* in the *larynx* (voicebox) into the *trachea* (windpipe), which leads to the lungs. Cilia lining these upper respiratory passages and mucus coating their surfaces generally prevent most small particles from entering the deep recesses of the respiratory system (Chap. 3).

However, the pathways leading to the lungs may accidentally be closed off by food and cause choking and death. This may occur when the epiglottis (a flap of cartilage normally covering the trachea) fails to block the entry of food into the glottis and food falls into the trachea rather than into the esophagus. Should mechanical means fail to dislodge the food, a *tracheotomy* (incision into the trachea) may be made below the point of the blockade to allow

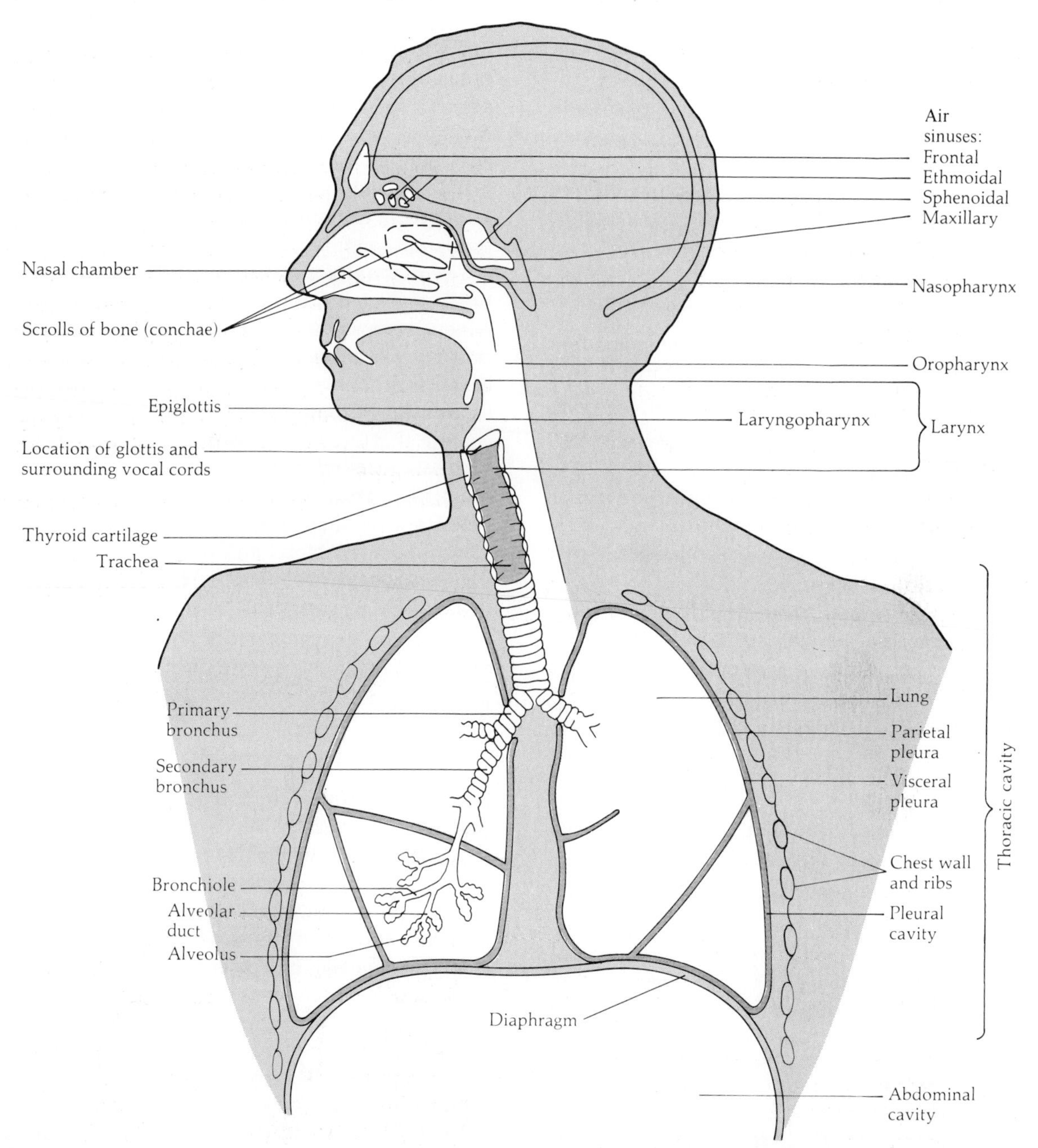

FIG. 20-1
Gross anatomy of the respiratory tract and its accessory structures.

breathing to resume. In some instances a tube is inserted to make the flow of air easier.

Hollowed, scroll-like passages in the nasal chambers are lined with ciliated mucous membranes, which warm the air to body temperature and saturate it with water vapor. The nasal passages also contain the olfactory receptors for smell (Chap. 8). These passages are connected with *air sinuses,* hollowed-out spaces in the skull which have a volume of about 60± 10 mL (Fig. 20-1). These spaces are rudimentary at birth, begin to grow at about 6 years, and are enlarged rapidly between adolescence and age 20 by the activity of osteoclasts. Changes in the dimensions of the sinuses alter the quality of sound generated by the larynx. Inflammation of the mucous membrane lining the sinuses is called *sinusitis*.

The larynx is formed by muscular folds of tissue attached to cartilages and is lined by mucous membranes. It is located between the base of the tongue and the trachea (Fig. 20-1). Contractions of various laryngeal muscles can cause different cartilages in the larynx to move and can stretch and relax the vocal cords. In this manner high-pitched or low-pitched sounds can be produced by air which vibrates the vocal cords. The movements are influenced by Broca's speech area in the brain (Fig. 11-10). Inflammation of the larynx, *laryngitis,* usually reduces the size of the laryngeal opening and the flow of air through it.

The trachea leads from the larynx toward the lungs and divides into two primary bronchi, which subdivide repeatedly to form *bronchioles* less than 1 mm in diameter. Except for the smallest bronchioles, rings or plates of cartilage in their walls keep all the tubes open. Bronchioles less than 1 mm in diameter contain smooth muscle. This muscle has little effect in normal respiration but may contract and decrease the diameter of bronchioles and increase resistance to air flow in an asthmatic attack. Grapelike clusters of *alveoli* extend from the ends of the bronchioles (Fig. 20-2).

The lungs are paired and contained within separate pleural cavities located within the chest cavity or thorax (Fig. 20-1). Each lung is surrounded by a double-layered *pleural membrane.* The outer layer is the *parietal pleura,* the inner one, the *visceral pleura.* A serous fluid secreted by cells of the membranes coats them and minimizes friction as the lungs expand and recoil during respiration. Inflammation of the membranes is called *pleurisy.* Their puncture causes collapse of a lung in a condition known as *pneumothorax.*

The lungs are pink at birth and darken with age as phagocytic cells, alveolar macrophages located in the lung's inner lining, ingest particulate material from inhaled air. Cigarette smoke and other pollutants accelerate the color change.

Alveoli are microscopic terminal sacs of the respiratory bronchioles and are the structures across which gas exchange takes place (Fig. 20-2). They are connected to the bronchioles by alveolar ducts (Fig. 20-3). The number of alveoli ranges from 24 million to 30 million in the newborn to a total of 300 million to 750 million, attained by about 8 years of age. The alveoli are closely enveloped by capillary beds which contain about 60 to 80 ml of blood. The surface area of all the capillary beds in the adult lungs is approximately 70 to 80 m^2 (754 to 861 ft^2), slightly less than half the size of a singles tennis court. Thus, because blood is spread in an extremely thin layer over the capillaries covering the surface of the alveoli, the rate of diffusion of gases between the alveoli and the capillaries is extremely high.

Breathing by the newborn is more difficult

FIG. 20-2
General surface structure of alveoli. The increase in color indicates elevation of oxygen levels in blood as it passes along the alveolar capillaries.

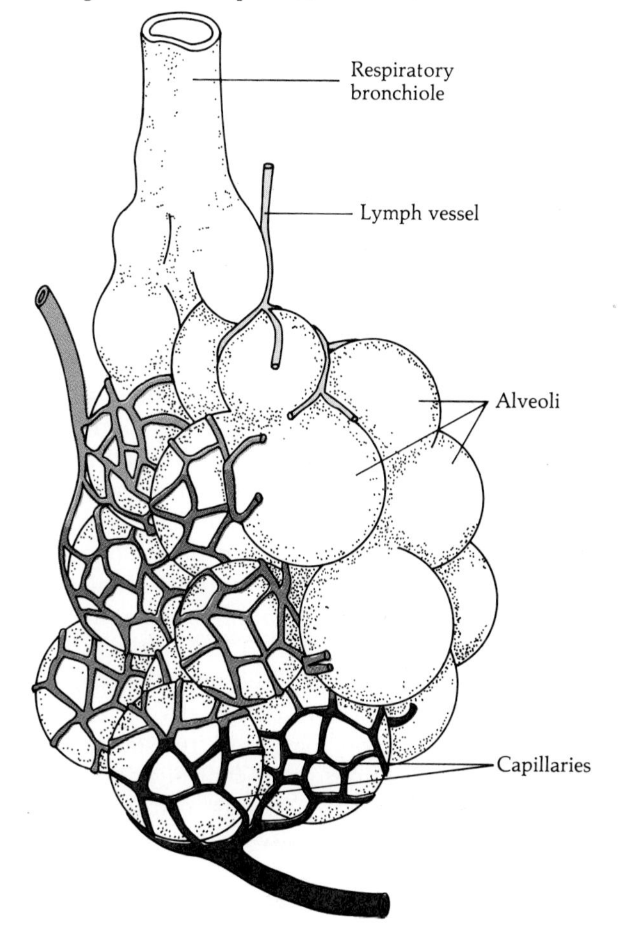

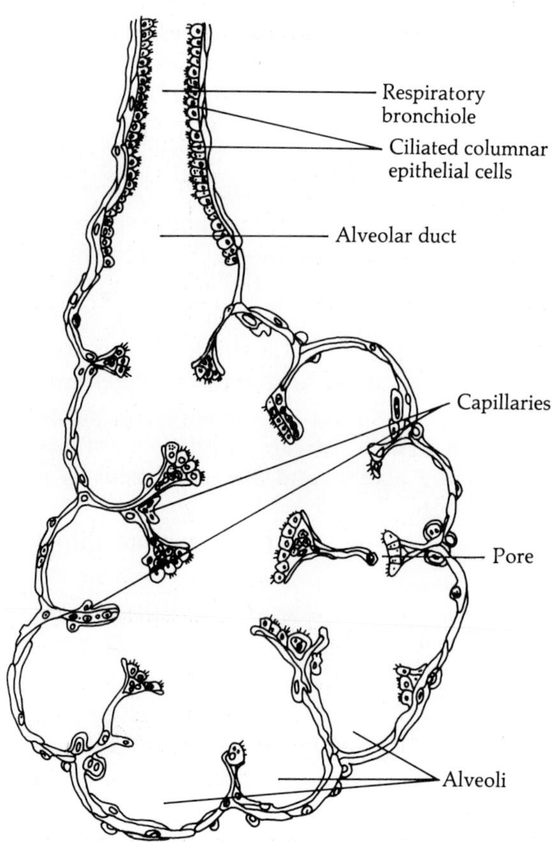

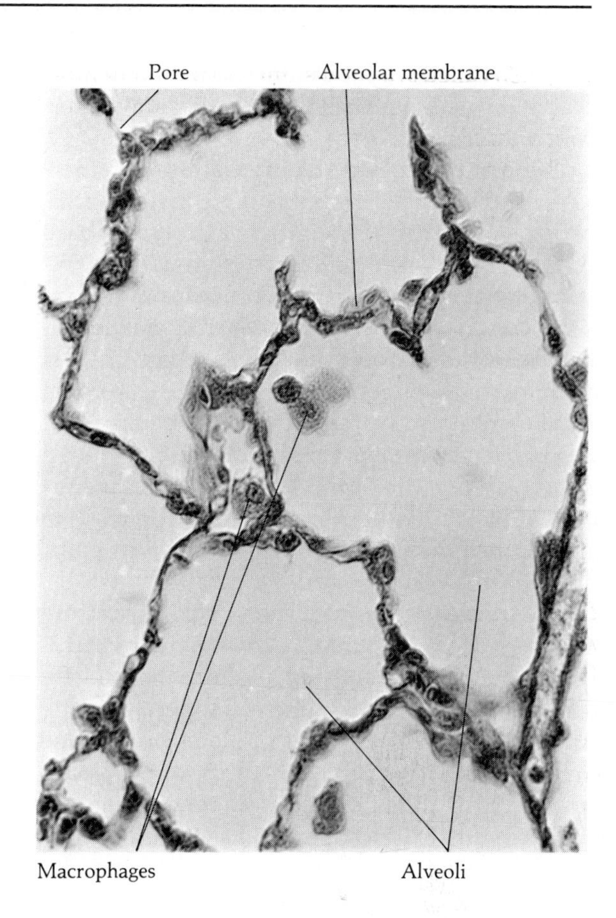

FIG. 20-3
Section through alveoli showing cellular detail. A section of the diagram on the left is enlarged in the photomicrograph of the alveoli on the right (×700).

than that by adults because the newborn child's lungs, are filled with fluid secreted by their cells prior to birth, are tiny, and have a small number of alveoli. There is approximately 50 mL of fluid in the alveoli prior to birth. A small proportion of it is excreted into the amniotic fluid bathing the unborn baby, about 25 mL exits through lymphatic vessels, and 15 mL is expelled through the nose and mouth. With his or her first few breaths the baby must exert 10 to 15 times the pressure required in adult breathing to expel the remaining fluid from the lungs. Furthermore, because of the relatively small size of the respiratory passages, resistance to air flow is about 5 to 6 times that of an adult. Consequently, newborn infants continue to breath with greater difficulty and 2 to 3 times more rapidly than do adults.

Cellular Anatomy and Physiology

The alveoli are generally larger in the upper part of the lung and have interconnections or pores which allow neighboring units to transmit air from one to the other (Fig. 20-3A).

Cell Types Lining the Respiratory Tract

Olfactory receptors, columnar epithelial cells with cilia, and mucus-secreting goblet cells are especially numerous in the lining of the upper respiratory tract. These cells are closely connected by tight junctions, which only allow passage of ions and small molecules (Fig. 3-1B).

Three types of epithelial cells and alveolar macrophages line the alveoli. Type I squamous cells cover 90 percent of the alveolar surfaces. Type II cuboidal cells seem to be the source of regenerating epithelium and are active in lipid metabolism. They secrete sheets of phospholipids that coat the inner alveolar lining. Type III cells, also called *brush cells,* may serve as chemoreceptors. The macrophages are scavenger cells, which rid the lung of foreign particles and are often shed with mucus by coughing.

Respiratory Membranes

There are several barriers between alveolar air and blood. Before oxygen passes into the blood, it must diffuse through the molecules of phospholipid that coat the alveoli. It then travels through the membranes of an alveolar epithelial cell, the basement membranes, and the interstitial fluid and finally moves through the endothelial cell membrane of a capillary to enter the plasma to confront a red blood cell (Fig. 20-4). The average distance involved is about 2 μm but it may be as little as 0.2 μm in areas where the alveolar membranes are very thin. The extremely large surface area available, coupled with the thinness of the membranes, offers optimal opportunity for gas exchange between air in the alveoli and the blood.

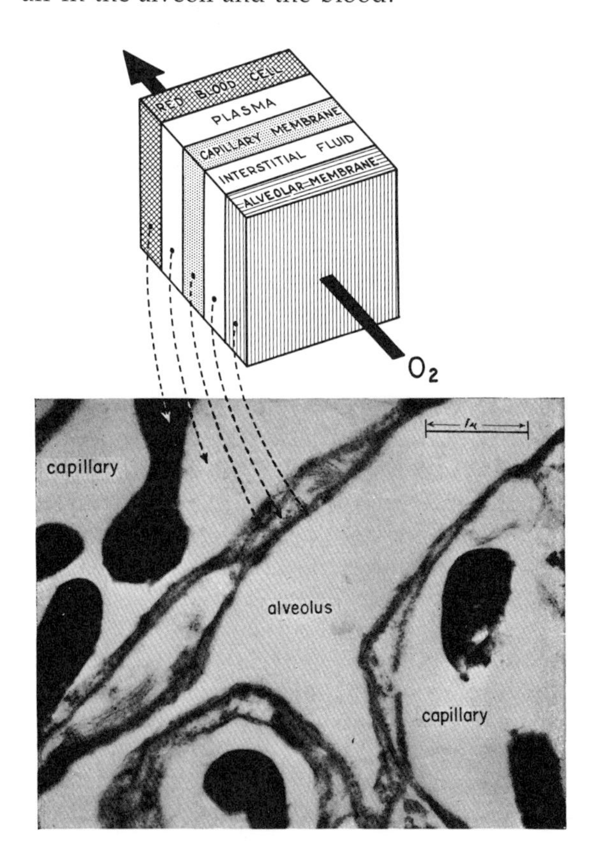

FIG. 20-4
Alveolar-capillary membranes. The arrow in the sketch indicates the direction of diffusion and the tissues through which O_2 must pass to reach the red blood cells in the capillaries. The electron micrograph depicts the tissues involved (× 20,000).

Surfactants and Respiratory Distress Syndrome

Surfactants containing a mixture of phospholipids, some protein, and a small percentage of carbohydrate coat the inner alveolar lining with a 10 to 20 nm thick layer of molecules. The phospholipids include the phosphoglycerides phosphatidylcholine (lecithin) and sphingomyelin, which are normal components of myelin and cell membranes. Acting as detergents, surfactants reduce the normal surface tension between gas molecules and the alveolar lining by a factor of 7 to 14. When there is too little surfactant, the force of attraction between gases within air and the alveolar lining pulls the alveolar walls together to make inhalation more difficult (Fig. 20-5).

The *respiratory distress syndrome* (RDS) is a disorder in which acute respiratory difficulty develops unexpectedly. It may occur in adults owing to a variety of causes, including trauma, infections, and drugs. The survival of an individual with RDS may require mechanical respiratory assistance, as described at the end of this chapter.

RDS also may occur in newborn children because of a deficiency in surfactant. It affects about 40,000 infants per year in the United States and may be a significant cause of infant deaths killing as many as several thousand of them each year, especially between 1 month and 1 year of age. With RDS alveolar membranes become disrupted and the alveoli accumulate protein, fluid, and fibrin. The RDS syndrome of the newborn was formerly referred to as *hyaline membrane disease.* In early prenatal development surfactants are secreted and released to the amniotic fluid bathing the fetus. Normally by the thirty-fourth week of pregnancy the secretion of lecithin has increased and become greater than that of sphingomyelin. The ratio of lecithin to sphingomyelin (L/S

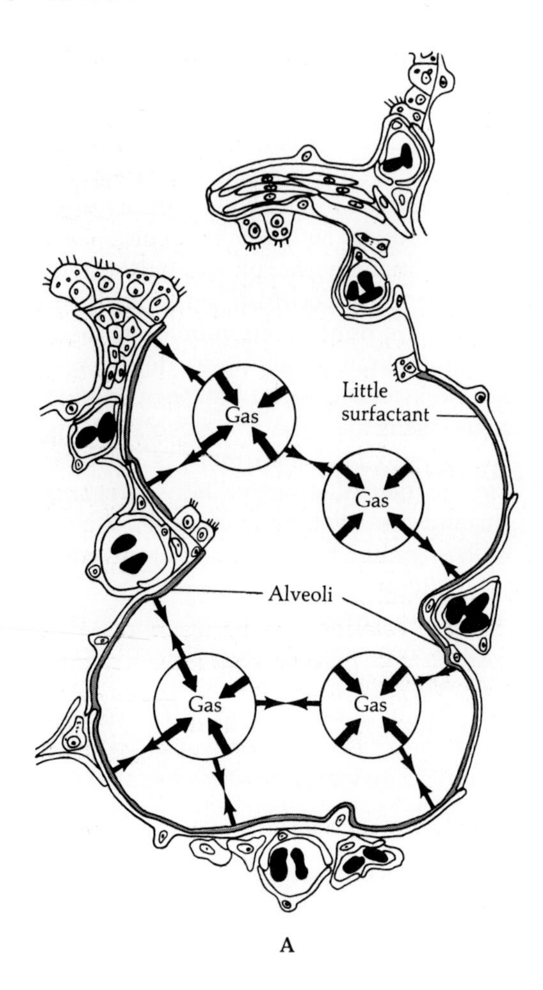

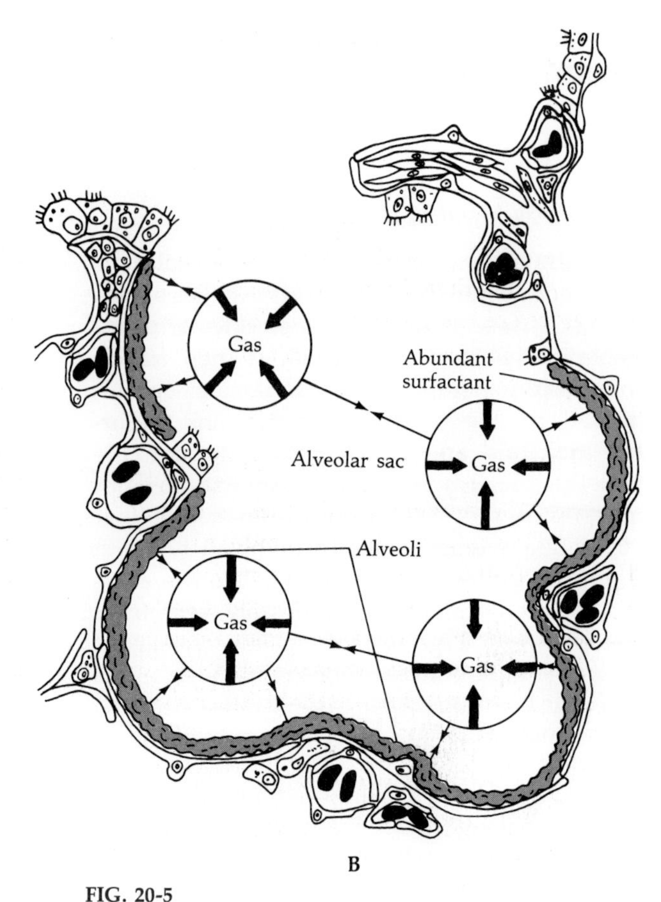

FIG. 20-5
The effects of surfactant on surface tension on the walls of the alveoli. The net forces of molecular attraction pulling the alveoli inward are indicated by the relative thickness and direction of the arrows. (A) Narrowed alveoli due to a deficiency of surfactant. (B) Widened alveoli with a normal amount of surfactant.

ratio) in a sample of amniotic fluid may be used as an index of lung development. Ratios below the mature lung level of 2.0 predict risk of RDS in the newborn, which rises with prematurity. Because phosphatidylcholine levels in the unborn baby's lungs increase sharply during the last few weeks of pregnancy and because saturated phosphatidylcholine (SPC) is also the major surfactant component, it has been recommended as an indicator of fetal lung maturity. A low level of SPC in the amniotic fluid seems to be an even more accurate predictor of RDS in the newborn than is the L/S ratio.

A deficiency in surfactant is associated with collapse of the alveoli (*atelactasis*) in sections of the lungs. Once babies get RDS, too little oxygen is delivered to blood as it passes through the lungs. The blood levels of oxygen decrease in a condition called *hypoxia*. Cortisone may be administered to premature infants to hasten fetal lung development and correct the disorder in some instances.

Another disorder, sudden infant death, may result from deficiency in respiratory control in the brain or carotid body. Factors associated with the disorder may include the presence of bacterial infection, anemia in the mother, and barbiturate use or smoking during pregnancy. The barbiturates from the mother retard respiration and ventilation of her lungs (Table 21-9) and the exchange of respiratory gases and impede the development of the baby's brain; smoking decreases blood flow to the placenta and the availability of oxygen to the fetus (Chap. 24) and slows its growth.

20-2 THE PHYSICS OF BREATHING

Dry air at sea level contains approximately 78.09 percent N_2, 20.94 percent O_2, 0.03 percent CO_2, and less than 1 percent of other gases. The movement of the gases from air through the alveoli into the plasma is governed by physical laws.

Gas Laws and Variations in Pressure

Much of the movement of gases is influenced by the physical principles embodied in the gas laws, including those that govern diffusion (Chap. 4). When temperature is constant, the diffusion rate of a substance across a barrier depends on differences in concentration of the substance and the properties of the barrier.

The concentration of gas is measured in terms of pressure it exerts. According to *Dalton's law,* the partial pressure (tension) of a gas is proportional to its concentration and the total pressure of a mixture of which it is a part. The average total pressure of air at sea level, defined as 1 atm (atmosphere) will support a column of mercury (Hg) 760 mm high in a narrow glass tube. It produces a pressure equal to 14.7 lb/in^2. The partial pressure of O_2 (Po_2) is equal to its percentage in air times the atmospheric pressure. At sea level it is 159 mmHg (20.94 percent × 760 mmHg).

The gases in air surrounding a fluid dissolve separately in that fluid and ultimately attain an equilibrium. The partial pressure of each gas in the fluid and the atmosphere around it may then become the same. It is important to note, however, that the ability of a gas to dissolve in plasma is directly related to the gas's solubility in it. All other factors being equal, gases which are readily soluble in a fluid move between it and the external environment more readily than those which are not very soluble in the fluid. The relative solubilities of CO_2, O_2, and N_2 in plasma are 24.0, 1.0, and 0.55, respectively. These figures indicate the comparative ease or difficulty with which these gases enter the plasma.

The effect of partial pressure on the availability of a gas can be illustrated by ascent to high altitude. With increasing altitude, the height of the column of air in the atmosphere decreases, that is, the atmospheric pressure declines (Fig. 20-6), as does the partial pressure of gases in air. As a result, although the percentages of the gases in air do not change significantly, less oxygen is available to the tissues of the body at high altitudes (Table 20-1). Furthermore, as atmospheric pressure declines, water vapor occupies more and more space in the lungs, reducing the room for other gases. Thus the presence of water vapor also can be significant in reducing gas exchange and the availability of O_2 at high altitudes.

The inverse relationship between pressure and volume of a gas (Boyle's law) has a significant effect on the movement of gases. This

FIG. 20-6
The relationship between altitude and atmospheric pressure.

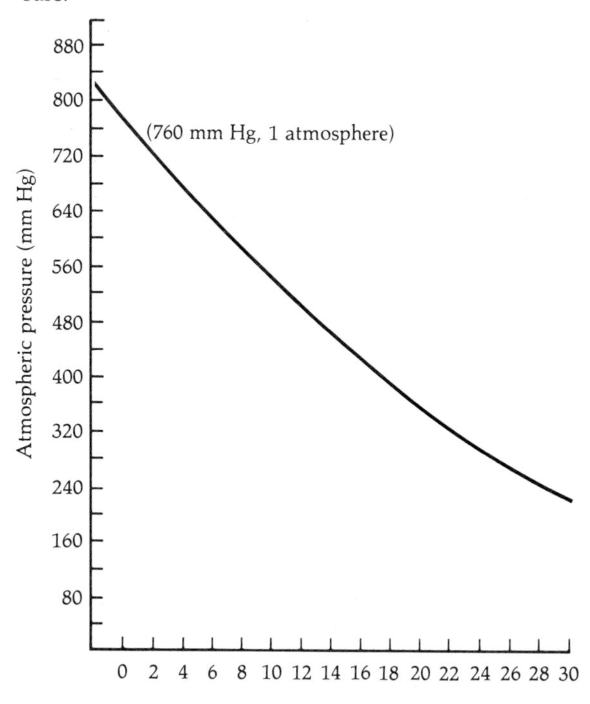

TABLE 20-1
Effects of altitude on respiratory gases*

			PARTIAL PRESSURE OF O_2		PARTIAL PRESSURE OF CO_2	
Elevation, ft	*Atmospheric Pressure*	*Approximate Equivalent % O_2 in Atmosphere*†	*Atmosphere*	*Alveoli*	*Atmosphere*	*Alveoli*
0	760	21	159	103	0.238	40.0
1,000	733	20	154	98	0.220	39.4
10,000	523	15	110	61	0.157	35.0
20,000	350	10	73	34	0.105	29.4

* The figures are rounded off and are given in mmHg unless stated otherwise.
† These values represent a decline in the partial pressure of oxygen rather than a true appreciable change in the percent of oxygen in air.

principle may be illustrated by transferring a volume of air which maximally inflates a small balloon to a larger balloon whose walls will be under less inflation and pressure. So too, as the volume of the chest cavity enlarges during inspiration, the pressure within the thoracic cavity and the lungs, the intrapulmonic pressure, decreases. Air in the atmosphere outside the lungs is then at higher pressure and rushes in. The opposite reaction takes place during expiration as the size of the thoracic cavity diminishes and intrapulmonic pressure increases.

Boyle's law applies to an ideal system. However, pressures in the lung are not uniform and flow rates vary with the diameter of the respiratory tubules through which air moves (Poiseuille's law, page 416). The rate of flow decreases with decreasing inner diameter of the respiratory tubules as resistance to air flow increases. An example is the difficulty in breathing experienced in an asthmatic attack when the bronchioles constrict. During quiet normal breathing, resistance changes little. In shallow nasal breathing, nearly half the resistance to air flow occurs in the nasal passages.

Pressure Gradients in Inspiration and Expiration

Air moves into and out of the lungs as gases diffuse down gradients from higher to lower partial pressures. *Inspiration* occurs when the diaphragm flattens as it contracts and the external intercostal muscles between the ribs elevate the rib cage, causing the thorax to enlarge (Fig. 20-7C and D) and air to rush into the lungs.

Expiration is the movement of air out of the lungs. It occurs passively when the elastic fibers in the lungs and body walls recoil, the diaphragm relaxes and rises as it returns to its dome-shaped position, and the rib cage returns to its smaller size as its associated muscles relax (Fig. 20-7A and B). This response is enhanced during vigorous *active expiration* in exercise when the internal intercostal muscles actively pull the rib cage down and contractions of the abdominal muscles force the diaphragm upward.

Ventilation of the Lungs

When air rushes into the lungs, the alveolar ducts expand like the bellows of an accordion. The total amount of air that enters the respiratory tract each minute is the *total ventilation, pulmonary ventilation,* or *respiratory minute volume.* A percentage of it reaches the alveoli and the remainder fills the rest of the respiratory passageways. The quantity of fresh air that enters the alveoli per minute is the *alveolar ventilation rate,* which is a major factor in determining gas exchange. It may be altered by changes in physiology or by diseases affecting lung function. For example, in *silicosis* the accumulation of particles in the lungs restricts lung expansion and decreases alveolar ventilation. Other disorders will be described later in this section.

Spirometry

The volume of air moved can be measured with a *spirometer* (Fig. 20-8). In a simple recording

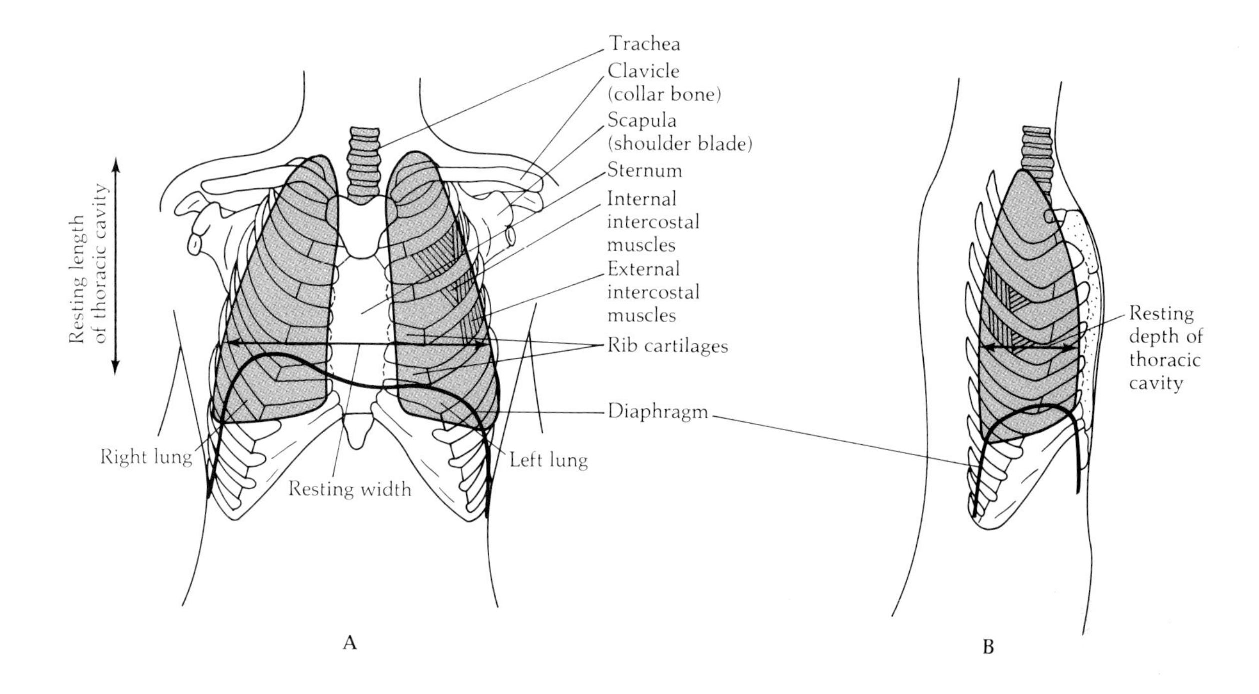

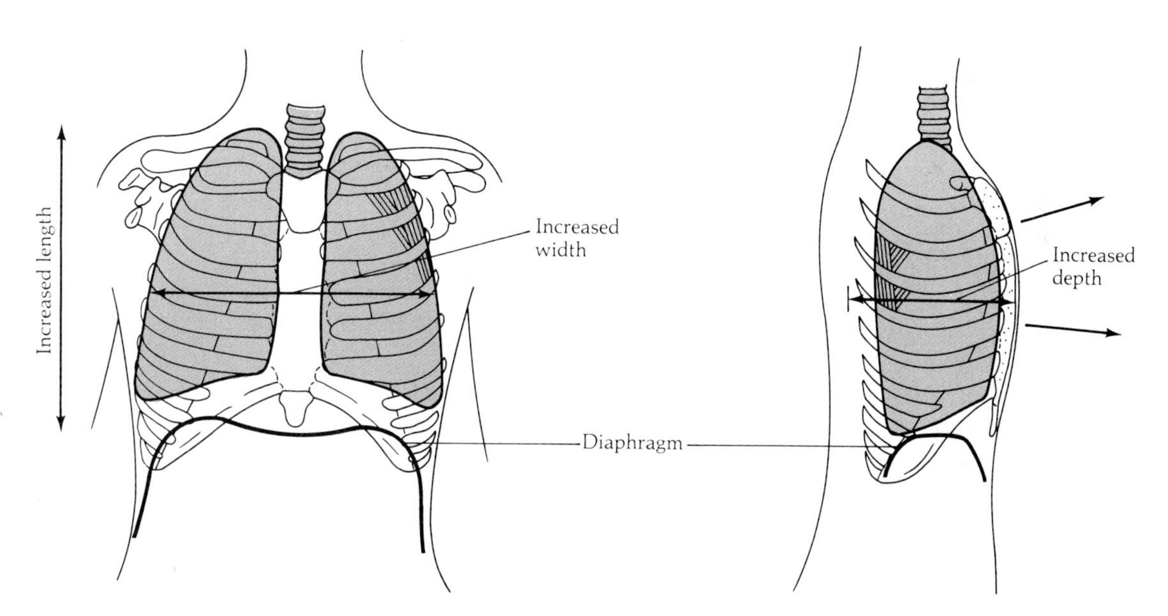

spirometer, into which an individual blows once, air enters the bottom of a tank filled with water and is trapped within a cylinder (the spirometer bell). This displaces the spirometer bell and the pen attached to it. The volume of air is thereby recorded on a moving calibrated

FIG. 20-7
Changes in the size of the thorax during respiration. (A and B) The thoracic cavity between respirations (A, anterior view; B, lateral view). (C and D) Increase in size of thorax during inspiration (C, anterior view; D, lateral view).

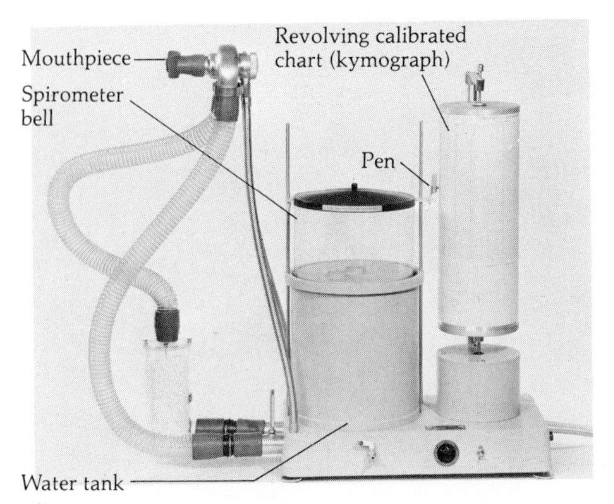

FIG. 20-8
A simple recording spirometer. Air blown into the spirometer displaces an inverted spirometer bell attached to a pen and recorder to make a graphic tracing of the volume of air expired.

chart. The time elapsed can be noted to determine the volume of air exhaled within a given time.

Lung Volumes

The volume of air exchanged by the lungs may be measured under different breathing conditions. The air volume which is inspired *or* expired during normal, quiet breathing is the *tidal volume* and is equivalent to about 500 mL, that is, about 1 fluid pint, in the average adult (Fig. 20-9).

The amount of air that can be forcibly expired by breathing actions in excess of the tidal volume is the *expiratory reserve volume* (or *supplemental air*). That which can be forcibly inspired beyond the tidal volume is the *inspiratory reserve volume* (or *complemental air*). Not all the air can be expelled from the two lungs. The *residual volume* is the air remaining in the lungs after a forcible expiration. It allows gas exchange to take place between the alveoli and the blood passing through the capillaries, between breaths.

The approximate functional air capacity of lungs is designated as the *vital capacity* and is measured as the maximal amount of air that can be forcibly expired after the greatest possible inspiration. It is equal to the sum of the tidal volume, inspiratory and expiratory reserves. The vital capacity varies with the size of the thoracic cavity, the amount of blood in the lungs, the condition of the alveoli, and the age, sex, and height of the subject. Some average values are shown in Table 20-2. Vital capacity is closely correlated with the cube of the body height in centimeters and may be approximated by the following equations:

$$\text{V.C. (males)} = 0.052H - 0.022A - 3.60$$
$$\text{V.C. (females)} = 0.041H - 0.018A - 2.69$$
$$\text{where V.C.} = \text{vital capacity}$$
$$H = \text{height in centimeters}$$
$$A = \text{age in years}$$

The average adult breathes about 12 to 20 times per minute, which is a normal resting respiratory rate. Under these conditions the volume of air exchanged per minute equals the respiratory minute volume and ranges from 6 to

FIG. 20-9
Respiratory volumes. The diagram illustrates the general range of volumes of air exhaled from the lungs of a young adult under varying degrees of effort.

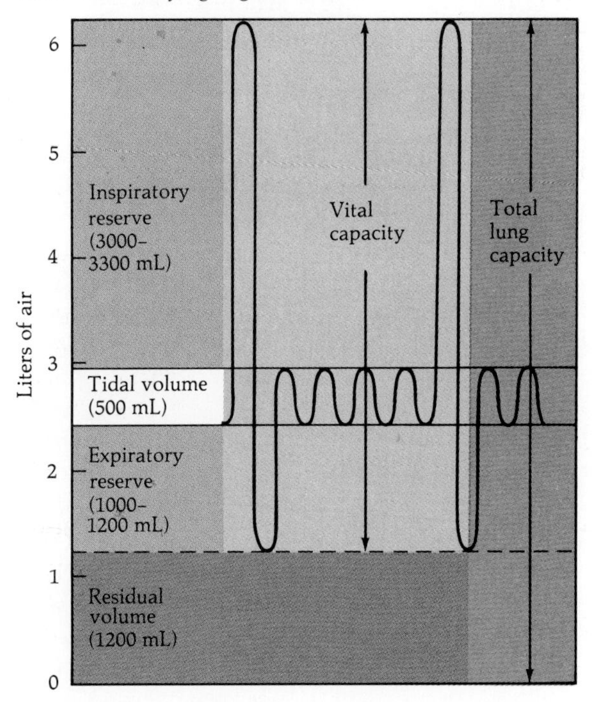

TABLE 20-2
Predicted vital capacities, L*

MALES								FEMALES							
Standing Height, in	AGE, YEARS 20	30	40	50	60	70	*Standing Height, cm*	*Standing Height, in*	AGE, YEARS 20	30	40	50	60	70	*Standing Height, cm*
61	3.97	3.65	3.35	3.04	2.73	2.42	155	57	2.81	2.63	2.45	2.27	2.09	1.91	145
63	4.30	4.00	3.70	3.40	3.10	2.80	160	59	3.08	2.89	2.71	2.53	2.35	2.17	150
65	4.62	4.32	4.02	3.72	3.42	3.12	165	61	3.34	3.15	2.97	2.79	2.61	2.43	155
67	4.94	4.64	4.35	4.05	3.74	3.44	170	63	3.60	3.41	3.22	3.05	2.87	2.69	160
69	5.26	4.96	4.66	4.36	4.06	3.76	175	65	3.88	3.68	3.50	3.32	3.14	2.96	165
71	5.58	5.28	4.98	4.68	4.38	4.08	180	67	4.13	3.94	3.76	3.58	3.40	3.22	170
73	5.90	5.60	5.30	5.00	4.70	4.40	185	69	4.38	4.20	4.02	3.84	3.66	3.38	175

* Variations 20% or more below predicted values are subnormal.

10 L/min. The respiratory minute volume varies with the functional capacity of the lungs, exercise, and disease and may be used as an index of respiratory physiology. Another measure of lung function is the FEV_1, the amount of air forcibly expired in the first second of a forcible expiration following a maximal inspiration. Alterations in the FEV_1 may be used as an index of the severity of certain lung disorders, described below.

Types of Respiration

NORMAL BREATHING *Normal breathing* (*eupnea*) can be either shallow or deep. *Shallow* or *costal breathing* exchanges the tidal air volume about 12 to 20 times per minute. However, of the tidal air, about 350 mL ventilates the alveoli while 150 mL remains in air passageways (the nose, mouth, pharynx, trachea, bronchi, bronchioles, and alveolar ducts), referred to as *anatomic dead space.* Because the alveoli receive a mixture of the air left in the passageways and that brought in from the atmosphere, the partial pressures of gases in alveolar and atmospheric air differ (Table 20-1). In *deep breathing* (*abdominal* or *diaphragmatic breathing*) the abdominal wall moves outward as the diaphragm descends. This type of respiration accounts for nearly 75 percent of gas exchange during quiet respiration.

ABNORMAL BREATHING AND EFFECTS OF DISEASE A variety of stimuli and diseases produce abnormal patterns of respiration. Response to sudden pain or exposure to cold temporarily stops breathing. The cessation of breathing, regardless of cause, is called *apnea.* Respiration then resumes and occurs at a faster rate and at greater depths if either type of stimulus persists. Psychological stimuli or exercise approaching maximal activity can cause *dyspnea* (air hunger), characterized by shortness of breath and/or difficult or painful breathing. The symptoms may be due to the increased metabolic requirements imposed on the body by the exertion of muscle used in vigorous expiration and/or by increased metabolic wastes from muscle exercise. Difficult breathing while in a horizontal position is called *orthopnea* and often follows congestive heart failure due to the accumulation of fluid in and around the alveoli (pulmonary edema). The effects on the lungs are minimized by propping the upper half of the body to the sitting position.

Cheyne-Stokes Breathing Rhythmic alternating periods of apnea and dyspnea occur in *Cheyne-Stokes* respiration. It may occur in otherwise healthy sleeping adults or it may result from exposure to high altitude or from disease of certain parts of the brain. It often precedes death.

The basis of this type of respiration is the accumulation of carbon dioxide, which causes hyperventilation which then lowers the Pco_2. As a result of the hyperventilation and the accompanying vasoconstriction of vessels in the brain, more time is required for Pco_2 to accumulate to levels that stimulate the respiratory center. Because CO_2 is the major signal to activate respiration, the prolonged low Pco_2 induces apnea. Blood pressure decreases during apnea and increases when breathing resumes.

Congestive Heart Disease Congestive heart disease causes the accumulation of interstitial fluid (Chap. 16) and may reduce air volume within the alveoli and decrease vital capacity and the FEV_1. Reflex acceleration of the heart usually occurs as arterial oxygen levels decline.

Emphysema *Emphysema* is a disorder in which there is an irreversible loss of elasticity in alveoli, which prevents them from recoiling after inspiration and leaves them patent (open). Consequently the residual air volume is increased and inspiration brings in less oxygen. The length of time required for expiration increases owing to loss of the elasticity in the lungs; the FEV_1 decreases as expiration becomes more difficult and prolonged; and the lungs become inflamed resulting in a decrease in effective respiratory surface area. Emphysema may be congenital; among its other causes is cigarette smoking. It often is accompanied by chronic bronchitis.

Asthma *Asthma* is a reversible functional disease caused by spasms of bronchiolar smooth muscle. The constriction of this muscle makes breathing difficult, especially obstructing outward air flow and diminishing the FEV_1. Sometimes asthma is induced by an allergic attack (Fig. 3-20, page 64).

Lung Infections *Pneumonia* is an inflammation of the lungs which restricts lung expansion and may be brought on by bacterial, viral, or fungal infections. It results when host defenses such as the alveolar macrophages and antibodies fail to eliminate invading microorganisms. Antibiotics can assist the natural defense mechanisms by killing or inhibiting bacteria. *Tuberculosis* commonly infects the lungs and bones. It is caused by a species of bacteria, *Mycobacterium tuberculosis,* which, in the absence of antibiotic therapy, often survives within macrophages as an intracellular parasite.

20-3 TRANSPORT AND EXCHANGE OF GASES

During respiration, oxygen diffuses from the atmosphere through lung tissues into the blood and CO_2 moves in the opposite direction, the exchange being greatly influenced by the physical and chemical properties of blood.

Oxygen Transport

About 19.5 mL of the 20 mL of oxygen usually carried in every 100 mL of blood is transported as oxyhemoglobin and a slight amount (0.5 mL) is transported as dissolved gas molecules in the plasma. The release of oxygen from oxyhemoglobin is promoted by the accumulation of hydrogen ions (low pH), carbon dioxide, and 2,3-diphosphoglycerate (2,3,-DPG) in the blood and also by an increase in body temperature, as occurs with exercise (Fig. 20-10). Association of oxygen with hemoglobin to form oxyhemoglobin is promoted mainly by an increase in PO_2 (and a decrease in the concentrations of hydrogen ions, CO_2, and 2,3-DPG as well as in temperature). Each of these agents can alter the shape and attraction of the four binding sites in hemoglobin, each of which loosely combines with a molecule of oxygen.

The dissociation of oxyhemoglobin may be depicted graphically by plotting the saturation of hemoglobin (percent HbO_2) against the Po_2 (in mmHg). As the Po_2 increases, the percent of hemoglobin found as oxyhemoglobin increases to 97 to 98 percent (Fig. 20-11). As the concentration of hydrogen ions increases in the blood (for example when the pH drops from 7.40 to 7.20), the percentage of oxyhemoglobin at a specific Po_2 declines (Fig. 20-11). This phenomenon is called the *Bohr effect.* Elevated levels of

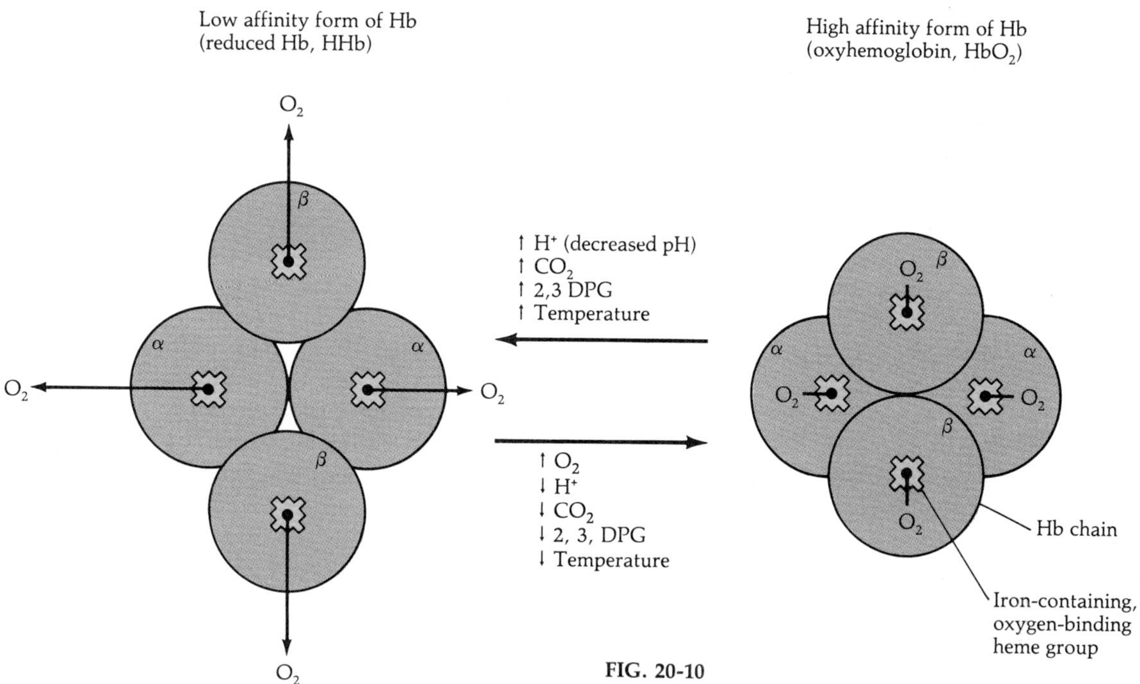

FIG. 20-10
Dissociation and association of oxygen and hemoglobin. As shown on the left, the β-chains move further apart when low-affinity Hb is formed and HbO_2 dissociates. As shown on the right, the β-chains move closer together when high-affinity Hb is formed and Hb and O_2 associate. (Also see Fig. 18-5, page 438.)

Pco_2 in the blood cause the effect by forming carbonic acid and then hydrogen ions, as shown in the following equation:

$$\underset{\text{Carbon dioxide}}{CO_2} + \underset{\text{Water}}{H_2O} \rightleftharpoons \underset{\text{Carbonic acid}}{H_2CO_3} \rightleftharpoons \underset{\text{Hydrogen ion}}{H^+} + \underset{\text{Bicarbonate ion}}{HCO_3^-}$$

2,3-DPG accumulation also causes oxyhemoglobin to dissociate. When too little oxygen is available to the tissues, 2,3-DPG is generated by the anaerobic breakdown of carbohydrate. It then binds to oxyhemoglobin and changes it to a form with low affinity for oxygen, thus releasing the gas for diffusion to the tissues (Fig. 20-10). The levels of 2,3-DPG increase within 60 min of exercise and in chronic hypoxia, anemia, and hyperthyroidism. The levels decrease when the red blood cell count is high, with aging of the cells (as occurs with storage of blood in blood banks), and in metabolic disorders that inhibit anaerobic metabolism in red blood cells. These changes render the red blood cells less able to provide oxygen to tissues.

FIG. 20-11
Effect of hydrogen ions (pH) on the dissociation of HbO_2. As the H^+ ion concentration increases (that is, as the pH decreases), the percent of HbO_2 decreases at a specific Po_2. From *Transport of Oxygen Carbon Dioxide and Inert Gases by the Blood.* C. J. Lambertsen. In *Medical Physiology,* by V. B. Mountcastle (ed.), 14th ed., copyright © C. V. Mosby Co., St. Louis, 1980.

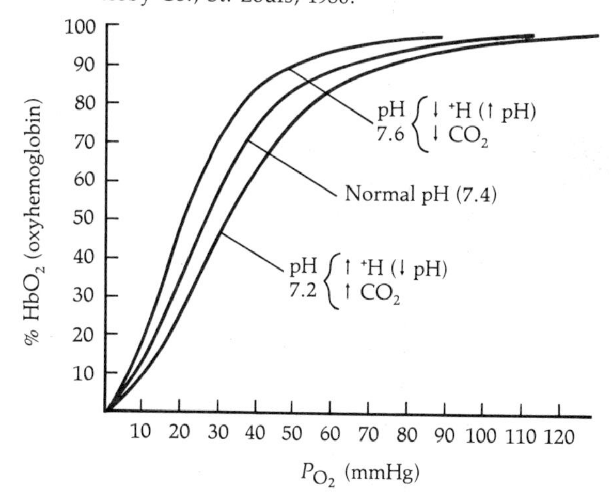

Because 2,3-DPG has a lower affinity for fetal hemoglobin than for adult hemoglobin, blood in the unborn and newborn infant has a high affinity for oxygen. This facilitates diffusion of oxygen across the placenta from the mother's hemoglobin to that of her unborn baby. However, the relative lack of responsiveness of fetal hemoglobin to 2,3-DPG leaves the newborn less able to release oxygen and to adapt to conditions that cause hypoxia, such as exercise, anemia, and exposure to high altitude.

Carbon Dioxide Transport

Carbon dioxide generally diffuses from tissues into plasma and into red blood cells. In the blood about one third of the carbon dioxide is in red blood cells and two thirds is in the plasma (Table 20-3). In both locations a small percentage of the gas combines with proteins to produce *carbamino proteins* such as *carbaminohemoglobin*, slightly more forms dissolved gas molecules, and most of it (over 60 percent in the plasma) is transported as bicarbonate ions.

Bicarbonate ions (HCO_3^-) are derived from carbonic acid (H_2CO_3), which is formed when carbon dioxide combines with water (H_2O) under the catalytic action of the enzyme carbonic anhydrase (Fig. 20-12A). The carbonic acid can dissociate into HCO_3^- and H^+. Bicarbonate ions diffuse out of the red blood cells into the plasma and are replaced by chloride ions, which diffuse into the cells in a reaction called the *chloride shift*. The chloride shift helps to maintain the isoosmolarity of the cells. Hydrogen ions from the carbonic acid displace sodium and potassium ions in hemoglobin, forming reduced hemoglobin and freeing the oxygen from hemoglobin to diffuse into the tissue.

As reduced hemoglobin (HHb) is formed, the blood becomes deoxygenated and is converted to the venous condition. The sodium and potassium ions released from hemoglobin cannot diffuse out of the red cells as readily as the negatively charged bicarbonate ions. This provides an opportunity for the negatively charged chloride ions to diffuse into the cells and replace the bicarbonate ions. The K^+ and Na^+ ions unite with Cl^- to form much KCl and some NaCl in the cell. They also combine with HCO_3^- to form some $KHCO_3$ and much $NaHCO_3$ in the plasma. As a result, venous blood attains higher levels of bicarbonates outside the cell, and a greater total number of K^+, Na^+, and Cl^- ions accumulate inside. The greater intracellular particle concentration causes an osmotic effect, so that water diffuses into the red blood cell and increases its diameter. These changes cause the hematocrit of venous blood to be 3 percent above that of arterial blood.

When venous blood arrives in the lungs, the reactions described above are reversed (Fig. 20-12B) and about 10 percent of the carbon dioxide is released to the alveolar air. The $P\text{CO}_2$ of venous blood declines from about 46 to 40 mmHg as the carbon dioxide diffuses into the alveoli and oxygen diffuses into the blood.

Exchange of Gases

Exchange in Tissues and the Lungs

The exchange of respiratory gases is governed physically by pressure gradients which favor the movement of oxygen into tissues from arterial capillaries and of carbon dioxide out of metabolically active tissues (Fig. 20-12A). Each gas diffuses in the reverse direction when the blood reaches the alveolar capillaries (Fig. 20-12B). Typical partial pressures of gases, which influence their movement at these sites, are shown in Table 20-4.

TABLE 20-3
Approximate percentage of modes of CO_2 transport in arterial blood*

Mechanism of Transport	*Plasma*	*Red Blood Cells*
Bicarbonate ions	62.7	27.3
Dissolved gas	3.2	3.6
Carbamino compounds	1.0	2.2
Total	66.9	33.1

* The average concentration of CO_2 in arterial blood ranges from 4.5 to 22.0 mmole. It is higher in venous blood, ranging between 23.2 and 23.9 mmole. The amounts of carbamino-Hb ($HbCO_2$) in the plasma and dissolved CO_2 in the red blood cells are the same in arterial and venous blood.

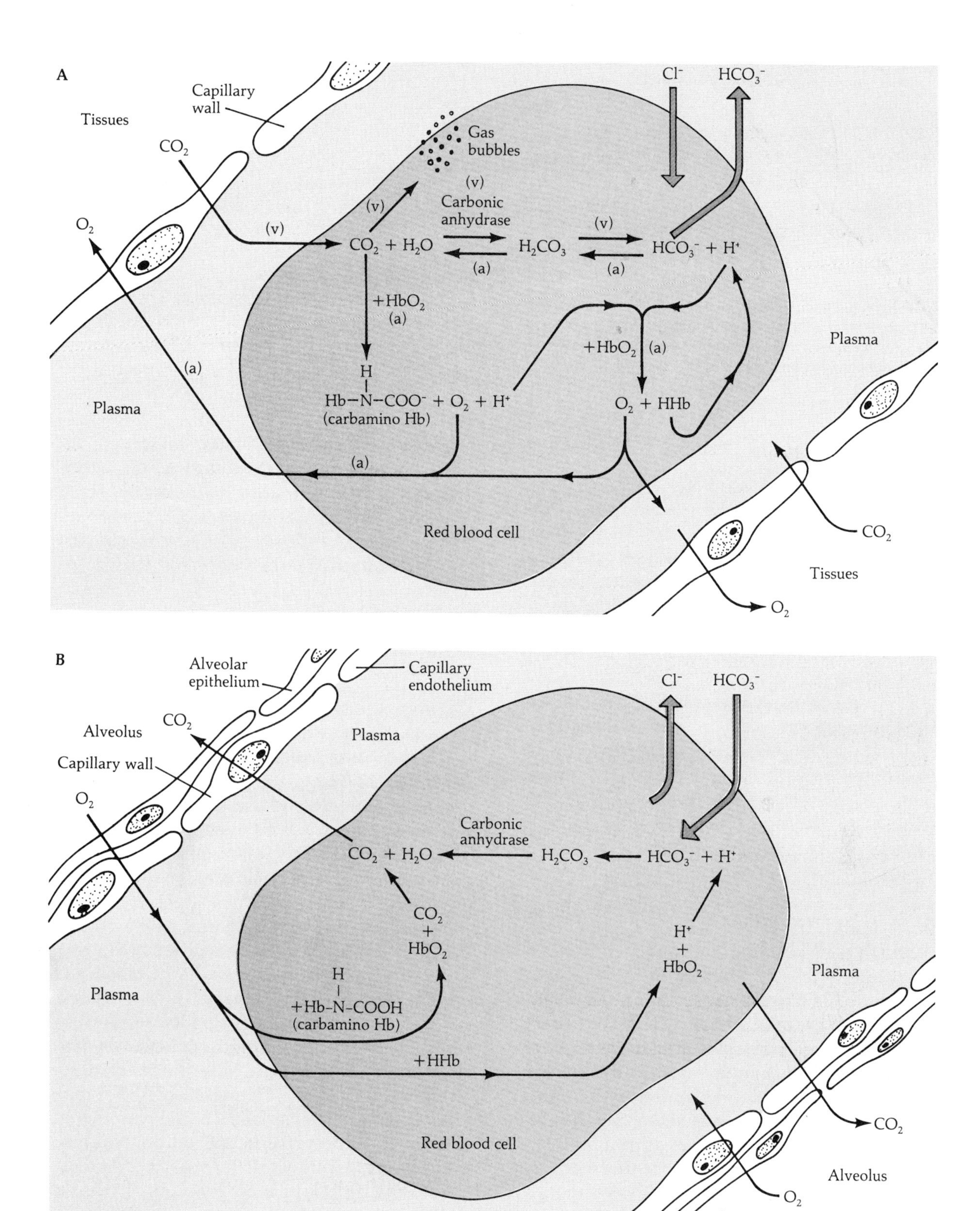

FIG. 20-12

Transport and exchange of CO_2 and O_2. CO_2 formed by metabolism in tissues enters a series of *reversible* reactions. Those favored in venous capillaries are labeled v. Those favored in arterial capillaries are labeled a. The reactions shown in the red blood cell proceed in the plasma but at a slower rate. (A) Gas exchange favored at tissue sites. (B) Gas exchange favored in alveoli in the lungs.

TABLE 20-4
Partial pressure of gases, mmHg, at sea level

Gas	Atmosphere Inspired	Alveoli	Arterial Blood	Tissues and Venous Blood	Expired Air
Oxygen	159	100	95	40	116†
Carbon dioxide	0.3	40	40	46+	28
Water vapor	5.7	47	47	47	47
Nitrogen	594	573	573	573	569
Totals	760*	760	755	706	760

* Inert gases other than N_2 produce a total pressure under 1.0 mmHg.
† It is important to note that expired air still contains substantial amounts of oxygen. This makes possible mouth-to-mouth ventilation in cardiopulmonary resuscitation (CPR).

Factors Affecting Gas Exchange

A number of factors affect gas exchange, some of which have been described above (Table 20-5). The ratio of alveolar ventilation to the flow of blood (*perfusion*) through the lungs must be optimal for gas exchange to occur efficiently. The optimal ratio is approximately 0.8 for average normal lungs. This is calculated by dividing alveolar ventilation (4 L/min) by cardiac output (5 L/min).

The ratio may be decreased by factors that lower ventilation, such as general anesthesia and asthma. Emphysema may increase or decrease the ratio in different sections of the lungs. If perfusion is too rapid, too little time exists for diffusion of gases, the ratio is lowered, and arterial $P\text{O}_2$ decreases.

20-4 REGULATION OF RESPIRATION

Respiration is rhythmically controlled by pathways in the nervous system which are responsive to changes in the chemistry and pressure of blood. Reactions to both variables are interrelated and are part of complex feedback mechanisms. Other stimuli may also modify the depth and rate of breathing through interneuronal connections in the central nervous system.

Respiratory reflex controls involve impulses carried to and from the respiratory center in the medulla (Fig. 20-13) and include chemoreceptors, which stimulate breathing, baroreceptors, which may inhibit it, and inflation (stretch) and deflation receptors in the lungs, which inhibit and stimulate inspiration, respectively. Interneurons between the respiratory center, the pons, and the cerebral cortex may modify impulses generated by the peripheral sensory receptors.

Nervous Control Mechanisms

Chemoreceptors

Chemoreceptors called the *carotid bodies* are located in discrete nerve endings in the origins of the wall of each external carotid artery. They stimulate the frequency and depth of respiration and a rise in blood pressure. In response to increases in $P\text{CO}_2$ and H^+ ions along with large decreases in blood $P\text{O}_2$, the carotid bodies generate excitatory impulses carried to the inspiratory center.

Blood flow through capillaries to the carotid bodies is rapid, allowing these chemoreceptors to respond promptly to changes in blood gas content. Their prompt response quickly reflects alterations in the composition of respiratory gases. Therefore, the carotid bodies act as *respiratory gas chemosensors*. In response to an increase in $P\text{CO}_2$, these chemosensors also generate impulses to the cardiac and vasomotor centers which decrease cardiac output and elevate blood pressure while increasing the blood flow to the heart and brain by dilating vessels in those organs. In coordination with these changes in circulation, impulses are generated which are carried to the inspiratory center to increase the frequency and depth of respiration

TABLE 20-5
Some factors affecting gas exchange in lungs

Component	*Property*	*Comment*
Alveolar and capillary membranes	Surface area	Less in the newborn
	Thickness	Varies within normal limits; may be significant in lung diseases
		Increased thickness due to accumulation of connective tissues in *pulmonary fibrosis* (may occur in response to poor circulation to lungs, which deprives them of oxygen)
	Permeability	Ranges from greater to lesser solubility of CO_2, O_2, and N_2
Gases	Partial pressure in alveoli and blood	Affected by environment (e.g., altitude, diving)
	Solubility in blood	Directly affects permeability and diffusion
	Affinity for Hb	Varies with gas and shape of Hb
Blood	Volume of blood in lungs	Increases in proportion to physical activity; decreases with significant fluid loss and hemorrhage
	Number and surface area of red blood cells	Altered by anemias, polycythemias
	Concentration and structure of Hb	Altered by anemias, hereditary disorders, normal genetic development, changes in Hb conformation, and nutritional deficiencies
	Time in alveolar capillaries	Optimal time required for efficient gas exchange

(Fig. 20-13). The aortic bodies also contain discrete chemosensors whose activities can be additive with those of the carotid bodies, but which are of lesser significance in humans. The slower rate of circulation through capillary beds to the aortic bodies make them more suited to serve as *circulatory gas chemosensors* and to better monitor O_2 in the circulatory system.

Baroreceptors

Baroreceptors (stretch receptors), located along the *aortic arch* and the *carotid sinus* in the vicinity of the origin of each carotid artery, may inhibit respiration but seem to exert little influence on it, although they do exert significant effects on blood pressure (Fig. 17-12, page 421).

Medulla

Clusters of neurons in the medulla form the *respiratory center,* which has *inspiratory* and *expiratory centers* that reflexively control the rhythm of breathing (Fig. 20-13). The groups of neurons affect each other by feedback signals, some of which are relayed by neurons in the pons. The impulses from the inspiratory center cause contraction of the respiratory muscles, which increases the size of the thorax (Fig. 20-7), causing intrapulmonic pressure to fall and air to rush into the lungs. The impulses from the expiratory center cause the opposite responses, and air to rush out of the lungs.

When breathing exceeds tidal volumes, pulmonary *inflation and deflation receptors* in the lungs inhibit and stimulate inspiration, respectively, in a response called the *Hering-Breuer reflex* (Fig. 20-13). The inflation receptors inhibit inspiration and relax the inspiratory muscles. As a result, blood P_{CO_2} eventually increases and expiration is thereby stimulated. Deflation receptors, located in *juxtacapillary cells* (*J cells*) within the alveoli next to capillaries, then generate impulses which relieve the inhibition of inspiration to complete the cycle. In labored breathing the deflation receptors seem to stimulate the inspiratory center directly.

Rhythmic respiration may be interrupted by hiccups, coughing, sneezing, sighing, and yawning. *Hiccups* are spasms of the diaphragm during which the glottis is forced closed during inspiration, causing the sound and sensations associated with the disorder. It has been hypothesized that stimulation of receptors in the diaphragm or upper abdomen may initiate hic-

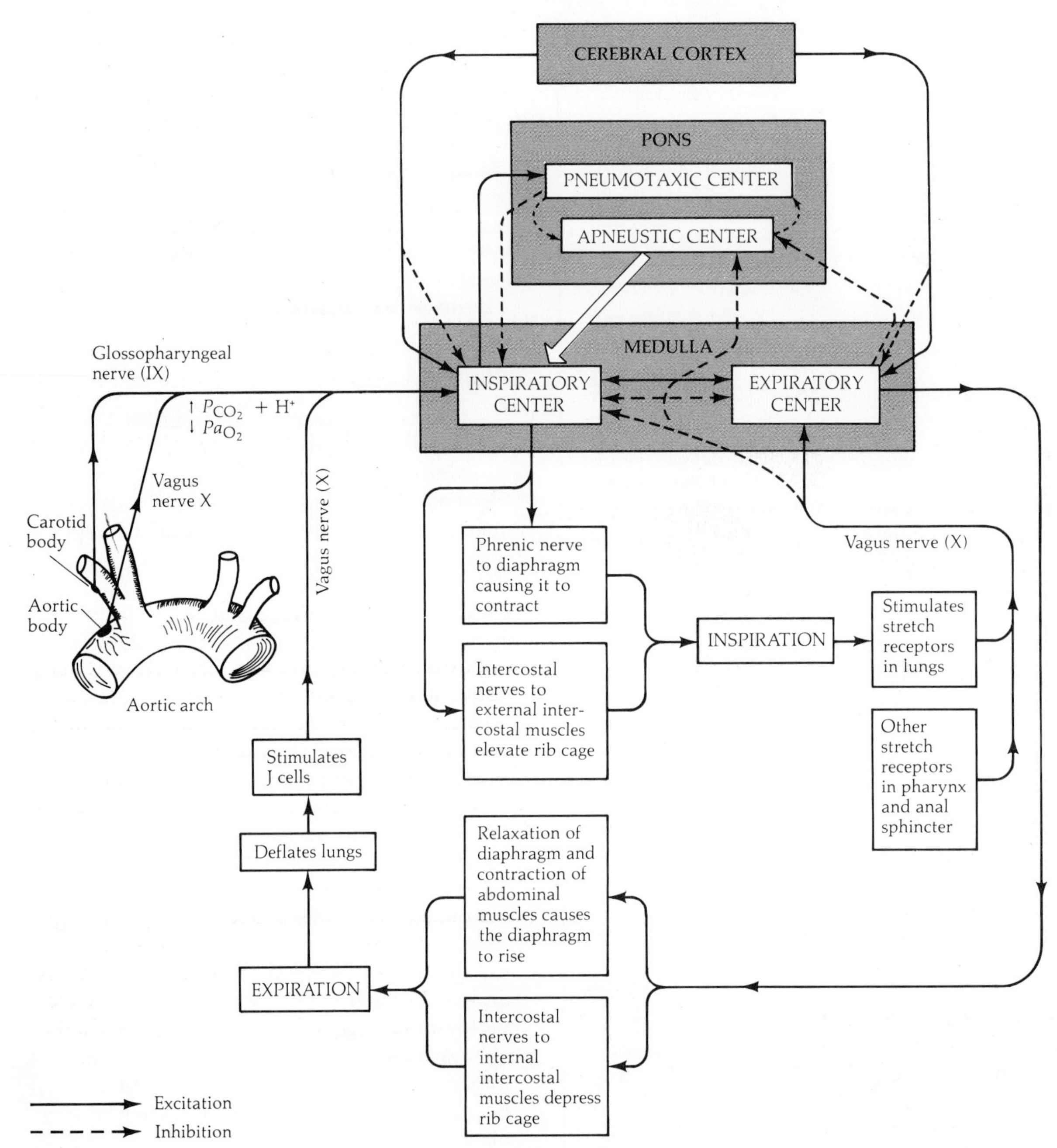

FIG. 20-13
Nervous control of respiration. The pathways for impulses illustrated on one side of the drawing are actually found on both sides of the nervous system. The figure simplifies a complex series of interactions. The lower portion of the diagram illustrates the Hering-Breuer reflex. It is a cycle in which inflation of the lungs stimulates stretch receptors that cause inhibition of the inspiratory center leading to deflation of the lungs which causes the opposite responses. During vigorous exercise the stretch receptors also may stimulate the expiratory center to produce more forceful expiration.

cups. The spasms may be eliminated in some instances by breathing air with a high $P\text{CO}_2$ (6 to 7 percent CO_2). This may be accomplished by breathing into and out of a small bag.

Coughing is a forceful reflex expiration often caused by foreign matter or fluid on the mucous membrane of the lower respiratory tract. Following inspiration, the epiglottis temporarily closes the glottis and air pressure builds in the lungs. The expiratory muscles forcefully contract, the glottis opens, and the air violently rushes out, usually carrying foreign material with it. *Sneezing* is a similar and even more forceful reflex response initiated by irritation of the mucous membrane in the upper tract and preceded by deep inspiration. *Sighing* and *yawning* are characterized by deep, prolonged inspiration. Their causes are unclear but they may be due to underventilation of the alveoli and may represent respiratory reflexes that prevent alveolar collapse (atelectasis). These involuntary acts periodically increase tidal volumes, resulting in stimuli that promote the formation of surfactant, which is required for full inflation of the lungs.

Pons

Clusters of neurons in the apneustic and pneumotaxic areas of the pons maintain the rhythmic respiration established by the medulla. The *apneustic center,* in the lower pons, is the driving force for the inspiratory center in the medulla. The apneustic center alone would stimulate continual inspiration. However, inhibitory signals from the pneumotaxic and expiratory centers, as well as some from the stretch receptors in the Hering-Breuer reflex, inhibit further inspiration (Fig. 20-13). The *pneumotaxic center* in the upper pons intermittently inhibits inspiration after it receives stimulatory feedback signals from the inspiratory center and inhibitory signals from the apneustic center.

Cerebral Cortex

Because impulses from the cerebrum may temporarily override those generated in the respiratory center, an individual may voluntarily increase or decrease the rate and strength of respiration. However, when the $P\text{CO}_2$ builds up to a certain level, the respiratory center is stimulated in spite of voluntary efforts to hold one's breath.

Chemical Regulators

Chemoreceptors in reflex centers are very sensitive to changes in blood $P\text{CO}_2$ and pH. The blood $P\text{O}_2$ seems to be effective primarily as an emergency control (Table 20-6). These chemical regulators cause the generation of impulses carried to and from the respiratory center.

Carbon Dioxide and Hydrogen Ions

A carbon dioxide level in arterial blood ($P a\text{CO}_2$) above 40 mmHg serves as the major stimulus of respiration. It reflexively increases the depth and rate of respiration (Table 20-6) by stimulating chemoreceptors in the medulla and, to a lesser extent, those in the carotid and aortic bodies. It may do so through its ability to combine with water to form carbonic acid and thereby lower blood pH. The change in pH is associated with the production of hydrogen ions whose effects on respiration are similar to a high blood $P\text{CO}_2$ and cumulative with it. These alterations have most marked effects when relayed through the cerebrospinal fluid to the medulla.

Oxygen

The partial pressure of oxygen in arterial blood ($P a\text{O}_2$) is an emergency control mechanism. A decrease in $P a\text{O}_2$ makes chemoreceptors more sensitive to changes in $P\text{CO}_2$ and reinforces their effects. Only when the $P a\text{O}_2$ declines below a level of 30 to 40 mmHg does it appear to

TABLE 20-6
Effects of chemical regulators on respiration

Regulator	*Alteration*	*Effect*
Carbon dioxide		
Pa_{CO_2} (arterial P_{CO_2})	Elevation above 40 mmHg	Increased respiratory minute volume
	Decrease	Decreased respiratory minute volume
CO_2 in air	Content 10%	Eightfold increase in respiratory rate
	Content 30%	Respiratory depression; death with exposure over 2 min
Hydrogen ions	Elevation in arterial blood (lower pH of blood)	Increased respiratory minute volume (e.g., a decline in pH from 7.4 to 7.1 causes a fourfold increase in respiratory rate)
	Decrease in arterial blood (higher pH of blood)	Decrease in respiratory minute volume (e.g., a rise in pH from 7.4 to 7.6 causes an 80% decrease in respiratory rate)
Oxygen		Emergency control
Pa_{O_2} (arterial P_{O_2})	Decrease below 30–40 mmHg	Stimulates ventilation, acting on carotid chemoreceptors
		Occurs with vigorous exercise and with severe lung disease at high altitude
PA_{O_2} (alveolar P_{O_2})	Decrease below 60 mmHg	Stimulates respiration
		Occurs at high altitude or when breath is held

be a significant factor. Such changes may occur with vigorous exercise or severe lung disease at high altitude.

20-5 EFFECTS OF EXERCISE

Exercise causes a number of responses, which in very complex ways brings about reflex acceleration of the respiratory rate. The increase in ventilation of the lungs is so rapid that the blood P_{CO_2} increases little, even with maximal exercise. No single stimulus can account for an increase in breathing rate and depth. Neural mechanisms may be fine-tuned by chemical stimulation of the respiratory center. The chemical stimuli may include a greater arteriovenous difference in O_2 and CO_2 levels due to exercise, that is, oxygen supply does not keep pace with CO_2 production (Fig. 16-1). Some evidence indicates that changes in blood temperature may play a role, as may those in blood pH, which may stimulate chemoreceptors in tendons and joints associated with exercising muscle.

During strenuous exercise, a respiratory minute volume of 100 to 120 L of air per minute may be maintained for long periods of time (Table 20-7). As a result, cardiac output rather than respiratory function becomes the limiting factor in endurance exercise (Figs. 16-1, 13-15, and 13-16). More vigorous rates of respiration cause the respiratory muscles to fatigue within seconds. There is a maximum ventilation volume beyond which all the surplus oxygen is utilized by the respiratory muscles. With prolonged, strenuous exercise, the accumulation of lactic acid, H^+ ions, and CO_2 stimulates the respiratory rate until the oxygen debt is repaid (Chap. 13). The demands for energy during prolonged strenuous exercise produce a shift to anaerobic metabolism. Subsequently, oxygen must be provided to utilize the lactic acid which has accumulated and to replenish high-energy phosphate stores of ATP and creatine phosphate. Following prolonged strenuous ex-

TABLE 20-7
Effects of strenuous exercise on pulmonary function

Property	*Resting Average*	*Attainable Value after Vigorous Exercise*
Respiratory rate (breaths/min)	12–20	40–50 (for 15–20 s)
Respiratory minute volume (L of air/min)	6–10	150–170 (for 15–20 s or 100–120 for long periods)
O_2 uptake (L/min)	0.25	4.0*
Cardiac output (L/min) (blood flow through lungs)	5.5	33.0

* Reports of higher values (up to 6.2 L/min) have been noted for superior champion athletes.

ercise the amount of oxygen that must be supplied to pay off the debt exceeds the preexercise basal level of oxygen consumption of 250 mL/min. In untrained individuals the debt may be 12 times the basal level, whereas in superior champion athletes it may reach a maximum of 25 times their preexercise level. Prior to fatigue (and sometimes nausea and vomiting) a total debt of 10 L and 18 L may be accumulated in untrained individuals and superior champion athletes respectively.

Ventilation affects P_{CO_2} in an inverse manner (as it increases, P_{CO_2} decreases), as long as CO_2 production is stable. Hyperventilation due to exercise, psychological stimuli, and low P_{O_2} at high altitude increases the rate and depth of respiration, resulting in a greater loss of carbon dioxide. Because blood P_{CO_2} also is a major determinant of blood flow to the brain, a decrease of P_{CO_2} following hyperventilation results in a corresponding cerebral vasoconstriction and a reduction in blood flow, which account for the dizziness that takes place. The decrease in alveolar P_{CO_2} diminishes the rate of carbonic acid formation in the blood causing a condition in which blood P_{CO_2} is diminished, termed *hypocapnia*. The kidneys normally compensate for the lesser acid production by excreting increased amounts of bicarbonate ions (see Chap. 21), which might otherwise accumulate and cause respiratory-induced alkalosis if uncorrected.

20-6 ENVIRONMENTAL EFFECTS

Exposure to High Altitude

The percent of oxyhemoglobin declines with increasing altitude from its maximal value of 97 to 98 percent at sea level to less than 85 percent at elevations of 10,000 ft or more. The adverse symptoms which may result vary with the degree of hypoxia, the rate of ascent, the duration of exposure, and differences among individuals. The symptoms become more severe at higher elevations. Consequently, cabin pressure must be raised in airplanes that fly above 10,000 ft to approximate 1 atm. In fact, persons with lung disease may experience difficulty at altitudes as low as 6000 ft. Sudden exposure of unacclimatized individuals to 20,000 ft or to higher altitudes for even a few minutes may be fatal.

Upon ascent to 10,000 ft or more and prior to adaptation, individuals who have been living at sea level may develop *mountain sickness* within 48 h of exposure. The individuals suffer from hypoxia and may experience dizziness, confusion, headaches, restlessness or insomnia, vomiting, diarrhea, lack of coordination, and an increase in heart and respiratory rates, which fail to keep pace with oxygen requirements. In a small percentage of individuals respiratory distress and pulmonary edema result. Adaptation to the elevated altitude, called *acclimatization,* is gradually lost with return to sea level. Natives of the Andes and Himalaya mountains live normal lives at altitudes of nearly 18,000 ft, and mountain climbers who are in excellent physical condition have been reported to have performed demanding physical work on Mt. Everest at an elevation above 26,000 ft while breathing that rarefied air.

However, the upper limit of permanent acclimatization at high altitude seems possible only at elevations around 0.5 atm (about 18,000 ft).

The symptoms of mountain sickness subside at rates inversely related to the altitude. Acclimatization to mountainous altitudes involves increases in ventilation, cardiac output, red blood cell count (polycythemia, Table 18-2), and hemoglobin levels and restoration of the acid-base balance. Under the influence of erythropoietin (Table 5-5 No. 10 and Fig. 18-4), the bone marrow produces more red blood cells within 4 days of exposure and their number rises above normal levels within 7 to 10 days or more. Many of the reversible changes of acclimatization requires weeks or months to become fully developed.

Underwater Diving

With the advent of relatively inexpensive equipment, scuba (self-contained underwater breathing apparatus) diving has become a popular sport. For each 10 m (about 33 ft) in depth of a dive the pressures of gases increase about 1 atm.

Nitrogen

The P_{N_2} (partial pressure of nitrogen) increases and more N_2 is driven into solution in body fluids in proportion to the depth of a dive. During rapid ascent the N_2 gas bubbles expand as the P_{N_2} of the body fluids becomes greater than that of the atmosphere. However, N_2 cannot diffuse rapidly enough across cell membranes to enter the blood. The expanding bubbles of N_2 are temporarily trapped within the tissues and damage them, resulting in the *bends,* that is, pain in the joints which occurs within 24 h of a dive. It also is called *decompression sickness* or *caisson disease.* The gas bubbles also may obstruct blood vessels, causing a lethal air embolism. For example, a rapid ascent from a depth of 66 ft to the surface causes a pressure change from 3 to 1 atm, which if uncontrolled may have serious consequences. In like manner, rapid ascent to high altitudes may produce *aviation sickness,* with similar results. For example, rapid ascent from sea level to an altitude of 28,000 ft represents a decrease of nearly 33 percent of atmospheric pressure (Fig. 20-6), allowing the rapid expansion of N_2 gas bubbles and their adverse effects.

Conversely, the extra nitrogen that moves into solution under the pressure of a deep dive initially may cause an anesthetic effect called *nitrogen narcosis.* This condition apparently occurs because of the relatively high solubility of nitrogen in fats such as myelin. Nitrogen narcosis is similar to the alcoholic effect of drinking a martini. According to the "martini rule," making a 100-ft dive is equivalent to drinking one martini; 200 ft, two to three martinis; 300 ft, four martinis; and 400 ft, more than four martinis.

Oxygen

Oxygen, a gas vital to life, can be toxic when the P_{O_2} exceeds certain limits. Breathing pure oxygen (100 percent oxygen) of itself does not produce toxic effects if the alveolar P_{O_2} does not exceed its normal level of nearly 100 mmHg (Table 20-1). Breathing oxygen under increased atmospheric pressure, called *hyperbaric oxygenation,* may produce toxic effects. The reactions may include tingling of the fingers and toes; disturbances in hearing, vision, thought processes, and muscle activity (including twitching of the lips); and dizziness, nausea, and convulsions. Loss of surfactant and pulmonary edema may cause respiratory distress. Although the exact causes of oxygen poisoning remain unclear, several events may be involved. Dehydrogenase enzymes may be oxidized irreversibly, adversely altering metabolism in many cells and tissues throughout the body. A deficiency may occur in hemoglobin available to transport carbon dioxide (Fig. 20-12A), causing the P_{CO_2} in the plasma to accumulate and stimulate the respiratory center.

The effect of hyperbaric oxygenation has serious consequences to divers. In order to equalize pressure in the body cavities with their environment, they must inhale air or other gas mixtures at a pressure equal to the water surrounding them to be able to breathe. This amounts to 1 atm for each 10 m of the dive. With descent below sea level, P_{O_2} increases dramatically, forcing more oxygen into the blood. For example, breathing 100 percent oxygen during a dive of 30 m (about 98 ft) produces a P_{O_2} of 3040 mmHg, which is about 19 times its value of 159 mmHg in air at sea level. The calculation is made as follows:

$$100\% \text{ oxygen} \times 4 \text{ atm} \times 760 \text{ mmHg} = 3040 \text{ mmHg}$$

(3 atm from the depth of the dive plus 1 atm at sea level)

After a few minutes of activity at this depth, inhalation of pure oxygen would cause convulsions and unconsciousness. Consequently, the use of pure oxygen is too hazardous for general use in diving.

Also, exposure of infants in incubators to a high P_{O_2} for more than 12 h at 40 percent O_2 may damage their retinas, causing *retrolental fibroplasia*. Blood vessels in the retina constrict and are damaged, ultimately pulling away from the retina, and permanent blindness may result. Because of its possible toxic effects, pure oxygen should be inhaled only briefly in medical emergencies.

Carbon Dioxide

Toxic reactions to CO_2 occur when the Pa_{CO_2} exceeds the normal range of 35 to 45 mmHg (Table 20-4, page 488). Accumulation of excessive CO_2 is called *hypercapnia*. Slight elevations of Pa_{CO_2} arouse and stimulate the respiratory center. Adverse symptoms generally begin as the CO_2 content of inspired air exceeds 4 percent (its normal level is 0.03 percent). These symptoms are related to the duration of exposure and become more severe with increasing levels of CO_2. Such ill effects may be experienced by divers, by people in submarines, closed aircraft, or space vehicles, and by workers where dry ice (solid CO_2) is manufactured. The excess CO_2 increases production of carbonic acid. The kidneys normally compensate for the greater acid production by excreting more hydrogen ions (Chap. 21). Failure to do so leads to uncompensated respiratory acidosis, mental impairment, and unconsciousness. The symptoms are generally reversible upon return to a normal atmosphere and a normal arterial P_{CO_2}.

Decompression

The consequences of decompression (a reduction in pressure) depend on the depth of a dive, its duration, and the rate of ascent. The responses to a dive become greater as depth increases because the partial pressure of gas rises and more gas goes into solution. As the time of a dive becomes longer, more gas diffuses from the blood into tissues. Conversely, ascent leads to expansion of gas according to Boyle's law and formation of gas bubbles, which expand and diffuse from the tissues into the blood as the diver rises toward the surface.

Generally, a dive to a depth of less than 30 ft does not require decompression. After descent to greater depths the bends may be avoided by a measured, slow rate of ascent, which gradually decreases the pressure of the atmosphere surrounding the diver. Pauses in ascent, or decompression stops, are described in literature on diving. Very prolonged or deep dives may require the use of a decompression chamber and breathing of special gas mixtures.

Pollutants

Carbon Monoxide

Carbon monoxide (CO) rapidly combines with hemoglobin to form *carbomonoxyhemoglobin* (*carboxyhemoglobin, COHb*). Since its affinity for Hb is 210 (200 to 300) times greater than that of O_2, as little as 0.02 to 0.04 percent CO in the atmosphere may cause carbon monoxide poisoning. Exposure in industrial jobs should not exceed an average of 0.0035 percent CO (35 parts per million of air) for 8 h. Prior to reexposure a worker's blood should return to normal.

Hypoxic symptoms result from the adverse effects of carbon monoxide on the CNS. Among them are severe frontal headache, decreased mental function, fainting, and nausea. The effects vary with the duration of exposure and the percent of CO in the air and can be fatal (Tables 20-8 and 20-9). The formation of COHb also leaves less Hb available for O_2 transport. The Hb molecules which do combine with O_2 do not release it readily, and the shape of the normal Hb dissociation curve shown in Fig. 20-11 becomes flattened. Treatment for carbon monoxide poisoning includes artificial respiration and temporary ventilation of the subject with pure oxygen under pressure (hyperbaric oxygenation) to facilitate rapid conversion of COHb to HbO_2.

TABLE 20-8
Effects of carbon monoxide

% COHb in Blood*	Probable Symptoms
5–10	Slight headache, breathlessness from slight exertion
20	Loss of appetite, nausea, headache
20–30	Drowsiness, lack of coordination, mucous membranes become cherry red (magenta)
40–50	Unconsciousness†
60 or more	Death

* Symptoms occur at lower blood levels in children.
† Exposure to as little as 0.03% CO (300 parts per million of air) for 10 h may cause unconsciousness.

TABLE 20-9
Sources of carbon monoxide exposure

Source	Generation of CO
Natural gas	Nearly 0% (but varies with source of natural gas)
Gasoline engines	6% or more CO in exhaust fumes
Cigarette smoke	Raises COHb 5% above normal in average smoker (COHb range is 1–16% above normal, depending on amount of smoking that is done)
Industrial propane furnaces	8% CO in air
Manufacture of gas from coke	10% CO in air

Cigarette Smoke

Cigarette smoke contains CO and raises its level in blood by 1 to 16 percent, depending on the amount the individual smokes (Table 20-9). Pregnant laboratory rats may be exposed to CO equivalent to that of humans who regularly smoke cigarettes. Such rats bear offspring that have lower birth weights, decreased weight gain, altered hormone levels, diminished behavioral activities, and less total brain protein. Similar research studies upon pregnant women confirm that those who smoke during pregnancy may retard the development of their babies.

Smoking stimulates mucus secretion and also may gradually cause respiratory columnar epithelium to acquire a more flattened shape. Smoking also increases *elastase* activity in alveolar macrophages, which degrades elastin. As a result, alveoli lose some of their natural recoil, which in turn increases the residual air volume and may cause a severe condition in the lungs of individuals with emphysema. Furthermore, the lungs of smokers show a decrease in surfactants, a change which makes breathing more difficult.

Smoking is also associated with a greater risk of lung cancer. Nearly 9 of 10 cases of primary lung cancer occur in cigarette smokers. The tars produced from the combustion of tobacco may be *carcinogenic* (cancer-producing). Radioactive isotopes in the air accumulate in higher concentration in the bronchial tissues of smokers than in those of nonsmokers. The rate of cellular mutation may be increased and the effect also may be related to the increased risk of cancer and atherosclerosis in smokers (Chap. 17). Furthermore, nicotine also constricts blood vessels, resulting in elevated peripheral resistance and blood pressure and thus causes the heart to work harder, which is a most significant risk factor in cardiovascular disease (Tables 16-1 and 16-4, pages 380 and 390).

Other Pollutants

In addition to CO, tars and nicotine, wastes from industrial combustions, and automobile exhaust pollute the air. *Nitrogen oxides* inflame the respiratory tract and eyes, *sulfur oxides* irritate the bronchial tree, and *ozone* promotes pulmonary edema. The adverse reactions to these gases depend upon the duration and intensity of exposure and on the general health and susceptibility of the subject. Particles emitted into the air may lodge in the upper respiratory tract and adhere to its mucous membrane. Particles less than 0.1 μm in diameter may eventually reach the alveoli. In this location, phagocytes ingest these particles and later are often coughed up and expectorated. If not, these cells laden with particles may enter the blood or lymph, pass into the digestive tract, and be excreted with wastes.

20-7 ARTIFICIAL RESPIRATION

A lack of gas exchange may result from the entry of fluid into the lungs (in drowning), from infections such as pneumonia, from paralysis of the respiratory muscles due to infections such as poliomyelitis or botulinal food poisoning, from trauma to the brain or spinal cord, or from drugs such as barbiturates which interfere with the nervous regulation of respiration. In these and other circumstances which interfere with normal breathing, artificial respiratory assistance may be required to sustain life.

Mechanical Devices: Advanced Life-Support Techniques

Mechanical devices are useful for long-term respiratory assistance in people with chronic respiratory problems. A *resuscitator* periodically moves air into and out of the lungs. It may consist of a simple bellowslike device which delivers gas from a cylinder through a face mask placed over the nose and mouth. These devices are positive-pressure ventilators and are most commonly used to assist respiration. They allow respiratory gases to escape from the lungs between intermittent delivery of air volumes which increase intrapulmonary pressure. The *Drinker artificial respirator* (iron lung) is both a positive- and a negative-pressure ventilator which consists of a chamber that surrounds all but the head of an individual. A motor-driven piston, adjusted to stimulate normal breathing rates, decreases and increases pressure within the chamber. These alterations enlarge and diminish the size of the chest cavity, which causes inflation and deflation of the lungs.

Mouth-to-Mouth and Mouth-to-Nose Resuscitation—Basic Life Support Techniques

A number of techniques have been developed to save lives by maintaining respiration and circulation in emergencies. *However, a person should have instruction and practice under the guidance of trained personnel before attempting these life-saving procedures.*

Mouth-to-mouth resuscitation may be used for

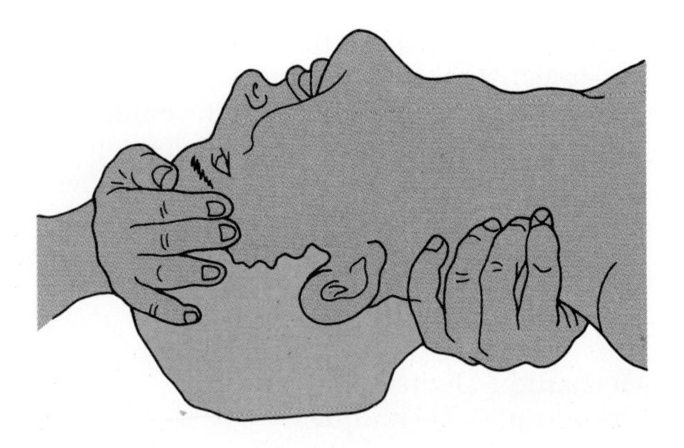

A Opening the airways

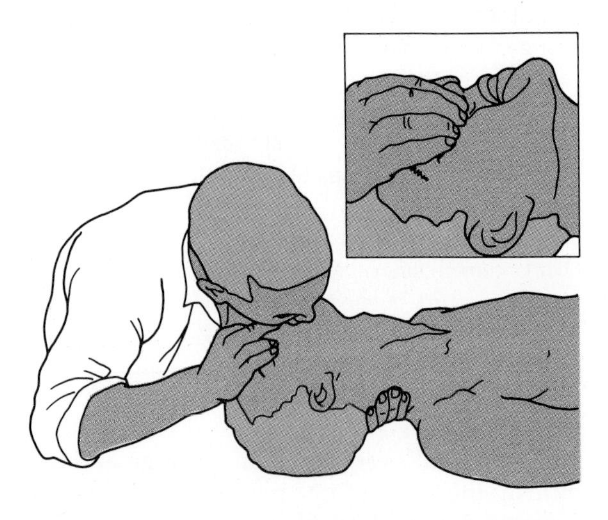

B Breathing

FIG. 20-14
Artificial mouth-to-mouth respiration. Individuals should be given instruction under the guidance of trained personnel prior to attempting this life-saving procedure.

victims of drowning, gas poisoning, or other respiratory distress. It is the method of choice except in accidents involving facial, neck, or head injuries which prevent its use. In these cases, the mouth-to-nose method should be used if possible.

For mouth-to-mouth respiration, the victim should be lying down, face up. Foreign material should be removed from the victim's mouth with the rescuer's index and/or middle finger. The victim's neck must be extended backward for maximum extension to clear the airways (Fig. 20-14A). One hand is placed under the neck to straighten it and keep the airways open. The jaw must be displaced forward as well. This position helps keep the tongue away from the back of the throat. The nostrils are pinched closed with the thumb and index finger of the other hand, whose palm is gently, but firmly, extending the victim's forehead in a backward tilt (Fig. 20-14B). The rescuer's mouth is placed in a wide circle over the subject's mouth, sealing it. The rescuer then blows air into the victim's mouth about every 5 s (12 to 15 times per minute). *When the victim's chest rises, the rescuer stops blowing and listens for exhalation from the victim's mouth. The rescuer also must observe the victim's chest fall at this time.* The blowing cycle is repeated until the victim breathes without assistance.

When the victim is an infant or small child, the rescuer's mouth should encircle both the mouth and nose. Much less air is needed to inflate the lungs. The blowing cycle should be increased and repeated every 3 s (20 times per minute).

In the event that *mouth-to-nose* resuscitation is required, the victim's head must be maintained in a fixed position without being extended or tilted. The mandible (lower jaw) is moved in a modified jaw thrust, as shown in Fig. 20-15A, to attempt to clear the airways. The rescuer takes a deep breath, holds the victim's lips together by hand, seals his or her mouth over the victim's nose, and blows in until the victim's chest expands. The rescuer then moves his or her mouth away from the victim's nose, allowing the air to escape and listening for it (Fig. 20-15B). The victim's mouth may have to be opened for this to occur. The rescuer repeats the cycle about every 5 s until the adult victim breathes independently.

When artificial respiration is used in conjunction with cardiac resuscitation, the technique is called *cardiopulmonary resuscitation*

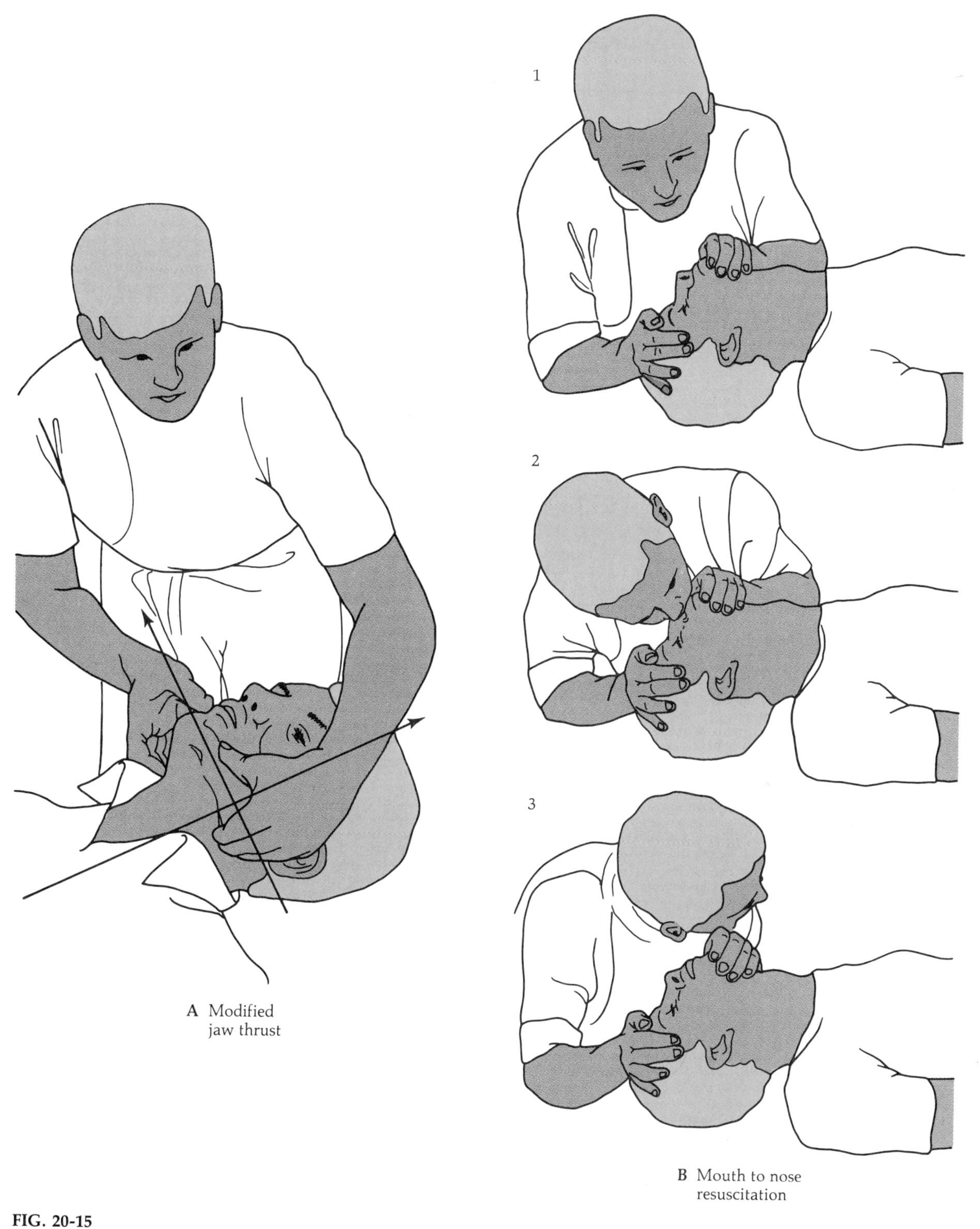

FIG. 20-15
Artificial mouth-to-nose respiration. Individuals should be given instruction under the guidance of trained personnel prior to attempting this life-saving procedure.

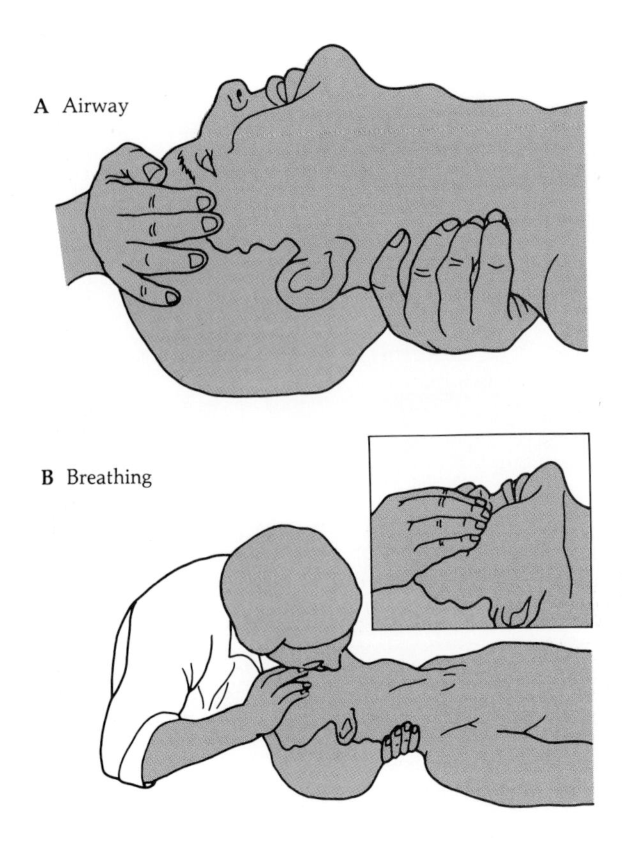

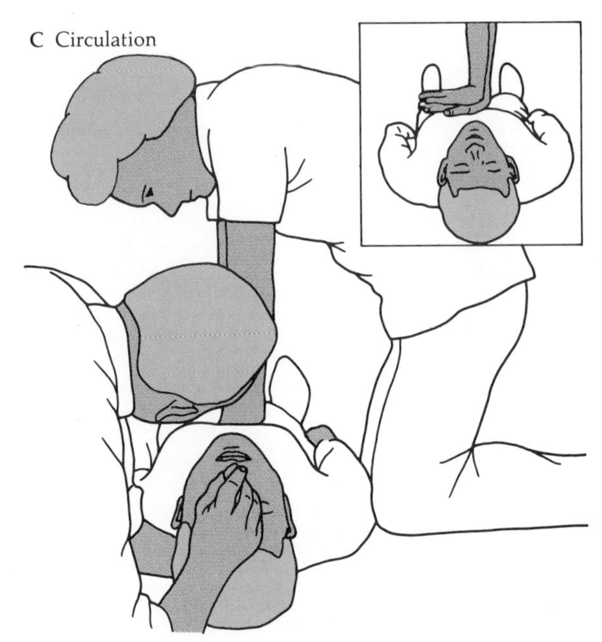

5:1 ratio = 5 chest compressions: 1 lung inflation
Compression rate = 60 per min
Lung inflations = interposed after each 5 compressions with no pause in compressions

FIG. 20-16
Steps of cardiopulmonary resuscitation (CPR). Individuals should be given instruction under the guidance of trained personnel prior to attempting this life-saving technique.

(*CPR*). External compressions of the heart are applied at the rate of 60 per minute. The sternum (breastbone) is compressed with the heel of the hand at a point about 1 to 1½ in from its tip near the abdomen and then allowed to return to its normal resting position (Fig. 20-16). If only one person is available for CPR, the rescuer should maintain four cycles per minute of 15 compressions of the heart and two full ventilations of the lungs. If two persons are available as rescuers, the compression rate should be 60 per minute and remain uninterrupted. Between every five compressions the second rescuer inflates the lungs, maintaining a cycle of five compressions to one ventilation, repeated 12 times per minute.

20-8 SUMMARY

1 External respiration is the exchange of gases between the lungs and the blood. It precedes cellular respiration, which uses oxygen in aerobic metabolism. Air passes from the nasal chambers and mouth through the pharynx and larynx into the trachea, bronchi, and bronchioles prior to delivery to the alveoli.

In addition to alveolar macrophages, three types of epithelial cells line the alveoli. Phosphoglycerides secreted by type II cuboidal cells act as detergent-like molecules called surfactants, which include lecithin and sphingomyelin. These compounds lessen surface tension within the alveoli and make inspiration easier. Immaturity of the fetal lung is reflected in a low lecithin to sphingomyelin ratio, which is an indicator of risk of a form of respiratory distress syndrome of the newborn formerly referred to as hyaline membrane disease.

2 At sea level air is a mixture containing 0.03 percent CO_2, 20.94 percent O_2, 78.09 percent N_2, and less than 1 percent of other gases. The direction of diffusion of each gas between the air in the alveoli and the blood in the surrounding alveolar capillaries is determined by the pressure gradients of the individual gases. A column of air in the atmosphere at sea level produces a pressure equivalent to that exerted by a column of mercury 760 mm high, is called 1 atm of pressure, and equals 14.7 lb/in^2.

During inspiration, elevation of the rib cage and contraction of the diaphragm enlarge the thorax and cause intrapulmonic pressure to decrease to less than 1 atm of pressure and air to rush into the lungs. The opposite sequence forces air out of the lungs during expiration as the diaphragm relaxes and the lungs recoil. Variations in the pressure produced by gases are governed by physical laws.

The total amount of air that enters the respiratory tract each minute is the total ventilation or respiratory minute volume, while alveolar ventilation is the volume that actually reaches the alveoli. Air volumes that may be measured with a recording spirometer include the tidal volume exchanged in quiet respiration, which, together with the inspiratory and expiratory reserve volumes, constitutes the vital capacity, a measure of functional lung capacity. The residual volume is air left in the lungs during respiration, and allows gas exchange to occur between breaths. The vital capacity is increased by exercise and is generally higher in males than in females. It is roughly a function of the cube of body height in centimeters. The vital capacity is decreased in diseases that alter the lung's elastic recoil, restrict their expansion, or obstruct air flow.

Breathing varies. It may be normal (eupnea), difficult while a person is lying down (orthopnea), temporarily suspended (apnea), painful (dyspnea), or it may alternate between apnea and dyspnea (Cheyne-Stokes respiration).

3 Approximately 20 mL of oxygen are carried in each 100 mL of blood, 19.5 mL as oxyhemoglobin (HbO_2) and 0.5 mL as dissolved O_2 gas. HbO_2 releases oxygen when combined blood levels of H^+ ions, CO_2, and 2,3-diphosphoglycerate increase and temperature increases. At a stated P_{O_2}, the percent of HbO_2 declines with decreasing pH (the Bohr effect).

Most carbon dioxide formed by metabolism is combined with water by carbonic anhydrase in red blood cells to make carbonic acid (H_2CO_3). Carbonic acid dissociates into bicarbonate ions (HCO_3^-) and hydrogen ions (H^+). Consequently, most of the carbon dioxide is transported in bicarbonate ions. Some is carried as dissolved gas in plasma and red blood cells and still less is combined with proteins called carbamino proteins, such as carbaminohemoglobin.

Bicarbonate ions diffuse from the red blood cells into the plasma and are replaced by chloride ions, which diffuse from plasma into the cells of metabolically active tissues in a reaction called the chloride shift. This reaction helps maintain the isoosmolarity of red blood cells. Hydrogen ions combine with HbO_2 to form reduced hemoglobin (HHb) and release oxygen, which diffuses into tissues.

The diffusion of oxygen and carbon dioxide is governed physically by pressure gradients which favor the movement of oxygen from arterial blood into tissues, and carbon dioxide out of tissues changing blood to a venous state. Each gas diffuses in the reverse direction when deoxygenated blood reaches the alveolar capillaries. The ratio of alveolar ventilation to the flow of blood (perfusion) through the lungs must be optimal for gas exchange to occur efficiently.

4 Respiration is rhythmically and reflexively controlled by chemoreceptors in the carotid and aortic bodies, baroreceptors in the aortic arch and carotid sinus, and pulmonary stretch and deflation receptors excited in the Hering-Breuer reflex to regulate normal breathing. The sensory receptors relay signals to the respiratory center in the medulla, which transmits impulses to the respiratory muscles and sends other impulses to centers in the pons. The apneustic center in the pons, which drives respiration, and the pneumotaxic center, which inhibits inspiration, are governed by feedback signals. The rhythm of respiration may be temporarily altered reflexively by interneurons in the medulla associated with hiccups, coughing and sneezing, sighing and yawning, and laughing and crying and voluntarily by neurons in the cerebrum.

Ventilation is primarily stimulated by an increase in P_{CO_2} and hydrogen ions, which act on chemoreceptors in the medulla and on those in the carotid body and to some extent on those in the aortic body. Ventilation also is influenced by the baroreceptors and stretch and deflation receptors described above.

5 Vigorous exercise increases the arteriovenous difference in O_2 and CO_2, using up O_2 and producing disproportionately more CO_2, and stimulates an increase in respiratory rate and cardiac output. The increase in cardiac output is limited and restricts the perfusion rate of the lungs so that strenuous exercise can be maintained only in proportion to cardiac output. The respiratory rate remains elevated until the oxygen debt built up during exercise is repaid.

6 Exposure to an altitude of 10,000 ft or more above sea level decreases P_{O_2} sufficiently to cause hypoxia. A decrease of HbO_2 below 85 percent may cause mountain sickness, the symptoms of which include restlessness, sleeplessness, increased heart, pulse, and respiratory rates, and respiratory distress in a small number of individuals.

Diving below a depth of 33 ft causes exposure to pressure which significantly exceeds that at sea level. Under the increased pressure, nitrogen gas dissolves in myelin of nerve fibers and may cause nitrogen narcosis. The greater the depth and duration of a dive, the greater the narcotic effect.

Nitrogen forced into the bloodstream and tissues escapes with difficulty upon rapid ascent to the surface. Nitrogen gas bubbles trapped in the joints expand to cause the bends and may induce the formation of embolisms in blood vessels. Rapid ascent to high altitude in unpressurized aircraft may also cause the bends and aviation sickness.

Hyperbaric oxygenation, the administration of oxygen under pressure, may cause oxygen toxicity, resulting in visual, auditory, mental, and vascular changes due to a lack of carbon dioxide. Adverse metabolic effects occur owing to oxidation of dehydrogenase enzymes.

Carbon monoxide is a pollutant produced by gasoline engines and propane furnaces. It forms carboxyhemoglobin, decreases the percent of oxyhemoglobin in the blood, induces the symptoms of hypoxia, and may be lethal. Cigarette smoke contains carbon monoxide, nicotine, and hydrocarbons, which may cause cardiovascular and pulmonary alterations associated with increased risk of atherosclerosis, lung cancer, and emphysema. Residues in cigarette smoke in a pregnant woman's blood may adversely affect her unborn baby's development.

7 Respiration may be maintained artificially by mouth-to-mouth or mouth-to-nose resuscitation procedures. The airways must be clear and open. Every 5 s air is breathed into the victim's lungs, which are allowed to deflate prior to each inflation. The cycle is continued until the victim breathes independently. Mouth-to-nose resuscitation is the method of choice for accident victims with facial, neck, or head injuries. Mechanical devices also may be used to help individuals who have chronic respiratory disease. A combination of artificial respiration with cardiac resuscitation is called cardiopulmonary resuscitation (CPR). A cycle of 15 compressions of the heart and two ventilations of the lungs is repeated 4 times each minute by one rescuer. With two rescuers, CPR involves a cycle of five compressions and one ventilation repeated 12 times per minute, providing rhythmical heart compressions at the rate of 60 per minute and a respiration rate of 12 per minute.

REVIEW QUESTIONS

1 Draw and label a sketch of the lungs, airways, and alveoli.
2 What are surfactants? What are their functions and their relations to respiratory distress?
3 Describe the structure of the respiratory membrane and cells that line the aveoli.
4 How do changes in pressure and pressure gradients influence inspiration and expiration?
5 How does one measure vital capacity? What is its relationship to tidal volume and to the inspiratory and expiratory reserves? What is its significance? How does the FEV_1 relate to respiratory function?
6 Describe congestive heart disease, emphysema, asthma, and their effects on vital capacity. What is the effect of respiratory infections on vital capacity?
7 Compare and give the terms used to describe normal and abnormal types of breathing.
8 What are the major ways in which O_2 and CO_2 are transported in the blood?
9 Describe the effects of PO_2, H^+, and PCO_2 on the formation and dissociation of HbO_2. What is the Bohr effect?
10 List the chemoreceptors and baroreceptors that influence respiration and describe their activities. Include an explanation of the Hering-Breur reflex and J cells in the lungs.
11 How do the apneustic and pneumotaxic centers interact with the respiratory centers to cause rhythmic respiration?
12 Why does cardiac function rather than respiratory function limit prolonged, vigorous exercise?

13 Why does the amount of O_2 available to the lungs decrease with increasing altitude? What effects result from exposure to high altitude?
14 What is nitrogen narcosis? Oxygen toxicity? Retrolental fibroplasia?
15 Compare mountain sickness and aviation sickness.
16 What causes the bends, and how can they be avoided?
17 What is CO poisoning? How does it occur? What are its effects?
18 Compare the likely condition, with respect to structure and function, of the lungs of Lisa M., who smokes heavily, with those of Rita D., who does not. What are the risks to Lisa?
19 Under what conditions would one use a mechanical respirator? Mouth-to-mouth artificial respiration? Mouth-to-nose artificial respiration?
20 Describe the technique of cardiopulmonary resuscitation (CPR).

THE EXCRETORY SYSTEM

21-1 ORGANS AND FUNCTIONS OF THE EXCRETORY SYSTEM

21-2 THE SKIN
The epidermis
Organization
Pigmentation (color)
Hair follicles
Sebaceous glands
Ceruminous glands
Sweat glands
Structure • Composition of sweat (perspiration) • Function of sweat • Regulation of sweat formation
The dermis (corium)

21-3 THE KIDNEYS
Anatomy
General functions
The nephrons: Functional units of the kidney
Physiology of the nephrons
Blood supply
Filtration and microscopic structure of the filtration membranes
Glomerular filtration rate (GFR) • Regulation of GFR
Reabsorption
Secretion
Effects of hormones
Antidiuretic hormone (ADH)
Aldosterone
Renin-angiotensin system
Countercurrent theory of nephron function
Regulation of acid-base balance
Measurement of kidney function
Effects of disease on kidney function
Kidney transplants and the artificial kidney

21-4 SUMMARY

OBJECTIVES

After completing this chapter the student should be able to:

1. Name the major excretory organs and describe the functions of each.
2. Compare the structure and function of the epidermis and dermis.
3. Describe the excretory functions of the skin, including the structure and function of sweat glands.
4. Sketch and label the macroscopic and microscopic structure of the kidneys, including the details of the nephron.
5. Describe the regulation of blood flow through the kidneys and the mechanisms affecting filtration in nephrons.
6. Compare the effects of antidiuretic hormone, aldosterone, and renin on kidney function. Include a detailed description of the functions of the renin-angiotensin system.
7. Explain the countercurrent theory of tubular function in the kidney and its relationship to filtration, reabsorption, and secretion in the formation of urine.
8. Name four buffer systems in the body. Describe the manner in which they influence the pH of body fluids and the way in which the kidneys regulate acid-base balance.
9. Describe some causes of kidney disease and their symptoms and effects.
10. Describe the principles of kidney transplants, the artificial kidney, and peritoneal dialysis.

The ability of excretory organs to eliminate wastes while conserving useful substances is essential to homeostasis and to life. Because the bulk of the human body is composed of water, fluid balance is critical to body function. Each day we lose about 1.5 to 3.0 L of fluid, which must be replaced. The volume of fluid loss is regulated by feedback mechanisms which normally exhibit the greatest effects in the kidneys. Although the fluid compartments of the body undergo constant exchange (Fig. 4-1), a relatively stable internal environment is maintained in spite of the diverse activities of cells, tissues, and organs.

21-1 ORGANS AND FUNCTIONS OF THE EXCRETORY SYSTEM

The excretory organs include the skin, which produces sweat; the kidneys, which form urine; the lungs, which eliminate carbon dioxide, water vapor, and heat (Chap. 20); and the digestive organs, which dispose of undigested foods, some digestive secretions, products of metabolism, water, and minor amounts of salts (Chaps. 22 and 23). This chapter will focus on the excretory functions of the skin and the primary organs of excretion, the kidneys.

21-2 THE SKIN

The skin, along with its outgrowths, the hair and nails, forms a protective *integumentary system*. It (1) provides a barrier that prevents the entry of many harmful agents and microorganisms; (2) contains in its inner dermal layers sensory receptors whose responses to various stimuli provide an awareness of the external environment (Chap. 8); (3) synthesizes forms of vitamin D (Chap. 14); (4) helps regulate to a limited extent the loss of fluid, salts, and wastes; and (5) helps to control body temperature.

The skin is a remarkable body covering, which in adults weighs between 2.7 and 3.6 kg and excretes approximately 1 L of sweat (perspiration) each day through nearly 2.5 million pores. The controlled evaporation of sweat helps to cool the skin, as well as the blood coursing through it, and thereby helps to regulate body temperature. The amount of salt lost in sweat influences the levels of electrolytes, such as sodium, potassium, and chloride ions, in body fluids.

Anatomically the skin is composed of two major subdivisions, the outer *epidermis* and the inner *dermis*, which are supported by the *hypo-*

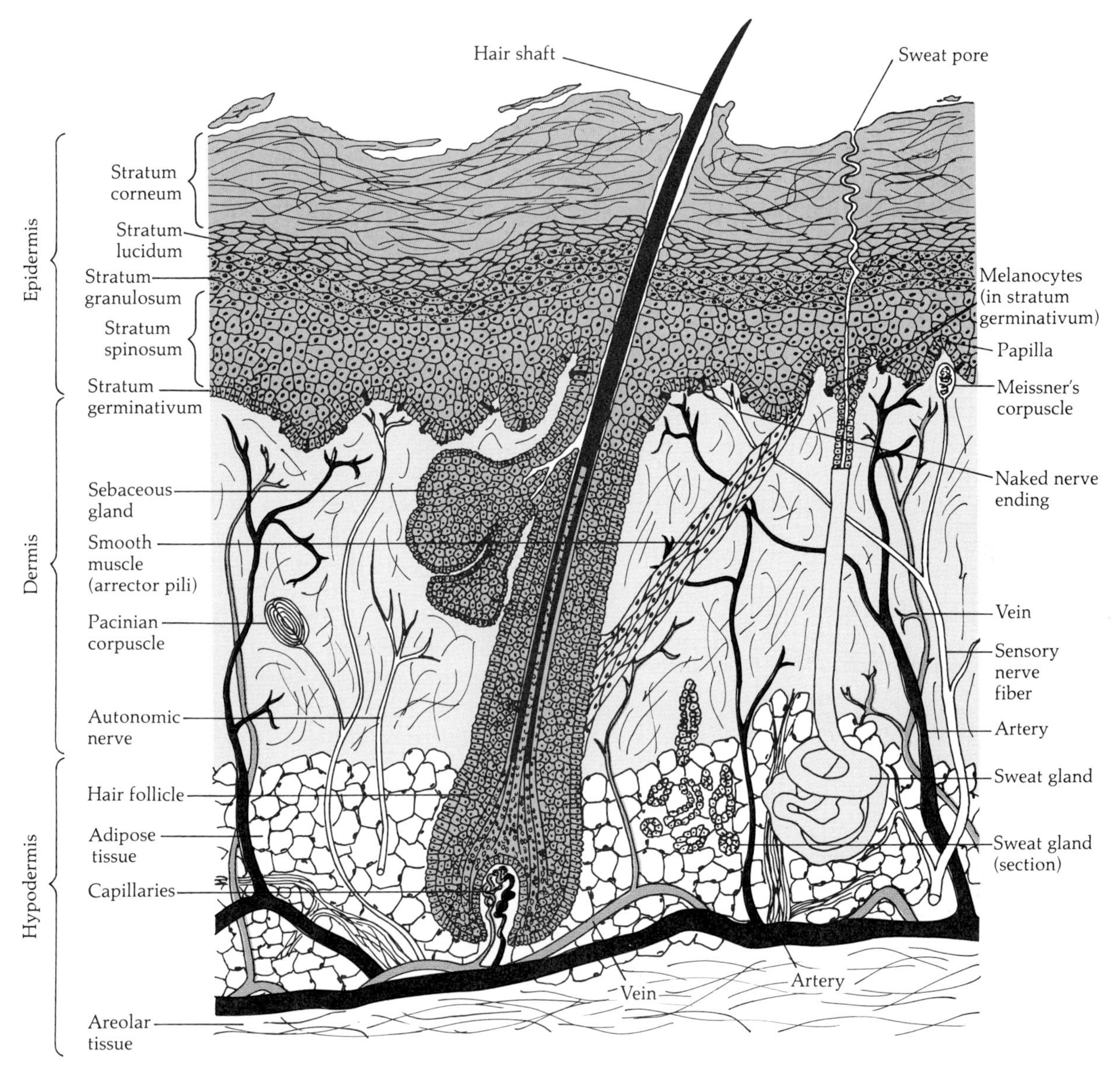

dermis, an underlying layer of loose connective tissue and fat. The *epidermis,* which originates from embryonic ectoderm, consists of several layers of epithelial cells plus pigment-forming cells, hair, nails, sweat glands, ceruminous (wax-producing) glands, and sebaceous (oil-producing) glands. Hair originates in follicles which extend from the epidermis into the dermis, as do the glands just mentioned. The *dermis,* which is derived from embryonic mesoderm, consists of dense connective tissue containing blood vessels, lymph vessels, nerve fibers, and sensory receptors (Fig. 21-1).

FIG. 21-1
Composite diagram of the skin and its tissue layers. The number of layers in the epidermis are illustrative of thick skin.

The Epidermis
Organization

The *epidermis* is made up of four to five strata (layers) of epithelial cells. It averages 0.1 mm in thickness but varies from 0.07 mm in the thin skin of the eyelids to 1.4 mm in the thick skin of the soles of the feet. The epidermis lacks blood vessels and relies on diffusion of substances from the dermis to exchange metabolic wastes for nutrients in the blood. Desmosomes (Fig. 3-1) tightly interconnect the cells of all but the outermost stratum of the epidermis. The outermost *stratum corneum* is about 20 cell layers deep (Fig. 21-1) and is formed by cells from underlying strata which have surfaced as the result of cell division. These cells lose their nuclei and cytoplasmic organelles, which are hydrolyzed by lysosomal enzymes and replaced by *keratin,* a water-resistant, sulfur-containing protein. Variations in keratin give the the fingernails and toenails their tough qualities. Formation, migration, and maturation of the cells of the stratum corneum require about 4 weeks. Since these events are continual, the outer surface of the skin is constantly being shed and replaced.

In thick skin, such as that of the palms of the hands and soles of the feet, the *stratum lucidum* lies immediately within the stratum corneum and consists of about four layers of thin, clear cells that lack nuclei.

The next inward layer is the *stratum granulosum,* often five cell layers deep and formed by flat spindle-shaped cells with granules containing keratin.

Next is the *stratum spinosum,* or prickle-cell layer, which is about 10 cell layers deep. It is composed of many-sided cells closely linked by interconnecting desmosomes. Small, prominent fibrils give the intercellular junctions a spiny or prickly appearance when viewed through a microscope.

The innermost layer of the epidermis is the *stratum germinativum* (or *stratum basale*). It is a single layer of columnar epithelial cells, which divide and push outward and change their appearance as they do so, to form the other layers.

Pigmentation (*Color*)

About 10 to 25 percent of the cells in the stratum germinativum are *melanocytes.* They originate in neuroectodermal cells and have cytoplasmic extensions that resemble the dendrites of neurons. The melanocytes behave as unicellular glands, which synthesize a black to brown pigment, *melanin,* in cytoplasmic granules, the melanosomes. The pigment is formed mainly under the influence of the melanocyte-stimulating hormone (MSH) and also the adrenocorticotropic hormone (ACTH) from the pituitary gland (Chap. 5 and Fig. 10-9). Melanosomes move in cytoplasmic extensions of melanocytes to reach keratinized cells in the epidermis. Melanin is also found in the choroid coat between the retina and sclera of the eye (Figs. 8-2 and 8-3), in the adrenal medulla, and in the brain.

Melanosomes scatter and absorb light to protect the lower layers of the skin from free chemical radicals (charged groups) formed by the interaction of ultraviolet light with substances in the fluids of the body. In another protective response epidermal cells proliferate and more melanosomes seem to be transferred to those cells when the skin is exposed to ultraviolet radiation.

Melanin, as well as reddish brown and yellow pigments and others, characteristic of human skin and hair color, is synthesized from tyrosine by variations of the following reaction:

$$\text{Tyrosine} \xrightarrow[\text{(oxidation)}]{\text{tyrosinase}} \text{L-dopa} \rightarrow \text{intermediates} \rightarrow \text{melanin}$$

(L-3,4—dihydroxyphenylalanine)

Alterations in tyrosinase activity cause differences in the skin color of white persons. Tanning of the skin by ultraviolet radiation in sunlight is initially due to darkening of existing melanin. Within a few days, however, an increase in tyrosinase activity leads to greater production of melanin and its release into epidermal cells.

Owing to genetic variations, whites generally have lower melanin levels than blacks, less well-developed melanocytes, smaller and less discrete melanosomes, and lower tyrosinase activity. *Albinism* is a recessive hereditary condition in which production of melanosomes, tyrosinase, and melanin is lacking. Color in albinos is a pinkish hue due to blood within capillaries in the skin.

Hair Follicles

Cells of the stratum germinativum produce hair follicles, which become embedded in the dermis and hypodermis, and hair which extends to the exterior of the epidermis. Smooth muscle cells (arrector pili, Fig. 21-1) are attached to each hair follicle and contract upon sympathetic autonomic nerve stimulation. The muscle fibers pull the hair from an oblique position to a more erect one, causing "goose pimples." This is analogous to fluffing of feathers by birds, which, in those animals, effectively reduces heat loss and makes them appear larger when threatened.

Follicular cells also synthesize melanin and other derivatives of tyrosine to give hair its color. A decline in tyrosinase activity in these cells leads to graying (loss of color) of the hair. Variations in hair color are controlled by specific genes which govern the number of melanosomes formed and which regulate the chemical conversion of melanin to other colored pigments.

As cells in the follicle divide, they migrate upward, lose their nuclei, become impregnated with melanin (or other pigments) and keratin, and die as they form the shaft of the hair (Fig. 21-2). The production of those cells that are transformed into hair appears to be cyclical, in that 50 percent or less within a follicle seem to be in the act of division at one time. Furthermore, some follicles rest while others are active. Cutting of the hair does not affect its rate of growth. Hair loss may be hereditary or may be caused by disease or other factors. An example of hair loss due to hereditary factors is the onset of pattern baldness in males approaching middle age, which is due to genes activated by the male sex hormone, testosterone.

Sebaceous Glands

The skin is lubricated by oils secreted from *sebaceous glands,* which are usually found in pairs around the upper portion of most hair follicles. These exocrine glands secrete a mixture of lipids through a duct into the follicle around the hair shaft (Fig. 21-1). Sebaceous glands are abundant in the face and scalp but are lacking in the palms of the hand and soles and upper surfaces of the feet.

The lipids are released as the entire epithelial cells are destroyed and shed in a process called *holocrine secretion* (page 51). The lipids include oils, which lubricate the hair and skin, and antimicrobial compounds, which inhibit or kill microorganisms.

The stimulation of oil production by the sex hormones becomes evident at puberty (Chap. 24). Follicular pores often become plugged with these oils and with keratin and may accumulate melanin to form blackheads. Within the glands bacteria may cause an inflammatory response that is recognized as *acne.*

Ceruminous Glands

Ceruminous glands are tubular and secrete a lubricating wax called *cerumen,* which is carried by oil in ducts shared with the sebaceous glands. In the external ear the wax may harden and cause pressure to be exerted on the tympanic membrane which separates the outer and

FIG. 21-2
Scanning electron micrograph of a section of a human hair (×1200).

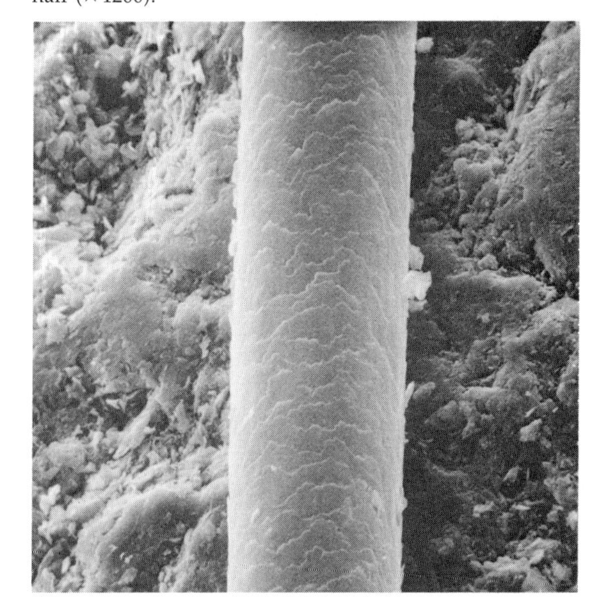

the middle ear (Fig. 8-8). The pressure created sometimes causes earaches and frequently impedes the conduction of sound, producing conduction deafness. The condition may be relieved and hearing regained after a physician removes the hardened wax. Ceruminous glands also are found in the periphery of the eyelids, where their secretions lubricate the conjunctiva and cornea (Table 8-2).

Sweat Glands

Perspiration, the primary excretion of the skin, is formed by 2 to 5 million *sweat glands* whose coiled ducts lead to pores on the surface of the skin (Fig. 21-1). Sweat glands are most densely packed in the palms of the hands, which may have up to 3000 pores per square inch.

STRUCTURE Sweat glands are lined with cuboidal or columnar epithelium, which actively secretes sweat. Most of them are *eccrine* glands, which are innervated by sympathetic cholinergic nerves, have no connections with hair follicles, and secrete a watery fluid in response to heat. *Apocrine* sweat glands are innervated by adrenergic nerves and secrete a more viscous milky secretion by shedding the tips of their epithelial cells in response to stressful stimuli but not to heat. These glands produce milky secretions in the armpit (axilla) and around the anus. The general structure of glands was described earlier in Chap. 3.

COMPOSITION OF SWEAT (PERSPIRATION) Sweat is about 99 percent water, while the remainder consists primarily of sodium chloride, potassium chloride, lactic acid, nitrogenous wastes (such as ammonia and uric acid), and traces of other substances. Odors come from the products of bacterial metabolism, which uses some of the components in sweat as nutrients.

FUNCTION OF SWEAT Evaporation of sweat from the surface of the skin results in a cooling effect, which helps lower body temperature. Active (sensible) perspiration produces an average of 0.5 to 1.0 L of sweat daily. Passive (insensible) sweating which occurs by evaporation, usually goes unnoticed but causes the loss of another 0.6 L of fluid each day.

REGULATION OF SWEAT FORMATION Neurons in the temperature regulation center of the hypothalamus balance heat loss and heat production to regulate the formation of sweat through reflex mechanisms to help maintain constant body temperature (page 267). The rate of formation and evaporation of sweat and its cooling effect are greater in hot, dry climates than in hot, humid ones. For short periods, sympathetic cholinergic stimuli may cause as much as 4 L of sweat to be formed each hour and up to 8 to 10 L daily in workers in summertime. Sweat formation increases by about 100 to 150 mL/day for each 1°C rise in body temperature. The number of active sweat glands increases with production of body heat and an elevation in body temperature. Because physical activity and an increase in environmental temperature require this response, persons native to tropical environments generally have a greater number of active sweat glands than those living in cooler climates.

When increased sweating occurs for long periods of time, salts and water must be replaced or an electrolyte imbalance may occur between the extracellular fluid (ECF) and intracellular fluid (ICF), and heat cramps (muscle cramps) may ensue. Heat loss mechanisms may fail to keep pace with heat production (Table 11-5 and page 594). Treatment for these disorders include fluid and electrolyte replacement and cooling of the body. Continued exposure to heat may cause sufficient loss of water and salts to induce *heat exhaustion,* which is characterized by weakness, dizziness, nausea, headache, a decrease in blood pressure, and fainting. With excessive exposure to heat, *heat stroke* may occur when, for reasons that are not clear, the sweating mechanism is inhibited and body temperature rises; brain damage and death may follow. Heat stroke is associated with the deaths of more than 200 Americans annually and even greater numbers in years with sustained heat waves. The elderly and infants under 1 year of age, both of whom have relatively ineffective temperature-regulating mechanisms, are at high risk, as are chronically ill people.

Under normal circumstances, with acclimatization to heat both the volume and salt content of sweat are reduced. The adrenal glands aid this process by releasing aldosterone, which promotes salt conservation. Aldosterone stimulates epithelial cells lining the ducts of the sweat glands to transport sodium ions (and water) from the ducts into the interstitial fluid for reabsorption (return to the blood).

A disorder which drastically increases sweat formation is *cystic fibrosis,* a heritable disease of glandular epithelium in the skin, lungs, and digestive tract. Abnormal secretions, which alter not only sweat but also tears and pulmonary and gastrointestinal secretions, are produced. Respiratory infections also are common as the secretions accumulate in the air passageways. The small intestine may become obstructed by excessive secretions and the intestine may fail to absorb nutrients normally. The average life expectancy of the victims of this disease is about 19 years, although some individuals live into their thirties. In 99 percent of those afflicted, sodium and chloride ion secretions are abnormally high.

The Dermis (Corium)

The *dermis,* the deepest segment of the skin, is composed of dense connective tissue, which can be divided into outer denser and inner looser areas (the *stratum compactum* and *spongiosum,* respectively). The outer portion of the dermis (the stratum compactum) forms ridges or *papillae* which project into the epidermis (Fig. 21-1). They produce unique patterns of hills and valleys, which appear on the under surface of the toes and on the fingertips as *fingerprints.* Within the papillae are capillaries, naked nerve endings and other sensory receptors such as Meissner's corpuscles and more deeply located Pacinian corpuscles (Figs. 21-1 and 8-1). The hypodermis (subcutaneous tissue beneath the skin) attaches the dermis to underlying structures such as muscles and is composed of areolar and adipose tissues. The subcutaneous tissues form a convenient site for the injection of medication (a subcutaneous injection) because of the relatively large amount of free space between its loose connective tissue cells.

Dermatitis (inflammation of the skin) may result from microbial infections caused by bacteria, fungi, or viruses and also may follow immunologic injury to the tissue in allergic reactions (see Chaps. 3 and 19). Pressure applied to the skin may spread the bacteria into the surrounding dermal tissues. Responses to poison ivy and poison oak are common examples of allergy-induced dermatitis. Cellular immune reactions cause inflammation at sites penetrated by allergens and produce conditions called eczema. *Eczemas* are noncontagious inflammations of the skin accompanied by redness, itching, loss of serous fluid, and a scaly, crusty appearance. Such reactions may impede the spread of infection but also bring with them the symptoms of dermatitis, namely, redness, swelling, and itching of the skin. Other types of eczema may involve humoral immunity and the formation of IgE antibodies (see page 63).

Therapy for various skin disorders may be preventative (avoidance of the causative agent, such as the oils of poison ivy), involve the administration of steroids to lessen the inflammatory response, or utilize antimicrobial drugs if the dermatitis is caused by bacteria or fungi. Certain severe skin diseases may require the use of specialized drugs to suppress the immune response.

21-3 THE KIDNEYS

Anatomy

The *kidneys* (renal organs) are paired, bean-shaped structures weighing about 150 g each and are about the size of the area covered by an individual's fingers folded against the palm of the hand. The kidneys are located posterior to the parietal peritoneum of the abdominal cavity (Fig. 5-1) and are surrounded by a connective tissue capsule. Each kidney is separated into an outer zone, the *cortex,* which is composed largely of spherical masses called *glomeruli,* and an inner zone, the *medulla,* which is formed

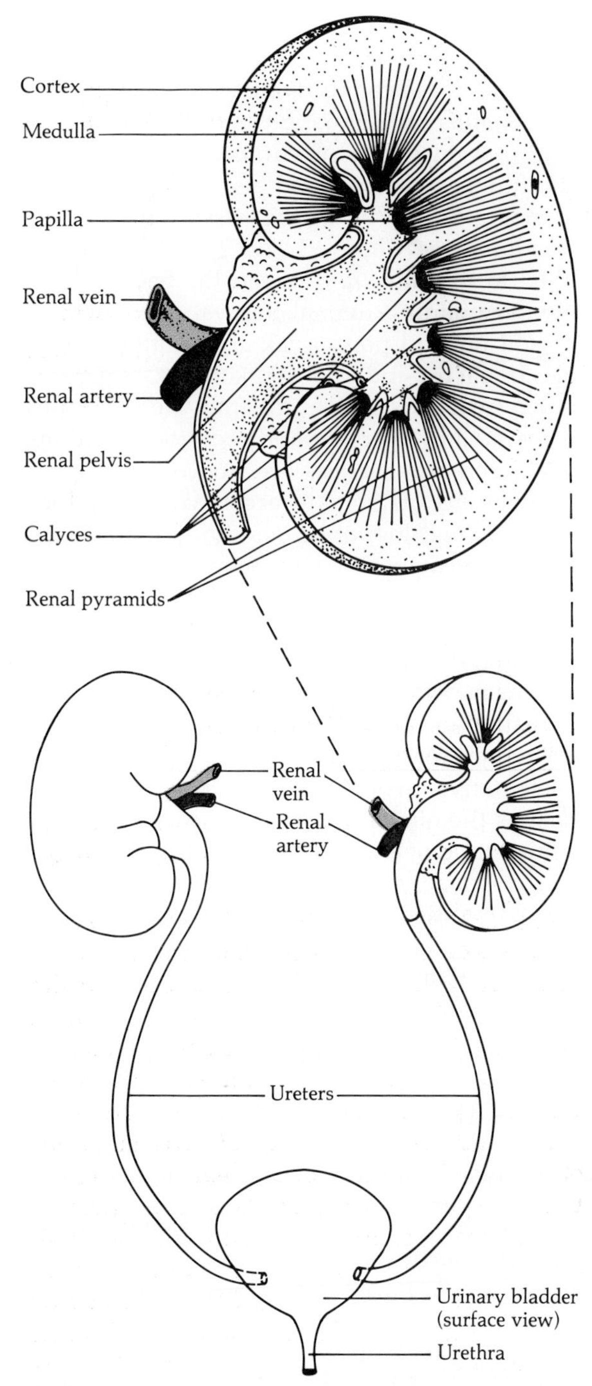

FIG. 21-3
The gross anatomy of the urinary system.

mostly of collecting structures (tubules and ducts) bound by connective tissues. The medulla is divided into 6 to 18 triangular *renal pyramids* (Fig. 21-3), whose tips, the *renal papillae*, extend into the central pelvis region of the kidney.

Urine formed in the outer cortical area of the pyramids flows into expanded cavities, the *renal calyces* (s. calyx) which join to form a single large, funnel-shaped collecting cavity, the *renal pelvis*. The pelvis continues from the body of the kidney as the *ureter*, a tube with smooth muscle in its wall that propels urine for storage into a muscular sac, the *urinary bladder*. The smooth muscle in the walls of the urinary bladder contracts in reflex responses initiated by stretch receptors and by impulses generated by the voluntary contractions of abdominal and pelvic cavity muscles. Urine is forced into the *urethra*, a tube whose external sphincter relaxes to allow the excretion of fluid to the exterior of the body in the process of urination (micturition).

General Functions

The kidneys are essential for the maintenance of homeostasis. These remarkable organs are the major means by which the body regulates the contents of the body fluids and eliminates wastes. The kidneys possess ultrafine filters and can selectively retrieve useful substances. In so doing they (1) regulate the osmotic pressure of body fluids; (2) help maintain the electrolyte equilibrium; (3) adjust the balance of acid and base in these fluids; (4) secrete hormones that influence red blood cell formation, blood pressure, and calcium storage and release; and (5) concentrate and excrete water-soluble wastes and substances that can otherwise accumulate to toxic levels in the blood. Mechanisms used by the kidneys to perform some of these functions include (1) filtration, (2) reabsorption (retrieval), and (3) selective secretion.

Substances filter from the blood through the capillaries of the kidneys and their surrounding basement membranes, which act as molecular sieves in a process called *ultrafiltration*. The substances are transported in a fluid, the filtrate, which flows through tubules in the kidney. Cells lining the walls of the tubules may return materials back to the blood by reab-

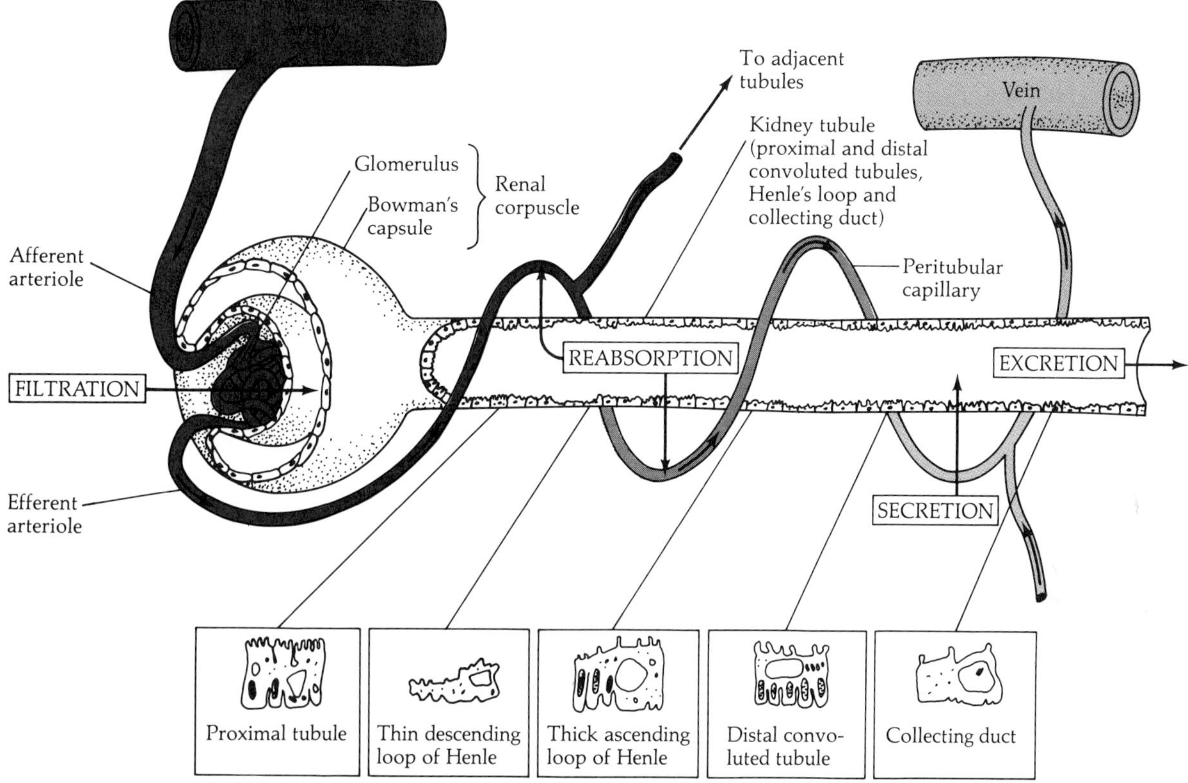

FIG. 21-4
Diagrammatic sketch of the major functions and cellular detail of the tubules of the kidneys. Arrows indicate the directions of filtration, reabsorption, secretion, and excretion in the kidney tubules. Note for simplicity that the complex coiling of the tubules is not shown.

sorption or add more of them to the filtrate by secretion. These events are illustrated in Fig. 21-4. As a result of these activities, urine is formed and excreted as a filtrate with a composition different from that of the plasma from which it is derived (Table 21-1).

The Nephrons: Functional Units of the Kidney

The filtration capacity of the kidneys is extraordinary and easily surpasses that of a swimming-pool filter. They provide nearly 50,000 times greater a ratio of filter bed area to total volume of fluid to be filtered than does a good swimming-pool filter. For example, the kidneys filter a total of about 180 L of fluid per day, which represents 60 times the average plasma volume, while the water in a pool is filtered about 5 times a day.

Each kidney has about 1 to 1.5 million *nephrons,* whose tubules form functional units approximately 12 mm in length. End to end the nephrons of the two kidneys would extend more than 22 mi, and if opened and spread out, cover an area of 12 m^2 (about 11.4 × 11.4 ft). The filtration component of the nephron is selective and restricts passage of substances out of the blood into the filtrate. Furthermore, active transport enzymes are located on numerous microscopic folds, the *microvilli* or *brush borders,* which line the inner surfaces of the tubules of each nephron (Fig. 21-5). The microvilli increase the surface area of the tubules and the opportunity for exchange between the urinary filtrate and the interstitial fluid.

The major parts of each nephron are made of two types of components. One is tubular, the other is vascular. Five subdivisions of the tubules are: (1) Bowman's capsule; (2) the proxi-

TABLE 21-1
Composition of urine*

Component		Amount	Urine/Plasma Ratio
Acid		pH Ave. 6.0 (range ~4.7–8.0)	
Osmolarity (mosmoles/L)		750–1400	
Specific gravity		1.015–1.038	
Sodium	2–4 g	100–200 mEq	0.8–1.5
Potassium	1.5–2.0 g	50–70 mEq	10–15
Magnesium	0.1–0.2 g	8–16 mEq	
Calcium	0.1–0.3 g	2.5–7.5 mEq	
Iron	0.2 mg		
Ammonia	0.4–1.0 g N	30–75 mEq	
H^+		$4 \times 10^{-8} - 4 \times 10^{-6}$ mEq/L	1–100
Uric acid	0.80–0.2 g N		20
Amino acids	0.08–0.15 g N		
Hippuric acid	0.04–0.08 g N		
Chloride		100–250 mEq	0.8–2
Bicarbonate		0–50 mEq	0–2
Phosphate	0.7–1.6 g P	20–50 mmol	25
Inorganic sulfate	0.6–1.8 g S	40–120 mEq	50
Organic sulfate	0.06–0.2 g S		
Urea	6–18 g N		35
Creatinine	0.3–0.8 g N		70
Peptides	0.3–0.7 g N		

* The values indicate the average contents of adult human urine collected for 24 h.
Adapted from *Principles of Biochemistry*, 6th ed., A. White et al. Copyright © 1977 McGraw-Hill Book Company. Used with the permission of McGraw-Hill Book Company.

mal convoluted tubule; (3) the loop of Henle; (4) the distal convoluted tubule; and (5) the collecting duct (Fig. 21-6).

Bowman's capsule is an epithelial chamber whose double wall appears as though it had been pushed in on itself like a partially compressed tennis ball. Its concavity is filled with a bed of capillaries, the *glomerulus* (plural, glomeruli). Bowman's capsule and its resident glomerulus are called the *renal* corpuscle and are located in the cortex of the kidney (Fig. 21-6). The *proximal convoluted tubule* is twisted and coiled next to Bowman's capsule and extends from it toward the medulla. The *loop of Henle* is shaped like a hairpin. It first courses toward the medullary area and then reverses toward the cortex. It has three segments: a thin descending limb, a thin loop, and a thick ascending limb (Fig. 21-6). The *distal convoluted tubule* extends from the ascending limb. Some of the cells in its first segment form a structure called the *macula densa,* located between the afferent (and efferent?) arterioles of the glomerulus (Fig. 21-7A). The distal tubule then forms a hairpin turn and extends toward the medullary area, where it merges with similar tubules to form *collecting ducts* which convey urine toward the papilla and renal calyces (Fig. 21-6).

There are two contrasting types of nephrons, designated according to location. *Cortical nephrons,* which are located in the cortex, lack the thin segment of the loop of Henle or have only a rudimentary one, and have shorter loops which extend at most to the outer area of the medulla next to the cortex. *Juxtamedullary nephrons* are those with renal corpuscles in the inner cortex nearest the medulla and have longer loops extending to the renal papillae.

FIG. 21-5
Microvilli in epithelial cells of the proximal convoluted tubules. (A) Diagram showing microvilli and intercellular connections between epithelial cells in the tubule. (B) Scanning electron micrograph of microvilli (× 14,900). From *Histology*, 4th ed., by L. Weiss and R. O. Greep. Copyright © 1977 McGraw-Hill Book Company. Used with permission of McGraw-Hill Book Company.

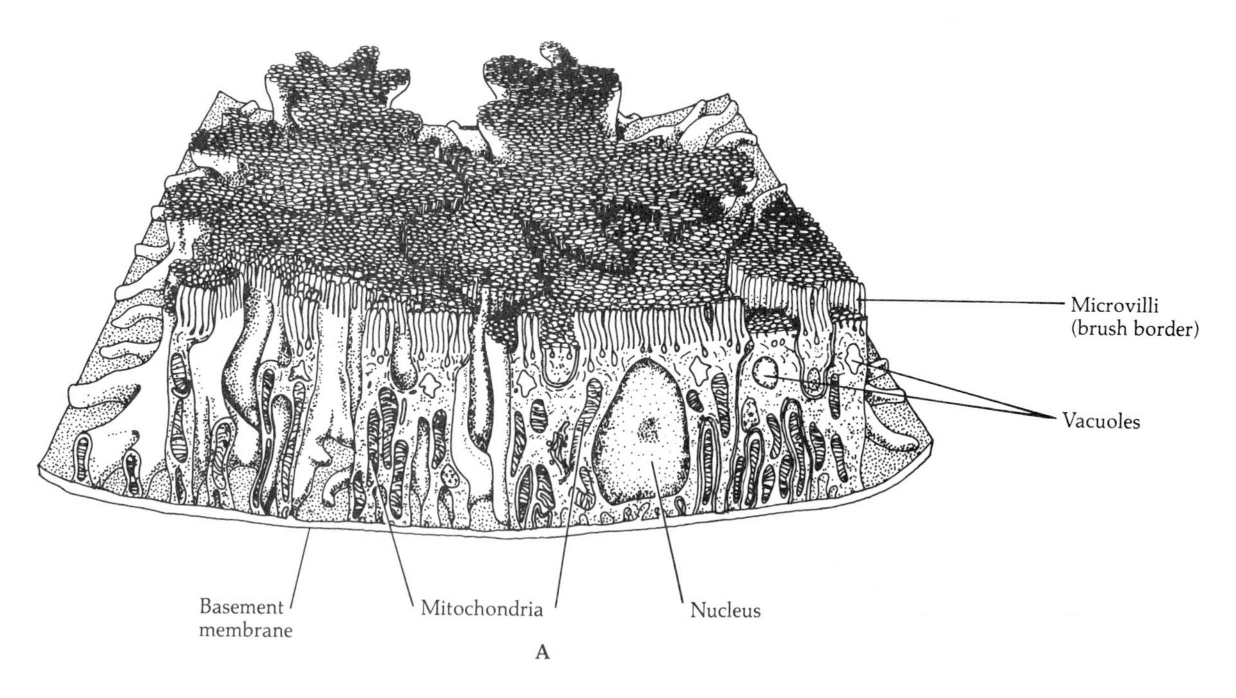
Microvilli
(brush border)
Vacuoles
Basement
membrane
Mitochondria
Nucleus

A

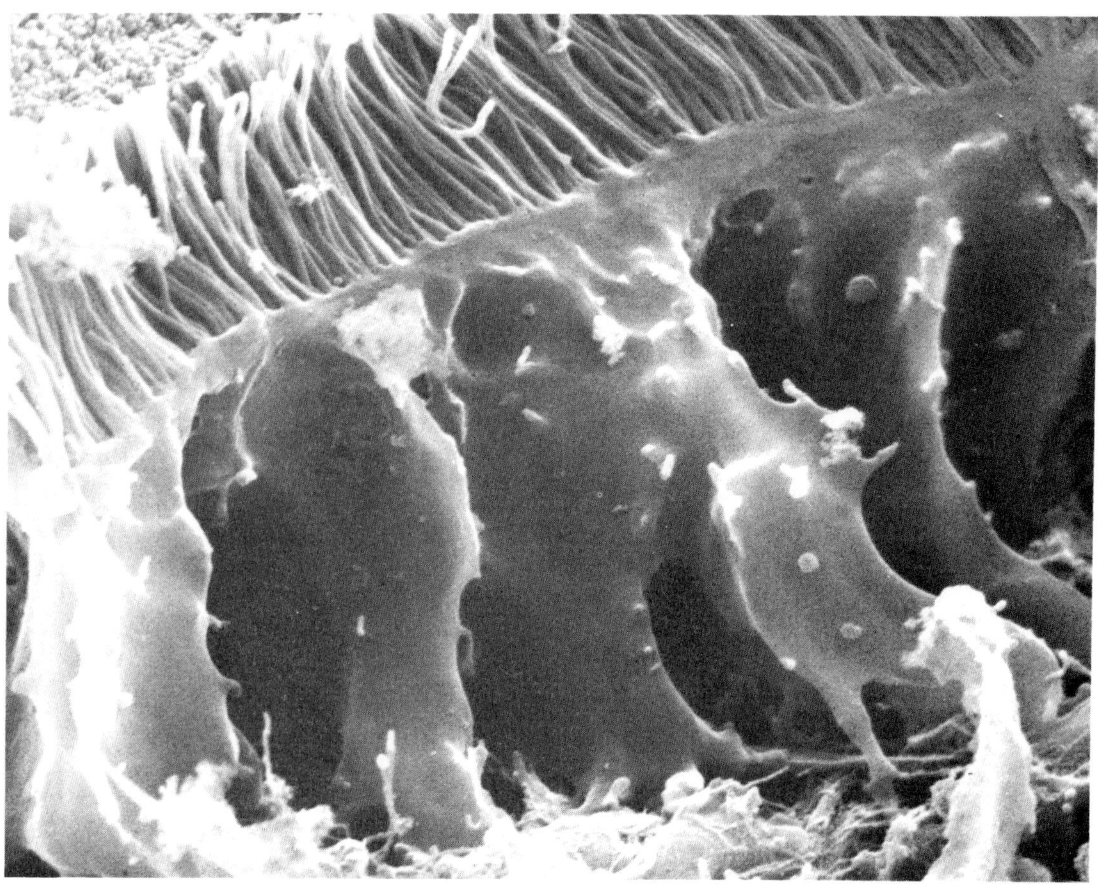

B

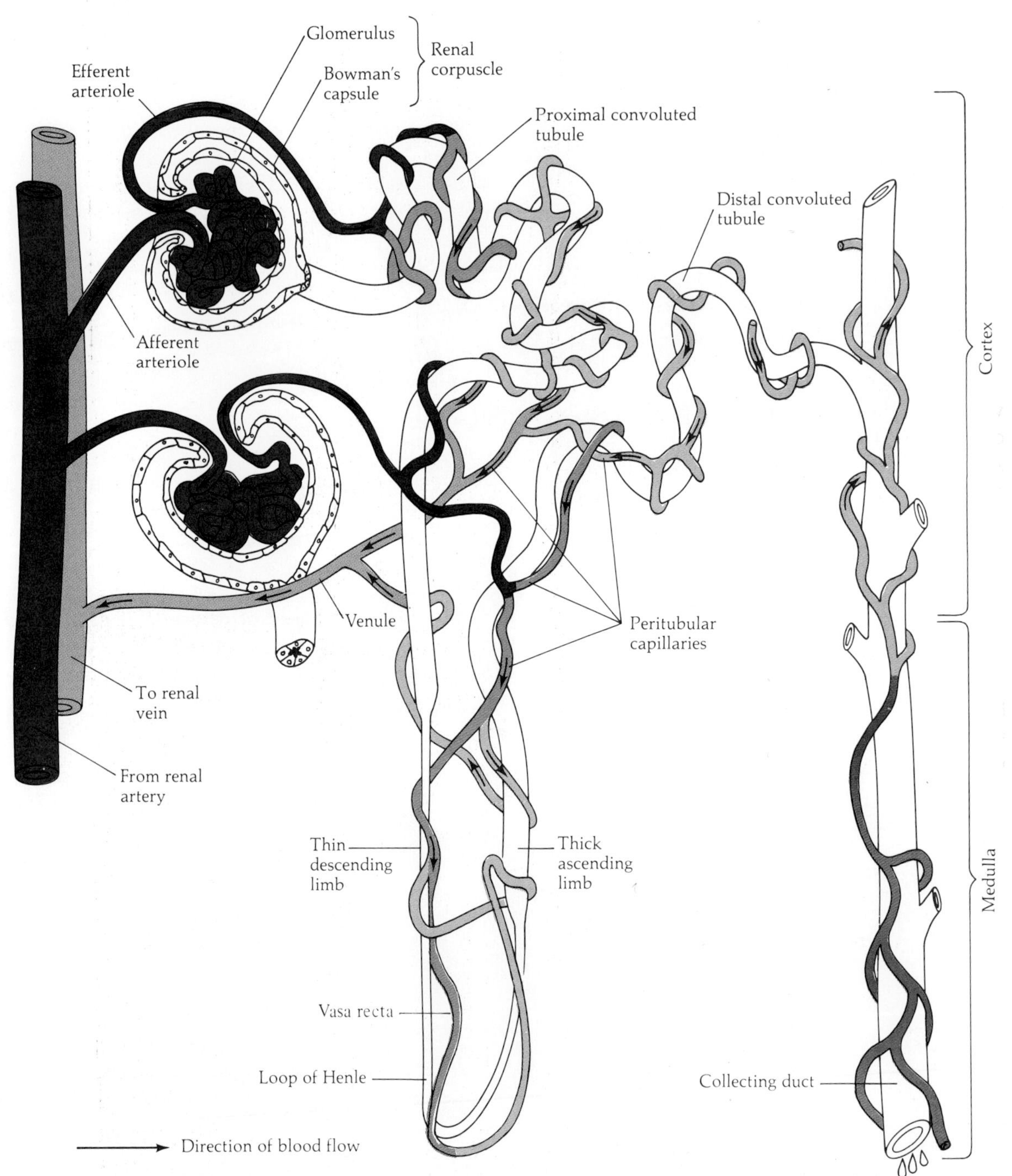

Physiology of the Nephrons

The nephrons function with extraordinary precision. The rates of filtration, reabsorption, and secretion of substances are closely regulated according to the needs of the body and critically related to the blood supply to the kidney.

FIG. 21-6
Structure of the nephrons and their blood supply. For simplicity, the contortion of these tubules is not illustrated fully in this diagram or the succeeding ones. In the kidney the first part of the distal tubule is actually located between the afferent and efferent arterioles of the glomerulus.

Blood Supply

Remarkably, the flow rate of blood to the kidney is nearly 100 times the rate to skeletal muscle. The kidneys account for less than 0.5 percent of body weight, yet they receive nearly 20 percent of the resting cardiac output. In 4 to 5 min the volume of blood that passes through the kidneys almost equals the entire blood volume of the body.

Each kidney is amply provided with blood by the *renal artery*, which subdivides within each organ to form *afferent arterioles* which transport blood to the *glomeruli* (Fig. 21-6). The glomerular capillaries merge to form an *efferent arteriole*, which is narrower than the afferent one and carries blood away from the glomerulus. A resistance to blood flow develops owing to the difference in vessel size and causes hydrostatic pressure (filtration pressure) to build up in the glomerulus.

The efferent arterioles freely merge with each other and form *peritubular capillaries*, which surround the tubules of one or more nephrons. Some of the peritubular capillaries form hairpin loops, vasa recta, around the loops of Henle in the medulla. The peritubular capillaries lead to venules, which eventually merge to form the renal vein, which delivers blood to the inferior vena cava.

Filtration and Microscopic Structure of the Filtration Membranes

Blood is filtered as it passes through the glomerulus, where water, ions, and small molecules flow into Bowman's capsule. There are three physical barriers that may impede the escape of substances from the blood. From the inside of a glomerular capillary outward they are: (1) the endothelial cell membrane; (2) the basement membranes of the capillary and Bowman's capsule; and (3) the epithelial cells of the inner (visceral) layer of Bowman's capsule (Fig. 21-7).

The size of the openings differ in these barriers (Table 21-2). The largest ones are open slits (fenestrations) in the endothelial cells of the glomeruli, while the spaces in the basement membranes which surround the glomerular capillaries are the smallest. Molecules about 6 nm in diameter and with molecular weights below 66,500 penetrate the basement membranes slightly. Substances of lesser size and molecular weight can pass more readily out of the capillaries to enter the tubules through the larger pore slits of the visceral epithelium of Bowman's capsule. The visceral epithelium is composed of octopuslike cells called *podocytes* (foot cells), whose footlike cytoplasmic extensions, or pedicels, surround the endothelial cells (Fig. 21-8). The pedicels are separated by *slit-pores* about 25 nm in diameter.

Because of the fine molecular sieve formed by the basement membranes, normally the filtrate and urine are nearly protein-free. Small protein molecules which cross this ultrafilter account for the bulk of the 34 to 144 mg of protein excreted in the urine each day. An average of about 0.008 g/dL of protein (0.008 percent) appears in urine collected during a 24-h period.

TABLE 21-2
Openings in the filtration barriers in the kidney

Site	*Characteristics*	*Approximate Width of Opening, nm*	PASSAGE OF MOLECULES *Type*	*Mol Wt*	*Passage*
Endothelial cells in glomerulus	Open slits (fenestrations)	50–100	Proteins (Ferritin	Variable 450,000	Easily Yes)
Basement membrane	Gel-like (collagen fibrils and mucopolysaccharide in matrix of gel)	6	Protein (Hemoglobin	40,000 66,500	Yes Slight)
Visceral epithelium of podocytes in Bowman's capsule	Slit pores (filtration slits bridged by thin membranes)	25	Protein	90,000	Slight

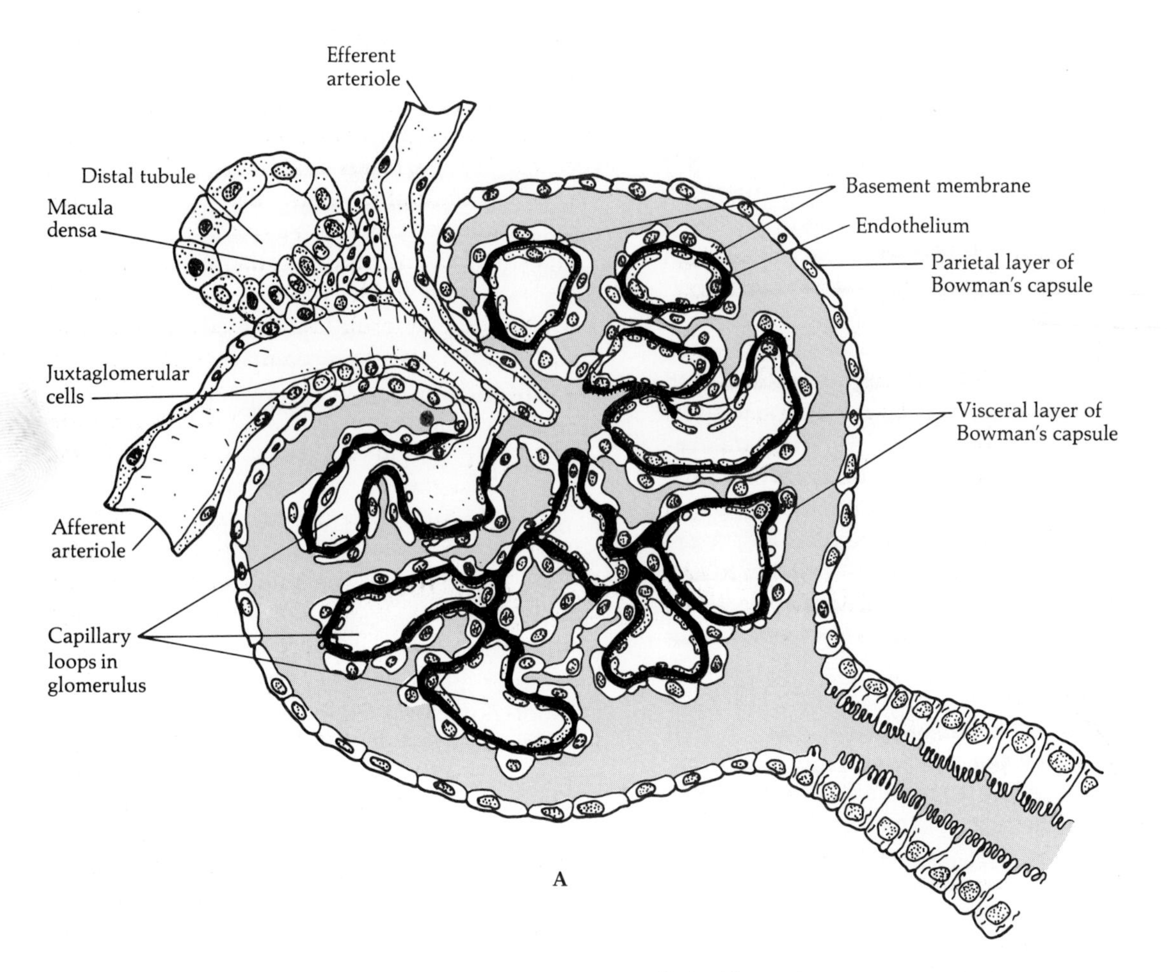

This value compares with a usual range of 6.0 to 8.4 g/dL of protein (6 to 8.4 percent) in plasma. Damage to the basement membranes leads to inflammation of the glomerular structure, *glomerulonephritis,* and to *proteinuria* (leakage of variable amounts of protein into the urine).

GLOMERULAR FILTRATION RATE (GFR) Nearly one-fifth of the plasma delivered to the nephron filters through the glomeruli while four-fifths continues on through the efferent arteriole and peritubular capillaries. The amount of plasma filtered through all the glomeruli each minute is the *glomerular filtration rate* (*GFR*). It averages 115 mL/min in females and 125 mL/min in males and totals about 165 to 180 L/day, respectively. About 0.8 percent of the body's plasma is excreted as urine each day, the remainder being

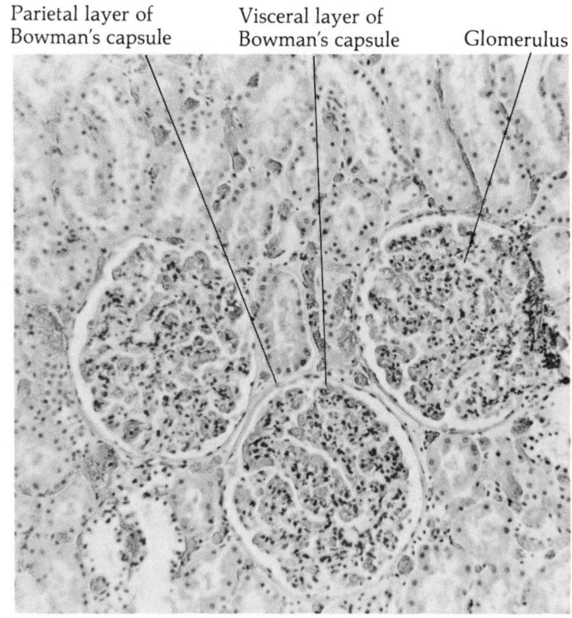

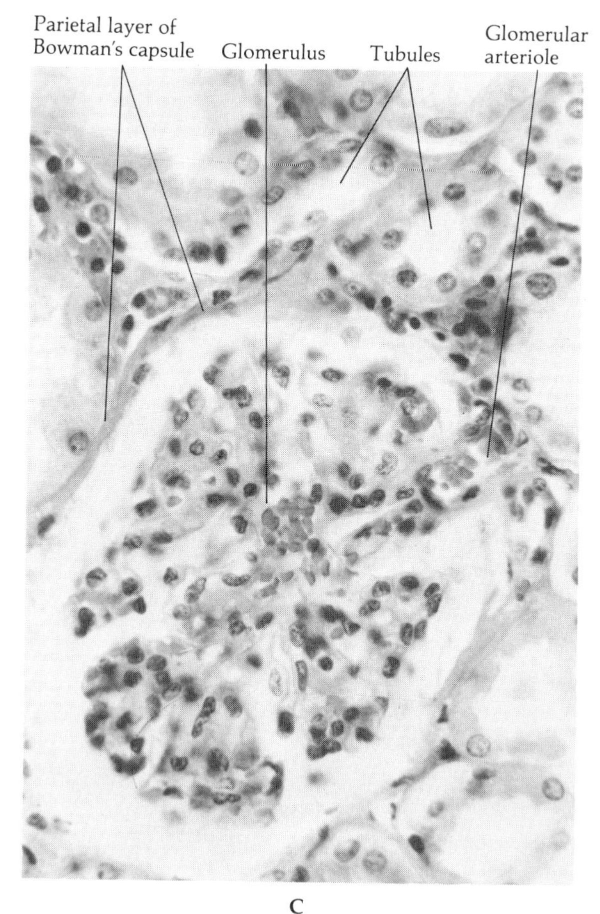

FIG. 21-7
Bowman's capsule and the glomerulus: the renal corpuscle. (A) Diagram illustrating the cellular detail of a renal corpuscle. (B) Photomicrograph of renal corpuscles (×170). The glomeruli and parietal and visceral layers in Bowman's capsule are evident. Cross sections of tubules also appear in the photomicrograph. (C) Photomicrograph of a renal corpuscle (×700).

reabsorbed. Differences in the GFR are related to the mass of nephrons and the sex of the individual. The mass of nephrons is associated with the surface area of the body. The amount of urine formed ranges between 600 and 1600 mL/day in the average adult, owing to variations in diet, the GFR, reabsorption, and other aspects of kidney function.

REGULATION OF GFR The principles which govern the GFR are the same as those that govern filtration in capillaries of other organs (Chap. 17 and Fig. 17-7). Glomerular filtration occurs by bulk flow, the process in which water and low-molecular-weight solutes move together. Increases in the GFR parallel those that cause greater resistance to develop in the efferent arterioles. The narrower diameter of the efferent arterioles compared with the afferent ones causes hydrostatic pressure to be higher in the glomeruli than in other capillaries in the kidneys. Filtration and loss of water from the blood through the glomeruli are favored, resulting in concentration and an accompanying elevation of osmotic pressure of the blood delivered to the peritubular capillaries. When hydrostatic pressure in the glomeruli decreases, the filtration rate decreases.

Owing to the effects of gravity, movement of the body from a horizontal to a vertical position increases pressure in the long columns of blood in the lower part of the body such as the legs (page 414). Therefore, more blood pools in the long veins, more fluid escapes into the interstitial spaces, and circulating blood volume decreases. As a result, the GFR declines and so does urine output. Conversely, during sleeping hours and recumbency in a horizontal position, pressure in the long blood vessels decreases and blood volume increases, as does the GFR. Urine output then becomes proportionately higher during sleeping hours when horizontal body posture is maintained.

Four prominent regulatory mechanisms that alter renal blood flow and pressure have been studied extensively. They are autoregulation and sympathetic regulation, both of which will be described directly below, and hormonal effects and the renin-angiotensin system, which will be discussed later in this chapter.

Autoregulation Nerves to the kidney may be severed, yet blood flow to these organs is still regulated in a self-control mechanism called *autoregulation.* The diameter of renal vessels is influenced by local feedback mechanisms, which cause smooth muscle in the walls of the arterioles to relax or contract in order to maintain a constant blood supply within a fairly narrow range.

Cells in the macula densa respond through feedback pathways to changes in the blood flowing through the afferent (and efferent?) ar-

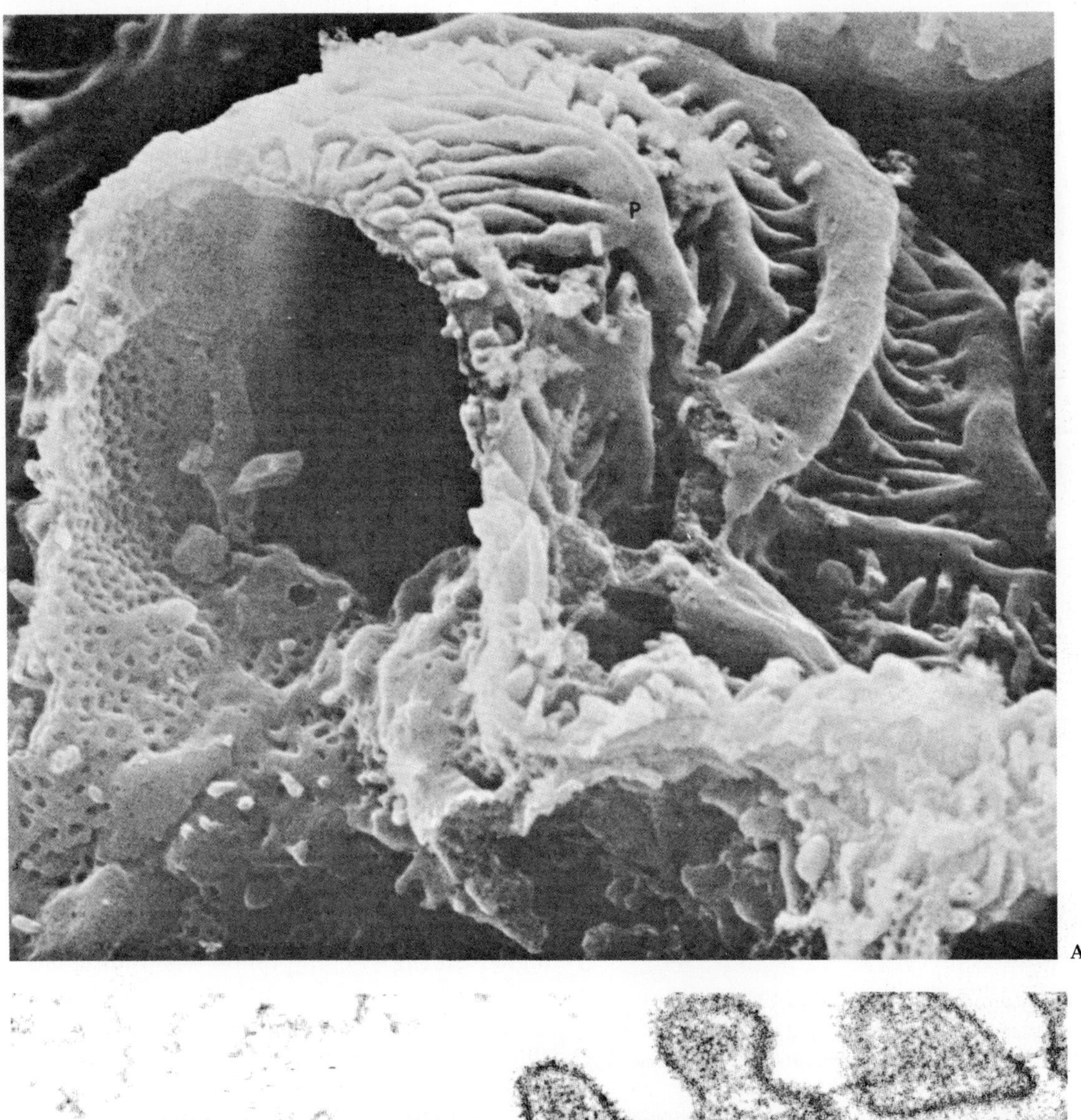

A

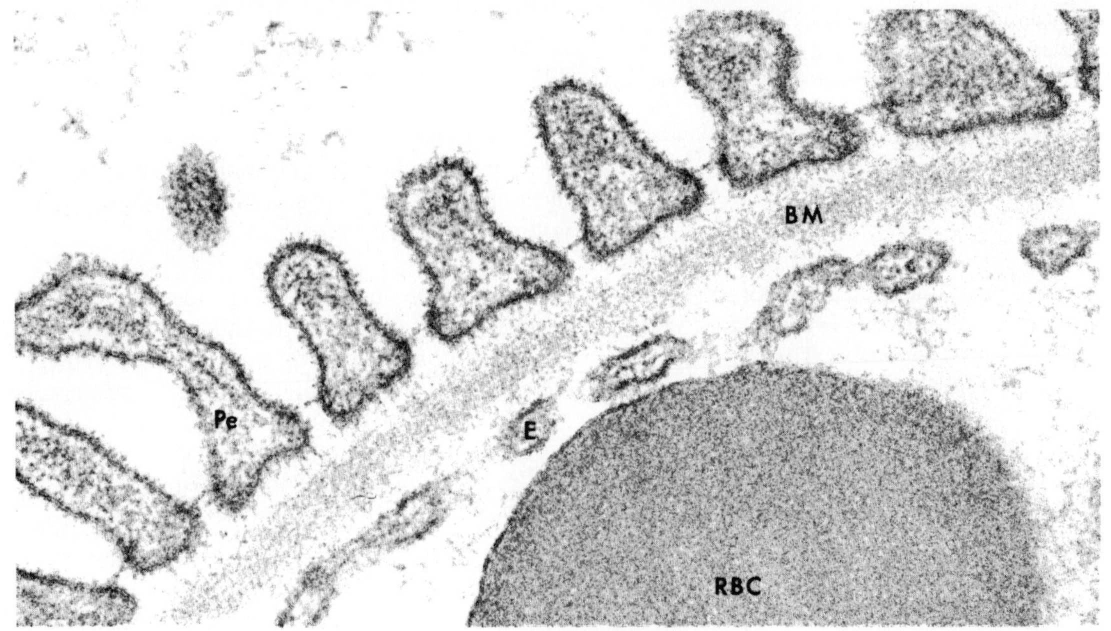

B

terioles of the glomerulus. The exact mechanism(s) by which this occurs are not clear. However, autoregulation may involve responses to physical stimuli (such as stretching of the vessels when blood pressure increases) or chemical stimuli (due to chemicals in the distal tubular fluid). Local reactions may include the intrinsic response of smooth muscle to stretch (page 302) and the release of vasoactive substances such as renin or prostaglandins. When these factors elevate renal blood flow, the condition is called *hyperemia*. Feedback mechanisms which lower renal blood flow and pressure may then follow.

A constant flow rate is maintained by alteration of pressure within the arterioles. The physical factors involved are described in a derivative of Poiseuille's law ($F = P/R$, or flow rate = pressure/resistance, page 419). When blood pressure P increases, autoregulatory factors cause the glomerular arterioles to constrict and resistance (R) increases. The flow rate F remains the same. Conversely, when P decreases, autoregulatory factors cause the glomerular arterioles to dilate and R decreases. Again, the flow rate remains constant. In this manner, the kidney is normally assured sufficient blood flow to maintain an adequate GFR.

Sympathetic Regulation Sympathetic nerve impulses from renal nerves cause constriction of renal blood vessels, which greatly decreases blood flow to the kidneys and shunts blood to other organs. Norepinephrine is released by sympathetic nerve terminals to cause vasoconstriction of renal arterioles. However, the effects are less on afferent than efferent arterioles and glomerular pressure, and the GFR are maintained.

FIG. 21-8
(A) Scanning electron micrograph of a glomerular capillary (×17,200). The capillary is surrounded by the footlike extensions of an epithelial cell, the podocyte, labeled P. Intercellular fenestrations between the endothelial cells appear as pox marks in the lower left side of the figure. (B) Transmission electron micrograph of epithelial cells of Bowman's capsule and a glomerular capillary (×71,700). The capillary is surrounded by footlike extensions (pedicels, Pe) of podocytes, epithelial cells which have spaces between them and which lie above the basement membrane (BM) in this figure. The latter acts as an ultrafilter for passage of substances into Bowman's capsule from the capillary loop through the fenestrations of the endothelial cells (E) which form a capillary that contains a red blood cell (RBC). From *Histology*, 4th ed., by L. Weiss and R. O. Greep. Copyright © 1977, McGraw-Hill Book Company. Used with permission of McGraw-Hill Book Company.

TABLE 21-3
Reabsorption of some components in the glomerular filtrate

Substance	*Primary Mode of Reabsorption*	*Approximate % Reabsorbed**
Glucose	Active	100
Bicarbonate (HCO_3^-)	Passive	99.9†
Sodium (Na^+)	Active and passive	99.6
Chloride (Cl^-)	Active and passive	99.5
Water (H_2O)	Passive	99.1
Potassium (K^+)	Active	92.6
Urea ($H_2N-C(=O)-NH_2$)	Passive	53.0
Total solute	Active and passive	98.9

* Varies with diet.
† Varies with pH of body fluids.

Reabsorption

The mechanisms involved in reabsorption of the filtrate include diffusion, facilitated diffusion, osmosis, and active transport (Chap. 4). Because most substances in the glomerular filtrate are useful to the body, about 99 percent of many of them are ultimately returned to the blood (Table 21-3).

The concentration of specific substances in various parts of the tubules may be determined by micropuncture techniques. With the aid of a microscope, a capillary pipette may be used to withdraw fluids from discrete areas of the nephron for analysis. The pipette also may be connected to a manometer to measure hydrostatic pressures (page 419).

Because of the ease of passage of water across the glomeruli into the proximal convoluted tubules, the filtrate there is very dilute and the blood which then flows to the peritubular capil-

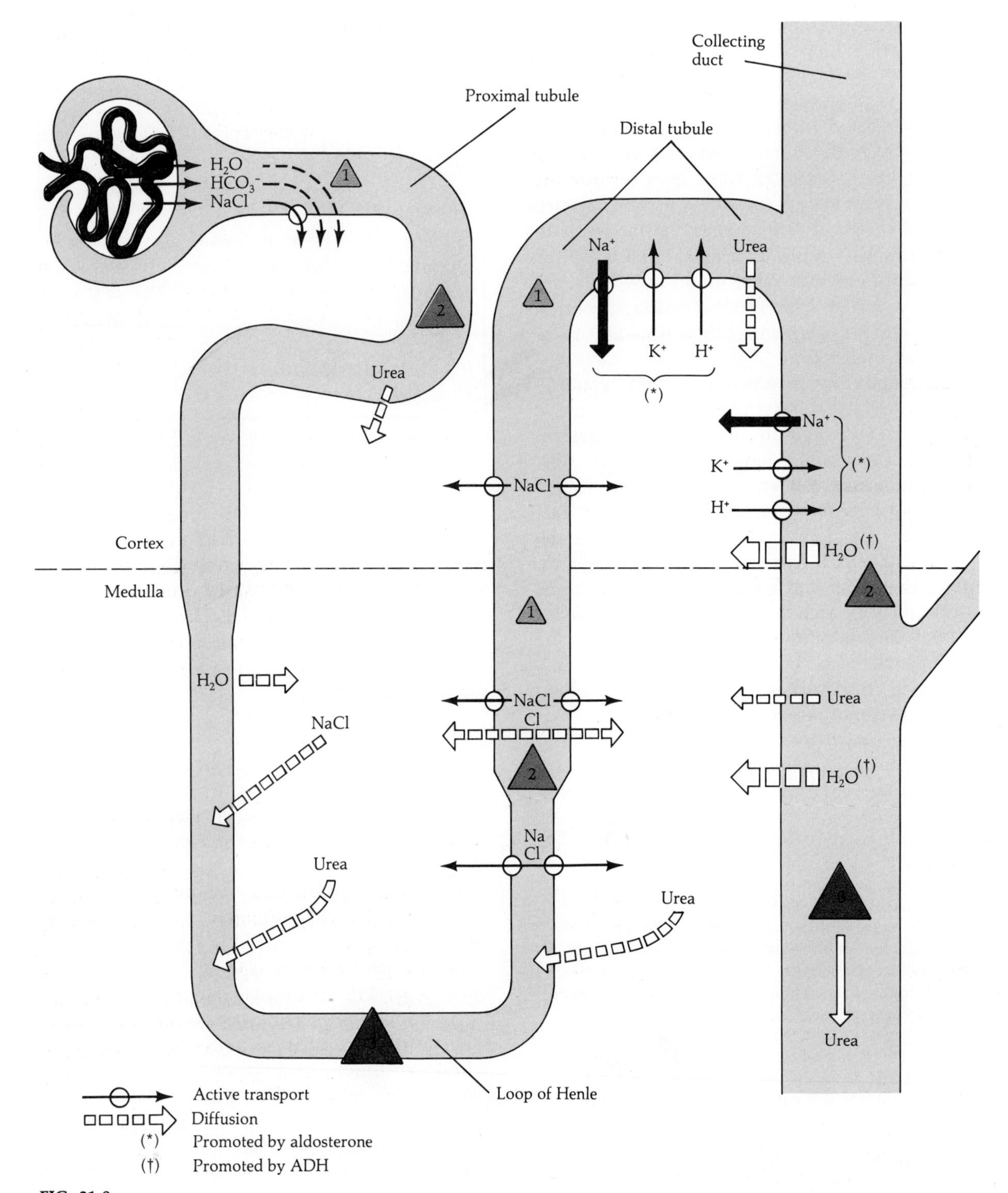

FIG. 21-9
Net direction of transport of major ions and water across the tubules of the kidneys. For simplicity, the peritubular capillaries adjacent to the tubules are not shown, and ions and molecules are shown only on one side of a specific area of a tubule rather than on all sides of it. The osmolarity of the tubular fluid is indicated by the relative color and size of the triangles: 1 is hypoosmotic, 2 is isoosmotic, 3 is hyperosmotic. The interstitial fluid is located outside the tubules.

laries is concentrated. As a result, water diffuses out of the proximal convoluted tubules into the peritubular capillaries in the reaction called *obligatory reabsorption* (Fig. 21-9).

Epithelial cells lining the tubules also reabsorb specific substances by active transport. The transporting enzymes pump useful solutes such as sodium ions, chloride ions, and glucose toward the blood. This process helps restore the filtrate to an isosmotic condition in the proximal tubules (Fig. 21-9).

If any one of the steps that move substances across the membranes of tubular cells requires the expenditure of energy, the entire sequence is generally referred to as *active* or *transepithelial transport.* Passive transport accounts for movement of substances in directions dictated by their concentration and electrical or pressure gradients. *Reabsorption* into the blood requires passage through the cytoplasm and membranes of the tubular epithelial cells, interstitial fluid, basement membranes, and the endothelial cell membrane of the capillaries. *Secretion,* net movement in the opposite direction, adds to urine formation (Fig. 21-4).

The ability of the kidneys to retrieve substances from the filtrate is limited. Different maximal rates of reabsorption and a *maximal tubular reabsorptive capacity,* or T_m (transport maximum), exist for individual substances. Should the filtration rate of a substance exceed its T_m, then the substance will appear in the urine. When this occurs, the kidney's capacity for reabsorption is overloaded, as its threshold for reabsorption is exceeded. As the T_m of a substance decreases, the percentage of it reabsorbed from the filtrate declines. Substances that are useful to body function, such as glucose and amino acids, have high T_m's and are reabsorbed extremely well.

For example, the T_m of glucose averages between 300 and 375 mg glucose per minute per 1.73 m^2 of body surface area in women and men, respectively. The normal plasma levels of glucose range between 70 and 110 mg/dL. The T_m is usually exceeded when excessive glucose is ingested and its levels rise to about 180 mg/dL of plasma. Glucose enters the filtrate faster than it can be reabsorbed and appears in the urine in the condition of *glucosuria* (the presence of glucose in the urine). Glycosuria is a general term that indicates elevation of sugar in the urine but does not specify the exact type of carbohydrate. After a big meal rich in carbohydrates, about 50 percent of the adults in the United States will excrete 2 to 3 mg glucose per deciliter of urine.

Glucosuria also can be a sign of body dysfunction. For example, excess glucagon secretion causes rapid hydrolysis of liver glycogen so that glucose blood levels exceed the T_m and glucosuria results. Insulin deficiency (as in diabetes mellitus) produces glucosuria by a different mechanism. The lack of insulin prevents adequate transport of glucose into muscle and liver cells, and the resulting high levels of blood glucose may cause the loss of up to 100 g glucose in the urine each day. Excess glucose in the urine produces an osmotic effect, which results in the diffusion of water into the tubules. The increase in urine volume is called *diuresis.* In this instance glucose acts as a *diuretic,* and one of the disabilities of diabetes mellitus is dehydration, which leads to low blood volume (hypovolemia). These responses sometimes develop as a consequence of massive osmotic diuresis, which accompanies the appearance of high levels of glucose in the urine. Significant electrolyte loss, mainly of sodium and sometimes of potassium, may also accompany the massive diuresis.

There are many other causes of diuresis. For example, drugs, poisons, and diseases may inhibit specific transport enzymes in the tubules. So too, caffeine in coffee and theophylline in tea act as diuretics, apparently by decreasing the active reabsorption of sodium and increasing the GFR. Thus, the tubular levels of sodium remain high, so that water moves into the filtrate by osmosis, and diuresis results. Disorders of the hypothalamus also may alter ADH secretion and cause diuresis, as described below.

Secretion

A number of substances are secreted (added to the filtrate) by active or passive mechanisms. Potassium and hydrogen ions and creatinine are actively transported into the filtrate. Water and urea enter or leave the filtrate by diffusion. Substances which are secreted and not reabsorbed are ultimately excreted in the urine.

The control of tubular secretion may be finely regulated. The transport of sodium ions and that of potassium ions are, to some extent, reciprocals of each other (Figs. 21-9 and 4-11); however, excretion of sodium is subject to finer control than that of potassium. As a consequence, an individual placed on a salt-free diet will exhibit a rapid decrease in sodium secretion several days before potassium secretion reaches a minimal level.

Effects of Hormones

The kidneys: (1) degrade and/or excrete many hormones or their metabolic by-products; (2) synthesize some; and (3) respond indirectly or directly to others. For example, the kidneys degrade insulin and glucagon and excrete human chorionic gonadotropin and the metabolic by-products of sex steroids in urine. They synthesize active vitamin D (1,25-dihydroxycholecalciferol), erythropoietic factor, and prostaglandins and respond indirectly to some hormones (such as insulin and glucagon as described above). Other hormones have direct effects on the kidneys. For example, parathormone and steroids alter electrolyte transport across tubular cells in the kidneys (Chap. 14) and prostaglandins increase renal blood flow. Two other hormones with notable effects are antidiuretic hormone (ADH) and aldosterone.

Antidiuretic Hormone (ADH)

ADH, which is formed in the hypothalamus and released into the blood, travels through the capillaries and binds to cell receptors on the surface of collecting ducts. ADH promotes synthesis of cAMP, which phosphorylates proteins to increase the permeability of the ducts to water. Water diffuses from the dilute filtrate in the collecting ducts into the more concentrated medullary fluid and blood (Figs. 21-9 and 21-12). The response diminishes urine volume, which accounts for the hormone's name *antidiuretic.* ADH is sometimes referred to as *vasopressin,* because it also elevates blood pressure, most probably by increasing blood volume.

Elevated blood levels of alcohol may inhibit osmoreceptors in the hypothalamus, block ADH secretion, and result in excretion of greater urine volumes. ADH secretion also may be reduced by trauma or disease of the hypothalamus in a disorder called *diabetes insipidus.* Diabetes insipidus produces as much as 20 L of dilute urine per day. Corrective therapy may include injections of ADH.

Aldosterone

Aldosterone is a steroid hormone secreted by the adrenal cortex which has marked effects on mineral metabolism and is classified as a *mineralocorticoid.* It is particularly effective in the distal tubules and in the collecting ducts (Fig. 21-9) and is frequently referred to as the *sodium-retaining hormone.* Aldosterone promotes reabsorption of sodium and chloride ions in exchange for secretion of potassium and hydrogen ions. More ions are retained than secreted and the resulting osmotic effect causes the reabsorption of water, which may result in an increase in extracellular fluid (ECF) and formation of edema. Aldosterone has similar effects on ion transport in epithelia of sweat glands and salivary glands and of the intestine.

Aldosterone secretion is regulated by negative feedback mechanisms involving sodium and potassium. More hormone is secreted when plasma levels of sodium ions are low and/or those of potassium ions are high. Salt retention results and the ECF volume increases. These changes inhibit the renin-angiotensin system (described below), which is the major stimulus for aldosterone secretion. Consequently, aldosterone levels decline and the cycle is completed.

Stress, kidney damage, and a number of diseases increase the secretion of aldosterone and glucocorticoids, which in turn regulate salt and water balance (Fig. 5-7). For example, the stress of menstruation stimulates secretion of these hormones and causes sodium and water retention in some females during the menstrual phase of each reproductive cycle.

Elevated blood levels of aldosterone may cause enough potassium loss to produce low blood potassium (*hypokalemia*). About 90 percent of potassium loss occurs through the kidneys and most of the remaining potassium exits from the digestive tract. When the loss of potassium and hydrogen ions in the urine is coupled with the retention of sodium (Fig. 21-9), the pH of the blood may increase and alkalosis may occur. If the change in blood pH remains uncorrected, the consequences to vital organs may be fatal owing to failure of critical enzymes in cells which results in accumulation of toxic levels of metabolites.

Renin-Angiotensin System

The major stimulus for aldosterone secretion is produced by the activity of the enzyme *renin*. Renin is a glycoprotein whose secretion by the kidney is increased when renal arteriolar pressure and blood flow decrease and when blood sodium and ECF levels decline. Renin causes the production of angiotensin II, an octapeptide, which is the most potent vasoconstrictor in the body. Angiotensin II raises blood pressure and thus elevates the GFR. Were it not for this response, waste products could accumulate to toxic and life-threatening levels when blood pressure decreases.

Renin, secreted by epithelial cells of the macula densa, acts as an enzyme and catalyzes the conversion of a bloodborne globular protein, angiotensinogen, from the liver, into angiotensin I (Fig. 21-10). Angiotensin I is then altered by converting enzymes in the lungs and blood into angiotensin II (hypertensin). Further enzyme activity converts angiotensin II into angiotensin III, which has lesser physiologic activity and, like angiotensin II, is rapidly destroyed by peptidase enzymes and converted into inactive compounds.

Angiotensin II constricts arterioles, increases cardiac output, promotes aldosterone secretion, and elevates blood pressure. In laboratory animals angiotensin II stimulates thirst and drinking, possibly by a direct effect on the brain. The resulting increase in water intake favors a rise in blood volume and blood pressure. Consequently, the production of renin by the kidney and the synthesis of angiotensin II can be a cause of hypertension and cardiovascular disease.

Effects contrary to those of the renin-angiotensin system are exerted by a similar but opposing group of compounds of the *kallikrein system*. Kallikrein, secreted by a number of tissues including those of the kidneys, converts plasma globulins into specific decapeptides called *kinins*, which oppose the vascular effects of angiotensins. Feedback pathways regulate the levels and net effects of these two opposing systems to help maintain kidney function.

Countercurrent Theory of Nephron Function

One hypothesis that has been proposed to explain how the kidneys concentrate urine is the *countercurrent theory*. Countercurrents are produced when fluids or gases flow close to each other in parallel but opposite directions. An example is a hot-water system in which the heater is placed in the middle of a tube bent like a U with its two limbs close to each other (Fig. 21-11A). Heat from outgoing water is used to warm the incoming water. As a result, the difference in temperature between the adjacent parts of the tubes is not great. In fact, the variations in temperature between incoming and outgoing water may be the same as in a conventional heating system (Fig. 21-11B). In a countercurrent system, however, although slight temperature gradients exist across its horizontal components, the temperature differences between the ends of the U and its bend are considerable.

Two countercurrent systems exist in the kidneys. One is formed by the hairpin-shaped loops of Henle and the other by the corresponding peritubular capillaries, the vasa recta. In both systems the gradients are osmotic rather than thermal and two different mechanisms, one passive and the other active, help to concentrate the urine. *Countercurrent exchange* involves *passive diffusion* between adjacent vessels and tubules and resembles the countercurrent exchanger noted above. The slow flow of

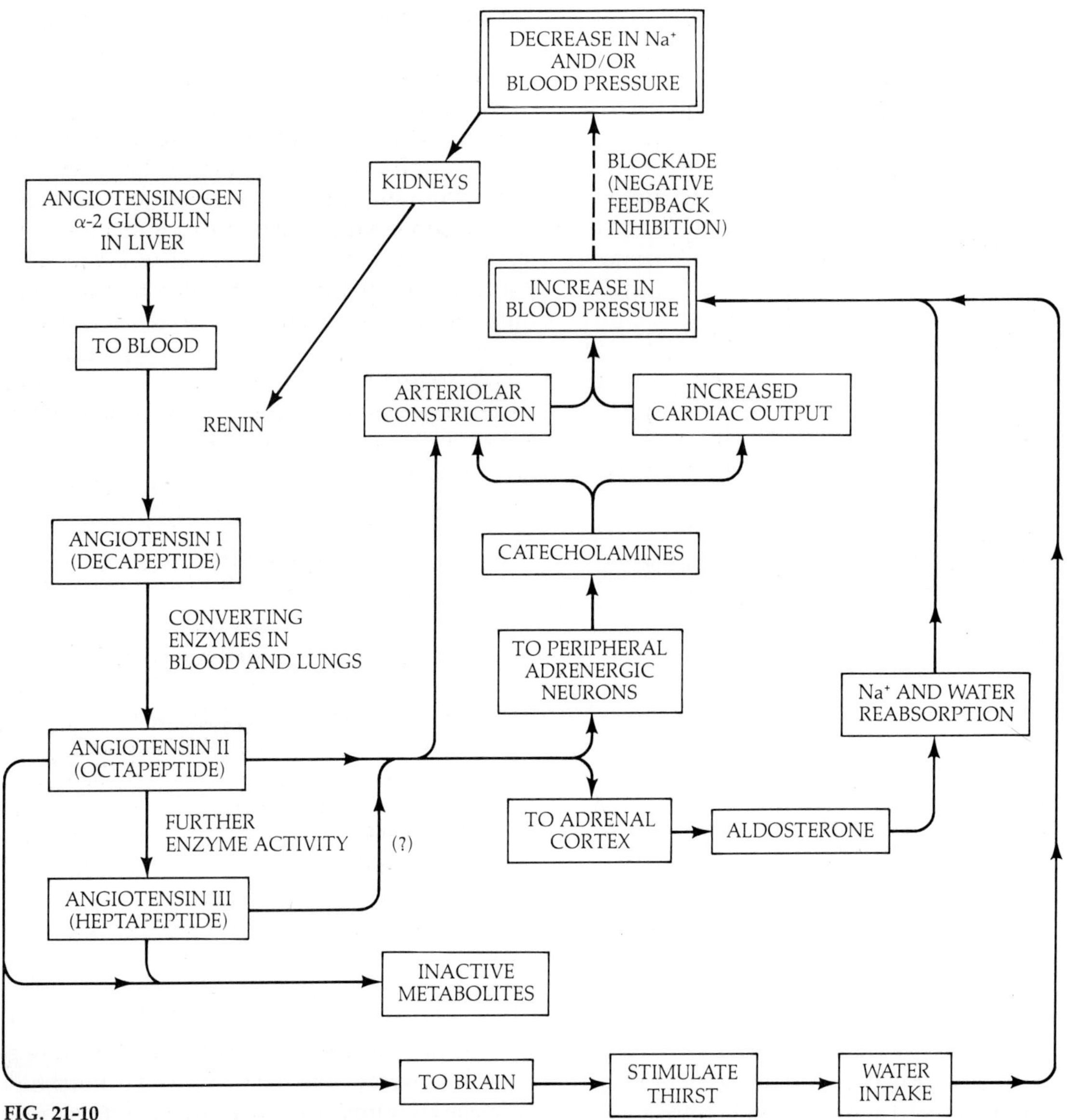

FIG. 21-10
The physiological effects of the renin-angiotensin system. The activity of renin produced by the kidneys may cause the changes noted in the diagram.

blood allows diffusion to occur between adjacent segments of descending (incoming) and ascending (outgoing) limbs of the blood vessels while a steep osmotic gradient exists at opposite ends of the system. Also, the osmolarity (concentration) of the filtrate increases significantly as fluid passes from cortical to medullary segments of the tubules (Fig. 21-12).

The kidneys also possess active mechanisms known as *countercurrent multipliers,* which resemble *pumps* (which in a heat exchanger might pump warm outgoing water to cooler incoming water). In the kidneys sodium chloride is actively transported out of the ascending limb and then diffuses through the interstitial fluid into the descending limb and vasa recta, whose contents become more concentrated (Fig. 21-12). Water diffuses in the reverse direction, from the descending limb into the vasa recta. Because the thick ascending limb of the tubule

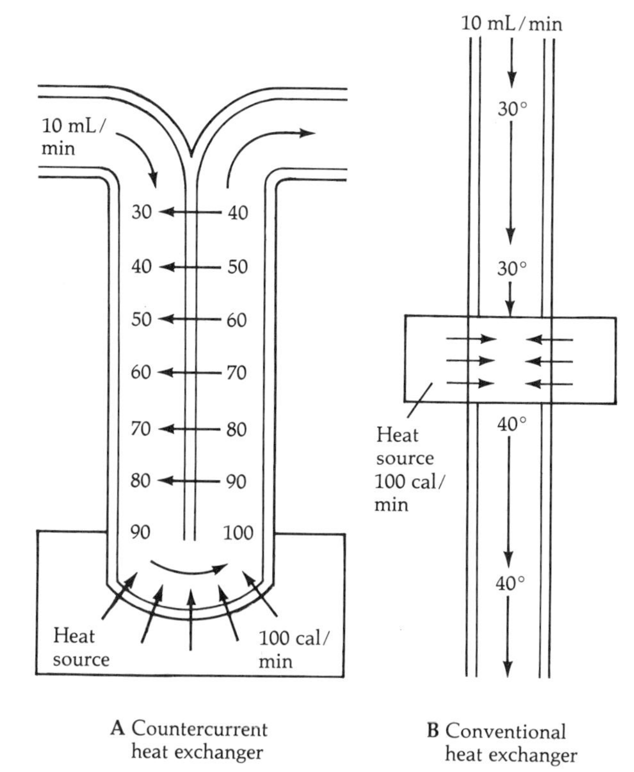

FIG. 21-11
Example of a countercurrent exchange. (A) The countercurrent heat exchanger warms outgoing water 10°C above its incoming temperature. However, a pronounced temperature difference of 60°C is produced between the ends of the tube and its U bend or loop. (B) The conventional heat exchanger illustrated consists of a straight tube in which incoming water passes through a heater and is warmed prior to its exit. Here the water temperature is also raised 10°C but, unlike the case of the countercurrent system, the temperature gradient does not exceed 10°C.

is relatively impermeable to water, the loss of salt from this segment of the tubule is *not* followed by the outward diffusion of water. Consequently, the fluid in the ascending limb becomes dilute (hypoosmotic).

Urea ($H_2N—\overset{\overset{\displaystyle O}{\|}}{C}—NH_2$), a breakdown product of protein metabolism, diffuses from the collecting ducts and accumulates in the medullary interstitial fluid (Fig. 21-9). This compound, along with sodium chloride, contributes greatly to the formation of a concentrated interstitial medullary fluid. The diffusion of water out of the collecting ducts is aided by ADH, and the volume of the urinary filtrate usually decreases as the urine becomes highly concentrated.

Owing to the continual formation of urea, urine is excreted even when no water is ingested. When metabolized, proteins form more solutes such as urea and less water than equivalent amounts of fat or carbohydrate. As a result a high-protein diet causes urinary solutes to accumulate and leads to osmotic withdrawal of water from blood into the urinary filtrate, limiting the ability of an individual to conserve water.

With normal solute production, a minimum amount of water is excreted daily; this is called *obligatory water loss.* About 500 mL is eliminated in urine. Another 400 mL is lost through the respiratory tract and skin, and 100 mL is lost by the digestive tract in the feces. The functions of the kidneys, lungs, skin, and digestive tract require the obligatory loss of some water. Consequently, the ability of the kidneys to regulate fluid loss is critical to homeostasis. The total obligatory loss of water from the body amounts to about 1 L/day; this is increased another 300 mL by water produced in the metabolism of a normal diet. When all routes are considered, the total water lost from the adult body ranges from 1500 to 3000 mL/day (Table 21-4). The volume varies with several factors, including the functions described, hormonal levels, fluid intake, environmental temperature, and exercise.

Regulation of Acid-Base Balance

The pH of the blood is maintained within normal limits of about 7.35 to 7.45 by means of buffers and by the capacity of the body to excrete excess acid or base. *Buffers* act as chemical sponges which can soak up or release hydrogen ions (protons). Those substances which donate hydrogen ions to other compounds are acids and those which accept hydrogen ions are bases (Chap. 2).

Acid-base homeostasis is influenced by dietary intake and excretion of ions in sweat and in the feces, the exchange of ions between bone and the blood (Chap. 14), the loss of carbon dioxide through respiration (Chap. 20), and especially by regulatory mechanisms related to

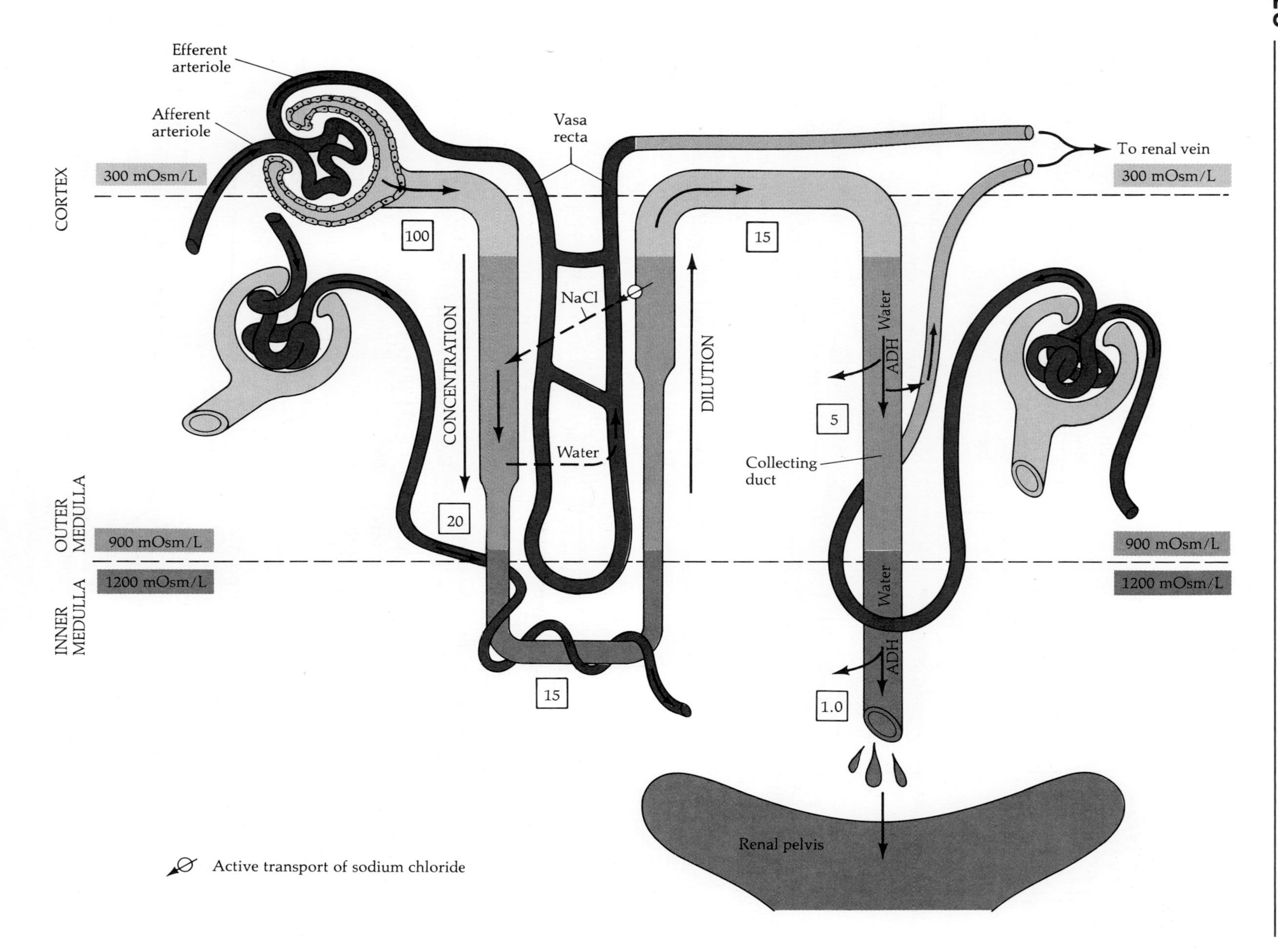

Efferent arteriole
Afferent arteriole
Vasa recta
To renal vein
CORTEX
300 mOsm/L
300 mOsm/L
100
15
NaCl
CONCENTRATION
DILUTION
Water
ADH
5
Water
Collecting duct
20
OUTER MEDULLA
900 mOsm/L
900 mOsm/L
INNER MEDULLA
1200 mOsm/L
1200 mOsm/L
Water
ADH
15
1.0
Renal pelvis
Active transport of sodium chloride

FIG. 21-12
Outline of changes in the osmotic pressure of renal fluids. The *osmolarity* of the filtrate is indicated in milliosmoles/per liter and the *percent* of filtrate remaining in different parts of the tubules is shown in the squares. The effects of sodium chloride transport, the diffusion of water, and the influence of ADH on the permeability of the collecting ducts to water are illustrated. The intensity of color in the blood vessels reflects in a general way the level of oxygen in the blood. The shade of color in the tubules represents in a general way the concentration of the filtrate.

the formation of urine by the kidneys. The respiratory tract can affect acid-base conditions on an acute, minute-to-minute basis. An increased respiratory rate may compensate for acidosis by causing excretion of more CO_2 (Fig. 20-12B). This effect reduces the levels of H^+ ions by shifting the equilibrium of the following reactions to the right:

$$\underset{\text{Hydrogen ions}}{H^+} + \underset{\text{Bicarbonate ions}}{HCO_3^-} \rightleftharpoons \underset{\text{Carbonic acid}}{H_2CO_3} \rightleftharpoons \underset{\text{Water}}{H_2O} + \underset{\text{Carbon dioxide}}{CO_2\uparrow}$$

A decrease in respiratory rate achieves the opposite effect.

The kidneys are not as rapid in response as the lungs but are the major organs of control of acid-base balance. They regulate the pH of body fluids by (1) excreting excess acid or base and (2) influencing the levels of buffers. In these processes the kidneys significantly affect acid-base equilibrium by exchanging hydrogen ions for alkali ions such as sodium and potassium (Fig. 21-9). The exchange is made within three major buffering systems, inorganic bicarbonates and phosphates (Table 21-5) and proteins, which can donate or accept hydrogen ions (as can hemoglobin, as described in Chap. 20 and Fig. 20-12B, page 487). Buffering occurs in both ICF and ECF compartments (Chap. 4). Consequently, a shift in pH of one compartment and/or buffering system eventually causes a shift in all of them.

In the reactions shown in Table 21-5 each inorganic buffer on the left acts as a hydrogen ion donor and behaves as a buffer acid while that on the right behaves as a hydrogen acceptor or buffer base. Each member of a buffer pair is normally in equilibrium. Since the buffer pairs donate or accept hydrogen ions, all the buffer systems are ultimately in equilibrium with one another.

The bicarbonate-carbonic acid buffering system is the most important in terms of buffering capacity and can be adjusted by activities of the kidneys and the lungs. The kidneys may excrete excess bicarbonate ions or hydrogen ions; the lungs can alter the level of carbonic acid by excreting carbon dioxide. In body fluids the normal ratio of bicarbonate base to carbonic acid is 20:1. As the ratio increases, the pH rises and body fluids become more basic (alkaline); as the ratio decreases, the pH drops and the fluids become more acidic. Usually excess bicarbonate ions are available which can combine with hydrogen ions after their release from strong acids (such as hydrochloric acid) to form carbonic acid, which is a weak acid.

The carbonic acid is converted to water and carbon dioxide, whose excretion by respiration

TABLE 21-4
Normal water intake and loss in adults*

Intake	*mL*	*Output*	*mL*
Drinking (water and beverages)	1200 (500–1700)	Urine	1400 (600–1600)
Water in food (ranges from 60 to 97% water)	900 (800–1000)	Evaporation (skin and lungs)	900 (850–1200)
Water derived from oxidation of foods	300 (200–400)	Digestive tract (feces)	100 (50–200)
Totals	2400 (1500–3000)		2400 (1500–3000)

Average volumes are illustrated for which the normal ranges are listed in parentheses.
* From *Water and Electrolyte Metabolism and Acid-Base Balance,* E. Muntwyler. Copyright © 1968. C. V. Mosby, St. Louis.

TABLE 21-5
Major inorganic buffering systems in the body

System	*Mechanism*
Bicarbonate buffers:*	H_2CO_3 (Carbonic acid) $\rightleftharpoons$ HCO_3^- (Bicarbonate ion) + H^+ (Hydrogen ion)
Phosphate buffers:	
Sodium phosphate buffers	NaH_2PO_4 (Monosodium phosphate) + Na^+ (Sodium ion) $\rightleftharpoons$ Na_2HPO_4 (Disodium phosphate) + H^+ (Hydrogen ion)
Potassium phosphate buffers	KH_2PO_4 (Monopotassium phosphate) + K^+ (Potassium ion) $\rightleftharpoons$ K_2HPO_4 (Dipotassium phosphate) + H^+ (Hydrogen ion)

* The kidneys can exchange Na^+ ions for H^+ ions to form sodium bicarbonate ($NaHCO_3$), a buffer base.

(Fig. 20-12) effectively decreases the availability of hydrogen ions and thereby increases the pH of body fluids. An increase in hydrogen ions in the blood (a lower pH) promotes such reactions as it stimulates an increased rate and depth of respiration.

The exchange of hydrogen ions for alkaline ions such as sodium is a major means by which the kidneys buffer the blood. Hydrogen ions enter the filtrate while sodium ions are reabsorbed into the blood (Fig. 21-9). The sodium ions then combine with bicarbonate ions to form sodium bicarbonate (Fig. 21-13A). The alkalinity of the blood is maintained while urine becomes more acidic as hydrogen ions continue to be secreted into the filtrate in the distal tubules and collecting ducts. Because the concentration of bicarbonate is much greater than that of phosphate, most of the exchange of sodium for hydrogen ions involves the bicarbonate buffering system (Fig. 21-13A).

The activity of the phosphate buffering system (Fig. 21-13B) increases when bicarbonate availability is low or when the rate of hydrogen ion formation is excessive. The phosphate buffering system is active in the distal and collecting tubules in which inorganic phosphates have accumulated. Hydrogen ions absorbed by phosphate buffers are the source of measurable acid in the urine.

The hydrogen ions also combine with ammonia (NH_3), derived from the amino acid building blocks of proteins, to form ammonium (NH_4^+) ions, which with chloride ions form ammonium chloride (NH_4Cl, Fig. 21-13C). This reaction also is prevalent in the distal and collecting tubules.

Measurement of Kidney Function

Four common means of measuring kidney function are estimation of: the renal plasma flow; the GFR, already discussed in connection with the physiology of the nephrons; the filtration fraction; and the clearance rate (Table 21-6). The GFR is related to the rate of flow of plasma through the kidneys, termed the *renal plasma flow* (RPF). The RPF is about 55 percent of the total blood volume delivered to the kidneys. Each minute about 1 mL of the 125 mL of the GFR is excreted as urine (Table 21-6A).

The *filtration fraction* (FF) measures the percentage of plasma filtered through the glomeruli and normally ranges from 16 to 20 per-

FIG. 21-13
The tubular secretion of acid and base. The exchange of hydrogen ions for sodium ions to maintain the alkalinity of the blood is illustrated in three different buffering systems. The net movement of substances is indicated by the direction of the arrows. Substances which appear in the tubular filtrate and are not reabsorbed are excreted in the urine. (A) The carbonic acid–bicarbonate buffering system. (B) The sodium (or potassium) monophosphate-diphosphate buffering system and hydrogen secretion. (C) The protein buffering system and secretion of ammonia.

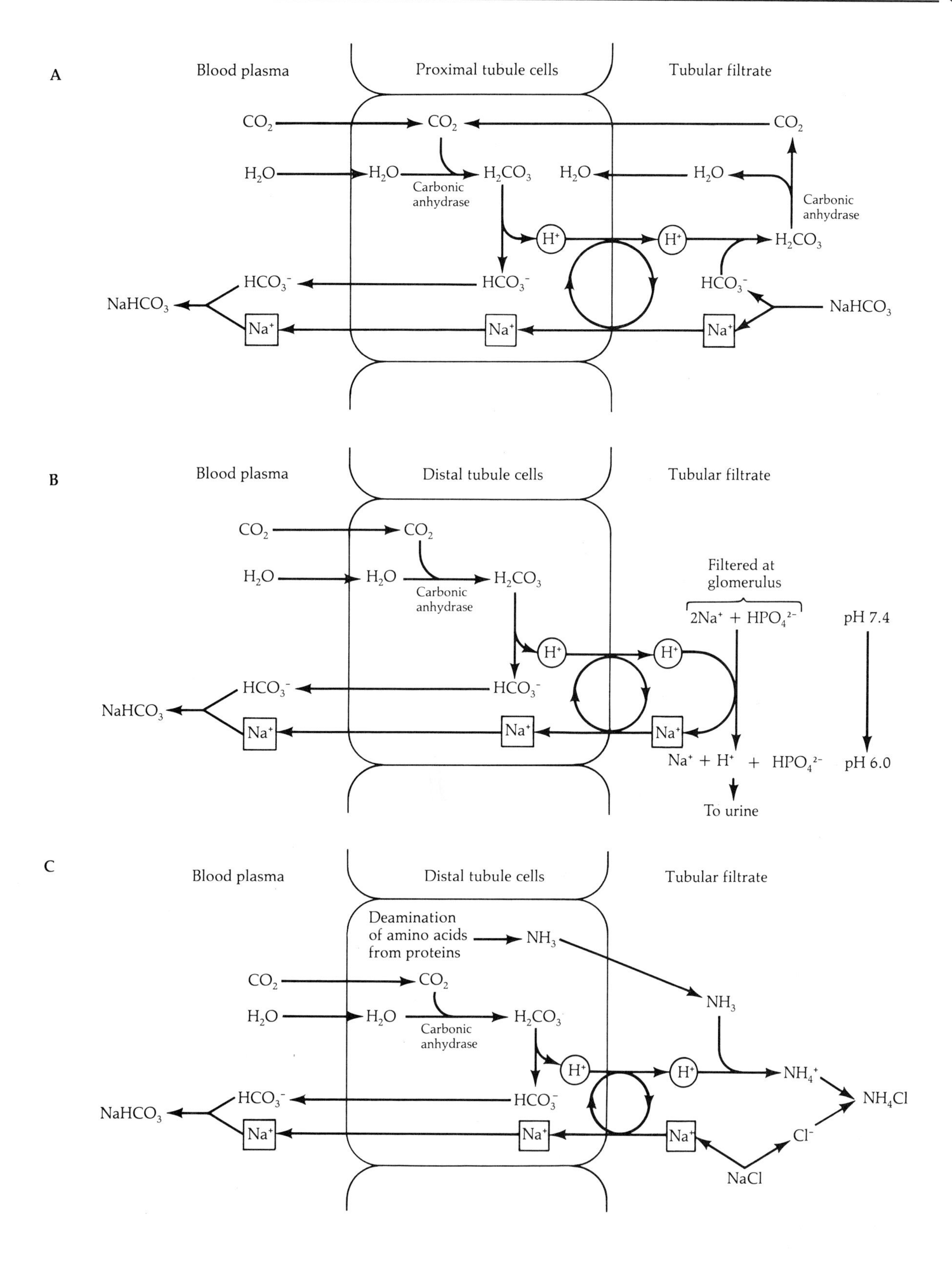
A
Blood plasma
Proximal tubule cells
Tubular filtrate
CO_2
CO_2
CO_2
H_2O
H_2O
H_2CO_3
H_2O
H_2O
Carbonic anhydrase
Carbonic anhydrase
H^+
H^+
H_2CO_3
HCO_3^-
HCO_3^-
HCO_3^-
$NaHCO_3$
$NaHCO_3$
Na^+
Na^+
Na^+
B
Blood plasma
Distal tubule cells
Tubular filtrate
CO_2
CO_2
H_2O
H_2O
H_2CO_3
Carbonic anhydrase
Filtered at glomerulus
$2Na^+ + HPO_4^{2-}$
pH 7.4
H^+
H^+
HCO_3^-
HCO_3^-
$NaHCO_3$
Na^+
Na^+
Na^+
$Na^+ + H^+ + HPO_4^{2-}$
pH 6.0
To urine
C
Blood plasma
Distal tubule cells
Tubular filtrate
Deamination of amino acids from proteins
NH_3
NH_3
CO_2
CO_2
H_2O
H_2O
H_2CO_3
Carbonic anhydrase
H^+
H^+
NH_4^+
NH_4Cl
HCO_3^-
HCO_3^-
$NaHCO_3$
Na^+
Na^+
Na^+
Cl^-
NaCl

TABLE 21-6
Measures of kidney function

A. Renal plasma flow (RPF) and glomerular filtration rate (GFR)
Assuming a cardiac output of 5 L/min, delivery of 20% of cardiac output to the kidneys and filtration of 20% of the plasma that flows to the glomeruli, the RPF and GFR can be calculated as shown:

(1) Daily cardiac output = cardiac output (L/min) × 60 min/h × 24 h/day
5 × 60 × 24 = 7200 L/day

(2) Renal blood flow (RBF), daily blood flow to the kidneys = % cardiac output to kidney × daily cardiac output
20% × 7200 L/day = 1640 L/day

(3) Renal plasma flow (RPF), liters of plasma to the kidneys daily = % plasma in blood × renal blood flow
55% × 1640 L/day = 902 L/day

(4) Glomerular filtration rate (GFR), liters of plasma filtered daily = % plasma filtering through glomeruli × liters of plasma to the kidneys daily
~20% × 902 L/day = 180 L/day
÷ 1440 min/day
= 0.125 L/min
= 125 mL/min

B. Filtration fraction (FF)

FF = GFR/RPF/
Filtration fraction = glomerular filtration rate/renal plasma flow
= 125 mL/min/902L/day
= 125 mL/min/626 mL/min = 0.1997
= 19.97%

C. Clearance

$$C = \frac{U}{P} \times V$$

where C = clearance of plasma, mL/min
U = concentration of the substance in the urine, mg/mL
P = concentration of the substance in the plasma, mg/mL
V = volume of urine formed, mL/min

cent. It can be calculated by dividing the GFR by the RPF (Table 21-6B).

The volume of plasma emptied of a substance each minute is the *clearance rate* of that substance. If a substance is freely filterable through the glomerulus, not reabsorbed, and not secreted by the renal tubules, then its clearance rate is equal to the GFR and is considered a good representation of glomerular function. Under these conditions, clearance is a measure of the average volume of plasma filtered through the glomeruli and is independent of the concentration of that substance. When the concentration of such a substance in the blood is high, more of it normally enters the filtrate. When the concentration is low, less enters the filtrate. However, under either condition, the volume of plasma cleared remains unchanged. Clearance is generally proportional to the surface area of the body, is lower in females than in males, and declines with aging. It also may be altered by changes in the diet. Values below normal for a person's body size, sex, and age indicate a decrease in the GFR and suggest the possibility of a kidney disorder. Conversely, elevated levels of certain substances in the serum also may be indicative of kidney dysfunction.

To determine clearance, the rate of disappearance of *creatinine* from the plasma is often measured. Creatinine is a product of metabolism of creatine phosphate in muscle (Fig. 13-12), and the amount excreted is related to the muscle mass of an individual. Although creatinine is secreted into the renal tubules, some is reabsorbed so that the amount excreted by an individual each day is rather constant and is closely related to the GFR.

Inulin is a polysaccharide made of fructose which most closely meets the three criteria necessary for measurement of the GFR. Its low molecular weight of 5000 and small diameter of 3 nm are well below the respective cutoff points of 40,000 MW and 6 nm pore size (Table 21-2), which allow its easy passage through glomeruli. However, inulin not only is a synthetic compound, it is also digestible. Therefore, in order to use it to measure clearance and GFR, inulin must be administered intravenously prior to and during the time of urine collection, a requirement which detracts from its use.

Clearance values above the glomerular filtration rate (about 125 mL/min) indicate secretion of a substance by the renal tubules into the filtrate. Conversely, clearance values below the GFR indicate reabsorption of a substance from the renal tubules into the blood. Representative clearance values are given in Table 21-7.

TABLE 21-7
Renal clearances of some substances in young adults*

Substance	*Clearance*
Glucose	0
Urea	75 (with urine flow of 2 mL/min)
Uric acid (metabolite of purine bases)	14
Inulin	125
Creatinine	104–125
p-aminohippuric acid	630

* These values represent figures typical of a young adult. Values above 125 indicate tubular secretion while those below that level occur with tubular reabsorption of the substance.

Effects of Disease on Kidney Function

The regulation of salt and water balance by the kidneys may be influenced by a number of diseases. Cirrhosis of the liver (Chap. 22) may result in elevated levels of aldosterone because more hormone remains undegraded as it passes through the diseased liver. The adrenal glands continue to secrete aldosterone and its levels increase above normal. Congestive heart failure also allows aldosterone levels to rise because less blood flows through the liver and less aldosterone is degraded.

Liver and kidney function also affect fluid balance in other ways. For example, when the liver is diseased, albumin synthesis is reduced, resulting in *hypoproteinemia* (low blood protein). Because albumin is the major protein in the plasma, hypoproteinemia is primarily due to *hypoalbuminemia* (low blood albumin). A low plasma osmotic pressure (a dilute plasma) results, and as blood flows in capillaries by the tissues, more water than usual remains in or enters tissue spaces, causing edema. If water accumulates in the abdominal cavity, the condition is called *ascites*. Hypoalbuminemia can also be due to excess excretion of albumin in the urine (*albuminuria*). It is the most common form of *proteinuria*, appearance of protein in the urine. The disorder may be caused by damage to the basement membranes of the kidneys and their inflammation in *glomerulonephritis*. The collective symptoms of edema, hypoproteinemia, proteinuria, and albuminuria, which may accompany structural changes in the kidney, are called the *nephrotic syndrome*.

A number of other diseases also affect kidney function. Some are heritable while others are hormonal (Chap. 5) or may be induced by physical trauma, drugs, infections, or immune reactions (Table 21-8). In various diseases urine formation may be increased (*polyuria*), diminished (*oliguria*), or absent (*anuria*). Elevated amounts of urea appear in the blood in *uremia*, an indication of severe kidney disease in which nitrogenous wastes also accumulate in the kidney. General symptoms of uremia are headache, nausea, and vomiting. Chronic (long-

TABLE 21-8
Some diseases which affect kidney function

Disease	*Mechanism*	*Effect(s)*
Hormonal:		
Diabetes mellitus	Insulin deficiency	High blood and urine glucose Diuresis Loss of kidney function occurs with chronic diabetes owing to adverse effects on blood vessels
Diabetes insipidus	Hypothalamic disorder; ADH secretion diminished; may be a dominant hereditary disorder, or induced by trauma	Marked diuresis of up to 20 L/day of urine
Conn's syndrome	Hyperactive adrenal cortex causing hypersecretion of aldosterone	Sodium retention Excess potassium excretion Low plasma potassium Increased ECF volume Elevated blood pressure
Infection and/or inflammation:	Bacterial or viral infections or immune mechanisms may damage kidney tubules and membranes	Protein filtration, proteinuria, especially albuminuria, may occur Protein may precipitate in small masses called *casts*
Pyelonephritis	Obstruction of urinary flow (for example following catheterization) or during pregnancy or with neurological disease may allow bacteria to ascend through the urethra and infect and inflame the kidney. *E. coli*, streptococci, or staphylococci are commonly involved	May cause destruction of tubules and glomeruli and lead to hypertension
Glomerulonephritis	Infection leading to autoimmune attack of glomeruli (streptococcal bacteria are often involved)	Decreased glomerular filtration rate Hypertension Edema
Metabolic:		
Kidney stones (urinary stones, urolithiasis)	Often formed by crystallization of calcium salts which narrow or block tubules and cause elevated hydrostatic pressure in them	Lowers glomerular filtration rate Predisposes kidney to infection and further derangements of kidney function
Gout	Produced when cations cause crystallization of uric acid (from nucleoprotein catabolism)	Accumulation of crystals in joints causes pain Stones and crystals may form in kidneys and thereby interfere with renal function
Cystinuria	Hereditary disease. Impairs reabsorption of amino acids cystine, arginine, lysine, and ornithine	Damages kidney by the formation of cystine stones

term) uremia is associated with progressive reduction in renal function. In addition to the symptoms already described, acidosis, electrolyte imbalance, dehydration, polyuria, diarrhea, disorientation, coma, and death may result.

The buffering capacity of the body is exceeded in respiratory and metabolic acidosis and alkalosis. In *acidosis,* the pH of the blood is lowered (made more acidic). In *alkalosis* the pH is raised (made more alkaline). If carbonic acid is in excess owing to a decrease in alveolar ventilation, the resulting condition is *respiratory acidosis* (Table 21-9). This type of acidosis is often induced by use of barbiturates or narcotics such as morphine. These drugs may depress respiratory rates and cause carbon dioxide to accumulate in *hypercapnia.* Conversely, hyperventilation, which causes excessive loss of carbon dioxide and decreases carbonic acid levels,

TABLE 21-9
Some causes of acidosis and alkalosis

Disorder	*Cause*	*Comments*
Respiratory acidosis		Occurs with accumulation of carbon dioxide and carbonic acid in plasma due to hypoventilation
	Barbiturates, morphine, and other narcotics	Inhibition of respiratory center in medulla
	Emphysema may cause hypoventilation. Pneumonia is a "common" cause of hypoventilation	Disease in alveoli or airways to the lungs may produce conditions leading to acidosis
Metabolic acidosis		Occurs with loss of bicarbonate ions from, or accumulation of hydrogen ions in ECF
	Diabetes mellitus	The inability to use glucose promotes the metabolism of fats to ketones, such as acetoacetic acid. The products of metabolism behave as negatively charged ions (anions) similar to bicarbonate. As a result, excess bicarbonate may be excreted and the pH of body fluids may decrease
	Ingestion of ammonium salts (such as ammonium chloride)	The ammonium ion (NH_4^+) is acid and can donate hydrogen ions to increase the acid load on buffers
	Diarrhea—often caused by viral or bacterial infection of the digestive tract	Excessive loss of bile, as well as alkaline potassium, sodium, and bicarbonate ions from the pancreas and intestine, are associated with excretion of watery feces and dehydration
Respiratory alkalosis		Hyperventilation causes excess loss of carbon dioxide and decreases carbonic acid levels
	High altitude exposure	Prolonged exposure to low partial pressure of oxygen for 3–4 weeks at 13,000 ft (4000 m) may stimulate hyperventilation
	Emotional disorders or aspirin (salicylate) poisoning	Stimulates hyperventilation, causing loss of carbon dioxide and decrease in carbonic acid levels
Metabolic alkalosis		Loss of hydrogen ions or accumulation of bicarbonate ions are most common causes
	Ingestion or administration of excess alkali (such as sodium bicarbonate in antacids)	May cause an excess of bicarbonate ions in body fluids
	Prolonged vomiting of gastric contents	The loss of chloride from the stomach decreases the quantity of acid leaving behind a relative excess of alkaline bicarbonate ions
	Some diuretics, such as thiazides, furosemide, and ethacrynic acid	The stimulation of chloride and sodium ion excretion and inhibition of their reabsorption in the ascending loop leaves a relative excess of alkaline bicarbonate in the blood

may lead to *respiratory alkalosis.* Variations in acid or base levels by causes other than respiratory alterations are called *metabolic acidosis* or *metabolic alkalosis* (Table 21-9).

When the body corrects pH imbalances, the condition is said to be *compensated* (corrected). If imbalances remain *uncompensated,* kidney failure and fatal damage to vital organs such as the brain and heart may occur.

Kidney Transplants and the Artificial Kidney

When one kidney is damaged or diseased, the number of nephrons and mass of the surviving healthy kidney increases in a response called *compensatory hypertrophy.* The larger number of nephrons in the healthy kidney assume the functions of both organs. When both organs are

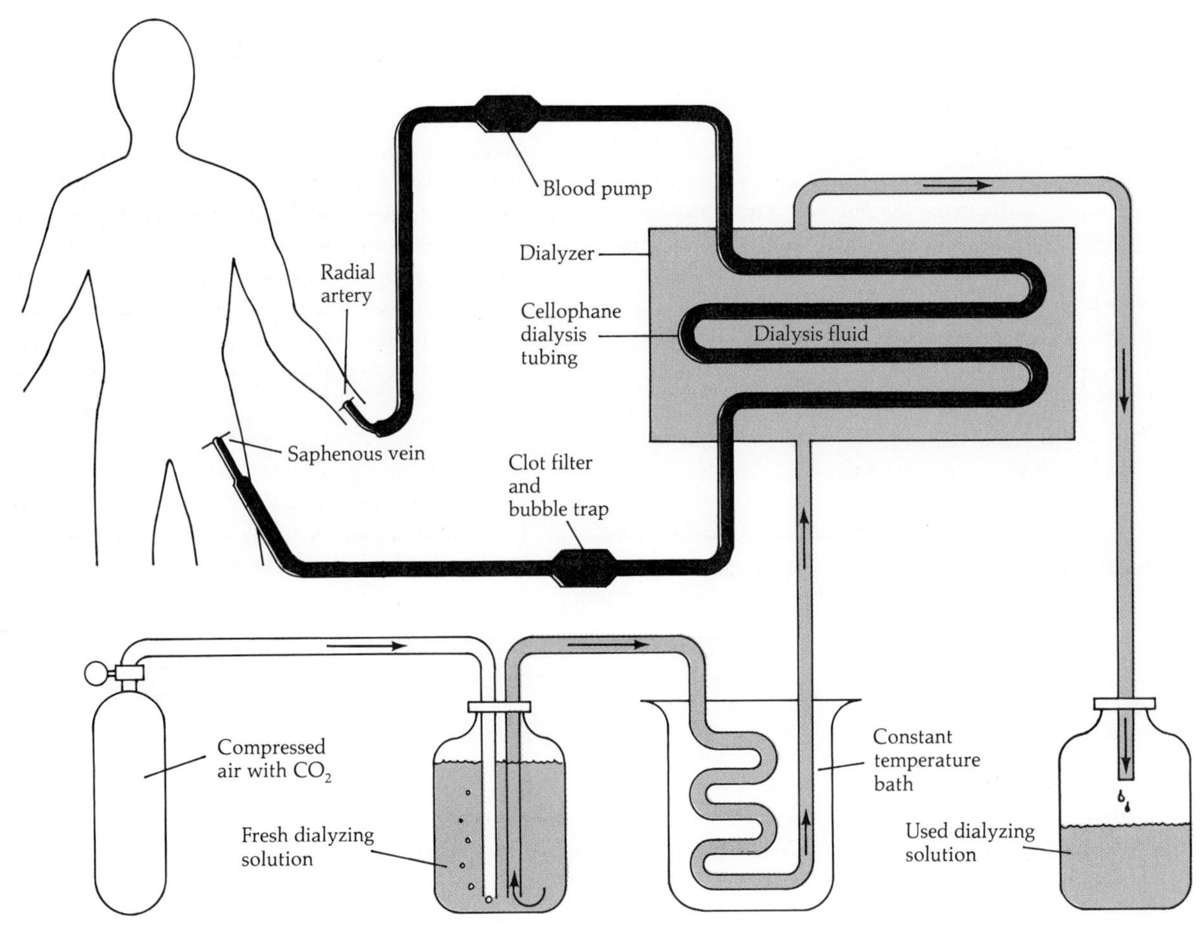

FIG. 21-14
Diagram of major components of an artificial kidney. The arrows indicate the direction of flow of blood in the darkly colored tubes and that of isotonic salt solution (dialysis fluid) in the lightly colored areas of the diagram.

diseased and life is threatened, other means must be found to maintain homeostasis.

A *kidney transplant,* which involves transfer of a healthy kidney from a donor to a recipient, may be considered. The survival of the transplant and the patient requires that the major histocompatibility antigens of donor and recipient be matched closely (pages 86 and 460). Survival for 2 or more years after the transplant is currently about 80 percent when the donor and recipient are unrelated. Immunosuppressive drugs are administered to prevent immune rejection of the transplanted organ. However, such immunosuppression makes the patient more susceptible to infectious diseases.

An *artificial kidney* may be used to filter the blood of an individual awaiting a transplant or to substitute temporarily for malfunctioning kidneys to clear wastes from the blood. The artificial kidney works on the principles governing diffusion, and permits substances to move down gradients through a selectively permeable dialysis membrane (in the same general way that nephrons function). Since blood is filtered by the membrane, the process is referred to as *hemodialysis* (blood dialysis). The membrane is made of a cellophane tubing whose pores retain large molecules such as proteins and allow the diffusion of small molecules and ions down their concentration and electrical gradients. The tubing is immersed in an isotonic salt solution maintained at body temperature. The salt solution resembles plasma but lacks metabolic wastes.

Blood which has been heparinized to inhibit clotting flows from an artery of the patient into the dialysis tubing (Fig. 21-14). Small molecules and ions in excess in the blood diffuse into the bath. Those in excess in the salt solution of the bath flow into the tubing and enter the blood. The blood is pumped through a filter and bubble trap to remove any air bubbles or clots which might form. After an antiheparin agent is added to the blood to neutralize the effects of the heparin, then the blood is returned to the patient through a large superficial vein. The procedure requires 4 to 5 h and must be repeated 2 or more times a week. The duration and frequency of dialysis depend on the severity of kidney disease and availability of the equipment. The use of hemodialysis has become much more widespread with increased experience, and the treatment now may be administered to the patient at home.

Dialysis also may be accomplished by utilizing the peritoneal membrane in the abdominal cavity of the patient. *Peritoneal dialysis* involves introduction of approximately 2 L of dialyzing fluid by gravity flow from a bag through a tube inserted into an artificial opening in the abdominal cavity. After the fluid flows into the abdominal cavity, toxic materials and wastes diffuse from blood across the peritoneal membrane into the fluid. The bag is held in place around the waist for 4 to 5 h in order to retain the fluid in the abdomen. When the bag is lowered, the fluid, along with excretions which have diffused into the abdominal cavity, flows from the abdominal cavity back into the bag. The process is repeated 4 times a day at 4 to 8 h intervals.

21-4 SUMMARY

1 The organs of the excretory system, which include the kidneys, skin, intestine, and lungs, eliminate substances from the body. The kidneys deliver water-soluble wastes in urine. The skin forms sweat. The large intestine excretes feces and the lungs exhale water vapor and carbon dioxide.

2 The skin is a protective barrier with sensory nerve endings. It synthesizes a form of vitamin D and helps regulate body temperature, fluid levels, and salt levels. The outer skin, or epidermis, is composed of as many as five distinct layers, which, from outside to inside, include the stratum corneum, stratum lucidum, stratum granulosum, stratum spinosum, and stratum germinativum. The more superficial layers originate by the proliferation and upward migration of cells from the stratum germinativum. As the cells migrate toward the epidermal surface, they become impregnated with keratin and are ultimately shed, to be replaced by underlying cells.

Melanin, a brown to black pigment, is primarily responsible for skin color. It is produced in cells called melanocytes, which are located in the stratum germinativum. Melanin is formed by the activity of the enzyme tyrosinase in cytoplasmic granules called melanosomes. Melanin is transported through extensions of the cytoplasmic membrane of melanocytes that cross the stratum germinativum.

Nearly 2.5 million sweat glands release about 1 L of sweat each day through pores on the surface of the epidermis. Sweat is 99 percent water and about 1 percent Na^+, K^+, and Cl^- ions, lactic acid, and nitrogenous wastes. The excretion of sweat increases with exercise and exposure to heat. Aldosterone enhances reabsorption of salts and water back into the blood from epithelial cells lining the ducts of the sweat glands.

Dense connective tissue forms the deepest layers of the skin and produces ridges (papillae), which are responsible for fingerprint patterns. Capillaries, naked nerve endings, Meissner's corpuscles, and Pacinian corpuscles are located in the dermis. Sebaceous glands, sweat glands, and hair follicles reside at least partially in this area. Inflammation of the dermis is called dermatitis and may be due to a noncontagious reaction which causes redness, itching, loss of serous fluid, and the formation of a scaly, crusty skin in a disorder known as eczema.

3 Each of the two kidneys has 6 to 18 renal pyramids, which project from the outer cortex to the central medulla. The kidneys are located retroperitoneally in the abdominal cavity and supplied by the renal artery with blood, which courses through a complex system of vessels and capillary beds prior to exiting in the renal veins. Urine formed in tubules of the kidney is collected in ducts, delivered to the central pelvis of each kidney, and then passed to the ureter and to the bladder. The excretion of urine from the bladder to the exterior through the urethra is reflexively controlled.

The kidneys regulate plasma osmotic pressure; balance electrolytes, acids, and bases; synthesize hormones; and concentrate and excrete wastes. They filter, reabsorb, secrete, and excrete a wide variety of substances. Up to 1.5 million microscopic tubular units, the nephrons, perform excretory functions in each kidney. Each nephron contains two capillary beds in series. The first, the glomerulus, is wedged into a C-shaped, double-walled cup called Bowman's capsule. Bowman's capsule leads to a tubule, the first and last segments of which are twisted and interconnected by a hairpin loop. The final twisted loops of each nephron lead to common collecting ducts, which deliver urine droplets to calyces in the renal pelvis. The second portion of the capillary bed surrounding the tubules and ducts are called peritubular capillaries.

Blood pressure forces small molecules and ions out of glomerular capillaries through their intercellular openings. In order to enter the tubules of the nephron, the substances must filter through intermediary basement membranes, which act as ultrafine filters. Their narrow intercellular slits measure approximately 6 nm and generally exclude passage of substances with molecular weight about 66,500 or higher.

The kidneys receive about 20 percent of cardiac output, which is delivered to them by the renal arteries. The blood is returned to the general circulation through the renal veins. The blood pressure generated in each glomerulus is related to the greater diameter of its incoming arteriole compared with that of its outgoing arteriole. The outgoing arterioles merge and form the peritubular capillaries, which often supply more than one nephron and overlap the nephrons which they serve.

The glomerular filtration rate (GFR) is the amount of plasma filtered through all the glomeruli each minute. It averages 115 to 125 mL/min and totals 165 to 180 L/day, which equals 60 times the average plasma volume. Alterations in blood pressure and resistance to flow markedly alter these values.

When the concentration of a substance in the filtrate exceeds the maximal capacity for its tubular reabsorption (T_m), the excess substance appears in the urine. The loss of water is increased by diuretics, which increase urine volume. Some diuretics used to treat edema decrease sodium and chloride ion reabsorption to promote the osmotic loss of water and diuresis. The diuretic activity of caffeine and alcohol are due to decreased sodium reabsorption and decreased ADH secretion, respectively.

Water retention is increased by ADH, aldosterone, glucocorticoids, and the renin-angiotensin system. ADH promotes the return of water to the blood from the collecting ducts. Stimuli which elevate aldosterone and glucocorticoid secretion enhance sodium reabsorption and the osmotic retrieval of water.

Renin, a glycoprotein with enzymatic properties, is secreted by the kidneys in response to decreased blood pressure and sodium. It converts globular proteins in the plasma into vasoactive molecules called angiotensin I, which in turn are transformed into angiotensin II by converting enzymes in the lungs and blood. Angiotensin II is a powerful vasoconstrictor which stimulates the secretion of aldosterone and raises fluid retention, blood volume, and blood pressure. These responses to the renin-angiotensin system are opposed in a feedback cycle by the kallikrein-kinin system of bloodborne polypeptides which alter vascular function to help maintain homeostasis.

The tubules and peritubular capillaries form countercurrent systems. Although slight osmotic differences are maintained in fluids between their adjacent segments, major differences develop elsewhere. Steep concentration gradients exist from the outer cortical to the central medullary segments of the tubules and capillaries. The capacity to maintain these gradients depends on diffusion (passive countercurrent exchange) and active transport (countercurrent multiplication) of substances between the tubules and capillaries. The maintenance of the gradient is assisted by differing permeabilities of the segments of the loops of Henle. The descending limbs are relatively water-permeable. The thick ascending limbs are relatively impermeable to water but actively transport chloride and sodium ions.

The kidneys influence acid-base balance by the selective tubular secretion of hydrogen ions and the secretion and reabsorption of the components of inorganic and protein buffer systems. A 20:1 bicarbonate/carbonic acid ratio in that buffering system is maintained by regulation of bicarbonate and hydrogen ion excretion by the kidneys and CO_2 excretion by the lungs. The buffering system is replenished by the activity of the enzyme carbonic anhydrase, which forms carbonic acid from available carbon dioxide and water. Failure to maintain the blood pH between 7.35 and 7.45 may result in acidosis or alkalosis, with lethal effects on the kidney, brain, and heart. Imbalances in pH can be respiratory-induced or metabolically induced.

Kidney function can be assessed by measurement of the renal plasma flow (RPF), glomerular filtration rate (GFR), filtration fraction (FF), and clearance of substances from the plasma. About 16 to 20 percent of the plasma in blood is filtered as it passes through a glomerulus. This volume, the FF, is a function of the ratio of the glomerular filtration rate to the RPF. The RPF equals about 55 percent of the total renal blood flow. Clearance tests measure the volume of plasma emptied of a substance each minute. The clearance rate of freely filtered substances that are not reabsorbed or secreted is a measure of the GFR.

When kidneys are diseased, the clearance rates may be too low to remove potentially toxic accumulations of wastes. In such cases, the organs may be assisted by an artificial kidney in which the patient's blood is passed through selectively permeable cellophane tubing bathed in isotonic salt solution. Wastes diffuse out of the blood prior to its reentry into the body. Peritoneal dialysis may be effectively employed in selected cases. Diseased kidneys also may be replaced by transplantation of a healthy kidney from donor to recipient. The major tissue antigens of recipient and donor must match if the patient is to survive. Drugs which suppress the immune response are administered to the recipient of the transplant in order to reduce the chances of immunological rejection and destruction of the donated organ.

REVIEW QUESTIONS

1 List five functions of the skin.
2 Sketch and label the layers in the epidermis and the structures in the dermis. What is the role of the stratum germinativum? The melanocytes? The papillae?
3 Describe the formation of melanin and its relation to skin structure and function.
4 Discuss the structure and function of hair follicles, sebaceous glands, and ceruminous glands.
5 What is dermatitis? Eczema? Cystic fibrosis?
6 Describe the physiology of sweat glands and sweating. What control mechanisms in sweat glands influence their excretory functions and body fluid composition?
7 Describe the anatomy of the kidneys and the route by which urine exits from the body.
8 Describe the major characteristics of urine, including its pH, volume formed, and major constituents.
9 List five major functions of the kidneys.
10 Draw and label a nephron, including its glomerulus and the origin of its peritubular capillaries, and describe the filtration functions of the renal corpuscle.
11 What is the route of blood flow through the kidney? Include a detailed description of the blood vessels associated with the nephron.
12 What is the GFR and how is it related to filtration? How does the anatomy of the glomerular arterioles influence the GFR? What is its relationship to blood pressure? What control mechanisms regulate the GFR?
13 What are reabsorption, transepithelial transport, and T_m? How are they related? How are they influenced by diabetes mellitus? By diuretics? By ADH?
14 What is tubular secretion? Describe this process for K^+, H^+, and creatine in the kidney's tubules. Compare the secretion of Na^+ and K^+.
15 How do ADH and aldosterone influence kidney function?
16 What is renin, and how does the renin-angiotensin system influence kidney function? How do its effects relate to the kallikrein-kinin system?
17 What is a countercurrent system? What is the difference between a countercurrent exchange and a countercurrent multiplication effect? Give examples of each in the kidney. What is the function of countercurrent systems in the kidney? How do they relate to changes in osmolarity in the tubular filtrate? How do the descending and ascending loops behave in this system? What is the role of the peritubular capillaries?
18 List the three major inorganic buffer systems in body fluids and explain how the kidney influences them. Give an example of a protein buffering system and explain how it works.
19 Distinguish metabolically induced and respiratory-induced acidosis and alkalosis. What effects do they have on body function and why? Name some causes of each.
20 Matthew Johnson's renal blood flow is 1200 mL/min. What is his renal plasma flow? If his glomerular filtration rate is 125 mL/min, what is his glomerular filtration fraction?
21 Joseph June's cardiac output is 4 L. Assuming that his renal blood flow represents 20 percent of the cardiac output, what is his GFR?

22 Describe clearance and tell how it aids in diagnosis of the extent of kidney disease.

23 What is a kidney transplant? What are some of the problems associated with it?

24 What is an artificial kidney? By what principles does it work? When is it used and what are its limitations? Compare it with peritoneal dialysis.

THE DIGESTIVE SYSTEM

Regulatory Mechanisms and Defenses

22

22-1 STRUCTURE
General gross anatomy and physiology
Microanatomy
Accessory digestive organs
Salivary glands
Exocrine portions of the pancreas
Liver
Gallbladder

22-2 SECRETIONS
Gastrointestinal hormones
Exocrine secretions
Saliva
Secretions of the stomach
Exocrine secretions of the pancreas
Exocrine secretions of the intestine
Secretion of bile

22-3 REGULATORY MECHANISMS
Control of secretions
Hormonal factors
Neural influences
Control of motility
Hormonal factors
Neural influences

22-4 DEFENSES AND INFLAMMATORY RESPONSES
Inflammatory reactions
Ulcers
Liver disease

22-5 SUMMARY

OBJECTIVES

After completing this chapter the student should be able to:

1. Describe the general structure of the digestive system.
2. Identify the four major layers of the wall of the digestive tract, listing their components and functions.
3. Name the accessory digestive organs and describe their general structure and functions.
4. Identify four major gastrointestinal hormones and describe their cellular origins and principal effects.
5. Discuss the properties and functions of exocrine secretions in the digestive tract.
6. Explain the regulatory influences of the nervous system and hormones on the motility and secretions of the digestive tract.
7. Describe in general the defenses of the digestive system and their relationships to inflammatory stimuli.

The digestive system forms an elaborate production line in which incoming materials are converted into building blocks that can enter the rest of the body through the blood or lymph. The substances which enter the circulation are carried to cells throughout the body to be used to produce energy or to synthesize a variety of molecules essential to the repair, replacement, or growth of cells, tissues, and organs. These molecules range from inorganic compounds such as sodium chloride and small organic compounds such as glucose and amino acids to large proteins such as collagen, myosin, hemoglobin, and nucleic acids, including deoxyribonucleic acid, (DNA, Chap. 2).

The digestive system consists of the mouth, parts of the pharynx, esophagus, stomach, small and large intestines, and accessory digestive organs, including the teeth, tongue, salivary glands, exocrine portions of the pancreas, liver, and gallbladder (Fig. 22-1). These organs digest and absorb foods and then excrete unused residues as wastes called *feces*. *Digestion* is the physical and chemical conversion of food molecules into smaller units which can be absorbed. *Absorption* is the passage of substances across selective epithelial barriers into the blood and lymph for transport to cells throughout the body. The chemical changes that nutrients further undergo in all types of cells are called *metabolism* (see Chaps. 13 and 23). Specific enzymes participate in each of these functions of the digestive system.

The digestive tract is also known as the *gastrointestinal* or *alimentary canal*. Epithelial cells with differing degrees of secretory capacity line its tubular segments and chambers. Its walls have variously arranged layers of smooth muscle, which reflexively mix and move food. Some of the muscle is circularly arranged, forming valves or *sphincters* which control the flow of food from one part of the tract to the next. Gastrointestinal hormones, internal nerve plexuses, and extrinsic fibers from the autonomic nervous system influence the activity of smooth muscle and the secretory activity of the epithelial cells. Exocrine secretions, which originate in accessory digestive organs, are delivered by ducts to the tract to perform important gastrointestinal functions.

22-1 STRUCTURE

General Gross Anatomy and Physiology

Digestive processes begin in the mouth, where 32 teeth in an adult bite, tear, and chew food to physically break it into smaller pieces that are more easily swallowed and attacked by digestive enzymes. During embryological development 20 temporary teeth are formed, which erupt through the gums between 6 and 24 months after birth. The formation of permanent teeth begins at about 16 weeks of intrauterine life and continues until a few years after birth. Some are embedded beneath the temporary ones; others are located posterior to them. From about 6 to 12 years of age the temporary teeth are shed as the permanent ones erupt through

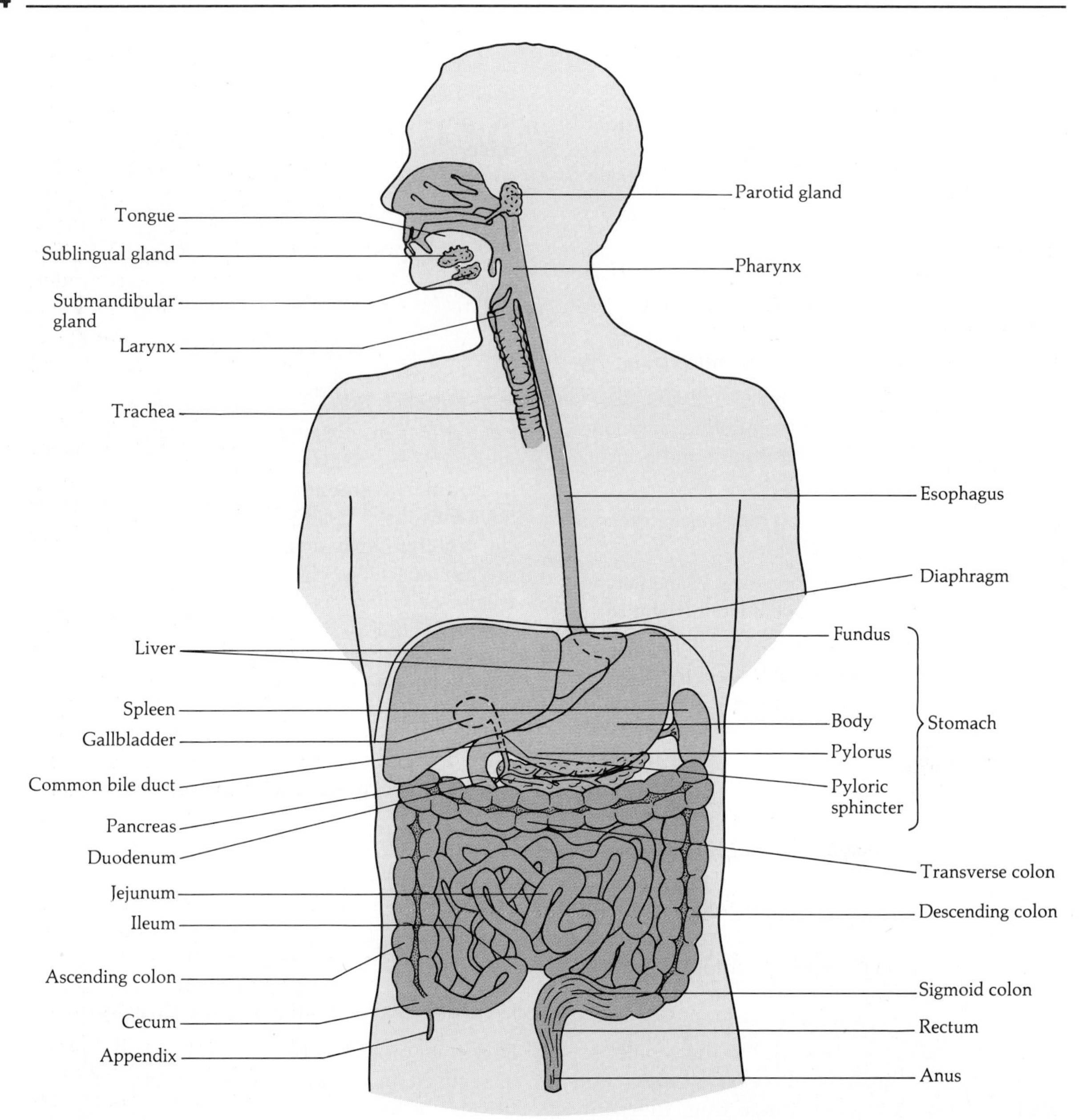

FIG. 22-1
General anatomy of the digestive tract.

the gums in a developmental sequence that is generally completed by age 25, when the skeletal system has matured. Because of this developmental pattern, nutrition of the mother, as well as that of the infant and the child, can markedly affect tooth and jaw structure (Chap. 14).

The *tongue* is a muscular structure within the mouth whose surface is composed of a mucous membrane that secretes fluid. Contractions of skeletal muscles of the tongue move food to the chewing areas of the teeth and to the back of the throat for swallowing. The tongue also participates in the modification of sound waves to produce speech. Taste buds formed of clusters of epithelial cells on the sides of elevations on the tongue's surface house taste receptors (Fig. 8-12, page 201).

Although chewing begins as a conscious voluntary process as the tongue initially propels food backward, chewing then becomes a subconscious reflex activity. Thereafter the skeletal muscles of the pharynx respond reflexively and push the food into the esophagus in the nonstop act of swallowing. The pharynx is a three-part muscular passageway composed of the *oropharynx* at the back of the mouth, the *nasopharynx* posterior to the nasal chambers, and the *laryngopharynx* superior to the trachea of the respiratory tract (Fig. 20-1). Masses of lymphoid tissues beneath the mucosa of the nasopharynx form the *pharyngeal tonsils* (or adenoids); others beneath the tongue are the *lingual tonsils;* and those at the arches at the back of the mouth anterior to the pharynx make up the *palatine tonsils.* When inflammation of the palatine tonsils makes swallowing and breathing difficult, the lymphoid structures sometimes are removed by surgical *tonsillectomy.*

The *esophagus* is a muscular tube about 9½ in long which extends from the laryngopharynx through the diaphragm to the stomach. The wall of its upper segment contains skeletal muscle and that of its lower two-thirds has smooth muscle, which provides the involuntary, reflexive movement (*peristalsis*) of food into the stomach.

Circular skeletal muscle in the wall of the esophagus (the cricopharyngeal muscle) forms the *upper esophageal sphincter,* which behaves as a control valve that relaxes in coordination with contraction of pharyngeal muscles to allow food to enter the esophagus during swallowing. The final 1½ in of the esophagus has a control valve of circular smooth muscle, the *gastroesophageal* or *cardiac sphincter,* so named because of its location near the heart. It relaxes to allow food to enter the stomach. As food moves down the pharynx toward the esophagus, the epiglottis is normally positioned down and backwards closing the opening (glottis) to the larynx (Fig. 20-1). During swallowing the movement of the larynx up to the epiglottis and the closure of the glottis usually prevent food from entering the airway and direct food into the posteriorly located esophagus. At the same time respiration is reflexively inhibited.

The *stomach* is a sac whose convoluted muscular wall, with folds called *rugae,* mixes food with digestive juices in a mixture called *chyme.* The stomach also regulates the entry of chyme into the small intestine. The upper part of the stomach attached to the esophagus and just beneath the diaphragm is the *fundus,* the main segment is the *body,* and that portion nearest the small intestine is the *pylorus* (Fig. 22-1). The terminal portion of the pylorus is surrounded by circular smooth muscle, which forms the *pyloric sphincter* in the distal region of the stomach. The sphincter regulates the entry of food into the small intestine.

The *small intestine* is about 1 to 1.5 in in diameter and 23 ft long when completely relaxed. During life it is partially contracted and only about 9 ft long. Thus, the adjective *small* refers to its diameter, which is less than that of the large intestine (about 2.5 in).

The small intestine consists of three parts, the duodenum, the jejunum, and the ileum (Fig. 22-1). The *duodenum* lies beneath the stomach and is 8 to 10 in long (*duodecim* is Latin for twelve, which indicates that the organ's length equaled the width of 12 fingers). The *pancreas* is cradled in the curvature of the duodenum. The *pancreatic duct* merges with the *common bile duct* from the liver and gallbladder at a common entry into the duodenum 2 to 3 in from the pylorus of the stomach (Figs. 22-1 and 22-5).

The middle segment of the small intestine, the *jejunum,* when relaxed is approximately 8 to 9 ft long and constitutes about 40 percent of the small intestine. By the time food has reached the middle of the jejunum, most of it has been digested and absorbed. The *ileum* when relaxed is nearly 12 to 14 ft long and constitutes almost 60 percent of the small intestine. Most of the remainder of food absorption takes place here.

The *ileocecal* (*ileocolic*) *sphincter* regulates the passage of food from the ileum into the *cecum,* which is a blind pouch situated at the first part of the *colon* (large intestine). The *appendix,* a narrow tube about 3 in long with little apparent digestive function, projects from the cecum. The outer muscular coat of the colon contains three longitudinal bands of smooth muscle (*taeniae*), which cause the wall to be puckered into pouches or *haustra.* The colon is about 4 to

6 ft long and forms sequentially, *ascending, transverse,* and *descending* segments (Fig. 22-1). They lead to a *sigmoid* (S-shaped) colon in the lower and central abdominal cavity. Much water and small amounts of other substances are absorbed in the colon.

The *rectum,* a terminal tube nearly 5 in long, serves to reduce the water content of waste materials and compact them. Two sphincters, one wrapped around the other, surround the terminal portions of the rectum, the *anal canal* and *anus,* and control the excretion of feces. The internal sphincter is composed of smooth muscle and is reflexively controlled. The external sphincter is composed of striated muscle and is under voluntary control.

Microanatomy

In most areas the wall of the digestive tract is composed of four layers of tissues. From the innermost to the outermost they are the mucosa, submucosa, external muscular coat, and adventitia (Table 22-1 and Figs. 22-2A and 3-8). The innermost layer, the *mucosa,* is made up of an epithelial lining, a lamina propria composed of loose connective tissue, and thin layers of inner circular and outer longitudinal smooth muscle (the *muscularis mucosa*). The *submucosa* is a supporting layer for the mucosa and consists of connective tissues with blood and lymph vessels and Meissner's nerve plexus. The *external muscular coat* is composed of an inner circular and outer longitudinal layer of smooth muscle, between which are nerves that form the myenteric plexus. *Meissner's nerve plexus,* which is sensory in function, and the *myenteric nerve plexus* (Auerbach's plexus), which has motor activity, form local reflex arcs which help regulate gastrointestinal function. The outermost layer, the *adventitia,* is composed of loose connective tissue that binds the digestive tract in place. In the abdominal cavity it is covered by a mesothelial membrane, the *serosa* or *peritoneum,* that somewhat loosely suspends the stomach and intestine.

In the small intestine the mucosa and submucosa are normally folded inward in circular elevations called *plicae,* which are most obvious in

TABLE 22-1
General structure of the wall of the digestive tract

Layer	*Tissue Components*
Mucosa (mucous membrane of inner lining)	Epithelium: some absorptive, some exocrine and endocrine cells
	Reticular connective tissue with blood vessels (*lamina propria*)
	Thin, usually double, muscular layer called the muscularis mucosa: inner circular smooth muscle, outer longitudinal smooth muscle
	The mucosa forms fingerlike folds called villi in the small intestine
Submucosa	Connective tissues: fibrous, elastic and loose
	Blood and lymph vessels (and lymph nodes which in the ileum are called *Peyer's patches*)
	Exocrine glands (Brunner's glands of the duodenum)
	Sensory nerve plexus (Meissner's plexus)
	The mucosa and submucosa form permanent circular folds, the plicae circulares in the small intestine
Muscularis externa (external muscular coat)	Circular smooth muscle adjacent to submucosa
	Motor nerve plexus (myenteric or Auerbach's plexus)
	Peripheral longitudinal smooth muscle
Adventitia	Loose connective tissue covered by mesothelial layer which forms the serosa or peritoneum that covers the stomach and intestines

FIG. 22-2
Microscopic structure of the small intestine. (A) Diagram of the microanatomy of the wall of the small intestine. (B) Photomicrograph of villi (×700). (C) Photomicrograph of the tip of a villus (×1700).

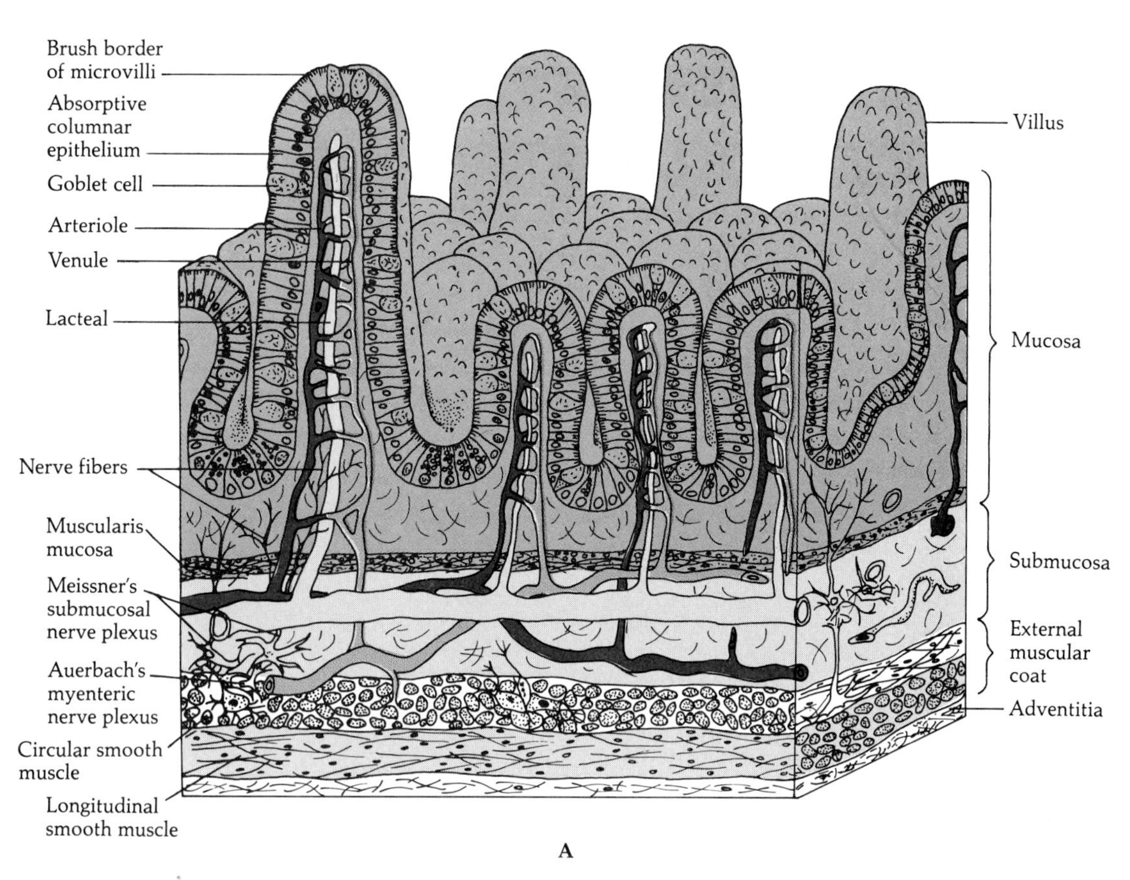
Brush border of microvilli
Absorptive columnar epithelium
Goblet cell
Arteriole
Venule
Lacteal
Nerve fibers
Muscularis mucosa
Meissner's submucosal nerve plexus
Auerbach's myenteric nerve plexus
Circular smooth muscle
Longitudinal smooth muscle
Villus
Mucosa
Submucosa
External muscular coat
Adventitia

A

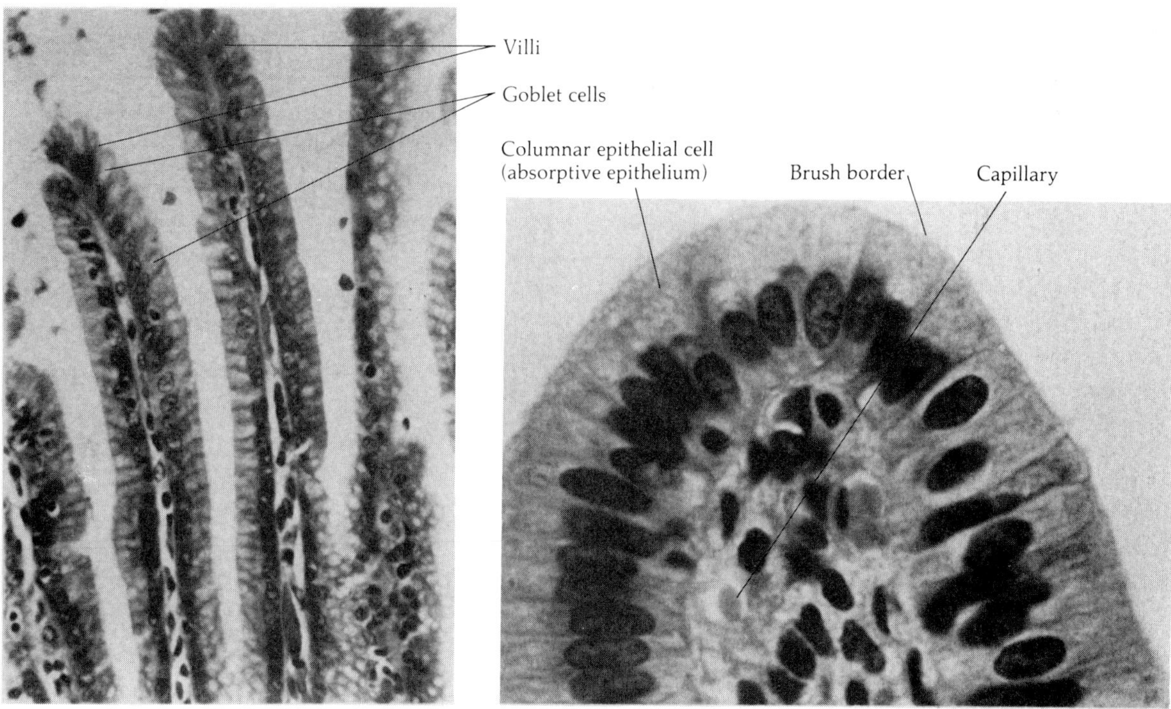
Villi
Goblet cells
Columnar epithelial cell (absorptive epithelium)
Brush border
Capillary

B

C

the jejunum. The mucosa of the small intestine is further pleated into very small, fingerlike projections known as *villi* (Figs. 22-2 and 22-3). They form absorptive and secretory units that protrude about 0.5 to 1.5 mm above the surface.

There are 4 to 5 million villi in the intestinal lining packed about 20 to 40 per mm^2. Furthermore, each of the superficial cells of the villi is studded with 2000 to 4000 microscopic folds of the cell membrane called *microvilli* or a *brush border* (Fig. 22-3B) and are packed at a density of 50,000 to 200,000/mm^2 of the intestinal lining. Because of the microvilli, the entire small intestine if spread out would cover an area of 3000 ft^2, which is slightly smaller than that of a full-size basketball court. All these convolutions increase the surface area of the small intestine about 600-fold.

A surface layer of glycoprotein about 0.1 to 0.5 μm thick covers the microvilli. It is a filamentous covering, called the "fuzzy coat" or *glycocalyx,* and lines the lumen of the digestive tract. The filaments provide sites of attachment for digestive enzymes while simultaneously protecting the mucosa from degradation by enzymes, acid, bile or such ingested substances as alcohol, aspirin, or other drugs.

The villi, which are composed of all of the tissues that make up the mucosa, change in length, sway, and mix food with digestive juices. They do so by means of strands of smooth muscle fibers arranged lengthwise and originating from the muscularis mucosa. The properties of absorptive and glandular epithelial cells which line the villi are described below. Intestinal glands, the *crypts of Lieberkükn* (Table 22-2), are located in depressions in the mucosa and extend toward the muscularis mucosae. These glands are prevalent in the small intestine but are also found in the large intestine.

Immature columnar cells, absorptive cells, goblet cells, and enterochromaffin cells are located on the surface of the villi and in the crypts. A fifth type, the Paneth cell, is found only in the crypts. *Immature columnar cells* are the principal epithelial cells within the villi. They rapidly divide by mitotic cell division in

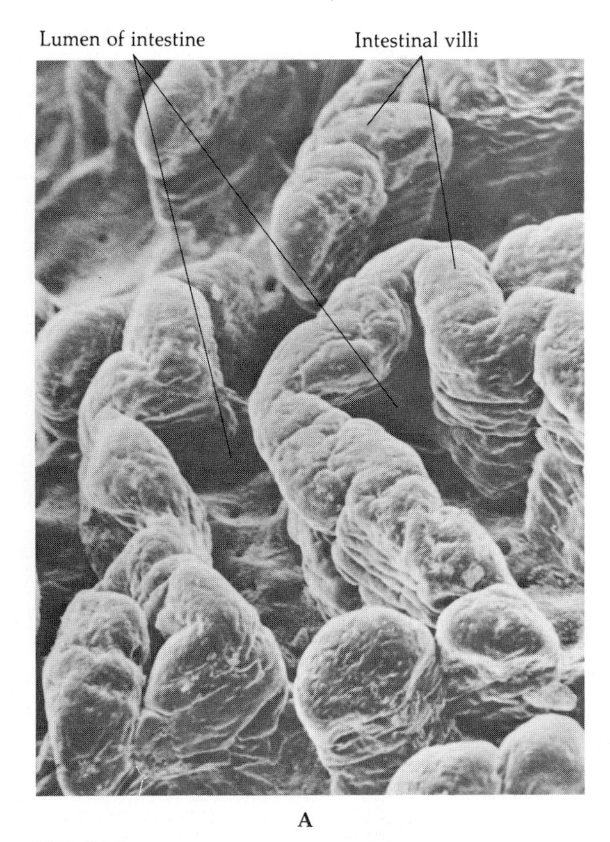

FIG. 22-3
Scanning electron micrographs of villi and microvilli. (A) Variations in appearance of human intestinal villi in a scanning electron micrograph (× 192). (B) Appearance of microvilli is shown in a transmission electron micrograph on the next page.

the crypts and mature as they push their way to the tips of the villi and replace the entire digestive and absorptive epithelium about every 2 to 6 days. New cells are formed at the rate of up to 50 million per minute. *Absorptive columnar cells* digest many nutrients prior to entry of the building blocks of food into the blood or lymph. They do so by means of digestive enzymes attached to the glycocalyx and thereby facilitate nutrient transport. *Goblet cells,* so called because of their appearance, (Fig. 22-2) secrete mucus. *Enterochromaffin cells* (enteroendocrine or argentaffin cells) produce gastrointestinal hormones and are concentrated in the duodenum and upper jejunum. The granular *Paneth cells* are located deep within the crypts, where they seem to make an antibacterial protein, ly-

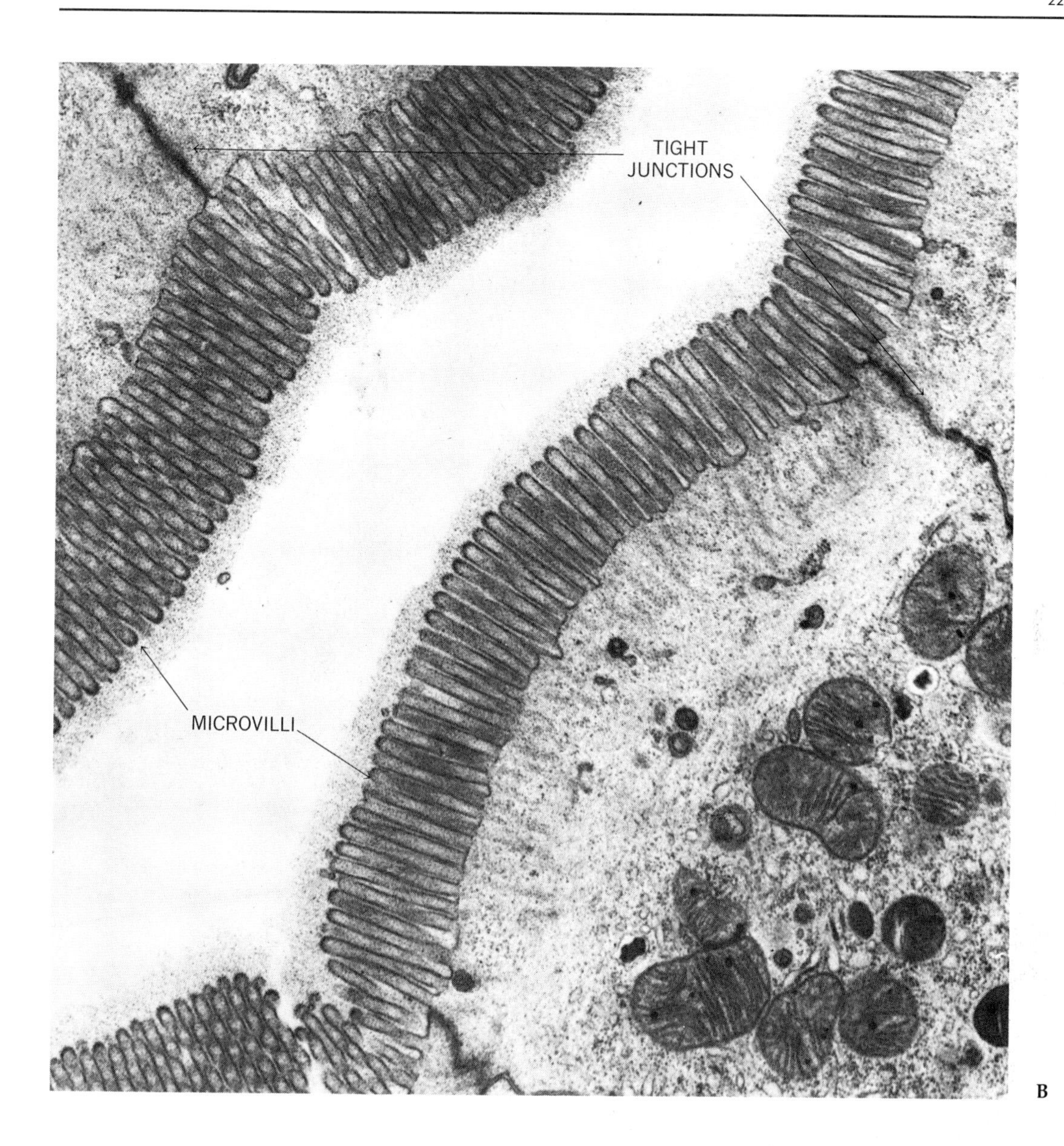

B

sozyme. It attacks mucopeptides in bacterial cell walls, killing some bacterial species and thereby influencing the ecology of the bacterial population in the intestine.

Accessory Digestive Organs

The salivary glands, liver, gallbladder, and pancreas are accessory digestive organs. They secrete juices that aid the digestion and absorption of foods (Table 22-3).

Salivary Glands

Three pairs of salivary glands secrete droplets of saliva, which are transported through ducts into the mouth (Fig. 22-1). They are the *parotid* glands, the *submaxillary* (*submandibular*) glands, and the *sublingual* glands. Saliva dissolves food, digests polysaccharides, and acts as a protective coating in the digestive tract. The parotid glands become quite evident when swollen during infection with mumps virus.

TABLE 22-2
Principal glands in the walls of the digestive tract*

Location	*Types of Cells*	*Product*
Mouth	Mucosal epithelium (buccal glands)	Mucus
Esophagus	Esophageal	Mucus
Stomach	Gastric glands	Mucus
	Mucous neck cells	Mucus
	Chief cells	Pepsinogen
	Parietal (oxyntic) cells	Hydrochloric acid and intrinsic factor
	Cardiac glands (in the cardiac segment of the stomach)	Mucus
	Pyloric glands (in pylorus of stomach)	Mucus
Small intestine	Intestinal glands in mucosa (the crypts of Lieberkühn)	Enzymes, hormones, and glycoproteins as described below
	Immature columnar cells (main cells in crypts)	As cells mature, they move toward the mucosal surface and ultimately are shed
	Absorptive cells (main cell type on surface of villi)	Glycoprotein and some digestive enzymes released to microvilli surfaces to facilitate digestion
	Goblet cells	Mucus
	Paneth cells (deep within the crypts)	Lysozyme
	Enterochromaffin cells*	Gastrointestinal hormones
	Brunner's glands (duodenal glands in submucosa of duodenum with ducts leading to crypts)	Alkaline mucus
Large intestine (colon)	Goblet cells (in the crypts of Lieberkühn)	Mucus

* All glands listed are exocrines except the enterochromaffin cells, which are endocrines. All are located in the mucosa, except Brunner's glands which are in the submucosa in the duodenum.

Exocrine Portions of the Pancreas

The endocrine functions of the pancreas and its hormones insulin and glucagon were described earlier (Chap. 5). The exocrine cells are clusters of epithelial cells called *acini,* which pour their secretions into ducts (Fig. 22-4). The ducts

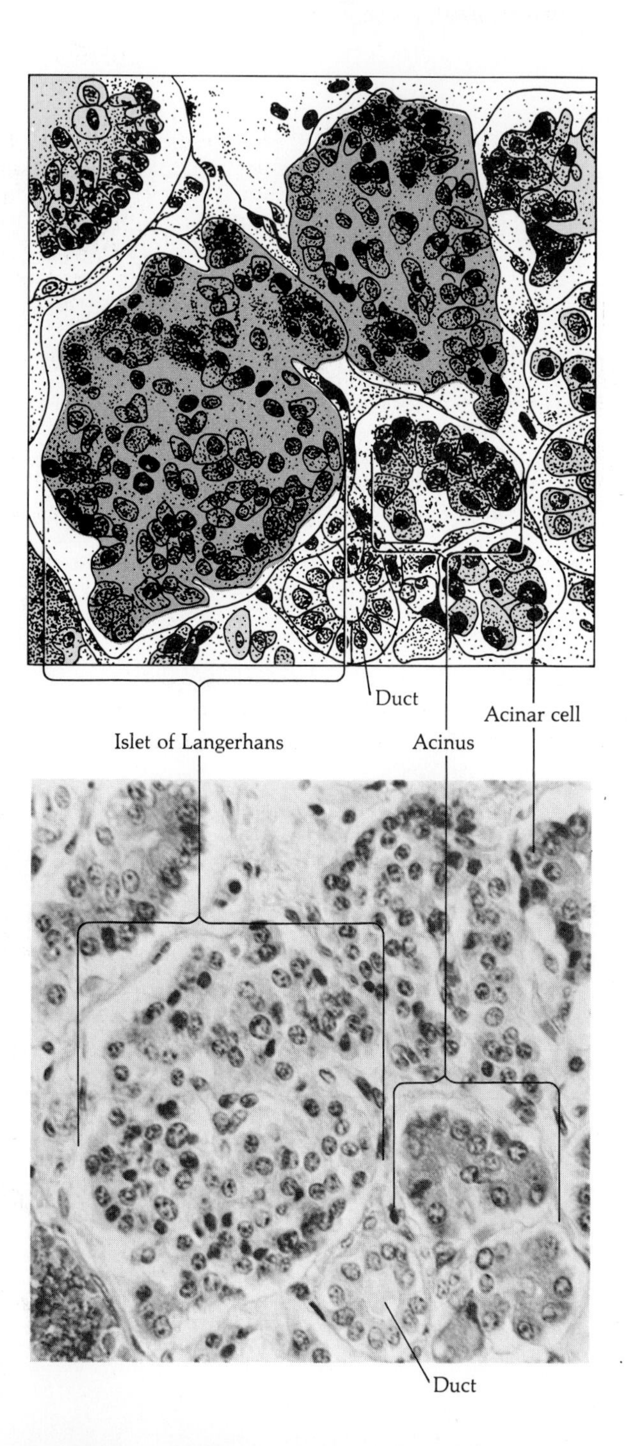

FIG. 22-4
Exocrine and endocrine segments of the pancreas. The diagram of pancreatic tissue is illustrated above the photomicrograph of pancreatic tissue (×700).

TABLE 22-3
Accessory digestive organs

Organ	*Location*	*Product*	*Destination of Product*	*Effect*
Parotid, submandibular, and sublingual salivary glands	In front of ears, lower rear jaw, and under base of tongue in front of mouth	Saliva Immunoglobulin A	Mucosal surface	Digestive Antimicrobial
		Amylase		Initiates digestion of polysaccharides in stomach and intestine
Pancreas	Curve of duodenum in abdominal cavity	Digestive enzymes such as amylase and lipase and protein-cleaving enzymes such as trypsin and chymotrypsin	Duodenum (then moves through intestine with food)	Degrade carbohydrates, fats and proteins to absorbable forms
		Bicarbonates	As above	Neutralize acid from stomach
Liver	Abdominal cavity inferior to the diaphragm	Bile	As above	Solubilizes fats; dissolves fat-soluble vitamins (A, D, E, and K)
		Products of carbohydrate lipid and protein metabolism	Enter bloodstream	
Gallbladder	Inferior surface of liver	Concentrated dehydrated bile	Duodenum	As above

FIG. 22-5
Secretory routes from the liver, gallbladder, and pancreas to the duodenum.

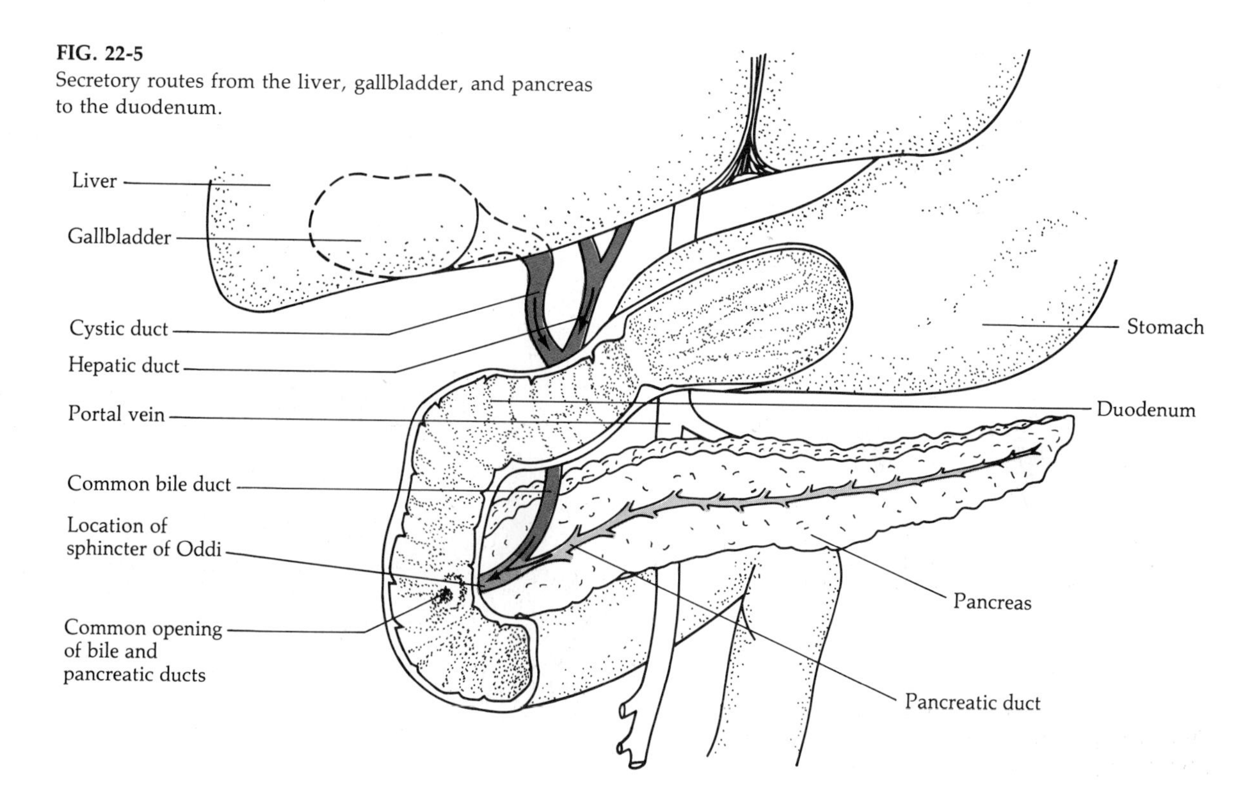

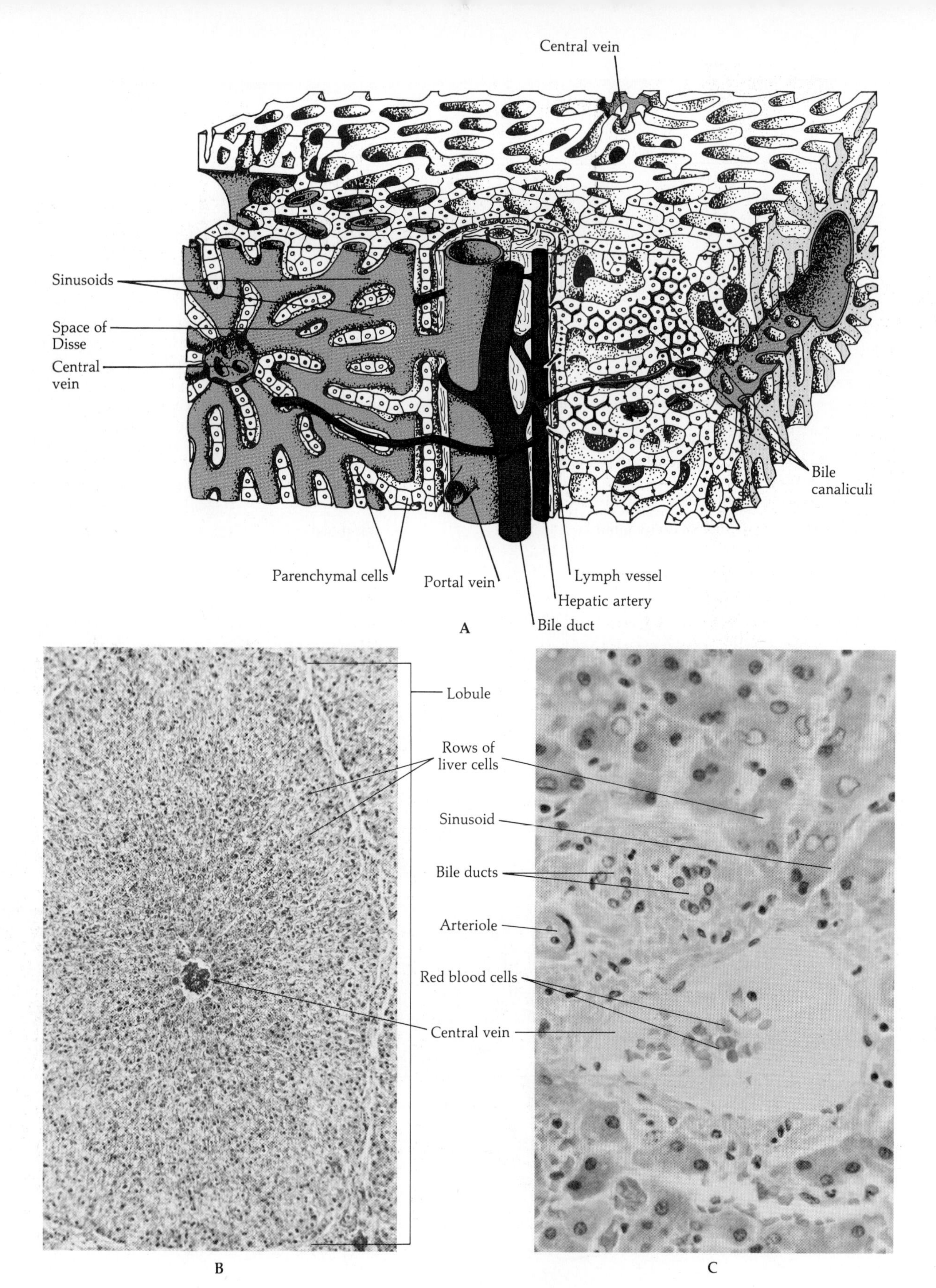

Central vein
Sinusoids
Space of Disse
Central vein
Bile canaliculi
Parenchymal cells
Portal vein
Lymph vessel
Hepatic artery
Bile duct
A
Lobule
Rows of liver cells
Sinusoid
Bile ducts
Arteriole
Red blood cells
Central vein
B
C

FIG. 22-6
Microanatomy of a lobule, the functional unit of the liver. Each row of parenchymal cells is flanked by a bile duct and a venous sinusoid. The flow of blood is toward the central vein in each lobule. (A) Diagram of liver lobule. (B) Photomicrograph of liver lobule (×70). (C) Photomicrograph of liver lobule (×700).

merge to form the pancreatic duct, which leads to the duodenum (Fig. 22-5). The cells of the acini secrete several different digestive enzymes; the duct cells produce bicarbonates.

Liver

The *liver,* the largest gland in the body, weighs 1.4 ± 0.2 kg, is separated into four lobes, and is located beneath the right side of the diaphragm in the upper abdominal cavity (Fig. 22-1). A small sac, the *gallbladder,* projects from its underside.

The liver is composed of tissue units termed *lobules.* Columnar epithelial cells, called *parenchymal cells* (*hepatocytes*) are aligned in rows like the spokes of a bicycle wheel and radiate toward a central vein in each lobule (Fig. 22-6). Each row of cells is flanked by a bile duct on one side and a highly permeable capillary, a venous sinusoid, on the other (Fig. 17-6C). The parenchymal cells are separated from the venous sinusoid by a space in which interstitial fluid collects (the *space of Disse,* Fig. 22-6A). The sinusoids are lined with endothelial cells and liver macrophages (*Küppfer cells*).

Substances carried to the liver by the blood may be transported into liver parenchymal cells and metabolized there. The metabolic products are secreted from the cells into the sinusoids on one side or into the bile ducts on the other. The bile ducts converge into one large *hepatic duct,* which joins the *cystic duct* from the gallbladder to form the *common bile duct* that leads into the duodenum (Fig. 22-5).

Because of its importance in body function, the liver receives nearly 30 percent of the cardiac output. Of this blood, 75 percent is brought to liver sinusoids by the hepatic portal vein, which collects blood rich in nutrients from the stomach, the small and large intestines, and the spleen and pancreas. As a result of this circulatory pattern, nutrients absorbed into blood from the digestive tract are carried by the hepatic portal vein directly into the liver. This circulatory route protects the rest of the body from direct exposure to ingested molecules, except for most fats, which are absorbed into the lymph stream.

The remainder of blood carried to the liver comes to the sinusoids by way of the abdominal aorta through the hepatic arteries. The sinusoids filter and exchange substances between the blood and the adjacent parenchymal cells (Fig. 22-6A). In each lobule the blood is then delivered to the central hepatic veins, which unite to make the three large hepatic veins that transport blood to the inferior vena cava and back to the right atrium of the heart.

Although the liver performs storage, secretory, and circulatory functions, it is primarily a huge conglomerate of metabolic factories whose activities affect all other body functions. Many of the metabolic reactions that occur in muscle are also carried out in the liver (see Fig. 13-11). The liver can synthesize carbohydrates, proteins, and fats, convert each one to the other, and degrade them to produce energy second only in quantity to that formed in skeletal muscle (Chap. 13). Some proteins formed in the liver are used to transport hormones (see Chap. 5); others convey lipids and cholesterol in lipoprotein complexes (Figs. 16-7 and 16-8); while still others are plasma proteins (Tables 10-2 and 18-1), several of which are critical in blood clotting (Table 19-3). The liver performs critical functions in lipid metabolism and synthesizes or degrades cholesterol and other steroids. It also secretes approximately 0.25 to 1.0 L of bile daily to the duodenum.

Potentially harmful compounds such as drugs, toxins, and ammonia from protein catabolism are detoxified by liver cells. Ammonia (NH_3), a strong base, is formed by the removal of amino groups (NH_2) from proteins in reactions called *deaminations.* The liver neutralizes ammonia by combining it with carbon dioxide to produce urea ($H_2N{-}\overset{\overset{\displaystyle O}{\|}}{C}{-}NH_2$) or uses it to synthesize glutamine and glutamic acid. Urea and other degradative products of liver metabolism may enter the blood to be excreted in urine by the kidneys (Chap. 21).

The liver also is a reservoir for blood and nutrients. For example, during exercise blood stored in the liver is released to the general circulation. Prominent nutrients stored in the liver are glycogen, vitamins, and minerals. Glycogen is packaged in granules in liver cells and converted by liver enzymes into glucose for release into the blood. The liver of an average 70-kg adult male stores 70 g of glycogen, which may provide 280 kcal of energy.[1] This energy reserve is second only to the skeletal muscle mass of the body which stores an average of 120 g of glycogen and 480 kcal of energy. The liver's storage capacity is also essential to vitamin metabolism. It contains vitamins A (retinol), B_2 (riboflavin), B_6 (pyridoxine), B_{12} (cobalamin), C (ascorbic acid), and D (cholecalciferol), as well as some of their precursors. The liver also conserves the minerals iron and copper, both of which are essential to normal red blood cell formation.

Many liver functions are related to the circulatory system's physiology. These functions include the storage of blood, as well as copper, iron, and vitamins necessary for blood cell formation; the conversion of hemoglobin into bile pigments (Fig. 18-6); and the synthesis of clotting factors (Table 19-3) and other plasma proteins. Macrophages in the liver sinusoids also perform defense functions. Each day they help remove hundreds of billions of aged red blood cells and other abnormal or foreign cells and particles in the blood. In so doing they prepare some antigens for interaction with lymphocytes to cause an immune response. In the fetus the liver is a source of red blood cells and of lymphocytes used in body defenses (Fig. 18-1).

[1] A calorie (cal) is the amount of heat needed to raise the temperature of 1 g of water by 1°C. Since metabolic reactions involve thousands of calories of energy, the kilocalorie (kcal = 1000 cal) is used to describe these reactions. This unit is sometimes called the large calorie or Calorie; the latter term is often used in discussions of nutrition.

Gallbladder

The *gallbladder,* a small muscular sac located on the inferior surface of the liver, is lined with mucosal epithelium, which concentrates bile 4 to 10 times and stores it in a fluid reserve of about 33 to 36 mL. The mucosal cells of the gallbladder actively transport sodium ions from bile into the blood. The osmotic effect of the sodium in the blood draws water from the bile, concentrating it.

When food distends the duodenum, bile normally flows from the liver down the common bile duct through the sphincter of Oddi to the duodenum (Fig. 22-5). However, the sphincter contracts when the duodenum is empty and bile backs up the common bile duct and cystic duct into the gallbladder for storage. When food rich in fats and fatty acids enters the small intestine, the sphincter of Oddi relaxes and the gallbladder contracts under the influence of a hormone called CCK-PZ (see below). As a result, concentrated bile is delivered from the gallbladder to the duodenum.

22-2 SECRETIONS

Mucosal epithelial cells include those of the endocrine glands which secrete gastrointestinal hormones and those of the unicellular and multicellular exocrine glands whose secretions influence a wide variety of critical digestive tract functions (Table 22-2).

Gastrointestinal Hormones

Hormones produced in the small intestine and the stomach have profound regulatory effects on the motility and secretions of the digestive tract. These activities are interrelated by feedback cycles (Chap. 1) in which the physical and chemical properties of food influence hormonal secretions and they in turn modify gastrointestinal motility and secretions. Changes in the composition of the food-fluid mixture then affect the secretion of specific hormones to complete the cycle.

Enterochromaffin cells in the mucosa secrete hormones whose targets are digestive organs. Some evidence indicates that these cells may be sources of numerous transmitter agents, including the enkephalins, substance P, somatostatin, catecholamines, and serotonin (Chaps. 5 and 10).

Four polypeptides ranging in size from 17 to 33 amino acid residues have been clearly identified as originating in enterochromaffin cells in the duodenum and upper jejunum (Table 22-4). The established hormones include *gastrin, cholecystokinin-pancreozymin* (CCK-PZ), *secretin,* and *gastric inhibitory polypeptide* (GIP). Gastrin is also secreted by cells in the *antrum* (dilated portion of the pylorus) of the stomach and in the pancreas. A fifth hormone, *glucagon,* primarily of pancreatic endocrine origin (Chap. 5), is also formed by cells in the stomach and duodenum. Gastrin and CCK-PZ resemble each other in their structure and effects on the digestive tract, as do secretin and glucagon.

The principal effect of gastrin is to stimulate the secretion of hydrochloric acid by the stomach and of enzymes by the pancreas; that of CCK-PZ is to contract the gallbladder and release enzymes from the pancreas. The main response to secretin is the secretion of bicarbonates from pancreatic duct cells and inhibition of stomach acid secretion, while GIP also inhibits stomach acid production and promotes insulin release.

Because the activities of the hormones overlap and interact, responses to them can be complex. For example, CCK-PZ and secretin augment each other's effects on pancreatic enzyme and bicarbonate production. Conversely, the stimulatory effect of CCK-PZ on HCl production becomes inhibitory in the presence of gastrin, the principal hormonal regulator of HCl secretion.

Exocrine Secretions

Exocrine secretions originate in the glands in the mucosa and submucosa and the accessory digestive organs (Tables 22-2 and 22-3). Secretory products often accumulate in *zymogen granules,* which are storage areas for inactive forms of enzymes. As the substances are secreted, the granules decrease in number and often disappear.

Saliva

The salivary glands, along with other secretory epithelia in the mouth, release about 1 mL of saliva per minute. The flow rate ranges from 0.5 to 7.0 mL/minute and totals 1 to 2 L/day. The presence of foods, especially acids such as lemon juice, reflexively promote higher rates of secretion through parasympathetic nerve stimulation.

As mentioned earlier, saliva dissolves food, digests polysaccharides, and lubricates and protects the lining of the digestive tract. Saliva has a pH of about 6.9 and consists of 99 percent water and 1 percent salts and other substances, including a number of proteins. Among the proteins is *salivary amylase* (ptyalin), an enzyme which is stored in zymogen granules. When secreted and mixed with starch, amylase hydrolyzes (divides) polysaccharides into smaller units such as dextrins, which on the average are composed of eight sugar units, and small amounts of the disaccharide maltose. The acid pH of the stomach inhibits the continued activity of amylase, which otherwise would degrade most of the starch to maltose.

Glandular epithelial cells in the mucosa, including those of the salivary glands, produce glycoproteins, which combine with IgA antibody. The glycoproteins are secretory components which promote transport of the antibody across mucosal surfaces. As a result, external secretions lining the digestive and respiratory systems, the tear ducts in the eyes, and other mucosal tissues are rich in protective IgA antibody.

Secretions of the Stomach

The stomach contains up to 30 million glands composed of cells arranged around deep gastric pits in the mucosal lining (Fig. 22-7). These glands release 2 to 3 L of fluid secretions daily. Those near the neck of the pit, the *mucous neck cells,* secrete mucus; the most common cells, the

TABLE 22-4
Some properties of gastrointestinal hormones

Hormone	*Origin*	*Source of Stimuli*
		Nervous reflex, or hormonal feedback controls
Gastrin	G cells of stomach and duodenum	Distention of stomach and stimulation by peptides and amino acids of meal in stomach
Cholecystokinin-pancreozymin (CCK-PZ)	Duodenum and jejunum	Presence of fat, protein, and their digestion products in the intestine
Secretin	S cells of duodenum	Presence of acid in the duodenum
Gastric inhibitory polypeptide (GIP)	Duodenum	Presence of fatty acids or simple sugars in duodenum

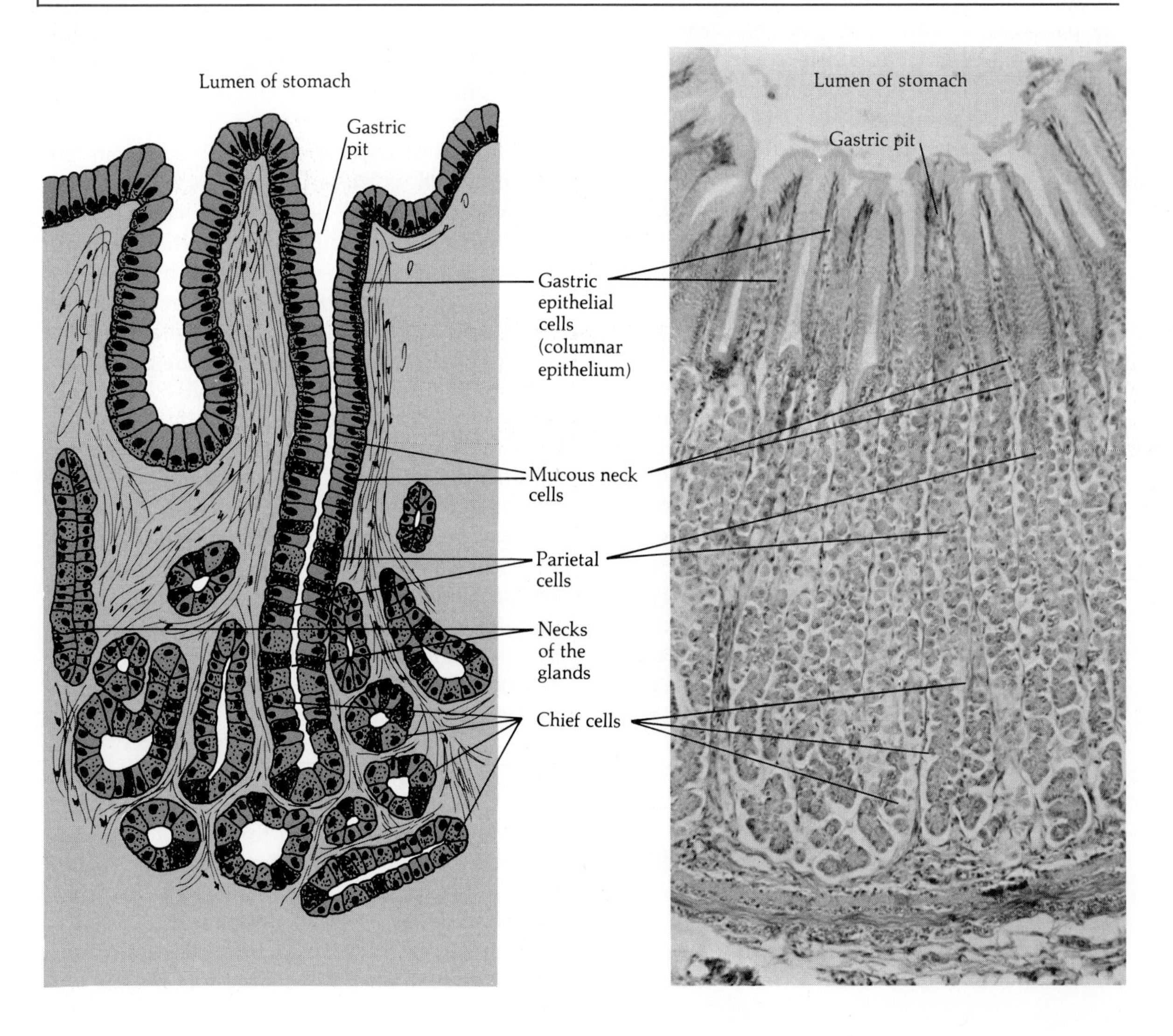

EFFECTS ON ORGANS AND TISSUES							
Stomach		*Liver*	*Gallbladder*	*Pancreas*			
HCl Secretion	*Pepsinogen Secretion*	*Bile Secretion*	*Contraction*	*Bicarbonate Secretion*	*Enzyme Secretion*	*Insulin Release*	*Mucosal Growth*
(+)	(+)	+	+	+	(+)	+	+
+	+		(+)	+	(+)	+	+*
(−)	+	+	+	(+)	+	+	−
−						(+)	

Legend: ( ) Major effect; + Stimulation; − Inhibition.
* Also enhances growth of exocrine pancreatic tissue.
Adapted from L. R. Johnson, "Gastrointestinal Hormones and Their Function," *Ann. Rev. Physiol.* **39**:135–158 (1977), p. 139, table 2; V. B. Mountcastle (ed.), *Medical Physiology*, vol. 2, 14th ed., C. V. Mosby, St. Louis, 1980, p. 1316, table 54.1.

chief cells, secrete pepsinogen; *parietal cells,* also called *oxyntic cells* (from Greek *oxnein,* to make acid) secrete hydrochloric acid and the intrinsic factor, a glycoprotein required for the absorption of vitamin B_{12}.

Mucus from the stomach, like that from the salivary glands, lubricates and protects epithelial cells. It contains enzymes and is viscous, alkaline, and rich in a variety of mucins (glycoproteins) that coat the cells in the lumen (inside space) of the digestive tract. *Pepsinogen* is an inactive molecule which, in the presence of hydrochloric acid, is converted to the active enzyme pepsin, which hydrolyzes proteins to polypeptides.

Hydrochloric acid denatures proteins (unfolds them) and kills bacteria, reducing their numbers in foods. The acid activates pepsin and inactivates salivary amylase by altering hydrogen bonds and other weak forces that hold parts of molecules together. It also exerts these effects on other enzymes. When hydrogen bonds are ruptured, the structure of proteins may be rearranged and their activity may be altered.

FIG. 22-7
Glandular cells in the mucosa of the stomach. A gastric pit and its major cell types are illustrated in a diagram on the left and a photomicrograph (× 170) on the right.

The formation of hydrochloric acid involves reactions similar to those that take place in the epithelia of the lungs and kidneys (Chaps. 20 and 21). Carbon dioxide and water combine to form carbonic acid, which dissociates into hydrogen ions, and into bicarbonate ions which diffuse into the blood in exchange for chloride ions (Fig. 22-8). The parietal cells rapidly and actively transport the hydrogen and chloride ions to little channels, *canaliculi,* that open into the gastric pits in the lumen of the stomach. The active transport occurs against concentration gradients and requires the participation of cell membrane–bound transport enzymes for both types of ions. The combination of hydrogen and chloride ions forms an acid with a pH of nearly 1.0 and a concentration of 150 m*M* (2 to 4 million times greater than the concentration in the plasma). When food and secretions mix with stomach acid, the chyme formed has a pH that ranges from 1.0 to 3.5.

Up to 2.0 L of hydrochloric acid is produced each day, performs its functions, and is then neutralized by intestinal juices as it enters the duodenum. Increased abdominal pressure (such as occurs in the later stages of pregnancy)

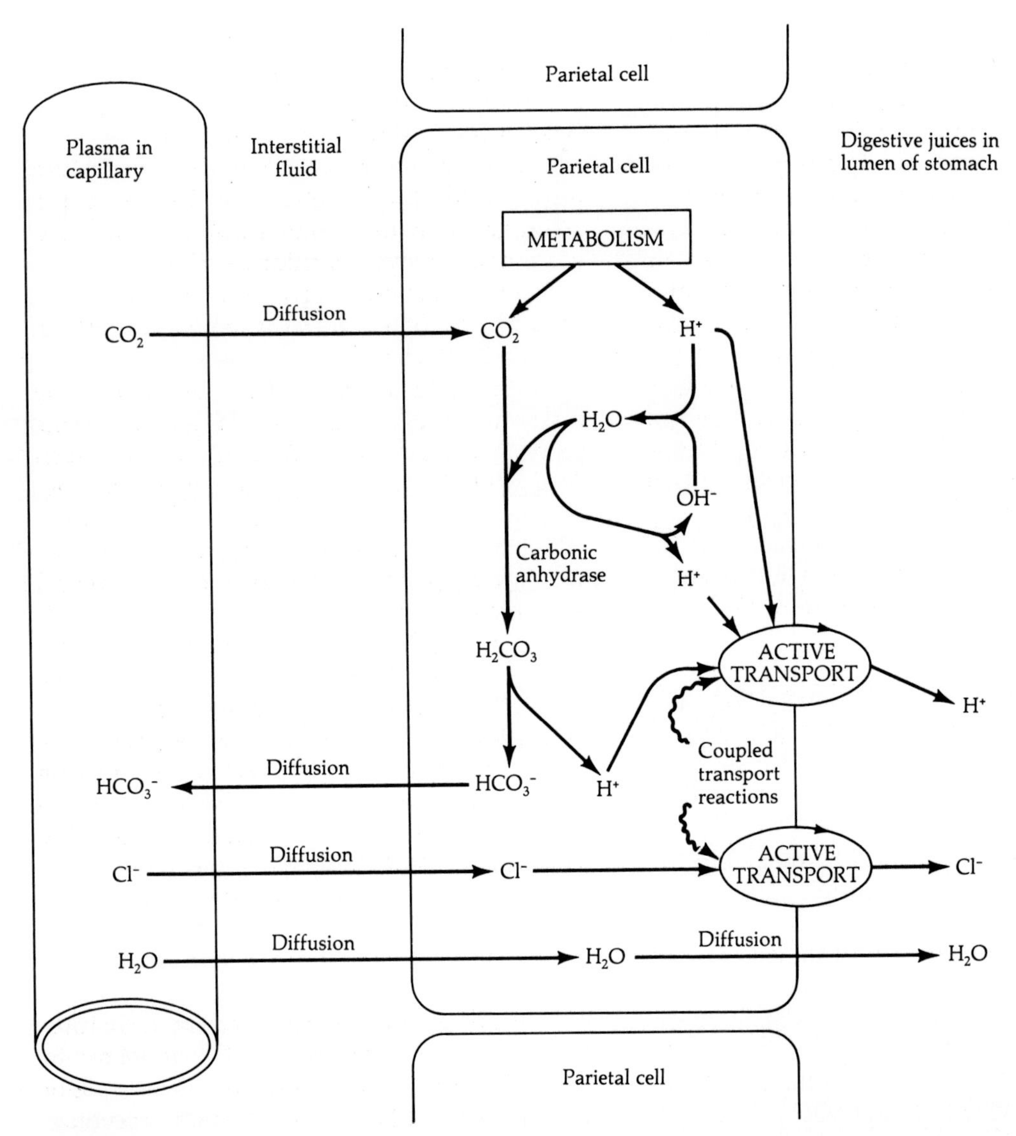

FIG. 22-8
Hydrochloric acid production by parietal cells in the stomach. The direction of net transport of substances is indicated by the arrows.

may cause the cardiac sphincter in the esophagus to remain open longer than usual and allow contractions of the stomach to force some of its acid contents into the esophagus. The esophageal mucosa is irritated in a condition called *heartburn,* which also may occur following overeating of foods which stimulate acid secretion by the stomach.

The parietal cells secrete a glycoprotein, the *intrinsic factor,* which has receptors for two molecules of vitamin B_{12}. In the presence of calcium ions the intrinsic factor–vitamin B_{12} complex binds to receptors on the mucosal cells in the lumen of the ileum. Absorption of the vitamin is stimulated by a mechanism that has yet to be clarified. Vitamin B_{12} is a coenzyme in nucleic acid metabolism and is essential for red blood cell formation (Chap. 18). Because vitamin B_{12} enters the body with food from the outside, it is sometimes called the *extrinsic factor.* The failure to secrete intrinsic factor is accompanied by inability to absorb vitamin B_{12}, and results in *pernicious anemia,* which is characterized by increasing weakness, fatigability, and loss of skin color, appetite, and weight (Table 18-8, p. 447).

Oral administration of vitamin B_{12} does not reverse pernicious anemia because of the deficiency in the intrinsic factor. However, intramuscular injection of the vitamin bypasses the defective absorption route and allows its entry into blood for delivery to the tissues. Some forms of pernicious anemia may result from autoimmune destruction of glands in the stomach by cell-mediated defense reactions. The cells in the fundus of the stomach degenerate, diminishing the production of intrinsic factor as well as that of hydrochloric acid and pepsinogen.

Exocrine Secretions of the Pancreas

The pancreas secretes about 1.5 L daily of fluid which contains abundant quantities of digestive enzymes and bicarbonate ions. The acini of the pancreas secrete: *pancreatic amylase,* which continues the breakdown of polysaccharides into disaccharides; the protein-cleaving enzymes *trypsin, chymotrypsin,* and *carboxypeptidase; RNAse* and *DNAse,* which cleave RNA and DNA into nucleotides; and *pancreatic lipase,* which degrades fats into fatty acids, glycerol, and monoglycerides.

Cells lining ducts of the acini produce carbonic acid, which dissociates into *bicarbonate ions,* which are transported into the ducts by mechanisms that are not fully understood. The bicarbonate flows with digestive enzymes through the pancreatic duct to the duodenum, where it aids in neutralizing stomach acid. It raises the pH of the fluid-food mixture toward neutrality (pH 7.0) but leaves it in the pH range 5.5 to 7.5. This pH range is optimal for the activity of pancreatic enzymes and bile salts.

Exocrine Secretions of the Intestine

Under normal conditions the secretory activity of the intestine produces about 1 L of fluid each day. Exocrine secretions in the small intestine are derived from goblet cells, which are simple glands in the villi, from the crypts of Lieberkühn, and from *Brunner's glands.*

Intestinal juice contains water and electrolytes at levels similar to those in plasma, with which it is isoosmotic. However, compared with plasma the secretions in the ileum and colon generally have higher levels of bicarbonate and lower levels of chloride. Higher concentrations of secretory IgA antibody compared with plasma levels are also found in fluids in the small intestine.

A variety of digestive enzymes secreted by different glands and accessory digestive organs seem to perform their functions in the brush border of the mucosa of the small intestine. Among them are enzymes such as *dextrinases,* which act on glucose linkages; *sucrase, maltase,* and *lactase,* which hydrolyze sucrose, maltose, and lactose respectively; *peptidases,* which cleave peptide bonds; *lipases,* which hydrolyze lipids; and *nucleases,* which degrade nucleic acids. Two additional enzymes, *enterokinase,* which activates proteolytic enzymes, and *amylase,* which degrades polysaccharides, may be the only ones to perform their functions in the lumen unattached to the brush border. A more complete description of the digestive enzymes will be given in Chap. 23 (Table 23-1).

Secretion of Bile

The cells of the liver secrete metabolic products of hemoglobin and cholesterol metabolism in *bile.* The liver of an adult secretes about 0.6 L of bile each day (although the volume ranges from 0.25 to 1 L). In composition bile resembles plasma (Tables 10-2 and 18-1) to which additional bile salts, pigments, cholesterol, and lecithin have been added. Bile is usually slightly alkaline, contains 97 percent water, and has a salt content similar to that of extracellular fluid (ECF, Fig. 4-2). In a route called enterohepatic circulation some of the components of bile are reabsorbed by the intestine and recirculated to the liver for secretion once again (Fig. 18-6).

Variations in the ratio of cholesterol, bile salts, and lecithin may cause the cholesterol to precipitate and form *gallstones.* Cholesterol accounts for 70 to 95 percent of the weight of

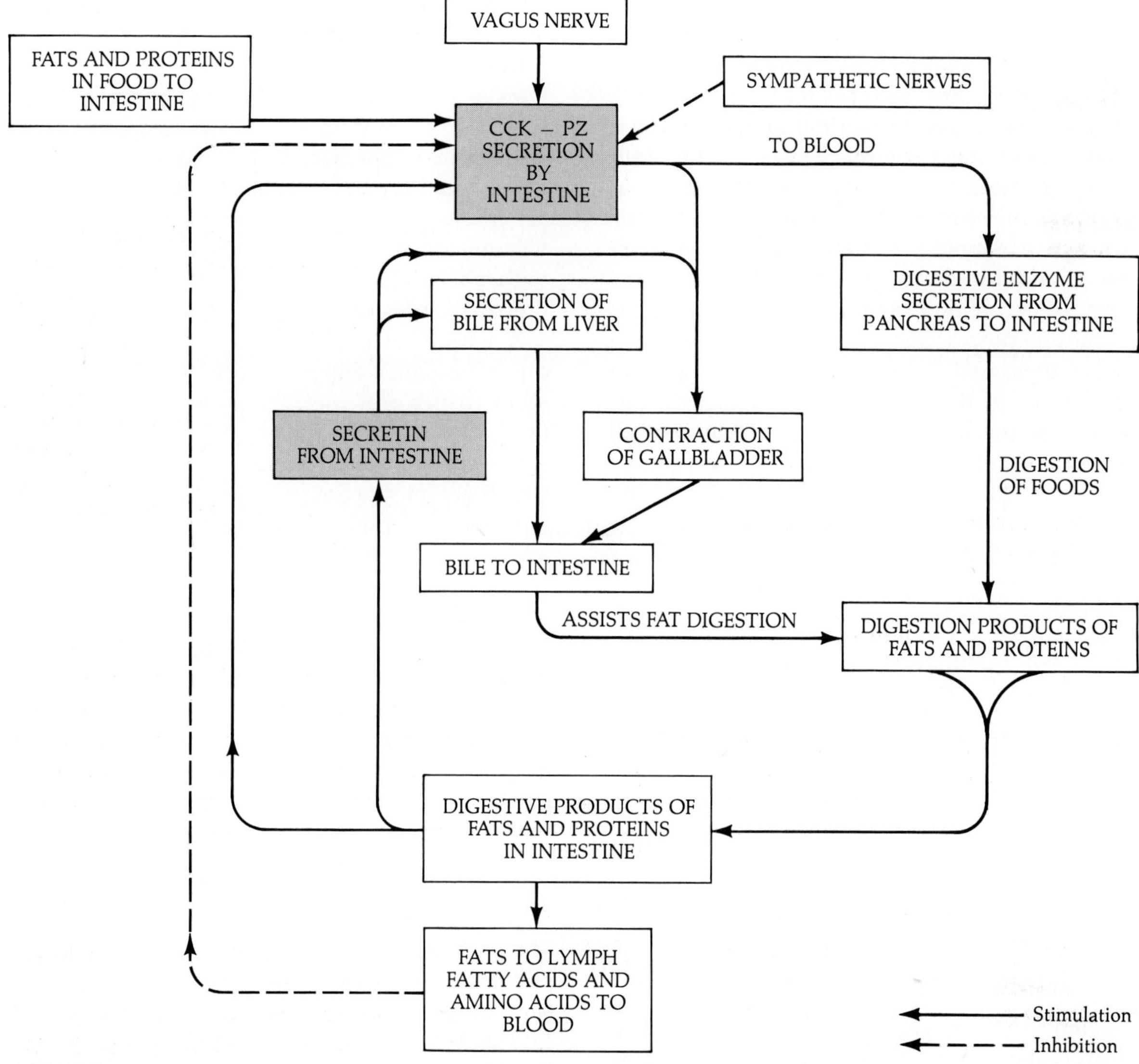

FIG. 22-9
Regulation of CCK-PZ secretion. Note that secretion of the hormones CCK-PZ and secretin promotes the digestion of fats. Secretion of the hormones in turn is stimulated by the presence of fats in the duodenum and inhibited when the levels of fats in the intestine decrease.

solids in most gallstones. Gallstones may lodge in ducts of the liver or in the gallbladder and cause the products of bile salt metabolism to accumulate in the blood and urine. These products may cause yellow discoloration of the mucous membranes, the skin, and the sclera (white of the eye) in the condition called jaundice (Chap. 18).

22-3 REGULATORY MECHANISMS

Gastrointestinal functions are regulated primarily by hormonal and neural mechanisms (Tables 22-4 and 22-5). *Cephalic* neural stimuli originate in the brain and are transmitted by the vagus nerve. *Gastric* stimuli arise in the stomach as a result of local reflex arcs and by the activity of gastrin and other hormones. *Intestinal stimuli* originate in the intestine and are caused by local reflex arcs and the activity of such hormones as secretin and CCK-PZ. Cephalic, gastric, and intestinal stimuli and the responses to them overlap. As a result, environmental stimuli, reflex activities, and hormonal feedback cycles that are influenced by the physical and chemical properties of food all affect the functions of the digestive system.

Control of Secretions

Hormonal Factors

Specific foods initiate feedback cycles between hormones and the exocrine secretions that influence digestion and absorption. For example, the entry of fats into the duodenum stimulates CCK-PZ secretion, which causes the gallbladder to contract and release bile (Table 22-4) and the pancreas to secrete lipases that digest fats (Fig. 22-9). As the products of fat digestion are absorbed, the secretion of CCK-PZ declines.

In another example, protein in a meal stimulates secretion of acid by the stomach. The proteins neutralize acid in the intestine. The decline in acidity stimulates gastrin secretion. Gastrin in turn causes acid production by the stomach and enzyme secretion by the pancreas (Fig. 22-10). The acid and enzymes then participate in the breakdown of proteins into amino acid building blocks, which can be absorbed. Thus, acid secretion by the stomach is proportional to the protein content of a meal.

Neural Influences

The sensory fibers of Meissner's plexus and the motor fibers of the myenteric plexus form reflex arcs which influence *localized segments* of the tract. In contrast, the central nervous system (CNS) delivers efferent impulses from the thalamus and hypothalamus through autonomic fibers that may influence *long segments* of the digestive system (Fig. 9-13). Parasympathetic stimuli, carried mostly through the vagus nerve, stimulate glandular secretions (Table 22-5) and promote motility of the tract. Sympathetic impulses usually are inhibitory but secondary to the dominant excitatory effects of the parasympathetic fibers.

Control of Motility

Hormonal Factors

Gastrointestinal motility, like gastrointestinal secretion, is regulated by hormonal and neural mechanisms. For example, gastrin promotes mixing of food in the stomach while CCK-PZ and secretin inhibit emptying (Table 22-5). Gastrin and CCK-PZ stimulate intestinal motility, which is opposed by secretin. These hormones and others interact in complex and as yet incompletely understood ways to influence movement of food along the digestive tract.

Neural Influences

Vagal stimulation of smooth muscle in the digestive tract maintains *muscle tone,* causes *segmental movement,* and produces waves of contraction called *peristalsis.* The parasympathetic impulses generate a low-grade *basic electrical rhythm* (BER) in smooth muscle, which in turn produces a low-grade contraction or slow waves of contraction called *tonus* (*muscle tone*). Parasympathetic impulses also may stimulate strong contractions and motility. The smooth muscle contracts as a single unit in defined segments of the tract (Fig. 12-13, page 302).

The chemical and physical properties of foods initiate a variety of reflexes (Table 22-5), whose names generally have dual origins. The first part of the name usually refers to the source of the stimulus and the second part to the site of the response. For example, the gastrocolic reflex has its origins in stimuli occurring in the stomach and results in mass peristalsis in the colon.

Among the reflex activities are those which influence thirst, feeding, and a feeling of fullness through groups of neurons in the hypothalamus (Table 11-5). Neurons which act as osmostats and glucostats respond respectively to changes in the osmotic pressure and glucose

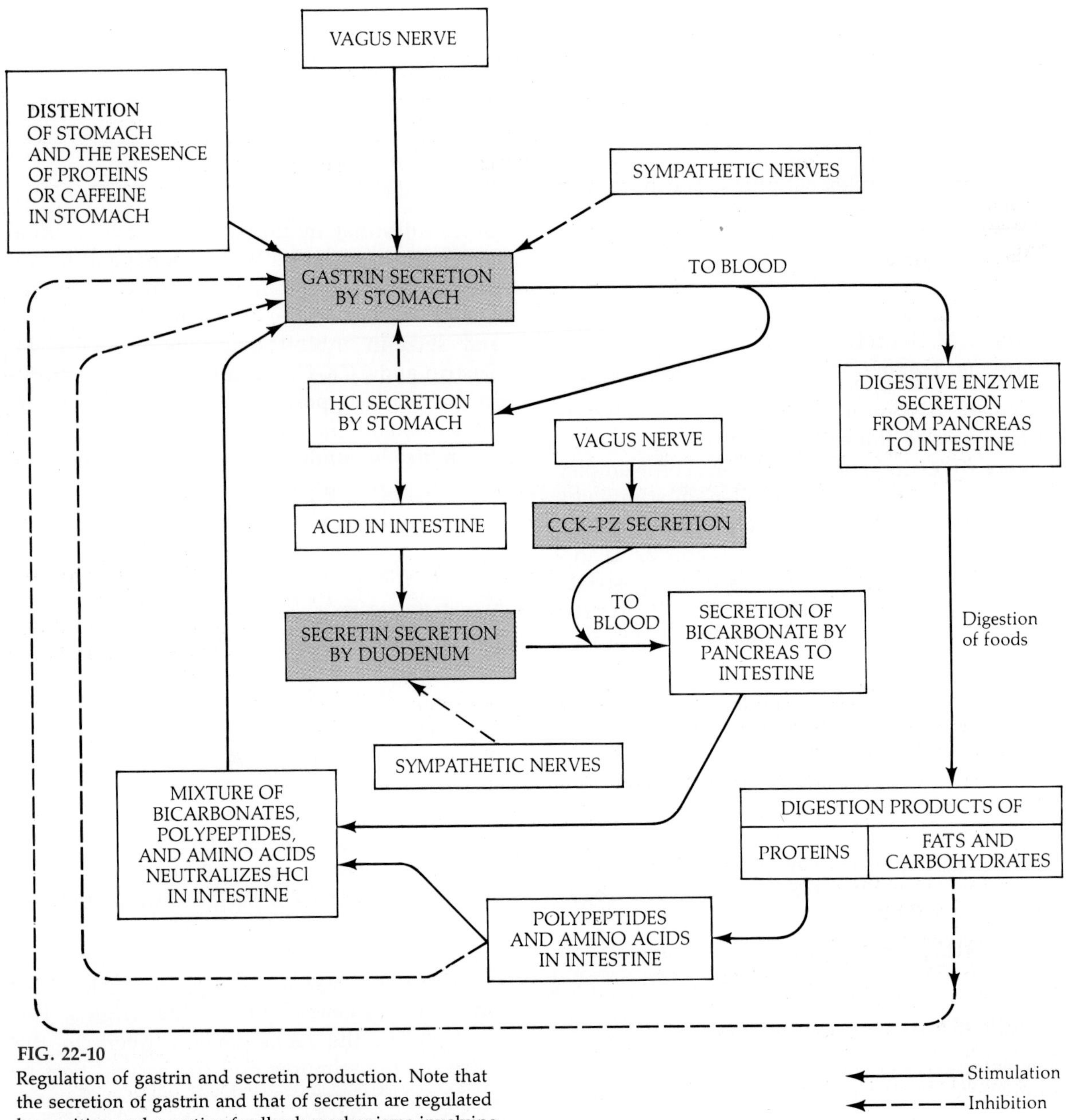

FIG. 22-10
Regulation of gastrin and secretin production. Note that the secretion of gastrin and that of secretin are regulated by positive and negative feedback mechanisms involving substances whose fates are in turn influenced by the hormones.

levels in the blood. These neurons make up *thirst, feeding,* and *satiety centers* in the hypothalamus. The exact mechanisms by which they function are not without debate. A desire to drink and a sense of thirst may develop when the osmotic pressure of the body fluids increases or when a decrease occurs in the ECF volume (Figs. 18-8 and 21-10). When glucose utilization by the satiety center is high, its neurons may inhibit the feeding center to produce a feeling of fullness. Conversely, when levels of glucose are low, its utilization by neurons in

TABLE 22-5
Motility of the digestive tract

	MOTILITY OF AFFECTED SEGMENT		
	Stomach		*Intestine*
Mechanisms	*Motility*	*Emptying**	
Hormonal			
Gastrin	+		+
Enterogastrones			
Secretin	−	−	−
CCK-PZ	+	−	+
Neural			
Central nervous system			
Stimuli of the CNS			
Smell, sight, thought of food	+		
Excitement, aggression (usually)	+		
Pain, sadness, fear, depression (usually)	−		
Parasympathetic fibers of the vagus nerve	+ (primarily)†	+	+
Sympathetic fibers	− (primarily)	−	−
Reflexes			
Increased food volume in stomach	+ (gastroenteric reflex)		+ (gastroileal reflex relaxes ileocecal sphincter) + (gastrocolic reflex causes mass peristalsis in the colon)
Stimulation of duodenal osmoreceptors (?)	− (enterogastric reflex)		
Presence of concentrated mixture of food and digestive juices (chyme)			
High fat content of meal (12- to 18-carbon fatty acids have the greatest inhibitory effect)			
High levels of polypeptides and amino acids from protein digestion			
Hypertonic salt solutions			
Acids (pH below 4.0)			
Carbohydrates (varying inhibitory effects but usually slight)			
Small to moderate distention of segments of the intestine			+ (myenteric reflex)
Small to moderate distention of the duodenum			+ (duodenocolic reflex causes vigorous mass peristalsis in the colon)
Distention of the rectum			+ (defecation reflex)
Large distention of specific segments of the intestine	− (ileogastric reflex)		− (intestino-intestinal reflex)

Key: + = stimulation; − = inhibition
* Emptying of the stomach and intestine is inhibited or delayed by stimuli which cause contraction of sphincters.
† Distention of the esophagus reflexively activates some vagal fibers which inhibit motility of the stomach.

the satiety center decreases and the inhibitory effects of the neurons on the feeding center may be diminished. Consequently, neurons in the feeding center may then be able to cause hunger pains by stimulating motility of the empty stomach.

Gastrointestinal smooth muscle cells act as mechanoreceptors when the walls of the digestive tract are stretched by food. Pacemaker cells in the inner circular layer stimulate *segmental movement,* which mixes the contents of a section in motions like those of a clock pendulum and like those used to knead dough. Segmental mixing occurs about 6 to 11 times per minute and decreases in frequency along the tract.

These oscillations can be overriden by peristaltic activity caused by the *myenteric reflex* (Table 22-5). It is initiated by mechanoreceptors or chemoreceptors in the wall of the intestine through Meissner's plexus and relayed by the myenteric plexus. As a result of impulses which are generated locally, longitudinal smooth muscle fibers in the muscularis externa in the small intestine alternately contract and relax. Waves of contractions, generally in proportion to the volume of food in the tract, propel the food through 60- to 70-cm segments of the small intestine at rates ranging from 2 to 25 cm/s. Abnormally strong or irritating stimuli may cause *mass peristalsis,* by which food is transported over greater distances even more rapidly. However, extremely large volumes of food inhibit intestinal motility, thereby affording greater opportunity for digestion and absorption.

The entry of fats into the duodenum causes a strong inhibitory *enterogastric reflex* of the stomach (Table 22-5). Eating hors d'oeuvres or other foods which contain fat slows the passage of food to the intestine. Absorption is slower and less extensive in the stomach than in the small intestine. As a result of the enterogastric reflex, food (and alcohol mixed with it) are more slowly absorbed. Conversely, carbohydrates generally have little inhibitory effect on the emptying of the stomach, and large quantities of carbohydrates rapidly enter the intestine after their ingestion. The osmotic effect of the carbohydrates draws water into the intestine, swelling it and producing afferent nerve impulses that may cause the sensation of nausea. In severe cases the vomiting center in the medulla is stimulated so that the stomach and abdominal muscles contract, forcing food through the relaxed gastroesophageal sphincter and out the mouth (regurgitation). Such responses may follow ingestion of large quantities of sweets.

Three to four times each day the *gastrocolic reflex,* initiated in the stomach, and the even stronger *duodenocolic reflex,* arising in the duodenum, cause food to move by mass peristalsis within long segments of the colon (Table 22-5). The reflex may be due either to neural activity or to the release of gastrin following a meal. The food ultimately enters the rectum. *Defecation reflexes* then stimulate the walls of the colon and rectum to contract. At the same time the internal anal sphincter reflexively relaxes while the external sphincter contracts. Then voluntary inhibition of the external anal sphincter allows the feces to exit. Spinal reflexes enhance these responses. Furthermore, the expulsion of feces is usually assisted by voluntary contractions of the abdominal muscles, or straining (the Valsalva maneuver).

22-4 DEFENSES AND INFLAMMATORY RESPONSES

The cells of the digestive tract are exposed to billions of different molecules and ions each second. In addition to nutrients, the lining is exposed to acids, alkalis, enzymes, hydrocarbons, and detergentlike molecules. The wall of the digestive tract is able to defend the body despite the potential for assault by these molecules. The mucosa sorts out, digests, and absorbs useful substances and eliminates undesirable or excess materials as waste. In some instances inflammatory reactions, which occur as part of the defense responses, cause gastrointestinal disorders. Some of them will be described here; others will be discussed in Chap. 23.

Among the defenses provided by the mucosa are its tight intercellular junctions, which prevent unauthorized entry of most substances, including hydrogen ions. The rapid and continual replacement of the mucosal barrier provides new cells at the front line of defense. Proliferation by the immature columnar cells in the mucosa pushes new cells to the surface every 2 to 6 days. Furthermore, the cells are shielded by the glycocalyx, which usually resists the attacks of bile, acids, and alcohol. Additionally, stomach acid is rapidly neutralized as it enters the small intestine and is mixed with bicarbonates.

Antimicrobial molecules such as IgA and lysozyme are delivered to the mucosal surface, as described earlier in this chapter. The lymphoid cells also produce other antimicrobial substances, which activate the cellular "militiamen," the macrophages, and which attract white blood cells to the sites of combat. The Küpffer cells, macrophages lining the liver sinusoids, are able to ingest large particulate materials, undesirable cells, and microbes. Liver enzymes also digest and neutralize potentially toxic substances and the colon eliminates them with the feces.

In spite of these defenses, disorders of the digestive tract occasionally occur. Bacteria attack the oral mucosa in periodontal disease (Fig. 14-9), irritate the walls of the intestine in bacterial food poisoning (Table 23-7), and even penetrate blood vessels to infect other organs in diseases such as typhoid fever. Poisons and foreign macromolecules which behave as allergens may penetrate the lining, causing spasm of smooth muscle, pain and abnormal function of segments of the tract (Fig. 3-20). Neural stimuli may cause similar responses. Because the number of gastrointestinal diseases is too extensive for enumeration in an introductory physiology text, only a few representative inflammatory disorders will be mentioned.

Inflammatory Reactions

Many gastrointestinal disorders may result from inflammatory reactions (Chap. 3). The diseases are usually named after the site of the inflammation coupled with the suffix *-itis,* which indicates an inflammatory response (Table 22-6). The reactions may be acute (severe but short) or chronic (protracted). They may cause abdominal pain, fever, localized inflammation, and other symptoms of varying severity. In the most serious disorders, tissue erosion and bleeding may result and can be fatal. Inflammation of the intestinal mucosa also can promote conditions which allow bacteria to penetrate to the blood, causing infection of the peritoneal membrane (the serosa) in *peritonitis,* which can lead to widespread infection of other organs and death.

Ulcers

Despite protective barriers, hydrogen ions and pepsin formed in the stomach may erode the mucosal lining and the underlying tissues of the digestive tract. *Ulcers* are erosions of the mucosa and are accompanied by inflammatory responses in the underlying tissues. They occur in about 1 in 10 people in the United States. More than 80 percent of the cases develop in males, with the incidence of the disorder increasing with age in both sexes. Three types of so-called peptic ulcers may be distinguished: *duodenal ulcers,* which occur with greatest frequency; *gastric* (stomach) *ulcers,* which are about one-fourth as common, and *esophageal ulcers,* which are the least common of the three. The causes of ulcers are complex. Hereditary and environmental factors may be involved, with a variety of contributing causes (Table 22-7).

Ulcers are more prevalent in some families than in others. Duodenal ulcers are more common in people with O type blood than in those with other blood types. Emotional stress, smoking, and alcohol consumption also increase the probability of ulcer formation.

The importance of acid production in certain ulcers is illustrated in *Zollinger-Ellison syndrome.* In this disease tumors of the pancreas (or stomach or duodenum) produce gastrin, which causes excessive hydrochloric acid secretion and may result in severe gastric, duodenal, or even jejunal ulcers.

Traditional ulcer therapy is somewhat controversial. It involves behavior modification to teach the patient to cope with stress, stop

TABLE 22-6
Examples of gastrointestinal inflammation

Name of Disorder	*Site*	*Comments*
Gastritis	Stomach	May progress to resemble ulcers
Pancreatitis	Pancreas	May involve disease of bile tract and effects of alcohol; may cause elevated serum enzyme activity (such as amylases, lipases)
Hepatitis	Liver	May be caused by viral infection, excess drugs, and alcohol Type A viral hepatitis (infectious hepatitis) transmitted by contaminated foods or water. Incidence: about 30,000 cases annually in United States Type B viral hepatitis (serum hepatitis) may be transmitted by a contaminated syringe needle, by contaminated fluid secretions, and by blood in transfusion*; may be transmitted by feces, urine, or saliva. Incidence: about 15,000 cases annually in United States†
Cirrhosis	Liver	Inflammation in which connective tissue cells replace parenchymal cells and disrupt the organization of the liver; often results in jaundice; may be caused by infections (such as hepatitis), toxins, drugs, and alcohol
Ileitis	Ileum	Causes pain in lower right quadrant of abdomen, fever, and leukocytosis
Appendicitis	Appendix	Symptoms resembling those of ileitis
Colitis	Colon	Rectal bleeding and fever may accompany this inflammatory response; most common between ages of 20 and 40; less common in blacks than in whites; may involve food allergies or hereditary and psychological factors
Diverticulitis	Along the length of the alimentary canal from the esophagus through the colon	Diverticula and herniated pouches often formed in the colon are found in over half of those over age 50, often as a result of increased pressure accompanying difficult bowel movements; often go unnoticed; incidence starts to increase in persons beyond age 35; inflammation of diverticula may lead to erosion of the lining and bacterial infection of the peritoneal cavity (peritonitis)

* Transmission of type B hepatitis by blood transfusions has almost been eliminated by screening of blood for the virus prior to its use.
† In addition, nearly 10,000 cases of hepatitis occur annually and are of unspecified origins, some being neither the A nor B type.

smoking, avoid spicy foods, caffeine, and alcohol, and eat small rather than large meals. Histamine antagonists (such as cimetidine, which blocks H_2 receptors) are commonly used to treat ulcers. They inhibit histamine-induced acid secretion by parietal cells. Consumption of antacids temporarily neutralizes hydrochloric acid but does not cure ulcers. Anticholinergic drugs (such as atropine, Table 9-4) inhibit vagal stimulation of acid secretion. In some cases surgical removal of the ulcerated tissue and/or cutting of the vagal fibers to the stomach (vagotomy) are recommended. Severing the vagus removes the extrinsic stimulus for acid production and allows intrinsic reflex arcs to regulate this activity. In some cases the antrum of the stomach (a source of gastrin) is also removed to further diminish acid secretion.

Liver Disease

Numerous agents, including drugs, toxins, and microorganisms may cause liver disease and alter bile pigment and protein metabolism. Products of bile metabolism such as conjugated bilirubin (Fig. 18-6) appear in elevated amounts in the blood and may cause jaundice when gallstones are formed or when other liver diseases occur. Transaminase enzymes are released from the liver into the blood during stress in proportion to liver cell damage. Some of these changes in the blood may be used to diagnose various diseases (Chap. 16, page 393).

Two common forms of liver disease, both of which are inflammatory, are *hepatitis* and *cirrhosis of the liver* (Table 22-6). Hepatitis is often caused by specific viral infections which afflict tens of thousands of people each year in the

United States. The more common of two varieties is type A virus, which causes *infectious hepatitis* (epidemic hepatitis) and is transmitted by contaminated food and water. Another prevalent form is type B virus, which causes *serum hepatitis* and is transmitted by contaminated syringe needles or contaminated fluids, including saliva, urine, feces, or blood. The resulting infections produce inflammatory reactions which disrupt liver organization and metabolism, cause jaundice and extreme fatigue, and require long periods of rehabilitation.

Cirrhosis of the liver is a diffuse inflammatory disease which destroys liver structure. It may be caused by viral infection (chronic hepatitis), by the toxic effects of alcohol or excess iron, by obstruction of the bile ducts, and by poor circulation resulting from congestive heart failure. When damaged, the liver has tremendous powers of regeneration, although cell division is infrequently observed in the healthy organ. Following damage, not only do surviving parenchymal cells proliferate but so do fat and other connective tissue cells, which causes a condition called *fibrosis.* The connective tissue growth disrupts the lobular organization of the liver and interferes with its blood flow and overall function. The diseased liver fails to metabolize bile normally and products of bile metabolism such as bilirubin accumulate in the blood, causing jaundice as described above. Thus, the name cirrhosis originated from the Greek word *kirrhos,* orange-colored.

Blood pressure increases in the hepatic portal vessels, so that portal hypertension results. As the disease progresses, marked fatigue, jaundice, increased susceptibility to infections, and enlargement of the spleen may occur. Ultimately, abnormal fluid retention, ascites, or peripheral edema may be observed (pages 389 and 412). The accumulation of fluid is due to combinations of elevated portal venous pressure, low serum protein levels, and/or excessive aldosterone secretion, occurring as a result of the stress of the disease and diminished capacity of the liver to degrade this hormone. Gastrointestinal bleeding, coma, and death may result.

TABLE 22-7
Comparison of gastric and duodenal ulcers

Property	*Gastric Ulcer*	*Duodenal Ulcer*
Usual time of occurrence of pain	Within 1 h after a meal	Within 2.5 to 3 h after a meal
Possible contributing physiological factors (in addition to genetics and stress)	Defective mucosa Poor blood supply Delay in gastric emptying Mucosal damage by the entry of bile salts into the stomach. They act as detergents, disorganizing the lipids of the epithelial cells Alcohol, aspirin, or other drugs	Rapid gastric emptying Excessive gastrin secretion Diminished bicarbonate secretion

22-5 SUMMARY

1 The digestive system digests and absorbs food. It is composed of the mouth, parts of the pharynx, esophagus, stomach, small and large intestines, and accessory digestive organs. The mouth of an adult usually contains 32 teeth, which bite, tear, and grind food, and the tongue, which moves food, receives and transmits sensory stimuli, and modifies sound. Cells on the surface of the tongue act as chemoreceptors and generate tastes. Although chewing and swallowing initially are voluntary, they become reflexive as the activities proceed.

Food moves through various segments of the tract by contraction of smooth muscle fibers arranged circularly and longitudinally. Certain circular muscle fibers form sphincters in the esophagus, stomach, ileum, and rectum, which regulate passage of food by narrowing and closing off segments of the tract or by relaxing to open them up.

Food leaves the mouth and passes through the pharynx and the esophagus to the stomach. From there it passes into the duodenum, jejunum, and ileum of the small intestine. The combination of food and wastes travels through the cecum and the ascending, transverse, descending, and sigmoid colon prior to exiting through the rectum by way of the anus. On this route it passes a narrow projection from the cecum, the appendix, a structure with little apparent digestive function.

Most of the wall along the digestive tract is composed of four layers of tissues. They are, from inside to outside, the mucosa, submucosa, muscularis externa, and adventitia. The mucosa is folded inward in the small intestine and forms fingerlike villi with microvilli on the epithelilial cells, giving a striated appearance referred to as a brush border. The cells are coated with glycoproteins that form a protective glycocalyx at their apexes which project into the interior of the gastrointestinal lumen.

Five different cell types—immature columnar cells, absorptive cells, goblet cells, enterochromaffin cells, and Paneth cells—are located in depressions of the mucosa, the crypts of Lieberkühn. All these cells, with the exception of intestinal Paneth cells, are also present in the mucosa of the villi.

The accessory digestive organs are glandular and release a variety of secretions into the tract. The salivary glands secrete saliva; the pancreatic acini synthesize enzymes; the pancreatic ducts manufacture bicarbonates; the liver produces bile; and the gallbladder concentrates the bile. Except for saliva, which is delivered into the mouth by ducts from the salivary glands, the secretions of the other accessory digestive organs are transported by the pancreatic and common bile ducts, which usually merge to deliver their contents through a common opening into the duodenum.

The liver is made up of functional units, lobules, that consist of rows of cells radiating toward central veins. Each row is flanked by a bile duct on one side and a venous sinusoid on the other. Bile secreted by the liver is carried in bile ducts to hepatic ducts, where it is transported to the common bile duct and delivered through the sphincter of Oddi to the duodenum.

The gallbladder is a muscular sac on the underside of the liver. The cystic duct from the gallbladder joins the hepatic duct from the liver to form the common bile duct. When the sphincter of Oddi is closed, bile released from the liver may back up through the common bile duct to enter the cystic duct and the gallbladder. The mucosal lining of the gallbladder concentrates bile for storage. When stimulated, the muscular wall of the gallbladder contracts to force bile into the duodenum.

Blood is delivered to liver sinusoids and adjacent parenchymal cells from digestive organs and the spleen by portal veins and from the abdominal aorta by hepatic arteries. The flow of blood from the digestive tract transports most foods directly into the liver prior to entry into the general circulation. Fats are an exception. Most of them enter the lymph stream from the small intestine. Because of the liver's critical metabolic, circulatory, and storage functions, it receives up to 30 percent of the blood pumped from the heart.

2 Digestive secretions are both endocrine and exocrine in nature. The endocrine secretions of specific enterochromaffin cells include four polypeptide hormones with some overlapping influences on gastrointestinal physiology. Gastrin causes secretion of stomach hydrochloric acid and pepsinogen, and pancreatic enzymes. Cholecystokinin-pancreozymin enhances pancreatic

enzyme secretion and the release of bile from the gallbladder. Secretin promotes the secretion of bicarbonates from the pancreas. Gastric inhibitory polypeptide (GIP) inhibits the secretion of stomach acid and causes insulin release from the pancreas.

Exocrine secretions of the digestive system include saliva, mucus, digestive enzymes, bicarbonates, and bile. Saliva is 99 percent water but also contains salts, amylase, immunoglobulin A, and glycoproteins. Exocrine glands in the stomach include mucous neck cells which secrete a viscous, alkaline mucus that contains enzymes and is rich in glycoproteins; chief cells which release pepsinogen; and parietal cells which produce hydrochloric acid and intrinsic factor. Up to 2 L of acid at a pH near 1.0 is produced daily. The acid kills bacteria and alters digestive enzymes by disrupting their hydrogen bonds.

Cells on the surface of the intestinal mucosa and in the crypts of Lieberkühn secrete mucus, digestive enzymes, and antimicrobial compounds such as lysozyme. Brunner's or duodenal glands in the submucosa also release alkaline mucus into the crypts in the duodenum.

The pancreas secretes enzymes which digest polysaccharides, proteins, fats, and nucleic acids. Bicarbonates produced in the pancreas and delivered to the duodenum aid other alkalies and bile to neutralize the acidic contents of the duodenum and provide an environment in which the pH is optimal for the activity of pancreatic enzymes and bile salts. The liver forms bile, an exocrine secretion resembling plasma to which bile salts, cholesterol, and lecithin have been added. When the ratio of these three components is altered, cholesterol may precipitate and gallstones may be formed.

3 Cephalic, gastric, and intestinal stimuli regulate gastrointestinal secretions. They do so through extrinsic nervous impulses from the autonomic nervous system, gastrointestinal hormone feedback mechanisms, and local neural reflexes involving the sensory fibers of Meissner's plexus in the submucosa and the motor fibers in the myenteric plexus in the muscularis externa.

Autonomic nerve fibers to the digestive tract transmit parasympathetic impulses, which modify digestive functions, through the vagus and usually stimulate motility and glandular secretions and dominate the opposing sympathetic impulses. Gastrointestinal functions are further influenced by local neural and hormonal feedback mechanisms. In general, small to moderate volumes of food in the tract increase motility, whereas large volumes produce osmotic effects and physical distension that inhibit it. For example, stimuli generated by the presence of moderate volumes of food in a segment of the tract generate local reflexes, causing myenteric motor impulses that increase motility. Also, the presence of fats in the duodenum stimulates the secretion of CCK-PZ, which inhibits emptying of the stomach and causes bile and lipases to be released into the duodenum. These responses enhance the opportunities for the absorption of fat. The decline of the fat content in the fluid-food mixture in the duodenum turns off CCK-PZ secretion.

4 The digestive tract's defenses include a glycocalyx shield, tight intercellular junctions, rapid replacement of its cellular lining, secretion of antimicrobial agents, and neutralization of acid by bicarbonates. Furthermore, lymphoid cells, macrophages, and white blood cells in the submucosa assist defenses through their cellular and molecular activities. Even so, external factors such as bacteria, viruses, poisons, and foodborne allergens may cause digestive diseases. In many instances they cause inflammatory reactions in a disease whose name includes that of the organ plus the suffix *-itis* (as in colitis or appendicitis). Ulcers, hepatitis, and cirrhosis of the liver are three diseases

in which inflammatory reactions occur. Ulcers are erosions of the stomach and duodenum caused by hydrogen ions and pepsin. Hepatitis is often caused by type A or type B hepatitis viruses or other viruses of unspecified type. Cirrhosis is an inflammatory disease of the liver in which damaged parenchymal cells are replaced by connective tissue cells. It may be caused by toxins, alcohol, or viruses. The proliferation of the connective tissue disrupts the organ's lobular organization, disturbs its function, and upsets bile metabolism to cause jaundice.

REVIEW QUESTIONS

1 How do digestion, absorption, and metabolism differ?
2 Draw and label a diagram outlining the major digestive organs.
3 Pretend that you and three of your friends are protein, starch, fat, and nondigestible molecules respectively, in a hamburger. Describe your routes of passage through the digestive tract. Note the geographic points of interest and natural stops along the way.
4 What is the general microscopic appearance of the layers of the wall of the digestive tract? What are villi, microvilli, the brush border, and the glycocalyx?
5 What are the principal glands in the digestive tract? Explain their functions. Be sure to describe the cells in the gastric glands, as well as those in the crypts of Lieberkühn.
6 List the accessory digestive organs and give their general anatomical relationships to the digestive tract.
7 Of what value is saliva? Hydrochloric acid? The intrinsic factor?
8 Why does the digestive tract not digest itself?
9 Why is the pancreas important to digestive function? Where do its exocrine secretions originate?
10 Describe the structure and function of the liver. Why is the liver so important to the body?
11 What is bile? How is it formed? What are its functions? What is its relationship to jaundice and gallstones?
12 Compare cephalic, gastric, and intestinal mechanisms of control of digestive tract motility. Be sure to describe local neural reflex mechanisms and differences between the effects of parasympathetic and sympathetic impulses.
13 Describe three types of smooth muscle activity in the digestive tract and their relationship to the movement of food.
14 Explain the enterogastric, gastrocolic, and myenteric reflexes. What are their effects?
15 Al Salvi eats a hamburger (which contains protein, fat, and carbohydrate) and drinks two glasses of strong lemonade. Describe the likely effects of this meal on his gastrointestinal hormones and the effects of the hormones on his digestive system.
16 Identify the major defenses provided by the digestive tract.
17 Describe the general relationship between the inflammatory response and gastrointestinal disorders.
18 What are ulcers and how do they occur?
19 Describe hepatitis and its causes and effects. What is cirrhosis of the liver? Describe some possible causes of this condition and its effects.

THE DIGESTIVE SYSTEM

Food Processing

23

23-1 DIGESTION

23-2 ABSORPTION
Role of various segments of the digestive tract
Absorption of specific substances
Carbohydrates
Proteins
Lipids
Vitamins
Ions and trace elements
Water
Digestive tract excretions
Malabsorption
Gastroenteritis and food poisoning
The role of the intestinal bacterial flora
Therapy

23-3 INTERMEDIARY METABOLISM
Carbohydrates
Proteins
Lipids

23-4 METABOLISM AND NUTRITION
Caloric values of foods
Basal metabolic rate
Metabolism and body temperature
Variations in body temperature
Effects of food and exercise
Obesity

23-5 SUMMARY

OBJECTIVES

After completing this chapter, the student should be able to:

1. Describe the sources and activities of the various digestive enzymes.
2. Discuss the absorption mechanisms for specific nutrients in various segments of the digestive tract.
3. Explain the origins and the nature of feces and identify the factors which regulate their formation and release.
4. Describe malabsorption, its principal causes and effects, and means of appropriate therapy.
5. Outline the steps in the intermediary metabolism of carbohydrates, proteins, and lipids and explain the transfer of energy to ATP.
6. Relate metabolism and metabolic rate to nutrition and caloric expenditure of energy and discuss factors which affect the basal metabolic rate, including specific foods and exercise.

For a substance to exhibit nutrient value it must: (1) be ingested; (2) in some instances be digested into its building blocks as it is moved along the digestive tract; (3) cross the mucosa (be absorbed); and (4) be incorporated (assimilated) into the body. Chapter 22 described the first two activities and the production of secretions that enter the digestive tract. This chapter will concentrate on the absorption of foods and their nutritional value.

23-1 DIGESTION

Digestion is the physical and chemical conversion of food molecules into smaller units which can be absorbed (Chap. 22). The smaller building blocks are used to provide energy or to synthesize compounds essential to body structure and function. Three of the major foodstuffs—carbohydrates, proteins, and fats—often require digestion prior to *absorption,* their passage across the gastrointestinal mucosa. During digestion, enzymes along the inner lining of the tract cleave specific molecular bonds of nutrients in the presence of water in a reaction called *hydrolysis.* Because most of the digestive enzymes cling to the microvilli in the small intestine, the products of digestion are near the surfaces through which they may be absorbed. Small molecules and ions move across the mucosa and may then enter the blood or lymph. Carbohydrates are converted into simple sugars, fats into fatty acids and glycerol, and proteins into amino acids. The digestive enzymes and their products are summarized in Table 23-1. Their properties were also described in Chaps. 2 and 22.

Among the enzymes which digest carbohydrates are salivary and pancreatic amylase and intestinal dextrinases, sucrase, maltase, and lactase. Proteins are hydrolyzed by peptidases, including gastric pepsin; pancreatic trypsin, chymotrypsin, and carboxypeptidases; and intestinal peptidases. Lipids are cleaved by pancreatic and intestinal lipases. Nucleic acids and their building blocks are broken down by intestinal nucleases, nucleotidases, and nucleosidases.

23-2 ABSORPTION

Absorption, the passage of substances across the mucosa, is usually followed by their entry into blood or lymph. The mechanisms, although not completely understood in each instance, involve active transport, diffusion, facilitated diffusion, and pinocytosis, as described in Chap. 4. Substances may pass through cell membranes and intracellular compartments of the epithelial cells and enter interstitial spaces to cross the membranes of the endothelial cells and enter a capillary or lymph vessel. Abnormalities in any of these locations can interfere with absorption.

Role of Various Segments of the Digestive Tract

The mucosa on the tongue demonstrates limited, though rapid, absorption of some substances. This route reduces the time usually

TABLE 23-1
Summary of principal digestive enzymes* (continued on next page)

(continued on next page)

Source	Enzyme	Requirements for Optimal Activity	Substrate(s)	Product(s) of Hydrolytic Action
Mouth				
Salivary glands	Salivary amylase (ptyalin)	pH 6.6–6.8, Cl^- ions	Glucose linkages in starch and glycogen	Dextrins (oligosaccharides) Maltotriose Maltose (a disaccharide) Other disaccharides
Stomach				
Chief cells	Pepsin (gastric protease)	pH 1.0–2.0	Peptide bonds next to aromatic amino acids in proteins, polypeptides	Proteoses Polypeptides
	Rennin (in infants)	pH 4.0, Ca^{2+} ions	Casein (milk protein)	Coagulates milk
	Lipase (slight activity in adults, more activity in infants)	pH 5.5	Fats composed of short-chain fatty acids	Fatty acids and glycerol
Pancreas				
Acinar cells	Pancreatic amylase	pH 6–7.1, Cl^- ions	Same as salivary amylase	Same as salivary amylase
	Trypsin	pH 5.2–6.0 and enteropeptidase (enterokinase) (see below) or pH 7.9 alone	Peptide bonds of basic amino acids in proteins, proteoses, and polypeptides	Polypeptides Dipeptides
	Chymotrypsins	pH 8.0 and trypsin	Same as trypsin	Same as trypsin
	Carboxypeptidase A	Trypsin	Carboxyamino acids with ring structure or branched chains at ends of proteins and polypeptides	Smaller peptides, free amino acids
	Carboxypeptidase B	Trypsin	Carboxyamino acids with basic side chains at ends of proteins and polypeptides	Same as carboxypeptidase A
	Pancreatic lipase	pH 8.0, emulsifiers (such as bile)	Ester bonds of triglycerides	Fatty acids Glycerol Diglycerides Monoglycerides
	Phospholipase A	Trypsin	Lecithin	Lysolecithin
	Cholesterol ester hydrolase	Bile salts	Ester linkages in cholesterol	Free cholesterol, fatty acids
	Ribonuclease		RNA	Nucleotides
	Deoxyribonuclease		DNA	Nucleotides
Small Intestine				
Brunner's glands in duodenal submucosa, and crypts of Lieberkühn at base of villi	Dextrinase	pH 6–9	Glucose linkages in dextrins (oligosaccharides)	Glucose

TABLE 23-1
Summary of principal digestive enzymes* (continued)

Source	*Enzyme*	*Requirements for Optimal Activity*	*Substrate(s)*	*Product(s) of Hydrolytic Action*
These enzymes are located on membranes of cells on the inner lining of the intestine	Sucrase (invertase)		Sucrose	Fructose and glucose
	Maltase		Maltose Maltotriose	Glucose Glucose
	Lactase		Lactose	Glucose and galactose
	Isomaltase (1,6 glucosidase)		Glucose linkages	Glucose
	Enteropeptidase (enterokinase)		Trypsinogen	Trypsin
	Aminopeptidases		Peptide bond of amino acids at ends of poly-peptides	Smaller peptides, amino acids
	Dipeptidases		Dipeptides	Amino acids
	Intestinal lipase		Triglycerides Monoglycerides	Glycerol Fatty acids
	Phospholipase		Phospholipids	Glycerol Fatty acids Phosphoric acid Bases from the phospho-lipid
	Nucleases		RNA, DNA	
	Polynucleotidases		RNA, DNA	Nucleotides
	Nucleosidases		Purine or pyrimidine	Purines or pyrimidines and pentose phosphates
			Nucleosides derived from nucleotides of DNA and RNA	
	Phosphatase	pH 8.6	Organic phosphates	Free phosphate

* Enzymes which cleave particular types of molecules are generally listed in the sequence in which they act on specific foods. Large molecules are thereby converted to their smaller building blocks as they pass from the stomach through the small intestine.

required for substances to pass along the tract prior to reaching a site of transport into the blood. For example, nitroglycerine (a lipid-soluble molecule) when placed under the tongue can pass directly into the blood to relieve quickly angina due to oxygen deficiency in heart muscle (Chap. 16). Few substances travel through tight junctions of the stomach's mucosa. However, low-molecular-weight lipid-soluble substances such as alcohol, weak acids, aspirin, and small amounts of iron can cross the stomach's mucosa.

The bulk of absorption occurs through the first part of the small intestine (Fig. 23-1) by means of transport enzymes incorporated into the large surface area of membranes of the microvilli. Intestinal absorptive capacity is so great that removal of most of the distal half of the small intestine can leave this function essentially intact. The cells of the jejunum behave as though their pore size were about twice that of cells of the ileum (0.75 nm versus 0.35 nm). Monosaccharides, amino acids, and fatty acid chains (up to 10 to 12 carbon atoms in length) enter the blood through a circulatory capillary within a villus. Longer chains of fatty acids and other lipids directly enter *lacteals,* the lymph capillaries within the villi.

Although the colon acts as though its pores were the smallest in the intestine (about

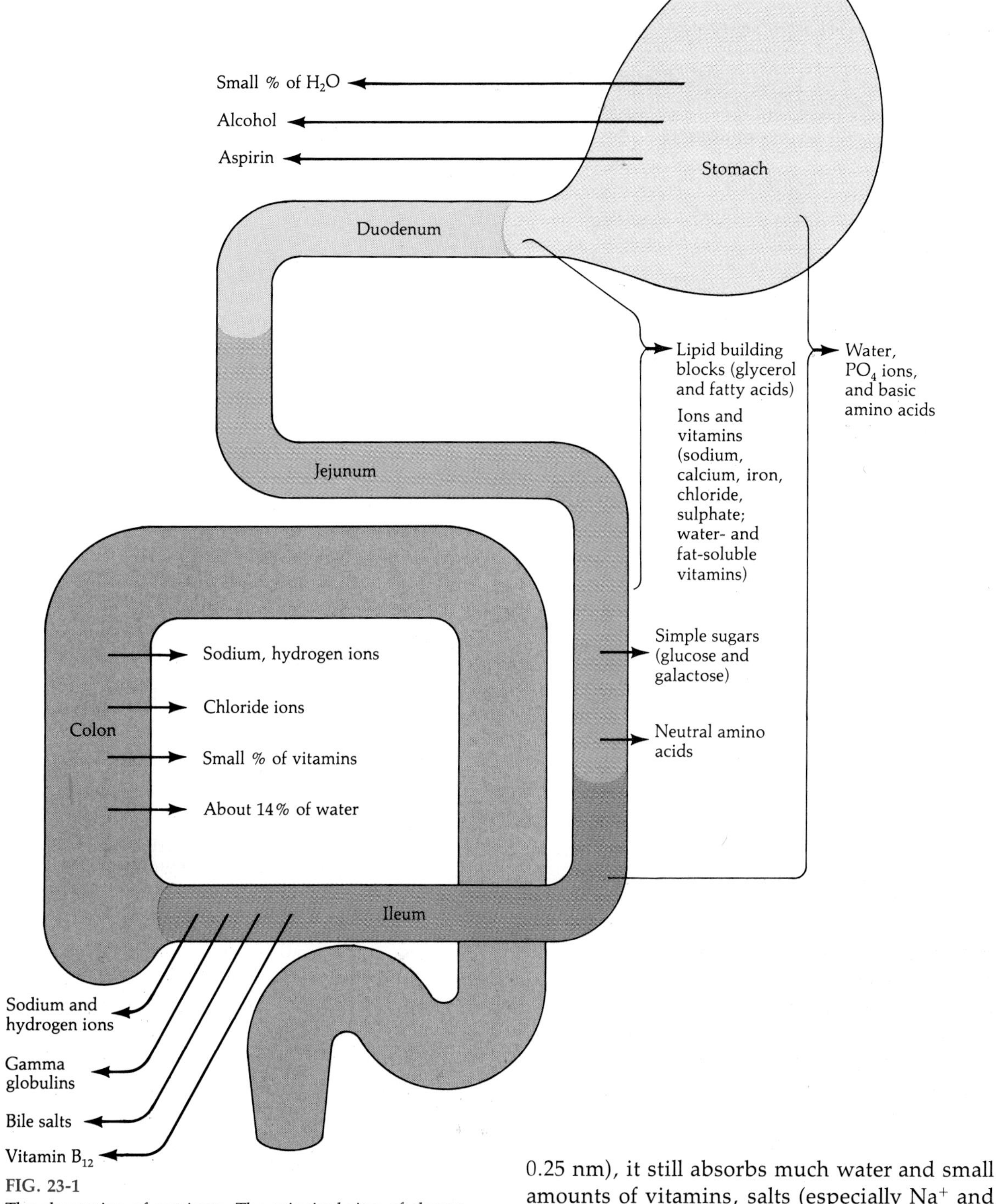

FIG. 23-1
The absorption of nutrients. The principal sites of absorption are illustrated. For most substances the process begins in the duodenum and is completed in the first third to first half of the small intestine.

0.25 nm), it still absorbs much water and small amounts of vitamins, salts (especially Na^+ and Cl^- ions), and drugs. This capacity allows the administration of medications in a *suppository* (a solid body containing medication, administered through one of the body openings such as that of the rectum or vagina).

TABLE 23-2
Major mechanisms of absorption of foods in the small intestine **(continued on next page)**

Food	*Mechanisms of Absorption*	*Comments*
Carbohydrates		
Glucose and galactose	Active transport	Glucose forms nearly 80% of the products of carbohydrate digestion. Its carrier actively transports glucose (or galactose, but less effectively). The active transport process is linked to that of sodium and requires energy. Insulin stimulates the glucose carrier in skeletal muscle and in the liver but not in the intestine
Fructose	Diffusion	The hydrolysis of sucrose produces one molecule of glucose and one of fructose. Fructose diffuses into the cell faster than does glucose but at a rate much slower than the active transport of glucose
	Facilitated diffusion	Slower than active transport
Other monosaccharides	Diffusion	The comparative rates of diffusion are fructose > mannose > xylose > arabinose
Proteins	Pinocytosis	Gamma globulins and other proteins in newborn May result in allergy (see Chaps. 3 and 4)
Peptides	Active transport	A single carrier system seems to exist for different di- and tripeptides The process is linked to active transport of sodium The rate of absorption is much more rapid than the rate of digestion of proteins and polypeptides The hydrolysis of the peptides is completed in the epithelial cells and releases amino acids, which can diffuse through the interstitial fluid into capillaries
Amino acids	Active transport	Specific carriers metabolically linked to sodium transport are responsible for movement of amino acids across the mucosal epithelium.
Lipids	Diffuse in chylomicrons into lymph unless noted otherwise.	Nearly 50% of lipid molecules are absorbed by the time they reach the end of the duodenum and 95% are absorbed by the middle of the jejunum
Neutral fats (triglycerides)		Fatty derivatives, phosphates, and bile salts combine in charged droplets called micelles. Lipases hydrolyze components in micelles, allowing diffusion and absorption of these derivatives to occur across the mucosa of the intestine. Short to medium-chain fatty acids move through mucosal cells to enter the blood. Other derivatives from the micelle may be used to resynthesize lipids and may be packaged with cholesterol, phospholipid, and protein to form particles called chylomicrons, which enter the lymph stream through lacteals
Fatty acids (up to 8–12 C atoms in length)	Diffusion into blood	
Monoglycerides	Diffusion into blood	
Fatty acids over 12 C atoms in length		
Glycerol		Absorption from micelles or direct diffusion into mucosal cell
Cholesterol esters (from cholesterol and fatty acids)		Free cholesterol diffuses from micelles into mucosal cell. Cholesterol is derived from food, bile, and intestinal cells which are shed into the lumen of the digestive tract
Vitamins		
Water-soluble vitamins (except vitamin B_{12})	Diffusion or facilitated diffusion	Maximal absorption into blood in first segments of small intestine through jejunum
Vitamin B_{12}	Not determined (may be active or passive)	Requires intrinsic factor Occurs in ileum (Table 18-8 and page 558)
Fat-soluble vitamins (A, D, E, and K)	Diffusion	Maximal absorption into lacteal after incorporation into a micelle; occurs primarily in first segments of small intestine

TABLE 23-2
Major mechanisms of absorption of foods in the small intestine (continued)

Food	*Mechanisms of Absorption*	*Comments*
(Fat-soluble vitamins)		Release and diffusion following hydrolysis of the micelle by pancreatic lipase
Nucleic acids		
Nucleosides, purines, pyrimidines	Diffusion	Purines are metabolized to uric acid. The rate of formation can exceed the rate of excretion by the kidney in a heritable disease called gout (Table 14-7). In pneumonia and leukemia excessive degradation of nucleic acids in white blood cells destroys these cells and elevates uric acid production

Absorption of Specific Substances

Carbohydrates

The absorption of nutrients is illustrated in Fig. 23-1 and that of some major foodstuffs is summarized in Table 23-2. Carbohydrates are absorbed as simple sugars, which may be actively transported across epithelial cells in the mucosa. The similarity of glucose to galactose allows the same carrier to bring one or the other sugar into the cell although it more effectively carries glucose. The carrier also seems to move sodium simultaneously into the cell in an active transport activity that is energy-requiring. Ultimately, the sodium pump expels sodium into the interstitial fluids (Fig. 23-2). Because the two systems are metabolically linked, the presence of sodium in the intestine stimulates the simultaneous active transport (cotransport) and absorption of glucose molecules into the mucosa and their eventual entry into the blood. "Downhill" movement of monosaccharides also may be assisted by carriers which facilitate their diffusion.

Proteins

Protein is digested by peptidases on the surface of the microvilli of intestinal mucosal cells and is converted into tri- and dipeptides and amino acids (Fig. 23-3). These products are actively transported into cells in a process that is metabolically linked to that of sodium and is similar to that described for glucose (Fig. 23-2). Four independent transport systems have been discovered: one for neutral amino acids, a second for basic amino acids, a third for the amino acids glycine and proline, and a fourth for glutamic and aspartic acids. Another transport system exists for di- and tripeptides. In the epithelial cell, peptides are further digested and then their ultimate products, amino acids and perhaps to a lesser extent some undigested di- or tripeptides, enter the blood.

About 20 amino acids are used to synthesize a variety of human proteins (or polypeptides). In adults eight are *essential amino acids,* those which cannot be synthesized by cells in the body and must be provided directly in proteins in the diet; they include isoleucine, leucine, lysine, methionine, phenylalanine, threonine, tryptophan, and valine (App. A). The 12 nonessential amino acids are those which can be synthesized in the body of an adult if the essential amino acids are provided in adequate quantities; the 12 nonessential amino acids are arginine, alanine, asparagine, aspartic acid, cystine, glutamic acid, glutamine, glycine, histidine, proline, serine, and tyrosine. The dietary amino acid requirements are listed in App. 23A at the end of this chapter.

While foods provide *exogenous* (external) sources of nutrients, cells shed from the mucosa also provide *endogenous* (internal) sources of nutrients. About 33 to 50 percent of the amino acid pool used to synthesize body protein is derived from endogenous sources. Fats, carbohydrates, and other nutrients are conserved in a similar fashion through endogenous recycling.

Slight amounts of protein may be absorbed intact by pinocytosis in the newborn for the first 72 h or so. Pinocytosis of protein diminishes when the amount of IgA antibody formed

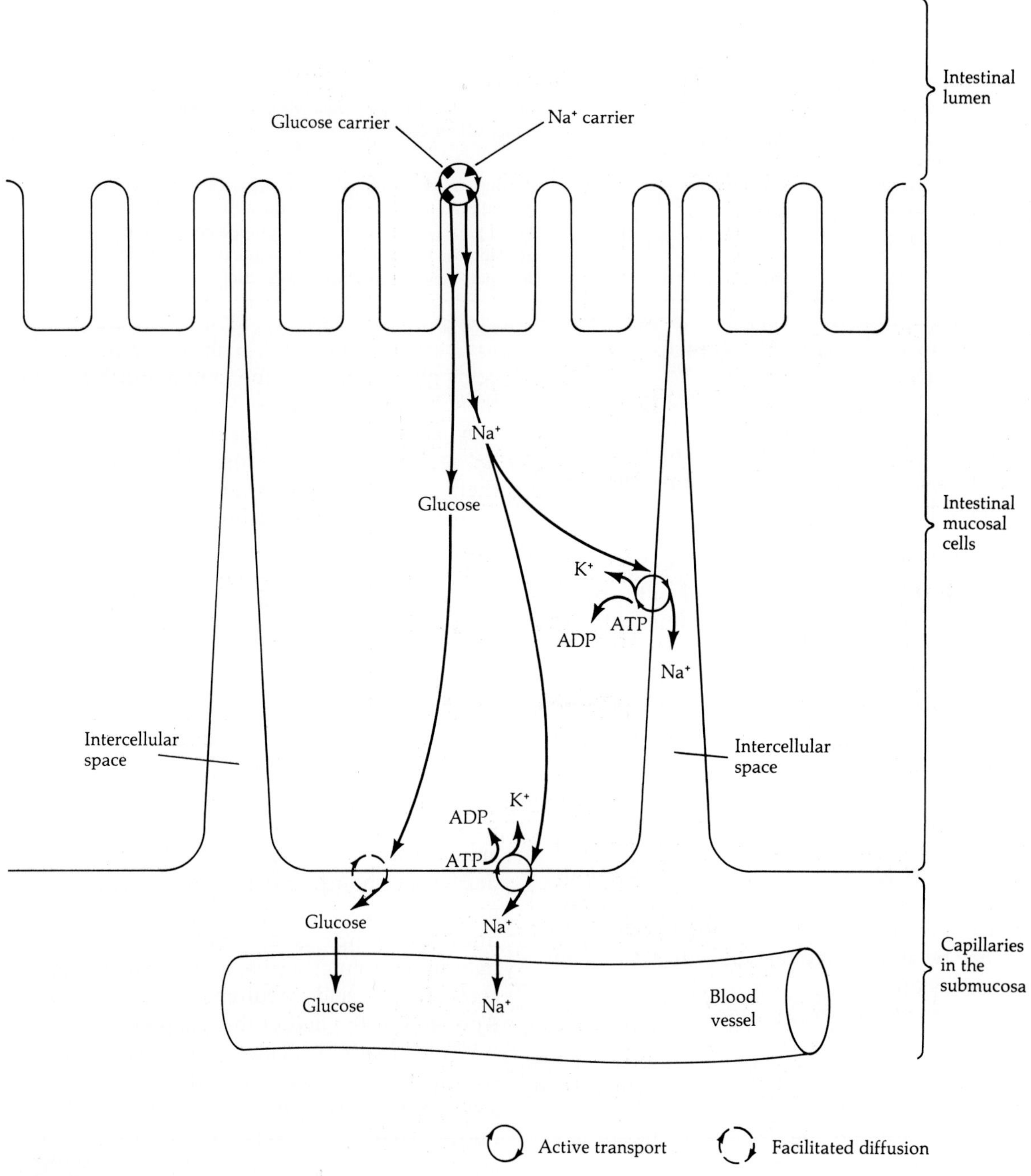

FIG. 23-2
The absorption of glucose across a mucosal epithelial cell in the small intestine. The movement of glucose from the lumen of the small intestine into the cell is linked to that of sodium. After entry into the cell, sodium is actively transported into the intercellular space by the sodium pump in an ATP-dependent reaction. Glucose enters the interstitial fluid by facilitated diffusion. Both substances may diffuse into the blood from the interstitial fluid.

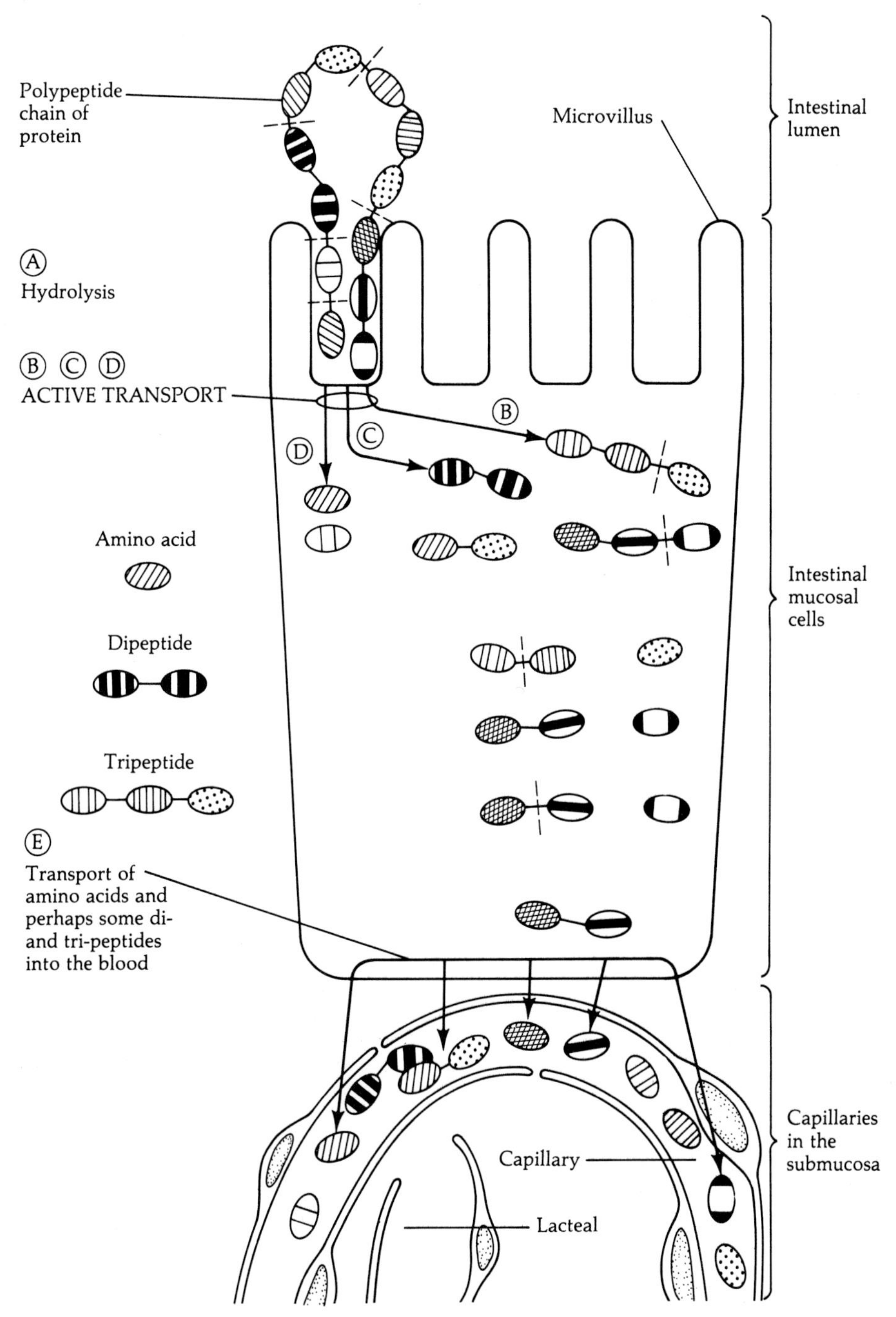

FIG. 23-3
The absorption of proteins. Proteins are composed of polypeptide chains which are cleaved by enzymes on the microvilli. The points of cleavage are illustrated by broken lines. Peptide bonds are (A) broken to release (B) tripeptides (C) dipeptides and (D) amino acids, which are actively transported into mucosal cells. Most of the peptides are further hydrolyzed within the cells to produce amino acids. (E) Amino acids and perhaps some di- and tripeptides are transported into the blood and carried to the liver.

against the specific protein increases. IgA secreted on the mucosal surface binds to the protein and enhances the opportunity for digestive enzyme activity. Digestion of the protein abolishes its ability to stimulate an immune response. Conversely, whole foreign protein molecules which penetrate the mucosa may act as antigens and stimulate an immune reaction that results in a foodborne allergy (Fig. 3-20). One example is an allergy to eggs caused by the entry of intact egg albumin.

Lipids

Dietary lipids are usually a mixture of neutral fats and compound and derived lipids (Chap. 2). Neutral fats, also called *triglycerides* or triacylglycerols, are the most common lipids in food and account for 95 percent or more of the fat in adipose tissue. Glycerol may bind one to three fatty acids to its carbon chain to form mono-, di-, and triglycerides. The fatty acids may be saturated or unsaturated in various fats. The dietary requirement for fatty acids may be fulfilled by a single unsaturated fatty acid, linoleic acid ($C_{18}H_{32}O_2$), if a total of 15 to 25 g of dietary fat is available daily to serve as a carrier of fat-soluble vitamins. Linoleic acid is also used for the synthesis of phospholipids, cholesterol, and some parahormones (such as the prostaglandins). Deficiency of this essential fatty acid is one of several causes of dry flaky skin.

Triglycerides are water-insoluble but form oil droplets which do not dissolve in water. Consequently, they do not diffuse through extracelluar fluid. However, bile salts and certain other molecules, such as protein, lecithin, and cholesterol, have apolar (noncharged) parts which dissolve in lipid droplets, as well as polar (charged) parts which are attracted to water. Molecules such as bile salts behave as ionic detergents and solubilize lipids. As fat droplets move through the digestive tract, they break down into smaller ones in which the nonpolar steroid rings of bile salts dissolve. Charged groups from the bile salts project outward from the mixture of molecules and keep lipid droplets dispersed in a process called *emulsification* (Fig. 23-4).

Lipases have greater opportunity to digest lipids in the mixture because the total surface area of the daughter droplets is greater than that of their parent. Microscopic water-soluble complexes called *micelles* are formed when the concentration of bile salts mixed with the lipids is high. Small micelles can nestle in the microvilli of the small intestine (Fig. 23-5). The size of the micelles range from 0.003 to 0.01 μm in diameter. Lipases bind to the micellar surface, digest its lipid contents, and allow rapid diffusion of molecules in and out of the aggregate in as little as one hundreth of a second. The close contact of the micelle and the mucosal surface encourages direct entry of some lipids through the epithelial cell membrane (Fig. 23-5).

Monoglycerides, free fatty acids, and cholesterol diffuse into the cell from the micelle. Glycerol is produced very slowly by the activity of lipases on triglycerides. As a result, the absorption of glycerol, independent of the micelle, is very slow.

Most dietary fat enters the lymph stream directly through the lacteals in the upper part of the small intestine. Notable exceptions are bile salts and short- and medium-length fatty acids (with chains of less than 8 to 12 carbon atoms), which enter blood through intestinal capillaries. About 90 percent of the bile salts absorbed in the ileum circulate back to the liver, where they were originally formed (Fig. 18-6, page 439).

In the mucosal cell longer fatty acid chains are linked to glycerol to form triglycerides. The triglycerides are packaged with phospholipids, cholesterol, and protein in a lipoprotein coat in particles called *chylomicrons.* Chylomicrons average 0.1 μm in diameter (Table 23-3) and are transported from the surfaces of epithelial cells at the innermost lining of the intestine toward their basal surfaces, which lie adjacent to capillaries and lacteals. Chylomicrons are too large to penetrate gastrointestinal capillaries whose fenestrated junctions are 0.08 μm wide at maximum. However, when provided with a lipoprotein coating, chylomicrons may enter lacteals, whose fenestrae can accommodate particles up to 25 μm in diameter (Chap. 17). Contrarily, individuals who lack β-lipoproteins in a rare disease called *abetalipoproteinemia*

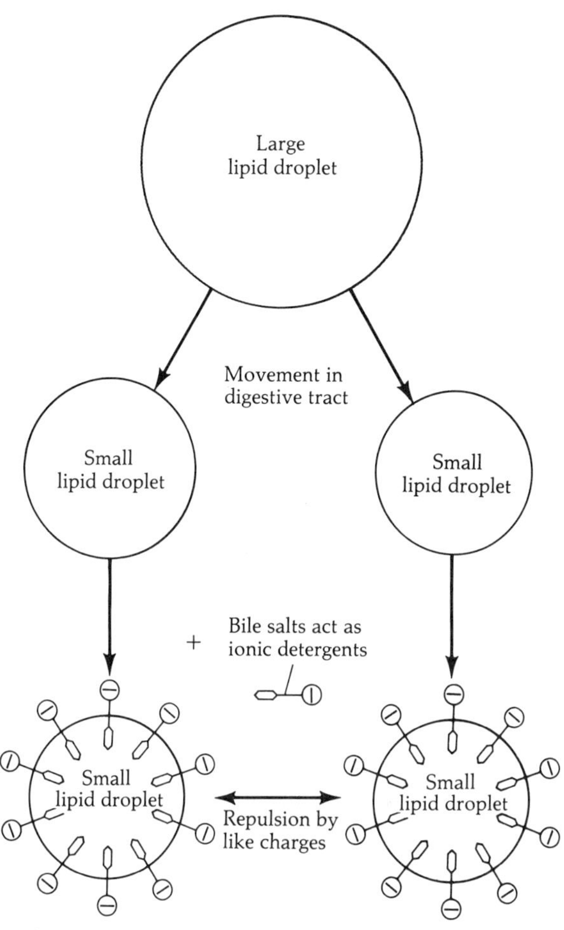

FIG. 23-4
The emulsification of lipids. Lipid droplets are broken into smaller ones as they are moved along the digestive tract. Molecules which behave as ionic detergents (such as bile salts) dissolve in the droplets, the repulsion by like charges on these molecules prevent the coalescence of the droplets.

cannot do so, and therefore accumulate fats in their intestinal mucosal cells.

Generally, chylomicrons in lymph are carried through the thoracic duct and then to the blood, where they may turn plasma a milky color within 1 to 8 h following ingestion of a meal with a high fat content. The particles are cleared from the blood by the activity of lipoprotein lipases found on endothelial cell membranes. These enzymes are released from several cell types under the influence of heparin. The relationship of dietary fats and chylomicrons to cardiovascular disease was noted in Chap. 16.

Vitamins

Vitamins are organic compounds which must be ingested in small amounts for the regulation of metabolism. The absence of one or more of them may result in deficiency diseases as described in App. 23B. Various vitamins act as enzymes, as helpers (cofactors) of enzymes, or as hormones. Many of the B vitamins are cofactors or coenzymes (Fig. 2-6B, page 23). Others, such as vitamin K, behave as enzymes, which are essential for the synthesis of blood clotting factors (Table 19-3). Vitamin D acts as a steroid hormone to influence mineral transport (Chap. 14). Some vitamins may play an essential role in tissue function by mechanisms which remain to be defined. For instance, although vitamin A forms the visual pigment rhodopsin (Chap. 8), it also participates in the development and maintenance of healthy epithelial tissue (Chap. 3) by mechanisms which remain to be identified.

Vitamins A, D, E, and K are fat-soluble. They dissolve in the steroid portion of bile salts, enter micelles, are absorbed by diffusion through the intestinal epithelium, and enter lacteals. All other vitamins are water-soluble, are rapidly absorbed by diffusion, and enter the blood via intestinal capillaries. The water-soluble vitamins are primarily absorbed in the first part of the small intestine (Fig. 23-1). Some, such as vitamin B_1 (thiamine), may also be transported by carriers in facilitated diffusion. Vitamin B_{12} (cyanocobalamin) combines with the intrinsic factor of the stomach prior to absorption in the ileum (Table 18-8 and p. 558

TABLE 23-3
Characteristics of chylomicrons

Property	*Average*	*Range**
Diameter (μm)	0.1	0.03–1.0
Concentration (mg/dL plasma)	95	100–250
Approximate composition (%)		
Triglyceride	80	80–95
Phosphoglyceride	5	3–9
Cholesterol	5	3–10
Protein	2	1–2

* The figures in this table represent a range of reported values. The electrophoretic properties and densities of chylomicrons relative to other fats are illustrated in Fig. 16-7, page 394.

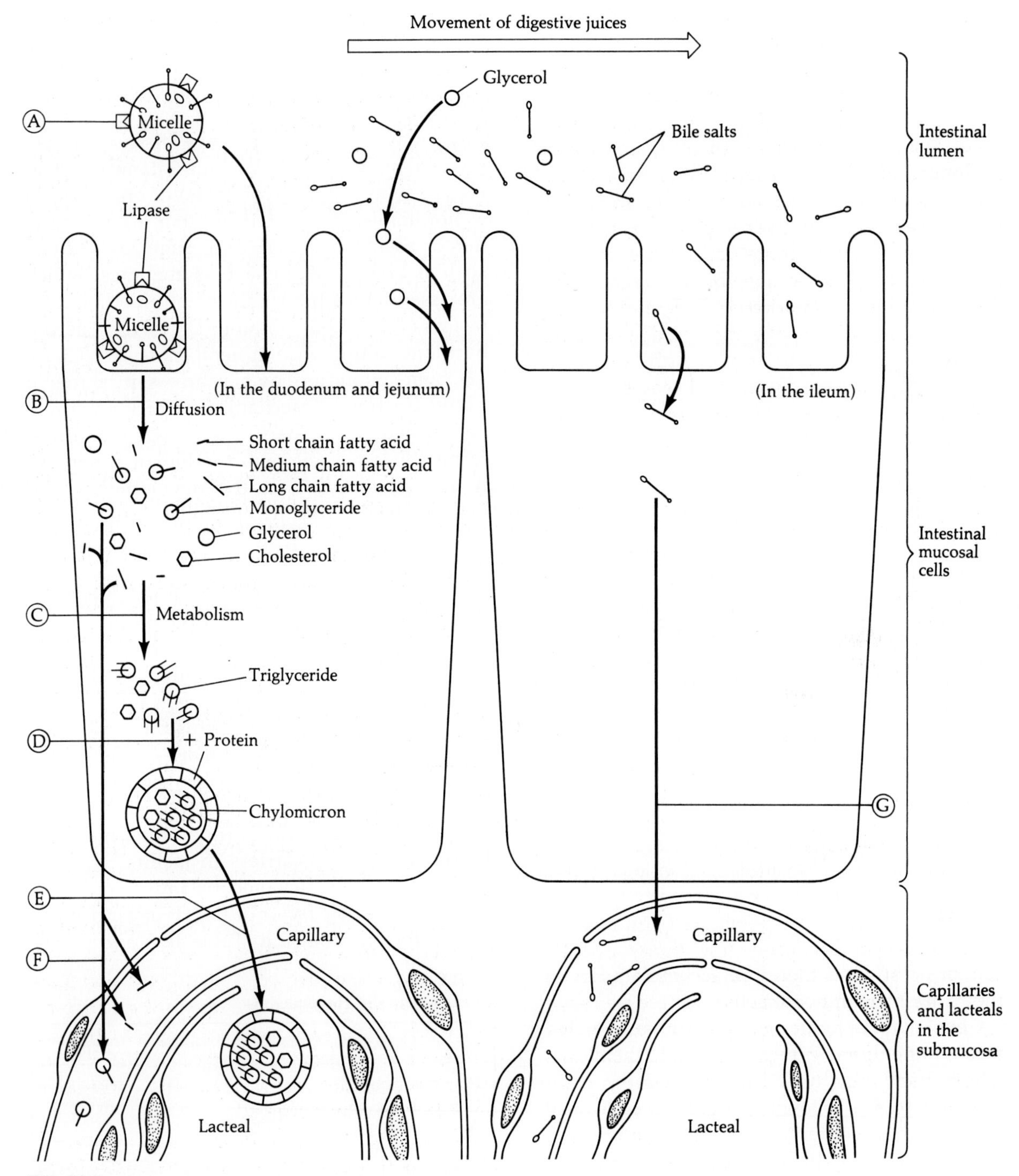

FIG. 23-5
The absorption of lipids. The absorption of glycerol and the contents of a micelle are illustrated. (A) Owing to action by lipase on the micelle, (B) the building blocks of lipids diffuse into the mucosal cell. (C) They are metabolized there and (D) packaged in protein to form chylomicrons, which (E) enter lacteals in the duodenum and jejunum. (F) Short- and medium-chain fatty acids and monoglycerides enter the blood directly. (G) Further down the intestine in the ileum, bile salts return through the mucosal cells to the blood.

and Fig. 23-1). Small amounts of B vitamins and vitamin K synthesized by intestinal bacteria are absorbed in the colon.

Vitamin deficiency may result from inadequate intake, poorly designed diets, and metabolic disorders (especially those which cause prolonged diarrhea or vomiting). Because many vitamins or their precursors are stored in the liver, diseases of the liver such as hepatitis or cirrhosis also can promote deficiencies of these essential nutrients. Because bile is essential for the absorption of vitamins A, D, E, and K, biliary diseases (such as gallstones) which alter the production and/or flow of bile to the small intestine may cause deficiency diseases due to malabsorption of fat-soluble vitamins. However, these diseases seem to occur infrequently.

Ions and Trace Elements

Ions are absorbed primarily in the duodenum and jejunum (Fig. 23-1). The ileum generally has a reserve but nonessential function in ion transport. Consequently, the osmolarity of the intestinal contents decreases rapidly along the first part of the small intestine and begins to resemble that of plasma before reaching the jejunum.

Eleven naturally occurring elements make up 99.99 percent of the weight of the body. They include hydrogen, oxygen, carbon, nitrogen, calcium, phosphorus, chlorine, potassium, sulfur, sodium, and magnesium. Fourteen other elements are indispensible to the body and constitute 0.01 percent of body weight. They are *essential trace elements* and are often present in concentrations ranging from parts per million to parts per billion (micrograms to nanograms per liter) in body tissues. These elements are iron, copper, zinc, manganese, cobalt, iodine, molybdenum, selenium, chromium, nickel, vanadium, silicon, tin, and fluorine. The functions of essential trace elements are summarized in App. 23C and the nutritional requirements for them are listed in App. 23D.

Iron is an interesting example of a trace element whose absorption is regulated by the needs of the body. Hydrochloric and ascorbic acids and other substances reduce the ferric form of iron (Fe^{3+}) to the more readily absorbed ferrous (Fe^{2+}) state to aid iron transport across the intestinal mucosa. Substances which form insoluble complexes with iron (such as phytic acid in cereals) hinder its absorption.

Mucosal epithelium in the first part of the small intestine actively transports iron in proportion to the body's needs. Slight excesses of dietary iron remain in the mucosa bound to protein in a complex called *ferritin,* which is shed with cells as they age and are excreted in the feces. The ability of the mucosal cells to retard entry of iron into the blood, referred to as the *mucosal block,* helps to regulate iron absorption. The importance of the duodenum to iron absorption is demonstrated following *gastrectomy* (partial or total surgical removal of the stomach, with the esophagus often being reattached to the jejunum). Gastrectomy thereby diverts food away from the duodenum and consequently may result in iron deficiency anemia.

When the total iron-binding capacity of mucosal proteins is saturated, excesses of iron can accumulate and attain toxic levels in the blood. This condition may result from the chronic overuse of iron tablets, repeated blood transfusions, liver disease, or hereditary factors that lead to an abnormally high rate of iron absorption. Iron storage diseases may occur as deposits of ferritin and another iron-containing complex, *hemosiderin,* accumulate and color cells in a condition called *hemochromatosis.* The individual shows weakness and sluggishness, increased amounts of melanin, which are deposited in the skin, and (in white persons) a bronze hue or a metallic gray color when iron deposits are also present. Weight loss also may occur with the onset of damage to the pancreas in a disorder called *bronzed diabetes.* The liver is damaged and becomes enlarged and cirrhotic, and abnormalities in heartbeat and congestive heart failure may develop.

Under normal conditions, iron is absorbed after it is combined with a carrier protein to form *apoferritin.* Apoferritin serves as a carrier in the intestinal lining and can release ferrous and ferric iron to the cell. Each molecule combines with up to 4000 ferrous ions to form *ferri-*

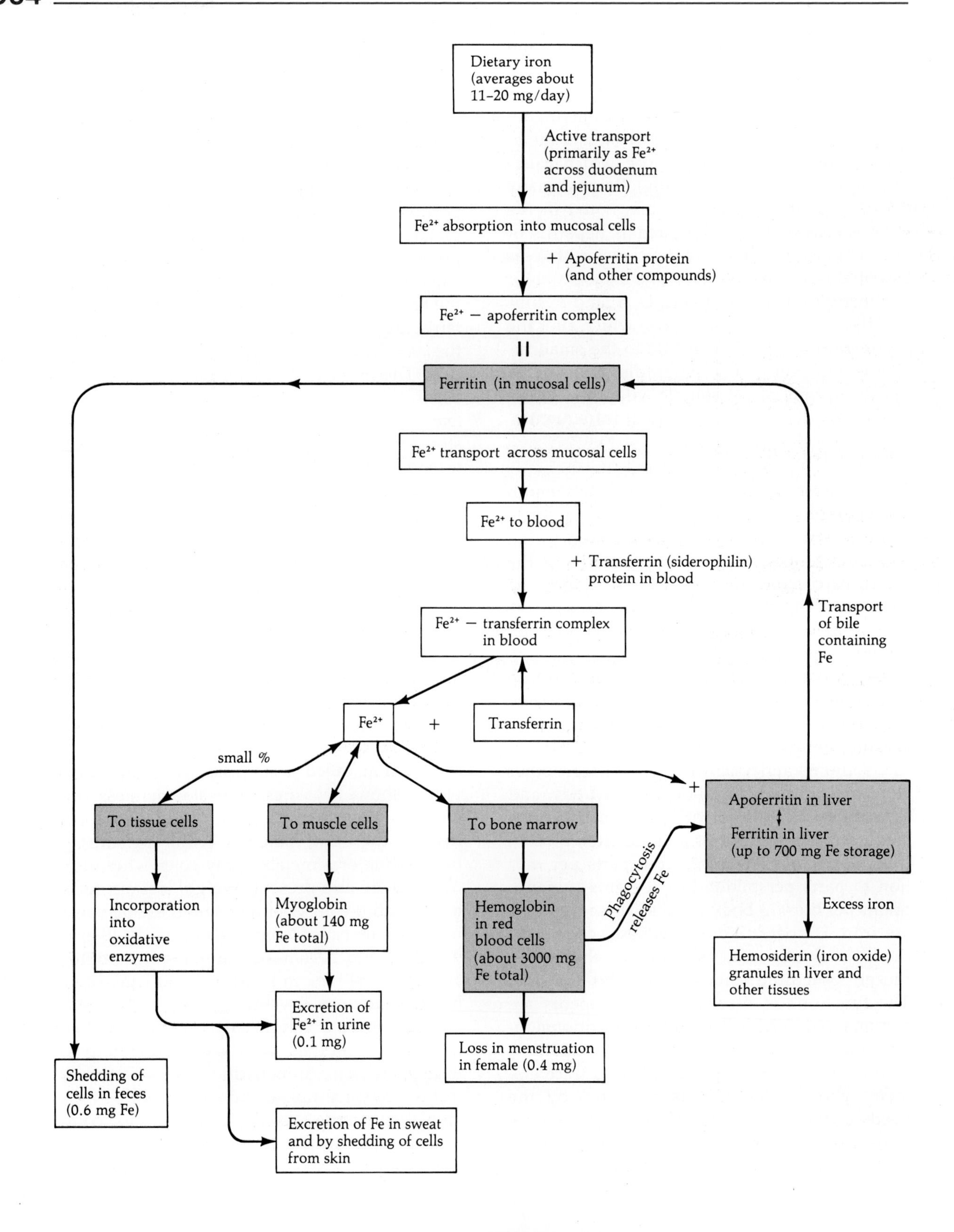
Dietary iron (averages about 11–20 mg/day)
Active transport (primarily as Fe²⁺ across duodenum and jejunum)
Fe²⁺ absorption into mucosal cells
+ Apoferritin protein (and other compounds)
Fe²⁺ — apoferritin complex
Ferritin (in mucosal cells)
Fe²⁺ transport across mucosal cells
Fe²⁺ to blood
+ Transferrin (siderophilin) protein in blood
Fe²⁺ — transferrin complex in blood
Transport of bile containing Fe
Fe²⁺
+
Transferrin
small %
+
To tissue cells
To muscle cells
To bone marrow
Apoferritin in liver
Ferritin in liver (up to 700 mg Fe storage)
Phagocytosis releases Fe
Incorporation into oxidative enzymes
Myoglobin (about 140 mg Fe total)
Hemoglobin in red blood cells (about 3000 mg Fe total)
Excess iron
Hemosiderin (iron oxide) granules in liver and other tissues
Excretion of Fe²⁺ in urine (0.1 mg)
Loss in menstruation in female (0.4 mg)
Shedding of cells in feces (0.6 mg Fe)
Excretion of Fe in sweat and by shedding of cells from skin

tin (Fig. 23-6). The ferritin complex not only transports iron across mucosal cells but releases it for entry into the blood. Ferrous ions released to the plasma combine with another carrier protein, *transferrin* (*siderophilin*). As the blood circulates, iron is released from transferrin in decreasing amounts to reticuloendothelial cells in the liver and to the bone marrow, skeletal muscle, and other tissues. However, the bone marrow, red blood cells, and spleen contain the major pool of iron and store about two-thirds of the body's content of this mineral.

The liver is critical to iron absorption and metabolism. It synthesizes apoferritin and transferrin, retrieves iron from red blood cells through its macrophages, and stores large quantities of the mineral. Iron is released from plasma transferrin to the liver, binds to apoferritin, and accumulates in it to form ferritin. Iron first accumulates in the liver's reticuloendothelial cells (Küpffer cells) and much later in its parenchymal cells. Excess iron is stored in the form of hemosiderin granules (iron oxide granules), from which it can be released to recirculate in the blood to other tissues. The spleen is the principal site of destruction of aged red blood cells from the circulation. Like the liver, it also stores iron as ferritin and hemosiderin.

Water

Most water is absorbed as it diffuses freely across the intestinal epithelium by osmosis. The entry of ions, and especially the active transport of sodium from the lumen of the digestive tract into mucosal cells produces an osmotic gradient (Fig. 23-2). As ions and molecules are transported into cells, water passively follows, moving from hypotonic to hypertonic fluids. The movement of ions and molecules to the lateral and basal surfaces across the mucosal cells and then into the capillaries produces an osmotic effect which causes water to rapidly enter the blood.

Although a total of about 2 to 3 L of fluid is ingested daily, the stomach absorbs only small amounts of water. Gastrointestinal secretions produce nearly 7 to 8 L of fluid and 98 percent of the 9 to 10 L of fluid entering the intestine is absorbed (over 61 percent in the jejunum, 22 percent in the ileum, and 14 percent in the colon). About 2 percent of the fluid is excreted, producing about 150 to 200 mL of water, mixed daily with solid wastes (Table 21-4).

FIG. 23-6
The absorption and destiny of iron. Iron is primarily absorbed in the ferrous state (Fe^{2+}) and transported across cell membranes in complexes with proteins such as ferritin. In the blood, iron is complexed with transferrin and carried to bone marrow, liver, muscle, and tissue cells. In those respective locations, iron becomes a critical part of protein-containing complexes such as hemoglobin, myoglobin, ferritin, and oxidative enzymes.

Digestive Tract Excretions

Excretions formed by the digestive tract are called *feces*. They contain exogenous substances derived from indigestible residues of food (such as cellulose from vegetables) and endogenous materials, which consist of cells shed from the mucosa, bile pigments, salts (especially potassium secreted by the colon), mucus, and bacterial cells which proliferate in the terminal segments of the intestine. The feces are about 25 percent solid and 75 percent water and are usually excreted within 24 h of entry into the colon.

Water is reabsorbed from the feces in the colon into the blood. Therefore, a prolonged stay in the colon can cause dehydration and hardening of the fecal mass. This results in *constipation*, difficult or delayed evacuation of the bowel. Feces also may accumulate, stretch the colon and rectum and generate afferent nerve impulses that result in the sensations of physical discomfort, abdominal pain or *colic*, nausea, headache, and loss of appetite called *anorexia*. These symptoms also may be induced by insertion and inflation of a balloon, which mimics distension of the colon by the fecal mass.

Motility of the digestive tract may be enhanced by *cathartics*, agents which evacuate the bowel, including *laxatives*, the least powerful cathartics. Many of these agents stimulate motility by promoting responses which stretch or

TABLE 23-4
Agents which promote intestinal motility and defecation

Agent	*Active Ingredient*	*Mechanism of Action*
Epsom salts	Magnesium sulfate	A nonabsorbable solute which causes osmotic entry of water into the colon and stimulates motility
Milk of magnesia	Magnesium hydroxide	As above
Plant cells	Cellulose	Indigestible polysaccharide constitutes bulk, which stimulates the colon; it also produces a hypertonic effect, which draws water into the bowel and lubricates it
Castor beans	Castor oil	Hydrolyzed in intestine to ricinoleic acid, which irritates mucosal lining and stimulates motility
Mineral oil	Mixture of hydrocarbons from petroleum	Nonabsorbable oil which lubricates hardened feces, easing their evacuation *Note:* Fat-soluble vitamins dissolve in mineral oil and are lost with the feces

irritate the walls of the colon and rectum (Table 23-4). For example, nonabsorbable solutes such as magnesium sulfate (in Epsom salt) or magnesium hydroxide (in milk of magnesia) cause an osmotic effect, drawing water into the colon. The feces are moistened and softened, the colon is distended, and its motility is enhanced. Castor oil and a component in prunes irritate the mucosa to increase colonic motility.

Malabsorption

Malabsorption is any disorder in which the passage of nutrients across the mucosa of the digestive tract is defective. It may result in loss of appetite; nausea; diarrhea; abdominal pain; foul-smelling, loose, fatty feces, called *steatorrhea;* and an upset in acid-base balance of body fluids (Table 21-9). Rapid transit of chyme through the intestine does not allow for sufficient absorption of water, sodium, and bicarbonates into the blood. Prolonged diarrhea can lead to acidosis (page 535) and such serious consequences that their reversal may require intravenous administration of fluid and electrolytes. Prolonged malabsorption can cause malnutrition, vitamin deficiency diseases, bleeding, anemia, and weight loss.

The causes of malabsorption may be broadly separated into those that create abnormalities in: (1) the lumen of the digestive tract; (2) mucosal cell transport; and (3) circulation of blood and lymph to the tissues of the digestive tract (Table 23-5). Specific examples of some disorders which may damage and destroy the microvilli and villi are noted in Table 23-6.

Primary malabsorption is a congenital (present at birth) condition. Intolerance of specific disaccharides can result from lack of disaccharidases required for their digestion. Disaccharidases are enzymes which split 12-carbon sugars such as lactose, maltose, or sucrose into 6-carbon subunits. The undigested disaccharides are converted by intestinal bacteria into organic acids, which irritate the digestive tract, stimulate its motility, reduce the opportunity for absorption of water, and result in *diarrhea.* Lactose intolerance is due to a decline in the activity of lactase, which normally breaks down milk sugar, or lactose, into glucose and galactose. This disorder occurs in 70 percent of blacks and 20 percent of whites in the United States. Interestingly, the enzymes lactase, maltase and sucrase have life-spans of only a few hours. Their activities at sustained levels is dependent on their continuous synthesis in mucosal cells, coupled with fragmentation of the microvilli, which sheds the cells and the enzymes into the lumen of the intestine. In most people the amount of lactose that can be digested usually diminishes with age.

Gastroenteritis and Food Poisoning

At times diarrhea may be *neurogenic* in origin. That is, it may be due to stress, anxiety, or other types of parasympathetic nerve stimulation. However, the most common cause of diarrhea is *gastroenteritis,* an inflammation of the

TABLE 23-5
Some causes of malabsorption

(1) *Intraluminal Defects*	(2) *Defective Mucosal Transport*	(3) *Abnormal Circulation of Blood or Lymph*
Alterations in intestinal microorganisms	Hereditary defects in the production of digestive enzymes such as disaccharidases	Atherosclerosis
Gastroenteritis	Abetalipoproteinemia	Thrombosis
Obstructions in intestine	Tropical sprue	Obstruction of lymph vessels
Diverticula	Protein-calorie malnutrition Kwashiorkor	Dilation of lymph vessels (intestinal lymphangiectasia)
	Immunological disorders:	
	Pernicious anemia	
	Food allergies, including gluten-induced allergy in celiac disease	
	Drug damage (from cancer therapeutic drugs)	
	Radiation damage	
	Endocrine disease (such as disease of the parathyroids or adrenals)	
Inadequacy of digestive juices		
Pancreatic insufficiency due to pancreatitis, cystic fibrosis, cancer		
Bile insufficiency due to hepatitis, cirrhosis, obstruction of bile ducts, inadequate resorption of bile salts (which may follow surgical removal of 100 cm or more of the ileum)		

gastric and intestinal mucosa, which usually results from the presence of pathogenic viruses or bacteria. When ingested with food or water, many pathogenic microorganisms or their toxins combine directly with cell receptors and remain localized in the mucosa to cause food poisoning. The symptoms produced are nausea, vomiting, fever, abdominal pain, and diarrhea. *Enteric organisms* are those which primarily affect the gastrointestinal system. They include the Coxsackie and Echo viruses, which are common sources of viral food poisoning, and *Salmonella, Staphylococcus* species, and *Clostridium perfringens,* which are prevalent causes of bacterial food poisoning (Table 23-7).

Specific products released from bacterial metabolism may cause food poisoning. Some of them are *exotoxins,* proteins secreted by bacterial cells. Others are *enterotoxins,* bacterial poisons which bind to intestinal mucosal cells to cause toxic effects. For example, *cholera* enterotoxin accumulates with large numbers of bacterial cells, inhibits intestinal sodium absorption, and stimulates sodium and chloride ion secretion. Sodium and chloride ions accumulate in the intestinal lumen, causing mass osmosis of water and excretion of these substances. Up to 10 to 15 L of fluid may be lost daily. If untreated, the severe fluid loss and electrolyte imbalance may be fatal. The mechanism of action of cholera toxin resembles that of some hormones (Chap. 5) in that it stimulates adenyl cyclase activity and greatly elevates intracellular levels of cAMP to produce its devastating effects.

Gastroenteritis caused by *Salmonella* species or *Clostridium perfringens,* like cholera, occurs when large numbers of whole cells are present. The toxins which elicit salmonella and perfringens food poisoning have yet to be clearly identified. On the other hand, the protein entero-

TABLE 23-6
Specific examples of malabsorption diseases

Disease	*Comment*
Lactase deficiency	Hereditary deficiency in intestinal lactase Allows bacteria to ferment lactose Metabolic products of bacteria cause increased motility of digestive tract and decreased time for absorption, resulting in diarrhea Common in adults of southern European, Oriental, and African ancestry
Sucrase-maltase deficiency	Hereditary failure to produce intestinal sucrase and maltase Symptoms as above
Glucose-galactose malabsorption	Hereditary lack of active transport enzymes for glucose and galactose Symptoms as above
Hartnup disease	Hereditary malabsorption of the amino acid tryptophan
Cystinuria	Hereditary malabsorption of dibasic amino acids (such as cystine)
Bile insufficiency	Interferes with fat absorption
Immunodeficiency diseases	Insufficient IgA antibody synthesis Overgrowth of microorganisms results Damage and destruction of villi occur May be hereditary
Tropical sprue	Occurs in tropical and subtropical climates Overgrowth of bacteria occurs in small intestine Excess of unconjugated bile acids produced as a result of bacterial activity Steatorrhea results Severe malabsorption of folic acid and vitamin B_{12} occurs
Celiac disease (celiac sprue)	Allergic reaction to gluten which may damage the villi and reduce their number (gluten is a polypeptide found in wheat and common in cereals)
Protein-calorie malnutrition	Severe deficiency in amino acids results in decrease in the surface area of the microvilli An example is the disease called kwashiorkor

TABLE 23-7
Examples of bacterial-induced gastroenteritis

Disease	*Organism*	*Toxin(s)**	*Common Food Source*	*Time of Onset of Symptoms after Ingestion*
Staphylococcal food poisoning	*Staphylococcus aureus*	Enterotoxin released from cells	Toxin released by bacteria in contaminated dairy products; especially cream products	1–12 h
Cholera	*Vibrio cholerae*	Enterotoxin	Toxin secreted from large number of bacteria in water or food contaminated with feces	2–5 days
Salmonellosis	*Salmonella* species (e.g., *Salmonella typhimurium*)	Lipopolysaccharide endotoxin; neurotoxin; requires large numbers of bacteria	Bacterial cells in contaminated poultry, eggs	8–48 h
Perfringens food poisoning	*Clostridium perfringens*	Enterotoxin; exotoxins; disease associated with whole bacteria	Bacterial cells which grow in reheated contaminated meats	about 12 h
Botulism (see Chap. 12)	*Clostridium botulinum*	Botulinal exotoxin secreted by bacterial cells	Toxin released from bacteria in improperly canned foods	18–36 h

* Protein unless noted otherwise.

toxins from staphylococci and the exotoxins of *Clostridium botulinum* have been well characterized. The mechanism of botulinal poisoning was noted earlier (Chap. 12, page 295).

The Role of the Intestinal Bacterial Flora

The mixture of bacterial species in the intestine is known as the *intestinal bacterial flora.* Among the usual species are a wide variety of aerobic and anaerobic bacteria, including *Escherichia coli, Enterobacter aerogenes,* and various species of *Lactobacillus.* In the normal course of digestion some microorganisms survive passage through the stomach's acid. As they move through the remainder of the digestive tract, the bacteria proliferate and attain numbers as high as 10^9 to 10^{10} per gram of feces in the colon.

Bacteria are essential to normal gastrointestinal function. In the large intestine they synthesize vitamins, stimulate antibody production, and produce about 0.55 ± 0.15 L of gas which is excreted daily. Animals born and raised in bacteria-free ("germ-free") environments have low levels of circulating antibody and a larger cecum than those reared in conventional environments.

A change in the bacterial flora can result from disease or from antibiotic administration. Antibiotics inhibit or kill microorganisms and may alter the ratio of numbers among surviving bacterial species. Attempts have been made, with limited success, to regulate the ratio. For example, the ingestion of certain species of lactobacilli which seem to be beneficial for gastrointestinal function has been advocated by some investigators. Lactobacilli are prevalent in foods such as yogurt and have even been added to special lactobacilli "sodas" for experimental nutritional purposes.

Abnormal bacterial residents of the small intestine may in some instances interfere with intestinal absorption. In addition to specific pathogens, anaerobic species may enzymatically degrade proteins to produce organic acids, gases such as ammonia, and potentially toxic amines such as putrescine and cadaverine, which are odorous and are called *ptomaines.* Although these compounds may induce abdominal discomfort, the term *ptomaine food poisoning* arose from the mistaken notion that these molecules rather than specific pathogens caused gastroenteritis.

Ammonia (NH_3) is a potentially toxic gas formed by intestinal bacteria and is absorbed from the colon and carried by the portal circulation to the liver. In the liver, ammonia is rapidly converted into urea [$H_2N(C{=}O)NH_2$], which is nontoxic. Urea is then transported by the blood to the kidneys, where it is excreted (Fig. 21-13). Failure to eliminate ammonia by this route can intoxicate the liver, elevate blood levels of ammonia, alter acid-base balance, and cause coma and death.

Abnormally high quantities of bacteria or viruses may accumulate in pockets (diverticula) in the small intestine. Overgrowth of bacteria may result, which elevates their numbers from less than 10^4 to 10^9 per gram of contents in the small intestine. The utilization of intestinal nutrients by the excess bacteria diminishes the availability of food to the person. Products of bacterial metabolism in these pockets also may alter digestive tract function. *Tropical sprue,* which interferes with mucosal cell transport, is an example of malabsorption caused by bacterial overgrowth (Table 23-6).

Abnormal growth of mucosal cells may result in masses, or *polyps,* that project inward, and in unusual circumstances interfere with motility of the intestinal contents. Limited evidence indicates that this interference may give bacteria greater opportunity to act on undigested food and produce carcinogens. Furthermore, the incidence of colonic cancer has been related directly to the size of the polyp. Some evidence indicates that a lack of bulk in the intestine may minimize motility and yield the same effect. As a result, it has been proposed that low-bulk diets, which are low in plant fibers such as cellulose and high in fat and protein, may dispose people to cancer of the colon, with associated symptoms of rectal bleeding, vomiting, and weight loss. Although the frequency of cancer of the sigmoid colon rises in both sexes between ages 40 and 70, its cause(s) remain to be clearly established.

Therapy

Therapy for malabsorption varies with the disorder. Prevention is the best therapy against gastroenteritis, the most common cause of malabsorption. Good personal and community hygiene applied to the handling, preservation, and storage of foods can markedly reduce the transmission of foodborne pathogens. If malabsorption causes severe fluid loss, an isoosmolar 5 percent dextrose solution may be administered intravenously. This solution is used to restore fluid balance but provides little nutrient value. Malnutrition characteristic of usual prolonged intravenous feeding may be avoided by *total intravenous feeding,* the administration of a hyperosmolar solution of amino acids, simple sugars, and medium-chain-length triglycerides. The fluid is administered through a large vessel (such as the inferior vena cava), in which the blood flow rapidly dilutes the nutrients, to avoid lysing blood cells. In some instances the hyperosmolar nutrient solution is administered directly through a surgical opening through the skin into the stomach (a *gastrostomy*).

Malabsorption therapy may involve the administration of antibiotics to reduce the number of bacteria in the small intestine and/or to eliminate pathogens. Other drugs may be given to decrease motility of the tract. Additionally, diets designed to avoid food allergies or reactions such as lactose intolerance may be used. In certain disorders surgery may be necessary to remove diseased segments of the digestive tract.

23-3 INTERMEDIARY METABOLISM

Metabolism of foods begins once their chemical building blocks enter cells. *Metabolism* is the totality of all chemical reactions in the body. It includes *catabolism,* in which molecules are degraded, and *anabolism,* in which they are synthesized. The chemical reactions by which nutrients are utilized are called *intermediary metabolism.* They are more properly the subject of biochemistry and thus merely will be outlined here. The general relationships between the chemical building blocks were described in Chap. 2 and their relationships to energy production in muscle were explained (page 318). A review of these concepts would be most helpful to the student at this time.

Metabolic reactions occur in steps catalyzed by specific enzymes. In some reactions compounds are synthesized; in others they are degraded as chemical bonds are broken. Some of the energy derived from the breakdown of molecules is transferred to high-energy phosphate compounds such as adenosine triphosphate (ATP) and creatine phosphate (C ~ P) and later released to drive other reactions.

The intermediary metabolism of carbohydrates, proteins, and fats forms some products that are shared in metabolic pools (Fig. 23-7). Among the common metabolic products are pyruvic acid and acetyl coenzyme A (acetyl CoA). These two substances can be derived from each of the three major types of food, and they also may be used in their synthesis. Furthermore, pyruvic acid and acetyl CoA can be funneled into the citric acid cycle and be degraded in a series of oxidative reactions that release hydrogen, carbon dioxide, and energy.

Chemical bonds are broken and energy released by the removal of pairs of hydrogen atoms from the carbon chains of compounds in the citric acid cycle. This energy is used to form high-energy phosphate bonds in ATP and C ~ P. The removal of hydrogen is controlled by enzymes so that maximal energy is transferred to ATP. The energy release is gradual in order to optimize its transfer to chemical bonds, lessening the amount that escapes as heat. The reactions may be likened to the combustion of coal or wood in a carefully stoked stove as compared with the sudden release of energy which might occur with the explosion of an oil tank. Pairs of hydrogen atoms are transferred to hydrogen acceptors located in an "assembly line" in the mitochondria. The downhill reactions degrade the carbon chains, strip off pairs of hydrogen atoms and an electron from each atom, and release energy to form ATP. Hydrogen and electron acceptors include NAD (nicotinamide adenine dinucleotide), coenzyme Q (CoQ), FAD (flavine adenine di-

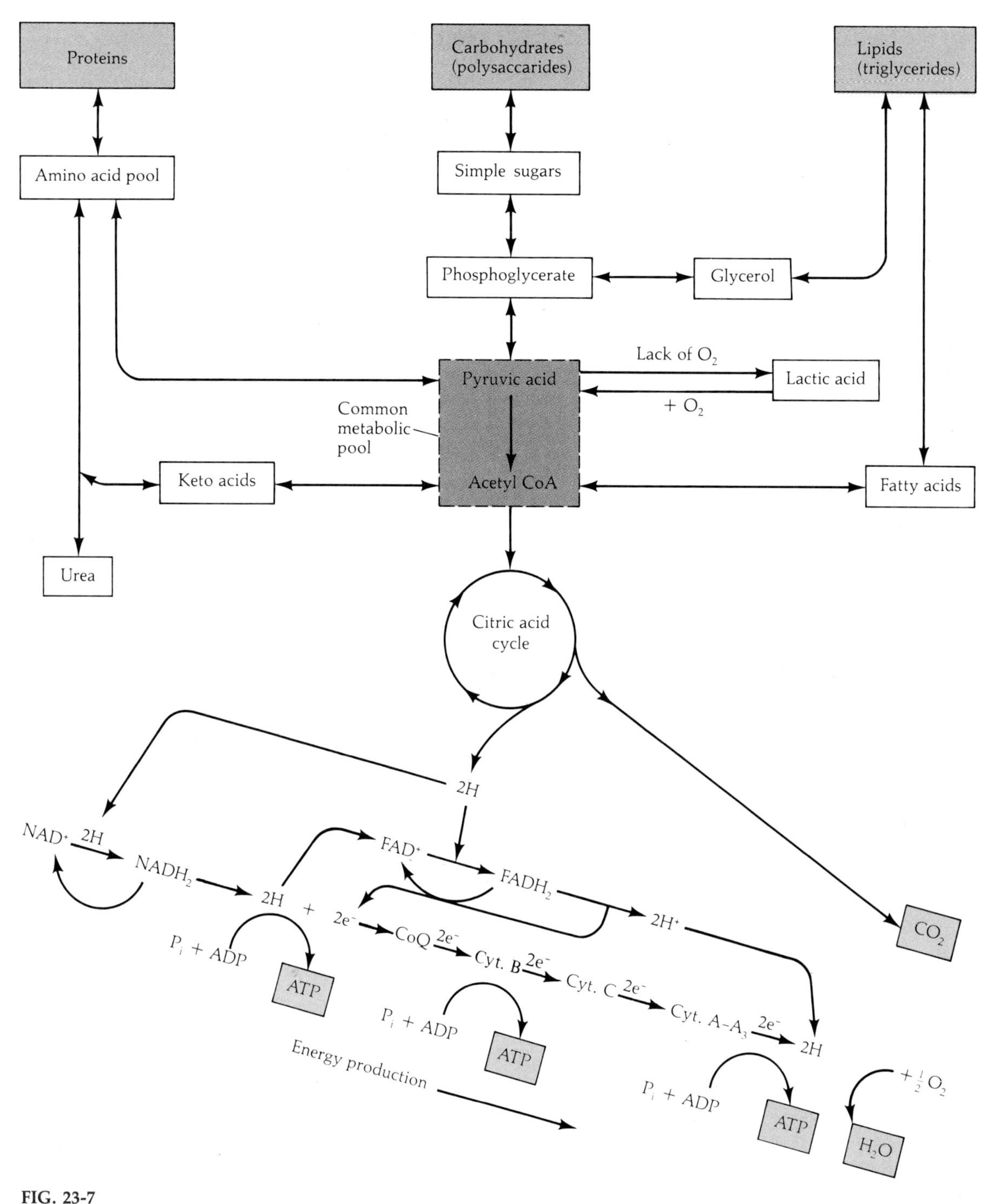

FIG. 23-7
Metabolic interconversions of major foods. This simplified scheme illustrates the main pathways of intermediary metabolism. Ultimately CO_2, H_2O, and energy are produced. The energy is derived from a series of enzymatically catalyzed, gradual, step-by-step reactions. They involve the degradation of carbon-to-carbon bonds and the transfer of hydrogen and its electrons to acceptors. The last in a series of acceptors is oxygen, whose union with hydrogen forms H_2O.

nucleotide), cytochromes b, c, and a (iron-containing oxidative enzymes), and oxygen (Fig. 23-7). The degradation of carbon chains and the ultimate transfer of the two hydrogen atoms (two hydrogen ions and their electrons) to an atom of oxygen yields the ultimate products of intermediary metabolism—H_2O, CO_2, and energy. The process is called *hydrogen* or *electron transport* and is coupled (linked) to oxidative phosphorylation reactions that lead to the synthesis of ATP. Of the energy released from the organic compounds, 40 percent is transferred to ATP and 60 percent forms heat.

Carbohydrates

When food intake exceeds energy expenditure, carbohydrates may be stored as glycogen, a polysaccharide which is abundant in the liver and skeletal muscle. The metabolism of glycogen requires its degradation to glucose, which is further cleaved to pyruvic acid. Pyruvic acid enters a metabolic pool in which its carbon chain can be used to synthesize amino acids or fats or be degraded in the citric acid cycle to carbon dioxide, water, and energy (Fig. 23-7). The breakdown of glucose (glycolysis) and the energy transferred to ATP were described in detail earlier (see Fig. 13-11, page 321).

Proteins

Proteins may be cleaved into amino acids, the amino groups of which may be removed, exchanged, and added to other compounds in enzymatically catalyzed reactions called *deamination, transamination,* and *amination,* respectively. These reactions are common in the liver and are involved in amino acid synthesis and also in amino acid degradation to form nitrogen-containing compounds such as urea and ammonia, mentioned previously. When protein catabolism exceeds protein anabolism, more nitrogen is lost than ingested and the body is in *negative nitrogen balance.* The opposite condition is *positive nitrogen balance* and is indicative of net protein synthesis. The assembly of amino acids in protein synthesis is mediated by mRNA, tRNA, and rRNA under the control of DNA (Fig. 2-14, page 35).

Lipids

Most of the lipids in the body are stored as triglycerides, which produce a fat depot found as a liquid (oil) in adipose tissue and in the liver. Triglycerides may be hydrolyzed to glycerol and fatty acids which may be converted to acetyl groups (2-carbon fragments) that are combined with CoA to enter the metabolic pool common to the three major foods (Fig. 23-7). The glycerol is converted to phosphoglycerate, which is then transformed into pyruvic acid, which may be converted to acetyl CoA. When caloric intake is high, reversal of the pathway funnels 2-carbon fragments (acetate) into reactions that result in the synthesis of fatty acid chains up to 16 carbon atoms long or other reactions in which the 2-carbon fragments are combined to form aromatic rings and steroids such as cholesterol.

Fats oxidized to acetyl CoA may be converted to ketoacids (ketone bodies) to provide energy. Ketoacids formed at normal rates are totally oxidized. However, when caloric intake is low, more fats are broken down and ketone levels may rise in the blood (*ketonemia*) and appear in excess in the urine (*ketonuria*). Ketogenic diets are sometimes prescribed to induce weight loss. However, uncontrolled accumulation of two major ketone bodies, acetoacetic acid and β-hydroxybutyric acid, may cause acidosis. Their excess in the blood is referred to as *ketosis* and may result in derangements of metabolism characteristic of a shift in the pH of body fluids (see Chaps. 20 and 21). Ketosis also may occur with gastrointestinal disorders which diminish carbohydrate levels and which lead to excess oxidation of fats. Prolonged diarrhea and insulin deficiency in untreated diabetes mellitus (Chap. 5) are two such disorders.

23-4 METABOLISM AND NUTRITION

Energy is released in large quantities, but in a controlled fashion, mostly by intermediary metabolism in skeletal muscle and in the liver. It is converted to heat and to chemical, mechanical, or electrical work, as shown in the following equation:

$$\underset{\text{Energy expenditure}}{E_E} = \underset{\text{Potential energy, stored energy}}{E_P} + \underset{\text{Work}}{W} + \underset{\text{Heat}}{H}$$

Energy in the body may be used chemically (in intermediary metabolism), electrically (by active transport pumps that carry ions to produce potentials across cell membranes, such as those of neurons and muscle cells), mechanically (to contract muscle), and thermally (to warm body fluids and tissues).

Caloric Values of Foods

The units of energy currency in intermediary metabolism are the high-energy phosphate bonds in ATP. The conversion of ATP to ADP releases 7 kcal of energy.[1] The complete oxidation of 1 g of fat, carbohydrate, and protein produces 9.3, 4.1, and 4.1 kcal, respectively. With allowance for some caloric loss due to incomplete digestion and absorption, the ingestion of 1 g of each of these foods provides about 9, 4, and 4 kcal, respectively. Therefore, the average caloric value of fat is a little over twice that of carbohydrates or proteins and reaches its highest level when the degree of saturation (hydrogenation) of fat is maximal.

The average American diet provides about 46 percent of its kilocalories from carbohydrate, 42 percent from fat, and 11 to 12% from protein. Proteins are critical to body and enzyme structure and are used as energy sources usually when in excess in the diet or, as a last resort, when carbohydrate and fat stores have been severely depleted.

Carbohydrate stores are not extensive. The average total carbohydrate reserve in a 70-kg adult provides about 1452 kcal (page 322). This source can be depleted rapidly during prolonged exercise while fats continue to be burned to provide energy. As a result, with training (Chap. 13) body fat decreases while protein is synthesized and a "lean" body mass may be maintained without a change in body weight. Excessive strenuous exercise can cause the degradation of proteins such as hemoglobin and myoglobin and the appearance of their metabolic products in urine. The average expenditures of energy by adult men and women in a variety of activities are listed in Apps. 23E and 23G at the end of this chapter.

The pathways by which the three major foods are utilized vary. They produce differing amounts of CO_2 and consume varying amounts of O_2. The ratio of CO_2 produced to O_2 consumed is the *respiratory quotient* (RQ), which is 1.0 for pure carbohydrate, 0.8 for pure protein, and 0.7 for pure fat. The average RQ is 0.82 for a mixed diet. Because of common metabolic pools, the RQ is usually measured to reflect in a general way the type of food that is being utilized in metabolism.

Basal Metabolic Rate

The *basal metabolic rate* (*BMR*) is the minimal level of metabolism to sustain life. It is the metabolic cost of just being alive. The BMR is the measurement of energy expenditure of a person lying down but awake in a comfortable environment, after a restful night's sleep and 12 to 14 h after a meal. It is directly related to the total body surface area of an individual, measured or calculated in square meters, and is a function of height and weight. The BMR of an average 70-kg young adult male about 25 years old is about 1620 kcal/day and that for an average 58-kg young adult female is approximately 1320 kcal/day. With activity, the respective total energy expenditures are 2740 and 2030 kcal/day (App. 23E). Energy expenditure increases with body mass as is indicated in the following equation:

$$\underset{\text{Energy expenditure in watts, or kcal/kg/h}}{\text{Log } E} = 0.75 \log \underset{\text{Body mass in kg}}{M} + 7.84$$

Metabolic rate may be altered by a variety of factors. In addition to body mass and physical activity, it is usually higher in males than fe-

[1] A calorie (cal) is the amount of heat needed to raise the temperature of 1 g of water 1°C. Since metabolic and chemical reactions involve thousands of calories of energy, the unit of measure used in describing these reactions is the kilocalorie (abbreviated kcal), which equals 1000 cal and is sometimes called the large calorie or Calorie. The term Calorie is often used in discussions of nutrition.

males and lower after young adulthood (Apps. 23E, 23F, and 23G). It is generally higher during periods of growth, assimilation of food, exposure to cold, fever, stress, and secretion of certain hormones such as epinephrine, norepinephrine, and thyroxine (Chap. 5).

In spite of these variations the number of kilocalories expended can be approximated when the amount of oxygen consumed is known. The caloric expenditure of the body averages 4.825 kcal per liter of oxygen consumed. For example if 400 L of oxygen were consumed in a 24-h period, the total energy expenditure for the day would be about 1930 kcal (400 L oxygen consumed times 4.825 kcal/L equals 1930 kcal).

Metabolism and Body Temperature

Variations in Body Temperature

Although heat production and heat loss are normally balanced by the hypothalamus to maintain body temperature (Fig. 1-2 and page 267), its normal flucuations may be considerable. Body temperature averages 37°C (98.6°F) orally but ranges from 36.3°C to 37.1°C among different individuals. Body temperature rises by about 0.5°C during the day, is 0.5°C higher when measured rectally than orally, and goes up by as much as 3°C with strenuous exercise. Because the temperature regulation center in the hypothalamus is less effective in the newborn and the elderly, changes in environmental temperature can alter body temperature and functions in these groups more readily than in other people.

Heat production occurs as a result of oxidation of foods in muscle and in liver cells (Figs. 23-7 and 13-9 to 13-11). Heat loss results from *evaporation* of sweat from the skin, *convection* to currents of cool air, *conduction* by contact with cool substances such as cold water, and *radiation* from the body by transfer of heat without direct contact in a manner similar to radiation of energy from the sun. A small amount of heat is also lost with warm bodily *excretions* in the form of water droplets with warm air from the lungs, urine from the kidneys, and feces from the digestive tract.

Body temperature may be altered by a number of factors, including hormones, exposure to heat or cold, ingestion of specific foods, exercise, and infections. The effects of hormones on metabolism and body temperature are evident during the female reproductive cycle. Within 24 h of the release of an egg from the ovary, a woman's body temperature generally rises to its highest point of the month (Fig. 24-12).

Infectious agents cause white blood cells to release proteins called *endogenous pyrogens* which stimulate an excess of heat production and result in *fever*, an abnormal increase in body temperature. For each 0.5°C rise in body temperature, energy production increases 7 percent. As a result, kilocalories are consumed rapidly and weight loss often occurs during prolonged infectious disease. The benefits of fever are debatable, although some evidence indicates that certain viruses replicate less often at elevated body temperature.

Under normal conditions temperature regulation takes precedence over water and salt balance. These priorities are reflected by changes in body function which occur when rectal temperature rises above 39.5°C in the condition of *hyperthermia*. An environmental temperature increase (or fever due to infectious disease) warms the body and results in perspiration. As sweat evaporates, blood flowing through the tissues of the skin is cooled. Excessive loss of sodium and chloride ions in sweat may result in muscle cramps, a decrease in blood pressure, general weakness, fainting, nausea, and vomiting. All these symptoms are indications of *heat exhaustion*. Should water loss cause plasma volume to decrease significantly, cardiac output will decrease, less blood will flow to the skin, less sweat will be formed, and a sudden, explosive rise in body temperature may result in *heat stroke* (heat shock). Many of the symptoms resemble those of hemorrhagic shock (Fig. 18-8). Convulsions usually result when failure to balance heat production and heat loss allows rectal temperature to rise to 41°C (106 to 107°F). If the temperature increase is prolonged, brain damage may occur. A rise above 43°C (109°F) denatures proteins and is often fatal.

Prolonged exposure to low environmental temperature can result in a decrease in body temperature, or *hypothermia*. Body functions slow measurably when rectal temperature reaches 33°C (91°F). At 28°C (82°F) irregularities in heart rate are evident and ventricular fibrillation and death may follow.

Effects of Food and Exercise

After consumption of a meal the BMR rises and body temperature is elevated in a response called the *specific dynamic action* (SDA) of food. The increase in temperature seems to be mostly due to the assimilation of foods by the liver and stimulation of sympathetic nerves. For foods to become assimilated, kilocalories must be expended which may be provided by any of the energy reserves in the body. The number of kilocalories consumed for the SDA of proteins, carbohydrates, and fats, represent 30, 6, and 4 percent of their respective energy content.

As noted above, the metabolic rate increases in direct proportion to the expenditure of energy in exercise. The number of kilocalories utilized varies with different types of physical activity (App. 23F) and can be related to the caloric value of specific foods (App. 23H). A significant amount of exercise is required to burn off excess kilocalories. For example, 18 min of bicycling or 8 min of running is needed to use up the 151 kilocalories provided by a single doughnut (App. 23H).

Obesity

Obesity may be defined as an increase in the ratio of fat to lean body mass which causes weight to rise 10 percent or more above the normal range for an individual's height, sex, and age. It occurs when the total number of kilocalories ingested exceeds those expended. Among multiple factors which may cause this imbalance are: physiological alterations in hormonal secretion, rate of growth, metabolism, and exercise; genetic factors, which are still obscure, but are indicated by the tendency of obesity to run in some families; and psychological factors, manifested in overeating due to stress or emotions.

Some evidence in mice indicates that *metabolic obesity* results from changes in metabolism which cause fats to accumulate. Although long suspected, specific metabolic mechanisms that cause obesity have yet to be clearly documented in humans. *Regulatory* obesity due to changes in the activity of nuclei in the hypothalamus has been demonstrated. Lesions in the satiety center (in the ventromedial hypothalamus) may cause overeating, hyperphagia (a condition involving extreme overeating), and a decrease in metabolic rate (with an increase in conversion of carbohydrate to fat), resulting in regulatory obesity. Conversely, lesions in the appetite center (in the lateral hypothalamus) cause undereating, hypophagia, and effects opposite to those cited above. Whether a relationship exists between regulatory and metabolic obesity is unclear.

Interestingly, the accuracy of weight tables (App. 23G) has been called into question recently. Some investigators believe the tables, published originally in 1959, underestimate the average weight of adults by 10 to 15 lb since the original figures were obtained from insurance studies in which the data were not random. New weight tables are currently being constructed. Even with such revisions in mind it is estimated that about 50 percent of middle-aged Americans are overweight. It is generally agreed that the longer a person is obese, the greater will be the adverse effects.

Weight loss and a reduction in obesity occur when the expenditure of energy exceeds caloric intake. Diets should be based on the normal recommended daily dietary allowances of proteins, vitamins, and minerals, as shown in App. 23I. The requirements for carbohydrates and fats may be altered according to the individual's need, as described in general terms in this chapter and in Table 16-5 on page 390.

Weight loss should be undertaken only under medical supervision and by following a prescribed diet and exercise routine. To do otherwise may be hazardous and even fatal. For example, diets high in proteins and fats and low in carbohydrates may cause ketosis. High-fat diets increase cardiovascular risk. One vegetarian diet that allows consumption of nothing of animal origin (even dairy products or eggs) results in Vitamin B_{12} deficiency, nervous maladies, and stunted growth in children.

23-5 SUMMARY

1 Digestion is a prerequisite for the absorption of polysaccharides, proteins, and fats. Digestion occurs when enzymes in the lumen of the gastrointestinal tract cleave molecular bonds of nutrients in the presence of water. Specific enzymes convert carbohydrates, proteins, and fats into simple sugars, amino acids, and fatty acids and glycerol, respectively. Because many of the enzymes are located on the intestinal microvilli, the products of digestion are close to the surfaces through which they may be absorbed. Definite requirements must be met for each enzyme to exert its optimal effects. These limits include specific ranges of pH and/or the presence of substances which convert inactive enzyme molecules into active ones. Examples of such enzyme activators are Ca^{2+} ions, enterokinase, and bile salts.

Among the enzymes which digest carbohydrates are salivary and pancreatic amylase and intestinal dextrinases, sucrase, maltase, and lactase. Proteins are hydrolyzed by peptidases, including gastric pepsin, pancreatic trypsin, chymotrypsin, and carboxypeptidase, and intestinal peptidases. Lipids are cleaved by pancreatic and intestinal lipases. Nucleic acids and their building blocks are broken down by intestinal nucleases, nucleotidases and nucleosidases.

2 Most absorption of foods occurs in the first half of the small intestine owing to the tremendous surface area afforded by the microvilli and the presence of digestive enzymes. Molecules and ions move into the mucosal surfaces of cells and out of their basal surfaces by the usual cellular mechanisms of transport. Substances mainly enter the intestinal capillaries, most lipids being the exceptions.

Bile salts act as ionic detergents to solubilize lipids and keep them dispersed in the fluids of the digestive tract in a process called emulsification. Triglycerides, fatty acids, cholesterol, and other lipids are packaged into small, water-soluble aggregates called micelles. Lipases bind to the surface of each micelle, digesting its contents, which then diffuse through the mucosal cell membrane. The lipids are metabolized in the cell and repackaged in a lipoprotein coat, producing microscopic chylomicrons, which are transported into lacteals. Fat-soluble vitamins follow this same route of absorption. Water-soluble vitamins diffuse into the mucosa and enter the intestinal capillaries.

Ions are absorbed primarily by the processes of active transport and diffusion. They include trace elements, which are minerals present in tissues in parts per million or lesser amounts. The transport of iron provides an interesting model for study of ion transport. Ferrous ions (Fe^{2+}) combine with a mucosal cell receptor protein, apoferritin, to form ferritin, which transports the ion across the mucosal cell and into the blood. The ferrous ion is released to another protein carrier in the blood, transferrin, which delivers the element to the liver, bone marrow, and other tissues.

Following the active transport of ions (especially sodium) into mucosal cells, water is absorbed by osmosis. From that location, water diffuses from hypotonic to hypertonic sites and most of it enters the intestinal capillaries. About 98 percent of the water entering the intestine from exogenous and endogenous sources is absorbed in the intestine. Only 2 percent (150 to 200 mL) exits from the colon with the feces. More fluid is lost with the feces when colonic motility is stimulated. Cathartics and laxatives often increase colonic motility by distending or irritating the colon.

Malabsorption results when mucosal passage of nutrients is defective. It may be initiated by abnormalities in the lumen of the tract, defective transport across the mucosa, and abnormal circulation of blood or lymph to its tissues. Disorders which alter the functional capacity of the microvilli most often are induced by specific pathogenic microorganisms or their toxins. These microbial species, which cause gastroenteritis, differ from the usual microorganisms in the intestine, which assist body function. For example, the normal bacterial flora stimulate antibody formation and provide some vitamins for absorption. Changes in the species of microorganisms among these flora can lead to bacterial overgrowth in intestinal diverticula and cause malabsorption.

3 The chemical reactions by which nutrients are utilized are called intermediary metabolism. Metabolic pools of substances such as pyruvic acid and acetyl CoA are formed, which may enter into anabolic or into catabolic reactions. The latter include oxidative reactions in which hydrogen and its electrons are stripped from carbon chains and transferred to hydrogen acceptors, the last of which is oxygen. The carbon chain is degraded and as a result, CO_2, H_2O, and energy are released. Forty percent of the energy is transferred to ATP; the remaining portion forms heat to warm body fluids. Each step in intermediary metabolism is catalyzed by a specific enzyme, so that the release of energy is gradual. This optimizes the transfer of energy to chemical bonds, reducing the amount that escapes as heat.

4 The energy expended by metabolism is the sum of that stored and converted to work and heat. The physical measure of energy expenditure is the kilocalorie (Calorie) and the unit of energy currency in metabolism is ATP. The complete oxidation of 1 g each of fat, carbohydrate, and protein generally makes available about 9, 4, and 4 kcal, respectively. Degradation of 1 mol of ATP to ADP releases 7 kcal, of energy, which may be expended in various ways.

The minimal metabolic cost of living is the basal metabolic rate (BMR), which is about 1620 and 1320 kcal/day for average young adult males and females, respectively. The metabolic rate increases with body surface area, with physical activity, during growth, with the consumption of food, and with fever, and it declines with aging. The caloric expenditure of the body averages 4.825 kcal per liter of oxygen consumed. It averages 2740 and 2030 kcal/day, respectively, for young adult males and females.

Heat production and heat loss are normally balanced by the hypothalamus to maintain body temperature near 37°C. Temperature regulation is so critical to body function that it takes precedence over water and salt balance. As a result sweat formation is used to cool the body as body temperature rises. If continued, the excessive loss of fluid and salts can cause muscle cramps, acidosis, heat exhaustion, and heat stroke, which may be fatal.

Fat depots are the greatest energy reservoirs in the body. Because carbohydrates provide limited caloric reserves, prolonged vigorous exercise burns up lipids, while proteins are usually spared. Obesity is an increase in the ratio of fat to lean body mass which causes body weight to rise 10 percent or more above normal for an individual's height, sex, and age. It may be caused by alterations in hypothalamic activity that stimulate overeating and a decrease in metabolic rate. Physiological, genetic, and psychological factors may help cause excess consumption of calories compared with those expended. Weight loss and a reduction in obesity occurs when caloric intake is less than caloric expenditure. A dietary and exercise regimen to cause weight loss should be based on medically accepted and recommended dietary allowances and carried out under qualified medical supervision.

REVIEW QUESTIONS

1 Name the enzymes secreted by and describe the functions of the salivary glands, stomach, pancreas, and small intestine.
2 How do the absorptive capacities of the stomach, small intestine, and large intestine compare?
3 Why is the surface of the microvilli critical to absorption?
4 How are the absorption of glucose and that of sodium related? Those of amino acids and of sodium?
5 Assume you are a protein molecule and three of your friends are fat, carbohydrate, and water molecules in a lunch consisting of a hamburger on a bun and a glass of water consumed by Bob Williams. Describe your voyage and its effect on you and your companions as you proceed through Bob's digestive tract.
6 How is the term *essential* used in connection with amino acids, fats, and vitamins?
7 Compare a micelle with a chylomicron and explain the functions of each in the absorption of lipids.
8 By what mechanisms is the body protected against iron toxicity? How is iron transported to tissues and recycled?
9 Chris Josephson ingests large quantities of nonabsorbable salt. What are its likely effects on his digestive tract?
10 Compare the effects of cathartics and laxatives. Compare constipation and diarrhea.
11 What are the three main causes of malabsorption? Give specific examples of each.
12 What is gastroenteritis? In what respects are bacterial overgrowth and bacterial food poisoning alike? How do they differ as causes of malabsorption?
13 Rachel Dee, a pregnant female astronaut, is persuaded to have her baby delivered "germ-free" during space flight and maintained in that state. What effects might a lack of intestinal bacteria have on the baby?
14 A starved, malnourished, underweight person is found unconscious. What alternative means are available to provide this person with nourishment? Why is intravenous feeding through a superficial arm vein insufficient for long-term survival?
15 What are metabolic pools? Name some. Of what nutritional significance are they? How are they interrelated?
16 What are the relationships between hydrogen transport and energy production?
17 How are carbohydrate, lipid, and protein stores used in energy production? Under what conditions is ketosis likely to develop and what is its significance?

18 What are the general ways in which the body uses energy derived from intermediary metabolism? What are the units of energy measurement? Of energy currency? What percentages of energy are supplied by the three major types of food?

19 Compare the metabolic rates of an average young adult female and an average young adult male. Compare the influence of fever and running a 26-mi marathon on a young adult's metabolism. Are an individual's carbohydrate reserves likely to provide enough energy for the marathon? Why or why not?

20 Stephen Jameson consumes an average of 3.0 L of O_2 per minute during 20 min of exercise, and his respiratory quotient is 0.82. Calculate his approximate caloric expenditure during this period. During this period, what is his most likely major source of energy? Why?

21 What are some of the factors that may be related to obesity?

APPENDIX 23A
Estimated amino acid requirements of humans

Amino Acid	REQUIREMENT, MG/KG BODY WEIGHT/DAY			Amino Acid Pattern for High-Quality Proteins, mg/g of protein*
	Infant (4–6 months)	*Child (10–12 years)*	*Adult*	
Histidine	33	?	?	17
Isoleucine	83	28	12	42
Leucine	135	42	16	70
Lysine	99	44	12	51
Total sulfur-containing amino acids (methionine) and cystine)†	49	22	10	26
Total aromatic amino acids (phenylalanine and tyrosine)†	141	22	16	73
Threonine	68	28	8	35
Tryptophan	21	4	3	11
Valine	92	25	14	48

* Two grams per kilogram of body weight per day of protein of the quality listed in this column would meet the amino acid needs of an infant.

† Cystine can in part substitute for some of the methionine requirement, as can tyrosine for some of the necessary phenylalanine.

APPENDIX 23B
Vitamins required in human nutrition

I. FAT-SOLUBLE VITAMINS			
Nomenclature	*Functions*	*Metabolism*	*Important Sources*
Vitamin A (retinol)	Maintenance of the integrity of epithelial tissue. Constituent of visual purple (rhodopsin) of retinal cells. Essential to growth, particularly of skeleton and other connective tissues	Absorption: From gastrointestinal tract, vitamin A follows pathway of fats; consequently, any impairment in fat absorption impairs absorption of vitamin A. Storage: In liver (95%). Deficiency: Dry, keratinized epithelium; night blindness; xerophthalmia; arrested growth	Fish liver oils, liver, butter, cream, whole milk, whole-milk cheese, egg yolk. Dark green leafy vegetables, yellow vegetables and fruits, fortified margarine. Principal source in food is as the provitamin beta-carotene. The vitamin itself (retinol) occurs relatively rarely in foods and is confined to lipids of animal tissues
Vitamin D Vitamin D_2: activated ergosterol (ergocalciferol) Vitamin D_3: activated 7-dehydrocholesterol (cholecalciferol)	Increases absorption of calcium and phosphorus from the intestine. Essential to ossification. Influences handling of phosphate by kidneys. Active form in tissues: 1,25-dihydroxycholecalciferol	Absorption: From intestine with fats, bile salts essential; can be synthesized in skin by activity of ultraviolet light on provitamin (D_3). Storage: Chiefly in liver. Deficiency: Rickets in children; tetanic convulsions in infants with severe deficiency	Fish-liver oils, fortified milk, activated sterols, exposure to sunlight. Very small amounts in butter, liver, egg yolks
Vitamin E Mainly alpha-tocopherol; also beta-, gamma-, delta-tocopherol	Exerts antioxidant effect to protect other vitamins in foods. May have auxiliary function in tissue respiration. In experimental animals, in association with other factors, prevents certain types of liver necrosis. No demonstrated function in human nutrition except in some infants immediately postpartum, particularly prematures. Need may be related to unsaturated fatty acid intake	Absorption: Similar to other fat-soluble vitamins. Transfer via placenta is limited; mammary gland transfer better; hence, breast milk is effective source for infants. Deficiency: May occur in malabsorptive states associated with lipid malabsorption	Plant tissues: oils of wheat germ, rice germ, cottonseed; green leafy vegetables, nuts, legumes, milk, eggs; muscle meats, fish

I. FAT-SOLUBLE VITAMINS	
Stability	*Chemistry*
Insoluble in water; fat-soluble. Associated with lipid in foods. Stable to heat by usual cooking methods. Destroyed by oxidation, drying, and very high temperatures	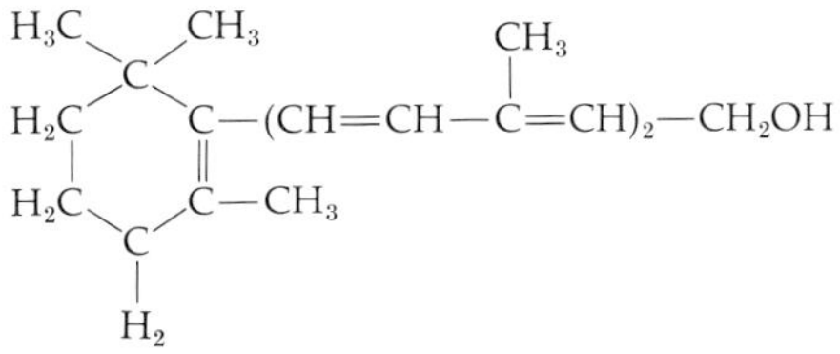 Retinol
Fat-soluble. Relatively stable to heat and oxidation	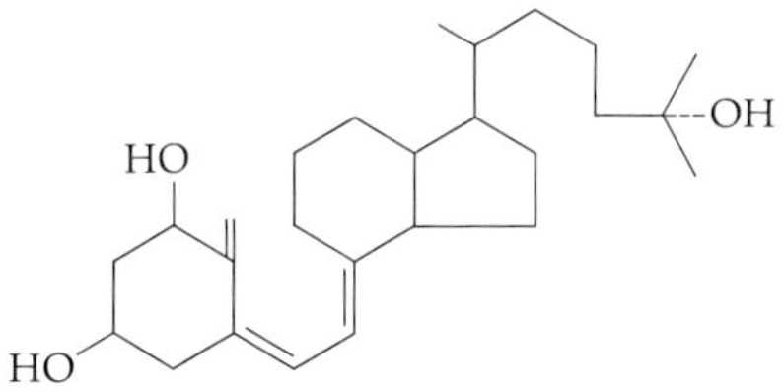 1,25-Hydroxycholecalciferol (1,25-HCC)
Fat soluble. Not affected by heat or acid. Oxidized in rancid fats and in the presence of lead and iron salts, alkali, ultraviolet light	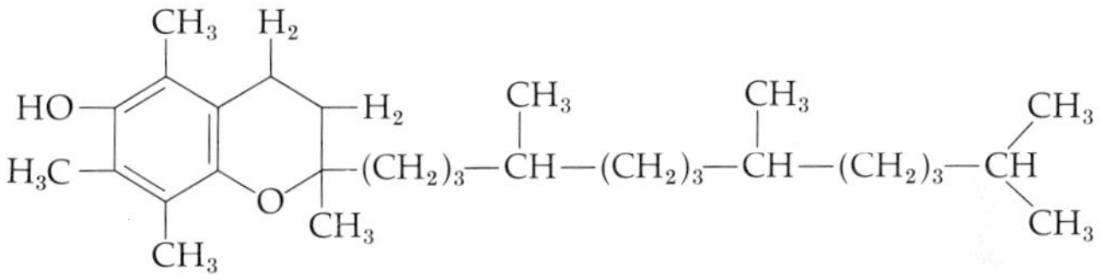 α-Tocopherol; β- and γ-tocopherol also active

APPENDIX 23B
Vitamins required in human nutrition (continued)

Nomenclature	*Functions*	*Metabolism*	*Important Sources*
Vitamin K Antihemorrhagic vitamin; coagulation vitamin. (Many compounds related to 2-methyl-1,4-naphthoquinone have some K activity)	Catalyzes synthesis of prothrombin by the liver. Required for activity of some blood clotting (thromboplastic) factors. May also participate in tissue respiration	Absorption: Similar to other fat-soluble vitamins. Utilization: Presence of bile needed; impairment in fat absorption seriously affects vitamin K absorption. Produced by intestinal microorganisms; therefore, antibiotic suppressive therapy may induce vitamin K deficiency. Storage: Limited amount in liver. Deficiency: Hypoprothrombinemia with resultant prolonged blood clotting; uncontrollable hemorrhage in newborn	Green leaves, such as alfalfa, spinach, cabbage; liver. Synthesis in intestine by activity of microorganisms is probably most important source. Dietary deficiency unlikely
II. WATER-SOLUBLE VITAMINS			
Ascorbic acid Vitamin C Antiscorbutic vitamin	Maintains normal intercellular material of cartilage, dentin, and bone. Probably has specific role in collagen synthesis by activity on proline hydroxylation. Association with oxidation-reduction systems in tissues. Metabolism of some amino acids, e.g., tyrosine, proline	Storage: Large quantity in adrenal cortex. With exception of muscle, tissues with high metabolic activity have increased concentrations. Excretion: Urine. Deficiency: Mild—petechial hemorrhages. Severe—loosening of teeth, lesions of gums, poor wound healing, easily fractured bones, scurvy	Citrus fruits, tomatoes, strawberries, cantaloupe, cabbage, broccoli, kale, potatoes, green peppers, salad greens
Thiamin Vitamin B_1 Antiberiberi vitamin	Constituent of enzyme systems of tissues, particularly in connection with decarboxylation (e.g., pyruvic and ketoglutaric acids). Deficiency mainly affects peripheral nervous system, gastrointestinal tract, and cardiovascular system	Absorption: Readily absorbed in aqueous solutions from both small and large intestines. Storage: Limited; hence, day-to-day supply needed. Excretion: Excess excreted to some extent in perspiration; also in urine. Deficiency: Anorexia, gastrointestinal atony and constipation, beriberi (including polyneuritis, cardiac failure, and edema). Requirement increased by high carbohydrate intake; also by fever, hyperthyroidism, pregnancy, lactation	Lean pork; liver, heart, kidney; brewer's yeast, wheat germ, whole-grain or enriched cereals and breads; soybeans, legumes, peanuts, milk

Stability	*Chemistry*
Fat soluble. Unstable to alkali and light. Fairly stable to heat	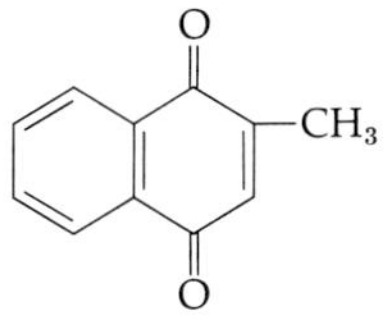 Vitamin K_3; other forms of vitamin K exist
II. WATER-SOLUBLE VITAMINS	
Soluble in water. Most easily destroyed of all vitamins—by heat, air, alkali, enzymes. Acid inhibits destruction. Copper accelerates destruction. Cooking generally reduces vitamin C content of food. Consumption of some uncooked foods essential to ensure adequate intake	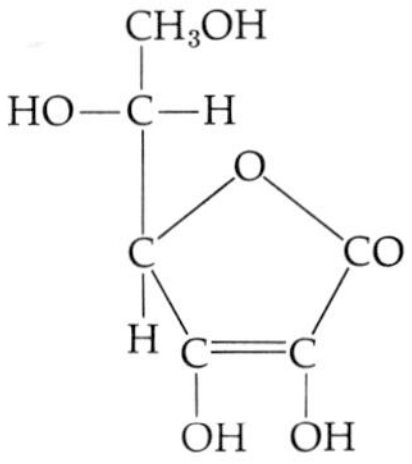 Ascorbic acid; reduced form
Soluble in water. Stable in slightly acid solution. Quickly destroyed by heat in neutral or alkaline solution. Sulfite quickly destroys thiamin	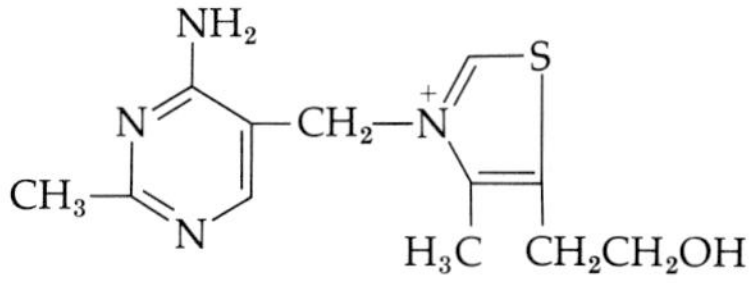 Thiamin

APPENDIX 23B
Vitamins required in human nutrition (continued)

Nomenclature	*Functions*	*Metabolism*	*Important Sources*
Riboflavin Vitamin B_2 (formerly lactoflavin, vitamin G	Constituent of tissue respiratory enzyme systems, as well as some enzymes (flavoproteins) involved in amino acid and lipid metabolism	Absorption: May require phosphorylation in intestinal mucosa. Storage: Limited in the body. Excretion: Excess excreted in urine. Deficiency: Cheilosis, seborrheic dermatitis of face, magenta tongue, certain functional and organic eye disorders	Milk, powdered whey; liver, kidney, heart, meats; eggs; green leafy vegetables; dried yeast, enriched foods (flour, bread). Cereals low; germination of oats, wheat, barley, and corn increases content
Niacin Nicotinic acid Niacinamide Antipellagra vitamin (pellagra-preventive [P-P] factor)	Constituent of 2 coenzymes (NAD, NADH) which operate as hydrogen and electron transfer agents in respiration. Tryptophan normally contributes to niacin supply (60 mg tryptophan equivalent to 1 mg niacin)	Storage: Limited in the body. Excretion: Urine, mainly as methylated derivatives. Deficiency: Pellagra with gastrointestinal, skin, and neurologic changes	Liver, kidney, lean meat, fish (salmon), poultry; whole-grain or enriched cereals, and breads; some leafy green vegetables, tomatoes; peanuts, brewer's yeast, tryptophan in proteins. Most fruits and vegetables are poor sources of niacin
Vitamin B_6 Pyridoxine Pyridoxal Pyridoxamine	Pyridoxal phosphate is a prosthetic group of enzymes which decarboxylate tyrosine, arginine, glutamic acid, and certain other amino acids. Essential to sulfur transfer and conversion of tryptophan to niacin; also as a coenzyme in transamination. Participates in metabolism of essential fatty acids. Essential in synthesis of porphyrins (e.g., heme for hemoglobin and cytochromes)	Absorption: Intestinal bacteria synthesize some pyridoxine, and it is absorbed from the intestine. Storage: Limited in the body. Excretion: Urine. Deficiency: Hypochromic macrocytic anemia, lesions of central nervous system evidenced by epileptiform seizures and encephalographic changes, particularly in in infants	Wheat germ; meat, liver, kidney; whole-grain cereals, soybeans, peanuts, corn, yams; brewer's yeast. Synthesis by activity of microorganisms
Panthothenic acid	Constituent of coenzyme A which participates in synthesis and breakdown of fatty acids, synthesis of cholesterol and steroid hormones, utilization of pyruvate and acetate, reactions of acetylation, metabolism of some amino acids, synthesis of heme for hemoglobin and cytochromes	Storage: Limited in the body. Excretion: Urine. Deficiency: Gastrointestinal symptoms, skin symptoms, anemia, and impairment in functions of adrenal cortex in experimental animals	Organ meats (liver, kidney), lean beef; egg yolk; peanuts; broccoli, cauliflower, cabbage; whole grains, cereal bran; skim milk; fruits, sweet potatoes

Stability	*Chemistry*
Sparingly soluble in water. Quickly decomposed by ultraviolet or visible light; very sensitive to alkali. Relatively resistant to heat in acid media	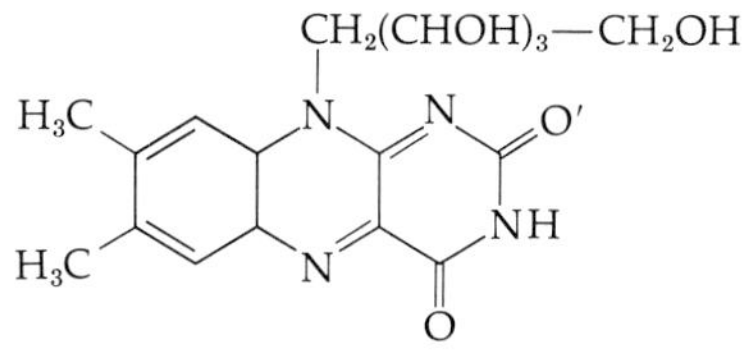 $CH_2(CHOH)_3-CH_2OH$ ... Riboflavin
Soluble in water. Relatively stable to heat, oxidation, and light. Relatively stable to acid and alkali	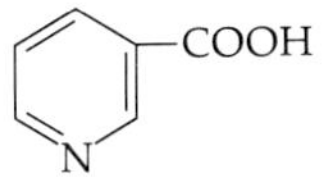 —COOH ... Niacin
Fairly stable to heat, but sensitive to ultraviolet light and oxidation	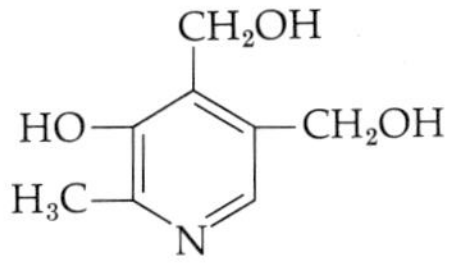 CH_2OH, HO—, $-CH_2OH$, H_3C- ... Pyridoxine
Easily destroyed by heat and alkali. Stable in neutral solution	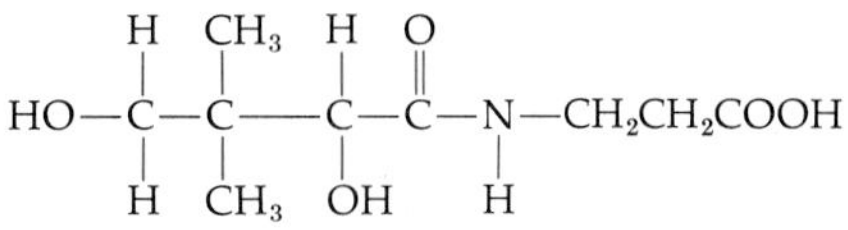 $HO-CH_2-C(CH_3)_2-CH(OH)-C(=O)-NH-CH_2CH_2COOH$ ... Pantothenic acid

APPENDIX 23B
Vitamins needed in human nutrition (continued)

Nomenclature	*Functions*	*Metabolism*	*Important Sources*
Folic acid Folacin Pteroylglutamic acid	Involved in transfer and utilization of the single-carbon moiety; participates in synthesis of purines, thymine, and methyl groups; has specific role in metabolism of histidine and well-demonstrated role in hematopoiesis	Excretion: Excess in both urine and feces. Deficiency: May produce macrocytic anemia with concurrent glossitis, gastrointestinal lesions, diarrhea, and intestinal malabsorption (sprue). Deficiency in pregnancy not uncommon	Liver, kidney; yeast; fresh green leafy vegetables, cauliflower. Synthesis by activity of intestinal microorganisms
Vitamin B_{12} Antipernicious anemia factor Cobalamin	Involved in purine and pyrimidine metabolism, synthesis of nucleic acid (DNA), methionine metabolism, transmethylation, and maturation of red blood cells. Contains cobalt, which is the only known function for this element	Absorption: From ileum, but requiring intrinsic factor and hydrochloric acid contributed by stomach. Storage: Principally in liver, for long periods. Excretion: In feces (represents unabsorbed vitamin). Deficiency: Macrocytic anemia or pernicious anemia with degenerative changes in gastric mucosa, characteristic lesions in nervous system (combined system disease)	Foods of animal origin: liver, kidney, muscle meat, eggs, milk, cheese. No significant amounts in higher plants. Synthesis within intestine by activity of microorganisms
Biotin Inositol Choline	Required by various animal species but of questionable need for humans. If they are in fact needed, the amounts required are very small and are probably synthesizable in tissues or provided by intestinal flora		

Stability	*Chemistry*
Easily oxidized in acid medium and sunlight. Labile to heat (Similar to thiamin)	Folic acid; monoglutamate form
Labile to heat, acids, alkali, and light	Vitamin B_{12}

APPENDIX 23C
Summary of essential trace elements

	FUNCTION		
Element	*Biochemical*	*Physiologic*	*Metabolism*
Iron	Integral part of hemoglobin, myoglobin Component of catalase, succinic dehydrogenase and other enzymes	Transport and utilization of oxygen	Absorption: 5–10% of that in diet. Regulation according to body needs Transport: Bound to serum protein (transferrin) Storage: In heme proteins, ferritin in liver, spleen, bone marrow Distribution: Hemosiderin, ⅔ as heme proteins, remainder as ferritin and hemosiderin Turnover: 0.8–1.5 mg/day. Iron is used over and over Excretion: Men, about 1 mg/day; women 1–2 mg/day in urine, perspiration, menstrual flow. Fecal excretion is iron unabsorbed from diet Deficiency: Anemia, frequent in infants, preschool children, teenaged girls, pregnant women
Iodine	Constituent of monoidotyrosine, diiodotyrosine, triiodothyronine (T_3), and thyroxine (T_4)	Regulates energy metabolism; promotes growth of young; maintains neuromuscular function	Absorption: Inorganic iodides, almost completely absorbed. Iodinated amino acids, absorbed more slowly and less completely Transport: As T_4 bound to thyroxine-binding alpha-globulin, albumin, and pre-albumin. Also as unbound iodide Storage: In combination with thyroglobulin in thyroid gland, storage regulated by thyroid stimulating hormone (TSH) Turnover: About 60 μg/day Excretion: Mainly in urine Deficiency: Goiter, cretinism, if deficiency is severe (rarely seen in U.S.)
Copper	Component of cytochrome oxidase and other metalloenzymes	Involved in the absorption and utilization of iron in the synthesis of hemoglobin and myoglobin. Involved in formation of melanin and myelin. Constituent of connective tissue. Has role in maintenance of neurologic function	Absorption: From the stomach and all portions of the small intestine. Extent is dependent on chemical forms of copper ingested, on other metals, and on organic substances in diet Transport: Mainly in loose binding with albumin. Small amount bound to amino acids and some as component of ceruloplasmin (a copper-binding plasma protein) Distribution: Concentrations in descending order: liver, brain, lung, kidney, ovary, testis, lymph nodes, muscle, pituitary, thymus, prostate Turnover: 0.6–1.6 mg/day Excretion: Chiefly in feces, less than 5% in urine Deficiency: In infants: anemia, hypoproteinemia, low serum iron and copper; anemia resulting from rehabilitation from malnourishment with high-calorie, low-copper diet; Menke's kinky hair syndrome

Element	FUNCTION		Metabolism
	Biochemical	*Physiologic*	
Manganese	Component of metalloenzymes, including pyruvate carboxylase, and mitochondrial superoxide dismutase. Activates glucosyltransferase	Involved in synthesis of polysaccharides, cholesterol	Absorption: Level of dietary manganese and the excretion of this element regulate absorption. Absorbed throughout length of small intestine. Inhibited by calcium and phosphorus in diet Transport: Bound to a β_1-globulin Distribution: Liver, pancreas, brain, heart, muscle, spleen, lung Turnover: 2.5 mg/day Excretion: In feces Deficiency: In animals: Impaired growth, reproductory function, skeletal abnormalities, ataxia in the newborn. In humans: not well-defined
Zinc	Component of at least 30 metalloenzymes, including carbonic anhydrase, alcohol dehydrogenase, alkaline phosphatase, leucine aminopeptidase	Involved in synthesis of DNA and RNA, glutathione, connective tissue, mobilization of vitamin A from liver	Absorption: Mainly from duodenum, ileum, jejunum. Regulated by needs, through zinc-binding protein in intestinal lumen, influenced by level of zinc in diet and with other dietary components Transport: 60–70 percent loosely bound to plasma albumin; remainder to α_2-macroglobulin and transferrin Distribution: In descending order of concentration: eye, prostate, bone, muscle, kidney, liver, pancreas, heart, lung Turnover: 0.5–2.0 mg/day Storage: Pancreas, liver, kidney, and spleen Excretion: Urine, perspiration, and menstrual fluid. Zinc leaving body in feces is mostly that unabsorbed from diet. Deficiency: Growth retardation, impaired sense of taste, hypogonadism in extreme deficiency, slow wound healing, acrodermatitis enteropathica
Cobalt	Integral part of vitamin B_{12}, which forms the cofactor 5,6-dimethylbenzimidazole ribonucleotide, required for the activation of methylmalonyl-CoA. No other function known	Involved in hematopoiesis	Absorption: Well absorbed, in proportion to concentration in diet. Is increased in iron deficiency. Is depressed by high iron concentrations in diet. (Cobalt must be absorbed as vitamin B_{12} by humans, since this cannot be synthesized from cobalt) Distribution: Kidney, liver, heart, spleen, pancreas Excretion: Chiefly in the urine, small amounts lost in feces, sweat, and hair Deficiency: Not known in humans or in laboratory animals. Vitamin B_{12} deficiency results in pernicious anemia

APPENDIX 23C
Summary of essential trace elements (continued)

	FUNCTION		
Element	*Biochemical*	*Physiologic*	*Metabolism*
Molybdenum	Constituent of a number of enzymes, xanthine oxidase, aldehyde oxidase, sulfite oxidase	Not established	Absorption: Rapidly from most diets and inorganic compounds of the element Transport: Bound to erythrocytes and to plasma proteins Distribution: Liver, kidney, spleen, lung, brain, muscle Turnover: May be 0.1–0.15 mg/day Excretion: Unknown Deficiency: Equivocal in animals, not known in humans
Selenium	Integral part of glutathione peroxidase	Protects against accumulation of hydrogen peroxide (has antioxidant function). Experimental and epidemiologic evidence indicate an inhibitory effect on carcinogenesis. This action may be secondary to its essential roles	Absorption: In form of selenite 40–70%. No evidence of a control mechanism. May depend on dietary form and amount Distribution: Liver, pancreas, blood, skin, muscle Excretion: Urine and feces Deficiency: Has been produced in animals. Reported hemolytic anemias in humans
Chromium	Integral part of glucose tolerance factor (GTF)	Involved in maintaining normal glucose tolerance	Absorption: Chromium salts absorbed poorly (1–3%); organic chromium in form of GTF absorbed more completely Transport: Bound to plasma albumin and transferrin Storage: Probably in liver Excretion: Mainly in urine Deficiency: Impaired glucose metabolism
Tin	May function as an oxidation-reduction catalyst. Also, as a catalyst in transesterification and polymerization reactions	Is required for growth in rat	
Vanadium	Probably acts as an oxidation-reduction catalyst. Catalyzes nonenzymatic oxidation of catecholamines. Controls synthesis of cholesterol	Required in rats for growth and reproduction; in chick for proper bone and feather development. Lowers cholesterol and phospholipids in blood of animals and humans	
Fluorine	Activates citrulline synthesis by liver. Activates adenyl cyclase	Needed in rats for growth and proper development. Inhibits dental decay in humans and in animals	Storage: Bones and teeth
Silicon	Acts as crosslinking agent in mucopolysaccharides and other components of connective tissue		Distribution: High in connective tissue and bone (especially in region of active calcification) and in skin

	FUNCTION		
Element	*Biochemical*	*Physiologic*	*Metabolism*
Nickel		Required by rats for optimal growth and reproduction. Required by chicks to maintain optimal liver function	Absorption: Poorly absorbed from ordinary diets (1–10%) Distribution: Widely distributed in body tissues but in low concentration Transport: In serum as ultrafiltrable, albumin-bound compound and as component of a metalloprotein Excretion: Mostly in feces but some in sweat and urine

APPENDIX 23D
Nutritional aspects of trace elements

Element	*Food Sources*	*Recommended Daily Allowances*	*Normal Serum Concentration*	*Laboratory Aids to Assessment of Nutriture*
Iron	Organ meats (liver, kidney), other meats, fish, poultry, egg yolk, cocoa, dark molasses; enriched cereals, nuts, legumes, green vegetables	Infants: 10–15 mg Children: 10–15 mg Men: 10–18 mg Women: 10–18 mg Pregnant women: 30–60 mg	Men: 60–150 μg/dL Women: 50–130 μg/dL	Measurement of blood hemoglobin, hematocrit, erythrocyte count, serum iron, iron binding capacity, and serum ferritin
Iodine	Seafood; iodized table salt,	Infants: 40–50 μg Children: 70–120 μg Men: 150 μg Women: 150 μg Pregnant women: 175 μg Lactating women: 200 μg	As inorganic iodine: 0.08–0.60 μg/dL As PBI: 4.8 μg/dL	
Copper	Shellfish, especially oysters; organ meats, nuts, legumes, cocoa, whole grain cereals, drinking water	Infants: 0.08 mg/kg Adults: 2–5 mg	73–199 μg/dL	Measurement of serum copper
Zinc	Seafoods, especially oysters; muscle meats; nuts, cereals	Infants: 3.5 mg Children: 1–10 years: 10 mg Men: 10–51+ : 15 mg Women: 11–50 : 15 mg Pregnant women: 20 mg Lactating women: 25 mg	69–121 μg/dL	Measurement of serum zinc
Manganese	Nuts, whole cereals, fruits, tubers, nonleafy vegetables; meats, poultry, fish, seafood	2–3 mg (estimated safe and adequate)	38–104 μg/dL	
Cobalt	Meats or pure vitamin B_{12}	Not established	Not known	Measurement of serum vitamin B_{12}. Measurement of urinary excretion of methylmalonic acid
Molybdenum	Plant foods. Content varies greatly with soils	Not established	1.5 + 0.12 μg/dL in whole blood	Not established
Selenium	Fish and other seafoods; plant foods rich in protein. Selenium is highly variable depending on soils	0.05–0.2 mg (estimated safe and adequate)	10–34 μg/dL in whole blood; may reflect concentration in diet	Not established but studies indicate that plasma levels may reflect status. Also, erythrocyte glutathione levels may be helpful
Chromium	Brewer's yeast, seafoods, meats, whole grains, oils, fat condiments	0.05–0.2 mg (estimated safe and adequate)	1–2 ng/mL (1–2 ppb)	Not established but some evidence indicates that urinary excretion may be helpful
Tin	Widespread in plant foods and in meats	3.6–17 mg (estimated)		
Vanadium	Fish, meats, grains, and nuts. Other plant foods vary greatly in vanadium content. Oxidation state may be important	2 mg (estimated)		

Element	*Food Sources*	*Recommended Daily Allowances*	*Normal Serum Concentration*	*Laboratory Aids to Assessment of Nutriture*
Fluorine	Abundant in many foods. Also often supplied in drinking water	4.4 mg	10 μg/dL	
Silicon	Present in all plant and animal foods; however, nothing is known of bioavailability		Has been reported as 200 μg/dL	
Nickel			Mean: Men: 258 μg/dL; Women: 270 μg/dL	

APPENDIX 23E

Examples of daily energy expenditures of mature women and men in light occupations

		MAN, 70 KG		WOMAN, 58 KG	
Activity Category	*Time, h*	*Rate, kcal/min*	*Total kcal*	*Rate, kcal/min*	*Total kcal*
Sleeping, reclining	8	1.0–1.2	540	0.9–1.1	440
Very light Seated and standing activities, painting trades, auto and truck driving, laboratory work, typing, playing musical instruments, sewing, ironing	12	up to 2.5	1300	up to 2.0	900
Light Walking on level, 2.5–3 mi/h, tailoring, pressing, garage work, electrical trades, carpentry, restaurant trades, cannery workers, washing clothes, shopping with light load, golf, sailing, table tennis, volleyball	3	2.5–4.9	600	2.0–3.9	450
Moderate Walking 3.5–4 mi/h, plastering, weeding and hoeing, loading and stacking bales, scrubbing floors, shopping with heavy load, cycling, skiing, tennis, dancing	1	5.0–7.4	300	4.0–5.9	240
Heavy Walking with load uphill, tree felling, work with pick and shovel, basketball, swimming, climbing, football	—	7.5–12.0		6.0–10.0	—
TOTAL	24		2740		2030

APPENDIX 23F
Approximate energy expenditure, kcal/h*

Activity	kcal/h
Restful Activity:	
Sleeping	70
Lying quietly	80
Sitting	100
Standing	110
Singing	120
Work:	
Mental work (seated)	105
Driving a car	140
Office work	145
Housekeeping	150
Bricklaying	205
Housepainting	210
Carpentry	230
Pick and shovel	400
Shoveling sand	405
Chopping wood	450
Sawing wood	480
Exercise:	
Bicycling	
At 5.5 mi/h	190
Rapidly	415
Rowing	
For pleasure	300
At 3.5 mi/h	660
At 11 mi/h	970
At 11.3 mi/h	1130
At 12 mi/h	1500
Running	
On a level surface	
5.7 mi/h	720
7 mi/h	870
11.4 mi/h	1300
13.2 mi/h	2330
14.8 mi/h	2880
15.8 mi/h	3910
17.2 mi/h	4740
18.6 mi/h	7790
18.9 mi/h	9480
Marathon running	990
Horizontally at 5 mi/h carrying a 43-lb load	820
Up an 8.6% grade at 7 mi/h	950
Skating	
at 9 mi/h	470
at 11 mi/h	640
at 13 mi/h	780
Swimming	
Backstroke at 1 mi/h	500
1.6 mi/h	800
2.2 mi/h	2000
2.4 mi/h	2530
Breaststroke at 1 mi/h	410
1.6 mi/h	490
2.2 mi/h	1850
2.7 mi/h	3690
Crawl at 1 mi/h	420
1.6 mi/h	700
2.0 mi/h	1600
Sidestroke at 1 mi/h	550
1.6 mi/h	1200
2.2 mi/h	3000
Walking	
Downstairs at 2 mi/h	200
Upstairs at 1 mi/h	180
2 mi/h	590
Horizontally at 2 mi/h	170
3.5 mi/h	290
Horizontally at 3 mi/h carrying a 43-lb load	350
Up a 3% grade at 3.5 mi/h	370
an 8.6% grade at 3.5 mi/h	560
a 10% grade at 3.5 mi/h	580
a 14.4% grade at 3.5 mi/h	740
In 12–18 in of snow	760
Up a 36% grade at 1 mi/h carrying a 43-lb load	680
Up a 36% grade at 1.5 mi/h carrying a 43-lb load	890
Other Sports	
Baseball (except pitcher)	280
Baseball (pitcher)	390
Basketball	395
Billiards	235
Bowling	215
Fencing	630
Football	1000
Mountain climbing	600
Parallel bar work	710
Pitching horseshoes	240
Skiing at 3 mi/h	540
Snowshoeing with trailshoes at 2.5 mi/h	620
Table tennis	345
Wrestling	790

* Because energy expenditure (and metabolic rates) are altered by a number of variables, including body size, environmental temperature, and training, the figures in these tables are only approximations. The calculations are based on a 70-kg (154-lb) person. To determine your approximate energy expenditure *per minute*, divide the above figures by 4200, then multiply the answer by your weight in kilograms (1 kg = 2.2 lb). Adapted from *Physiology of Exercise*, by L. E. Morehouse and A. T. Miller, Jr., 7th ed., copyright © 1976, C. V. Mosby. Compiled from data obtained chiefly from the Harvard Fatigue Library.

APPENDIX 23G
Mean heights and weights and recommended energy intake*

		WEIGHT		HEIGHT		ENERGY NEEDS (WITH RANGE)	
Category	*Age, years*	*kg*	*lb*	*cm*	*in*	*kcal*	
Infants	0.0–0.5	6	13	60	24	kg × 115	(95–145)
	0.5–1.0	9	20	71	28	kg × 105	(80–135)
Children	1–3	13	29	90	35	1300	(900–1800)
	4–6	20	44	112	44	1700	(1300–2300)
	7–10	28	62	132	52	2400	(1650–3300)
Males	11–14	45	99	157	62	2700	(2000–3700)
	15–18	66	145	176	69	2800	(2100–3900)
	19–22	70	154	177	70	2900	(2500–3300)
	23–50	70	154	178	70	2700	(2300–3100)
	51–75	70	154	178	70	2400	(2000–2800)
	76+	70	154	178	70	2050	(1650–2450)
Females	11–14	46	101	157	62	2200	(1500–3000)
	15–18	55	120	163	64	2100	(1200–3000)
	19–22	55	120	163	64	2100	(1700–2500)
	23–50	55	120	163	64	2000	(1600–2400)
	51–75	55	120	163	64	1800	(1400–2200)
	76+	55	120	163	64	1600	(1200–2000)
Pregnancy						+300	
Lactation						+500	

* The data in this table have been assembled from the observed median heights and weights of children, together with desirable weights for adults, for the mean heights of men (70 in) and women (64 in) between the ages of 18 and 34 years as surveyed in the U.S. population (HEW/NCHS data).

The energy allowances for the young adults are for men and women doing light work. The allowances for the two older age groups represent mean energy needs over these age spans, allowing for a 2 percent decrease in basal (resting) metabolic rate per decade and a reduction in activity of 200 kcal/day for men and women between 51 and 75 years, 500 kcal for men over 75 years, and 400 kcal for women over 75 years.

The customary range of daily energy output is shown in parentheses for adults and is based on a variation in energy needs of ±400 kcal at any one age, emphasizing the wide range of energy intakes appropriate for any group of people.

Energy allowances for children through age 18 are based on median energy intakes of children of these ages followed in longitudinal growth studies. The values in parentheses are 10th and 90th percentiles of energy intake, to indicate the range of energy consumption among children of these ages.

APPENDIX 23H
Energy equivalents chart. Number of minutes required at the activities listed to expend the caloric energy of food items shown below*

Type of Food	*Caloric Content*	*Reclining*	*Walking*	*Bicycle Riding*	*Swimming*	*Running*
Beverages, alcoholic						
Beer, 8 oz glass	114	88	22	14	10	6
Gin, $1\frac{1}{2}$ oz jigger	105	81	20	13	9	5
Manhattan, $3\frac{1}{2}$ oz cocktail	164	126	32	20	15	8
Martini, $3\frac{1}{2}$ oz cocktail	140	108	27	17	13	7
Old-fashioned, 4 oz glass	179	138	34	22	16	9
Rye whiskey, $1\frac{1}{2}$ oz jigger	119	92	23	15	11	6
Scotch whiskey, $1\frac{1}{2}$ oz jigger	105	81	20	13	9	5
Tom Collins, 10 oz glass	180	138	35	20	16	9
Wine, $3\frac{1}{2}$ oz glass	84	65	16	10	8	4
Beverages, nonalcoholic						
Carbonated, 8 oz glass	106	82	20	13	9	5
Ice cream soda, chocolate	255	196	49	31	23	13
Malted milk shake, chocolate	502	386	97	61	45	26
Milk, 8 oz glass	166	128	32	20	15	9
Milk, skim, 8 oz glass	81	62	16	10	7	4
Milk shake, chocolate	421	324	81	51	38	22
Desserts						
Cake, 2-layer, $\frac{1}{12}$	356	274	68	43	32	18
Cookie, chocolate chip	51	39	10	6	5	3
Doughnut	151	116	29	18	13	8
Ice cream, $\frac{1}{6}$ qt	193	148	37	24	17	10
Gelatin, with cream	117	90	23	14	10	6
Pie, apple, $\frac{1}{6}$	377	290	73	46	34	19
Sherbet, $\frac{1}{6}$ qt	177	136	34	22	16	9
Strawberry shortcake	400	308	77	49	36	21
Fruits and fruit juices						
Apple, large	101	78	19	12	9	5
Banana, small	88	68	17	11	8	4
Orange, medium	68	52	13	8	6	4
Peach, medium	46	35	9	6	4	2
Apple juice, 8 oz glass	118	91	23	14	10	6
Orange juice, 8 oz glass	120	92	23	15	11	6
Tomato juice, 8 oz glass	48	37	9	6	4	2
Meats						
Bacon, 2 strips	96	74	18	12	9	5
Ham, 2 slices	167	128	32	20	15	9
Pork chop, loin	314	242	60	38	28	16
Steak, T-bone	235	181	45	29	21	12
Miscellaneous						
Bread and butter, 1 slice	78	60	15	10	7	4
Cereal, dry, $\frac{1}{2}$ cup, with milk and sugar	200	154	38	24	18	10
French dressing, 1 tbsp	99	45	11	7	5	3
Mayonnaise	92	71	18	11	8	5
Pancake, with syrup	124	95	24	15	11	6
Spaghetti, 1 serving	396	305	76	48	35	20
Cottage cheese, 1 tbsp	27	21	5	3	2	1

APPENDIX 23H
Energy equivalents chart. Number of minutes required at the activities listed to expend the caloric energy of food items shown below* (continued)

Type of Food	*Caloric Content*	*Reclining*	*Walking*	*Bicycle Riding*	*Swimming*	*Running*
Poultry and eggs						
Chicken, fried, 1/2 breast	232	178	45	28	21	12
Chicken, "TV dinner"	542	217	104	66	48	28
Turkey, 1 slice	130	100	25	16	12	7
Egg, fried	110	85	21	13	10	6
Egg, boiled	77	59	15	9	7	4
Sandwiches and snacks						
Club	590	454	113	72	53	30
Hamburger	300	269	67	43	31	18
Roast beef, with gravy	430	331	83	52	38	22
Tuna salad	278	214	53	34	25	14
Pizza, with cheese, 1/8	180	138	35	22	16	9
Potato chips, 1 serving	108	83	21	13	10	6
Cheddar cheese, 1 oz	111	85	21	14	10	6
Seafood						
Clams, 6 medium	109	77	19	12	9	5
Cod, steamed, 1 piece	80	62	15	10	7	4
Crabmeat, 1/2 cup	68	52	13	8	6	4
Haddock, 1 piece	71	55	14	9	6	4
Halibut steak, 1/4 lb	205	158	39	25	18	11
Lobster, 1 medium	55	38	10	6	4	3
Shrimp, French fried, 1 serving	180	136	35	22	16	9
Vegetables						
Beans, green, 1 cup	27	21	5	3	2	1
Beets, canned, 1/2 cup	38	29	7	5	3	2
Carrot, raw	42	32	8	5	4	2
Lettuce, 3 large leaves	30	23	6	4	3	2
Peas, green, 1/2 cup	55	43	11	7	5	3
Potato, boiled, 1 medium	100	77	19	12	9	5
Spinach, fresh, 1/2 cup	20	15	4	2	2	1

* The caloric value of foods and portions vary. So too, energy expenditures (and metabolic rates) are altered by a number of variables, including body size, environmental temperature, training and nature of the exercise (as noted in App. 23F). Therefore, the figures in these tables are only approximations.

APPENDIX 23I

Food and Nutrition Board, National Academy of Sciences–National Research Council Recommended Daily Dietary Allowances,[a] revised 1980
Designed for the maintenance of good nutrition of practically all healthy people in the United States

		WEIGHT		HEIGHT			FAT-SOLUBLE VITAMINS			WATER-SOLUBLE VITAMINS							MINERALS					
	Age, *years*	*kg*	*lb*	*cm*	*in*	Protein, *g*	Vitamin A (*μg* RE)[b]	Vitamin D (*μg*)[c]	Vitamin E (*mg* α-TE)[d]	Vitamin C *mg*	Thiamin *mg*	Riboflavin *mg*	Niacin *mg* NE[e]	Vitamin B-6 *mg*	Folacin[f] *μg*	Vitamin B-12 *μg*	Calcium *mg*	Phosphorus *mg*	Magnesium *mg*	Iron *mg*	Zinc *mg*	Iodine *μg*
Infants	0.0–0.5	6	13	60	24	kg × 2.2	420	10	3	35	0.3	0.4	6	0.3	30	0.5[g]	360	240	50	10	3	40
	0.5–1.0	9	20	71	28	kg × 2.0	400	10	4	35	0.5	0.6	8	0.6	45	1.5	540	360	70	15	5	50
Children	1–3	13	29	90	35	23	400	10	5	45	0.7	0.8	9	0.9	100	2.0	800	800	150	15	10	70
	4–6	20	44	112	44	30	500	10	6	45	0.9	1.0	11	1.3	200	2.5	800	800	200	10	10	90
	7–10	28	62	132	52	34	700	10	7	45	1.2	1.4	16	1.6	300	3.0	800	800	250	10	10	120
Males	11–14	45	99	157	62	45	1000	10	8	50	1.4	1.6	18	1.8	400	3.0	1200	1200	350	18	15	150
	15–18	66	145	176	69	56	1000	10	10	60	1.4	1.7	18	2.0	400	3.0	1200	1200	400	18	15	150
	19–22	70	154	177	70	56	1000	7.5	10	60	1.5	1.7	19	2.2	400	3.0	800	800	350	10	15	150
	23–50	70	154	178	70	56	1000	5	10	60	1.4	1.6	18	2.2	400	3.0	800	800	350	10	15	150
	51+	70	154	178	70	56	1000	5	10	60	1.2	1.4	16	2.2	400	3.0	800	800	350	10	15	150
Females	11–14	46	101	157	62	46	800	10	8	50	1.1	1.3	15	1.8	400	3.0	1200	1200	300	18	15	150
	15–18	55	120	163	64	46	800	10	8	60	1.1	1.3	14	2.0	400	3.0	1200	1200	300	18	15	150
	19–22	55	120	163	64	44	800	7.5	8	60	1.1	1.3	14	2.0	400	3.0	800	800	300	18	15	150
	23–50	55	120	163	64	44	800	5	8	60	1.0	1.2	13	2.0	400	3.0	800	800	300	18	15	150
	51+	55	120	163	64	44	800	5	8	60	1.0	1.2	13	2.0	400	3.0	800	800	300	10	15	150
Pregnant						+30	+200	+5	+2	+20	+0.4	+0.3	+2	+0.6	+400	+1.0	+400	+400	+150	*h*	+5	+25
Lactating						+20	+400	+5	+3	+40	+0.5	+0.5	+5	+0.5	+100	+1.0	+400	+400	+150	*h*	+10	+50

[a] The allowances are intended to provide for individual variations among most normal persons as they live in the United States under usual environmental stresses. Diets should be based on a variety of common foods in order to provide other nutrients for which human requirements have been less well defined. See text and appendixes in this chapter for nutrients not tabulated and for weights and heights by year of age and for suggested average energy intakes.

[b] Retinol equivalents. 1 retinol equivalent = 1 μg retinol of 6 μg β-carotene.

[c] As cholecalciferol. 10 μg cholecalciferol = 400 IU of vitamin D.

[d] α-tocopherol equivalents. 1 mg *d*-α-tocopherol = 1 α-TE.

[e] 1 NE (niacin equivalent) is equal to 1 mg of niacin or 60 mg of dietary tryptophan.

[f] The folacin allowances refer to dietary sources as determined by *Lactobacillus casei* assay.

[g] The recommended dietary allowance for vitamin B-12 in infants is based on average concentration of the vitamin in human milk. The allowances after weaning are based on energy intake (as recommended by the American Academy of Pediatrics) and consideration of other factors, such as intestinal absorption.

[h] The increased requirement during pregnancy cannot be met by the iron content of habitual American diets nor by the existing iron stores of many women; therefore the use of 30–60 mg of supplemental iron is recommended. Iron needs during lactation are not substantially different from those of nonpregnant women, but continued supplementation of the mother for 2–3 months after birth of the baby is advisable in order to replenish stores depleted by pregnancy.

THE PHYSIOLOGY OF REPRODUCTION

24

24-1 THE MALE REPRODUCTIVE SYSTEM
The testes
Accessory sex organs
Seminal vesicles
Prostate gland
Bulbourethral (Cowper's) glands
The production and release of sperm
Sperm cell production (spermatogenesis)
Erection, emission, and ejaculation
Composition of seminal fluid (semen)
Orgasm
The influence of hormones

24-2 THE FEMALE REPRODUCTIVE SYSTEM
The ovaries
Accessory sex organs
The production of eggs (oogenesis)
Orgasm
The influence of hormones
Effects on the ovaries
Effects on the uterus and vagina
Effects on the mammary glands

24-3 CONCEPTION AND PREGNANCY
Early embryonic development
The placenta
Human chorionic gonadotropin (HCG) and pregnancy tests
The pattern of development from the third week
Maternal hormonal changes during pregnancy
Amniocentesis

24-4 REPRODUCTIVE CAPACITY
Fertility and infertility
Effects of sexually transmitted diseases
Control of conception
Rhythm
The pill
Physical methods
Surgical methods
Miscellaneous techniques
Other methods of birth control
Spontaneous and induced abortion • Intrauterine devices • Morning-after pill

24-5 CHILDBIRTH (PARTURITION)

24-6 SUMMARY

OBJECTIVES

After completing this chapter the student should be able to:

1. Describe the general structure and functions of the male reproductive system.
2. Discuss spermatogenesis, orgasm, and the influence of hormones on male reproductive function.
3. Describe the general structure and functions of the female reproductive system.
4. Discuss oogenesis, orgasm, and the influence of hormones on female reproductive function.
5. Explain the relationships among follicle development, the corpus luteum, the uterus, and cyclic changes in ovarian and pituitary hormones.
6. Discuss the events surrounding conception and early embryonic development.
7. Describe embryonic and fetal development.
8. Describe the basis of fertility and infertility in males and females.
9. Describe and compare various methods of control of conception.
10. Describe abortion and some procedures that are used to induce it. Discuss the mechanisms of action of intrauterine devices and the morning-after pill.
11. Discuss the major events in each stage of childbirth.

The human race is perpetuated by *sexual reproduction,* a remarkable physiological activity greatly influenced by psychological and social factors and one in which human beings give rise to more human beings. A genetic legacy, whose origins go back to the beginning of the human race, is transferred through coded information in genes. They are carried in a sperm and an egg, whose union begins each human life in an event called *fertilization* or *conception.* Human development continues through pregnancy, infancy, early childhood, adolescence, maturity, and old age. The remarkable cycle which begins for an individual at conception may be continued by means of reproduction when sexual maturity is attained. This chapter describes the physiology of human reproduction and the events surrounding conception and early development.

24-1 THE MALE REPRODUCTIVE SYSTEM

The Testes

The primary male reproductive organs are the paired *testes* (singular, *testis*) each of which lies in a sac called the *scrotum* located outside and beneath the pelvic cavity. Each testis is divided into over 200 lobules containing small, highly coiled *seminiferous tubules,* which produce sperm. The tubules merge to form a network that leads to a long, coiled duct, the *epididymis,* found on the posterior-superior surface of each testis (Fig. 24-1). Another duct, the *vas deferens* (seminal duct), originates from the epididymis. The vas deferens ascends from each scrotal sac and courses with the spermatic cord through the inguinal canal in the groin to the abdomen to join ejaculatory ducts, which lead into the urethra. The *urethra* is a tubular structure which extends from the bladder through the penis to the exterior. The spermatic cord contains the vas deferens, lymph vessels, blood vessels, and nerve fibers to the testes, all encased in connective tissue and surrounded by circular loops of muscle (cremaster muscle), which can pull the testes up toward the abdominal cavity. An *inguinal hernia* occurs when a loop of intestine penetrates either inguinal canal, these being weak points in the wall between each scrotal sac and the abdominal cavity.

The *penis,* the external reproductive organ in the male, has a dual function. It can transmit sperm through the urethra during sexual union or at separate times carry urine during urina-

FIG. 24-1
The male reproductive organs. (A) Sagittal view. (B) Posterior view, partly sectioned.

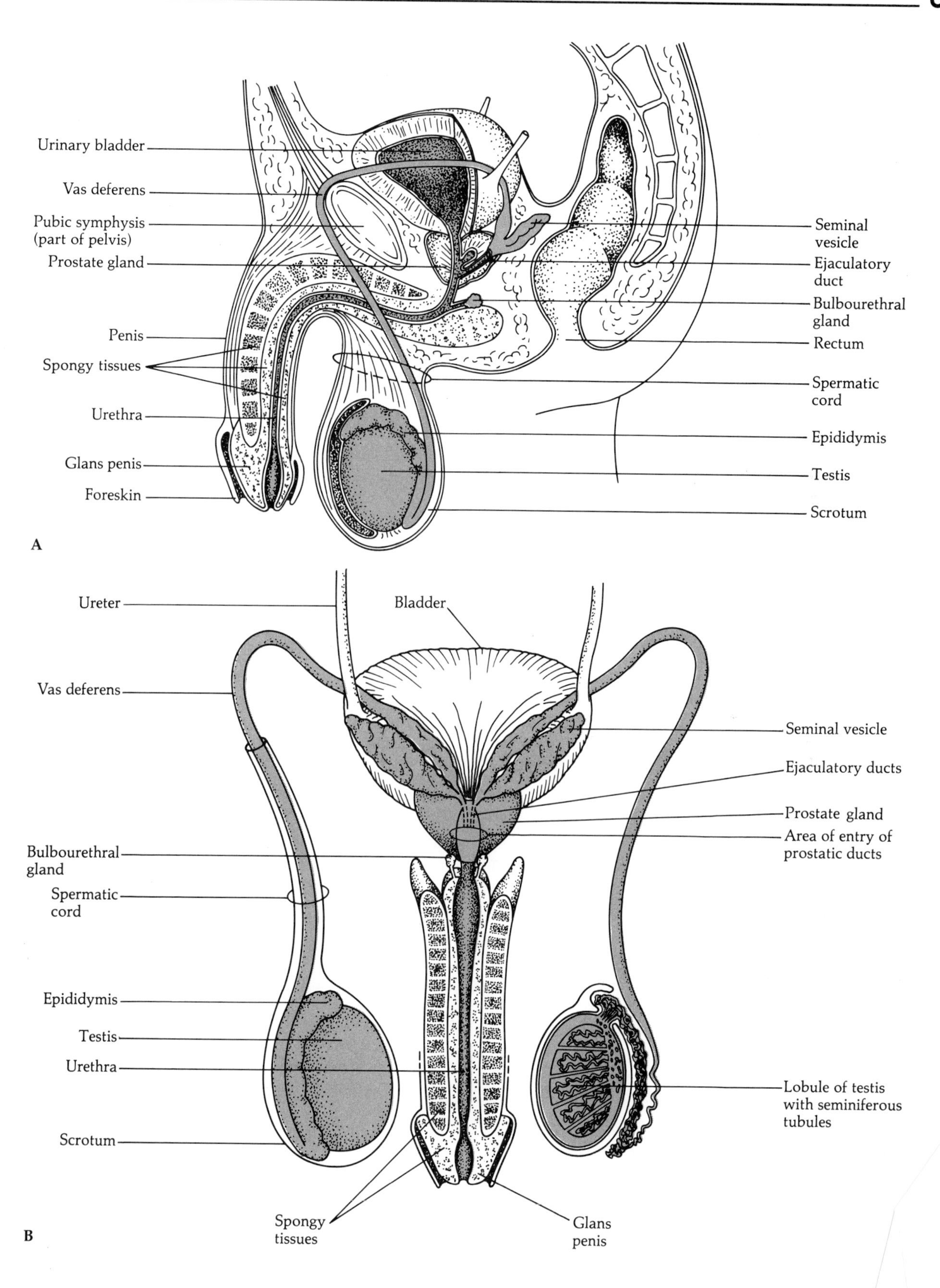
Urinary bladder
Vas deferens
Pubic symphysis (part of pelvis)
Prostate gland
Penis
Spongy tissues
Urethra
Glans penis
Foreskin
Seminal vesicle
Ejaculatory duct
Bulbourethral gland
Rectum
Spermatic cord
Epididymis
Testis
Scrotum
A
Ureter
Bladder
Vas deferens
Seminal vesicle
Ejaculatory ducts
Prostate gland
Area of entry of prostatic ducts
Bulbourethral gland
Spermatic cord
Epididymis
Testis
Urethra
Scrotum
Lobule of testis with seminiferous tubules
Spongy tissues
Glans penis
B

tion. The penis contains three parallel cylinders of spongy tissue (venous sinusoids) that can become swollen with blood. The urethra passes through the center of one of them (Fig. 24-1B). The tip of the penis has a high density of sensory nerve endings in an expanded tip, the *glans*, which is covered by a cuff of skin, the foreskin (prepuce). The rim of the foreskin is often removed in the surgical procedure of *circumcision*.

Accessory Sex Organs

Three other organs involved with male reproductive physiology are the seminal vesicles, prostate gland, and bulbourethral (Cowper's) glands. Their secretions aid the reproductive process, as described below.

Seminal Vesicles

The *seminal vesicles* are paired, compartmentalized glands, whose ducts join the vas deferens to form two ejaculatory ducts that lead to the urethra (Fig. 24-1B). The thick, yellow, alkaline secretions of the seminal vesicles contain abundant fructose and provide metabolic fuel for the transport of sperm. The fluid can neutralize acid secretions of the female, which otherwise rapidly immobilize sperm. The seminal vesicles also produce large quantities of prostaglandins, which may help stimulate sperm activity. The prostaglandins also may stimulate contractions of the oviducts after transmission in seminal fluid to the female in sexual intercourse. Both these hormonal effects aid sperm transport up the oviducts.

Prostate Gland

The *prostate*, a doughnut-shaped gland about the size of a chestnut, surrounds the neck of the bladder, the urethra, and the junction of the two ejaculatory ducts (Fig. 24-1). Up to 30 ducts from the prostate enter the urethra directly. The prostate produces proteolytic enzymes, such as fibrinolysin and acid phosphatases, which make the male's reproductive secretions less viscous and thereby increase the motility of the sperm.

Infection and/or inflammation of the prostate with accompanying edema may narrow the urethra, render urination difficult and painful, and at times totally prevent it. Under these conditions and also owing to enlargement of the prostate, which may occur with aging, surgical removal of the gland may be advisable. Sexual activity is usually normal following this type of surgery, but adverse effects such as impotency can occur.

Bulbourethral (Cowper's) Glands

The *bulbourethral glands* are paired, pea-shaped structures which lie below the prostate and secrete small quantities of fluid during sexual stimulation. Their thick, clear, mucuslike alkaline secretions are delivered through a pair of ducts to the urethra to neutralize its otherwise acidic contents.

The Production and Release of Sperm

A sperm (spermatozoon) is a flagellated cell resembling a tadpole with a head, middle-piece, and tail and measures about 60 μm in total length (Fig. 24-2). The head measures 3×5 μm and includes the nucleus, which enters the egg (at fertilization) and contributes the sperm's chromosomes to it.

The head of the sperm also contains an organelle, the *acrosome*, which develops while the sperm is in the testis and matures in the epididymis. During this period of maturation the acrosome acquires a high content of carbohydrate and lysosomal enzymes such as hyaluronidase. Many sperm are needed to release enough hyaluronidase to degrade the hyaluronic acid around the egg to allow a sperm to penetrate the egg.

The sperm's tail is composed of microfibrils and microtubules, which are similar in structure to flagella and cilia found in other cell types (Fig. 2-12). They form a flagellum, which is used like a whip to propel the sperm at speeds up to 4 mm/min.

Sperm Cell Production (Spermatogenesis)

Mitosis, equal cell division, is a process in

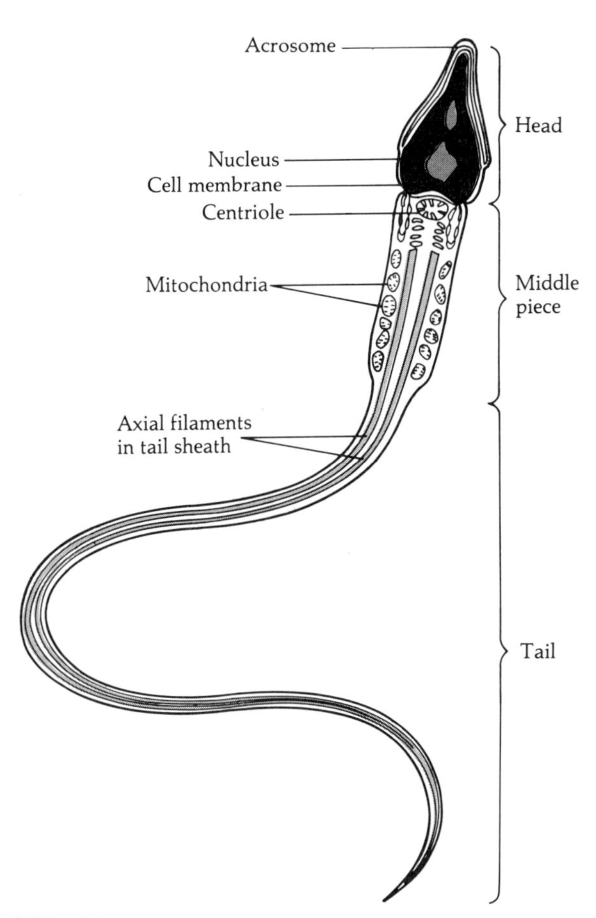

FIG. 24-2
The male reproductive cell: the sperm. Diagrammatic sketch of a sperm in longitudinal section.

which duplication of chromosomes (and DNA within them) occurs and which usually produces two chromosomally identical cells. Immature sperm cells (spermatogonia) divide repeatedly by mitosis during the reproductive life of the male. They develop into spermatocytes, which divide by *meiosis* (reductive or unequal cell division, Fig. 24-3). In this activity chromosomes are duplicated only once in two successive divisions in which one cell ultimately forms four, each of which receives half the original number of chromosomes in the parent cell ($\frac{1}{2} \times 46 = 23$, Fig. 24-3).

Body cells normally contain 46 chromosomes, half of which are derived from the mother and half from the father of an individual. Prior to meiosis the 46 *single-stranded chromosomes* duplicate. In the early phases of meiosis the duplicated strands pair up, members of the pairs may swap parts, and 23 *four-stranded* chromosomes are formed (Fig. 24-3C).

In the first meiotic cell division two of the four strands from each chromosome pass to two cells. Each of these cells receive 23 *double-stranded chromosomes.* In the second division, the double strands separate into four cells, each of which acquires 23 *single-stranded* chromosomes. The X and Y chromosomes are among the original 23 pairs in the male and they also separate, so that each of the immature sperm, the spermatids, receives an X chromosome or a Y chromosome.

Sperm mature in 70 to 74 days. For nearly 60 days they reside in the seminiferous tubules, and for an average of 12 days (1 to 21 days) they are stored in the epididymis prior to their release. Sperm may retain their viability for weeks in the epididymis.

The descent of the testes into the scrotal sacs, which usually occurs during fetal development, allows sperm to survive. This is so because the temperature of the tissues in the scrotal sacs is below that of the tissues in the abdominal cavity. Spermatogenic cells, the seminiferous tubules, and sperm are damaged by prolonged exposure to normal body temperature. So too, because low temperatures prolong their survival, sperm which are properly frozen may remain viable for years. Such procedures have been used to preserve male reproductive cells for animal breeding for some time and have more recently been used to store human sperm in sperm banks.

Sertoli cells are sandwiched between sperm-forming cells in the basement membrane of the seminiferous tubules. Sertoli cells assist movement of immature sperm by the activity of fibrous strands of the contractile protein *actin.* Sertoli cells also are phagocytic and ingest aged, damaged, or degenerate sperm. The rich glycogen reserves of Sertoli cells can be used to provide energy to maturing sperm.

Although gap junctions allow communication between Sertoli cells, their tight cell connections (Fig. 3-1B) form a barrier between them and capillaries to the testis. This is called

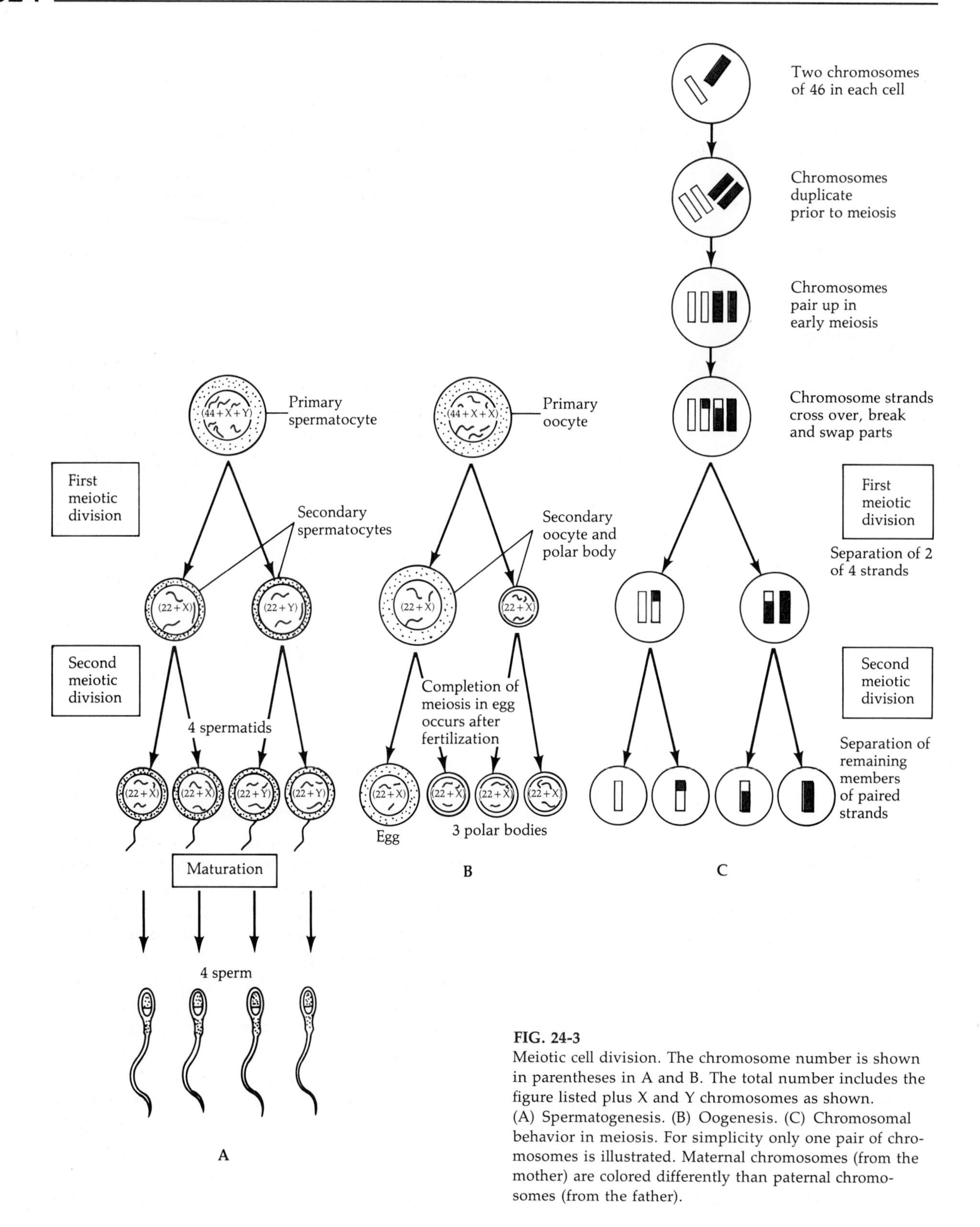

FIG. 24-3
Meiotic cell division. The chromosome number is shown in parentheses in A and B. The total number includes the figure listed plus X and Y chromosomes as shown. (A) Spermatogenesis. (B) Oogenesis. (C) Chromosomal behavior in meiosis. For simplicity only one pair of chromosomes is illustrated. Maternal chromosomes (from the mother) are colored differently than paternal chromosomes (from the father).

the *blood-testis barrier* and prevents passage of proteins between the general circulation and the testis. The tight cell junctions prevent sperm antigens from gaining access to an individual's own immune system and thereby normally prevent the male from forming antisperm antibodies. Such molecules if formed could attack the testis and cause its autoimmune destruction. The tight cell junctions also retain androgen-binding protein secreted by the Sertoli cells.

Interstitial cells (Leydig cells) are endocrine components found between the seminiferous tubules. They secrete male sex hormones called *androgens,* of which the most active form in cells is *dihydrotestosterone* (DHT). DHT is produced in some target cells from its precursor, testosterone, which acts as its prohormone. High concentrations of DHT are required for spermatogenesis, the maturation and the maintenance of the male's accessory reproductive glands, and his secondary sex characteristics.

Erection, Emission, and Ejaculation

Sensory signals from touch receptors (especially those in the glans of the penis) and those of sight, sound, and smell, as well as psychological stimuli, may generate outgoing involuntary impulses that initiate erection, emission, and ejaculation. *Erection* of the penis is caused by involuntary, autonomic, parasympathetic impulses that increase blood flow into its spongy tissue spaces (Fig. 24-1). The local rise in blood pressure enlarges the penis and stiffens it to produce an erection. Simultaneously, the veins in the penis are compressed, inhibiting blood outflow and adding to the distension of the organ.

Seminal fluid is secreted and its active propulsion occurs in two stages: *emission,* or movement to the urethra, and *ejaculation,* or transport from the urethra to the exterior. The latter usually occurs during the climax of sexual excitement, *orgasm,* mediated by involuntary, autonomic, sympathetic impulses, which cause smooth muscle contractions in the walls of the vas deferens and seminal vesicles.

Ejaculation occurs as the result of contraction of muscle in the ejaculatory ducts, urethra, and penis. Erection subsides when smooth muscle in the walls of the arteries supplying the penis contract, thereby decreasing the rate of blood flow and pressure within its spongy tissues. In adult males and with some frequency in teenagers, spontaneous emission and ejaculation of naturally stored semen may occur occasionally during sleep (nocturnal emission).

Composition of Seminal Fluid (Semen)

Seminal fluid is white and opalescent, with a slightly alkaline pH (7.35 to 7.50), and is produced by secretions of the testes and their accessory glands. The fluid volume (ejaculate) averages between 2.5 and 3.5 mL when released after several days of abstinence from sexual activity. It decreases, as does its sperm content, with repeated ejaculations.

The average sperm count is 100 million to 120 million/ml, ranging from 20 million to 120 million, per milliliter. Sperm account for about 5 percent of the volume of semen. The remainder is composed of secretions from the seminal vesicles (60 percent), prostate (20 percent), epididymis (5 percent), and other structures.

Orgasm

Orgasm is an intense climax of sexual excitement in males and females generated by sexual intercourse or manipulation of the genital organs. It occurs as a series of responses which may be divided into excitement, plateau, orgasmic, and resolution phases. It primarily is due to sensations which are pelvic in origin and arise mostly from the penis in the male and the clitoris, vagina, and uterus in the female.

In the male, during the *excitement phase* the penis is swollen with blood and enlarges in width and length in erection. Muscles in the scrotal sac elevate the testes. Reflexes begin to cause an increase in heart rate, blood pressure, respiration, blood flow to the skin, and contractions of muscles in the legs, thighs, and back. During the *plateau phase* these alterations are enhanced, and two or three drops of mucus secretion from the bulbourethral glands (and

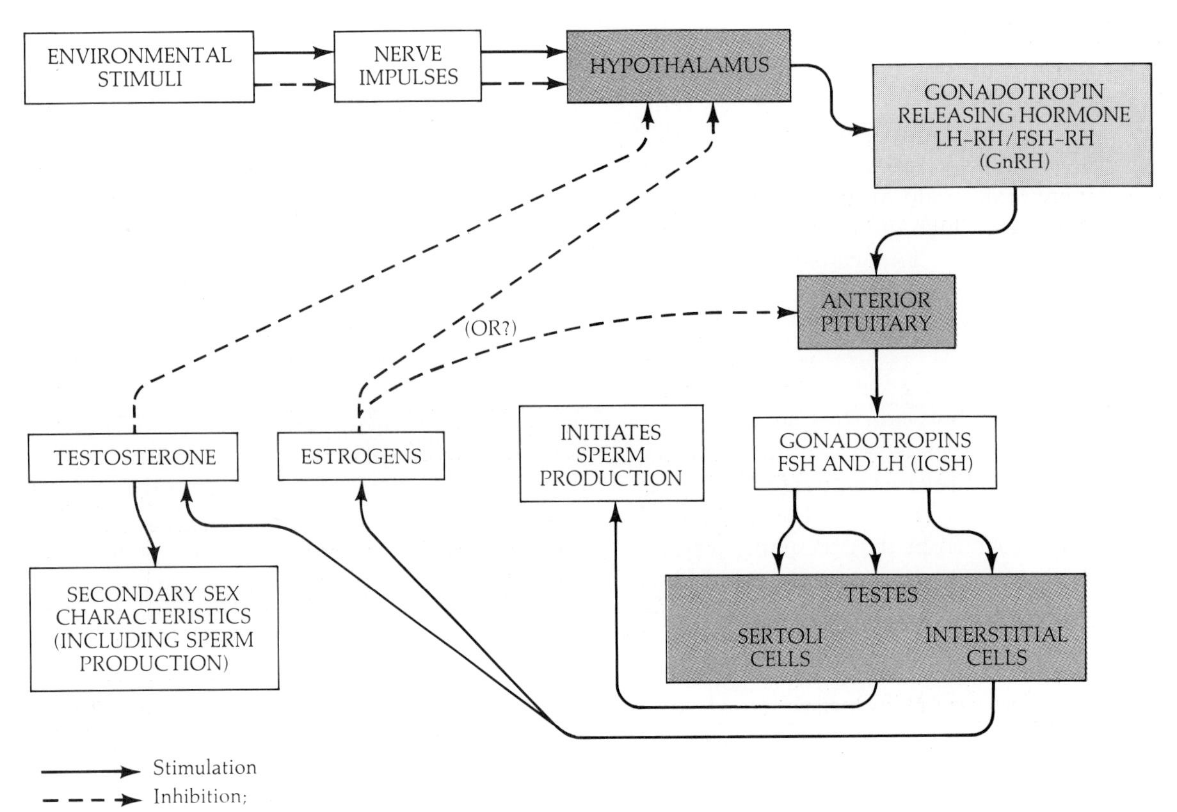

FIG. 24-4
Regulation of sex hormone secretion in the male. FSH = follicle-stimulating hormone; LH = luteinizing hormone (also called ICSH, the interstitial cell-stimulating hormone in the male) RH = releasing hormone. The pathways of regulation of gonadotropin secretion have not yet been completely clarified. Some evidence indicates that inhibin, a protein hormone from the testes and ovaries may inhibit the secretion of FSH by the anterior pituitary.

some sperm) may exit from the penis. The *orgasmic phase* occurs during ejaculation and a sensation of physical pleasure radiates from the pelvic area throughout the body. About four rhythmic contractions occur in the pelvic organs at intervals of 0.8 s. The heart and respiratory rates may double or triple. In the *resolution phase* erection subsides rather rapidly and the body begins to return toward the levels of activity prior to sexual arousal in a process that may require up to 2 h.

The Influence of Hormones

Reproductive physiology is influenced by sex hormones, whose levels are controlled by feedback mechanisms and complex hormonal interactions. Their effects on specific cells may be determined by the nature of cellular receptors (Chaps. 4, 5, and 7). Furthermore, environmental influences of learning may modify behavioral aspects of human reproduction to complicate our understanding of this basic body function. In addition to higher brain functions and conscious activity, the hypothalamus, anterior pituitary, and the testes have important influences on sexual development and behavior in the male (Fig. 24-4). The hypothalamus secretes luteinizing hormone-releasing hormone/follicle-stimulating hormone-releasing hormone (LH-RH/FSH-RH), which causes the release of LH and FSH from the pituitary (Tables 5-1 and 5-5). LH-RH/FSH-RH also may be referred to as gonadotropin releasing hormone (GnRH), since it causes secretion of follicle-stimulating hor-

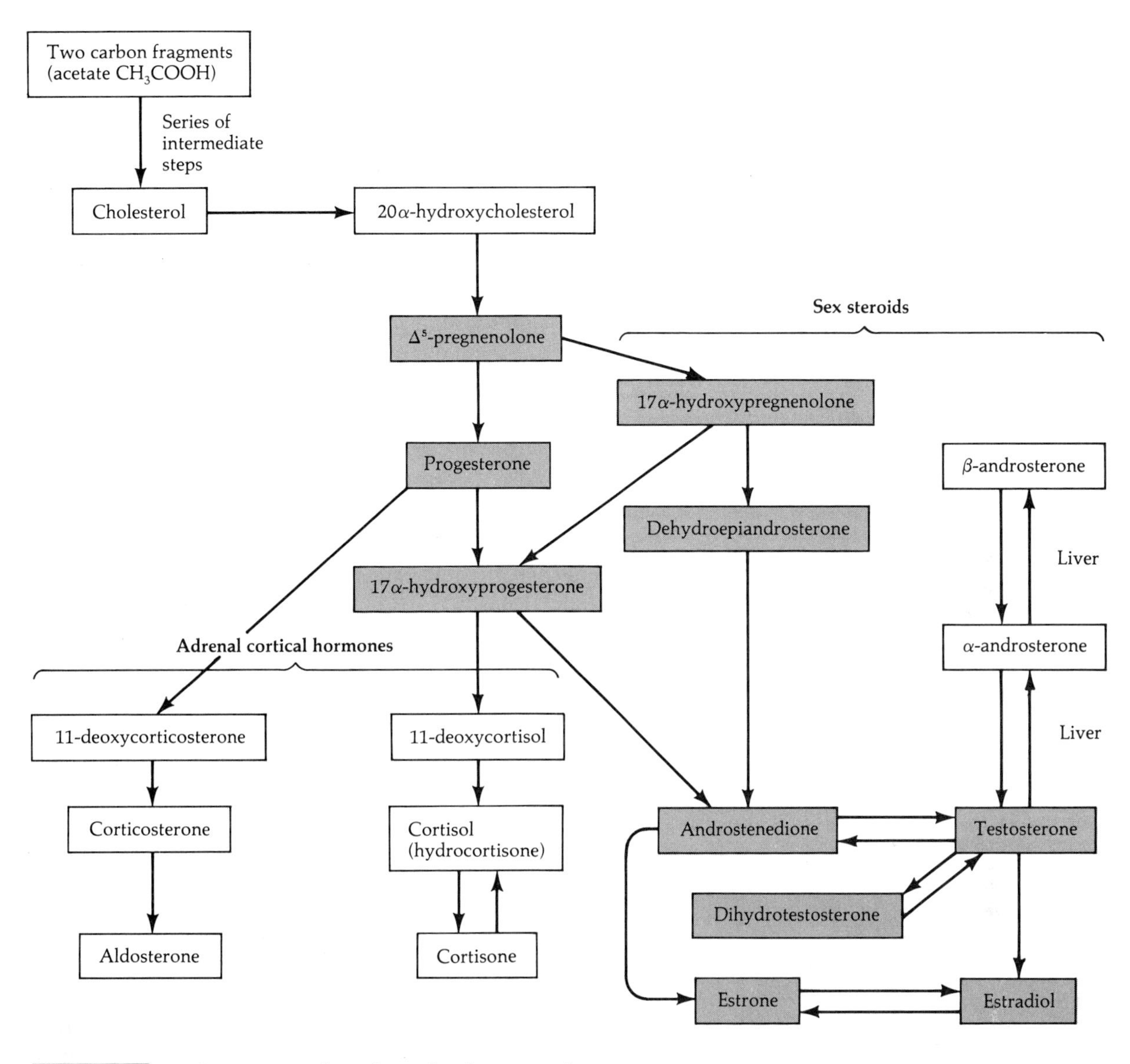

FIG. 24-5
Synthesis of sex steroids and adrenal steroid hormones.

mone (FSH) and luteinizing hormone (LH), which are gonadotropins. FSH causes the epithelial cells to multiply and form sperm (spermatogenesis). LH, which causes the interstitial cells in the testes to secrete male sex hormones, is sometimes termed ICSH, the interstitial cell-stimulating hormone.

In a series of reactions, androgens and estrogens are produced from 2-carbon fragments (acetate) and combined to form cholesterol, which in turn is converted into steroid hormones (Fig. 24-5). These reactions occur in the testes, ovaries, and adrenal cortex. Compared with other tissues the testes and the ovaries synthesize much greater quantities of sex steroids.

Like other steroids, DHT influences gene expression (Fig. 5-10) and is effective in nano-

TABLE 24-1
Major properties of the androgens

Property	*Comments*
Structure	Nineteen carbon atoms forming steroid compounds
Most active form	Dihydrotestosterone
Other forms	Testosterone, dehydroepiandrosterone, androstenedione, hydroxyprogesterone
Major source	Testes (interstitial or Leydig cells)
Other sources	Adrenal cortex (and ovaries in females)
Immediate precursors	17-OH-Pregnenolone, 17-OH-progesterone
Mode of action	Combine with cytoplasmic receptor proteins and enter nucleus of cell to influence transcription of DNA
General effects	Generally anabolic
	Promote growth, maturation, and maintenance of male reproductive tract
	Promote protein synthesis (with notable effects on muscle and bone)
	Increase secretory activity of sebaceous glands
	Influence brain development and behavior
	Promote the development of secondary sex characteristics in males

gram quantities per milliliter of serum. The concentration of testosterone averages about 20 times higher in males than in females. Most of it circulates bound to a globulin protein. The hormone is degraded in the liver and the rate of appearance of its metabolic by-products in the urine is an indirect measure of testosterone production.

Testosterone is a general anabolic hormone which in males promotes the growth and maturation of the sex organs and accessory sex glands and causes the development of secondary sex characteristics at puberty (Table 24-1). Under the influence of genes on the Y chromosomes, the testes develop and secrete testosterone, which regulates the development of male external genital organs, which are evident by 4 to 6 weeks of intrauterine life.

Recent research with laboratory animals indicates the androgens may bind to receptors in brain cells within the first few weeks after conception to influence the development of behavior. Further studies indicate that testosterone may activate human aggressive behavior and, by its action on the brain, contribute to heightened self-assertiveness in boys and men as compared with girls and women. However, its roles in these regards are not clear. In some areas of the brain testosterone may be converted into estrogens such as estradiol (Fig. 24-5) prior to interacting with receptors in brain neurons to trigger behavioral changes. This discovery highlights the fact that estrogens, although described as female sex hormones, have important functions in normal male physiology.

Testosterone synthesis by the testes of the unborn male is stimulated during embryonic development by human chorionic gonadotropin (HCG). HCG is secreted by cells of the chorion, an embryonic membrane that penetrates the walls of the mother's uterus and forms the fetal part of the placenta. Following birth, when the placenta detaches and is shed, this source of gonadotropin and its stimulatory effect on testosterone secretion and sexual development is lost.

Consequently, sexual development remains dormant during infancy and early childhood owing to steroids from the adrenal cortex that inhibit the secretion of LH-RH/FSH-RH by negative feedback pathways. With the onset of adolescence (puberty) at about 10 to 14 years of age, the hypothalamus becomes less sensitive to inhibition by the low levels of steroids. LH-RH/FSH-RH is secreted, testosterone secretion increases, and secondary sex characteristics begin to appear. The individual gradually acquires the capability of sexual reproduction. Increasing amounts of testosterone promote protein synthesis, including the growth of muscular and skeletal tissues, as well as the appearance of body hair on the face, chest, axilla (armpits), abdomen, and pubic region. Growth of the thyroid cartilage produces a significant enlargement in the throat (the "Adam's apple") and changes in the larynx that deepen the pitch of the voice. The activity of the sex and adrenal steroids promote enlargement of the sinuses in the skull, which increases the resonance of the voice. The penis and accessory sex organs also increase in size and become sexually functional. The sebaceous glands become more active and produce thicker, oily secretions. As a result, the skin acquires greater susceptibility to inflammatory reactions in the oil glands, which may result in *acne.*

A gradual decrease in male sex hormones occurs in a phase of reproductive life that begins about age 40, termed the *male climacteric.* Less testosterone is formed, and, over a period of years, its levels decline. This occurs in spite of the lesser feedback inhibition of gonadotropins by testosterone (Fig. 24-4), the resulting higher levels of the gonadotropins, and their stimulatory effects on the testes. With aging, the decrease in testosterone may be the basis for a decline in sexual activity, which peaks at about 18 to 21 years of age. However, social conditioning of behavior may lead to more rather than less sexual activity by males later in life. Furthermore, about 20 percent of the peak value of testosterone may still remain at 80 years of age, an amount sufficient for sexual reproductive function.

24-2 THE FEMALE REPRODUCTIVE SYSTEM

Although the origins of male and female reproductive organs are from similar embryonic tissues in the abdominal cavity, female body cells possess two X chromosomes, while male body cells possess one X and one Y chromosome as the paired sex chromosomes (Fig. 24-3). Specific and separate genes on the Y chromosome regulate the development of the testis and the sperm, respectively. Genes on the paired X chromosomes, in the *absence* of a Y chromosome, normally lead to development of ovaries, eggs, and female characteristics. In abnormal situations, when two X chromosomes and a Y chromosome appear in a fertilized egg, a congenital abnormality, *Klinefelter's syndrome,* produces male characteristics but undeveloped sexual organs and sterility in the individual. Conversely, absence of a second sex chromosome causes an XO chromosomal constitution and *Turner's syndrome,* in which the ovaries are missing, the breasts are underdeveloped, and the skeleton is deformed. Further, an extra Y chromosome in males yields an XYY chromosomal pattern, whose effects are unknown (although it was initially proposed that it produced aggressive, violent, "supermales").

In both sexes, genes on the autosomes (chromosomes other than the X or Y) also influence the development of sexual traits. Consequently, the sex organs, which are very similar early in development, mature in different ways. For example, the ovaries do not migrate from the abdominal cavity as do the testes, nor does the clitoris attain the size of the penis. The pathways for sex steroid synthesis are similar, but females normally produce more estrogens and less androgens than do males. Furthermore, once the ovaries or testes are established, the net balance of sex steroids produced by these organs seems to have profound effects on further sexual development.

The Ovaries

The paired ovaries are the primary female reproductive organs and are similar in origin (homologous) to the male's testes. The ovaries are almond-shaped and situated on either side of the uterus (womb) in the pelvic cavity (Fig. 24-6). Each ovary is divided into an inner medulla and an outer cortex, in which clusters of cells, the *follicles,* are located (Fig. 24-9). One cell in each follicle is the female reproductive cell, or *egg* (ovum).

Accessory Sex Organs

The accessory sex organs in the female are the uterus and its tubes (the fallopian tubes or oviducts), the vagina, the external genital organs (Fig. 24-6A and 24-7), and the mammary glands or breasts. The *uterus,* or womb, is a pear-

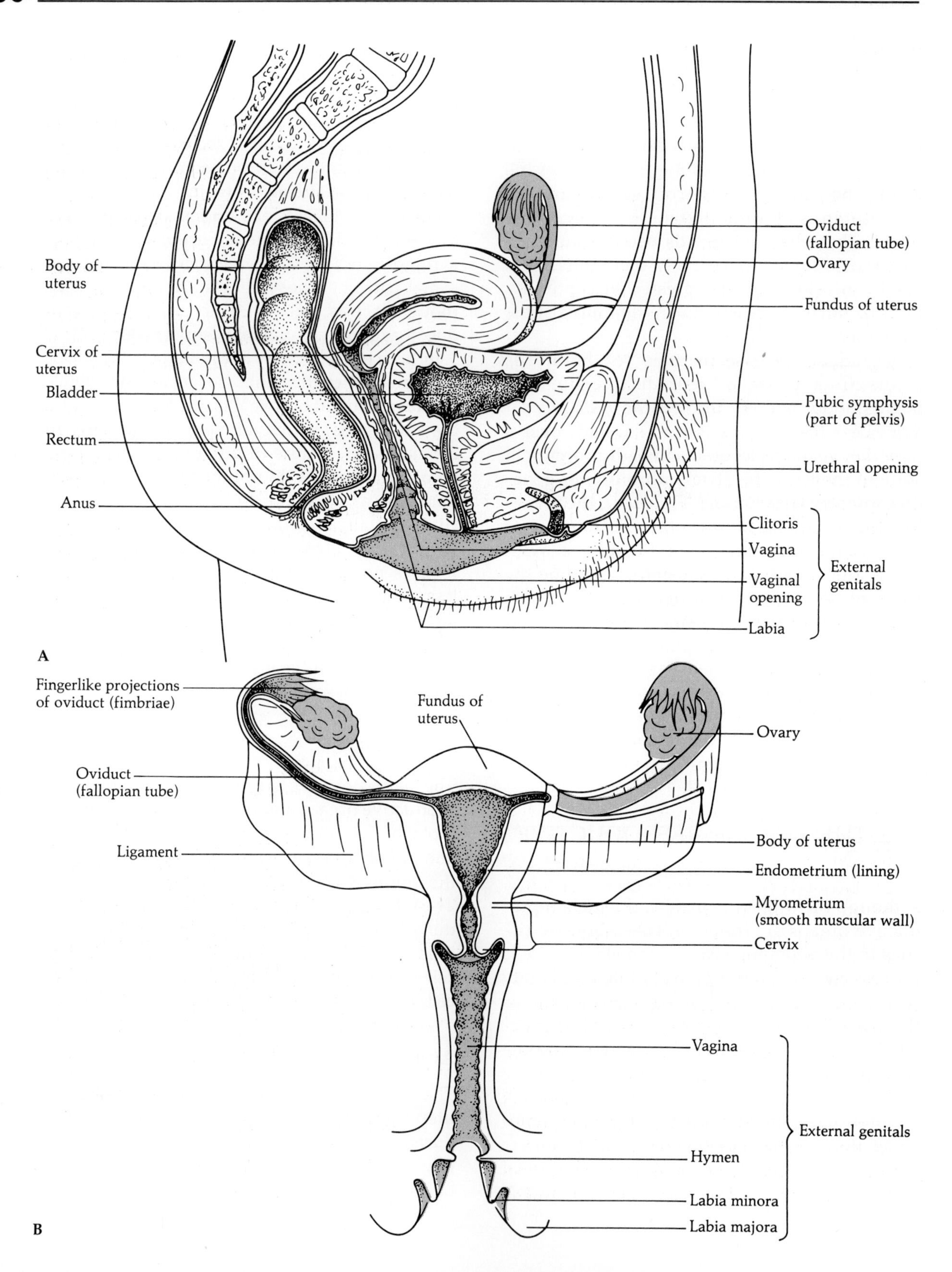

Oviduct (fallopian tube)
Body of uterus
Ovary
Fundus of uterus
Cervix of uterus
Bladder
Pubic symphysis (part of pelvis)
Rectum
Urethral opening
Anus
Clitoris
Vagina
Vaginal opening
Labia
External genitals
A
Fingerlike projections of oviduct (fimbriae)
Fundus of uterus
Ovary
Oviduct (fallopian tube)
Ligament
Body of uterus
Endometrium (lining)
Myometrium (smooth muscular wall)
Cervix
Vagina
External genitals
Hymen
Labia minora
Labia majora
B

FIG. 24-6
The female reproductive organs. (A) Sagittal view. (B) Frontal view, partly sectioned.

shaped, hollow, muscular sac located between the urinary bladder and rectum in the pelvic cavity (Fig. 24-6A). It is a remarkable organ, which undergoes monthly cyclic changes during the reproductive life of a woman. Its lining is prepared to support the development of a fetus, but is shed each month in a menstrual cycle if pregnancy does not occur. Nine months after the start of pregnancy, the muscular walls of this organ propel the newborn child from this warm, moist supportive chamber to the outside world. The inner mucosal uterine lining, the *endometrium,* is a target tissue whose cyclic proliferation and glandular activity are regulated by the female sex hormones estradiol and progesterone. Abnormal growth of the endometrium is called *endometriosis* and may interfere with reproductive capacity. In fact, the tissue may invade the ovaries and other sites outside of the uterus and interfere with reproductive function. The endometrium is normally joined by connective tissue to the *myometrium,* a thick wall of smooth muscle. The myometrium is composed of three layers of smooth muscle interwoven with connective tissues and an abundance of elastic fibers.

The upper portion of the uterus, the *fundus,* is covered by a serous membrane, the *parietal peritoneum,* which suspends the ovaries and oviducts loosely in place. The *body* or main part of the uterus leads to a narrow, constricted segment, the *cervix* (neck), which connects the uterus to the vagina.

The oviducts project laterally from each side of the fundus of the uterus and terminate in fingerlike projections called *fimbriae* (Fig. 24-6B). The oviducts, like the uterus, have mucosal linings and walls composed of smooth muscle.

The *vagina* is a 6- to 9-cm-long channel that connects the cervix to the exterior of the body in an area called the vestibule (Fig. 24-7). The vagina is lined with mucosal epithelium that produces secretions, which are released mostly during sexual excitement. Its muscular walls widen during sexual stimulation. To facilitate intercourse, glandular ducts which are lateral to the urethra and the posterior portion of the vagina also release secretions, which lubricate the vagina to a limited extent. The secretions are produced by the *lesser vestibular glands* (*Skene's glands*) and the *greater vestibular glands* (*Bartholin's glands*), respectively. The *hymen,* a mucous membrane of varying thickness and prominence, may narrow or block the vaginal opening. The hymen may not be apparent and may be broken during numerous activities, including horseback riding, bicycling, a fall, and sexual intercourse.

The opening of the vagina is within a narrow chamber, the *vestibule,* which is immediately surrounded by two small folds of skin and fatty tissue, the *labia minora* (Fig. 24-7). The labia minora are folds of mucous membrane homologous with the tissues in the male which fuse within the penis. Within the vestibule and from its anterior to its posterior borders are the openings of the urethra and vagina, and the hymen. The openings of the lesser and greater vestibular glands are also within this area.

FIG. 24-7
The female external genital organs.

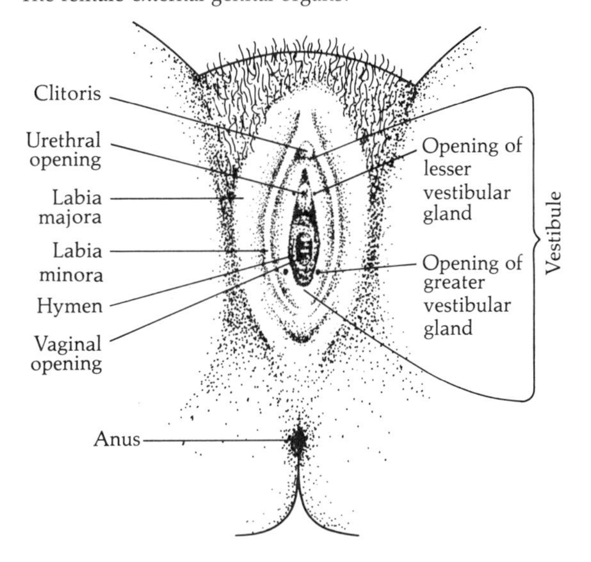

The *labia majora* are two prominent folds of skin and fatty tissue lateral to the labia minora and vestibule. They are homologous with the scrotal sacs in the male. The *clitoris,* an organ homologous with the penis but having more limited growth, is at the anterior border of the vestibule. The clitoris may be unique in that it seems to be the only organ in either sex with a single function, namely pleasure.

The tissues around and within the vestibule are susceptible to ragged tears and injury during childbirth. Therefore, in the United States but rarely elsewhere, a surgical incision, an *episiotomy,* is often made in the area between the vagina and anus just prior to childbirth. The incision enlarges the vestibule and thereby minimizes stretching and damage, substituting a clean surgical incision for a ragged tear.

The *mammary glands* (breasts) are exocrines which are considered accessory sex organs because they provide milk for the newborn infant. However, their embryological origin and structure is more closely related to that of complex sweat glands than to that of other female sex organs.

The breasts are highly susceptible to the development of cysts and tumors, which often can be detected early by periodic self-examination. Successful therapy and prevention of cancer of the breast, as with other cancers, are dependent on early detection. X-ray examination may reveal whether cysts or tumors in the breast are cancerous. One x-ray technique is *mammography,* in which two right-angle x-ray images of each breast are made on film. Another one is *xeroradiography,* which requires less radiation and produces an x-ray image on paper. A heat-detecting technique, *thermography,* measures heat produced by tumor cells and records it graphically. Because tumor cells produce more heat, they may be distinguished from normal ones. A highly successful test for the early detection of cancer of the uterus (cervical cancer) is the *Pap test* or *Papanicolaou smear.* Cells are removed from the upper part of the vagina and cervix with a swab, stained, and observed on a glass slide under the microscope. Changes in the appearance of the stained cells provide opportunity for early diagnosis and treatment, which improve the chances of survival significantly.

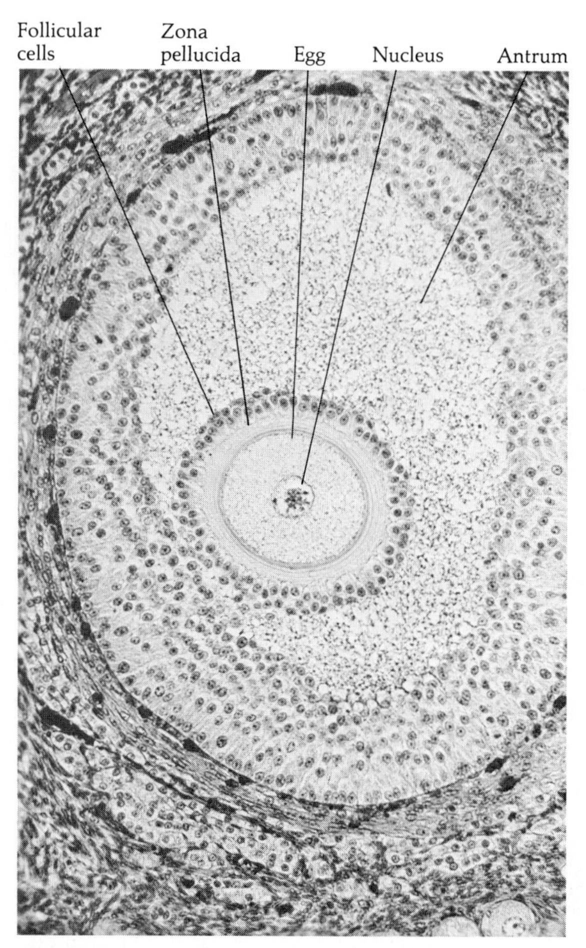

FIG. 24-8
The female reproductive cell; the egg within a mature (secondary) follicle (×250).

The Production of Eggs (Oogenesis)

The mature egg (ovum) is about 140 to 150 μm (0.006 in) in diameter and is enclosed in a follicle (Graafian follicle), which measures about 10 mm in diameter (Fig. 24-8). In its early development, the ovum is surrounded by layers of cells in a structure called a *primary follicle* (Fig. 24-9). Generally within 10 to 14 days of the beginning of its growth under the influence of FSH, the cells in the follicle multiply, the whole mass becomes oval, and the egg moves toward the surface of the ovary. The follicular cells se-

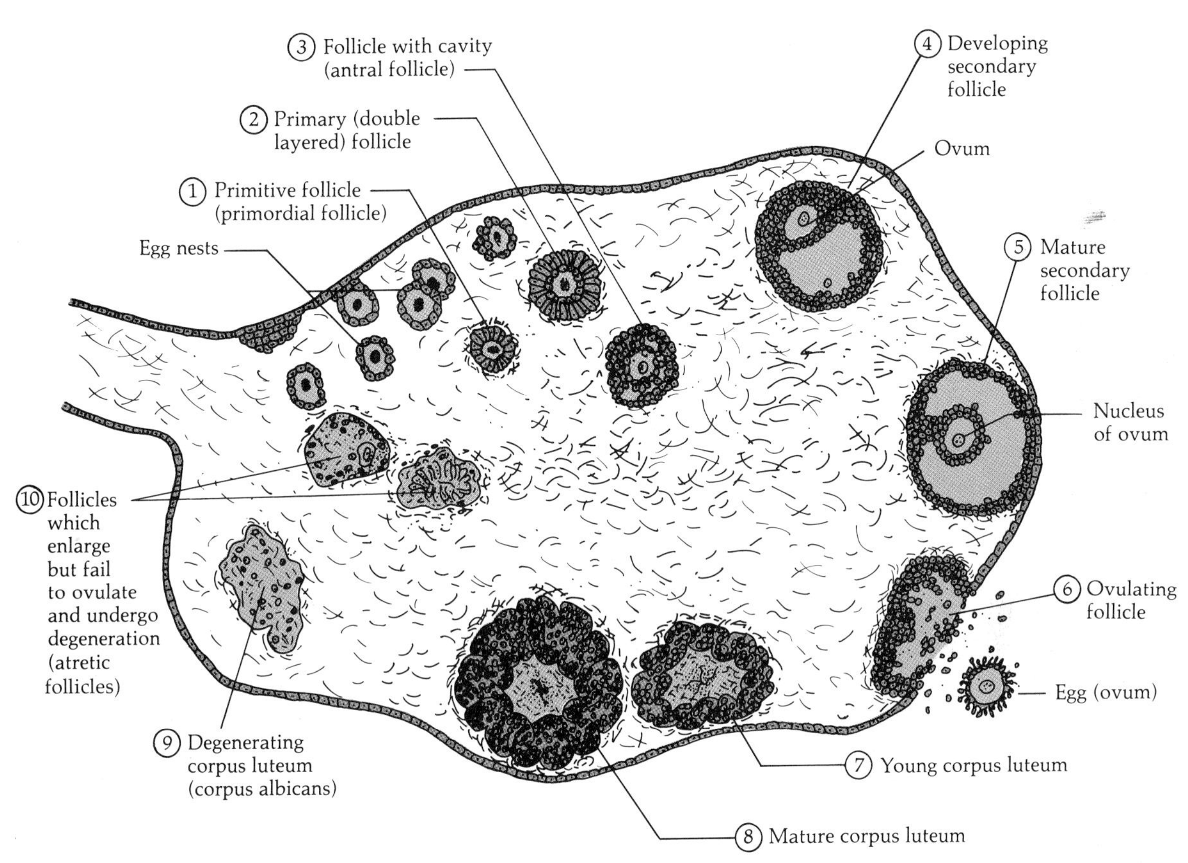

FIG. 24-9
Diagram of development of a follicle and formation of an ovum in the ovary. Note the various structures schematically illustrated in the ovary. The maturation of a follicle may be followed in sequence by the numbers on the diagram. However, in actuality, development occurs at a single site in the ovary and in various locations during different cycles. The clockwise scheme is for illustrative purposes only. The mature secondary follicle (5) is shown in Fig. 24-8. The atretic follicle (10) is one which does not fully mature but eventually degenerates and then shrivels up. Modified from B. M. Patten, *Human Embryology*. Copyright © 1953. McGraw-Hill Book Company, used with permission of the McGraw-Hill Book Company.

crete estrogens. At the completion of cell proliferation, the egg and its surrounding cells constitute a mature *secondary follicle*. The fully mature follicle ruptures at *ovulation*, and the egg and its surrounding cells are released from the ovary and normally move toward the opening of the oviduct. Follicles which start but do not complete maturation at this time shrivel; these are called *atretic follicles*.

The number of eggs that reach maturity is extremely small. The 7 million immature oocytes of a 20-week female fetus decline to 300,000 to 400,000 by puberty. Only 400 to 500 mature eggs are released in ovulations that occur during a woman's reproductive life. In each cycle of ovulation, one ovary normally releases one egg. Meiotic cell division, which forms an ovum in a follicle, begins before birth but is interrupted until after fertilization. Ultimately the chromosome number is reduced from 46 to 23 in each of the four cells that are formed by two successive stages of meiotic division (Fig. 24-3B). The two X chromosomes, like all other chromosomes, are duplicated only once prior to the two successive cell divisions of meiosis, so that normally each daughter cell acquires one X chromosome. During meiosis most of the cytoplasm is enclosed in the cell which ultimately matures to form the egg. The other three smaller cells, *polar bodies*, formed in each division die prematurely. It is interesting to note that completion of the second stage of meiotic cell division occurs only after a sperm penetrates the egg.

Orgasm

Orgasm in the female, like that in the male, occurs in identifiable phases. Sights, sounds, smells, and psychological and touch stimuli generate an *excitement phase* in which the vagina and vestibular glands produce secretions and the clitoris undergoes erection. This leads to a *plateau phase,* in which the walls of the vagina, the labia, and the nipples of the breasts become engorged with blood and the clitoris retracts. An increase also may occur in blood flow to the thighs and back, and heart and respiratory rates increase. A short but intense *orgasmic phase* may then occur, which involves a sensual awareness of the pelvic region and a "warmth" which spreads from that area to the rest of the body followed by "pelvic throbbing." During orgasm the vagina's walls dilate and rhythmic waves of smooth muscle contraction occur in the uterus, oviducts, and breasts. The first contraction may last 2 to 4 s, but successive ones occur at intervals of 0.8 s for as few as 3 times in a mild orgasm to as many as 15 times in intense ones. A woman, unlike a man, can sustain a plateau and have two or more orgasms in a brief period prior to entering the *resolution* phase, in which she gradually returns to an unexcited state.

The Influence of Hormones

The hypothalamus, anterior pituitary, and ovaries regulate female reproductive physiology by mechanisms which are generally similar to those observed in males but which seem to involve a much more complex rhythm of events during the years of reproductive capability. At puberty the hypothalamus overcomes its extreme sensitivity to inhibition by low levels of sex steroids, and LH-RH/FSH-RH is released, stimulating the secretion of FSH by the pituitary. Some follicles develop which secrete estrogens in increasing amounts, which reach high levels and a peak at ovulation and by positive feedback cause secretion of LH (Fig. 24-10).

After a follicle matures, it releases an egg and the cells left behind in the follicle become fat-laden and form the *corpus luteum.* This is the start of the *luteal* or *progestational phase* of the female's reproductive cycle. Under the influence of LH the corpus luteum secretes increasing amounts of progesterone, which stimulates maturation of the uterine lining (Fig. 24-12). Cells in the corpus luteum also secrete moderate amounts of estrogens, which inhibit FSH and LH secretion by negative feedback mechanisms. The estrogens also inhibit LH-RH/FSH-RH secretion by negative feedback mechanisms. The decrease in LH-RH/FSH-RH is followed by a corresponding decrease in estrogen and progesterone and atrophy of the corpus luteum, the uterine lining, and other follicles which had begun to mature. Should the egg be fertilized, these responses are altered by the release of human chorionic gonadotropic hormone (as discussed below).

Of the three major forms of estrogens produced in the body, β-estradiol is the most active female sex hormone (Table 24-2). The estrogens are generally anabolic and, along with adrenal steroids, cause development of female secondary sex characteristics during puberty. The latter include growth of fat and glandular tissue in the breasts; growth of hair in the axillae (armpits) and pubic region; rapid bone development, causing widening of the pelvis and leading to early cessation of the growth of long bones (Chap. 14); and growth of smooth muscle in the uterus, which increases the size of this organ significantly.

Cycles of alternating peak secretions of FSH and LH characterize female reproductive life. The first sign of these changes in girls usually is the appearance of "breast buds" or pubic hair. These and other female secondary sex characteristics appear with the advent of sexual maturity during *puberty,* which usually begins about age 13 but may start between 9 and 14. During puberty cyclic changes called menstrual cycles begin to occur in the ovaries and uterus. After a spurt in total body growth the first menstrual cycle or *menarche* occurs. The initial cycles are irregular and nonovulatory for about 1 to 1.5 years.

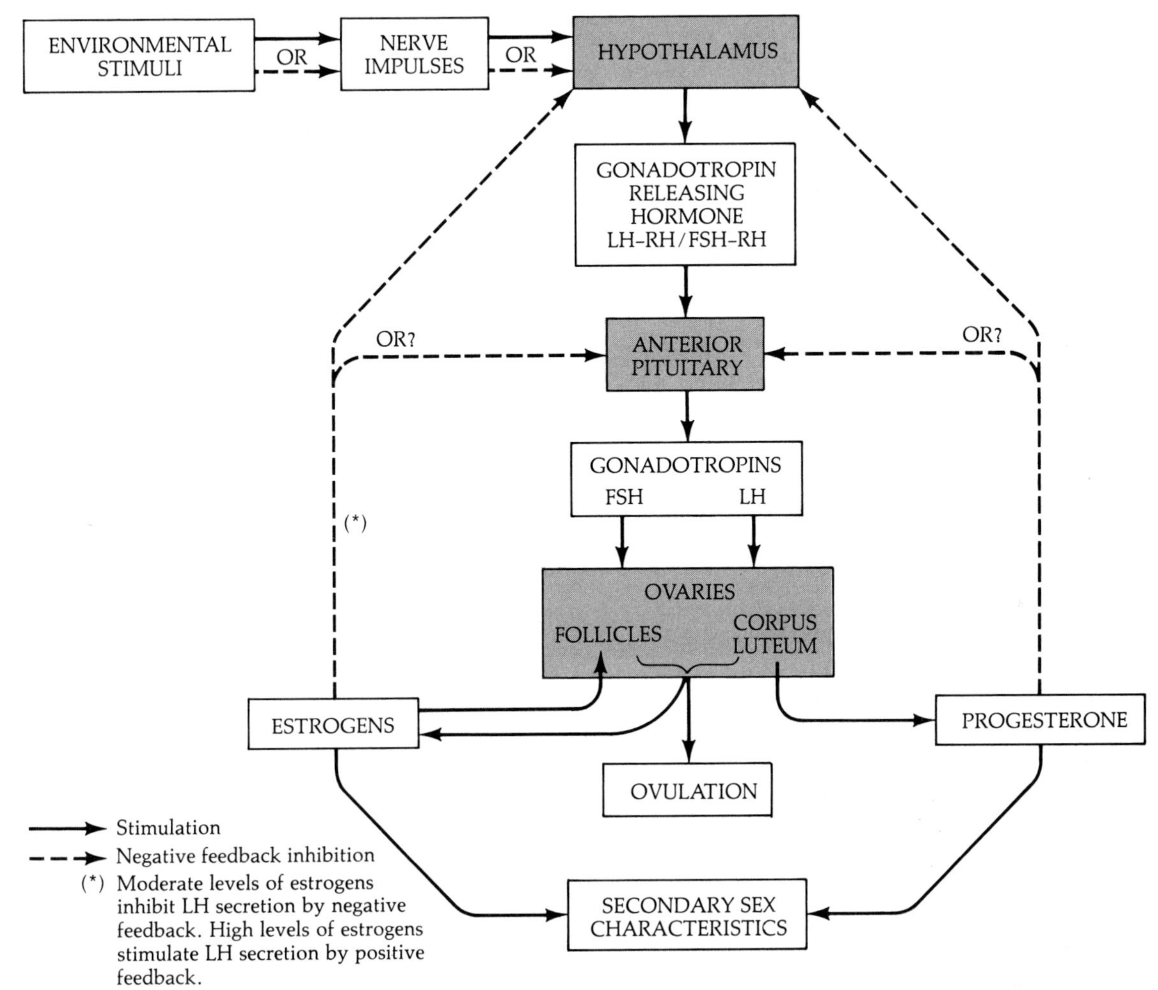

FIG. 24-10
Regulation of sex hormone secretion in the female. As in the male, some regulatory mechanisms have not been documented completely, hence the question marks.

Regular cycles last about 28 days (but may range from 21 to 45 days) and consist of three phases, *menstrual, follicular,* and *luteal* (Fig. 24-11). During the *menstrual phase,* which lasts about 3 to 5 days (but may last as long as 7) the uterine lining is shed and new follicles start to mature in the ovary. As a follicle grows, it produces more estrogens during the *follicular* or *estrogenic phase,* which lasts approximately 9 days. Variations in the length of the menstrual and follicular phase account for differences in the duration of individual cycles. The term *follicular phase* is sometimes used to describe the first two parts of the cycle because the follicle develops then. The *luteal* or *progestational phase* begins at ovulation and is rather constant. It usually lasts and precedes the start of the next menstrual phase by 14 days. The luteal phase is promoted by the secretion of increasing amounts of progesterone from the corpus luteum, which further prepares the uterine lining for reception of a fertilized egg.

Effects on the Ovaries

During the follicular phase FSH causes proliferation of the ovarian epithelium and maturation of the follicle. Under the influence of FSH (and LH) the follicle increases in size, glycogen and fat are stored in its cells, and the egg matures. Fluid is secreted, causing some cells and the

TABLE 24-2
Major properties of the estrogens

Property	*Comments*
Structure	Eighteen carbon atoms forming steroid compounds
Most active form	β-Estradiol
Other forms	Estrone, estriol
Major source	Ovaries (cells of follicles and corpus luteum)
Other sources	Adrenal cortex (and testes in males)
Immediate precursor	Testosterone
Mode of action	Combine with cytoplasmic receptor proteins and enter nucleus of cell to influence transcription of DNA
General effects	Generally anabolic
	Promote growth maturation and maintenance of the female reproductive tract
	Promote protein synthesis; decrease circulating blood lipids; elevate serum calcium and inorganic phosphate
	Decrease secretory activity of sebaceous glands
	Influence brain development and behavior
	Promote the development and maintenance of secondary sex characteristics

egg to be segregated in a stalk in the follicle as it grows and migrates toward the surface of the ovary (Fig. 24-9).

By positive feedback, high levels of estrogens stimulate peak secretion of LH near the fourteenth day of a 28-day cycle. LH, with the help of FSH, causes *ovulation,* the escape of the egg and its surrounding cells from the follicle (and the ovarian surface, Fig. 24-12). As mentioned earlier, this event signals the start of the luteal phase of the cycle and usually precedes the start of the next menstrual phase by 14 days. Generally, within 24 h of ovulation normal early morning body temperature rises 0.56°C (1°F, Fig. 24-12) under the influence of progesterone.

Ovulation is made possible by the increased activity of cellular enzymes, which hydrolyze the connective tissue capsule of the ovary near the mature follicle. The egg and the cells surrounding it are released into the fimbriae of the oviducts, which have positioned themselves to "catch" the cell mass. Motions of the cilia lining the oviducts create a current that draws the ovum into the tube and propel it toward the uterus. Other follicles which had started to mature shrivel and become atretic (Fig. 24-12).

Effects on the Uterus and Vagina

The cyclic secretions of the ovaries cause corresponding changes in the uterus, which undergoes a *proliferative* phase corresponding in time with the follicular phase of the ovarian cycle, a *secretory* phase which parallels the luteal part of the ovarian cycle, and a *menstrual phase* in which the uterine lining is shed (Fig. 24-11).

In the *proliferative phase,* under the influence of increasing amounts of estrogens the endometrial cells multiply and form tubular glands. Branches of the uterine artery proliferate, forming coiled spiral arteries whose arterioles and capillary beds produce vessels just beneath the surface of the endometrium (Fig. 24-12). The estrogens also cause the cervical epithelium to proliferate beyond the capacity of the blood vessels to supply nutrients. The surface cells die and fall off as the remaining cells secrete a slight amount of acidic mucus (pH about 4 to 5). At ovulation, mucus is thin, watery, and acidic, under the influence of high levels of estrogen.

As mentioned above, the high estrogen levels stimulate LH secretion by positive feedback (Fig. 24-10). LH causes the corpus luteum to produce increasing quantities of progesterone (Fig. 24-12), which stimulate the uterus to enter its *secretory phase.* The nutrient supply of the cells in the lining of the uterus improves. The stores of glycogen, fat, salt, and water increase. The lining attains a 5-mm average thickness 24 days into the cycle. After ovulation, under the influence of progesterone the epithelial cells of the cervix secrete less mucus which is more viscous and alkaline. The more acidic, watery

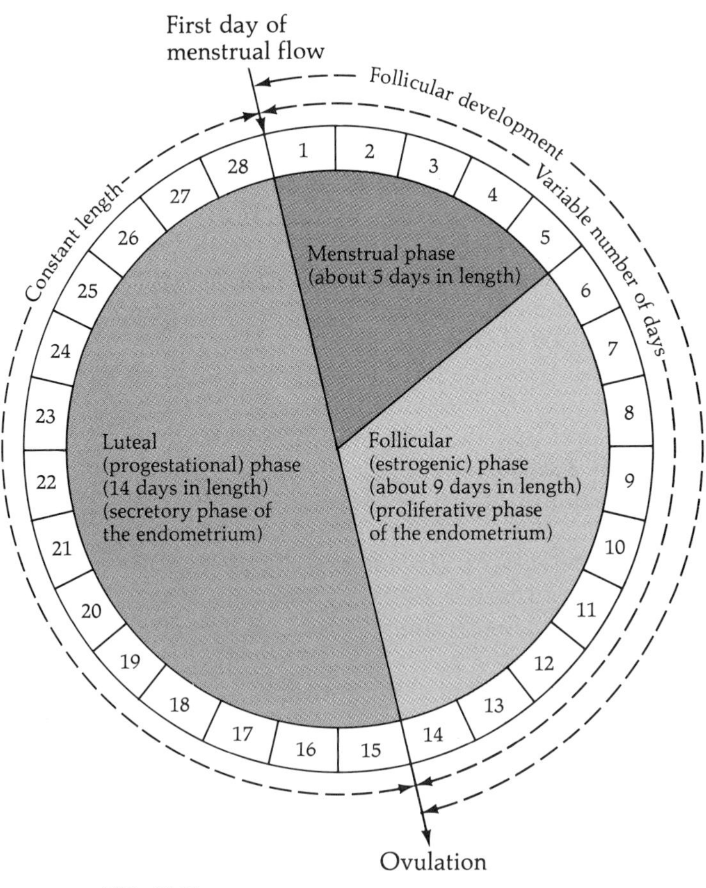

FIG. 24-11
Phases of the menstrual cycle. Adapted from Searle Pharmaceutical Company. *Research in the Service to Medicine,* Vol. 55. Copyright © 1962 and Allen Lein, *The Cycling Female: Her Menstrual Rhythm.* Copyright © 1979. W. H. Freeman and Company.

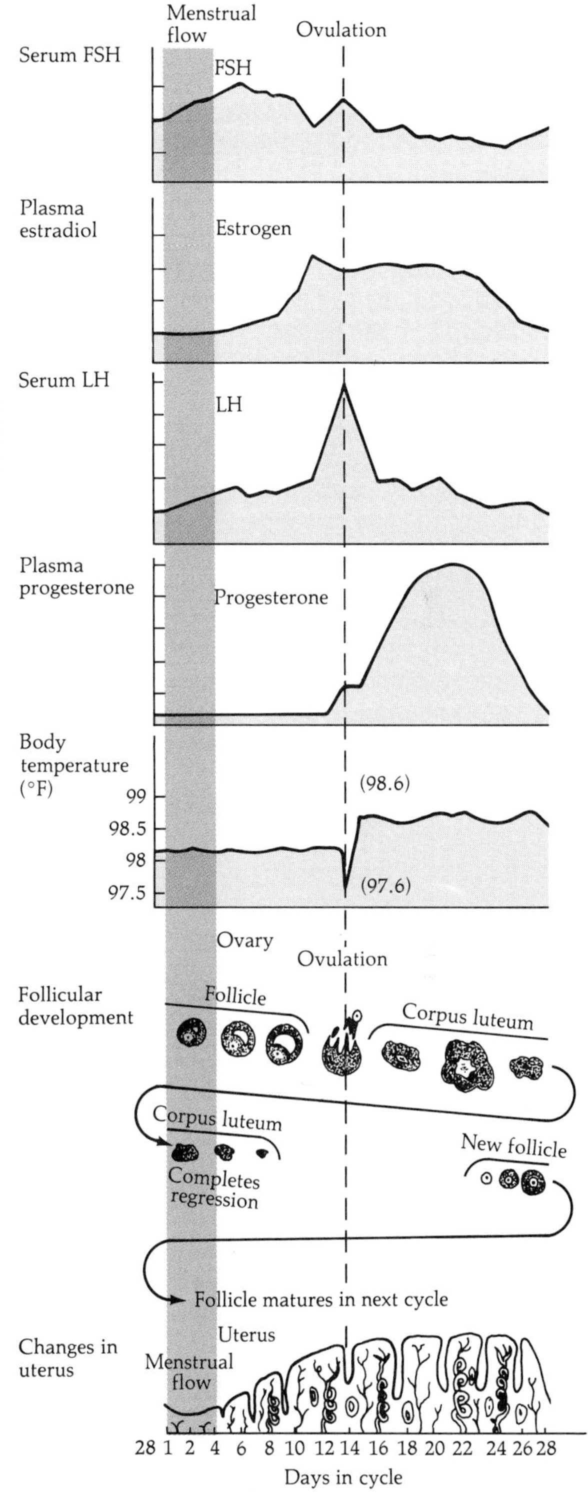

FIG. 24-12
Hormonal, body temperature, and tissue changes that occur in the menstrual cycle.

mucus secreted earlier aids the transport and survival of the sperm.

If the egg is fertilized, gonadotropins produced by the developing embryo cause the corpus luteum to continue to secrete estrogens and progesterone. The progesterone helps to maintain the thick, vascularized lining of the uterus and sustain pregnancy. This hormone also inhibits contractions of the uterus and stimulates the mammary glands, preparing them for milk production. Because of these effects, progesterone is named the *progestational, procarrying,* or *propregnancy* hormone.

If fertilization does not take place, gonadotropin levels decline, the corpus luteum begins to shrivel, and the levels of estrogens and progesterone decrease, beginning about day 24. The corpus luteum is gradually infiltrated by collagenous connective scar tissue at the surface of the ovary, producing a white mass called the *corpus albicans* (Fig. 24-9). The uterine lining loses its accumulated fluid and atrophies, and its blood supply diminishes. Its cells shed lysosomal enzymes, which cause spasm of the spiral arteries in the walls to further reduce the blood supply to the lining. The loss of localized areas of tissue, which die and fall into the uterine cavity, is accompanied by bleeding called *menstruation*. It signals the start of a new cycle.

About 30 to 100 mL of blood, but as much as 300 mL in severe cases, is lost during the first 2 to 7 days of the menstrual period. However, the average volume of menstrual flow is less than 85 mL, of which 35 to 50 mL is blood. The remainder consists of dead cells and serous fluid that flows from the denuded uterine tissue. During menstruation the low levels of estrogen and progesterone cause negative feedback stimulation of LH and FSH secretion. These gonadotropins cause more follicles to mature and begin a new cycle. The replacement of the endometrial lining is initiated from its remaining viable cells (Fig. 24-12).

Various factors may cause the pain and cramps that sometimes occur prior to or during menstruation. Overproduction of prostaglandins may trigger strong uterine contractions when progesterone levels are low. The fact that inhibitors of prostaglandin synthesis relieve menstrual cramps supports this possibility. Other factors that may explain menstrual pain include: *ischemia*, a deficiency in blood and oxygen supply to the uterus and reproductive organs; an imbalance in the ratio of estrogens to progesterone; failure of the uterine lining to break down; and mucus blockage of the canal leading from the cervix of the uterus to the vagina.

As stated previously, the first menstrual period or *menarche* generally occurs as early as 9 years of age or as late as 14. A *reproductive peak* lasts for nearly 40 more years (until about age 45 to 52 for most women). Then the number of cycles slows down during the *female climacteric*, which lasts from several months to a few years. Menstrual cycles cease in a phase of life called the *menopause*. As menopause approaches, nearly all the follicles in the ovaries have matured, ruptured, and/or atrophied. Women are still fully capable of normal sexual activity after the menopause but they are incapable of reproduction. The average age of women who reach menopause is 47. However, 30 percent do so by age 45, 60 percent by age 50, 98 percent by age 55, and 100 percent by age 60.

At this time, the lack of maturation of follicles and the accompanying decline in female sex hormones upset hormonal balance. Sensations of hot flashes, sweating, irritability, fatigue, anxiety, and difficult or labored breathing are not uncommon. The physiological effects may be severe enough to warrant professional treatment, and the psychological impact may promote a need for counseling. Because the administration of estrogens may have harmful effects, their use to minimize the symptoms should be considered only upon professional advice and in the most severe instances.

Effects on the Mammary Glands

The mammary glands are essentially modified sweat glands which are designed to produce milk in a process called *lactation*. They are embedded in fatty tissues which form the contours of the breasts. Lactation may be likened to an endocrine "symphony," which follows a score conducted by the hypothalamus and in which parts are played by hormones secreted by the hypothalamus, the anterior and posterior pituitary, and several other endocrines. Although the exact mechanisms of lactation are not fully understood, evidence indicates that the following general series of events is likely.

During pregnancy high levels of estrogens, prolactin, and a placental hormone, chorionic *somatomammatropin* (*placental lactogen*), stimulate growth of the breasts, whose full development is completed under the influence of progesterone. Through the activity of these hormones some milk is produced during pregnancy. After childbirth, when the placenta is shed, estrogen levels decline and no longer effectively block prolactin's stimulatory action on milk secretion by the ducts of the breast. Increases in adrenal glucocorticoid activity also seem to assist the secretion of milk. Although milk production is now markedly greater, it is stored in the breasts until the nipple of a breast is stimulated by suckling. Sensory signals from the nipple travel to the hypothalamus in the brain. The hypothalamus causes the anterior pituitary to secrete prolactin and the posterior pituitary to release oxytocin within 1 min of suckling. Prolactin stimulates ducts in the breast to secrete milk and oxytocin causes the walls of the ducts to contract and eject the milk from the nipples (the milk is "let down").

During the first 3 days after childbirth, the breasts release *colostrum,* a watery fluid rich in protein (including IgA antibody), but lower in milk sugar (lactose) and fat than milk, which is released later. Human milk contains nearly 88 percent water, 7 percent lactose, 4 percent fat, slightly more than 1 percent protein, and a variety of minerals and vitamins. Nursing stimulates continued prolactin secretion and milk production. It also inhibits LH-RH and thereby blocks ovulation in about 50 percent of nursing mothers.

24-3 CONCEPTION AND PREGNANCY

The union of the sperm and egg, *fertilization* or *conception,* begins a new life. Conception is the beginning of pregnancy. The word conception comes from Latin *concipere,* which literally means to take together. Yet this new self is not a clone but is a unique human being. The improbability of an individual's conception is staggering. One specific egg and one specific sperm are the only two cells that will produce a particular individual. Over 4 billion people are available to provide alternate parental combinations for different offspring. Furthermore 7 million immature eggs are originally present in the ovaries of one woman before her birth and countless sperm in excess of hundreds of billions are formed in the lifetime of one man. These computations prompted one biologist to write:

> Statistically, the probability of any one of us being here is so small that you'd think the mere fact of existing would keep us all in a contented dazzlement of surprise. We are alive against stupendous odds of genetics, infinitely outnumbered by all the alternates who might, except for luck, be in our places.[1]

Under natural conditions, conception takes place after the release of sperm into the vagina. Only a few hundred sperm usually reach the egg in the upper part of the oviduct. During their vigorous race up the uterus and oviducts, the sperm undergo a final stage of maturation, termed *capacitation.* Rapid rhythmical motions of cilia in the fimbriae and oviducts, along with those of muscles in the walls of the tubes, cause fluid secretions and the egg to flow toward the uterus. These muscular contractions, along with some in the uterus, may be induced by prostaglandins from the seminal fluid of the male and by oxytocin released from the posterior pituitary of the female. The contractions move the egg toward the sperm while producing a current against which the sperm must swim in order to meet the egg.

The egg is surrounded by a covering consisting of a few thousand follicular cells, which must be penetrated by a sperm. Lysosomal enzymes such as hyaluronidase are released from acrosomes of the sperm and provide a point of entry. The egg and a sperm normally fuse in the upper third of the oviduct, and the egg's final meiotic division then results. Granules in the egg's cytoplasm disappear, indicating the occurrence of functional changes, and the cell membrane of the egg is altered to prevent the entry of another sperm. The nuclei of the two cells unite, each contributing one member to each of the 23 pairs of chromosomes, in a new single cell, the *zygote.*

In unusual instances, the ovaries may release two or more eggs in the same cycle. If fertilized, the two eggs produce offspring which are not

[1] L. Thomas, *The Lives of a Cell: Notes of a Biology Watcher,* Viking Press, New York, 1974, p. 165.

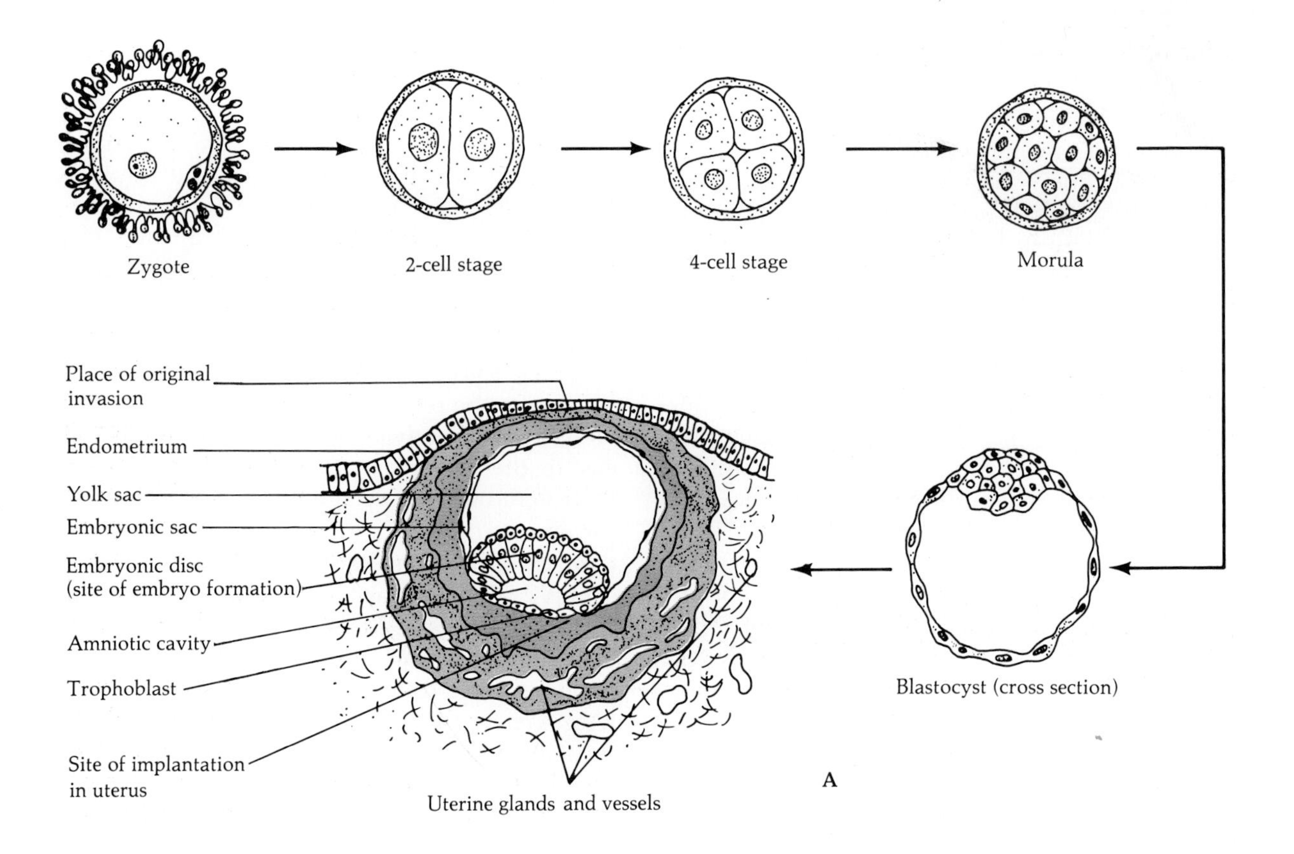

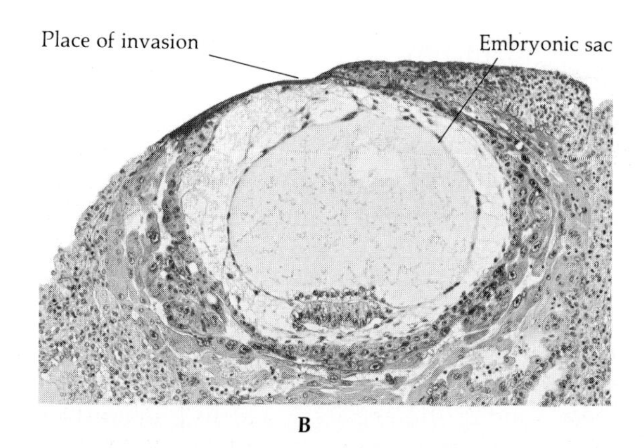

FIG. 24-13
Early development after fertilization of the egg. (A) Diagrammatic sketch of the development of a blastocyst from a fertilized egg (zygote). (B) Photomicrograph of the human embryo as a blastocyst which is burrowed deep in the uterine lining 12 days after fertilization of the egg.

genetically identical (they are dizygotic or *fraternal* twins). In other cases, for reasons which are not clear, the cells formed by division of the fertilized egg separate into individual entities which retain the capacity to develop into two, three, or more genetically identical (monozygotic) individuals. A combination of these events can produce offspring who are born at the same time of whom some are monozygotic and others dizygotic.

Early Embryonic Development

A fascinating and complex process of development follows conception. All the genetic information necessary to produce a unique human being is now in the fertilized egg. Within hours of early development, cells derived from the zygote are predestined to form specific organs. Although not visibly apparent, various types of cells with different potentials are sorted out and develop according to an exquisite plan stored in their genetic code and executed with extreme precision and according to an exact timetable.

The abundant cytoplasm in the egg and secretions of the oviducts provide early nourishment to sustain the initial very rapid mitotic cell divisions (cleavage) characteristic of early development. In the first 30 days of life a fertilized egg about 140 μm in diameter increases about 40-fold in size and 3000-fold in weight, developing into an embryo which is only about 5 to 6 mm (about 0.25 in) long.

Within the third to fourth day after fertilization, a solid ball of cells, the *morula,* is transformed into a hollow ball, the *blastocyst,* which produces an outer layer of cells that later erode the uterine lining (Fig. 24-13). About 7 to 10 days after fertilization the blastocyst burrows into the lining in a process called *implantation.* The embryo itself then forms membranes which invade the uterine lining (endometrium) to produce the *placenta,* an organ of communication between embryo and mother.

The cells of the blastocyst multiply to form the *embryonic disc* (blastoderm) and cells which spread to each side to produce the *trophoblast.* Cells of the trophoblast multiply and ultimately separate into an outer layer, the *chorion,* which erodes the uterine lining, beginning placental formation. An inner mass of cells produces a membrane, the *amnion,* which surrounds the embryonic disc and in later stages of development encases the embryo in a fluid-filled chamber called the amniotic cavity (Fig. 24-13). While the blastocyst burrows into the uterine lining, the embryonic disc forms three germinal tissue layers, an outer *ectoderm,* an inner *endoderm,* and later a middle *mesoderm.*

In some cases, embryonic development occurs outside the uterus. This condition, called *ectopic* or out-of-place pregnancy may occur in the oviduct (a tubal pregnancy), the ovary (an ovarian pregnancy), the abdominal cavity (an abdominal pregnancy), or the cervix (a cervical pregnancy). Development in all these unusual locations is hazardous to the mother, is unlikely to be completed, and often results in early loss of the fetus. Ectopic pregnancies also carry a high risk of rupturing blood vessels both of the mother and of the chorion of the embryo, and may result in shock and death of the mother and fetus.

The Placenta

The chorion produces fingerlike projections, or *villi,* which are invaded by fetal blood vessels and which penetrate the uterine lining to form the *placenta* (Fig. 24-14). The placenta is a selective barrier which controls the exchange of ions and molecules between fetus and mother. Removing nutrients and other compounds from the mother's blood, the placenta nurtures the fetus and also functions as the fetal respiratory, digestive, and endocrine systems.

In order to develop a mechanism for fetal-maternal exchange, capillaries supplied by the fetal umbilical arteries proliferate within the chorionic villi. The blood vessels of the fetus, separated by the selective placental barrier, are located in a stalk, the *umbilical cord,* which originates in its abdomen. At the same time spiral blood vessels from the mother's uterine artery multiply in the uterine lining, delivering blood to spaces between the villi (intervillous spaces) which are drained by maternal veins (Fig. 24-14). Blood flows from the fetus to the placenta by way of two umbilical arteries and returns to it by an umbilical vein (Fig. 16-2). The blood remains within a closed system independent of and separated from the maternal circulation.

Human Chorionic Gonadotropin (HCG) and Pregnancy Tests

The cells of the trophoblast, and later those of the placenta, secrete a number of hormones, including *human chorionic gonadotropin* (*HCG*). These hormones are transported through the placenta into the uterine lining, and are eventually excreted in the urine of the pregnant female. HCG can be identified by radioimmunoassay (Fig. 5-9) a few days after conception and may be used to diagnose pregnancy. HCG also may be identified later in pregnancy by immunologic assays in which the activity of antibodies to the hormone are neutralized (inhibited) when mixed with urine which contains the hormone from a pregnant female. Other less sensitive but effective assays are based on the reaction of the gonads of animals to the injection of urine containing HCG from a pregnant female. HCG causes the formation of hemorrhagic follicles (bloody follicles) in immature mice (the mouse test), the induction of ovula-

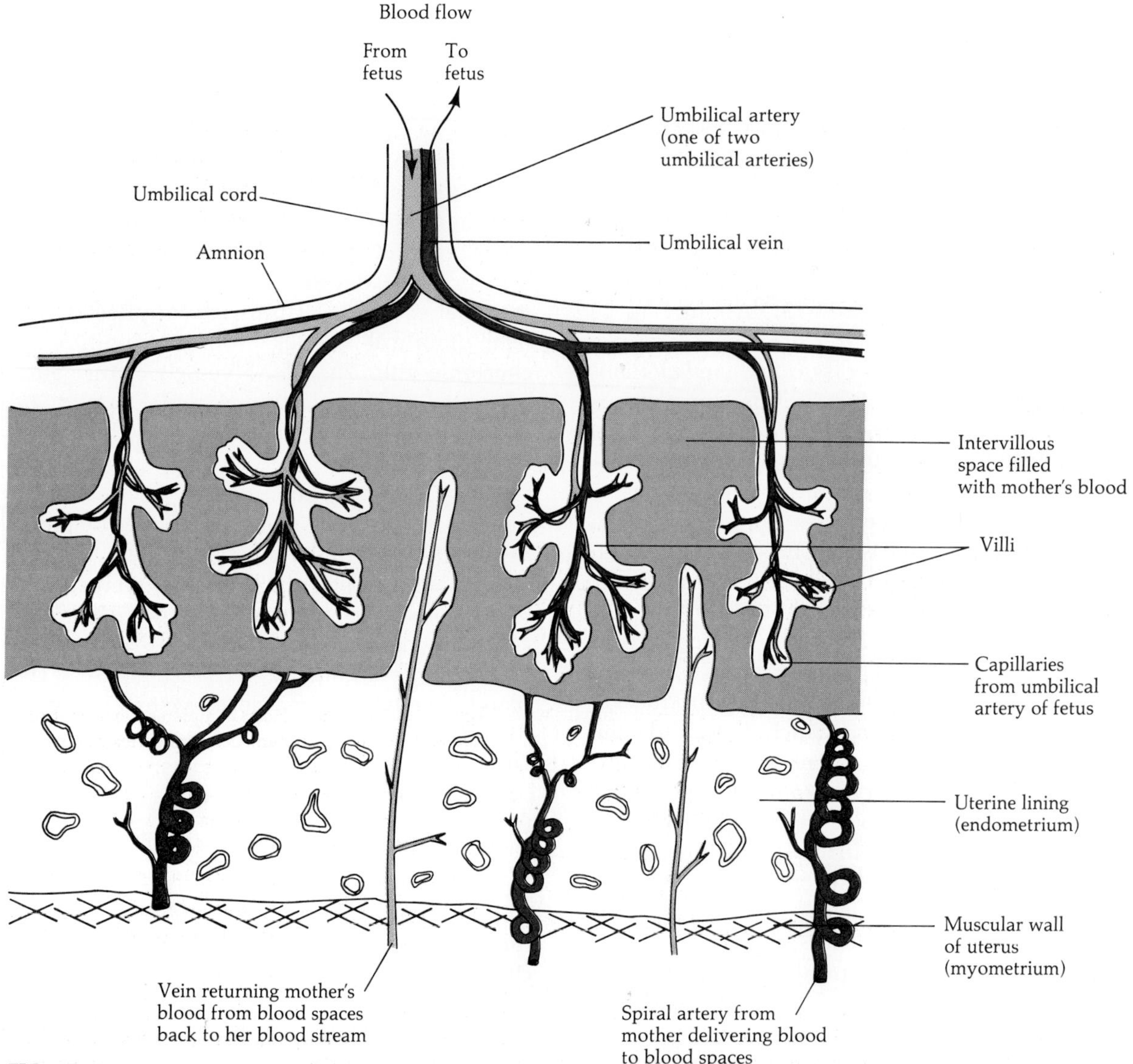

FIG. 24-14
Enlarged detailed diagram of the placenta.

tion in isolated female rabbits (the rabbit test), and the release of sperm from male frogs or toads (the frog test).

The Pattern of Development from the Third Week

The continued growth and development of the embryonic germinal tissue layers produces specific tissues and organs. The ectoderm primarily forms the outer layers of the body, the sense organs and nervous system, and the posterior pituitary gland. Nervous tissue is produced so rapidly that the size and maturation of the brain dominates that of other organs (Figs. 10-1 to 10-3). The mesoderm forms the moving, supporting, and connecting tissues of the body (muscle, bone, blood) and the tissues lining the blood and lymph vessels, kidneys, and ducts of glands. The endoderm forms tissues lining the digestive, respiratory, and urinary tracts and parts of the reproductive tract.

The timetable of human development before birth is well documented (Table 24-3). For ex-

TABLE 24-3
Timetable of human development before birth

Age	*What Takes Place*
1 day	The union of egg and sperm forms the zygote, the beginning of the new individual. The first cell division of the zygote, the first step in its growth, is completed within 36 h. By future cell divisions (mitosis) all the cells and tissues of the new individual will arise from the zygote.
4 days	Morula stage—special techniques can tell the sex of the new individual at this early stage.
7 to 9 days	Blastocyst stage—embryo reaches cavity of uterus and attaches to the lining of the uterine wall, burying itself in its glands.
2.5 to 4 weeks	Neurula stage—by three weeks the foundation for brain, spinal cord, and entire nervous system are established. Blood vessels start forming at 2.5 weeks, the heart a day later. At 3.5 weeks, the heart, a simple tube, starts to pulsate. From 3 weeks, the primitive digestive system and the forerunner of the kidney form.
4.5 weeks	The three main parts of the brain are present. Eyes, ears, nasal organs, digestive tract, liver, gallbladder, and arm and leg buds are forming.
5 weeks	Embryo is 1/3-in long and weighs 1/1000 oz. The early differentiation of the cerebral cortex is seen. Pituitary gland begins to form.
5.5 weeks	All muscle blocks present. Embryo may begin to move, but mother does not feel this for another 6 to 10 weeks. The heart begins to subdivide into its four chambers.
6 weeks	Embryo is ½ in long. Earliest reflexes can be elicited. Electrocardiogram (ECG) and electroencephalogram (EEG) can be recorded. Fingers, then toes, begin to form. Especially during the first 6 to 8 weeks of embryonic life, the embryo is most vulnerable to the effects of drugs, radiations, infections (particularly viral), noxious substances (such as alcohol and nicotine), and nutritional deficiencies of the mother.
8 weeks	Embryo is 1.5 in long and weighs 1/30 oz. The face appears quite human. Heart completes the formation of its four chambers. Hands and feet are well-formed and distinctly human. Cerebral cortex begins to acquire typical cells. At the end of 8 weeks all organs, facial features and limb structures have begun to form. Everything is present that will be found in the newborn baby. The fundamental plan of the human body is completely mapped out by the end of the second month. During the remainder of pregnancy the various organs will mature in structure and function.
9 weeks	The growing child is now called a fetus. When the eyelids or palms of the hand are touched, they both respond by closing; this indicates that both nerves and muscle are functioning.
10 weeks	Except for refinements, the brain is much as it will be at birth. If the forehead is touched, the fetus turns the head away.
12 weeks (3 months)	Fetus is 3–4 in crown-rump length, and weighs about ½ oz. The thumb can now be opposed to the forefinger (a characteristic of all the primates). Fetuses of this age begin to show individual variations, probably based on behavioral patterns inherited from the parents. By the end of the twelfth week, the fetus has developed all organ systems and is virtually a functioning organism. The fetal organs become more and more like what they will be in the newborn infant.
4 months	Fingerprints, unique to the individual, are formed. The fetus responds to touch and spontaneously stretches and exercises both arms and legs.
5 months	Fetus measures 8 in crown-rump length, and weighs 8 to 10 oz. The fetus exhibits a firm hand grip, good muscular strength, coordination and reflex action, and kicks, moves, turns in the womb, hiccups, develops patterns of sleep and wakefulness, and reacts in an individual way to loud noise, music, or jarring or tapping the abdomen.
6 months	During this month, the eyes become sensitive to varying intensities of light and darkness but not to objects.
7 months	Fetus measures 12 in crown-rump length and weighs 2 to 3 lb. The fetus (now called a premature baby if born) continues growing and maturing. Every added day spent in the uterus until birth prepares the baby all the better to assume an independent role.

ample, the heart is pulsating before the fourth week and the cerebral cortex and pituitary gland are evident by the fifth week. By the sixth week some reflexes can be elicited. The brain generates electrical waves which can be recorded, as can electrical activity of the heart. By the end of the eighth week the fundamental plan of the human body is mapped out and the embryo is now called a *fetus.* It is recognizable as human (Fig. 24-15) with the facial features of eyes and ears, hands with fingers, feet with toes, and a four-chambered pulsating heart. About this time the hormonal activity of the corpus luteum begins to diminish and is taken over by the placenta, although some of the function of the corpus luteum is maintained throughout pregnancy.

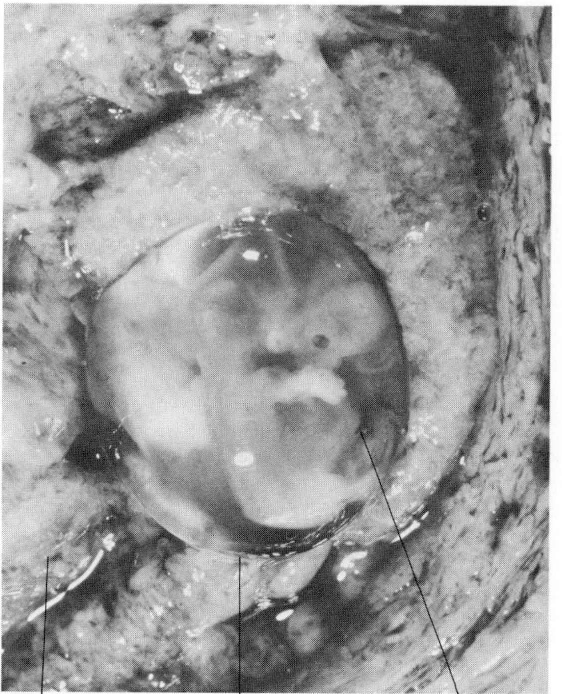

FIG. 24-15
The human fetus at the end of 8 weeks. Note the facial characteristics, hands with fingers, feet with toes, the amniotic sac, umbilical cord and placenta.

Maternal Hormonal Changes during Pregnancy

The requirements for fetal development impose demands on the mother's respiratory, circulatory, and excretory physiology, along with an extra load on her dietary intake. To adjust to these new requirements the size of some of her endocrines increases and the levels of many hormones are altered. Estrogens and progesterone produced by the placenta enter the maternal bloodstream and in moderate levels inhibit maternal secretion of FSH and LH. The placenta secretes somatomammotropin, which stimulates development of the woman's breasts and which converts a fetal adrenal steroid, *estriol,* into the major fetal estrogen which sustains pregnancy. The presence of fetal estriol in the mother's blood may be used to monitor pregnancy.

The elevated levels of certain hormones stimulate maternal appetite and metabolism to accommodate fetal needs. Greater quantities of salt and water-retaining hormones, such as estrogens, aldosterone, and antidiuretic hormone, cause extracellular fluid volume to increase. As a sign of excessive demands made on her kidneys, the mother's wrists and ankles may swell. Toxic levels of substances may accumulate in her blood and cause high blood pressure, and even convulsions if untreated, in a disorder termed the *toxemia of pregnancy.*

Amniocentesis

As mentioned above, the fetus floats in a salty bath of amniotic fluid suspended by an umbilical cord attached to the uterus but separated from the uterine wall by the amniotic sac (Fig. 24-15). The fluid accumulates rapidly and amounts to about 175 to 225 mL (6 to 8 oz) by the fourteenth to fifteenth week of pregnancy. Amniotic fluid originates from several fetal sources, principally from fetal urine but also from the respiratory tract, skin, and amniotic membranes.

Cells from the amnion and fetal skin slough off into the amniotic fluid. The diagnosis of many genetic diseases is made possible by analysis of the chromosomal patterns observed in these cells at the fourteenth to fifteenth week of pregnancy (about the sixteenth to seventeenth week after the last menstrual cycle). Biochemical analyses may reveal errors in metabolism induced by genetic defects and allow treatment to be instituted immediately after birth (see phenylketonuria, Chap. 6).

The amniotic fluid may be sampled by insertion of a syringe needle through the mother's abdominal wall and into the amniotic cavity. The procedure, called *amniocentesis,* allows up to 25 mL of amniotic fluid to be withdrawn for analysis. Risks of the technique include rupture of the amniotic sac (usually followed by spontaneous termination of pregnancy), bleeding, and an immune reaction by the mother against the fetus if fetal cells from damaged tissues enter her bloodstream. Other reactions include lower abdominal pain, leakage of amniotic fluid, bleeding, and fetal death. The risk of miscarriage (spontaneous abortion) caused by amniocentesis is about 1 to 2 percent higher than that which naturally occurs at the same stage of pregnancy. Thus, the diagnostic procedure is considered to involve a low risk, which should be weighed against its potential benefit prior to its use.

24-4 REPRODUCTIVE CAPACITY

Fertility and Infertility

Fertility refers to the capability of reproduction; *infertility* or sterility is the inability to reproduce. About 10 to 15 percent of marriages in the United States are infertile and nearly twice as many women as men are sterile.

The *fertile period* during the menstrual cycle in women, the time when the egg is capable of being impregnated by a sperm, lasts about 6 to 24 h following ovulation. Sperm retain their capacity to fertilize the egg for about 48 to 72 h after sexual intercourse and must be present in sufficient numbers while the egg is still fertile. The egg is made accessible to the sperm by ovulation, which occurs in the luteal phase of the cycle about 14 $\pm$ 2 days prior to the start of the next menstrual phase (Fig. 24-11). The probability of conception increases as the time of ovulation approaches. As previously stated, female reproductive cycles average 27 to 30 days in length, with a range of 21 to 45 days. In spite of these variations the day of ovulation can be estimated by subtracting 15 and 13 from the total length of the cycle. For example, in a 40-day cycle, ovulation occurs between days 25 and 27 from the beginning of menstruation ($40 - 15 = 25$ and $40 - 13 = 27$). In a 28-day cycle it takes place between days 13 and 15 ($28 - 15 = 13$ and $25 - 13 = 15$). The time of ovulation may be more accurately approximated by using changes in body temperature and in the viscosity and pH of cervical mucus, as indicated below in a discussion of the rhythm method of conception control.

The major causes of female infertility are closure of the oviducts, alterations in the cervix, and hormonal changes which inhibit ovulation. Blockage of the oviducts accounts for 30 percent or more of the cases of infertility and often is caused by infection and inflammation of the tubes in a condition called *salpingitis.* About 20 percent of infertility is due to functional blockade of the cervix by secretions which are too viscous or too acid and hostile to the sperm. Approximately 15 percent of infertility is due to hormone abnormalities. Deficiencies in the secretion of gonadotropins by the anterior pituitary often result in failure to ovulate.

The prime cause of infertility in males, accounting for about 30 percent of all cases, is due to the formation of insufficient or abnormal sperm. A low sperm count may result from damage to the testes by infectious disease, from exposure to irradiation, or from a congenital disorder in which the testes fail to descend into the scrotal sac (cryptorchidism). Furthermore, some males have endocrine disorders which cause the testes to produce too few sperm or a high percentage of abnormal ones.

A male is usually fertile if his sperm count exceeds 50 million per milliliter of semen, and if 60 percent or more of his sperm are highly motile and fewer than 25 percent are abnormal in appearance. In general, for minimal fertility, more than 20 million sperm per milliliter of semen should be present in 3 to 5 mL of ejaculate. About 50 percent of those males whose sperm count is 20 million to 40 million per milliliter are sterile. This figure approaches 100 percent among those who produce fewer than 20 million sperm per milliliter. Distinct from sterility, *impotent* males are unable to achieve or maintain an erection and cannot transfer sperm into the vagina. Such males are not necessarily sterile. Furthermore, their impotence may be a temporary condition and usually is related to psychological, dietary, or physical disorders.

Effects of Sexually Transmitted Diseases

The term *venereal disease* was originally used to emphasize the role of sex in the transmission of syphilis and gonorrhea, which were erroneously thought to be a single disease. Since that time a number of infectious microorganisms have been shown to be capable of infecting reproductive organs and the phrase, "sexually transmitted disease" is now used to designate common contagious diseases generally transmitted through sexual contact. Some, such as syphilis and gonorrhea, may cause sterility in both sexes. Infectious agents spread by other routes may also damage the reproductive tract. For example, mumps virus spread by respiratory droplets may infect and destroy seminiferous tubules to cause infertility.

Although a variety of microorganisms can cause sexually transmitted diseases, medical case reports indicate that annually over 1 million cases are due to *Neisseria gonorrhea,* which causes gonorrhea (or "clap"), and *Treponema pallidium,* which produces syphilis. These diseases are epidemic among sexually active young adults in the United States. More than 900,000 cases of gonorrhea are reported annually. Gonorrhea may go undetected in females although it often causes abnormal secretions and irritation of the reproductive tract openings. If detected early, gonorrhea and syphilis usually can be treated successfully with antibiotics. However, strains of gonorrhea have recently developed which are extremely resistant to penicillin, the drug most commonly used to combat the infection.

Type II *herpes simplex virus* (genital herpes) may infect the genital organs. This disorder is a sexually transmitted disease for which there is no cure. It is perhaps second only to gonorrhea in incidence. Like gonorrhea and syphilis, it can be transferred from a mother to her unborn child during its passage through an infected birth canal. Type II herpes is life-threatening to the newborn if it infects the nervous system during the first 4 weeks of life. A related form of the virus, type I herpes simplex, is most often associated with fever blisters or "cold sores" and is usually transmitted orally, but occasionally infects the genital tract.

Group A *Chlamydia* are intracellular parasitic microorganisms that depend on the cells in which they reside for their energy requirements. They also may be transmitted sexually and cause an infectious disease termed *lymphogranuloma venereum.* The microorganisms damage mucous membranes. However, unlike type II herpes, these organisms can be eliminated by antibiotic therapy.

A wide variety of other sexually transmitted diseases occur, including infections by a protozoan (*Trichomonas*), yeast (*Candida*), papilloma viruses (which cause venereal warts), lice (called "crabs," which cause pediculosis), mites (which cause scabies), and a number of disorders of unknown origins grouped together as nonspecific urethritis.

Control of Conception

Among several methods of controlling conception are abstinence from sexual intercourse during the fertile period (the rhythm method), the administration of hormones to regulate ovulation, and the use of physical and surgical preventive techniques. Owing to variations among individuals, differences in effectiveness of the methods, and differences in their application by individuals, the incidence of pregnancy varies with the use of alternative conception control techniques (Table 24-4).

Rhythm

The *rhythm method* requires abstinence from sexual intercourse near the time of ovulation during the *fertile period* of the female. This pro-

TABLE 24-4
Relative effectiveness of birth control methods

Method	*Pregnancies per 100 Women per Year*
Combination pill	
1. At least 50 μg estrogen plus some progestin	Fewer than 1
2. 25–35 μg estrogen plus some progestin	Fewer than 1–2
Minipill (progestin only)	About 3
Intrauterine devide (IUD)	Fewer than 1–6
Diaphragm with spermicidal cream or jelly	2–20
Condom (rubber)	3–36
Spermicidal aerosol foams	2–29
Spermicidal jellies or creams	4–36
Rhythm (periodic abstinence from sexual intercourse)	Less than 1–47
1. Calendar method	14–47
2. Temperature method	1–20
3. Temperature method (intercourse only in postovulatory phase)	Fewer than 1–7
4. Mucus method	1–25
Surgical sterilization	
1. Males (vasectomy)	Fewer than 1
2. Females (tubal ligation)	0.4–2.8 (or more)
Interrupted intercourse (withdrawal)	2.1–16
No contraception	60–80

cedure is most effective when intercourse is restricted to the postovulatory period. The rhythm method is based on studies which indicate that the sperm do not remain viable more than 2 to 3 days after entry into the female reproductive tract, that the egg retains its fertility for 24 h or less after ovulation, and that ovulation occurs 14 ± 2 days before the next menstrual period.

On the basis of this information, the fertile period in a 28-day cycle is considered most likely to fall between the eleventh and the eighteenth day of the menstrual cycle. These 8 days take into account 2 days for survival of the sperm, 1 for that of the egg, 3 possible days on which ovulation might occur (days 13 to 15) and 2 days (one on each side of the fertile period) for good measure.

The fertile period may then be calculated as follows:

1. Record the length of cycles in days over a period of at least 6 months, preferably 1 year. The shortest and longest cycles are noted.
2. Subtract 18 days from the shortest menstrual period.
3. Subtract 11 days from the longest one.

The result is the calculated fertile period and is the time span during which sexual intercourse is most likely to result in conception. It is the "unsafe period" if conception control is desired. Conversely, the days outside this period are the "safe" or "infertile periods" indicating that conception is unlikely. For example, if a woman's cycles vary between 25 and 28 days the fertile period is calculated as follows:

1. 25 and 28 days are the shortest and longest cycles
2. $25 - 18 = 7$
3. $28 - 11 = 17$

Days 7 to 17 of the cycle would be most likely to include the fertile period, where day 1 is the first day of menstrual flow of the current cycle.

Variations among individuals, longevity of the fertility of the sperm or egg, and factors that change the cycle and cause early or late ovulation (such as emotional stress, dietary factors, or illness) may, in some cases, alter the "safe" period. Therefore, to ascertain the time of ovulation, early morning temperature and the properties of cervical mucus delivered to the vagina should be monitored. As noted above, early morning temperature rises 0.56°C (1°F) from its low point of the month within 24 h of ovulation (Fig. 24-12). However, infections often elevate body temperature and may mask this hormonal effect. As noted earlier, the cervical mucus, which is thin and watery at ovulation, becomes more viscous and alkaline after ovulation. Research currently underway indi-

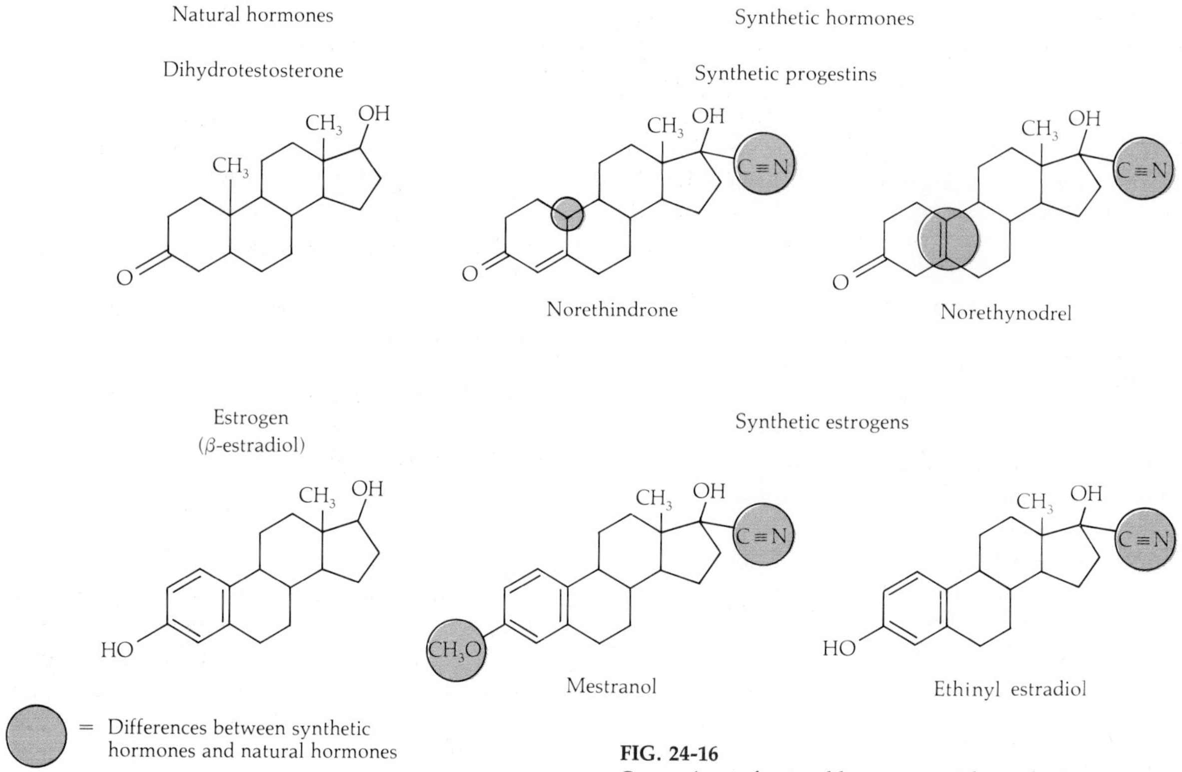

Natural hormones

Dihydrotestosterone

Synthetic hormones

Synthetic progestins

Norethindrone

Norethynodrel

Estrogen
(β-estradiol)

Synthetic estrogens

Mestranol

Ethinyl estradiol

= Differences between synthetic hormones and natural hormones

FIG. 24-16
Comparison of natural hormones with synthetic counterparts used in various forms of the pill.

cates that ovulation may be predicted very accurately by observing changes in the cervical mucus. When sexual intercourse is restricted to the postovulatory phase based on changes in cervical mucus or body temperature, then the rhythm method significantly reduces the probability of pregnancy and approaches that attained by use of the pill (Table 24-4).

The Pill

Ovulation may be regulated by the administration of synthetic female sex hormones taken orally as pills. Synthetic hormones are more effective than their natural counterparts because they are less susceptible to degradation when administered orally. Oral contraceptives contain synthetic estrogens and progesterones (progestins or progestogens) which suppress gonadotropins and thereby inhibit ovulation. The progestins structurally resemble testosterone (Fig. 24-16). Under the influence of the pill, follicles which have begun to develop shrivel up, and consequently an egg is not released. The follicles resemble atretic ones which form in natural cycles and during pregnancy (Fig. 24-9).

Pills with progestins and synthetic estrogens are called *combination pills* and are the most commonly used birth control pills in the United States. They are usually taken orally for a specified number of days in each cycle: daily between the fifth and twenty-fifth days of a cycle, followed by 7 days in which no hormones are taken. Menstrual flow occurs during this latter time frame. Each day that the pill is not taken (other than the 7 days noted) increases the possibility of pregnancy.

Although the pills are the most effective conception control agents (Table 24-4), they may cause a variety of adverse side effects. The less serious consequences which may occur include bleeding between menstrual periods, weight gain, tenderness associated with swelling of the breasts, irritation of the mucosa of the stomach (which may cause nausea and vomiting), diminished excretion of sodium by the kidney (which causes edema, generalized bloating, and headaches), reactions in the skin (including acne and the appearance of brown spots), central nervous system effects (increases or decreases in sexual desire, irritability, and depression), and migraine headaches. Some debatable evidence indicates that a predisposition to fungal and yeast infections also may occur. Precise estimates of the relative risk of incidence of these disorders in users of the pill compared with nonusers remain to be established.

An increased risk of thrombophlebitis (inflammation of the walls of blood vessels, especially veins) has been associated with the use of the combined pills. Therefore, *minipills* have been developed which contain only progestins. However, these pills have been associated with increased possibility of pregnancy, intermenstrual bleeding, and irregularity of cycles.

A slight but significant increase in probability of more serious side effects in the mother, such as circulatory disorders (Table 24-5), liver tumors, gallbladder disease, and birth defects occur if the pill is used immediately before or during pregnancy. The risks increase with age (especially beyond 35) and with heavy smoking (15 or more cigarettes daily) and depend on the type of pill and the individual. Because pills vary in dose and the chemistry of the synthetic estrogens and progestins they contain, their side effects also may differ.

Development of birth control pills for males is still experimental. Some steroid hormones and nonsteroid compounds have been tried experimentally, which, through negative feedback inhibition, suppress gonadotropin secretion and sperm production. Some of the hormones tested cause a number of undesirable side effects in males, such as feminization or loss of sexual drive. Drugs such as serotonin and reserpine suppress sperm production but also decrease sexual desire and activity.

An antifertility organic compound, gossypol, found in the cotton plant, has been studied chemically as a potential male contraceptive. It reduces sperm count by blocking the enzyme lactic dehydrogenase X, found only in sperm and testis cells. Unfortunately, gossypol also affects enzymes in other cells, and in extremely high doses it may cause serious side effects. The significance of the reactions must be established before gossypol can be cleared for use as an antifertility agent in men.

TABLE 24-5
Relative risks of vascular problems in users of oral contraception

Disorder	*Risk Factor Compared with Nonusers*
Thromboembolic disease (formation, dislodgement, and migration of a clot which blocks a blood vessel)	2–11
Thrombotic stroke (clot blocking circulation to part of the brain)	4–9.5
Hemorrhagic stroke (bursting of a blood vessel in the brain)	2–2.3
Heart attack (myocardial infarction)	2–12

Physical Methods

The *diaphragm* is a round, rubber, cuplike device placed in the upper vagina to seal off the cervix in another method of birth control. The diaphragm is formed of a circular metal spring covered with smooth latex rubber, and in practice is covered with a spermicidal jelly and is inserted prior to intercourse. It usually is removed 6 to 8 h later. The efficacy of the diaphragm in conception control varies widely with its fit and consistency of use.

A *condom* is a thin, latex rubber covering that encloses the penis to trap seminal fluid and prevent conception. Condoms of different shapes are available and are often covered with a lubricating fluid to simulate that which is normally released by the male. The use of condoms reduces the transmission of sexually transmitted diseases. However, in addition to the risk of pregnancy (Table 24-4), the use of the condom reduces the pleasure of sexual intercourse for some couples.

Interruption of sexual intercourse (*coitus interruptus* or *withdrawal* prior to sperm release is another means of controlling conception (Table 24-4). However, since sperm are extremely motile, pregnancy may result from those that may spill on the external genitals. Furthermore, some sperm released prior to ejaculation in the secretions of the lubricating fluids of the penis may cause pregnancy.

Surgical Methods

Vasectomy and tubal ligation are highly effective methods of conception control (Table 24-4). *Vasectomy* is a surgical procedure in which each vas deferens is cut and tied after incisions in each of the scrotal sacs have been made. Sperm are then unable to pass from the testis to the exterior, but the remainder of the semen still is released upon ejaculation. The simplicity of the operation has been widely reported. Although the vas deferens may be rejoined, atrophy of the tissues prevents successful surgical reversal of vasectomy in more than 50 percent of the cases.

Spontaneous reunion of the cut ends of the vas deferens has been reported in less than 1.0 percent of the males who have had vasectomy, which of course nullified the desired effects. Perhaps of greater significance are studies which indicate that up to 50 percent of vasectomized males form antibodies against sperm antigens. The inability to release sperm with the ejaculate causes the male sex cells to accumulate in the testis. White blood cells, including lymphocytes and monocytes, migrate into the area and an inflammatory response results (Chap. 3) as the sperm are removed by phagocytosis. Sperm antigens, which normally do not enter the circulation, now do so as they are ingested by wandering macrophages. The molecules from sperm, which are alien to the circulatory system, elicit an immune reaction against the individual's own tissues. It is an autoimmune response which may be associated with long-term side effects, including an increased susceptibility to atherosclerosis. Alterations in the urinary excretion of metabolites of androgens of the vasectomized male indicate that some changes also occur in hormone metabolism. The long-term results of these changes have yet to be clearly established.

Castration, removal of the testes, has serious effects. It produces androgen deficiency and results in profound hormonal changes. The negative effects of castration are most marked in the development of secondary sex characteristics, body structure, and function in preadolescent males (Chaps. 5 and 14). Significant diminution of these traits also may occur in fully mature males.

Tubal ligation is surgery in which the oviducts are cut and tied. The oviducts also may be closed by cauterization (burned). Tubal ligation blocks the entry of sperm to the egg and the passage of the egg to the uterus. The failure rate of this surgery in terms of subsequent pregnancy ranges from 0.4 to 2.8 percent or more (Table 24-4). Tubal ligation requires entry through the abdomen. It is irreversible and occasionally alters blood flow to the ovaries, causing abnormal cycles, and may necessitate hysterectomy later in life. Therefore, it is not to be considered lightly.

Hysterectomy is a major surgical procedure in which the entire uterus is usually removed. Since the other reproductive organs remain, hormonal secretion and sexual functions continue in the absence of reproductive capacity. On the other hand, *ovariectomy,* removal of the ovaries, causes profound and undesirable changes in hormone balance. It, too, is a major surgical operation with the risks such surgery involves.

Miscellaneous Techniques

Other methods of conception control with varying degrees of effectiveness include the *douche* and *spermicidal agents* (Table 24-4). A douche cleanses the female reproductive tract with a stream of fluid delivered into the vagina by a syringe. The wash reduces the number of sperm remaining there after intercourse but results in a rather high rate of pregnancy (about

31 per 100 women per year). Spermicidal (sperm-killing) foams, jellies, or creams may be inserted into the vagina prior to sexual intercourse, with wide ranges in degree of effectiveness.

A variety of methods have been tested to control conception by using immune reactions. For example, female laboratory animals immunized with sperm antigens produce sperm antibodies, which attack sperm that gain access to the female's reproductive tract after sexual intercourse. This method and other immunological approaches to conception control in humans are still experimental.

Other Methods of Birth Control

Some birth control techniques do not prevent conception. Rather, they prevent birth by removal or expulsion of the embryo or fetus. Among these techniques are abortion, intrauterine devices, and the morning-after pill.

SPONTANEOUS AND INDUCED ABORTION *Abortion* is the termination of pregnancy. When it occurs naturally, it is called a *miscarriage* or *spontaneous abortion.* Miscarriages often are the result of errors in chromosomal replication and division. More than 1 million *induced abortions* are reported annually in the United States. The procedures used in induced abortions are described in Table 24-6. The side effects of abortion include psychological reactions, bleeding, infection, and the risk of the woman's death. The risks generally increase with the duration of the pregnancy and vary with the method and the conditions under which an abortion is performed.

TABLE 24-6
Methods of abortion*

Technique	
Dilatation and evacuation (D and E)	The cervix is dilated and a suction aspirator (like a small vacuum cleaner) sucks out the uterine lining and the fetus. This is the most common current form of abortion and has replaced the second technique which was previously most widely used
Dilatation and curettage (D and C)	The cervix is dilated (widened) with an instrument that wedges it open so that a scraping instrument, a curette, may be used to scrape the uterine lining and the fetus away
Instillation of saline	The uterus is flooded with hypertonic saline, which induces muscular contractions (labor), sheds the lining, and usually destroys the fetus
Instillation of prostaglandins	These hormones induce muscular contractions (labor), which expel the fetus
Hysterotomy	A hysterotomy or Caesarean section is an incision made in the uterus. The fetus may then be removed

* The first two techniques are used to terminate pregnancy during the first trimester (3 months) of pregnancy. The other procedures are used at later stages.

INTRAUTERINE DEVICES Several types of intrauterine devices (IUDs) are available which also may terminate pregnancy. Some are plastic coils or loops of various sizes and shapes which may be inserted into the uterus. They may also contain hormones or have metal wrappings such as copper. The IUD usually causes an inflammatory response in the uterine lining, making it an unsuitable location for implantation and development of the embryo. The IUD usually does not prevent union of the sperm and egg. Side effects include an inflammatory response in the uterus, risk of infection, excessive menstrual bleeding, and pain in a significant percentage of users (5 to 20 percent). Furthermore, the IUD is expelled without knowledge of the user in 3 to 20 percent of cases. The highest expulsion rate occurs in women during the first year of use.

MORNING-AFTER PILL *"Morning-after" pills* (postcoital pills) are composed of synthetic hormones, which when taken orally for 4 to 6 days after sexual intercourse, terminate pregnancy. Some morning-after pills contain the synthetic estrogen ethinyl estradiol, while others contain sodium estrone sulfate. Morning-after pills do not prevent union of the sperm and egg. Rather, they block implantation of the embryo and its continued development. Their side effects are linked to those of estrogenic agents. Earlier versions of morning-after pills con-

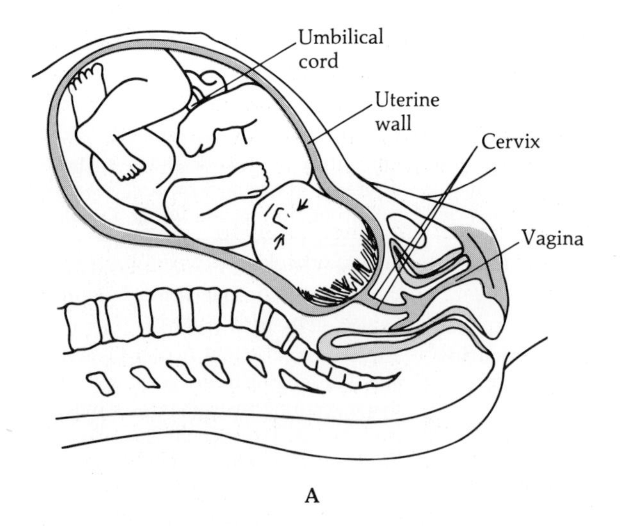

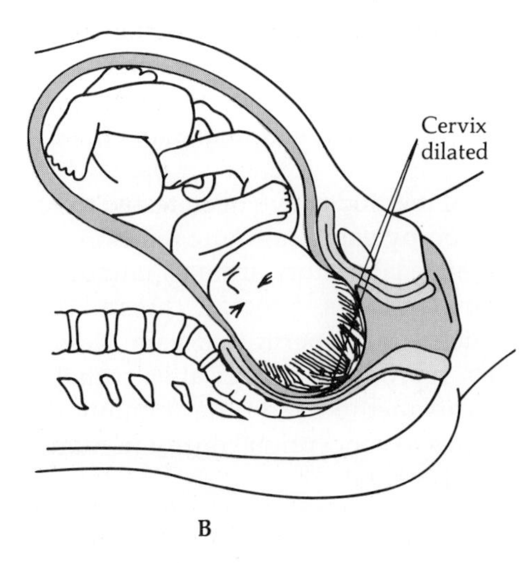

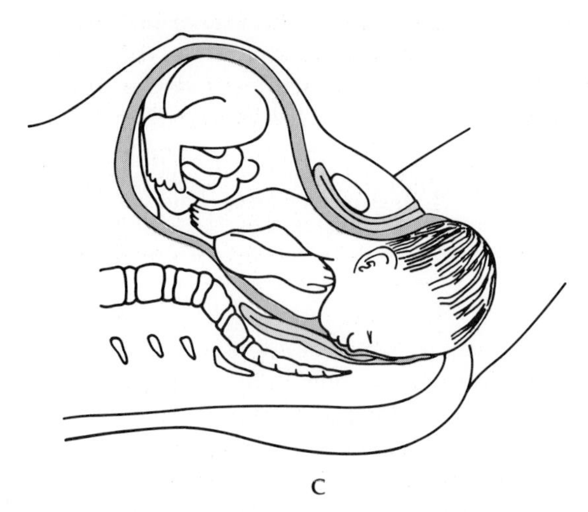

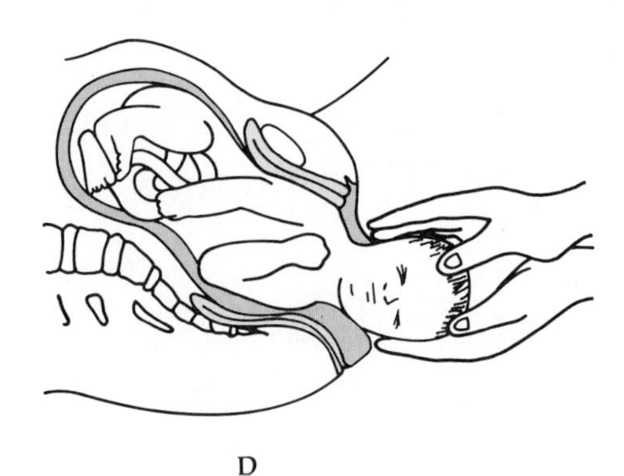

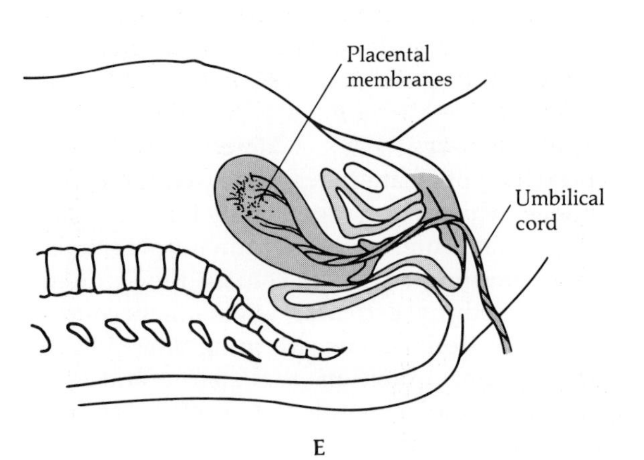

FIG. 24-17
Childbirth. The movement of the baby from the uterus through the cervix and birth canal and out of the vagina are shown in A to D. The expulsion of the placental membranes as the afterbirth completes the last stage of normal labor is shown in E. The movements of the baby in successive stages of delivery are noted. (A) Flexion: The narrow birth canal causes the baby's chin and head to bend toward its chest. (B) Internal rotation: The baby's head rotates from a sideways position with the chin turned towards its shoulder to one with the chin straight ahead. (C) Extension: The baby's head meets resistance of the floor of the mother's pelvic cavity, causing the head to be extended backward and upward toward the opening of the vagina. (D) External rotation: As the baby's body comes out of the vagina, the attendant supports the baby's head and helps rotate it back to its original sideways position to line up the chin with the shoulders and make delivery easier. (E) Expulsion of the afterbirth: The placental membranes are expelled from the uterus and out the cervix by uterine contractions which occur a few minutes to an hour after birth.

tained diethylstilbestrol (DES), which increases the risk of vaginal and cervical cancer in female offspring of women who use it during pregnancy. DES has been implicated in congenital malformations in embryos and in babies born to mothers who used it during pregnancy.

24-5 CHILDBIRTH (PARTURITION)

In the natural course of events following conception, development usually continues so that "A fertilized egg generates a baby in 9 months, an adult in 15 years."[2] In about the seventh or eighth month of development the baby descends, usually with the head moving toward the mother's cervix and the arms and legs folded close to its abdomen. The descent is commonly called *lightening*. In about 5 percent of pregnancies, breech presentations occur, in which parts of the baby other than the head are nearest the cervix.

In the last few months prior to childbirth, painless muscular spasms, *Braxton Hicks contractions,* alter the size of the uterus to accommodate the enlarging fetus. Other intermittent contractions in the abdomen cause *false labor* and *false labor contractions* (or pains). *True labor* and *labor contractions* are uterine in origin, occur generally within 24 h of childbirth and 266 days from the time of conception, and average 280 days after the start of the last regular menstrual period. The normal time of childbirth can range from 260 to 315 days after conception.

Intense rhythmical uterine contractions begin the *first stage of labor*. Each contraction initially lasts about 45 to 60 s and occurs at 10-min intervals. The initial contractions may be stimulated by prostaglandins and later ones by oxytocin. The feedback signals that initiate labor and delivery are unknown.

[2] S. A. Luria, *36 Lectures in Biology*, MIT Press, Cambridge, 1975.

During the first stage of labor the contractions become more intense, rhythmic, and closely spaced in time. Their force widens the opening of the cervix, which is normally 0.5 cm in diameter, to 5 cm in the beginning of the first stage and 10 cm by the end of it. In animals a polypeptide hormone, *relaxin,* secreted by the corpus luteum, softens and widens the cervix, relaxes the pelvic joints, and inhibits uterine contractions. Its role in humans remains to be established.

During the *second stage of labor* the rhythmic muscular contractions of the uterus last nearly 90 s and occur at intervals of 90 to 120 s. The amniotic sac usually ruptures ("the water breaks") and fluid flows out of the vagina early in this stage. In 10 percent of pregnancies the sac ruptures in the first stages of labor. The contractions in the second stage of labor usually push the baby headfirst through the cervix and out through the vagina (Fig. 24-17).

The person assisting childbirth removes mucus from the baby's mouth and ties and cuts the umbilical cord. The baby's back may be tapped gently to stimulate his or her respiratory tract and breathing. An antibiotic solution (penicillin) or a silver nitrate solution is applied to the baby's eyes to prevent infection from gonorrhea. Up to an hour after delivery, more rhythmic muscular uterine contractions occur, which cause the placental tissues still adhering to the uterine wall to break away and be expelled as the *afterbirth*. A new and uniquely wondrous human being has been born against odds that defy comprehension. The human life cycle continues.

24-6 SUMMARY

1 Human reproduction is a sexual process in which people generate people. The primary male reproductive organs, the testes, contain seminiferous tubules whose epithelial cells produce sperm. The sperm are delivered from the epididymis to the vas deferens to the urethra and to the exterior through the penis.

On their route to the exterior sperm are mixed with secretions from the accessory sex organs, the seminal vesicles, the prostate gland, and the bulbourethral glands. The secretions, along with the sperm from the testes, constitute seminal fluid (semen). Semen is an alkaline fluid high in fructose, prostaglandins, and proteolytic enzymes.

The testes contain three functionally important and distinct cell types. Spermatogenic cells produce two types of sperm, each with 23 chromosomes. Mature sperm are produced by meiosis, in which chromosomes are duplicated once but each cell divides twice. In this manner a single cell with 46 chromosomes produces four cells, each of which contains 23 chromosomes. In half of the sperm one of the 23 chromosomes is of the X type and in the remaining sperm one of the 23 chromosomes is of the Y type. Sertoli cells are polyfunctional phagocytic cells which nourish sperm, assist their maturation, and contain contractile proteins which may move sperm. Sertoli cells also have tight junctions which form a blood-testis barrier. Interstitial connective tissue cells are the major source of the major male sex hormone testosterone, whose most active form is dihydrotestosterone.

Orgasm occurs in phases of excitement, plateau, orgasm, and resolution. Sensory receptors cause reflex responses and a surge of blood that swells the spongy tissues of the penis, enlarges it, and stiffens it in an erection. Semen is released in two stages, emission, and then ejaculation which occurs at the time of the orgasmic phase. These responses are associated with rhythmic muscular contractions of the pelvic organs and elevation of the heart rate and blood pressure, which are also characteristics of orgasm. Resolution, or return to the sexually unexcited state, may take up to 2 h.

The male reproductive cycle is regulated by LH-RH/FSH-RH from the hypothalamus, which stimulates the secretion of the gonadotropins LH and FSH from the anterior pituitary. LH causes secretion of androgens, including testosterone, and some estrogens by the interstitial cells while FSH promotes the formation of sperm. The levels of androgens and estrogens are controlled by negative feedback inhibition of the hypothalamus and/or the anterior pituitary. Testosterone, the major androgen, is generally anabolic in its activity and promotes the development of the reproductive system and secondary sex characteristics. Testosterone levels increase markedly during puberty, resulting in attainment of reproductive capability, which usually lasts for the lifetime of the male.

2 The primary female reproductive organs, the ovaries, form eggs by meiosis within masses of cells in follicles. Beginning at puberty and continuing until menopause, the eggs are released from the surface of an ovary into the oviduct about once each month. The egg is propelled by cilia lining the walls of the oviduct and enters the muscular, pear-shaped uterus.

The oviducts, uterus, cervix of the uterus, vagina, external reproductive organs, and mammary glands are accessory female reproductive organs. The uterus, or womb, is a hollow sac composed mostly of smooth muscle lined with tissues that undergo monthly cyclical changes during the reproductive life of a woman. It leads through a cervical region to the vagina, a channel which opens to the exterior of the body.

The external female reproductive organs include the labia majora and minora and the clitoris; the last-named organ is located in the vestibule, a narrow chamber which also contains the exterior openings of the urethra, the vagina, and vestibular glands. The clitoris, an erectile organ homologous to the male penis, is at the anterior border of the vestibule. The mammary glands are modified sweat glands embedded in fatty tissue that forms the contours of the breasts. They develop initially during puberty under the influence of estrogens and progesterone, and in pregnancy under the further influence of prolactin and chorionic somatomammatropin. The mammary glands secrete milk when the suckling reflex stimulates prolactin secretion from the anterior pituitary and eject milk under the influence of oxytocin from the posterior pituitary.

Orgasm in the female may be divided into phases of excitement, plateau, orgasm, and resolution. It usually begins with glandular secretions in the vagina and vestibule, followed by reflex responses involving sensory receptors that produce erection of the clitoris and dilation of the vagina; these are succeeded by rhythmic smooth muscle contractions in the uterus, oviducts, and breasts and terminated by a return to the unexcited state.

In the female reproductive cycle changes promoted by FSH and LH cause the follicular cells to secrete increasing amounts of estrogens in the follicular or estrogenic phase. The production of an egg takes place as the follicle matures and increases in size. The egg is segregated within layers of nourishing and protective cells of the follicle as the whole structure moves toward the ovary's surface. The estrogens are generally anabolic and reach a peak near the time of ovulation, when the egg is released.

Of the 7 million eggs in the ovaries of the 20-week-old female embryo, only 400 to 500 mature in the 30 or more years of female reproductive life. This compares with the almost unlimited number of sperm that can be produced by the male.

Ovulation usually occurs 14 days before the start of the next menstrual flow, when the LH levels reach a peak. The cells left behind in the follicle constitute the corpus luteum and continue to secrete estrogens and increasing amounts of progesterone under the influence of LH in the luteal or progestational phase of the cycle. Progesterone causes arteries to form a coiled, rich vascular bed within a thickened uterine lining whose cells store fat and glycogen. Sustained secretion of estrogens and progesterone prevent further ovulation.

If the egg is not fertilized, the corpus luteum degenerates, estrogen and progesterone levels decline, and the uterine lining is shed during the menstrual phase of the cycle. A complete reproductive cycle lasts 21 to 45 days and averages 28 days in length.

3 The union of sperm and egg is called fertilization or conception and is the beginning of a new individual. Fertilization occurs in the oviduct within 6 to 24 h after ovulation and within 72 h of release of the sperm. Fertilization stimulates the final stage of meiosis in the egg, which reduces its chromosome number from 46 to 23. The union of the 23 chromosomes from the nucleus of the sperm and 23 from the nucleus of the egg form a zygote with 23 pairs of chromosomes. When a female is to be produced, an egg, which contains an X chromosome, and an X-chromosome-bearing sperm unite at conception. A fertilized egg destined to produce a male is formed when the X-chromosome-containing egg and a sperm bearing a Y chromosome unite to produce a zygote.

The embryologic origins of the male and female reproductive organs are similar. However, the presence of genes on the Y chromosome in the male seem to inhibit those on the X chromosome which otherwise cause female sexual development. Furthermore, lack of a Y chromosome, coupled with a lesser supply of testosterone, allows genes on the X chromosome to develop and maintain the female reproductive tract. Genes on autosomes (chromosomes other than the X and Y chromosomes) also influence the development of sexual characteristics.

Pregnancy starts at conception and is followed by rapid cell division.

Within 10 days the embryo is a blastocyst, which nestles into the uterine lining and begins to form the placenta. The placenta houses the embryonic and fetal blood vessels, which penetrate caverns in the uterine lining that are filled with the mother's blood. The placenta serves as the respiratory, digestive, and endocrine systems for the fetus. Three embryonic tissue layers, namely the ectoderm, mesoderm, and endoderm, are formed within 21 days after conception. The brain and nervous system are derived from the ectoderm and dominate early growth. The ectoderm forms the outer layers of the body, the mesoderm produces the muscle and connective tissues, and the endoderm develops the tissues lining the digestive, respiratory, urinary, and reproductive systems. By the end of the eighth week, development has proceeded so that the embryo is easily recognizable as human and is now called a fetus.

The outer layer of the placenta, the chorion, secretes gonadotropins which foster embryonic development and which cross the placenta and appear in the mother's blood and urine. Human chorionic gonadotropin (HCG) may be detected in the mother's urine a few days after conception by radioimmunoassay or other immunological tests as a basis of early diagnosis of pregnancy. Fetal hormones which enter the maternal circulation cause profound hormonal changes in the mother that increase her metabolism and appetite and make other adjustments related to the demands of pregnancy.

The innermost layer of the placenta, the amnion, is a fluid-filled sac derived from the fetus, which it surrounds and bathes. The fluid and cells may be withdrawn to assess fetal abnormalities after the fourteenth to fifteenth week of development in a technique called amniocentesis.

4 Fertility, the ability to reproduce, is dependent on the production and union of a viable sperm and egg. It usually requires 20 million or more sperm per milliliter of semen to provide enzymes that peel away the protective follicular cells and thus to make the union possible. Infertility (sterility) may result from a number of causes. It may be due to infection by gonorrhea, syphilis, or other microorganisms which damage reproductive organs and/or to hormonal disorders or other diseases which may produce similar effects. Males are usually considered sterile if the number of sperm ejaculated is less than 20 million per milliliter of semen. Obstruction of the oviducts or cervix and failure to ovulate are common causes of infertility in females.

Conception control may be achieved by a variety of means, none of which is 100 percent effective. Techniques include periodic abstinence from sexual intercourse (rhythm), the use of synthetic hormones (the pill) that inhibit gonadotropins and ovulation, surgical sterilization (vasectomy in the male and tubal ligation in the female), methods that chemically kill sperm (jellies, foams, creams), physical devices that prevent fertilization of the egg (the condom and diaphragm), and interruption of intercourse (withdrawal). All the methods have specific advantages and disadvantages.

Abortion is the termination of pregnancy. Abortions may occur spontaneously or be induced by surgical or chemical techniques. The IUD and morning-after pill are birth control methods that abort pregnancy.

5 Childbirth is preceded by labor, the rhythmic contractions of the uterus, some 260 to 315 days after conception. Labor seems to be initiated by prostaglandins and later stimulated by oxytocin. In animals the hormone relaxin, secreted by the corpus luteum, relaxes the pelvic organs, facilitating birth. Intense rhythmic uterine contractions increase in duration and occur more rapidly as childbirth nears. They last about 90 s and occur at intervals of 90 to 120 s just before they expel the baby out of the vagina at childbirth. As a re-

sult of conception and completion of pregnancy and childbirth, the human life cycle is continued as a new and unique human being enters the universe.

REVIEW QUESTIONS

1 Describe the anatomical relationships between the primary and accessory male reproductive organs.
2 Name the functions of the seminal vesicles, prostate glands, and bulbourethral glands.
3 Describe the formation of sperm by meiosis.
4 Describe the structure of sperm and compare the functions of spermatogenic cells, Sertoli cells, and interstitial cells.
5 Describe the composition of seminal fluid and its origins.
6 Define and explain emission, ejaculation, and orgasm.
7 Describe in general terms the synthesis of sex hormones and their origins and interrelationships.
8 How do releasing hormones from the hypothalamus and gonadotropins regulate male sex hormone levels? What feedback mechanisms assist homeostasis of male sex hormone levels?
9 Do males form female sex hormones and vice versa? If so, how? Name the male and female sex hormones.
10 Discuss the role of testosterone in the development of male characteristics and their relationship to the changes which occur at puberty. What prevents these changes before puberty?
11 List the major properties of the androgens.
12 What is the relationship of the adrenal cortex and adrenal steroids to sex hormones?
13 Compare the origins of female with male reproductive organs.
14 Describe the anatomical relationships between the primary and accessory female sex organs.
15 Describe the production of an egg and polar bodies by meiosis.
16 Describe the formation of a follicle and its relationship to the corpus luteum.
17 What is the uterus? Illustrate with a sketch and describe the functional relationships among the uterus, oviducts, and vagina.
18 Describe the changes in the ovaries and uterus that occur as an egg is released from a follicle and unites with a sperm in the oviduct.
19 Describe the phases and changes that occur during orgasm in the female.
20 Describe the effects of LH and FSH on the ovaries.
21 Name the hormones that originate from the corpus luteum and describe their interactions with the hypothalamus, anterior pituitary, and uterus.
22 What are the three major phases of the female's reproductive cycle and what general changes occur in the uterus and ovaries during the cycle?
23 What changes signal the fertile period in the sexually mature female?
24 If Janet Bee's menstrual cycle varied between 26 and 29 days during the last year, during what time of each cycle is her fertile period? During what time of each cycle is her "unsafe" period?
25 How do hormones influence the mammary glands and lactation?
26 What is conception? Where does it take place? Why is the conception of a specific individual an event that defies tremendous odds?
27 What are the blastocyst, the placenta, the amnion, and the chorion? Describe their formation and functions.

28 What are the bases of radioimmunoassays, other immunological tests, and mouse, frog, and rabbit tests for pregnancy?
29 What is amniocentesis? Why is it used? What are its risks?
30 What factors are required for fertility in males and females and what are some causes of sterility?
31 Discuss the relative effectiveness of various birth control methods, including a description of the rhythm method. How do the combination pill, minipill, and morning-after pill differ? What side effects and risks do they entail?
32 Is surgical sterilization a foolproof method of birth control? Compare the side effects, advantages, and disadvantages of vasectomy and tubal ligation.
33 What is abortion? Describe five methods used to induce abortion. What are the side effects of abortion? Why may the IUD and morning-after pill be considered abortion-inducing?
34 What causes labor and childbirth? When does labor begin? Outline the major changes that occur during the various stages of labor. What is the afterbirth?

APPENDIX: NORMAL REFERENCE LABORATORY VALUES*

Blood, plasma, or serum values

Determination	Reference range: Conventional	Reference range: SI	Method
Acetoacetate plus acetone	0.3–2.0 mg/100 mL	3–20 mg/L	Behre: *J Lab Clin Med* **13:**770, 1928 (modified)
Aldolase	1.3–8.2 mU/mL	12–75 nmol·s^{-1}/L	Beisenherz, et al.: *Z Naturforsch* **8b:**555 (1953)
Alpha amino nitrogen	3.0–5.5 mg/100 mL	2.1–3.9 mmol/L	Szentirmai, et al.: *Clin Chim Acta* **7:**459 (1962)
Ammonia	80–110 μg/100 mL	47–65 μmol/L	Seligson, Hirahara: *J Lab Clin Med* **49:**962 (1957)
Amylase	4–25 U/mL	4–25 arb. unit	Huggins, Russell: *Ann Surg* **128:**668 (1948)
Ascorbic acid	0.4–1.5 mg/100 mL	23–85 μmol/L	Roe, Kuether: *J Biol Chem* **147:**399 (1943)
Barbiturate	0 Coma level: phenobarbital, approximately 10 mg/100 mL; most other drugs, 1–3 mg per 100 mL	0 μmol/L	Goldbaum: *Anal Chem* **24:**1604 (1952)
Bilirubin (van den Bergh test)	One minute: 0.4 mg/100 mL Direct: 0.4 mg/100 mL. Total: 1.0 mg/100 mL. Indirect is total minus direct	up to 7 μmol/L up to 17 μmol/L	Malloy, Evelyn: *J Biol Chem* **119:**481 (1937)
Blood volume	8.5–9.0% of body weight in kg	80–85 mL/kg	Isotope dilution technic with ^{131}I albumin
Bromide	0 Toxic level: 17 meq/L	0 mmol/L	Adapted from Wuth: *JAMA* **88:**2013 (1927)
Bromsulfalein (BSP)	Less than 5% retention	<0.05 L	Goebler: *Am J Clin Pathol* **15:**452 (1945)
Calcium	8.5–10.5 mg/100 mL (slightly higher in children)	2.1–2.6 mmol/L	Bett, Fraser: *Clin Chim Acta* **4:**346 (1959). Kessler, Wolfman: *Clin Chem* **10:**686 (1964) (modified)

* Abbreviations used: SI, Système international d'Unités; d = 24 h; P = plasma; S = serum; B = blood; U = urine; L = liter; h = hour; s = second

Printed by permission from *The New England Journal of Medicine* **302:**37–48 (1980).

Blood, plasma, or serum values (continued)

Determination	Reference range: Conventional	Reference range: SI	Method
Carbon dioxide content	24–30 meq/L 20–26 meq/L in infants (as HCO_3^-)	24–30 mmol/L	Van Slyke, Neill: *J Biol Chem* **61**:253 (1924) Tech AutoAnalyzer Meth
Carbon monoxide	Symptoms with over 20% saturation	0 (1)	Bruchner, Desmond: *Clin Chim Acta* **3**:173 (1958)
Carotenoids	0.8–4.0 μg/mL	1.5–7.4 μmol/L	Natelson: *Microtechniques of Clinical Chemistry*, 2d ed, 1961, p 454
Ceruloplasmin	27–37 mg/100 mL	1.8–2.5 μmol/L	Ravin: *J Lab Clin Med* **58**:161 (1961)
Chloride	100–106 meq/L	100–106 mmol/L	Modification of Schales: *J Biol Chem* **140**:879 (1941) Tech AutoAnalyzer Meth
Cholinesterase (pseudocholinesterase)	0.5 pH U or more/h 0.7 pH U or more/h for packed cells	0.5 or more arb. unit	Michel: *J Lab Clin Med* **34**:1564 (1949)
Copper	Total: 100–200 μg/100 mL	16–31 μmol/L	MGH Methodology; atomic absorption
Creatine phosphokinase (CPK)	Female 5–35 mU/mL Male 5–55 mU/mL	0.08–0.58 $\mu mol \cdot s^{-1}/L$	Rosalki: *J Lab Clin Med* **69**:696 (1967) (modified)
Creatinine	0.6–1.5 mg/100 mL	60–130 μmol/L	Fabriny and Ertingshausen: *Clin Chem* **17**:696 (1971)
Doriden (glutethimide)	0 mg/100 mL	0 μmol/L	Rieder, Zervas: *Am J Clin Pathol* **44**:520 (1965)
Ethanol	0.3–0.4%, marked intoxication; 0.4–0.5% alcoholic stupor; 0.5% or over, alcoholic coma	65–87 mmol/L 87–109 mmol/L >109 mmol/L	Natelson: *Microtechniques of Clinical Chemistry* 2d ed, 1961, p 208
Glucose	Fasting: 70–110 mg/100 mL	3.9–5.6 mmol/L	Suguira and Huano: *Clin Chim Acta* **75**:387 (1977) Stein in Bergmeyer: *Methods of Enz Anal* p 117 (1965)
Iron	50–150 μg/100 mL (higher in males)	9.0–26.9 μmol/L	Tech AutoAnalyzer Meth (modified)
Iron-binding capacity	250–410 μg/100 mL	44.8–73.4 μmol/L	Scalata, Moore: *Clin Chem* **8**:360 (1962) Tech AutoAnalyzer Meth
Lactic acid	0.6–1.8 meq/L	0.6–1.8 mmol/L	Hadjivassiliou, Rieder: *Clin Chim Acta* **19**:357 (1968)
Lactic dehydrogenase	60–120 U/mL	1.00–2.00 $\mu mol \cdot s^{-1}/L$	Wacker et al.: *N Engl J Med* **255**:449 (1956)
Lead	50 μg/100 mL or less	up to 2.4 μmol/L	Berman: *Atom Absorp Newsl* **3**:9 (1964 modified)
Lipase	2U/mL or less	up to 2 arb. unit	Comfort, Osterberg: *J Lab Clin Med* **20**:271 (1934)
Lipids			
Cholesterol	120–220 mg/100 mL	3.10–5.69 mmol/L	Tech AutoAnalyzer Meth
Cholesterol esters	60–75% of cholesterol		Creech, Sewell: *Anal Biochem* **3**:119 (1962) Tech AutoAnalyzer Meth

Blood, plasma, or serum values (continued)

Determination	REFERENCE RANGE		Method
	Conventional	*SI*	
Phospholipids	9–16 mg/100 mL as lipid phosphorus	2.9–5.2 mmol/L	Fiske, SubbaRow: *J Biol Chem* **66:**2 (1925)
Total fatty acids	190–420 mg/100 mL	1.9–4.2 g/L	Stoddard, Drury: *J Biol Chem* **84:**741 (1929)
Total lipids	450–1000 mg/100 mL	4.5–10.0 g/L	Freedman: *Clin Chim Acta* **19:**291 (1968)
Triglycerides	40–150 mg/100 mL	0.4–1.5 g/L	Tech AutoAnalyzer Meth
Lipoprotein electrophoresis (LEP)			Lees, Hatch: *J Lab Clin Med* **61:**518 (1963)
Lithium	Toxic level 2 meq/L	2 mmol/L	Flame Photometry
Magnesium	1.5–2.0 meq/L	0.8–1.3 mmol/L	Willis: *Clin Chem* **11:**251 (1965) (modified)
Methanol	0		Natelson: *Microtechniques of Clinical Chemistry* 2d ed, 1961, p 298
5′Nucleotidase	0.3–3.2 Bodansky U	30–290 nmol·s^{-1}/L	Rieder, Otero: *Clin Chem* **8:**727 (1969)
Osmolality	285–295 mOsm/kg water	285–295 mmol/kg	Crawford, Nicosia: *J Lab Clin Med* **40:**907, 1952
Oxygen saturation (arterial)	96–100%	0.96–1.00 L	Gordy, Drabkin: *J Biol Chem* **227:**285 (1957)
Pco_2	35–45 mmHg	4.7–6.0 kPa	By CO_2 electrode
pH	7.35–7.45	same	Glass electrode
Po_2	75–100 mmHg (dependent on age) while breathing room air Above 500 mmHg while on 100% O_2	10.0–13.3 kPa	Oxygen electrode
Phenylalanine	0–2 mg/100 mL	0–120 μmol/L	Cullay, et al.: *Clin Chem* **8:**266 (1962 modified)
Phenytoin (Dilantin)	Therapeutic level, 5–20 μg/mL	19.8–79.5 μmol/L	Gas-liquid chromatography
Phosphatase (acid)	Male—Total: 0.13–0.63 Sigma U/mL Female—Total: 0.01–0.56 Sigma U/mL Prostatic: 0–0.7 Fishman-Lerner U/100 mL	36–175 nmol·s^{-1}/L 2.8–156 nmol·s^{-1}/L	Bessey et al.: *J Biol Chem* **164:**321 (1946) Babson et al.: *Clin Chim Acta* **13:**264 (1966)
Phosphatase (alkaline)	13–39 IU/L; infants and adolescents up to 104 IU/L	0.22–0.65 μmol·s^{-1}/L up to 1.26 μmol·s^{-1}/L	Bessey et al.: *J. Biol Chem* **164:**321 (1946)
Phosphorus (inorganic)	3.0–4.5 mg/100 mL (infants in 1st year up to 6.0 mg/100 mL)	1.0–1.5 mmol/L	Fiske, SubbaRow: *J Biol Chem* **66:**375 (1925) Adapted for Tech AutoAnalyzer Meth
Potassium	3.5–5.0 meq/L	3.5–5.0 mmol/L	Flame photometry
Primidone (Mysoline)	Therapeutic level 4–12 μg/mL	18–55 μmol/L	Gas-liquid chromatography
Protein: Total	6.0–8.4 g/100 mL	60–84 g/L	Refractometry (American Optical Co)
Albumin	3.5–5.0 g/100 mL	35–50 g/L	Doumas et al.: *Clin Chim Acta* **31:**87 (1971)

Blood, plasma, or serum values (continued)

Determination	Reference Range: Conventional	Reference Range: SI	Method
Globulin	2.3–3.5 g/100 mL	23–35 g/L	Gornall et al.: *J Biol Chem* **177:** 751 (1949 modified)
Electrophoresis	% of total protein		Kunkel, Tiselius: *J Gen Physiol* **35:**89 (1951)
Albumin	52–68	0.52–0.68 L	
Globulin:			Durrum: *J Am Chem Soc* **72:**2943 (1950)
$Alpha_1$	4.2–7.2	0.042–0.072 L	
$Alpha_2$	6.8–12	0.068–0.12 L	
Beta	9.3–15	0.093–0.15 L	
Gamma	13–23	0.13–0.23 L	
Pyruvic acid	0–0.11 meq/L	0–0.11 mmol/L	Hadjivassiliou, Rieder: *Clin Chim Acta* **19:**357 (1968)
Quinidine	Therapeutic: 1.5–3 μg/mL Toxic: 5–6 μg/ml	4.6–9.2 μmol/L 15.4–18.5 μmol/L	Fluorometry after extraction
Salicylate:	0		Keller: *Am J Clin Pathol* **17:**415 (1947)
Therapeutic	20–25 mg/100 mL; 25–30 mg/100 mL to age 10 yrs. 3 h post dose	1.4–1.8 mmol/L 1.8–2.2 mmol/L	
Toxic	Over 30 mg/100 mL over 20 mg/100 mL after age 60	over 2.2 mmol/L over 1.4 mmol/L	
Sodium	135–145 meq/L	135–145 mmol/L	Flame photometry
Sulfate	0.5–1.5 mg/100 mL	0.05–1.2 mmol/L	Letonoff, Reinhold: *J Biol Chem* **114:**147 (1936)
Sulfonamide	0 mg/100 mL Therapeutic: 5–15 mg 100/mL	0 mmol/L	Bratton, Marshall: *J Biol Chem* **128:**537 (1939)
Thymol:			
Flocculation	Up to 1+ in 24 h	up to 1+ arb. unit	Maclagen: *Nature* **154:**670 (1944)
Turbidity	0–4 U	0–4 arb. unit	
Transaminase (SGOT) (aspartate aminotransferase)	10–40 U/mL	0.08–0.32 $\mu mol \cdot s^{-1}/L$	Karmen et al.: *J Clin Invest* **34:**126 (1955)
Urea nitrogen (BUN)	8–25 mg/100 mL	2.9–8.9 mmol/L	Bretaudiere et al.: *Clin Chem* **22:**1614 (1976) Weatherburn: *Anal Chem* **39:** 901 (1967)
Uric acid	3.0–7.0 mg/100 mL	0.18–0.42 mmol/L	Practorius: *Scand J Clin Lab Invest* **1:**22 (1949)
Vitamin A	0.15–0.6 μg/mL	0.5–2.1 μmol/L	Natelson: *Microtechniques of Clinical Chemistry* 2d ed, 1961, p 451
Vitamin A tolerance test	Rise to twice fasting level in 3 to 5 h		Josephs: *Bull Johns Hopkins Hosp* **65:**112 (1939)

Hematologic values

Determination	Reference Range: Conventional	Reference Range: SI	Method
Coagulation factors:			
Factor I (fibrinogen)	0.15–0.35 g/100 mL	4.0–10.0 μmol/1	Ratnoff, Menzies: *J Lab Clin Med* **37:**316 (1951)
Factor II (prothrombin)	60–140%	0.60–1.40	Owren, Aas: *Scand J Clin Lab Invest* **3:**201 (1951)

Hematologic values (continued)

Determination	Reference Range: Conventional	Reference Range: SI	Method
Factor V (acceleratory globulin)	60–140%	0.60–1.40	Lewis, Ware: *Proc Soc Exp Biol Med* **84**:640 (1953)
Factor VII-X (proconvertin-Stuart)	70–130%	0.70–1.30	Same as factor II
Factor X (Stuart factor)	70–130%	0.70–1.30	Bachman et al.: *Thromb Diath Haemorrh* **2**:29 (1958)
Factor VIII (antihemophilic globulin)	50–200%	0.50–2.0	Tocantins, Kazal: *Blood Coag, Hemorrh Thrombosis* 2d ed, 1964
Factor IX (plasma thromboplastic cofactor)	60–140%	0.60–1.40	*Idem:*
Factor XI (plasma thromboplastic antecedent)	60–140%	0.60–1.40	*Idem:*
Factor XII (Hageman factor)	60–140%	0.60–1.40	*Idem:*
Coagulation screening tests:			
Bleeding time (Simplate)	3–9 min	180–540 s	Simplate Bleeding Time Device (General Diagnostics)
Prothrombin time	Less than 2-s deviation from control	Less than 2-s deviation from control	Colman et al.: *Am J Clin Pathol* **64**:108 (1975)
Partial thromboplastin time (activated)	25–37 s	25–37 s	Babson, Babson: *Am J Clin Pathol* **62**:856 (1974)
Whole-blood clot lysis	No clot lysis in 24 h	0/d	Page and Culver: *Syllabus Lab Exam Clinical Diagnosis* 1960, p 207
Fibrinolytic studies:			
Euglobin lysis	No lysis in 2 h	0 (in 2 h)	Sherry et al.: *J Clin Invest* **38**:810 (1959)
Fibrinogen split products:	Negative reaction at greater than 1:4 dilution	0 (at >1:4 dilution)	Carvalho: *Am J Clin Pathol* **62**:107 (1974)
Thrombin time	control ±5 s	control ±5 s	Stefanini, Dameshek: *Hemorrhage Disorders* 1962, p 492
"Complete" blood count:			
Hematocrit	Male: 45–52% Female: 37–48%	Male: 0.42–0.52 Female: 0.37–0.48	
Hemoglobin	Male: 13–18 g/100 mL Female: 12–16 g/100 mL	Male: 8.1–11.2 mmol/L Female: 7.4–9.9 mmol/l	
Leukocyte count	4300–10,800/mm^3	4.3–10.8 × 10^9/L	
Erythrocyte count	4.2–5.9 million/mm^3	4.2–5.9 × 10^{12}/L	
Mean corpuscular volume (MCV)	80–94 μm^3	80–94 fl	
Mean corpuscular hemoglobin (MCH)	27–32 pg	1.7–2.0 fmol	
Mean corpuscular hemoglobin concentration (MCHC)	32–36%	19–22.8 mmol/L	
Erythrocyte sedimentation rate	Male: 1–13 mm/h Female: 1–20 mm/h	Male: 1–13 mm/h Female: 1–20 mm/h	Modified Westergren method. Gambino et al.: *Am J Clin Pathol* **35**:173 (1965)

Hematologic values (continued)

Determination	Reference Range: Conventional	Reference Range: SI	Method
Erythrocyte enzymes:			
Glucose-6-phosphate dehydrogenase	5–15 U/g Hb	5–15 U/g	Beck: *J Biol Chem* **232:**251 (1958)
Pyruvate kinase	13–17 U/g Hb	13–17 U/g	Beutler, *Red Cell Metabolism,* 2d ed. 1975, p 60
Ferritin (serum)			
Iron deficiency	0–20 ng/ml	0–20 μg/L	Addison et al: *J Clin Pathol* **25:**326 (1972)
Iron excess	Greater than 400 ng/L	>400 μg/L	
Folic acid			
Normal	Greater than 1.9 ng/mL	>4.3 mmol/L	Waxman, Schreiber: *Blood* **42:** 281 (1973)
Borderline	1.0–1.9 ng/mL	2.3–4.3 mmol/L	
Haptoglobin	100–300 mg/100 mL	1.0–3.0 g/L	Behring Diagnostic Reagent Kit
Hemoglobin studies:			
Electrophoresis for abnormal hemoglobin			Singer: *Am J Med* **18:**633 (1955)
Electrophoresis for A_2 hemoglobin	1.5–3.5%	0.015–0.035	Abraham: *Hemoglobin* **1:**27 (1976)
Hemoglobin F (fetal hemoglobin)	Less than 2%	<0.02	Maile: *Lab Med-Hemat.* 2d ed, 1962, p 845
Hemoglobin, met- and sulf-	0	0	Michel, Harris: *J Lab Clin Med* **25:**445 (1940)
Serum hemoglobin	2–3 mg/100 mL	1.2–1.9 μmol/L	Hunter et al.: *Am J Clin Pathol* **20:**429 (1950)
Thermolabile hemoglobin	0	0	Dacie et al.: *Br J Haematol* **10:**388 (1964)
Lupus anticoagulant	0	0	Boxer et al.: *Arthritis Rheum* **19:**1244 (1976)
L.E. (lupus erythematosus) preparation:			
Method I	0	0	Hargraves et al.: *Proc Staff Meet Mayo Clin* **24:**234 (1949)
Method II	0	0	Barnes et al.: *J Invest Dermatol* **14:**397 (1950)
Leukocyte alkaline phosphatase:			
Quantitative method	15–40 mg of phosphorus liberated/h/10^{10} cells	15–40 mg/h	Valentine, Beck: *J Lab Clin Med* **38:**39 (1951)
Qualitative method	Males: 33–188 U Females (off contraceptive pill): 30–160 U	33–188 U 30–160 U	Kaplow: *Am J Clin Pathol* **39:**439 (1963)
Muramidase	Serum, 3–7 μg/mL Urine, 0–2 μg/mL	3–7 mg/L 0–2 mg/L	Osserman, Lawlor: *J Exp Med* **124:**921 (1966)
Osmotic fragility of erythrocytes	Increased if hemolysis occurs in over 0.5% NaCl; decreased if hemolysis is incomplete in 0.3% NaCl		Beutler, in Williams et al., eds. *Hematology,* McGraw-Hill, 1972, p 1375

Hematologic values (continued)

Determination	Reference range: Conventional	Reference range: SI	Method
Peroxide hemolysis	Less than 10%	<0.10	Gordon et al.: *Am J Dis Child* **90:**669 (1955)
Platelet count	150,000–350,000/mm^3	150–350 × 10^9/L	(Hand count): Brecher et al.: *Am J Clin Pathol* **23:**15 (1955)
Platelet function tests:			
Clot retraction	50–100%/2 h	0.50–1.00/2 h	Benthaus: *Thromb Diath Haemorrh* **3:**311 (1959)
Platelet aggregation	Full response to ADP, epinephrine and collagen	1.0	Born: *Nature* **194:**927 (1962)
Platelet factor 3	33–57 s	33–57 s	Rabiner, Hrodek: *J Clin Invest* **47:**901 (1968)
Reticulocyte count	0.5–1.5% red cells	0.005–.015	Brecher: *Am J Clin Pathol* **19:**895 (1949)
Vitamin B_{12}	90–280 pg/mL (borderline: 70–90)	66–207 pmol/L (borderline: 52–66)	*Difco Manual,* 9th ed, 1953, p 221 (modified)

Cerebrospinal-fluid values

Determination	Reference range: Conventional	Reference range: SI	Method
Bilirubin	0	0 μmol/L	*See* blood bilirubin (adapted)
Cell count	0–5 mononuclear cells		
Chloride	120–130 meq/L		*See* blood chloride
Colloidal gold	0000000000 to 0001222111	Same readings in 10 tubes of doubling dilutions of CSF in which precipitation of gold by protein changes a red solution to a deep blue one and finally to a colorless precipitate; rated on a scale of 0 to 5.	Wuth, Faupel: *Bull Johns Hopkins Hosp* **40:**297 (1927)
Albumin	Mean: 29.5 mg/100 mL ±2 SD: 11–48 mg/100 mL	0.295 g/L ±2 SD: 0.11–0.48	Mancini et al.: *Immunochem* **2:**235 (1965)
IgG	Mean: 4.3 mg/100 mL ±2 SD: 0–8.6 mg/100 mL	0.043 g/L ±2 SD: 0–0.086	
Glucose	50–75 mg/100 mL	2.8–4.2 mmol/L	*See* blood glucose
Pressure (initial)	70–180 mm of water	70–180 arb. u.	
Protein:			
Lumbar	15–45 mg/100 mL	0.15–0.45 g/L	Meulmans: *Clin Chim Acta* **5:**757 (1960)
Cisternal	15–25 mg/100 mL	0.15–0.25 g/L	
Ventricular	5–15 mg/100 mL	0.05–0.15 g/L	

Urine values

Determination	REFERENCE RANGE Conventional	SI	Method
Acetone plus acetoacetate (quantitative)	0	0 mg/L	Behre: *J Lab Clin Med* **13:**770 (1928)
Alpha amino nitrogen	64–199 mg/d; not over 1.5% of total nitrogen	4.6–14.2 mmol/d	Hamilton, Van Slyke: *J Biol Chem* **150:**231 (1943)
Amylase	24–76 U/ml	24–76 arb. unit	Huggins, Russell: *Ann Surg* **128:**668 (1948)
Calcium	150 mg/d or less	3.8 or less mmol/d	Atomic absorption
Catecholamines	Epinephrine: under 20 μg/d Norepinephrine: under 100 μg/d	<55 nmol/d <590 nmol/d	DuToit: *WADC Tech Report no. 59-175,* 1959
Chorionic gonadotropin	0	0 arb. unit	Immunologic technic
Copper	0–100 μg/d	0–1.6 μmol/d	MGH methodology Atomic absorption
Coproporphyrin	50–250 μg/d	80–380 nmol/d	Schwartz: *J Lab Clin Med* **37:**843 (1951)
	Children under 80 lb 0–75 μg/d	0–115 nmol/d	With: *Scand J Clin Lab Invest* **7:**193 (1955)
Creatine	Under 100 mg/d or less than 6% of creatinine. In pregnancy: up to 12%. In children under 1 yr.: may equal creatinine. In older children: up to 30% of creatinine	<0.75 mmol/d	Folin: *Lab Manual Biol Chem,* 5th ed, 1933, p 163
Creatinine	15–25 mg/kg of body weight/d	0.13–0.22 mmol·kg^{-1}/d	*Idem:*
Creatinine clearance	150–180 L/d (104–125 mL/min) per 1.73 m^2 of body surface	1.7–2.1 mL/s	Brod, Sirota: *J Clin Invest* **27:**645 (1948)
Cystine or cysteine	0	0	Hawk et al.: *Practical Physiological Chemistry* 13th ed, 1954, p 141
Follicle-stimulating hormone:			Radioimmunoassay
Follicular phase	5–20 IU/d	same	
Mid-cycle	15–60 IU/d		
Luteal phase	5–15 IU/d		
Menopausal	50–100 IU/d		
Men	5–25 IU/d		
Hemoglobin and myoglobin	0		Spectroscopy
Homogentistic acid	0		Neuberger: *Biochem J* **41:**431 (1947)
5-Hydroxyindole acetic acid	2–9 mg/d (women lower than men)	10–45 μmol/d	Sjoerdsma et al.: *JAMA* **159:** 397 (1955)
Lead	0.08 μg/ml or 120 μg or less/d	0.39 μmol/L or less	Willis: *Anal Chem* **34:**614 (1962) (modified)
Phenolsulfonphthalein (PSP)	At least 25% excreted by 15 min; 40% by 30 min; 60% by 120 min	0.25 L	Chapman: *N Engl J Med* **214:** 16 (1936)
Phenylpyruvic acid	0	0	Penrose, Quastel: *Biochem J* **31:**266 (1937)
Phosphorus (inorganic)	Varies with intake; average 1 g/d	32 mmol/d	Tech AutoAnalyzer Meth
Porphobilinogen	0	0	Watson, Schwartz: *Proc Soc Exp Biol Med* **47:**393 (1941)

Urine values (continued)

Determination	REFERENCE RANGE Conventional			SI		Method
Protein:						
Quantitative	<150 mg/24 h			<0.15 g/d		Meulmans: *Clin Chim Acta* **5:**757 (1951)
Electrophoresis						*See* blood protein
Steroids:						
17-Ketosteroids (per day)	Age	Male	Female	*μmol/d*	*μmol/d*	Vestergaard: *Acta Endocrinol* **8:** 193 (1951)
	10	1–4 mg	1–4 mg	3–14	3–14	Normal values taken from Hamburger: *Acta Endocrinol* **1:**19 (1948)
	20	6–21	4–16	21–73	14–56	
	30	8–26	4–14	28–90	14–49	
	50	5–18	3–9	17–62	10–31	
	70	2–10	1–7	7–35	3–24	
17-Hydroxysteroids	3–8 mg/d (women lower than men)			8–22 μmol/d as hydrocortisone		Epstein: *Clin Chim Acta* **7:**735 (1962)
Sugar:						
Quantitative glucose	0			0 mmol/L		Slein, in Bergmeyer: *Methods of Enzyme Analysis,* p 117, 1965
Identification of reducing substances						
Fructose	0			0 mmol/L		Roe et al: *J Biol Chem* **178:**839 (1948)
Pentose	0			0 mmol/L		Roe, Rice: *J Biol Chem* **173:**507 (1948)
Titratable acidity	20–40 meq/d			20–40 mmol/d		Henderson, Palmer: *J Biol Chem* **17:**305 (1914)
Urobilinogen	Up to 1.0 Ehrlich U			to 1.0 arb. unit		Watson et al.: *Am J Clin Pathol* **15:**605 (1944)
Uroporphyrin	0			0 nmol/d		Schwartz et al.: *Proc Soc Exp Biol Med* **79:**463 (1952)
Vanilmandelic acid (VMA)	Up to 9 mg/24 h			up to 45 μmol/d		Pisano et al.: *Clin Chim Acta* **7:**285 (1962)

Special endocrine tests

Determination	REFERENCE RANGE Conventional	SI	Method
Steroid hormones			
Aldosterone	Excretion: 5–19 μg/24 h	14–53 nmol/d	Bayard et al.: *J Clin Endocrinol* **31:**507 (1970)
	Supine: 48 ± 29 pg/ml	133 ± 80 pmol/L	
	Fasting, at rest, 210 meq sodium diet		Poulson et al.: *Clin Immunol Immunopathol* **2:**373 (1974)
	Upright: (2 h) 65 ± 23 pg/mL	180 ± 64 pmol/L	
	Upright, 2h, 210 meq sodium diet		
	Supine: 107 ± 45 pg/mL	279 ± 125 pmol/L	
	Fasting, at rest, 110 meq sodium diet		
	Upright: (2 h) 239 ± 123 pg/mL	663 ± 341 pmol/L	
	Upright, 2h, 110 meq sodium diet		
	Supine: 175 ± 75 pg/mL	485 ± 208 pmol/L	
	Fasting, at rest, 10 meq sodium diet		
	Upright: (2 h) 532 ± 228 pg/mL	1476 ± 632 pmol/L	
	Upright, 2h, 10 meq sodium diet		

Special endocrine tests (continued)

	REFERENCE RANGE		
Determination	*Conventional*	*SI*	*Method*
Cortisol	8 a.m. (fasting): 5–25μg/100mL	0.14–0.69 μmol/L	Murphy: *J Clin Endocr* **27:**973 (1967)
	8 p.m. (at rest): Below 10 μg/100 mL	0–0.28 μmol/L	
	4 h ACTH test: 30–45 μg/100 mL	0.83–1.24 μmol/L	
	Overnight suppression test: Below 5 μg/100 mL	<0.14 nmol/L	
	Excretion: 20–70 μg/24 h	55–193 nmol/d	
11-Deoxycortisol	Responsive: Over 7.5 μg/100 mL	>0.22 μmol/L	Kliman: *Advan Tracer Meth* **4:**227 (1968)
Testosterone	Adult male: 300–1100 ng/100 mL	10.4–38.1 nmol/L	Chen et al.: *Clin Chem* **17:**581 (1971)
	Adolescent male: Over 100 ng/100 mL	>3.5 nmol/L	
	Female: 25–90 ng/100 mL	0.87–3.12 nmol/L	
Unbound testosterone	Adult male: 3.06–24.0 ng/100 mL	106–832 pmol/L	Forest et al: *Steroids* **12:**323 (1968)
	Adult female: 0.09–1.28 ng/100 mL	3.1–44.4 pmol/L	
Polypeptide hormones			
Adrenocorticotropin (ACTH)	15–70 pg/mL	3.3–15.4 pmol/L	Ratcliffe, Edwards: *Radioimmunoassay Methods,* Churchill Livingstone, Edinburgh, 1971
Calcitonin	Undetectable in normals	0	Deftos et al.: *Metabolism* **20:**1129 (1971)
	>100 pg/mL in medullary carcinoma	>29.3 pmol/L	Deftos et al.: *Metabolism* **20:**428 (1971)
Growth hormone	Below 5 ng/mL	<233 pmol/L	Glick et al.: *Nature* **199:**784 (1963)
		Fasting, at rest	
	Children: Over 10 ng/mL	>465 pmol/L	
		After exercise	
	Male: Below 5 ng/mL	<233 pmol/L	
		After glucose load	
	Female: Up to 30 ng/mL	0–1395 pmol/L	
	Male: Below 5 ng/mL	<233 pmol/L	
	Female: Below 10 ng/mL	0–465 pmol/L	
Insulin	6–26 μU/mL	43–187 pmol/L	Morgan, Lazarow: *Proc Soc Exp Biol Med* **110:**29 (1962)
		Fasting	
	Below 20 μU/mL	<144 pmol/L	
		During hypoglycemia	
	Up to 150 μU/mL	0–1078 pmol/L	
		After glucose load	
Luteinizing hormone	Male: 6–18 mU/mL	6–18 u/L	Odell et al.: *J Clin Invest* **46:**248 (1967)
	Female: 5–22 mU/mL	5–22 u/L	
		Pre- or postovulatory	
	30–250 mU/mL	30–250 u/L	
		Mid-cycle peak	

Determination	REFERENCE RANGE Conventional	SI	Method
Parathyroid hormone	<10 μL equiv/mL	<10 mL equiv/L	Segre et al.: *J Clin Invest* **51:** 3163 (1972)
Prolactin	2–15 ng/mL	0.08–6.0 nmol/L	Sinha et al.: *J Clin Endocrinol Metab* **36:**509 (1973)
	Normal diet		
Renin activity	Supine: 1.1 ± 0.8 ng/mL/h	0.9 ± 0.6 (nmol/L)h	Haber et al.: *J Clin Endocrinol* **29:**1349 (1969)
	Upright: 1.9 ± 1.7 ng/mL/h	1.5 ± 1.3 (nmol/L)h	
	Low sodium diet		
	Supine: 2.7 ± 1.8 ng/mL/h	2.1 ± 1.4 (nmol/L)h	
	Upright: 6.6 ± 2.5 ng/mL/h	5.1 ± 1.9 (nmol/L)h	
	Diuretics: 10.0 ± 3.7 ng/mL/h	7.7 ± 2.9 (nmol/L)h	
Thyroid hormones			
Thyroid-stimulating-hormone (TSH)	0.5–3.5 μU/mL	0.5–3.5 mU/L	Ridgway et al.: *Clin Invest* **52:**2785 (1973)
Thyroxine-binding globulin capacity	15–25 μg T_4/100 mL	193–322 nmol/L	Levy et al.: *J Clin Endocrinol Metab* **32:**372 (1971)
Total triiodothyronine by radioimmunoassay (T_3)	70–190 ng/100 mL	1.08–2.92 nmol/L	Larsen et al.: *J Clin Invest* **51:** 1939 (1972)
Total thyroxine by RIA (T_4)	4–12 μg/100 mL	52–154 nmol/L	Chopra: *J Clin Endocrinol Metab* **34:**938 (1972)
T_3 resin uptake	25–35%	0.25–0.35	Taybearn et al.: *J Nucl Med* **8:** 739 (1967)
Free thyroxine index (FT_4I)	1–4 ng/100 mL	12.8–51.2 pmol/L	Sarin, Anderson: *Arch Intern Med* **126:**631 (1970)

Miscellaneous values

Determination	REFERENCE RANGE Conventional	SI	Method
Ascorbic acid load test (after 500 mg orally)	0.2–2.0 mg/h in control sample 24–49 mg/h after loading	0.3–3.2 nmol/s 38–77 nmol/s	Harvard Fatigue Labs: *Laboratory Manual,* 1945
Autoantibodies			
Thyroid colloid and microsomal antigens	Absent		Doniach et al.: *Protocol of Autoimmunity,* Lab. Middlesex Medical School, London
Stomach parietal cells	Absent		Doniach et al.: *Protocol of Autoimmunity,* Lab. Middlesex Medical School, London
Smooth muscle Kidney mitochondria	Absent Absent		Doniach et al.: *Clin Exp Immunol* **1:**237 (1966)
Rabbit renal collecting ducts	Absent		Forbes et al.: *Clin Exp Immunol* **26:**426 (1976)
Cytoplasm of ova, theca cells, testicular interstitial cells	Absent		Forbes et al.: *Clin Exp Immunol* **26:**436 (1976)
Skeletal muscle	Absent		Osserman, Weiner: *Ann NY Acad Sci* **124:**730 (1965)

Miscellaneous values (continued)

Determination	REFERENCE RANGE *Conventional*	*SI*	*Method*
Adrenal gland	Absent		Doniach et al.: *Protocol of Autoimmunity,* Lab. Middlesex Medical School, London
Carcinoembryonic antigen (CEA)	0–2.5 ng/mL, 97% healthy nonsmokers	0–2.5 μg/L, 97% healthy nonsmokers	Hansen et al.: *J Clin Res* **19:** 143 (1971)
Chylous fluid			Todd et al.: *Clin Diag* 12th ed. 1953, p 624
Cryoprecipitable proteins	0	0 arb. unit	Barr et al.: *Ann Intern Med* **32:**6 (1950) (modified)
Digitoxin (after medication)	17 ± 6 ng/mL	22 ± 7.8 nmol/L	Smith, Butler, and Haber: *N Engl J Med* **281:**1212 (1969)
Digoxin, medication with:			
0.25 mg/day	1.2 ± 0.4 ng/mL	1.54 ± 0.5 nmol/L	Smith, Haber: *J Clin Invest* **49:**2377 (1970)
0.5 mg/day	1.5 ± 0.4 ng/mL	1.92 ± 0.5 nmol/L	
Duodenal drainage:			
pH	5.5–7.5	5.5–7.5	
Amylase	Over 1200 U/total sample	>1.2 arb. u	Huggins, Russell: *Ann Surg* **128:**668 (1948)
Trypsin	Values from 35 to 160% "normal"	0.35–1.60	Anderson, Early: *Am J Dis Child* **63:**891 (1942)
Viscosity	3 min or less	180 s or less	
Gastric analysis	Basal:		Marks: *Gastroenterology* **41:**599 (1961)
	Females: 2.0 ± 1.8 meq/h	0.6 ± 0.5	
	Males 3.0 ± 2.0 meq/h	0.8 ± 0.6 μmol/s	
	Maximal: (after histalog or gastrin)		
	Females 16 ± 5 meq/h	4.4 ± 1.4 μmol/s	
	Males 23 ± 5 meq/h	6.4 ± 1.4 μmol/s	
Gastrin-I	0–200 pg/mL	0–95 pmol/L	Dent et al.: *Ann Surg* **176:**360 (1972)
Immunologic tests:			
Alpha-fetoglobulin	Abnormal if present		
Alpha 1-Antitrypsin	200–400 mg/100 mL	2.0–4.0 g/L	
Antinuclear antibodies	Positive if detected with serum diluted 1:10		
Anti-DNA antibodies	Less than 15 units/mL		Pincus: *Arthritis Rheum* **14:**623 (1971)
Bence-Jones protein	Abnormal if present		
Complement, total hemolytic	150–250 U/mL		Hook, Muschel: *Proc Soc Exp*
C3	Range 55–120 mg/100 mL	0.55–1.2 g/L	*Biol Med* **117:**292 (1964)
C4	Range 20–50 mg/100 mL	0.2–0.5 g/L	
Immunoglobulins:			
IgG	1140 mg/100 mL	11.4 g/l	
	Range 540–1663	5.5–16.6 g/L	
IgA	214 mg/100 mL	2.14 g/l	
	Range 66–344	0.66–3.44 g/L	
IgM	168 mg/100 mL	1.68 g/l	
	Range 39–290	0.39–2.9 g/L	
Viscosity (of serum to water)	1.4–1.8		Barth: *Serum Proteins and Dysproteinemias,* Sunderman, Sunderman, 1964, p 102
Iontophoresis	Children: 0–40 meq sodium/L.	0–40 mmol/L	Gibson, Cooke: *Pediatrics* **23:** 545 (1959)
	Adults: 0–60 meq sodium/L	0–60 mmol/L	

Miscellaneous values (continued)

Determination	REFERENCE RANGE *Conventional*	*SI*	*Method*
Propranolol (includes bioactive 4-OH metabolite) (Blood sample 4 h after dose)	100–300 ng/mL	386–1158 nmol/L	M.G.H. method of Rockson, Homcy, Haber: by radio-immunoassay
Stool fat	Less than 5 g in 24 h or less than 4% of measured fat intake in 3-d period (24 h or 3 d specimen)	<5 g/d	Jover et al.: *J Lab Clin Med* **59:** 878 (1962)
Stool nitrogen	Less than 2 g/d or 10% of urinary nitrogen (24 h or 3 d specimen)	<2 g/d	Peters, Van Slyke: *Quant Clin Chem* vol 2: Methods, 1932, p 353
Synovial fluid:			
Glucose	Not less than 20 mg/100 mL lower than simultaneously drawn blood sugar	see blood glucose mmol/L	*See* blood glucose
Mucin	Type 1 or 2	1–2 arb. u.	Grade as: Type 1—tight clump Type 2—soft clump Type 3—soft clump breaks up Type 4—cloudy, no clump
D-Xylose absorption	5–8 g/5 h in urine 40 mg per 100 mL in blood 2 h after ingestion of 25 g of D-xylose	33–53 mmol 2.7 mmol/L	Roe, Rice: *J Biol Chem* **173:**507 (1948); for directions see Benson et al.: *N Engl J Med* **256:**335 (1957)

ADDITIONAL READINGS

INTRODUCTORY REFERENCES

CHAPTER 1

Hardy, R. W.: *Homeostasis* Studies in Biology, no. 63, Crane Russak, New York, 1976.

Kennedy, D. (introduction by): *Cellular and Organismal Biology. Readings from Scientific American.* W. H. Freeman & Company, Publishers, San Francisco, 1974.

Langley, L. L. (ed.): *Homeostasis: Origins of the Concept.* Benchmark Papers in Human Physiology Series, Halstead Press, John Wiley, New York, 1973.

CHAPTER 2

Abelson, J., and E. Butz: "Recombinant DNA," *Science* **209:**1317–1438, 1980.

American Society for Testing and Materials: *Metric Practice Guide,* American Society for Testing and Materials, Philadelphia, 1973.

Frieden, E.: "The Chemical Elements of Life," *Scientific American,* July 1972.

Hanawalt, P. C. (introduction): *Molecules to Living Cells, Readings from Scientific American,* W. H. Freeman & Company Publishers, San Francisco, 1980.

Hanawalt, P. C., and R. H. Haynes (introductions): *The Chemical Basis of Life. Readings from Scientific American.* W. H. Freeman & Company, Publishers, San Francisco, 1973.

Hollaway, M. R.: *The Mechanism of Enzyme Action,* Oxford Biology Readers, Oxford University Press, New York, 1978.

Horowitz, N. H., and E. Hutchings, Jr. (Eds.): *Genes, Cells, and Behavior,* W. H. Freeman & Company, Publishers, San Francisco, 1980.

Jackson, R. J.: *Protein Biosynthesis,* Oxford Biology Readers, Oxford University Press, New York, 1978.

Kornberg, A.: *DNA Replication,* W. H. Freeman & Company Publishers, San Francisco, 1980.

Mortimer, C. E.: *Chemistry: A Conceptual Approach.* 4th ed, D. Van Nostrand, New York, 1979.

Phelps, C. F.: *Polysaccharides,* Oxford Biology Readers, Oxford University Press, New York, 1972.

Phillips, D. C., and C. T. North: *Protein Structure,* Oxford Biology Readers, Oxford University Press, New York, 1973.

Ritchie-Calder, Lord: "Conversion to the Metric System," *Scientific American,* July 1970.

Watson, J. D., and J. Tooze: *The DNA Story,* W. H. Freeman & Company Publishers, San Francisco, 1981.

The Cell

Allison, A. C.: *Lysosomes,* 2d ed., Oxford Biology Readers, Oxford University Press, New York, 1977.
Chappell, J. B., and S. C. Rees: *Mitochondria,* Oxford Biology Readers, Oxford University Press, New York, 1972.
Claude, A.: "The Coming Age of the Cell," *Science,* **189**:433–435 (1975).
Cohen, S. N.: "The Manipulation of Genes," *Scientific American,* July 1975.
Cook, G. M. W.: *The Golgi Apparatus,* Oxford Biology Readers, Oxford University Press, New York, 1975.
Dustin, P.: "Microtubules," *Scientific American,* August 1980.
Engelman, D. M., and P. B. Moore: "Neutron-Scattering Studies of the Ribosome," *Scientific American,* October 1976.
Jordan, E. G.: *The Nucleolus,* 2d ed., Oxford Biology Readers, Oxford University Press, New York, 1978.
Kornberg, R. D., and A. Klug: "The Nucleosome," *Scientific American,* February 1981.
Lake, J. A.: "The Ribosome," *Scientific American,* August 1981.
Lucy, J. A.: *The Plasma Membrane,* Oxford Biology Readers, Oxford University Press, New York, 1975.
Maniatas, T., and M. Ptashne: "A DNA Operator-Repressor System," *Scientific American,* January 1976.
Miller, O. L., Jr.: "The Visualization of Genes in Action," *Scientific American,* March 1975.
Porter, K. R., and J. B. Tucker: "The Ground Substance of The Living Cell," *Scientific American,* March 1981.
Rich, A., and S. H. Kim: "The Three-Dimensional Structure of Transfer RNA," *Scientific American,* January 1978.
Stein, G. S., J. S. Stein, and L. J. Kleinsmith: "Chromosomal Proteins and Gene Regulation," *Scientific American,* February 1975.
Travers, A. A.: *"Transcription of DNA,"* Oxford Biology Readers, Oxford University Press, New York, 1975.

CHAPTER 3

Albrecht-Buehler, G.: "The Tracks of Moving Cells," *Scientific American,* April 1978.
Capra, D. and A. B. Edmundson: "Antibody-Combining Site," *Scientific American,* January 1977.
Cooper, M. D., and A. R. Lawton III: "The Development of the Immune System." *Scientific American,* November 1974.
Cunningham, B. A.: "The Structure and Function of Histocompatibility Antigens," *Scientific American,* October 1977.
Gorton, R., and A. G. Jacobson: "The Shaping of Tissues in Embryos," *Scientific American,* June 1978.
Notkins, A. L., and H. Kaprowski: "How the Immune Response to a Virus Can Cause Disease," *Scientific American,* January 1973.
Old, L. J.: "Cancer Immunology," *Scientific American,* May 1977.
Raff, M. C.: "Cell-Surface Immunology," *Scientific American,* May 1976.
Staehelin, L. A., and B. E. Hull: "Junctions between Living Cells," *Scientific American,* May 1978.
Verzar, F.: "The Aging of Collagen," *Scientific American,* April 1963.

CHAPTER 4

Capaldi, R. A.: "A Dynamic Model of Cell Membranes," *Scientific American,* March 1974.
Culliton, B. J.: "Cell Membranes: A New Look at How They Work," *Science* **175**:1348–1350 (1972).
Cunningham, B. A.: "The Structure and Function of Histocompatibility Antigens," *Scientific American,* October 1977.
Edelman, G. M.: "Surface Modulation in Cell Recognition and Cell Growth," *Science* **192**:218–226 (1976).
Fox, C. F.: "The Structure of Cell Membranes," *Scientific American,* February 1972.
Gregor, H. P. and C. D. Gregor.: "Synthetic-Membrane Technology," *Scientific American,* July 1978.
Kolata, G. B.: Cell Surface Protein: No Simple Cancer Mechanisms. *Science* **190**:39–40 (1975).
Lodish, H. F., and J. E. Rothman: "The Assembly of Cell Membranes," *Scientific American,* January 1979.
Marx, J. L.: "Biochemistry of Cancer Cells: Focus on the Cell Surface," *Science,* **183**:1279–1282 (1974).
Raff, M. C.: "Cell-Surface Immunology," *Scientific American,* May 1976.
Satir, B.: "The Final Steps in Secretion," *Scientific American,* October 1975.
Singer, S. J., and G. L. Nicolson: "The Fluid Mosaic Model of the Structure of Cell Membranes," *Science,* **175**:720–731 (1972).
Staehlin, L. A., and B. E. Hull: "Junctions Between Living Cells," *Scientific American,* May 1978.

CHAPTER 5

Bloom, F. E.: "Neuropeptides," *Scientific American,* October 1981.
Cheng, W. Y.: "Calmodulin," *Scientific American,* June 1982.
Guillemin, R., and R. Burgus: "The Hormones of the Hypothalamus," *Scientific American,* November 1972.
Hardy, R. N.: *Endocrine Physiology,* University Park Press, Baltimore, 1982.
Levine, S.: "Stress and Behavior," *Scientific American,* January 1971.
Martin, C. R.: *Textbook of Endocrine Physiology,* Oxford University Press, New York, 1976.
Nathanson, J. A., and P. Greengard: "'Second Messengers' in the Brain," *Scientific American,* August 1977.
Notkins, A. L.: "The Causes of Diabetes," *Scientific American,* November 1979.
O'Malley, B. W., and W. T. Schroder: "The Receptors of Steroid Hormones," *Scientific American,* February 1976.
Pastan, I.: "Cyclic AMP," *Scientific American,* August 1972.
Randle, P. J., and R. M. Denton: *Hormones and Cell Metabolism.* Oxford Biology Readers, Oxford University Press, London, 1974.
Rasmussen, H.: "The Parathyroid Hormone," *Scientific American,* April 1971.
Rubenstein, E.: "Diseases Caused by Impaired Communication Among Cells," *Scientific American,* March 1980.
Wilkens, L.: "The Thyroid Gland," *Scientific American,* March 1960.
Wurtman, R. J.: "The Effects of Light on the Human Body," *Scientific American,* July 1975.
Wurtman, R. J., and J. Axelrod: "The Pineal Gland," *Scientific American,* July 1965.

CHAPTER 6

Adrian, R. H.: *The Nerve Impulse,* Oxford Biology Readers, Oxford University Press, New York, 1974.

Brady, R. O.: "Hereditary Fat Metabolism Diseases," *Scientific American,* August 1973.

Bray, D.: "The Fibrillar Proteins of Nerve Cells," *Endeavour* **120:**131–136 (1976).

Dustin, P.: "Microtubules," *Scientific American,* August 1980.

Fernstrom, J. D., and R. J. Wurtman: "Nutrition and the Brain," *Scientific American,* February 1974.

Jacobson, M., and R. K. Hunt: "The Origins of Nerve Cell Specificity," *Scientific American,* February 1973.

Kandel, E. R.: "Small Systems of Neurons," *Scientific American,* September 1979.

Keynes, R. D.: "Ion Channels in the Nerve Cell Membrane," *Scientific American,* March 1979.

Levi-Montalcani, R., and P. Calissano: "The Nerve Growth Factor," *Scientific American,* June 1979.

Morell, P., and W. T. Norton: "Myelin," *Scientific American,* May 1980.

Patterson, P. H., D. D. Potter, and E. J. Furshpan: "The Chemical Differentiation of Nerve Cells," *Scientific American,* July 1978.

Schwartz, J. H.: "The Transport of Substances in Nerve Cells," *Scientific American,* April 1980.

Stevens, C. F.: "The Neuron," *Scientific American,* September 1979.

Winick, M.: *Early Nutrition and Brain Development,* Oxford Biology Readers. Oxford University Press, New York, 1978.

Wurtman, R. J.: "Nutrients that Modify Brain Function," *Scientific American,* April 1982.

Wurtman, R. J., and J. H. Growdon: "Dietary Enhancement of CNS Neurotransmitters," pp. 59–65, in D. T. Krieger and J. C. Hughes (eds.): *Neuroendocrinology,* Sinauer Associates, Sunderland, Mass., 1980.

CHAPTER 7

Development of Impulses

Adrian, R. H.: *The Nerve Impulse,* 2d ed., Oxford Biology Readers, Oxford University Press, New York, 1980.

Keynes, R. D.: "Ion Channels in the Nerve Cell Membrane," *Scientific American,* March 1979.

Lester, H. A.: "The Response of Acetylcholine," *Scientific American,* February 1978.

Morell. P., and W. T. Norton: "Myelin," *Scientific American,* May 1980.

Sheperd, G. M.: "Microcircuits in the Nervous System," *Scientific American,* February 1978.

Stevens, C. F.: "The Neuron," *Scientific American,* September 1979.

Synapses and Neurotransmitters

Axelrod, J.: "Neurotransmitters," *Scientific American,* June 1974.

Gray, E. G.: *The Synapse,* 2d ed., Oxford Biology Readers, Oxford University Press, New York, 1977.

Llinas, R. R.: "Calcium in Synaptic Transmission," *Scientific American,* October 1982.

CHAPTER 8

General References

Haagen-Smit, A. J.: "Smell and Taste," *Scientific American,* March 1952.
Kalmus, H.: "Inherited Sense Defects," *Scientific American,* May 1952.
Loewenstein, W. R.: "Biological Transducers," *Scientific American,* August 1960.

The Eye

Abrahamson, E. W., and S. E. Ostroy (eds.): *Benchmark Papers in Biochemistry,* vol. 3: *Molecular Processes in Vision,* Academic Press, New York, 1981.
Bizzi, E.: "The Coordination of Eye-Head Movements," *Scientific American,* October 1974.
Glickstein, M., and A. R. Gibson: "Visual Cells in the Pons of the Brain," *Scientific American,* November 1976.
Gregory, R. L.: *Visual Perception,* Oxford Biology Readers, Oxford University Press, New York, 1973.
Hubel, D. H., and T. Wiesel: "Brain Mechanisms of Vision," *Scientific American,* September 1979.
Land, E. H.: "The Retinex Theory of Color Vision," *Scientific American,* December 1977.
Pettigrew, J. D.: "The Neurophysiology of Binocular Vision," *Scientific American,* August 1972.
Regan, D., K. Beverley, and M. Cynader: "The Visual Perception of Motion in Depth," *Scientific American,* July 1979.
Rodieck, R. W.: *The Vertebrate Retina,* W. H. Freeman & Company, Publishers, San Francisco, 1973.
Ross, J.: "The Resources of Binocular Perception," *Scientific American,* March 1976.
Rushton, W. A. H.: "Visual Pigments and Color Blindness," *Scientific American,* March 1975.
Van Heyningen, R.: "What Happens to the Human Lens in Cataract?" *Scientific American,* December 1975.
Weale, R. A.: *The Vertebrate Eye,* Oxford Biology Readers, Oxford University Press, New York, 1978.
Werblin, F. S.: "The Control of Sensitivity in the Retina," *Scientific American,* January 1973.
Wurtz, R. H., M. E. Goldberg, and D. L. Robinson, "Brain Mechanisms of Visual Attention," *Scientific American,* June 1982.
Yellott, J. I., Jr.: "Binocular Depth Inversion," *Scientific American,* July 1981.

The Ear

Friedman, I.: *The Human Ear.* Oxford Biology Readers, Oxford University Press, New York, 1979.
Oster, G.: "Auditory Beats in the Brain," *Scientific American,* October 1973.
Von Bekesy, G.: "The Ear," *Scientific American,* August 1957.

CHAPTER 9

Evarts, E. V.: Brain Mechanisms of Movement, *Scientific American,* September 1979.
Kandel, E. R.: "Small Systems of Neurons," *Scientific American,* September 1979.

Lefkowitz, R. J.: "Adrenergic Receptors: Recognition and Regulation," *N. Engl. J. Med.* **295**(5): (1976).
Lester, H. A.: "The Response To Acetylcholine," *Scientific American*, February 1977.
Merton, P. A.: "How We Control the Contraction of Our Muscles," *Scientific American*, May 1972.

CHAPTER 10

"The Brain," *Scientific American*, September 1979.
Bloom, F. E.: "Neuropeptides," *Scientific American*, October 1981.
Devey, G. B., and P. N. T. Wells: "Ultrasound in Medical Diagnosis," *Scientific American*, May 1978.
Guillemin, R.: "Beta-Lipotropin and Endorphins: Implications of Current Knowledge," p. 67–74. in D. T. Krieger and J. C. Hughes (eds.), *Neuroendocrinology*, Sinauer Associates Inc., Sunderland, Mass., 1980.
Julien, R. M.: *A Primer of Drug Action*, W. H. Freeman & Company, Publishers, San Francisco, 1981.
Lassen, N. A., D. H. Ingvar, and E. Skinh: "Brain Function and Blood Flow," *Scientific American*, October 1978.
Luria, A. R.: "Functional Organization of the Brain," *Scientific American*, March 1970.
Nathanson, J. A., and P. Greengard: "Second Messengers in the Brain," *Scientific American*, August 1977.
Pappenheimer, J. R.: "The Sleep Factor," *Scientific American*, August 1976.
Pykett, I. L.: "NMR Imaging in Medicine, *Scientific American*, May 1982.
Regan, D.: "Electrical Responses Evoked from the Human Brain," *Scientific American*, December 1979.
Routtenberg, A.: "The Reward System of the Brain," *Scientific American*, November 1978.
Rubenstein, E.: "Diseases by Impaired Communication among Cells," *Scientific American*, March 1980.
Siegel, R. K.: "Hallucinations," *Scientific American*, October 1977.
Snyder, S. H.: "Opiate Receptors and Internal Opiates," *Scientific American*, March 1977.
Van Dyke, C., and R. Byck: "Cocaine," *Scientific American*, March 1982.
Wurtman, R. J.: "Nutrients that Modify Brain Function," *Scientific American*, April 1982.
Wurtman, R. J.: "The Effects of Light on the Human Body," *Scientific American*, July 1975.

CHAPTER 11

"The Brain," *Scientific American*, September 1979.
Atkinson, R. C., and R. M. Shiffrin: "The Control of Short-Term Memory, *Scientific American*, August 1971.
Binkley, S. A.: "Timekeeping Enzyme in the Pineal Gland," *Scientific American*, April 1979.
Constantine-Paton, M., and M. I. Law: "The Development of Maps and Stripes in the Brain," *Scientific American*, December 1982.
Fernstrom, J. D., and R. J. Wurtman: "Nutrition and the Brain," *Scientific American*, February 1974.

Gazzaniga, M. S.: "The Split Brain in Man," *Scientific American,* August 1967.
Geschwind, N.: "Specializations of the Human Brain," *Scientific American,* September 1979.
Heller, H. C., L. I. Crawshaw, and H. T. Hammel: "The Thermostat of Vertebrate Animals," *Scientific American,* August 1978.
Lassen, N. A., D. H. Ingvar, and E. Skinhoj: "Brain Function and Blood Flow," *Scientific American,* October 1978.
Llinas, R. R.: "The Cortex of the Cerebellum," *Scientific American,* January 1975.
McEwen, B. S.: "Interactions Between Hormones and Nerve Tissue," *Scientific American,* July 1976.
Olton, D.: "Spatial Memory," *Scientific American,* June 1977.
Pappenheimer, J. R.: "The Sleep Factor," *Scientific American,* August 1976.
Peterson, L. R.: "Short-Term Memory," *Scientific American,* July 1966.
Springer, Sally P.: *Left Brain, Right Brain,* W. H. Freeman & Company, Publishers, San Francisco, 1981.
Waltz, D. L.: "Artificial Intelligence," *Scientific American,* October 1982.
Young, J. Z.: *The Evolution of Memory,* Oxford Carolina Biology Readers, Oxford University Press, New York, 1976.

CHAPTER 12

Cohen, C.: "The Protein Switch of Muscle Contraction," *Scientific American,* November 1975.
Evarts, E. V.: "Brain Mechanisms in Movement," *Scientific American,* July 1973.
Harrington, W. F.: *Muscle Contraction,* Oxford Biology Readers, Oxford University Press, New York, 1981.
Hurwitz, L., D. Fitzpatrick, G. Debbas, and E. J. Landon: "Localization of Calcium Pump Activity in Smooth Muscle," *Science* **179**:384–386 (1973).
Lazarides, E., and J. P. Revel: "The Molecular Basis of Cell Movement," *Scientific American,* May 1979.
Lester, H. A.: "The Response to Acetylcholine," *Scientific American,* February 1977.
Morton, P. A.: "How We Control the Contraction of Our Muscles," *Scientific American,* May 1972.
Murray, J. M., and A. Weber: "The Cooperative Action of Muscle Proteins," *Scientific American,* February 1974.

CHAPTER 13

Buller, A. J.: *The Contractile Behavior of Mammalian Skeletal Muscle,* 2nd ed., Oxford Biology Readers, Oxford University Press, New York, 1980.
Carlson, F. D., and W. R. Wilkie: *Muscle Physiology,* Prentice-Hall, Englewood Cliffs, N.J., 1974.
Carter, L. C.: *Guide to Cellular Energetics,* W. H. Freeman & Company, Publishers, San Francisco, 1973.
Chappell, J. B.: "ATP," *Oxford Biology Readers,* Oxford University Press, New York, 1977.
Cohen, C.: "The Protein Switch of Muscle Contraction," *Scientific American,* November 1975.
Hinkle, P. C., and R. E. McCarthy: "How Cells Make ATP," *Scientific American,* March 1978.
Murray, J. M., and A. Weber: "The Cooperative Action of Muscle Proteins," *Scientific American,* February 1974.

CHAPTER 14

Engel, R. H.: "The Dynamic Role of Vitamin D in Calcium and Phosphate Metabolism. Part I. Biosynthesis and Effects of Vitamin D,"*Lab. Management,* March 1979.

Engel, R. H.: "The Dynamic Role of Vitamin D in Calcium and Phosphate Metabolism. Part II. Vitamin D in Calcium and Phosphate Regulation and the Use of Vitamin D Analogues," *Lab. Management,* April 1979.

National Dairy Council: "Calcium in Bone Health," *Dairy Council Digest,* pp. 31–36, November-December 1976.

Pritchard, J. J.: *Bones,* Oxford University Press, New York, 1974.

Sonstegard, D. A., L. S. Matthews, and H. Kaufer: "The Surgical Replacement of the Human Knee Joint," *Scientific American,* January 1978.

Wurtman, R. J.: "The Effects of Light on the Human Body," *Scientific American,* July 1975.

CHAPTER 15

American National Red Cross: *Cardiopulmonary Resuscitation,* American National Red Cross, Washington, 1974.

Muir, A. R.: *The Mammalian Heart,* Oxford Biology Readers, Oxford University Press, New York, 1971.

Neil, E.: *The Mammalian Circulation,* Oxford Biology Readers, Oxford Unversity Press, New York, 1975.

CHAPTER 16

American Heart Association: *Risk Factors and Coronary Heart Disease,* American Heart Association, New York, 1968.

American Heart Association: *Diet and Coronary Heart Disease,* American Heart Association, New York, 1978.

Benditt, E. P.: "The Origin of Atherosclerosis," *Scientific American,* February 1977.

DeVey, G. B., and P. N. T. Wells: "Ultra-Sound in Medical Diagnosis," *Scientific American,* May 1978.

Glass, D. C.: "Stress, Behavior Pattern and Coronary Disease," *Am. Scientist,* **65**:177–187 (1977).

Reiser, S. J.: "The Medical Influence of the Stethoscope," *Scientific American,* February 1979.

Stallones, R. A.: "The Rise and Fall of Ischemic Heart Disease," *Scientific American,* November 1980.

Zohman, L. R.: *Beyond Diet . . . Exercise Your Way to Fitness and Heart Health,"* Best Foods, CPC International Inc., Englewood Cliffs, N.J., 1979.

CHAPTER 17

Benditt, C. P.: "The Origin of Atherosclerosis," *Scientific American,* February 1977.

Brown, Helen B.: *Current Focus on Fat in the Diet,* The American Dietetic Association. Chicago, 1977.

Fein, J.: "Microvascular Surgery For Stroke," *Scientific American,* April 1978.

Johansen, K.: "Aneurysms." *Scientific American,* July 1982.

Neil, E.: *The Mammalian Circulation,* Oxford University Press, New York, 1975.

CHAPTER 18

Cerami, A., and C. M. Peterson: "Cyanate and Sickle-Cell Disease," *Scientific American,* April 1975.
Cunningham, B. A.: "The Structure and Function of Histocompatibility Antigens," *Scientific American,* October 1977.
Friedman, M. J., and W. Trager: "The Biochemistry of Resistance to Malaria," *Scientific American,* March 1981.
Kolata, G. B.: "Thalassemias: Models of Genetic Disease," *Science* **210**:300–302 (1980).
Maugh, T. H., III: "A New Understanding of Sickle Cell Emerges," *Science* **211**:265–267 (1981).
Perutz, M. F.: "Hemoglobin Structure and Respiratory Transport," *Scientific American,* December 1978.

CHAPTER 19

Buisseret, Paul D.: "Allergy," *Scientific American,* August 1982.
Burke, D. C.: "The Status of Interferon," *Scientific American,* March 1977.
Cooper, M. D., and A. R. Lawton III: "The Development of the Immune System," *Scientific American,* November 1974.
Doolittle, R. F.: "Fibrinogen and Fibrin," *Scientific American,* December 1981.
Leder, P.: "The Genetics of Antibody Diversity," *Scientific American,* May 1982.
Lerner, R. A., and F. J. Dixon: "The Human Lymphocyte as an Experimental Animal," *Scientific American,* June 1973.
Raff, M. C.: "Cell Surface Immunology," *Scientific American,* May 1976.
Sharon, N.: "Lectins," *Scientific American,* June 1977.
Stroud, R. M.: "A Family of Protein-Cutting Proteins," *Scientific American,* July 1974.
Zucker, M. B.: "The Functioning of Blood Platelets," *Scientific American,* June 1980.

CHAPTER 20

American National Red Cross, *First Aid For Foreign Obstruction of the Airway,* 1978.
Avery, M. E., N. S. Wang, and H. W. Taeusch, Jr.: "The Lung of the Newborn Infant," *Scientific American,* April 1973.
Hock, R. J.: "The Physiology of High Altitude," *Scientific American,* February 1970.
Hughes, G. M.: *The Vertebrate Lung,* Oxford Biology Readers, Oxford University Press, New York, 1973.
Naeye, R. L.: "Sudden Infant Death," *Scientific American,* April 1980.
Perutz, M. F.: "Hemoglobin Structure and Respiratory Transport," *Scientific American,* December 1978.

CHAPTER 21

Breathnach, A. S.: *Melanin Pigmentation in the Skin,* Oxford Carolina Biology Readers, Oxford University Press, New York, 1971.
Cannon, P. J.: "The Kidney in Heart Failure," *N. Engl. J. Med.* **296**:26–32 (1977).
Jamison, R. L., and R. H. Maffly: "The Urinary Concentrating Mechanism," *N. Engl. J. Med.* **295**:1059–1067 (1976).

Koffler, D.: "Systemic Lupus Erythematosus," *Scientific American*, July, 1980.
Marples, M. J.: "Life on the Human Skin," *Scientific American*, January 1969.
Moffat, D. B.: *The Control of Water Balance by the Kidney*, 2d ed., Oxford Carolina Readers, Oxford University Press, New York, 1978.
Montagna, W.: "The Skin," *Scientific American*, February 1965.
Rose, N. R.: "Autoimmune Diseases," *Scientific American*, February 1981.
Smith, H. W.: "The Kidney," *Scientific American*, January 1953.

CHAPTER 22

Davenport, H. W.: "Why the Stomach Does Not Digest Itself," *Scientific American*, January 1972.
Kappas, A., and A. P. Alvares: "How the Liver Metabolizes Foreign Substances," *Scientific American*, June 1975.
Lieber, C. S.: "The Metabolism of Alcohol," *Scientific American*, March 1976.
Mayer, J.: "The Dimensions of Human Hunger," *Scientific American*, September 1976.
McMinn, R. M. H.: *The Human Gut*, Oxford Biology Readers, Oxford University Press, New York 1974.
Moog, F.: "Lining of the Small Intestine," *Scientific American*, November 1981.
Pribor, H. C., and T. J. Duello: "Hepatitis. Diagnostic Profiles," *Lab. Management* **19**:8–11 (July 1981).
Sherlock, S. V.: *The Human Liver*, Oxford Biology Readers, Oxford University Press, New York, 1978.

CHAPTER 23

Carter, L. C.: *Guide to Cellular Energetics*, W. H. Freeman & Company, Publishers, San Francisco, 1973.
Dudrick, S. J., and J. E. Rhoades: "Total Intravenous Feeding," *Scientific American*, May 1972.
Fernstrom, J. D., and R. J. Wurtman: "Nutrition and the Brain," *Scientific American*, February 1974.
Frieden, E.: "The Chemical Elements of Life," *Scientific American*, July 1972.
Hanawalt, P. C. (introduction): *Molecules to Living Cells*, Readings from Scientific American. W. H. Freeman & Company Publishers, San Francisco, 1980.
Hollaway, M. R.: *The Mechanisms of Enzyme Action*, Oxford Biology Readers, Oxford University Press, New York, 1976.
Jackson, R. J.: *Protein Biosynthesis*, Oxford Biology Readers, Oxford University Press, New York, 1978.
Kappas, A., and A. P. Alvares: "How the Liver Metabolizes Foreign Substances," *Scientific American*, June 1975.
Kretchmer, N.: "Lactose and Lactase," *Scientific American*, October 1972.
Kretchmer, N., and W. van B. Robertson: *Human Nutrition*, Readings from Scientific American, W. H. Freeman & Company Publishers, San Francisco, 1978.
Moog, F.: "Lining of the Small Intestine," *Scientific American*, November 1981.
Scrimshaw, N. S., and V. R. Young: "Nutrition, The Requirements of Humans," *Scientific American*, September 1976.
Sherlock, S. V.: *The Human Liver*, Oxford Biology Readers, Oxford University Press, New York, 1978.
Young, V. R., and N. S. Scrimshaw: "The Physiology of Starvation," *Scientific American*, October 1971.

CHAPTER 24

Annis, L. F.: *The Child before Birth,* Cornell University Press, Ithaca, N. Y., 1978.
Beaconsfield, P., G. Budwood, and R. Beaconsfield: "The Placenta," *Scientific American,* August 1980.
Edwards, R. G.: *The Beginnings of Human Life,* Oxford Biology Readers, Oxford University Press, New York, 1981.
Epel, D.: "The Program of Fertilization," *Scientific American,* November 1977.
Fuchs, F.: "Genetic Amniocentesis," *Scientific American,* June 1980.
Gordon, R., and A. G. Jacobson: "The Shaping of Tissues in Embryos," *Scientific American,* June 1978.
Grobstein, C.: "External Human Fertilization," *Scientific American,* July 1979.
Hart, G.: *Sexually Transmitted Diseases,* Oxford Biology Readers, Oxford University Press, New York, 1976.
Head, J. J.: *The Biology of Human Reproduction,* J. J. Head, Publisher, Hampton, Middlesex, England, 1979.
Lein, A.: *The Cycling Female: Her Menstrual Rhythm,* W. H. Freeman & Company Publishers, San Francisco, 1979.
Nilsson, L.: *A Child is Born,* rev. ed., Delcorte Press, distributed by Dial Press, New York, 1977.
Rhodes, P.: *Birth Control,* Oxford Biology Readers, Oxford University Press, 1971.
Rhodes, P.: *Childbirth,* Oxford Biology Readers, Oxford University Press, New York, 1981.
Segal, S. S.: "The Physiology of Human Reproduction," *Scientific American,* September 1974.
Tietze, C., and S. Lewit: "Legal Abortion," *Scientific American,* January 1977.
Wolpert, L.: "Pattern Formation in Biological Development," *Scientific American,* October 1978.

GLOSSARY

A band: anisotropic band; that portion of a striated muscle cell which bends polarized light unequally; a mixture of thick and thin myofilaments

abdominal breathing: diaphragmatic breathing; deep breathing in which the abdominal wall moves outward as the diaphragm descends

abetalipoproteinemia (ay″bay″tuh·lip″o·pro·teen·ee′mee·uh): a rare disease in which the lack of β-lipoproteins results in an inability to transport lipids from the intestinal mucosa and causes their accumulation in that location

ABO grouping: determining the major blood group of a potential donor; performed by observing the reactions that occur when known types of blood cells and antibodies in serum are mixed in separate tests with the donor's antibodies and cells, respectively

abortion (uh·bor′shun): the termination of pregnancy

absolute refractory period: the period after a response, when an excitable membrane will not respond to a second stimulus under any condition

absorb: to transport a substance from the extracellular to the intracellular fluid, with its eventual entry into the blood; a function of transport across epithelia; to move substances across the digestive tract into the blood or lymph

absorption: the passage of substances into cells

accommodation: the ability of the ciliary muscles attached to the lens to change its shape, to bend light more or less, and to focus light on the retina

ACD: acid citrate dextrose solution, used to preserve whole blood stored under refrigeration

acetylcholine (ACh) (a·see″til·ko′leen): a neurotransmitter released from the terminals of axons of cholinergic nerves

acetylcholinesterase (AChE) (a·see″til·ko′leen·es′tur·ase): cholinesterase (ChE); an enzyme which degrades the neurotransmitter acetylcholine

acid (as′id): a substance which can act as a proton (H^+) donor

acid phosphatase: an enzyme that in an acid environment catalyzes reactions involving inorganic phosphate; lysosomal enzyme produced by osteoclasts which digests calcium phosphate in the matrix of bone

acidity (a·sid′i·tee): a measure of a substance's ability to donate hydrogen ions (H^+) to another substance; expressed as a pH below 7.0

acidosis (as″i·do′sis): condition caused when blood pH drops below 7.35

acinus (as′i·nus) (pl. **acini**): grapelike cluster of glandular epithelial cells; pancreatic acini secrete digestive enzymes

acne: an inflammation of sebaceous glands in the skin often caused by bacterial infection

acrosome (ack′ro·sohm): an organelle in the head of a sperm that contains hyaluronidase, an enzyme that helps sperm penetrate an egg at fertilization

actin (ack′tin): a protein found as a fibrous polymer in muscle and which during contraction slides toward the center of a sarcomere in striated muscle; the major component of thin myofilaments

action potential: a nerve impulse; see impulse; a self-propagating wave of current that travels along an entire membrane of nerve or muscle

activation heat: the liberation of energy during a period prior to and in the initial development of muscle tension in which 30 to 50 percent of the energy used by skeletal muscle is consumed; most of the energy is used to return Ca^{2+} to the sarcoplasmic reticulum prior to its release to the sarcoplasm to trigger the development of tension

active expiration: expelling air by the active engagement of intercostal (rib) and abdominal muscles to decrease the size of the thoracic cavity

active site: part of the three-dimensional structure of proteins which fits or binds to complementary molecules; found in enzymes and antibodies

active symport: *see* cotransport

active transport: the movement of specific substances across a cell membrane against concentration, electrical, or pressure gradients by active transport enzymes or pumps provided with energy by the cell

acute (uh·kewt): referring to conditions that are usually sudden at the onset, severe, and often quickly fatal

adaptation (in general): feedback-regulated alterations in cell, organ, or body functions in response to changes in environment

adaptation (of receptors): decline in the response of an afferent neuron to a prolonged constant stimulus

adenosine diphosphate (ADP) (a·den″o·seen, ·sin dye·fos″fate): formed by the cleavage of a high-energy phosphate bond from ATP

adenosine triphosphate (ATP) (a·den″o·seen, ·sin trye·fos″fate): the unit of currency of energy exchange, formation of its phosphate bonds is correlated with energy production and storage in cells

adenyl cyclase (ad′e·nil sigh′klace, ·klaze): an enzyme bound to cell membranes that converts ATP to cyclic AMP, a secondary messenger of certain hormones

ADH: antidiuretic hormone; vasopressin

adipocyte: fat cell

adrenal cortex: the outer area of the adrenal glands, whose cells secrete mineralocorticoids, glucocorticoids, and sex hormones

adrenal glands: paired, flattened yellowish endocrines, each of which rests on the superior portion of a kidney

adrenaline (a·dren′uh·lin, ·leen): see epinephrine

adrenal medulla (pl. **adrenal medullae**): the inner or central portions of the paired adrenal glands, that secrete norepinephrine and epinephrine

adrenal steroids: steroid hormones produced in the adrenal cortex

adrenergic (ad″re·nur′jick): a nerve fiber whose terminals secrete norepinephrine (noradrenaline)

adrenocorticotropic hormone (ACTH) (a·dree″no-kor″ti·ko·tro′fick, ·trof′ick): a hormone released by the anterior pituitary which stimulates the release of adrenal cortical hormones, the glucocorticoids

adsorb: to adhere to a surface without the formation of chemical bonds

adsorption: adherence of one substance to the surface of another without the formation of chemical bonds

adventitia (ad″ven·tish′ee·uh): an outer covering of an organ which is not an integral part of it; the loose connective tissue that binds the digestive tract in place; outer covering of blood vessels, composed of loose and fibrous connective tissues and in certain cases some longitudinal smooth muscle

aerobic (ay″ur·o′bick, air·): requiring oxygen

afferent (af′ur·unt): carrying toward; used in reference to impulses carried toward the central nervous system or fluids which move in vessels toward tissues or organs

afterbirth: the placental tissues expelled from the vagina after the birth of a baby

agglutination (a·gloo′ti·nay′shun): the clumping or aggregation of cells

agglutinin (a·gloo′ti·nin): agglutinating antibody; an antibody which with others will cross-link specific antigens on the surface of cells, clumping them together in a reaction called agglutination

agglutinogen (ag″leu·tin′o·jen, a·gloo′tin·o·jen): a molecule on the surface of a cell, which when abundant can be cross-linked by antibodies with identical molecules on other cells; this clumping of cells together is called agglutination

agnosia (ag·no·see·uh): inability to recognize familiar objects

agonist (ag·uh·nist): an agent which produces a response that enhances that of another agent; drugs or hormones whose effects are additive; muscles whose forces together cause the prime movement of the body

agranulocytes (ay·gran′yoo·lo·sites): white blood cells such as monocytes and lymphocytes which lack coarse cytoplasmic granules

albinism (al′bi·niz·um): lack of pigmentation with melanin due to a recessive hereditary condition

albumin (al·bew′min): a molecule that makes up about 55 percent of the protein in blood plasma and accounts for much of its osmotic pressure

albuminuria (al·bew″mi·new′ree·uh): the presence of albumin in the urine

aldosterone (al·dos′te·rone, al″do·ste·rone′): a mineralocorticoid that promotes the reabsorption of Na^+ in exchange for the secretion of K^+ and H^+ ions in the distal tubules of the kidneys; "sodium-retaining hormone"

alexia (a·leck′see·uh, ay): word blindness; an inability to read

algesireceptors (al·jee′zee·re·sep′turs): naked nerve

fibers which generate pain impulses in response to irritating or harmful stimuli; nociceptors

alimentary canal: the digestive tract

alkaline phosphatase: an enzyme which, in an alkaline environment, catalyzes reactions involving inorganic phosphate; an enzyme found on the plasma membrane of osteoblasts whose net effect aids in the formation of calcium phosphate and in the mineralization of bone

alkalinity (al″kuh·lin′i·tee): a measure of a substance's ability to accept hydrogen ions (H^+) from another substance; expressed as a pH from 7.0 up to 14.0; a measure of a substance's ability to accept hydrogen ions in a solution

alkalosis (al″kuh·lo′sis): elevation of blood pH above 7.45

allergy (al′ur·jee): one of two types of pathological effects of the immune response, due to the reactions caused by antibodies or T lymphocytes produced in response to exposure to a foreign substance; see delayed-type and immediate-type allergies.

all-or-none law: following a single stimulus, an excitable cell (muscle fiber or neuron) either responds or does not; a stronger single stimulus will not produce a greater response in that single cell; a response in which all the cells in the heart act as a single unit and generate the strongest possible contractions under a single set of conditions

alpha-adrenergic: (α-adrenergic); characteristic of cellular receptors which are mainly responsive to norepinephrine and to a lesser extent to epinephrine

alpha helices: right-handed coils in some protein molecules

alpha waves (al′fuh): brain waves characteristic of the relaxed state when awakening from restful sleep; high-voltage slow waves occurring about 8 to 12 times per second

alveolar ventilation rate (al·vee′uh·lur): the volume of fresh air that enters the alveoli each minute

alveolus (pl. **alveoli**): microscopic terminal air sacs in the lungs where gas exchange takes place

amination: addition of an amino group to a molecule

amniocentesis (am″nee·o·sen·tee″sis): a procedure in which amniotic fluid is withdrawn by inserting a syringe needle through the mother's abdominal wall into the amniotic cavity and analysis of the fluid to detect the presence of fetal abnormalities

amnion (am′nee·on): the innermost embryonic membrane which surrounds the embryo and fetus in a fluid-filled chamber called the amniotic cavity

AMP: adenosine monophosphate

amplifying circuit: a pathway in which an impulse from one neuron is spread by fibers to many other neurons and effector cells

amplitude: size; extent; the measure of voltage associated with electrical activity such as brain waves, and reflected by the height of the wave

ampulla (am·pull′uh, am·pool·uh) (pl. **ampullae**): widened end of a canal or duct; end of a semicircular canal in the ear

amylase (am″i·lace, ·laze): an enzyme which splits polysaccharides into smaller building blocks; see salivary amylase, pancreatic amylase

anabolism (a·nab′uh·liz·um): chemical reactions which synthesize molecules from smaller components in cells; a building up or constructive process

anaerobic (an″air·o′bick): occurring in the absence of oxygen

analgesic (an″al·jee·zick, je′sick): a drug that inhibits the perception of pain

analogous: similar in function

anastomosis (a·nas″tuh·mo′sis) (pl. **anastomoses**): a direct junction between tubules such as blood vessels

anatomic dead space: air passageways (in the nose, mouth, pharynx, trachea, bronchi, bronchioles, and alveolar ducts) which hold about 150 mL of air that does not reach the alveoli for gas exchange during respiration

anatomy (uh·nat′uh·mee): study of the structure of an organism and its parts

androgen (an′dro·jin): male sex hormone, the most active of which is dihydrotestosterone (DHT)

anemia (uh·nee′mee·uh): a decrease below normal levels in the quantity and/or function of red blood cells and/or hemoglobin

aneurysm (an′yoo·riz·um): formation of an outward bulge in a weakened blood vessel wall

angina (an′ji·nuh, an·jye′nuh): angina pectoris; referred pain over the chest or abdominal wall and/or arm, due to insufficient oxygen supply to the heart often as a result of coronary artery insufficiency

angiogram (an′jee·o·gram): *see* angiography

angiography (an″jee·og·ruh·fee): x-ray analysis of blood vessels following intravenous injection of a radiopaque dye

angiotensin (an″jee·o·ten′sin): specific polypeptide molecules termed angiotensins I, II, and III, some of which act as powerful vasoconstrictors that increase blood pressure; polypeptides derived from protein in the blood by the activity of renin and converting enzymes; the group of molecules is termed the renin-angiotensin system

angstrom (Å) (ang′strum, awng′strem): 10^{-10} m; 0.1 nm

anisotropic (an″eye″so·trop′ick): not the same in all directions, as when polarized light is bent (refracted) unequally; indicating molecular heterogeneity

anorexia (an″o·reck′see·uh): loss of appetite

antagonists (an·tag′uh·nists): describing agents which produce effects opposite to each other; drugs or hormones whose effects produce opposing responses; muscles which oppose the prime movement of the body and return it to its anatomical position

anterior pituitary (an·teer′ee·ur pi·tew′i·terr″ee): an endocrine gland located at the base of the center of the brain; sometimes called the master endocrine because it produces several hormones which stimulate activities of other endocrines

anterograde transport: movement of substances in an axon away from the cell body of a neuron

antibody (an′ti·bod″ee) (pl. **antibodies**): gamma globulin (globular protein) in the serum with components which are physically and chemically complementary to parts of antigens which stimulate their production

anticoagulant (an″tee·ko·ag′yoo·lunt): substance which inhibits blood clotting

anticodon (an″tee·ko′don): a nucleotide triplet in tRNA whose sequence forms a code complementary to the nucleotide triplet (codon) in mRNA

antidiuretic hormone (ADH) (an″tee·dye·yoo·ret′ick): vasopressin; a hormone produced in the hypothalamus and released by the posterior pituitary which promotes the reabsorption of water by the collecting tubules in the kidney, thereby minimizing urine formation

antigen (an′ti·jin): large molecule, usually foreign to the body, which stimulates the production of antibodies

antihistamine (an″tee·hiss′tuh·meen, ·min): drug that blocks the effects of histamine

antrum: a cavelike chamber; the portion of the stomach nearest the small intestine

anuria (an·yoo′ree·uh): lack of urine formation caused by kidney malfunction

aorta (ay·or′tuh): the largest artery, carrying blood to the body from the left ventricle and branching into arteries and arterioles

aortic bodies: location of chemoreceptors in the wall of the aorta which behave as chemosensors of gases in the circulatory system

apnea (ap′nee·uh): cessation of breathing; may be a response to sudden pain or exposure to cold

apneustic center (ap·new′stick): a cluster of neurons in the lower pons which is the driving force for the inspiratory center in the medulla and would cause continual inspiration if uninhibited

apocrine secretion (ap′o·krin): a process in which exocrine glands release products by shedding the tips of their cells; the mammary glands secrete products this way

apoferritin (ap″o·ferr′i·tin): a carrier protein whose combination with iron produces ferritin, enabling iron to be absorbed across the intestinal mucosa

appendix (a·pen′dicks): a narrow tube about 3 in long with much lymphoid tissue in its walls and projecting from the cecum but having little apparent digestive function

arachnoid (uh·rack′noid): a fibrous, weblike membrane between the dura mater and the pia mater on the surface of the brain and spinal cord

areolar tissue (a·ree′o·luh): loose connective tissue; composed of several cell types scattered among elastic and collagen fibers

arteriole (ahr·teer′ee·ole): smallest of arteries

arteriosclerosis (ahr·teer″ee·o·skle·ro′sis): any one of a number of disorders in which the arterial wall degenerates, some of its cells proliferate, and deposits accumulate in them and the lining of the vessel

arteriovenous oxygen difference (ahr·teer″ee·o·vee′nus): the difference in oxygen content of blood between the arterial supply to, and the venous drainage from, a tissue

artery (ahr′tur·ee): blood vessel that carries blood away from the heart to tissues

arthritis (ahr·thrigh′tis): any one of more than 100 disorders which cause progressive degeneration of joints; common in the elderly and may be due to mechanical stress, infection, metabolic disorders, genetic factors, or autoimmune reactions

articular cartilage (ahr·tick′yoo·lur): hyaline cartilage covering the ends of long bones and providing a smooth, resilient surface

ascending colon: the first major portion of the large intestine

ascending tract: a discrete bundle of nerve fibers which carries impulses up the spinal cord toward the brain

ascites (a·sigh′teez): the pooling of excess fluid in the abdominal cavity

aspartate aminotransferase (AST) (as·pahr′tate a·mee″no·trans′fur·ace): also SGOT or GOT, serum glutamic-oxalacetic transaminase; an enzyme whose activity in the serum is elevated by stressful conditions, including myocardial infarction; see transamination

associative neurons (a·so′shee·uh·tiv, a·so′see·ay″-

tiv): neurons with junctions which interconnect areas within a cerebral hemisphere (one half of the cerebrum)

asthma (az'muh): a reversible functional disease caused by spasms of smooth muscle constricting the bronchioles, making breathing difficult and especially obstructing outward air flow; may be caused by an immediate-type allergic reaction

astigmatism (a·stig'muh·tiz·um): visual blurring; uneven focusing of light on the retina

atelectasis (at"e·leck'tuh·sis): collapse of alveoli

atherosclerosis (ath"ur·o"skle·ro'sis): a disorder in which the walls of the arteries degenerate as they are infiltrated by fatty substances, forming patches called plaques; a specialized form of arteriosclerosis

atom (at'um): the smallest part of an element which cannot be subdivided without changing the nature of the substance

atomic weight: the weight of an element relative to a standard of 12 for carbon which is approximately twelve times the value assigned to an hydrogen atom

ATP: adenosine triphosphate

ATP-ase carriers: enzymes which cleave ATP and which simultaneously actively transport specific substances across cell membranes

atretic follicle (a·tret'ick): the shriveled remains of an ovarian follicle which has not completed maturation

atrium (ay'tree·um) (pl. **atria**): one of two chambers of the heart which receive blood from the body and lungs, respectively

atrioventricular node (ay"tree·o·ven·trick'yoo·lur) **A-V node:** a cluster of modified cardiac muscle cells in the right atrium near the right ventricle which conducts impulses from the S-A node to the bundle of His which delivers them to each ventricle by way of ventricular bundles

atrioventricular valve (A-V valve): valve located between the atrium and its corresponding ventricle; see bicuspid (mitral) and tricuspid valves

atrophy (at'ruh·fee): a reduction in size or bulk of cells, tissues, or organs

auditory (aw'di·tor"ee): referring to sound

auditory apparatus: mechanoreceptors in the ear which transduce sound waves to nerve impulses that are interpreted as sound in the brain

auscultatory method (aws·kul'tuh·to·ree): measuring blood pressure by listening to sounds with a stethoscope placed on a large artery while assessing arterial pressure with a sphygmomanometer

autoimmunity (aw'to·i·mew'ni·tee): immune reactions against one's own tissues; examples include rheumatoid arthritis and pernicious anemia

autonomic (aw"tuh·nom'ick): self-controlling

autonomic nervous system (ANS): the part of the nervous system which transmits involuntary impulses to the organs of the body

autoregulation (aw"to·reg·yoo·lay'shun): a self-control mechanism in which local feedback mechanisms in tissues influence the rate of blood flow to them

autorhythmicity: the ability of contractile tissue, such as smooth and cardiac muscle, to set its own rate of contraction independently of external stimuli; due to pacemaker cells in the tissue which generate their own electrical rhythm

autosome (aw'to·sohm): a chromosome in a cell other than the X or Y chromosomes, and usually one of 22 pair in each body cell

auxotonic contraction (awk'so·ton'ick): the development of force by a muscle using isotonic and isometric contractions

aviation sickness: a condition produced by rapid ascent to high altitude, resembling decompression sickness in its causes and symptoms. See the "bends."

Avogadro's number (ah"vo·gah'dro): the number of molecules (6.02×10^{23}) in a mole of any substance

axilla (ack·sil'uh): armpit

axon (acks'on): cytoplasmic extension of a neuron which usually carries impulses away from its cell body

axon hillock: thickened portion of the cell body of a neuron from which its axon emerges

backward failure: a type of congestive heart failure caused when blood does not return to the heart adequately but accumulates in vessels or tissues

bacterial flora (back'teer'ee·ul flo'ruh): the resident population of bacterial species in a location

Bainbridge reflex: reflex acceleration of the heart occurring after intravenous infusion of large quantities of blood or salt solution; initially proposed, but now discarded, as a reflex response to increased blood pressure due to exercise

baroreceptor (băr"o·re·sep'tur): pressoreceptor, mechanoreceptor; clusters of cells in the heart or major blood vessels which generate impulses in response to changes in pressure

basal ganglia (bay'sul gang'glee·uh): groups of neurons deep within the cerebral hemispheres above and lateral to the thalamus which coordinate sensory and motor activity; dysfunctions may cause muscular disorders such as Parkinson's disease and Huntington's chorea

basal lamina: basement membrane

basal metabolic rate (BMR): the minimal level of metabolism to sustain life; the energy expended per day by an individual lying down, awake, and comfortable after a restful sleep and 12 to 14 h after a meal

base: an alkali; a substance which can combine with or accept protons (H^+) from other substances

basement membrane: basal lamina; glycoproteins and connective tissue fibers which anchor epithelial cells to underlying tissues

basilar membrane (bas'i·lur): a membrane containing up to 30,000 strands of varying lengths between the middle and lower chambers of the cochlea in the inner ear; shorter to longer strands are responsive to specific sound waves, ranging from high to low pitch respectively; the energy is transmitted by the strands to hair cells of the organ of Corti

basophilic (bay"so·fil'ick): having an attraction for alkaline substances, such as the blue alkaline dye of Wright's or Giemsa stains for blood smears

basophils (bay'so·fils): granulocytic white blood cells; bloodborne equivalents of mast cells which release histamine and heparin in allergic reactions; the cell's cytoplasmic granules have an attraction for alkaline (basic) dyes, staining blue to purple in a classically stained blood smear

B cells (B lymphocytes): small lymphocytes originating in the bone marrow which contribute to humoral immunity by secreting antibodies into body fluids

bends ("the bends"): decompression sickness; caisson disease; the expansion of N_2 gas bubbles in body tissues and blood, which may cause a fatal embolism when a rapid ascent is made from a deep water dive

Bernouilli's principle (Ger. be^{h}r·noo'lee, F. be^{h}r·noo·yee'): the observation that the total energy of a moving fluid is constant and is the sum of the energy of its pressure and its flow, and gravity; Bernouilli's principle states that as a blood vessel divides, the area of its subdivisions increases, the rate of blood flow decreases, and the lateral pressure increases

beta-estradiol (bay'tuh es·truh·dye'ol): the most active of the female sex hormones found in the ovaries; important in the development and maintenance of female secondary sex characteristics during puberty

beta-adrenergic (β-adrenergic): cellular receptors which are mostly responsive to epinephrine and to a lesser extent to norepinephrine

beta-lipotropin (β-lipotropin) (bay'tuh lip"o·trop'-in): a polypeptide hormone in the anterior and intermediate pituitary gland which may stimulate lipid release from fat cells; shares structural sequences with the endorphins, enkephalins, the melanocyte-stimulating hormone (MSH), and the adrenocorticotropic hormone (ACTH)

beta sheets: folded sheets of paired polypeptide chains linked by hydrogen bonds

beta waves: brain waves predominant during mental activity and in the aroused state; low-voltage fast waves occurring about 13 to 32 times per second

bicuspid valve (bye·kus'pid): a valve composed of two flaps and located between the left atrium and left ventricle; mitral valve; the left atrioventricular (A-V) valve

bile: an alkaline secretion of the liver which emulsifies fats and contains 97 percent water, plus metabolic products of hemoglobin, salts, cholesterol, and the phospholipid lecithin

bilirubin (bil"i·roo"bin): a product from the degradation of hemoglobin

biofeedback (bye"o·feed'back): the modification of otherwise involuntary activities by the influence of learned voluntary nerve impulses on autonomic functions; an example is the learned voluntary modification of blood pressure

bipolar recording (bye·po'lur): measurement of voltage by comparison of potentials between two locations in the body at one time

-blast: a suffix indicating "formation of . . ."

blastocyst (blas'to·sist): an early stage in development, formed initially as a hollow ball of cells 3 to 4 days after fertilization of the egg and which upon continued maturation becomes implanted in the uterine lining 7 to 10 days after conception

bleeding time: usually 1 to 3 min, the time required for blood to stop flowing from a sharp standardized incision made on the finger or earlobe

blood: a specialized connective tissue composed of a fluid intercellular matrix called plasma and cells and cell-derived components called formed elements

blood-brain barrier: a protective barrier of limited permeability formed by phagocytic cells called astrocytes, which cover neurons in the brain and which are located between neurons and blood capillaries in the brain

blood pressure: the lateral pressure of blood (in excess of atmospheric pressure) against the walls of circulatory vessels

Bohr effect: alterations in the percent of oxyhemoglobin due to a change in the acidity of the blood at a specific partial pressure of oxygen

bond: a union of atoms formed by exchanging, sharing, or donating electrons

bone: connective tissue with cells trapped in a firm matrix of inorganic salts, such as calcium phosphate or calcium carbonate; osseous tissue

botulism (bot'yoo·liz·um): food poisoning due to a protein toxin released from *Clostridium botulinum,* which blocks the release of acetylcholine from motor neurons causing muscular paralysis that is often fatal

Bowman's capsule: a double-walled microscopic epithelial chamber in the kidney which collects substances that filter from glomerular capillaries

bradycardia (brad"ee·kahr'dee·uh): very slow heartbeat

brain waves: impulses from electrical neurons in the brain conducted to the surface of the scalp by body fluids and relayed to a recording instrument to produce a visual display called an electroencephalogram (EEG)

Braxton Hicks contractions: painless muscular spasms during the last few months of pregnancy

breasts: mammary glands

Broca's area (bro^h·kah'): a speech center in the brain; consists of neurons in the left side of the frontal lobe which receive impulses from Wernicke's area (for comprehension of language) and relay them to motor neurons to regulate respiratory muscles, as well as to those in the mouth to produce sound and speech

bronchus (pl. **bronchi**): two large main branches of the trachea, each going to a lung

bronchiole (bronk"ee·ole) (pl. **bronchioles**): very small branch of a bronchus

Brunner's glands (brŏon'ur): clusters of cells in the submucosa of the duodenum whose ducts lead to the crypts of Lieberkühn and which secrete alkaline mucus containing a wide variety of intestinal enzymes

brush border: a fuzzy surface on a cell due to the presence of microvilli present in the epithelial cells lining the intestine and the tubules of the kidney

buffers: organic and inorganic compounds that can neutralize excess acid or base; chemical sponges which can soak up or release hydrogen ions

bulbourethral glands (bul"bo·yoo·ree'thrul): Cowper's glands; paired pea-shaped structures which lie below the prostate and which, during sexual arousal in the male secrete a thick, clear, mucus-like, alkaline fluid that neutralizes acid

bundle of His (hiss): a group of fibers in the heart wall between the ventricles that conducts electrical impulses from the atrioventricular node through right and left branches to each ventricle

bursa (pl. **bursae**): a fluid-filled sac which surrounds and relieves pressure between moving body parts such as muscles, tendons, ligaments or bones

caisson disease (kay'sun): *see* "bends"

calcitonin (thyrocalcitonin) (kal"si·to'nin): a hormone that is synthesized by C cells in the thyroid gland; counteracts the effects of parathormone and vitamin D_3, thereby inhibiting the resorption of calcium and phosphate from bone and lowering their blood levels

calcium/phosphate ratio (kal'see·um fos'fate): the ratio of calcium to inorganic phosphate in a tissue; often measured as the product of the concentrations of calcium and phosphate ions in the blood

calmodulin: one of a number of calcium-binding proteins that may affect the activity of cAMP and many enzymes in cells

Calorie (kcal) (kal'uh·ree): *see* kilocalorie

canaliculus (kan"uh·lick'yoo·lus) (pl. **canaliculi**): small channel; in bone, microscopic intercellular canal

cancellous bone (kan'se·lus): spongy bone; a lacy network of bone cells found within long bones

capacitance (ka·pas'i·tance): the number of charges which maintain a voltage across a membrane

capillary (kap'i·lair"ee): smallest vessel of the circulatory and lymphatic systems; smallest of blood vessels and one through which the exchange of substances between blood, body fluids, and tissues takes place

capsule (kap'sool, ·syool): a dense fibrous connective tissue membrane which surrounds, protects, and anchors organs and body parts to each other; for example, capsules enclose the kidney, spleen, and other organs, as well as synovial joints, from which they extend to merge with ligaments

carbaminohemoglobin (kahr"buh·mee"no·hee'muh-glo"bin): a hemoglobin molecule with CO_2 bonded to its structure; $HbCO_2$

carbohydrate (kahr"bo·high'drate): simple sugars or their multiples with a formula ratio of $C_X(H_2O)_X$

carbohydrate loading: glycogen loading; a regimen of diet and exercise designed to theoretically increase glycogen stores in muscle and the length of time of endurance exercise prior to fatigue

carbomonoxyhemoglobin: carboxyhemoglobin; hemoglobin molecules bonded to carbon monoxide (CO); HbCO

carboxypeptidase (kahr·bock"see·pep'ti·dace, ·daze): a protein-cleaving enzyme secreted by the pancreas

cardiac catheterization (kahr'dee·ack, kath'e-tyr·i-zay'shun): the insertion of a tube into the heart through large blood vessels leading into its chambers

cardiac center: a group of neurons in the medulla oblongata that influences the rate and strength of contraction of the heart

cardiac conducting system: the various elements which form an electrical bridge that regulates the heart's electrical activity; they include the sinoatrial node, the atrioventricular node, the bundle of His and its branches, and the Purkinje fibers

cardiac cycle: the events occurring in systole and diastole of one heartbeat

cardiac fibrillation (kahr'dee·ack figh"bri·lay'shun): loss of coordination between the electrical activity and the mechanical activity of the heart

cardiac muscle: tissue composed of multinucleated, striated, branching cells in the heart, but not under voluntary control; forms the myocardium, the heart wall

cardiac output: the amount of blood pumped from either ventricle in 1 min; averages 5.25 L/min in adults; the minute volume of the heart

cardiac plexus (pleck'sus): groups of autonomic nerves surrounding the aortic arch and parts of the heart

cardiac sphincter (sfink'tur): *see* gastroesophageal sphincter

cardiac vector: an arrow derived from an ECG tracing; its size and angle reflect the sum of the electromotive force generated by the heart

cardio: pertaining to the heart

cardiopulmonary resuscitation (CPR) (kahr"-dee·o·pŏŏl'muh·nerr·ree re·sus"i·tay'shun): artificial respiration used in conjunction with cardiac resuscitation (restoration of the heartbeat)

cardiovascular system (kahr"dee·o·vas'kew·lur): the circulatory system; the heart and blood vessels

carotid (ka·rot'id): principal artery in the neck; divides into an external and internal branch

carotid body: location of respiratory gas chemosensors; chemoreceptors located in the walls of each external carotid artery at its origin, whose responses to increases in CO_2 and H^+ or to large decreases in O_2 increase the frequency and depth of respiration and raise blood pressure

carotid sinus (sigh'nus): site of baroreceptors in the walls of the carotid arteries; sensitive to changes in blood pressure

carrier molecule (mol·e·kyool): substance which assists the movement of specific chemicals across cell membranes

cartilage (kahr'ti·lij): connective tissue composed of groups of cells, chondrocytes, which are embedded in a firm but somewhat flexible matrix of fibers and proteoglycans (protein-carbohydrate complexes)

cascade (kas·kade): a shower of effects; a very rapidly spreading series of responses which may alter many aspects of cell function

castration (kas·tray'shun): surgical removal of the gonads, which causes profound and undesirable changes in hormonal balance; removal of testes in the male or ovaries in the female

catabolism (ka·tab'uh·lizz·um): a series of chemical reactions which degrade molecules into smaller components and usually release energy in cells

catalyst (kat'uh·list): a substance that alters the rates of chemical reactions without being altered in the process

cataract (kat'uh·rakt): accumulation of pigmented compounds in the aging lens of the eye, which decreases its transparency and impairs vision

catecholamine (kat"i·kol'uh·meen, ·min): chemical control agent with a ring structure; examples include dopamine, norepinephrine, and epinephrine

catechol-O-methyl-transferase (COMT): an enzyme which may inactivate catecholamines such as norepinephrine and epinephrine

cathartic (ka·thar'tick): substance which enhances the motility of the digestive tract

catheter (kath'e·tur): a hollow, flexible tube designed to be inserted into hollow body structures such as tubules, vessels, or cavities for the introduction or removal of fluids and/or the measurement of specific functions

CAT scan (computer-assisted tomography): x-ray images sorted out by computer analysis to construct three-dimensional images; their contrast is often increased by the intravenous injection of radiopaque substances such as iodine containing compounds prior to x-ray photography

cauda equina (kaw'duh e·kwye'nuh): the sacral and coccygeal spinal nerves which are located at the inferior (lower) end of the spinal cord

caudal: referring to the tail or inferior (lower) part of the human body when in the upright position

caudal anesthesia: administration of an anesthetic at the sacral segment of the spinal cord

cecum (see'kum): a blind pouch situated at the beginning of the large intestine

cell: the structural and functional building block of living organisms

cell body: the part of a cell which surrounds the nucleus

cell membrane: the plasma membrane; a selectively

permeable barrier surrounding the cell, composed primarily of proteins and lipids; the outermost boundary of the cell

cellular immunity (sel'yoo·lur i·mew'ni·tee): defenses provided by T lymphocytes and phagocytes

cellular receptors: molecules in the cell membrane and other cell parts to which substances, such as hormones, neurotransmitters, drugs, toxins, and microrganisms may bind specifically

cellular respiration (res"pi·ray'shun): process in which cells use oxygen and release energy, carbon dioxide, and water as products of aerobic metabolism

center: a cluster of neurons in the brain whose boundary has not been clearly delineated but which performs a specific function

centi- (c): a prefix ascribing a value of one-hundredth of the whole

central canal: a cavity running through the center of the length of the spinal cord which is continuous with the ventricles, the fluid-filled chambers in the brain

central nervous system (CNS): the brain and spinal cord

central sulcus (sul'kus): *see* fissure of Rolando

centrioles (sen'tree·ole): cell organelles found as paired cylindrical bodies which give rise to the fibers that separate chromosomes in cell division

cephalic (se·fal'ick): pertaining to the head

cerebellar (serr"e·bel'ur): referring to the cerebellum

cerebellum (serr'e·bel'um): a three-lobed segment of the hindbrain which fine-tunes motor activity and influences posture, equilibrium, and speech

cerebral cortex (serr'e·brul, se·ree·brul): a 2.5- to 4-mm-thick outer layer of gray matter on the surface of the cerebral hemispheres, which controls voluntary movement and the highest of brain functions such as thoughts, judgments, and memory

cerebral hemisphere: one-half of the cerebrum, the largest component of the human brain

cerebral hemorrhage (hem'uh·rij): rupture of and bleeding from a cerebral artery

cerebral sclerosis (skle·ro'sis): a disorder characterized by the replacement of normal tissues by fibrous connective tissue; hardening of tissues in the brain

cerebral thrombosis (throm·bo'sis): blockage of an artery in the brain

cerebrospinal fluid (CSF) (serr"e·bro·spye'nul): liquid which is formed mainly by filtration from the blood in the choroid plexus in each of the ventricles of the brain; CSF circulates through them and the central canal of the spinal cord; its composition resembles that of the extracellular fluid around the brain and differs slightly from that of plasma

cerebrovascular accident (CVA) (serr"e·bro·vas'-kew·lur): stroke; apoplexy; an interruption of normal blood flow to a part of the brain, often caused by an embolism, thrombosis, or hemorrhage in the middle cerebral artery

cerebrum (serr'e·brum, se·ree'brum): largest part of the human brain and divided into two halves called cerebral hemispheres

ceruminous glands (se·roo·mi·nus): tubular glands which secrete a lubricating wax called cerumen; found in the external ear and in the periphery of the eyelids

cervi-: referring to the neck

cervix (sur'vicks): the narrow, constricted neck of an organ; the lower cylindrical segment of the uterus leading to the vagina

chemo-: referring to chemical properties

chemoreceptor (kee"mo·re·sep"tur): sensory nerve ending responsive to chemical stimuli such as acid, carbon dioxide, glucose

chemotactic: referring to the response of cells to specific chemicals such as lymphokines produced by lymphocytes as part of the body's defense reactions

Cheyne-Stokes respiration (chain): rhythmic alternating periods of apnea and dyspnea

chief cell: cell in gastric pits of the stomach which secretes the enzymes pepsinogen, lipase in adults, and rennin in infants

cholecystokinin-pancreozymin (CCK-PZ) (kol"e·sis"-to·kigh'nin, ko"le· pan"kree·o·zye'min): a hormone secreted by cells of the duodenum and jejunum which stimulates the secretion of stomach acid and pancreatic enzymes and causes contraction of the gallbladder

cholera (kol'ur·uh): a disease caused by a bacterial species (*Vibrio cholera*) which secretes a powerful exotoxin; a form of gastroenteritis marked by severe diarrhea with fluid and electrolyte loss caused by cholera organisms

cholesterol (ko·les·tur·ol): a steroid constituting 25 percent of the cell membrane and usually present in plasma at levels of 120 to 220 mg/100 ml; a precursor to steroid hormones

cholinergic (ko"lin·ur'jick, kol"in): a nerve fiber whose terminals secrete acetylcholine

chordae tendineae (kor'dee ten·din'ee·ee): strands attached to the atrioventricular valves and which originate from the papillary muscles that extend to the walls of the ventricles and help keep the valves closed during systole; *also see* papillary muscle

chorion (ko'ree·on): outer embryonic and fetal portion of the placental membrane which erodes the uterine lining during implantation of the blastocyst into the uterine lining

choroid coat (kor'oid, ko'roid): vascular lining between the outer coat of the eye (the sclera) and the innermost layer, the retina

choroid plexus (pleck'sus): a formation of blood vessels which penetrate the pia mater in each of the four ventricles within the brain and is the main source of cerebrospinal fluid

chrom-: referring to color

chromophore (kro'mo·fore): colored pigment: oxidized forms of vitamin A in rods and cones which are responsible for sensitivity to light and colors

chronic (kron'ick): reference to diseases, gradual in development and long-standing

chronotropic effect (kron″o·trop'ick, ·tro'pick): an influence on the rate of heartbeat

chylomicron (kigh″lo·migh'kron): particle composed of triglycerides, phospholipids, cholesterol, and protein, enclosed in lipoprotein; the major form in which lipids cross the intestine into lacteals; chylomicrons appear as the largest and least dense of lipoproteins in the blood after a meal with a high fat content

chymotrypsin (kigh″mo·trip'sin): a protein-cleaving enzyme secreted by the pancreas

cilium (sil'ee·uhm) (pl. **cilia**): cellular hairlike extension which moves back and forth and which consists of nine pairs of microtubules surrounding a central pair; found in the lining of the respiratory tract and the oviducts

circadian rhythms (sur·kay'dee·un): biological rhythms that follow a cycle of about 1 day

circulatory gas chemosensor: chemoreceptor of the aortic body which, because of the relatively slow rate of delivery of blood to it, is suited to detect changes in the levels of gases in that fluid; compare with respiratory gas chemosensor

circumcision (sur″kum·sizh'un): surgical removal of the rim of the foreskin of the penis

cirrhosis of the liver (si·ro'sis): a diffuse, inflammatory disease which disrupts liver structure as functional liver cells (parenchyma) are replaced by connective tissue

citric acid cycle (Krebs cycle; tricarboxylic acid cycle): the aerobic metabolism of carbohydrates, fats, and proteins by which energy is released in small steps to form ATP, and carbon dioxide and water are produced as waste products; 2-carbon fragments of acetyl CoA are combined with citric acid in the early stages of the cycle; enzymatically catalyzed reactions remove hydrogen and electrons stepwise from molecules to release energy used to form ATP and to produce other organic acids some of which, like citric acid, have three carboxyl groups in their structure (hence the alternative name, tricarboxylic acid cycle)

-clast: a suffix indicating a process in which something is broken down

clearance rate (clearance): the number of milliliters of plasma from which a particular substance is completely removed each minute by glomerular filtration; an indication of filtration capacity and kidney function

click: heart murmur

climacteric (klye·mack'tur·ick, klye″mack·terr'ick): a period often between ages 40 to 50 in which a decrease in sexual activity and/or reproductive capability occurs; in females, a period of several months to a few years during which menstrual cycles slow down and cease, causing physical and emotional changes; in males, changes are characterized by a gradual decline in testosterone as well as a decline in sexual activity; unlike females, males retain reproductive capacity

clitoris (klit'o·ris, klye'to·ris): a female external genital organ, highly sensitive to touch and made of erectile tissues, homologous to the penis

clone: a group of cells derived from one cell by asexual reproduction so that all members of the group are genetically identical

clotting time: the measured time needed for blood to clot in a capillary tube or test tube at 37°C, ranging from 3 to 6 or 7 to 15 minutes, respectively

CNS: the central nervous system

coccyx (kock'sicks): "tailbone," the lower end of the vertebral column

cochlea (kock'lee·uh): a spiral tube in the inner ear, resembling a snail's shell, that houses the organ of hearing

codon (ko'don): a triplet of nucleotides within DNA or mRNA that directs the insertion of a specific amino acid into a polypeptide

coenzyme (ko·en'zyme): a substance which assists or enables an enzyme to perform its function; a cofactor

colic (kol'ick): abdominal pain often caused by smooth muscle contraction, due to accumulation of feces, gas, and stretching of the colon and rectum

collagen (kol'uh·jin): fibers in connective tissue formed of triple-helical cross-linked coils of protein

collagenase (kol'uh·je·nace, ·naze): an enzyme which hydrolyzes collagen

collateral (kuh·lat'ur·al): an alternate route; two or

more branches of an axon or blood vessel lying side by side

colon (ko'lun): the segment of the large intestine between the cecum and the rectum

colostrum (ko·los'trum): a watery fluid rich in protein and IgA antibody but lower in milk sugar and fat than is milk; released from the breasts during the first 1 to 3 days after childbirth

columnar epithelium (ko·lum'nur ep″i·theel'ee·um): column-shaped cells found lining specific ducts; lines parts of the digestive tract, respiratory tract, urethra, and bile ducts of the liver

combination pills: the most commonly used birth control pills in the United States, containing synthetic progestins and estrogens

commissural fibers (kom″i·shoor'ul; syoor'ul; ·soor'-ul): fibers carrying impulses between the two cerebral hemispheres

common bile duct: a vessel that is formed by the union of the hepatic duct of the liver and the cystic duct of the gallbladder which delivers secretions to the duodenum

compact bone: densely packed bony tissue, composed of orderly arranged building blocks called Haversian systems, in the ends of and outer layer of the shafts of long bones

compatibility: similarity of molecular structure of the major molecules at the surface of cells in different tissues; precludes the formation of antibodies by the recipient against tissues of the donor and allows the donor's transplanted tissues or transfused cells to survive in the recipient; *see* histocompatibility antigens

compensatory hypertrophy (kum·pen'suh·to″ree high·pur'truh·fee): formation of an increased mass of a tissue or an organ following damage by injury or disease

compensatory pause: a longer than usual relaxation period which follows a premature beat of the heart

complementary base pairs (kom″ple·men'tuh·ree): pairs of nucleotides that combine; each pair consists of a pyrimidine joined to a matching purine; the respective pairs in double strands of DNA are thymine: adenine and cytosine: guanine

compound: a complex molecule formed by the union of definite proportions of two or more dissimilar atoms to produce a substance with properties normally unlike those of the original atoms

compound glands: multicellular glands that have ducts branching into several lobules

compound lipids: substances formed of lipids and other groups; examples are phospholipids, glycolipids, and lipoproteins

conception: fertilization; the union of egg and sperm

conditioned reflex: an automatic involuntary response to a stimulus which is learned, that is, modified by experience, as opposed to an inborn or natural reflex; a learned response

condom (kon'dum): a thin latex rubber covering that is used to enclose the penis to trap seminal fluid and to prevent conception

cones: photoreceptors in the retina which contain opsins, allowing for sensitivity to color vision and vision in bright light

congestive heart failure: congestive heart disease; a disorder in which too little blood is pumped through the heart and cardiac output is low; results from failure of blood to flow back to the heart normally (backward failure) or from a condition in which the heart pumps too little blood out to peripheral tissues (forward failure); either condition may cause blood to back up in the lungs

Conn's syndrome: hyperactivity of the adrenal cortex causing oversecretion of aldosterone, sodium retention, and an increase in extracellular fluid, blood pressure, and potassium excretion

constipation (kon″sti'pay'shun): difficult or delayed emptying of the bowel producing hard, dry feces

continuous capillary: microscopic blood vessel whose cells have tight junctions 10 to 20 nm in width which prevent many substances from escaping from the blood

contraception (kon″truh·sep'shun): prevention of fertilization and pregnancy

contraction (kun·track'shun): the generation of force by muscle, occurring when myofilaments within each muscle cell slide toward each other

contraction period: the time between start and development of peak tension in a muscle fiber

contracture (kon·track'chur): failure of a muscle to relax fully between successive stimuli; sustained partial contraction of muscle between stimuli; clinically, the shortening of a muscle which limits mobility of a joint

contralateral (kon″truh·lat'ur·ul): on the opposite side; refers to reflexes which cross the spinal cord

converging circuit (kun·vur'jing): a pathway in which impulses from two or more neurons are funneled to a lesser number of neurons

Cooley's anemia (Mediterranean anemia): a form of thalessemia in which the synthesis of one or more chains of hemoglobin is defective and which is most common in persons of Mediterranean ancestry

Cori cycle: recycling of lactic acid derived from glycolysis in muscle and delivered by diffusion into the blood to the liver, where it is converted to glucose, which may recirculate to muscle for use in energy metabolism

cornea (kor'nee·uh): the transparent epithelium of the anterior portion of the eye

coronary arteries (kor'uh·nerr"ee ahr'tur·ees): the right and left branches of the aorta which supply the tissues of the heart with blood and which are immediately formed as the aorta leaves the left ventricle

coronary insufficiency: oxygen deprivation to tissue of the heart caused by insufficient circulation of blood through narrowed coronary vessels

coronary sinus (sigh'nus): a large vein that empties blood into the posterior surface of the right atrium

coronary veins: vessels carrying deoxygenated blood from the heart tissues to the coronary sinus, thence to the right atrium

corpus albicans (kor'pus al'bi·kanz): a white mass of cells produced by the infiltration of the corpus luteum by connective tissue cells and collagen fibers beginning the latter part of the menstrual cycle (about day 24 in a 28-day cycle) if conception does not occur

corpus callosum (kor'pus ka·lo'sum): largest bundle of fibers connecting the cerebral hemispheres; *see* commissural fibers

corpus luteum (kor'pus lew'tee·um): a mass of follicular cells that is left behind after ovulation and is the major source of estrogens and progesterone during the luteal (progestational) phase of the menstrual cycle

corpuscle of Ruffini (kor'pus·ul): sensory receptor that is encapsulated in connective tissue in the skin and in capsules of joints and ligaments; responsive to touch and tension

cortex (kor'tecks): outer portion of an organ

cortical nephron (kor'ti·kul nef'ron): microscopic tubule located in the cortex of the kidney which usually lacks a well-developed thin segment of the loop of Henle

corticotropin releasing factor (CRF) (kor"ti·ko·tro'-pin): corticotropin releasing hormone; a hypothalamic hormone which causes the anterior pituitary to release adrenocorticotropic hormone (ACTH) into the bloodstream

costal cartilage (kos'tul kahr'ti·lij): hyaline cartilage joining the ribs to the sternum (breastbone)

cotransport: active symport; process in which the active transport of one substance derives energy from and is linked to the transport of another substance

coughing: a forceful reflex expiration caused by foreign matter or fluid on the mucous membrane of the lower respiratory tract

Coumadin: trade name for warfarin, an anticoagulant derived from fermented sweet clover which prevents the synthesis of vitamin K–dependent clotting factors in the liver; used as a rodenticide

countercurrent: a stream of liquid or gas moving closely and in opposite direction to another stream

countercurrent exchange: passive diffusion of substances between adjacent vessels as in the tubules of the kidneys

countercurrent multiplication: the amplification of differences in concentration between two oppositely moving fluids by means of active transport pumps which assist movement of specific substances across epithelial cells

countercurrent theory: a hypothesis which proposes that the ability of the kidney to produce concentrated urine is enhanced by the formation of steep osmotic gradients along tubules, and is made possible by active transport and diffusion of substances between fluids, which move in opposite directions in hairpin-shaped loops of the organ's vessels

CPD: a citrate phosphate dextrose solution commonly used in blood banks to preserve whole blood stored under refrigeration

cramps: involuntary, intermittent muscular contractions, which usually produce pain

cranial nerves (kray'nee·ul): one of 12 pairs of nerves which originate in or close to the brain and mainly supply structures around the head and neck

creatine kinase (CK) (kree'uh·teen, ·tin kigh'nace, kin'ace, kin'aze): also CPK, creatine phosphokinase; an enzyme whose serum level reflects damage to skeletal and cardiac muscle

creatine phosphate (C~P): a high-energy phosphate compound which releases its stored energy to drive reactions that lead to the synthesis of ATP

creatinine (kree·at'i·neen, ·nin): a metabolic derivative of creatine phosphate produced in muscle; the clearance rate of creatinine by the kidneys is often used to assess kidney function

crista (kris·tuh) (pl. **cristae,** kris·tee): a fold or projection of the inner mitochondrial membrane; the location of respiratory enzymes that use oxygen to release energy from organic molecules

crista ampullaris (kris·tuh am'puh·lair'is): a sensory organ containing ciliated hair cells which form mechanoreceptors in the ampulla (enlargement) at the end of each semicircular canal of the ear; a dynamic sensor that detects the rotational acceleration as well as the acceleration and deceleration of the head

cross matching: assessment of compatibility of major

blood types of donor and recipient; cells and antibodies (serum) of the donor are mixed in separate tests with the recipient's antibodies (serum) and cells, respectively

crypts of Lieberkühn (lee'bur·kuᵉn"): intestinal glands; glands in depressions in the lining of the intestine composed of Paneth cells, immature columnar cells, absorptive columnar cells, goblet cells, and enterochromaffin cells

cuboidal epithelium (kew·boy'dul ep"i·theel·ee·um): cubical cells which line hollow structures such as ducts, tubules, and secretory cells

curare (kew·rahr·ee): in small doses, a muscle relaxant; in large doses, a poison causing muscle paralysis; may act by competing with acetylcholine

current: a flow of electrical charges

cyanosis (sigh·uh·no'sis): blueing of the surface tissues of the body due to a low level of oxygenation of the blood

cyclic AMP (cAMP) (sigh'click, sick'lick): cyclic adenosine monophosphate; a "secondary messenger" to certain hormones whose activities within the cell may cause effects generally opposite to those of cGMP

cyclic GMP (cGMP): cyclic guanosine phosphate; a "secondary messenger" to certain hormones which may cause overall effects generally opposite to those of cAMP

cyclic structure: a molecular structure in which the bonds between atoms unite them in the form of a ring

cyst (sist): accumulation of gas, liquids, or solids surrounded by a membrane, which forms a sac

cystic duct: a vessel which carries bile from the gallbladder and merges with the hepatic duct of the liver

cystic fibrosis (sis'tick figh·bro'sis): a heritable disease of the glandular epithelium in the lungs, the digestive tract, and the skin, characterized by abnormal secretions, usually containing excessive amounts of sodium and chloride ions; usually fatal by young adulthood

cystinuria (sis"ti·new'ree·uh): accumulation of cystine in the urine which may form stones to damage the kidneys; a hereditary disease which impairs the reabsorption of cystine, arginine, lysine, and ornithine

-cyte (sight): a suffix denoting a cell

cytochrome (sigh'to·krome): one of a group of iron-containing proteins which enzymatically transfer electrons released from the removal of hydrogen in hydrogen transport, the end result of which is the oxidation of molecules aerobically to produce water and energy; the cytochromes include cytochrome b, c, and a

cytoplasm (sigh'to·plaz·um): intracellular material surrounding the nucleus of a cell

Dalton's law: a gas law which states that the partial pressure of a gas in a mixture of gases is proportional to its concentration and to the total pressure of the mixture of which it is a part

deamination (dee·am"i·nay'shun): removal of an amino group from a protein molecule, polypeptide, or amino acid

decompression sickness (dee'kum·presh'un): *see* the "bends"

dedifferentiation (dee"dif·ur·en"shee·ay'shun): a process in which cells lose their characteristic properties

deep breathing: *see* abdominal breathing

defecation reflex (def·e·kay'shun): a response to distension of the rectum which causes its contraction and the excretion of feces

defibrillation (dee·fib"ri·lay'shun, ·figh"bri·): interruption of fibrillation by the application of a strong artificial electric shock; usually used to coordinate the mechanical activity of the ventricles with the electrical activity of the heart

dehydrogenated fat (dee·high'druh·je·nate·ed dee·high·droj'e·): *see* unsaturated fat

delayed heat: heat formed during the recovery of a muscle as it is readied for its next contraction; the greatest amounts of ATP hydrolysis and energy production take place at this time and the globular heads of myosin are moved from the perpendicular to the more angular rigor position

delayed-type allergy: a pathological cellular response by T lymphocytes, occurring hours to days after the second or subsequent exposures to the allergen; *also see* immediate-type allergy

delta waves (del'tuh): brain waves predominant during deep sleep or unconsciousness; slow, high-voltage waves occurring about 0.5 to 3.5 times per second

dendrite (den'drite): cytoplasmic extension of a neuron which usually carries impulses toward its cell body

dense connective tissue: aggregates of cells packed closely together in space filled with regularly or irregularly arranged elastic or collagenous fibers

dentin (den'tin): hard, bonelike, acellular substance which makes up the bulk of teeth

deoxyribonucleic acid (DNA) (dee·ock"see·rye"bo-new·klee'ick as'id): the genetic material of a cell; nucleotides and the sugar deoxyribose linked in a polymer by phosphate groups in two chains which are bound together as a pair of coils by hydrogen bonds

depolarized (dee·po'lur·ized): having lost polarity; decreased potential difference across a membrane; electrically neutral

deposition: the act of adding or depositing a substance

derived lipids (lip'ids, lye'pids): compounds formed from other lipids; they include cholesterol, steroid hormones, bile acids, and some vitamins

dermatitis (dur"muh·tye'tis): inflammation of the dermis or skin caused by bacterial, fungal, or viral infections as well as by immunologic reactions

dermatome (dur'muh·tome): a segment of skin innervated by sensory fibers from one pair of spinal nerves

dermis (dur'mis): the corium; inner layer of skin consisting of dense connective tissue

descending colon (de·sen'ding ko'lun): the portion of the large intestine which leads from a segment (the transverse colon across the abdominal cavity), down to an S-shaped portion (the sigmoid colon) in the lower part of that chamber

descending tract: fibers which carry impulses down the spinal cord away from the brain

desmosome (dez'mo·sohm): intercellular junction in which protein helps firmly bind cells together

diabetes insipidus (dye"uh·bee'teez, ·bee'tis in·sip'i·dus): a disease of the hypothalamus and/or posterior pituitary which diminishes the secretion of ADH and results in excretion of up to 20 liters of urine per day

diabetes mellitus (mel·eye'tus): a disorder in which the pancreas secretes too little insulin, resulting in a diminished ability to transport glucose into liver and skeletal muscle; may lead to glucosuria, increased urine volume, and metabolism of fats for energy, causing ketosis

dialysis (dye·al'i·sis): the separation of substances according to size by a selectively permeable membrane

diapedesis (dye"uh·pe·dee'sis): ability of cells to migrate between the blood and tissue spaces by squeezing through openings in capillaries

diaphragm (dye'uh·fram): a partition separating one area from another; the thin, dome-shaped muscle which separates the thoracic cavity above from the abdominal cavity below; a birth control device which is round, rubber, and cuplike and used to separate the cervix from the vagina and thereby prevent conception

diaphragmatic breathing (dye"uh·frag·mat'ick): *see* abdominal breathing

diaphysis (dye·af'i·sis); the shaft of a long bone

diarrhea (dye"uh·ree·uh): excretion of watery feces caused by rapid transport of digested food through the large intestine, resulting in too little water reabsorption

diarthrosis (dye"ahr·thro'sis,) (pl., **diarthroses**): synovial joint; freely moving joint

diastole (dye·as'tuh·lee): relaxation of the chambers of the heart between contractions

diastolic pressure: force produced by the recoil of the elastic tissue in arterial walls, occurring during ventricular diastole; measures 80 mmHg in the average adult

differential count (dif"ur·en'shul): determination of the percentage of each type of white blood cell among a population of 100 to 200 in a stained blood smear

diffusion (di·few'zhun): the random movement of ions and molecules along concentration, electrical, or pressure gradients in gases and fluids

diffusion rate: the net rate of random movement of substances down gradients from one area to another

digestion (di·jes'chun, dye·jes'chun): the physical and chemical conversion of food molecules into smaller units that can be absorbed

dihydrotestosterone (DHT): the most potent derivative of the major male sex hormone, testosterone; a male sex steroid active in promoting spermatogenesis, the maturation and maintenance of the accessory reproductive organs, and secondary sex characteristics at adolescence

1,25-dihydroxycholecalciferol (1,25-diHCC): vitamin D_3; the most active form of vitamin D

diploë (dip'lo·ee): a spongy central mass between an inner and an outer layer of compact bone in flat bone

direct arteriovenous anastomoses (ahr·teer"ee·o-vee'nus a·nas"tuh·mo'seez): direct connections between arteries and veins without intervening capillary beds

dislocation (dis"lo·kay'shun): a tear or stretch of tissues which displaces bone from a joint

dissociation (di·so"see·ay'shun): separation; the process in which certain molecules in solution separate into particles

distal convoluted tubule (dis'tul kon"vo·lew'tid tew'bewl): the coiled segment farthest along a kidney tubule which merges with similar structures to form collecting ducts

diuresis (dye"yoo·ree'sis): formation of urine

diuretic (dye"yoo·ret'ick): a substance which promotes the formation of urine

diverticulum (dye″ver·tick′yoo·lum): a saclike opening from an organ; a herniated pouch, often formed in the colon after age 50, and may become inflamed in diverticulitis

DNA: the substance of which genes are made; hereditary material; *see* deoxyribonucleic acid

DNAse: an enzyme that cleaves DNA into nucleotides

DOPA (do′puh): dihydroxyphenylalanine; a precursor of the CNS neurotransmitter dopamine

dopamine: dihydroxyphenylethylamine; a CNS neurotransmitter which may be converted to norepinephrine

dorsal horn (dor′sul): posterior segment of the gray matter of the spinal cord containing afferent (sensory) neurons

dorsal white columns: sensory fibers for proprioception, discriminative touch, and deep sensation; a chain of three neurons which relay impulses from the posterior part of the spinal cord to the thalamus and ultimately to the opposite side of the cerebral cortex

douche (doosh): a stream of fluid delivered into the vagina by a syringe to cleanse the female reproductive tract

Drinker artificial respirator: "iron lung;" a chamber enclosing all of the body except the head and whose mechanical piston causes ventilation of the lungs by alternating positive and negative pressures

drugs: medicines used to treat disease or otherwise alter body functions; artificially administered chemicals that affect the physiology of cells, tissues and organs, by mimicking, exaggerating, or blocking the actions of normal chemical control agents of the body

ductus arteriosus (duck′tus ahr·teer′ee·o·sus): a vessel in the fetus which carries blood from the pulmonary artery directly to the aorta, bypassing the lungs and the remainder of the heart

duodenocolic reflex (dew″o·dee″no·kol′ick, ·ko′lick): a response to small and moderate distention of the duodenum which results in vigorous mass peristalsis in the colon

duodenum (dew″o′·dee″num): the first 8 to 10 in of the small intestine

dura mater (dew′ruh may′tur, mah′tur): the outermost and double layer of fibrous connective tissue, forming a tough covering surrounding the brain and spinal cord

dynorphin (di″nor·fin): a tridecapeptide formed in the brain which may exhibit 30 to 200 times more pain-killing ability than morphine

dyslexia (dis·leck′see·uh): incomplete alexia; an inability to correctly organize letters and numbers, causing failure to read or understand more than a few lines at a time

dyspnea (disp·nee′uh, dis·nee′uh): air hunger, characterized by shortness of breath; difficult or painful breathing

dystrophy (dis′truh·fee): defective or abnormal development or degeneration of cells, tissues, or organs

eccrine glands (eck′rin, ·rine, ·reen): a group of cells which produce secretions or excretions, such as sweat glands

ECG: EKG; electrocardiogram

echocardiogram (eck″o·kahr″dee·o·gram): a graphic recording of ultrasound waves reflected off heart tissues

ectoderm (eck′to·durm): outermost of the three embryonic germinal layers from which develop the nervous system, mucous membranes, and outer layers of the body

ectopic (eck·top′ick): out of place

eczema (eck′se·muh, eg·zee′muh): noncontagious inflammation causing redness, itching, and scaling of the skin and loss of serous fluid

edema (e·dee′muh): swelling of tissue caused by accumulation of fluid in intercellular spaces

efferent (ef′ur·unt): carrying away or out of, as with neurons which transmit impulses away from the central nervous system or vessels which move fluids away from a tissue or organ

egg: the female reproductive cell; an ovum

Einthoven's triangle (aeynt′ho·vun, ·ho·vuh): a hypothetical pyramid formed by three electrodes whose placement parallels the frontal axis of the heart (on the right arm, left arm and left leg); used in an ECG to record changes in potential as the heart generates electricity during the cardiac cycle

ejaculation (e·jack′yoo·lay′shun): movement of seminal fluid from the urethra to the exterior of the penis during the second stage of the fluid's secretion

EKG: ECG; electrocardiogram

elastase (e·las′tace, ·taze): an enzyme which degrades the protein elastin in elastic fibers

elastic artery (e·las′tick ahr′tur·ee): conducting artery; vessel with a predominance of elastic tissue which conducts blood away from the heart

elastic cartilage (kahr′ti·lij): highly flexible tissue with a large amount of elastic fibers; located in the external ear, the eustachian tubes, the epiglottis, and parts of the larynx

elastic components: noncontractile elements which

are stretched and recoil in and around muscle cells, transmitting force to their surface and to the bone to which the muscle is attached; series elastic components are the connective tissue elements which form tendons; parallel elastic elements are the sarcolemma and the resistance to flow offered by the sarcoplasm

elastic fibers: threadlike elements composed of elastin, an amorphous protein which imparts pliability to connective tissue

electrocardiogram (ECG, EKG) (e·leck″tro·kahr′-dee·o·gram): a graphic display of the electrical activity of the heart and the rate of heartbeat

electrochemical equilibrium (ee″kwi·lib′ree·um): a balance of the electrical charges *and* the concentration of charged particles across each side of a selectively permeable membrane, maintained by the expenditure of energy by active transport enzymes

electrocorticogram: a visual display of the electrical activity of the cerebral cortex obtained by direct placement of electrodes on it

electroencephalogram (EEG) (e·leck″tro·en·sef′uh-·lo·gram): a visual display of electrical activity of the brain (its brain waves); often recorded on the screen of an oscilloscope from impulses relayed by electrodes placed on the scalp

electromyogram (EMG) (e·leck″tro·migh′o·gram): a visual display of the electrical activity of muscle; relayed from electrodes placed on the overlying skin or in the muscle and often recorded on an oscilloscope screen

electron (e·leck′tron): a negatively charged particle which orbits the nucleus of an atom

electron transport: *see* hydrogen transport

electrophoresis (e·leck″tro·fo·ree′sis): the separation of molecules in an electrical field utilizing their varying degrees of attraction and migration to the positive and negative electrodes (anode and cathode)

elephantiasis (el″e·fan·tye′uh·sis): a parasitic infection caused by filaria, a nematode worm, which stretches tissues, obstructs lymph vessels, and causes marked edema, especially in the legs and scrotum

embolism (em′bo·liz·um): a physical blockade of a vessel by an undissolved mass, caused by any bit of matter which may obstruct blood flow

embryonic disc (em′bree·on′ick): the blastoderm; a mass of cells from the blastocyst which produces the three germinal layers of the embryo, namely, the ectoderm, mesoderm, and endoderm

emission (e·mish′un): movement of seminal fluid to the urethra during the first stage of its secretion

emphysema (em′fi·see′muh): an irreversible loss of elasticity in alveolar membranes in the lungs, resulting in reduction of the efficiency of expiration due to alveoli which fail to recoil to their usual size

emulsification (e′mul″si·fi·kay′shun): the process of maintaining lipid droplets dispersed in watery fluids

enamel (e·nam′ul): a hard coating on the surface of teeth, secreted onto dentin by specialized epithelial cells known as the enamel organ

encephalomyelitis (en·sef″uh·lo·migh″e·lye′tis): acute inflammation in the central nervous system

end artery: an artery without collateral circulation, which is the sole source of blood to a tissue; examples are the coronary arteries and middle cerebral arteries

endo-: inner or innermost

endocarditis (en″do·kahr·dye′tis): inflammation of the inner lining of the heart and/or heart valves

endocardium (en″do·kahr′dee·um): a smooth inner lining of the heart composed of simple squamous epithelium

endochondral bone formation (en″do·kon′drul): *see* intracartilaginous bone formation

endocrine gland (en′do·krin, ·krine): ductless gland; clusters of epithelial cells which synthesize hormones and release them directly into the bloodstream

endocrine system: the ductless glands of the body, which secrete hormones directly into the bloodstream

endocytosis (en″do·sigh·to′sis): the movement of material into a cell due to infolding of its cell membrane

endoderm (en′do·durm): innermost of the three embryonic germinal layers, from which the respiratory, digestive, and urinary tracts and parts of the reproductive tract develop

endogenous (en·doj′e·nus): internal; originating from sources within the body

endogenous pyrogens (en·doj′e·nus pye′ro·jens): proteins released by phagocytic cells in response to infectious agents and which cause elevation of body temperature, or fever

endolymph (en′do·limf): fluid in the middle chamber of the inner ear; its vibrations transmit energy from sound waves to stimulate mechanoreceptors in the vestibular apparatus of the semicircular canals

endometriosis (en″do·mee″tree·o′sis): the growth of the uterine mucosa outside the uterine lining

endometrium (en″do·mee′tree·um): the inner mucosal lining of the uterus

endomysium (en″do·mis′ee·um, ·miz′ee·um): the connective tissue coat of an individual muscle cell

endoneurium (en″do·new′ree·um): the connective tissue coat surrounding each axon; the neurilemma

endoplasmic reticulum (ER) (en′do·plaz·mick re·tick′yoo·lum): a network of paired membranes found throughout the cytoplasm; includes the rough endoplasmic reticulum (RER), with ribosomal granules, along which protein synthesis occurs, and the smooth endoplasmic reticulum, a site of lipid synthesis

endorphin (en·dor′fin): a natural opiate; a generic name for polypeptide neurotransmitters in the brain which exhibit the pain-killing activity of morphine

endosteum (en·dos′tee·um): a condensed layer of marrow lining the inner surface of a bone and surrounding the marrow cavity

endothelium (en″do·theel′ee·um): simple squamous epithelium lining the circulatory and lymphatic systems

endotoxin: poison in the cell wall of certain bacteria and composed of lipid and polysaccharide

end-plate potential: an excitatory change in voltage measured across a part of the muscle cell membrane, termed the motor end plate, with which the motor neuron synapses and to which acetylcholine binds after its release from the axon's terminals

endurance: in muscle, the ability to maintain physical activity against less than maximal resistance

endurance exercise: *see* submaximal exercise

enkephalin (en″kef′a·lin): one of several pentapeptide neurotransmitters formed in the brain and pituitary with greater pain-killing activity than morphine

enteric organisms (en·terr′ick): microorganisms, such as viruses or bacteria, which reside in or primarily affect the gastrointestinal system

entero-: intestinal

enterochromaffin cells (en″tur·o·kro′muh·fin): argentaffin cells; enteroendocrine cells; endocrine cells concentrated in the duodenum and upper jejunum which produce gastrointestinal hormones

enterogastric reflex: a response in the duodenum to osmotic stimuli, fats, and products of protein digestion, which results in inhibition of motility of the stomach

enterogastrone (en″tur·o·gas′trone): phantom hormone(s) with gastric activity; may include secretin and CCK-PZ.

enterokinase (en″tur·o·kigh′nace, ·naze, ·kin′ace, ·aze): an intestinal enzyme which activates enzymes that cleave proteins

enterotoxin (en″tur·o·tock′sin): a toxin which poisons the gastrointestinal tract, secreted by certain species of microorganisms

enzyme (en′zime, ·zim): an organic catalyst; a protein which alters the speed of chemical reactions without being used up in the reaction

eosinophil (ee′o·sin′uh·fil): a granular white blood cell with oxidative enzyme activity important in defense reactions against allergens and parasites; the cells' cytoplasmic granules have an affinity for the red dye eosin in a classically stained blood smear

eosinophilic: pertaining to material which has an affinity for and stains red when combined with the red acid dye eosin

epi-: a prefix meaning upon

epicardium (ep″i·kahr′dee·um): *see* visceral pericardium

epidermis (ep″i·dur′mis): the outer part of skin, consisting of as many as five distinct layers of epithelial cells

epididymis (ep″i·did′i·mis): a long coiled duct on the posterior-medial surface of the testis in which sperm are stored and mature and which leads to the vas deferens

epiglottis (ep″i·glot′is): a flap of mucous membrane–covered cartilage, attached to the root of the tongue, which forms a seal over the opening of the trachea (the glottis) during swallowing and usually prevents food or liquids from entering airways to the lungs

epilepsy (ep″i·lep′see): a neurological disorder characterized by a temporary abnormal increase in electrical activity of the brain; petit mal epilepsy occurs in brief episodes, often lasting 1 min or less; grand mal epilepsy is more prolonged and is often preceded by a warning or aura, followed by periods of unconsciousness, muscular spasms, and deep sleep

epimysium (ep″i·miz′ee·um, ·mis′ee·um): the fascia; the connective tissue coat surrounding a whole muscle

epinephrine (ep″i·nef′rin, ·reen): adrenaline; a hormone released by the adrenal medullae which elevates blood sugar and initiates a series of reactions in the body known as the "fight or flight" syndrome

epineurium (ep″i·new′ree·um): the outermost connective tissue sheath surrounding a nerve

epiphyseal disc (plate) (e·pif″i·see′ul, ep″i·fiz′ee·ul): an area between the epiphysis and diaphysis of a long bone in which cartilage cells retain their capacity for growth and conversion to bone, thereby allowing the shaft to lengthen

epiphysis (e·pif'i·sis) (pl. **epiphyses**): one of the two extremities of a long bone

episiotomy (e·piz"ee·ot'uh·mee, e·pee"see, ep"i-sigh): a surgical incision made in the tissue around the vestibule between the vagina and the anus to prevent ragged tearing of tissues during childbirth

epithelium (ep"i·theel'ee·um): a tissue which lines surfaces of the body and functions in protection, absorption, secretion, and excretion

equilibrium (ee"kwi·lib'ree·um): the state in which two opposing activities equalize each other; a balanced condition

equilibrium potential: the number of millivolts required to oppose the natural direction and force created by the concentration gradient of a specific ion

equivalent weight (eq): the combining weight of an element or ionizable compound; its atomic or molecular weight in grams divided by its valence or the charge of its ions; thus for NaOH and HCl the equivalent weight equals the molecular weight but for H_2SO_4 it is half the molecular weight

erection (e·reck'shun): stiffening and enlargement of the penis or clitoris caused by involuntary nerve impulses that increase blood flow into their cavernous tissue spaces

erythroblastosis fetalis (e·rith"ro·blas·to'sis fee-tay'lis): a hemolytic disease destroying red blood cells in the fetus and the newborn, caused by incompatibility between the blood cells of the fetus and the mother; hemolytic disease of the newborn (HDN)

erythrocyte (e·rith'ro·site): mature red blood cell

erythropoietin: a glycoprotein hormone which stimulates red blood cell production and is produced by the activity of erythropoietic factor secreted by the kidneys which converts a globulin protein in the blood into the active hormone

esophagus (e·sof'uh·gus): a muscular tube about 9.5 in long that carries food from the laryngopharynx to the stomach

essential amino acids: eight amino acids that cannot be synthesized by the adult body and must be provided in proteins in the diet; include isoleucine, leucine, lysine, methionine, phenylalanine, threonine, tryptophan, and valine

essential hypertension (high"pur·ten'shun): high blood pressure in which the cause is not apparent

estriol (es'tree·ole, es'trye'ol): a major form of estrogen; used to monitor pregnancy since it is transported into the mother's blood from the adrenal gland of fetus

estrogenic phase (es"tro·jen'ick): follicular phase of the menstrual cycle

estrogens (es"tro·jenz): female sex hormones; β-estradiol, estrone, and estriol

eustachian canal (yoo·stay'kee·un, ·stay'shun): auditory tube; eustachian tube; channel extending from the middle ear to the nasopharynx

excitable: capable of responding to stimuli; able to conduct impulses in response to stimuli

excitation-contraction coupling: occurrence of excitation and contraction in sequence, the first event causing the second, namely, a change in muscle membrane potential (excitation) causing contraction of the cell

excitatory postsynaptic potential (EPSP) (eck·sight'-uh·to"ree pohst"si·nap'tick): a change in voltage which results when neurotransmitters depolarize the postsynaptic membrane and stimulate the development of local current flow

excitement phase: the first stage of orgasm, during which the penis in the male, or clitoris in the female, becomes erect and cardiovascular, respiratory, and skeletal muscle reflexes occur

excretion (eck·skree'shun): the elimination of substances not needed by the body

exercise (eck'sur·size): physical activity in which muscles perform work and/or produce heat

exocrine gland (eck'so·krin, ·krine): a cell or group of cells which release products through ducts

exocytosis (eck·so·sigh·to'sis): cell vomiting, emiocytosis; the delivery of substances packaged in membrane-enclosed sacs, termed secretory vesicles, to the exterior of the cell

exogenous (eck"soj'e·nus): external; originating outside the body

exotoxin (eck"so·tock'sin): protein poison secreted by bacterial cells

expiration (eck"spi·ray"shun): the movement of air out of the lungs caused by an elevation of pressure within the lungs above atmospheric pressure; occurs when the size of the thorax decreases, the diaphragm relaxes and ascends and the rib cage returns down to its resting position, and elastic fibers in the lungs recoil

expiratory center (eck'spye'ruh·tor·ee): a cluster of neurons in the medulla which stimulates outward breathing by causing the relaxation of external intercostal rib muscles and the diaphragm

expiratory reserve volume: supplemental air; the volume of air which can be forcibly expired in excess of the tidal volume

external muscular coat: muscularis externa; third layer of tissue from the interior of the wall of the digestive tract; it consists of inner circular muscle and outer longitudinal muscle and the myenteric nerve plexus between them

external respiration (res″pi·ray·shun): gas exchange between the lungs and blood

exteroceptor (eck″stur·o·sep′tur): nerve ending responding to stimuli on the outer surface of the body

extra-: outside of

extracellular fluid (ECF) (eck″struh·sel′yoo·lur): fluid located outside of cells

extrapyramidal tracts (ecks″truh·pi·ram′i·dul): descending motor tracts, originating in areas of the brain outside of the cerebral cortex, which coordinate muscular activity and influence instinctive behavior; reticulospinal, vestibulospinal, and rubrospinal tracts

extrinsic factor (eck·strin′zick): vitamin B_{12}

extrinsic system: a series of chemical reactions in which clotting is initiated by damaged tissues

facilitated diffusion: movement of a substance across a cell membrane assisted by a carrier that passively but specifically transports that substance

F-actin: fibrous actin; a fibrous molecule that is the major protein in thin filaments of muscle

fallopian tube (fa·lo′pee·un): oviduct; a duct between each ovary and the uterus

Fallot's tetrad: a congenital heart defect characterized by four abnormalities; a hole in the interventricular septum, hypertrophy of the right ventricle, movement of the aorta toward the right side of the heart, and narrowing of the pulmonary artery; the defects result in mixing of oxygenated and deoxygenated blood and severe abnormalities in cardiac function

false labor: intermittent abdominal contractions which do not lead to childbirth

fasciculus (fa·sick′yoo·lus) (pl. **fasciculi**): a bundle of cells; in skeletal muscle, usually groups of 12 or more cells

fast-twitch glycolytic fibers (glye″ko·lit·ick): A fibers; skeletal muscle fibers predominant near the body surface which primarily obtain their energy from anaerobic metabolism and which contract and fatigue rapidly under exercise conditions requiring the development of greater force for short periods

fast-twitch oxidative fibers: C fibers; skeletal muscle fibers predominant in muscles deep within the body of a muscle which obtain energy from aerobic metabolism and which contract rapidly, but fatigue at moderate rates under exercise conditions requiring long-term activity

fat cell (adipocyte): major cell type in adipose connective tissue and commonly filled with oil

fatigue (fa·teeg′): a condition in which muscle runs out of energy and can no longer respond to a stimulus

fatty acid: an organic compound which consists of a carbon chain ending with a carboxyl group (COOH)

feces (fee′seez): waste products of the digestive system

feedback mechanism: a reaction in which a stimulus provokes a response which in turn stimulates or inhibits the original reaction; *see* positive feedback and negative feedback signals

femto-(f) (fem·tow): a prefix ascribing a value of one-quadrillionth (10^{-15}) of a whole

fenestrated capillary (fen′e·stray·tid): capillary that has large, circular openings 60 to 80 mm in diameter which are highly penetrable

Fenn effect: the consumption of more energy and oxygen and the release of extra shortening heat by a muscle fiber in isotonic contraction compared to one in isometric contraction

ferritin (ferr′i·tin): a complex formed between iron and protein in the cells of the mucosal lining of the intestine and liver

fertile period (fur′til): the time during the menstrual cycle when the egg is capable of being penetrated by a sperm; lasts about 6 to 24 h after ovulation

fertility (fur·til′i·tee): capability of reproduction

fertilization (fur″ti·li·zay′shun): conception; union of an egg and a sperm, usually taking place in the oviducts

fetus (fee′tus): a human embryo after the end of its eighth week of development

FEV_1 (forced expiratory volume$_1$): the amount of air breathed out during the first second of forcible exhalation following a maximal inhalation; may be used as an index of the severity of lung disorders

fever (fee′ver): an abnormal increase above the usual range of body temperature

fiber (figh′bur): a threadlike strand; in connective tissue, a strand of collagen or elastin; in muscle, a cell; in nerve, an axon

fibril (figh′bril, fib′ril): a component of a fiber in a muscle cell which is called a *myofibril;* strands within collagen fibers

fibrillation (figh″bri·lay′shun): the lack of coordination of contraction of fibers within a muscle

fibrin (figh′brin): a polymerized insoluble protein which traps blood cells as it is produced in the formation of a blood clot

fibroblast (figh′bro·blast): a small, flat, often star-shaped cell which plays an active role in the synthesis of fibers in connective tissue

fibrosis (figh·bro'sis): a condition in which connective tissue cells proliferate to replace other cell types normal to an organ

fibrous cartilage (figh'brus kahr'ti·lij): slightly flexible cartilage, found in joints, intervertebral discs, and the symphysis pubis; allows for expansion of the pelvic cavity during childbirth

fibrous pericardium (figh'brus perr"i·kahr'dee·um): the tough connective tissue membrane surrounding and supporting the heart and its double serous membranes

filtration fraction (FF) (fil·tray'shun): the percentage of plasma filtered through the glomeruli; the glomerular filtration rate divided by the rate of renal plasma flow, normally ranging between 16 and 20 percent

fimbria (fim'bree·uh) (pl. **fimbriae**) (·bree·ee): a fringe; fingerlike projections of the ends of the oviducts, which seem to guide the ovum to the uterus after its release from the ovary

final common pathway: nerve fibers of lower motor neurons which transmit impulses from upper motor neurons to skeletal muscle

first pain: rapid and localized sensation transmitted by myelinated, A-delta fibers

first stage of labor: the initiation of rhythmical uterine contractions during childbirth, starting with 10-min intervals and lasting 45 to 60 s, and gradually becoming closer in time and more intense

fissure (fish'ur): a cleft or groove; a deep depression in the outer surface of the cerebral cortex

fissure of Rolando: the central sulcus; a groove in the cerebral cortex that separates the frontal lobe from the parietal lobe

fissure of Sylvius: the lateral fissure; a deep groove in the cerebral cortex which divides the frontal lobe from the temporal lobe beneath it

flagellum (fla·jel'um) (pl. **flagella**): an organelle similar to but longer than a cilium; the part which propels a spermatozoon (sperm); the tail of a sperm

fluoroscopy (floo"ur·os'kuh·pee): projection of x-ray images on a fluorescent screen

follicle (fol'i·kul): a cluster of cells forming a sac, pouch, or cavity; in the skin a hair follicle produces hair and a sebaceous follicle or gland produces oil. Groups of lymphoid cells in lymphoid follicles are abundant in the connective tissue of the submucosa of the respiratory and gastrointestinal tracts. A mass of cells in the ovaries forms a follicle which may mature into a Graafian follicle, one of whose cells, the oocyte, develops into an ovum

follicle-stimulating hormone: *see* FSH

follicular phase: estrogenic phase; part of the menstrual cycle during which the follicle-cell-stimulating hormone causes the maturation of follicles within an ovary, which in turn produce increasing amounts of estrogens

foodborne allergies: allergies caused by substances in specific foods such as strawberries, shellfish, and chocolate

foramen ovale (fo·ray'mun o·vay'lee): a small oval opening between the atria in the fetal heart which allows the flow of blood to bypass the lungs

forebrain: the most anterior segment of the brain in early development, which includes all structures surrounding the third ventricle. It encompasses the pineal body, thalamus, hypothalamus, and cerebral hemispheres

foreskin: a cuff of skin covering the glans of the penis; prepuce

formed elements: red blood cells, white blood cells, and platelets

forward failure: a type of congestive heart failure in which the heart pumps insufficient blood to peripheral tissues

fraternal (fra·tur'nul): referring to offspring developing from the fertilization of two or more eggs and therefore not identical genetically

frequency: rate of occurrence; number of events in a specific time

frequency code: a message created by the rate and rhythm of generation of impulses by neurons

FSH: follicle-stimulating hormone, secreted from the anterior pituitary and stimulating maturation of the ova and sperm

fundus (fun'dus): the portion of an organ farthest from its outlet; fundus of the stomach, fundus of the uterus

future growth zone: the epiphyseal plate and metaphysis; the part of bone where mesenchymal cells allow for lengthening

fuzzy coat: see glycocalyx

gallbladder (gawl'blad'ur): a small sac under the liver and composed of smooth muscle lined with a mucosal membrane which stores and concentrates bile secreted by the liver

gallstones: precipitates of cholesterol, bile salts, and lecithin, which may lodge in ducts of the liver or gallbladder

gamma-aminobutyric acid (GABA): an inhibitory neurotransmitter in the central nervous system formed by removal of CO_2 from glutamic acid; heritable GABA deficiency causes Huntington's chorea

gamma globulin (gam'uh glob'yoo·lin): immunoglobulin; globular protein in the plasma whose level increases after immunization and which acts as an antibody

ganglion (gang'glee·un) (pl. **ganglia**): clusters of cell bodies of neurons outside of the CNS

gap junction: intercellular connection which forms a pore that allows passage of substances with molecular weights under 1000

gastrectomy (gas·treck'tuh·mee): partial or total surgical removal of the stomach (in which the esophagus is often reattached to the jejunum)

gastric (gas'trick): pertaining to the stomach

gastric inhibitory polypeptide (GIP): a digestive hormone, produced by enterochromaffin cells in the intestinal lining, which inhibits stomach acid production while promoting insulin release from the pancreas

gastrin (gas'trin): a hormone secreted by the stomach and duodenum which stimulates the secretion of hydrochloric acid by the stomach and digestive enzymes by the pancreas

gastrocolic reflex (gas"tro·kol'ick, ·ko'lick): a response to large volumes of food in the stomach which causes mass peristalsis in the colon

gastroenteritis (gas"tro·en·tur·eye'tis): inflammation of the mucosa of the digestive tract; enteritis

gastroesophageal sphincter (gas"tro·e·sof"uh·jee'uhl sfinck'tur): circular smooth muscle between the esophagus and the stomach; also called cardiac sphincter

gastrointestinal (gas"tro·in·tes'ti·nul) **canal:** the digestive tract

gastrointestinal mucosa (mew·ko'suh, ·zuh): the inner lining of the digestive tract composed of an epithelial lining surrounded by connective tissue and layers of smooth muscle

gastrostomy (gas·tros'tuh·mee): a surgically acquired opening through the skin into the stomach

gate model: a hypothesis which proposes the existence of channels for the passage of ions across the cell membrane which are similar to specific gates that may open to lesser or greater degrees

generator potential: a small temporary voltage which may remain local, failing to spread along the entire length of an excitable membrane

-genesis (·jen'e·sis): suffix indicating the formation of—

genetic code (je·net'ick): the pattern of building blocks called nucleotides in DNA, which account for inherited characteristics; groups of nucleotides in triplets called codons, whose sequences are responsible for heritable traits

genetics (je·net'icks): the study of heredity

germinal center (jur'mi·nul): clusters of cells in lymphoid tissue which may be stimulated to actively proliferate to form large numbers of B lymphocytes

gestation (jes·tay'shun): pregnancy, from conception until birth

gingivitis (jin"ji·vye·tis): inflammation of the gums

gland: a cell or cluster of epithelial cells which synthesizes and secretes specific substances

glans (glanz): expanded tip of the penis containing a high density of sensory nerve endings

glia (glye'uh, glee'uh): cells which "glue" nerves together; connective tissue cells of several types which surround, support, or protect neurons and their extensions

glomerular filtration rate (GFR) (glom·err'yoo·lur): the number of milliliters of plasma filtered through all the glomeruli per minute; averages 115 and 125 mL/min in adult females and males, respectively

glomerulonephritis (glom·err"yoo·lo·ne·frye'tis): inflammation of glomerular membranes in the kidneys

glomerulus (glom·err"yoo·lus) (pl. **glomeruli**): the capillary bed enclosed within a Bowman's capsule in the kidney; about 1 million are found in each kidney

glottis (glot'is): an opening into the larynx formed by folds of tissue which control air flow from the trachea to modify vocal sounds

glucagon (gloo'kuh·gon): a hormone, secreted by the pancreas, which regulates the breakdown of glycogen to glucose

glucocorticoid (gloo"ko·kor'ti·koid): steroid hormone produced in the adrenal cortex which promotes the conversion of noncarbohydrates to carbohydrates and which affects salt and water metabolism; an example is cortisone

gluconeogenesis (gloo"ko·nee"o·jen'e·sis): the conversion of noncarbohydrates to carbohydrates

glucostat: neuron in the hypothalamus whose rate of metabolism of glucose generates impulses that influence the sensations of hunger and fullness

glucosuria (gloo"ko·syoo'ree·uh): the presence of glucose in the urine

glutamic acid (gloo·tam'ick): a probable excitatory neurotransmitter in the brain

glycocalyx (glye"ko·kal'licks): a "fuzzy coat"; a gelatinous coating of glycoproteins on the inner surface of the small intestine, which protects the mucosa from degradation by enzymes, acid, bile, or ingested substances

glycogen (glye'kuh·jin): a polymerized form of glucose, often abundant in liver and muscle cells

glycogen loading: *see* carbohydrate loading

glycolipid (glye″ko·lip′id): compound lipid which includes a sugar in its structure

glycolysis (glye·kol′i·sis): "glucose-lysis;" anaerobic reactions releasing energy by breaking chemical bonds of glucose, that often is derived from glycogen stored in liver or muscle

glycoprotein (glye″ko·pro′teen, ·pro′tee·in): a carbohydrate-protein complex; see mucopolysaccharide

glycosuria (glye″ko·syoor′ee·uh): a general term indicating the presence of sugar in the urine; *see* glucosuria

goblet cell: goblet-shaped cell which secretes mucus in the epithelial lining of the small intestine

Golgi apparatus (gohl′jee): aggregations of membrane-enclosed sacs or vesicles in the cytoplasm; may be a means of transport for the exit of substances synthesized in cells

Golgi tendon organ: sensory nerve fibers which monitor tension between tendons and muscle cells

gonads (go′nads); the primary sex organs, the testes in the male and the ovaries in the female

gonorrhea (gon″uh·ree″uh): a sexually transmitted disease caused by *Neisseria gonorrhea,* which can cause sterility in both sexes and may often go undetected in females

gout (gaowt): a disorder caused by uric acid crystals (that are derived from the breakdown of nucleoproteins) and which may accumulate in joints to cause pain and may form stones which interfere with kidney function

graafian follicle (graf′ee·un): a secondary (mature) follicle near the surface of an ovary

gradient (gray′dee·unt): gradually varying difference between two areas in temperature, pressure, concentration, or electrical charges of specific substances

granule cell (gran′yool): excitatory interneuron in the cerebellar cortex, each of which forms junctions with hundreds of Purkinje cells; granule cells are the most common type of neuron in the brain, approaching 500 billion in number

granulocyte (gran′yoo·lo·site): white blood cell that contains coarse cytoplasmic granules; neutrophils, eosinophils, and basophils

gray matter: regions of the CNS composed of cell bodies and nerve fibers which are naturally gray in color; unmyelinated neurons and their components

ground substance: a thin gel consisting mostly of proteoglycans (protein-carbohydrate complexes) in the intercellular substance of connective tissue

growth hormone (GH): somatotropin; a polypeptide hormone that is secreted by the anterior pituitary gland and promotes protein synthesis and multiplication of cartilage, bone, and blood cells

gustation (gus·tay′shun): sense of taste

gyrus (jye′rus) (pl. **gyri**) (·rye): a fold or convoluted ridge on the outer surface of the cerebral cortex

haversian system (ha·vur′zhun, hay·): a microscopic building block of compact bone, composed of concentric layers of cells with intercellular channels, embedded in crystalline calcium salts and that surround a central canal penetrated by blood vessels, lymph vessels, and nerves

hay fever: an immediate-type allergic inflammation sometimes caused by pollen from trees, grasses, and shrubs

Hb: *see* hemoglobin

HCG: *see* human chorionic gonadotropin

heartbeat: a single cycle of contraction (systole) and relaxation (diastole) of all four chambers of the heart, which lasts about 0.8 s in an individual at rest

heart block: inhibition of conduction of electrical impulses from the sinoatrial node through the atrioventricular node to the ventricles and, which may be fatal; a possible effect of overdose with digitalis (digoxin)

heartburn: a burning sensation caused by entry of the acidic contents of the stomach into the esophagus

heart murmur: click; abnormal heart sound often caused by stenosis or faulty heart valves or by whirlpools of blood

heart sound: noise produced by the flow of blood during the cardiac cycle and due to vibrations and closure of heart valves as well as to vibrations of heart muscle; sounds of Korotkoff

heat exhaustion: a condition characterized by weakness, dizziness, nausea, headache, a drop in blood pressure, and fainting; caused by loss of water and salts during prolonged heat exposure

heat stroke: heat shock; a condition that may follow heat exhaustion and occur during excessive exposure to heat; inhibition of the sweating mechanism causes body temperature to rise and may result in brain damage and death

hematocrit (he·mat′o·krit): the packed red blood cell volume; the percentage of the volume of blood occupied by red blood cells

hematopoietic tissue (hee″muh·to·poy·et′ick, hemat″o, hem″uh·to): blood-forming tissue; reticular tissue in bone marrow containing stem cells which produce blood cells

hemochromatosis (hee″mo·kro″muh·to′sis): an iron

storage disease due to iron toxicity which occurs as deposits of ferritin and hemosiderin accumulate in and color cells

hemodialysis (hee″mo·dye·al′i·sis): the filtration of blood through a selectively permeable membrane

hemoglobin (hee′muh·glo″bin): Hb; an iron containing globular protein that is composed of four chains each of which can bind oxygen and transport it through the blood

hemolysis (he·mol′i·sis): the bursting of red blood cells and leakage of hemoglobin from them

hemolytic disease of the newborn (hee″mo·lit′ick): HDN; *see* erythroblastosis fetalis

hemophilia (hee″mo·fil′ee·uh): disorders characterized by prolonged blood clotting and hemorrhage. Hemophilia A, the most common hereditary type, is due to a deficiency in factor VIII caused by a recessive gene on the X chromosome; hemophilia B is due to a similar hereditary pattern in which factor IX is deficient

hemopoiesis (hee″mo·poy·ee′sis): hematopoiesis; formation of blood cells

hemorrhage: loss of blood which may be external or internal; usually due to rupture of a blood vessel

hemosiderin (hee″mo·sid′ur·in): granules produced when excess iron oxide accumulates in the liver and other tissues

hemostasis (hee″mo·stay′sis): the stoppage of bleeding

heparin (hep′uh·rin): a sulfur-containing polysaccharide released by mast cells and liver cells which acts as an anticoagulant by inhibiting thrombin activity

hepatitis (hep″uh·tye′tis): an inflammation of the liver; often caused by specific viral infections. Type A (infectious) hepatitis is transmitted most often by contaminated food or water; type B (serum) hepatitis is transmitted most often by contaminated fluid secretions

hepatocyte (he·pat′o·site, hep′uh·to·): parenchymal cell; metabolically active columnar epithelial cells in rows radiating around a central vein in each lobule of the liver

hernia (hur′nee·uh): rupture; the abnormal protrusion of an organ or body part through the wall of a cavity; *see* inguinal hernia

herpes simplex virus (hur′peez): a virus which infects mucous membranes and may attack the nervous system to cause viral encephalitis; type I herpes simplex often causes fever blisters or cold sores and is usually transmitted orally, but occasionally infects the genital tract; type II herpes simplex (genital herpes) is a common sexually transmitted disease, which is incurable and can also be passed from a mother to her unborn child

herpes zoster virus (hur′peez zos′tur): a virus which may infect and damage dorsal root neurons in the spinal cord, causing pain and rashes in dermatomal areas associated with the corresponding nerves in a disease commonly called *shingles*

hiccup (hick′up): spasm of the diaphragm during which the glottis is forced closed during inspiration

hindbrain: the most posterior segment of the brain in early embryonic development, which forms the cerebellum, pons, and medulla oblongata

histamine (his′tuh·meen, ·min): a derivative of the amino acid histidine; released by mast cells, histamine dilates capillaries and arterioles and increases their permeability and the loss of fluid into intercellular spaces

histiocyte (his′tee·o·site): a stationary macrophage; a tissue macrophage

histocompatibility antigen (his″to·kom·pat″i·bil′i·tee): transplantation antigen; specific protein on the exterior surface of cells of certain individuals; the presence of such antigens on cells of a donor and their absence in the recipient can cause immune reactions that destroy transplanted cells, tissues, or organs in the recipient

hive: itching, elevated area of the skin (wheal) caused in some people by immediate-type allergic reactions to certain foods such as strawberries or shellfish (see wheal and flare reaction)

Hodgkin's disease: a lymphoma of T lymphocytes, occurring most often in individuals 15 to 35 and over 50 years of age; cancer of lymph nodes

holocrine secretion (hol′o·krin, ho′lo): a process by which exocrine glands shed whole cells and their contents; the sebaceous glands exhibit holocrine secretion

homeostasis (ho″mee·o·stay′sis): the ability of cells and the body to maintain a relatively constant internal environment

homeotherm (ho′mee·o·thurm): a warm-blooded animal; an animal that maintains its body temperature within a narrow range

homologous (ho·mol′uh·gus): corresponding in structure

hormone: chemical control agent synthesized by specific cells and released into the blood to be delivered to specific target cells upon which it exerts regulatory effects

human chorionic gonadotropin (kor″ee·on′ick

go·nad″o·tro″pin, gon″uh·do): HCG; a hormone that is secreted by the trophoblast of the embryo and later by the chorion, whose presence in the urine indicates pregnancy; stimulates the corpus luteum to continue progesterone secretion

humoral immunity: defense against disease provided by antibodies circulating through body fluids

Huntington's chorea: a condition caused by a dominant heritable lethal gene expressed by middle age (about 40 to 45); due to degeneration of cells in the brain and most probably the loss of gamma-aminobutyric acid, an inhibitory neurotransmitter; characterized by loss of voluntary motor control, progressive mental retardation, and death

hyaline cartilage (high′uh·lun): a flexible type of cartilage characterized by cells surrounded by a translucent matrix

hyaline membrane disease: HMD; a respiratory distress syndrome (RDS) in newborn children caused by a deficiency of surfactant; involves disruption of the alveolar membranes and accumulation of protein, fluid, and fibrin in the microscopic air sacs of the lungs

hyaluronic acid (high″uh·lew·ron·ick): an acid mucopolysaccharide in the intercellular matrix of connective tissue

hyaluronidase (high′uh·lew·ron′i·dace, ·daze): a lysosomal enzyme released by certain cells including sperm, and which degrades hyaluronic acid in the intercellular substances that help bind follicular cells together around the ovum; assists penetration of the egg by a sperm

hydro: pertaining to water

hydrocephalus (high″dro·sef′uh·lus): accumulation of cerebrospinal fluid compressing the brain against the skull

hydrogenated fat: saturated fat; fat to which hydrogen has been added; the chemical addition of hydrogen atoms to unsaturated fats raises their melting points and converts them from liquids to solids at room temperature

hydrogen bond: a force produced by the attraction of a weak positive charge around the polar covalent bond of hydrogen in a molecule to an electronegative element in another molecule; though individually weak, hydrogen bonds are numerous enough cumulatively to produce strong intermolecular forces

hydrogen transport: a series of metabolic reactions in which pairs of hydrogen atoms and an electron from each are removed by enzymes from molecules to gradually release energy in controlled steps

hydrostatic pressure (high″dro·stat′ick): fluid pressure; blood pressure in blood vessels

hydroxyapatite (high·druck″see·ap′uh·tite): the major calcium phosphate crystal in bone, whose ionized structure allows for exchange of many ions between bone substance and blood: $Ca_{10}(OH)_2(PO_4)_6$

hydroxybutyric dehydrogenase (high·drock″see-bew·tirr′ick): HBD; an enzyme whose activity is elevated in the serum after myocardial infarction and which seems to be an isozyme of lactic dehydrogenase, namely LD_1

hymen (high′mun): a membrane of varying thickness and prominence which often narrows the opening of the vagina

hyper-: prefix meaning excess, high-level, above normal

hyperbaric oxygenation (high″pur·băr′ick): an increase above the normal partial pressure of oxygen

hypercapnia (high″pur·kap′nee·uh): accumulation of excessive carbon dioxide in the blood; may lead to respiratory acidosis, mental impairment, and unconsciousness

hyperemia (high″pur·ee′mee·uh): elevated blood flow to tissues, an organ, or a region of the body

hyperfunction: overactivity of a cell, tissue, organ, or system

hyperglycemia (high″pur·glye·see′mee·uh): high blood sugar

hyperlipoproteinemia (high″pur·lip″o·pro·tee·nee′-mee·uh): above normal levels of lipoproteins in blood

hyperopia (high″pur·o′pee·uh): farsightedness; good distant vision, poor near vision; a condition in which the lens focuses light behind the retina

hyperosmolar (high″pur·oz″mo′lur): pertaining to a solution whose osmolarity exceeds that of another fluid with which it is compared

hyperpolarization: a condition in which more than the usual number of positive charges accumulate on the exterior of an excitable membrane and/or more negative charges accumulate on its interior, making the difference a greater than usual total charge or potential; the presence of a larger than usual number of oppositely charged ions on different sides of a membrane

hypertension (high″pur·ten′shun): high blood pressure; a systolic pressure over 160 mmHg or a diastolic pressure greater than 95 mmHg

hyperthermia (high″pur·thur′mee·uh): body temperature above the normal range; fever

hypertonic (high″pur·ton·ick): pertaining to a fluid whose osmolarity exceeds that of plasma

hypertrophy (high″pur·truh·fee): an increase in size or bulk of cells, tissues, or organs

hypo-: prefix meaning little, low-level, below normal

hypoalbuminemia (high″po″al·bew′min·ee′mee·uh): a low blood level of albumin

hypocapnia (high″po·kap·nee·uh): a reduction in CO_2 pressure in blood; may be caused by hyperventilation, exercise, or exposure to high altitude

hypodermis (high″po·dur′mis): subcutaneous tissue; layer of loose connective tissue and of fat that supports the dermis

hypofunction (high″po·funk′shun): underactivity of cells, tissues, organs, or systems

hypoglycemia (high″po·glye·see′mee·uh): low levels of blood sugar

hypokalemia (high″po·ka·lee′mee·uh): low levels of potassium in the blood

hypoosmolar (high″po·os·mow′lahr): a solution, having an osmolarity less than that of another liquid with which it is compared

hypoproteinemia (high″po·pro″tee·in·ee′mee·uh): low level of protein in the blood

hypotension (high″po·ten′shun): low blood pressure

hypothalamus (high″po·thal·uh·mus): portion of the forebrain on the floor of the third ventricle whose neurons influence a number of autonomic functions, including appetite and thirst, temperature regulation, behavior, and sleep, and whose neurons secrete releasing and inhibitory neurohormones, as well as the hormone oxytocin and antidiuretic hormone

hypothermia (high″po·thur′mee·uh): body temperature below the normal range

hypotonic (high″po·ton′ick): having an osmolarity less than that of plasma

hypoxia (high·pock′see·uh, hi·pock′): low levels of oxygen

hysterectomy (his′tur·eck′tuh·mee): a major surgical procedure in which the entire uterus is removed

I band: isotropic band; that portion of a striated muscle cell which bends polarized light equally in all directions

identical twins: individuals of the same sex and genetics developed from one fertilized egg

ileocecal sphincter (il″ee·o·see′kul): circular smooth muscle located in the wall of the digestive tract at the junction of the ileum and cecum

ileum (il′ee·um): a segment of the small intestine 12 to 14 ft in length which leads into the large intestine

immediate-type allergy: an immune reaction caused by interaction of allergens with antibodies which bind to basophils or mast cells, releasing vasoactive compounds such as histamine within minutes to hours of the second or subsequent exposure to the allergen

immune response: a dual defense mechanism which includes protection by antibodies and cells, called respectively, humoral and cellular immunity

immunoglobulin: gamma globulin; globular protein in the plasma which acts as an antibody and is produced against antigens in increasing amounts during the immune response

immunology (im″yoo·nol′uh·jee): the study of defense mechanisms in the body

implantation: process in which the blastocyst burrows into the uterine lining 7 to 10 days after conception

impotency (im′puh·tun·see): inability of males to achieve or maintain an erection to penetrate the vagina to transfer sperm

impulse: an action potential; a change in electrical properties transmitted by redistribution of ions along excitable membranes, such as those of neurons and muscle cells

induced abortion: an intentionally caused abortion

infarct (in′fahrkt, in·fahrkt′): death of tissue due to an obstruction of a vessel which supplies it with blood

infertility (in″fur·til′i·tee): inability to conceive children

inflammatory response: the migration of plasma molecules and blood cells (such as lymphocytes, monocytes, neutrophils, and others) into an irritated tissue site, which often becomes swollen, painful, and reddened

inguinal (ing′gwi·nul): referring to the groin

inguinal hernia (ing′gwi·nul hur′nee·uh): the penetration of the inguinal canal in the groin by a loop of intestine

inhibitory postsynaptic potential (IPSP): a change in voltage which results when neurotransmitters hyperpolarize the postsynaptic membrane and inhibit the transmission of impulses

inorganic (in″or·gan′ick): not organic; those molecules lacking carbon-to-hydrogen bonds

inotropic effect (in″o·trop′ick): alteration of the strength of muscle contraction

insertion: the part of a muscle attached to a movable bone or tendon

inspiration: the process whereby air rushes into the lungs owing to differences between atmospheric pressure and that within the lungs, caused by enlargement of the thorax as the diaphragm contracts and the rib cage is elevated

inspiratory center (in·spye·ruh'to·ree): a cluster of neurons in the medulla which stimulate inward breathing by causing contraction of rib muscles and the diaphragm

inspiratory reserve volume: complemental air; the volume of air which can be forcibly inspired beyond the tidal volume

insula (island of Reil) (in·sue·luh): internal lobe of the brain, formed by an island of tissue within the cerebral hemispheres and the fissure of Sylvius above the middle of the temporal lobe, which receives impulses for taste from the thalamus

insulin (in'suh·lin, ·sue): a polypeptide hormone secreted by beta cells in the pancreas and which promotes glucose transport into liver and skeletal muscle cells

insulin shock: reactions brought on when excess insulin promotes too rapid a loss of sugar from the blood to muscle and liver and thereby seriously affects brain function and respiratory control, causing unconsciousness, coma, and sometimes death

integument (in·teg'yoo·munt): a covering; the integumentary system of the body includes the skin, hair, and nails

inter: prefix meaning between

intercalated disc (in·tur'kuh·lay"tid): intercellular junction that facilitates passage of electrical current between cardiac muscle cells

intercellular fluid: liquid between cells; interstitial fluid

interferon (in"tur·feer'on): one of a number of low molecular weight molecules released from cells such as T lymphocytes or fibroblasts, which inhibit the replication of viruses in cells; a specific type of lymphokine

intermediary metabolism: chemical reactions, catalyzed by enzymes in cells; reactions by which molecules are degraded in catabolism or united in the synthesis of new compounds in anabolism

internal respiration: gas exchange between blood and tissues

interneuron (in"tur·new'ron): a neuron between others; an internuncial neuron; a neuron which transmits impulses between others

interoceptor: a nerve ending which responds to a stimulus from an internal organ

intersegmental (in"tur·seg·men·tul): between segments; may refer to an impulse or reflex transmitted between two or more levels of the spinal cord

interstitial (in"tur·stish·ul): pertaining to spaces between other components; intercellular

interstitial cells: also called Leydig cells in the male; endocrine cells located between the seminiferous tubules, which secrete the male sex hormones, the androgens; cells between the germinal cells in the reproductive organs

interstitial cell stimulating hormone (ICSH): *see* luteinizing hormone

interstitial fluid: *see* intercellular fluid

intestinal bacterial flora: the mixture of bacterial species in the intestine, essential to normal gastrointestinal function

intima (in'ti·muh): an innermost layer; innermost layer of tissues of a blood vessel

intra-: prefix meaning within

intracartilaginous bone formation (endochondral bone formation) (in"truh·kahr'ti·laj·i·nus): the development of bone through the multiplication, enlargement, and replacement of cartilage cells by bone cells as the tissue lengthens

intracellular (in"truh·sel'yo·lur): within a cell

intracellular fluid (ICF): fluid within cells

intramembranous (mesenchymal) bone formation (in"truh·mem'bruh·nus): bone development occurring between connective tissue membranes; the process whereby flat bones grow and the diameter of long bones increases

intrapleural pressure (in"truh·ploo'rul): subatmospheric pressure created between the layers of the membranes surrounding the lungs as the thoracic wall and the lungs pull away from each other

intrapulmonic pressure: the pressure of gases within the alveoli, the microscopic air sacs of the lungs

intrasegmental (in"truh·seg·men'tul): within a segment; may refer to a nerve impulse or reflex involving one level of the spinal cord

intrauterine device (IUD) (in"truh·yoo'tur·in): one of several forms of coil or loop, usually made of plastic, which may be inserted into the uterus to prevent implantation of the embryo; it may induce abortion by causing an inflammatory response in the uterine lining

intrinsic factor (in·trin'sick): a glycoprotein, secreted by parietal cells in the stomach, which is required for absorption of vitamin B_{12} in the ileum

intrinsic system: a series of chemical reactions by which clotting is initiated by factors in the blood

inulin (in'yoo·lin): a synthetic polymer of the sugar fructose; the appearance of inulin in the urine may be measured after its intravenous administration to assess kidney function via a clearance test

in vitro (in·vee'tro): literally, in glass; pertaining to use of glass or plastic tubes or containers to grow cells, tissues, or organs in a procedure called tissue culture

in vivo (in·vee'vo): in the living animal, in contrast to studies in artificial containers, which are termed in vitro

ion (eye'on): a negatively or positively charged atom or group of atoms

ionic (electrovalent) bond (eye·on'ick): a bond between oppositely charged ions

ipsilateral (ip"si·lat'ur·ul): on the same side; pertaining to reflexes occurring on the same side of the spinal cord

iron lung: *see* Drinker artificial respirator

irreversible shock: a likely reaction to loss of 30 percent or more of blood volume, in which tissue hypoxia causes cells to shift to anaerobic metabolism, to produce organic acids that exceed the body's buffering capacity, and to release vasoactive compounds which increase fluid loss from the capillaries and result in leakage of bacterial toxins from the intestine to the blood, resulting in death

ischemia (is·kee'mee·uh): reduction of blood (and oxygen) supply to a tissue, causing local anemia; usually temporary and due to narrowing of the arterial vessel to the area

iso-: prefix meaning the same, equal

isoenzymes (eye"so·en'zimes): isozymes; slightly different molecular species of the same enzyme; molecules which catalyze the same type of chemical reaction but at different rates. Examples are creatine phosphokinases 1, 2, and 3 and lactic dehydrogenases 1, 2, 3, 4, and 5

isometric contraction (eye"so·met'rick): isometric tension; development of increasing force by a muscle whose length remains unchanged

isoosmolar (eye"soz·mow'lahr): pertaining to two solutions of the same osmolarity

isoosmotic (eye"soz·mot'ick): equal in osmotic pressure

isotonic (eye"so·ton'ick): having the same osmolarity as plasma

isotonic contraction, isotonic tension (eye"so·ton'-ick): development of force which remains the same during the shortening of a muscle

isotropic (eye"so·tro'pick): having the same properties in all directions; specifically, bending (refracting) polarized light equally in all directions, a sign of molecular homogeneity

-itis: suffix meaning inflammation of

jaundice (jawn'dis): yellowing of the mucous membranes, the skin, and sclera caused by the accumulation of the products of bile salt metabolism in the blood; often accompanies diseases of the liver, gallbladder, and blood

J-cell: juxtacapillary cell; a cell located next to a capillary in a microscopic chamber in the lung and acts as an inflation and deflation receptor, generating impulses to the brainstem that stimulate expiration and inspiration, respectively

jejunum (je·joo'num): the middle segment of the small intestine, approximately 8 to 9 ft long

juxtaglomerular (jucks"tuh·glom·err'yoo·lur): next to the glomerulus; the capillaries in Bowman's capsule in the kidney

juxtamedullary nephrons: functional microscopic units in the kidney whose renal corpuscles are adjacent to the center of the kidney and which have long loops extending to the renal papillae

kallikrein (kal"i·kree'in): a proteolytic enzyme originally isolated in the pancreas and that produces kinins (polypeptides which are vasodilators) from globular proteins in the blood; the group of molecules is called the kallikrein system and its effects in part oppose those of the renin-angiotensin system

keratin (kerr'uh·tin): a water-resistant, sulfur-containing protein found in cells of the outer layer of skin as well as in hair, toenails, and fingernails

kernicterus (kair·nick'tur·us, kur·): a serious, sometimes fatal, complication of erythroblastosis fetalis, in which high levels of bilirubin derived from hemolyzed red blood cells cause the degeneration of neurons in an infants' brain

ketone (kee'tone): product of lipid metabolism, such as acetone or acetoacetic acid, which contains a ketonic carbonyl group ($-\overset{\displaystyle O}{\overset{\|}{C}}-$)

ketonemia (kee"to·nee'mee·uh): elevation of ketone levels in the blood; observed when excess lipids are metabolized; may occur in starvation, diabetes mellitus, or with strenuous prolonged exercise

ketonuria (kee"to·new'ree·uh): excessive level of ketones in the urine

ketosis (kee"to'sis): accumulation of excess ketone bodies (such as acetoacetic and β-hydroxybutyric acids) in body tissues and fluids

kidneys: renal organs; paired, bean-shaped structures posterior to the abdominal cavity which regulate plasma osmotic pressure; which balance electrolytes, acids, and bases; secrete hormones; and which concentrate and excrete wastes

kilo (k): a prefix ascribing a value 1000 times the value of what follows

kilocalorie (kcal) (kil"o·kal'o·ree): the amount of heat necessary to raise the temperature of 1 kg (1000 g, 1 L) of water 1°C; the equivalent of 1000 small calories; a measure of the energy content of food, often called large calorie or Calorie in this context

kinesiology (ki·nee″see·ol·uh·jee): the study of voluntary muscular movement

kinin (kin'in, kigh'nin): polypeptide formed by the action of kallikrein on globular protein in the plasma and whose vasodilator activity opposes the vasoconstrictor effects of the angiotensins

Klinefelter's syndrome (XXY): trisomy; a congenital abnormality in which body cells acquire two X chromosomes and a Y chromosome (instead of one X and one Y chromosome); produces male characteristics but undeveloped sexual organs and sterility

Krause's end bulb: cold receptor; sensory receptor encapsulated in connective tissue in the skin and responsive to a decrease in temperature

Krebs cycle: *see* citric acid cycle

Kuppfer cell (ko͝op'fur): a liver macrophage

kymograph (kigh'mo·graf): a recording device that registers graphic waves; an instrument used to measure the rates and extent of muscle contraction

labia majora (lay'bee·uh): two prominent folds of skin and fatty tissue which are lateral to the labia minora and vestibule in the female; homologous to male's scrotal sacs

labia minora (lay'bee·uh mi·no'ruh): two small folds of mucous membrane and fatty tissue which surround the vagina and are anterior to and surround the clitoris; these folds are medial to the labia majora in the female

labor: uterine contractions associated with birth; *also see* true and false labor, first and second stages of labor

lactase (lack'tace, ·taze): intestinal enzyme that cleaves milk sugar, lactose, into its two building blocks, glucose and galactose

lactation (lack·tay'shun): the production of milk by mammary glands influenced by chorionic somatomammatropin, estrogens, and prolactin during and after pregnancy

lactic acid: a 3-carbon acid derived from the breakdown of glucose in the absence of oxygen

lactic dehydrogenase (LD; also **LDH):** an enzyme which removes hydrogen from lactic acid to produce pyruvic acid; LD levels increase in serum when cells are damaged in certain disorders such as myocardial infarction

lacuna (la·kew'nuh): space or small hollow in cartilage or bone in which living cells are located

lamella (la·mel'uh) (pl. **lamellae**): layer or plate of bone composed of cells in concentric circles in a haversian system

lamina propria (lam'i·nuh pro'pree·uh): a connective tissue layer beneath the basement membrane of epithelial tissue

laryngitis (lar″in·jye'tis): inflammation of the larynx

laryngopharynx (la·ring″go·făr'inks): the lower part of the throat; a passageway leading to the esophagus and to the hallway of the larynx

larynx (lăr'inks): the voicebox; the upper end of the respiratory tract below the pharynx; contains vocal cords and cartilages used to produce sound

latent period (lay'tunt): the time between the application of a stimulus and the mechanical response of a muscle fiber

lateral fissure (lat'ur·ul fish'ur): *see* fissure of Sylvius

lateral ventricle (lat'ur·ul ven'tri·kul): one of two C-shaped chambers filled with cerebrospinal fluid, each in a cerebral hemisphere

laxative (lack'suh·tiv): the least powerful of substances which enhances motility of the digestive tract; a weak cathartic

leukemia (lew·kee'mee·uh): a disease in which the quantity of specific white blood cells increases permanently owing to cancerous proliferation of specific hematopoietic cells

leukocyte (lew'ko·site): white blood cell

leukocytosis (lew″ko·sigh·to'sis): a white blood cell count above 11,000/μl; often a sign of infection

leukopenia (lew″ko·pee'nee·uh): a white blood cell count below 4000/μl; occurs in some stages of infectious disease and after exposure to radiation or drugs that inhibit the activity of bone marrow cells

leukotriene (lew·ko'try·een): one of several lipid-type relatives of prostaglandins and whose activity slowly causes smooth muscle contraction and which may be responsible for some symptoms of certain types of allergies

Leydig cells (lye'di^{k}h): interstitial cells which secrete androgens in the testes

LH: *see* luteinizing hormone

ligament (lig'uh·munt): fibrous connective tissue which, in joints, connects bone to bone or bone to cartilage and which encloses muscle or other body parts

limbic system (lim'bick): segment of the forebrain that influences feelings and emotions; tissue surrounding the brainstem and some nuclei in the thalamus, hypothalamus, and reticular formation

lipase (lye'pace, lip'ace): an enzyme which degrades fats into fatty acids and glycerol; gastric lipase or pancreatic lipase

lipid (lip'id, lye'pid): a fat or fatlike substance that is insoluble in water but does dissolve in organic solvents such as ether, chloroform, and benzene

lipoprotein (lip″o·pro′tee·in, ·teen): combination of lipid and protein; compound lipid; molecules of varying densities that are found in blood plasma which are useful in predicting the risk of heart attack

liter (lee′tur): a unit equal to 1000 mL; a volume of 1000 cm^3; 1.056 quarts

liver: an accessory digestive organ which is the largest gland in the body; it is a metabolic factory which performs storage, secretory, and circulatory functions

lobe: a subdivision of an organ

loop of Henle: the hairpin-shaped segment of a kidney tubule; composed of a thin descending limb, a thick loop, and a thick ascending limb which course toward the medullary area and then reverses direction, extending toward the cortex

lower motor neuron: neuron in the ventral horn of the spinal cord that receives impulses from an upper motor neuron in the brain

lumbar (lum′bar, ·bahr): pertaining to the small of the back

lumen (lew′min): the cavity within a tube

luteal phase (lew′tee·ul): progestational phase; part of the menstrual cycle beginning with ovulation under the influence of the luteinizing hormone from the anterior pituitary and during which increased progesterone secretion from the corpus luteum prepares the uterine lining and the woman's body for pregnancy

luteinizing hormone (LH): interstitial cell-stimulating hormone, ICSH; a polypeptide hormone secreted by the anterior pituitary that stimulates testosterone secretion in the male and ovulation and maintenance of the corpus luteum in the female; the level of LH reaches its peak in the menstrual cycle just before ovulation

lymph (limf): fluid in lymph capillaries which had escaped from circulatory capillaries

lymphatic pump (lim′fat′ick): factors which move lymph through lymph vessels; they include passive squeezing by nearby skeletal muscle and smooth muscle, compression by tissues, contraction of smooth muscle in the walls of lymph vessels, and suction in the thoracic duct

lymphoblast (lim′fo·blast): cell in the bone marrow which migrates to lymphoid tissue and differentiates to form lymphocytes

lymphocyte (lim′fo·site): agranular white blood cell usually second in abundancy to neutrophils; lymphocytes derived from the thymus (T cells) and bone marrow (B cells) are involved in cellular and humoral defense mechanisms, respectively

lymphoid tissue (lim′foid): specialized connective tissue which contains lymphocytes and is found in lymph nodes, tonsils, spleen, and thymus

lymphokine (lim′fo·kine): one of several types of molecules secreted by T lymphocytes which regulate the activity of other cells (examples are interferons and chemotactic factors)

lymphoma (lim·fo′muh): cancer of lymphoid tissue which initially tends to remain localized but may become leukemic

lysosome (lye′so·sohm): membrane-enclosed sac in the cytoplasm containing enzymes which may digest molecules and even whole cells

lysozyme (lye′so·zyme): a protein whose enzymatic activity digests cell walls, killing some species of bacteria; secreted by Paneth cells and other cells

macro-: prefix meaning large

macrocircuits: axodendritic circuits; common electrical circuits formed by a synapse of long axons from one neuron with dendrites of another

macromolecule (mack″ro·mol′e·kyool): a large molecule composed of units called monomers linked together as a polymer; polysaccharides, proteins, and nucleic acids are macromolecules

macrophage (mack′ro·faij): a large phagocytic cell common in connective tissues

macula densa (mack′yoo·luh den′suh): a group of cells of the distal tubules of the kidneys and located adjacent to the afferent arterioles of the glomeruli, and which influence feedback mechanisms that control renal blood flow

macula utriculi, macula sacculi (mack′yoo·luh) (pl. **maculae,** mack′yoo·lay): ciliated receptors covered by calcium carbonate crystals termed otoliths located in the utricle and saccule of the semicircular canals of the ears; they act as gravity sensors

malabsorption (mal″ub·sorp′shun): any disorder in which the passage of nutrients across the mucosa of the digestive tract is defective

male climacteric: a phase in male reproductive life between the ages of 40 to 50, during which testosterone secretion gradually decreases

maltase (mawl·tace, ·taze): an intestinal enzyme which splits the disaccharide maltose into two molecules of glucose

mammary glands (breasts) (mam′uh·ree): exocrine glands and accessory sex organs similar in origin and structure to sweat glands and which in the female may secrete milk for the newborn infant

mammography (ma·mog′ruh·fee): x-ray analysis of the breast provided by right-angle images of each organ

marrow (măro): a specialized connective tissue in

hollow spaces in bone which forms blood cells and stores fat

mass peristalsis: rapid transportation of food over great distances of the gastrointestinal tract as a response to abnormally strong or irritating stimuli

mast cell: tissue basophil; round or polygonal connective tissue cell which contains large amounts of heparin and histamine; often found adjacent to capillaries

matrix (may'tricks): intercellular substance; material between connective tissue cells composed of a variety of fibers and chemicals

maximal exercise: a physical activity in which O_2 is used at rates beyond the rate of supply

maximal tubular reabsorptive capacity: *see* T_m

maximum heart rate: the highest heart rate recorded for an individual during a medically supervised graded exercise test

mechano-: referring to changes in pressure and tension

media (mee'dee·uh): the middle of three layers in a blood vessel, composed of circular smooth muscle and elastic and collagenous tissues

mediated transport: movement of substances with the assistance of components in the cell membrane

Mediterranean anemia: *see* Cooley's anemia

medulla (me·dul'uh): inner portion of an organ

medulla oblongata (me·dul'uh ob"long·gay'tuh ·gah'tuh): lowest part of the brain, and is connected to the spinal cord; contains regulatory centers for vital activities that exert respiratory, cardiac, and vasomotor control

medullary cavity (med'yoo·lerr·ee, med'uh, me·dul'-ur·ee): the marrow cavity of bone

mega- (M): a prefix ascribing a value one million times the value of what follows

megakaryocyte (meg"uh·kăr"ee·o·site): very large cell of the bone marrow which produces platelets by pinching off bits of membrane and cytoplasm

meiosis (migh·o'sis): a form of nuclear cell division in which the chromosome number of reproductive cells is reduced to half that found in most body cells, with the chromosomes duplicating only once as a cell undergoes two successive divisions to produce four cells; reductive cell division; unequal cell division

Meissner's corpuscle (mice'nurs): sensory receptor which is responsive to gentle touch and is encapsulated in connective tissue within the outer ridges of the dermis of the skin

melanin (mel'uh·nin): a black to brown pigment giving color to the skin, hair, eye, adrenal medulla, brain, and some tumors

melanocyte (mel'uh·no·site, me·lan'o·): neuroectodermal cells in the stratum germinativum of the skin which synthesize the pigment melanin

melanoma (mel"uh·no'muh): cancer of the melanin-producing layer of cells in the skin

menarche (me·nahr'kee): the first menstruation; the initial irregular, nonovulatory menstrual cycles of puberty

meninges (me·nin'jeez) (sing., **meninx**): three distinct layers of membranes surrounding the brain and spinal cord; they are called the dura mater, arachnoid, and pia mater

meniscus (me·nis'kus) (pl. **menisci,** me·nis'skye, ·sigh): crescent-shaped disc of fibrocartilage which separates the synovial cavity of a joint into two separate chambers

menopause (men'o·pawz): phase of a female's life in which menstrual cycles cease and she becomes incapable of reproduction

menstrual cycle (men'stroo·ul): a period in which alternating peaks of FSH, LH, and estrogens occur; it averages 28 days in length but ranges from 21 to 45 days; a period in which the ovaries produce an egg and the uterine lining is prepared for pregnancy and shed if conception does not occur as a new cycle begins

menstrual phase: a time of 3 to 5 days during which the uterine lining is shed and new follicles begin to mature

menstruation (men"stroo·ay"shun): shedding of the lining of the uterus with the loss of fluid and blood; a signal of the start of a new menstrual cycle

Merkel's touch corpuscle (mehr'k^{e}l): sensory receptor stimulated slowly by deep pressure which deflects nearby hair follicles in the skin

merocrine gland (merr'o·krin ·krine): a group of exocrine cells which release their secretions by exocytosis; salivary and sweat glands

mesenchymal bone formation (me·seng'ki·mul, mes"in·kye'mul, mez"): *see* intramembranous bone formation

mesenchyme (mes"in·kime, mez'): embryonic tissue derived from the middle embryonic germ layer, the mesoderm, which can develop into hematopoietic stem cells as well as into several different tissue types

mesentery (mes'un·terr"ee, mez'): serous membranes which support the small intestine in the abdominal cavity; intestinal peritoneal membrane

mesoderm (mez'o·durm, mes'): the middle of the three embryonic germinal layers from which the supporting and connective tissues, such as muscle, bone, and blood develop

mesothelium (mes″o·theel′ee·um, mez″): epithelium in serous membranes which line the body cavities

messenger RNA (mRNA): a form of ribonucleic acid that is encoded with genetic information from DNA in the nucleus, and that associates with ribosomes in the cytoplasm to direct the assembly of amino acids into polypeptides and proteins

metabolic acidosis (met″uh·bol′ick as″i·do′sis): abnormal decrease in pH of the blood caused by alterations other than respiratory ones

metabolic alkalosis (met″uh·bol′ick al″kuh·lo′sis): above normal increase in pH of the blood caused by alterations other than respiratory ones

metabolism (me′tab′o·liz·um): the totality of chemical reactions in living cells; anabolism and catabolism

metaphysis (me·taf′i·sis): area of spongy bone between the epiphysis and diaphysis which, with the epiphyseal disc, makes up the future growth zone in immature bones

metarteriole (met″ahr·teer′ee·ole): the smallest of the arterioles, whose walls have scattered smooth muscle cells and which deliver blood into capillaries

micelle (mi·sel, migh·sel′): water-soluble complexes of lipids; a combination of detergentlike bile salts whose hydrophilic groups mix with water and whose hydrophobic, lipid-soluble groups help maintain lipids in a dispersed state; may consist of monoglycerides, free fatty acids, and cholesterol mixed with bile salts

micro- (μ): prefix meaning very small; numerical prefix assigning a value of one-millionth (10^{-6}) of the value that follows

microcircuits: compact local electrical circuits formed by synapses between dendrites; dendrodendritic circuits; common in brain

microfilament (migh″kro·fil′uh·munt): protein molecule 4 to 6 nm in diameter found in axons and resembling actin; strand in cell cytoplasm performing various roles in different cell types; examples are centrioles, cilia and flagella, as well as actin and myosin in muscle cells

microphage (migh′kro·faij): small phagocyte; neutrophil

microtubule (migh″kro·tew′bewl): hollow tubule in cell cytoplasm which may be used to transport materials, maintain cell shape, and assist in movement

microvillus (migh″kro·vil′us) (pl. **microvilli,** ·eye): microscopic projection on the surface of a cell which increases its surface area; brush border

micturition (mick″tur·rish′un): urination

midbrain: the middle segment of the brain in which early development forms centers for hearing and vision and which contains tracts connecting anterior and posterior divisions of the central nervous system; the smallest subdivision of the brain; a heart-shaped mass of tissue nestled among the pons, cerebellum, and thalamus

milli- (m): a prefix ascribing value of one-thousandth (10^{-3}) of the value that follows

milliliter (mL) (mil′i·lee′tur): one-thousandth of a liter; a volume of 1 cm^3

mineralocorticoid (min″ur·uh·lo·kor′ti·koid): a steroid salt-retaining hormone produced in the adrenal cortex; an example is aldosterone

minipill: birth control pill which contains only progestins

minute volume: *see* cardiac output, respiratory minute volume

miscarriage (mis·kăr′ij): an abortion from natural causes; spontaneous abortion as opposed to artificially induced abortion

mitochondrion (migh″to·kon′dree·on, mit″o·) (pl. **mitochondria**): membrane-bound structure which is the major site of oxygen consumption within human cells; mitochondria contain oxidative (respiratory) enzymes and their own DNA

mitosis (migh·to′sis, mi·): equal nuclear cell division in which the chromosomes duplicate and separate into two genetically identical daughter cells

mitral valve (migh′trul): *see* bicuspid valve

molar solution (mo′lur): a solution made by measuring the molecular weight of a substance in grams and dissolving it in sufficient water to make a total volume of 1 L

mole (m): the number of grams of an element or compound equal to its molecular weight

molecular weight (MW) (mo·leck′yoo·lur): the total weight of the atoms in a molecule compared with the weight of an atom of the carbon-12 isotope

molecule (mol′e·kyool): two or more atoms joined together by bonds formed through gaining, donating, or sharing electrons

monoamine oxidase (MAO) (mon″o·am′een ock′si·dace, ·daze): an enzyme which degrades catecholamine-type neurotransmitters such as norepinephrine, serotonin, and dopamine; for example, it degrades the neurotransmitter dopamine by removing an amino group (NH_2) from it

monoblast (mon′o·blast): bone marrow cell which proliferates to form monocytes

monoclonal theory (mon″o·klo′nul): a hypothesis regarding the origin of atherosclerosis, which proposes that mutations may lead to tumor formation by proliferation of clones of smooth muscle cells in the walls of arteries, with accompanying accumulation of fat and protein

monocyte (mon'o·site): agranulocytic white blood cell which migrates to tissue spaces and matures to produce a macrophage of the reticuloendothelial system

mononuclear phagocytic system (MPS) (mon″o-new'klee·ur fag″o·sit'ick): a term applied to cells derived from monocytes

monosaccharide (mon″o·sack'uh·ride, ·rid): a simple sugar commonly containing six carbon atoms in a ring; examples are glucose and fructose

monosodium glutamate (MSG) (mon″o·so'dee·um gloo'tuh·mate): the sodium salt of glutamic acid, an excitatory neurotransmitter in the brain; used as a flavor enhancer

monosynaptic (mon″o·si·nap'tick): a reflex path involving a single junction between two neurons

"morning-after" pill: postcoital pill; pill taken daily for 4 to 6 days after intercourse whose combination of hormones prevents implantation of the embryo and induces its abortion

morula (mor'yoo·luh, mor'oo·luh): a stage in embryonic development in which a solid mass of cells is formed by divisions of the zygote within hours of fertilization

mossy fiber: nerve fiber that receives impulses from nuclei of the spinocerebellar tracts and the pons and transmits excitatory impulses to the granule cells of the cerebellar cortex and to neurons in the cerebellar medulla

motility: capacity for spontaneous movement

motor cortex: the thin outer layer of the cerebral hemispheres in the precentral gyrus and anterior to it in the frontal lobe, and whose neurons initiate voluntary muscular movement

motor end plate: specific region of the skeletal muscle cell membrane forming a junction with an axon terminal from a motor nerve

motor unit: an individual motor neuron, its fibers and their axon terminals which form synaptic junctions with skeletal muscle cells

mountain sickness: a condition affecting some individuals who move too rapidly from sea level to high altitudes; the low P_{O_2} in the atmosphere may cause dizziness, confusion, headaches, fever, restlessness, vomiting, diarrhea, and increases in heart and respiratory rates

mRNA: messenger RNA

mucopolysaccharide (mew″ko·pol″ee·sack'uh·ride, ·rid): a gummy carbohydrate macromolecule with an amino containing sugar; a material common to the intercellular substance of connective tissues

mucosa (mew·ko'suh, ·zuh): the inner lining of the walls of the digestive, respiratory, excretory, and reproductive tracts

mucous membrane (mew'kus): epithelial membrane which lines the inner surface of the body and which produces a viscous fluid secretion called mucus

mucous neck cell: cell in a gastric pit of the stomach which secretes mucus

mucus (mew'kus): viscous fluid rich in glycoproteins and enzymes secreted by epithelial cells

multiple sclerosis (skle·ro'sis): a neuromuscular disorder caused by degeneration of myelin surrounding nerve fibers in the brain and spinal cord

muscle fiber: muscle cell

muscle spindle: modified skeletal muscle cells with stretch receptors and associated afferent and efferent nerve fibers that regulate muscle tone

muscle wasting: reduction in the bulk of muscle due to degradation of myofibrillar protein; often due to prolonged protein malnutrition or to immobility of a muscle for long periods

muscular artery: distributing artery; artery with a predominance of muscle tissue which carries blood directly to tissues and organs

muscular dystrophy (dis'truh·fee): any one of a number of heritable disorders in which muscles degenerate as their cells are replaced by fibrous connective tissue and fat

mutation (mew·tay'shun): permanent heritable changes in the genetic material of a cell

myasthenia gravis (migh″as·theen'ee·uh grav'is): extreme muscular weakness due to an autoimmune disease in which an individual forms antibodies against his or her own ACh receptors, leading to destruction of the receptors

myelin (migh'e·lin): a lipoprotein, rich in the compound lipid, sphingomyelin, which is abundant in the connective tissue cells that surround and insulate certain axons

myelinated: wrapped in layers of myelin

myeloblast (migh'e·lo·blast): cell in the bone marrow which proliferates to form myelocytes that later mature to produce granulocytes

myeloperoxidase (migh″e·lo·pur·ock'si·dace, ·daze): bacteria-killing enzyme released by the lysosomal granules of the cytoplasm of polymorphonuclear leukocytes; the enzyme causes the production of substances such as hydrogen peroxide to oxidize susceptible molecules and kill microorganisms

myenteric plexus (Auerbach's plexus) (migh″-en·terr·ick): a group of autonomic nerve fibers in the external muscle layer of the digestive tract

myenteric reflex: a contractile response of longitudinal smooth muscle in the intestine that overrides segmental mixing actions and causes peristalsis

myo-: a prefix which refers to muscle

myoblasts: embryonic cells which fuse to form skeletal muscle cells

myocardial infarct (migh″o·kahr′dee·uhl): death of heart muscle due to oxygen deprivation caused by obstruction of a coronary artery

myocarditis (migh″o·kahr·dye′tis): inflammation of the myocardium

myocardium (migh″o·kahr″dee·um): the bulk of the heart wall composed of cardiac muscle

myofibril (migh″o·figh′bril, ·fib′ril): microscopic strand of protein in striated muscle formed by groups of protein molecules called myofilaments

myofilament (migh″o·fil′uh·mint): an aggregate of protein molecules whose movement causes contraction of muscle cells; includes thin and thick myofilaments

myoglobin (migh″o·glo″bin): a single-chained protein in striated muscle and which has a greater attraction for O_2 than does hemoglobin

myometrium (migh″o·mee′tree·um): the thick smooth muscle wall of the uterus

myopia (migh″o·pee·uh): nearsightedness; good near or close-point vision; poor distant vision; a condition in which the lens focuses light in front of the retina

myosin (migh′o·sin, migh′uh·sin): a two-headed globular protein with a double coiled tail; each globular head has two active sites, one of which has ATPase activity, the other of which binds to actin; thick myofilaments of skeletal muscle are composed of this protein

myositis ossificans (migh″o·sigh′tis os·if′i·kanz): a rare disease in which damage to bone causes the accumulation of calcium in muscle

myotome (migh′o·tome): an area of muscle innervated by motor fibers from one pair of spinal nerves

Na^+–K^+ ATPase: *see* sodium-potassium ATPase

nano- (n): a numerical prefix assigning a value of one-billionth (10^{-9}) of the value of what follows

nanometer (nm) (nay′no·mee·tur): 10^{-9} m; 10 Å (angstrom units)

narcotic (nahr·ko′tick): medically, a drug that induces drowsiness (acts as a sedative) and inhibits pain (acts as an analgesic)

nasopharynx (nay′zo·făr′inks): the upper section of the throat above the soft palate and posterior to the mouth

necrosis (ne·kro′sis): tissue death

negative feedback signal: a message received by a cell causing an activity that generates effects opposite to that which existed originally; an opposite response signal

negative nitrogen balance (nigh′truh·jin): the loss of nitrogen (which appears in the urine) when protein degradation exceeds protein synthesis in cells

negatively charged: having more negative than positive charges

nephron (nef′ron): the microscopic functional unit of the kidneys, composed of capillaries and tubules and numbering about 1 million per organ

nephrotic syndrome (ne·frot′ick): the collective symptoms of edema, hypoproteinemia, and proteinuria accompanying structural changes or damage to the kidneys

nerve: a bundle of axons, dendrites, or both, arranged in parallel and bound by connective tissue

nerve fiber: an axon

nerve growth factor: a protein secreted by cells which causes extensions of neurons to grow toward them

nerve impulse: an action potential; *see* impulse

nerve terminals: the end of an axon composed of tiny swellings, called terminal boutons or terminal knobs, whose vesicles contain neurotransmitters

neural crest (new′rul): ectodermal cells lateral to the neural tube which proliferate in early embryonic development to form the components of the peripheral nervous system

neural plate: a layer of ectodermal tissue in the middle of an early embryo, whose growth forms ridges, the neural folds; these merge to produce a cylinder, the neural tube, which develops distinct outgrowths to form the brain and spinal cord

neurilemma (new″ri·lem′uh): a thin membranous layer around an axon formed by Schwann cells in the peripheral nervous system; *see* endoneurium

neurofibril (new″ro·figh′bril): a strand composed of proteins in the cytoplasm of neurons

neurofilament (new″ro·fil′uh·munt): a fibrous protein about 10 nm in diameter that is found in bundles and which may aid the flow of cytoplasm in axons

neurogenic (new″ro·jen′ick): caused by neural stimuli, including stress and anxiety

neuroglia (new″rog′lee·uh): connective tissue cells which surround, support, protect, and nourish neurons in the CNS; astrocytes and oligocytes

neurohormone: chemical control agent synthesized by special neurons, and released from their terminals to enter the blood for delivery to target cells; examples are ADH and oxytocin

neuroma (new′ro·muh): a tumor of the nervous system characterized by proliferation of supporting connective tissue cells and a dense outgrowth of nerve fibers in a bulbous sprout

neuromuscular junction (new″ro·mus′kew·lur): the junction between a neuron and a muscle cell membrane

neuron (new′ron): nerve cell; excitable cell in the nervous system

neurosecretory cell (new″ro·se·kree′tuh·ree): special type of neuron that functions like an endocrine and synthesizes and releases neurohormones into the blood to affect distant target cells

neurotransmitter (new″ro·trans·mit′ur): chemical agent secreted from the terminals of neurons to affect nearby cells

neurotubule (new″ro·tew′byool): tubular element about 25 nm in diameter which may offer structural support to a neuron and convey cytoplasm through it; a neurotubule is composed of 13 parallel fibrous protein filaments about 5 nm in diameter arranged in a circle

neutral fat: triglyceride; glycerol combined with three fatty acids

neutron (new′tron): neutral particle found with protons in the nucleus of an atom

neutrophil (new″truh·fil): usually the most common white blood cell whose cytoplasmic granules stain equally well with acid and alkaline dyes and whose nucleus is often lobed; a microphage

neutrophilic (new″truh·fil·ick): describes material which will readily stain with either acid or alkaline dyes (absorbing both red and blue dyes in a stained blood smear)

nexus (neck′sus): a gap junction between involuntary muscle cells, measuring less than 2 nm and providing a low-resistance pathway for current flow

nicotinamide adenine dinucleotide (NAD) (nick′-uh·tin·uh·mide ad′i·neen dye·new′klee·o·tide): a coenzyme which accepts hydrogen atoms from compounds in oxidative reactions and transfers them to cytochrome B

Nissl granule: modifications of the rough endoplasmic reticulum the ribosomal granules of which synthesize proteins in neurons

nitrogen narcosis (nigh′truh·jin nahr·ko′sis): drunkenness of the deep; an anesthetic effect caused by increased nitrogen forced into body fluids during a deep-water dive

nociceptor (no″si·sep′tur): naked nerve fiber which generates pain impulses in response to irritating or harmful stimuli; algesireceptor

node of Ranvier: a gap between adjacent Schwann cells which are wrapped around an axon

noninvasive: nonhostile; the least aggressive means of analysis requiring no damaging entry into a tissue or organ

nonpolar covalent bond: a molecular bond in which the distribution of electrons is equal between the atoms

non–rapid eye movement sleep (NREM sleep): slow-wave sleep; a pattern of sleep that is characterized by slow delta waves and which accounts for about four-fifths of total sleeping time

norepinephrine (nor·ep″i·nef′reen, ·rin): noradrenaline; a hormone produced by the adrenal medullae which elevates blood pressure and blood sugar used for energy metabolism in skeletal muscle

normo-: prefix meaning normal

normoblast: *see* pronormoblast

nuclei: clusters of cell bodies of neurons in the central nervous system, grouped together to perform common functions. Also, the portions of cells which contain the genetic material located in strands called chromosomes; the cell nucleus is surrounded by a double-layered semipermeable membrane

nucleic acid (new·klee′ick): a large molecule formed of nucleotides; DNA and RNA

nucleolus (new·klee′uh·lus): a small, often spherical, structure within the nucleus where ribosomal RNA is synthesized

nucleotide (new′klee·o·tide): molecular building block of the nucleic acids, DNA and RNA; composed of a ring structure of nitrogen and carbon, attached to a 5-carbon sugar which is bonded to a phosphate group

nucleus (new′klee·us) (pl. **nuclei**): *see* nuclei

obligatory reabsorption: diffusion of water from the more dilute fluid in the proximal convoluted tubules of a nephron in the kidney and dictated by the more concentrated blood in the peritubular capillaries

obligatory water loss: the amount of water loss necessitated by the functions of the skin, lungs, and gastrointestinal tract, as well as that lost in the urine because of the osmotic effects of solutes in the tubules of the kidneys

oils: liquid fats

olfaction: sense of smell

oliguria: scanty urine; an abnormal decrease in urine formation

oncogenic (onk″o·jen′ick): tumor-causing

on-off receptor: phasic receptor; sensory nerve ending which produces an impulse only when a stimulus is applied or removed

oocyte (o′o·site): an immature ovum; a primary oocyte which initiates meiotic cell division to produce two secondary oocytes, which complete the process by dividing into an egg and three smaller polar bodies, each with 23 chromosomes

opsin (op'sin): protein in the rods and cones of the retina which is bound to colored pigments and which is a derivative of vitamin A

optic vesicle: outgrowth of the embryonic brain that forms the retina in each eye

organ: a structure composed of different tissues interacting to perform a specific function

organic: pertaining to molecules which contain carbon-to-hydrogen bonds

organ of Corti: the site of auditory (hearing) receptors; a spiral membrane in the inner ear whose ciliated hair cells vibrate in response to specific sound waves at frequencies of 20 to 20,000 per second thereby generating nerve impulses carried by the cochlear (statoacoustic) nerve to the auditory area of the brain

orgasm (or'gaz·um): the climax of sexual excitement

orgasmic phase: third phase of orgasm in which ejaculation occurs through the penis in the male and in which dilation of the vagina occurs in the female, while several rhythmic waves of smooth muscle contraction, warmth spreading from the pelvic area, and pelvic throbbing occur in both sexes

origin: the part of the muscle attached to a tendon and bone which remain stationary while their counterparts move

oropharynx (or″o·far″inks): the central portion of the throat, posterior to the mouth between the soft palate and the hallway leading to the larynx

orthopnea (or″thup·nee·uh, or·thop'nee·uh): difficulty in breathing while lying horizontally

oscilloscope (os·il'uh·skope): an instrument which displays electrical waves such as action potentials visually, on a fluorescent screen

osmo-: referring to osmotic properties, which are determined by the number of particles per unit volume of fluid

osmolarity (oz″mo·lări·tee): a measure of the osmotic pressure of a solution, dependent on the number of solute particles; a 1-osmolar solution of a substance contains Avogadro's number of solute particles in 1 L of water

osmoreceptor (oz″mo·re·sep'tur): an hypothalamic neuron which responds to changes in the osmotic pressure of the blood and influences the sense of thirst

osmosis (oz'mo'sis, os): the diffusion of water across a selectively permeable cell membrane

osmotic pressure (oz·mot'ick, os): the tendency of water to diffuse into a solution due to the total number of solute particles dissolved in the solution

osseous tissue (os'ee·us): bone

osteo-: pertaining to bone

osteoarthritis (os″tee·o·ahr·thrigh'tis): wear-and-tear arthritis; a disease in which articular cartilage degenerates owing to production of abnormal collagen which cannot withstand mechanical stress

osteoblast (os'tee·o·blast): bone-forming cell; a cell in bone which secretes protein, lays down collagen fibers and deposits calcium phosphate crystals between the fibers to form bone; a cell that is active during growth and repair of bone

osteoclast (os'tee·o·klast): a large, multinucleated phagocyte which is active in the reconstruction of bone

osteocyte (os'tee·o·site): a bone cell; an osteoblast which becomes trapped in its secretions in bone

osteogenesis (os″tee·o·jen'e·sis): bone formation

osteomalacia (os″tee·o·muh·lay'shee·uh): sick bones; an adult variation of rickets due to deficiency of calcium or vitamin D, causes soft, less dense bones; often occurs in chronically ill, bedridden persons

osteoporosis (os″tee·o·po·ro'sis): softened, porous bones due to widened haversian canals, decreased density, and malformed matrix in bone; may be due to inadequate calcium intake, and an imbalance of glucocorticoids in relationship to sex steroids, as well as vitamin C deficiency, and hyperthyroidism; onset occurs most often in women after age 50

otolith (statolith) (o'to·lith): calcium carbonate granules whose movement on the gelatinous layer in the utricles and saccules in the ear generates impulses from sensory hair cells that create an awareness of gravity and of forces while body movement is at rest (static forces)

oval window: an opening between the middle and inner ear chambers; the foot of a small bone, the stapes of the middle ear, fits in the oval window; the pistonlike action of the stapes produces a wave in fluid in the spiral bony labyrinth, the cochlea, of the inner ear

ovariectomy (o·văr″ee·eck'tum·ee): surgical removal of the ovaries

ovaries (o'vur·eez): primary female reproductive organs; paired almond-shaped organs on either side of the uterus, which produce the female reproductive cells, or eggs, and which are the chief source of female sex hormones, the estrogens

oviduct (o'vi·dukt): one of two tubes, on either side of the uterus, extending from the ovaries to the uterus; fallopian tubes; uterine tubes

ovulation (o″vyoo·lay'shun, ov″yoo): the release of an

egg and its surrounding cells from the surface of an ovary

ovum (o'vum): a female reproductive cell; an egg

oxidation (ock″si·day'shun): the loss of electrons by an atom which becomes more positively charged; a molecule usually becomes oxidized by the loss of two hydrogen atoms

oxidative phosphorylation (ock'si·day″tiv fos″-for·i·lay'shun): oxidation of organic molecules to convert ADP into ATP in an aerobic series of reactions catalyzed by respiratory enzymes located in the mitochondria; the extraction of energy through hydrogen and electron transport from organic compounds which are degraded to synthesize ATP

oxygen debt (ock'si·jen): the excess oxygen consumed during recovery from work or exercise; the amount of oxygen needed by cells to complete the metabolism of lactic acid, which accumulates during work or exercise

oxyhemoglobin (HbO_2) (ock·si·hee'muh·glo″bin): a globular protein which has a molecule of oxygen loosely bonded to iron in each of its four chains

oxyntic cells (ock″sin'tick): *see* parietal cells

oxytocin (ock″si·to'sin): a hormone secreted by the posterior pituitary gland which strengthens uterine contractions through positive feedback signals during childbirth and causes milk to be let down during suckling

ozone (o'zone): a form of oxygen which produces molecules composed of three oxygen atoms; a product of industrial combustions which can cause accumulation of fluid in the lungs (termed pulmonary edema)

pacemaker cells: cells in smooth and cardiac muscle which generate an electrical rhythm independent of external nerve impulses; pacemaker of the heart, the sinoatrial node

Pacinian corpuscle (pa·sin'ee·un): sensory receptor encapsulated in connective tissue far within the dermis of the skin and responsive to deep pressure

palpatory method (pal'puh·to″ree): a technique of measuring blood pressure by feeling the pulse in a major artery while observing pressure levels with a sphygmomanometer

pancreas (pan'kree·us, pang'kree·us): an organ with digestive and endocrine functions that is located in the curvature of the duodenum; the source of pancreatic enzymes, bicarbonates, and the hormones insulin and glucagon

pancreatic amylase (pan″kree·at'ick, am″i·lace, ·laze): an enzyme secreted by the pancreas which splits polysaccharides into disaccharides

pancreatic duct: a vessel which delivers exocrine secretions from the pancreas to the duodenum, in which they mix with fluids delivered through the common bile duct

pancreatic lipase (lye'pace, lip·ace): an enzyme secreted by the pancreas which cleaves fats into fatty acids, glycerol, and monoglycerides

Paneth cells (pa^h·net): granular cells that are located deep within the crypts of Lieberkühn of the intestine, and which may secrete the antibacterial protein, lysozyme

Papanicolaou (Pap) smear (pa^h·pa^h·nee·ko^h·laow'): a spreading on a slide of cells taken from the upper vagina and the cervix, followed by staining and microscopic observation of cell structure, to distinguish normal and abnormal cells; often used in order to detect cervical (uterine) cancer

papilla (pl. **papillae,** pa·pil'uh·ee): nipplelike projection, found, for example, on the tongue and in the kidney; fingerlike projection or ridge in the outer dermis which with others is responsible for a fingerprint

papillary muscle (pap'i·lerr·ee): small, nipple-shaped muscle attached to tendinous strands, the chordae tendineae, which hold the atrioventricular valves in place and prevent them from turning inside out during systole

paradoxical sleep (păr″uh·dock'si·kul): *see* rapid eye movement sleep (REM sleep)

parahormone (par″uh·hor'mone): chemical control agent which can be synthesized by a number of different cell types; examples are histamine and prostaglandins

parallel circuits: a series of neurons in pathways which lie side by side

parasympathetic nervous system (par″uh·sim″-puh·thet'ick): a subdivision of the autonomic nervous system which originates in the cranial and sacral parts of the central nervous system and whose ganglionic fibers release acetylcholine; craniosacral division of the ANS

parathormone: parathyroid hormone; a polypeptide secreted by the parathyroid glands; promotes synthesis of vitamin D_3 and elevates blood calcium by causing release of the mineral from bone, its absorption across the intestine, and its reabsorption by the kidneys

parenchymal cells (pa·renk'i·mul): metabolically active cells, the hepatocytes of the liver

parietal (puh·rye'e·tul): a term referring to the wall of a cavity in the body

parietal cells: cells found in the stomach lining which secrete hydrochloric acid and the intrinsic factor; oxyntic cells

parietal pericardium: the outer of two serous membranes which surrounds the heart but which allows room for its expansion and contraction

parietal pleura (ploor'uh): serous membrane outside of the lungs and lining the wall of the thoracic cavity

Parkinson's disease: loss of some voluntary muscular control due to deficiency in the formation of the neurotransmitter dopamine in the basal ganglia in the forebrain

parotid gland (pa·rot'id): one of a pair of large salivary glands that are located in front of and below the ear, and which secrete saliva through a duct into the mouth

partial pressure (P): the force exerted by a gas in a mixture of gases; a force which is proportional to the percent concentration of a gas in a mixture; the partial pressure, Po_2, of oxygen at sea level is equal to its percentage (20.94 percent) times the total pressure of air at sea level (760 mmHg) and is about 159 mmHg

parturition (pahr"tew·rish'un): childbirth

passive diffusion: *see* diffusion

passive loading: a condition produced when a resting muscle is moderately stretched causing tension in the elastic elements and resulting in a greater force of contraction in the next response to a stimulus

pathogenic (path"uh·jen'ick): causing disease

pathology (pa·thol'uh·jee): the study of disease

penis (pee'nis): the external male reproductive organ which transmits sperm or urine through its tubular urethra

pepsin (pep'sin): an enzyme that is secreted by the chief cells in the stomach and which degrades proteins into polypeptides

pepsinogen (pep·sin'o·jen): an inactive form of pepsin

peptidase (pep'ti·dace, ·daze): an enzyme which splits peptide bonds that link amino acids to each other

peptide bond (pep'tide): a linkage between carboxyl and amino groups of two adjacent amino acids

perfusion (pur·few'zhun): the movement of fluid through tissues or organs or spaces

peri-: a prefix meaning around, surrounding

pericardial cavity (perr"i·kahr·dee·uhl): space between the two serous membranes (the visceral and parietal pericardium) which form a double-layered sac around the heart and its blood vessels

pericardial membrane: a fibroserous membrane surrounding the heart and its blood vessels; pericardium

pericarditis (perr"i·kahr·dye'tis): inflammation of the pericardium

pericardium (perr"i·kahr'dee·um): the pericardial membrane

perimysium (perr"i·mis'eeum, ·miz'ee·um): connective tissue layer surrounding a bundle of muscle fibers

perineurium (perr"i·new'ree·um): connective tissue sheath surrounding a bundle of nerve fibers

periodontal disease (perr"ee·o·don'tul): disease around a tooth; *see* gingivitis, pyorrhea

periosteal band (collar) (perr"i·os·tee·ul): a ring of tissue which surrounds the center of the shaft of developing bone and the primary ossification center, and whose cells help form marrow spaces

periosteum (perr"ee·os'tee·um): a fibrous connective tissue which adheres to the outer surface of bone

peripheral edema (pe·rif'e·rul): accumulation of excess interstitial fluid in body tissues

peripheral nervous system (PNS): all parts of the nervous system outside the central nervous system

peripheral resistance: opposition to the flow of blood in vessels

peristalsis (perr"i·stal'sis, ·stahl'sis): involuntary reflex waves of contraction of smooth muscle which move food and waste along the gastrointestinal tract

peritoneal dialysis (perr"i·to·nee'ul dye·al'i·sis): filtration of the blood across the abdominal membrane; artificial exchange of peritoneal fluid

peritoneum (perr"i·to·nee'um): serous membrane that lines the abdominal cavity and supports the digestive organs

peritonitis (perr"i·to·nigh'tis): inflammation of the membrane which lines the abdominal cavity

peritubular capillaries: a network of capillaries in the kidneys between the renal arterioles and renal venules, and which surrounds the tubules of one or more nephrons

permissiveness: one event making possible the occurrence of another; potentiation

pernicious anemia (pur·nish'us uh·nee'mee·uh): a severe anemia caused by vitamin B_{12} deficiency due to lack of the vitamin's absorption for want of the intrinsic factor

pH: a measure of acidity or alkalinity; the reciprocal of the concentration of hydrogen ions expressed as a power of 10 (for example, a pH of 7 equals a hydrogen ion concentration of 10^{-7} mol/L). The pH scale ranges from 0 to 14; a solution with a pH less than 7.0 is acid, 7.0 is neutral, and greater than 7.0 is alkaline

phagocyte (fag'o·site): a defense cell which can ingest particulate material; a macrophage or microphage

phagocytosis (fag"o·sigh·to'sis): cell eating by endocytosis; the ingestion of particles or cells by cells

phagosome (phagolysosome) (fag'o·sohm): lysosome of a phagocytic cell which contains enzymes that can degrade foreign molecules

pharmacological level (făhr"muh·ko·loj'ick·ul): abnormally high concentrations of a substance in the body due to its artificial administration or to hyperfunction of an organ

pharmacology (fahr"muh·kol'uh·jee): the study of the action of drugs and their properties

pharynx (făr·inks): the throat; the cavity at the back of the mouth where the passages to the nose, lungs, and digestive tract meet

phasic receptor (fay'zick): velocity receptor; on-off receptor; sensory nerve ending which produces an impulse only when the stimulus changes

phenylketonuria (PKU) (fen"il·kee"to·new'ree·uh): heritable accumulation of phenylalanine and phenylpyruvic acid in body fluids; associated with a deficiency in the enzyme phenylalanine hydroxylase, which prevents normal maturation of the brain, causing mental retardation in young children

pheromone (ferr'o·mone): a chemical control agent that affects the physiology and behavior of animals in a population, and most often attracts males to females during reproductive cycles

phleb-: referring to veins

phlebitis (fle·bye'tis): inflammation of tissue in walls of a vein

phospholipid (fos"fo·lip'id, ·lye'pid): a compound lipid which includes some form of phosphorus

phosphorylation (fos"for·i·lay'shun): addition of phosphate to a molecule; for example, the conversion of ADP to ATP by the addition of inorganic phosphate

photo-: referring to light

photopic vision (fo·to'pick, ·top'ick): color vision; vision in bright light due to three types of retinal cone cells whose pigments show maximal sensitivity to red, green, and violet light, respectively

physiological level (fiz"ee·uh·loj'i·kul): the normal concentration of a substance in the body, such as a hormone level in body fluids

physiology (fiz"ee·ol'uh·jee): study of the function of the whole organism and its parts

pia mater (pye'uh may·tur, pee'uh mah'tur): a delicate fibrous membrane adhering to the nervous tissue of the brain and spinal cord

pico- (p): a numerical prefix ascribing a value of one-trillionth (10^{-12}) of the value of what follows

pill: *see* combination pills

pineal body (pineal gland) (pin'ee·ul, pye'nee·ul): a cluster of neurosecretory cells posterior to the thalamus and in the roof of the third ventricle; secretes melatonin, which may influence other hormones by feedback mechanisms; contains cells whose responses to changes in light may produce a biological timing mechanism

pinocytic vesicle: a membranous sac formed by the fusion of parts of the cell membrane that fold inward during pinocytosis (cell drinking)

pinocytosis (pin"o·sigh·to'sis, pye"no): cell-drinking; the movement of water and substances in solution into a cell due to infolding of the cell membrane stimulated by salts and/or proteins in the extracellular fluid

pituitary (pi·tew'i·terr"ee): hypophysis; the "master" endocrine gland located at the base of the brain and composed of anterior and posterior divisions

PKU: *see* phenylketonuria

placebo (pla·see'bo): a substance containing no medication given as a control or merely as a suggestive effect to a patient

placenta (pluh·sen'tuh): an organ of communication that is formed from fetal and maternal components by the invasion of the uterine wall by embryonic membranes and serving to nurture the fetus by functioning as its respiratory, digestive, and endocrine systems

placental lactogen: somatomammatropin; a hormone secreted by the placenta that assists the growth of the breasts during pregnancy

plaque (plack): a microscopic or macroscopic abnormal flat patch in the wall of an artery, characterized by a buildup of fatty substances and the accumulation of smooth muscle cells, which may narrow or close the vessel

plasma (plas'muh): the fluid component of blood

plasma cell: an antibody-producing cell in connective tissue which is derived from a B lymphocyte

plasmin (plaz'min): fibrinolysin; an enzyme which hydrolyzes proteins in blood clots

plateau phase (pla·tow): second stage of orgasm, during which a few drops of mucus and some sperm may be secreted through the penis in the male, and the vagina, labia, and nipples of the breasts become engorged with blood as the clitoris retracts in the female; an increase in blood flow may warm the thighs and back in either sex

platelet (plait"lit): thrombocyte; a formed element in the blood that helps plug up holes in damaged blood vessels and releases a number of factors, some of which are essential to clotting

pleural effusion (ploor'ul e·few'zhun): elevation in fluid content of the thoracic cavity
pleural membrane (ploor'ul): pleura; double-layered serous membrane surrounding each lung
plexus (pleck'sus): an interconnected network of nerves, blood vessels, or lymph vessels
pneumo-: referring to the terms lung, respiration, air, gas
pneumoencephalography (new"mo·en·sef"uh·log'-ruh·fee): x-ray analysis of tissues and chambers of the brain following removal of the cerebrospinal fluid and its replacement with air
pneumonia (new·mo'nyuh, ·nee·uh): inflammation of the lungs, which restricts their expansion and is brought on by bacterial, viral, or fungal infections
pneumotaxic center (new"mo·tack'sick): a cluster of neurons in the upper pons which intermittently inhibits the apneustic center, thereby temporarily blocking inspiration
pneumothorax (new"mo·tho'racks): air in the pleural cavity; collapse of the lung caused by a puncture of the pleural membranes which allows air to surround the lung(s) in the pleural cavity
PNS: the peripheral nervous system
podocyte (pod'o·site): octopuslike cell of the visceral epithelium of Bowman's capsule, whose footlike cytoplasmic extensions contain slit-pores about 25 nm in diameter that influence the entry of substances from the glomerular capillaries into the renal tubule
poikilotherm (poy"ki·lo·thurm): an animal whose body temperature rises and falls with that of the environment; a cold-blooded animal such as a frog or fish
Poiseuille's law (pwah·zoeys'): an equation which states that flow rate along a vessel is a direct function of pressure differences between the ends of the vessel multiplied by the fourth power of the radius of the vessel and is an inverse function of the length of the vessel and the viscosity of the fluid:

$$F = \frac{(P_1 - P_2)\pi r^4}{8Lv}$$

where F = flow rate
P_1 and P_2 are the pressures at the ends of the vessels
r = vessel radius
L = vessel length
v = viscosity

polar body: each of the three smaller cells derived from meiotic nuclear division of the oocytes in the ovaries and which die spontaneously, leaving the egg as the sole survivor
polar covalent bond: a molecular link connecting atoms in which electrons are unevenly distributed between them
polarized: maintaining an unequal number of charges across a membrane; having positively and negatively charged sides
polycythemia (pol"ee·sigh·theem'ee·uh): an increase in the number of red blood cells; occurs with exposure to high altitude or to low levels of oxygen
polymer (pol'i·mer): a large molecule formed by the chemical union of several smaller chemical building blocks called monomers
polymorphonuclear leukocyte (PMN) (pol"ee·mor'-fo·new'klee·ur lew'ko·site): one of a few types of granulocytic white blood cells whose nuclei have a variable appearance; neutrophil, eosinophil, basophil
polyneuritis (pol"ee·new·rye'tis): inflammation of several nerves often due to deficiencies of specific vitamins, such as thiamine and nicotonic acid and sometimes due to certain infections
polyp (pol'ip): an abnormal mass or growth of mucosal cells
polypeptide (pol'ee·pep'tide): a compound formed by the linkage of several amino acids; polymer of amino acids
polyribosome (pol"ee·rye'buh·sohm): a chain of ribosomes which may be active in protein synthesis
polysaccharide (pol"ee·sack'uh·ride, ·rid): a complex carbohydrate composed of chains of simple sugar molecules
polysynaptic pathway (pol"ee·si·nap'tick): a route of travel of nerve impulses involving junctions among several neurons
polyuria (pol"ee·yoo'ree·uh): an abnormal increase in urine formation
pons (ponz): portion of the hindbrain that contains neurons which bridge various parts of the brain and centers which influence respiration
pore: a small opening; a space between or within molecules in the cell membrane; opening of sweat glands
portal vessel: blood vessel which links two capillary networks to separate areas without the usual intervening transition from arterial to capillary to venous systems; the hypothalamic-pituitary portal system and the hepatic portal system are examples
positive feedback signal: a message received by a cell causing an increase in an activity of the original source of the message
positively charged: having more positive than negative charges
positive nitrogen balance: the accumulation of nitro-

gen in excess of its loss as net protein synthesis occurs in cells

posterior pituitary (pos·teer'ee·ur pi·tew'i·terr"ee): neurohypophysis; the portion of the pituitary whose origin is shared with that of the neurons at the base of the brain and which releases antidiuretic hormone (ADH) and oxytocin, secreted by the hypothalamus, with which it has neuronal connections

postganglionic fiber (pohst"gang"glee·on'ick): a cytoplasmic extension of the cell body of a neuron of the autonomic nervous system, which extends from a ganglion to an effector tissue or organ; postganglionic fibers of the sympathetic nervous system release norepinephrine, those of the parasympathetic system release acetylcholine

postsynaptic membrane (pohst"si·nap'tick): limiting membrane of a cell to which neurotransmitters from the presynaptic membrane diffuse and bind; the membrane located on the receiving end of a junction between cells

potential (po·ten'chul): voltage; possibility for ion flow caused by differences in concentration of electrical charges across a cell membrane

potentiation (po·ten"shee·ay'shun): one event making possible the occurrence of another; permissiveness

precapillary sphincter (pree·kap'i·lerr"ee sfinck'tur): autonomically controlled smooth muscle fibers in a metarteriole which can open and close and control the amount of blood flow to true capillaries

preganglionic fiber (pree"gang·glee·on'ick): cytoplasmic extension from the cell body of a neuron of the autonomic nervous system that extends from the brain or spinal cord to a ganglion to which it releases acetylcholine

prepuce (pre'pewce): foreskin of the penis

presbyopia (prez"bee·o'pee·uh): loss of power of accommodation of the lens with aging

presynaptic membrane (pree"si·nap'tick): terminal boundary of an axon from which neurotransmitters are released; the membrane which releases neurotransmitter across a junction between cells

primary effect: the first result; the initial event

primary follicle (fol'i·kul): an immature follicle containing an ovum surrounded by layers of cells in the ovary

primary malabsorption: congenital malabsorption; inability, present since birth, to absorb nutrients from the digestive tract

primary motor cortex: neurons that control voluntary motor activity, located anterior to the central sulcus in the outer layer of the cerebrum

primary ossification center (os"i·fi·kay'shun): an area in the center of the shaft of a long bone where the transformation of cartilage to bone begins

primary sensory cortex: an area of the outer layer of the cerebrum, posterior to the central sulcus, where conscious awareness of sensory information occurs

primary shock: the initial reaction to injury; with loss of blood symptoms resembling those of fainting and reflex responses are elicited which may restore low blood pressure which is one of the symptoms of shock

P-R interval: in an electrocardiogram, a measure of the time required for an impulse to travel from the S-A node to the ventricles

progestational phase (pro"jes·tay'shun·ul): the luteal phase of the menstrual cycle in which the secretion of progesterone helps prepare the uterine lining for pregnancy

progesterone (pro"ges'tur·ohn): a female sex hormone secreted by the ovary which promotes changes that favor maintenance of pregnancy

progestin (pro·jes'tin): a compound that has the properties of progesterone

prohormone: a substance released as an inactive hormone which is converted in the blood or in tissues to a fully active hormone

prolactin (pro·lack'tin): a hormone secreted by the anterior pituitary gland which stimulates secretion of milk by the mammary glands

proliferative phase (pro·lif'ur·uh·tiv): the stage of the uterine cycle during which estrogens influence the endometrial cells and uterine blood vessels to proliferate; corresponds to the follicular stage of the ovarian cycle

pronormoblast (pro·nor'mo·blast): cell in the bone marrow which develops into a normoblast that loses its nucleus as it matures into a reticulocyte, which later is transformed into a mature erythrocyte

proprioceptor (pro"pree·o·sep'tur): mechanoreceptor which responds to changes in length or tension of tissues in muscles, tendons, or joints

proprioception: awareness of position, balance, movement

prostacyclin: a derivative of prostaglandin H_2 that inhibits clotting and induces vasodilation

prostaglandin (pros"tuh·glan'din): one of a number of fatty acid derivatives, 20 carbon-long with a 5-carbon ring, which behave as parahormones; prostaglandins A_2, D_2, E_2, I_2, F_2, and H_2 have been discovered. The molecules have different effects which may be opposing: some cause contraction of smooth muscle, others cause its relaxation; some facilitate clotting, others inhibit it

prostate (pros'tate): a doughnut-shaped gland surrounding the urethra at the neck of the bladder that produces enzymes that make seminal fluid less viscous and increase sperm motility

protein (pro'tee·in, pro'teen): a large (polymeric) molecule composed of many amino acids linked in chains by peptide bonds; complex molecules of carbon, hydrogen, nitrogen, and often sulfur

proteinuria (pro'tee·new'ree·uh, pro"tee·i·new'-ree·uh): leakage of protein into the urine

proteoglycan: complex of protein and mucopolysaccharide; found in connective tissues, where it improves the tissue's pliability; examples are hyaluronic acid, chondroitin sulfate, and keratin sulfate

prothrombin (pro'throm'bin): an inactive precursor of the active enzyme thrombin, which enzymatically converts soluble fibrinogen protein into insoluble molecules in the third and final stage of blood clotting

prothrombin time (PT): a test for clotting based on the number of seconds required for plasma to clot when it is mixed with 3.8% citrate solution and standard amounts of thromboplastin and Ca^{2+} ions; normally 12 to 17 s

proton (pro'ton): positively charged particle found in the nucleus of an atom

protoplasm (pro'tuh·plaz·um): all the material within a cell; the matter of which living organisms are composed

proximal convoluted tubule (prock'si·mul kon"-vo·lew'tid tew'bewl): the first segment of a kidney tubule, a twisted and coiled vessel adjacent to Bowman's capsule and extending toward the inner kidney

psychotropic (sigh"ko·trop'ick): affecting behavior, mood, and perception

ptomaine (to'mane, to'may·een): one of a group of odorous and potentially toxic amine molecules which are produced by the anaerobic breakdown of proteins by bacteria in the intestine and which cause abdominal discomfort; examples are putrescine and cadaverine

ptyalin (tye'uh·lin): salivary amylase

pulmonary artery (pul'muh·nerr"ee, pŏol'): a vessel that transports deoxygenated blood from the right ventricle to the lungs

pulmonary edema (e·dee'muh): accumulation of interstitial fluid of the lungs

pulmonary fibrosis (figh·bro'sis): a respiratory disorder in which gas exchange is retarded when respiratory membranes thicken through accumulation of fibrous connective tissue

pulmonary veins: vessels carrying oxygenated blood from the lungs to the left atrium

pulmonary ventilation: see respiratory minute volume

pulse: the rhythmic change in the size of vessels as the heart beats; often is felt in arteries near the surface of the body

pulse deficit: an abnormal condition in which pulse rate is less than heart rate

pulse pressure: the difference between systolic and diastolic pressure; averages 40 mmHg in adults

pumps: active transport enzymes; proteins in the cell membrane which move substances against gradients across the cell membrane

purine base (pew'reen, ·rin): a molecule which consists of a pyrimidine attached to another ring formed by two nitrogen atoms and one carbon atom; purines include adenine and guanine, which are found in DNA and RNA

Purkinje cell (poor'kin·ye^h): large flask-shaped inhibitory neuron in the cerebellar cortex

Purkinje fibers (poor'kin·ye^h): a network of modified cardiac muscle cells which relay electrical impulses from the right and left bundles of His to the muscle cells in each ventricle

P wave: a deflection or tracing in an electrocardiogram which occurs as the atria depolarize

pyelonephritis (pye"e·lo·ne·frye'tis): inflammation of the kidney and especially the renal pelvis

pyloric sphincter (pye·lo'rick sfinck'tur): circular smooth muscle in the distal part of the stomach that regulates entry of food into the small intestine

pylorus (pye·lo'rus, pi·lor'us): the opening between the stomach and the small intestine

pyorrhea (pye"o·ree'uh): inflammatory reaction causing reabsorption of bony sockets around the teeth

pyramidal tract (pi·ram'i·dul): corticospinal tract; descending motor tract which originates in the cerebral cortex of the brain, and crosses over in the medulla prior to reaching the ventral horn of the spinal cord to relay impulses for the conscious control of motor activity

pyrimidine base (pye·rim'i·deen, pi·): a nitrogen-containing organic ring structure which makes up part of DNA and RNA; pyrimidines include thymine, cytosine, and uracil

QRS complex: a pattern in an electrocardiogram occurring during ventricular depolarization

radioimmunoassay (ray"dee·o·im"yoo·no·a·say'): an immunological assay for measuring the level of a hormone in body fluids; unlabeled hormone in body fluids competes with artificially labeled radioactive hormone for antihormone antibody molecules; the formation of radioactive hormone–antihormone antibody complex decreases in proportion to the concentration of unlabeled hormone in the body fluids which are added to the mixture

rapid eye movement sleep (REM sleep): paradoxical sleep; a type of sleep lasting 5 to 30 min which occurs during dreams, with rapid eye movement and intermittent muscular twitching, in which beta waves of low voltage replace slower delta waves of high voltage; REM sleep accounts for about one-fifth of total sleeping time

RAS: reticular activating system

reabsorption: retrieval; the act of returning substances to the bloodstream

receptor potential (re·sep'tur po·ten'chul): a local or generator potential which is relayed from a sensory receptor to a second neuron to reach the central nervous system

recruitment (re·kroot'munt): the involvement of more excitable cells (muscle fibers or neurons) in an activity due to an increase in the strength of the stimulus

rectum (reck'tum): the terminal segment of the intestinal tract, nearly 5 in long and surrounded by sphincters to regulate the excretion of feces

reduced hemoglobin (HHb): a hemoglobin molecule in which hydrogen binds to its structure and oxygen is displaced from it

reduction (re·duck'shun): the gain of electrons by an atom, which thereby becomes less positive in charge; a molecule becomes less positive via the uptake of hydrogen and its electron

referred pain: pain coming from an internal organ but mistakenly identified as originating from another part of the body; an example is pain from the heart identified as though it originated elsewhere, commonly the left arm

reflex (ree'flecks): a basic automatic involuntary response to a stimulus

reflex arc: the functional unit of the nervous system consisting of at least one afferent neuron and one efferent neuron but also usually containing at least one interneuron as well

refractory (ree·frack'tuh·ree): nonresponsive

regurgitation (re·gur"ji·tay'shun): backward flow of a liquid causing turbulent flow; may occur with blood in the circulatory system or food in the digestive system

relative refractory period: a recovery phase during which an excitable membrane may respond to a second stimulus only if it is stronger than the initial one

relaxation period: the time required for tension in a muscle fiber to decrease from its peak to zero

renal (ree'nul): referring to the kidneys

renal calyx (pl. **calyces,** kay'licks, kal'i·seez): expanded cavity which collects urine from the renal pyramids and joins with others to form a funnel-shaped collecting tube, the renal pelvis

renal corpuscle (kor'pus·ul): Bowman's capsule and its glomerulus

renal papilla (pl. **papillae,** pa·pil'uh): nipplelike projections making up the tip of the renal pyramid, from which urine is excreted into the central pelvic region of the kidney

renal pelvis (pel'vis): a large, funnel-shaped cavity in the center of each kidney which collects urine formed in the renal pyramids

renal plasma flow (RPF): the rate of flow of plasma through the kidneys; about 55 percent of the total blood volume delivered to the kidneys

renal pyramid (pirr'uh·mid): one of many conical segments of the medulla of the kidneys, each of which contains the openings of ducts through which urine travels to the renal pelvis

renin (ree·nin): an enzyme released by the kidney which converts specific globular proteins in the blood to active angiotensin II; the group of molecules is called the renin-angiotensin system

rennin (ren'in): an enzyme secreted by the chief cells in the infant's stomach and that curdles milk by its action on the protein casein

residual air volume (re·zid'yoo·ul): the quantity of air remaining in the lungs after a forcible expiration

resolution phase: final stage of orgasm during which sexual excitement subsides

resorption (re·sorp'shun, ·zorp): removal of a substance from a tissue followed by entry of the substance into the blood

respiration (res"pi·ray'shun): the exchange of gases between tissues of the body and their surrounding environment

respiratory acidosis: abnormal decrease in blood pH due to respiratory dysfunction such as hypoventilation; may be drug-induced

respiratory alkalosis: abnormal elevation of blood pH due to respiratory dysfunction such as hyperventilation; sometimes induced by anxiety

respiratory center: clusters of neurons in the medulla, including an inspiratory and expiratory center which control the rhythm of breathing

respiratory distress syndrome (RDS): one of several

disorders in which acute respiratory difficulty develops unexpectedly; an example is hyaline membrane disease in newborn children

respiratory gas chemosensor: one of a number of receptors in each of the carotid bodies whose rapid response to changes in the composition of respiratory gases is made possible by the rapid flow of blood to them

respiratory minute volume: the quantity of air exchanged by the lungs per minute; pulmonary ventilation rate; total ventilation

respiratory quotient: the ratio of CO_2 produced to O_2 consumed as food is metabolized; equal to 1.0, 0.8, and 0.7 for pure carbohydrate, protein, and fat, respectively, averaging 0.82 for a mixed diet

response to injury theory: a hypothesis proposing that atherosclerosis develops in blood vessels following injury to the endothelial lining

resting membrane potential (resting potential): the total measurable voltage across a membrane when it is not transmitting an impulse

resuscitation (re·sus″i·tay′shun): restoration to life

resuscitator (re·sus′i·tay″tur): a mechanical device which moves air into and out of the lungs

reticular activating systems (RAS) (re·tick′yoo·lur): reticular formation; neurons in the medulla, pons, and midbrain which collectively seem to stimulate arousal and may form a "waking center"

reticular tissue (re·tick′yoo·lur): connective tissue which forms the framework of many organs, such as the liver, spleen, and kidneys, as well as lymphoid, reticuloendothelial, and blood-forming tissues

reticulocyte (re·tick′yoo·lo·site): immature, nonnucleated red blood cell; young red blood cell with abundant ribosomal granules and endoplasmic reticulum

reticuloendothelial system (RES) (re·tick″yoo·lo-en″do·theel′ee·ul): collective name for all connective tissue cells which are phagocytic

retina (ret′i·nuh): layers of neurons in the inner surface of the eye containing photoreceptor cells

retrograde transport (ret′ro·grade): movement of substances in an axon toward the cell body of a neuron

retrolental fibroplasia (ret″ro·len″tul figh″bro·play′-zhuh, ·zee·uh): literally, the production of fibrous tissue behind the lens; a disorder which may occur owing to exposure to air with a high percent of O_2 at a high partial pressure of O_2, causing the blood vessels to constrict and pull away from the retina

reverberating circuits: pathways formed by loops of neurons in which impulses return to the cells of their origin until fatigue or inhibition stops the activity

Rh antigen (Rh factor): a molecule, first discovered in red blood cells of the rhesus monkey, which is present on most human red blood cells; its presence provokes antibody formation in recipients who lack the antigen (Rh-negative persons) and whose serum may clump cells containing the antigen after such a response

rheumatic fever (roo·mat′ick): an acute inflammatory condition characterized by fever and often caused by infection which leads to inflammation of the joints and sometimes of the heart valves as well; a response to streptococcal infection in which antistreptococcal antibodies may cause an inflammation of the heart valves termed endocarditis; rheumatism

rheumatism (roo′muh·tiz·um): a disease of muscle nerve and/or joints, causing stiffness and pain in associated tissues; rheumatic fever

rheumatoid arthritis: a disease caused by an immune reaction with the production of antibody (immunoglobulin M) against the connective tissue in joints

rhodopsin (ro·dop′sin): visual purple; a protein pigment which becomes more abundant in the retinal rods in the dark, accounting for their usefulness as photoreceptors in twilight or night vision. Rhodopsin is unstable in the light, which causes it to dissociate into its opsin and chromophore, releasing Ca^{2+} ions to hyperpolarize the external membranes of the rods

Rhogam: trade name of an anti-Rh antibody which is an anti-Rh gamma globulin; it may be given to an Rh-negative mother just after the birth of her Rh-positive baby to block the mother's reaction to Rh-positive antigens that may have seeped into her blood from the baby; Rh.(D) immunoglobulin

Rh-positive: indicative of the presence of the Rh antigen on red blood cells

rhythm method: method of conception control in which sexual intercourse is avoided during the fertile period of the female

ribonucleic acid (RNA) (rye″bo·new·klee′ick): polymer of nucleotides containing the sugar ribose; different forms include messenger RNA (mRNA), transfer RNA (tRNA), and ribosomal RNA (rRNA)

ribosomal RNA (rRNA) (rye″bo·so′mul): ribonucleic acid that is formed as a complement to the nucleotides of DNA and becomes incorporated into a ribosome in the cytoplasm to serve as a station in which amino acids are united in the assembly of polypeptides and proteins

ribosome (rye'bo·sohm): flat or spherical granule that contains RNA and protein, and is attached to the endoplasmic reticulum

rickets: softening of bones in children within 18 months of birth due to an abnormal calcium/phosphate ratio; often due to vitamin D deficiency

righting reflex: capacity to maintain or return to an upright position; mediated by neurons in the midbrain; results from integration of visual and auditory impulses with motor impulses carried by the pyramidal and extrapyramidal tracts

rigor mortis (rig'ur mor'tis): stiffened condition of muscles due to failure to dissociate bonds between actin and myosin which results from depletion of ATP following death

RNA: ribonucleic acid; three forms are mRNA (messenger), tRNA (transfer), and rRNA (ribosomal)

RNase: ribonuclease; an enzyme that cleaves RNA into nucleotides

rod: photoreceptor cell in the retina which contains large amounts of rhodopsin and is sensitive to low levels of light

rough endoplasmic reticulum (RER) (en"do·plaz'mick re·tick'yoo·lum): endoplasmic reticulum that contains ribosomes; site of protein synthesis in cytoplasm

saccharide (sack'uh·ride): a sugar molecule

saccule (sack'yool): the smaller of two bulges that occur in the membranous labyrinth of the semicircular canals in the inner ear; the saccule houses sensory receptors, termed maculae sacculi, which are static or gravity sensors

saliva (suh·lye'vuh): a watery, slightly acid secretion of the salivary glands which moistens and dissolves food, digests polysaccharides, and acts as a protective coating in the digestive tract

salivary amylase (sal'i·verr·ee am"i·lace, ·laze): an enzyme in saliva which degrades polysaccharides into smaller units such as dextrins; ptyalin

salivary gland: one of three pair of glands which secrete saliva into the mouth: the parotid, submaxillary (submandibular), and sublingual glands

salpingitis (sal"pin·jye'tis): infection and inflammation of the oviducts, a common cause of female infertility; inflammation of the eustachian canal(s), which often occurs with colds and ear infections

saltatory conduction (sal'tuh·to"ree): rapid passage of an impulse by skipping from one node of Ranvier to another along a myelinated nerve fiber

sarco-: referring to muscle, literally means "flesh"

sarcolemma (sahr"ko·lem'uh): cell membrane of muscle cell

sarcomere (sahr"ko·meer): the functional unit of contraction in a striated muscle cell; it extends from Z line to Z line

sarcoplasm (sahr"ko·plaz·um): the cytoplasm of a muscle cell

sarcoplasmic reticulum: the endoplasmic reticulum of muscle whose release of calcium ions triggers contraction and whose active transport enzymes retrieve the ions into its storage sites, causing muscle to relax

satiety center (sa·tye'uh·tee, say'shee·uh·tee): neurons in the hypothalamus whose activities are stimulated by utilization of available glucose and which inhibit the feeding center and hunger pains

saturated fat (satch'uh·ray"tid): a fat in which chains of adjacent carbon atoms are bonded to one another by single bonds while simultaneously being bonded to as much hydrogen as possible

Schwann cell (shua^{h}n): a connective tissue cell which may wrap itself around peripheral nerve fibers and whose phospholipids produce a myelinated insulation; gaps between adjacent Schwann cells are called the nodes of Ranvier

scintigram (sin'ti·gram): scintiscan; an image generated by radioactivity onto a film from short-lived radioactive isotopes incorporated into a tissue or organ

sclera (skleer'uh): white of the eye; connective tissue coat on the outer surface of the eyeball continuous anteriorly with the transparent cornea and posteriorly with the connective tissue around the optic nerve

scotopic vision (sko·to'pick, ·top'ick): night vision; twilight vision due to the abundance of unbleached rhodopsin in retinal rod cells

scrotum (skro'tum): a sac outside the abdominal cavity which normally contains the testes

scurvy (skur'vee): a dietary deficiency disease due to lack of vitamin C, causing the collagen in the matrix of connective tissue to be poorly formed or malformed

sebaceous gland (se·bay'shus): an oil-secreting exocrine gland located near the base of a hair follicle in the skin

secondary follicle: a mature ovarian follicle in which an ovum and its surrounding cells have migrated to the surface of the ovary; a graafian follicle

secondary messenger: an intracellular regulator; a substance whose activity is altered by hormones and which causes further alterations in cellular functions; examples include cAMP, cGMP, and Ca^{2+} ions

secondary shock: a state which may follow primary shock; when blood volume is not restored it is characterized by a marked drop in blood pressure and tissue hypoxia

second pain: dull, aching, diffuse sensation transmitted slowly by unmyelinated C nerve fibers

second stage of labor: rhythmic uterine contractions during childbirth lasting nearly 90 s and occurring at 90- to 120-s intervals until the baby is born

secretin (se·kree′tin): a hormone that is secreted by S cells in the duodenum and that promotes the secretion of bicarbonates from pancreatic duct cells

secretion (se·kree′shun): the synthesis and release of useful substances by cells

secretory phase (se·kree·tuh·ree): the stage of the uterine cycle during which the endometrial lining thickens and, following ovulation, in which the cervix secretes a more alkaline and viscous mucus; corresponds to the luteal phase of the ovarian cycle

secretory vesicle: membranous sac formed in the cytoplasm by the Golgi apparatus which fuses with the cell membrane and releases its contents through the cell membrane by exocytosis

sedative (sed′uh·tiv): a drug that induces drowsiness or quiets nervous excitement

segmental movement: contractions of the inner circular layer of smooth muscle in the gastrointestinal tract which cause mixing of food in a specific area

selective permeability: a membrane's capacity to control passage of substances across its borders, specifically admitting some while excluding others

semen (see′mun): *see* seminal fluid

semilunar valve (sem′ee·lew′nur): a membranous structure composed of three half-moon-shaped flaps that guides blood to flow in a single direction; the pulmonary semilunar valve is found between the right ventricle and the pulmonary artery and the aortic semilunar valve between the left ventricle and the aorta

seminal fluid (sem′i·nul): a yellowish white, opalescent, slightly alkaline, thick fluid containing fructose, sperm, and secretions from the seminal vesicles, prostate, epididymis, testes, and other structures in the male reproductive tract; semen

seminal vesicle: one of a pair of compartmentalized glands which produce the bulk of semen and whose ducts join the vas deferens to form ejaculatory ducts

seminiferous tubules (sem″i·nif′ur·us, see″mi): small, highly coiled tubules whose epithelial cells produce sperm

semipermeable membrane: a barrier which exhibits selective permeability

sensor: a cell which receives signals

sensory fiber: a nerve fiber which delivers impulses to parts of the brain to produce awareness of a stimulus

sensory receptor: a modified neuronal ending whose afferent fibers terminate in the CNS and which responds to specific stimuli such as touch, pressure, pain, light, and sound

septum (sep′tum): a wall; used in reference to the division of organs, as in the interventricular septum which separates the right and left ventricles

serosa (se·ro′suh): a serous membrane; the peritoneum, a mesothelial membrane which loosely suspends the stomach and intestine in the abdominal cavity

serotonin (5-hydroxytryptamine) (seer′o·to′nin, serr″): a neurotransmitter derived from the amino acid tryptophan and synthesized in the brain, especially in the hypothalamus; serotonin depletion seems to have calming effects on electrical activity in the brain; the effects of serotonin on smooth muscle are related to the location of that tissue; serotonin may be vasoactive and may participate in some allergic reactions

serous fluid (seer′us, serr′us): a watery lubricant secreted by serous membranes

serous membrane: a thin membrane covered with simple squamous epithelial cells, and strengthened by connective tissue, and which lines cavities, supports organs, and secretes a watery (serous) fluid to reduce friction between tissues and organs; examples are the pleural, pericardial, and peritoneal membranes

Sertoli cell (serr′to·lee): one of many cells sandwiched between sperm-forming cells in the basement membrane of the seminiferous tubules, which assist the movement of immature sperm, act as phagocytes, form a blood-testis barrier, and help concentrate androgens in the testis

serum (seer′um, serr′um): yellowish liquid part of the blood left behind after removal of the formed elements and clotting factors

sex-linked trait: X-linked trait; a trait determined by genes on that part of the X chromosome for which the corresponding Y chromosome has no genes

sex steroid: sex hormone; steroid produced in the gonads which affect sexual development and maintain the sex organs and secondary sex characteristics; androgens and estrogens are sex steroids

sexually transmitted disease: contagious diseases generally transmitted through sexual contact; common examples are gonorrhea, syphilis, and type II herpes simplex; venereal disease

shivering: weak rhythmic contractions of skeletal

muscle which generate heat but perform no useful work

shock (circulatory shock): a condition which may be caused by blood loss; characterized by decreased blood pressure and volume, rapid "thready" pulse, cool pale skin, dizziness, and fainting

shortening heat: heat released as the globular heads of myosin in muscle move to positions perpendicular to actin filaments with development of peak tension

sickle-cell anemia (sickle-cell disease): a heritable disorder due to a pair of recessive genes which cause an abnormality in the structure of the paired β chains in hemoglobin, producing red blood cells which become misshapen when oxygen levels in the blood are low

sickle-cell trait: condition of an individual who possesses one recessive gene for sickle-cell anemia and one normal gene and who can still produce a significant amount of normal hemoglobin

sigmoid colon (sig'moyd): the portion of the large intestine which is S-shaped and leads to the rectum in the lower central abdominal cavity

silicosis (sil″i·ko'sis): accumulation of particles in the lungs

simple epithelium: epithelial cells arranged in a single layer

simple exocrine gland: unicellular or multicellular glands with a single unbranched duct

simple lipid: neutral fat and wax

sinoatrial node (sigh″no·ay'tree·ul): S-A node or the pacemaker; a cluster of modified cardiac muscle cells that are located in the right atrium near the superior vena cava and generate the dominant rhythm of the heart

sinus (sigh'nus): a hollow, a cavity, or a chamber; air sinuses are hollow spaces in the bone of the skull interconnected with each other and with the nasal passages

sinusitis (sigh'nuh·sigh'tus): inflammation of the mucous membrane lining the sinuses

sinusoid (sigh'nuh·soid): a space or channel in a vein in which blood collects in pools; examples include liver and spleen sinusoids and those in the pregnant uterus

skeletal muscle: contractile tissue composed of multinucleated, striated cells, attached to the bony skeleton and under voluntary control

sleep center: neurons in the hypothalamus, pons, thalamus, and frontal lobes of the brain which influence sleep; *see* waking center

sliding filament theory: the hypothesis that skeletal muscle contraction occurs when thick and thin filaments slide past each other

slow-twitch fibers: B fibers; cells that are predominant in skeletal muscles used to maintain posture, which obtain energy from aerobic metabolism and which contract and fatigue slowly under long-term exercise conditions

slow-wave sleep: *see* NREM sleep

small intestine: a tubular segment of the gastrointestinal tract between the stomach and the large intestine, 1 to 1.5 in in diameter

smooth endoplasmic reticulum: paired membranes in the cytoplasm which lack ribosomal granules; they seem to be sites for the synthesis of steroids, carbohydrates, lipids, and other substances

smooth muscle: visceral muscle; a contractile tissue located in the walls of internal organs; a tissue composed of elongated, spindle-shaped, nonstriated, contractile, autonomically controlled cells, each containing a single nucleus

sneezing: a forceful expiratory reflex, preceded by deep inspiration and initiated by irritation of the mucous membrane of the upper respiratory tract

sodium-potassium ATPase (Na^+–K^+ ATPase): an active transport enzyme which pumps sodium out of and potassium into cells against their respective concentration gradients

solute (sol'yoot, so'lewt): a substance dissolved in a fluid

solution: a homogeneous mixture of a fluid dissolving medium (solvent) in which one or more substances (solutes) are dispersed

solvent: a fluid in which a substance is dissolved

somatic (so'mat'ick): pertaining to general body cells, particularly skeletal muscle cells

somatomammatropin (so″muh·to·mam'muh·tro″-pin): placental lactogen; a hormone secreted by the chorion which, with estrogens and prolactin, stimulates growth of the breasts of a woman during pregnancy

somatomedin (so″muh·to·mee'din): a polypeptide hormone which is produced in the liver, kidney, and muscle(?) under the influence of growth hormone and that stimulates the synthesis of proteoglycans and bone growth

somatotopic organization (so″muh·to·top'ick): orderly patterns of neurons in the cerebral cortex, correlated with the parts of the body with which they are associated; a map of the brain correlating structure with function

somatotropin (so″muh·to·tro'pin): *see* growth hormone

somesthetic sense (so″mes·thet′ick): body feeling (from Greek *soma,* body, *esthesis,* feeling)
sounds of Korotkov (kor′ut·kof): heart sounds transmitted to blood vessels, such as those noises created by cardiac vibrations, valve closure, and blood turbulence; sounds of Korotkoff or Korotkow
space of Disse: a fluid-filled interstitial space between liver parenchymal cells and liver sinusoids
specific dynamic action (SDA) of food: the elevation of body temperature and basal metabolic rate (BMR) after consumption of a meal; the energy expended to assimilate foods, amounting to 30, 6, and 4 percent of the energy content of proteins, carbohydrates and fats, respectively
specific gravity: the weight of a substance compared with that of an equal volume of another substance, which is usually distilled water for liquids and solids, and is hydrogen or air for gases
sperm: a male reproductive cell; a spermatozoon
spermatid (spur′muh·tid): an immature cell which is formed in the final stage of development prior to the production of mature sperm
spermatocyte (spur·mat′o·site, spur′muh·to): one of two cell types formed early in the production of sperm by meiosis; the first is a primary spermatocyte which initiates meiosis to form two secondary spermatocytes, which complete the process by the production of four spermatids, each with 23 chromosomes
spermatogonium (spur″muh·to·go′nee·um): a primitive cell type in the testis which undergoes mitosis and matures to form primary spermatocytes
spermicidal agents: sperm-killing substances
sphincter (sfinck′tur): a circular arrangement of muscle whose contraction narrows a passage and controls flow through it; located in the digestive, respiratory, and circulatory systems
sphincter of Oddi: circular smooth muscle in the walls of the pancreatic and common bile ducts that regulates the size of their opening in the duodenum
sphygmomanometer (sfig″mo·ma·nom′e·tur): an instrument used to measure arterial blood pressure
spicule (spick′yool): threadlike splinter of spongy (cancellous) bone formed by units of little beams or plates of bony connective tissue called trabeculae
spike potential: the peak of the action potential, which shows depolarization of an excitable membrane due to the rapid inward movement of positive ions; a maximal change in voltage and pattern of deflection of an electron beam displayed visually as an inverted V (a spike) on an oscilloscope screen
spinal anesthesia (spye′nul an″es·theezh′uh, ·theez-ee·uh): administration of an anesthetic by puncture of the connective tissue covering the spinal cord between the third and fourth lumbar vertebrae in adults and one to two vertebrae lower in young children
spinal nerve: one of a number of nerves which originate in pairs, one on each side of the spinal cord, as opposed to cranial nerves most of which originate in the brainstem
spinal shock: temporary loss of reflex activities controlled by lower segments of the spinal cord, due to damage to its upper levels, which removes the influence of descending pathways from segments above
spinocerebellar tract (spye″no·serr″e·bel′ur): fibers forming a relay between two neurons which transmit subconscious sensations for balance and proprioception from the spinal cord, mainly to the same side of the cerebellum
spinothalamic tract (spye″no·tha·lam′ick): sensory fibers for pain, pressure, touch, and temperature in a chain of three neurons which ascend from the spinal cord through the thalamus to ultimately reach the opposite side of the sensory portion of the cerebral cortex
spirometer (spye·rom′e·tur): a device used for measuring the volume of air exchanged by the lungs
split brain: division of the left brain from the right brain; refers to differences between the cerebral hemispheres in terms of their dominance over various functions; alludes to the ascendancy of right and left cerebral hemispheres with respect to particular functions
spontaneous abortion: a miscarriage
sprain: a stretch or tear of tissues, especially in a joint
squamous epithelium (skway′mus): an aggregation of flat, irregularly shaped epithelial cells
Starling's equilibrium of capillary exchange: the difference between the amount of fluid leaving arterial capillaries and that reabsorbed into the venous capillaries; a measure of the net filtration rate of the circulatory system
Starling's law of the heart: the Frank-Starling law of the heart: a principle which indicates that cardiac fibers generate greater energy during contraction if they are moderately stretched prior to shortening; it also reflects that the number of milliliters expelled from the heart with each beat is proportional to its blood volume during diastole
statolith (stat′o·lith): an otolith
steady state: a continual exchange of substances into

and out of cells with no apparent net change in their levels inside or outside of the cells

steatorrhea (stee'uh·to·ree'uh): flow of hard fats; in the intestine, the presence of fatty feces; the flow of fats caused by overactivity of the sebaceous glands

stem cell: undifferentiated cell, which in bone marrow may divide and mature to produce various types of blood cells and platelets

stenosis (ste·no'sis): abnormal narrowing of an opening of a valve, duct, or blood vessel

sterile (sterr'il): unable to reproduce; lacking living microorganisms and viruses capable of reproduction

steroid (steer'oid, sterr'oid): a derived lipid characterized by four interconnected rings of carbon atoms, as found in the structure of cholesterol, certain hormones and vitamin D_3

stethoscope (steth'uh·scope): an instrument used to listen to sounds, which it amplifies; a device most often used to hear lung or heart sounds from the thoracic cavity

stimulus (stim'yoo·lus): any event or substance which causes a response in a cell, tissue, organ, whole organism or their components

stomach (stum'uck): a muscular sac which mixes food with digestive juices to form a mixture called chyme and which regulates its entry into the small intestine

stratified epithelium (strat'i·fide): layered epithelial cells

stratum corneum (stray'tum, strah'tum, kor'nee-um): outermost segment of the epidermis of the skin consisting of about 20 layers of keratinized cells

stratum germinativum (jur"mi·nuh·tye'vum): innermost segment of the epidermis of the skin consisting of a single layer of columnar epithelial cells which produces all other epidermal layers; stratum basale

stratum granulosum (gran·yoo·lo'sum): portion of the epidermis of the skin consisting of about five layers of spindle-shaped cells which contain granules of keratin

stratum lucidum (lew'si·dum): the second segment of the epidermis in thick skin, consisting of about four layers of thin, clear cells which lack nuclei

stratum spinosum (spye'no·sum): the next-to-innermost segment of the epidermis of the skin, consisting of about 10 layers of prickly appearing cells with close interconnecting desmosomes

stress: any stimulus which prompts the release of corticotropin releasing factor (CRF) at a greater than normal rate; a stimulus that causes an imbalance in homeostasis; a mechanical force

striated (strye'ay·tid): striped; refers to the microscopic appearance of lines in skeletal and cardiac muscle caused by their myofilaments

stroke: *see* cerebrovascular accident

stroke volume: the number of milliliters of blood delivered from either ventricle per heartbeat

subarachnoid spaces (sub"uh·rack'noid): areas within the arachnoid membrane filled with cerebrospinal fluid

subcutaneous tissue (sub"kew·tay'nee·us): the hypodermis

subliminal (sub·lim'i·nul): a stimulus whose intensity is below that required to cause a response; subthreshold

submaximal exercise: endurance exercise; physical activity requiring less oxygen than can be supplied

submucosa (sub"mew·ko'suh): a layer of tissue inside the mucosa which in the digestive tract contains connective tissues, blood and lymph vessels, exocrine glands, and Meissner's nerve plexus

substantia gelatinosa (sub·stan'shee·uh je·lat·i·no'-suh): interneurons in the dorsal horn of the spinal cord whose release of inhibitory neurotransmitters, such as gamma-aminobutyric acid, may hyperpolarize and temporarily inhibit pain fibers

substrate (sub'strate): a substance acted on by an enzyme and converted to a product or products

succinylcholine (suck"si·nil·ko'leen): in low doses a muscle relaxant; promotes the generation of motor end-plate potentials, causing muscle contractions which lead to temporary paralysis

sulcus (sul'kus) (pl. **sulci**): a shallow depression in the outer surface of the cerebral cortex

summation (sum·ay'shun): the addition of the electrical effects of multiple stimuli; may be either spatial or temporal

summation of contractions: an accumulation of tension caused by repeated stimulation of muscle whose contractions occur before complete relaxation between responses

superficial fascia (sue"pur·fish'ul fash'uh, ee·uh): loose connective tissue attaching the dermis of the skin to underlying muscle

suppository (suh·poz'i·tor"ee): a solid body containing medication administered through body openings such as the rectum or vagina

suprasegmental: above the segment; impulses which proceed between the spinal cord and the brain

surfactant (sur·fack'tunt): a substance which reduces surface tension; phospholipids secreted on the lining of the lungs and which reduce the attraction between gas molecules and the organs' microscopic sacs, thereby increasing the ease with which the sacs widen during inspiration

sweat gland: an exocrine gland with a coiled duct leading to the skin surface, which excretes fluid as perspiration

sympathetic nervous system (sim″puh·thet′ick): a subdivision of the autonomic nervous system whose fibers originate in the thoracic and lumbar segments of the central nervous system and whose postganglionic fibers usually release norepinephrine; thoracolumbar division of the ANS

synapse (sin·aps, si·naps′): a junction between two excitable cells; examples include neuron-to-neuron junctions and neuron-to-muscle junctions (neuromuscular junctions)

synaptic cleft (si·nap′tick): an intercellular space between two excitable cells; a space between the presynaptic and postsynaptic membranes of two cells

synaptic vesicle (ves′i·kul): a membrane-enclosed sac which contains neurotransmitters released by exocytosis from terminals of neurons

synarthrosis (sin″ahr·thro′sis) (pl. **synarthroses**): a joint having little or no movement, such as those in the skull

syncytium (sin·sish′ee·um, sin·sit′ee·um): a multinucleated mass surrounded by a single cell membrane, such as that found in skeletal muscle cells

synergistic (sin″ur·jis′tick): cooperative interactions; ascribed to different hormones whose effects enhance one another, also to muscles whose effects are additive

synovial fluid (si·no′vee·ul): a viscous fluid secreted by cells of the synovial membrane and which minimizes friction between tissues in freely movable joints, bursae, and tendon sheaths

synovial joint (si·no′vee·ul): *see* diarthrosis

syphilis (sif′i·lus): a sexually transmitted disease caused by the bacterial species *Treponema pallidium,* which, if untreated, can cause sterility in both sexes as well as other ill effects; the disease also may be transmitted from mother to fetus

system: a group of organs which cooperate to perform specific functions in the body; examples are the organs of the circulatory system, which includes the heart and blood vessels

systole (sis′tuh·lee): contraction of the chambers of the heart; generally used in reference to ventricular contraction unless noted otherwise

systolic pressure (sis·tol′ick): blood pressure produced when the ventricles contract; averages nearly 120 mmHg in adults

T_3: Triiodothyronine, an active form of thyroid hormone; *see* thyroxine

T_4: Tetraiodothyronine, an active form of thyroid hormone; *see* thyroxine

tabes dorsalis (tay′beez dor·say′lis): progressive destruction of the dorsal part of the spinal cord and its peripheral nerves, disrupting touch and proprioceptive fibers, while leaving small pain fibers intact; gradual loss of sensation, muscular coordination, and intellectual faculties occurs; most often due to damage by syphilitic infection

tachycardia (tack″i·kahr′dee·uh): very rapid heartbeat

target cell: a cell that responds to a particular hormone

target heart rate: training heart rate; the maximum heart rate permissible for safe exercise by an individual; a heart rate to be achieved during exercise which is a specified percentage of the individual's maximum heart rate

T cell (T lymphocyte): small lymphocyte originating in the thymus gland which contributes to cellular immunity

teloreceptor: nerve ending that responds to stimuli originating from distant sources such as light and sound

tendon sheath: a sleeve of connective tissue that forms a tunnel in which tendons glide over bones or ligaments

testis (tes′tis) (pl. **testes,** tes·teez): one of a pair of the male reproductive organs containing small, highly coiled tubules that produce sperm, and interstitial cells that secrete the male sex hormones, the androgens

testosterone (tes·tos′tur·ohn): the principal male sex hormone secreted primarily as a sex steroid; promotes the growth and maturation of the sex organs and accessory sex glands and the development and maintenance of secondary sex characteristics in males; a precursor of dihydrotestosterone which is its most active form

tetanus (tet′uh·nus): a continuous sustained maximal contraction of a muscle due to extremely rapid repetition of stimuli

thalamus (thal′uh·mus): that portion of the forebrain located above the hypothalamus which serves as the "switchboard" of the brain; it processes sensory information and influences perceptions of environmental stimuli, and the formation of primitive feelings such as hunger and discomfort

thalassemia (thal″uh·see′mee·uh): one of several types of hemolytic anemia in which red blood cells with low levels of hemoglobin are formed, due to heritable defects in synthesis of one or more alpha or beta chains; may cause jaundice, enlargement of the spleen, and stunted growth and may be fatal

thermo-: referring to heat

thermography (thur·mog′ruh·fee): graphical display of heat production from cells; used to distinguish tumor cells which produce more heat than normal ones

theta waves (thay′tuh): θ waves; brain waves produced with a frequency about 4 to 7 times per second in times of disappointment or frustration in young children and adolescents

thirst center: neurons in the hypothalamus that generate a sensation of thirst when the osmotic pressure of the blood is high

thoroughfare capillary: a vessel that provides a direct connection between an arteriole and a venule

threshold: the level of minimal intensity required for a stimulus to cause a response, such as an action potential in an excitable membrane

thrombosthenin (throm·bo·sthee′nin): a contractile protein released from blood platelets that causes blood clots to retract or shrivel

thrombin: a molecule with enzymatic properties that converts fibrinogen, a soluble blood protein in the blood, into fibrin monomers, which are polymerized to entrap blood cells in a meshwork of protein fibers

thrombocyte (throm′bo·site): *see* platelet

thrombocythemia (throm′bo·sigh·theem′ee·uh): a rare disease of bone marrow which causes proliferation of platelets, as well as abnormal platelet activity; the disorder may be accompanied by thrombosis and hemorrhage

thrombocytopenia (throm″bo·sigh″to·pee′nee·uh): a reduction in number of platelets below normal levels in the blood

thrombocytopenic purpura (throm″bo·sigh″to·pee′nick pur′pew·ruh): a condition characterized by purple patches appearing in mucous membranes and skin and by prolonged bleeding time; may be caused by the destruction of platelets by antiplatelet antibodies

thrombosis (throm·bo′sis): formation of a clot in a blood vessel

thromboxane (TXA_2): a derivative of prostaglandin H_2 which causes platelet aggregation and vasoconstriction

thyrocalcitonin: *see* calcitonin

thyroxine (thigh·rock′seen, sin): active thyroid hormone; triiodothyronine and tetraiodothyronine; T_3 and T_4

tidal volume: the quantity of air which is inspired or expired during normal, quiet breathing

tight junction: relatively impermeable discrete intercellular union

tissue: group of similar cells which work together to perform a specialized function

tissue culture: growth of cells, tissues, or organs in glass or plastic tubes or other containers

tissue hypoxia: insufficient oxygen supply to tissues

T_m (transport maximum): the maximal rate of reabsorption of a substance from the kidney filtrate; the maximal tubular reabsorptive capacity

tolerance: a diminished response due to repeated exposure

tone: the degree of tension in a muscle; a state of partial contraction

tongue: a muscular structure, covered with a mucous membrane which secretes fluid into the mouth; the tongue aids in mixing and swallowing food, tasting food, and modifying sounds made by the vocal cords

tonicity (to·nis′i·tee): the measure of the effective osmotic pressure of a solution compared with that of plasma

tonic receptor (ton′ick): a nerve ending which produces action potentials as long as the stimulus is applied, generating impulses which reach a peak and decline to a lower steady level

tonsil (ton′sil): one of various masses of lymphoid tissue located in the nasopharynx, under the tongue, and at the arches in the back of the mouth anterior to the pharynx; the pharyngeal, lingual, and palatine tonsils, respectively

tonsillectomy: the surgical removal of the tonsils

toxemia of pregnancy (tock·see′mee·yuh): accumulation of toxic levels of substances in the blood of a pregnant woman, which may cause high blood pressure and convulsions

trabecula (tra·beck′yoo·luh) (pl. **trabeculae,** tra·beck′yoo·laye): the unit of which spongy bone is composed; a connective tissue strand which, with others, forms plates or spicules of spongy bone; a strand of connective tissue which supports or anchors a tissue or organ, often extending from a capsule into the organ

trace element: one of 14 elements which are found in parts per million to parts per billion concentrations in tissues and which together constitute only 0.01 percent of total body weight but are indispensable to the body; iron, copper, zinc, manganese, cobalt, iodine, molybdenum, selenium, chromium, nickel, vanadium, silicon, tin, and fluorine are trace elements

tracheotomy (tray″kee·ot′uh·mee): an incision made in the trachea

tract: a well-defined group of nerve fibers in the white matter of the central nervous system, usually of a common origin and/or termination

training: adaptation to repeated exercise

training heart rate: *see* target heart rate

transamination (trans·am″i·nay′shun): a chemical transfer of an amino group (NH_2) from one organic compound to another; *see* aspartate aminotransferase (AST)

transcellular fluid: liquid secreted by cells; includes saliva, mucus, and cerebrospinal fluid

transcription (tran·skrip′shun): direction of the synthesis of mRNA, tRNA, and rRNA by the code within DNA

transepithelial transport: active transport of a substance across epithelial cells, which may occur in steps, at least one of which requires energy expenditure by the cell

transferrin (trans·ferr′in): a carrier protein whose combination with iron in the blood aids transport of the mineral to body tissues; siderophilin

transfer RNA (tRNA): a form of ribonucleic acid whose nucleotide triplets are complementary to (are anticodons of) the codons of mRNA; activated amino acids are attached to tRNA and then linked to each other by peptide bonds after tRNA adheres to a complementary segment of mRNA in a ribosome

translation: the direction of protein synthesis by the code in mRNA

transplantation antigen: *see* histocompatibility antigen

transport maximum: *see* T_m

transverse colon: the portion of the large intestine which crosses from one side of the abdominal cavity to the other between the ascending and descending colon

treppe (trep′eh): a stepwise increase of muscular force caused by repeated maximal stimuli at a rate below that which will induce tetanus

triacylglycerol (trye·as″il·glis′ur·ole): *see* triglyceride

triad (trye′ad): a three-membered structure; in skeletal muscle composed of a transverse tubule and, on either side and perpendicular to it, cisterns of the endoplasmic reticulum

tricarboxylic acid (TCA) cycle (trye·kahr″bock·sil′-ick): *see* citric acid cycle

tricuspid valve (trye·kus′pid): a membranous structure composed of three flaps located between, and directing blood from the right atrium to the right ventricle of the heart; the right atrioventricular (A-V) valve

triglyceride (trye·glis′ur·ide): neutral fat; compound of glycerol and fatty acids which forms the most common lipids in foods and accounts for 95 percent of the fat in adipose tissue

trophoblast (trofo·blast, tro′fo): the lateral and peripheral cells from the blastocyst and the embryonic disc; later in development some of its cells produce the fetal membrane, the chorion

tropic hormone: a chemical control agent secreted by an endocrine gland which stimulates the functions of or has an affinity for other endocrine glands; several tropic hormones are secreted by the anterior pituitary

tropical sprue (trop′i·kul sproo): intestinal malabsorption occurring in tropical and subtropical climates and probably caused by bacterial overgrowth interfering with mucosal cell transport

tropomyosin (tro″po·migh′o·sin): a cablelike muscle protein found twisted around actin in thin myofilaments

troponin (tro′po·nin): a complex of three globular polypeptides in thin myofilaments which binds to tropomyosin and actin and whose attraction for calcium ions changes its shape in the initial stages of skeletal muscle contraction

true capillary: a side branch of a metarteriole which does not directly connect to a venule

true labor: uterine contractions which occur generally within 24 h of childbirth and about 266 days after conception and which lead to propulsion of the baby from the uterus through the cervix and vagina to the exterior of the mother′s body

trypsin (trip′sin): a protein-cleaving enzyme secreted by the pancreas

T-tubule (tew′byul): a small vessel which projects inward as a perpendicular extension of the sarcolemma in skeletal and cardiac muscle to provide a path of low electrical resistance to the interior of the cells

tubal ligation: a term often applied to a procedure whereby small tubes are tied and cut; a conception control procedure in which the oviducts are tied and surgically cut, following an incision made in the abdominal cavity

Turner′s syndrome (XO): a congenital abnormality in which body cells are missing a second sex chromosome, and possess only one X chromosome; this results in an individual with female characteristics but the ovaries are missing and the breasts are underdeveloped and the skeleton is deformed

T wave: the pattern in an electrocardiogram occurring during recovery as the ventricles repolarize following contraction

twitch: the mechanical response of a muscle fiber to a single stimulus

tympanic membrane (tim·pan'ick): eardrum; a membrane that separates the outer and middle ear and vibrates in response to sound waves

tyrosinase (tye'ro·sin·ace, ·aze, tirr'o): an enzyme which oxidizes the amino acid tyrosine, an essential step in the formation of the neurotransmitter L-dopa and the pigment melanin; a decline in tyrosinase activity in hair follicles causes graying of the hair

ulcer (ul'sur): erosion of the skin, the mucous membrane, or other epithelia leading to an inflammatory response in underlying tissues lining the digestive tract

ultrafiltration (ul″truh·fil·tray'shun): the separation of substances from the blood by passage of fluid through renal capillaries, and their surrounding basement membranes which act as molecular sieves

ultrasound (ul'truh·saownd): beyond sound; sound waves beyond the frequencies to which the human ear responds, namely, about 20,000 cycles per second (20,000 hertz)

umbilical cord (um·bil'i·kul): a stalk containing fetal blood vessels which connects the abdomen of the fetus to the placenta

unipolar recording (yoo″ni·po'lur): measurement of voltage in an ECG by an active electrode at a single location by comparing its potential with that of a standard or indifferent electrode which shows no major changes during the cardiac cycle

universal donor: a term referring to a person with type O blood cells, which have neither A nor B antigens

universal recipient: an individual who has type AB blood cells, but neither anti-A nor anti-B antibodies, and may therefore in an emergency receive washed red blood cells from certain A, B, or O donors

unmyelinated (un·migh'e·li·nay·tid): lacking significant amounts of myelin

unsaturated fat (un·sath'uh·ray″tid): a fat in whose chains some adjacent carbon atoms have double bonds between them and thus do not contain all the hydrogen which they might

upper gastroesophageal sphincter: the circular skeletal muscle between the pharynx and first part of the esophagus which regulates the entry of food into the esophagus during swallowing

upper motor neuron: a nerve cell within the brain that causes voluntary motor activity and relays its impulses to a lower motor neuron in the ventral horn of the spinal cord

uracil (yoor'uh·sil): a nitrogen-containing pyrimidine base in RNA found in place of thymine which is its equivalent in DNA; uracil is complementary in structure to cytosine of DNA and pairs with it when DNA forms RNA

urea (yoo·ree'uh): a product of protein catabolism carried by the blood to the kidney where it is excreted in urine: $H_2N(C{=}O)NH_2$

uremia (yoo·ree'mee·uh): elevated levels of urea in the blood; an indication of severe kidney disease

ureter (yoo·ree'tur, yoor'e·tur): a tube carrying urine from the renal pelvis to the urinary bladder

urethra (yoo·ree'thruh): a tube leading from the urinary bladder to the exterior of the body; it carries urine in both sexes as well as sperm in males

urinary bladder (yoor'i·nerr·ee): a muscular sac that stores urine delivered to it from the kidneys through the ureters and expels urine to the exterior through the urethra

urolithiasis (yoor″o·li·thigh'uh·sis): urinary stones; kidney stones formed by calcium salt crystals often containing calcium

uterus (yoo'tur·us): a pear-shaped, hollow, smooth muscular sac in the female pelvic cavity in whose lining the developing embryo may become embedded, nourished, mature, and grow; the womb

utricle (yoo'tri·kul): the larger of two bulges in the membranous labyrinth of the semicircular canals in the inner ear; it houses sensory receptors, the maculae utriculi, which are static or gravity sensors

U wave: a pattern sometimes present in an electrocardiogram which may occur because of slow depolarization of the papillary muscles

vagina (va·jye'nuh): a channel about 2.3 to 3.5 inches long which leads from the cervix of the uterus to the vestibule at the exterior of the body

vagus (vay'gus): the tenth cranial nerve which contains most of the body's parasympathetic fibers

valence (vay'lunce): the number of electrons which an atom donates, accepts, or shares in forming a chemical bond

valve: a membranous fold which prevents backward flow of materials in a passage

varicose vein (văr'i·koce, ·koze): twisted, dilated vein caused by forces which wear out venous valves, thereby putting great pressure on the vessel

vasa vasorum (vay'suh, zuh, vay·so'rum): blood vessels which supply blood to the thick walls of large arteries and veins

vascular (vas'kew·lur): pertaining to blood vessels

vas deferens (vas def″ur·enz): a long duct leading from the epididymis of each testis through the inguinal canal to the urethra

vasectomy (vas·eck'tuh·mee): a surgical procedure used for conception control in the male in which each vas deferens is cut and tied to prevent the passage of sperm to the urethra in the penis

vaso-: relating to blood vessels

vasoactive (vay″zo·ack'tiv): a term referring to an effect on blood vessels, such as increased permeability, widening (vasodilation), or narrowing (vasoconstriction)

vasoconstriction (vay″zo·kun·strick'shun, vas″o·): narrowing of a blood vessel by contraction of circular smooth muscle in the vessel wall

vasodilation (vay″zo·dye·lay'shun, vas″o): vasodilatation; widening of a blood vessel by inhibition of circular smooth muscle in the vessel wall

vasomotor (vay″zo·mo'tur, vas″o): referring to narrowing or widening of the inner diameter of blood vessels

vasopressin (vay″zo·pres'in, vas″o): antidiuretic hormone, ADH

vectorcardiogram (veck″tur·kahr'dee·o·gram): a graphic illustration formed by a line drawn as a loop that interconnects the ends of the arrows of cardiac vectors of the P wave, QRS complex, and T wave in an ECG of one cardiac cycle

vein: a blood vessel which carries blood toward the heart

vena cava (vee'nah kay'vah) (pl. **venae cavae,** vee'-nee kay'vee): the two large veins which deliver blood from the upper and lower parts of the body to the right atrium, the superior and inferior vena cava, respectively

venereal disease (ve'neer'ee·ul): a term applied to syphilis and gonorrhea when they were considered to be one disease; sexually transmitted disease

venous baroreceptor (ven'ous băr″o·re·sep″tur): receptor in veins which is sensitive to changes in blood pressure; notable in walls of venae cavae at the entry to the right atrium

ventilation: movement of air into and out of the lungs to aerate the blood

ventral horn: ventral segment of the gray matter of the spinal cord, composed of efferent (motor) neurons

ventricle (ven'tri·kul): one of two chambers of the heart which pump blood out of the organ to the lungs and to the body, respectively; also, one of four interconnected, hollow chambers in the brain filled with cerebrospinal fluid

venule (ven'yool): the smallest of veins which receives blood from capillaries

vertebral canal (vur'te·brul): a channel in the center of the vertebrae which houses the spinal cord

vesicle (ves'i·kul): a membrane-enclosed sac; a synaptic vesicle; a pinocytic vesicle

vestibular apparatus (ves·tib'yoo·lur): a membranous labyrinth in the inner ear containing sensory organs responsive to gravitational changes, acceleration, and movement of the head

vestibule (ves'ti·bewl): a small chamber or space at the entry of a canal; a narrow chamber, between the external female genital organs, into which the urethra and the vagina open to the exterior

villus (vil'us) (pl. **villi,** vil'eye): a fingerlike projection that increases the absorptive surface area of tissues; intestinal villi are located in the small intestine

virus (vye'rus): a submicroscopic, infectious microorganism composed of an outer protein coat which surrounds a core of DNA or RNA; viruses are microorganisms which can reproduce only after gaining entry into living cells

visceral (vis'ur·ul): pertaining to inner organs, smooth muscle, cardiac muscle, and glands

visceral pericardium: the inner of two serous membranous layers directly attached to, and surrounding the heart; the epicardium

visceral pleura: a serous membrane which adheres to the outer surface of the lungs

viscosity (vis·kos'i·tee): resistance to liquid flow due to intermolecular attraction

visual acuity: clarity of retinal focus; the minimal distance at which two lines can be distinguished as separate from each other; the distance from which a person must view an object for clear vision divided by that required for a person with average eyesight; 20/20 vision represents normal visual acuity

vital capacity (vye'tul): the approximate functional air capacity of the lungs, measured as the maximal amount of air that can be forcibly breathed out after the greatest possible inward breathing

vitamin B_{12}: a nutrient required in the diet whose action as a coenzyme in nucleic acid metabolism is essential for red blood cell formation; the extrinsic factor; cobalamin

vitamin D_3: ergocalciferol and cholicalciferol and its active metabolite 25-hydroxycholicalciferol (25-HCC) and its most active derivative (1,25-di-HCC); steroids which promote intestinal absorption of calcium and phosphate and the mobilization of calcium from old to new bone

vitamin (vye'tuh·min): an organic compound which must be present in small amounts for the normal regulation of metabolism

$Vo_{2_{max}}$: the largest amount of oxygen a person can deliver to muscle during the most strenuous exercise

volt: a unit of electromotive force (electrical pressure)

voltage: the potential across a membrane expressed in volts; the possibility of ion flow to do work

waking center: neurons of the reticular activating system which override those of the sleep center to maintain the wakened state; *see* sleep center

Wallerian degeneration (wah·lirr'ee·un): deterioration of a neuron moving along the axon away from its cell body

warfarin: *see* Coumadin

Wernicke's area (ve^{h}r'ni·ke^{h}, vur'ni·kee, wur'): a center for the comprehension of language; neurons in the left temporal lobe which receive and integrate impulses from spoken words relayed through the auditory center as well as written words transmitted from the visual center to produce thoughts

wheal and flare reaction: local allergic response associated with swelling, pain or itching, and redness (*see* hive)

white matter: region of the CNS composed of myelinated (white) fibers

womb: *see* uterus

xeroradiography (zeer″o·ray″dee·og'ruh·fee): reproduction of x-ray images on paper as opposed to their conventional display on film

X-linked trait: *see* sex-linked trait

Z-line: regularly repeating transverse lines in skeletal and cardiac muscle, which mark the boundaries of the sarcomere and which consist solely of thin myofilaments

Zollinger-Ellison syndrome: a condition in which tumors of the pancreas, stomach, or duodenum elevate gastrin production, thus causing excessive hydrochloric acid secretion and possibly leading to severe ulcers

zygote (zye'gote): the single cell formed by the union of egg and sperm, each contributing one member of each pair of 23 sets of chromosomes

zymogen granule (zye'mo·jen): a minute cytoplasmic particle in which inactive forms of enzymes are stored prior to their secretion from a cell

ACKNOWLEDGMENTS

The following figures are used by permission or adapted from the sources cited.

2-6 Martin, D. W. Jr., P. A. Mayes, and V. W. Rodwell: *Harper's Review of Biochemistry*, 18th ed, Copyright © 1981, Lange Medical Publications, Los Altos, California.

2-10 Weiss, L., and R. O. Greep: *Histology*, 4th ed. Copyright © 1977. McGraw-Hill Book Company, New York.

2-11 A Courtesy of Dr. Wilbert Bowers, U.S. Army Research Institute for Environmental Medicine, Natick, Mass.

2-12 Courtesy of R. M. Brenner, from *Progress in Gynecology*, vol. V, p. 82, S. H. Sturgis and M. L. Taymor (eds.). Copyright © 1970, Grune and Stratton, Inc., New York.

3-1 A and B (left) Hull, B. E., and L. A. Staehelin, *J. Cell Biol.* **81**:67. Copyright © 1979, Rockefeller University Press, New York.

3-1 C (left) Perrachia, C., and A. F. Dulhunty, *J. Cell Biol.* **70**:426. Copyright © 1976, Rockefeller University Press, New York.

3-1 A (right) Pappas, G. D., "Junctions between Cells," *Hospital Practice* **8**:40. Copyright © 1973, Rockefeller University Press, New York.

3-1 B and C (right) Staehelin, L. A., *Intl. Rev. Cytol.* **39**:207, 249. Copyright © 1974, Rockefeller University Press, New York.

3-3 Mueller, J. C., A. L. Jones, and J. A. Long, *Gastroenterology* **63**:860. Copyright © 1972, The Williams and Wilkins Co., Baltimore.

3-5 B Port, C., and I. Corvin, *Science* **177**: Cover (September 22, 1972). Copyright © 1972 by the American Association for the Advancement of Science, Washington, D.C.

3-8, 3-9, and 3-22 Adapted from Ham A. W., *Histology*, 8th ed. Copyright © 1979, J. B. Lippincott Co., Philadelphia.

3-10 A White, A, P. Handler, E. L. Smith, R. L. Hill, and I. R. Lehman, *Principles of Biochemistry*, 6th ed. Copyright © 1978, McGraw-Hill Book Company, New York.

3-10 B and 3-11 Weiss, L., and R. O. Greep: *Histology*, 2d ed. Copyright © 1966, McGraw-Hill Book Company, New York.

3-12 and 3-16 Bloom, W. and D. W. Fawcett, *A Textbook of Histology*, 10th ed. Copyright © 1975, W. B. Saunders Company, Philadelphia.

3-13, 3-14, and 3-24 Weiss, L., and R. O. Greep: *Histology*, 4th ed. Copyright © 1977, McGraw-Hill Book Company, New York.

3-17 Weiss, L., *J. Immunol.* **101**:1356. Copyright © 1968, The Williams and Wilkins Co., Baltimore.

3-18 Courtesy of D. Zucker-Franklin from L. Weiss, *Cells and Tissues of the Immune System*. Copyright © 1972, Prentice-Hall, Inc., Englewood Cliffs, New Jersey.

3-19 Courtesy of R. J. North from L. Weiss, *Cells and Tissues of the Immune System*. Copyright © 1972, Prentice-Hall, Inc., Englewood Cliffs, New Jersey.

3-20 Adapted from H. H. Fudenberg, D. P. Stites, J. L. Caldwell, and J. V. Wells, *Basic and Clinical Immunology*. Copyright © 1980, Lange Medical Publications, Los Altos, California.

4-1 Adapted from I. S. Edelman, and J. Leibman, *Amer. J. Med.* **27**:261. Copyright © 1959, Technical Publishing Co., New York.

4-4 DeRobertis, E., *Science* **171**:963. Copyright © 1971 by the American Association for the Advancement of Science.

4-10 A Singer, S. J., *Hospital Practice* **8**:88. Copyright © 1973, *Hospital Practice*, New York.

4-10 B Dutton, A., E. D. Rees, and S. J. Singer, *Proc. National Academy of Sciences* **73**:1535. Copyright © 1976, National Academy of Sciences, Washington, D.C.

5-2 A, D, and F From the Fuller Albright Teaching Slide Collection. Courtesy of the Endocrine Unit of the Massachusetts General Hospital, Boston, Mass.

5-2 B, C, and E Courtesy of the Armed Forces Institute of Pathology, Washington, D.C., Negative Numbers 57, 10583-1, 56-20-699-1, and 60-10307-1.

5-9 Adapted from A. N. Delaat, *Primer of Serology*. Copyright © 1976, Lippincott, Harper and Row, New York.

6-3 Bray, D., *Endeavour*, XXIII (120):132–135; and courtesy of John Hopkins, Kings College, London. Copyright © 1974, Pergamon Press Ltd., Elmsford, New York.

6-7 B Bunge, M. B., R. P. Bunge, and H. Ris, *J. Biophysical and Biochemical Cytology* **10**:67. Copyright © 1961, Rockefeller University Press, New York.

6-8 Courtesy of Dr. K. A. Seigesmund, Medical College of Wisconsin, Milwaukee.

6-9 Ganong, W. F., *Review of Medical Physiology*, 9th ed. Copyright © 1979, Lange Medical Publications, Los Altos, California.

6-10 Patton, H. D., J. W. Sundsten, W. E. Crill, and P. D. Swanson, *Introduction to Basic Neurology*. Copyright © 1976, W. B. Saunders Co. Philadelphia; and Maynard, C. A., R. L. Schultz, and C. D. Pease: *Americam J. Anatomy* **100**:423. Copyright © Alan R. Liss, Inc., New York.

7-8 A Tharp, G. D., *Experiments in Physiology*, 3d ed. Copyright © 1976, Burgess Publishing Company, Minneapolis.

7-8 B Hodgkin, A. L., and A. F. Huxley, *J. Physiology* **117**:530 (1952). Physiological Laboratory, Cambridge, England; and Eccles, J. C.: *The Understanding of the Brain*, Copyright © 1977. McGraw-Hill Book Company, New York.

7-9 Selkurt, C. E. (ed.), *Basic Physiology for the Health Sciences*. Copyright © 1975, Little, Brown and Company, Boston.

7-11 B Peters, A. S., L. Palay, and H. DeF. Webster, *The Fine Structure of the Nervous System*. Copyright © 1970, J. B. Lippincott Company, Philadelphia.

7-13 Adapted from W. M. Copenhaver, R. M. Bunge, and M. B. Bunge, *Bailey's Textbook of Histology*. Copyright © 1971,

Williams and Wilkins Company, Baltimore.

7-15 B Rockel, A. J., and E. G. Jones, *J. Compar. Neurology* **147**:85. Copyright © 1973, Alan R. Liss, Inc., New York.

8-3 Adapted from J. E. Dowling, and B. B. Boycott, *Proc. Roy. Soc. of Biol.* (London) **166**:104. Copyright © 1966, The Royal Society, London.

8-9 Myers, D., et al., *Clinical Symposia* **21**(22):42. Copyright © 1969, CIBA Pharmaceutical Company, Division of Ciba-Geigy Corporation. Reprinted with permission from *Clinical Symposia*, illustrated by Frank H. Netter, M.D. All rights reserved.

8-10 Davis, H., *J. Acoustic Soc. America* **34**: 1379. Copyright © 1962, American Institute of Physics, New York.

8-13 Zotterman, Y. (ed.), *Olfaction and Taste.* Courtesy of A. D. J. DeLorenzo. Copyright © 1963, Pergamon Press Inc. Elmsford, New York.

9-4 A Keegan, J. J., and F. D. Garrett, *Anatomical Record* **102**:411. Copyright © 1948, Alan R. Liss, Inc., New York.

9-4 B House, E. L., B. Pansky, and A. Siegel, *A Systematic Approach to Neuroscience.* Copyright © 1980, McGraw-Hill Book Company, New York.

9-6 Patton, H. D., H. W. Sundsten, W. C. Crill, and P. D. Swanson. *Introduction to Basic Neurology.* Copyright © 1976, W. B. Saunders Company, Philadelphia.

9-7 and 9-11 Elliot, A. C., *Textbook of Neuroanatomy.* Copyright © 1969, J. B. Lippincott Company, Philadelphia.

10-2 House, E. L., B. Pansky, and A. Siegel, *A Systematic Approach to Neuroscience.* Copyright © 1979, McGraw-Hill Book Company, New York.

10-7 Carpenter, N. B., *Human Neuroanatomy*, 7th ed. Copyright © 1976, Williams and Wilkins Company, Baltimore.

10-8 Courtesy of the Radiology Department of the Framingham Union Hospital. Framingham, Mass.

11-8 A Ransom, S. W., and S. L. Clark, *The Anatomy of the Nervous System.* Copyright © 1953, W. B. Saunders Company, Philadelphia.

11-8 B and C Penfield, W., and T. Rasmussen, *The Cerebral Cortex of Man.* Copyright © 1950, 1978, Macmillan Publishing Company, Inc., New York.

12-3 Schmalbruch, A., *Advances in Anatomy, Embryology and Cell Biology* **43**: Fig. 1. Copyright © 1979, Springer-Verlag, Inc., New York.

12-4 and 12-7 Bloom, W., and D. W. Fawcett, *A Textbook of Histology*, 10th ed. Copyright © 1975, W. B. Saunders Company, Philadelphia.

12-5 and 12-6A Martin, D. W. Jr., P. A. Mayes, and V. W. Rodwell, *Harper's Review of Biochemistry*, 18th ed. Copyright © 1981, Lange Medical Publications, Los Altos, California.

12-9 A and B Elliot, H. C., *Textbook of Neuroanatomy.* Copyright © 1969, J. B. Lippincott Company, Philadelphia.

12-9 C Courtesy of Dr. John E. Heuser, Washington University Medical School, St. Louis, Missouri.

12-10 Staendaert, F., and K. L. Dretchen, *Fed. Proc.* **38**:2183. Copyright © 1979, Federation of American Societies for Experimental Biology, Bethesda, Maryland.

12-12 Fawcett, D. W., and N. S. McNutt, *J. Cell Biol.* **42**:24. Copyright © 1969, Rockefeller University Press, New York.

13-2, 13-3, and 13-4 Vander, A. J., J. H. Sherman, and D. W. Luciano, *Human Physiology: The Mechanisms of Body Function*, 3d ed. Copyright © 1980, McGraw-Hill Book Company, New York.

13-9 Edington, D. W., and V. R. Edgerton, *The Biology of Physical Activity.* Copyright © 1976, Houghton Mifflin Company, Boston.

13-15 Pollock, M. L., J. H. Wilmore, and S. M. Fox III, *Health and Fitness through Physical Activity.* Copyright © 1978, John Wiley and Sons, New York.

13-16 Zohman, L. R., *Beyond Diet: Exercise Your Way to Fitness and Heart Health.* Copyright © 1979, CPC International, Inc., Englewood Cliffs, N.J.

14-1 B Courtesy of Radiology Department, Framingham Union Hospital, Framingham, Mass.

15-3 and 15-7 Hurst, J. S., and R. B. Logue (eds.), *The Heart*, 2d ed. Copyright © 1970, McGraw-Hill Book Company, New York.

15-5 Adapted from E. E. Selkurt (ed.), *Basic Physiology for the Health Sciences.* Copyright © 1975, Little, Brown and Company, Boston.

15-8 Burch, G. E., and T. Winsor, *A Primer of Electrocardiography.* Copyright © 1966, Lea and Febiger, Philadelphia.

15-9 Goldman, M. J., *Principles of Clinical Electrocardiography*, 10th ed. Copyright © 1979, Lange Medical Publications, Los Altos, California.

15-10 Selkurt, E. E., *Physiology*, 4th ed. Copyright © 1976, Little, Brown and Company, Boston.

16-1 Berne, R. M., and M. N. Levy, *Cardiovascular Physiology*, 4th ed. Copyright © 1981, C. V. Mosby Co., St. Louis; data from A. Carlsten and G. Grimby, *The Circulatory Response to Muscular Exercise in Man.* Copyright © 1966, Charles C. Thomas, Publisher, Springfield, Ill.

16-5 *Clinical Electrophoresis: Gelman Procedures for Special Electrophoresis.* Copyright © 1969, Gelman Instrument Co., Ann Arbor, Mich.

16-6 Henry, J. B., *Clinical Diagnosis and Management by Laboratory Methods*, 16th ed., Chap. 12, "Clinical Enzymology" by H. J. Zimmerman and H. B. Henry. Copyright © 1979, W. B. Saunders Company, Philadelphia; and courtesy of Worthington Diagnostics, Inc., Freehold, N.J.

16-9 Courtesy of Dr. Daniel Savage, Framingham Heart Study, Framingham Union Hospital, Framingham, Mass., and *The Middlesex News*, Framingham, Mass.

17-1, 17-3, and 17-10 (two upper figures) Shepherd, J. T., and P. M. Vanhoutte, *The Human Cardiovascular System, Facts and Concepts.* Copyright © 1979, Raven Press Publishers, New York.

17-4 A and B Courtesy of Dr. Toichi Kuwabara, National Eye Institute, Bethesda, Maryland.

17-5 and 17-6A, B, and C Weiss, L., and R. O. Greep, *Histology*, 4th ed. Copyright © 1977, McGraw-Hill Book Company, New York.

17-10 (lower third) Smith, J. J., and J. P. Kampine, *Circulatory Physiology—The Essentials.* Copyright © 1980, Williams and Wilkins Company, Baltimore.

17-13 Stare, M. D. (ed.), *Atherosclerosis*: Copyright © 1974, Medcom, Inc., New York.

18-2 Courtesy of Dr. L. Weiss. Weiss, L., and R. O. Greep, *Histology*, 4th ed. Copyright © 1977, McGraw-Hill Book Company, New York.

18-3 Morrel, F. M. M., R. F. Baker, and H. Wayland, *J. Cell Biol.* **48**:98. Copyright © 1971, Rockefeller University Press, New York.

18-5 Dickerson, R. E., and I. Geis, *Hemoglobin: Structure, Function, Evolution, and Pathology.* Copyright © 1982, Benjamin Cummings Publishing Company, Menlo Park, California. Illustration copyright, Irving Geis.

18-7 *ABO and Rh Systems.* Copyright © 1969, Ortho Diagnostics Systems, Inc., New York.

19-1 and 19-6 Weiss, L., and R. O. Greep, *Histology*, 4th ed. Copyright © 1977, McGraw-Hill Book Company, New York.

19-2 Courtesy of Dr. Vincent Marchesi, Yale University School of Medicine, New Haven, Conn. *Q. J. Exp. Physiol.* **45**:343 (1960).

19-3 Courtesy of Dr. Aaron Polliack, Hadassah Medical Organization, Jerusalem, Israel.

19-5 Courtesy of M. J. Cline. From Hanifin, J. M. and M. J. Cline, *J. Cell Biol.* **46**:97. Copyright © 1970, Rockefeller University Press, New York.

19-8 Hovig, T., et al, *Experimental Hemostasis in Normal Dogs and Dogs with Congenital Disorders of Blood Coagulation, Blood* **30**: 636. Copyright © 1967, Grune and Stratton Inc., New York.

20-4 (diagram) Comroe, J. H. Jr., et al, *The Lung—Clinical Physiology and Pulmonary Function Tests*, 2d ed. Copyright © 1962, Year Book Medical Publishers, Chicago.

20-4 (photograph) Low, F. N., *Anatomical Record* **117**:241. Copyright © 1953, American Association of Anatomists, Wistar Institute Press, c/o Alan R. Liss, Inc. N.Y.

20-8 Collins Stead-Wells Spirometer, courtesy of W. C. Collins Inc., Braintree, Mass.

20-10 Huehns, E. R., "Oxygen Delivery to Tissues," *Lab Lore* **8** (10):575. Copyright © 1979, Wallace Reagents Div., Burroughs Wellcome Co., Research Triangle Park, North Carolina.

20-11 Lambertsen, C. J., "Transport of Oxygen, *Carbon Dioxide, and Inert Gas by the Blood," in V. B. Mountcastle (ed.), Medical Physiology*, 14th ed. Copyright © 1980, C. V. Mosby Company, St. Louis.

20-14 to 20-16 Adapted from American Red Cross, *Cardiopulmonary Resuscitation.*

Copyright © 1974, The American National Red Cross, Washington, D.C.

21-2 Courtesy of Dr. K. A. Seigesmund, Medical College of Wisconsin, Milwaukee.

21-4 Adapted from E. E. Selkurt (ed.), *Basic Physiology for the Health Sciences.* Copyright © 1975, Little, Brown and Company, Boston.

21-5 A Bulger, R., *Amer. J. of Anat.* **116**:237. Copyright © 1965, Alan R. Liss, Inc., New York.

21-5 B and 21-8 Weiss, L., and R. O. Greep, *Histology,* 4th ed. Copyright © 1977, McGraw-Hill Book Company, New York.

21-9 Adapted from P. J. Cannon, *N.E.J. Med.* **296**:26. Copyright © 1977, *New England Journal of Medicine,* Boston.

21-11 Berliner, R. W., *N.E.J. Med.* **24**:730. Copyright © 1958, *New England Journal of Medicine,* Boston.

21-13 Harper, H. A., V. W. Rodwell, and P. A. Mayes, *Review of Physiological Chemistry,* 17th ed. Copyright © 1979, Lange Medical Publications, Los Altos, California.

23-3 A Carr, K. E., *International Rev. of Cytol.* **30**:213. Copyright © 1971, Academic Press, New York.

22-3 B Fawcett, D. W., *J. Histochem. and Cytochem.* **13**:75. Copyright © 1965, Elsevier, North Holland, Inc., New York.

22-6 A Elias, H. *Functional Morphology of the Liver.* Research in the Service of Medicine, vol. 37. Copyright © 1953, Searle Laboratories, Division of Searle Pharmaceuticals, Inc., Chicago.

24-8 Courtesy of Peter Arnold, Inc., Copyright © Manfred Kage.

24-9 Adapted from B. M. Patten, *Human Embryology.* Copyright © 1953, McGraw-Hill Book Company, New York.

24-11 Adapted from G. D. Searle and Co., "Envoid: The Pattern and the People," *Research in the Service of Medicine* **55**:9. Copyright © 1962, Searle Laboratories, Skokie, Ill.; and W. H. Lein, *The Cycling Female: Her Menstrual Rhythm.* Copyright © 1979, W. H. Freeman and Company, San Francisco.

24-12 Adapted from Taymor, L., in D. P. Lauler and R. W. Kistner (eds.), *Reproductive Endocrinology,* p. 28. Copyright © 1973, Wyeth Laboratories, Philadelphia; and I. H. Thorneycroft et al., "The Relation of Serum Hydroxyprogesterone and Estradiol-17 β Levels during the Human Menstrual Cycle," *Amer. J. Obstetrics and Gynecol.* **111**:947–951. Copyright © 1971, C. V. Mosby Company, St. Louis.

24-13 B Hertig, A. T., and J. Rock, *Contributions to Embryology* **29**:Plate 6. Copyright © 1941, Carnegie Laboratories of Embryology, Davis, California.

24-15 Courtesy of Taurus Photos, New York. Copyright © 1981, Martin M. Rotker.

CREDITS FOR TABLES

The following tables are used by permission or adapted from the sources cited.

10-2 and Appendix Rieder, S. V., L. Ellman, B. Kliman and K. J. Bloch: *Case Records of the Massachusetts General Hospital.* **302**: 37–48. Copyright © 1980. Printed by permission from *The New England Journal of Medicine,* Boston.

16-7 and 16-8 Levy, R. I., *J. Cardiovascular Medicine* **5**(5):439, 443. Copyright © 1980, Group Medicine Publications, Inc., New York.

17-3 Lasser, R. P., and A. M. Master, *Geriatrics* **14**(6):347. Copyright © 1959, Modern Medical Publishers, Division of Harcourt Brace Jovanovich Publishers, New York.

18-3 Adapted from P. L. Altman and D. D. Katz, *Human Health and Disease Biological Handbooks,* vol. 2, Federations of American Societies for Experimental Biology, Bethesda, Maryland.

18-5 Williams, W. J., C. Beutler, A. J. Ersley, and R. W. Rundles, *Hematology,* 2d ed. Copyright © 1977, McGraw-Hill Book Company, New York.

20-2 Bates, D. V., et al., *Respiratory Function in Disease,* 2d ed. Copyright © 1971, W. B. Saunders Company, Philadelphia.

21-1 White, A. P. Handler, C. L., R. L. Hill, and I. R. Lehman, *Principles of Biochemistry,* 6th ed. Copyright © 1978, McGraw-Hill Book Company, New York.

21-4 Muntwyler, E., *Water and Electrolyte Metabolism and Acid-Base Balance.* Copyright © 1968, C. V. Mosby Company, St. Louis.

23- A, E, G, and I Committee on Dietary Allowances, Food and Nutrition Board, *Recommended Dietary Allowances.* Copyright © 1980, National Research Council, National Academy of Sciences, Washington, D.C.

23- B and H Harper, H. A., V. W. Rodwell, and P. A. Mayes, *Review of Physiological Chemistry,* 17th ed. Copyright © 1979, Lange Medical Publications, Los Altos, California.

23- H Konishi, F., "Food Energy Equivalents of Various Activities," *J. Amer. Dietetic Assoc.* **46**:186. Copyright © The American Dietetic Association, Chicago.

23- C and D Faulkner, W. R., "The Trace Elements for Laboratory Medicine," *Laboratory Management,* July, 1981. Copyright © 1981, United Business Publications, New York.

23- F Morehouse, L. E., and A. T. Miller Jr., *Physiology of Exercise,* 7th ed. Copyright © 1976, C. V. Mosby Co., St. Louis.

24-3 Physician Members of the Value of Life Committee, Inc., *Timetable of a Human Individual's Development before Birth.* Copyright © 1980, Value of Life Committee, Inc., Brighton, Mass.

24-4 and 24-5 Searle Pharmaceutical Company, Publication A05472-5. Copyright © 1980. Searle Laboratories, Division of Searle Pharmaceuticals, Chicago.

Glossary's phonetic guides: *Blakiston's Gould Medical Dictionary,* 4th ed. Copyright © 1979, McGraw-Hill Book Company, New York.

INDEX

A antigens, 440–441
A bands, 287–290
AB blood type, 440–441
A,B,C fibers
  of nerve, 147–148
  of skeletal muscle 297–298, 322, 325
Abdominal breathing, 483
Abdominal cavity, 49
Abdominal pregnancy, 641
Abetalipoproteinemia, 579, 586
A blood type, 440–441
Abnormal breathing, 483–484
ABO blood grouping, 441–442
Abortion, 651–653
Absolute refractory period, 163, 164, 312, 367
Absorption, 42, 543, 572, 575–580
  of amino acids, 574–576, 580
  bile salts and, 575
  of ions, 575, 583–585
  of lipids, 395, 575–579, 581, 582
  pressure in capillaries, 413
  of proteins, 59, 575–577, 580
  of trace elements, 583–585
  of vitamins, 577, 581, 582
  of water, 575, 585
  (*See also* Malabsorption)
Absorptive epithelium, 547, 548, 550
Accelerator globulin (ACG), 464, 465
ACD (*see* Acid-citrate dextrose solution)
Acceleration, perception of (*see* Sense organs, ear; Sensor receptors, vestibular receptors)
Acceleration of heart, 369
  (*See also* Heart rate)
Accessory sex organs, 621–622, 629–632
  (*See also* Reproductive system)
Acclimatization
  to altitude, 493–494
  to heat, 510–511
Accommodation, 192–193
A cells of pancreas, 129
Acetoacetic acid, 321, 535, 592
Acetylcholine (ACh), 104, 122, 175, 176, 189, 230, 273, 379, 380
  anticholinergic drugs and, 252, 253
  fate of, 294–295
  receptors and, 84, 85, 97, 294
  skeletal muscle and, 292, 294–297, 311, 314
  smooth muscle and, 302–304
  (*See also* Autonomic nervous system; Cholinergic nerves; Nervous system; Neurons; Neurotransmitters)
Acetylcholinesterase (AChE), 84, 87, 175, 294, 296, 297
Acetyl coenzyme A (CoA), 320, 321, 393, 590–592
AC globulin (*see* Accelerator globulin)
Acid, 21, 521, 527
  measurable in urine, 530
  production and ulcers, 565
Acid-base balance, 512
  kidneys and, 527–531, 572
  respiratory tract and, 487, 529–530
  (*See also* Buffers)
Acid citrate dextrose solution (ACD solution), 466
Acidity, 16, 18
  effects on bone, 342, 345, 349, 350
  effects on hemoglobin, 485
Acidophils, 457
Acidosis, 19, 22, 433, 534–535, 586, 592
  metabolic, 535
  respiratory, 534–535
Acid phosphatases, 342, 351, 622
Acinar glands, 51
Acini, 550
Acne, 509, 629, 649
Acromegaly, 106, 353
Acrosome, 623, 624, 639
ACTH (*see* Adrenocorticotropic hormone)
Actin, 34, 139, 287–289, 321, 324, 623
  (*See also* Skeletal muscle, myofilaments)
Action potential (*see* Impulses; Potentials, action)
Activation:
  of amino acids, 35–36
  of membrane to sodium permeability, 164
Activation heat, 323–324
Active expiration, 480
Active site of enzyme, 23
Active symport, 95, 558, 576, 577, 579
Active transport, 87, 93–95, 576, 577, 579
  of amino acids, 92, 95, 576
  of calcium, 295–296, 340, 347
  of chloride, 522, 528, 558, 575
  of glucose, 95, 523, 558, 576, 577, 579
  of hydrogen ions, 522, 531, 557, 558
  of potassium, 93–95, 522–524
  of sodium, 93–95, 149, 160–162, 375, 522–524, 531, 577, 579
  of sodium chloride, 522, 526, 528
  (*See also* Impulses)
Actomyosin, 464
  content of muscle, 285, 290, 311, 314
  velocity of contraction and, 311
  warming up, 314
Actomyosin ATPase, 314, 323
Acuity, visual, 189, 191, 194
Acupuncture, 222, 251
Acute leukemia, 468
Acute reactions, defined, 565
Adam's apple, 629
Adaptation
  bright, 189
  dark, 189
  general adaptation syndrome (GAS) of, 109
  of receptors, 186
  to stress, 111
Addison's anemia, 447
Addison's disease, 106, 131
Adenine, 26–28
Adenohypophysis (*see* Pituitary, anterior)
Adenosine diphosphate (ADP):
  platelets and, 462
  preservation of blood and, 466
  (*See also* adenosine triphosphate)
Adenosine monophosphate (AMP), 27
  (*See also* Cyclic adenosine monophosphate)
Adenosine triphosphate (ATP), 87, 93, 139, 149, 292, 295, 392, 590, 592, 593
Adenosine triphosphatase (ATPase), 87, 228, 294–296, 310, 314, 320–324
  (*See also* Active transport, calcium; Active transport, sodium; Metabolism; Myosin ATPase)
Adenyl cyclase, 120, 121, 123, 232
ADH (*see* Antidiuretic hormone)
Adhesion(s), 361
Adiadochokinesia, 264
Adipocyte (*see* Fat cells)
Adipose tissue, 56, 116, 349, 507
  (*See also* Fat cells)
ADP (*see* Adenosine triphosphate)
Adrenal(s), 6, 8, 56, 98, 104, 105, 130–132
  cortex, 52, 126, 130–131, 526, 628, 636
  glucocorticoids, 52, 106, 130, 131, 350, 351, 457, 467, 627, 638
  medullae, 52, 107, 111, 113, 131–132, 176
    autonomic nervous system and, 226–228, 231
    blood pressure and, 423
    melanin and, 508
    stress and, 109–112
    sweat glands and, 510
  steroids, 130, 627, 634
    (*See also* Aldosterone; Adrenal glucocorticoids)
  stress and, 109–112
    (*See also* Stress)

Adrenaline (*see* Epinephrine)
Adrenergic nerves, 176, 224–233
biological properties, 231–233
blood pressure and, 420–421, 526
drug effects on, 231, 252, 253
effects of, 228–229
on smooth muscle, 302–303
neurotransmitters and, 227–228, 230
origins of, 225–227
regulation of heat production and heat loss by, 267–268
secondary messengers and, 232
vasoconstrictor effects, 266–268
vasodilator effects, 266, 268
Adrenocortical steroids (*see* Adrenal steroids)
Adrenocorticotropic hormone (ACTH), 52, 108, 109, 112, 122, 126, 127, 269, 508
relation to endorphins, 249, 250
stress and, 109–112
Adsorption of minerals to bone, 339
Adult hemoglobin (HbA), 439
Adult rickets, 354
Adventitia, 403, 404, 546
Aerobic metabolic pathways (*see* Metabolism; Skeletal muscle, energy metabolism)
Afferent arterioles of kidney, 514, 516–518
Afferent nerve fibers (*see* Nerves, afferent; Sensory fibers)
A fibers
of nerve, 147–148, 221
of skeletal muscle, 297–298, 322, 325
Afterbirth, 652, 653
Aggregation of platelets, 462–463
Agglutination, 440, 441
Agglutinins, 440, 459
Agglutinogens, 440
Aging, 61, 72
atherosclerosis and, 389
autoimmune reactions and (*see* Autoimmunity)
blood pressure and, 417
body defenses and, 459
bone and, 341, 343, 344, 351–352
cell receptors and, 115, 351
collagen and, 54–55
heart and, 300, 329, 381, 382, 390
hematocrit and, 433, 435
hormones and, 115, 351
(*See also* Steroids, sex steroids)
lactose intolerance and, 586
lens and, 194
lungs and, 475
memory and, 351
metabolic rate and, 593–594, 615
prostate and, 622
sex hormones and, 629, 638
T cells and, 351
temperature regulation and, 510–511
thymus and, 351
water content of body and, 80
Agnosia, 278
Agonist:
to drugs, 252, 253
of skeletal muscles, 312
Agranulocytes, 454
(*See also* Leukocytes; Lymphocytes; Monocytes)
AHF (*see* Antihemophilic factor)
AHG (*see* Antihemophilic globulin)
Air:
composition of, 479
high altitude and, 479
pressure of, 479
Air sinuses, 474–475, 629
Albinism, 508
Albumin, 113, 412
elcetrophoresis of, 394
Albuminuria, 533, 534
Alcohol
absorption of, 564, 574, 575
effects:
on antidiuretic hormone secretion, 524
on blood vessels, 379
on heart, 379, 380
on nervous system, 149
lipoproteins and, 395, 396
penetration of blood-brain barrier, 147
temperature regulation and, 510–511
ulcers and, 565–567
Alcoholism, 225
Alexia, 278
Algesireceptors (*see* Sensory receptors, mechanoreceptors)
Alimentary canal, 543
(*See also* Digestive system)
Alkaline phosphatase, 342, 351
Alkalinity, 16, 18
of venous blood, 486
Alkalosis, 19, 22, 433, 525, 534–535
metabolic, 535
respiratory-induced, 494, 534–535
Allergen, 63, 64, 511, 565
(*See also* Allergy)
Allergic encephalomyelitis, 149
Allergy, 63–65, 98, 177, 467
autonomic nervous system and, 229, 232
cortisol and, 117
delayed-type, 63
immediate-type, 59, 63
digestive tract and, 565, 576, 588
histamine and, 65
leukotrienes and, 462–463
mast cells and, 64, 65
nervous system disorders and, 147, 149–150
in respiratory tract, 63–64, 484
in skin, 64–65, 511
(*See also* Allergen)
All or none law, 162–163, 311, 366
Alpha adrenergic fibers (*see* Adrenergic nerves)
Alpha cells in pancreas, 111
Alpha chains, 83
of hemoglobin, 431, 438, 439
of proteins, 394
Alpha globulin, 465
Alpha-2 globulin, 526
Alpha helices of proteins, 83
Alpha nerve fibers, 148
Alpha receptors (*see* Receptors, cellular)
Alpha rhythm (*see* Alpha waves)
Alpha thalassemia, 449
Alpha waves, 247, 248
Altitude exposure, 480, 492–494, 535
2,3 diphosphoglycerate and, 486, 493–494
effects on erythrocytes, 432, 436, 437, 448
erythropoietin and, 437
sickle cell anemia and, 448
stress and, 109
Alveolar air
composition of, 480, 482, 486–488
Alveolar duct(s), 474, 475
Alveolar glands, 51
Alveolar macrophages, 66, 477, 484, 496, 497
Alveolar membranes, 475–477
Alveolar ventilation, 480, 488, 534
Alveolus, 402, 474–478
atelectasis of, 478
anatomy of, 475–477, 487
capillary membranes and, 477
respiratory distress syndrome and, 477–478
surfactant and, 476–478
p-Amino hippuric acid, 533
Aminotransferase (*see* Aspartate aminotransferase)
Amino acid(s), 21, 366
absorption of, 57, 61, 575–577, 580
active transport of, 92, 95, 576
chemistry of, 21–23
deamination of, 322
deficiencies of, 150
essential, 576
as neurotransmitters, 150–151
in proteins, 28
requirements for, 599
Amination of proteins, 592
Amino groups in proteins, 28
Aminopeptidases, 574
Ammonia, 21, 510, 530, 553, 592
detoxification by liver, 589
Ammonium chloride, 53
Ammonium salts, 535
Amniocentesis, 644–645
Amnion, 641, 642
Amniotic cavity, 640, 641
Amniotic fluid, 476, 477, 644–645
Amniotic sac, 644–645, 653
AMP (*see* Adenosine monophosphate)
Amphetamines, 104, 249, 252, 253
Amplifying circuit, 214, 215
Amplitude, 172, 185, 247
Ampulla, 195, 197, 199
(*See also* Sense organs, ear; Sensory receptors, vestibular receptors)
Amygdala, 251, 272, 273
Amylase, 551, 559, 566, 572, 573
pancreatic, 551, 573
salivary, 551, 555, 572
Amyl nitrite, 380, 389
Anabolism, 318, 590, 628
(*See also* Metabolism)
Anaerobic bacterias, 589
Anaerobic glycolysis, 318–323
(*See also* Metabolism)
Anaerobic reactions, 318
shock and, 443, 444, 445
(*See also* Anaerobic glycolysis; Metabolism)
Anal canal, 546
Analgesic, 249, 250, 252, 253
Anal sphincter, 490, 546, 564
Anaphylaxis, 64
Anastomosis, 407
Anatomic dead space, 483
Anatomy, 2
Androgens, 127, 131, 133, 625–629
binding protein of, 625
effects on bone, 350, 351
secretion after vasectomy, 650
(*See also* Dehydrotestosterone; Reproductive system; Testosterone)
Anemia, 432, 446–449
hematocrit and, 432
iron deficiency, 447, 583
oxyhemoglobin and, 486
sickle cell, 446–449
thalassemia, 446–449
Anesthesia, 251
Aneurysm, 246, 422

Angina pectoris, 221, 370, 389, 574
Angiogram, 398
Angiography, 245, 246
Angiotensin, 524–526
  blood pressure and, 423, 424, 444
  renin-angiotensin system and, 525
Angiotensinogen, 526
Angstrom unit (Å), 16
Anorexia, 585
Antacids, 535
Antagonist:
  to drugs, 252, 253
  of hormones, 116, 122
  skeletal muscle and, 312
Anisotropic bands of muscle (*see* A bands)
Anode, 391, 392
Anterior pituitary (*see* Pituitary, anterior)
Anterior root (*see* Ventral root)
Anterograde transport, 139, 140
Anterograde Wallerian degeneration, 145
Antiacetylcholinesterases, 297
Antiobiotics, 457, 589, 590, 646
  (*See also* Penicillin)
Antibodies, 59, 60, 61, 85, 459
  aging and, 351
    (*See also* Autoimmunity)
  as analytical reagents, 117, 118, 246, 641
  to endocrines, 115–118
  fluorescent, 246
  in germ-free animals, 589
  as immunoglobulins:
    immunoglobulin A (IgA), 59, 551, 555, 565, 576, 588, 639
    immunoglobulin D (IgD), 59
    immunoglobulin E (IgE), 59, 63, 229, 232
    immunoglobulin G (IgG), 59, 354
    immunoglobulin M (IgM), 59, 354
  interaction with antigens, 439–440
  to major blood groups, 440–443
  to platelets, 469
  radioimmunoassay and, 117, 118, 246, 641
  rheumatic fever and, 361
  to Rh factor, 446
  to sperm, 625, 650
  structure of, 64, 459
  tagged, 246
  vasectomy and, 650
  (*See also* Agglutinins; Allergy; Antigens; Autoimmunity; Defense cells; Histocompatibility antigens; Leukocytes, B lymphocytes; Plasma cells)
Anticholinergic drugs, 252, 253, 566
Anticoagulants, 465–466
Anticodon (*see* Genetics; Transcription)
Antidiuretic hormone (ADH), 52, 91, 103, 107, 108, 111, 122, 267, 269
  effects on kidney, 91, 523, 524, 644
  shock and, 444
Antigen(s), 59, 60, 87
  of blood cells, 439–443
  histocompatibility, 85, 86, 440, 536
  interaction with antibodies, 63, 64, 459, 460
  (*See also* Antibodies)
Antihemophilic factor A (AHF), 464, 465
Antihemophilic factor B, 465
Antihemophilic globulin (AHG), 464
Antihistamines, 64, 566
Antimicrobial compounds, 509, 511, 551, 565
  (*See also* Antibiotics; Antibodies; Interferon; Lymphokines; Penicillin)
Antiserotonergic drugs, 252
Antithrombin III, 465
Antral follicle, 633
Antrum of stomach, 555, 566
Anuria, 533
Anus, 546, 630, 631
Anxiety, 129, 253
Aorta, 362, 384, 388
  semilunar valve of, 362, 363
Aortic arch, 364, 369
Aortic bodies (*see* Sensory receptors, baroreceptors, chemoreceptors)
Aplastic anemia, 448
Aphasia, 278
Apnea, 483
Apneustic center, 260, 490, 491
Apocrine glands, 51, 510
Apoferritin, 583–585
Apomorphine, 253
Aponeurosis, 66
Apoplexy, 426
Appendicitis, 566
Appendix, 544, 545
  referred pain and, 210
Appetite, 131, 151, 233
  (*See also* Hypothalamus, appetite center)
Aqueous humor, 187, 188, 191, 192
Arabinose, 577
Arachidonic acid
  leukotrienes and, 462–463
  prostaglandins and, 462
Arachnoid membrane, 242
Arachnoid villi, 242, 244
Areolar tissue (*see* Connective tissue, loose)
Argentaffin cells, 548
Arousal, 112, 177, 247, 248, 253
Arrector pili, 507, 509
Arrythmias of heart, 381
  (*See also* Fibrillation)
Arsenic poisoning, 225, 447
Arterial pressure (*see* Blood pressure)
Arteries, 54, 402–404, 415
  atherosclerosis and, 425–426
  composition of walls of, 403, 404
  coronary (*see* Coronary arteries)
  end, 407, 426
  middle cerebral, 407, 426
Arterioles, 363, 402, 403, 408
Arteriosclerosis, 339, 422, 425–426
Arteriovenous anastomoses, 407
Arteriovenous oxygen differences, 382, 492
Arthritis, 63, 354
  (*See also* Osteoarthritis, Rheumatoid arthritis)
Articular cartilage, 68, 73, 336–338, 343
Articulation(s), 336–338
Artificial kidney, 535–537
Artificial respiration, 497–500
Ascending colon, 544, 546
Ascending tract (*see* Spinal cord)
Ascites, 389, 533, 567
Ascorbic acid (*see* Vitamins, vitamin C)
Aspartate aminotransferase (AST), 23, 391–393, 432, 566
Aspartic acid, 23
Asphyxiation, 314
Aspirin, 89, 463, 535, 567, 574
Associative fibers of neurons of brain, 237, 270, 274
Asthenia, 64
Asthma, 63, 64, 229, 232, 467, 475, 484
  (*See also* Allergies)
AST (*see* Aspartate aminotransferase)
Astigmatism, 194
Astrocytes, 145, 146
Asynergia, 264
Ataxia, 264
Atelectasis, 478, 491
Atherosclerosis, 25, 111, 363, 422, 425–426, 587
  aging and, 389
  defined, 425
  blood vessels in eyes and, 187
  platelets and, 426, 462
  prostaglandins and, 462–463
  prostacyclin and, 462–463
  thromboxane and, 462–463
  vasectomy and, 650
  (*See also* Heart, disease of)
Atmospheric pressure, 479
Atom(s), 15, 87
ATP (*see* Adenosine triphosphate)
ATPase, 87, 295–296, 323
  carrier of calcium, 295–296, 347
  carrier of sodium and potassium, 93–95, 149, 160–162, 375
  secondary messengers and, 232
  (*See also* Active transport; Metabolism)
Atretic follicle, 633, 636
  birth control pills and, 648
Atria, 361, 362, 365, 371
Atrial fibrillation, 375
Atrial flutter, 375
Atrioventricular (A-V) block, 375
Atrioventricular node, 367–369, 371, 380
Atrioventricular valves, 354, 361–363
Atrophy of skeletal muscle, 314
Atropine, 104, 231, 252, 253, 273, 566
Attack reactions
  behavior and, 266
Auditory apparatus (*see* Sense organs, ear)
Auditory area of brain, 277
Auditory canal, 69
Auditory nerve fibers (*see* Cranial nerves, statoacoustic nerve)
Auditory receptors (*see* Sense organs, ear; Sensory receptors, auditory receptors)
Auditory reflexes (*see* Reflexes)
Auditory tube, 196
Auerbach's plexus, 546, 547
Augmented limbleads (aVR, aVL, aVF), 372, 373
Aura, 249
Auscultatory method of measuring blood pressure, 416–419
Autoimmunity, 129, 149, 354, 447, 460, 467, 469, 534, 559, 650
  (*See also* Grave's disease; Myasthenia gravis; Multiple sclerosis; Pernicious anemia; Rheumatoid arthritis; Thrombocytopenic purpura)
Autonomic nervous system, 225–233
  adrenals and, 226–228
  allergies and, 229, 232
    (*See also* Allergies)
  anatomy of, 226–227, 231
  brain and, 229, 233
    (*See also* Brain; Brain, medulla oblongata; Brain, pons; Hypothalamus; Medulla oblongata)
    centers of, 177
    (*See also* limbic system, hypothalamus, medulla oblongata, pons below)
    cranial nerves and, 226, 227, 229, 333
    hypothalamus and, 233, 266–269
    limbic system and, 233
    medulla oblongata and, 233, 260, 369, 407, 420, 421, 424, 444, 488, 490, 491
    pons and, 233, 260, 490, 491

Autonomic nervous system (*Cont.*):
catechol-O-methyltransferase (COMT), 228, 230
circulatory system and, 226, 419–421
heart, 73, 225, 227–229, 231, 233, 368–370
vascular smooth muscle, 226–229, 231, 233, 260, 407
digestive system and, 225–227, 231, 233, 302–304
(*See also* Digestive system)
drugs and;
antiadrenergic, 252, 253
anticholinergic, 253
atropine, 231
carbachol, 231
dibenzyline, 231
isoproterenol, 229, 231
neosynephrine, 231
phenylbenzamine, 231
phenylephrine, 231
propranalol, 231
excretory system and, 226, 227, 521
eye and, 226, 227, 233
fibers of, 225–227, 233
adrenergic (*see* Adrenergic nerves)
cholinergic (*see* Cholinergic nerves)
origins of, 226, 227
postganglionic, 226, 227, 229
preganglionic, 226, 227, 229
ratio of fiber types, 227
ganglia of, 141, 145, 225–227, 380
hair follicles and, 226
kidneys and, 226, 521
monoamine oxidase (MAO), 228, 230
neurotransmitters of:
acetylcholine, 227, 228, 230–232
epinephrine, 228, 229, 232
metabolism of, 228–230
receptors to, 123, 227–232
phenylethanolamine-N-methyl transferase, 226
respiratory system and, 226, 625
bronchioles, 226, 229, 230, 232, 233
lungs, 226, 229
salivary glands, 226, 227, 233
secondary messengers and, 228–229, 231
small intestine and, 226, 231
sphincters of the digestive tract and, 231
stomach and, 226, 231
subdivisions of:
parasympathetic, 225, 226, 229, 368–370, 419–421, 561, 625
sympathetic, 225, 226, 229, 368–370, 419–421, 509, 560–563, 625
(*See also* Adrenergic nerves; Cholinergic nerves; Brain)
Autoprothrombin I, II, III, 464
Autoregulation of blood flow, 383, 407, 408
exercise and, 383
kidney and, 519–521
Autorhytmicity, 367
Autosomes, 629
Auxotonic contraction, 308–309
AV node (*see* Atrioventricular node)
aVF, aVL, aVR (*see* Augmented limb leads)
Aviation sickness, 494
Avogadro's number, 89, 91
Axilla, 510, 634
Axoaxonic synapse, 169, 171
Axodendritic synapse, 169, 171
Axolemma, 143
Axon, 75, 135, 137, 139–140, 147
collaterals of (*see* Collaterals)
Axon (*Cont.*):
conduction of impulses, 141
(*See also* Impulses)
hillock, 137, 139, 143
terminals, 168–171, 291–295
transport in, 139–140
(*See also* Synapse)
Axoplasm, 143
Axosomatic synapse, 169, 171

Babinski sign, 213
Backward heart failure, 388, 389
Bacteria
digestive tract mucosa and, 565, 588, 589
effects on bile pigments, 439
food poisoning and, 587
gastroenteritis and, 586–590
ingestion of, by macrophages, 457
(*See also* Macrophages; Phagocytosis)
Bactericidal chemicals, 457
Bacteriostatic chemicals, 457
Bainbridge reflex, 370
Balance, sense of, 194, 259–261 (*See also* Sense organs, ear)
Band cell, 455, 466
Banded appearance of skeletal muscle, 284, 287, 288, 290
B antigens (*see* Red blood cells, major blood groups)
Barbiturates, 89, 268, 478, 535
Baroreceptors (*see* Sensory receptors, baroreceptors)
Bartholin's glands, 631
Basal body complex, 331
Basal ganglia, 177, 265, 272, 445
Basal lamina (*see* Basement membrane)
Basal metabolic rate (BMR), 593–595
Basal propria (*see* Lamina propria)
Base, 527
Basement membrane, 44–47, 409–411, 515, 518, 520
Base pairs in DNA, 27, 28
Basic electric rhythm of smooth muscle, 561
Basilar membrane, 195, 197–199
Basket cells, 262
Basophil, 63, 229, 434, 435, 455–458, 465
Basophilic, 455
B blood type, 441–442
(*See also* Red blood cells, major blood groups)
B cells (*see* Leukocytes, B lymphocytes)
B cells of pancreas, 129
Becker form of muscular dystrophy, 315
Behavior
(*See also,* Emotions; Brain, hypothalamus, limbic system; Hypothalamus)
centers and, 226–229
chromosomes and, 629
drugs and, 251, 252, 272
and hormones, 104, 628, 634
(*See also* Endocrines)
type and heart attack risk, 389–390
Belladona, 253
Benadryl, 64
Bend, 494
Benign tumor, 425
Benzene, 447
Benzodiazepines, 252
Beri-beri, 225
Bernoulli's principle, 414–415
Beta adrenergic stimulation (*see* Adrenergic nerves; Sensory receptors, cellular, beta)
Beta cells, 98, 112
Beta chains of hemoglobin, 431, 438, 439, 446, 447
Beta-estradiol (*see* Estradiol)
Beta helices, 83
Beta lipotropin, 127, 249, 394
Beta receptors (*see* Adrenergic nerves; Receptors, cellular, beta)
Beta rhythm, 247, 248
Beta sheets, 83
Beta thalassemia, 449
Beta waves, 247, 248
Betz cells, 270
B fibers
of skeletal muscle, 297, 298, 300, 322
of nerve, 148, 227
Bicarbonate
buffering system, 433, 486, 487, 529–531
excretion, 493
in extracellular fluids, 81, 82
filtration, 521
pancreas and, 551
secretion, 555, 558, 562
ulcers and, 567
Bicarbonate-carbonic acid ratio, 529
Biconcave lens (*see* Sense organs, eye, lens)
Bicuspid valve, 362
Bile
cholesterol and, 394, 395
composition of, 559
gallbladder and, 551, 554
hemoglobin metabolism and, 438–439
insufficiency of, 586, 588
liver and, 551, 554, 566, 567
mucosa of digestive tract and, 565, 567
recycling of, 439
secretion of, 557, 560
vitamin absorption and, 465
Bile canaliculi, 552
Bile ducts, 552
Bile pigments, 438–439, 445
Bilirubin, 439
Biochemistry, 3
Biofeedback, 75
Biological Abstracts, 34
Biological clock (*see* Biological rhythyms)
Biological rhythyms, 128, 177, 266, 269–270
Bipolar recordings, 373
Bipolar neuron, 141, 187, 190, 202
(*See also* Neurons, types of)
Birefringence of skeletal muscle, 288
Birth, 9, 133, 632, 652–653
Birth canal, 133, 646
(*See also* Cervix; Vagina)
Birth control (*see* Conception control)
Birth control pills, 647–649
Bitter taste, 201
Blackheads, 509
Bladder (*see* Urinary bladder)
Blastocyst, 640, 641
Blastoderm, 641
Bleeding, 468
(*See also* Blood; Hemorrhage; Thrombocytes, clotting, inhibition of)
Bleeding time, 466
Blood, 55, 66, 431, 436–452, 453–471
blood-brain barrier, 146, 177, 269
blood-testis barrier, 625
buffers and, 433–434
(*See also* Buffers)
cell storage, 91
centrifugation of, 433
dialysis of, 536–537
disease of
erythrocytes, 435–449

Blood, disease of (*Cont.*):
leukocytes, 466–468
thrombocytes, 469
flow of, 414–416
to kidneys, 517, 521
to skeletal muscle, 517
(*See also* Flow rate of blood)
gas chemosensors and, 370
(*See also*, Sensory receptors, chemoreceptors)
gases, 486, 488
general functions of, 431–432
general properties of, 432–435
group antigens, 439–443
(*See also* Antigens; Erythrocytes)
leukocytes (*See* Leukocytes)
loss in hemorhage (*see* Hemorhage)
loss in menstruation, 638
origins of, 434–435
plasma, 245, 431–432
pulse and, 404
smear, 454–456
specific gravity of, 432
storage of, 466, 553
thrombocytes, 432–433
(*See also* Thrombocytes)
transfusion reactions, 91, 441–443, 446, 469, 566, 583
viscosity of, 419, 432
Blood pressure, 111, 366, 406, 416–421, 484
age and, 417
angiotensin and, 423
area of vessels and, 414–416
average values of, 415, 417
autonomic nervous system and (*see* Autonomic nervous system, circulatory system)
central nervous system and, 421
(*See also* Brain, medulla oblongata)
defined, 416
erection and, 625
exercise and, 383
heart and, 369, 380, 381, 420–421
hormonal effects on, 128, 131, 422–423
measurement of, 417–419
receptors and, 184, 185, 420
(*See also* Receptors, cellular; Sensory receptors, chemoreceptors, mechanoreceptors)
regulation of, 419–422
renin-angiotensin system and, 525, 526
risk of heart attack and, 390
shock and, 443, 444
stress and, 423
(*See also* Blood vessels)
Blood bank
storage of blood, 466, 485
Blood vessels, 401–429
autonomic nervous system and (*see* Autonomic nervous system, circulatory system)
cross-sectional areas of, 414–415
disorders of, 421–426
hormonal effects on, 131–133
neurotransmitter effects on
norepinephrine and, 131, 227–228, 231, 300, 304, 422, 423
serotonin, 177
platelets and, 462
prostaglandins and, 462–463
reflex control of, 419–421
smooth muscle and, 301, 304
total length of, 431
B nerve fibers, 148, 227

Body building blocks, 12–39
Body, elemental composition of, 583
Body temperature, 267–268, 510
alcohol and, 379
basal metabolic rate and, 594–595
fever and, 594
heart rate and, 381
normal range of, 268, 594
ovulation and, 594, 636, 637
water (*see* Water)
weight (*see* Weight, body)
(*See also* Hypothalamus, temperature regulation center)
Body of stomach, 544, 545
Body of uterus, 630, 631
Bohr effect, 484–485
Bond, 16, 17, 19, 22
(*See also* individual names of types of bonds)
Bone, 6, 54, 67, 69–72, 81, 334–358
aging and, 341, 351–352
anatomy of, 335–338
cancellous, 70, 72, 336, 342, 581
compact, 70, 72
calcium and, 335, 338–341, 345–349, 351
(*See also* Calcium)
cancer of, 352, 353
depolymerization of, 349
dynamic exchanges in, 341
electrolyte regulation and, 128
formation and growth of, 68, 126, 341–352
acidity and, 339, 348, 349
cells and, 341–345, 348
exercise and, 351, 352
intracartilaginous, 343–345
intramembranous, 342–343
mechanical stress and, 342, 351–352
functions of, 335
homeostasis and, 345
hormones and, 348–351
adrenal steroids, 350
calcitonin, 128, 348, 349
estrogens, 348, 350, 351, 634
glucocorticoids, 130, 348, 350, 351
growth hormone, 348, 349
mineralocorticoids, 350
parathormone, 128, 347–349
somatomedins, 349, 350
testosterone, 348, 350, 628
thyroxine, 349, 350
vitamin D_3, 132, 345–350
(*See also* Endocrines)
joints and, 336–337
marrow, 65, 66, 132, 337–338, 352, 460, 461, 466–469, 584
sinusoids of, 411
transplant of, 448, 449
vitamins and, 345–348
vitamin A, 347
vitamin C, 347–348
vitamin D, 345–350
water content of, 80, 81, 338
Bonyy labyrinth of inner ear, 195, 196
Books in Print, 34
Botulinal toxin, 295
Botulism, 295, 588
Boutons (*see* Axon terminals)
Bowel, 91
(*See also* Colon; Small intestine)
Bowman's capsule, 513, 514, 516, 517
Boyle's Law, 479, 480, 495
Brachial plexus, 211, 370
Brachiocephalic vein, 413
Bradycardia, 406

Brain, 6, 7, 135, 136, 236–255, 255–281
anatomy, 237
blood-brain barrier, 146
brainstem, 241, 262–264
cerebellum, 141, 241, 258, 260–265, 312
(*See also* Cerebellum)
cerebrospinal fluid (CSF), 145, 242
blood-brain barrier and, 243–244
circulation of, 242–244
composition of, 244, 245
disorders and, 243
enkephalins and, 250
formation of, 242–244
functions of, 244
ventricles of brain and, 145, 208, 241–244, 260
volume of, 243
withdrawal by spinal tap, 208, 243, 244
cerebrum, 241, 268, 270–279
(*See also* Cerebral hemispheres)
cranial nerves (*see* Cranial nerves)
death of, 247
development of, 237–241, 247, 643
influence of estradiol, 257
influence of stimulation, 257
diagnostic tests, 244–249
cerebral angiography, 245, 246
computer assisted tomography (CATscan), 245, 246
electrical activity of, 246–249
brain waves and, 247–249
epilepsy and, 248–249
immunological studies, 246
pneumoencephalography, 245
staining and microscopy, 244
x-ray analysis, 245
(*See also* X-rays)
divisions of, 239, 241, 265–266
dominance in, 274–278
drugs and, 249–253
(*See also* Drugs)
energy consumption by, 237
forebrain, 177, 239, 241, 251, 265–279
hindbrain, 177, 239, 241, 259–265
hypothalamus (*see* Hypothalamus)
limbic system (*see* Limbic lobe; Limbic system)
lobes, 202
medulla oblongata, 241, 258–261, 420–421
cardiac center, 260, 369, 420, 421, 444
expiratory center, 490
inspiratory center, 490, 491
respiratory center, 260, 444, 488, 490, 491
vasomotor center, 260, 407, 420, 421, 444, 488, 535, 564
(*See also* Vasoconstriction; Vasodilation)
vomiting center, 564
membranes, 242
midbrain, 239, 241, 251, 259, 265
neurons
associative, 237
centers and, 257
(*See also* centers in specific areas of brain)
lifespan of, 257
nuclei of, 257
numbers of, 237, 262, 263
synapses and, 237
neurotransmitters and, 177, 249–253
drugs and, 249, 251–253
dynorphins, 250
endorphins, 249, 250
enkephalins, 249–250
(*See also* Neurotransmitters)

Brain (*Cont.*):
nutrition and, 148–151
(*See also* Nutrition)
olfactory bulb (*see* Olfactory bulb)
olfactory lobe (*see* Olfactory lobe)
pineal body, 6, 7, 52, 105, 28, 50, 177, 260, 266, 269–270
pons, 241, 253, 260, 261, 490, 491
apneustic center, 260, 490, 491
pneumotaxic center, 260, 490, 491
thalamus, 216, 237, 241, 266, 369
cerebellum and, 260–264
functions of, 269
limbic system and, 271, 272
opiate receptors of, 251
weight of, 237, 244
Brainstem, 241, 262–264, 271
Brain waves (*see* Brain, electrical activity of)
Braxton Hicks contractions, 653
Breast buds, 634
Breasts, 632, 634, 638–639, 644
Breathing
of newborn, 476
reflexes of, 489, 490
(*See also* Respiration)
Bright adaptation, 189
Broca's area, 274, 277, 475
Brodmann's areas, 274, 276
Bronchioles, 474, 475
allergies and, 63, 64
(*See also* Asthma)
hormones and, 132, 133
Bronchus, 46, 68, 301, 474
Bronzed diabetes, 583
Brown fat, 56
Brunner's glands, 546, 550, 559, 573
Brush border, 513, 515, 547, 548, 559
Brush cells, 477
Buccal glands, 550
Buffers, 22, 433, 527, 529–531
Building blocks of body, 12–39
Bulbourethral glands, 621, 622, 625
Bulk effect on colon, 586, 589
Bulk flow, 412, 518
Bundle of His, 367–368
Burkitt's lymphoma, 468
Bursa, 337, 338
Bursitis, 354

Cadaverine, 589
Caesarean section, 651
Caffeine, 26, 27, 91, 367, 380, 381, 523, 562, 566
Caisson disease, 494
Calcification (*see* Bone; Calcium)
Calcitonin, 8, 52, 122, 128, 348, 349
Calcium
absorption of, 339–342, 345–347
adsorption of, 339
ATPases, 295, 296, 347
bone and, 335, 338–342, 345–347
clotting and, 461, 464, 465
daily dietary requirement of, 618
disorders of metabolism and, 352–355
(*See also* Calcitonin; Endocrines, parathyroids, thyroid; Parathormone)
exchange, 341
exocytosis and, 98, 170–172
extracellular fluids and, 81, 82
heart and, 368, 379, 380
hormonal regulation and, 8–10, 113, 115, 636
calcitonin, 8, 128
Calcium, hormonal regulation and (*Cont.*):
parathormone, 8, 10, 128
as a secondary messenger, 120–123
intrinsic factor absorption and, 558
neurotransmission and, 169, 170–172, 230
to phosphate ratio, 339–340, 347, 352, 353
pores in cell membrane and, 88, 89
rennin and, 573
skeletal muscle and, 288, 291, 292, 294–296, 310, 314, 323–324
smooth muscle and, 301, 302
transport, 95
(*See also* ATPases)
visual impulses and, 189, 191
Calcium carbonate, 69, 338
crystals in inner ear, 197
Calcium phosphate, 69, 338–340
Calmodulin, 123
Caloric value of foods, 593, 616–617
Calorie (kcal), 14–16
expenditure in exercise, 593–595, 613–617
(*See also* Exercise)
heart attack risk and, 390, 394–396
hyperlipoproteinemia and, 394–396
storage in liver glycogen, 554
storage in skeletal muscle, 554
cAMP (*see* Cyclic adenosine monophosphate)
Canal:
of root in tooth, 73
of Schlemm, 187, 188
semicircular in inner ear, 195, 197, 199–201, 259
Canaliculi
in bone, 69–71
in liver, 552
Cancellous bone, 70, 72, 336, 342, 581
Cancer, 63, 86, 467–469
of bone, 352, 353, 435
of breast, 632
of cervix, 632, 653
colonic, 589
interferon and, 460
leukemias, 146, 340, 467–468
of lungs, 496–497
lymphomas, 467, 468
macrophage defense and, 61, 63, 146
prostacyclins and, 463
smoking and, 496–497
T cell defense and, 459–460
of uterus, 632
of vagina, 632, 653
Candida infection, 646
Cannabis, 115, 252, 253
C antigen, 442–443, 446
Capacitance, 167
Capacitation, 639
Capillaries, 49, 81, 403, 407–412, 415
alveolar, 475, 487
beds of, 407, 408
in brain, 145, 146
of digestive tract, 578, 579
exchange of materials by, 412–413
in heart, 362, 364
junctions of, 408–412, 517
in kidneys, 517
permeability of, 49, 444–445, 457, 578, 579
types of, 407, 412
Capsule, 67, 69, 336, 337, 636
Captopril, 424
Carbachol, 231
Carbaminohemoglobin ($HbCO_2$), 437, 486, 487
Carbaminoproteins, 486
Carbohydrate(s), 20
absorption of, 575, 576
caloric value of, 593
chemistry of, 20
digestion of, 572–574
(*See also* Digestive system)
effects on gastrointestinal motility, 564
excretion of, 523
loading, 327, 593
metabolism, 320, 321, 328, 329, 366, 591, 592
(*See also* Metabolism)
nutrition and, 149
(*See also* Digestive system)
respiratory quotient of, 593
storage of, 593
Carbon dioxide
in alveoli, 486
in blood, 486
partial pressure of, 480, 488
poisoning, 149, 495
as regulator of respiration, 485–486, 488, 490–492
transport of, 486–487
(*See also* Respiratory system)
Carbonic acid, 491
bicarbonate buffering system and, 433, 487
anhydrase, 487, 531, 558
Carbon monoxide
combination with hemoglobin, 438, 496
poisoning, 496
respiration and, 438, 496
sources of, 496
Carbonmonoxyhemoglobin (CoHb), 438, 496
Carboxyhemoglobin (COHb), 438, 496
Carboxyl group, 21
Carboxypeptidase, 559, 572, 573
Carcinogens, 86, 497, 589
(*See also* Cancer)
Cardiac catherization, 398
Cardiac center, 260, 380, 381, 444
Cardiac circulation in fetus, 384–387
Cardiac cycle, 364–365
Cardiac disease, 360, 373–377
atherosclerosis and, 363
coronary arteries, and, 363–364, 387, 389–398
myocardial infarction, 364
(*See also* Heart disease)
Cardiac glands, 550
Cardiac insufficiency, 363
Cardiac muscle (*see* Muscle, cardiac)
Cardiac output, 360–363, 365–366, 379–383, 444
blood pressure and, 419–421
excercise and, 492
heart disease and, 387–389
(*See also* Heart disease)
kidney and, 517, 525–526, 532
liver and, 553
respiration and, 488, 492–493
Cardiac pacemaker, 300, 367, 375
Cardiac plexus, 368, 369
Cardiac sphincter, 545, 558
Cardiac vector, 373
Cardioinhibitory center, 369, 370
Cardiopulmonary resuscitation (CPR), 374, 498–500
Cardiovascular system, 360
disease of, 422, 649
(*See also* Atherosclerosis; Coronary arteries; Heart Disease)
shock and, 443-445

Cardiovascular system (*Cont.*):
smoking and (*see* Nicotine; Smoking)
(*See also* Blood vessels; Heart)
Carotid bodies (*see* Sensory receptors, chemoreceptors)
Carotid sinuses (*see* Sensory receptors, baroreceptors)
Carrier, 81
mediated transport and, 87, 92–94
Carrier molecules, 92, 113, 114, 120
Cartilage, 54, 67, 81
articular, 68, 73, 336–338, 343
elastic, 68, 69
fibrous, 68, 69
hyaline, 68, 73, 336, 337, 343, 475
Cascade effect
of clotting, 464
of hormones, 116
Casein, 573
Castor oil, 586
Castration, 350, 650
Casts, 534
Catabolism, 318, 590
(*See also* Metabolism)
Catalyst, 21–23
(*See also* Enzymes)
Cataract, 194
Catecholamines
degradation of, 230
effects of drugs on, 253, 424
as neurotransmitters, 176–177, 228, 230, 252, 424, 526
receptors and aging, 115, 380, 381
Catechol-O-methyltransferase (COMT), 228, 230
Cathartics, 585–586
Catheter, 374, 398, 419
Catheterization, 398
Cathode, 391, 392
Cathode-ray oscilloscope, 163, 371
CAT scan (*see* Computer assisted tomography)
Cauda equina, 208
Caudal anesthesia, 208
Caudate nucleus, 272–273
Cauterization, 650
C cells, 128, 349
CCK-PZ (*see* Cholecystokinin-pancreozymin)
Cecum, 545, 589
Celiac disease, 587
Cell body, 135, 137, 209
Cell, 2, 5, 29–36
of body, 75
division of 34, 623–624
(*See also* Meiosis; Mitosis)
Cell membrane, 29, 30, 82–87, 120, 292, 458
hormone receptors of, 114–123
(*See also* Endocrines; Receptors; Neurotransmitters; Receptors, cellular)
Cellular immunity, 59, 86, 458, 511
(*See also* Leukocytes, T lymphocytes)
Cellular respiration, 473
(*See also* Metabolism)
Cellulose, 20, 585, 586, 589
Cellulose acetate electrophorersis, 392
Celsius scale of temperature, 13, 16
Cement substance, 70, 73
Cementum, 355
Centers of neurons, 257
(*See also* Brain, medulla oblongata, centers of; Hypothalamus, centers of)
Centrigrade to Fahrenheit conversion, 13, 16
Centimeter, 16
Central canal of spinal cord, 145, 208, 209, 242, 244
Central nervous system (CNS), (*See* Brain; Spinal Cord)
Central sulcus, 270, 273, 274
Central veins of liver, 552
Centrioles, 24, 29, 623
Centrosome, 31
Cephalic neural stimuli, 560
Cerebellar peduncles, 260–262
Cerebellum, 141, 241, 258, 260–265, 312
cellular anatomy, 261–262
computer activities, 261, 263
coordination of skeletal muscle (*see* Skeletal muscle)
cortex of, 261, 262
disorders and, 246, 265
excitatory cells of, 263
functions of, 261
inhibitory cells of, 263
medulla of, 261, 262
nuclei of, 261, 262
pathways and, 261–263
thalamus and, 260–264
Cerebral angiography, 245, 246
Cerebral aqueduct, 241, 244
Cerebral artery, 242, 407, 426
Cerebral cortex (*see* Cerebral hemispheres)
Cerebral dominance, 274–278
Cerebral hemispheres, 237, 241, 268, 270–279
anatomy, 270–273
conditioned reflexes and, 278
cortex of, 237, 241, 242, 263, 270
learning and, 278
lobes of, 270–273
major functions of, 270–273
memory and, 270–273
organization of, 186–187, 273–278
primary motor areas, 273–275
primary sensory areas, 273–275
split brain, 274–278
speech center and, 277–278
respiration and, 488–491
(*See also* Brain)
Cerebral hemorrhage, 426
Cerebral peduncles, 265
Cerebral sclerosis, 147
Cerebral thrombosis, 426
Cerebral vein, 242
Cerebrosides, 25
Cerebrospinal fluid (CSF), (*see* Brain, cerebrospinal fluid)
Cerebrovascular accident (CVA), 426
Cerebrum (*see* Brain; Cerebral hemispheres)
Cerumen, 509
Ceruminous glands, 196, 507, 509
Cervical epithelium, 636
Cervical cancer, 632, 653
Cervical mucus, 636–638, 645, 647–648
Cervical pregnancy, 641
Cervical nerves, 206, 207
Cervical vertebrae, 69, 206, 207
Cervix, 630, 631
childbirth and, 652, 653
C fibers
of skeletal muscle, 297, 298, 322
of nerve, 148
Chambers of eye, 188, 191–193
Channels in cell membrane (*see* Pores)
Chemical control agents, 103–104
Chemoreceptors (*see* Sensory receptors, chemoreceptors)
Chemotaxis, 64, 457, 459, 467
Chest cavity (see Thoracic cavity)
Chewing, 259, 554–555
Cheyne-Stokes respiration, 388, 483–484
Chickenpox, 467
Chief cells, 550, 556, 557, 573
Childbirth, 632, 652–653
Chinese food syndrome, 177
Chlamydia infection, 646
Chloride
absorption, 575
channels (*see* Potentials, inhibitory)
excretion (*see* Kidney, excretion, secretion)
in extracellular fluids, 81, 82
heart and, 364
hydrated size of, 83, 84, 89
hydrochloric acid and (*see* Hydrochloric acid)
as index of meningitis, 245
ion permeability, 89
(*See also* Potentials, inhibitory)
loss, effects of, 535
shift, 486, 481
transport in digestive tract, 558
(*See also* Digestive system; Malabsorption)
Chlorophyll, 437
Chlorpromazine, 252, 253
Chlortrimeton, 64
Cholecalciferol, 346, 352
Cholecystokinin-pancreozymin (CCK-PZ), 52, 129, 554–557, 560–563
Cholera, 587, 588
Cholesterol, 25, 265, 573, 578, 581
in bile, 558–560
in cell membrane, 87
esters of, 577
gallstones and, 559–560
heart attack risk and, 390
metabolism of, 346, 395, 397
sex steroids and, 627
Cholesterol ester hydrolase, 573
Cholinergic nerves, 176, 225–233, 510
biological properties, 231–233, 268
blood pressure and, 420–421
drug effects on, 231, 253
effects of, 228–229
on smooth muscle, 304
neurotransmitters of, 227–228, 230
origins of, 225, 226
secondary messengers and, 232
Cholinesterase (ChE), 297
Chondrocyte, 67, 344
Chondroitin sulfate, 53, 67, 339, 347
Chordae tendineae, 361, 362
Chorion, 641
Chorionic gonadotropin, 123, 634, 638
Chorionic villi, 641, 642
Choroid coat, 187, 189–192, 508
Choroid plexus, 242–244
Christmas disease, 466
Christmas factor, 464, 466
Chromatin, 455, 456, 458, 459
Chromium, 583, 610, 612
Chromophore, 187, 189
Chromosome(s), 27, 30, 34, 137, 622–624
analysis in amniocentesis, 623–624
division of, 622–624
errors and miscarriage, 651
X and Y, 30, 149, 315, 353, 466, 622–624, 628–629, 633
Chronic hepatitis, 567
Chronic leukemia, 468
Chronic reactions, defined, 565

Chronotropic effect, 368
Chylomicrons, 394–397, 577, 578, 581, 582
Chyme, 545, 586
Chymotrypsin, 551, 559, 572, 573
Cigarette smoke (*see* Nicotine; Smoking)
Cilia, 33, 34, 44, 45, 48, 53, 139, 622
  in respiratory tract, 48, 476
  of oviducts, 636, 639
  in sensory receptors of ear, 194, 197, 199–201
  in sensory receptors of nasal chambers, 202
  in sensory receptors of taste buds, 201
Ciliary body, 181
Ciliary muscle, 181, 192, 233
Cimetidine, 566
Circadian rhythms, 128, 177, 266, 269–270
Circuits, 214–216
  amplifying, 214–215
  in brain, 237, 246, 270, 271
  of cerebral cortex, 270, 271
  converging, 191, 214, 215
  diverging, 214–215
  pathways and, 186–187
  reverberating, 215, 263
    memory and, 278–279
  synaptic, 169–171
Circulatory system, 6, 49, 81
  aging and, 351, 360
  autonomic nervous system and (*see* Autonomic nervous system, circulatory system)
  of blood, 402–412, 430–452, 453–471
  blood vessels, lymph vessels and, 401–429
  exercise and, 325, 326, 328, 329, 382–385, 406
  heart and, 359–377, 378–400
  (*See also* Arteries; Blood; Blood vessels; Capillaries; Heart; Lymph vessels; Veins)
Circulatory gas chemosensors, 489
Circumcision, 622
Circus movement, 374
Cirrhosis of the liver, 435, 533, 566–567
Cisterns of striated muscle, 286, 295, 299
Citrate
  blood clotting and, 465–467
  phosphate dextrose (CPD) solution, 466
Citric acid cycle, 320–322, 590–592
Clap, 646
Clavicle(s), 338, 481
Clearance, 532–533
  of glucose, 523
Cleavage, 640
Clicks, 363
Climacteric, 629, 638
Climbing fibers, 262
Clitoris, 630–632, 634
Clonic phase of epilepsy, 249
Clostridium bolulinum, 285, 588, 589
Clostridium perfringens, 587, 588
Clostridium tetani, 89
Clotting of blood, 464–467
  factors of, 464–466
  inhibition of, 465–467
  pathways of, 464–465
  tests for, 466–467
  time required, 466
  (*See also* Thrombocytes)
CNS (central nervous system, *see* Brain; Spinal cord)
CoA (*see* Acetyl coenzyme A)
Cobalt, 583, 609, 612
Cocaine, 221, 249, 252, 253
Coccygeal nerves, 206, 207
Cochlea, 195–201
  (*See also*, Sense organs, ear)
Cochlear duct, 195, 200
Cochlear nerve (*see* Cranial nerves, statoacoustic nerve)
Codes, 182, 185–187
  patterns and, 185–186
  frequency, 186, 198, 201
  number of receptors and, 186
  on-off, 185–186
  for taste, 201–202
Codon, 34–36
Coenzyme, 23, 581
Coenzyme A (CoA), 320, 321, 393, 590–592
Coenzyme Q (CoQ), 591
CoHb (*see* Carbonmonoxyhemoglobin)
Coitus interruptus, 650
Cold
  response to, 109, 233, 266–268, 325, 380, 448
  (*See also* Sensory receptors, thermoreceptors; Stress)
Cold-blood animals, 268
Cold sores, 641
  (*See also* Herpes simplex virus)
Colic, 585
Colitis, 566
Collagen, 53–56, 66–68, 70–73, 409, 410
  in bone, 335, 338, 339, 341, 342, 347–350
  clotting of blood and, 462–463
Collagenase, 63, 355
Collateral(s) 135, 137, 139, 141, 213, 214, 223
  in cerebellum, 262, 263
  ganglia, 227
Collateral circulation, 364, 407
Collecting ducts, 513, 514, 522, 527, 528
Colloidal osmotic pressure, 412
Colon, 544, 574–575, 585
  absorption by, 575
  ascending, 544, 546
  cathartics and, 585–586
  cancer of, 589
  cecum and, 544, 546
  descending, 544, 546
  haustra of, 545
  motility of, 226, 561–564
  reflexes in, 561, 563
  sigmoidal, 544, 546
  taeniae of, 545
  transverse, 544
  (*See also* Bowel, Digestive system; Small intestine)
Colony forming cells, 434
Color (*see* Sense organs, eye; Sensory receptors, photoreceptors)
Color blindness, 191
Color vision, 189–191
Colostrum, 639
Columnar epithelium, 547
Coma, 130, 251, 534, 567
Combination pills, 647, 648
Commissural fibers, 270, 276
Common bile duct, 544, 551, 553
Common metabolic pool, 590–592
Compact bone, 336
  (*See also*, Bone)
Compartments, body fluid, 80–82
Compatibility (*See* Blood, transfusion reactions; Histocompatibility antigens, Artificial kidney)
Compensated acidosis, 534
Compensated alkalosis, 534
Compensatory hypertropy of kidney, 534–535
Compensatory pause of heart, 367
Complementary base pairs of DNA, 26–28, 34–36
Compound, 17, 19
Compound glands, 51
Compound lipids, 24
Computer assisted tomography (CAT scan), 245–246
COMT (*see* Catechol-O-methyltransferase)
Concentration, 16
Concentration gradient, 88, 89, 92
Conception, 620, 637, 639–640
Conception control, 646–651
  effectiveness of, 647
Conchae, 474
Conditioned reflex,213, 278
Condom, 647, 649
Conducting artery, 403
Conduction
  of impulses in heart, 594
  of sound waves, 194–198
Cones (*see* Sense organs, eye)
Congenital malformations, 385–387
  chromosomes and, 651
  diethylstilbesterol and, 653
    (*See also* Chromosomes; Fetus; X and Y chromosomes)
Congestive heart failure, 387–389, 397, 483, 484, 533, 567
  effects on respiration, 483, 484
  (*See also* Heart disease)
Conjunctiva, 187, 188, 510
Connective tissue(s), 5, 53–73, 347
  areolar, 55, 59, 507
  blood (*see* Blood)
  bone (*see* Bone)
  cartilage (*see* Cartilage)
  cell types in, 55, 142–147
  dense, 66–68
  diseases of, 53, 64
  (*See also* Blood; Bone; Cartilage)
  loose, 55, 59, 507
  (*See also* Collagen; Elastic fibers)
Conn's syndrome, 130, 423, 534
Conscious activity, 216
Constipation, 585
Contact factor in clotting, 464, 465
Continuous capillaries, 411, 412
Contraception, 646–651
Contraction of muscle (*see* Muscle, cardiac, smooth; Skeletal muscle)
Contracture, 313
Contralateral reflex, 212, 214
Convection of heat, 594
Converging circuit, 214, 215
  of retinal rods, 191
Converting enzymes, 525, 526
Convoluted tubules
  distal, 513, 514, 516, 518, 522, 531
  proximal, 513, 514, 516, 522, 531
Convulsions, 130, 249, 594
Cooley's anemia, 446, 448–449
Coordination of skeletal muscle contraction (*see* Cerebellum; Sensory receptors, Golgi tendon organ, muscle spindle; Skeletal muscle)
Cooper, 583, 608, 612
Cordotomy, 250
Cori cycle, 319
Corium, 506–507, 509, 511

Cornea, 187, 188, 191, 194, 510
Coronary arteries, 363, 407
  autonomic nervous system and, 227, 380, 389
    (*See also* Autonomic nervous system, circulatory system)
  disease of (*see* Heart, disease of)
Coronary circulation, 363–365
Coronary insufficiency, 363
Coronary sinus, 363
Coronary surgery and prostacyclin, 463
Coronary veins, 363
Corpora quadrigemina, 241, 261, 265
Corpus albicans, 633, 638
Corpus callosum, 241, 260, 270, 273
Corpus luteum, 52, 127, 133, 633–635, 637–638, 644, 653
Corpus striatum, 272
Corpuscles of Ruffini (see Sensory receptors, thermoreceptors)
Cortex
  cerebral, 237, 241, 242, 263, 270
    (*See also* Cerebral hemispheres)
  of kidneys, 511, 512, 516
  motor, of brain, 273–275
  sensory, of brain, 273–275
  visual, 273, 274, 276
Corti, organ of, 195–199
Cortical nephrons, 514
Corticospinal tract, 216, 219, 220, 514
  (*See also* Spinal cord, tracts of)
Corticosteroids, 122, 297
  (*See also* Glucocorticoids)
Corticosterone, 351, 627
  (*See also* Glucocorticoids)
Corticothalamic fibers, 269
Corticotropin (*see* Adrenocorticotropic hormone)
Corticotropin releasing factor (CRF), 109, 112
Cortisol, 65, 112, 117, 627
  (*See also* Glucocorticoids)
Cortisone, 26, 52, 478, 627
Costal breathing, 483
Costal cartilage, 68
Cotransport, 95, 558, 576, 577, 579
Coughing, 491
Coumadin, 466
Countercurrent theory, 525–527
Coupled transport, 95, 558, 576, 577, 579
Covalent bonds, 17, 19
Cowper's glands, 621, 622
Coxsackie virus, 587
CPD (*see* Citrate phosphate dextrose)
CPK (*see* Creatine kinase)
CPR (*see* Cardiopulmonary resuscitation)
Crabs, 646
Cramps
  premenstrual, 638
  of skeletal muscle, 326, 510, 594
Cranial cavity, 49
Cranial nerves, 135, 140, 145, 147, 257, 258, 259
  abducens (VI), 257–259
  accessory (XI), 257–259
  facial (VII), 201, 257–259
  glossopharyngeal (IX), 201, 257–259, 490
  hypoglossal (XII), 257–259
  mixed, 257–259
  nuclei of, 226, 227, 229, 233, 260, 265
  oculomotor (III), 257–259
  olfactory (I), 202, 257–259, 271, 272
  optic (II), 187, 188, 190, 257, 258, 273, 274, 277
Cranial nerves, optic (II) (*Cont.*):
    (*See also* Sense organs, eye)
  parasympathetic fibers of, 258
  reflexes, 211
    (*See also* Reflexes)
  sensory fibers of, 257, 259
  statoacoustic (VIII), 197, 199, 200, 257, 258, 277
    auditory division, 195, 197–200, 259
    vestibular division, 200, 259
    (*See also* Sense organs, ear; Sensory receptors, auditory receptors)
  trigeminal (V), 257–259
  trochlear (IV), 257–259
  vagus (X), 201, 257–259, 490, 560
    abdominal organs and, 258, 259, 560–563
    digestive tract and, 258, 259, 560–563, 567
    heart and, 368, 369
    thoracic organs and, 258, 259
Creatine kinase (CK, CPK), 391–393
Creatine phosphate (CP), 318–320, 322–324, 392, 590
Creatinine, 523, 533
Cremaster muscle, 620
Crenation of cells, 91
Cretinism, 102, 113, 128, 353
Crista ampullaris, 197, 199, 200
Cristae of mitochondria, 33
Crossbridges, 290, 295, 296, 301, 310, 323–324
Crossmatching of blood, 441–442
Crown of tooth, 355
Cryoglobulinemia, 435
Cryptorchidism, 645
Crypts of Lieberkühn, 549–500, 559, 573
Cuboidal epithelium, 476
  (*See also* Epithelium)
Curare, 297, 311
Current, 160, 166, 247, 309
  (*See also* Impulses; Potentials; Synapse)
Current Contents, 3, 4
Cushing's syndrome, 106, 126, 131, 467
Cyanate, 449
Cyanocobalamin, 98
  (*See also* Vitamin B_{12})
Cyanosis, 387, 438
Cyclic adenosine monophosphate cAMP), 116, 119–123, 231, 232
  cholera and, 587
  memory and, 279, 294, 349
Cyclic guanosine monophosphate cGMP), 119–123, 231, 232
Cyclic ring structure, 20
Cyst, 128, 222, 223, 632
Cystic duct, 551, 553, 554
Cystic fibrosis, 511, 587
Cystinuria, 534, 588
Cytochromes, 437, 591, 592
Cytoplasm, 30–33, 137–139
  receptors in, 115, 119
  vesicles in, (*see* Vesicles)
Cytosine, 26, 28

Dalton's law, 479
D antigen, 442–443, 445
Dark adaptation, 189
D cells, 130
DDT (*see* Dichlordiphenyl-trichlorethane)
Deamination of proteins, 322, 553, 592
Death and brain, 247
Decibels and perception of sound, 199
Decomposition of movement, 264
Decompression, 495
Decompression sickness, 494, 495
Decussation (*see* Spinal cord tracts, decussation)
Dedifferentiation, 53
Deep breathing, 483
Defecation reflex, 563, 564
Defense cells, 56, 128, 459
  (*See also* Immune system; Leukocytes; Mast cells, Macrophages; Mononuclear phagocytic system; Phagocytes, Plasma cells; Reticuloendothelial system)
Defibrillation, 374
Deflation of lungs, 481–483, 489
Degeneration
  of nerve, 22, 146, 314
  of skeletal muscle, 314
  Wallerian, 145, 222
Degree:
  Celsius, 16
  centigrade, 13, 16
  Fahrenheit, 13, 14, 16
Dehydrogenated fat, 25
Dehydrogenases:
  hyperbaric oxygenation and, 494
  (*See also* Hydroxybutyric dehydrogenase; Lactic dehydrogenase Metabolism)
Delayed heat, 323–324
Delayed-type allergy, 63
Delta chains of hemoglobin, 438, 439
Delta nerve fibers, 148, 221
Delta waves in EEG, 247
Dementia, 225
  (*See also* Mental retardation)
Demerol, 252, 253
Demylenation, 147
Dendrite(s), 75, 135, 137, 140, 147
Dendrodendritic synapse, 169, 170
Denervation atrophy, 314
Dense connective tissue, 66, 68, 81
  irregular, 66, 67
  regular, 66, 67, 337
Dental plaque, 355
Dentin, 72, 73, 347, 355
Deoxycorticosterone, 627
Deoxyribonucleic acid (DNA), 26–28, 34, 574
  in chromosomes, 30
  in mitochondria, 32
  transcription of, 34, 592
  (See also Chromosomes)
Deoxyribonuclease (DNAse), 573
Depolarization, 160, 161, 163, 164, 166, 167, 172–175, 292
  of smooth muscle, 301, 303
  of skeletal muscle, 310
Depressant, 177, 252
Derived lipids, 24
Dermatitis, 151, 511
Dermatome(s), 210, 211, 221, 225, 370
Dermis, 507–507, 509
DES (*see* Diethylstilbesterol)
Descending colon, 544, 546
Descending tract (*see* Spinal cord, tracts)
Desmosomes, 42, 43, 73, 86, 508
Development, 385–387, 640–645
  of alveoli, 475–478
  of air sinuses, 474–475
  of blood cells, 434, 435
  of bone, 343–344
  of heart, 367, 381
  hemoglobin and, 439
  of lungs, 475–476
  of nervous system, 238–241

Development (*Cont.*):
of teeth, 543–544
(*See also* Fetus)
Dexedrine, 252, 253
Dextran polymers in plaque, 355
Dextrinases, 559, 572, 573
Dextrins, 573
Dextro- (d-), 20
Dextrose in blood preservation, 466
Diabetes
bronzed, 583
insipidus, 128, 267, 534
mellitus, 115, 130, 389, 523, 534–535, 592
hemoglobin IAc (HbIAc) and, 439
Dialysis, 536–537
Diapedesis, 435, 436, 454, 457
Diaphragm, 360, 474, 481, 483
effects in respiration, 479–480, 489, 490
Diaphragm of capillary junction, 410, 411
Diaphragm for birth control, 647, 649
Diaphragmatic breathing, 483
Diaphysis, 336
Diarrhea, 534, 535, 586–588, 592
Diarthrosis, 336, 337
Diastole, 364–365, 367
Diastolic pressure, 416, 417
Dibenzyline, 103, 231
Dichlordiphenyl-trichlorethane (DDT), 447
Diencephalon, 239, 241, 265
Diet
exercise and, 327
foods and caloric value, 593, 616–617
heart disease and, 390
hyperlipoproteinemia and, 395–397
salt-free, and response to, 524
water loss and, 523, 527
Diethylstilbesterol (DES), 653
Differential count, 455
Differentiation of tissues, 53
Diffusion, 87–89, 156, 157, 412, 576, 577
facilitated, 87, 577
of ions across nerve cell membrane, 158
(*See also* Impulses; Potentials; Synapse)
of water in kidney, 526–527
Diffusion rate
calculation of, 88
Digestion, 543, 572–574
(*See also* Digestive enzymes; Digestive system)
Digestive enzymes, 572–573
aminopeptidase, 574
amylases, 559, 572, 573
carboxypeptidases, 559, 572, 573
cholesterol ester hydrolases, 573
chymotrypsin, 559, 572, 573
deoxyribonuclease (DNAse), 573
dextrinases, 559, 572, 573
dipeptidases, 574
disaccharidases, 586–587
enterokinase, 559
enteropeptidase, 559
invertase, 574
isomaltase, 574
lactase, 559, 586, 588
lipases, 559, 561, 572, 577, 578, 581
of lysosomes, 32
maltase, 559, 572, 573, 586, 588
nucleases, 559, 572, 574
nucleosidases, 572, 574
nucleotidases, 572
pepsin, 481, 557, 572, 573
peptidases, 559, 572, 576
phospholipases, 573, 574
polynucleotidases, 574
Digestive enzymes (*Cont.*):
rennin, 573
ribonuclease (RNAse), 573
sucrase, 572, 573, 586, 588
Digestive system, 6, 542–570, 571–618
absorption and, 572–585
of carbohydrates, 576, 579
defined, 543
of ions and trace elements, 583–585
of lipids, 576–582
of proteins, 576–578, 580
of vitamins, 581, 582
of water, 585
(*See also* Water)
anatomy, 50, 543–554
of accessory digestive organs, 549–554
of capillaries, 578
cellular, 47, 50, 184, 301, 546–549
of whole organs, 544–546
defenses and inflammatory responses, 564–567
gastroenteritis, 586–589
inflammatory reactions, 564–566
(*See also* Inflammatory responses)
intestinal flora and, 589–590
liver disease and, 566–567
malabsorption and, 586–590
ulcers, 565–567
digestion, 551, 555–559, 572–574
defined, 543
(*See also* Digestive enzymes)
excretions, 585–586
malabsorption, 586–590
metabolism and nutrition, 590–595
amino acid requirements, 599
energy expenditure and, 593–595, 613–615
energy expenditure and caloric values of foods, 593, 595, 616–617
recommended daily dietary allowances, 618
trace elements, 608–613, 618
vitamins, 600–607
regulatory mechanisms, 129, 130, 132, 133, 560–564
hormonal influence on motility, 561, 563
hormonal influence on secretions, 561, 562
neural influences on motility, 225, 267, 561–564
neural influences on secretions, 52, 104, 128, 561–562
secretions, 554–560
by exocrine glands, 555–560
of gastrointestinal hormones, 550, 554–556
Digitalis, 95, 375, 380, 389
Dihydrotestosterone (DHT), 133, 625, 627, 648
(*See also* Reproductive system; Testosterone)
1,25-Dihydroxycholecalciferol (*see* Vitamin D)
Dihydroxyphenylalanine (L-dopa), 131, 150, 176, 230, 248, 273, 508
Dihydroxyphenylethylamine (dopamine), 104, 131, 150, 176–177, 230, 250, 253, 273
antidopaminergic drugs, 253
Diisopropylfluorophosphate (DFP), 297
Dilation and curettage, 651
Dilation and evacuation, 651
Dim light vision, 189
Dipeptidases, 574
Dipeptides, 22, 574, 580
2,3, Diphosphoglycerate, effects on hemoglobin, 484, 485
Diplöe, 342
Direct arteriovenous anastomosis, 407
Dissacharidases, 586–587
Dissacharides, 20, 339, 586
Disc
intercalated, 73, 74
intervertebral, 69, 224–225
Discontinuous capillaries, 411
Dislocation
of intervertebral disc, 69, 224–225
of joints, 351
Displaced foci, 374
Dissociation
of oxyhemoglobin, 484–485
of molecules, 91
Distal convoluted tubules (*see* Kidneys)
Distributing arteries, 403
Disuse atrophy, 314–315
Diuresis, 91, 122, 130, 523
Diuretic(s), 91, 423–424, 534, 535
activity of glucose, 523
Diurnal rhythm, 128, 177, 266, 269–270
Divergence, 214–215
Diverticula, 566, 587, 589
Diverticulitis, 566
Diving, 494–496
Dizygotic twins, 639–640
DNA (*see* Deoxyribonucleic acid)
DNAse (*see* Deoxyribonuclease)
Dominance in brain, 274–278
Dominant alpha rhythms, 247
1-Dopa (*see* Dihydroxyphenylalanine)
DOPA decarboxylase, 176
Dopamine (*see* Dihydroxyphenylethylamine)
Dorsal columns (*see* Spinal cord, pathways)
Dorsal horn, 208, 209
Dorsal root, 208, 209, 211
Dorsal root ganglion, 140, 145, 209, 211
Doublets, 249
Douche, 650
DPG (*see* 2,3 Diphosphoglycerate)
Dramamine, 64
Drinker artificial respirator, 497
Drinking (*see* Antidiuretic hormone; Hemorrhage; Hypothalamus; Thirst; Thirst center)
Drugs
addiction to, 251
antibiotics, 63, 145, 361, 457
(*See also* Penicillin)
antihypertensive, 424
blood-brain barrier and, 145
chemotherapeutic, 467
defined, 104
depressants, 252
diuretic (*see* Diuretics)
effects on clotting, 469
fate of, 252
mechanisms of action, 251–253
neurotransmitters and, 89, 249, 251–253
(*See also* Neurotransmitters)
receptors and, 85, 115, 251, 253
secondary messengers and, 251
sickle cell anemia and, 449
sites of action, 252
timing and biological rhythms, 270
tolerance to, 251
white blood cells and, 467
(*See also* Autonomic nervous system, drugs; individual names of drugs)
Duchenne-type muscular dystrophy, 315
Ductus arteriosus, 384–387

Ductus venosus, 384, 385
Duffy antigen, 442, 446
Duodenal glands, 550
Duodenal ulcers, 565–567
Duodenocolic reflex, 563, 564
Duodenum (*see* Small intestine)
Dura mater, 242
D^u variant of Rh antigen, 443
Dwarfism, 126, 353
(*See also* Cretinism)
Dynorphin, 250
Dysarthria, 264
Dysdiakochokinesia, 264
Dysergia, 269
Dyslexia, 278
Dysmetria, 264
Dyspnea, 483
Dystrophy, 314

E antigen, 442–443, 446
Ear (*see* Sense organs, ear; Sensory receptors, auditory and vestibular receptors)
Eccrine glands, 510
ECF (*see* Body fluids; Extracellular fluids)
ECG (*see* Electrocardiogram)
Echocardiogram, 397, 398
Echo virus, 587
Ectoderm, 237–239, 641, 642
Ectopic foci, 374
Ectopic pregnancy, 641
Eczema, 63, 571
Edema, 63, 149, 380, 414, 462, 524, 533, 534, 567
(*See also* Ascites; Peripheral edema; Pulmonary edema)
EDTA (*see* Ethylenediaminetetraacetic acid)
EEG (*see* Electroencephalogram)
Efferent arterioles in kidneys, 514, 516–518
Efferent nerves, 141, 147
Efficiency of metabolism, 323–324, 366
Egg, 30, 34, 46, 126, 133, 629–633, 666, 639, 641
ovulation and, 632–637
production of, 632–634
viability of, 636, 647
Egg albumin, 576
Einthoven's triangle, 372, 373
Ejaculate, 625
Ejaculation, 625
Ejaculatory ducts, 621, 622
EKG (*see* Electrocardiogram)
Elastase activity, 496
Elastic arteries, 403
Elastic cartilage (*see* Cartilage)
Elastic elements in skeletal muscle, 308, 314
Elastic fibers, 53–55, 66, 67
Elastin, 54
Electrical activity of cells, tissues and organs (*see* Impulses; Potentials; Synapses)
Electrical synapses, 169
(*See also* Gap junctions)
Electrical vector, 373
(*See also* Heart, electrical activity)
Electrocardiogram (EEG, EKG), 310, 370–375, 391
(*See also* Heart, electrical activity)
Electrochemical equilibrium, 88, 156–158, 160
Electrocorticogram, 247
Electrodes, 156, 158
active or exploring, 158, 370
Electrodes (*Cont.*):
bipolar, 373
reference, 156, 158
unipolar, 371, 373
Electroencephalogram (EEG), 247
Electrolyte(s)
equilibrium of and potassium, 512
loss and diuresis, 523
(*See also* Adrenals; Bone; Digestive system; Diuretics; Endocrines, glucocorticoids and mineralocorticoids; Lungs; Respiratory system; Salts; Skin)
Electromagnetic force, 373
Electromyogram (EMG), 310
Electron(s), 15, 87
acceptors of, 590, 591
Electron transport, 591, 592
Electrophoresis, 391–392
of hemoglobin, 448, 449
of lipoproteins, 393
of serum enzymes, 391–392
Electrovalent bond, 17, 18
Element, 15
Elephantiasis, 414
Embolism, 422, 426, 463
EMG (*see* Electromyogram)
Embryo, 385, 637
Embryonic development, 640–645
of heart, 384–387
human chorionic gonadotropin (HCG) and, 133, 628, 634, 641
of muscle, 238
of nervous system, 238–241, 642
autonomic nervous system, 238, 239
dorsal root ganglion, 238
forebrain, 239, 241
hindbrain, 239, 241
midbrain, 239, 241
optic nerve, 240
optic vesicles, 241
peripheral nervous system, 239, 241
retina, 240, 241
of spinal cord, 238, 239, 241
of spinal nerves, 238
of skeletal system, 341–345
somites and, 238
(*See also* Fetal development)
Embryonic disc, 640, 641
Embryonic sac, 640
Emiocytosis (*see* Exocytosis)
Emission, 625
Emotions, 109, 233, 270–272
brain waves and, 247–248
(*See also* Behavior; Limbic system)
Emphysema, 484, 488, 496, 535
Emulsification, 578, 581
Enamel, 72, 73, 355
Enamel organ, 72
Encephalitis, 225, 245
Encephalomyelitis, 149
End arteries, 407
Endocardium, 49, 361
Endochondral bone formation, 341–344
Endocrines, 5–7, 102–133
(*See also* Digestive system, regulatory mechanisms; Hormones; names of individual endocrine glands and hormones)
Endocytosis, 95–98, 457
(*See also* Phagocytosis)
Endoderm, 237, 238, 641, 642
Endogenous nutrients, 576
Endogenous pyrogens, 594
Endolymph, 187–200
Endometriosis, 631
Endometrium, 630, 631, 636–638, 640, 642
Endomysium, 283, 285
Endoneurium, 142
Endoplasmic reticulum, 31, 97, 292
of reticulocytes, 435
rough, 29–31, 59, 60, 137, 140
smooth, 29, 286
Endorphins, 127, 222, 249–250
alpha, 249–250
beta, 249–250
gamma, 249–250
Endosteum, 336
Endothelium, 46, 49, 403–405, 517–519
cells of, 409, 410, 457
damage to in atherosclerosis, 425–426
of lymph capillaries, 412
Endotoxin, 588
End-plate potential (EPP), 292–295, 311
Endurance exercise, 318, 327–330
Energy:
content of foods, 593, 616–617
daily requirements, 615
expenditure of
body mass and, 593
exercise and, 613–615
forms of, 592–593
metabolism and,
basal metabolic rate (BMR), 593–595
(*See also* Metabolism)
Enkephalins, 127, 155
effect on pain, 222, 326
as neurotransmitters, 249, 250
schizophrenia and, 251
Enteric organisms, 587
Enteritis (*see* Digestive system, defenses and inflammatory responses)
Enterobacter aerogenes, 589
Enterochromaffin cells, 548, 550, 555
Enteroendocrine cells, 548
Enterogastric reflex, 563, 564
Enterogastrone, 563
Enterohepatic circulation, 559
Enterokinase, 559
Enteropeptidase, 574
Enterotoxins, 587, 588
Environmental effects
of altitude, 480, 492–494
of diving, 494–495
of heat, 56, 233, 267–268, 283, 308, 323–325, 509–511, 594
of pollutants, 475, 496–497
Enzymes, 21–24, 36, 85, 87, 122, 590
active site of, 23
converting, 525, 526
digestive (*see* Digestive enzymes)
general functions, 22–24, 119, 121–123
hormones and, 119–121
as indicators of disease, 391–393
in lysosomes, 32, 59, 60, 96
metabolism and (*see* Metabolism)
protein kinases, 121, 122
respiratory, 32–33
deficiencies, 149, 150
(*See also* Malabsorption)
Eosinophilic, 455
Eosinophilic chemotactic factor, 64
Eosinophils, 434, 435, 455–457
Ependymal cells, 145, 241, 243
enkephalins and, 250
Epicardium, 361
Epidemic hepatitis, 567

Epidermis, 506–507
(*See also* Skin, epidermis)
Epididymis, 620–623, 625
Epiglottis, 69, 201, 473, 474, 491
Epilepsy, 248–249, 252, 276–277
Epimysium, 283
Epinephrine, 52, 122, 132, 176
beta-adrenergic effect of, 303
blood pressure and, 423
heart rate and, 368
inhibition of, 115
secretion of, 11, 113
stress and, 112
vasoconstrictive effects, 423
(*See also* Endocrines; Sensory receptors, cellular)
Epineurium, 142
Epiphyseal disc, 335, 343, 345
Epiphyseal plate, 335, 343, 345
Epiphysis, 336
Episiotomy, 632
Epithalamus, 241, 269
Epithelium, 5, 42–53
absorptive, 547
alveolar, 476
characteristics of, 42, 44–49
columnar, 44–46, 53
cuboidal, 44, 45, 47
pseudostratified, 45–48
simple, 46
squamous, 44, 45, 53, 476
stratified, 44, 46, 201, 507
transitional, 45, 46
dynamics of, 53
EPP (*see* End-plate potential)
Epsilon chains of hemoglobin, 438, 439
Epsom salts, 586
EPSP (*see* Potentials, excitatory postsynaptic)
Equilibrium, 89, 90, 113
electrochemical, 88, 156, 157
potential, 159–162
sense of, 194, 259–261
(*See also* Balance; Sense organs, ear)
Equivalent weight, 16, 18
ER (*see* Endoplasmic reticulum)
Erection, 625
Ergosterol, 346
Erythroblast, 436
Erythroblastosis fetalis, 445–446
Erythrocytes, 405, 437, 430–452, 456
(*See also* Red blood cells)
Erythropoietic factor, 435–437
Erythropoiesis, 434–437
Erythropoietin, 132, 435
altitude exposure and, 436, 437, 494
Escape reaction, 266
Escherichia coli, 534, 589
Esophageal sphincters, 545
Esophageal ulcers, 565
Esophagus, 544, 545
Essential amino acids, 576
Essential fatty acids, 578
Essential hypertension, 423
β-Estradiol, 627, 628, 631, 634, 636, 637, 648
(*See also* Estrogens)
neuronal development and, 257
Estriol, 604
Estrogenic phase of reproductive cycle, 634–638
Estrogens, 52, 131, 132–133, 348, 350, 351, 353, 614, 625, 626–628, 631, 634, 636–639, 644, 648
chemistry, 627, 633–638, 648
Estrogens (*Cont.*):
early epiphyseal closure and, 350, 634
receptors, 115
side effects, 638
Estrone, 627, 636
Estrone sulfate, 651
Ethacrynic acid, 535
Ethmoidal sinus, 474
Ethylenediaminetetraacetic acid (EDTA), 465
Ethinyl estradiol, 648, 651
Eustachian canal, 69
Evaporation of water, 529
Excerpta Medica, 3, 4
Exchange transfusion, 446
Excitability, 135, 155, 156, 367
(*See also* Impulses; Skeletal muscle, neuromuscular junction; Potential)
Excitation-contraction coupling, 292, 294
Excitatory postsynaptic potential (EPSP), 171–175, 214, 221, 292, 294–296
Excitement phase of orgasm, 625, 634
Excretion, 7, 42
digestive system and, 585–586
heat loss and, 594
kidneys and, 512, 513, 522–524, 527–531
lungs and, 506
Excretory system 6, 7, 505–541
(*See also* Digestive system; Kidneys; Lungs; Respiratory system; Skin)
Exercise, 325–330
Bainbridge reflex, and, 370
circulatory system and, 325–327, 329, 363, 364–366, 382–385
cramps and, 326
diet and, 327
fatigue and, 325, 327, 492, 493
(*See also* Fatigue; Skeletal muscle, fatigue)
heart and (see Heart; training, target heart rate, below)
high density liproteins and, 397
joints and, 351, 352
maximal, 326, 328–330
metabolism
carbohydrates, 327–329
2,3, diphosphoglycerate and, 484, 485
energy, 318–325, 327, 329, 553–554, 613–617
fats, 321–322, 329, 397
oxygen consumption, 326–327, 330, 382, 383, 492–493
proteins, 593
rate of, 593–595
second wind, 330
muscle and, 326, 328–330
pain, 326
respiratory system and, 325, 329, 483, 492–493
stroke and, 389
submaximal, 327–328
endurance and, 325
training, 326–330, 383, 385, 492
target heart rate and, 328–329
V_{O_2} max, 326–327
(*See also* Metabolism)
Exocrine gland, 51
Exocytosis, 32, 51, 95, 97, 457
of hormones, 97, 113
of neurotransmitters, 140, 168, 292
Exogenous nutrients, 576
Exopthalmic goiter, 128
(*See also* Grave's disease)
Exotoxins, 587, 588
Expiration, 480–481, 490
Expiratory center, 489–491
Expiratory muscles, 480–481
Expiratory reserve volume, 482
Exploring electrode, 371
External ear, 69, 195, 196, 509–510
External ear canal, 196
External female genital organs, 630
External intercostal muscles, 486, 490
External muscularis, 546, 547
External respiration, 473
Exteroceptors, 182
Extracellular compartments, 80
Extracellular fluid (ECF), 6, 80, 82, 130, 155, 322, 412, 644
Extraocular muscles, 189
Extrapyramidal tracts (*see* Spinal cord, tracts)
Extra shortening heat, 323–324
Extrinsic factor, 558
Extrinsic clotting system, 464–465
Eye (*see* Sense organs, eye)

Facial nerve (*see* Cranial nerves)
Facilitated diffusion, 87, 92, 577
Facilitation of subthreshold stimuli, 221
F-actin, 288
Factors in clotting, 464–466
Fahrenheit to centigrade conversion, 13, 14, 16
Fainting, 416, 444
Fallopian tubes, 34, 622, 629–631, 634, 639, 640, 650
Fallot's tetrad, 387
False labor, 653
Falx cerebelli, 242
Falx cerebri, 242
Familial hypophosphatemic rickets, 352, 353
Fascia, 285
Fasciculus
of nerves, 142
of skeletal muscle, 283, 287
Fasciculus cuneatus, 216
(*See also* Spinal cord pathways, dorsal columns)
Fasting, 322
Fast-red fibers, 297, 298
Fast-twitch glycolytic fibers, 297, 298, 328
Fast-twitch oxidative fibers, 297, 298, 329
Fast-white fibers, 297, 298,
Fat (*see* Fats; Lipids)
Fat cells
functions, 56, 148–149
hormones and, 116, 127, 129, 130, 132, 133
in marrow, 335
number of, 56, 58
(*See also* Adipose tissue; Connective tissues)
Fatigue
of muscle, 297, 298, 312, 314, 318, 319, 322
(*See also* Exercise; Skeletal muscle, fatigue)
Fats
brown, 56
caloric value of, 593
heat production and, 56
metabolism of (*see* Metabolism)
neutral, 24
respiratory quotient of, 593
saturated, 576
saturated fatty acids of, 24–25
white, 56
(*See also* Fatty acids; Lipids)
Fat-soluble vitamins (*see* Vitamins)

Fatty acid(s), 24, 25, 87, 103, 133, 573 576–578, 592
prostaglandin synthesis and, 462
release by growth hormone, 349
synthesis of steroid hormones and, 627
F cells, 130
Feces, 91, 543, 564, 565, 584, 585
Feedback mechanisms, 8–10, 75, 117, 147, 312
negative, 8, 9, 108, 110, 111, 215, 347, 420–421, 626, 635
positive, 8, 9, 16, 164, 635
Feeding center, 267, 562, 564
(*See also* Hypothalamus, centers of control)
Female climacteric, 638
Female reproductive system (*see* Reproductive system)
Femto-, 16
Fenestrated capillaries, 410–412, 517, 519, 579
(*See also* Capillaries; Intercellular junctions)
Fenn effect, 314
Ferric ion (Fe^{3+}), 583
hemoglobin and, 437
Ferritin, 519, 583–585
Ferrous ion (Fe^{2+}), 583
hemoglobin and, 437
Fertile period, 645–648
Fertility, 645–648
Fertilization, 620, 637, 639–640
development and, 237
Fetal circulation, 384–387, 434, 641–645
Fetal development, 640–645
Fetal hemoglobin, 438, 439, 448, 449, 486
Fetal respiration, 384, 475–476, 641–644
Fetus, 631, 633
change in type of fat, 56
circulation of, 384–387, 434, 641–645
liver of, 384, 385, 434
lungs of, 384, 475–476, 641–644
reproductive system development, 623, 628
smoking of mother, effects on, 478
(*See also* Development)
FEV_1 (*see* Forced expiratory volume)
Fever, 594
Fever blisters, 641
Fibers:
muscle, 283, 284
nerve
associative, 270
commissural, 270, 276
(*See also* Adrenergic nerves; Collagen; Cholinergic nerves; Nerve fibers; Skeletal muscle, fibers)
Fibril, 53
of cardiac muscle, 299
of skeletal muscle, 285–287, 308, 314
Fibrillation, 374–375, 380, 595
Fibrin, 436, 464
Fibrinase, 464
Fibrinogen, 464
Fibrinolysin, 622
Fibrin serum factor, 464
Fibrin stabilizing factor, 464, 465
Fibroblast, 55–57, 73, 355
interferon and, 460
Fibrosis, 387, 567
respiratory distress syndrome and, 477–478
Fibrositis, 354
Ficrous cartilage, 68, 69
Fibrous pericardium, 360
Filaments, 33, 34
in muscle fibrils, 285–290, 311, 314, 323, 324
Filarial parasite in elephantiasis, 414, 467
Filtration barriers in kidney, 519
Filtration of blood, 412
Filtration fraction, 530, 532
Filtration in kidneys, 512, 513, 518, 519
Filtration pressure, 413, 517
Fimbriae, 630, 631, 636
Final common pathway, 220, 223, 263
(*See also* Neuron, lower motor)
Fingerprints, 511
First pain, 221
First stage of labor, 652–653
Fissure, 270
Fissure of Rolando, 270
Fissure of Sylvius, 270, 272, 274
Flagella, 34, 139, 622
Flavine adenine dinucleotide (FAD), 590–592
Flavoprotein-cytochrome system, 591
Flight or fight syndrome, 111
Flow rate of blood, 402, 406, 407, 414–416, 419
area of vessels, pressure and, 415
to kidney, 517, 521
(*See also* Blood flow; Blood pressure)
Fluid(s), 80–82, 155–156
blood, 81
cerebrospinal fluid (*see* Brain, cerebrospinal fluid)
compartments of body and, 81, 529
composition of, 82, 155
exchange, 81
extracellular, 6, 80–82, 155
interstitial, 6, 81
intracellular, 6, 80–82, 155
lymph, 81
(*See also* Lymph; Lymphoid tissue)
plasma, 80, 81, 244, 245, 391, 431, 432, 454
(*See also* Plasma)
regulation of movement, 82, 404–421
water and (*see* Water)
Fluorescent antibodies, 176, 246
Fluorescent microscope, 246
Fluoride, 339
Fluorine, 583, 610, 613
Fluoroscopy, 397, 398
Flutter, 375, 406
Folic acid (*see* Vitamins)
Follicle, 46, 126, 629, 632–638
atretic, 633, 636
cells of, 632–633, 639
changes during menstrual cycle, 634–638
Graafian, 632
Follicle stimulating hormone (FSH), 52, 103, 108, 109, 126, 127, 133, 626, 632, 634–638, 644
(*See also* Gonadotropins)
Follicle cell stimulating hormone-releasing hormone (FSH/RH)
(*see* Gonadotropic releasing hormone)
Follicular phase of ovarian cycle, 635–638
Fontanelle, 337
Food-borne allergies, 63, 64
Food-borne hepatitis, 566–567
Food posioning, 565, 586–589
Foramen
intervertebral, 244
of Luschka, 244
of Magendie, 244
Foramen magnum, 259
Foramen ovale, 384–386
Force of skeletal muscle contraction, 315–318, 326
Forced expiratory volume$_1$ (FEV_1), 483, 484
Forebrain, 177, 239, 241, 265–269
Foreskin of penis, 621, 622
Formed elements, 431, 432
Forward heart failure, 388, 389
Fovea centralis, 187, 188
(*See also* Sense organs, eye)
Fragility test, 432
Frank-Starling law, 366
Fraternal twins, 639–640
Frequency of blood groups, 440
Frequency code, 186, 198, 201
Frog test, 642
Frontal lobe of brain, 270, 271
Frostbite, 380
(*See also* Cold exposure)
Fructose, 533, 622
FSH (*see* Follicle stimulating hormone)
Fulcrum, 315
Fundus of stomach, 544, 545
Fundus of uterus, 630, 631
Furosemide, 535
Fusiform cells of cerebral cortex, 270
Future growth zone, 345
Fuzzy coat, 548

GABA (*see* Gamma amino butyric acid)
G-actin, 288
Galactose, 574, 576, 577
Gallbladder, 544, 545, 551, 554–557, 560
autonomic nervous system and, 226, 560
function of, 554
hormonal influence on, 129, 556–557, 560
referred pain and, 210
Gallstones, 339, 559–560, 566, 583
Gammaaminobutyric acid (GABA), 104, 177–178, 273, 276
sleep and, 248
Gamma globulins, 575, 577
(*See also* Antibodies)
Gamma nerve fibers, 148
Gamma receptors (*see* Autonomic nervous system, receptors; Receptors, cellular, gamma)
Ganglion, 135, 140, 141, 145, 227
cells of retina, 187, 190
of parasympathetic nervous system (*see* Autonomic nervous system)
of sympathetic nervous system (*see* Autonomic nervous system)
Gangliosides, 25
Gangrene, 448
Gap junctions, 42, 43, 73, 86, 166, 299, 302, 623
(*See also* Intercellular junctions; Synapses)
Gas
alveolar, 486–488
chemosensors for, 488, 489
content of air, 479
content in blood, 488
exchange
heart failure and, 388, 389
in lungs and tissues, 486–488
laws, 479–480
partial pressures of, 488
production by intestinal bacteria, 589
solubility of, 479
transport in body, 484–487
volumes in lungs, 480–483
Gastrectomy, 583

Gastric emptying, 561–564
Gastric glands, 550
  hydrochloric acid and, 555–558, 560
Gastric inhibitory polypeptide (GIP), 52, 129, 556–557
Gastric motility, 561–564
Gastric protease, 573
Gastric secretions, 555–560
Gastric stimulation of digestive tract, 560
Gastric ulcers, 565
Gastrin, 52, 122, 129, 555–557, 561–563, 567
  ulcers and, 565, 567
Gastritis, 566
Gastrocolic reflex, 561, 563, 564
Gastrocnemius muscle, 292
Gastroenteric reflex, 563
Gastroenteritis, 586–590
  (*See also* Food poisoning; Infections)
Gastroesophageal sphincter, 564
Gastroileal reflex, 563
Gastrointestinal canal, 543
  (*See also* Digestive system)
Gastrointestinal hormones, 550, 554–557
Gastrostomy, 589
Gate model, 164–165, 167, 169, 174, 221, 251, 292, 294, 296
Gaucher's disease, 150
G cells, 556–557
Gelatinous layer of maculae in inner ear, 200
Gemmules, 140
Gene (*see* Genetics)
General adaptation syndrome (GAS), 109
Generator potentials, 167
  (*See also* Generator potentials)
Genetics, 327
  anticodon, 34–36
  of blood groups, 439–441
  code of, 27, 34, 640
  codons of, 34–36
  deficiency diseases and anemias, 446–449
    of nervous system, 149–150
    of skeletal muscle, 314
  determination of sex and, 629
  genes and, 53, 85
  lipoproteins and, 394, 397
Genital herpes, 646
Germ-free animals, 589
Germinal center, 65, 66, 460
Germ layers, 55, 237, 238
GFR (*see* Glomerular filtration rate)
GH (*see* Growth hormone)
GH-RF (*see* Growth hormone releasing factor)
Giemsa stain, 454, 455
Gigantism, 106, 126, 353
Gingiva, 355
Gingivitis, 354
GIP (*see* Gastric inhibitory polypeptide)
Gland(s), 47, 50
  endocrine, 51, 52
    (*See also* Endocrine system)
  exocrine, 51, 550–560
    mammary, 51, 127, 132, 133, 632, 638–639
    salivary, 51, 130, 544, 549, 551, 573
    sebaceous, 46, 51, 506, 509
    sweat, 51, 510–511
Glandular hypophysis (*see* Pituitary, anterior)
Glans penis, 621, 622, 625
Glass factor, 464
Glia, 142–146, 175, 242
  of central nervous system, 145–146, 270, 273
  of peripheral nervous system, 142–145
Globulin(s)
  in cerebrospinal fluid, 245
  (*See also* Antibodies)
Globus pallidus, 272, 273
Glomerular capillaries, 411
Glomerular filtration rate (GFR), 517–521
  measurement of, 530, 532
  obligatory water loss and, 527, 529
  regulation of, 518–521
Glomerular filtrate
  composition of, 521
Glomerulonephritis, 517, 531, 533
Glomerulus, 412, 511, 512, 514, 516–519, 531
Glottis, 473, 479, 489, 491
Glucagon, 52, 111, 116, 122, 129, 130, 523, 555
Glucocorticoids, 52, 126, 130, 131, 457, 467, 627, 638
  adrenal secretion of (*see* Adrenals)
  anti-inflammatory effects of, 117
  blood pressure and, 423
  effects of
    on basophils and eosinophils, 457
    on bone, 130, 348–351, 353
    on milk secretion, 638
  stress and, 11, 112, 457
  water and electrolyte metabolism and, 524
Gluconeogenesis, 11, 112, 131, 349–351
Glucose
  absorption of, 573, 574, 577, 579
  chemistry of, 18, 20
  diuretic effect of, 523
  as energy source, 116, 149, 318, 319
  in extracellular fluids, 81, 82
  metabolism of, 23, 320–322, 590–592
  reabsorption of, 523
  regulation of, 111
  transport of, 92, 95, 439, 577, 579
  in urine, 523
  (*See also* Endocrines, glucagon, insulin; Glycogen; Liver; Metabolism, carbohydrates)
Glucostats, 9, 184, 267
Glucosuria, 523
Glucuronide:
  conjugation to bilirubin, 439
Glutamate:
  monosodium (MSG), effects of, 177
Glutamic acid, 23
  as a neurotransmitter, 177, 273
Glutamic oxalacetic transaminase (GOT) (*see* Aspartate aminotransferase)
Glutamine in hemoglobin, 447
Gluten:
  allergic reaction to, 588
Glycerol, 24, 25, 573, 574, 576–578, 581, 591, 592
Glycine, 22
Glycocalyx, 548, 565
Glycogen, 20, 21, 23, 111, 122, 623, 626
  in liver, 322, 554
  loading, 327
  in skeletal muscle, 286, 318, 320, 322, 323, 325, 327, 554
Glycogenesis, 116, 129, 349
Glycogenolysis, 116, 122, 129, 130
Glycolipids, 25, 87
Glycolysis, 318, 592
  (*See also* Metabolism, carbohydrates)
Glycoprotein(s), 87, 525, 550
  blood group antigens and, 441
  as receptors, 85, 98, 114
  as hormones, 126–128
Glycosuria, 523
GnRH (*see* Gonadotropic releasing hormone)
Goblet cells, 45, 51, 547, 548, 550, 559
Goiter, 113
Golgi apparatus, 29, 31, 32, 51, 97, 98, 137, 342
Golgi cells, 262
Golgi tendon organ, 183–185, 260, 263, 312
Gonadotropins, 52, 126, 127, 133, 626–629, 635–638, 648
  (*See also* Follicle stimulating hormone; Luteinizing hormone)
Gonadotropic releasing hormone (GRH), 52, 108, 109, 126, 626, 628, 634, 635
  (*See also* Luteinizing releasing hormone)
Gonads, 64, 104, 105
  (*See also* Ovaries; Testes)
Gonorrhea, 646
Goose pimples, 509
Gossypol, 649
GOT (Glutamic oxalacetic transaminase) (*see* Aspartate amino-transferase)
Gout, 354, 534, 578
Graafian follicle, 632
Gradient(s), 91, 92
  concentration, 88, 89, 156, 160, 161
  electrical, 88, 156, 160, 161
Gram (g), 13, 16
Grand mal epilepsy, 249
Granular layer of cerebellum, 261, 262
Granule cells, 136, 261, 262, 270
Granules
  metachromatic, 63
  zymogen, 555
Granulocytes, 455–458, 468
  (*See also* Leukocytes, types of)
Grave's disease, 106, 116, 128
Gravity, 194, 200, 201, 311
  blood pressure and, 414, 416, 519
  filtration in kidney and, 519
  urine formation and, 519
  (*See also* Sense organs, ear; Sensory receptors, vestibular receptors)
Gray matter, 270
  (*See also* Unmyelinated fibers)
Graying of hair, 509
Greater vestibular glands, 631
GRH (*see* Growth hormone)
Ground substance, 53, 55
Growth
  of bone (*see* Bone formation and growth; Bone, hormones and)
  puberty and, 628–629, 634
Growth factor
  nerves and, 139
Growth hormone (GH), 105–109, 126, 130, 133, 348, 349
  disorders and, 353, 355
Growth hormone releasing factor (GH-RF), 107, 109
Growth hormone release inhibiting hormone (GH-RIH), 107, 109
Guanethidine, 424
Guanine, 26–28
Guanosine triphosphate (GTP), 232
Guanyl cyclase, 121, 232
Guanyl phosphodiesterase, 121, 232
Gums, 72
Gustation (*see* Sense organs, taste; Sensory receptors, gustator receptors)
Gyrus, 270

Hageman factor, 464, 465

Hair, 509
Hair cells
  in ear, 197–200
  in nasal chamber, 202
  in taste buds, 201
  (*See also* Sense organs, ear, tongue; Sensory receptors, auditory receptors, gustatory receptors)
Hair follicles, 226, 507, 509
  smooth muscle and, 301
  (*See also* Skin)
Hallucinogen, 177, 253
Hardening of arteries, 425
Hartnup disease, 588
Haustra of colon, 545
Haversian canals, 352
Haversian system, 70, 71
Hay fever, 63, 64
Hb (*see* Hemoglobin)
H ban, 288
HCG (*see* Human chorionic gonadotropin)
HDL (*see* High density lipoprotein)
HDN (*see* Hemolytic disease of newborn)
Headache, 245, 422, 649
Hearing (*see* Sense organs, ear)
Heart, 49, 359–377, 378–400
  age and, 381, 382
  anatomy, 360–364
    blood supply, 227, 363–364
    chambers, 361–362
    lining, 360–361
    valves, 354, 361–363
    wall, 361
  blood flow and, 361–362, 364–365
  blood pressure and, 366, 381
  body size and, 381
  body temperature and heart rate, 381
  cardiac cycle, 364–365
  cardiac output, 327–329, 360, 363, 365–366, 368–370, 379–383
    calculation of, 366
    minute volume and, 366
    Starling's law and, 366
    work of heart and, 365–366
  chemical agent effects on, 379–381
    alcohol, 379–380
    caffeine, 367
    hormones, 379
      (*See also* Epinephrine; Norepinephrine)
    ions, 379, 380
    neurotransmitters, 379, 380
    nicotine, 367, 380
    prostacyclin, 463
  chemoreceptors and, 369–370
  development of, 384–387
  diagnostic tests and, 390–398
    images and, 397–398
    electrocardiogram (ECG) and, 370–375
    physical examination and, 391
    plasma lipids and lipoproteins, 383–397
    serum enzymes, 391–393
    vectorcardiogram, 373
  disease of, 360, 373–377, 381, 387–398, 463
    abnormal valves, 362–363
    angina pectoris, 370
    atheroscleoris, 389
    congestive heart failure, 387–389, 397, 483, 484, 533
    coronary artery disease and, 381, 389–390
    electrical activity and, 373–375
    fibrillation and, 374–375
    lipoproteins and, 393–397
Heart, disease (*Cont.*):
    murmurs and, 363
    myocardial infarction, 389–390
    pain and, 210, 221
    regurgitation and, 363
    risk factors and, 389–390, 397, 497
  electrical activity of, 366–375
    elecrocardiogram (ECG, EKG), 370–374
    extrinsic regulation of, 260, 368–370
    intrinsic regulation of, 367–368
  energy production and, 363, 366
    (*See also* Metabolism)
  exercise and, 327–329, 363–366, 382–385, 406
    Bainbridge reflex and, 370
    (*See also* Exercise)
  oxygen deprivation and, 363, 366, 370
  pain, 221, 370
    (*See also* Pain)
  reflexes of, 369, 370, 375, 420–421
  sensory receptors and, 369–370
    (*See also* Sensory receptors, chemoreceptors, aortic bodies, carotid bodies; baroreceptors, aortic bodies, carotid sinus)
  sex and, 381
  sounds of, 363, 365, 417–418
  stress and, 112, 365, 368, 379
  stretching of, 366
Heartbeat, 364, 365, 368
Heartburn, 557–558
Heart clicks, 363
Heart murmurs, 363, 385
Heart rate, 327–329, 364–365, 368–370, 380, 420–421
  age and, 381
  body temperature and, 595
  exercise and, 327–330, 383
  maximal, 327–330, 383
  orgasm and, 626, 634
  training and, 327–330
Heat:
  cramps, 510
  exhaustion, 510, 594
  loss, 233, 266–268, 509, 594
  production, 56, 233, 267–268, 594
    response to, 266
    (*See also* Liver, metabolism; Metabolism; Skeletal muscle, metabolism)
  stress, 109
  sweating and, 510
Height-weight tables, 615
Helper T cells, 460
  (*See also* Leukocytes, T lymphocytes)
Hematocrit, 432, 433
  of venous blood, 486
Hematopoiesis, 335, 436
Hematopoietic tissue, 66, 335, 345, 435, 436, 467, 468
Heme group of hemoglobin, 437–438 485
Hemochromatosis, 583
Hemochromic anemia, 447
Hemocytoblast, 434
Hemodialysis, 536–537
Hemoglobin (Hb), 31, 36, 338, 436–439, 446–449
  altitude effects on, 403
  anaerobic metabolism, effects on, 485
  anemias and, 446–449, 486
  carbon monoxide and, 496
  chains of, 438–439, 446–449, 486
  2,3,diphosphoglycerate and, 485
  dissociation of, 485, 487
  electrophoresis of, 448, 449
Hemoglobin (*Cont.*):
  filtration in kidneys, 519
  general properties, 437–438
  iron content of, 584
  levels of, 432, 433, 437, 438
  metabolism of, 438–439, 554, 559, 583–585
  myoglobin and, 437
  structure of, 437–439
  transport of gases by
    carbon dioxide, 437, 486–487
    carbon monoxide, 438, 496
    oxygen, 436–438
    pH effects on, 436
  types of, 438–439
  adult (HbA), 438, 439
  Bart's, 439, 449
  carbaminohemoglobin ($HbCO_2$), 437, 484–487
  carboxyhemoglobin (HbCO), 431, 438, 496
  fetal (HbF), 438–439, 448, 486
  Gower, 439
  H (HbH), 449
  oxyhemoglobin (HbO_2), 436, 484–487
  potassium hemoglobinate (KHb), 486
  reduced (HHb), 436
  sickle cell (HbS), 436
  sodium hemoglobinate (NaHb), 486
Hemolysis, 438
Hemolytic anemia, 432, 446
Hemolytic disease of the newborn (HDN), 445–446
Hemophilia A and B, 466
Hemopoiesis (*see* Hematopoiesis; Hematopoietic tissue)
Hemorrhage, 422, 425, 426, 432, 443–445, 466, 469
Hemorrhagic shock, 443–445, 594
Hemosiderin, 583, 584
Hemostasis, 462
Heparin, 465, 581
  basophils and, 465, 581
  mast cells and, 63, 459
Hepatic artery, 553
Hepatic circulation, 553–554
Hepatic duct, 551, 553
Hepatic vein, 384
Hepatitis, 435, 467, 566–567, 583
Hepatocytes, 552, 553
Hereditary anemias, 446–449
Hereditary immune deficiency, 467
Hereditary lipid storage diseases, 149–150
Hering-Breuer reflex, 489–491
Hernia, 620
Heroin, 251–253
Herpes simplex virus, 225, 646
  (*See also* Sexually transmitted disease)
Herpes zoster, 225
Hertz (Hz), defined, 397
Hexachlorophene, 149
H gene, 441
H substance, 440–441
Hiccups, 489
High altitude (*see* Altitude exposure; Respiratory system, environmental effects)
High blood pressure (*see* Blood pressure; Hypertension)
High density lipoproteins (HDL), 390, 394, 395, 397
High oxidative fibers, 298, 300
High voltage slow waves, (HVS), 247
Hindbrain, 177, 239, 241, 259–265
Hippocampus, 272, 273

Histamine, 185, 458, 459, 463
  allergic reactions and, 63–65
  basophils and, 229, 458
  mast cells and, 63, 64, 229, 458
  as parahormone, 103, 122
  receptors to, 85, 229, 566
  vasodilator activity, 458
Histamine antagonists, 566
Histiocyte, 59, 459, 468
  (*See also* Defense cells; Macrophages)
Histocompatibility antigens, 85, 86, 440, 441 460, 536
Hives, 63
HMD (*see* Hyaline membrane disease)
Hodgkin cycle, 161
Hodgkin's disease, 467, 468
Holocrine glands, 51
Holocrine secretion, 509
Homeostasis, 2, 5–10, 35, 308, 345, 431, 506
  autonomic nervous system and, 225
    (*See also* Autonomic nervous system)
  endocrine system and, 105–111
  heart and, 402
  kidney and, 512–517
Homeotherms, 268
Homozygous, 466
Hormones, 7, 87, 102–133
  aging and, 115, 351
    (*See also* Androgens; Estrogens; Steroids)
  blood pressure and, 422
  cascade effects, 116
  cellular permeability and, 120
  chemical composition of, 105, 126–133
  defined, 7, 103
  degradation, 113, 114, 524, 533
  effects of, 126–133
  endorphins and, 249
  excretion of, 113, 114
  feedback mechanisms and, 108–111
    (*See also* Feedback mechanisms)
  of females (*see* Estrogens; reproduction, below; Steroids)
  gastrointestinal, 550, 554–557
  hypothalamic (*see* Hypothalamus)
  interactions of, 116, 123, 350, 555
  life span, 114
  of males (*see* Androgens; reproduction, below; Steroids)
  maternal changes during pregnancy, 644–645
  measurement by radioimmunoassay, 117–119
  mechanisms of action, 119–123
  pharmacological levels of, 117
  physiological levels, of, 117
  pituitary, 126–128, 133
    (*See also* Pituitary)
  primary effects of, 116
  properties of, 105, 126–133
  receptors for, 114–123, 397
    (*See also* Receptors, cellular)
  reproduction and, 626, 634–639
  as secondary messengers, 119–123
  secretion of, 98, 113, 114
  stress and, 109–111, 112, 115
  synthesis of, 111
  target cells and, 103
  transport of, 113, 114
  transcription and, 119–121
  translation and, 119–121
  (*See also* Endocrines; names of individual hormones)
Housemaid's knee, 354
Human chorionic gonadotropin (HCG), 133, 628, 634, 641
Human chorionic somatomammatropin (HCS), 133
Human placental lactogen (HPL), 133, 638
Humoral immunity, 59, 86, 458, 511
  (*See also* Antibodies)
Hunger, 9, 267
  (*See also* Hypothalamus)
Huntington's chorea, 178, 272–273
Hyaline cartilage, 68, 73, 336, 337, 343, 475
Hyaline membrane disease, 477–478
Hyaluronic acid, 53, 622
Hyaluronidase, 622, 639
Hydralazine, 424
Hydrocarbon, 19
Hydrocephalus, 243
Hydrochloric acid, 48, 85, 122, 550, 565
  functions, 557
  secretion, 555, 557–558, 560–562
Hydrocortisone, 351
Hydrogen acceptors, 590–592
Hydrogenated fat, 24
Hydrogen bonds, 17, 557
Hydrogen ions, 17, 564
  hemoglobin dissociation and, 415, 487
  pH and, 16
  respiration and, 488, 490, 492
  secretion of, 555–560
    by kidneys, 94, 522–524, 527–530
    by stomach, 94, 122, 555, 557–558, 560–562
Hydrogen transport, 590–592
Hydrolysis, 572
Hydrophilic properties
  of glutamine in hemoglobin, 447
  of hormones, 105
  of phospholipids in cell membranes, 87
Hydrophobic properties
  of hormones, 105
  of phospholipids in cell membranes, 87
  of valine in hemoglobin, 447
Hydrops fetalis, 449
Hydrostatic pressure, 403, 412, 413, 517, 518, 521
Hydroxyapatite, 339
β-Hydroxybutyrate, 322
β-Hydroxybutyric acid, 592
Hydroxybutyric dehydrogenase (HBD), 391–393
  (*See also* Lactic dehydrogenase)
Hydrocholecalciferol (*see* Vitamin D)
Hydroxyergocalciferol (*see* Vitamin D)
Hydroxyproline, 347–348
5-Hydroxytryptamine (5-HT) (*see* Serotonin)
Hymen, 630, 631
Hyperaldersteronism, 130, 534
Hyperbaric oxygenation, 494–496
Hypercalcemia, 349
Hypercapnia, 495, 534
Hyperchromic anemia, 446, 448
Hyperemia, 521
Hyperfunction of endocrines, 105, 106, 126–133
Hyperglycemia, 267
Hyperlipidemia, 425–426
Hyperlipoproteinemia, 393–397
  atherosclerosis and, 425–426
Hyperopia, 193
Hyperosmolar, 91
Hyperparathyroidism, 128, 268, 352, 353, 381, 594
Hyperphagia, 595
Hyperpolarization, 163, 164, 172–175, 221
  of heart, 368
  of smooth muscle, 303
  visual impulses and, 189, 191
  (*See also* Potentials, inhibitory postsynaptic)
Hypersensitivity (*see* Allergy)
Hypertensin, 525
Hypertension, 420, 421, 534, 644
  Conn's syndrome and, 130, 423, 534
  defined, 422
  essential, 423
  heart attack risk and, 388, 390
  kidneys and, 423
  nervous factors, 422–423
  renin-angiotensin system and, 111, 423, 525
  therapy for, 423–424
Hyperthermia, 26, 381, 594
Hyperthyroidism, 106, 113, 116, 352, 467, 485
Hypertonic solution, 90, 91
Hypertrophy
  of cardiac muscle, 300, 364, 385, 387, 389
  of skeletal muscle, 300, 314, 315, 326, 328
Hyperventilation, 493, 534, 535
Hypnosis, 250, 251
Hypoalbuminemia, 533
Hypocapnia, 493
Hypochromic anemia, 446
Hypodermis, 506–507, 509, 511
Hypofunction of endocrines, 105, 106, 126–133
Hypoglycemia, 116–117, 267
Hypokalemia, 525
Hypoparathyroidism, 353
Hypophosphatemic rickets, 353
Hypophyseal portal vessels, 107–109
Hypophysis (*see* Pituitary)
Hypoproteinemia, 533
Hypoosmolar solution, 91
Hypophagia, 595
Hypotension, 420, 421
Hypothalamus, 91, 104–111, 177, 241, 260, 265–269
  anatomy, 105–108
  autonomic nervous system and, 233
  behavioral responses and, 267
  centers of control:
    appetite, 266–268
    behavior, 266–268
    feeding, 267, 562, 564
    hunger, 9, 233, 249, 266–268, 325, 510, 562, 564
    satiety, 267, 562, 564, 595
    temperature regulation, 233, 407, 594–595
      (*See also* Cold exposure; Heat; Sweat; Sweating)
    thirst, 233, 244, 266–269, 562
    (*See also* Sensory receptors, chemoreceptors, osmoreceptors, thermoreceptors)
  hormones and, 52, 104–111, 126, 266–269
    (*See also* Antidiuretic hormone; Oxytocin; Releasing hormones)
  nerve tracts and, 267
  neuroendocrine functions of (*see* hormones above)
  pituitary and, 107–110
  portal system of, 107–108, 266
  reproductive system and, 626–629, 634, 635, 638
    (*See also* hormones, above)
  sleep and, 266, 268
    (*See also* Sleep)

Hypothermia, 268, 380, 594
Hypothrombinemia, 466
Hypothyroidism, 116
Hypotonia, 264
Hypotonic solution, 90, 91
Hypoventilation, 535
Hypovolemia, 523
Hypoxia, 366, 437, 443–444, 478, 485, 493, 496
Hysterectomy, 650
Hysterotomy, 651
H zone, 288

I band, 278–290
ICF (*see* intracellular fluid)
ICSH (*see* Interstitial cell stimulating hormone)
Identical twins, 639
IDL (*see* Intermediate density lipoprotein)
Ileitis, 566
Ileocecal sphincter, 545
Ileocolic sphincter, 545
Ileogastric reflex, 563
Ileum (*see* Small intestine, ileum)
Immature cells of digestive tract, 548, 550, 565
Immediate-type allergy (*see* Allergy)
Immune response, 56, 59, 246, 454, 460, 467, 650, 651
  (*See also* Allergy; Autoimmunity; Cellular immunity; Defense cells; Histocompatibility antigens; Humoral immunity; Leukocytes; Inflammatory response)
Immunodeficiency disease, 588
Immunological assays, 117–119, 176, 246, 641
Immunosuppression, 536
Implantation, 113, 640, 641, 651
Impotency, 622, 646
Impulses, 7, 135–136, 140, 155–168, 311
  all or none law and, 162–163
  amplitude and, 175, 185, 247
  brain and, 246–249
  destination and direction, 147
  duration of, 164, 165, 214
  electrochemical equilibrium and, 156–157
  excitatory, 171–175, 214, 221, 292, 294–296
  frequency of, 185–186
    hearing and, 198
    taste and, 201
  gate model and, 164–165
  inhibitory, 171, 172, 174, 178
  intensity of stimulus and, 186
  memory and, 279
  muscle fatigue and, 326
  Nernst equation and, 159
  potentials and (*see* Potentials)
  summation of stimuli and, 167–168, 172–175, 189, 221
  threshold for, 160
  velocity of, 148, 166–167, 186
  visual, 189, 191
  (*See also* Potentials; Neurotransmitters; Synapses)
Incus, 195, 196
Index Medicus, 3, 4
Induced abortion, 651, 653
Inducer T cells, 460
Infant death syndrome, sudden, 477–478
Infarct, 364, 389
Infection(s)
  blood-brain barrier and, 145
  epilepsy and, 248
  of heart, 360, 361
  in kidneys, 533–534
  leukocyte responses to, 459, 467
    (*See also* Leukocytes)
  of liver, 566, 567
  of lungs, 484
  of nervous system, 149, 225
  sexually transmitted, 646
  sickle cell anemia and, 448
  skin and, 509, 511, 629, 649
  (*See also* Gastroenteritis; names of individual infectious agents
Infectious hepatitis, 566, 567
Infectious mononucleosis, 467, 469
Inferior colliculi, 241, 265
Inferior vena cava, 362, 517
  (*See also* Vena cava)
Infertility, 645
Inflammatory response, 61, 62, 72, 117, 352–355, 517, 534, 564, 565
  intrauterine device and, 651
  leukotrienes and, 463
  prostate and, 622
  vasectomy and, 650
Inflation of lungs, 481–483, 489
Inguinal hernia, 620
Inhibiting factors (*see* Inhibiting hormones)
Inhibiting hormones, 107–109
Inhibition by negative feedback (*see* Feedback mechanisms, negative)
Inhibition of pain, 221–222, 251
Inhibitory postsynaptic potential (IPSP), 171, 172, 174, 178
Inner ear, 141, 194–199
  (*See also* Sensory organs, ear)
Inorganic compounds, 19
Inotropic effect, 368
Insensible perspiration, 510
Insertion of muscle, 285
Inspiration, 480, 481, 490
Inspiratory center, 488–491
Inspiratory reserve volume, 482
Instillation induced abortion
  by prostaglandins, 651
  by saline, 651
Insufficiency
  cardiac, 363
Insula, 270, 272, 273
Insulin, 52, 105, 111, 116, 129, 523, 577
  cAMP and, 122
  in diabetes mellitus, 115, 130, 389, 523, 534, 535
  release of, 555–557
  receptors to, 115
  resistance to, 115
  secretion of, 98, 113
Integumentary system, 506
Intention tremor, 264
Interaction of hormones, 116, 123, 350, 555
Intercalated discs, 73, 74, 284, 299
Intercellular fluid, 81, 412
  (*See also* Extracellular fluid; Insterstitial fluid)
Intercellular junctions
  of capillaries, 408–412, 517
  desmosomes, 42, 43, 73, 86, 508
  gap, 42, 43, 73, 86, 166, 299–302, 623
  intercalated discs, 73, 74, 284, 299
  neuromuscular, 291–298
  nexuses, 42, 43, 73, 299
Intercellular junctions (*Cont.*):
  tight, 42, 43, 73, 86, 411, 549, 564, 623, 624
Intercostal muscles, 490
Intercostal nerves, 490
Interferons, 460
  (*See also* Lymphokines)
Intermediary metabolism, 590
  (*See also* Metabolism)
Intermediate density lipoprotein (IDL), 394, 395, 397
Intermediate-term memory, 278
Intermedins, 127
Internal intercostal muscles, 481
Internal respiration, 473
Interneuron, 141, 209
Interoceptors, 182
Intersegmental reflex, 212–214
Interstitial cells of Leydig, 127, 625–626, 628
Interstitial cell stimulating hormone (ICSH)
  (*see* Luteinizing hormone)
Intervertebral disc, 69
Intervertebral foramen, 244
Interventricular septum, 362, 368, 386, 388, 398
Intervillous spaces, 641, 642
Intestinal bacterial flora, 589
Intestinal cell pore size of, 574
Intestinal colic, 585
Intestinal effects of hormones, 123, 128, 550, 554–557
Intestinal juice:
  chyme, 545, 586
  compared to plasma, 559
Intestinal lymphangiectasia, 587
Intestinal malabsorption, 586–590
Intestinal motility
  regulation of, 561–564
Intestinal mucosa, 50, 513, 515, 546–550
Intestinal polyps, 589
Intestinal secretions
  regulation of, 561
Intestinal smooth muscle, 304, 561–564
Intestinal stimuli, 561
Intestine (*see* Colon; Small intestine)
Intestino-intestinal reflexes, 563
Intima, 403–405
Intracartilaginous bone formation, 341–344
Intracellular compartments, 80–82
Intracellular fluid (ICF), 6, 80–82, 155, 412
Intracellular regulators, 119–123
Intrafusal fibers, 185
Intramembranous bone formation, 341, 342
Intrapulmonic pressure, 480, 489
Intrasegmental reflex, 212–213
Intrauterine device (IUD), 647, 651
Intravenous feeding, 590
Intrinsic clotting system, 464–465
Intrinsic conduction system of heart, 300, 367–368
Intrinsic factor, 98, 447, 550, 557–559, 581, 584, 585
Inulin, 533
Invasive diagnostic techniques, 390
Invertase, 574
Involuntary muscle (*see* Muscle, cardiac, smooth)
Involuntary nerve fibers, 147
(*See also* Adrenergic nerves; Autonomic nervous system; Cholinergic nerves)
Iodine, 94, 583, 608, 612, 618
Ion, 5, 81
  absorption of, 583–585

Ion (*Cont.*):
  channels (*see* Gate model; Pores)
  size, 83, 84, 158
Ionic bond, 17
Ionic composition of body fluids, 82, 245, 431–432
  (*See also* Intracellular fluid; Extracellular fluid)
Ipsilateral reflex, 212
IPSP (*see* Inhibitory postsynaptic potential)
Iris, 181, 188, 301
  (*See also* Sense organs, eye)
Iron, 457, 574, 583, 608, 612
  absorption of, 583–585
  daily requirements for, 618
  deficiency diseases, 432, 447, 469, 583
  in hemoglobin, 437, 438
  induced cirrhosis, 567
  in oxidative enzymes, 592
  storage disease, 584, 585
  in tissues, 583
  transport and recycling, 439, 583–585
Iron lung, 497
Iron oxide granules, 585
Irradiation effects (*see* Radiation effects)
Irreversible shock, 445
Ischemia, 389, 638
Island of Reil, 270, 272, 273
Islets of Langerhans, 98, 104, 550
Isoenzymes, 391–392
Isomaltase, 574
Isometric contraction, 308, 311, 324
Isometric tension (*see* Skeletal muscle, contraction)
Isometric ventricular contraction, 365
Isometric ventricular relaxation, 365
Isoosmotic solution, 91
  therapy for malabsorption, 590
Isoproniazid, 252
Isoproterenol, 229, 231
Isotonic contraction, 308, 310, 313, 324
Isotonic exercise, 382–383
  (*See also* Exercise; Skeletal muscle, exercise; Submaximal exercise)
Isotonic solution, 90, 91
Isotropic appearance of skeletal muscle, 286
Isotropic band (*see* I band)
Isozymes, 391–392
Itch, 184, 185

Jaundice, 438, 448, 566, 567
Jaw, 337, 355, 498–499
  resorption of, 351, 354, 355
J cells, 489–490
Jejunal ulcers, 565
Jejunum (*see* Small intestine, jejunum)
Journals in the life sciences, 3, 4
Junctions of cells (*see* Intercellular junctions; Neuromuscular junction; Synapses)
Juxtaglomerular cells, 518
Juxtamedullary nephrons, 514

Kallikrein system, 525
  (*See also* Kinins)
Kell antigen, 442, 446
Keratin, 508, 509
Keratin sulfate, 339
Kernicterus, 445
Keto acids, 591, 592
Ketogenic diets, 592
Ketoglutaric acid, 23
Ketone bodies, 592
Ketonemia, 592
Ketonuria, 592
Ketosis, 592, 595
Kidneys, 6, 7, 65, 506, 511–537
  acid-base balance and, 495, 527–530
    bicarbonate excretion, 493, 529–530
    regulation of, 7
  anatomy of, 46, 511–512
  artificial, 536–537
  autonomic nervous system and, 226, 521
  blood supply to, 517, 518–521
    autoregulation of, 519–521
    capillaries of, 517–521
  calcium excretion and, 348
  cells of monkey kidney in tissue culture, 56
  collecting ducts of, 513, 514, 516, 522, 528
  countercurrent theory, 525–527
  dialysis and, 536–537
  disease and, 533–535
    effects on bone, 352, 353
    effects of leukemias, 468
  erythropoietin and, 435–437
  excretion, 513, 529
    of hormones, 113–114
    of phosphate, 348, 349
  filtration by, 513, 517–522, 528
  general functions of, 512–513
  glomerular filtration rate (GFR), 517–521, 530, 532
  heart failure and, 387
  homeostasis and, 512, 527
  hormonal effects on, 132, 524–525
    aldosterone, 130, 131, 524–525
    antidiuretic hormone, 91, 267, 524, 528
    glucocorticoids, 130, 131
    parathormone, 128
    prostaglandins, 133
  hypertension and, 525
  measurement of function, 530–533
  nephrons of, 513–524
    anatomy, 513–516
    physiology, 516–524
  reabsorption by, 513, 521–523, 528
  referred pain and, 210
  renin-angiotensin system and, 525–526
  secretion by, 512, 513, 522–524, 528
    of hormones, 104, 132
    of hydrogen, 495, 522, 528–531
    of phosphate, 348, 349
    of potassium, 94–95, 522–524, 529
  stones, 339, 347–348, 534
  transplants, 536
  vitamin D synthesis and, 346–347
  water
    content, 80
    loss by, 527, 529
    regulation and, 128
    (*See also* Excretory system)
Killer T cells, 460
Kilo-, 15
Kilocalorie, 15, 16
Kilogram, 16
Kinases
  protein, 121, 122
Kinesiology, 316
Kinins, 64, 185, 326, 525
Klinefelter's syndrome (XXY), 629
Kneecap, 338
Knee jerk reflex, 213
Korotkoff's sounds, 417
Krause's end bulbs, 182–185
Kreb's cycle, 320–322, 590–592
Küppfer cells, 66, 553, 565, 588
Kwashiorkor, 587, 588
Kymogram, 309
Kymograph, 309, 482

Labia majora, 630–632
Labia minora, 630–632
Labile factor, 464
Labor, 652–653
Laboratory values (*see* Appendix to text)
Lacrimal glands, 188
Lactase, 559, 586, 588
Lactase deficiency, 586–588
Lactation, 638–639
Lacteals, 547, 574, 577–582
Lactic acid, 318–319, 321, 591
  muscle fatigue and, 326, 492
Lactic dehydrogenase, 391–393
Lactic dehydrogenase X, 649
Lactobacillus bacteria, 345, 589
Lactoferrin, 457
Lactogen, 638
Lactogenic hormone (*see* Luteotropic hormone, LTH)
Lactose, 574, 639
  calcium absorption and, 345
  intolerance to, 586–588
Lacunae, 67, 69
Laki-Lorand factor, 464
Lamellae, 70
Lamina propria, 47, 50, 201, 546
Langerhans, islets of, 98, 104, 550
Language, 270, 273, 276, 278
Large external transformation sensitive protein (LETS), 86
Large lymphocyte, 456, 458, 460
  (*See also* Leukocytes, lymphocytes)
Laryngitis, 475
Laryngopharynx, 473, 474, 545
Larynx, 69, 128, 473–475, 544
Latent period, 310, 311
Lateral corticospinal tract (*see* Spinal cord, tracts)
Lateral fissure, 274
Lateral spinothalamic tract (*see* Spinal cord, tracts)
Lateral sulcus, 270, 272, 274
Lateral ventricles of brain, 241, 243, 244
  (*See also* Brain, cerebrospinal fluid)
Laxative, 91, 585, 586
LD (*see* Lactic dehydrogenase)
LDH (*see* Lactic dehydrogenase)
LDL (*see* Low density lipoprotein)
Lead poisoning, 339–340
Leads I, II, III of electrocardiogram, 373
Learning, 278
Lecithin, 87, 477–478
  to sphingomyelin ratio (L/S ratio), 477–478
Lens (*see* Sense organs, eye)
Lenticular nucleus, 272
Lesser vestibular glands, 631
Leu-enkephalin, 249
Leukemias, 146, 340, 435, 467–478, 578
Leukocytes, 98, 431, 454–460, 468
  appearance, 454–456
  differentiation of, 454–456
  general functions, 454
    body defenses and, 454, 459–461
    (*See also* Defense cells; Immune response; Macrophages)
  life span of, 454

Leukocytes (*Cont.*):
  numbers of, 269, 431, 454, 455
  origins of, 433–435, 454–455
  pathophysiology and, 467–468
  types of
    basophils, 63, 229, 43, 435, 455–458, 465
    eosinophils, 434, 435, 455–457
    lymphocytes, 59, 60, 62, 65, 85, 129, 454, 455, 456, 458, 459, 460
  B lymphocytes, 59, 434, 435, 458, 460
    (*See also* Humoral immunity; Immune response; Plasma cells)
  T lymphocytes, 59, 61, 63, 85, 297, 351, 354–355, 434, 435, 458, 460
    (*See also* Cellular immunity; Lymphokines; Thymus)
  monocytes, 60, 62, 146, 434, 455, 456, 458–459
  neutrophils, 434, 435, 455–457
  (*See also* Autoimmunity; Cellular immunity; Defense cells; Immune response; Inflammatory response; Lymphokines; Macrophages; Thymus)
Leukocytosis, 468, 566
Leukopenia, 468
Leukotrienes, 64, 462–463
  (*See also* Slow reacting substance anaphylaxis)
Levers
  skeletal muscle action and, 315–318
Levo- (l), 20
Leydig cells, 127, 625–626, 628
LH (*see* Luteinizing hormone)
Librium, 252, 253
Lice, 646
Lieberkühn
  crypts of, 548–550, 559, 573
Ligament, 66, 69, 224, 336
Light, 184
  (*See also* Sense organs, eye)
Light chains of myosin, 289
Lightening, 653
Limb lead recordings of ECG, 373
Limbic lobe, 270, 271
  brain waves and, 248
Limbic system, 271, 272
  autonomic nervous system and, 233
  behavior and, 266–268
  hypothalamus and, 266–268
  reticular formation and, 271
  thalamus and, 271, 272
Lingual tonsils, 545
Linoleic acid, 578
Lipases, 559, 561, 566, 572, 573, 577, 578, 581
  lipoprotein, 394, 395, 426, 581
  pancreatic, 551
Lipid, 56
  absorption of, 575, 577, 582
  atherosclerosis and, 394–395, 425–426
  in cell membranes, 30, 82, 83, 87, 144
  chemistry of, 24–26
  digestion of, 572–574
  disorders and, 149–150
  droplet, 58
  emulsification of, 578, 581
  metabolism of, 321–322, 393–395, 591–593
  nutrition and, 148–150, 395–397
  from sebaceous glands, 509
  secretion of, 127
  in skeletal muscle, 286
  (*See also* Fats)
Lipid storage diseases, 149–150
Lipoprotein, 25, 393–397
  atherosclerosis and, 394–395, 425–426
  in cell membrane, 87, 144
  electrophoresis of, 393–394
  groups of, 394–396
  lipases, 394–395, 426, 581
  metabolism of, 393, 395, 426
    (*See also* Metabolism)
  ultracentrifugation of, 394
Lipopolysaccharide endotoxin, 588
Lipotropins, 127
Liter (L), 13, 16
Liver, 50, 65, 150, 551–554
  anatomy, 544, 551–553
  ascites and, 389, 567
  bile secretion by, 557, 559–561
  capillaries of, 410, 411, 552–553
  cells, 552–553
  cirrhosis of, 533, 567
  clotting factors and, 464–465
  congestive heart failure and, 533
  Cori cycle and, 319
  detoxification of ammonia by, 589
  disease, 566–567
  function of, 553–554
  hemoglobin and, 438–439, 446, 554, 559, 583–585
  heparin and, 465, 466
  hormones and, 116, 126, 129–131, 350, 556–567
    (*See also* Glucagon; Insulin)
  leukemia effects on, 468
  lobules of, 552, 553
  metabolism and:
    glycogen, 122, 322
      (*See also* Glucagon; Glycogenesis; Glycogenolysis; Insulin)
    hemoglobin, 438, 446, 583–585
    hormones and, 113, 114, 533, 553, 591–593, 627, 628
    lipoproteins, 394, 397
  portal circulation of, 553
  referred pain and, 210
  regeneration of, 567
  sinusoids of, 411, 565
  space of Disse and, 410, 411
  steroid metabolism and 627, 628
  storage functions, 553–554
  vitamins and, 346–347, 465, 466
    (*See also* Vitamins)
Lobes of brain, 270–273
Local potentials, 167
Lockjaw, 89
Longitudinal cerebral fissure, 270
Long-term memory, 278
Loop of Henle, 513, 514, 516, 522–525
Loose connective tissue, 55, 59, 507
Low blood pressure, 416, 420–421
Low density lipoprotein (LDL), 394, 395
Lower motor neurons, 220, 223
Low voltage fast waves (LVS), 248
LRH (*see* Gonadotropin releasing hormone; Luteinizing releasing hormone)
LSD (*see* Lysergic acid diethylamide)
LTH (*see* Luteotropic hormone)
Lumbar nerves, 206–208
Lumbar vertebrae, 207, 208
Lumbosacral plexuses, 211
Lumen, 410, 55, 557
Lung, 6, 34, 46, 261, 298
  aging and, 475
  alveoli of, 475–478
    capillaries, 475
    macrophages, 477
Lung, alveoli of (*Cont.*):
    ventilation, 480
  anatomic dead space, 483
  anatomy, 474–477
  artificial respiration and, 497–500
  blood gases and, 486
  breathing and, 488–492
  cancer of, 496–497
  capillaries, 475
  carbonic acid of blood and, 433, 487
  disorders and (*see* Respiratory system, disorders of)
  environmental effects on, 493–497
  excretion by, 506
  exercise and, 483, 492–493
  of fetus, 384, 475–478
  forced expiratory volume of, 483, 484
  gas exchange and transport in, 484–488
  heart and, 362
    (*See also* Heart)
  infections of, 484
  inflation and deflation of, 481–483, 489
  intrapulmonic pressure, 489
  metabolism and gas exchange, 484–487
  nervous control and, 226, 229, 488–491
  oxygen diffusion and absorption by, 484–487
  perfusion, 488
  pleura of, 49, 475
  pleural cavity and, 49, 474–475
  pleural effusion, 389
  pressures of gases in, 480
  receptors in, 369, 488–490
    (*See also* Sensory receptors, mechanoreceptors)
  reflexes and, 370, 488–490
  spirometry and, 480–483
  surfactants of, 477
  ventilation of, 473, 480–484, 493
    artificial, 497–500
  ventilation-perfusion ratio, 488
  volumes of, 480–484
    exercise and, 484, 492–493
    respiratory minute volume, 480, 482
  water loss from, 529
  (*See also* Alveoli; Respiratory system)
Luteal phase of reproductive cycle, 634–638
Luteinizing hormone (LH), 52, 108, 109, 122, 126, 127, 133, 266, 626, 627, 644
  female reproductive cycle and, 634–638
Luteinizing hormone releasing hormone (LH-RH/FSH-RH), 108, 109, 626, 628, 634, 635, 639
  (*See also* Gonadotropic releasing hormone)
Luteotropic hormone (LTH), 52, 108, 109, 127, 133, 639
Luteotropin (*see* Luteotropic hormone)
Lymph, 81, 412–414, 577
Lymph capillaries, 412–413, 574
  (*See also* Lacteals)
Lymph nodes, 59, 65, 66
  (*See also* Lymph tissue)
Lymph vessels, 402, 412–413, 476
  obstruction of, 414, 587
Lymph volume, 412
Lymphatic organs, 413
Lymphatic pump, 413
Lymphatic system, 6, 49, 81
Lymphoblasts, 434, 468
Lymphocytic leukemia, 468
Lymphocytes, 59, 60, 62, 65, 85, 129, 455, 456, 458, 460
  atypical, 459
  B cells, 59, 434, 435, 458, 460

Lymphocytes (*Cont.*):
  large, 456, 458, 460
  memory, 454, 460
  small, 456, 458, 460
  T cells, 59, 61, 63, 85, 297, 351, 354–355, 434, 435, 458, 460, 468
  (*See also* Autoimmunity; Cellular immunity; Defense cells; Immune response; Inflammatory response; Lymphokines; Macrophages; Thymus)
Lymphogranuloma venereum, 646
Lymphoid organs, 459
  (*See also* Lymphoid tissue)
Lymphoid tissue, 65, 129, 460, 468
  as source of lymphocytes, 434
  (*See also*, Appendix; Peyer's patches; Lymph nodes; Spleen; Thymus; Tonsils)
Lymphokines, 60, 61, 64, 354–355, 360
  activation of macrophages by, 63, 355
  defined, 460
  as interferons, 460
  as lymphotoxins, 355
  (*See also* Chemotaxis)
Lymphotoxins, 355
Lymph vessels and blood vessels, 401–429
  space of Disse and, 552, 553
Lysergic acid diethylamide (LSD), 177, 252, 253
Lysis of cells, 91
Lysolecithin, 573
Lysosomal enzymes, 59, 61, 69, 508, 622, 638, 639
  (*See also* Lysosomes)
Lysosomal storage diseases, 149–150
Lysosomes, 29, 32, 96, 98, 394, 457
  enzymes of, 622, 638, 639
Lysozyme, 549, 550, 565

Magnesium, 449
Macrocircuits, 169
Macrocytic anemia, 446–447
Macromolecule, 20
Macrophage:
  activation of, 61, 62, 354–355
  alveolar, 477, 497
  in bone, 354–355
  as defense cells, 56, 61, 63, 650
    in mononuclear phagocytic system (MPS), 65
    in reticuloendothelial system (RES), 65
  endocytosis and, 61, 98
  general properties of, 59–61, 66
  interaction with lymphocytes, 61–62, 354–355, 459–460
  liver, 66, 553, 565
  in nervous system, 146
  transformation from monocytes, 354, 460
  (*See also* Defense cells; Immune response)
Macula densa, 514, 518, 519
Macula lutea, 187
Maclae, 197, 200–201
  (*See also* Sense organs, ear; Sensory receptors, vestibular receptors)
Magnesium, 121, 618
Magnesium hydroxide, 586
Magnesium phosphate, 338
Magnesium sulfate, 586
Malabsorption, 586–590
Male climacteric, 629
Male reproductive system, 620–628
  (*See also* Reproductive system)
Malleus, 195, 196
Maltase, 559, 572, 573, 586, 588
Maltose, 573, 574
Maltotriose, 573, 574
Mammary glands, 51, 127, 132, 133, 632, 638–639
Mammography, 632
Mandible, 337, 355, 498–499
  (*See also* Jaw)
Manganese, 583, 609, 612
Mannose, 577
Manometer, 419, 521
MAO (*see* Monoamine oxidase)
Marijuana, 115, 252, 253
Marrow, 72, 335, 337–338, 340, 342, 343, 345
Marsilid, 252
Martini's rule, 494
Masculinization, 131
Mass peristalsis, 564
Mast cells, 63–65, 232, 279, 457, 465
Mathematical computations, 270, 276
Mating behavior, 104, 266
Matrix, 53, 68, 69
  of cell membrane, 87
Mature follicle, 633
Maxillary sinus, 474
Maximal exercise, 326–327
Maximum heart rate, 326, 328, 383
Maximum tubular reabsorptive capacity (T_m), 523
  of glucose, 523
M band, 287–289
Measles, 149, 461
Measured resting potential, 156, 158
Mechanical stress
  bone and, 351–352, 354
Mechanics of skeletal muscle contraction, 308–318
  (*See also* Skeletal muscle)
Mechanoreceptors (*see* Sensory receptors, mechanoreceptors)
Media, 403–405
Mediastinum, 360
Mediated transport, 92
Mediterranean anemia, 446, 448–449
Medulla
  of adrenal glands, 52, 107, 111, 113, 131, 132, 176
  of kidney, 511, 512, 516
Medulla oblongata, 241, 258–261, 369, 420–421, 444, 488, 490, 491
  (*See also* Brain, Medulla oblongata)
Medullary cavity, 72, 336
Megakaryoblast, 434
Megakaryocytes, 435, 461
Meiosis, 623–624, 633, 639
Meissner's corpuscles, 182–185, 507, 511
Meissner's plexus, 546, 547, 561, 564
Melanin, 187, 508, 509, 583
Melanocyte, 507, 508
Melanocyte stimulating hormone (MSH), 108, 109, 122, 127, 508
  endorphins and, 249, 250, 252
Melanoma, 53
Melanophores, 127, 128
Melanosomes, 508, 509
Melatonin, 52, 128, 150, 177, 266, 269
Membranes:
  abdominal (*see* peritoneal, below)
  alveolar, 477, 478
    (*See also* Respiratory system)
  basement (*see* Basement membrane)
  cell (*see* Cell membrane)
  of digestive tract, 50, 59
    (*See also* Digestive system, absorption; Mucosa)
Membranes (*Cont.*):
  mesothelial, 48–49, 546
  motor end-plate, 285, 291, 314
  mucous (*see* Digestive system; Mucosa; Respiratory system)
  nuclear, 30
  pericardial, 49, 360, 361
  peritoneal, 49, 51, 537, 546
  pleural, 49, 474–475
  of respiratory system, 59, 474, 477, 478
    (*See also* Respiratory system)
  synovial, 63, 336–338, 354
  (*See also* names of individual membranes)
Membranous labyrinth of ear, 195, 196
Memory, 151, 270, 272, 278–279
Memory cells in immune system, 454, 460
Menarche, 634–635, 638
Meninges, 242
Meningitis, 242
  chloride ion level as an index of, 245
Meniscal cartilage, 337
Meniscus, 336
Menopause, 638
Menstrual cycle, 634–638
  birth control pills and, 648–649
  body temperature during, 594, 636, 637, 647
  follicular phase of, 635–638
  flow of blood during, 637–638
  hormonal influences, 126, 634–638
  loss of iron in, 584
Menstrual phase of reproductive cycle, 635, 637
  salt retention and, 524
Mental activity
  brain waves and, 247
  energy consumption and, 237
Mental deterioration, 273, 279
Mental disorders and enkephalins, 250
Mental retardation, 149–151, 178, 225
  amino acid deficiencies and, 150–151
  carbohydrate deficiencies and, 149
  lipid storage diseases and, 150–151
  vitamin deficiencies and, 151
  (*See also* Vitamins)
Meperidine, 253
Meprobromate, 252, 253
Merkel's touch corpuscles, 184
Merocrine glands, 51
Meromyosin, 289–290
Mesencephalon, 239, 241
Mesenchymal bone formation, 341–342
Mesenchyme, 55, 338, 341–343, 345
Mesentery, 49, 50
Mesoderm, 55, 237, 238, 341, 507, 641, 642
Mesothelium, 46, 48, 49, 546
Messenger RNA (mRNA) (*see* Ribonucleic acid, messenger RNA)
Messengers (*see* Secondary messengers)
Mestranol, 648
Metabolic acidosis, 534
Metabolic alkalosis, 534
Metabolic obesity, 595
Metabolic paths, 321, 591
  (*See also* Metabolism)
Metabolic pool, 590
Metabolic rate, 593–595
  basal, 593–595
  body weight and, 593, 615
  exercise and, 595, 613–615
    (*See also* exercise)
  sex and, 593, 594, 615
Metabolism, 318–325, 590–595
  aerobic reactions and, 297, 298, 318, 321

Metabolism (*Cont.*):
aerobic oxidative phosphorylation (*see* oxidative phosphorylation below)
anabolism, 590
anaerobic glycolysis, 297, 298, 318–323, 592
of carbohydrates, 320–322, 591–592
catabolism, 590
citric acid cycle, 320–322, 590–592
Cori cycle, 319
efficiency of, 324–325, 592
energy production and, 122, 286, 292, 318–325, 366, 553–554, 578, 592–593
exercise and, 325–330, 485
fat metabolism and, 320–322, 591–593
heat production, 320–325, 554, 592–595
high energy phosphate compounds, 322–323, 590
and adenosine triphosphate, 318–323, 590–591
(*See also* Adenosine triphosphate)
and creatine phosphate, 318–323, 590
(*See also* Creatine phosphate)
hydrogen transport and, 590–592
lactic acid formation, 318–319, 321, 591
of lipids, 321–322, 393, 395, 591–592
of lipoproteins, 395
(*See also* Lipoprotein lipases)
nutrition and, 592–595, 599–618
oxidative phosphorylation and, 28, 320–325, 591, 592
pathways of, 318–321, 591
of proteins, 321–322, 591–592
rate of, 593–595
storage of energy, 554
Metacarpal tunnel syndrome, 337
Metachromatic granules, 63
Metamyelocyte(s), 435
Metaphysis, 345
Metarteriole, 403, 408
Metastasis, 463
Metencephalon, 239, 241
Met-enkephalins, 249, 250
Meter (m), 13, 16
Methane, 14
Methemoglobin, 437
Methyldopa, 424
Methylphenylethanolamine transferase, 176
Metric system, 13–26
Micelles, 577, 578, 582
Micro-, 15
Microbial toxins, 285, 587–589
shock and, 444–445
Microcircuits, 170, 171
Microcytic anemia, 446–448
Microfilaments, 29, 34, 86, 113, 138, 139, 461, 622
Microglia, 66, 146
Microgram (μg), 16
Microliter (μL), 16
Micrometer (μm), 16
Microorganisms, 61, 587–589
(*See also* names of individual organisms; Infections; Viruses)
Microphages, 62, 457
(*See also* Leukocytes, neutrophils; Phagocytes)
Micropuncture techniques, 521
Microtubules, 29, 33, 34, 86, 112, 139, 461, 622
Microvilli, 513, 515, 547–549, 572
absorption and, 574, 578
Micturition, 512
Midbrain, 239, 241, 266, 274
corpora quadrigemina and, 261
opiate receptors and, 251
sound and, 265
vision and, 259, 265
Middle cerebral artery, 407, 426
Middle ear, 195, 196, 198
Migraine headaches, 649
Milk, 573, 632
secretion of, 27, 133, 638–639
Milk of magnesia, 91, 586
Milk sugar, 639
Milli- (m), 15
Milliequivalent (mEq), 16, 18
Milligram (mg), 16
Milliliter (mL), 16
Millimeter (mm), 16
Miltown, 252, 253
Mineralocorticoids, 52, 350, 524
Mineral, 335, 338, 340–341
absorption of ions and trace elements, 583–585
adsorption to bone, 339, 340
functions and metabolism, 610–611
nutritional aspects of, 611–613
Minipill, 647, 649
Minoxidil, 424
Minute volume
of heart, 366
of respiratory tract, 492, 493
Miscarriage, 645, 651
Mites, 646
Mitogen, 426, 462
Mitochondria, 29, 31, 32, 515, 590
in cardiac muscle, 299, 300
in neuromuscular junction, 292, 293
in neurons, 137, 168, 170, 173
in skeletal muscle, 286, 292, 293, 326, 328, 329
Mitosis, 622, 623, 639–640
Mitral valve, 363, 398
Mixed nerves, 147
M line, 287–289
Modified jaw thrust, 498–499
Molar (M), 16
Molar solution, 16
Mole, 18, 89, 91
Molecule, 16
Molecular layer
of cerebellum, 261, 262
of cerebral cortex, 271
Molecular weight, 18
diffusion rate and, 88
Molecule, 15–16
Molybdenum, 583, 610, 612
Monoamine oxidase, (MAO), 175, 177, 228, 230, 253
Monoblast, 436
Monoclonal theory of atherosclerosis, 425
Monocyte, 60, 62, 146, 434, 455, 456, 458–459
(*See also* Macrophages: Mononuclear phagocytic system)
Monocytic leukemia, 468
Monoglycerides, 573, 574, 577, 578
Monomer, 53, 73
Mononuclear cells, 341
(*See also* Leukocytes; Lymphocytes; Monocytes; Mononuclear phagocytic system)
Mononuclear phagocytic system (MPS), 65, 66
Monosaccharides, 20
Monosodium glutamate, 177
Monosynaptic reflex arc, 212, 213
Monozygotic twins, 640
Morning-after pill, 651
Morphine, 249, 251, 252, 253, 535
Morula, 640, 641
Mossy fibers, 262, 264
Motility of digestive tract, 561–564
Motion
sense of, 194, 197
Motor cortex of brain, 273–276
Motor end-plate, 285, 291, 314
Motor nerve fibers, 147, 211
impulses of, 274
(*See also* Lower motor neurons)
Motor neurons, 141, 208
lower, 220, 223
upper, 220
Motor unit, 291, 311, 312, 314, 317
Mountain sickness, 493–494
Mouse test, 641
Mouth, 46, 201, 543–545, 550
Mouth-to-mouth resuscitation, 497–500
Mouth-to-nose resuscitation, 498–499
Movement, decomposition of, 264
M protein of skeletal muscle, 287–289
mRNA (*see* Ribonucleic acid, messenger RNA)
MSH (*see* Melanocyte stimulating hormone)
MSG (*see* Monosodium glutamate)
Mucopeptide(s), 549
Mucosa, 47, 50, 546, 549, 564, 631
absorption and, 572
exocrine secretions of, 548, 557
growth of, 548, 557, 565
iron absorption by, 583, 584
(*See also* Digestive system, anatomy; Respiratory system, anatomy)
Mucosal block, 583
Mucous membrane (*see* Mucosa; Respiratory system, anatomy)
Mucous neck cells, 550, 556, 557
changes during female reproductive cycle, 636–638
Multicellular glands, 51
Multiple myeloma, 435
Multiple sclerosis, 147, 149, 245, 460
Multipolar neurons, 141
Multi-unit smooth muscle, 301, 303
Murmurs, 363
Muscle, 30, 55, 74, 282–286, 307–333
cardiac, 73, 74, 122, 283, 299–300
actomyosin content, 285, 299
anatomy, 283–285
blood supply, 361–365
cell membranes of, 284, 299
contraction phases of, 367
electrical activity of, 167, 300
energy production by, 300, 366
exercise and, 327–329, 363–366
fat metabolism and, 322
general properties, 285
hormones and, 132
(*see also* Epinephrine; Norepinephrine)
hypertrophy of, 300, 383
intercalated discs of, 284, 299
intercellular junctions of, 299
metabolic properties, 300, 366
molecular architecture, 299–300
multiplication of components, 300
myotomes and, 211
nerve fibers and, 147
neurotransmitters and, 300
nexuses of, 299

Muscle, cardiac (*Cont.*):
regeneration of components, 300
relative refractory period of, 312
ultrastructure of, 299
(*See also* Heart)
skeletal, 283–298, 307–333
(*See also* Skeletal muscle)
smooth, 300–304
actomyosin content, 285, 301
anatomy of, 283–285, 300
atherosclerosis and, 301, 425–426
autonomic nervous system and, 302–304
in bronchioles, 475
cell junctions of, 300–302
cell membranes of, 74, 300
contraction mechanism, 301–304
multi-unit, 301–304
single-unit, 301–304, 561
electrical activity of, 167, 301–304, 366
energy production and, 301
general properties, 285
in hair follicle, 507, 509
hormones and, 127, 130–132
intercellular junctions of, 300, 302
metabolic properties of, 301
myofibrils and, 301
nerve fibers to, 147, 302–304
neurotransmitters, and, 127, 130–132, 301
nexuses and, 300, 302
pacemaker activity of, 301–302
prostaglandins and, 132–133, 622, 638, 639, 651, 653
receptors of, 300, 303–304
regeneration of, 301
in reproductive tract, 301, 622, 625, 631, 634, 639, 651, 653
(*See also* names of individual muscles)
Muscle cramps, 510, 594
Muscle spindle, 148, 183–185, 260, 263, 512
Muscle tone, 561
Muscle wasting, 322
Muscular arteries, 403
Muscular dystrophy, 314, 315, 393
Muscular hypertrophy, 315
Muscular system, 6
Muscularis externa, 546, 547
Muscularis mucosa, 546–548
Musical creativity and brain, 276, 278
Mustard plaster, 222
Mutation, 340, 425
Myasthenia gravis, 297
Mycobacterium tuberculosis, 245, 484
Myelencephalon, 239, 241
Myelin, 137, 138, 140, 142–145
composition of, 144, 149
degeneration of, 147, 149–150
regeneration of, 145, 148
(*See also* Myelinated nerve fibers; Nerves)
Myelinated nerve fibers, 139, 140, 143, 144, 146, 147, 149, 209, 221
autonomic nervous system and, 227
brain and, 270
transmission of first pain and, 221
velocity of nerve impulses and, 148, 166–167
Myeloblast, 434, 468
Myelocyte, 435
Myelogenous leukemia, 468
Myeloperoxidase, 457, 459
Myenteric nerve plexus, 546, 547
Myenteric reflex, 564
Myoblast, 285
Myocardial infarct, 389
birth control pill and susceptibility to, 649
serum enzyme indicators of, 392–393
(*See also* Heart, disease; Infarct)
Myocarditis, 361
Myocardium, 361, 398
Myofibrillar protein, 314
(See also Muscle, cardiac; Myofibril; Myofilaments; Skeletal muscle, myofibrils, myofiflaments)
Myofibril, 285, 293, 318
exercise and, 326, 328
hypertrophy of, 314, 315
(*See also* Hypertropy; Skeletal muscle, myofibrils, myofilaments)
Myofilaments, 285–288, 311, 314
(*See also* Skeletal muscle, myofibrils, myofilaments)
Myoglobin
exercise and, 326, 328, 329
function of, 297, 437–438
iron content of, 584
in skeletal muscle, 297
Myometrium, 630, 631, 642
Myopia, 193
Myosin, 34, 287, 288, 295, 296, 323, 324
Myosin-ATPase, 288, 291, 295, 310, 311, 314, 323
Myositis ossificans, 353
Myotome, 211
Myxedema, 113, 128

NAD (*see* Nicotinamide adenine dinucleotide)
Na^+-K^+ ATPase (*see* Sodium-potassium ATPase)
Naked nerve ending(s), 183–185, 507
Nanogram (ng), 16
Nanometer (nm), 16
Narcosis, nitrogen, 494
Narcotic, 249, 251, 535
Nasal cavity, 46
Nasal chamber, 474
sense of smell and, 201
Nasopharynx, 473, 474, 545
Nausea, 564
Nearsightedness, 193
Negative feedback signals (*see* Feedback mechanisms)
Negative nitrogen blance, 592
Neisseria gonorrhea, 646
Neoplastic disease, 468
Neosynephrine, 104, 231
Nephron, 513–524
anatomy, 513–516
physiology, 516–524
(*See also* Kidney)
Nephrotic syndrome, 533
Nernst equation, 159
Nerve, 141–151, 162
anatomy, 141–142
autonomic (*see* Autonomic nervous system)
cardiac, 368, 369
cranial (*see* Cranial nerves)
degeneration of, 144–146, 151, 222–225
(*See also* Spinal cord, damage to)
endings, 168–171
in neuromuscular junction, 291–295
fibers, 139
adrenergic, 176, 225–233
(*See also* Adrenergic nerves)
afferent, 141, 147, 182, 208, 209, 211
of autonomic nervous system (*see* Autonomic nervous system)
Nerve, fibers (*Cont.*):
A, B, and C, 147, 148, 221, 225
cholinergic, 176, 225–233, 510
(*See also* Cholinergic nerves)
collaterals of, 135, 137, 139, 141, 213, 214, 223, 262, 263
efferent, 147, 211
sensory, 148, 182, 211
growth factor and, 139, 222
of heart, 368–370
impulses and, (*see* Impulses; Potentials)
mixed, 147
(*See also* Cranial nerves; Spinal nerves, below)
motor, 147, 211
muscle cell properties and, 298, 326
myelin content of (*see* Myelinated fibers; Unmyelinated fibers)
nutrients and, 148–151
parasympathetic (*see* Autonomic nervous system, subdivisions)
peripheral (*see* spinal nerves below)
regeneration of, 139, 145, 158
sensory, 147, 182
somatic, 147, 211–212
spinal, 135, 136, 140, 147, 206–211
sympathetic (*see* Autonomic nervous system, subdivisions)
tissue, 5, 134–151
visceral, 212
(*See also* Autonomic nervous system)
Nervous system, 7, 135, 136, 206, 207
central, 135, 136
development of, 238–241, 642
nutrition of, 148–151
peripheral, 135, 136, 206, 207
(*See also* Autonomic nervous system; Brain; Nerves, Neurons; Spinal cord)
Neural crest, 238, 239
Neural plate, 237, 238
Neural stimulation of digestive tract, 560–563
Neural tube, 237, 238, 241
Neurilemma, 142–145, 272
Neuritis, 222
Neuroencephalitis, 245
Neuroendocrinology, 107
(*See also* Hypothalamus; Pituitary)
Neuroepithelial cell, 201
Neurofibrils, 137, 138
Neurofilaments, 138
Neurogenic diarrhea, 586
Neuroglia, 76
(*See also* Glia)
Neurohormones, 104, 107
(*See also* Antidiuretic hormone, Hypothalamus; Oxytocin)
Neurohypophysis (*see* Pituitary, posterior)
Neuroma, 222
Neuromuscular junction, 291–298
Neuron, 7, 75, 105, 134–141, 239
afferent, 141, 208, 209
associative, 208
axons of (*see* Axons)
cell body of, 135–136, 209
in cerebellum, 261–262
in cerebral cortex, 270–271
collaterals of, 135, 137, 141
dendrites of, 135, 137, 140
efferent, 141, 208
interneurons, 208, 209, 215
membranes of (*see* Impulses; Myelin; Potentials; Synapses)

Neuron (*Cont.*):
motor, 141, 208, 209, 211, 220, 223
neurotransmitters and (*see* Neurotransmitters)
nuclei of,
autonomic nervous system and, 266
cerebellum and, 261–264
cranial nerves and, 261, 265
defined, 267
septal, 272, 273
subthalamic, 272
sensory, 209
shapes, 140–141
synapses of, (*see* Synapses)
types of, 147–148, 187, 202
(*See also* Brain; Spinal cord; individual components of nervous system)
Neurosecretory cell, 104
Neurosyphylis, 245
Neurotoxin, 588
Neurotransmitters, 104, 169, 249–253
activity of, 170
deficiency diseases and, 148–151
degradation of, 113, 175, 214, 252
drugs and, 249, 253
examples of:
acetylcholine, 227–229, 230–232, 253, 300
catecholamines, 176–177, 228–230, 253, 380, 381
dihydroxyphenylalanine (L-dopa), 131, 150, 176, 230, 248, 273, 508
dopamine, 104, 131, 150, 176–177, 230, 250, 253, 273
dynorphin, 250
endorphins, 127, 249, 250
enkephalins, 127, 249–250, 253
epinephrine, 176, 228, 229, 232
gamma-amino butyric acid (GABA), 104, 177–178, 221
glutamic acid, 177
norepinephrine, 176, 177, 227–228, 230–232, 300
P substance, 249
serotonin, 177, 252, 253
(*See also* individual names of neurotransmitters)
excitatory, 171–175
fate of, 113, 175, 176, 214, 230, 252
functions of, 171–178
inhibitory, 171–175
nutrition and, 150–151
pain and, 249, 250
(*See also* Endorphins; Pain)
receptors for, 114, 175, 380, 381
(*See also* Receptors, cellular)
reuptake, 252
secondary messengers, and, 176
secretion of, 98, 168, 170, 173, 174, 252
synthesis of, 149, 252
transport of, 17, 252
Neurotubules, 138, 139
Neutral fats, 576, 577
Neutron, 15
Neutrophilic, 454–455
Neutrophils, 434–435, 455–457
(*See also* Defense cells; Immune response; Microphages; Phagocytes)
Nexuses, 42, 43, 73, 299
(*See also* Intercellular junctions)
NFG (*see* Nerve growth factor)
Niacin (*see* Vitamins)
Nickel, 583, 611, 613
Nicotinamide adenine dinucleotide (NAD), 590, 591
Nicotine, 253
atherosclerosis and, 425
cancer and, 496–497
heart disease and, 380
respiratory system and, 497
(*See also* Smoking)
Niemann-Pick disease, 150
Night vision, 189
Nissl granules, 137, 139
Nitrilosides, 449
Nitrogen:
balance, 592
diving and, 494
gas bubble effects, 494
narcosis, 494
partial pressure of (P_{N_2}), 488
Nitrogen oxides, 497
Nitroglycerin, 380, 389, 574
Nobel laureates in physiology and medicine, 13, 14
Nociceptors, 148, 184, 185
Nocturnal emission, 625
Node of Ranvier, 137, 143, 144, 166
Noninvasive diagnostic techniques, 390
Nonmediated transport, 87–91
Non-polar covalent bond, 17
Non-rapid eye movement sleep (NREM sleep), 248, 268
Nonspecific urethritis, 646
Norepinephrine, 104, 176, 269
blood vessels and, 131, 227–228, 231, 303, 304, 422, 423
fate of, 253
heart and, 368, 379, 380, 422, 423
hormonal properties, 52, 122, 131–132
kidneys and, 521
metabolism and, 175, 228, 230
secretion of, 98, 111
smooth muscle and, 123, 131, 423
synergism with cortisol, 116
Norethindrone, 648
Norethynodrel, 648
Normal breathing, 483
Normal laboratory values, 659–671
Normoblast, 435, 436
Normochromic anemia, 446–447
Nose (*see* Nasal cavity; Nasal chamber; Olfactory receptors; Respiratory system)
NREM sleep (*see* Non-rapid eye movement sleep)
Nucleases, 559, 572, 574
Nuclei (*see* Neurons)
Nucleic acid, 26, 27
absorption of, 577
Nucleolus, 30, 31
Nucleosidases, 572, 574
Nucleosides, 27, 573, 574
Nucleotidases, 572
Nucleotide, 26–28, 34, 36, 573
Nucleus
of atom, 15, 87
of cell, 29–31, 136
membrane of, 30, 31
receptors in, 115, 120, 121
Nucleus pulposus, 69, 224–225
Nursing effects on reproductive capacity, 639
Nutrient
criteria for, 572
Nutrition
hormones and, 113
metabolism and, 592–595, 599–618
Nutrition (*Cont.*):
nervous system and, 148–151
summary tabulation of data on, 599–618
Nystagmus, 264

Obesity, 80, 595
cardiovascular disease and, 390
hormones and, 131
hypothalamus and, 267
metabolism and, 595
Obligatory reabsorption of water, 523
Obligatory water loss, 527
O blood type, 440–442
duodenal ulcers and, 565
Occipital fissure, 274
Occipital lobe, 270, 273, 274, 277
Odontoblast, 73, 355
Odor
detection of, 184, 202
(*See also* Sense organs, Nose, smell and)
Oil, 25
puberty and, 509
sex hormones and, 509
Oil producing glands (*see* Sebaceous glands)
Olfaction, 184, 202
(*See also* Sense organs, nose)
Olfactory bulb, 202, 260, 271–273
(*See also* Sense organs, nose)
Olfactory lobe, 237, 259
Olfactory mucosa, 259
Olfactory nerve fibers, 141
(*See also* Sense organs, nose)
Olfactory receptors, 202, 474, 476
Oligocyte, 145
Oligosaccharides, 573
Oliguria, 533
Oncotic pressure, 412
On-off receptors, 185–186
Oocytes, 634, 635
Oogenesis, 624, 632–638
Operant conditioning, 278
Opiate receptors, 251–252
(*See also* Brain, neurotransmitters)
Opiate-like molecules, 251
Opiates, 222
Opioid activity, 249
Opium, 249
Opsin, 187, 189
Opthalmoscope, 187
Optic disc, 189
Optic nerves (*see* Cranial nerves, optic nerve)
Optic tracts, 274
Oral contraceptives, 647–649
Organ, 5
Organelles, 30, 137
Organic compounds, 19
Organic phosphates, 574
(*See also* Adenosine diphosphate; Adenosine triphosphate; Creatine phosphate)
Organ of Corti, 195, 197–199
(*See also* Sense organs, ear; Sensory receptors, auditory receptors)
Orgasm, 625–626, 631, 634
Orgasmic phase of orgasm, 626, 634
Origin of skeletal muscle, 285
Oropharynx, 473, 474, 545
Orthopnea, 483
Oscilloscope, 163, 371, 397
Osmolarity, 89
maintenance of in body fluids, 512, 522, 528
of urinary filtrate, 522, 526, 528
(*See also* Kidneys)

Osmoreceptors, 182, 184, 267, 562, 563
Osmosis, 87, 89–91, 585, 587
Osmotic diuresis, 523
Osmotic effects
  on blood cells, 486
  in gastrointestinal tract, 91, 586
Osmotic fragility
  of blood cells, 432
Osmotic pressure, 89, 512, 517
  lymph formation and, 412, 413
Osseous tissue (*see* Bone)
Ossicles of ear, 195, 196, 198
  muscles of, 196
Osteitis fibrosa cystica, 128, 353
Osteoarthritis, 73, 354
Osteoblast, 72, 336, 341, 342, 344, 348, 350, 351, 353
Osteoclast, 66, 341, 342, 345, 347, 348, 350, 353, 474
  activating factor of, 355
Osteocyte, 69–72, 341, 342
Osteogenesis, 341
Osteomalacia, 352, 353
Osteomyelitis, 354
Osteoporosis, 352
Osteosclerosis, 353
Otolith, 197, 200–201
Oval window of ear, 194, 195, 198
Ovarian pregnancy, 641
Ovariectomy, 650
Ovary, 6, 7, 46, 52, 104, 105, 126
  anatomy of, 132, 629–631
  contraception and, 639–640
  cycles of, 634–638
  enzymes and, 127, 636
  hormones and, 103, 132–133, 634–638
  ovulation and, 633–636
  production of eggs in, 632–634
Oviduct, 34, 622, 629–631, 634, 639, 640
  prostaglandins and, 622
  tubal ligation of, 650
Ovulation, 633–636
  body temperature and, 594, 636, 637
  events surrounding, 632–638
  nursing and, 639
  timing of, 645, 647
Ovum (*see* Egg)
Oxalacetic acid, 23
Oxidation, 17
Oxidative enzymes, 437, 584, 591–592
  (*See also* Metabolism, aerobic; Metabolism, oxidative phosphorylation and; Mitochondria; Oxidative phosphorylation)
Oxidative metabolism, 590, 591
  (*See also* Metabolism, aerobic; Metabolism, oxidative phosphorylation and; Mitochondria; Oxidative phosphorylation)
Oxidative phosphorylation, 28, 320–325, 591, 592
Oxygen
  in air, 479
  at altitude, 480
  consumption, 329, 492–493
    energy production and, 594
    metabolic rate and, 593–594
  debt, 319–320, 327, 366, 492–493
  diving and, 494–495
  general functions of, 473
  hemoglobin dissociation curve and, 485, 487
  hypoxia and, 366, 437, 443–444, 478, 485, 493, 496
  partial pressure (P_{O_2}) of, 488
Oxygen (*Cont.*):
  respiratory signals and, 490–492
  toxicity of, 494–495
  transport of, 484–485
    myoglobin and, 297
  utilization of, 366
  oxygenation and,
    hyperbaric, 494–496
Oxyhemoglobin (HbO_2), 436, 484–487
  (*See also* Hemoglobin)
Oxylate
  inhibition of clotting by, 465, 466
Oxyntic cells, 550, 557
Oxytocin, 10, 52, 103, 107, 108, 122, 127, 255–269, 639, 653
Ozone, 494

Pacemaker:
  cells:
    in heart, 300, 367, 368
    in intestine, 564
    of smooth muscle, 301, 564
  of heart, 368, 375
Pacinian corpuscle, 182–185, 507
Pain, 148, 220–222, 354
  absence in brain, 247
  acupuncture and, 222, 251
  angina pectoris and, 221, 370, 389
  cordotomy and, 250
  fast and slow, 147–148, 221
  fibers, 147–148, 221
  first, 147–148, 221
  of heart, 221, 370, 389
  inhibition of, 221–222, 249–251
  labor and, 653
  of muscle, 319, 326, 510, 594
  neurotransmitters and, 249–251
  placebo and, 250
  P substance and, 249
  referred, 221, 370
  second, 221
  substantia gelatinosa and, 221, 251
  temperature and, 223
    (*See also* Spinal cord, spinothalamic tracts)
  types of, 221
Palatine tonsils, 545
Palpatory method of measuring blood pressure, 417
Pancreas, 6, 8, 52, 544, 545
  acini of, 550, 559
  anatomy of, 544, 550–551, 553
  bicarbonate secretion and, 129, 556–557, 559–562
  ducts of, 551
  enzyme secretion by, 129, 551–557, 559–562, 572–573
  hormonal functions of, 98, 104, 105, 111, 115, 129–130, 550
  islets of Langerhans of, 550
Pancreatic amylase, 551
Pancreatic duct, 545, 551
Pancreatic lipase, 551
Pancreatic polypeptide (PP), 130
Pancreatitis, 566
Pancreozymin (*see* Cholecystokinin-Pancreaozymin, CCK-PZ)
Paneth cells, 548–550
Papanicolaou smear, 632
Pap test, 632
Papilla
  of kidney, 512, 514
Papilla (*Cont.*):
  of skin, 507, 511
  of tongue, 201
Papillary muscles, 361, 362, 371
Paradoxical sleep, 248
Parahormones, 103, 578
Parallel circuits, 214, 215
Parallel elastic elements, 308–309
Parallel fibers of cerebellum, 262, 263
Paralysis, 224, 225, 295, 297, 314, 426
Parasites:
  defenses against, 414, 459, 467
  (*See also* Cellular immunity; Defense cells; Humoral immunity; Immune response; Macrophages; Phagocytes)
Parasympathetic division of autonomic nervous system, 225, 226, 229, 368–370, 419–421, 625
  (*See also* Autonomic nervous system)
Parathion, 297
Parathormone, 52, 122, 128
  calcium metabolism and, 8, 342, 347
Parathyroids, 6, 8, 52, 104, 105
Parathyroid hormone, 52, 122, 128
Paravertebral ganglia, 227
Parenchymal cells, 552, 553, 566
  hemoglobin metabolism and, 439
Parietal cells, 550–552, 556–558, 566
Parietal epithelium, 518, 519
Parietal lobe, 270, 273, 275
Parietal membrane(s) (*see* Membranes, pericardial, peritoneal, pleural)
Parietal pericardium, 360, 361
Parietal peritoneum, 631
Parietal pleura, 474, 475
Parietal-occipital area of brain, 247
Parkinson's disease, 177, 272, 273
Parotid gland, 554
Partial pressure
  defined, 479
  of gases, 480
    at altitude, 479, 480
    at sea level, 488
Particles, subatomic, 15
Pars nervosa (*see* Pituitary, posterior)
Parturition, 652–653
Passive diffusion, 87, 525
Passive loading, 312
Passive transport, 87, 525
Patent, 484
Patent ductus arteriosus, 386
Pathology, 3
Pathophysiology
  of leukocytes, 467–468
  of red blood cells, 443–449
  of thrombocytes, 469
  (*See also* names of specific cells, diseases, organs, or systems)
Pathways (*see* Associative fibers of neurons of brain; Circuits; Spinal cord, tracts of)
Pavlov, 278
PBI (*see* Protein-bound iodine)
Pedicels, 517, 520
Pediculosis, 646
Pellagra, 151, 604
Pelvic cavity, 631
Pelvis, 337, 338, 621, 630, 631
Penicillin, 145, 361, 646, 653
  allergy to, 63
Penis, 620–623, 625, 632
Pepsin, 48, 557, 572, 573
Pepsinogen, 550, 557
Peptic ulcers, 565–566

Peptidases, 525, 559, 572, 576
Peptide bonds, 22, 573
Peptide hormones, 107, 126–133
 (*See also* individual names of hormones)
Peptides, 574
 absorption of, 577
Perfringens food poisoning, 588
Perfusion, 326, 488
 of heart, 326
 of lungs, 488
Pericapillary space, 409, 410
Pericardial cavity, 49, 360
Pericardial membrane, 49, 360, 361
Pericarditis, 360
Perichondrium, 67
Pericytes, 408
Perilymph, 194, 196, 198, 199
Perimysium, 284
Perineurium, 142
Periodontal disease, 355, 565
Periodontal membrane, 355
Periosteal band, 345
Periosteal collar, 345
Periosteium, 67, 242, 335, 336, 343
Peripheral edema, 389, 567
Peripheral nervous system, 135, 136
 (*See also* Autonomic nervous system; Nerves)
Peripheral resistance, 382, 383, 410–421
Peristalsis, 301, 561, 564
Peritoneal dialysis, 537
Peritoneal membrane, 49, 51, 537, 546
Peritoneum (*see* Peritoneal membrane)
Peritonitis, 565, 566
Peritubular capillaries, 516–518, 521, 523, 525
Permeability
 of blood-brain barrier, 145, 146, 177, 269
 of capillaries, 403, 411
  (*See also* Capillaries)
 hormones and, 120
  (*See also* Antidiuretic hormone)
 of intercellular junctions (*see* Intercellular junctions)
 of membranes, 87–89, 160, 164
Permissiveness of hormones, 116, 133
Pernicious anemia, 447, 467, 557, 587
Perpendicular crossbridge position, 290
Perspiration, 510–511
Petechiae, 469
Petit mal epilepsy, 249
Peyer's patches, 546
 (*See also* Lymphoid tissue)
PGA, PGE, PGF (*see* Prostaglandins)
pH, 16, 18, 19
 of blood, 19, 524
 of cerebrospinal fluid, 244–245
 of cervical secretions, 636
 effect on enzymes, 319
 effect on hemoglobin dissociation, 485, 487
 effect on respiration, 484–496, 491, 492
Phagocyte, 59, 62, 65, 145, 146, 341, 342, 457, 623
 (*See also* Defense cells; Macrophages; Microglia; Mononuclear phagocyte system; Neutrophils; Osteoclasts)
Phagocytosis, 60–62, 95–98, 448, 459, 497, 623
 (*See also* Endocytosis)
Phagolysosomes, 459
Phagosomes, 459
Pharmacological levels of hormones, 116–117
Pharmacology, 3
Pharynx, 201, 474, 490, 544, 545
Phasic receptors, 185–186
Phenoxybenzamine, 231
Phenylalanine, 150
Phenylethanolamine N-methyltransferase, 228
Phenylephrine, 231
Phenylketonuria, 150, 645
Phenylpyruvic acid, 150
Pheochromocytoma, 113, 131
Pheromone, 104
Phlebitis, 422
Phonoreceptors, 184, 195, 198–199
Phosphatase, 342, 351, 622, 636
Phosphate buffers, 530–531
Phosphatidylcholine, 87, 477–478
Phosphatidylethanolamine, 87
Phosphodiesterase, 121, 123, 232
Phosphoglycerides, 477–478, 591
 in chylomicrons, 581
Phospholipids, 25, 574
 in alveolar membrane, 476–478
 in cell membrane, 30, 84, 87, 97
Phosphoric acid, 574
Phosphorous
 absorption as inorganic phosphate, 345
 excretion of, 340, 348, 349
 functions, 340
 ratio to calcium, 340
 recommended daily dietary requirements, 618
 storage of, 335
Photoelectric transducers, 189
Photopic vision, 189
Photoreceptors, 184, 186
 (*See also* Sense organs, eye)
Phototherapy of hemolytic disease of newborn, 446
Phrenic nerve, 490
Physical exam, 391
Physical training
 cardiovascular system and, 382–385
 food and, 595
 respiratory system and, 392–393
 skeletal muscle and, 325–330
 (*See also* Exercise)
Physiological levels of hormones, 116
Physiology, 2
Physiology journals, 4
Phytic acid, 583
Pia mater, 242, 271
Pico- (p), 15
Picoliter (pL), 16
Pigmentation, 126–128, 131, 508–509
 accumulation in lens, 194
Pill, birth control and, 647–649
Pineal body, 6, 7, 52, 105, 128, 150, 177, 260, 266, 269–270
Pineal gland (*see* Pineal body)
Pinna, 195
Pinocytic vesicle, 96, 98
Pinocytosis, 96, 98, 576, 577
 by nerve terminals, 140
 (*See also* Endocytosis)
Pituitary, 6, 7, 52, 104, 105, 108, 130, 260
 anterior, 52, 107, 108, 634, 635
 endorphins and, 249
 enkephalins and, 249, 250
 hormones of, 126–128, 133, 269
 hypothalamus and, 107–110
 posterior, 52, 107, 108
PKU (*see* Phenylketonuria)
Placebo, 250
Placenta, 52, 104, 384, 386, 628, 638, 641, 642, 644
Placenta (*Cont.*):
 hemolytic disease of newborn and, 444–446
 hormones of, 132–133, 638
 membranes of, 651–652
 oxygen exchange and, 439
 Rh incompatibility and, 445
 as selective barrier, 641
Placental lactogen, 133, 638
Plaque
 atherosclerotic, 389, 425–426
 dental, 355
Plasma, 391, 431, 432, 454
 chemistry, 432
  (*See also* Normal reference laboratory values, 659–671)
 composition, 244, 245, 431, 432
  compared to bile, 559
  compared to intestinal juice, 559
 as extracellular fluid, 80, 81
 normal values (*see* 659–671)
 volume of, 431
Plasma cells, 56, 59, 60, 65, 460
Plasma membrane (*see* Cell membrane)
Plasma protein, formation of, 411, 412
 (*See also* Plasma chemistry; Liver)
Plasma thromboplastin antecedent (PTA), 464, 465
Plasma thromboplastin component (PTC), 464–466
Plasmin, 464
Plasminogen, 464
Plateau phase of orgasm, 625, 634
Platelet, 432–433, 456, 461–467
 atherosclerosis and, 425–426
 (*See also* Blood; Thrombocytes)
Platelet lipoprotein factor, 465
Pleural cavity, 49, 474–475
Pleural effusion, 389
Pleural membrane, 49, 474–475
Plexus, 211
Plicae, 546
PMN (*see* Polymorphonuclear leukocyte)
Pneumoencephalography, 245
Pneumonia, 484, 578
Pneumotoaxic center, 260, 490, 491
Pneumothorax, 475
PNS (*see* Peripheral nervous system)
Podocyte, 517, 520
Poikilotherm, 268
Poiseuille's law, 416, 419, 521
Poison ivy, 511
Poison oak, 511
Polar bodies, 624, 633–634
Polar covalent bond, 17
Polarization of membranes, 156, 159–160, 166, 172–175
Polarized light
 skeletal muscle and, 286–288
Polio, 137, 223
Pollutants
 respiratory tract and, 475, 496–497
Polycythemia, 432, 494
Polycythemia vera, 432
Polymer, 20, 21, 53
 of fibrin in clotting, 464–465
Polymorphonuclear leukocyte (PMN), 457
 (*See also* Leukocytes)
Polyneuritis, 151
Polynucleotidases, 574
Polynucleotide, 28
Polyp, 589
Polypeptides, 22, 573
Polyribosomes, 29, 31, 36

Polysaccharide, 20, 23
(*See also* Glycogen)
Polysome, 29, 31, 36
Polysynapses, 171
(*See also* Circuits; Synapses)
Polyuria, 533
Pons, 241, 253, 260, 261, 490, 491
Pores, 83–85, 88–89
cell membrane and, 29, 83, 84, 88, 91, 93, 94, 517, 519
as channels in retinal cells, 189
in endothelial cells, 412
in nerve fibers, 158, 160–161, 164, 165, 169, 171–172, 174
(*See also* Synapses)
of intestinal cells, 574
intercellular, 42, 43
of alveoli, 476
(*See also* Intercellular junctions)
nuclear membrane and, 29, 31
in kidney cells, 517, 519
in skeletal muscle cells, 292
slit-pores, 519
in sweat glands, 506, 507, 510
in taste buds, 201
Porphyrin ring, 437
Portal hypertension, 567
Portal vessels
hepatic, 384, 553, 589
hypophyseal, 107–109
Positive feedback signals, 8, 9, 161, 164, 635
Positive nitrogen balance, 592
Postcentral gyrus, 273, 274
Posterior pituitary, 52, 107, 108
(*See also* Pituitary)
Posterior root, 208–211
Posterior root ganglion, 140, 145, 209, 211
Postganglionic fibers, 226, 227, 229
Postsynaptic membrane, 168, 169, 292–294
Posture, 260, 261, 265, 311
(*See also* Cerebellum; Equilibrium, perception of; Sensory receptors, proprioceptors)
Potassium
in bicarbonate buffer, 433
(*See also* Buffers)
bone and, 338
in extracellular fluids, 81, 82, 155
heart and, 368, 379, 380
hemoglobin and (KHb), 486
hydrated size of, 83, 84, 89
ion permeability, 89
nerve impulses and (*see* Impulses; Potentials)
Potentials, 88, 156, 247
action, 160–167, 186, 286, 300, 302, 310
amplitude of, 172, 175, 185, 247
electrocardiogram and, 370–375
electrochemical equilibrium and, 156–157
end-plate (EPP), 292, 294–296
equilibrium and, 159–162
excitatory postsynaptic potential (EPSP), 171–175, 214, 221, 294–296
generator, 185, 186, 189, 202
inhibitory postsynaptic potential (IPSP), 171, 172, 174, 178
Nernst equation and, 159
pacemaker, 167, 301, 380
receptor, 164, 185, 189
stimulus intensity and, 186
resting, 156–160
in heart, 367, 368
in skeletal muscle, 292
in smooth muscle, 301
spike, 164
Potentials, spike (*Cont.*):
in skeletal muscle, 312
in smooth muscle, 301, 302
transduced, 167
types of, 167–168
(*See also* Impulses; Neurotransmitters; Synapses; Voltage)
Potentiation, 116, 279
Power
defined, 317
muscle and, 316–318
work and, 317–318
p position of crossbridge, 323–324
Pre-beta lipoproteins, 394
Precapillary sphincter, 408
Precentral gyrus, 273
Precocious pseudopuberty, 131
Prednisone, 467
Preganglionic fibers, 226, 227, 229
Pregnancy, 639–645, 652–653
amniocentesis and, 644–645
corpus luteum and, 127, 633–635, 637–638, 644, 653
(*See also* Corpus luteum)
hormonal changes of, 644
human chorionic gonadotropin and, 133, 628, 634, 641
smoking and, 496
tests for, 641–642
toxemia of, 393, 644
varicose veins and, 422
Premature heartbeat, 367, 374
Premenstrual cramps, 638
Prepuce of penis, 622
Presbyopia, 194
Pressoreceptors (*see* Sensory receptors, baroreceptors, mechanoreceptors)
Pressure
of blood, 111, 416–421, 484
(*See also* Blood pressure)
of pulse, 416
of respiratory gases, 488
(*See* Sensory receptors, baroreceptors, mechanoreceptors)
Presynaptic membrane, 168, 169, 292–294
(*See also* Synapses, membranes of)
Prevertebral ganglia, 227
Primary bronchus, 474, 475
Primary effect of hormones, 116, 120
Primary follicle, 632–633
Primary malabsorption, 586
Primary motor cortex, 273–276
Primary ossification center, 343, 344
Primary sensory cortex, 273–276
Primary shock, 443, 444
Primitive follicle, 633
Primordial follicle, 633
P-R segment, 371
Proaccelerin, 464–466
Proconvertin, 464
Product
of enzyme activity, 22, 23
Progestational hormone (*see* Progesterone)
Progestational phase of reproductive cycle, 634–638
Progesterone, 26, 52, 133, 631, 634–638, 644
metabolism and, 627, 628
(*See also* Progestins)
Progestins, 647–649
(*See also* Progesterone)
Progestogens, 648
(*See also* Progesterone; Progestins)
Prohormones, 105
Prolactin (PRL) (*see* Luteotropic hormone)
Proliferative phase of uterine cycle, 636–638
Promyelocyte, 435
Pronormoblast, 434, 435
Propranolol, 115, 231, 424
Propregnancy hormone (*see* Progesterone)
Proprioception, 148, 216, 225
(*See also* Cerebellum; Equilibrium, perception of; Sense organs, ear, vestibular functions; Sensory receptors, mechanoreceptors; Skeletal muscle, coordination)
Proprioceptors, 184, 312
Prosencephalon, 238, 241
Prostacyclin (PGI_2), 462–463
(*See also* Prostaglandins)
Prostaglandins, 52, 103, 105, 120, 122, 132, 133, 462–463, 578, 622, 638, 639, 651, 653
Prostate gland, 621, 622, 625
Prostatic ducts, 621, 622
Proteases, 573
Protein:
absorption, 98, 576, 577, 580
acid secretion by stomach and, 561, 562
amino acid building blocks and, 21–22
calorie malnutrition, 587, 588
caloric value of, 593
in cell membrane, 82–87, 93, 94, 144, 458
(*See also* Cell membrane)
in cerebrospinal fluid, 245
chains of, 83
in actin, 287, 290
in antibody, 459
in collagen, 53, 54
in creatine isozymes, 393
in hemoglobin, 437–439
in myosin, 287, 289, 291
chemistry of, 21–24, 28, 54, 83, 437–439
in chylomicrons, 581
daily dietary requirements for, 576, 599
denaturation by hydrochloric acid, 557
digestion of, 557, 572–574, 576
as enzymes, 22–24
exercise and, 593
(*See also* Exercise; Muscle, cardiac; Skeletal muscle)
filtration in urine, 517, 519, 533
hormone chemistry and, 126–133
hormone transport and, 113, 114
in muscle (*see* Muscle, cardiac, smooth; Skeletal muscle)
memory and, 279
metabolism of, 320–322, 591–594
in bone, 348, 349
synthesis of, 30, 35, 36, 120, 122, 133, 149
(*See also* metabolism *above*)
neurotransmitters, and, 150–151
(*See also* Neurotransmitters)
in plasma, 245, 412
as pumps, 87
(*See also* Active transport; ATPases)
respiratory quotient of, 593
in urine, 517, 519, 533
Protein-bound iodine (PBI), 117, 128
Protein kinases, 121, 122, 172
Protein wasting, 117
Proteinuria, 517, 519, 533
Proteoglycans, 53, 67
in bone, 336–339, 349, 350
heparin and, 465
Proteolytic enzymes, 572–573, 622
(*See also* Digestive enzymes)
Prothrombin, 464–466

Prothrombin time (PT) test, 466–467
Proton, 15
Protoplasm, 30
Protozoan infections, 646
  monocytes and, 467
Proximal convoluted tubules, 513, 514, 516, 522, 531
Pseudohermaphroditism, 131
Pseudopolar neurons, 140, 141
Pseudostratified epithelium, 45–48
P substance, 249, 370
Psychological stress, 109
Psychotropic drugs, 251–253
  (*See also* names of individual drugs)
PT (*see* Prothrombin time test)
PTA (*see* Plasma thomboplastin antecedent)
PTC (*see* Plasma thromboplastin component)
Ptomaine food poisoning, 589
Ptomaines, 589
Ptyalin, 551, 553, 573
Puberty, 127, 128, 509, 628–629, 634
Pubic bones, 337, 621
Pubic hair, 628, 634
Pubic symphysis, 621
Pulmonary artery(ies), 362, 363, 384, 386
Pulmonary receptors, 184, 369, 420–421
Pumps (*see* Active transport)
Pupil, 187, 188
Pupillary reflex, 259, 265
Purines, 26, 574
  gout and, 354
Purkinje cells, 141, 262, 263
Purkinje fibers, 300, 367–368
Purkinje layer of cerebellum, 261–264
Pus, 62, 457
Putamen, 272, 273
Putrescine, 589
P wave, 371, 374
Pyelonephritis, 534
Pyloric glands, 550
Pyloric sphincter, 544, 545
Pylorus, 550, 544, 545
Pyorrhea, 354, 355
Pyramidal cells, 270, 271
Pyramidal tracts, 260
Pyrimidine, 26, 28, 574
Pyrogens, 594
Pyronine, 59
Pyruvic acid, 150, 318–321, 590–592

QRS complex, 371, 374
QT interval, 371

Rabbit test, 641–642
Radial artery, 417
Radiation
  accumulation in bone, 340
  anemia and, 447
  effects on clotting, 466
  effects on platelets, 469
  gastrointestinal malabsorption and, 587
  heat loss by, 594
  leukocyte responses to, 467, 468
  skin and, 508
  sterility and, 645
  x-ray analysis and, 184, 244–246, 336, 369, 397, 398, 632
Radioimmunoassay, 117–119, 246, 641
Radionuclides, 340
Radiopaque dye
  use of, 245, 398
Rapid eye movement sleep (REM), 248, 268
RAS (*see* Reticular activating system)
RBF (*see* Renal blood flow)
RBC (*see* Red blood cells)
RDS (*see* Respiratory distress syndrome)
Reabsorption, 42, 341, 412, 512, 513, 521–523
  (*See also* Antidiuretic hormone; Kidneys; Parathormone; Vitamin D)
Reading, 278
Receptors, cellular, 87, 175, 225–232, 457
  aging and, 115, 351
  agonists to, 253
  alpha, 176, 225, 227–232, 303, 420
  antagonists to, 253
  for antibodies, 59, 63, 459
    immunoglobulin A and, 59
    immunoglobulin E and, 63
  beta, 123, 148, 176, 227–232, 303, 304
    β_1, 228, 424
    β_2, 228, 420, 424
  blood vessels and, 420–421
  of cerebral cortex, 270
  drugs and, 249–253
  gamma, 176, 228, 230–232, 294, 311
    (*See also* Acetylcholine)
  histamine, 85, 229, 566
  hormones and, 114–123, 397
    aging and, 115
    calcium and, 115
  killer T cells and, 460
  lipoproteins and, 394, 395, 397
  opiate, 249–253
  in skeletal muscle, 292, 294, 296
  in smooth muscle, 303, 304
Receptor potential, 167, 185, 189
Recommended daily dietary allowances, 618
Recordings
  bipolar, 373
  of electrocardiogram, 370–375
  of electrocortigram, 247
  of electroencephalogram, 247
  of electromyogram, 310
  in kymograph, 309, 482
  unipolar, 371, 373
Recovery
  of resting membrane potential in nerve, 163–165
  of skeletal muscle, 323
Recruitment, 311
Rectum, 544, 546, 563, 575, 621, 630, 631
Red blood cells, 405, 430–452, 456
  anatomy of, 435–437
  antigens of, 440–443
  carbon dioxide transport and, 484–487
  cell membrane composition, 82, 435
  chemistry of major blood groups and, 440–441
  in clotting, 463
  disorders of, 435, 443–449
    anemias, 446–449
    hemolytic disease of the newborn, 445–446
    sickle cell anemia, 446–448
    thalassemia, 448–449
  erythropoietin and, 132, 435–437, 512
  fragility of, 432
  groups of, 437–443
  hematocrit, and, 432, 433
  hemoglobin and, 436–439, 484–487
    properties of, 437–448, 484–487
    types of, 436–439, 486
  hemorrhage, shock and, 443–445
  kidneys and, 132, 435, 512
  life cycle of, 435, 436
Red blood cells (*Cont.*):
  liver and, 554
  major blood groups and, 440–442
  numbers of, 431, 435
  nutrient requirements for formation of, 98
  origin of, 434, 435
  oxygen transport and, 436–437, 484–487
  pathophysiology and, 435, 443–449
  viscosity of blood and, 419
  (*See also* Blood)
Red bone marrow, 132
Red muscle, 325
Redifferentiate, 53
Red nucleus, 265
Reduced hemoglobin (HHb), 436, 485–487
Reduction, 17
Referred pain, 221, 370
Reflex arc, 212, 546
Reflex, 211–213, 563
  anal, 546
  auditory, 265
  Babinski, 213
  Bainbridge, 213, 370
  blood vessels and, 419–421
  in breathing, 489–490
  chewing and, 545
  conditioned, 213, 278
  contralateral, 212, 214
  cranial, 211
  defined, 211
  in digestive tract, 546, 561, 563
  eye and,
    movements of, 197
    pupillary response and, 211, 233, 259, 265
    visual response and, 265, 269
  heart and, 369, 370, 420
  Hering-Breuer, 489–490
  intersegmental, 212–214
  intrasegmental, 212–214
  ipsilateral, 212
  knee-jerk, 213, 264
  monosynaptic, 212, 213
  myenteric, 546, 564
  polysynaptic, 212
  pupillary, 211, 233, 259, 265
  righting, 265
  skeletal muscle, and, 312
  spinal, 211, 564
  stretch, 213
  suckling effects of, 639
  suprasegmental, 213, 214
  swallowing, 545
  unconditioned, 278
Refraction, 286
  of polarized light by skeletal muscle, 286–289
Refractory period
  in excitation of nerve, 163–165
  in heart, 367
  in skeletal muscle, 312
Regeneration
  of liver, 567
  of myofibrillar protein
    in cardiac muscle, 300, 389
    in skeletal muscle, 314–315
  of nerve fibers, 139, 145, 148
  of smooth muscle, 301
Regulatory obesity, 595
Regurgitation of blood in heart, 363
Regurgitation of food, 564
Reissner's membrane, 194, 197–201
Rejection of organs and tissues, 449
  (*See also* Histocompatibility antigens; Kidney, transplants)

Relative refractory period, 312, 313, 367
Relaxed awakening
  brain waves and, 247
Relaxation period, 310
Relaxin, 133, 653
Releasing factors, 107–109
Releasing hormones, 107–109
Remission
  of leukemia, 468
  of multiple sclerosis, 147
REM sleep (*see* Rapid eye movement sleep)
Renal blood flow (RBF), 532
Renal calyces, 512, 514
Renal circulation, 512, 516, 517, 532
Renal erythropoietin factor, 132, 435
  (*See also* Erythropoiesis; Erythropoietin)
Renal functions, 512–513
  (*See also* Kidneys)
Renal hypertension, 524
Renal nerves, 519–520
Renal papillae, 512
Renal pelvis, 512
Renal plasma flow (RPF), 530, 532
Renal pyramids, 512
Renin-angiotensin system, 111, 524, 526
  in hypertension, 423, 525
  in shock, 444
Rennin, 573
Repolarization of nerve, 163–165
Reproductive system, 6, 47, 619–658
  accessory sex organs:
    of female, 629–632
    of male, 621–622
  amniocentesis and, 644–645
  autonomic nervous system and, 226, 233, 625–626, 635, 639
  birth and, 652–653
  birth control and, 646–653
    abortion, 651
    intrauterine device, 651
    miscellaneous techniques, 650–651
    morning-after pill, 651–652
    pill, 648–649
    rhythm, 646–648
    surgical techniques, 650
    tubal ligation, 650
    vasectomy, 650
  bulbourethral glands, 621, 622, 625
  childbirth, 632, 652–653
  conception, 620, 637, 639–640
    control of, 646–651
  Cowper's glands, 621, 622
  eggs:
    ovulation and, 632–637
    production of, 632–634
    viability of, 636, 637
  embryonic and fetal development, 640–645
    timetable of, 642–644
  of female, 629–645
  fertility and infertility, 645–648
  fetal development, 644–645
  hormones and
    androgens, 627–629
    estrogens, 634–639
    human chorionic gonadotropin (HCG), 133, 628, 641
    (*See also* Endocrines; Hormones; individual names of hormones)
  mammary glands, 51, 127, 132, 133, 632, 638–639
  meiosis and, 623–624, 633
  menstrual cycle and, 634–638
  mitosis and, 622–623, 639–640
  oogenesis, 624, 632–638
Reproductive system (*Cont.*):
  orgasm
    in female, 631, 634
    in male, 625–626
  ovaries, 6, 8, 52, 103, 105, 126, 127, 132–133, 629–638
  penis, 620–623, 625, 632
  placenta, 52, 104, 384, 386, 439, 444–446, 628, 638, 642, 644, 652
  pregnancy, 639–645, 652–653
  prostate, 621, 622, 625
  seminal fluid, 625, 649, 650
  seminal vesicles, 133, 621, 622, 625
  sexually transmitted diseases, 646, 649
  spermatogenesis, 30, 34, 127, 621–625, 639
  sperm ejaculation, 625
  sperm emission, 625
  sperm production, 621, 645–646
  testes, 6, 8, 52, 104, 105, 127, 133, 620–629, 636
  uterus, 34, 132, 385, 629–632, 634, 636, 639, 652–653
  vagina, 575, 629–632, 650, 652
  (*See also* Endocrines; Androgens; Estrogens; Hormones; names of individual components)
RER (*see* Rough endoplasmic reticulum)
RES (*see* Reticuloendothelial system)
Reserpine, 177, 424, 629
Residual air volume, 482
Resistance:
  to air flow, 474, 480–484
    (*See also* Asthma; Emphysema; names of other respiratory diseases)
  to blood flow, 382, 383, 419–421
Resolution phase of orgasm, 626, 634
Resorption of minerals from bone, 341
Respiration, 473
  (*See also* Breathing; Cellular respiration; Lungs; Pulmonary functions; Respiratory system)
Respiratory acidosis, 534–535
Respiratory alkalosis, 534–535
Respiratory allergies (*see* Allergies; Asthma)
Respiratory center, 488–491
  (*See also* Brain, medulla oblongata, centers)
Respiratory distress syndrome (RDS), 477–478
Respiratory gas chemosensors, 488
  (*See also* Sensory receptors, chemoreceptors)
Respiratory membranes, 475–477
Respiratory minute volume, 480, 482–483 492, 493
Respiratory paralysis, 297, 314
Respiratory quotient, 593
Respiratory rate, 483, 493, 626
Respiratory system, 6, 34, 36, 37, 472–504
  acid-base balance and, 487, 529–530
  alveoli, 475–478
  anatomy, 473–477
  artificial respiration, 497–500
  disorders of:
    abnormal breathing, 483–484
    asthma, 475, 484
    atelectasis, 478
    barbiturate effects, 478
    choking, 473
    congestive heart disease and, 388, 483, 484
    emphysema, 484, 488
    hyaline membrane disease, 477–478
Respiratory system, hyaline membrane disease (*Cont.*):
    hypoxia and, 478
    infections and, 484
    pneumothorax, 475
    pollutants and, 496–497
    respiratory distress syndrome, 477–478
    silicosis, 480
    sinusitus, 475
    smoking and, 484, 486, 496–497
    (*See also* Smoking)
  diving effects and, 494–495
  exercise and, 483, 492–493
  environmental effects, 493–497
  gas exchange and transport by, 484–489
  measurement of breathing of, 480–482, 484
  physical laws and, 479–480
  regulators of, 484–492
    chemical, 491–492
    nervous control, 184, 215, 225, 226, 229, 230, 232, 488–491
  types of, 483–484
  vocal activity and, 277–278, 474–475
  volumes of, 480–482, 484
  water loss by, 527, 529
  (*See also* Alveoli; Hemoglobin; Lungs)
Response to injury theory of atherosclerosis, 425–426
Resting heart rate, 383
Resting membrane potential, 156, 158
Resuscitation:
  artificial, 497–500
  cardiopulmonary, 374, 498–500
  mechanical, 497
  mouth-to-mouth, 497–498
  mouth-to-nose, 497–498
Resuscitator, 497
Reticular activating system, 248, 268
Reticular formation, 248, 265, 268, 271
Reticular tissue, 59, 65, 413, 459
Reticulospinal tract, 216, 219, 220, 264, 420
  (*See also* Spinal cord, extrapyramidal tracts)
Reticulocyte, 434–436
Reticuloendothelial system (RES), 60, 65, 459, 464, 468, 585
Retina, 181, 185, 187–191, 193, 259, 274, 446
  (*See also* Sense organs, eye, retina)
Retinal detachment, 187
Retinol (*see* Vitamin A)
Retrograde transport, 139, 140
Retrolental fibroplasia, 495
Reuptake of neurotransmitter by neurons, 113, 175, 230, 252
Reverberating circuits, 215, 263
Reversible shock, 446
Rh antigen, 442–443
Rh factor, 442–443
Rh incompatibility, 442–443, 445–446
Rh positive blood, 442–443
Rheumatic fever, 354, 361, 362
Rheumatism, 354
Rheumatoid arthritis, 63, 354, 435
Rheumatoid factors, 442, 446
Rhodopsin, 189, 191, 347
Rhogam, 446
Rhombencephalon, 239, 241
Rhythm:
  biological, 128, 177, 266, 269–270
  circadian, 269–270
    (*See also* Rhythms, biological)
  in electroencephalogram, 247
  in female reproductive cycles, 634–638

Rhythm (*Cont.*):
method of conception control, 646–648
in respiration, 479–484, 488–492
theta, of brain waves, 247
Riboflavin (*see* Vitamin B_2)
Ribonuclease (RNAse), 573
Ribonucleic acid, 26–28, 34, 119, 574
memory and, 279
messenger RNA (mRNA), 28, 34, 35, 119, 121, 449, 460
ribosomal RNA (rRNA), 28, 30, 119, 121
transfer RNA (tRNA), 28, 119
translation of, 35, 36, 119–121
Ribose, 28
Ribosomal RNA (*see* Ribonucleic acid, rRNA)
Ribosome, 28, 30, 35, 36
Ribcage, 474, 481, 490
Rickets, 352, 353
familial hypophosphatemic, 352, 353
vitamin D resistant, 353
Righting reflex, 265
Rigor crossbridge position, 290, 323–324
Rigor mortis, 322
RNA (*see* Ribonucleic acid)
RNAse (*see* Ribonuclease)
Risk factors of coronary artery disease, 389–390
Rodenticides, 466
Rods, 187–191
Root canal, 355
Root of tooth, 355
Rough endoplasmic reticulum (RER), 29–31, 59, 60, 137, 140
Rouleau effect, 435
Round window of ear, 195–198
RPF (*see* Renal plasma flow)
r position of crossbridge, 290, 323–324
r RNA (*see* ribosomal RNA)
Rubber, 647
Rubrospinal tract, 216, 220, 264
(*See also* Spinal cord, extrapyramidal tracts)
Ruffini nerve endings, 182–184
Rugae, 545

Sabin vaccine, 225
SA node (*see* Sino-atrial node)
Saccule, 195, 197, 199–201
Sacral nerves, 206–208
Sacral segment of parasympathetic division, 226–227
(*See also* Autonomic nervous system, subdivisions)
Safe period, 647
Salicylate poisoning, 535
Saliva, 549, 551, 555
Salivary amylase, 551, 553, 573
Salivary glands, 51, 544, 549, 551, 573
amylase secretion by, 551, 553
autonomic nervous system and, 226, 227, 233, 259, 535
hormonal influences on, 130
parotid, 544, 549, 551
ptyalin, 573
saliva from, 551, 555, 559
sublingual, 544, 549, 551
submandibular, 544, 549, 551
Salmonella, 587, 588
Salmonella typhimurium, 588
Salmonellosis, 588
Salpinigitis, 645
Salt
blood pressure and, 423
coronary artery disease and, 390

Salt (*Cont.*):
heart and, 379, 380
loss in sweat, 326, 506, 510, 594
mineralocorticoids and, 52, 350, 524
(*See also* Aldosterone; Glucocorticoids)
taste of, 201
Saltatory conduction, 148, 166–167
Saralasin, 424
Sarcolemma, 74, 292, 295, 299, 308–309, 383–385
Sarcomere, 286, 287, 296, 299
Sarcoplasm, 286
Sarcoplasmic reticulum, 31
of cardiac muscle, 299
of skeletal muscle, 286, 295, 296, 310, 323, 324
Satellite cell, 145
Satiety center, 562, 564, 595
(*See also* Hypothalamus)
Saturated fats, 576
Saturated fatty acids, 24–25
Saturated phosphatidylcholine, 478
Scala media, 195, 197–199
Scala tympani, 196, 198
Scala vestibuli, 196, 198, 199
Scanning speech, 264
Scapula, 481
S cells, 556
Schizophrenia, 251
Schlemm canal of, 187, 188
Schwann cell, 142–145, 147, 148, 166, 222
Scientific journals, 3, 4
Scintillation counter, 117–119, 246
Scintigram, 397
Sclera, 187, 189, 560
Sclerosis
cerebral, 147
multiple, 147, 149, 245, 460
Scopolamine, 252, 253
Scotopic vision, 189
Scrotal sac, 133, 620, 621, 623, 632, 645, 650
Scrotum, 133, 620, 621, 623, 632, 645, 650
SCUBA diving (*see* Self-contained underwater breathing apparatus)
Scurvy, 73, 602
SDA (*see* Specific dynamic action of foods)
Sebaceous glands, 46, 51, 506, 509, 628, 629
Secondary bronchus, 474
Secondary follicle, 633
Secondary shock, 443–444
Secondary messenger, 120–123, 172
autonomic nervous system and, 228–229, 231, 232
caffeine and, 381
(*See also* Caffeine)
calcium as, 120–123
(*See also* Calcium)
cyclic adenosine monophosphate (cAMP), 120–123, 171
cyclic guanosine monophosphate (cGMP) 120–123, 171
drugs and, 251
neurotransmitters and, 175, 176
(*See also* names of individual secondary messengers)
Secondary ossification center, 343, 344
Secondary sex characteristics, 132, 133, 625, 626, 628–629, 634–636
Secondary shock, 443–445
Second pain, 221
Second stage of labor, 653
Second wind, 330

Secretin, 52, 122, 128, 555–557, 560–563
Secretion, 43
of hormones, 98, 113, 114
of hydrochloric acid, 555, 557–558, 560–562
of hydrogen ions (*see* Hydrogen ions, secretion of)
of neurotransmitters, 98, 168, 170, 173, 174, 252
tubular, in kidney (*see* Kidneys, secretion)
Secretory antibody (*see* Antibodies, Immunoglobulin A)
Secretory phase of uterine cycle, 636–638
Secretory vesicles, 97, 98
(*See also* Exocytosis; Neurotransmitters, secretion)
Sedative, 249, 251, 253
Sedimentation rate, 432
Segmental contraction, 301
Segmental movement, 564
Seizures, 151, 248–249
Selective permeability, 87–89, 158
Selenium, 583, 610, 612
Self-contained underwater breathing apparatus (SCUBA), 494
diving and, 494
Semen, 625, 649, 650
Semicircular canals, 195, 197, 199–201, 259
(*See also* Sense organs, ear)
Semilunar valves, 362, 363
Seminal fluid, 625, 649, 650
Seminal vesicles, 621, 622, 625
Seminiferous epithelium, 46
Seminiferous tubules, 46, 620, 623
Sensation, 186, 222, 223, 259
(*See also* Pain; Sense organs; Sensory receptors)
Sense organs, 187–202
ear, 194–201
acoustic functions, 181, 184, 194, 196–199, 259, 270, 274, 276, 277
auditory apparatus, 194–199, 265, 277
organ of Corti, 195–199
sound, amplitude of, 197–199
sound, pitch of, 195, 197, 198
vestibular functions, 151, 184, 194, 195, 197–200
(*See also* Sensory receptors, auditory and vestibular receptors)
eye, 187–194
adaptation to light, 186, 189, 259, 265
anatomy, 188–190, 193
autonomic nervous system and, 226, 227, 233
blind spot of, 189
extraocular muscles and, 151, 189, 259
functions of components, 187–189, 191–193
lens
accommodation by, 192–193, 233, 259, 301
artificial lenses, 193, 194
astigmatism and, 194
cataract and, 194
focusing by, 191–194
hyperopia and, 193
myopia and, 193
muscles of, 189, 259, 264, 265
reflex movements, 197, 211, 233, 259, 265
(*See also* Reflexes)
retina, 181, 185, 187–191, 193, 259, 274, 446
anatomy of, 190

Sense organs,
retina (*Cont.*):
bipolar neurons of, 187
choroid coat of, 187, 189, 190
cones of, 187–191
fovea centralis and, 189
detachment of, 187
ganglia cells of, 187
macula lutea and, 189
oxygen toxicity and, 495
retinal detachment, 187
retrolental fibroplasia and, 494
rhodopsin in, 189, 191, 347
rods of, 187–191
vision and, 141, 150, 186–187, 189–191, 259
bright, 189
dark adaptation, 189
farsightedness, 193
impulses, 189, 191
midbrain and, 259, 265
nearsightedness, 193
night, 189
physiology of, 187–191
responses to color, 189–191
thalamus and, 259, 265
vitamin deficiencies and, 151
(*See also* Vitamin A)
Young-Helmholtz theory and, 189–191
visual acuity, 189, 191, 194
visual cortex, 276, 277
visual pathways, 186–187, 190, 276, 277
visual reflexes, 265, 269
visual signals
destination of, 273, 274, 276, 277
nose
smell, 141, 148, 202, 259, 270, 271, 274, 276, 474, 476
molecular requirements for generation of odor, 202
(*See also* Sensory receptors, olfactory receptors; Olfactory bulb; Olfactory lobe)
tongue
taste and, 44, 148, 184, 201–202, 258, 259, 276
taste buds and, 44, 201–202, 259, 544
(*See also* Sensory receptors, gustatory receptors; Tongue)
(*See also* Sensory receptors)
Sensible perspiration, 510
Sensitization, 63, 64
Sensors, 8
Sensory fibers, 148, 182, 211
Sensory impulses, 274
Sensory information
coding of, 185–187
(*See also* Codes; Sense organs; Sensory receptors; Somatotopic organization)
destination of, 273–274
Sensory neurons, 209
Sensory receptor potential, 167
thalamus and, 269
Sensory receptors, 140, 148, 182–185, 217
algesiceptors, 148, 184, 185
auditory, 184, 195, 198–199
(*See also* Sense organs, ear)
baroreceptors, 184, 369–370, 381, 407, 420–421
aortic bodies, 184, 381, 420–421, 488–490
carotid sinus, 184, 185, 259, 369–370, 375, 381, 420–421, 488–490

Sensory receptors,
baroreceptors (*Cont.*):
naked nerves, 183–185
(*See also* Naked nerve endings)
chemoreceptors, 184, 201, 370, 407, 488, 490
aortic bodies, 184, 369, 370, 489, 490
brush cells, 477
carotid bodies, 184, 259, 478, 488–490
glucostats, 9, 184, 267
(*See also* Hypothalamus)
gustatory receptors, 184, 201–202
(*See also* Sense organs, tongue)
olfactory receptors, 184, 202
(*See also* Sense organs, nose)
osmoreceptors, 182, 184, 267, 562, 563
exteroreceptors, 182
intensity, 185
interoreceptors, 182
mechanoreceptors, 184–186, 197, 369
acceleration detectors, 197
afferent fibers and, 184
deceleration detectors, 197
deflation receptors, 489–490
Golgi tendon organ, 183–185, 260, 312
gravity detectors, 194, 197
(*See also* Sense organs, ear, vestublar functions)
inflation receptors, 488–490
maculae, 197
Meissner's corpuscle, 182–185, 507, 511
Merkel's touch corpuscles, 184
muscle spindle, 148, 183–185, 260, 263, 312
nociceptors, 148, 184, 185
Pacinian corpuscles, 182–185, 507
proprioceptors, 184, 312
stretch receptors, 488–489, 491, 521
naked nerve endings, 183–185, 507, 511
nociceptors (*see* chemoreceptors above)
olfactory receptors (*see* chemoreceptors, olfactory receptors above; Olfactory bulb; Olfactory lobe; Sense organs, nose)
osmoreceptors (*see* chemoreceptors above)
pain receptors, 148, 184, 185, 221–223
(*See also* Pain; nociceptors above)
phasic receptors, 186
photoreceptors, 184, 186
(*See also* Sense organs, eye)
pressure receptors (see baroreceptors, mechanoreceptors above)
proprioceptors, 184, 312
smell and (*see* Sense organs, nose)
taste (*see* chemoreceptors, gustatory receptors above; sense organs, tongue)
teloreceptors, 182
thermoreceptors, 148, 182, 184, 185, 223
corpuscles of Ruffini and, 182–184
Krause's end bulbs and, 182–185
tonic receptors, 185–186
touch, 148, 182, 185, 186, 223, 276
(*See* also mechanoreceptors above; Spinal cord, spinothalamic tracts)
velocity receptors, 185
(*See also* mechanoreceptors, acceleration detectors, above)
vestibular receptors
cristae ampullaris, 184
maculae, 184
(*See also* Sense organs, ear)
Sensory unit, 182
Septal nuclei, 272, 273
Septum of heart, 362

SER (see Smooth endoplasmic reticulum)
Series eleastic elements, 308–309, 312
Serosa, 549, 565
(*See also* Serous membranes)
Serotonin, 122, 128, 269, 555
allergies and, 64
antiserotonergic drugs, and, 252, 253
as neurotransmitter, 150, 151
platelets and, 462
sleep and, 248
Serous fluid, 49, 51, 360, 475, 638
Serous membrane(s), 48–49, 639
Sertoli cells, 623, 625, 626
Serum, 340, 431
Serum enzymes
diagnosis of disease and, 392
electrophoresis of, 391
Serum glutamic-oxalacetic transaminase (*see* Aspartate aminotransferase)
Serum transaminase (*see* Aspartate aminotransferase)
Serum hepatitis, 566
Serum hormones, 117
(*See also* Normal reference laboratory values, 659–671)
Serum prothrombin conversion acceleerator (SPCA), 464–466
Sex
characteristics of 132, 133, 625, 626, 628–629, 634–636
determination of, 629
heart rate and, 381
Sex chromosomes, 30
(*See also* X and Y chromosomes)
Sex hormones (*see* Androgens; Estrogens; Reproductive system, hormones and; Steroids)
Sex-linked traits, 30, 191, 466
(*See also* X and Y chromosomes)
Sex steroids (*see* Steroids, sex steroids)
Sexual development and maturation, 126, 127, 132, 266, 269, 628–629, 634–636
Sexual excitement, 625–626, 631, 634
Sexual reproduction, 620
(*See also* Reproductive system)
Sexually transmitted disease, 646, 649
SGOT (*see* Serum glutamic oxalacetic transaminase)
Shaft of bone, 72, 336, 343, 344
Shallow breathing, 483
Shingles, 225
Shivering, 233, 266, 268, 325
Shock
anaphylactic, 64
circulatory, 42
hemorrhagic, 422, 425, 426, 443, 444
Shortening heat, 323
Short-term memory, 278
Sickle cell anemia, 432, 446–449
Sickle cell disease, 447
Sickle cell trait, 448
Siderophilin, 583–585
Sigmoid colon, 544, 546, 589
Sighing, 491
SI unit (see Système international units)
Silicon, 583, 610, 613
Silicosis, 480
Silver nitrate, 653
Simple epithelium, 46
Simple exocrine gland, 51
Simple lipids, 24
Single-unit smooth muscle, 301–30
Sinoatrial node (SA node), 300, 3
(*see also* Heart, electrical activi

Sinus (*see* Air sinus)
Sinusitis, 475
Sinusoid, 411, 622
  of liver, 552, 553
Skeletal muscle, 73, 74, 283–298, 307–333
  acetylcholine receptors of, 311
  actomyosin content, 285, 290, 311, 314
  atrophy and dystrophy, 314–315
  bands, 284, 287–290
  blood flow to
    autonomic nervous system and, 227
    autoregulation of, 383, 407
    capillaries in, 410
  calcium and, 288, 291, 292, 294–296, 310, 314, 323–324
  cell structure, 30, 283, 284–288
    types of, 297–298, 322, 328
  Cori cycle and, 319
  contraction, 288–291, 296, 308–314
    all or none law and, 311
    defined, 311
    efficiency of, 323–324
    Fenn effect and, 324
    isometric, 308, 311, 318, 324
    isotonic, 295, 308, 310, 311, 324
    mechanics of, 308–318
    molecular events in, 295–296
    period, 310, 312
    phases of, 310, 312
    relation to load, 310
    sliding filament theory of, 289–290
    strength of, 311–312
    summation of stimuli and, 311
    tetanus and, 312–314
    velocity of, 310, 311, 316–318, 325
  coordination of, 217, 220, 260–264, 273, 312
    vision and, 261
    (*See also* Cerebellum; Sensory receptors, Golgi tendon organs, muscle spindles; Spinal cord, tracts)
  cramps and, 319, 326, 510, 594
  crossbridges of, 290, 295, 296, 310, 323–324
  development of, 238, 285
  disorders and,
    cerebellum and, 264–265
    dystrophies and, 314–315
  elastic elements in, 308–309, 314
  end-plate potential (EEPs), 292–295, 311
  energy metabolism and, 122, 286, 292, 295, 297, 298, 308, 318–325
    (*See also* Metabolism; heat production below)
  exercise, 297, 298, 318, 319, 325–330
    (*See also* Exercise)
  extraocular, 189, 291
  fatigue and, 312, 318, 319, 322, 492, 493
  fiber types, 297–298, 322, 325
  force and, 315–318
  general properties, 285
  glycogen, 318, 320, 322, 323
  growth of, 126, 315
  heat production and, 283, 323–325
  (*See also* Hypothalamus; energy production above; Metabolism; Shivering)
  high energy phosphate compounds and, 318–325
  hormonal effects on, 116, 126
    insulin and, 116
  hypertrophy, 300, 314, 315, 326, 328
  kymograph and, 309
  lactic acid metabolism, 319, 321

Skeletal muscle (*Cont.*):
  levers and, 315–318
  metabolism, 318–325
    exercise and, 325–330
  molecular architecture, 286–291
  motor unit, 311–314, 317
  muscle spindles of, 148, 183–185, 260, 263, 312
  myofibrils of, 285–287, 308, 314
  myofilaments, 285–290
    actin in, 34, 287, 288, 311, 314, 323, 324
    myosin in, 34, 287, 288, 323, 324
    thick, 287–289, 290
    thin, 287–290
  nerves to, 258, 259, 292–295, 311–314
    cell properties and, 298
  neuromuscular junction and, 291–297, 311–314
  oxygen debt and, 319–320, 327, 366, 492–493
  pain and, 319, 326, 510, 594
  passive loading, 312
  poisons of, 295, 297, 311, 314
  polarized light and, 286–288
  power of, 315–318
  reflexes and, 312
    (*See also* Reflexes)
  regeneration of components, 300
  regulation of motor activity and strength of contraction, 311–312
    (*See also* Brain; neuromuscular junction above)
  resting length of, 290, 312
  rhythmical activity of, 215
  rigor mortis and, 290, 322
  sarcolemma, 283, 284, 286, 308–309
  sarcomere, 286, 287, 290
  sliding filament theory of contraction, 289–290
  spindle, 148, 183–185, 260, 263, 312
  stimulation of, 309
    frequency of, 131
    maximal, 314
    strength of, 311
    summation and, 311–314
  stretching of, 290, 312
  system, 6, 68
  tissue, 5, 73–75
  tremors, 129, 177, 272, 273, 325
  triads of, 286
  T-tubules of, 286, 295
  twitch, 210, 313
  ultrastructure of, 286
  warming up, 314
  wasting of, 322
  water content of, 80
  work and, 315–318
Skin, 54, 55, 66, 506–511
  anatomy, 506–511
  blood flow, 132, 507–508
  ceruminous glands in, 196, 507, 509–510
  dermatomes and, 210, 211, 221
  epidermis of, 46, 506–508
  hair follicles in, 509
  keratin and, 508
  pigmentation of, 126–128, 131, 508–509
  sebaceous glands in, 509
  sensory receptors of, 181–185, 507
    (*See also* Sensory receptors)
  stratum basale of, 508
  stratum compactum of, 511
  stratum corneum of, 507–508
  stratum germinativum of, 507–508
  stratum lucidum of, 507–508

Skin (*Cont.*):
  stratum spinosum of, 507–508
  stratum spongiosum of, 511
Sleep, 177, 233, 247–248, 266, 268
  breathing and, 483
  center, 248, 268
  cycle, 248
  epilepsy and, 249
  neurotransmitters and, 248
  non-rapid eye movement (NREM) during, 248, 268
  paradoxical, 248
  rapid eye movement (REM) during, 248, 268
  urine production during, 519
Sliding filament theory of muscle contraction, 288–290
Slit pores, 517
Slow reacting substance anaphylaxis (SRS-A), 64, 462–463
Slow red fibers, 298
Slow twitch fibers, 297, 298, 301
Slow wave, 301
Slow wave sleep, 248
Small intestine, 544–545, 547–551, 555–567, 572–574
  absorption by, 575–580, 585
  anatomy of
    gross, 543–546
    microanatomy, 546–549
    surface area, 548
    villi, 547–548
  autonomic nervous system and, 226, 231, 560–564
  Brunner's glands and, 546, 550, 559, 573
  brush border of, 513, 515, 547, 548, 559
  crypts of Lieberkühn in, 548–550, 559, 573
  duodenum of, 104, 544, 545, 550, 551, 555–557, 563–567, 575, 577, 583
  exocrine secretions of, 549, 550
  enzymes, 559, 572–574
    (*See also* Digestive enzymes)
  fuzzy coat of, 548
  glands of, 550
  glycocalyx of, 548
  hormones and, 555–557
  ileum of, 544, 545, 577, 578, 581–585
    absorption of bile salts by, 575
    aborption of vitamin B_{12} by, 575
  jejunum of, 544, 545, 548, 577, 584, 585
  motility of, 561–564
  reflexes of, 563–564
    (*See also* Reflexes)
Small lymphocytes, 456, 458, 460
  (*See also* Leukocytes, lymphocytes)
Smear of blood, 454–455
Smell (*see* Sense organs, nose; Sensory receptors, olfactory receptors)
Smoking
  physiological effects of, 367, 380, 390, 425, 475, 478, 484, 496–497, 565–566, 649
Smooth endoplasmic reticulum (SER), 29, 286
Smooth muscle (*see* Muscle, smooth)
Sneezing, 491
Sodium
  absorption of, 575, 577, 579
  activation, 163
  active transport of, 93–95, 149, 160–162, 375, 522–524, 531, 577, 579
  excretion, 522, 524, 528
    in cystic fibrosis, 511
  filtration of, 521

gate, 164–165, 167
heart and, 379, 380
Sodium (*Cont.*):
hemoglobinate (NaHb), 486
hydrated size of, 83, 84
inactivation, 163
pump (*see* Active transport)
reabsorption, 526
storage, 335, 338, 340
Sodium bicarbonate buffer, 433, 530–531, 535
(*See also* Buffers)
Sodium chloride, 17, 18
Sodium-potassium ATPases (Na-K ATPases) 93–95, 149, 160–162, 375
(*See also* Impulses)
Sodium phosphate buffer, 530–531
(*See also* Buffers)
Sodium-retaining hormone, 524
(*See also* Aldosterone)
Solubility of respiratory gases, 479
Solute, 18, 80, 89
Solution:
hyperosmolar, 91
hypertonic, 90, 91
hypoosmotic, 91
isoosmotic, 91, 590
isotonic, 90, 91
Solvent, 18, 24, 80
Soma, 135
Somatic motor impulses, 274
Somatic nerve fibers, 147, 211–212
(*See also* Nerve, fibers)
Somatic sensory impulses, 274
Somatomammatropin, 133, 638, 644
Somatomedin, 126, 349, 350
Somatostatin, 52, 130, 555
Somatotopic organization, 273–278
Somatotropic hormone (*see* Growth hormone)
Somatotropin (*see* Growth hormone)
Somasthetic sense, 182
Sound:
ear and (*see* Sense organs, ear, acoustic functions; Sensory receptors, auditory receptors)
heart and, 363–365, 417–418
of Korotkoff, 417–418
Sour taste, 201
Space of Disse, 410, 411, 552, 553
Spasticity, 151
Spastic paralysis, 314
Spatial summation of stimuli, 168
SPCA (*see* Serum prothrombin conversion accelerator)
Special sense organs (*see* Sense organs, ear, eye, nose, tongue)
Specific gravity, 432
Spectrin, 435
Spectrophotometry, 391–392
Speech, 260, 264, 270, 273, 274, 276–278, 475
Speech center, 274, 277, 475
Sperm, 30, 34, 634, 639
anatomy of, 622–623
antibody to, 629, 650
count of, 625, 645–646
hormonal effects on, 127
maturation of, 623, 624
sterility and, 624
storage of, 623
viability of, 645, 647
Spermatic cord, 621
Spermatid, 624
Spermatocyte, 624
Spermatogenic cells, 623
Spermatogenesis, 624–627
Spermatogonia, 623
Spermicidal agents, 647, 649, 650
Sphenoidal sinus, 474
Sphincter, 543
anal, 231, 543, 564
cardiac, 559
esophageal, 555
gastroesophageal, 564
ileocecal, 545
of Oddi, 551, 554
precapillary, 408
pyloric, 544, 545
upper esophageal, 545
Sphingolipids, 148, 150
Sphingomyelin, 87, 477–478
Sphygmomanometer, 417
Spicule, 72, 342
Spike potential, 163, 164, 301, 302, 312
Spinal anesthesia, 208
Spinal cord, 6, 69, 135, 136, 147, 206–225, 241, 258, 260
anatomy, 208–211
cordotomy, 250
damage to, 222–225
ascending tracts and, 223–224
descending tracts and, 223–224
by infections, 225
by physical factors, 224–225
syringomyelia, 223
vitamin deficiencies and, 225
effects of damage 222, 223
Wallerian degeneration, 222
(*See also* Nerves, degeneration of)
growth of, 206–207
membranes of, 242
nerves of (*see* Nerves; Spinal nerves)
neurotransmitters and, 177
(*See also* Neurotransmitters)
pain and (*see* Pain)
pathways, 185–186, 214
defined, 214
dorsal columns, 213, 216–218, 223
(*See also* Circuits; tracts below)
regeneration of, 222
(*See also* Nerves, growth factor, regeneration)
tracts, 75
ascending, 214, 216–218, 260, 265
corticospinal (*see* pyramidal below)
corticothalamic, 269
descending, 214, 263, 266
decussation of, 216
defined, 214
extrapyramidal, 216, 219, 220, 264, 312
location of, 216–220
pyramidal, 216, 219, 220, 223
reticulospinal, 216, 219–220, 264, 420
rubrospinal, 216, 220, 264
spinocerebellar, 213, 216, 217, 223
spinothalamic, 213, 216–220, 223
vestibulospinal, 216, 220, 264
Spinal nerves, 69, 135, 136, 140, 147–211
neurotransmitters and, 177
(*See also* specific names of neurotransmitters)
plexuses of, 211
segmental distribution of, 210, 211
Spinal pathways (*see* Spinal cord, pathways)
Spinal reflexes, 211, 564
(*See also* Reflexes)
Spinal shock, 222
Spinal tap, 208, 243, 244
Spindle apparatus, 34
Spines of dendrites, 140
Spinocerebellar tract, 213, 216, 217, 262, 263
Spinothalamic tract, 213, 216, 217
Spiral arteries, 636–638, 641, 642
Spironolactone, 424
Spirometer, 480–482
Spirometry, 480–483
Spleen, 65, 150, 544, 585
ascites and, 389
anemia and, 448
capillaries of, 411
Splenomegaly, 448
Split brain, 274–278
Spongy bone, 72, 336, 342
Spontaneous abortion, 645, 651
Sprain, 351
Sprue, tropical, 587–589
Squamous epithelium, 44, 45, 53, 476
SRS-A (*see* Slow reacting substance anaphylaxis)
Stable factor, 464
Staircase effect, 314
Stapedius muscle, 196
Stapes, 195, 198
Staphylococci, 534, 587–589
Staphylococcus aureus, 588
Starch:
animal (*see* Glycogen)
plant, 20
Starling's equilibrium of capillary exchange, 412–413
Starling's law of the heart, 366, 383
Static sensors, 199–201
(*See also* Sense organs, ear, vestibular functions)
Statolith, 197, 200–201
Steady state, 6
Steatorrhea, 586, 588
Stellate cells, 262
Stem cell, 66, 340, 434, 435, 437, 449, 454, 455, 467, 468
Stenosis, 362–363, 387, 388
Stercobilin, 439
Stercobilinogen, 439
Sterility, 645–646, 650
Sterilization, 647
Sternum, 128, 338, 360, 481, 500
Steroid, 65
bone physiology and, 132–133
chemistry, 25, 627, 648
electrolyte metabolism and (*see* Adrenals; Kidneys; names of individual hormones; Mineralocorticoids)
(*See also* Adrenals; Glucocorticoids; Mineralocorticoids; Sex steroids)
Sex steroids:
reproductive tract effects, 52, 105, 627–629, 634, 649
physiology of bone and, 351, 353
synthesis, of, 627
(*See also* Adrenals; Reproductive system, ovaries, testes)
Stethoscope, 363, 391, 417, 418
Stimulator, 309
Stimulus:
intensity, 186
nerve and, 161, 164, 165
skeletal muscle and, 309, 311, 313–314
smooth muscle and, 302
subthreshold, 160, 167, 172, 189
summation of, 167–168, 172–175, 189, 221

Stimulus (*Cont.*):
threshold, 160, 161, 172, 292
(*See also* Impulses; Potentials)
S-T interval, 371
Stomach, 48, 89, 104, 544, 551, 575
absorption by, 98, 575
anatomy of, 302, 544, 545
antrum of, 555
body of, 544, 545
fundus of, 544, 545, 549
pylorus of, 544, 545, 555
chief cells of, 550, 556, 557, 573
enzymes from
gastric protease, 573
lipase, 573
pepsin, 557, 572, 573
rennin, 573
(*See also* Digestive enzymes)
gastric pits of, 555–557
glands of, 550
hormonal effects on, 556–557
hormonal secretion by, 555, 556
hydrochloric acid secretion by, 48, 85, 94, 122, 129, 555–557, 559
intrinsic factor and, 550, 557–559
motility of, 226, 231, 561–564
mucous neck cells of, 550, 556, 557
mucus of, 550, 557
parietal cells of, 550–552, 556–558, 566
referred pain and, 210
secretions of, 555–559
(*See also* Digestive system)
Stratified epithelium, 44, 46, 201, 507
Stratum basale, 508
Stratum compactum, 511
Stratum corneum, 507–508
Stratum germinativum, 507–508
Stratum granulosum, 507–508
Stratum lucidum, 507–508
Stratum spinosum, 507–508
Stratum spongiosum, 511
Streptococcal infection, 361, 534
Stress:
basophils and, 457
blood pressure and, 112, 423
defined, 109
eosinophils and, 457
general adaptation syndrome (GAS) and, 109
general responses to, 112
glucocorticoids and, 112, 457
heart and, 65, 112, 368, 378, 390, 392
hormones and, 55–56, 109–112, 457
hypothalamus and, 109–112, 266, 268–269
leukocytes and, 457, 467
mechanical:
effects on bone, 73, 342
receptors and, 115
sweating and, 510
ulcers and, 565
Stretching
response of cardiac muscle, 366
response of skeletal muscle, 290, 312
response of smooth muscle, 302
Stretch receptors, 488–489, 491, 521
Striated muscle (*see* Muscle, cardiac; Skeletal muscle)
Stroke, 389, 426, 649
Stroke volume of heart, 366, 382, 406
Strontium, 340
Structural proteins in cell membrane, 84, 86
Strychnine poisoning, 314
S-T segment, 371
Stuart-Prower factor, 464–466
Subarachnoid space, 242–244
Subatomic particles, 15
Subcutaneous injection, 511
Subcutaneous tissue, 511
Subdural space, 242
Subliminal stimulus (*see* Subthreshold stimulus)
Submaximal exercise, 325, 327–328
Submucosa, 50, 546, 547
Substance P, 555
Substantia gelatinosa, 221, 251
Substantia nigra, 265, 272
Substrate, 22, 23
Subthalamic nuclei, 272
Subthreshold stimulus, 160, 167, 172, 189
Succinylcholine, 297
Suckling reflex, 639
Sucrase, 572, 573, 586–588
Sucrose, 577
polymers of in dental plaque, 355
Suction of lymph, 414
Sudden infant death syndrome, 477–478
Sudoriferous glands (*see* Sweat glands)
Sulcus, 270
Sulfur oxides, 497
Summation of contractions:
in skeletal muscle, 311–314
in smooth muscle, 301
Summation of stimuli, 167–168, 172–175, 221, 292, 301, 311–314
spatial, 168
temporal, 167–168
Superior colliculi, 241
Superior sagittal sinus, 242
Superior vena cava, 362, 364, 384, 386, 388, 413
(*See also* Vena cava)
Superficial fascia, 55
(*See also* Fascia)
Supplemental air volume, 482
Suppository, 575
Suppressor T cells, 460
Suprasegmental paths, 260
(*See also* Spinal cord, paths and tracts)
Surfactant, 447–478, 494, 496
Surgical sterilization, 647
Suspensory ligaments, 187, 192
Suture, 337
Swallowing, 259, 545
Sweat, 51, 510–511
composition of, 510, 584
function of, 510
odor of, 510
regulation of, 130, 510
autonomic nervous system and, 226, 227, 233, 266, 268
shock and, 444
volume of, 510
(*See also* Fluids; Glands)
Sweat glands, 46, 51, 444, 524, 632
Sweet taste, 201
Sylvian fissure, 270, 272, 274
Sympathetic nerves
to heart, 368–370
in hemmorhagic shock, 443
stress and, 111, 112
vasoconstriction and, 420–421
(*See also* Nerve fibers, adrenergic, cholinergic; Autonomic nervous system, subdivisions)
Symphysis, 337, 621
Symphysis pubis, 69, 133, 627
Symport, 95, 558, 576, 577, 579
Synapse, 168–178
anatomy, 168–169
calcium activity and, 169–172, 230
circuits and, 169–171, 229
cleft of, 169, 291, 293
defined, 169
excitatory, 171, 174
(*See also* Potentials, excitatory)
inhibitory, 171, 174
(*See also* Potentials, inhibitory)
memory and, 279
membranes of, 168, 169, 292–294
neuromuscular junctions and, 291–298
neurotransmitters and, 175–178
(*See also* Neurotransmitters; names of specific neurotransmitters)
numbers of
in brain, 237
in cerebellum of brain, 270
in spinal cord, 211, 214
terminals in, 169
types of, 169–171
Synarthrosis, 336, 337
Syncytium, 285
Syndrome of inappropriate ADH (SIADH), 128
Synergist, 116
Synostosis, 337
Synthetic estrogens, 648
Synthetic progestins, 648
Synovial cavity, 337, 338
Synovial fluid, 336
Synovial joint, 336, 337
Synovial membrane, 63, 336–338, 354
Syphilis, 225, 245, 646
Syringomyelia, 223
System, 5, 6
(*See also* names of specific systems)
Système international units, 659–671
Systole, 364–365, 404
Systolic pressure, 415–417

T_3 (*see* Triiodothyronine)
T_4 (*see* Tetraiodothyronine)
Tabes dorsalis, 225
Tachycardia, 406
Taeniae in colon, 545
Tagged antibody, 246
Target cell, 103, 109, 110, 114, 139
Target heart rate, 328–329
Taste, 44, 148, 184, 201–202, 258, 259, 270, 276
Taste buds, 44, 201–202, 259, 544
Taste receptors, 184, 201–202
Tay-Sach's disease, 150
TCA (*see* Tricarboxylic acid cycle)
T cells, 59, 61, 63, 85, 297, 351, 354–355, 435, 458, 460, 468
(*See also* Cellular Immunity; Leukocytes; Lymphokines; Thymus)
Tears:
antibody in, 59, 555
cranial nerve VII and, 233
in cystic fibrosis, 506
lacrimal gland in, 188
Tectorial membrane, 194, 195, 198, 199
Teeth, 72, 73, 339, 347, 354–355, 542–544
Telencephalon, 239, 241, 266, 270–279
Teloceptor, 182
Temperature
body, 267–268, 510
metabolism and, 594
variations, 268, 594

diffusion rate and, 88
fever and, 594
ovulation and, 494, 647
regulation of, 233, 407, 594–595
cold exposure and, 109, 233, 266–268, 325, 380, 448
heat and, 233, 266–268, 509, 594
(*See also* Heat; Sweat; Sweat glands)
respiration and, 492
skin and, 506, 510
Temporal lobe of brain, 270, 274, 277
Temporal summation of stimuli, 167–168
Tendon, 54, 66, 338
muscle and, 283–285
Tendon organ, 183–184, 260, 263, 312
Tendon sheath, 337–338
Tennis elbow, 354
Tensile strength of bone, 335, 338, 340
Tension, muscle and, (*see* Skeletal muscle, contraction)
Tentorium cerebelli, 242
Terminal boutons, 137, 140, 149
Terminal ganglia, 227
Terminal knobs, 137, 139, 140
Testis, 6, 8, 52, 104, 105, 620–625, 628, 636
Testosterone, 26, 52, 625–629
chemistry of, 627, 648
effects, 133, 348, 350, 353, 623–629, 636
hair loss and, 509
Tetanus, 89, 312, 314, 324, 367
Tetany, 89, 128
Tetraiodothyronine (T_4), 117, 128
9-Tetrahydrocannabinol, 115
Tetralogy of Fallot, 387, 388
Tetrodoxin, 165
Thalamus, 216, 237, 241, 266, 369
(*See also* Brain, thalamus)
Thalassemia, 432, 446–449
Theobromine, 26, 27
Theophylline, 26, 27, 91, 523
Thermoreceptor, 148, 182–185, 223
Thermography, 632
Theta waves, 247, 248
Thiamine (*see* Vitamin B_1)
Thiazides, 424, 535
Thick myofilaments, 289–290
Thin myofilaments, 287–290
Thirst, 233, 244, 266–269, 562
angiotensin and, 444, 525–526
Thoracic cavity, 49, 68, 474, 475, 481
Thoracic duct, 413
Thoracolumbar division of autonomic nervous system, 227
Thorazine, 252, 253
Thorax, 49, 68, 474, 475, 481
Thoroughfare capillaries, 408
Thoughts and brain, 277
Threshold stimulus, 160, 172, 292, 310
Thrombin, 464, 465
Thrombocytes, 431–433, 456, 461–469
appearance, 456, 461
atherosclerosis and, 425–426
clotting and, 464–467
inhibition of, 465–466
sequence of, 464–465
tests for, 466–467
general functions, 454, 462–463
life span, 461
Thrombocytes (*Cont.*):
numbers of, 461, 466
origins of, 434, 435, 461
pathophysiology and, 422, 425, 426, 469
prostacyclin and, 462–463
prostaglandins and, 462–463
thromboxane and, 462–463
(*See also* Blood)
Thrombocythemia, 469
Thrombocytopenia, 469
Thrombocytopenic purpura, 463, 469
Thromboembolic disease
birth control pills and, 649
Thromboplastin, 464, 465
Thrombosis, 422, 488, 469, 587
Thrombotic stroke
birth control pills and, 649
Thrombosthenin, 464
Thromboxane (TXA_2), 462–463
Thymine, 26
Thymosin, 52, 129
Thymus, 6, 8, 52, 104, 105, 129, 355
aging and, 351
inhibition by stress, 112
hormones and, 131
T cells and, 434, 460
Thyrocalcitonin, 8, 52, 122, 128, 348, 349
Thyroid cartilage, 474, 629
Thyroid gland, 6, 7, 52, 94, 104, 105, 128, 423
Thyroid stimulating hormone, 52, 108, 109, 116, 122, 126
Thyrotropin (*see* Thyroid stimulating hormone)
Thyroxine, 8, 52, 94, 113, 116, 122, 126, 128, 349
blood pressure and, 423
measurement of, 117, 118
(*See also* Thyroid gland; triiodothyronine; tetraiodothyronine)
Tibia, 338
Tidal volume, 482
Tin, 583, 610, 612
Tight junction, 42, 43, 73, 86, 411, 549, 564, 574, 623, 624
Tissue basophils, 457
(*See also* Mast cells)
Tissues, 5, 41–78
(*See also* names of specific tissue types)
Tissue culture, 56
Tissue factor, 464
Tissue hypoxia, 443–444
Titubation, 264
T lymphocetes (*see* T cells)
T_m (*See* transport maximum)
Tocopherol (*see* Vitamin E)
Tolerance to drugs, 251
Tone of muscle, 264–265, 561
Tongue, 44, 46, 148, 184, 258, 259, 276, 475, 544, 572
(*See also* Taste; Taste buds)
Tonic phase in epilepsy, 249
Tonicity of solutions, 91
Tonic receptors, 185–186
Tonsils, 545
Tonsillectomy, 545
Tonus, 265, 561
Tooth (*see* Teeth)
Total intravenous feeding, 589
Total ventilation of lungs, 480
Touch, 216, 217, 223
pathways for (*See* Spinal cord, dorsal columns; spinothalamic tracts)
Touch (*Cont.*):
sensory receptors and, 148, 182, 185, 186, 223, 276
(*See also* Sensory receptors, mechanoreceptors)
skin and, 506–507
Toxemia of pregnancy, 393, 644
Toxins, 587–589
of black widow spider venom, 295–296
blood-brain barrier and, 145
botulinal, 295
nervous system and, 145, 149, 165
shock and, 444–445
(*See also* Gastroenteritis; Skeletal muscle, toxins)
Trabecula, 342, 344
Trace elements, 340–341, 583, 608–613
absorption of, 583–585
nutritional aspects of, 611–613, 618
Trachea, 46, 53, 54, 68, 69, 481, 544
Tracheotomy, 473
Tract (*see* Spinal cord, pathways, tracts)
Training
exercise and, 491, 492
heart rate and, 328–329
muscle and, 297, 298, 318, 319, 325–330
(*See also* Exercise; Physical training)
Tranquilizer, 177, 252, 253
Transaminase, 566
(*See also* Aspartate aminotransferase)
Transamination, 392, 592
Transcellular fluid, 81
Transcription, 34–36
hormones and, 119–121, 628, 636
Transducers, 189, 247, 397
(*See also* Sensory receptors)
Transepithelial transport, 523
Transferrin, 447, 584, 585
Transfer RNA (*see* Ribonucleic acid, transfer RNA)
Transfusion (*see* Blood, transfusions)
Translation, 35, 36
hormones and, 119–121
(*See also* Ribonucleic acid, messenger RNA)
Transmission of impulses (*see* Impulses; Neuromuscular junction; Neurotransmitters; Potentials)
Transmitters, 8
(*See also* Hormones; Neurotransmitters; Secondary messengers)
Transplantation antigens (*see* Histocompatibility antigens)
Transplant(s)
of bone marrow, 448, 449
of kidney, 535–536
Transport
active, 97, 93–95
anterograde, 139, 130
carrier mediated, 92–94
of hormones, 113, 114
maximum (T_m), 523
nonmediated, 87–91
diffusion, 87–89
osmosis, 89–91
pumps and, 87, 95
retrograde, 139, 140
Transverse colon, 544
Trauma
nervous system and, 148, 149, 222
(*See also* Regeneration)
physical, 353
Tremors, 129, 177, 272, 273, 325
Treponema pallidum, 646
Treppe, 314
TRH (*see* Thyrotropin releasing hormone)
Triacylgycerols (*see* Triglycerides)
Triad, 286, 300
Tricarboxylic acid cycle (TCA cycle), 320–322, 590–592

Trichomonas, 646
Tricuspid valve, 362, 363
Triglyceride, 24, 25, 573, 574, 576–578, 581, 592
(*See also* Fats; Lipids)
Tripeptides
absorption of, 580
Triplet, 34
Trisaccharides, 20
Triiodothyronine (T_3), 113, 117, 128
tRNA (*see* Ribonucleic acid, transfer RNA)
Trophoblast, 640, 641
Tropical sprue, 587–589
Tropic hormones, 107, 109, 120, 126–127
(*See also* Pituitary, anterior)
Tropocollagen, 53, 54
Tropomyosin, 289, 295, 296, 324
Troponin, 289, 295, 296, 324
in smooth muscle, 301
True capillaries, 408
True labor, 652–653
Trypsin, 551
Trypsinogen, 574
Tryptophan, 64, 128, 150, 177
TSH (*see* Thyroid stimulating hormone)
T-tubules
in cardiac muscle, 299, 300
in skeletal muscle, 286, 295
in smooth muscle, 301
Tubal ligation, 647, 650
Tubal pregnancy, 641
Tuberculosis, 245, 459, 467, 484
Tubule, 33
(*See also* Kidneys, nephrons of; Seminiferous tubules; T-tubules)
Tubulin, 139
Tumor, 63, 85, 113, 116, 131, 150, 223, 632
benign, 425
epilepsy and, 248
interferon and, 460
melanoma, 53
of nervous system, 243, 245, 246
pheochromocytoma and, 113, 132
of pineal gland, 128, 269
ulcers and, 565
vitamin A and, 53, 347
Tunic of blood vessels, 403
T-wave, 371, 374
Twighlight vision, 189
Twitch, 310, 313, 323
Twins, 639
Tympanic membrane, 195, 196, 198, 509
(*See also* Sense organs, ear)
Type A and B behavior, 389–390
Type I and II cells of alveoli, 476
Type I and II herpes, 646
Type I and II synapses, 169, 171
Typhoid fever, 467, 565
(*See also* Gastroenteritis)
Typing of blood, 441–443
Tyrosinase, 508
Tyrosine, 128, 131, 150, 176, 508
Tyrosine hydroxylase, 150, 176, 230

Ulcers, 111, 565–567
Ultracentrifugation of lipoproteins, 394
Ultrafiltration, 512
Ulstrasound, 397, 398
Ultraviolet radiation:
as phototherapy in erythroblastosis fetalis, 446
skin and, 508
Umbilical artery, 384, 385, 641, 642
Umbilical cord, 641, 642, 652, 653
Umbilical vein, 384, 385, 541, 642
Uncompensated acidosis, 534
Uncompensated alkalosis, 534
Unconditioned reflex, 278
Unfused tetanus, 313
Unicellular glands, 51
Unipolar neurons, 140, 141
Unipolar recording, 371, 373
Units:
international, 659–671
of measure, 16, 18, 19
of metric system, 16
Système international (*see* Appendix of text)
Universal donor of blood, 442
Universal recipient of blood, 442
Unmyelinated nerve fibers, 139, 140, 144–146, 166, 177, 209
transmission of second pain and, 22[illegible]
Unsafe period, 647
Unsaturated fats, 25, 576
Upper gastroesophageal sphincter, 555
Upper motor neuron, 220
Uracil, 26, 28
Urea, 449, 527
metabolism of, 591–592
production in liver, 589
role in excretion of water, 521, 522, 524, 527, 533
Uremia, 533–534
Ureter, 47, 301, 512
Urethra, 512, 620, 621, 630, 631
Urethritis, 646
Uric acid, 533, 578
in gout, 354, 534, 578
in sweat, 510
Urinary bladder, 47, 512, 621, 622, 630, 631
Urinary stones, 534
Urinary system, 47
Urination, 512
Urine:
chemistry of, 514, 517, 578, 584
concentration of, 522, 527, 528
formation, 512–537
hormone excretion and, 113, 114, 117
regulation of, 91, 122, 128, 130, 512–537
volume, 517, 519, 529
Urobilinogen, 439
Urolithiasis, 534
Uterine arteries, 636
Uterine cycle, 636–638
Uterine glands, 640
Uterine lining, 630, 631, 636–639, 640, 642
Uterus, 34, 132, 385, 629–632, 634, 636, 639, 652–653
hormones and, 127, 133, 634–638
hysterectomy and, 650
hysterotomy and, 651
smooth muscle of, 301, 631, 634
Utricle, 195, 199–201
(*See also* Sense organs ear; Sensory receptors, vestibular receptors)
U-wave, 371, 374

Vacuoles, 32, 56
Vagina, 575, 629–632, 650, 652
Vaginal cancer, 632, 653
Vagus nerve
heart and, 368–369
respiration and, 490
(*See also* Cranial nerves; Brain, medulla oblongata, expiratory center, inspiratory center, respiratory center; Brain, pons, apneustic center, pneumotaxic center)
Valence (*see* Equivalent weight)
Valine, in hemoglobin, 447
Valium, 252, 253
Valsalva maneuver, 564
Valve:
of heart, 316–363
echocardiogram and, 398
of lymph vessels, 413
of veins, 403
Vanadium, 583, 610, 612
Varicose veins, 414, 422
Vasa recta, 517, 525, 526
Vasa vasorum, 403
Vascular side effects of birth control bills, 649
Vas deferens, 620–622, 650
Vasectomy, 647, 650
Vasoactive amines, 444–445
(*See also* Histamine)
Vasoconstriction, 266–268, 420–424, 462
(*See also* Adrenergic nerves; Cholinergic nerves; Vasomotor center)
Vasodilation, 266, 268, 407, 420–421, 457
(*See also* Adrenergic nerves; Cholinergic nerves; Vasomoter center)
Vasomotor center, 260, 407, 420–421, 444, 488
Vasopressin (*see* Antidiuretic hormone)
Vectorcardiogram, 373
Vein:
anatomy, 403, 404
capacity, 414–416
pressure and, 414
regulation of blood flow and, 414–416
valves of, 414
Velocity of blood flow, 415, 416
Velocity of muscle contraction, 316–317
Velocity of conduction of nerve impulses, 148
Venae cava, 403
receptors in, 184, 369–370
(*See also* Inferior vena cava; Superior vena cava)
Venereal disease, 646
Venous baroreceptors, 184, 369–370, 420–421
Venous return to heart, 366, 383
Venous sinus, 242
Venous sinusoids (*see* Sinusoids)
Ventilation of lungs, 480–484
artificial ventilation, 497–500
exercise and, 493
Ventral corticospinal tract, 216, 219–220
(*See also* Spinal cord, tracts)
Ventral horn, 209, 211
Ventral spinothalamic tract, 216–218
(*See also* Spinal cord, tracts)
Ventricles of brain, 145, 208, 241–244, 260
Ventricles of heart, 361–366
fibrillation of, 374
premature beats of, 367
systole of, 364–365
tachycardia of, 406
Venule, 403, 415, 516
Vertebra, 69
Vertebral canal, 69, 206
Vertebral column, 206–208
growth compared to spinal cord, 206–207
Very low density lipoprotein (VLDL), 394, 395
Vesicle, 29, 32, 87
calcium and, 169, 170
(*See also* Calcium)
cytoplasmic, 139, 140

of endothelial cells, 409, 410
in mediated transport, 87, 95–98
(*See also* Endocytosis; Exocytosis)
in neuromuscular junction, 291–295
pinocytic, 96, 98
secretory, 97, 98, 113
synaptic, 169–171
Vessels
blood (*see* Blood vessels)
lymph (*see* Lymph vessels)
Vestibular apparatus (*see* Sense organs, ear; Sensory receptors, vestibular receptors)
Vestibular glands, 631
Vestibular membrane, 194, 197–201
Vestibule
of ear, 195, 259
of female reproductive tract, 631, 632
Vestibulospinal tract, 216, 220, 264
(*See also* Spinal cord, extrapyramidal tracts of)
Vibration, detection of, 197, 201
Villus, 50
cells of, 548–550
chorionic, 641, 642
number, 548
structure of, 547, 548
Viral infections (*see* Infections; Virus; Sexually transmitted diseases)
Virilization, 131
Virus:
cancer and, 86
cell receptors and, 85
fever and, 594
food poisoning and, 566, 587
hepatitis and, 566
infections by, 225, 459, 467, 469, 566, 587
interferon and, 460
multiple sclerosis and, 147
mumps, 549, 646
papilloma, 646
polio, 137, 223
sexually transmitted disease and, 646
Viscera, 75, 206
autonomic regulation of, 225
(*See also* Autonomic nervous system; Digestive system)
Visceral epithelium, 517–519
Visceral muscle, 75
Visceral nerve fibers, 147, 212
(*See also* Autonomic nervous system)
Visceral pericardium, 361
Visceral pleura, 474, 475
Viscosity of blood, 419
Viscosity of cervical secretions, 636, 637
Vision (*see* Sense organs, eye)
Visual acuity, 189, 191, 194
Visual cortex, 274, 276, 277
Visual purple, 189
Visual reflexes, 261, 265, 269
eye movement and, 197
lens and, 191–194, 233, 259
pupil and, 211, 233, 259, 265
(*See also* Sense organs, eye; Reflexes)
Vital capacity, 482–483
calculation of, 482
Vitamins, 600–607
absorption of, 98, 577, 581, 582
deficiency diseases and, 151, 225, 352, 353, 444, 600–607
defined, 581
fat soluble, 537, 581
A (retinol), 53, 187, 347, 581, 600–601, 618
D (1,25 dihydroxycholecalciferol), 132, 345–350, 352, 581, 600–601, 618
E (tocopherol), 347, 581, 600–601, 618
K, 465, 466, 681, 602–603
intestinal bacteria and, 589
liver and, 554
recommended daily dietary allowances, 618
transport of, 92
water soluble
B_1 (thiamine), 151, 225, 575, 581, 602–603, 618
B_2 (riboflavin), 604–605, 618
B_6 (pyridoxine), 151, 447, 581, 604–605, 618
B_{12} (cobalamin), 98, 151, 225, 447, 466, 469, 557–559, 575, 581, 588, 595, 606–607, 618
Vitamin, water soluble (*Cont.*):
B_{17} (nitrilosides), 449
biotin, 606–607
C (ascorbic acid), 73, 347–348, 353, 602–603, 618
choline, 606–607
folic acid, 447, 448, 466, 469, 588, 606–607, 618
inositol, 606–607
niacin (nicotinic acid), 151, 604–605, 618
pantothenic acid, 604–605
Vitreous humor, 187, 188, 191, 192
VLDL (*see* Very low density lipoprotein)
V lead recordings, 371–373
V_{O_2} max, 326, 329, 330, 383
Vocal cords, 474, 475
Voice:
cranial nerve XI and, 233
(*See also* Cranial nerves)
speech area in brain and, 274–278
Voice box, 69, 473–475
Volkmann canals, 71, 72
Voltage, 88, 156, 163
brain waves and, 247
electrocardiogram and, 370–371
measurement of, 158, 163
(*See also* Potentials)
Voluntary activity, 270
(*See also* Cerebral hemispheres; Upper motor neurons)
Voluntary fibers, 147
Vomiting center, 535, 564

Waking center, 266, 268
Wallerian degeneration, 145, 222
Warm-blooded animal, 268
Warfarin, 466
Warming-up, 314
Warts, 646
Water:
absorption of, 575, 585
in childbirth, 653
compartments in body, 80–82, 412
content of body, 80
diuresis by kidneys, 517, 522, 585
endogenous secretion of, 585
intake and loss, 510, 526–529, 585
loss in shock, 444–445
molecular size of, 83
reabsorption of, 351
regulation of, 82, 111, 131, 133
(*See also* Aldosterone, Antidiuretic hormone, Glucocorticoids; Hypothalamus)
stress and, 524
sweating and loss of, 506, 510–511
water-borne hepatitis and, 566–567
water-soluble vitamins and (*see* Vitamins, water soluble)
vapor:
partial pressure of, 488
respiration and, 479
Waves:
in electrocardiogram (ECG), 371–375
in electroencephalogram (EEG), 247
Wax in ear, 509–510
WBC (White blood cells) (*see* Leukocytes)
Wear and tear arthritis, 354
Weight:
body:
average by age, height and sex, 615
blood pressure and, 423
bone and, 335, 341
bone marrow and, 338
calcium and, 339
heart and, 360
heart disease risk and, 390
kidneys and, 517
loss of, 592, 594, 595
skeletal muscle and, 283
(*See also* Obesity)
mass and, 16
molecular, 18
Wernicke's area, 277
Wernicke-Korsakoff syndrome, 277
Wheal and flare reaction, 63
White blood cells (*see* Leukocytes)
White matter (*see* Myelinated nerve fibers)
White muscle fibers, 297–298, 325
Whooping cough, 467
Wintrobe tube, 433
Withdrawal method of conception control, 647
Womb (*see* Uterus)
Work, skeletal muscle activity and, 315–318, 327
Worm infections, 414, 467
Wound healing, 73, 462–463
Wright's stain, 454, 456
Writing, brain and, 276

Xanthines, 91
X chromosome, 30
in meiosis, 633
Xeroradiography, 632
X-linked traits, 30, 149, 191, 315, 353, 466, 623–624, 628, 629
(*See also* Sex-linked traits)
XO genotype, 629
X-ray, 184, 244–246, 336, 369, 397, 398, 632
(*See also* Radiation)
XXY genotype, 629
Xylose, 588

Yawning, 491
Y chromosome, 30, 623–624, 628–629
in meiosis, 633
(*See also* Sex-linked traits)
Yeast infections, 646, 649
Yellow bone marrow, 337–338
Yolk sac, as a source of blood cells, 434
Young-Helmholtz theory of color vision, 189–191

Z-disc (*see* Z-line)
Zinc, 583, 609, 612, 618
Z-line, 286, 300, 312
Zollinger-Ellison syndrome, 565
Zygote, 639–640
Zymogen granules, 555